AF696307

TRANSLATIONAL CONTROL OF GENE EXPRESSION

COLD SPRING HARBOR MONOGRAPH SERIES

The Lactose Operon
The Bacteriophage Lambda
The Molecular Biology of Tumour Viruses
Ribosomes
RNA Phages
RNA Polymerase
The Operon
The Single-Stranded DNA Phages
Transfer RNA:
    Structure, Properties, and Recognition
    Biological Aspects
Molecular Biology of Tumor Viruses, Second Edition:
    DNA Tumor Viruses
    RNA Tumor Viruses
The Molecular Biology of the Yeast *Saccharomyces:*
    Life Cycle and Inheritance
    Metabolism and Gene Expression
Mitochondrial Genes
Lambda II
Nucleases
Gene Function in Prokaryotes
Microbial Development
The Nematode *Caenorhabditis elegans*
Oncogenes and the Molecular Origins of Cancer
Stress Proteins in Biology and Medicine
DNA Topology and Its Biological Effects
The Molecular and Cellular Biology of the Yeast *Saccharomyces:*
    Genome Dynamics, Protein Synthesis, and Energetics
    Gene Expression
    Cell Cycle and Cell Biology
Transcriptional Regulation
Reverse Transcriptase
The RNA World
Nucleases, Second Edition
The Biology of Heat Shock Proteins and Molecular Chaperones
Arabidopsis
Cellular Receptors for Animal Viruses
Telomeres
Translational Control
DNA Replication in Eukaryotic Cells
Epigenetic Mechanisms of Gene Regulation
C. elegans II
Oxidative Stress and the Molecular Biology of Antioxidant Defenses
RNA Structure and Function
The Development of Human Gene Therapy
The RNA World, Second Edition
Prion Biology and Diseases
Translational Control of Gene Expression

TRANSLATIONAL CONTROL OF GENE EXPRESSION

Edited by

Nahum Sonenberg
McGill University, Montreal

John W.B. Hershey
University of California, Davis

Michael B. Mathews
New Jersey Medical School
University of Medicine and Dentistry of New Jersey

COLD SPRING HARBOR LABORATORY PRESS
Cold Spring Harbor, New York

LOYOLA UNIVERSITY MEDICAL CENTER LIBRARY

QH 450.5 T772 2000

TRANSLATIONAL CONTROL OF GENE EXPRESSION

Monograph 39
© 2000 by Cold Spring Harbor Laboratory Press,
Cold Spring Harbor, New York
All rights reserved
Printed in the United States of America

Project Coordinator: Joan Ebert
Production Editor: Patricia Barker
Desktop Editor: Danny deBruin
Interior Book Designer: Emily Harste

Library of Congress Cataloging-in-Publication Data

Translational control of gene expression / edited by Nahum Sonenberg, John W.B. Hershey, Michael B. Mathews--
p. cm. -- (Monograph, ISSN 0270-1847 ; 39)
Includes bibliographical references and index.
ISBN 0-87969-568-4 (alk. paper)
1. Genetic translation. I. Sonenberg, Nahum. II. Hershey, John W.B. III. Mathews, Michael B. IV. Cold Spring Harbor monograph series 39.

QH450.5 .T725 2000
572'.645--dc21

00-055481

10 9 8 7 6 5 4 3 2 1

Authorization to photocopy items for internal or personal use, or the internal or personal use of specific clients, is granted by Cold Spring Harbor Laboratory Press, provided that the appropriate fee is paid directly to the Copyright Clearance Center (CCC). Write or call CCC at 222 Rosewood Drive, Danvers, MA 01923 (508-750-8400) for information about fees and regulations. Prior to photocopying items for educational classroom use, contact CCC at the above address. Additional information on CCC can be obtained at CCC Online at http://www.copyright.com/

All Cold Spring Harbor Laboratory Press publications may be ordered directly from Cold Spring Harbor Laboratory Press, 10 Skyline Drive, Plainview, New York 11803-2500. Phone: 1-800-843-4388 in Continental U.S. and Canada. All other locations: (516) 349-1930. FAX: (516) 349-1946. E-mail: cshpress@cshl.org. For a complete catalog of Cold Spring Harbor Laboratory Press publications, visit our World Wide Web Site http://www.cshl.org

For orders from Europe, the Middle East, and Africa, British pound pricing is provided. Orders are fulfilled and shipped from Cold Spring Harbor Laboratory Press–Europe c/o Lavis Marketing, 73 Lime Walk, Headington, Oxford OX3 7AD, U.K. Phone: +44 (0) 1865 741541. FAX: +44 (0)1865 750079. E-mail: cshl.press.europe@cshl.org. WWW site: WWW: cshl.press.co.uk

Contents

Preface

The major reason for the publication of a second edition of *Translational Control*, under the augmented title *Translational Control of Gene Expression*, is the remarkable progress that has been made in the field in the past five years, since the appearance of the first edition. During this time there has been a new wave of interest in protein synthesis, which is abundantly reflected in the many different chapters of this book. There has been excellent progress in understanding the mechanisms of initiation, elongation, and termination of translation; the mechanisms of translational control during development, in response to extracellular stimuli (including signal transduction pathways that control translation rates); and how the translational machinery is affected during virus infection and in disease. The idea that translational control plays a larger role than generally appreciated is further bolstered by recent studies using genomics and proteomics techniques, which show a large discrepancy between mRNA and protein levels in cells.

The first edition of *Translational Control* was well received by translation researchers and by the scientific community at large. Reviews of the book were very favorable and chapters have been widely cited. In fact, there were two hardcover printings, and after the second printing ran out, Cold Spring Harbor Laboratory Press decided to publish the book in a paper cover format. Then, John Inglis from Cold Spring Harbor Laboratory Press approached us at the CSH Translational Control Meeting in 1998 and requested that we prepare a second edition of this treatise. Somewhat reluctantly, recalling all of the hard work we put into the first edition, we agreed to undertake this important task. In addition to the intellectual satisfaction achieved by editing the book, we fondly remembered the fun and pleasure in interacting among ourselves and with the many contributors. However, we were not interested in simply updating the content of the first edition, but rather wished to take a fresh look at the field. As a result, the book has been extensively revised, a number of new topics were added, and topics that did not change greatly over the past 5 years were omitted. Our hard work was well worth the effort in this second edition as witnessed by all the new discoveries and exciting developments in the field.

The format of the current book differs from that of the first edition. The first 8 chapters are broad and are intended to provide an overview of major themes in translational control, including the basic mechanisms and factors involved in translation initiation and elongation, the mechanism and regulation of Met-tRNA binding, the regulation of mRNA binding to ribosomes, translational control of developmental decisions, and virus–cell interactions. The next 28 chapters are focused reviews on a wide range of research topics in translation and translational control.

In the preface to the first edition of this book, we expressed our optimism that studies on translational control would highlight its importance in regulating gene expression in cell proliferation, development, and differentiation and for integrating the various metabolic pathways in the cell. Our optimism was fully justified, as attested by the many chapters in this book that document numerous new and exciting examples of translational control in a wide range of biological systems.

We assembled this monograph in the hope that it would be helpful to students entering the field, as well as to researchers working on the regulation of gene expression who come to realize that translational control plays a key role in their systems. As a way of emphasizing the growing recognition of this role, the second edition has been retitled *Translational Control of Gene Expression*. We are confident that continued research in the translational field will yield a wealth of information and many surprises, and that it will increase our understanding of the function of many biological systems.

We are grateful to all the authors for their thoughtful reviews and for their patience and good humor in dealing with our numerous editorial requests. We thank the staff of the Cold Spring Harbor Laboratory Press, John Inglis, and Patricia Barker. We would be remiss if we did not single out Joan Ebert for her cheerful support and unflagging encouragement in the preparation and completion of this monograph.

N. Sonenberg
J.W.B. Hershey
M.B. Mathews

June, 2000

1

Origins and Principles of Translational Control

Michael B. Mathews
Department of Biochemistry and Molecular Biology
New Jersey Medical School
University of Medicine and Dentistry of New Jersey
Newark, New Jersey 07103

Nahum Sonenberg
Department of Biochemistry
McGill University, Montreal
Quebec H3G 1Y6, Canada

John W.B. Hershey
Department of Biological Chemistry
University of California, School of Medicine
Davis, California 95616

Proteins occupy a position high on the list of molecules important for life processes. They account for a large fraction of biological macromolecules—about 44% of the human body's dry weight, for example (Davidson et al. 1973); they catalyze most of the reactions on which life depends; and they serve numerous structural, transport, regulatory, and other roles in all organisms. Accordingly, a large proportion of the cell's resources is devoted to translation. The magnitude of this commitment can be appreciated in both genetic and biochemical terms.

Translation is a sophisticated process requiring extensive biological machinery. One way to estimate the minimal amount of genetic information needed to assemble the protein synthetic machinery would be to compile a "parts list" of essential proteins and RNAs and add up their sizes. However, this approach entails several questionable assumptions about the identity of the essential components and their minimal sizes. An alternative approach is to examine the genomes of simple living organisms. The smallest known cellular genome, that of the parasitic bacterium

© 2000 Cold Spring Harbor Laboratory Press 0-87969-568-4/00 $5 + .00

Mycoplasma genitalium, encodes 480 proteins, of which no fewer than 101 have been ascribed a function in protein synthesis (Fraser et al. 1995; Hutchison et al. 1999). Excluding genes that are less directly involved in translation per se (e.g., those for proteases and peptidases), *M. genitalium* has about 90 genes encoding proteins in the translation system. Additionally, 37 genes specify RNA molecules, chiefly ribosomal and transfer RNAs (rRNA and tRNAs), which fill critical translational roles. Thus, some 127 genes, a quarter of the *M. genitalium* complement, are involved in protein synthesis. Nearly all of these are held in common with *M. pneumoniae*, which has a somewhat larger genome (Himmelreich et al. 1996), and have been shown by transposon mutagenesis to be essential for growth under laboratory conditions (Hutchison et al. 1999). Discounting genes that are dispensable for mycoplasma growth in the laboratory, it can be calculated that the fraction of genes in a theoretical minimal genome that is devoted to translation may be as high as 35–45%.

Such a heavy genomic investment is not surprising in view of the high proportion of a cell's resources and energy budget that is allotted to translation. Protein synthesis consumes 5% of the human caloric intake but as much as 30–50% of the energy generated by rapidly growing *Escherichia coli* (Meisenberg and Simmons 1998). A portion of this is accounted for by the substantial input of energy required during translation itself (4 high-energy bonds per peptide bond, or ~25 kcal/mole, plus additional consumption for initiation and termination). Extensive resources are invested in the translation system—the ribosomes, tRNAs, and enzymes that constitute the physical plant for making proteins. A rapidly growing yeast cell, for example, contains nearly 200,000 ribosomes occupying some 30–40% of its cytoplasmic volume (Warner 1999). Growth alone demands that the yeast cell produce 2000 ribosomes per minute, an operation that absorbs ~60% of its transcriptional activity in manufacturing rRNA, as well as a large fraction of its translational capacity, since ribosomal protein messenger RNAs (mRNAs) account for almost one-third of the cell's mRNA population (for review, see Warner 1999). It would be surprising if a biological process of this importance were not closely monitored and regulated.

ORIGINS OF TRANSLATIONAL CONTROL

The central idea of translational control is that gene expression can be regulated by the efficiency of utilization of mRNA in specifying protein synthesis. This notion emerged only a few years after the articulation of the central dogma of molecular biology (Crick 1958) and very soon after the formulation of the messenger hypothesis. In 1961, Jacob and Monod perceived that "the synthesis of individual proteins may be provoked or suppressed within a cell, under the influence of specific external agents, and . . . the relative rates at which different proteins are synthesized may be profoundly altered, depending on external conditions." They pointed out that such regulation "is absolutely essential to the survival of the cell," and went on to advance the concept of an unstable RNA intermediary between gene and protein as a key feature of their elegant model for transcriptional control (Jacob and Monod 1961). The idea that this mRNA could be subject to differential utilization depending on the circumstances was accorded scant attention in the bacterial culture of the time, but it was taken up enthusiastically by workers in other fields, to the extent that 10 years later, one writer could allude to the "now classical conclusion" that eggs contain translationally silent mRNA that is activated upon fertilization (Humphreys 1971).

The term Translational Control was certainly in use as early as 1968, by which date at least four clearly distinct exemplars had been recognized and were already coming under mechanistic scrutiny. The groundwork for these four paradigms—developing embryos, reticulocytes, virus- and phage-infected cells, and higher cells responding to stimuli ranging from heat to hormones and starvation to mitosis—had all been laid by the middle of the 1960s. They founded a thriving and expanding field of study that has advanced from its largely eukaryotic origins to embrace prokaryotes (although not yet the archaebacteria, as far as we are aware).

The Early History of Translation

The genesis of the translational control field took place at a time when translation itself was in its infancy; many (although not all) of the reactions had been observed, but most of the components were not yet characterized and mechanistic details were essentially unknown. To place the origins of translational control in context, we briefly outline the development of protein synthesis.

Biochemical investigations of the process began in the latter half of the 1950s, at the same time as the view of proteins as unique, nonrandom linear arrays of just 20 amino acid residues was solidifying. Enabled by the availability of radioactive amino acids as tracers, biochemistry ran ahead of genetics, as it continued to do in this field until the advent of cloning and the systematic exploitation of the yeast system, which began to make their mark in the 1980s. Siekevitz and Zamecnik produced a cell-free preparation from rat liver that incorporated amino acids into protein, showing that energy was required in the form of ATP and GTP (Siekevitz 1951). The system was refined by stages and resolved into subfractions including a microsomal fraction that included ribosomes attached to fragments of intracellular membrane (for review, see Zamecnik 1960). It is salutary to recall that this was accomplished in advance of an understanding of the central role of RNA in the flow of genetic information to protein and in an era when theories of protein synthesis via enzyme assembly and peptide intermediates were entertained along with template theories (Campbell and Work 1953). Further biochemical work demonstrated that the ribonucleoprotein particles later called ribosomes comprise the site of protein synthesis, but it was not until the early 1960s that polysomes were observed and their function appreciated in light of the messenger hypothesis (Marks et al. 1962; Warner et al. 1963).

At much the same time, the role of aminoacyl-tRNA was being established. The existence of an intermediate, activated amino acid state was detected (Hultin and Beskow 1956) and characterized (Hoagland et al. 1959), then understood as the physical manifestation of the adapter RNA predicted on theoretical grounds (Crick 1958). Once its function had been realized, the name transfer RNA rapidly displaced the original descriptive term, "soluble" RNA. Later, chemical modification of the amino acid moiety of a charged tRNA confirmed that it is the RNA component which decodes the template (Chapeville et al. 1962). Thus, responsibility for the fidelity of information transfer from nucleic acid to protein rests in part on the aminoacyl-tRNA synthetases, which became the first macromolecular component of the protein synthetic apparatus to be purified (Berg and Ofengand 1958). These, together with the other enzymes, or protein "factors" as they became known, were steadily characterized and purified such that nearly all of the protein components have been known for more than 20 years. Yet, new ones continue to be reported (e.g., eIF5B; see Chapters 2 and 9), and even today there is no certainty that the full complement of protein factors involved in translation has been identified.

It was genetics rather than biochemistry that supplied the missing cornerstone of the protein synthetic system, mRNA. According to the messenger hypothesis, the ribosomes and other components of the protein synthesis machinery constitute a relatively stable decoding and synthetic apparatus that is programmed by an unstable template (Jacob and Monod 1961). This insight soon received confirmation in bacteria (Brenner et al. 1961; Gros et al. 1961) and in bacterial cell-free systems. The discovery that poly(U) can direct the synthesis of polyphenylalanine (Nirenberg and Matthaei 1961) was particularly fruitful, greatly speeding the elucidation of the genetic code by the mid 1960s. Because of the greater stability of most eukaryotic mRNAs, the applicability of the messenger hypothesis to higher cells was less readily apparent. Nonetheless, the existence of a class of rapidly labeled RNA, heterogeneous in size and with distinct chromatographic properties, was recognized. Its essential features as informational intermediary were confirmed and it was universally accepted several years before the discovery (in the early 1970s) of 5′ caps and 3′ poly(A) tails, the modern hallmarks of most eukaryotic mRNAs. The mRNA concept immediately revolutionized thinking about gene expression in all cells.

To appreciate the pace at which protein synthesis advanced during the decade of the 1960s, it is instructive to compare the Cold Spring Harbor Symposium volume of 1962 (on Cellular Regulatory Mechanisms) with that of 1970, a much thicker book devoted to a narrower topic (the Mechanism of Protein Synthesis). By the end of the decade, much of the translational apparatus had been characterized (although much also remained to be done), many problems of regulation had been laid out, and translational control came to receive increasing attention.

General Features of Translational Control

In a multistep, multifactorial pathway like that of protein synthesis, regulation can be exerted at many levels. Examples of translational control are indeed found at different levels, but the overwhelming preponderance of known instances—including all of the earliest cases recognized—is at the level of initiation. This empirical observation conforms to the biological (and logical) principle that it is more efficient to govern a pathway at its outset than to interrupt it in midstream and have to deal with the resultant logjam of recyclable components and the accumulation of intermediates as by-products. Nevertheless, well-characterized cases do occur at later steps in the translational pathway, especially at the elongation level, where it seems that a translational block may be imposed as a safety measure to halt further peptide bond formation.

One of the chief virtues of translation as a site of regulation is that it offers the possibility of rapid response to external stimuli without invoking nuclear pathways for mRNA synthesis and transport. Predictably, the first cases to be recognized were those in which it was simplest to establish, if it was not self-evident, that transcription and other nuclear events were not responsible. By the same token, the relative scarcity of prokaryotic examples and their generally later recognition can be largely attributed to the lack of a nuclear barrier between the sites of mRNA synthesis and translation. The greater speed of macromolecular synthesis in bacteria and their lesser dependence on mRNA processing are other factors. These circumstances allow a coupling of transcription and translation that all but obviates the need for translational control. That it occurs at all in prokaryotes is due to the exigencies of particular circumstances and to the potency of translational control mechanisms.

The earliest cases of translational control to be explored in depth, in fertilized invertebrate eggs and mammalian reticulocytes, were those in which the departure from the transcription-based regulatory model was the most obvious and extreme. Protein synthesis is abruptly turned on (in fertilized eggs) and off (in iron-starved reticulocytes) in the absence of ongoing transcription. A further distinction that made it easier to define and study these two particular cases is that the regulation is apparently indiscriminate in that it affects protein synthesis generically, rather than the synthesis of specific proteins. Not all translational controls are of this type, however. A distinction is often drawn between global and selective controls (sometimes referred to, rather misleadingly, as quantitative and qualitative controls). Global controls, such as those operating in eggs and reticulocytes, impact the entire complement of mRNAs within a cell, switching their translation on or off or modulating it by degrees in unison. This kind of regulation is usually implemented by substantial alteration in the activity of general components of the protein synthesis machinery that act in a nonspecific manner. Selective controls, on the other hand, affect a subset of the mRNAs within a cell, in the extreme case a single species only. This can be accomplished through mechanisms that target ligands to individual mRNAs or classes of mRNA, but it is achieved more commonly by exploiting the differential sensitivity of mRNAs to more subtle changes in the activity of general components of the translation system, e.g., eIF4E (Chapter 6) or eIF2 (Chapter 5). Although examples of all these exist and are discussed at length in this monograph, in the context of the historical origins of translational control, it should come as no surprise that the earliest examples were mainly of the global variety and that (with notable exceptions) definitive evidence in favor of selective translational control accumulated more slowly.

PARADIGMS OF TRANSLATIONAL CONTROL

In large part, the origins of translational control can be traced to the confluence of four early streams of investigation, which still continue to flow. Their early courses are described below, followed by an example involving elongation control.

Sea Urchin Eggs

The eggs of sea urchins and other invertebrates provide a striking example of regulated gene expression which, it was quickly realized, did not harmonize with the emerging theme of transcriptional control. These cells are essentially quiescent until they are galvanized into action by fertilization. Egg ribosomes synthesize protein at a very low rate but are triggered to incorporate amino acids within a few minutes of fertilization (Hultin 1961). Although the rate of protein synthesis accelerates rapidly after fertilization, there is little or no concomitant RNA synthesis (Hultin 1961; Nemer 1962; Gross and Cousineau 1963). Translation in enucleated eggs can be activated parthenogenetically (Denny and Tyler 1964). Moreover, actinomycin D fails to block the first wave of increased translation, which lasts for several hours, and both cell division and many morphogenetic events proceed unimpeded by the transcriptional inhibition. A second wave of increased protein synthesis is prevented by actinomycin D, however, presumably because this wave does depend on new mRNA synthesis (Gross et al. 1964). Such observations are explained by the fact that the eggs contain preexisting mRNA in a form that is not translated until some stimulus dependent on fertilization is received. In principle, the limitation could be due to a deficiency in the translational machinery, but unequivocal evidence in this direction has been more difficult to obtain. For example, a comparison of polysome sizes and translation rates in eggs and embryos did not disclose any defect in the apparatus itself (Humphreys 1969). On the other hand, a good deal of evidence points to a defect in the availability of mRNA. Consistent with the conclusion that mRNA is largely sequestered in eggs, deproteinized egg RNA can be translated in a cell-free system (Maggio et al. 1964). The ribosomes from eggs—unlike those from embryos—display little intrinsic protein synthetic activity, although they are able to translate added poly(U) (Nemer 1962; Wilt and Hultin 1962), suggesting that they possess latent translational capacity. Egg mRNA exists in a masked form: Cytoplasmic messenger ribonucleoprotein (mRNP) particles have been observed (Spirin and Nemer 1965), and some studies even indicated that the template could be activated by trypsin treatment, presumably by removing masking proteins (Monroy et al. 1965). Since the assembly of masked mRNP complexes must take place during oogenesis, the sea urchin system exemplifies a reversible process of mRNA repression and activation. Developments in this arena are discussed in Chapters 7 and 27.

Reticulocytes

These immature red cells have endowed researchers with a unique and especially dynamic system for studying the mechanism and control of translation. Because mammalian reticulocytes are enucleate, unlike those of most vertebrates, it was taken for granted that the regulation of protein production would be exercised at the translational level, an assumption that has been borne out in numerous studies. More than 90% of the protein made in the reticulocyte is hemoglobin, which consists of two α-globin and two β-globin chains together with four molecules of heme, an iron-containing porphyrin. In the intact rabbit reticulocyte, the synthesis of heme parallels that of globin (Kruh and Borsook 1956), and subsequent work showed that globin synthesis is controlled by the availability of heme or of ferrous ions (Bruns and London 1965). The phenomenon was made experimentally accessible by the development of the highly active unfractionated reticulocyte lysate translation system (Lamfrom and Knopf 1964), which became the forerunner of the widely used messenger-dependent system of Pelham and Jackson (1976). Regulation by heme is reversible in intact cells, and, to a limited extent, the repression of protein synthesis that ensues in the reticulocyte lysate soon after heme deprivation can also be rescued by restoring the heme level. When globin synthesis is inhibited in cells or extracts, the polysomes dissociate to monosomes (Hardesty et al. 1963; Waxman and Rabinovitz 1966), arguing that heme is involved in regulating translation initiation. Contrary to intuitive expectation, there is no necessary linkage between the role of heme as the prosthetic group of globin and its role as translational regulator. The effects of heme deprivation on protein synthesis in the reticulocyte or its lysate are mimicked by unrelated stimuli such as double-stranded RNA (dsRNA) and oxidized glutathione (Ehrenfeld and Hunt 1971; Kosower et al. 1971) and extend to all mRNAs in the reticulocyte lysate (Mathews et al. 1973). Such observations imply that a general mechanism of translational control is being invoked: In each of the conditions under which protein synthesis is down-regulated, inhibitors—now known to be the eIF2 kinases HRI and PKR—are activated (for reviews, see Chapters 13 and 14). By 1977, a unifying scheme could be advanced (Farrell et al. 1977), centering on the phosphorylation of the α-subunit of initiation factor eIF2 and the loading of the 40S ribosomal subunit with Met-$tRNA_i$. This mechanism has been found to have wide applicability in cells and tissues responding to a range of stimuli (see also Chapters 5 and 15).

Virus-infected Cells

During the 1960s, it came to be appreciated that cellular protein synthesis is suppressed during infection with many viruses (see Chapters 8, 31–35). This inhibition may begin before the onset of viral protein synthesis and without any apparent interference with cellular mRNA production or stability. In poliovirus infection, an early example, the shutoff of host-cell translation can be complete within 2 hours after infection and is followed by a wave of viral protein synthesis (Summers et al. 1965). The first phase is accompanied by the reduction of polysomes to monosomes without any effect on the elongation or termination phases of protein synthesis (Penman and Summers 1965; Summers and Maizel 1967). In the second phase, virus-specific polysomes form (Penman et al. 1963), evidence that initiation has become selective for a class of mRNA—in this case viral, rather than cellular. Later studies showed that cellular mRNA remains intact in the infected cell (Leibowitz and Penman 1971) and is translatable in a cell-free system, although it is not translated in the infected cell. Furthermore, the inhibition extends to the mRNAs of several other viruses introduced together with poliovirus in a double infection (Ehrenfeld and Lund 1977), indicative of a general effect. Although circumstantial evidence aroused suspicions that viral dsRNA and PKR might be responsible for the phenomenon, later work incriminated a modification of the cap-binding complex, eIF4F. Cleavage of the eIF4G subunit of this complex prevents cap-dependent initiation on cellular mRNAs but does not interfere with initiation on the viral mRNA, which occurs by internal ribosome entry (see Chapters 4, 6, 8, and 31).

Bacteriophage f2 provided the first evidence for translational control in a prokaryotic system, as well as the first clear case of mechanisms specific for the synthesis of individual protein species. The phage RNA genome encodes four polypeptides, the maturation protein, coat protein, lysis protein, and replicase, that are initiated individually but produced at dissimilar rates. Several regulatory interactions among them are now known. One was revealed by the observation that a nonsense mutation early in the cistron coding for the viral coat protein down-regulates replicase synthesis (Lodish and Zinder 1966). Apparently, passage of ribosomes through a critical region of the coat protein cistron is required to melt RNA structure and allow replicase translation. In contrast, a second nonsense mutation leads to overproduction of the replicase, suggesting that the coat protein acts as a repressor of replicase translation. This inference has been amply confirmed, and the binding of the coat protein to the hairpin structure containing the replicase AUG has become one of the best-characterized cases of RNA–protein interaction (Witherell et al. 1991). Subsequent studies have disclosed translational control mechanisms in the DNA phages as well as in bacterial genes themselves (see Chapter 8), but it was eukaryotic systems that made most of the early running.

Physiological Stimuli

The cells and tissues of higher organisms have been reported to regulate the expression of individual genes or of whole classes of genes at the translational level in response to a wide variety of stimuli or conditions. These include cell state changes, such as mitosis (Steward et al. 1968; Hodge et al. 1969; Fan and Penman 1970) and differentiation (Heywood 1970); stress resulting from heat shock (McCormick and Penman 1969), treatment with noxious substances or the incorporation of amino acid analogs (Thomas and Mathews 1984); and normal cellular responses to ions (Drysdale and Munro 1965) and hormones (Eboué-Bonis et al. 1963; Garren et al. 1964; Martin and Young 1965; Tomkins et al. 1965). Not in every case was the evidence for regulation at the translational level complete, and in a few instances, the trail has gone cold or been erased upon more detailed examination, but the accumulated volume of information added conviction to the view that translational control is both widespread and important. One of the chief stumbling blocks in this arena lay in determining that the level at which control was exerted was indeed translational. This can be a difficult task in nucleated cells, let alone in a tissue or whole animal (or plant), and it was addressed in various ways. A popular approach exploited selective inhibitors of transcription or translation, such as actinomycin D and cycloheximide, but the results were liable to be complicated (if not confounded) by side effects of the drugs or their indirect sequelae in complex systems. Another argument that could be made for an effect at the translational level, although not without some reservations, came from its rapidity (see below). Timing alone cannot provide a definite assignment, however, and the most convincing proofs often came from subsequent investigations of the underlying biochemical processes—for example, by demonstrating changes in polysome profiles or initiation factor phosphorylation states as discussed below and in Chapters 6, 13–17, 20, and 23. The ultimate goal is to achieve an understanding of the regulatory mechanisms set in train by the stimuli applied, and within this wide array of phenomena lie many of the challenges for the future.

Secretory Proteins

No overview of the principal themes of translational control would be complete if it dwelt exclusively on the initiation phase. One of the best-studied examples of regulation during the elongation phase is found in the synthesis of proteins that are destined for secretion. These are made on polysomes that are attached to the endoplasmic reticulum, isolated from cellular homogenates in the form of microsomes. In the early 1970s, it began to seem likely that ribosomes become associated with cell membranes only after protein synthesis has been initiated (Lisowska-Bernstein et al. 1970; Rosbash 1972). Contemporaneously, the existence of what came to be called a signal peptide was reported on an immunoglobulin light chain (Milstein et al. 1972) and other secreted proteins (Devillers-Thiery et al. 1975). These findings lent substance to the signal hypothesis (Blobel and Sabatini 1971), which suggested that an amino-terminal sequence might ensure secretion, and prompted the development of cell-free systems that enabled the biochemical dissection of the secretory pathway (Blobel and Dobberstein 1975). One of the surprising discoveries to emerge was the involvement in secretion of a ribonucleoprotein particle, the signal recognition particle (SRP), which interacts with the signal peptide, the ribosome, and the endoplasmic reticulum. Remarkably, binding of the SRP to a nascent signal peptide protruding from the ribosome causes translational arrest in the absence of cell membranes (Walter and Blobel 1981). This elongation block is relieved when the ribosome docks with the endoplasmic reticulum, allowing the protein chain to be completed and simultaneously translocated across the lipid bilayer. It has been speculated that this mechanism serves to ensure cotranslational protein export and to prevent the accumulation of secretory proteins in an improper subcellular compartment (the cytosol). Interestingly, a similar rationale has been offered to account for control at the elongation level during heat shock (for review, see Chapter 17). In this situation, it has been proposed that a translational arrest is imposed to prevent the synthesis of proteins that might be folded abnormally. Thus, elongation blocks might be used under exceptional circumstances to preserve cellular integrity when it is threatened by the production of protein at the wrong time or in the wrong place, or perhaps in the event of a sudden shortage of energy or an essential metabolite.

WHAT LIMITS PROTEIN SYNTHESIS IN PRINCIPLE?

Given that translational controls are so widespread in eukaryotic cells, it is appropriate to examine the fundamental principles on which these controls are based. Translational control is defined as a change in the rate (efficiency) of translation of one or more mRNAs, i.e., the number of completed protein products changes per mRNA per unit time. It is generally believed that during protein synthesis, the number of protein chains initiated is about the same as the number of proteins completed; in other words, few nascent polypeptides abort and fall off the ribosome (Tsung et al. 1989). Therefore, under steady-state conditions, the number of initiation events per unit time approximates the number of protein products produced during the same time interval. It follows logically that the rate of protein synthesis is determined by the number of initiation events, i.e., the rate of initiation. What determines the number of initiation events per unit time? Four major parameters may influence or define the rate of protein synthesis. Each is considered briefly below.

The Activity of the Protein Synthesis Machinery

Numerous examples exist of cells that possess ribosomes and mRNAs in excess of those actively engaged in protein synthesis. This may occur if a single translational component (e.g., a soluble factor) is limiting in amount or if one or more components have reduced specific activities. Such regulation frequently involves the phosphorylation status of translational components, as detailed in numerous chapters in this monograph. Regulation of the overall activity of the translational apparatus is expected to affect the translation of essentially all mRNAs. As argued earlier by Lodish (1976), down-regulation of the initiation steps that occur prior to the binding of mRNAs is expected to lead to greater inhibition of those mRNAs whose initiation rate constants are relatively low ("weak" mRNAs), as compared to "strong" mRNAs. Reciprocally, activation of such steps may stimulate more greatly the translation of weak mRNAs. Alteration of the activities of components that interact with mRNAs and affect their binding to ribosomes also would be expected to generate differential effects on the translation of the mRNA population (Godefroy-Colburn and Thach 1981). The mechanisms affecting mRNA binding and differences in the translational efficiency of specific mRNAs are reviewed in Chapter 6.

The Rate of Elongation

The initiation rate on an mRNA can be inhibited if a ribosome, having already initiated, vacates the initiation region too slowly. A ribosome bound at the AUG initiation codon occupies about 12–15 nucleotides (4–5 codons) downstream from the AUG and about 20 nucleotides upstream. Another ribosome can occupy the initiation site only after the first ribosome has moved about 7 codons down the mRNA. When the time needed to vacate the initiation region approaches or exceeds the time required for initiation, the elongation rate becomes limiting. In general, it is believed that the elongation rate is about the same for all mRNAs (3–8 amino acids per second per ribosome in eukaryotes, faster in prokaryotes), because measurements of a few specific examples gave similar results in this range (Lodish and Jacobsen 1972; Palmiter 1974). Nevertheless, the rate of elongation is not uniform throughout the coding region of an mRNA, as pausing may occur at specific locations, possibly due to the occurrence of rare codons or RNA secondary structure (Wolin and Walter 1988). If ribosome pausing occurs such that it impedes initiation, mRNA efficiency is decreased. The question of which translation phase is rate-limiting, initiation or elongation/termination, is addressed in greater detail below.

The Amount or Efficiency of mRNAs

The level of mRNA in the cytoplasm is determined by the rate of transcription, the proportion of primary transcripts that are processed and transported into the cytoplasm, and the degradation rate of cytoplasmic mRNAs. In actively translating mammalian cells, mRNAs often are found entirely in polysomes, as shown for actin (Endo and Nadal-Ginard 1987); thus, the rates of synthesis of such specific proteins are mRNA-limited. However, total mRNA in the cytoplasm frequently appears to be present in excess, with about 30% of the mRNA in cultured cells present as free mRNP particles (Geoghegan et al. 1979; Kinniburgh et al. 1979; Ouellette et al. 1982). Therefore, the level of mRNA appears not to limit the overall number of translational initiation events in these cells. In cells exhibiting low translational activity, many mRNAs are repressed and apparently unavailable to the translational apparatus (masked), as seen most dramatically in oocytes and unfertilized eggs as described above, but also in somatic cells in culture (Lee and Engelhardt 1979). Such repression sometimes appears to be all or none, as some mRNAs are distributed bimodally in polysome profiles; a fraction of the specific mRNA is completely repressed (nontranslating mRNP particles), whereas a portion is actively translated as large polysomes (Yenofsky et al. 1982; Agrawal and Bowman 1987). In instances of specific regulation of protein synthesis, mRNA repression and availability to the translational apparatus likely have a dominant role, for example, in the translation of ferritin mRNA (see Chapter 21) and ribosomal protein mRNAs (see Chapter 22). Furthermore, individual activated mRNAs differ greatly in their efficiencies of translation as deduced from polysome sizes, thereby contributing to regulation of gene expression. These innate efficiencies are determined in large part by the primary and higher-order structures of the mRNAs (for reviews, see Chapters 2, 4, and 8).

The Abundance of Ribosomes

The cellular levels of ribosomes may be rate-limiting under some circumstances. Cells active in protein synthesis, for example, liver cells from fed rats, engage 90–95% of their ribosomes in protein synthesis (Henshaw et al. 1971), suggesting that still higher rates of protein synthesis might have been possible were there a greater number of ribosomes. On the other hand, in translationally repressed cells, such as liver cells from fasted rats (Henshaw et al. 1971) or in quiescent cells in culture (Duncan and McConkey 1982; Meyuhas et al. 1987), fewer than half of the ribosomes may be actively translating mRNAs. The level of ribosomes surely is not limiting in these cells, since a rapid increase in the rate of protein synthesis can be induced within 20 minutes, before the assembly of more ribosomes is possible (Duncan and McConkey 1982). Translation also may be limited by the levels (as opposed to specific activities) of other components of the translational apparatus, e.g., eIF2 and eIF4F, the latter likely through its eIF4E subunit (see Chapter 6). When amino acids become limiting, global protein synthesis is rapidly repressed by inhibiting the activity of initiation factors (Clemens et al. 1987; Chapter 16).

WHICH PHASE OF PROTEIN SYNTHESIS IS RATE-LIMITING AND REGULATED?

The analysis above identifies three ways in which the rate of protein synthesis may be limited and thus regulated over a relatively short time period (on the order of minutes): the rate of initiation, the rate of elongation/termination, and the repression/activation of mRNAs/mRNPs. How is the rate of protein synthesis measured and how is the rate-limiting step identified? The overall rate of protein synthesis can be measured by assaying the time course of incorporation into protein of radioactively labeled amino acids added to the culture medium. The method is complicated only by the uncertainty of the specific radioactivities of the precursors within cells, as intracellular de novo synthesis of amino acids and degradation of proteins may influence these values. A second method measures the absolute number of active ribosomes and the elongation rate, from which the number of amino acids incorporated per unit time can be calculated. The elongation rate is obtained by dividing the number of amino acids in the average protein by the ribosome transit time (Fan and Penman 1970), the time it takes to translate an average-sized mRNA. The fraction of total ribosomes that is active is assessed by high-salt sucrose gradient centrifugation of cell lysates to generate a polysome profile: Active ribosomes in polysomes are separated from nonactive, free ribosomal subunits. The second method, although more laborious than the measurement of labeled amino acid incorporation, is not complicated by uncertainties of amino-acid-specific radioactivities. Both methods serve to analyze global rates of protein synthesis. The relative synthesis rates of specific proteins can be measured by radioactively labeling proteins, followed by immunoprecipitation or fractionation of proteins by high-resolution two-dimensional gel electrophoresis.

Which phase of protein synthesis is rate-limiting, initiation or elongation/termination? Although most mRNAs are thought to be limited by their initiation rate, others are limited by the rate of elongation/termination. Therefore, the question is best addressed to specific mRNAs rather than to the whole population. Insight into which phase is rate-limiting is gained by an examination of polysome profiles, where the specific mRNA is located in the sucrose gradient fractions by hybridization techniques. The rate of initiation, i.e., the number of initiation events per minute, can be calculated from the number of ribosomes translating an mRNA (polysome size) and the ribosome transit time (the time required for the ribosome to traverse the mRNA). As elegantly determined for ovalbumin mRNA in chick oviducts (Palmiter 1975), ovalbumin polysomes average 12 ribosomes and the ribosome transit time is 1.3 minutes, giving a rate of initiation of 9.2 events per minute (or one initiation every 6.5 seconds). Since the elongating ribosome requires only about 2 seconds to vacate the initiation site, it is clear that the initiation rate is slower than potentially possible and thus is rate-limiting. Parenthetically, if the number of mRNA molecules in the polysomes is known, an absolute rate of specific protein synthesis can be calculated.

A second way to determine whether initiation or elongation/termination is rate-limiting for an mRNA is to treat cells with low concentrations of an elongation (e.g., cycloheximide and sparsomycin) or initiation (e.g., pactamycin) inhibitor. If translation of the specific mRNA is limited by the elongation rate, its synthesis will be sensitive to the inhibitors of elongation. Conversely, if initiation is rate-limiting, such mRNAs will be insensitive to elongation inhibitors but sensitive to initiation inhibitors. For example, when mRNAs encoding α- and β-globin (Lodish and Jacobsen 1972) or reovirus proteins (Walden et al. 1981) were analyzed, initiation was the sensitive step. Because the majority of mRNAs in cells are resistant to low concentrations of cycloheximide, it is thought that the translation of most mRNAs is limited at the initiation phase.

Further evidence that the rate of initiation limits the translation of most mRNAs is obtained by examining polysome sizes from sucrose gradients. On the average, ribosomes in polysomes occur once every 80–100 nucleotides. For example, the average polysome size for globins is about 5 ribosomes per mRNA, or 1 ribosome per 90 nucleotides. When protein synthesis is inhibited by cycloheximide such that elongation becomes rate-limiting, polysomes increase in size (to more than 12 ribosomes per globin mRNA, for example). Therefore, polysome densities of one ribosome per 30–40 nucleotides are possible. This approaches the limit for close packing, since a ribosome occupies about 30 nucleotides of mRNA. That average polysome densities are much less is due to the relatively low rate of initiation.

Changes in the size (number of ribosomes per mRNA) or amounts (amplitude) of polysomes may be diagnostic of the phase of global protein synthesis that is being modulated. If the size of polysomes decreases, either initiation is inhibited or elongation/termination is stimulated, or a combination of both occurs. Conversely, an increase in polysome size can be caused by an increased rate of initiation and/or a decreased rate of elongation/termination. To interpret polysome profiles unambiguously, it is advisable to measure the elongation rate by determining the ribosome transit time and average length of mRNAs being translated. In cases where the overall rate of protein synthesis is repressed and polysomes are smaller, initiation has clearly been inhibited. Regulation of a specific mRNA is readily evaluated by these methods, since the average size of its polysomes is readily determined by hybridization techniques with cloned probes. Repression or activation of protein synthesis need not always affect polysome size. Instead, the number of translating mRNAs may be affected by masking mRNAs or mobilizing them into polysomes. In this case, there is a change in the *amount* (i.e., amplitude) of polysomes, but the average *size* of the polysomes may remain the same.

Are there cases where the elongation rate is regulated? Examination of a number of specific mRNAs shows that rather modest changes in the rate of elongation are found following treatment of cells with hormones and other agents (Chapter 24). A dramatic example is the fivefold stimulation of the rate of elongation of tyrosine aminotransferase seen when rat hepatoma cells are treated with dibutyryl-cAMP (Roper and Wicks 1978). Similarly, the elongation rate on vitellogenin mRNA drops about fourfold when cockerel liver explants are treated with 17β-estradiol (Gehrke and Ilan 1987). Even small changes in the elongation rate will affect the efficiencies of those mRNAs that are elongation-limited; whether or not moderate inhibition of elongation affects initiation-limited mRNA expression depends on the degree that initiation is limiting.

TARGETS AND MECHANISMS OF TRANSLATIONAL CONTROL

Having defined the rate-limiting steps in the protein synthesis pathway, we now turn to the means by which its regulation is accomplished in the cell. Translational control is realized through multiple mechanisms that target structural features of the mRNA and *trans*-acting components; the latter may be either protein or (less commonly) RNA in nature. The survey that follows takes stock of the principal targets of translational control and the mechanisms which they coordinate, giving reference to chapters in this monograph where these topics are considered in greater detail.

mRNA

The intrinsic translational efficiency of an mRNA is dependent on several *cis*-acting elements, which also have critical roles in the regulation of mRNA utilization, as discussed in many chapters of this work. It is convenient to divide the *cis*-acting elements into two categories: those that act alone or with general translation factors; and those whose actions are mediated by specific *trans*-acting factors.

In prokaryotes, the first category is of overriding importance. Translational efficiency is heavily influenced by mRNA primary structure, especially the Shine-Dalgarno sequence, as well as by the degree of secondary structure that can be modulated by various mechanisms (Chapters 2, 4, and 8). In eukaryotes, *cis*-acting elements distributed along the length of the mRNA modulate translational efficiency. Primary structure, notably the 5′ cap, the sequence flanking the initiator AUG (its "context"), and the presence of upstream AUG triplets all determine translational efficiency (Chapters 2 and 4). Secondary structure, particularly in the 5′ -untranslated region (5′UTR), can also have a determinative role. Upstream open reading frames (uORFs) participate in translational control in yeast and higher eukaryotes. Regulation of the translation of uORF-containing mRNAs is dependent on many factors, including the amino acid sequence encoded by the uORF, the length of intercistronic regions, and the sequence context of the termination codon of the uORFs (Chapters 5 and 18).

Within the coding sequence of some mRNAs are elements that signal ribosome frameshifting, hopping, termination codon readthrough, and the incorporation of selenocysteine (for review, see Chapters 11, 25, and 26). Some of these processes are known to be regulated. For example, ribosomal frameshifting is regulated in both eukaryotes (in antizyme) and prokaryotes (in RF2 and tryptophanase).

cis-Acting elements belonging to the second category also occur throughout the mRNA. The iron-responsive element (IRE) is a sequence- and structure-specific negative regulatory element, found in the 5′UTR of ferritin mRNA (and subsequently in other mRNAs), that modulates its translation in accordance with the level of cellular iron. This regulation is mediated by a *trans*-acting iron repressor protein (IRP) that binds to the IRE and inhibits translation (Chapter 21). It is reasonable to expect that other such *negative* mRNA-specific *trans*-acting regulators of translation are awaiting discovery. *Positive* mRNA-specific regulators of translation have been described in bacteriophages. For example, the Com protein of bacteriophage Mu activates translation of *mom* mRNA by binding near its initiation site and altering its secondary structure (Chapter 8). Although no factor with similar activity has yet been reported in eukaryotes, several proteins interact with the internal ribosome entry site (IRES) of picornavirus RNAs and stimulate their translation (Chapters 4 and 31).

The past decade has seen the surprising discovery that the 3′UTR is a rich repository of *cis*-acting elements that determine mRNA stability and localization in the cytoplasm and also serve to regulate translation initiation. These controls are most likely mediated by *trans*-acting factors (Chapters 7, 27, and 29). Most such examples of translational control occur during early development, but some cases have been described in somatic cells. An unusual case is seen in the developmentally regulated *Caenorhabditis elegans* gene *lin-14*. Translation of this mRNA is inhibited by a short (22-nucleotide) RNA transcribed from the *lin-4* gene, which can base-pair with sequences in the 3′UTR of the *lin-4* mRNA. At the 3′ end of eukaryotic mRNAs, the poly(A) tail also has an important role as an enhancer of translation. Intriguingly, the poly(A) tail acts in synergy with the mRNA 5′ cap structure, and the translational activity of the poly(A) tail may be mediated by the poly(A)-binding protein (Chapter 10).

mRNA stability is an important determinant of cytoplasmic mRNA levels and therefore of protein synthesis. In many instances, translation has a direct role in determining mRNA stability, as mRNA degradation may be coupled to translation (Chapter 29). Most but not all of the *cis*-acting elements that trigger mRNA degradation are localized to the 3′UTR; the poly(A) tail influences the degradation of mRNAs via the poly(A)-binding protein, and short-lived mRNAs possess sequence-specific elements that mediate mRNA degradation. A separate pathway exists to degrade mRNAs that contain premature termination codons (nonsense-mediated decay) (Chapters 29 and 30). This pathway has most probably evolved to prevent the synthesis of truncated proteins that might function in a dominant-negative manner. It is puzzling that this degradative pathway seems to operate in the cytoplasm in yeast, whereas it is nuclear in mammals. The nuclear mode of nonsense-mediated decay poses intriguing questions concerning the mechanism whereby nonsense codons are recognized in the nucleus, and the possible coupling between translation and nuclear-cytoplasmic mRNA transport.

Initiation Factors

The effects of the various *cis*-acting elements in the mRNA 5′UTR are modulated through the activity of initiation factors and other *trans*-acting factors. Phosphorylation of initiation factors provides the chief means to control the rate of mRNA binding. Several factors that promote mRNA binding to ribosomes (eIF4E, eIF4G, eIF4B, and eIF3 in mammalian cells; also eIF4A in plants) are phosphorylated, and the phosphorylation status of these proteins correlates positively with both translational and growth rates of the cell (Chapter 6). The phosphorylation state of these initiation factors is modulated in a wide variety of circumstances and affects translation during the cell cycle, during infection with viruses, after heat shock, or in response to growth factors and hormones (Chapters 6, 8, 17, and 20). Although there is some biochemical evidence that the phosphorylation of eIF4E potentiates its cap-binding activity, for eIF4B and eIF4G, the consequences of phosphorylation are not yet established (Chapter 6).

Phosphorylation of eIF2 also has a central role in regulating translation by affecting the binding of Met-tRNA$_i$. In contrast to the eIF4 group, phosphorylation of eIF2 inactivates its ability to recycle, as the exchange of GDP for GTP on the factor is blocked, leading to inhibition of translation (Chapter 5). Phosphorylation of eIF2, like that of the eIF4 proteins, has a role in differentiation (Chapter 7) and occurs under conditions of stress, including heat shock (Chapter 17), viral infection (Chapters 8, 32–35), and serum deprivation (Chapters 16 and 17). Extensive analyses of the mechanisms of eIF2 phosphorylation led to the identification and characterization of four mammalian protein kinases, PKR, HRI, PERK, and GCN2 (Chapters 5, 13–15), the first having a key role in the antiviral host defense mechanism that is mediated by interferons (Chapter 8). GCN2 in yeast regulates translation reinitiation on the 5′UTR of GCN4 mRNA and mediates the response to amino acid deprivation (Chapter 5). It would be of interest to know whether GCN2 plays a similar role in vertebrates. Thus, phosphorylation of eIF2 controls the rate of reinitiation of translation on mRNAs that contain uORFs. Phosphorylation also controls the activity of eIF2B, the eIF2 guanine nucleotide exchange factor (Chapters 5 and 16).

Apart from phosphorylation, translation initiation factor activity can be modulated in principle by other reversible or irreversible modifications. One important example that occurs as a result of infection with certain picornaviruses is the cleavage of eIF4G. This cleavage is responsible in part for the shutoff of host-protein synthesis after viral infection (Chapters 8 and 31) and a different cleavage pattern occurs in cells undergoing apoptosis (Bushell et al. 2000).

An important recent development is the discovery that initiation factor activity can be modulated by proteins that interact with initiation factors. For example, polypeptides (4E-BPs) that bind eIF4E and inhibit cap-dependent translation initiation have been identified; their activity is modulated by phosphorylation under the control of growth factors and hormones (Chapter 6). Also, proteins exhibiting homology with eIF4G (p97/DAP5/NAT1 and Paip1) have been described (Chapter 6). These proteins modulate translation most likely via their interaction with eIF4G-binding proteins. Similarly, eIF2 activity may be modulated by an accessory protein, p67, which binds to eIF2 and prevents its phosphorylation by eIF2 kinases (Chapter 5).

Elongation Factors

Elongation rates are also modulated by phosphorylation, particularly through the activity of the translation elongation factor eEF2. This factor undergoes phosphorylation in response to growth-promoting stimuli, calcium ion fluxes, and other agents, to affect translation (Chapter 24). eEF2 and the other elongation factors are also altered posttranslationally by other modifications. For example, eEF2 is a substrate for ADP-ribosylation by diphtheria toxin on the unique diphthamide residue (derived from histidine). There is evidence that diphthamide has a role in polypeptide chain elongation (Chapter 3). Both bacterial EF1A and eukaryotic eEF1A also contain modifications, but their functions are not yet clear.

Ribosomes

Phosphorylation of ribosomal proteins may also affect translational initiation. Of these, ribosomal protein S6 provides the best-studied example: Its phosphorylation promotes the initiation of translation on mRNAs encoding ribosomal proteins and elongation factors. Recent studies have revealed that the mechanism underlying this selectivity involves an oligopyrimidine tract in the 5′UTR of the target mRNAs and have shed light on the signal transduction pathways that link growth-promoting stimuli to S6 phosphorylation (Chapters 22 and 23).

WHY CONTROL TRANSLATION?

Thus far, we have considered the basis and principles of translational control. As mentioned above, there is a clear-cut rationale for regulating a biochemical pathway at its first step; this principle holds true, by and large, for protein synthesis, in that regulation is most often exercised at the initiation phase. From a broader perspective, however, matters become less clear-cut. Viewing gene expression in totality, translation occupies a position somewhere in the middle of a complex pathway that begins with transcription, continues with RNA processing and transport, and ends with protein translocation, modification, folding, assembly, and degradation. Each of these steps is known to be regulated in one or another biological system. Yet, two of the steps in this grand scheme, transcription and translation, are especially critical for the cell. Both are biosynthetic steps in which the cell makes large investments of energy. Consequently, both are steps at which the cell's expenditure of resources is checked. Indeed, transcription is subject to a multitude of controls. So, why control translation, too? And where and when is this option exercised?

To these frequently asked questions there is no single answer. Rather, there are several compelling reasons for cells to deploy translational control in their arsenal of regulatory mechanisms. Some of the advantages offered by translational control are considered briefly below. Evidently, the benefits more than compensate for the energetic and other penalties paid for the privilege of exerting regulation over a downstream reaction in a long pathway.

Directness and Rapidity

Immediacy is the most conspicuous advantage of translational control over transcriptional and other nuclear control mechanisms. Whereas transcriptional control affects the first step in the flow of genetic information, translational control affects the last step. When control is applied at a step prior to translation, the cell has to confront subsequent biochemical reactions (splicing, nuclear transport, etc.) that might be rate-limiting and inevitably entail a delay in implementing changes in protein synthesis. No such time lag applies in the case of translational control.

Reversibility

Most translational controls are effected by reversible modifications of translation factors, chiefly through phosphorylation. The readily reversible nature of translational control mechanisms is economical in energetic terms, a feature that is of particular biological significance in energy-deprived cells.

Fine Control

There are numerous examples of genes that are under both transcriptional and translational control (e.g., TNF-α, C/EBPB, VEGF, ornithine decarboxylase). In most instances, but not all, the changes in transcription rates are considerably greater in magnitude than the changes in translation rates. Thus, regulation of gene expression at the translational level provides a means for fine control.

Regulation of Large Genes

Some genes are extremely long (e.g., dystrophin, >2000 kb), and their transcription is estimated to take an extended period of time (>24 hours for dystrophin). It is reasonable to assume that if their expression needs to be regulated in a relatively short period, it is likely to be accomplished at the level of translation.

Systems That Lack Transcriptional Control

In some systems (e.g., reticulocytes, oocytes, and RNA viruses), there is little or no opportunity for transcriptional control, and gene expression is modulated mostly at the translational level. The widespread use of translational controls to regulate gene expression during development suggests that this mode of control preceded transcriptional control in evolution. Such a hypothesis is consistent with the notion of the existence of an RNA world prior to the emergence of DNA. Is it therefore possible that translational control was more prevalent early in evolution and that we are now witnessing only the relics of such control mechanisms?

Spatial Control

Regulation of the site of protein synthesis within the cell can generate concentration gradients of proteins. Such gradients are known to affect the translational efficiency of other specific mRNAs that determine patterning in early development (Chapter 7). Similar mechanisms are likely to explain synaptic plasticity (see section below).

Flexibility

Because of the wide variety of mechanisms for translational control, it can be focused by specific effector mechanisms on a single or a few gene(s) or cistrons, such as the coat protein and replicase of RNA phages, antizyme, and ferritin (see Chapters 8, 21, and 25); alternatively, by influencing general factors, it can encompass whole classes of mRNAs, as in heat shock and virus-induced host-cell shutoff (see Chapters 8, 17, and 22). Such flexibility affords the cell a powerful and adaptable means to regulate gene expression.

FUTURE TRENDS

Applied Genomic Approaches to Translational Control Studies

In the past two years, the development of cDNA microarray technology has provided a powerful means to explore the control of gene expression at a genome-wide level (Iyer et al. 1999). This technology has been applied primarily to studies of global expression profiles at the transcriptional level, but has recently been adapted for studies at the translational level (Johannes et al. 1999; Zong et al. 1999). The basis of this modification is the fact that the number of ribosomes associated with an mRNA reflects, under most circumstances, the rate of translation initiation, the rate-limiting step in translation as described above. Thus, mRNAs associated with a small number of ribosomes (light polysomes) are translated inefficiently, whereas those associated with a large number of ribosomes (heavy polysomes) are translated efficiently (see Chapter 19). The cDNA microarray technology has already been used, albeit on a small scale, to identify mRNAs that are translationally regulated in response to mitogens (Zong et al. 1999). Another study identified mRNAs, which are likely to translate via an IRES-dependent mechanism, because they could be translated in poliovirus-infected cells (Johannes et al. 1999). It is certain that this approach will be extended to identify translationally controlled mRNAs during development, differentiation, proliferation, and through the cell cycle, with the prospect of exciting findings. The results will be of importance in the understanding of translational control in diseases such as cancer and virus infection where there are clear indications that normal translation patterns are disrupted (see Chapters 8 and 20).

mRNA 5′–3′ Interactions

In the past four years, it has become abundantly clear that the 5′ end of the eukaryotic mRNA physically interacts with the 3′end. This interaction, which was first discovered in yeast (Tarun and Sachs 1996), is phylogenetically conserved and is mediated primarily by the interaction of eIF4G with the poly(A)-binding protein. The interaction of the mRNA ends brings about circularization of the mRNA, a phenomenon observed occasionally as polysome circles or spirals by electron microscopy by several investigators during the past four decades (see, e.g., Christensen et al. 1987). mRNA circularization could explain the synergistic activation of translation by the mRNA 5′ and 3′ ends (Gallie 1991). Although the mechanism of translational activation is not clear, it may involve the direct shunting of ribosomes (following termination) from the 3′ to the 5′ end of the mRNA. What is tantalizing about the circularization model is that it holds much promise to explain how sequences in the mRNA 3′UTR affect translation initiation from the 5′UTR. Such examples of translational control abound in development and in response to extracellular stimuli (see Chapters 7 and 27). One attractive hypothesis is that proteins, which interact with the 3′UTR positively or negatively, affect mRNA circularization. A likely mechanism is that 3′UTR-binding proteins interact with PABP or eIF4G or their partners to modulate their binding affinity. The circularization model could well explain the difference between initiation and re-initiation (recycling) with respect to their dependence on the cap structure (see Chapter 2). Future work is bound to shed light on these mechanisms.

Synaptic Plasticity

Synaptic plasticity is the mechanism that leads to changes in neurons in response to experience. Local translation of specific proteins in individual synapses plays a key role in effecting synaptic plasticity. This translation occurs on mRNAs that are localized at synapses, and on mRNAs that are transported after learning to synapses. Regulation also occurs at the level of the localization of mRNAs to synapses. For example, the mRNA for Arc, which is inducible, is specifically localized to previously activated synapses (Steward et al. 1998)

Translation inhibitors block long-term facilitation (L-LTP) in the snail, *Aplysia*, by specific blockade of synaptic translation. Translation at synapses in *Aplysia* is increased after treatment with serotonin, a response that is partially blocked by rapamycin. This suggests a role for the

FRAP/TOR rapamycin-sensitive pathway in synaptic plasticity (Casadio et al. 1999). Indeed, local application of rapamycin to synapses blocks the retention of long-term facilitation in this system.

The mechanism by which translation is activated in synapses is not understood, but some clues have recently been obtained. For example, polyadenylation of α-CaMKII, which is localized to synapses and is important in synaptic plasticity, is increased in dark-reared rats that are exposed to light. This is accompanied by enhanced translation (Chapter 27). CaMKII synthesis is also increased at synapses after NMDA stimulation. Interestingly, this is coupled to a decrease in general translation mediated by calcium-dependent phosphorylation of eEF2 (Sheetz et al. 2000). It is possible that this decrease is required to facilitate the enhancement of the translation of specific mRNAs.

mRNPs and mRNA Localization

Proteins that interact with mRNA and mediate its genesis, transport, activity, and destruction continue to be characterized in profusion. Elucidation of their interactions with one another and with mRNA, and the determination of their precise functions in the cell, remain formidable challenges, as discussed in many chapters of this monograph. The central roles played by such RNA–protein interactions are illustrated by recent work on mRNA localization in developing embryos and neural tissue. For example, the *cis*-acting elements known as "zip codes," which specify the sorting of some mRNAs within the cytoplasm, are recognized by a variety of *trans*-acting proteins such as ZBP and Vg1 (Ross et al. 1997; Deshler et al. 1998; Havin et al. 1998); similarly, Staufen recognizes structured RNA elements in the 3′UTR of mRNA species and targets them to specific locations (St Johnston et al. 1991; Broadus et al. 1998; Kiebler et al. 1999). How the resulting mRNP complexes are transported through the cytoplasm is not well understood. Evidence suggests that the mRNPs are assembled into large granules or "locusomes," which may contain ribosomes and other components of the translation apparatus, and implicates cytoplasmic structures such as microfilaments, microtubules, and the endoplasmic reticulum as migration pathways (for review, see Bassell et al. 1999; Kiebler and DesGroseillers 2000). Presumably, additional factors mark the final destinations of the mRNAs and govern their translational activity once they have been delivered there. The full elucidation of these mechanisms presents an exciting prospect as well as a technical challenge.

CONCLUDING REMARKS

The recognition of translational control formally requires the measurement of two parameters—the rate of protein synthesis and the concentration of the corresponding mRNA—so its rigorous demonstration can be demanding. Nevertheless, appreciation of the range of biological processes that entail translational control is expanding rapidly. At the same time, our understanding of the underlying protein synthetic apparatus is well advanced and provides a solid platform to address the mechanisms exploited by cells to control gene expression at this level. Goals for the future lie in many directions: to identify and characterize the *cis*- and *trans*-acting elements that mediate translational control, to visualize the interactions at the atomic level, and to integrate this information within the framework of the physiology and evolution of intact cells and organisms. The next few years will undoubtedly see progress toward all of these goals, as well as insights and unlooked-for discoveries that will open further vistas in this dynamic field.

ACKNOWLEDGMENTS

The authors' work has been supported by grants from the National Institutes of Health (M.B.M. and J.W.B.H.) and from the Medical Research Council, National Cancer Institute of Canada, and Howard Hughes Medical Institute (N.S.).

REFERENCES

Agrawal M.G. and Bowman L.H. 1987. Transcriptional and translational regulation of ribosome protein formation during mouse myoblast differentiation. *J. Biol. Chem.* **262:** 4868–4875.

Bassell G.J., Oleynikov Y., and Singer R.H. 1999. The travels of mRNAs through all cells large and small (comments) *FASEB J.* **13:** 447–54.

Berg P. and Ofengand E.J. 1958. An enzymatic mechanism for linking amino acids to RNA. *Proc. Natl. Acad. Sci.* **44:** 78–86.

Blobel G. and Dobberstein B. 1975. Transfer of proteins across membranes. II. Reconstitution of functional rough microsomes from heterologous components. *J. Cell Biol.* **67:** 852–862.

Blobel G. and Sabatini D.D. 1971. Ribosome-membrane interaction in eukaryotic cells. In *Biomembranes* (ed. L.A. Mason), vol. 2, pp. 193–195. Plenum Press, New York.

Brenner S., Jacob F., and Meselson M. 1961. An unstable intermediate carrying information from genes to ribosomes for protein synthesis. *Nature* **190:** 576–581.

Broadus J., Fuerstenberg S., and Doe C.Q. 1998. Staufen-dependent localization of prospero mRNA contributes to neuroblast daughter-cell fate. *Nature* **391:** 792–795.

Bruns G.P. and London I.M. 1965. The effect of hemin on the synthesis of globin. *Biochem. Biophys. Res. Commun.* **18:** 236–242.

Bushell M., Wood W., Clemens M.J., and Morley S.J. 2000. Changes in integrity and association of eukaryotic protein synthesis initiation factors during apoptosis. *Eur. J. Biochem.* **267:** 1083–1091.

Campbell P.N. and Work T.S. 1953 Biosynthesis of proteins. *Nature* **171:** 997–1001.

Casadio A., Martin K.C., Giustetto M., Zhu H., Chen M., Bartsch D., Bailey C.H., and Kandel E.R. 1999. A transient, neuron-wide form of CREB-mediated long-term facilitation can be stabilized at specific synapses by local protein synthesis. *Cell* **99:** 221–237.

Chapeville F., Lipmann F., von Ehrenstein G., Weisblum B., Ray W.J., and Benzer S. 1962. On the role of soluble ribonucleic acid in coding for amino acids. *Proc. Natl. Acad. Sci.* **48:** 1086–1092.

Christensen A.K., Kahn L.E., and Bourne C.M. 1987. Circular polysomes predominate on the rough endoplasmic reticulum of somatotropes and mammotropes in the rat anterior pituitary. *Am. J. Anat.* **178:** 1–10.

Clemens M.J., Galpine A., Austin S.A., Panniers R., Henshaw E.C., Duncan R., Hershey J.W.B., and Pollard J. 1987. Regulation of polypeptide chain initiation in CHO cells with a temperature-sensitive leucyl-tRNA synthetase: Changes in phosphorylation of initiation factor eIF-2 and in the activity of the guanine nucleotide exchange factor GEF. *J. Biol. Chem.* **262:** 767–771.

Crick F.H.C. 1958. On protein synthesis. *Symp. Soc. Exp. Biol.* **12:** 138–163.

Davidson S.D., Passmore R., and Brock J.F. 1973. *Human nutrition and dietetics,* 5th edition. Churchhill Livingstone, London, United Kingdom.

Denny P.C. and Tyler A. 1964. Activation of protein biosynthesis in non-nucleate fragments of sea urchin eggs. *Biochem. Biophys. Res. Commun.* **14:** 245–249.

Deshler J.O., Highett M.I., Abramson T., and Schnapp B.J. 1998. A highly conserved RNA-binding protein for cytoplasmic mRNA localization in vertebrates *Curr. Biol.* **8:** 489–496.

Devillers-Thiery A., Kindt T., Scheele G., and Blobel G. 1975. Homology in amino-terminal sequences of precursors to pancreatic secretory proteins. *Proc. Natl. Acad. Sci.* **72:** 5016–5020.

Drysdale J.W. and Munro H.N. 1965. Failure of actinomycin D to prevent induction of liver apoferritin after iron administration. *Biochem. Biophys. Acta* **103:** 185–188.

Duncan R. and McConkey E.H. 1982. Rapid alterations in initiation rate and recruitment of inactive RNA are temporally correlated with S6 phosphorylation. *Eur. J. Biochem.* **123:** 539–544.

Eboué-Bonis D., Chambaut A.M., Volfin P., and Clauser H. 1963. Action of insulin on the isolated rat diaphragm in the presence of actinomycin D and puromycin. *Nature* **199:** 1183–1184.

Ehrenfeld E. and Hunt T. 1971. Double-stranded poliovirus RNA inhibits initiation of protein synthesis by reticulocyte lysates. *Proc. Natl. Acad. Sci.* **68:** 1075–1078.

Ehrenfeld E. and Lund H. 1977. Untranslated vesicular stomatitis virus messenger RNA after poliovirus infection. *Virology* **80:** 297–308.

Endo T. and Nadal-Ginard B. 1987. Three types of muscle-specific gene expression in fusion-blocked rat skeletal muscle cells: Translational control in EGTA-treated cells. *Cell* **49:** 515–526.

Fan H. and Penman S. 1970. Regulation of protein synthesis in mammalian cells. *J. Mol. Biol.* **50:** 655–670.

Farrell P.J., Balkow K., Hunt T., and Jackson R.J. 1977. Phosphorylation of initiation factor eIF-2 and the control of reticulocyte protein synthesis. *Cell* **11:** 187–200.

Fraser C.M., Gocayne J.D., White O.I., Adams M.D., Clayton R.A, Fleischmann R.D., Bult C.J., Kerlavage A.R., Sutton G., Kelley J.M. et al. 1995. The minimal gene complement of *Mycoplasma genitalium*. *Science* **270:** 397–403.

Gallie D. 1991. The cap and poly(A) tail function synergistically to regulate mRNA translational efficiency. *Genes Dev.* **5:** 2108–2116.

Garren L.D., Howell R.R., Tomkins G.M., and Crocco R.M. 1964. A paradoxical effect of actinomycin D: The mechanism of regulation of enzyme synthesis by hydrocortisone. *Proc. Natl. Acad. Sci.* **52:** 1121–1129.

Gehrke L. and Ilan J. 1987. Regulation of messenger RNA translation at the elongation step during estradiol-induced vitellogenin synthesis in avian liver. In *Translational regulation of gene expression* (ed. J. Ilan), pp. 165–186. Plenum Press, New York.

Geoghegan T., Cereghini S., and Brawerman G. 1979. Inactive mRNA-protein complexes from mouse sarcoma-180 ascites cells. *Proc. Natl. Acad. Sci.* **76:** 5587–5591.

Godefroy-Colburn T. and Thach R.E. 1981. The role of mRNA competition in regulating translation. IV. Kinetic model. *J. Biol. Chem.* **256:** 11762–11773.

Gros F., Hiatt H., and Gilbert W. 1961. Unstable ribonucleic acid revealed by pulse labelling of *Escherichia coli*. *Nature* **190:** 581–585.

Gross P.R. and Cousineau G.H. 1963. Effects of actinomycin D on macromolecular synthesis and early development in sea urchin eggs. *Biochem. Biophys. Res. Commun.* **10:** 321–326.

Gross P.R., Malkin L.I., and Moyer W.A. 1964. Templates for the first proteins of embryonic development. *Proc. Natl. Acad. Sci.* **51:** 407–414.

Hardesty B., Miller R., and Schweet R. 1963. Polyribosome breakdown and hemoglobin synthesis. *Proc. Natl. Acad. Sci.* **50:** 924–931.

Havin L., Git A., Elisha Z., Oberman F., Yaniv K., Schwartz S.P., Standart N., and Yisraeli J.K. 1998. RNA-binding protein conserved in both microtubule- and microfilament-based RNA localization. *Genes Dev* **12:** 1593–1598.

Henshaw E.C., Hirsch C.A., Morton B.E., and Hiatt H.H. 1971. Control of protein synthesis in mammalian tissues through changes in ribosome activity. *J. Biol. Chem.* **246:** 436–446.

Heywood S.M. 1970. Specificity of mRNA binding factor in eukaryotes. *Proc. Natl. Acad. Sci.* **67:** 1782–1788.

Himmelreich R., Hilbert H., Plagens H., Prikl E., Li B.C., and Hermann R. 1996. Complete sequence analysis of the genome of the bacterium *Mycoplasma pneumoniae*. *Nucleic Acids Res.* **24:** 4420–4449.

Hoagland M.B., Zamecnik P.C., and Stephenson M.L. 1959. A hypothesis concerning the roles of particulate and soluble ribonucleic acids in protein synthesis. In: *A symposium on molecular biology* (ed. R.E. Zirkle), pp. 105–114. Univ. Chicago Press, Chicago, Illinois.

Hodge L.D., Robbins E., and Scharff M.D. 1969. Persistence of messenger RNA through mitosis in HeLa cells. *J. Cell Biol.* **40:** 497–507.

Hultin T. 1961. Activation of ribosomes in sea urchin eggs in response to fertilization. *Exp. Cell Res.* **25:** 405–417.

Hultin T. and Beskow G. 1956. The incorporation of ^{14}C-L-leucine into rat liver proteins in vitro visualized as a two-step reaction. *Exp. Cell Res.* **11:** 664–666.

Humphreys T. 1969. Efficiency of translation of messenger-RNA before and after fertilization in sea urchins. *Dev. Biol.* **20:** 435–458.

———. 1971. Measurements of messenger RNA entering polysomes upon fertilization of sea urchin eggs. *Dev. Biol.* **20:** 201–208.

Hutchison C.A., Peterson S.N., Gill S.R., Cline R.T., White O., Fraser C.M., Smith H.O., and Venter J.C. 1999. Global transposon mutagenesis and a minimal Mycoplasma genome. *Science* **286:** 2165–2169.

Iyer V.R., Eisen M.B., Ross D.T., Schuler G., Moore T., Lee J.C.F., Trent J.M., Staudt L.M., Hudson J. Jr., Boguski M.S., Lashkari D., Shalon D., Botstein D., and Brown P.O. 1999. The transcriptional program in the response of human fibroblasts to serum [see comments]. *Science* **283:** 83–87.

Jacob F.C. and Monod J. 1961. Genetic regulatory mechanisms in the synthesis of proteins. *J. Mol. Biol.* **3:** 318–356.

Johannes G., Carter M.S., Eisen M.B., Brown P.O., and Sarnow P. 1999. Identification of eukaryotic mRNAs that are translated at reduced cap binding complex eIF4F concentrations using a cDNA microarray. *Proc. Natl. Acad. Sci.* **96:** 13118–13123.

Kiebler M.A. and DesGroseillers L. 2000. Molecular insights into mRNA transport and local translation in the mammalian nervous system. *Neuron* **25:** 19–28.

Kiebler M.A., Hemraj I., Verkade P., Köhrmann M., Fortes P., Marion R.M., Ortin J., and Dotti C.G. 1999. The mammalian staufen protein localizes to the somatodendritic domain of cultured hippocampal neurons: Implications for its involvement in mRNA transport. *J. Neurosci.* **19:** 288–297.

Kinniburgh A.J., McMullen M.D., and Martin T.E. 1979. Distribution of cytoplasmic poly(A^+)RNA sequences in free messenger ribonucleoprotein and polysomes of mouse ascites cells. *J. Mol. Biol.* **132:** 695–708.

Kosower N.S., Vanderhoff G.A., Benerofe B., Hunt T., and Kosower E.M. 1971. Inhibition of protein synthesis by glutathione disulfide in the presence of glutathione. *Biochem. Biophys. Res. Commun.* **45:** 816–821.

Kruh J. and Borsook H. 1956. Hemoglobin synthesis in rabbit reticulocytes *in vitro*. *J. Biol. Chem.* **220:** 905–915.

Lamfrom H. and Knopf P.M. 1964. Initiation of hemoglobin synthesis in cell-free systems. *J. Mol. Biol.* **9:** 558–572.

Lee G.T.Y. and Engelhardt D.L. 1979. Peptide coding capacity of polysomal and nonpolysomal messenger RNA during growth of animal cells. *J. Mol. Biol.* **129:** 221–233.

Leibowitz R. and Penman S. 1971. Regulation of protein synthesis in HeLa cells. III. Inhibition during poliovirus infection. *J. Virol.* **8:** 661–668.

Lisowska-Bernstein B., Lamm M.E., and Vassalli P. 1970. Synthesis of immunoglobulin heavy and light chains by the free ribosomes of a mouse plasma cell tumor. *Proc. Natl. Acad. Sci.* **66:** 425–432.

Lodish H.F. 1976. Translational control of protein synthesis. *Annu. Rev. Biochem.* **45:** 39–72.

Lodish H.F. and Jacobsen M. 1972. Regulation of hemoglobin synthesis. Equal rates of translation and termation of α- and β-globin chains. *J. Biol. Chem.* **247:** 3622–3629.

Lodish H.F. and Zinder N.D. 1966. Mutants of the bacteriophage f2 VIII, control mechanisms for phage-specific syntheses. *J. Mol. Biol.* **19:** 333–348.

Maggio R., Vittorelli M.L., Rinaldi A.M., and Monroy A. 1964. *In vitro* incorporation of amino acids into proteins stimulated by RNA from unfertilized sea urchin eggs. *Biochem. Biophys. Res. Commun.* **15:** 436–441.

Marks P.A., Burka E.R., and Schlessinger D. 1962. Protein synthesis in erythroid cells. I. Reticulocyte ribosomes active in stimulating amino acid incorporation. *Proc. Natl. Acad. Sci.* **48:** 2163–2171.

Martin T.E. and Young F.G. 1965. An *in vitro* action of human growth hormone in the presence of actinomycin D. *Nature* **208:** 684–685.

Mathews M.B., Hunt T., and Brayley A. 1973. Specificity of the control of protein synthesis by haemin. *Nat. New Biol.* **243:** 230–233.

McCormick W. and Penman S. 1969. Regulation of protein synthesis in HeLa cells: Translation at elevated temperatures. *J. Mol. Biol.* **39:** 315–333.

Meisenberg G. and Simmons W.H. 1998. *Principles of medical biochemistry.* Mosby, St. Louis, Missouri.

Meyuhas O., Thompson E.A., and Perry R.P. 1987. Glucocorticoids selectively inhibit translation of ribosomal protein mRNA in P1798 lymphosarcoma cells. *Mol. Cell. Biol.* **7:** 2691–2699.

Milstein C., Brownlee G.G., Harrison T.M., and Mathews M.B. 1972. A possible precursor of immunoglobin light chains. *Nat. New Biol.* **239:** 117–120.

Monroy A., Maggio R., and Rinaldi A.M. 1965. Experimentally induced activation of the ribosomes of the unfertilized sea urchin egg. *Proc. Natl. Acad. Sci.* **54:** 107–111.

Nemer M. 1962. Interrelation of messenger ribonucleotides and ribosomes in the sea urchin egg during embryonic development. *Biochem. Biophys. Res. Commun.* **8:** 511–515.

Nirenberg M.W. and Matthaei J.H. 1961. The dependence of cell-free protein synthesis in *E. coli* upon naturally occurring or synthetic polyribonucleotides. *Proc. Natl. Acad. Sci.* **47:** 1588–1602.

Ouellette A.J., Ordahl C.P., Van Ness J., and Malt R.A. 1982. Mouse kidney nonpolysomal messenger ribonucleic acid: Metabolism, coding function, and translational activity. *Biochemistry* **21:** 1169–1177.

Palmiter R.D. 1974. Differential rates of initiation on conalbumin and ovalbumin messenger ribonucleic acid in reticulocyte lysates. *J. Biol. Chem.* **249:** 6779–6787.

———. 1975. Quantitation of parameters that determine the rate of ovalbumin synthesis. *Cell* **4:** 189–197.

Pelham H.R.B. and Jackson R.J. 1976. An efficient mRNA-dependent translation system from reticulocyte lysates. *Eur. J. Biochem.* **67:** 247–256.

Penman S. and Summers D. 1965. Effects on host cell metabolism following synchronous infection with poliovirus. *Virology* **27:** 614–620.

Penman S., Scherrer K., Becker Y., and Darnell J.E. 1963. Polyribosomes in normal and poliovirus-infected HeLa cells and their relationship to messenger RNA. *Proc. Natl. Acad. Sci.* **49:** 654–662.

Roper M.D. and Wicks W.D. 1978. Evidence for acceleration of the rate of elongation of tyrosine aminotransferase nascent chains by dibutyryl cyclic AMP. *Proc. Natl. Acad. Sci.* **75:** 140–144.

Rosbash M. 1972. Formation of membrane-bound polysomes. *J. Mol. Biol.* **65:** 413–422.

Ross A.F., Oleynikov Y., Kislauskis E.H., Taneja K.L., and Singer R.H. 1997. Characterization of a beta-actin mRNA zipcode-binding protein. *Mol. Cell. Biol.* **17:** 2158–2165.

Scheetz A.J., Nairn A.C., and Constantine-Paton M. 2000. NMDA receptor-mediated control of protein synthesis at developing synapses. *Nat. Neurosci.* **3:** 211–216.

Siekevitz P. 1951. In vitro incorporation of 1-^{14}C-dl-alanine into protein of rat liver granular fractions. *Fed. Proc.* **10:** 246–247.

Spirin A.S. and Nemer M. 1965. Messenger RNA in early sea-urchin embryos: Cytoplasmic particles. *Science* **150:** 214–217.

Steward D.L., Schaeffer J.R., and Humphrey R.M. 1968. Breakdown and assembly of polyribosomes in synchronized Chinese hamster cells. *Science* **161:** 791–793.

Steward O., Wallace C.S., Lyford G.L., and Worley P.F. 1998. Synaptic activation causes the mRNA for the IEG Arc to localize selectively near activated postsynaptic sites on dendrites. *Neuron*. **21:** 741–751.

St Johnston D., Beuchle D., and Nüsslein-Volhard C. 1991. Staufen, a gene required to localize maternal RNAs in the Drosophila egg. *Cell* **66:** 51–63.

Summers D.F. and Maizel J.V. 1967. Disaggregation of HeLa cell polysomes after infection with poliovirus. *Virology* **31:** 550–552.

Summers D.F., Maizel J.V., and Darnell J.E. 1965. Evidence for virus-specific noncapsid proteins in poliovirus-infected cells. *Proc. Natl. Acad. Sci.* **54:** 505–513.

Tarun S.Z., Jr. and Sachs A.B. 1996. Association of the yeast poly(A) tail binding protein with translation initiation factor eIF-4G. *EMBO J.* **15:** 7168–7177.

Thomas G.P. and Mathews M.B. 1984. Alterations of transcription and translation in HeLa cells exposed to amino acid analogues. *Mol. Cell. Biol.* **4:** 1063–1072.

Tomkins G.M., Garren L.D., Howell R.R., and Peterkofsky B. 1965. The regulation of enzyme synthesis by steroid hormones: The role of translation. *J. Cell. Comp. Physiol.* **66:** 137–151.

Tsung K., Inouye S., and Inouye M. 1989. Factors affecting the efficiency of protein synthesis in *Escherichia coli:* Production of a polypeptide of more than 6000 amino acid residues. *J. Biol. Chem.* **264:** 4428–4433.

Walden W.E., Godefroy-Colburn T., and Thach R.E. 1981. The role of mRNA competition in regulating translation. I. Demonstration of competition in vivo. *J. Biol. Chem.* **256:** 11739–11746.

Walter P. and Blobel G. 1981. Translocation of proteins across the endoplasmic reticulum. III. Signal recognition protein (SRP) causes signal sequence-dependent and site-specific arrest of chain elongation that is released by microsomal membranes. *J. Cell Biol.* **9:** 557–561.

Warner J.R. 1999. The economics of ribosome biosynthesis in yeast. *Trends Biochem. Sci.* **24:** 437–440.

Warner J.R., Knopf P.M., and Rich A. 1963. A multiple ribosome structure in protein synthesis. *Proc. Natl. Acad. Sci.* **49:** 122–129.

Waxman H.S. and Rabinovitz M. 1966. Control of reticulocyte polyribosome content and hemoglobin synthesis by heme. *Biochem. Biophys. Acta* **129:** 369–379.

Wilt F.H. and Hultin T. 1962. Stimulation of phenylalanine incorporation by polyuridylic acid in homogenates of sea urchin eggs. *Biochem. Biophys. Res. Commun.* **9:** 313–317.

Witherell W.G., Gott J.M., and Uhlenbeck O.C. 1991. Specific interaction between RNA phage coat proteins and RNA. *Prog. Nucleic Acid Res. Mol. Biol.* **40:** 185–220.

Wolin S.L. and Walter P. 1988. Ribosome pausing and stacking during translation of a eukaryotic mRNA. *EMBO J.* **7:** 3559–3569.

Yenofsky R., Bergmann I., and Brawerman G. 1982. Messenger RNA species partially in a repressed state in mouse sarcoma ascites cells. *Proc. Natl. Acad. Sci.* **79:** 5876–5880.

Zamecnik P.C. 1960. Historical and current aspects of the problem of protein synthesis. *Harvey Lect.* **54:** 256–281.

Zong Q., Schummer M., Hood L., and Morris D.R. 1999. Messenger RNA translation state. The second dimension of high-throughput expression screening. *Proc. Natl. Acad. Sci.* **96:** 10632–10636.

2

The Pathway and Mechanism of Initiation of Protein Synthesis

John W. B. Hershey
Department of Biological Chemistry
School of Medicine
University of California
Davis, California 95616

William C. Merrick
Department of Biochemistry
School of Medicine
Case Western Reserve University
Cleveland, Ohio 44106-4935

Elucidation of the detailed molecular mechanism of protein synthesis is essential for understanding translational controls. This chapter is concerned with the prokaryotic and eukaryotic pathways of initiation, where most translational controls are found (Chapter 1). In general terms, it focuses on (1) how the initiation factors catalyze the binding of the initiator tRNA and mRNA to the small ribosomal subunit; (2) how the initiation codon is recognized; and (3) how the large ribosomal subunit joins to form an initiation complex capable of elongation. Considerable progress has been made during the past 5 years in refining our knowledge of the pathway and in determining the three-dimensional structures of some of the macromolecular components of initiation. Emphasis is placed on the structures of the initiation factors and how the factors function to promote and regulate the pathway. The reader is directed to other chapters in this volume for descriptions of elongation (Chapter 3) and termination (Chapter 11).

The process of translation initiation was elucidated during the late 1960s through the 1970s primarily by biochemical studies that utilized radiolabeled amino acids and fractionated lysates derived from bacterial or mammalian cells. The major macromolecular components were identified by purifying proteins and nucleic acids required to reconstitute

Translational Control of Gene Expression
© 2000 Cold Spring Harbor Laboratory Press 0-87969-568-4/00 $5 + .00

translation in vitro. The ribosome itself was viewed essentially as a "black box" that appeared to provide a rigid surface onto which the other translational components bind. Surprisingly, genetic approaches contributed only modestly to the identification of the 200 or more macromolecular components that comprise the translational apparatus. Because the biochemical approach was so fruitful, subsequent in vitro studies on how these molecules interact proceeded rapidly. It is only recently that genetic studies, especially with the yeast *Saccharomyces cerevisiae* (see Chapters 5 and 12), or experiments using recombinant DNA techniques have enabled researchers to examine the mechanism of protein synthesis in vivo. Thus, in the sections that follow, the described pathways and mechanisms of initiation are based almost entirely on in vitro biochemical studies, although in several instances the views have been confirmed by in vivo experiments. High-resolution three-dimensional structures of the initiation factors and the ribosome also are contributing to a better understanding of the molecular mechanisms involved. Nevertheless, the reader is cautioned that the pathways proposed here are working models and that corrections and fine tuning of these pathways are anticipated in the future.

INITIATION IN BACTERIA

In bacteria, translation is coupled temporally and spatially to transcription, allowing protein synthesis to begin on mRNAs that are still being transcribed. These mRNAs are usually polycistronic, possessing multiple signals for the initiation and termination of protein synthesis. The translational apparatus therefore must recognize and initiate protein synthesis at specific start signals at several different locations in the same mRNA. The recognition process must be precise, because an error in phasing of only a single nucleotide results in translation of the mRNA in the wrong reading frame. This chapter describes briefly the steps in the pathway whereby the ribosome binds the unique initiator, formyl-methionyl-$tRNA_f$ (fMet-$tRNA_f$), and mRNA and recognizes the correct initiation codon. It also focuses on how the three initiation factors, IF1, IF2, and IF3 (Table 1), promote the rate and fidelity of these reactions.

The Pathway

The goal of the initiation pathway in prokaryotes is to assemble a 70S initiation complex with fMet-$tRNA_f$ in the ribosomal P site, interacting with the initiation codon in the mRNA. The pathway has been studied primari-

Table 1 Prokaryotic initiation factors

Name	Mass (kD)		Accession number	Function
	SDS	sequence		
IF1	8	7.1	Y00373	Rb subunit dissociation; assists IF2 and IF3
IF2-1	120	97.3	X00513	binds fMet-tRNA$_f$; GTPase
IF2-2	90	79.7	X00513	binds fMet-tRNA$_f$; GTPase
IF3	20	20.7	K02844	ribosome subunit anti-association; tRNA$_f$-codon interaction
W2		~71	AAC76196	ATPase, helicase, eIF4A homolog
EF-P		~21	X61676	eIF5A homolog

Data are for initiation factors from *Escherichia coli*. The two initiation factors in the lower part of the table are described in Lu et al. (1999); their roles in the initiation phase of protein synthesis are not yet clearly defined.

ly in *Escherichia coli* and is depicted in Figure 1. Characteristics of the initiation factors are reported in Table 1. The 70S ribosome is in dynamic equilibrium with its 30S and 50S subunits. The extent of association into 70S ribosomes is influenced by the free Mg^{++} concentration; at physiological concentrations, estimated to be around 5 mM, ribosomal subunits are mostly associated into 70S particles (Godefroy-Colburn et al. 1975). IF3 binds stably to 30S ribosomal subunits and prevents their association with 50S subunits. The "native" 30S subunits thus generated contain bound IF3, and subsequently may bind IF1 and IF2•GTP (Zucker and Hershey 1986). Next, either fMet-tRNA$_f$ or mRNA binds to the 30S subunit to form a relatively unstable bimolecular complex, followed by binding of the other component (for review, see Gualerzi and Pon 1990). The ternary complex thus formed undergoes a rate-limiting conformational change to generate a more stable 30S initiation complex (Gualerzi et al. 1977; Gualerzi and Pon 1990). Although the binding is not ordered, specific mRNAs may favor one pathway over the other for complex assembly. The binding of fMet-tRNA$_f$ is promoted and accelerated by IF2 and IF3, which recognize different aspects of the initiator tRNA. IF2 detects the unique formylmethionyl group, and therefore forms a complex only with fMet-tRNA$_f$, not with aminoacyl-tRNAs involved in elongation. The ternary complex, IF2•GTP•fMet-tRNA$_f$, is rather unstable and is not readily isolated, but nevertheless appears to be an intermediate in the binding of the tRNA to

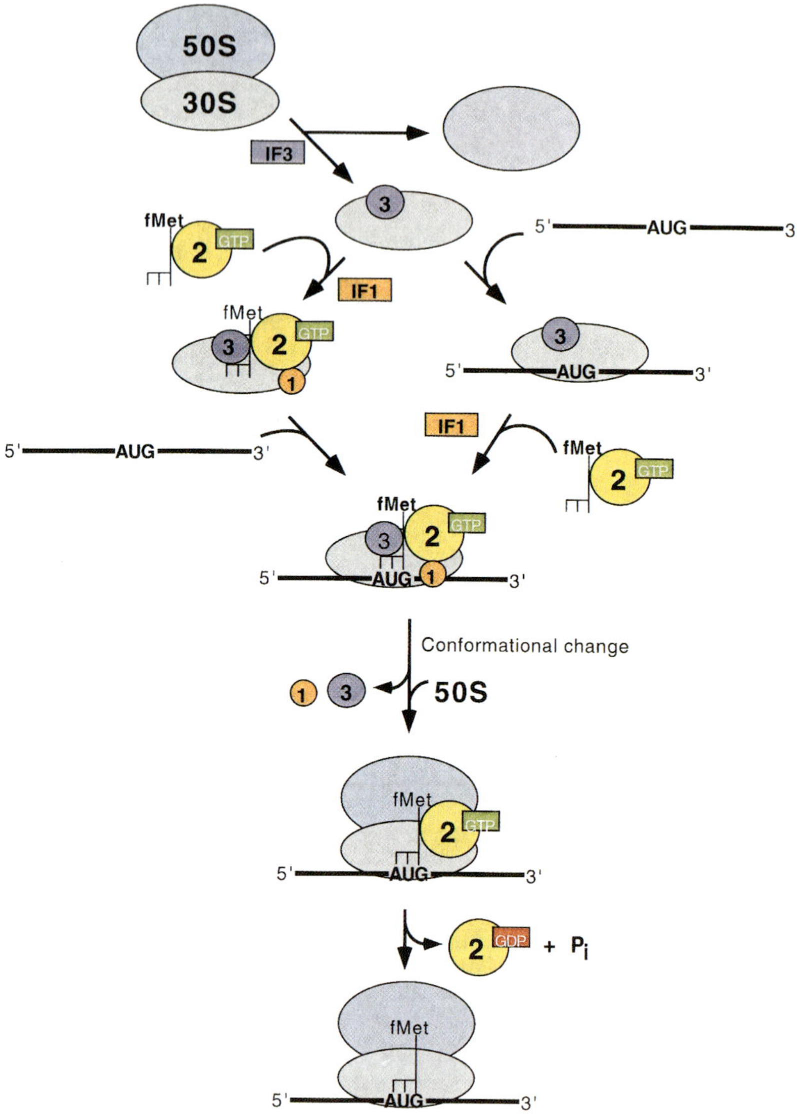

Figure 1 Initiation pathway in prokaryotes. The figure represents a working model of the pathway and should not be taken literally. The placement of initiation factors IF1, IF2, and IF3 (shown as color-coded labeled circles) on the 30S ribosomal subunit relative to other components and to one another is hypothetical and is not based on actual structural information. Reaction arrows point in the productive direction, although most of the reactions are likely reversible.

the 30S subunit (Wu et al. 1996; Wu and RajBhandary 1997). Additional kinetic studies are needed to determine whether fMet-tRNA$_f$ also may interact with IF2 already bound to the 30S subunit.

The role of IF3 is to stabilize fMet-tRNA$_f$ binding in the 30S ribosomal P site (Mangroo et al. 1995) by examining the structure of the anticodon stem containing three G-C base pairs, and to destabilize tRNAs lacking this feature (Hartz et al. 1989). IF3 also distinguishes between matched and mismatched codon–anticodon interactions (see below). Thus, IF2 and IF3 assure that it is fMet-tRNA$_f$ that is bound to the 30S ribosomal subunit, not an elongator aminoacyl-tRNA, especially Met-tRNA$_m$. Following mRNA binding to form a 30S initiation complex, it is thought that both IF1 and IF3 dissociate. The 30S initiation complex joins with the 50S ribosomal subunit to form a 70S complex that contains mRNA, fMet-tRNA$_f$ bound in the P site, and IF2•GTP. The 70S ribosome catalyzes a GTPase reaction that results in the rapid release of IF2•GDP. The resulting 70S initiation complex is competent to enter the elongation phase. When initiation complexes are constructed with a nonhydrolyzable GTP analog, 70S complexes are formed, but IF2 is not released and the complexes are not able to begin protein synthesis. More detailed descriptions of the pathway of prokaryotic initiation are available elsewhere (Gualerzi and Pon 1990; McCarthy and Brimacombe 1994; Voorma 1996).

mRNA Binding and Selection of the Initiator Codon

About 90% of bacterial mRNAs initiate protein synthesis at an AUG codon; other codons are GUG (8%) and UUG (1%). These three codons are called *canonical*, based on their not being discriminated against by IF3, (Sussman et al. 1996). Other codons such as AUU are called *non-canonical*, are discriminated against by IF3, and are used much less frequently. The translational apparatus must distinguish the correct initiator AUG from other AUGs that encode internal methionines or that occur out of phase. Two important mechanisms determine this selection: (1) mRNA secondary structure, which masks AUGs not serving as start signals and thereby prevents the binding of 30S ribosomal subunits and (2) RNA–RNA interactions between the mRNA and both ribosomal RNA and the anticodon of the ribosome-bound fMet-tRNA$_f$.

A cistron whose ribosome-binding site lies in an unstructured region is translated efficiently, whereas a cistron whose ribosome-binding site is masked by secondary structure is inefficiently translated. Since translation is coupled to transcription and the ribosome can bind to the mRNA soon after it emerges from the RNA polymerase, transiently unstructured initiation sites may occur in nascent transcripts that later may be occluded. The elegant experiments of de Smit and van Duin (1990) with the MS2 phage coat cistron showed quantitatively the influence of secondary structure on translational efficiency. Structures with stabilities up to –5 to –6 kcal/mole had little or no influence on protein synthesis, but more stable structures exhibited a tenfold inhibition for every increase in stability of –1.4 kcal/mole (de Smit and van Duin 1990). There is no in vivo evidence that bacteria employ RNA helicases to remove mRNA secondary structure at initiation sites. However, factor W2, a bacterial homolog of eIF4A (the prototypical RNA helicase of eukaryotes), stimulates in vitro the translation of a cistron whose ribosome-binding site is masked by secondary structure (Lu et al. 1999). How W2 might specifically recognize ribosome-binding sites and thereby stimulate translation is not easily explained. The stabilization/destabilization of mRNA secondary structure near the initiation site frequently is employed as a regulatory mechanism in many examples of translational control in prokaryotes (for descriptions of such control mechanisms, see Simons and Grunberg-Manago 1998; Chapter 8).

The initial binding of the mRNA's unstructured region to the 30S ribosomal subunit occurs in a cleft or channel between the head and platform regions of the subunit (Fig. 2B). One of the RNA–RNA interactions that stabilizes mRNA binding involves the purine-rich Shine-Dalgarno (SD) region in the mRNA and a complementary pyrimidine-rich sequence at the 3′ terminus of 16S rRNA called the anti-Shine-Dalgarno (ASD) region. The SD sequence lies 5–7 nucleotides upstream of the initiator AUG; the consensus sequence is 5′-UA**AGGAG**GU-3′, with the nucleotides shown in bold being most frequently present. The SD region binds in an anti-parallel manner with a portion of the ASD sequence, namely 5′-ACCUCCUUA-3′, at the 3′ terminus of the 16S rRNA. The mRNA–rRNA interaction was first demonstrated by Steitz and Jakes (1975) and later proven to occur in vivo by using a mutated SD sequence and compensatory mutations in the ASD of the 16S rRNA (Hui and de Boer 1987; Jacob et al. 1987). The interaction stabilizes mRNA binding sufficiently to enable detection by primer extension inhibition assays of the bound 30S ribosome (without fMet-$tRNA_f$) in the correct initiation region (Hartz et al. 1991). Interestingly, such binding does not require, nor is it influenced by, the initiation factors. mRNAs with a "poor" SD sequence, i.e., those with little complementarity to the ASD in the 16S rRNA, are expected to be less efficiently translated. mRNA binding also

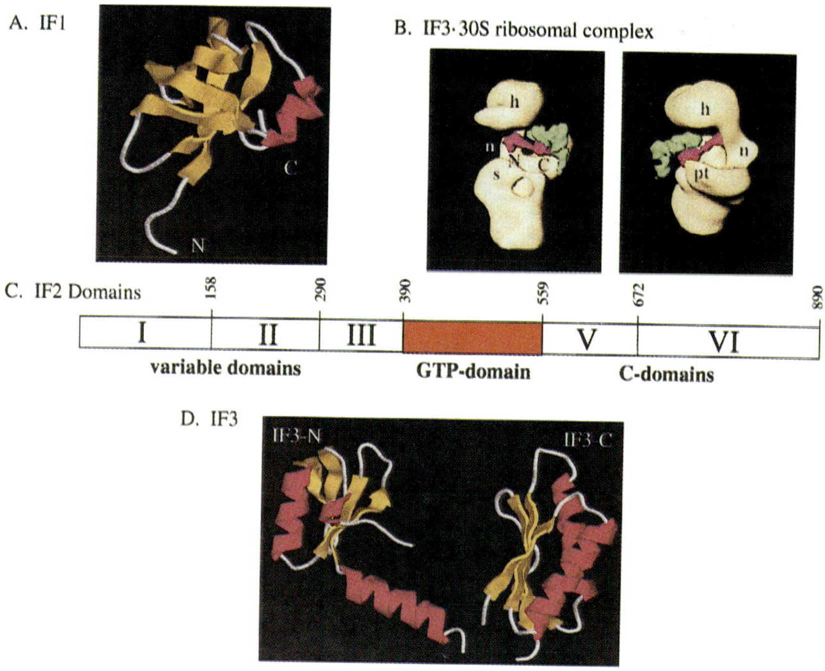

Figure 2 Structures of bacterial components of translation. (*A*) The three-dimensional solution structure of IF1 solved by NMR spectroscopy (Sette et al. 1997). (*B*) Two views of a cryo-EM-derived model of the 30S ribosomal subunit complexed with fMet-tRNA$_f$ and IF3. IF3 is shown in magenta; fMet-tRNA$_f$ is shown in green. (*C*) The domain structure of IF2. The GTP-binding domain is shown in red. (*D*) The three-dimensional structures of the amino- and carboxy-terminal domains of IF3 determined by X-ray crystallography (Biou et al. 1995). (The structures shown in *A* and *D* were taken from the Brookhaven protein structure data base; *B*, reprinted, with permission, from McCutcheon et al. 1999 [copyright National Academy of Sciences]; *C*, adapted from Moreno et al. 1999.)

may be stabilized by other interactions with the 16S rRNA. The occurrence of so-called translational enhancers and statistically preferred nucleotides in ribosomal binding sites suggests that such interactions may occur (for review, see Jackson 1996; Voorma 1996), although they are thought to be less important. Initial binding of ribosomal protein S1 to U-rich elements in some mRNAs also may promote mRNA binding (Boni et al. 1991). The stabilizing RNA–RNA interactions are especially important when the ribosome-binding site is occluded by weak secondary structure (de Smit and van Duin 1994).

A second stabilizing RNA–RNA interaction occurs between the initiation codon and the anticodon of the bound fMet-tRNA$_f$. When mRNAs employ GUG, UUG, or non-canonical initiation codons, there is less stabilization by the codon–anticodon interaction, and the efficiency of initiation is reduced. IF3 examines the match between the anticodon and codon, and labilizes fMet-tRNA$_f$ binding when the match is imperfect. The discrimination by IF3 does not depend on the actual sequence of the initiation codon, but rather only detects mismatches between the codon and anticodon (Meinnel et al. 1999). This activity is important for the autoregulation of IF3 expression, because IF3 mRNA initiates with an AUU codon and therefore is translated inefficiently when IF3 levels are high (Butler et al. 1987).

Structure of Initiation Complexes

Exciting advances in our understanding of the structure of the 70S bacterial ribosome have begun to shed light on the molecular mechanisms of initiation of protein synthesis. High-resolution models of the 30S, 50S, and 70S ribosomes have been constructed based on X-ray crystallographic analyses (Ban et al. 1999; Cate et al. 1999; Clemons et al. 1999) and cryo-electron microscopic (cryo-EM) reconstructions (Frank et al. 1995; Stark et al. 1995; Gabashvili et al. 2000). Visualization by cryo-EM of the binding sites for tRNAs in the A, P, and E sites (Agrawal et al. 1996; Stark et al. 1997) and localization of fMet-tRNA$_f$ in the P site (Malhotra et al. 1998) have been reported (see Fig. 2B). At the same time, atomic resolution structures for IF1 and IF3 have been solved.

The solution structure of IF1 was obtained by nuclear magnetic resonance (NMR) spectroscopy (Sette et al. 1997); the structure (Fig. 2A) resembles proteins in the OB family (Murzin 1993), of which the major cold shock protein, CspA, is an example. Residues whose NMR spectra were altered by binding to a 30S ribosomal subunit lie over a broad area on the viewer's side of the IF1 structure shown in Figure 2A, suggesting that this region binds to the ribosome. IF1 binding protects the same 16S rRNA nucleotides from chemical modification as tRNA binding to the A site (Moazed et al. 1995).

The structure of IF2 has not yet been solved, because it has not been possible to crystallize the protein. However, partial protease fragmentation experiments suggest a 6-domain structure (Fig. 2B) (Spurio et al. 1993; Vornlocher et al. 1997; Moreno et al. 1999). Domains 1–3 (amino-terminal) are not highly conserved and are not essential for activity in vivo (Laalami et al. 1994). Domain 4 is the GTP-binding domain whose structure has been modeled on the basis of its homology with elongation factor EF1A (formerly EF-Tu) (Cenatiempo et al. 1987). Domains 5 and 6 are responsible for binding to fMet-tRNA$_f$ and to the 30S ribosomal subunit (Spurio et al. 2000). Footprinting experiments demonstrate that IF2 interacts primarily with the T loop and T stem minor groove of the fMet-tRNA$_f$. It also protects the fMet ester linkage from hydrolysis. Comparison of the IF2, IF1, and EF2 (formerly EF-G) structures led to the suggestion that a complex of IF1 and IF2 contains regions that resemble domains IV and V in EF2 (Brock et al. 1998). The EF2 domains have been localized near the A site of the ribosome (Agrawal et al. 1998) and are thought to mimic the tRNA structure (see Chapter 3 for detailed discussions of EF2 structure and molecular mimicry). Sprinzl and coworkers (Brook et al. 1998) make the interesting hypothesis that a portion of the IF1–IF2 complex binds to the A site, thereby blocking access of fMet-tRNA$_f$ to this site and guiding it into the P site. The presence of IF2 in the A site of the 30S ribosome might also help to align the 50S subunit during the junction reaction. Confirmation of these ideas requires a better description of the structure of IF2 and its complex with IF1.

The solution structure of IF3 was solved by NMR analysis of the individual amino- and carboxy-terminal halves of the molecule (Garcia et al. 1995a,b). A model based on X-ray crystallography also was developed independently (Biou et al. 1995). The structure of IF3 consists of two globular α/β domains joined by a flexible α-helical linker (Fig. 2D). The carboxy-terminal domain binds to the 30S ribosomal subunit and possesses the ribosome anti-association activity of IF3. The structure of the amino-terminal domain suggests that it also may have an RNA-binding capability.

Additional insight into IF3 function derives from cryo-EM analyses of 30S ribosomal complexes containing fMet-tRNA$_f$ and IF3 (McCutcheon et al. 1999) (Fig. 2B). In the model, the carboxy-terminal domain is bound to the 30S platform, adjacent to the anticodon stem of the initiator tRNA. The structure therefore accounts for the ribosome anti-association activity by the C-domain alone (Garcia et al. 1995b), as the platform makes contact with the 50S subunit in the 70S ribosome. It also helps explain how IF3 distinguishes the presence of the anticodon arm of the initiator tRNA versus that of other aminoacyl-tRNAs (Hartz et al. 1989). The amino-terminal domain is situated very near the codon decoding site, deep in the cleft of the 30S subunit, and therefore is likely to provide the IF3 activity that destabilizes tRNAs with imperfect codon-anticodon matches. It is anticipated that cryo-EM analyses will determine whether IF1 and IF2 bind to the A site of the 30S ribosome, and whether IF2 helps align the fMet-tRNA$_f$ in the P site of the 50S subunit upon its junction with the 30S initiation complex.

INITIATION IN EUKARYOTES

General Features and Pathways

Initiation of protein synthesis in the cytoplasm of eukaryotes is in some ways similar to prokaryotic initiation. Both involve dissociation of ribosomes into ribosomal subunits, the use of a unique initiator $tRNA^{Met}$, usually recognition of the same initiation codon (AUG), formation of a preinitiation complex on the small ribosomal subunit, and promotion by initiation factors. However, there are a number of important mechanistic differences that derive from the use of altered strategies for mRNA binding and initiation codon recognition. The strategy in eukaryotes employs many more initiation factors and involves primarily protein–RNA and protein–protein interactions, whereas in bacteria mostly RNA–RNA interactions are used.

In eukaryotes, protein synthesis is uncoupled both temporally and spatially from transcription. Before translation can begin, the mRNA transcript must be synthesized in the nucleus, then processed by capping, splicing, and polyadenylation. The mature mRNA is exported into the cytoplasm through nuclear pores, where it emerges as a messenger ribonucleoprotein (mRNP) complex. Initiation of protein synthesis on an mRNP results in mobilization of the mRNA into a monosome, and additional initiation events convert the mRNA into a polysome.

It is thought that the mRNA in mRNP particles is highly structured, ruling out recognition of initiation sites by a mechanism similar to that employed by prokaryotes. Instead, eukaryotes have evolved an entirely different mechanism for recognition of the initiation site, commonly called the scanning model (Kozak and Shatkin 1978). Key features include the recognition of the 5′ terminus of the mRNA and its m^7G-cap structure, followed by binding of the 40S ribosomal subunit and scanning downstream to the initiation codon. It is noteworthy that an mRNA–rRNA interaction, comparable to the SD–ASD interaction in prokaryotes, is not employed. A consequence of m^7G-cap recognition is that eukaryotic mRNAs are monocistronic, since an mRNA contains only a single 5′ terminus. A second strategy, similar to that in prokaryotes but used more rarely in eukaryotes, is 5′ terminus-independent and involves direct binding of the 40S ribosomal subunit to an internal ribosome entry site (IRES) on the mRNA at, or just upstream of, the initiation codon. A priori, this mechanism could allow the use of polycistronic mRNAs, but no such cellular mRNA has been found to date. We focus primarily on the scanning model; detailed treatment of the internal initiation mechanism is found in Chapters 4 and 31.

Another aspect of eukaryotic initiation that differs from the prokaryotic mechanism is the involvement of numerous initiation factors. Eleven or more initiation factors have been identified, comprising over 25 polypeptide chains (Table 2). No compelling explanation has been given for the need of so many initiation factors. However, the reliance on protein–RNA and protein–protein interactions, rather than RNA–RNA interactions, may have contributed to this need. Another common notion is that regulation of the initiation phase requires a more complex process. The fact that most eukaryotic initiation factors are phosphoproteins supports this idea. All of the initiation factor cDNAs/genes have been cloned from a variety of species (Tables 2 and 3) and the three-dimensional structures of some have been determined (see below). The functions of each of these factors are described in detail in the sections below, and the step in the pathway where each appears to function first is depicted in Figure 3.

Features of mRNA Structure Recognized during the Scanning Mechanism

The rate of initiation on different mRNAs varies enormously, and innate efficiencies are influenzed by the mRNA's primary and secondary structures. We provide here a brief sketch of the structural elements most important for determining innate initiation rates. The reader is referred elsewhere for a more detailed discussion and for references to the literature (Kozak 1989, 1999; Merrick and Hershey 1996; Chapter 12).

The presence and availability of the m^7G-cap structure is important, although not absolutely essential for translation (Palmer et al. 1993; Gunnery and Mathews 1995). All cytoplasmic mRNAs are thought to be capped, but some cap structures are "hidden" by secondary structure and cannot be recognized readily by the m^7G-cap-binding initiation factor, eIF4F. Thus, m^7G-cap accessibility correlates with high mRNA efficiency.

The length and secondary structure of the 5′UTR influence translational efficiency, the latter being far more important. The 5′UTRs of most cellular mRNAs are 50–70 nucleotides in length, although much shorter or longer mRNAs translate efficiently. A systematic shortening of a 5′UTR from 32 to 3 nucleotides led to a progressive decrease in recognition of the first AUG (Kozak 1991b), whereas lengthening it from 17 to 77 nucleotides increased efficiency (Kozak 1991a). In yeast, the average length of the 5′UTR is a bit shorter than that in mammalian cells, and cases are known where the AUG is adjacent to the m^7G-cap (Ellis et al. 1987).

The presence of strong secondary structure (–50 kcal/mole) inhibits initiation (presumably scanning), whereas less stable structures can be

Table 2 Initiation factors from mammalian, plant, and yeast cells

	Mammals[a]		Plants[a]		Yeast[a]				
Name	mass (kD)	acc. no.	mass (kD)	acc. no.	gene	mass (kD)	acc. no.	%ID[b]	Function
eIF1	12.6	L26247	12.6	AC005287	*SUI1*	12.3	M77514	58	AUG recognition
eIF1A	16.5	L18960	16.6	AC006951	*TIF11*	17.4	U11585	65	Met-tRNA$_i$ binding to 40S subunit
eIF2α	36.2	J02646	41.6	AF085279	*SUI2*	34.7	M25552	58	affects eIF2B binding by phosphorylation
eIF2β	39.0	M29536	26.6	AL162351	*SUI3*	31.6	M21813	42	binds to eIF2B, eIF5
eIF2γ	51.8	L19161	50.9	AC002411	*GCD11*	57.9	L04268	71	binds GTP, Met-tRNA$_i$; GTPase
eIF2Bα	33.7	U05821	39.8	AC016529	*GCN3*	34.0	M23356	42	nonessential; helps recognize P-eIF2
eIF2Bβ	39.0	U31880	43.6	AC012395	*GCD7*	42.6	L07116	36	binds GTP, helps recognizes P-eIF2
eIF2Bγ	50.4	U38253			*GCD1*	65.7	X07846		GEF activity
eIF2Bδ	57.8	Z48225	29.4	AC016041	*GCD2*	70.9	X15658	36	binds ATP, helps recognizes P-eIF2
eIF2Bε	80.2	U19511	81.9	AC004238	*GCD6*	81.2	L07115	30	GEF activity
eIF3	*See Table 3*								
eIF4AI	44.4	X03039.1	46.7	AB019229	*TIF1*	45.1	X12813	65	ATPase, RNA helicase
eIF4AII	46.3	X12507.1	46.8	AC005287	*TIF2*	44.6	X12814		ATPase, RNA helicase
eIF4B	69.2	S12566	57.6	AF021805	*TIF3*	48.5	X71996	26	binds RNA; stimulates helicase
eIF4E	25.1	M15353	26.5	AL110123	*CDC33*	24.3	M21620	33	binds m^{7}G-cap of mRNA
		(eIFiso4E)	22.5	AB013393					
eIF4GI	171.6	AF104913	153.2		*TIF4631*	107.1	L16923	22	binds eIF4E, 4A, 3, PABP•RNA
eIF4GII	176.5	AF012072	176.5		*TIF4632*	103.9	L16924	21	binds eIF4E, 4A, 3, PABP•RNA
		(eIFiso4G)	87.0	AB013396					
eIF5	48.9	L11651	48.6	AC007576	*TIF5*	45.2	L10840	39	stimulates eIF2 GTPase
eIF5B	139.0	AF078035			*FUN12*	97.0	L29389	70	GTPase; promotes junction reaction

[a] The masses (kD) and accession numbers pertain to human or rat, *Arabidopsis thaliana,* and *S. cerevisiae* proteins. Only one of the numerous isoforms of the plant proteins was included arbitrarily. A complete listing of *A. thaliana* initiation factors is found in Browning (1996).

[b] %ID, percent sequence identity shared by yeast and human proteins.

Table 3 eIF3 subunits from mammalian, plant, and yeast cells

Human			*Arabidopsis*			Yeast				
name	mass	acc. no.	name	mass	acc. no.	name	gene	mass	acc. no.	Comments
p170	166.5	D50929	eIF3a	114.3	AL050399	p110	*TIF32*	111.1	AF004912	(*RPG1*) binds RNA, PCI family
p116	98.9	U78525	eIF3b	81.9	AC007478	p93	*NIP1*	88.1	J02674	binds eIF1, eIF5
p110	105.3	U46025	eIF3c	103.0	AF040102	p90	*PRT1*	93.4	L02899	PCI family
p66	64.0	U54558	eIF3d	66.2	AB001475					binds RNA
p48	52.2	U54562	eIF3e	51.8	A1137080					PCI family, Int-6
p47	37.6	U94855	eIF3f	31.9	AF002109					MDN motif
p44	35.4	U96074	eIF3g	32.7	AC008153	p36	*TIF35*	30.5	AF004913	binds RNA, eIF4B
p40	39.9	U54559	eIF3h	38.4	AC007354					MDN motif
p36	36.5	U39067	eIF3i	36.4	AC005397	p39	*TIF34*	38.8	U56937	WD repeats, TRIP1
p35	29.0	U97670					*HCR1*	29.6	U14913	
p28	25.1	N/A	eIF3k	25.7	AL61583					
						p135	*TIF31*	145.2	AF004911	(also called *CLU1*)

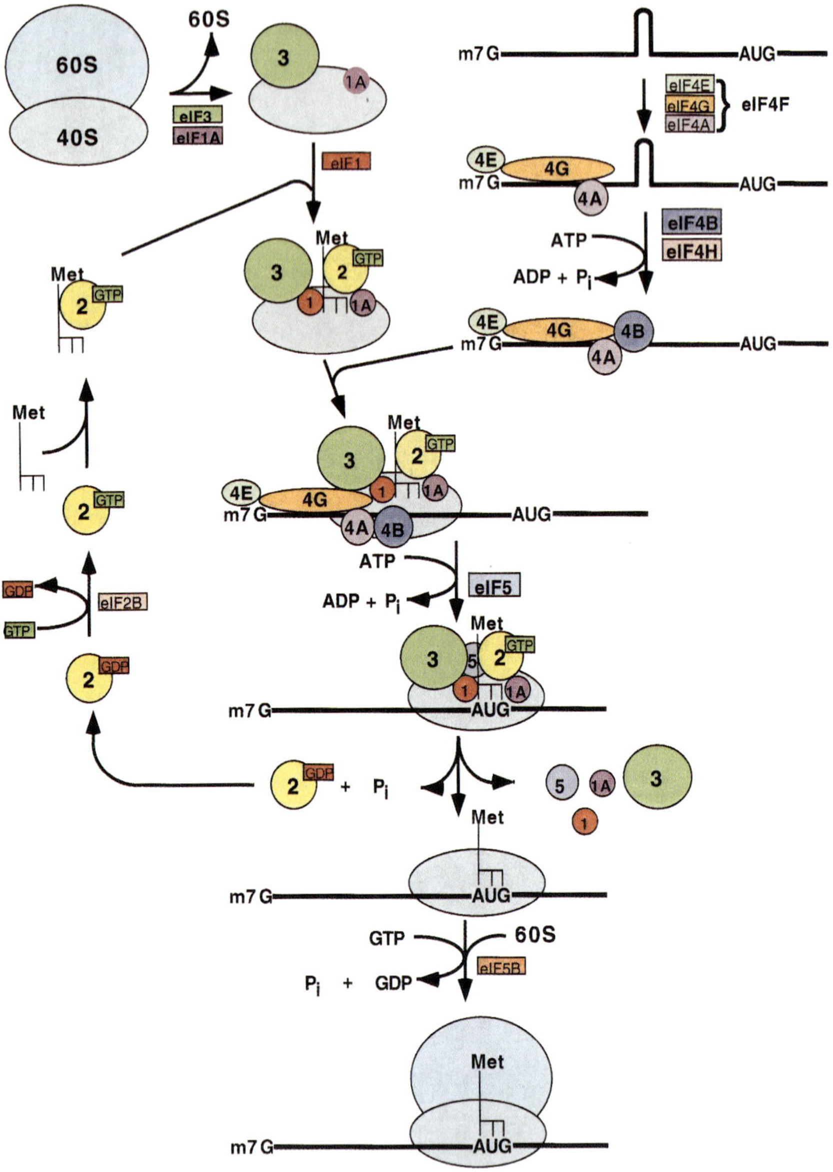

Figure 3 Initiation pathway in eukaryotes. The pathway is a working model of the individual reactions in the initiation pathway. Initiation factors are shown as color-coded labeled circles, and appear when first implicated in the pathway. The symbols for ribosomal subunits, Met-tRNA$_i$, and mRNA are self-evident. Higher-order complexes are hypothetical and may not be accurate representations of native structures.

removed by the RNA helicase activity of the initiation factors (see below). However, such less stable structures, when close to the m^7G cap, inhibit, presumably by reducing the cap's accessibility to the cap-binding factor (Lawson et al. 1988). It is noteworthy that a stable hairpin structure about 12 nucleotides downstream from the AUG initiation codon enhances initiation efficiency, likely by causing a pause in scanning at the initiation codon. Such secondary structure is likely encountered just as the ribosome leaves the 5′UTR and enters the coding region (Merrick 1992). It is important to realize that no detailed three-dimensional structure for any mRNA has been determined, much less the structure of an mRNA complexed with protein (mRNP). An obvious caveat is that a computer-derived structure of the 5′UTR, or a naked mRNA prepared with denaturing solutions such as phenol, may not reflect or possess the native structure of the mRNA.

The nature and placement of the initiation codon are critical. AUG is the predominant initiation codon in mammalian cells, and appears to be used exclusively in yeast. The 5′-proximal AUG serves as the initiation codon in more than 90% of mRNAs, suggesting that its position in the 5′UTR plays a dominant role in its selection. The sequence surrounding the AUG is critical in mammalian cells (but much less so in yeast): Efficient initiation codons lie in a context with purines at positions –3 and +4 (where the A of the AUG is +1). Scanning ribosomes that encounter an AUG with a poor match to the consensus sequence may pass over the AUG and initiate downstream at another AUG in a more favorable context. This phenomenon, called "leaky scanning," leads to two protein products, either with different amino termini or with entirely different sequences if translated in different reading frames. The presence of an AUG upstream from the major initiation codon generally reduces the latter's efficiency of initiation. Thus, upstream AUGs and open reading frames are used to regulate initiation, as described in Chapters 5 and 18.

Finally, the poly(A) tail and elements in the 3′UTR affect the efficiency of translation initiation. A positive synergy between the m^7G-cap and poly(A) tail structures has been shown that depends on the poly(A)-binding protein (Chapter 10). In yeast, the poly(A)-binding protein complexed with poly(A) stimulates the binding of 40S ribosomal subunits to the 5′ end of mRNAs. An intriguing possibility is that the poly(A) tail enables ribosomes to efficiently recycle on the same mRNA, thereby maintaining large polysomes (discussed below in the section Recycling and Reinitiation).

Building the 40S Ribosomal Subunit Pool

80S ribosomes are the most prominent species at the presumed physiological free magnesium concentration (~1–2 mM), and must dissociate into 40S and 60S subunits for initiation to proceed. Dissociation (or anti-association) is thought to be promoted by two initiation factors, eIF3 and eIF1A, although the dissociation reaction is poorly characterized. One view is that eIF3 and eIF1A bind to the 40S ribosomal subunit, preventing its association with the 60S ribosomal subunit by steric hindrance (Goumans et al. 1980). Indeed, eIF3 binds to the 40S ribosomal subunit quite avidly in the absence of other components of initiation (Benne and Hershey 1976). Furthermore, native 40S subunits containing eIF3, but devoid of eIF2, Met-tRNA$_i$, and mRNA, are detected in reticulocyte lysates (Ayuso-Parilla et al. 1973). Early studies of complexes of eIF3 (and eIF2) on the 40S ribosomal subunit by electron microscopy led to a model where the eIF3 touches eIF2, but binds somewhat removed from the decoding site (Bommer et al. 1991). In contrast, eIF3 visualized on the native 40S subunit by cryo-EM at 48 Å resolution appears to be oriented away from the subunit–subunit interface, and thus, the decoding site (Srivastava et al. 1992). The latter study appears to rule out a simple steric hindrance mechanism; an allosteric effect due to a change in the structure of the 40S subunit upon eIF3 binding remains possible. Resolution of the issue of eIF3 placement on the 40S subunit awaits studies at higher resolution. An alternative explanation for how eIF3 functions as an anti-association factor has been proposed where eIF3 does not dissociate 80S ribosomes directly, but rather prevents 60S subunits from displacing the eIF2•GTP•Met-tRNA$_i$ ternary complex from the 40S preinitiation complex (Merrick et al. 1973; Chaudhuri et al. 1999). Finally, a third protein, called eIF6, has been shown to bind to the 60S subunit and prevent its association with the 40S subunit (Russell and Spremulli 1979; Raychaudhuri et al. 1984), although its role in the initiation pathway has been questioned (Si and Maitra 1999).

eIF1A

eIF1A is a small, stable protein (17–22 kD) that is one of the most highly conserved of the initiation factors (Table 2). It is an essential protein in *S. cerevisiae*, and mammalian eIF1A cDNA can substitute for the yeast gene in vivo (Wei et al. 1995). Depletion of eIF1A in yeast cells results in polysome runoff and thus an inhibition of initiation (M. Kainuma and J.W.B. Hershey, unpubl.). Yeast eIF1A exhibits 21% sequence identity with *E. coli* initiation factor IF1 (Kyrpides and Woese 1998b). It is very polar, with 10 of the first 22 amino-terminal residues being basic and 13 of the last 20 carboxy-terminal residues being acidic. The three-dimensional solution structure of the human factor (Fig. 4A) has been determined by NMR spectroscopy (Battiste et al. 2000). The protein contains two structural domains, and like its bacterial homolog IF1, it also contains an OB domain. eIF1A binds RNA in a non-sequence-specific manner, either to mRNA or rRNA, although the actual physiological target is not known. The role of eIF1A in initiation is pleiotropic, as it not only may affect ribosome dissociation, but also is involved in the binding of Met-$tRNA_i$ to 40S ribosomes and in mRNA binding and scanning (see below). eIF1A interacts with eIF5B (Schreier et al. 1977; Chapter 9), thereby mimicking the binding of IF1 to IF2 (the prokaryotic homolog of eIF5B) (Boileau et al. 1983). On the basis of the possible function of IF1 and IF2 (see above), eIF1A and eIF5B may occupy the tRNA-binding A site on the eukaryotic 40S ribosome.

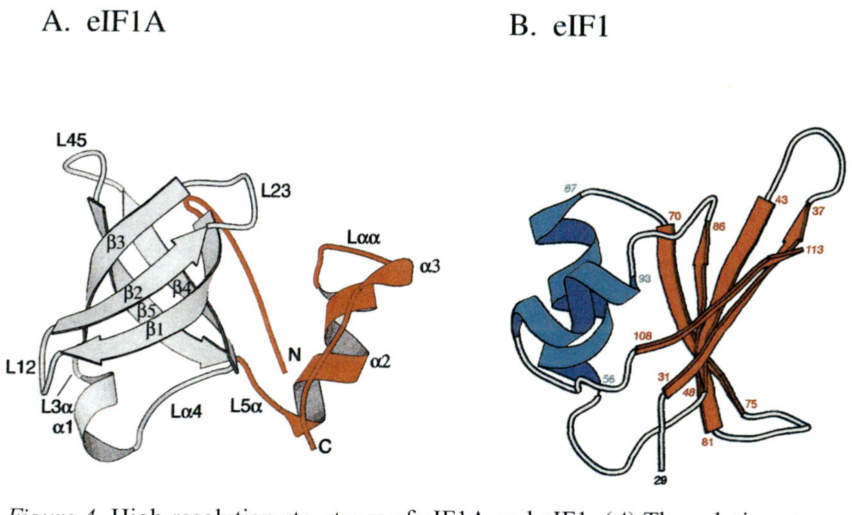

Figure 4 High-resolution structures of eIF1A and eIF1. (*A*) The solution structure of human eIF1A solved by NMR spectroscopy. The small carboxy-terminal domain is shown in red. Portions of the amino and carboxyl termini are not shown as they lack structure. (*B*) The solution structure of human eIF1 solved by NMR spectroscopy. The β-sheet domain is shown in red and the α-helical domain is shown in blue. The unstructured 28 amino-terminal residues are not shown. (*A*, reprinted, with permission, from Battiste et al. 2000 [copyright Cell Press]; *B*, reprinted, with permission, from Fletcher et al. 1999 [copyright Oxford University Press].)

eIF3

eIF3 was first isolated and purified from rabbit reticulocytes as a high-molecular-weight complex (Benne and Hershey 1976; Safer et al. 1976; Schreier et al. 1977). The mammalian factor, by far the largest of the initiation factors, possesses a molecular mass of about 600,000 daltons and contains at least eleven different subunits: p170, p116, p110, p66, p48, p47, p44, p40, p36, p35, and p28 (Table 3). The subunits are thought to be present in stoichiometric amounts in the complex, although this fact is not well established. A similar complex has been characterized from plants (Table 3) (Browning 1996), whereas eIF3 in yeast (*S. cerevisiae*) is related, but smaller (see below). Mammalian eIF3 has been implicated not only in 80S ribosome dissociation, but also in Met-tRNA$_i$ and mRNA binding to 40S ribosomal subunits (Table 3). It also interacts with numerous other initiation factors (see below) and likely helps to organize higher-order initiation complexes on the 40S ribosomal surface. At least four of its subunits, p170, p116 or p110, p66, and p44, bind to RNA.

The cDNAs/genes encoding the eIF3 subunits have been cloned from mammals, plants, and yeast (Table 3), but a detailed structure of the mammalian or plant eIF3 complex has not yet been determined. Attempts to identify specific mammalian eIF3 subunits that bind to other initiation factors have been carried out. NMR spectroscopy was used to demonstrate the binding of eIF1 to p110 (Fletcher et al. 1999). eIF5 copurifies with oligohistidine-tagged eIF3 (Bandyopadhyay and Maitra 1999), and a fragment of eIF4B binds in vitro to the p170 subunit, which had been subjected to SDS-PAGE, transferred to a membrane, and renatured prior to probing (Far-Western blot analysis) (Méthot et al. 1996a). However, full-length eIF4B appears not to interact with p170 in this analysis, but rather binds to p44 (F. Peiretti and J.W.B. Hershey, unpubl.), a result consistent with the yeast eIF4B–p33 interaction described below. The central domain of mammalian eIF4G interacts with eIF3 (Lamphear et al. 1995; Mader et al. 1995), but the eIF3 subunit(s) responsible for the interaction has not been identified.

eIF3 was purified from *S. cerevisiae* by employing either of two assay systems: stimulation of methionyl-puromycin synthesis based on mammalian assay components (Naranda et al. 1994a) and stimulation of protein synthesis in a heat-inactivated yeast lysate derived from a conditional mutant, *prt1*-1 (Danaie et al. 1995). Complexes of six to eight subunits were isolated, and the genes encoding the eIF3 subunits were cloned (Table 3). Rapid isolation of eIF3 containing an oligohistidine-tagged p90 (PRT1) subunit identified a core of five subunits in a complex with eIF5 (Phan et al. 1998). The five core subunits, p110 (TIF32), p93 (NIP1), p90

(PRT1), p39 (TIF34), and p33 (TIF35), are all essential for yeast growth and have orthologs in mammalian eIF3 (Table 3). Other yeast proteins appear to associate less tightly with the core complex, and their possible roles in eIF3 activity are not known. GCD10 (p62) coimmunoprecipitates with eIF3 (Garcia-Barrio et al. 1995) but is not required for eIF3 activity (Anderson et al. 1998). eIF1 (SUI1, p16) also coimmunoprecipitates and, furthermore, interacts specifically with the p93 subunit (Asano et al. 1998) but is only loosely associated with eIF3. A 135-kD protein (TIF31) copurifies with eIF3, but its depletion does not affect initiation or cell growth (Vornlocher et al. 1999). Curiously, the 135-kD protein is identical to CLUA, a protein implicated in mitochondrial morphology (Fields et al. 1998). Finally, HCR1 (see below) when overexpressed represses a temperature-sensitive mutant form of the p110 subunit (Valasek et al. 1999), but its role in eIF3 remains unclear.

Studies in yeast using conditional mutant forms of the p110 (Valasek et al. 1998), p93 (Greenberg et al. 1998), p90 (Evans et al. 1995), and p39 (Verlhac et al. 1997; Asano et al. 1998) subunits, and in vivo depletion of p110 (Vornlocher et al. 1999), p93 (Greenberg et al. 1998), p39 (Naranda et al. 1997), and p33 (Hanachi et al. 1999), show inhibition of initiation (i.e., polysomes are reduced) whenever one of the subunits is inactivated or removed. The p33 subunit contains an RRM and binds RNA (Hanachi et al. 1999); p90 possesses a degenerate RRM (Evans et al. 1995), but RNA binding has not been shown with certainty. Identification of protein–protein interactions by yeast two-hybrid analyses and GST pull-down experiments (Asano et al. 1998; Phan et al. 1998), together with genetic interactions (for review, see Chapter 5), has generated a working model of the core complex (Asano et al. 1998). Similar techniques were used to demonstrate that p93 binds to eIF1 and to eIF5 (Asano et al. 1998; Phan et al. 1998), whereas p33 binds to eIF4B (Vornlocher et al. 1999). The possible binding of yeast eIF4G to an eIF3 subunit has not yet been reported. A challenge for future research is to obtain high-resolution three-dimensional models of eIF3 and its complexes with other initiation factors.

Given the similarities between other yeast and mammalian initiation factors (Table 2), the structural differences seen with eIF3 are surprising. An example is the mammalian p170 subunit, which contains a repeat region that is lacking in the yeast ortholog, p110. It appears that mammalian and plant eIF3 have evolved to incorporate extra subunits in addition to the five subunits homologous with the yeast core eIF3. Three of the mammalian subunits (p48, p110, and p170) contain the PCI homology domain (a domain present in the "lid" regulatory complex of the 26S

*P*roteasome, the plant photomorphogenic regulatory complex *C*OP9, and initiation factor e*I*F3); the p40 and p47 subunits contain the MPN motif, first observed at the *N*-terminus of the yeast protein *M*pr1 and *P*ad1 (Aravind and Ponting 1998; Glickman et al. 1998; Hofmann and Bucher 1998). Since the PCI and MPN domains are found in components of large protein complexes, it was suggested that the five eIF3 subunits may serve as a structural scaffold or as docking sites for other proteins.

The p48 subunit is the product of the *Int-6* gene, which in mouse is the site of frequent integration by the mouse mammary tumor virus (MMTV). This surprising finding suggests that eIF3 may play a role in the regulation of the cell cycle (see Chapter 20). Mammalian p35 is related to the yeast *HCR1* gene product, whose overexpression suppresses the phenotype of a temperature-sensitive p110 mutant (Valasek et al. 1999). The p35 subunit appears to be absent in plant eIF3. Given the striking similarities of the initiation process in all eukaryotic species, the apparent discrepancies above may be due in part to subtle differences in the strengths of various protein–protein interactions, as noted also for eIF4F in the next section. Other differences likely reflect the early stage of research in this area and may disappear as better structural evidence is generated.

Met-tRNA$_i$ Binding to 40S Ribosomal Subunits

A specific tRNA derivative is used to initiate protein synthesis: methionyl-tRNA$_i$ (Met-tRNA$_i$). Met-tRNA$_i$ binds to eIF2•GTP to form a ternary complex that is an obligate intermediate in its binding to ribosomes. eIF2 distinguishes the initiator tRNA from elongator tRNAs by recognizing the methionyl residue and the A-U base pair at the end of the acceptor stem. Initiator tRNAs also uniquely possess three G-C base pairs in their anticodon stems, but this feature is important for ribosome binding, not eIF2 recognition. A detailed description of initiator tRNAs is provided in Chapter 5. eIF2 comprises three non-identical subunits: α, β, and γ (Table 2; Fig. 5). Formation of the ternary complex requires GTP (or a nonhydrolyzable GTP analog) and is inhibited by GDP. The γ subunit is implicated in both GTP and Met-tRNA$_i$ binding (see below). Ternary complexes can be prepared in vitro at physiological Mg^{++} concentration in the absence of other translational components and are readily detected by filtration through nitrocellulose membranes. At dilute concentrations the complex dissociates, but the equilibrium shifts toward the ternary complex in the presence of either eIF3 or eIF2C (Gupta et al. 1990). The role of eIF2C in protein synthesis is not well established, although its cDNA has been cloned (Zou et al. 1998; see below).

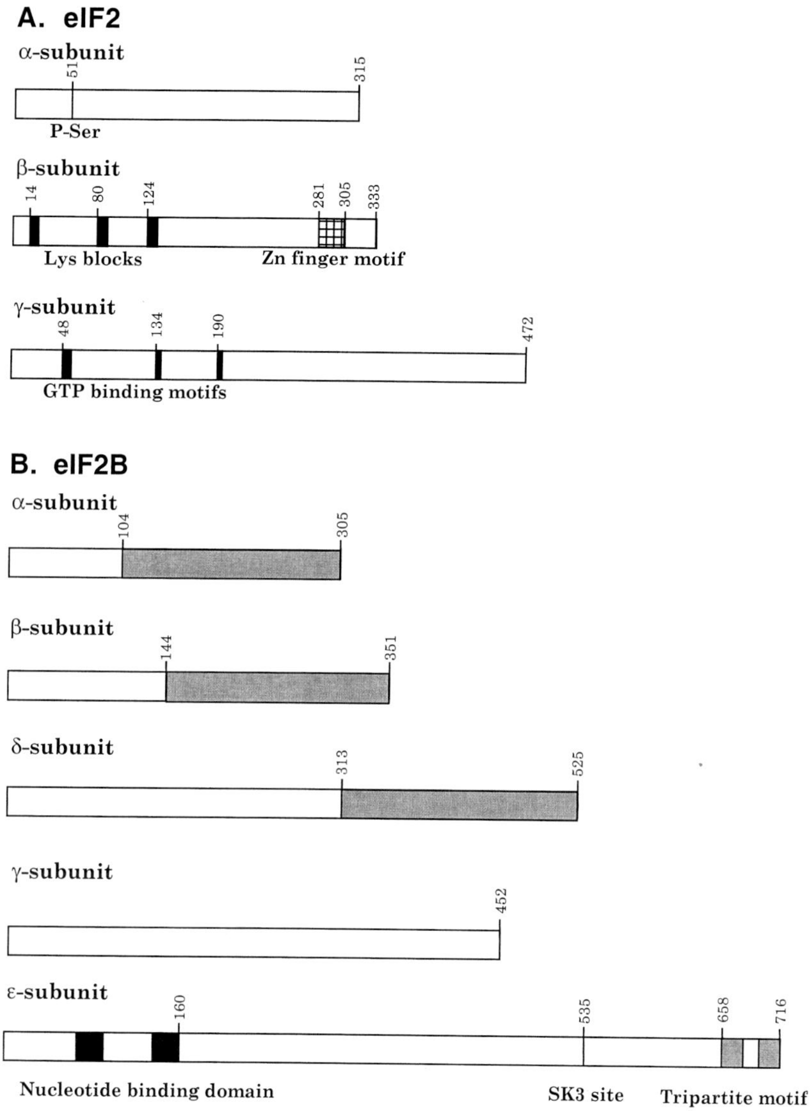

Figure 5 Structural motifs in the subunits of eIF2 and eIF2B. (*A*) eIF2 subunits (*human*). The site of phosphorylation in the α subunit is shown. For the β subunit, the three Lys blocks (*black*) and the Zn finger motif (*hatch*) are identified. The three GTP-binding motifs in the γ subunit are shown in black. (*B*) eIF2B subunits (rat). The homologous regions of the α, β, and δ subunits are shown in gray. In the ε subunit, the nucleotide-binding site (*black boxes*), the tripartite motif involved in binding to eIF2β (*gray*), and the glycogen synthase kinase 3 phosphorylation site (SK3 site) are identified (Welsh et al. 1998; Anthony et al. 2000). Domains for the γ subunit have not yet been identified for the mammalian protein. For a detailed description of yeast eIF2B subunits, see Chapter 5.

eIF2, as it recycles after a round of initiation, leaves the ribosome as a binary complex with GDP. In order to bind another Met-tRNA$_i$, it must be converted from its inactive eIF2•GDP form to its active eIF2•GTP form by a guanylate exchange reaction. The nonenzymatic exchange reaction is slow and requires catalysis by eIF2B. eIF2B contains five nonidentical subunits (Table 2; Fig. 5) and forms a complex with eIF2, GDP, and GTP. Evaluation of mutant forms of yeast eIF2 and the five subunits of eIF2B have shed light on the molecular interactions involved (see Chapter 5), but the detailed catalytic mechanism remains controversial (Manchester 1997). The interaction of eIF2B with eIF2 involves the lysine blocks found at the amino terminus of eIF2β (Asano et al. 1999). A complex of the ε and γ subunits of yeast eIF2B possesses strong guanylate exchange activity, whereas the remaining three subunits form a subcomplex that distinguishes between the phosphorylated and nonphosphorylated forms of eIF2 (described in detail in Chapter 5).

The ternary complex binds to the 40S ribosomal subunit to form a 40S preinitiation complex (sometimes called the 43S initiation complex). The complex in the absence of mRNA and other translational components can be detected by sucrose gradient centrifugation and is stabilized by the presence of eIF3 and eIF1A. That Met-tRNA$_i$ binding can precede mRNA binding is indicated by the detection of 40S preinitiation complexes lacking mRNA in rabbit reticulocytes (Smith and Henshaw 1975). Recent experiments indicate that eIF1A acts catalytically to promote ternary complex binding to the 40S ribosomal subunit in the absence of 60S subunits (Chaudhuri et al. 1997, 1999). The presence of AUG or mRNA also stabilizes Met-tRNA$_i$•40S complexes when analyzed by sucrose gradient centrifugation (see next pathway step below).

The binding of Met-tRNA$_i$ to 40S ribosomal subunits is a common step in the translation of all mRNAs. Regulation of this step is frequently used to control global rates of protein synthesis. Ternary complex formation and Met-tRNA$_i$ binding to ribosomes are inhibited indirectly by phosphorylation of the α subunit of eIF2 by highly specific protein kinases. Four different eIF2α kinases have been identified that in general are activated by cell stress; their function and regulation are described in detail elsewhere in this volume: HRI (Chapter 14); PKR (Chapter 13); PERK (Chapter 15); and GCN2 (Chapter 5). Phosphorylated eIF2 is inactive in the eIF2B-catalyzed guanylate exchange reaction and therefore cannot form a ternary complex after completing a round of initiation. Phosphorylation converts eIF2 from an exchange substrate to a competitive inhibitor of eIF2B activity. By binding more tightly to eIF2B, it in effect sequesters eIF2B, whose level in cells is two- to fivefold lower than that of eIF2 (Oldfield et al. 1994). Therefore, only partial phosphorylation of eIF2 is sufficient to inhibit all of the eIF2B and to prevent the recycling of eIF2. eIF2B itself may be regulated by phosphorylation of the ε subunit (Welsh and Proud 1993; Singh et al. 1994). Detailed descriptions of the regulation of the Met-$tRNA_i$ binding step are found in Chapters 5, 8, and 16.

eIF2

eIF2 has been purified from animals, plants, insects, and eukaryotic microorganisms and is highly conserved (Tables 2 and 4), with homologs found in archaea, but not in eubacteria. All three eIF2 subunits are essential in yeast. Although normally isolated as a heterotrimeric complex, an active α–γ dimeric complex is sometimes purified. Thus, an intact β subunit appears not to be essential for ternary complex formation in vitro; however, fragments of the β subunit may remain in such preparations, which could confer essential activities. The concentration of eIF2 in HeLa cells has been estimated to be about 1 μM, with 0.5 eIF2 molecule per ribosome (Duncan and Hershey 1983).

There has been no report of the crystallization of eIF2 or its individual subunits, nor has a high-resolution three-dimensional structural model been proposed. However, insight into the function of eIF2 comes from an examination of the subunits' primary sequences and from analyses of mutant forms of the yeast factor (Fig. 5). Mammalian eIF2α is phosphorylated on Ser-51, which lies in a highly conserved region (except in archaea). The amino-terminal region of the α subunit is implicated in AUG recognition, as *sui2*$^-$ mutations occur there (Chapter 12).

The β subunit contains three lysine blocks near its amino terminus that have been implicated in binding to eIF2B and to eIF5 (Asano et al. 1999) and also to mRNA (Flynn et al. 1994; Laurino et al. 1999). eIF2β also contains a zinc-finger motif (but no Zn^{++}) near its carboxyl terminus that plays an important role in the recognition of the initiation codon (Chapter 12).

eIF2γ is homologous to bacterial SelB and EF1A, as well as to other G-proteins. Except for the GTP-binding domain, eIF2γ (or the other eIF2 subunits) does not resemble bacterial IF2, despite the apparent similarity in their function. eIF2γ crosslinks to GTP and to Met-$tRNA_i$ (Gaspar et al. 1994). Mutant forms of yeast eIF2γ have been characterized that show reduced binding of either Met-$tRNA_i$ or GTP (Harashima and Hinnebusch 1986; Erickson and Hannig 1996). More detailed descriptions of eIF2 subunits and their interactions with eIF2B are provided in Chapter 5.

eIF2B

Mammalian eIF2B comprises five subunits with masses of 26 kD (α), 39 kD (β), 58 kD (γ), 67 kD (δ), and 82 kD (ε) (Table 2), although small mass differences are seen in various species (Kimball et al. 1994). Comparable subunits are found in yeast, where all but the α subunit are essential for growth. Although eIF2 homologs are found in archaebacteria, no protein comparable to the eIF2B subunits is found encoded in the complete genome sequence of the archaebacteria. The cellular level of eIF2B varies in different cell types, but generally is lower than that of eIF2 (Oldfield et al. 1994).

Neither eIF2B nor any individual subunit has been crystallized, and no high-resolution structure is available to date. However, the five subunits have been expressed in Sf921 insect cells, allowing reconstitution and purification of active eIF2B (Fabian et al. 1998). The ability to reconstitute recombinant eIF2B should allow rapid advances in our understanding of its structure and function. Computer-assisted analysis of amino acid sequences has identified a complex multidomain organization of yeast and human eIF2B subunits (Fig. 5).

eIF2B binds weakly to GTP, with photoaffinity labeling implicating the β subunit (Dholakia et al. 1989). The binding of ATP to the γ and δ subunits, and NADPH to the β subunit, also have been reported (Dholakia et al. 1986; Oldfield and Proud 1992). Only the γ and ε subunits contain potential nucleotide-binding motifs in their sequences (Koonin 1995). ATP and NADPH binding may sense the energy and redox state of the cell, but their functional roles remain to be better established.

In yeast, a subcomplex of the γ and ε subunits possesses strong guanylate exchange activity, even with phosphorylated eIF2 (Pavitt et al. 1998), whereas a subcomplex of the α, β, and δ subunits has no such activity but distinguishes between the phosphorylated and nonphosphorylated forms of eIF2. The α, β, and δ subunits show sequence similarities between one another, yet all three subunits are required for binding to eIF2 (Pavitt et al. 1998). The subcomplexes, together with mutational analyses of both yeast and mammalian eIF2B subunits, are described in detail in Chapter 5.

Binding of the 40S Preinitiation Complex to mRNA

We restrict our discussion of mRNA binding to the 5′-end-dependent mechanism. The m^7G-cap structure is recognized by eIF4E, likely in the form of eIF4F. eIF4F is a heterotrimeric complex comprising eIF4E, eIF4G, and eIF4A (Table 2). The affinity of eIF4F for capped mRNAs varies, with cap accessibility playing an important role. If stable secondary structure is present close to the m^7G cap, eIF4F is not able to bind efficiently. The influence of RNA-binding proteins present in free mRNPs and polysomes on eIF4F binding to the m^7G cap has not been determined. For a discussion of mRNA masking by such proteins, see Chapter 7. The binding of eIF4F to an m^7G cap commits the translational apparatus to the translation of that mRNA. eIF4F is recognized as the key factor in selecting mRNAs for translation, and its activity therefore is regulated by several important mechanisms (Chapter 6). The level of the eIF4F complex is affected by a family of proteins, called 4E-BPs, that bind to eIF4E and prevent its association with eIF4G. Their affinity for eIF4E is greatly reduced by 4E-BP phosphorylation. Furthermore, eIF4E and eIF4G phosphorylation regulates the activity of the cap-binding complex. The presence of proteins structurally related to eIF4G, such as PAIP and p97, further complicates the situation. Conditions that down-regulate eIF4F activity increase the competition between mRNAs. Consequently, regulation of eIF4F activity influences not only the level of total protein synthesis, but also the class of mRNAs being translated. The proteins and regulatory mechanisms are described in detail elsewhere (Gingras et al. 1999; Chapter 6).

An unstructured mRNA region appears to be essential for the binding of the 40S preinitiation complex to the 5′-terminal region of the mRNA. To accomplish this, eIF4F, together with eIF4B, possesses ATP-dependent RNA helicase activity that presumably melts out secondary structure in the 5′-proximal region of the mRNA (Rozen et al. 1990). The eIF4A subunit is responsible for the RNA helicase activity and can catalyze RNA unwinding in the absence of eIF4E and eIF4G. However, the eIF4F complex possesses even stronger helicase activity and is likely the physiologically relevant form of the activity. The helicase activity is further increased by eIF4B and eIF4H, which cause a switch in the eIF4F activity from nonprocessive to processive (Rogers et al. 1999; G. Rogers, Jr. and W.C. Merrick, unpubl.). By coupling the action of eIF4A and eIF4B or eIF4H to eIF4E through eIF4G, the helicase activity is directed to duplexes near the 5′ terminus of the mRNA. The activity is rather weak, as high concentrations of the factors are required in vitro. A weak activity may be essential in the context of initiation, where too much unwinding of the mRNA and removal of associated proteins may be deleterious.

Once secondary structure is removed from the 5′-terminal region, an interaction between eIF4G and eIF3 (bound to the 40S subunit) brings the ribosome to the mRNA. Thus, a protein bridge is constructed between the mRNA m^7G cap and the ribosome: m^7G cap – eIF4E – eIF4G – eIF3 – 40S ribosomal subunit. The initial binding of the 40S subunit to an mRNA differs substantially from the prokaryotic process where two RNA–RNA interactions promote mRNA binding to ribosomes. By employing an m^7G-cap-dependent mechanism, ribosome binding necessarily occurs at the 5′ terminus of the mRNA.

eIF4E

eIF4E is conserved from yeast to man (Table 2), with the mammalian protein capable of substituting for the yeast factor (Altmann et al. 1989). A number of eIF4E forms have been detected in various organisms; e.g., two in humans (Rom et al. 1998) and five in *Caenorhabditis elegans* (Keiper et al. 2000). The functional implications of multiple forms is not clear, however. eIF4E alone binds to m^7G cap analogs (e.g., m^7GDP) with high specificity and moderate affinity, allowing the facile purification of eIF4E and its complexes by affinity chromatography with m^7G columns. When eIF4G associates with eIF4E, the complex binds capped mRNA tenfold more tightly than eIF4E alone (Haghighat and Sonenberg 1997). A study using cap analogs to inhibit protein synthesis in vitro reported an apparent inhibitor constant of about 4 μM for m^7GTP (Cai et al. 1999); this value presumably reflects its affinity for eIF4F. Recently, the binding of eIF4E to the m^7G cap was visualized in three dimensions with the solution of the eIF4E•m^7GDP structure (Fig. 6) by both X-ray crystallography and high-field NMR spectroscopy (Marcotrigiano et al. 1997; Matsuo et al. 1997). m^7GDP binding occurs in a pocket on the concave side of the protein (Fig. 6). It is achieved by intercalating the m^7G ring into a stack of two highly conserved tryptophan residues (56 and 102). The specific recognition of G occurs through hydrogen bonds to the side-chain oxygens of Glu-103 and the main-chain nitrogen of Trp-102 (this mimics the recognition of G by C in a normal Watson-Crick base pair: two hydrogen-bond acceptors and one donor). Ser-209, the site of regulated phosphorylation (Chapter 6), lies on the same concave face and is in the vicinity of Lys-159. It has been suggested that phosphorylation of this serine enables formation of a salt bridge, thereby clamping the protein to the m^7G structure.

eIF4G binds to eIF4E at a convex surface on the side of the protein opposite that which binds m^7G. The same surface of eIF4E binds to 4E-BP1, as shown by NMR analysis of complexes of eIF4E and peptide fragments of 4E-BP1 and 4E-BP2 (Matsuo et al. 1997). A striking feature of eIF4E binding to the 4E-BPs is that the 4E-BP proteins are unfolded (Matsuo et al. 1997; Fletcher et al. 1998), and binding occurs by an

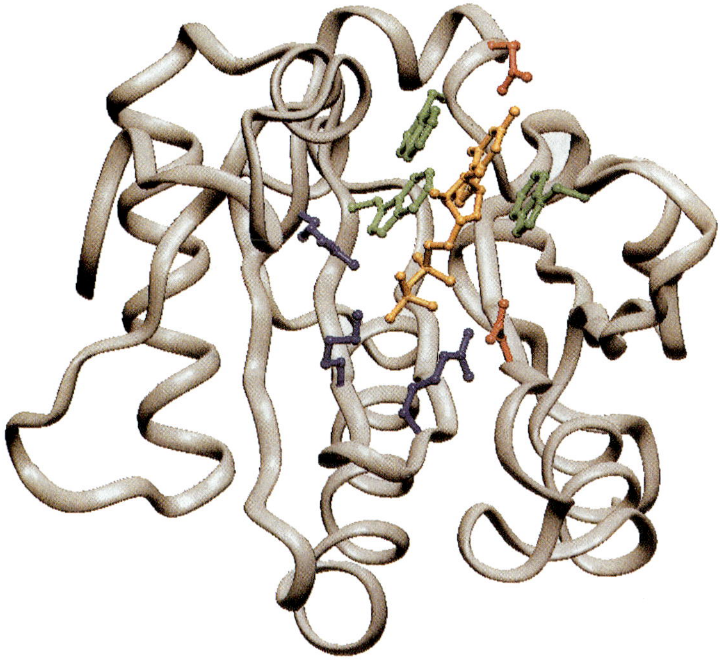

Figure 6 The high-resolution structure of the eIF4E•m^{7}GDP complex. The structure is based on X-ray crystallographic analysis of the murine eIF4E•m^{7}GDP complex (Marcotrigiano et al. 1997), kindly provided by A.-C. Gingras, J. Marcotrigiano, S. Burley, and N. Sonenberg. The m^{7}GDP cap analog, shown in gold, lies in a pocket; the three tryptophan residues (W56, W102, W166) that interact with the m^{7}GDP are green; the R112, K162, and R157 residues are blue (counterclockwise in the structure); E103 (*top of figure*) and D90 (*middle*) are red.

induced fit mechanism. Further details of the structure of the 4E-BPs and their interaction with eIF4E are provided in Chapter 6.

eIF4E is thought to be a limiting initiation factor in cells, one involved in discriminating between mRNAs. Measurements of the cellular level of eIF4E vary, with estimates in rabbit reticulocyte lysates ranging from 0.02 copies (Hiremath et al. 1985) to 1 copy (Rau et al. 1996) per ribosome. However, since eIF4E functions as an eIF4F complex, and the amount of eIF4F is regulated by the 4E-BP family of proteins, knowledge of the actual cellular level of eIF4E cannot predict its activity. Since the levels of the 4E-BPs have not yet been defined, and furthermore, their activities are regulated by phosphorylation (Chapter 6), it has been difficult to assess the actual activity of eIF4E.

eIF4G

eIF4G was first recognized as a high-molecular-mass protein (formerly called p220) found in eluates of m^7G affinity columns in a complex with eIF4E and eIF4A, called eIF4F. Its involvement in the initiation of cap-dependent translation was further confirmed by the finding that eIF4G is cleaved following poliovirus infection (Chapter 31). Two forms of mammalian and yeast eIF4G have been characterized (called eIF4GI and eIF4GII) and their cDNAs (genes) have been cloned from mammalian (Yan et al. 1992; Gradi et al. 1998; Imataka et al. 1998) and yeast cells (Goyer et al. 1993). Two forms also exist in plants, called eIF4G and eIFiso4G (Browning 1996).

Regions of mammalian and yeast eIF4GI that interact with other proteins have been identified and mapped (Fig. 7A), indicating that eIF4G serves as a scaffolding protein that brings together other components of the initiation pathway. Human eIF4GI may be divided into three distinct domains of roughly similar size. The amino-terminal third (residues 1–634) binds the poly(A)-binding protein (PABP) and eIF4E and is required for cap-dependent translation (Lamphear et al. 1995; Mader et al. 1995; Imataka et al. 1998). Picornaviral proteases cleave the amino-terminal third from the rest of the protein (Lamphear et al. 1993), in effect separating the m^7G-cap-recognition region from downstream functions. The central domain (residues 635–1039) binds eIF3 and eIF4A (Imataka and Sonenberg 1997) and possesses an RNA-binding site (Pestova et al. 1996). It alone promotes 40S initiation complex formation with encephalomyocarditis virus (EMCV) IRES RNA (Pestova et al. 1996) and stimulates the translation of uncapped mRNAs (De Gregorio et al. 1998). The carboxy-terminal third contains a second eIF4A-binding site (Imataka and Sonenberg 1997) and binds to the protein kinase, Mnk1 (Pyronnet et al. 1999). The minimal region required for cap-dependent translation has been mapped to residues 550–1090 (Morino et al. 2000); this fragment includes the eIF4E-binding site and the central domain. A point mutation in eIF4GI that abolishes eIF4A binding to the central domain does not support translation, whereas eIF4GI with a comparable mutation in the eIF4A-binding site of the carboxy-terminal domain is active, although about sixfold less than wild-type eIF4GI (Morino et al. 2000). Thus, the carboxy-terminal domain (which is absent in yeast eIF4G) is not absolutely required, but plays a modulatory role. A more detailed description of the structure, function, and regulation of eIF4G is presented in Chapter 6.

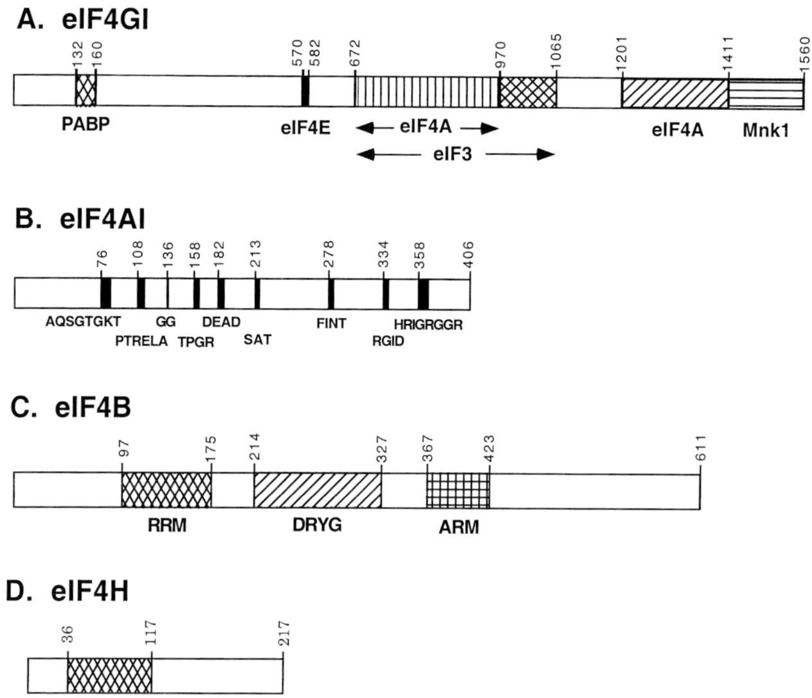

Figure 7 Domain structures of eIF4G, eIF4A, and eIF4B. (*A*) eIF4G domains for binding to other proteins are identified and labeled. (*B*) eIF4A motifs shared by the DEAD-box family of proteins. (*C*) eIF4B functional domains: (RRM) RNA recognition motif; (DRYG) aspartate, arginine, tyrosine, and glycine-rich region; (ARM) arginine-rich motif. (*D*) eIF4H contains an RRM similar to that in eIF4B (*cross-hatched*). (*A*, adapted from Morino et al. 2000; *B*, adapted from Pause et al. 1993; *C*, Méthot et al. [1994] and Naranda et al. [1994b].)

eIF4A

eIF4A has been characterized from a variety of sources, most notably from yeast and mammalian cells. Unlike many of the initiation factors that are present at about 0.2–0.5 copy per ribosome, eIF4A is more abundant, at about 3 copies per ribosome. eIF4A is unusual in that it appears to participate in protein synthesis both as an individual polypeptide and as a subunit of eIF4F, since translation in highly fractionated systems requires both factors (Conroy et al. 1990). In both the plant and yeast systems, the association of eIF4A with the other two subunits of eIF4F is less stable, and usually little or no eIF4A is associated with eIF4F when purified from these sources.

eIF4A binds ATP and is a helicase capable of bidirectional unwinding of RNA duplexes. It is the prototypic member of the DEAD box family of proteins, a family of helicases named after one of the eight conserved sequence motifs found in this family (Fig. 7B). Current models for the mechanism of RNA duplex unwinding suggest that eIF4A is a nonprocessive helicase due to the rapid off-rate of either eIF4A•ADP or eIF4A•ATP from the duplex (Lorsch and Herschlag 1998a,b; Rogers et al. 1999). Thus, eIF4A by itself is capable of unwinding only 3–5 base pairs. If this is not sufficient to destabilize the duplex, unwinding is not seen. However, in the presence of eIF4B, single-stranded RNA may be prevented from reassociating, leading to the unwinding of larger duplexes. Because eIF4A is the only ATPase identified in the initiation pathway, it likely serves as the "motor" for scanning.

Recently, a crystallographic structure for the ATP-binding domain (residues 9–232) of eIF4A was obtained (Benz et al. 1999; Johnson and McKay 1999). The structure is essentially the same as those seen with two other helicases, PcrA and HCV NS3 (Subramanya et al. 1996; Yao et al. 1997; Kim et al. 1998). PcrA is a DNA helicase, HCV NS3 unwinds either RNA/RNA or DNA/DNA duplexes, and eIF4A only unwinds duplexes with an RNA strand (either RNA/RNA or RNA/DNA). However, the conserved sequence motifs of the three helicases appear to be in the same positions in the three-dimensional structures, even though there are considerable differences in their spacing in the primary sequences. This suggests that all helicases may share a common ATP-binding domain.

Mutations directed to the consensus elements have helped to elucidate their various functions (Pause and Sonenberg 1992; Pause et al. 1993), as identified in Figure 7B. The studies indicate that ATP binding induces a conformational change in eIF4A which allows RNA binding to the HRIGRXXR motif, and RNA binding in turn induces ATP hydrolysis followed by more stable RNA binding. In vitro studies with a dominant negative inhibitor mutant of eIF4A (a change in the HRIGRXXR region) suggest that eIF4A may interact with the translational machinery primarily through its association with and cycling through eIF4F (Pause et al. 1994).

There are several isoforms of eIF4A (eIF4AI, eIF4AII, and eIF4AIII) that possess similar characteristics in vitro (e.g., RNA-dependent ATPase activity, RNA duplex unwinding), but only eIF4AI and eIF4AII appear to function in vivo in protein synthesis (Li et al. 1999). The failure of eIF4AIII to promote protein synthesis is reminiscent of the observation that mammalian eIF4A does not function in yeast protein synthesis (Prat et al. 1990). This is surprising because most other mammalian translation factors substitute for their yeast counterparts in intact cells or in cell lysates, even those that share less sequence identity than the eIF4As.

eIF4B

Human eIF4B is a 69-kD protein that appears to function as a homodimer. It is an RNA-binding protein that promotes the recruitment of ribosomes to mRNA and stimulates the RNA helicase activities of eIF4A and eIF4F. However, 40S initiation complexes can form in the absence of eIF4B, albeit less efficiently, and yeast lacking eIF4B are viable but grow more slowly (Altmann et al. 1993; Coppolecchia et al. 1993). The precise mechanisms whereby eIF4B stimulates eIF4A RNA helicase activity and the initiation process are not yet known.

A number of functional regions of eIF4B have been mapped by deletion and point mutation analyses (Fig. 7C). A central 99-amino-acid region called the DRYG domain is responsible for the dimerization of eIF4B (Méthot et al. 1996a). Two sequence nonspecific RNA-binding domains were identified (Méthot et al. 1994; Naranda et al. 1994b): an RRM comprising two RNP motifs near the amino terminus, and two arginine-rich motifs (ARM) in the carboxy-terminal half of the protein. The ARMs bind RNA more strongly than the RRM and are essential for promoting RNA helicase activity (Méthot et al. 1994). An in vitro RNA selection analysis (SELEX) with the RRM generated an RNA that binds with high affinity and inhibits RRM binding to 18S rRNA (Méthot et al. 1996b). It has been proposed that eIF4B may recognize the junction between single-stranded and double-stranded RNA (Méthot et al. 1994) or may bridge the mRNA and rRNA (Méthot et al. 1996a). eIF4B also may play a role in remodeling RNA–RNA interactions between rRNA and mRNA (Altmann et al. 1995).

Human eIF4B comprises 611 amino acid residues, whereas in other species, the initiation factor is smaller (Table 2): Plant eIF4B has 531 residues and *S. cerevisiae* eIF4B has only 436 residues. In the case of the yeast protein, about 50 residues are missing from the carboxyl terminus, and substantial sequence gaps occur in the DRYG domain. Mammalian eIF4B is homologous to another mammalian initiation factor of 25 kD, eIF4H (see below). The homology occurs primarily in the RRM domain; the DRYG and carboxy-terminal regions are missing in eIF4H.

Human eIF4B occurs in cells as multiple isoelectric forms when subjected to IEF/SDS-PAGE (Duncan and Hershey 1984). The isoelectric variants are due in part to phosphorylation. Two sites (Ser-406 and Ser-422) appear to be phosphorylated by the p70 S6 kinase (S6K1), and Ser-

442 is phosphorylated by PKA (F. Peiretti and J.W.B. Hershey, in prep.). Although hyperphosphorylation of eIF4B correlates with activation of protein synthesis, recombinant human eIF4B purified from *E. coli* and lacking phosphates is as active in the in vitro RNA helicase assay as eIF4B purified from mammalian cells (Goyer et al. 1993).

eIF4H

eIF4H has only recently been discovered (Richter-Cook et al. 1998), in large part due to its instability and a requirement for 25% glycerol in buffers during its purification. eIF4H is homologous with the amino-terminal region of eIF4B (39% sequence identity), especially in the RRM region shared by the two proteins (Fig. 7D). However, unlike eIF4B which functions as a dimer, eIF4H appears to function as a monomer of 25 kD, consistent with the fact that the DRYG domain for dimerization of eIF4B is absent in eIF4H. The factor stimulates β-globin synthesis in a highly fractionated rabbit reticulocyte lysate system. Like eIF4B, it binds RNA weakly and stimulates the ATPase activities of eIF4A and eIF4F. Its stimulation of protein synthesis in vitro is most apparent when eIF4B is present in subsaturating amounts, suggesting that the two proteins may perform similar functions.

Bacterially expressed recombinant eIF4H has the same biological properties as eIF4H isolated from mammalian cells, allowing the preparation of large amounts of the protein for structural analysis. A number of folding programs predict an average α-helix and β-sheet content of about 20% and 15%, respectively, but surprisingly, the actual values determined by CD are 50% β-sheet and about 5% α-helix (Richter et al. 1999). No eIF4H homolog has been identified in *S. cerevisiae* (other than *TIF3* for eIF4B), but the *Schizosaccharomyces pombe* gene *Sce3* (Schmidt et al. 1997) may encode either an eIF4B or eIF4H homolog. Given the similarity in structure and function of the two mammalian initiation factors, it is not clear whether Sce3 is more closely related to eIF4B or to eIF4H.

mRNA Scanning and AUG Recognition

Following the binding of the 40S preinitiation complex to the m^{7}G-cap-proximal region of the mRNA, the ribosome seeks the initiation codon and binds there. The mechanism believed to be used for most mRNAs is called "scanning." The hypothesis states that the 40S ribosomal subunit scans or migrates downstream along the mRNA from the 5′ terminus toward the initiation codon by a process that consumes energy in the form of ATP. The ribosome stops when it binds stably at the initiation codon, primarily through the RNA–RNA interaction of the AUG and the CAU anticodon of the bound Met-tRNA$_i$. In rarer instances, the 40S ribosome does not simply migrate along the 5′UTR until it encounters the initiation codon, but rather "hops" from one region to another by a process called shunting. The shunting mechanism is described in Chapters 4 and 8.

Evidence supporting the scanning model is summarized briefly below and is reviewed in greater detail in Chapter 4 and Kozak (1999). (1) The 5′-proximal AUG is utilized in about 90% of mRNAs that employ the scanning mechanism. If an AUG within a good context is placed artificially between the 5′-m^{7}G cap and the natural AUG, the upstream AUG is recognized instead. (2) RNA secondary structure in the 5′UTR blocks scanning and prevents the 40S subunit from binding at the AUG; rather, the ribosome may be found bound to the mRNA just upstream of the secondary structure. (3) Multiple 40S preinitiation complexes may bind to an mRNA in the presence of edeine which interferes with AUG recognition. (4) Regulation of GCN4 translation is most easily explained by a scanning mechanism (described in detail in Chapter 5). Nevertheless, a ribosome in the process of scanning has never been visualized directly on a native mRNA. Furthermore, the rate of scanning and the probability that a scanning ribosome may dissociate from the mRNA before reaching the initiation codon have not been measured. It also is not known if or when the eIF4F dissociates from the m^{7}G cap and eIF3 during scanning. It is possible that eIF4G remains bound to both eIF4E and eIF3 during scanning, at least until the initiation codon is recognized. If so, the 40S ribosomal subunit would be bound simultaneously to the m^{7}G cap and to the AUG, with single-stranded 5′UTR RNA looping out from the ribosome. A figure depicting this model of scanning is shown in Figure 4 of Chapter 4.

How does the scanning 40S initiation complex recognize the initiation codon? The dominant interaction is the codon/anticodon interaction, as mutation of the Met-tRNA$_i$ anticodon leads to recognition of a new cognate codon, not the AUG, as the initiation codon (Cigan et al. 1988). Thus, to recognize the initiation codon, it is essential that Met-tRNA$_i$ be bound to the 40S subunit before or very soon after scanning begins. In addition, the context of the AUG (see above) plays a role, as AUG codons in a poor context are bypassed and scanning continues until a more suitable AUG is encountered. Thus, a whole range of probabilities for recognizing an AUG exists, leading to variable extents of "leaky" scanning. How are the context nucleotides recognized? The answer to this question is not known. It is unlikely that initiation factors such as eIF2 are directly responsible. A possible explanation is that the structure of the ribosome in the decoding region of the tRNA P-binding site contributes to the recognition. The impressive advances in elucidating ribosome structure may soon provide insight into this issue.

An important aspect of AUG recognition appears to be the time that a ribosome "stalls" over the initiation codon. Initiation at a given site can be enhanced by the presence of RNA secondary structure appropriately placed downstream from the codon (Kozak 1990). However, initiation factors do play an important role in AUG recognition. Mutations in the yeast genes encoding eIF1, eIF5, or any of the three subunits of eIF2 allow ribosomes to initiate at UUG instead of AUG (for a detailed review, see Chapter 12). The eIF5-promoted hydrolysis of GTP bound to eIF2 also plays a role, as mutant forms of eIF2 or eIF5 that result in more rapid GTP hydrolysis allow initiation at sites that normally would be bypassed in a wild-type background (for review, see Chapter 12). If the GTP hydrolysis reaction proceeds more rapidly than normal, it may occur while the ribosome stalls briefly at a non-AUG codon such as UUG. Such stalling would be due to a weak interaction between the UUG and the Met-$tRNA_i$ anticodon. Therefore, the rate of the eIF5-promoted GTPase reaction on eIF2 plays an important role in the fidelity of initiation codon recognition. GTP hydrolysis results in GDP-bound eIF2, whose affinity for the 40S ribosomal subunit is reduced, resulting in its ejection. By analogy with EF1A (Chapter 3), dissociation is likely due to a conformational change in eIF2 dictated by the bound guanine nucleotide. Ejection of eIF2 and associated factors such as eIF3 prepares the 40S initiation complex for the subsequent junction reaction as described below.

Analysis of 40S preinitiation complexes bound to β-globin mRNA by primer extension inhibition has provided important new insights into the process of scanning and AUG recognition (Pestova et al. 1998). A 40S initiation complex formed with eIF2, eIF3, eIF4F, and eIF4B is found exclusively at or near the $5'm^7G$ cap and does not scan downstream to the AUG. However, when eIF1 and eIF1A are included in the incubation, mRNA binding, subsequent scanning, and AUG recognition occur. The complex at the 5′ terminus formed in the absence of eIF1 and eIF1A cannot be chased to the AUG by the late addition of the two factors, suggesting that the complex at the 5′ terminus is defective. Instead, it must first dissociate, then bind again in order to begin scanning. A caveat is that these effects of eIF1 and eIF1A have been shown for only a single mRNA, that encoding β-globin; other mRNAs may generate different results.

Since both eIF1 and eIF5 bind to eIF3 on the same subunit (yeast p93), both are likely present on the scanning ribosome. Thus, a super complex of nearly all of the initiation factors may be present on the 40S ribosome during initial binding to a mRNA and subsequent scanning (Fig. 3). This may account for the fact that all of these factors are ribosome-bound in cell lysates and are released only in the presence of high-salt buffers. Knowledge of the detailed structure of the 40S initiation complex before and after scanning is expected to provide insights into the mechanism of scanning and AUG recognition.

eIF1

eIF1 is the smallest of the initiation factors (12.7 and 12.3 kD in human and yeast, respectively). Yeast eIF1, also called SUI1, is essential for cell viability (Yoon and Donahue 1992), and a portion is found loosely associated with eIF3 (Naranda et al. 1996; Phan et al. 1998), where it binds to the p93 (NIP1) subunit (Phan et al. 1998). Although mammalian eIF1 is not found in preparations of eIF3, an interaction between eIF1 and the eIF3-p110 subunit (the ortholog of yeast p93) was detected by NMR (Fletcher et al. 1999). The solution structure of human eIF1, determined by NMR techniques (Fletcher et al. 1999), contains a tightly packed domain (Fig. 4B) that resembles several ribosomal proteins and RNA-binding domains. A number of mutations in yeast eIF1 (SUI1) have been identified that affect initiation codon selection by allowing initiation at UUG (Chapter 12). The residues in human eIF1 that correspond to these yeast mutations map close together on the surface of the three-dimensional structure. They and the surrounding residues (residues 69 and 86–90) are conserved between eukaryotes, bacteria, and archaea (Kyrpides and Woese 1998b). The results suggest that this surface comprises a binding site, although its target has not yet been identified.

The precise function of eIF1 remains unclear. As described above, it only weakly promotes the binding of 40S ribosomes to the initiation codon in a system that contains eIF2, eIF3, eIF4A, eIF4B, and eIF4F (Pestova et al. 1998). However, the activity is strongly stimulated by eIF1A, which without eIF1 does not stimulate at all. It remains unresolved whether eIF1 destabilizes 40S complexes on mRNA near the m^7G cap, is required for scanning, or is involved in recognition of the AUG. The genetic analyses of yeast eIF1 strongly support the view that recognition of the initiation codon is at least a part of the function of eIF1 (described in detail in Chapter 12). Since both yeast eIF1 and eIF5 bind to the p93 subunit of eIF3, an attractive hypothesis is that eIF1 influences the eIF2 GTPase activity stimulated by eIF5. Involvement of eIF1 in the decoding site of the ribosome also is indicated by a mutant form of eIF1 (*mof2*) that increases the frameshifting frequency of elongating ribosomes (Cui et al. 1998). eIF1 in the *mof2* strain carries a substitution at residue G-112 that is situated somewhat removed from the cluster of *sui* mutations described above (Fig. 4B), and therefore may interact with a different component of the initiation complex. This surprising result indicates that eIF1 also functions during the elongation phase of protein synthesis.

eIF5

eIF5 activates the GTPase center in eIF2γ (i.e., GAP activity) following establishment of the codon/anticodon interaction between the Met-tRNA$_i$ and the initiation codon in the mRNA. The mammalian initiation factor has a molecular mass of 48.9 kD and can replace its yeast ortholog which is slightly smaller (45.2 kD). eIF5 should not be confused with a larger initiation factor, now called eIF5B. eIF5B had been isolated earlier as a 150–160 kD protein, then called eIF5, that appeared to possess GAP activity for eIF2 and was required for the 60S subunit joining reaction. This protein (eIF5B) is described in the next section. The currently named eIF5 binds to eIF2 in crude mixtures of initiation factors (Chaudhuri et al. 1994). The interaction involves the oligo-lysine blocks of eIF2β and a highly conserved carboxy-terminal bipartite motif in eIF5 that contains aromatic and acidic residues, called AA boxes (Asano et al. 1999; for details, see Chapter 5). eIF5 also copurifies with eIF3, and the same AA boxes interact with the p93 (NIP1) subunit of eIF3 in yeast. It is not yet clear whether eIF5 can bind simultaneously to both eIF2β and p93. A high-resolution structure of eIF5 has not yet been achieved, and the molecular details of how eIF5 senses the codon/anticodon interaction are not known. However, mutations in eIF5 (*sui5*) that exhibit a gain of GAP function (faster GTPase reaction) result in initiation at non-AUG codons (for review, see Chapter 12).

Junction with the 60S Ribosomal Subunit

Once the initiation factors that were bound to the 40S initiation complex have dissociated, the 60S subunit can bind. The junction reaction has long been thought to be a passive reaction, but recent experiments suggest that the reaction requires the activity of eIF5B•GTP. eIF5B was identified over 20 years ago as a 150–160 kD protein, called eIF5 at the time (Schreier et al. 1977; Benne et al. 1978; Merrick 1979). Its function was detectable only after 40S initiation complex formation, and it was thought to promote the hydrolysis of GTP bound to eIF2. However, when the currently named eIF5 was cloned and shown to stimulate GTP hydrolysis on eIF2, the high-molecular-mass protein was ignored until very recently. Using purified initiation factors (including eIF5) to form the 40S initiation complex on β-globin mRNA, Pestova and coworkers showed that an 80S initiation complex fails to form with the 60S subunit unless eIF5B and GTP are included in the reaction (Pestova et al. 2000 and Chapter 9). This work established a role for eIF5B, even in the presence of eIF5.

eIF5B acts catalytically with GTP to convert preformed 40S initiation complexes into 80S initiation complexes. However, if GTP hydrolysis is prevented by the use of GMP-PNP, 80S complexes form but require stoichiometric amounts of eIF5B. This indicates that GTP hydrolysis by eIF5B is not required for 80S complex formation, but may be needed to promote the dissociation of eIF5B and its efficient recycling in the junction reaction. In this respect, eIF5B resembles its prokaryotic homolog, IF2, which also hydrolyzes GTP to effect rapid dissociation from 70S ribosomes. In effect, two GTP hydrolysis reactions are required for initiation in eukaryotes: one, with GTP bound to eIF2; the other, with GTP bound to eIF5B. In a kinetic study of initiation complex formation with low amounts of eIF2 and Met-tRNA$_i$, the eIF5B-catalyzed GTP hydrolysis step was one of the rate-limiting steps in the system (Lorsch and Herschlag 1999). That the junction of 40S initiation complexes with 60S subunits may be slow is indicated by the occasional appearance of "half-mers" in polysome profiles, where a half-mer comprises a polysome plus a 40S ribosomal subunit not yet complexed with the 60S subunit.

How might eIF5B function in the junction reaction? Insight is derived from our understanding of IF2 function in bacteria. IF2 binds to the initiator tRNA, fMet-tRNA$_f$, and to the 30S subunit, as described above. Together with IF1, it is thought to bind in the ribosomal A site, perhaps thereby guiding the initiator tRNA into the P site. By analogy, eIF5B (and eIF1A) might also bind to the A site of the 40S and 60S ribosomal subunits, helping proper alignment of the 60S subunit on the 40S initiation complex and ensuring placement of the Met-tRNA$_i$ in the P site. This hypothesis should be amenable to testing by cryo-EM techniques.

eIF5B

Yeast eIF5B (also called yIF2) (Choi et al. 1998) and human eIF5B (Wilson et al. 1999; Pestova et al. 2000) are homologous to bacterial IF2. Like bacterial IF2, eIF5B binds GTP and is a ribosome-dependent GTPase, as its sequence contains the three consensus motifs found in GTP-binding proteins. It is possible that eIF5B (in the absence of eIF5) also can stimulate the GTPase activity of eIF2 (GEF activity), as eIF5B added to 40S initiation complexes causes the rapid dissociation of Met-$tRNA_i$ (Peterson et al. 1979). Yeast eIF5B is not required for cell viability, but deletion of its gene results in a severe slow-growth phenotype (Choi et al. 1998). There is a report that mammalian eIF5B is phosphorylated (Traugh et al. 1976), but the effect of this modification on function has not yet been studied. A detailed comparison of the structures and mechanisms of action of IF2 and eIF5B is described in Chapter 9.

Other Proteins Implicated in the Initiation Pathway

A number of other proteins have been implicated in the process of initiation, but their functions are uncertain and poorly characterized. Nevertheless, future work may establish important roles for some of them, and so these proteins are listed and described briefly below.

The dissociation of 80S ribosomes into subunits is promoted in vitro by a 25-kD protein called eIF6 that binds to the 60S subunit (Russell and Spremulli 1979; Valenzuela et al. 1982). The human cDNA (Si et al. 1997) and yeast gene (Si and Maitra 1999) encoding eIF6 have been cloned. The yeast factor, which shares 72% sequence identity with the human protein, is essential for cell growth. Depletion of eIF6 inhibits protein synthesis in vivo (Si and Maitra 1999), but lysates from depleted cells retain their capacity to translate mRNAs in vitro. Depleted cells show greatly reduced levels of 60S subunits, and polysome profiles show half-mers (polysomes with an extra 40S subunit). The authors conclude that eIF6 affects protein synthesis indirectly, through its maintenance of 60S ribosomal subunit levels.

At low concentrations, the ternary complex, eIF2•GTP•Met-$tRNA_i$, is stabilized in the presence of RNA by a 94-kD protein called eIF2C (Gupta et al. 1990). A rabbit cDNA encoding eIF2C has been cloned (Zou et al. 1998), but the protein's involvement in initiation has not been studied further. The eIF2C gene belongs to a large family (the piwi/sting/argonaute/zwille/eIF2C gene family) that is conserved from plants to vertebrates. One family member, the *C. elegans* rde-1 gene, is involved in the phenomenon of RNA interference (Hunter 2000) by double-stranded RNA (Tabara et al. 1999). Whether eIF2C functions directly in translation initiation or during targeted mRNA degradation by dsRNA remains to be determined.

Another protein implicated in Met-$tRNA_i$ binding to ribosomes is eIF2A. It was first identified in the early 1970s on the basis of its stimulation of Met-$tRNA_i$ binding to 40S ribosomal subunits dependent on the triplet AUG (Merrick and Anderson 1975). It also stimulated poly(Phe) synthesis with ribosomes programed with poly(U) and *E. coli* Phe-tRNA. The recent cloning of the mammalian cDNA (W. Zoll and W.C. Merrick, unpubl.) unfortunately has not shed light on the function of eIF2A. Disruption of a homologous yeast gene displayed no phenotype, but both transformation and sporulation appeared compromised (L. Horton and W.C. Merrick, unpubl.).

eIF5A stimulates the synthesis of methionyl-puromycin in an AUG-based assay with purified eIF2, eIF3, eIF1A, and eIF5B (Merrick et al. 1975; Schreier et al. 1977; Benne et al. 1978). However, severe depletion in yeast causes little effect on protein synthesis or polysome profiles (Kang and Hershey 1994), suggesting that eIF5A does not play a significant role in the translation of most mRNAs. The results do not rule out a requirement for the translation of a small subset of mRNAs. It may be significant that a bacterial homolog of eIF5A, called elongation factor EF-P, also affects formation of the first few peptide bonds in vitro (Aoki et al. 1997). eIF5A may play a role in other cell processes unrelated to translation (Chapter 36). It has been implicated in transcription (Morehouse et al. 1999) and mRNA turnover (Zuk and Jacobson 1998). It also is a cellular cofactor of HIV-1 Rev and HTLV-I Rex involved in the nuclear export of incompletely spliced or unspliced mRNAs (Ruhl et al. 1993; Katahira et al. 1995) and is a retinoic-acid-stimulated binding partner for tissue transglutaminase II (Singh et al. 1998). Thus, the primary function of eIF5A remains to be identified.

A mammalian protein associated with mRNP particles, called p50, inhibits translation. p50 may play a dual role in cells, both as a transcription factor that binds to a DNA element called the Y box, and as a translation factor (Evdokimova et al. 1995). High levels of p50 inhibit protein synthesis in vitro and in vivo (Davydova et al. 1997; Evdokimova et al. 1998). The physiological role of this protein in regulating protein synthesis is not well understood.

The *DED1* gene encodes another member (besides eIF4A) of the DEAD box family of putative RNA helicases that may be required for translation initiation in yeast (Chuang et al. 1997). Although *DED1* mutant proteins were first implicated in mRNA splicing (Jamieson et al. 1991) and Pol III transcription (Thuillier et al. 1995), a number of cold-sensitive *DED1* mutants, when shifted to 15°C, exhibited ribosome run-off from polysomes indicative of a defect in initiation (Chuang et al. 1997). A mouse homolog, *PL10*, can substitute for *DED1*, but a precise function for the protein in initiation has not been shown.

Recycling and Reinitiation

Most biochemical experiments designed to elucidate the pathway of initiation in eukaryotes (Fig. 3) employ either crude lysates or purified components to which is added an mRNA. Such systems are quite inefficient, and mRNAs are usually translated only once or a few times. As a result, the reaction measured this way reflects primarily the first initiation event on a naked mRNA or one likely associated with fewer proteins. In contrast, most initiation events in vivo occur on polysomes, i.e., mRNAs already being translated. The 40S ribosomal subunit that initiates translation on a polysomal mRNA may come either from the pool of nontranslating 40S subunits ("native" 40S subunits) or from a ribosome that has just terminated protein synthesis on that mRNA. In the latter case, we call this event "recycling," as the initiation mechanism may differ in some detail from that which involves free ribosomal subunits.

The phenomenon of ribosome recycling has not been demonstrated unambiguously, and the mechanism enabling ribosomes to recycle is not known. How might recycling occur? The synergistic action of the m^7G cap and the poly(A) tail, apparently through an interaction between the eIF4G and the poly(A)-binding protein (PABP), may contribute. The ability for the mRNA to circularize and the enhanced recruitment of 40S ribosomal subunits to capped mRNAs are described in Chapter 10. Involvement of a circularized mRNA implies that the 40S subunit following termination would have to scan down through the 3′UTR to reach the poly(A) tail and thereby arrive near to the m^7G cap. This seems unlikely, especially since many mRNAs have exceedingly long 3′UTRs. However, the recently described interaction between eRF3 and PABP might bring the terminating ribosome close to the poly(A) tail and the m^7G cap (Hoshino et al. 1999). Although this is plausible, there is as yet no experimental evidence that the eRF3–PABP interaction affects initiation or translational efficiency. Elucidating how ribosomes may recycle and contribute to translational efficiency is one of the major challenges for the future. If recycling occurs, its stimulation would result in an increase in polysome size, whereas its inhibition would cause the runoff of polysomes. Thus, recycling may be an important target of translational control.

It is possible that a terminating ribosome might initiate protein synthesis at an initiation codon near the site of termination, leading to the synthesis of a different protein from the same mRNA. This type of initiation event is called "reinitiation." The phenomenon of reinitiation has been shown to occur after the translation of short open reading frames, the best examples being the yeast GCN4 mRNA (Chapter 5) and the cauliflower mosaic virus 35S mRNA (Ryabova and Hohn 2000), the latter complicated by the phenomenon of "shunting." Other examples of reinitiation are described in Chapter 18, and a discussion of the mechanism involved is provided in Chapter 4. Reinitiation following the translation of a long open reading frame occurs extremely rarely, if at all, and no natural example of a dicistronic mRNA producing two functional polypeptides is known in mammalian cells.

EVOLUTION OF TRANSLATION FACTORS

Examination of sequence relationships of proteins from eubacteria, archaebacteria, and eukaryotes provides insight into how the translational machinery evolved and how some of the factors function (Kyrpides and Woese 1998b). Table 4 lists such relationships, many having been recognized only after the sequencing of several bacterial and archaebacterial genomes. The relationships are not always easily discerned, because the degree of sequence identity between the bacterial and eukaryotic proteins is small, just above what would be predicted as a random match.

Some of the related proteins appear to perform very similar functions: EF1A and eEF1A, EF2 and eEF2, RF1/2 and eRF1, and RF3 and eRF3. A number of others have shed light on the function of the proteins. For example, the homology between bacterial IF1 and eukaryotic eIF1A suggests that eIF1A may function by binding to the ribosomal A site, helping to position the initiator tRNA in the P site. Bacterial IF2 also may mimic tRNA and bind to the A site, suggesting that its homolog, eIF5B, functions similarly in this regard. Other homologs appear to perform related functions: Bacterial SelB and eukaryotic eIF2γ each bind a specific tRNA and recognize the charged amino acid, while excluding other aminoacyl-tRNAs. On the other hand, some functions apparently common to bacteria and eukaryotes are peformed by entirely unrelated proteins. The binding of the initiator tRNA is an example, with bacterial IF2 being unrelated to eukaryotic eIF2. For catalyzing guanylate exchange on EF1A/eEF1A, bacterial EF1B is not related to any of the subunits of eEF1B.

The failure to find homologs in bacteria for some of the eukaryotic initiation factor proteins indicates how the basic mechanism of initiation differs between these phyla. The most striking difference concerns the eukaryotic eIF4 group of initiation factors, which are almost entirely missing in eubacteria and archaebacteria. An interesting exception is bac-

Table 4 Evolutionary relationships of translation factors

Eubacteria	Archaebacteria	Eukaryotes	Reference
(+/-)	*	eIF1	Kyrpides and Woese (1998b)
IF1	*	eIF1A	Kyrpides and Woese (1998b)
	*	eIF2α	Kyrpides and Woese (1998a)
	*	eIF2β	Kyrpides and Woese (1998a)
SelB	*	eIF2γ	Keeling et al. (1998)
	(none?)	eIF2A	W. Zoll and W.C, Merrick (unpubl.)
	I, II, III	eIF2Bα,β,δ	Kyrpides and Woese (1998a)
	no	eIF3	
W2	*	eIF4A	Lu et al. (1999)
	no	eIF4B	
	no	eIF4E	
	no	eIF4G	
	no	eIF4H	
	no	eIF5	
EF-P	*	eIF5A	Kyrpides and Woese (1998b)
IF2	*	eIF5B	Kyrpides and Woese (1998b); Lee et al. (1999)
EF1A	*	eEF1A	Chapter 3; Creti et al. (1994)
	*	eEF1B	
EF2	*	eEF2	Creti et al. (1994)
		eEF3 (yeast, fungi only)	
RF1/RF2	no	no	
	*	eRF1	
RF3	no	no	
(EF1A)	(*)	eRF3	

(+/-) indicates that this protein is present in some bacterial species.
* indicates the presence of an archaebacterial gene related to the eukaryotic gene.

terial W2 and eukaryotic eIF4A, the RNA helicase factor (Lu et al. 1999). The homology suggests that W2 may function as a helicase in bacteria, perhaps by unwinding RNA secondary structure at initiation sites, but such a function has yet to be demonstrated in vivo. Eukaryotes possess eIF1, trimeric eIF2, pentameric eIF2B, and eIF3 subunits, but no eubacterial equivalents exist. On the other hand, archaebacteria contain homologs of eIF1 and eIF2. Whether or not bacterial EF-P and eukaryotic eIF5A function in protein synthesis in vivo remains to be proven.

PERSPECTIVES

Since publication of the first edition of this monograph about five years ago, significant advances have been made in our understanding of the pathway and mechanism of initiation. In the bacterial system, dramatic advances have occurred in our knowledge of the structure of the ribosome and the solving of detailed three-dimensional structures of many of the soluble factors. The concept of protein mimicry of RNA has emerged, which sheds light on how the factors function. Noteworthy in eukaryotic systems is the elucidation of the functions of the initiation factors, especially how eIF1, eIF1A, and eIF5B act; how eIF2B catalyzes the guanylate exchange reaction on eIF2; and how eIF4G links many other components involved in mRNA binding to ribosomes. All of the known initiation factor cDNAs/genes have been cloned, providing tools to better study their functions, and the three-dimensional structures of many of the smaller factors have been solved. In addition, the importance of the 3′UTR in promoting initiation is recognized, especially interactions involving the poly(A)-binding protein.

What might we expect during the coming five years? In the bacterial system, atomic-level structures of the 70S ribosome and its various complexes will help elucidate the molecular mechanism of protein synthesis and show how the ribosome itself undergoes conformational changes and provides catalysis. Solving the structure of IF2 also is likely. Detailed structural information on the numerous conformational changes that occur in the translation factors and ribosome during the discrete phases of initiation, elongation, and termination, combined with kinetic experiments, should provide a rather complete molecular mechanism of translation in bacteria.

In eukaryotic systems, one can anticipate solving the structures of many more of the initiation factors. In particular, a detailed three-dimensional structure of eIF2, eIF2B, and eIF3 should emerge. The application of cryo-EM methods toward solving high-resolution structures of the ribosome and various initiation factor supercomplexes also is practical. The continued use of genetic approaches in yeast will provide additional insights into mechanism and factor function. It is anticipated that additional initiation factors or interacting proteins that influence factor activity will be discovered via genetic screens. Kinetic analyses of the reaction steps in the pathway also will be important. Major challenges are to elucidate how the 3′UTR contributes to initiation, how ribosomes may recycle, and how scanning occurs. Another challenge is to understand how mRNPs packaged in the nucleus and exported into the cytoplasm are either translated efficiently or repressed. As our understanding of the molecular details of initiation increase, it should be possible to explain better the mechanisms and kinetics of translational control.

REFERENCES

Agrawal R.K., Penczek P., Grassucci R.A., and Frank J. 1998. Visualization of elongation factor G on the *Escherichia coli* 70S ribosome: The mechanism of translocation. *Proc. Natl. Acad. Sci.* **95:** 6134–6138.

Agrawal R.K., Penczek P., Grassucci R.A., Li Y., Leith A., Nierhaus K.H., and Frank J. 1996. Direct visualization of A-, P-, and E-site transfer RNAs in the *Escherichia coli* ribosome. *Science* **271:** 1000–1002.

Altmann M., Müller P.P., Pelletier J., Sonenberg N., and Trachsel H. 1989. The mammalian translation initiation factor 4E substitutes for its yeast homologue in vivo. *J. Biol. Chem.* **264:** 12145-12147.

Altmann M., Wittmer B., Méthot N., Sonenberg N., and Trachsel H. 1995. The *Saccharomyces cerevisiae* translation initiation factor Tif3 and its mammalian homologue, eIF-4B, have RNA annealing activity. *EMBO J.* **14:** 3820–3827.

Altmann M., Müller P.P., Wittmer B., Ruchti F., Lanker S., and Trachsel H. 1993. A *Saccharomyces cerevisiae* homologue of mammalian translation initiation factor 4B contributes to RNA helicase activity. *EMBO J.* **12:** 3997–4003.

Anderson J., Phan L., Cuesta R., Carlson B.A., Pak M., Asano K., Bjork G.R., Tamame M., and Hinnebusch A.G. 1998. The essential Gcd10p-Gcd14p nuclear complex is required for 1-methyladenosine modification and maturation of initiator methionyl-tRNA. *Genes Dev.* **12:** 3650–3662.

Anthony T.G., Fabian J.R., Kimball S.R., and Jefferson L.S. 2000. Identification of domains within the ε-subunit of the translation initiation factor eIF2B that are necessary for guanine nucleotide exchange activity and eIF2B holoprotein formation. *Biochim. Biophys. Acta* **1492:** 56–62.

Aoki H., Dekany K., Adams S.L., and Ganoza M.C. 1997. The gene encoding the elongation factor P protein is essential for viability and is required for protein synthesis. *J. Biol. Chem.* **272:** 32254–32259.

Aravind L. and Ponting C.P. 1998. Homologues of 26S proteasome subunits are regulators of transcription and translation. *Protein Sci.* **7:** 1250–1254.

Asano K., Anderson J., Phan L., and Hinnebusch A.G. 1998. Complex formation by all five homologues of mammalian translation initiation factor 3 subunits from yeast *Saccharomyces cerevisiae*. *J. Biol. Chem.* **273:** 18573–18585.

Asano K., Krishnamoorthy T., Phan L., and Hinnebusch A.G. 1999. Conserved bipartite motifs in yeast eIF5 and eIF2Bε, GTPase-activating and GDP-GTP exchange factors in translation initiation, mediate binding to their common substrate eIF2. *EMBO J.* **18:** 1673–1688.

Ayuso-Parilla M., Henshaw E.C., and Hirsch C.A. 1973. The ribosome cycle in mammalian protein synthesis. III. Evidence that the nonribosomal proteins bound to the native smaller subunit are initiation factors. *J. Biol. Chem.* **248:** 4386–4393.

Ban N., Nissen P., Jansen J.C., Capel M.., Moore P.B., and Steitz T.A. 1999. Placement of protein and RNA structures into a 5Å-resolution map of the 50S ribosomal subunit. *Nature* **400:** 841–847.

Bandyopadhyay A. and Maitra U. 1999. Cloning and characterization of the p42 subunit of mammalian translation initiation factor 3 (eIF3): Demonstration that eIF3 interacts with eIF5 in mammalian cells. *Nucleic Acids Res.* **27:** 1331–1337.

Battiste J.L., Pestova T.V., Hellen C.U.T., and Wagner G. 2000. The eIF1A solution structure reveals a large RNA-binding surface important for scanning function. *Mol. Cell* **5:** 109–119.

Benne R. and Hershey J.W.B. 1976. Purification and characterization of initiation factor IF-E3 from rabbit reticulocytes. *Proc. Natl. Acad. Sci.* **73:** 3005–3009.

Benne R., Brown-Luedi M.L., and Hershey J.W.B. 1978. The purification and characterization of protein synthesis initiation factors eIF-1, eIF-4C, eIF-4D and eIF-5 from rabbit reticulocytes. *J. Biol. Chem.* **253:** 3070–3077.

Benz H., Jr., Trachsel H., and Baumann U. 1999. Crystal structure of the ATPase domain of translation initiation factor 4A from *Saccharomyces cerevisiae* — The prototype of the DEAD box protein family. *Structure* **7:** 671–679.

Biou V., Shu F., and Ramakrishnan V. 1995. X-ray crystallography shows that translational initiation factor IF3 consists of two compact α/β domains linked by an α-helix. *EMBO J.* **14:** 4056–4064.

Boileau G., Butler P., Hershey J.W.B., and Traut R.R. 1983. Direct cross-links between initiation factors 1, 2 and 3 and ribosomal proteins promoted by 2-iminothiolane. *Biochemistry* **22:** 3162–3170.

Bommer U.A., Lutsch G., Stahl J., and Bielka H. 1991. Eukaryotic initiation factors eIF-2 and eIF-3: Interactions, structure and localization in ribosomal initiation complexes. *Biochimie* **73:** 1007–1019.

Boni I.V., Isaeva D.M., Musychenko M.L., and Tzarera N.V. 1991. Ribosome-messenger recognition: mRNA target sites for ribosomal protein S1. *Nucleic Acids Res.* **19:** 155–162.

Brock S., Szkaradkiewicz K., and Sprinzl M. 1998. Initiation factors of protein biosynthesis in bacteria and their structural relationship to elongation and termination factors. *Mol. Microbiol.* **29:** 409–417.

Browning K.S. 1996. The plant translational apparatus. *Plant Mol. Biol.* **32:** 107–144.

Butler J.S., Springer M., and Grunberg-Manago M. 1987. AUU-to-AUG mutation in the initiator codon of the translation initiation factor IF3 abolishes translational autocontrol of its own gene (*infC*) in vivo. *Proc. Natl. Acad. Sci.* **84:** 4022–4025.

Cai A., Jankowska-Anyszka M., Centers A., Chlebicka L., Stepinski J., Stolarski R., Darzynkiewicz E., and Rhoads R.E. 1999. Quantitative assessment of mRNA cap analogs as inhibitors of in vitro translation. *Biochemistry* **38:** 8538–8547.

Cate J.H., Yusupov M.M., Yusupova G.Z., Earnest T.M., and Noller H.F. 1999. X-ray crystal structures of 70S ribosome functional complexes. *Science* **285:** 2095–2104.

Cenatiempo Y., Deville F., Dondon J., Grunberg-Manago M., Sacerdot C., Hershey J.W.B., Hansen H.F., Petersen H.U., Clark B.F.C., Kjeldgaard M., la Cour T.F.M., Mortensen K.K., and Nyborg J. 1987. The protein synthesis initiation factor 2 G-domain. Study of a functionally active C-terminal 65 kilodalton fragment of IF2 from *Escherichia coli*. *Biochemistry* **26:** 5070–5076.

Chaudhuri J., Chowdhury D., and Maitra U. 1999. Distinct functions of eukaryotic translation initiation factors eIF1A and eIF3 in the formation of the 40S ribosomal preinitiation complex. *J. Biol. Chem.* **274:** 17975–17980.

Chaudhuri J., Das K., and Maitra U. 1994. Purification and characterization of bacterially expressed mammalian translation initiation factor 5 (eIF-5): Demonstration that eIF-5 forms a specific complex with eIF-2. *Biochemistry* **33:** 4794–4799.

Chaudhuri J., Si K., and Maitra U. 1997. Function of eukaryotic translation initiation factor 1A (eIF1A) (formerly called eIF-4C) in initiation of protein synthesis. *J. Biol. Chem.* **272:** 7883–7891.

Choi S.K., Lee J.H., Zoll W.L., Merrick W.C., and Dever T.E. 1998. Promotion of binding Met-tRNA$^{Met}_{i}$ to ribosomes by yIF2, a bacterial IF2 homolog in yeast. *Science* **280:**

1757–1760.

Chuang R.-Y., Weaver P.L., Liu Z., and Chang T.-H. 1997. Requirement of the DEAD-box protein Ded1p for messenger RNA translation. *Science* **275:** 1468–1471.

Cigan A.M., Pabich E.K., and Donahue T.F. 1988. Mutational analysis of the *HIS4* translational initiator region in *Saccharomyces cerevisiae*. *Mol. Cell. Biol.* **8:** 2964–2975.

Clemons W.M., Jr., May J.L.C., Wimberly B.T., McCutcheon J.P., Capel M., and Ramakrishnan V. 1999. Structure of a bacterial 30S ribosomal subunit at 5.5Å resolution. *Nature* **400:** 833–840.

Conroy S.C., Dever T.E., Owens C.L., and Merrick W.C. 1990. Characterization of the 46,000 dalton subunit of eIF-4F. *Arch. Biochem. Biophys.* **282:** 363–371.

Coppolecchia R., Buser P., Stotz A., and Linder P. 1993. A new yeast translation initiation factor suppresses a mutation in the eIF-4A RNA helicase. *EMBO J.* **12:** 4005–4011.

Creti R., Ceccarelli E., Bocchetta M., Sanangelantoni A.M., Tiboni O., Palm P., and Cammarano P. 1994. Evolution of translational elongation factor (EF) sequences — Reliability of global phylogenies inferred from EF-1α (Tu) and EF-2 (G) proteins. *Proc. Natl. Acad. Sci.* **91:** 3255–3259.

Cui Y., Dinman J.D., Kinzy T.G., and Peltz S.W. 1998. The mof2/sui1 protein is a general monitor of translational accuracy. *Mol. Cell. Biol.* **18:** 1506–1516.

Danaie P., Wittmer B., Altmann M., and Trachsel H. 1995. Isolation of a protein complex containing translation initiation factor Prt1 from *Saccharomyces cerevisiae*. *J. Biol. Chem.* **270:** 4288–4292.

Davydova E.K., Evdokimova V.M., Ovchinnikov L.P., and Hershey J.W.B. 1997. Overexpression in COS cells of p50, the major core protein associated with mRNA, results in translation inhibition. *Nucleic Acids Res.* **25:** 2911–2916.

De Gregorio E., Preiss T., and Hentze M.W. 1998. Translational activation of uncapped mRNAs by the central part of human eIF4G is 5′ end-dependent. *RNA* **4:** 828–836.

de Smit M.H. and van Duin J. 1990. Secondary structure of the ribosome binding site determines translational efficiency: A quantitative analysis. *Proc. Natl. Acad. Sci.* **87:** 7668–7672.

———. 1994. Translational initiation on structured messengers — Another role for the Shine-Dalgarno interaction. *J. Mol. Biol.* **235:** 173–184.

Dholakia J.N., Francis B.R., Haley B.E., and Wahba A.J. 1989. Photoaffinity labeling of the rabbit reticulocyte guanine nucleotide exchange factor and eukaryotic initiation factor 2 with 8-azidopurine nucleotides: Identification of GTP and ATP-binding domains. *J. Biol. Chem.* **264:** 20638–20642.

Dholakia J.N., Mueser T.C., Woodley C.L., Parkhurst L.J., and Wahba A.J. 1986. The association of NADPH with the guanine nucleotide exchange factor from rabbit reticulocytes: A role of pyridine dinucleotides in eukaryotic polypeptide chain initiation. *Proc. Natl. Acad. Sci.* **83:** 6746–6750.

Duncan R. and Hershey J.W.B. 1983. Identification and quantitation of levels of protein synthesis initiation factors in crude HeLa cell lysates by two-dimensional polyacrylamide gel electrophoresis. *J. Biol. Chem.* **258:** 7228–7235.

———. 1984. Heat shock-induced translational alterations in HeLa cells. *J. Biol. Chem.* **259:** 11882–11889.

Ellis S.R., Hopper A.K., and Martin N.C. 1987. Amino-terminal extension generated from an upstream AUG codon is not required for mitochondrial import of yeast N^2,N^2-dimethylguanosine-specific tRNA methyl transferase. *Proc. Natl. Acad. Sci.* **84:** 5172–5176.

Erickson F.L. and Hannig E. 1996. Ligand interactions with eukaryotic translation initiation factor 2: Role of the gamma-subunit. *EMBO J.* **15:** 6311–6320.

Evans D.R.H., Rasmussen C., Hanic-Joyce P.J., Johnston G.C., Singer R.A., and Barnes C.A. 1995. Mutational analysis of the Prt1 protein subunit of yeast translation initiation factor 3. *Mol. Cell. Biol.* **15:** 4525–4535.

Evdokimova V.M., Kovrigina E.A., Nashchekin D.V., Davydova E.K., Hershey J.W.B., and Ovchinnikov L.P. 1998. The major core protein of messenger ribonucleoprotein particles (p50) promotes initiation of protein biosynthesis in vitro. *J. Biol. Chem.* **273:** 3574–3581.

Evdokimova V.M., Wei C.-L., Sitikov A.S., Simonenko P.N., Lazarev O.A., Vasilenko K.S., Ustinov V.A., Hershey J.W.B., and Ovchinnikov L.P. 1995. The major protein of messenger ribonucleoprotein particles in somatic cells is a member of the Y-box binding transcription factor family. *J. Biol. Chem.* **270:** 3186–3192.

Fabian J.R., Kimball S.R., and Jefferson L.S. 1998. Reconstitution and purification of eukaryotic initiation factor 2B (eIF2B) expressed in Sf21 insect cells. *Protein Expr. Purif.* **13:** 16–22.

Fields S.D., Conrad M.N., and Clarke M. 1998. The *S. cerevisiae* and *D. discoideum cluA* genes are functional homologues that influence mitochondrial morphology and distribution. *J. Cell Sci.* **111:** 1717–1727.

Fletcher C.M., Pestova T.V., Hellen C.U.T., and Wagner G. 1999. Structure and interactions of the translation initiation factor eIF1. *EMBO J.* **18:** 2631–2637.

Fletcher C.M., McGuire A.M., Gingras A.-C., Li H., Matsuo H., Sonenberg N., and Wagner G. 1998. 4E binding proteins inhibit the translation factor eIF4E without folded structure. *Biochemistry* **37:** 9–15.

Flynn A., Shatsky I.N., Proud C.G., and Kaminski A. 1994. The RNA-binding properties of protein synthesis initiation factor eIF-2. *Biochim. Biophys. Acta* **1219:** 293–301.

Frank J., Zhu J., Penczek P., Li Y., Srivastava S., Verschoor A., Radermacher M., Grassucci R., Lata R.K., and Agrawal R.K. 1995. A model of protein synthesis based on cryo-electron microscopy of the *E. coli* ribosome. *Nature* **376:** 441–444.

Gabashvili I.S., Agrawal R.K., Spahn C.M.T., Grassucci R.A., Svergun D.I., and Frank J. 2000. Solution structure of the *E. coli* 70S ribosome at 11.5 Å resolution. *Cell* **100:** 537–549.

Garcia C., Fortier P.L., Blanquet S., Lallemand J.Y., and Dardel F. 1995a. ^{1}H and ^{15}N resonance assignments and structure of the N-terminal domain of *Escherichia coli* initiation factor 3. *Eur. J. Biochem.* **228:** 395–402.

———. 1995b. Solution structure of the ribosome-binding domain of *E. coli* translation initiation factor IF3. Homology with the U1A protein of the eukaryotic spliceosome. *J. Mol. Biol.* **254:** 247–259.

Garcia-Barrio M.T., Naranda T., Vazquez de Aldana C.R., Cuesta R., Hinnebusch A.G., Hershey J.W.B., and Tamame M. 1995. GCD10, a translational repressor of GCN4, is the RNA-binding subunit of eukaryotic translation initiation factor-3. *Genes Dev.* **9:** 1781–1796.

Gaspar N.J., Goss-Kinzy T., Scherer B.J., Hümbelin M., Hershey J.W.B., and Merrick W.C. 1994. Translation initiation factor eIF-2. Cloning and expression of the human cDNA encoding the γ-subunit. *J. Biol. Chem.* **269:** 3415–3422.

Gingras A.-C., Raught B., and Sonenberg N. 1999. eIF4 initiation factors: Effectors of mRNA recruitment to ribosomes and regulators of translation. *Annu. Rev. Biochem.* **68:** 913–963.

Glickman M.H., Rubin D.M., Coux O., Wefes I., Pfeifer G., Cjeka Z., Baumeister W., Fried V.A., and Finley D. 1998. A subcomplex of the proteasome regulatory particle required for ubiquitin-conjugate degradation and related to the COP9-signalosome and eIF3. *Cell* **94:** 615–623.

Godefroy-Colburn T., Wolfe A.D., Dondon J., and Grunberg-Manago M. 1975. Light-scattering studies showing the effect of initiation factors on the reversible dissociation of *Escherichia coli* ribosomes. *J. Mol. Biol.* **94:** 461–478.

Goumans H., Thomas A., Verhoeven A., Voorma H.O., and Benne R. 1980. The role of eIF-4C in protein synthesis initiation complex formation. *Biochim. Biophys. Acta* **608:** 39–46.

Goyer C., Altmann M., Lee H.S., Blanc A., Deshmukh M., Woolford J.L., Jr., Trachsel H., and Sonenberg N. 1993. *TIF4631* and *TIF4632:* Two yeast genes encoding the high-molecular-weight subunits of the cap-binding protein complex (eukaryotic initiation factor 4F) contain an RNA recognition motif-like sequence and carry out an essential function. *Mol. Cell. Biol.* **18:** 334–342.

Gradi A., Imataka H., Svitkin Y.V., Rom E., Raught B., Morino S., and Sonenberg N. 1998. A novel functional human eukaryotic translation initiation factor 4G. *Mol. Cell. Biol.* **18:** 334–342.

Greenberg J.R., Phan L., Gu Z., deSilva A., Apolito C., Sherman F., Hinnebusch A.G., and Goldfarb D.S. 1998. Nip1p associates with 40S ribosomes and the Prt1p subunit of eIF3 and is required for efficient translation initiation. *J. Biol. Chem.* **273:** 23485–23494.

Gualerzi C.O. and Pon C.L. 1990. Initiation of mRNA translation in prokaryotes. *Biochemistry* **29:** 5881–5889.

Gualerzi C.O., Risuleo G., and Pon C.L. 1977. Initial rate kinetic analysis of the mechanism of initiation complex formation and the role of initiation factor IF-3. *Biochemistry* **16:** 1684–1689.

Gunnery S. and Mathews M.B. 1995. Functional mRNA can be generated by RNA polymerase III. *Mol. Cell. Biol.* **15:** 3597–3607.

Gupta N.K., Roy A.L., Nag M.K., Kinzy T.G., MacMillan S.E., Hileman R.E., Dever T.E., Wu S., Merrick W.C., and Hershey J.W.B. 1990. New insights into an old problem: Ternary complex (Met-tRNA$_i$•eIF2•GTP) formation in animal cells. In *Transcriptional control of gene expression* (ed. J.E.G. McCarthy and M.F. Tuite), pp. 521–526, Springer Verlag, Berlin.

Haghighat A. and Sonenberg N. 1997. eIF4G dramatically enhances the binding of eIF4E to the mRNA 5′-cap structure. *J. Biol. Chem.* **272:** 21677–21680.

Hanachi P., Hershey J.W.B., and Vornlocher H.-P. 1999. Characterization of the p33 subunit of translation initiation factor eIF3 from *Saccharomyces cerevisiae*. *J. Biol. Chem.* **274:** 8546–8553.

Harashima S. and Hinnebusch A.G. 1986. Multiple *GCD* genes required for repression of *GCN4*, a transcriptional activator of amino acid biosynthetic genes in *Saccharomyces cerevisiae*. *Mol. Cell. Biol.* **6:** 3990–3998.

Hartz D., McPheeters D.S., and Gold L. 1989. Selection of the initiator tRNA by *Escherichia coli* initiation factors. *Genes Dev.* **3:** 1899–1912.

Hartz D., McPheeters D.S., Green L., and Gold L. 1991. Detection of *Escherichia coli* ribosome binding at translation initiation sites in the absence of tRNA. *J. Mol. Biol.* **218:** 99–105.

Hiremath L.S., Webb N.R., and Rhoads R.E. 1985. Immunological detection of the messenger RNA cap-binding protein. *J. Biol. Chem.* **260:** 7843–7849.

Hofmann K. and Bucher P. 1998. The PCI domain: A common theme in three multi-protein complexes. *Trends Biochem. Sci.* **23:** 204–205.

Hoshino S., Imai M., Kobayashi T., Uchida N., and Katada T. 1999. The eukaryotic polypeptide chain releasing factor (eRF3/GSPT) carrying the translation termination signal to the 3′-poly(A) tail of mRNA. *J. Biol. Chem.* **274:** 16677–16680.

Hui A. and de Boer H.A. 1987. Specialized ribosome system: Preferential translation of a single mRNA species by a subpopulation of mutated ribosomes in *Escherichia coli. Proc. Natl. Acad. Sci.* **84:** 4762–4766.

Hunter C. 2000. Gene silencing: Shrinking the black box of RNAi. *Curr. Biol.* **10:** R137–R140.

Imataka H. and Sonenberg N. 1997. Human eukaryotic translation initiation factor 4G (eIF4G) possesses two separate and independent binding sites for eIF4A. *Mol. Cell. Biol.* **17:** 6940–6947.

Imataka H., Gradi A., and Sonenberg N. 1998. A newly identified N-terminal amino acid sequence of human eIF4G binds poly(A)-binding protein and functions in poly(A)-dependent translation. *EMBO J.* **17:** 7480–7489.

Jackson R.J. 1996. A comparative view of initiation site selection mechanisms. In *Translational control* (ed. J.W.B. Hershey et al.), pp. 71–112. Cold Spring Harbor Laboratory Press, Cold Spring Harbor, New York.

Jacob W.F., Santer M., and Dahlberg A.E. 1987. A single base change in the Shine-Dalgarno region of 16S rRNA of *Escherichia coli* affects translation of many proteins. *Proc. Natl. Acad. Sci.* **84:** 4757–4761.

Jamieson D.J., Rahe B., Pringle J., and Beggs J.D. 1991. A suppressor of a yeast splicing mutation (*prp8*-1) encodes a putative ATP-dependent RNA helicase. *Nature* **349:** 715–717.

Johnson E.R. and McKay D.B. 1999. Yeast initiation factor 4A N-terminal domain. *RNA* **5:** 1526–1534.

Kang H.A. and Hershey J.W.B. 1994. Effect of initiation factor eIF-5A depletion on protein synthesis and proliferation of *Saccharomyces cerevisiae. J. Biol. Chem.* **269:** 3934–3940.

Katahira J., Ishizaki T., Sakai H., Adachi A., Yamamoto K., and Shida H. 1995. Effects of translation initiation factor eIF-5A on the functioning of human T-cell leukemia virus type I Rex and human immunodeficiency virus Rev inhibited *trans* dominantly by a Rex mutant deficient in RNA binding. *J. Virol.* **69:** 3125–3133.

Keeling P.J., Fast N.M., and McFadden G.I. 1998. Evolutionary relationship between translation initiation factor eIF2γ and selenocysteine-specific elongation factor SELB: Change of function in translation factors. *J. Mol. Evol.* **47:** 649–655.

Keiper B.D., Lamphear B.J., Deshpande A.M., Jankowska-Anyszka M., Aamodt E.J., Blumenthal T., and Rhoads R.E. 2000. Functional characterization of five eIF4E isoforms in *Caenorhabditis elegans. J. Biol. Chem.* **275:** 10590–10596.

Kim J.L., Morgenstern K.A., Griffith J.P., Dwyer M.D., Thompson J.A., Murcko M.A., Lin C., and Caron P.R. 1998. Hepatitis C virus NS3 RNA helicase domain with a bound oligonucleotide: The crystal structure provides insights into the mode of unwinding. *Structure* **6:** 89–100.

Kimball S.R., Karnich A.M., Feldhoff R.C., Mellor H., and Jefferson L.S. 1994. Purification and characterization of eukaryotic translational initiation factor eIF-2B from liver. *Biochim. Biophys. Acta* **1201:** 473–481.

Koonin E.V. 1995. Multidomain organization of eukaryotic guanine nucleotide exchange translation initiation factor eIF-2B subunits revealed by analysis of conserved sequence motifs. *Protein Sci.* **4:** 1608–1617.

Kozak M. 1989. The scanning model for translation: An update. *J. Cell Biol.* **108:** 229–240.

———. 1990. Downstream secondary structure facilitates recognition of initiator codons by eukaryotic ribosomes. *Proc. Natl. Acad. Sci.* **87:** 8301–8305.

———. 1991a. Effects of long 5′ leader sequences on initiation by eukaryotic ribosomes *in vitro*. *Gene Expr.* **1:** 117–125.

———. 1991b. A short leader sequence impairs the fidelity of initiation by eukaryotic ribosomes. *Gene Expr.* **1:** 111–115.

———. 1999. Initiation of translation in prokaryotes and eukaryotes. *Gene* **234:** 187–208.

Kozak M. and Shatkin A.J. 1978. Identification of features in the 5′ terminal fragments from reovirus mRNA which are important for ribosome binding. *Cell* **13:** 201–212.

Kyrpides N.C. and Woese C.R. 1998a. Archaeal translation initiation revisited: The initiation factor 2 and eukaryotic initiation factor 2B α-β-δ subunit families. *Proc. Natl. Acad. Sci.* **95:** 3726–3730.

———. 1998b. Universally conserved translation initiation factors. *Proc. Natl. Acad. Sci.* **95:** 224–228.

Laalami S., Timofeev A.V., Putzer H., Leautey J., and Grunberg-Manago M. 1994. *In vivo* study of engineered G-domain mutants of *Escherichia coli* translation initiation factor IF2. *Mol. Microbiol.* **11:** 293–302.

Lamphear B.J., Kirchweger R., Skern T., and Rhoads R.E. 1995. Mapping of functional domains in eukaryotic protein synthesis initiation factor 4G (eIF4G) with picornaviral proteases. *J. Biol. Chem.* **270:** 21975–21983.

Lamphear R.J., Yan R., Yang F., Waters D., Liebig H.D., Klump H., Kuechler E., Skern T., and Rhoads R.E. 1993. Mapping of the cleavage site in protein synthesis initiation factor eIF-4γ of the 2A proteases from human coxsackievirus and rhinovirus. *J. Biol. Chem.* **268:** 19200–19203.

Laurino J.P., Thompson G.M., Pacheco E., and Castilho B.A. 1999. The β subunit of eukaryotic translation initiation factor 2 binds mRNA through the lysine repeats and a region comprising the C_2-C_2 motif. *Mol. Cell. Biol.* **19:** 173–181.

Lawson T.G., Cladaras M.H., Ray B.K., Lee K.A., Abramson R.D., Merrick W.C., and Thach R.E. 1988. Discriminatory interaction of purified eukaryotic initiation factors 4F plus 4A with the 5′ ends of reovirus messenger RNAs. *J. Biol. Chem.* **263:** 7266–7276.

Lee J.H., Choi S.K., Roll-Mecak A., Burley S.K., and Dever T.E. 1999. Universal conservation in translation initiation revealed by human and archaeal homologs of bacterial initiation factor IF2. *Proc. Natl. Acad. Sci.* **96:** 4342–4347.

Li Q., Imataka H., Morino S., Rogers G.W., Jr., Richter N.J., Merrick W.C., and Sonenberg N. 1999. Eukaryotic translation initiation factor eIF4AIII is functionally distinct from eIF4AI and eIF4AII. *Mol. Cell. Biol.* **19:** 7336–7346.

Lorsch J.R. and Herschlag D. 1998a. The DEAD box protein eIF4A. II. A cycle of nucleotide and RNA-dependent conformation changes. *Biochemistry* **37:** 2194–2206.

———. 1998b. The DEAD box protein eIF4A. I. A minimal kinetic and thermodynamic framework reveals coupled binding of RNA and nucleotide. *Biochemistry* **37:** 2180–2193.

———. 1999. Kinetic dissection of fundamental processes of eukaryotic translation initiation *in vitro. EMBO J.* **18:** 6705–6717.

Lu J., Aoki H., and Ganoza M.C. 1999. Molecular characterization of a prokaryotic translation factor homologous to the eukaryotic initiation factor eIF4A. *Int. J. Biochem. Cell Biol.* **31:** 215–229.

Mader S., Lee H., Pause A., and Sonenberg N. 1995. The translation initiation factor eIF-4E binds to a common motif shared by the translation factor eIF-4γ and the translational repressor 4E-binding proteins. *Mol. Cell. Biol.* **15:** 4990–4997.

Malhotra A., Penczek P.,. Agrawal R.K., Gabashvili I.S., Grassucci R.A., Jünemann R., Burkhardt N., Nierhaus K.H., and Frank J. 1998. *Escherichia coli* 70S ribosome at 15 Å resolution by cryo-electron microscopy: Localization of fMet-$tRNA_f$ and fitting of L1 protein. *J. Mol. Biol.* **280:** 103–116.

Manchester K.L. 1997. Catalysis of guanine nucleotide exchange on eIF-2 by eIF-2B: Is it a sequential or substituted enzyme mechanism? *Biochem. Biophys. Res. Commun.* **239:** 223–227.

Mangroo D., X.Q. Wu, and RajBhandary U.L. 1995. *Escherichia coli* initiator tRNA: Structure-function relationships and interactions with the translational machinery. *Biochem. Cell Biol.* **73:** 1023–1031.

Marcotrigiano J., Gingras A.-C., Sonenberg N., and Burley S.K. 1997. Cocrystal structure of the messenger RNA 5' cap-binding protein (eIF4E) bound to 7-methyl-GDP. *Cell* **89:** 951–961.

Matsuo H., Li H., McGwire A.M., Fletcher C.M., Gingras A.-C., Sonenberg N., and Wagner G. 1997. Structure of translation factor eIF4E bound to m7GDP and interaction with 4E binding protein. *Nat. Struct. Biol.* **4:** 717–724.

McCarthy J.E.G. and Brimacombe R. 1994. Prokaryotic translation: The interactive pathway leading to initiation. *Trends Genet.* **10:** 402–407.

McCutcheon J.P., Agrawal R.K., Philips S.M., Grassucci R.A., Gerchman S.E., Clemons W.M.J., Ramakrishnan V., and Frank J. 1999. Location of translational initiation factor IF3 on the small ribosomal subunit. *Proc. Natl. Acad. Sci.* **96:** 4301–4306.

Meinnel T., Sacerdot C., Graffe M., Blanquet S., and Springer M. 1999. Discrimination by *Escherichia coli* initiation factor IF3 against initiation on non-canonical codons relies on complementarity rules. *J. Mol. Biol.* **290:** 825–837.

Merrick W.C. 1979. Purification of protein synthesis initiation factors from rabbit reticulocytes. *Methods Enzymol.* **60:** 100–101.

———. 1992. Mechanism and regulation of eukaryotic protein synthesis. *Microbiol.Rev.* **56:** 291–315.

Merrick W.C. and Anderson W.F. 1975. Purification and characterization of protein synthesis initiation factor M1 from rabbit reticulocytes. *J. Biol. Chem.* **250:** 1197–1206.

Merrick W.C. and Hershey J.W.B. 1996. The pathway and mechanism of eukaryotic protein synthesis. In *Translational control* (ed. J.W.B. Hershey et al.), pp. 31–69. Cold Spring Harbor Laboratory Press, Cold Spring Harbor, New York.

Merrick W.C., Kemper W.M., and Anderson W.F. 1975. Purification and characterization of homogeneous initiation factor M2A from rabbit reticulocytes. *J. Biol. Chem.* **250:** 5556–5562.

Merrick W.C., Lubson N.H., and Anderson W.F. 1973. A ribosome dissociation factor from rabbit reticulocytes distinct from initiation factor M3. *Proc. Natl. Acad. Sci.* **70:** 2220–2223.

Méthot N., Song M.S., and Sonenberg N. 1996a. A region rich in aspartic acid, arginine, tyrosine and glycine (DRYG) mediates eIF4B self association and interaction with eIF3. *Mol. Cell. Biol.* **16:** 5328–5334.

Méthot N., Pause A., Hershey J.W.B., and Sonenberg N. 1994. The translation initiation factor eIF-4B contains an RNA-binding region that is distinct and independent from its ribonucleoprotein consensus sequence. *Mol. Cell. Biol.* **14:** 2307–2326.

Méthot N., Pickett G., Keene J., and Sonenberg N. 1996b. In vitro RNA selection identifies RNA ligands that specifically bind to eukaryotic translation initiation factor 4B: The role of the RNA recognition motif. *RNA* **2:** 38–50.

Moazed D., Samaha R.R., Gualerzi C.O., and Noller H.F. 1995. Specific protection of 16S rRNA by translational initiation factors. *J. Mol. Biol.* **248:** 207–210.

Morehouse H., Buratowski R.M., Silver P.A., and Buratowski S. 1999. The importin/karyopherin kap114 mediates the nuclear import of TATA-binding protein. *Proc. Natl. Acad. Sci.* **96:** 12542–12547.

Moreno J.M.P., Dyrskjotersen L., Kristensen J.E., Mortensen K.K., and Sperling-Petersen H.U. 1999. Characterization of the domains of *E. coli* initiation factor IF2 responsible for recognition of the ribosome. *FEBS Lett.* **455:** 130–134.

Morino S., Imataka H., Svitkin Y.V., Pestova T.V., and Sonenberg N. 2000. Eukaryotic translation initiation factor 4E (eIF4E) binding site and the middle one-third of eIF4GI constitute the core domain for cap-dependent translation, and the C-terminal one-third functions as a modulatory region. *Mol. Cell. Biol.* **20:** 468–477.

Murzin A.G. 1993. OB (oligonucleotide/oligosaccharide binding) fold: Common structural and functional solution for non-homologous sequences. *EMBO J.* **12:** 521–526.

Naranda T., MacMillan S.E., and Hershey J.W.B. 1994a. Purified yeast translational initiation factor eIF-3 is an RNA-binding protein complex that contains the PRT1 protein. *J. Biol. Chem.* **269:** 32286–32292.

Naranda T., Kainuma M., MacMillan S.E., and Hershey J.W.B. 1997. The 39-kilodalton subunit of eukaryotic translation initiation factor 3 is essential for the complex's integrity and for cell viability in *Saccharomyces cerevisiae*. *Mol. Cell. Biol.* **17:** 145–153.

Naranda T., MacMillan S.E., Donahue T.F., and Hershey J.W.B. 1996. SUI1/p16 is required for the activity of eukaryotic translation initiation factor 3 in *Saccharomyces cerevisiae*. *Mol. Cell. Biol.* **16:** 2307–2313.

Naranda T., Strong W.B., Menaya J., Fabbri B.J., and Hershey J.W.B. 1994b. Two structural domains of initiation factor eIF-4B are involved in binding to RNA. *J. Biol. Chem.* **269:** 14465–14472.

Oldfield S. and Proud C.G. 1992. Purification, phosphorylation and control of the guanine nucleotide exchange factor from rabbit reticulocyte lysates. *Eur. J. Biochem.* **208:** 73–81.

Oldfield S., Jones B.L., Tanton D., and Proud C.G. 1994. Use of monoclonal antibodies to study the structure and function of eukaryotic protein synthesis initiation factor eIF-2B. *Eur. J. Biochem.* **221:** 399–410.

Palmer T.D., Miller A.D., Reeder R.H., and McStay B. 1993. Efficient expression of a protein coding gene under control of an RNA polymerase I promoter. *Nucleic Acids Res.* **15:** 3451–3457.

Pause A. and Sonenberg N. 1992. Mutational analysis of a DEAD box RNA helicase: The mammalian translation initiation factor eIF4A. *EMBO J.* **11:** 2643–2654.

Pause A., Méthot N., and Sonenberg N. 1993. The HRIGRR region of the DEAD box RNA helicase eukaryotic translation initiation factor eIF4A is required for RNA binding and ATP hydrolysis. *Mol. Cell. Biol.* **13:** 6789–6798.

Pause A., Méthot N., Svitkin Y., Merrick W.C., and Sonenberg N. 1994. Dominant negative mutants of mammalian translation initiation factor eIF-4A define a critical role for eIF-4F in cap-dependent and cap-independent initiation of translation. *EMBO J.* **13:** 1205–1215.

Pavitt G.D., Ramaiah K.V.A., Kimball S.R., and Hinnebusch A.G. 1998. eIF2 independently binds two distinct eIF2B subcomplexes that catalyze and regulate guanine-nucleotide exchange. *Genes Dev.* **12:** 514–526.

Pestova T.V., Borukhov S.I., and Hellen C.U.T. 1998. Eukaryotic ribosomes require initiation factors 1 and 1A to locate initiation codons. *Nature* **394:** 854–859.

Pestova T.V., Shatsky I.N., and Hellen C.U.T. 1996. Functional dissection of eukaryotic initiation factor 4F: The 4A subunit and the central domain of the 4G subunit are sufficient to mediate internal entry of 43S preinitiation complexes. *Mol. Cell. Biol.* **16:** 6870–6878.

Pestova T.V., Lomakin I.B., Lee J.H., Choi S.K., Dever T.E., and Hellen C.U.T. 2000. The ribosomal subunit joining reaction in eukaryotes requires eIF5B. *Nature* **403:** 332–335.

Peterson D.T., Safer B., and Merrick W.C. 1979. Role of eukaryotic initiation factor 5 in the formation of 80S initiation complexes. *J. Biol. Chem.* **254:** 7730–7735.

Phan L., Zhang X., Asano K., Anderson J., Vornlocher H.-P., Greenberg J.R., Goldfarb D.S., Qin J., and Hinnebusch A.G. 1998. Identification of a translation initiation factor 3 (eIF3) core complex, conserved in yeast and mammals, that interacts with eIF5. *Mol. Cell. Biol.* **18:** 4935–4946.

Prat A., Schmid S.R., Buser P., Blum S., Trachsel H., Nielsen P.J., and Linder P. 1990. Expression of translation initiation factor 4A from yeast and mouse in *Saccharomyces cerevisiae*. *Biochim. Biophys. Acta* **1050:** 140–145.

Pyronnet S., Imataka H., Gingras A.-C., Fukunaga R., Hunter T., and Sonenberg N. 1999. Human eukaryotic translation initiation factor 4G (eIF4G) recruits Mnk1 to phosphorylate eIF4E. *EMBO J.* **18:** 270–279.

Rau M., Ohlmann T., Morley S.J., and Pain V.M. 1996. A reevaluation of the cap-binding protein, eIF4E, as a rate-limiting factor for initiation of translation in reticulocyte lysate. *J. Biol. Chem.* **271:** 8983–8990.

Raychaudhuri P., Stringer E.A., Valenzuela D.M., and Maitra U. 1984. Ribosomal subunit anti-association activity in rabbit reticulocytes. *J. Biol. Chem.* **259:** 11930–11935.

Richter N.J., Rogers G.W., Jr., Hensold J.O., and Merrick W.C. 1999. Further biochemical and kinetic characterization of human eukaryotic initiation factor 4H. *J. Biol. Chem.* **274:** 35415–35424.

Richter-Cook N.J., Dever T.E., Hensold J.O., and Merrick W.C. 1998. Purification and characterization of a new eukaryotic protein translation factor: Eukaryotic initiation factor 4H. *J. Biol. Chem.* **273:** 7579–7587.

Rogers G.W., Jr., Richter N.J., and Merrick W.C. 1999. Biochemical and kinetic characterization of the RNA helicase activity of eukaryotic initiation factor 4A. *J. Biol. Chem.* **274:** 12236–12244.

Rom E., Kim H.C., Gingras A.-C., Marcotrigiano J., Favre D., Olsen H., Burley S.K., and Sonenberg N. 1998. Cloning and charcterization of 4EHP, a novel mammalian eIF4E-related cap-binding protein. *J. Biol. Chem.* **273:** 13104–13109.

Rozen R., Edery I., Meerovitch K., Dever T.E., Merrick W.C., and Sonenberg N. 1990. Bidirectional RNA helicase activity of eucaryotic translation initiation factors 4A and 4F. *Mol. Cell. Biol.* **10:** 1134–1144.

Ruhl M., Himmelspach M., Bahr G.M., Hammerschmid F., Jaksche H., Wolff B., Aschauer H., Farrington G.K., Probst H., Bevec D., and Hauber J. 1993. Eukaryotic initiation factor 5A is a cellular target of the human immunodeficiency virus type I Rev activation domain mediating trans-activation. *J. Cell Biol.* **123:** 1309–1320.

Russell D.W. and Spremulli L.L. 1979. Purification and characterization of a ribosome dissociation factor (eukaryotic initiation factor) from wheat germ. *J. Biol. Chem.* **254:** 8796–8800.

Ryabova L.A. and Hohn T. 2000. Ribosome shunting in the cauliflower mosaic virus 35S RNA leader is a special case of reinitiation of translation functioning in plant and animal systems. *Genes Dev.* **14:** 817–829.

Safer B., Adams S.L., Kemper W.M., Berry K.W., Lloyd M., and Merrick W.C. 1976. Purification and characterization of two initiation factors required for maximal activity of a highly fractionated globin mRNA translation system. *Proc. Natl. Acad. Sci.* **73:** 2584–2588.

Schmidt S., Hofmann K., and Simanis V. 1997. Sce3, a suppressor of the *Schizosaccharomyces pombe* septation mutant cdc11, encodes a putative RNA-binding protein. *Nucleic Acids Res.* **25:** 3433–3439.

Schreier M.H., Erni B., and Staehelin T. 1977. Initiation in mammalian protein synthesis. I. Purification and characterization of seven initiation factors. *J. Mol. Biol.* **116:** 727–753.

Sette M., van Tilborg P., Spurio R., Kaptein R., Paci M., Gualerzi C.O., and Boelens R. 1997. The structure of the translational initiation factor IF1 from *E. coli* contains an oligomer-binding motif. *EMBO J.* **16:** 1436–1443.

Si K. and Maitra U. 1999. The *Saccharomyces cerevisiae* homologue of mammalian translation initiation factor 6 does not function as a translation initiation factor. *Mol. Cell. Biol.* **19:** 1416–1426.

Si K., Chaudhuri J., Chevesich J., and Maitra U. 1997. Molecular cloning and functional expression of a human cDNA encoding translation initiation factor 6. *Proc. Natl. Acad. Sci.* **94:** 14285–14290.

Simons R.W. and Grunberg-Manago M., Eds. 1998. *RNA structure and function.* Cold Spring Harbor Laboratory Press, Cold Spring Harbor, New York.

Singh L.P., Aroor A.R., and Wahba A.J. 1994. Phosphorylation of the guanine nucleotide exchange factor and eukaryotic initiation factor 2 by casein kinase II regulates guanine nucleotide binding and GDP/GTP exchange. *Biochemistry* **33:** 9152–9157.

Singh U.S., Li Q., and Cerione R. 1998. Identification of the eukaryotic initiation factor 5A as a retinoic acid-stimulated cellular binding partner for tissue transglutaminase II. *J. Biol. Chem.* **273:** 1946–1950.

Smith K.E. and Henshaw E.C. 1975. Binding of Met-$tRNA_f$ to native ribosomal subunits in Ehrlich ascites tumor cells. *J. Biol. Chem.* **250:** 6880–6884.

Spurio R., Severini M., La Teana A., Canonaco M.A., Pawlik R.T., Gualerzi C.O., and Pon C.L. 1993. Novel structural and functional aspects of translational initiation factor IF2. In *The translational apparatus: Structure, function, regulation, evolution* (ed. K.H. Nierhaus et al.), pp. 241–252. Plenum Press, New York.

Spurio R., Brandi L., Caserta E., Pon C.L., Gualerzi C.A., Misselwitz R., Krafft C., Welfle K., and Welfle H. 2000. The C-terminal subdomain (IF-C2) contains the entire fMet-tRNA binding site of initiation factor IF2. *J. Biol. Chem.* **275:** 2447–2454.

Srivastava S., Verschoor A., and Frank J. 1992. Eukaryotic initiation factor 3 does not prevent association through physical blockage of the ribosomal subunit-subunit interface.

J. Mol. Biol. **226:** 301–304.

Stark H., Müller F., Orlova E.V., Schatz M., Dube P., Erdemir T., Zemlin F., Brimacombe R., and van Heel M. 1995. The 70S *Escherichia coli* ribosome at 23 Å resolution: Fitting the ribosomal RNA. *Structure* **3:** 815–821.

Stark H., Orlova E.V., Rinke-Appel J., Junke N., Müller F., Rodnina M., Wintermeyer W., Brimacombe R., and van Heel M. 1997. Arrangement of tRNAs in pre- and post-translocational ribosomes revealed by electron cryomicroscopy. *Cell* **88:** 19–28.

Steitz J.A. and Jakes K. 1975. How ribosomes select initiator regions in mRNA: Base-pair formation between the 3′ terminus of 16S rRNA and mRNA during initiation of protein synthesis in *Escherichia coli*. *Proc. Natl. Acad. Sci.* **72:** 4734–4738.

Subramanya H.S., Bird L.E., Brannigan J.A., and Wigley D.B. 1996. Crystal structure of a DExx box helicase. *Nature* **384:** 379–383.

Sussman J.K., Simons E.L., and Simons R.W. 1996. *Escherichia coli* translation initiation factor 3 discriminates the initiation codon *in vivo*. *Mol. Microbiol.* **21:** 347–360.

Tabara H., Sarkissian M., Kelly W.G., Fleenor J., Grishok A., Timmons L., Fire A., and Mello C.C. 1999. The rde-1 gene, RNA interference, and transposon silencing in *C. elegans*. *Cell* **99:** 123–132.

Thuillier V., Stettler S., Sentenac A., Thuriaux P., and Werner M. 1995. A mutation in the C31 subunit of *Saccharomyces cerevisiae* RNA polymerase III affects transcription initiation. *EMBO J.* **14:** 351–359.

Traugh J.A., Tahara S.M., Sharp S.B., Safer B., and Merrick W.C. 1976. Factors involved in initiation of haemoglobin synthesis can be phosphorylated in vitro. *Nature* **263:** 163–165.

Valasek L., Trachsel H., Hasek J., and Ruis H. 1998. Rpg1, the *Saccharomyces cerevisiae* Homologue of the largest subunit of mammalian translation initiation factor 3, is required for translational activity. *J. Biol. Chem.* **273:** 21253–21260.

Valasek L., Hasek J., Trachsel H., Imre E.M., and Ruis H. 1999. The *Saccharomyces cerevisiae HCRI* gene encoding a homologue of the p35 subunit of human translation eukaryotic initiation factor 3 (eIF3) is a high copy suppressor of a temperature-sensitive mutation in the Rpglp subunit of yeast eIF3. *J. Biol. Chem.* **275:** 27567–27572.

Valenzuela D.M., Chaudhuri A., and Maitra U. 1982. Eukaryotic ribosomal subunit anti-association activity of calf liver is contained in a single polypeptide chain protein of M_r = 25,500 (eukaryotic initiation factor 6). *Biochim. Biophys. Acta* **565:** 305–314.

Verlhac M.H., Chen R.H., Hanachi P., Hershey J.W.B., and Derynck R. 1997. Identification of partners of TIF34, a component of the yeast eIF3 complex, required for cell proliferation and translation initiation. *EMBO J.* **16:** 6812–6822.

Voorma H.O. 1996. Control of translation initiation in prokaryotes. In *Translational control* (ed. J.W.B. Hershey et al.), pp. 759–777, Cold Spring Harbor Laboratory Press, Cold Spring Harbor, New York.

Vornlocher H.-P., Hanachi P., Ribeiro S., and Hershey J.W.B. 1999. A 110-kilodalton subunit of translation initiation factor eIF3 and an associated 135-kilodalton protein are encoded by the *Saccharomyces cerevisiae TIF32* and *TIF31* genes. *J. Biol. Chem.* **274:** 16802–16812.

Vornlocher H.-P., Scheible J.W.-R., Faulhammer H.G., and Sprinzl M. 1997. Identification and purification of translation initiation factor 2 (IF2) from *Thermus thermophilus*. *Eur. J. Biochem.* **243:** 66–71.

Wei C.-L., Kainuma M., and Hershey J.W.B. 1995. Characterization of yeast translation initiation factor 1A and cloning of its essential gene. *J. Biol. Chem.* **270:** 22788–22794.

Welsh G.I. and Proud C.G. 1993. Glycogen synthase kinase-3 is rapidly inactivated in response to insulin and phosphorylates eukaryotic initiation factor eIF–2B. *Biochem. J.* **294:** 625–629.

Welsh G.I., Miller C.M., Loughlin A.J., Price N.T., and Proud C.G. 1998. Regulation of eukaryotic initiation factor eIF2B: Glycogen synthase kinase-3 phosphorylates a conserved serine which undergoes dephosphorylation in response to insulin. *FEBS Lett.* **421:** 125–130.

Wilson S.A., Sieiro-Vazquez C., Edwards N.J., Iourin O., Byles E.D., Kotsopoulou E., Adamson C.S., Kingsman S.M., Kingsman A.J., and Martin–Rendon E. 1999. Cloning and characterization of hIF2, a human homologue of bacterial translation initiation factor 2, and its interaction with HIV-1 matrix. *Biochem. J.* **342:** 97–103.

Wu X.-Q. and RajBhandary U.L. 1997. Effect of the amino acid attached to *Escherichia coli* initiator tRNA on its affinity for the initiation factor IF2 and on the IF2 dependence of its binding to the ribosome. *J. Biol. Chem.* **272:** 1891–1895.

Wu X.-Q., Iyengar P., and RajBhandary U.L. 1996. Ribosome-initiator tRNA complex as an intermediate in translation initiation in *Escherichia coli* revealed by use of mutant initiator tRNAs and specialized ribosomes. *EMBO J.* **15:** 4734–4739.

Yan R., Rychlik W., Etchison D., and Rhoads R.E. 1992. Amino acid sequence of the human protein synthesis initiation factor eIF-4γ. *J. Biol. Chem.* **267:** 23226-23231.

Yao N.H., Hesson T., Cable M., Hong Z., Kwong A.D., Le H.V., and Weber P.C. 1997. Structure of the hepatitis C virus RNA helicase domain. *Nat. Struct. Biol.* **4:** 463–467.

Yoon H.J. and Donahue T.F. 1992. The *suil* suppressor locus in *Saccharomyces cerevisiae* encodes a translation factor that functions during $tRNA_i^{Met}$ recognition of the start codon. *Mol. Cell. Biol.* **12:** 248–260.

Zou C., Zhang Z., Wu S., and Osterman J.C. 1998. Molecular cloning and characterization of a rabbit eIF2C protein. *Gene* **211:** 187–194.

Zucker F.H. and Hershey J.W.B. 1986. Binding of *Escherichia coli* protein synthesis initiation factor IF1 to 30S ribosomal subunits by fluorescence polarization. *Biochemistry* **25:** 3682–3690.

Zuk D. and Jacobson A. 1998. A single amino acid substitution in yeast eIF-5A results in mRNA stabilization. *EMBO J.* **17:** 2914–2925.

3

The Protein Biosynthesis Elongation Cycle

William C. Merrick
Department of Biochemistry
School of Medicine
Case Western Reserve University
Cleveland, Ohio 44106

Jens Nyborg
Department of Molecular and Structural Biology
University of Aarhus
DK-8000 Århus C, Denmark

The detailed mechanism of protein biosynthesis has been studied for many years, especially in prokaryotic systems, but in recent years also increasingly in eukaryotes. The reader is referred to recent books on various aspects of protein biosynthesis in general (Hill et al. 1990; Nierhaus et al. 1993a; Söll and RajBhandary 1995). Protein biosynthesis as it happens on the ribosome is conveniently divided into three phases: initiation, elongation, and termination. Here we are primarily concerned with the description of the elongation cycle from a functional and structural point of view (for discussion of the regulation of the elongation cycle, see Chapter 24. Aspects of initiation and termination are only dealt with to illustrate some important concepts of elongation, which seem to be general for all three phases. Aminoacylation of tRNA catalyzed by tRNA-synthetases is not described nor discussed in this review. Readers are referred to recent reviews on this subject (Arnez and Moras 1997; Cusack 1997). Because the amount of work performed over many years is enormous, many important biochemical studies on the function of elongation factors are not mentioned in detail, but a discussion of some of these can be found in a recent book chapter (Clark et al. 1995).

During the last six years, major advances have been made in generating a structural description of the prokaryotic elongation cycle. Currently available is an essentially complete structural picture of most of the major functional forms of the elongation factors (Krab and Parmeggiani 1998).

Translational Control of Gene Expression
© 2000 Cold Spring Harbor Laboratory Press 0-87969-568-4/00 $5 + .00

Many of the structural principles found in prokaryotic factors can be used as models for the structures of eukaryotic elongation factors. The structural studies of the ribosome particle itself are progressing at a rapid rate. For some years, important image reconstructions of the prokaryotic ribosome obtained by cryo-electron microscopy (cryo-EM) on randomly oriented ribosome particles in vitreous ice have been published by two groups of researchers (Frank et al. 1995; Stark et al. 1995). A reconstruction of a eukaryotic ribosome has also been published recently (Dube et al. 1998). These reconstructions not only give views of the ribosome particle alone, which are essentially identical from the two groups, but also give pictures of the ribosome in various functional states, interacting with tRNAs and with elongation factors (Agrawal et al. 1996, 1998, 1999; Stark et al. 1995; 1997a,b). Some of these reconstructions are important for an understanding of the functional and structural details of the interaction between the elongation factors and the ribosome that are discussed in this review.

Crystallographic investigations of ribosome particles and of ribosomal subunits have been under way for many years (Yonath and Franceschi 1998). Recently, significant progress has been obtained in the phasing of crystallographic data such that electron densities of both subunits have been obtained with resolution to 5.5 Å for the 30S subunit (Clemons et al. 1999), to 5 Å for the 50S subunit (Ban et al. 1999), and to 7.8 Å for the complete particle (Cate et al. 1999). The structures of many ribosomal proteins from both subunits are known (Liljas and Garber 1995; Ramakrishnan and White 1998). Structures of some fragments of ribosomal RNA are also available (Szewczak and Moore 1995; Fourmy et al. 1996; Correll et al. 1997; 1998; 1999; Dallas and Moore 1997; Puglisi et al. 1997; Woodson and Leontis 1998), as well as a complex of a ribosomal protein and a fragment of ribosomal RNA (Conn et al. 1999; Wimberly et al. 1999). All of this structural information, together with many years of biochemical studies, is changing our view on the detailed mechanism of protein biosynthesis. Undoubtedly, our perception of ribosome function will change even more dramatically in the very near future when higher-resolution structural information will be available for the many well-defined states of the functional cycle of the ribosome, and then these structures will be the basis for the design of many new biochemical and functional experiments.

PROKARYOTIC ELONGATION

During the elongation phase of protein biosynthesis, the ribosome selects aminoacylated-tRNA (aa-tRNA) according to the sequence of codons in the mRNA and catalyzes the formation of a peptide bond between a growing peptide and the incoming amino acid. This ribosomal process occurs in a cyclical manner assisted by three protein elongation factors. The factors act as catalysts, mediating the speed and accuracy of translation for the production of protein macromolecules in the cell. The accuracy is partly in the aminoacylation process on tRNA-synthetases, where energy in the form of ATP is used to ensure that the amino acid coupled by an amino acid ester bond to the 3′ CCA end of tRNA corresponds to the anticodon of the tRNA. The second part relies on the correct recognition between the codon in the mRNA and the anticodon of the tRNA as it is being bound to the ribosome. Another form of accuracy applies during translocation of tRNAs and mRNA, where the ribosome advances exactly one codon on the mRNA. Errors in translocation can lead to frameshifting, and thus to erroneous protein products (Chapter 25). Both correct codon/anticodon recognition and translocation rely on the use of energy in the form of GTP. The trade-off between speed and accuracy has been an essential parameter during evolution of the ribosomal machinery, and is of utmost importance for the efficient functioning of the cell. The detailed interaction of elongation factors with tRNA, with GTP, and with ribosomal RNAs and proteins is thus of crucial importance for the survival of any organism.

The Elongation Factors

The prokaryotic (and mitochondrial) protein biosynthesis elongation cycle is catalyzed by three polypeptide *e*longation *f*actors (EFs): EF1A (formerly called EF-Tu), EF1B (formerly EF-Ts), and EF2 (formerly EF-G). Two of these, EF1A and EF2, are GDP/GTP-binding proteins (G proteins), which are active when complexed to GTP and inactive in their GDP form. The third, EF1B, is a *g*uanosine nucleotide *e*xchange *f*actor (GEF) for EF1A. In bacterial systems, the K_d of EF1A for GTP is about 10^{-6} M and that for GDP is 10^{-8} M. This poses two problems. First, the off-rate for GDP is quite slow. Second, given an average GTP/GDP ratio of 10/1, normally the EF1A would exist in the GDP-bound state (an inactive form). To overcome these problems, a nucleotide exchange factor exists, EF1B, that enhances the off-rate for GDP. The net result is the favored formation of EF1A•GTP. There is no known GEF for EF2.

EF1A•B

EF1A is a protein of about 400 amino acids and is very well conserved among prokaryotes and mitochondria (see Fig. 1). The molecular mass of

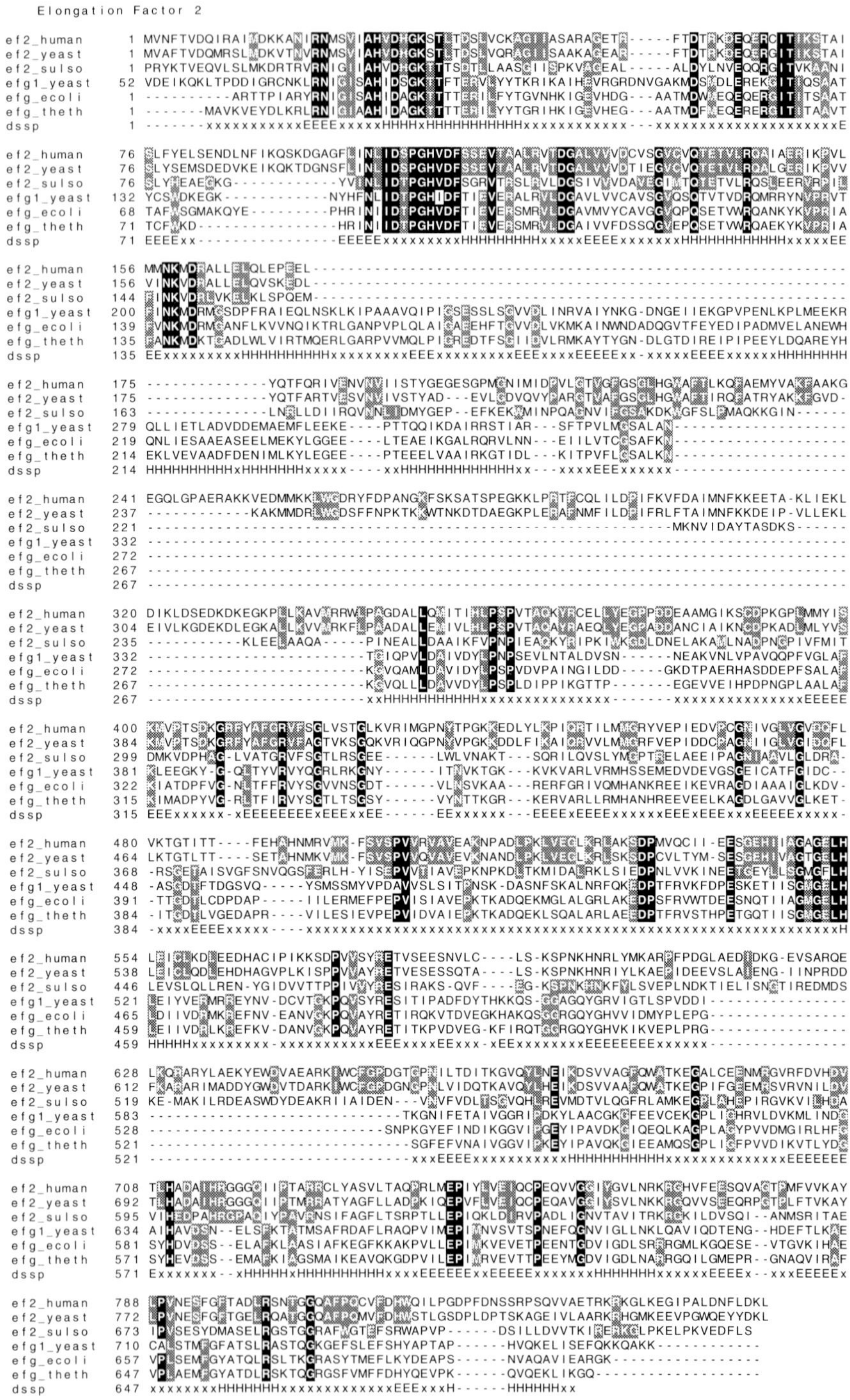

Figure 1 (*See facing page for legend.*).

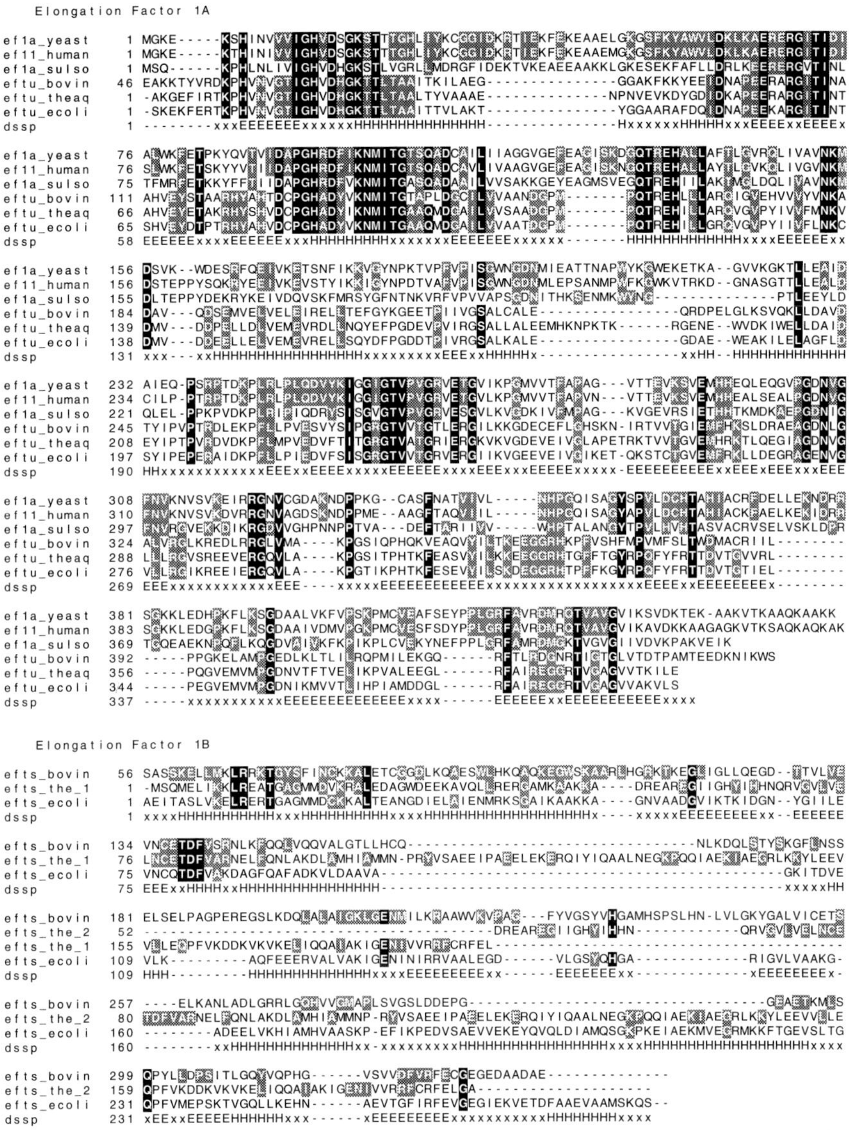

Figure 1 Sequence alignments of EF2, EF1A, and EF1B. Selected sequences of eukaryotes, archaebacteria, and eubacteria from an alignment of all sequences in the SWISSPROT database are shown (Bairoch and Boeckmann 1994). The line "dssp" shows secondary structures, H is helix, and E is β strand based on known structures (Kabsch and Sander 1983). In black background are universally conserved residues, whereas a gray background shows residues conserved within each kingdom. The sequence alignments are produced with the program ALMA (Thirup and Larsen 1990).

EF1A is approximately 44 kD. It is composed of three structural domains, of which domain 1 is about 200 amino acids, and domains 2 and 3 are about 100 amino acids each. Prokaryotic EF1As are shorter than the ones from archaea or eukaryotes (see also Fig. 6). EF1A from *Escherichia coli* has 393 amino acids, whereas human eEF1A has 463 amino acids. The *Thermus* family has an insertion in domain 1, sometimes referred to as the "thermophilic loop," although it is in reality an extension of an α helix. Some mitochondrial EF1As have a carboxy-terminal extension of about 10 amino acids similar to that found in eukaryotic eEF1A. Many of the inserts of the longer proteins occur at loop regions of the structure. In prokaryotes (but not in mitochondria) the GDP is bound more tightly (in the nM range) than GTP by two orders of magnitude (Louie and Jurnak 1985). In mitochondria (and in eukaryotes) the binding of the two nucleotides is not as tight (μM range), and both have a similar affinity for EF1A (Y.-C. Cai and L. Spremulli, in prep.). The function of EF1A in its GTP form is to bind aa-tRNA and to protect the amino ester bond from hydrolysis. Furthermore, this so-called ternary complex of EF1A (EF1A•GTP•aa-tRNA) will assist in the binding of aa-tRNA to the ribosome.

In *E. coli*, EF1B is a protein of 282 amino acids and thus has a molecular mass of approximately 30 kD. The sequence alignment between prokaryotic and mitochondrial EF1B has been very difficult (see Fig. 1). The structural determinations of EF1B from *E. coli* and from *Thermus thermophilus* in complex with EF1A (see later) showed that the basic folds are very similar but that the *E. coli* factor has rudiments of an internal repeat giving a pseudo twofold symmetry within the molecule. In Figure 1 this is shown by letting the sequence from *T. thermophilus* (efts_the_1 and efts_the_2) align with both the amino-terminal part and the carboxy-terminal part of the sequence from *E. coli*. The *T. thermophilus* EF1B is functionally a dimer. The function of EF1B is to catalyze the nucleotide exchange of EF1A•GDP to EF1A•GTP. This is needed because there is a large structural change between the two forms of EF1A (see below).

EF2

EF2 is a protein of about 700 amino acids. The molecular mass is thus on the order of 77 kD. Sequence comparisons reveal extensive similarities in the amino-terminal half of the molecule between all organisms with large inserts in the eukaryotic proteins. The sequences are much more diverse in the carboxy-terminal half (see Fig. 1). GDP and GTP are bound with similar affinity (μM range) to EF2. The lack of a GEF for EF2 is explained by the higher off-rate for GDP than GTP (thus, poorer binding affinity) and the relatively higher level of GTP over GDP in the cell, thus shifting the physiologic equilibrium in favor of EF2•GTP (Bourne et al. 1991). The function of EF2 in its GTP form is to catalyze translocation on the ribosome.

Description of the Elongation Cycle

The elongation cycle of protein biosynthesis on the ribosome will add one amino acid at a time to a growing polypeptide according to the sequence of codons found in the mRNA (Fig. 2). The next available codon on the mRNA is exposed in the aa-tRNA binding site (A site) on the 30S subunit. Ternary complexes of aa-tRNA•EF1A•GTP enter the ribosome in a "testing phase," where the anticodon of the tRNA attempts to make a codon/anticodon interaction with the A-site codon of the mRNA. Upon cognate recognition, the EF1A•GTP is brought into the GTPase activating center of the ribosome, GTP is hydrolyzed, and EF1A•GDP leaves the ribosome (Pape et al. 1998). The 3′ CCA end of aa-tRNA enters the A site on the 50S subunit, and the peptidyl transferase center of the ribosome quickly catalyzes the formation of a peptide bond between the incoming amino acid and the peptide found in the peptidyl-tRNA binding site (P site). This leaves the tRNAs in so-called mixed hybrid states (Moazed and Noller 1989) with the newly formed peptidyl-tRNA in the A/P state, where the anticodon is still in contact with the codon of the A site of the 30S subunit, but the CCA end is at the P site of the 50S subunit. The newly deacylated tRNA is left in a similar P/E state, with its CCA end in an exit site (E site) on the 50S subunit. In this pre-translocation state of the ribosome, the EF2•GTP enters a site that is similar to the testing site of the ternary complex, physically forcing the peptidyl-tRNA out of the A site on the 30S subunit, and possibly preventing it from rebinding to the A site. Thus, the peptidyl-tRNA is brought fully into the P-site, and the deacylated-tRNA fully into the E site. During this process, GTP bound to EF2 will be hydrolyzed at the GTPase center, and EF2•GDP leaves the ribosome (Rodnina et al. 1997). The action of EF2 that "forces" the peptidyl-tRNA into the P site also accounts for the precise movement of the mRNA by 3 nucleotides. By moving the peptidyl-tRNA and preserving the tRNA/mRNA contacts, the mRNA is moved precisely 3 nucleotides. In this model, the peptidyl-tRNA is "dragging" the codon of the mRNA from the A site into the P site. In its post-translocation state, the ribosome is ready to receive a new ternary complex. The EF1B catalyzes the nucleotide exchange of EF1A, so that inactive EF1A•GDP is converted into active EF1A•GTP.

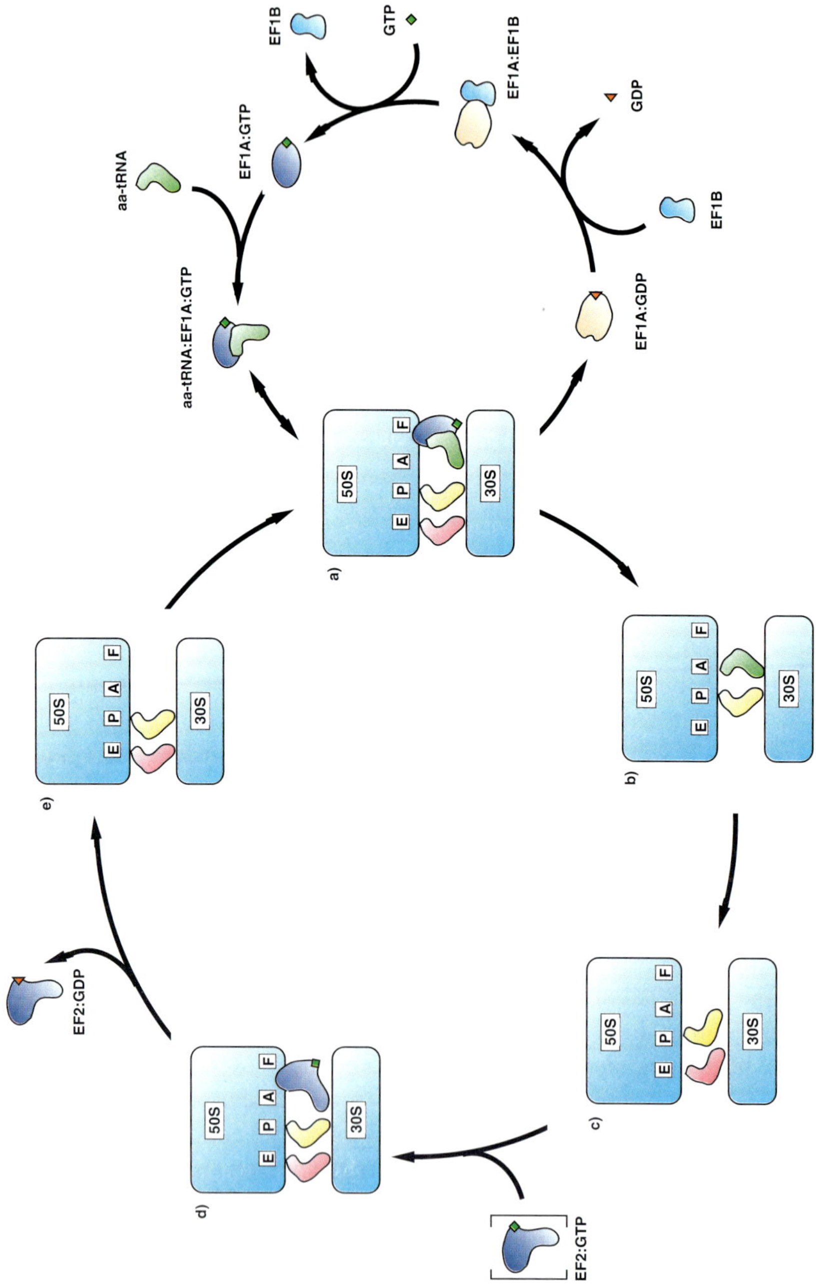

Figure 2 Schematic drawing of the elongation cycle.

The overall functional cycle of elongation as outlined above is generally accepted and supported by many experiments. This chapter addresses only some of the work performed in recent years and tries to identify some concepts that are still under debate at the moment.

By using pre-steady-state kinetic experiments, six steps were identified in the kinetic mechanism of EF1A-dependent binding of cognate aa-tRNA to the ribosome (Pape et al. 1998). The initial binding of the ternary complex of EF1A is rapid and readily reversible. This binding is independent of the codon, and involves a preorientation of the ternary complex possibly by interaction with the L7/L12 stalk. This binding site will be referred to as the factor-binding site (F site). The codon/anticodon interaction is rapid and the binding of a ternary complex with a cognate tRNA, with the tRNA in the A site of the 30S subunit and EF1A at the F site (hybrid A/F site), is greatly stabilized. Subsequently, a rapid induction of the GTPase conformation of EF1A occurs, which is instantaneously followed by GTP hydrolysis. Ternary complexes with noncognate tRNA dissociate at a fast rate from the ribosome before the induction of the GTPase (Bilgin et al. 1992). The conformation of EF1A switches from the GTP form to the GDP form and aa-tRNA is rapidly released from EF1A (Pape et al. 1998). The following accommodation of an aa-tRNA into the A site is relatively slow, but is immediately followed by peptide-bond formation (Bilgin et al. 1992; Pape et al. 1998). The slowest step is dissociation of EF1A•GDP from the ribosome. For ternary complexes with near-cognate aa-tRNA, it has been found that the GTPase activation of EF1A (preceding GTP hydrolysis) and A-site accommodation of aa-tRNA (preceding peptide-bond formation) are significantly slower (Pape et al. 1999). Conformational coupling between the ribosomal subunits or induced fit are thus important contributions to aa-tRNA discrimination.

Some of the kinetic parameters appear to depend on the exact buffer conditions, noticeably on Mg^{++}-ion concentration and on temperature. This may explain some of the differences in the values of these parameters obtained in various laboratories. The accuracy of aa-tRNA selection by the bacterial ribosome has been proposed to be due to a proofreading step (Hopfield 1974; Thompson 1988), which should discriminate between cognate and near-cognate tRNA. The exact nature of this step has not been established. One simple, most likely too simple, explanation for a conformational coupling between subunits could be the large distance (about 70Å) between the anticodon and the CCA-aa end of tRNA. Even a slight mismatch on the order of a fraction of 1 Å at the codon/anticodon interaction over such a distance could result in suboptimal interactions of EF1A with the GTPase center and of the CCA-aa end of the aa-tRNA with the peptidyl transferase center of the 50S subunit. Optimal interactions could, of course, further result in an induced fit of the ribosome or EF1A at these reaction centers (Rodnina et al. 1995).

The concept of hybrid sites (Moazed and Noller 1989) is now generally accepted and also supported by a number of protection experiments and hydroxyl radical probings (Joseph and Noller 1996; Wilson and Noller 1998a,b). The existence of an E site is likewise accepted and supported by small-angle neutron scattering and cryo-EM (Spahn and Nierhaus 1998). However, the allosteric three-site model (Nierhaus et al. 1993b), which proposes that two tRNAs are always interacting with mRNA and that there is allosteric negative regulation between the E and A sites, has been the subject of much discussion (Rodnina et al. 1994a; Semenkov et al. 1996; Spahn and Nierhaus 1998). This discussion has not been resolved due to the fact that deacylated tRNA has been found in different sites in different cryo-EM reconstructions (Agrawal et al. 1996, 1999; Stark et al. 1997b), although the possibility exists that there are a number of subsites of the E site. However, studies in yeast that have a unique elongation factor (eEF3) are very supportive of this model (Triana-Alonso et al., 1995; see also Eukaryotic Protein Biosynthesis Elongation, below).

It is interesting that the ribosomal contact pattern of the two tRNAs at the A and P sites, although strikingly different from each other, hardly changes during the EF2-catalyzed translocation to the P and E sites. This suggests that there is a movable domain of the ribosome, which tightly binds two tRNAs and the mRNA during the translocation reaction (Dabrowski et al. 1998) and has led to the proposal of a modified allosteric three-site model, the α–ε model (Spahn and Nierhaus 1998). Kinetic investigations of the translocation reaction indicate that also during this reaction a conformational coupling exists between the binding of domain 4 of EF2•GTP to the A site of the 30S subunit and the GTPase center of the 50S subunit. EF2 lacking domain 4 does not function as a translocase (Martemyanov and Gudkov 1999). The observation that the induced GTPase activity of EF2 precedes the actual translocation of tRNAs and mRNA is contrary to previous views on this reaction (Rodnina et al. 1997). EF2 may also play a non-elongation function in translation. Recent data from studies in *E. coli* indicate that whereas mimicry of the elongation cycle may take place with the RFs (either RF1•3 or RF2•3) to allow for both reading of the stop code word in the ribosomal A site and the "activation" of water for attack at the peptidyl-tRNA bond, there are several additional steps as well. These additional steps require RF4, EF2, IF3, and GTP where the EF2-driven hydrolysis of GTP provides the energy for ribosome dissociation. If the proposed model is correct, the termination process brings the terminated ribosome back to an early step in initiation where the mRNA and IF3 are bound to the 30S subunit (Karimi et al. 1999).

Structural Studies of Elongation Factors

From structural studies of the last six years, a fairly complete picture of the various functional states of the elongation factors is now available in the prokaryotic system. The structural results have been compared to functional studies in a recent comprehensive review (Krab and Parmeggiani 1998).

EF1A

The structure of EF1A•GDP (see Fig. 3) has been obtained from *E. coli* (Abel et al. 1996; Polekhina et al. 1996; Song et al. 1999) and from *Thermus aquaticus* (Polekhina et al. 1996). Recently the structure of EF1A•GDP from bovine liver mitochondria (see Fig. 3) has also been determined to high resolution (Andersen et al. 2000). The structure of EF1A•GDPNP (Fig. 3), where GDPNP is a nonhydrolyzable GTP analog, has been determined from *T. thermophilus* (Berchtold et al. 1993) and from *T. aquaticus* (Kjeldgaard et al. 1993). A model for the structure of EF1A•GTP from *Bacillus stearothermophilus* has been put forward based on these structures (Krásny et al. 1998).

The structures reveal that EF1A consists of two structural units. One unit is the G domain (or domain 1) found in many other G proteins (Kjeldgaard et al. 1996). This domain consists of about 200 amino acids and is responsible for the binding of the GDP/GTP nucleotides. The domain is a typical nucleotide-binding domain having a central mostly parallel β sheet surrounded by α helices. However, one β strand is antiparallel and is preceded by the so-called effector loop or switch I region of the G domain. The switch II region consists of an α helix (helix B or helix α2). The local structures of the switch regions depend critically on the nature of the bound nucleotide. In all G proteins, these regions signal the state of the nucleotide to interacting partners (Kjeldgaard et al. 1996). During activation of EF1A, by the exchange of GDP for GTP, the switch I region changes structure from a β hairpin to a short α helix (Abel et al. 1996; Polekhina et al. 1996; Song et al. 1999). The α helix of the switch II region is shifted along the amino acid sequence by about four residues, thereby rotating the axis of the helix by approximately 45°

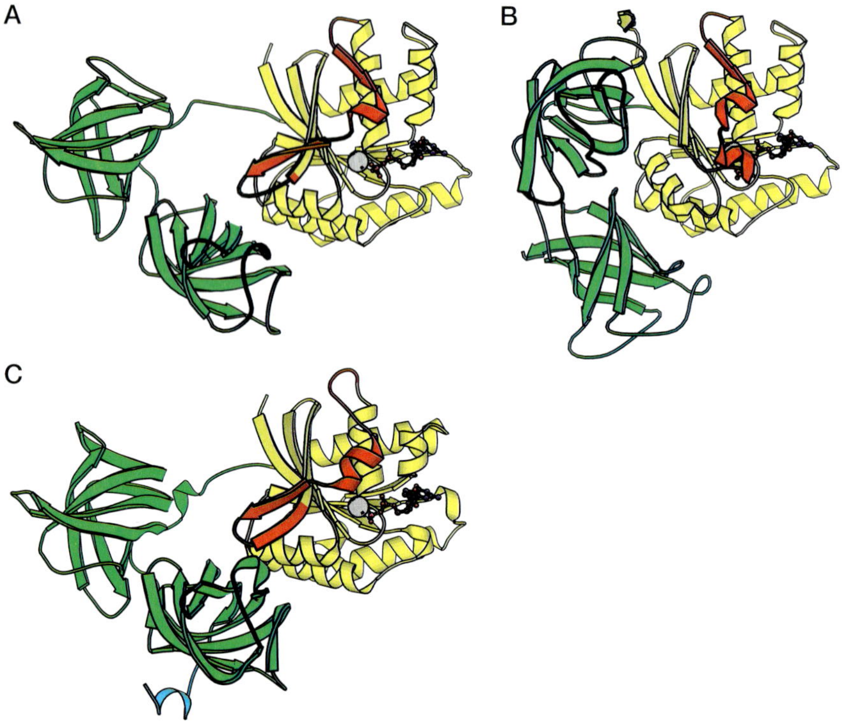

Figure 3 Structures of EF1A•GDP (*A*) and of EF1A•GDPNP (*B*) from *T. aquaticus* and the structure of bovine mitochondrial EF1A•GDP (*C*). Domain 1 is yellow, and domains 2 and 3 are green. Switch region 1 is shown in red, and the special carboxy-terminal extension in mitochondrial EF1A, in blue. Nucleotides are drawn as a ball-and-stick model, and the Mg^{++} ion is a gray ball.

(Berchtold et al. 1993; Kjeldgaard et al. 1993). As both switch regions are contact areas to the second structural unit, consisting of two β barrels of about 100 amino acids each, the overall conformational change of EF1A upon activation is dramatic. Domain 1 is rotated by approximately 90° relative to domains 2 and 3 (Kjeldgaard et al. 1993).

The nucleotides are bound in a nucleotide-binding pocket formed by a number of loops containing highly conserved consensus sequences common to all G proteins (Dever et al. 1987; Bourne et al. 1990, 1991). One of these is a phosphate-binding or so-called P loop (Walker et al. 1982; Wierenga et al. 1985) having a sequence of G/AXXXXGKS/T, while two others, NKXD and SAL/K, are involved in specific recognition of the G base. The last two loops, DXXG just before the switch II region and a T within the switch I region, are involved in signaling the nucleotide state (GTP versus GDP). Of these, the D fixes a water ligand to a Mg^{++} ion and the T is a ligand in the GTP state, but far removed from Mg^{++} in the GDP state (Polekhina et al. 1996). The peptide bond before the G flips by about 150° such that the main-chain NH can make contact with the γ phosphate of GTP. This very local structural change induces the shift of helix B along the sequence of switch region II.

A comparison of the high-resolution structures of *E. coli* EF1A•GDP (Song et al. 1999) and of bovine mitochondrial EF1A•GDP (Andersen et al. 2000) should in principle provide some insight into the reasons for the tighter binding of GDP in the prokaryotic elongation factor. The well-conserved residues in the nucleotide-binding site provide the same direct and specific binding interactions with GDP or GTP. Therefore, the difference in nucleotide affinity (GDP>>GTP) must be found in the next shell of residues, where variations are found. It has been observed that a number of larger side chains are found in *E. coli* EF1A in this shell when compared to mitochondrial EF1A (Andersen et al. 2000). As a prominent example, W184 of *E. coli* EF1A is in a position which is structurally similar to G232 in mitochondrial EF1A. A glycine residue is also found at this position in mammalian and fungal EF1A. Although this large change in size of side chain is partly compensated for by L231 in the mitochondrial structure, it is possible that this and other changes allow greater flexibility of the protein around the GDP/GTP-binding site, thereby accounting for some of the differences in affinities for the nucleotides (Andersen et al. 2000). The interactions between domain 1 and domains 2 and 3 also seem to modulate the affinity of the nucleotides. The cloning and analysis of domain 1 alone from *E. coli* revealed an alteration of the affinity of GDP to the level of that of GTP (Parmeggiani et al. 1987).

Complex of EF1A•EF1B

Crystal structures of the EF1A•EF1B complex (see Fig. 4) have been determined from *E. coli* (Kawashima et al. 1996) and from *T. thermophilus* (Wang et al. 1997b). The structures show that EF1B is contacting most of the nucleotide-binding loops of EF1A, thus modifying many parts of the binding pocket. It has been suggested that removal of the Mg^{++} ion is an essential first step of dissociation (Kawashima et al. 1996). However, the extensive contacts over large regions between EF1A and EF1B, and the many alterations of the local structures of loops in the GDP/GTP-binding site, seem to be as important for the release of GDP. Details of how EF1B catalyzes the formation of EF1A•GTP are not known. From the structure, it is seen that domains 2 and 3 are moved away

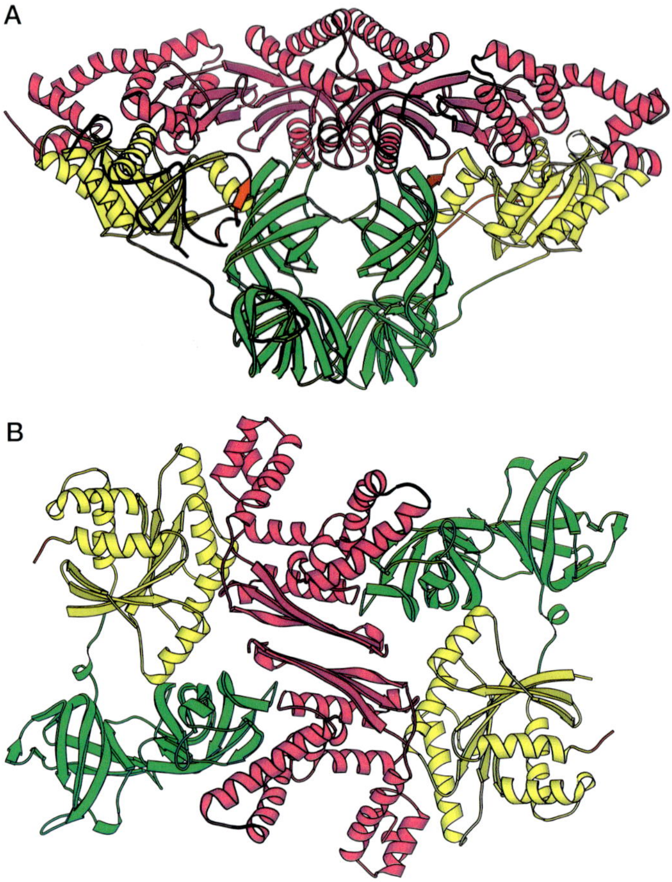

Figure 4 Structures of EF1A•EF1B from *E. coli* (*A*) and *T. thermophilus* (*B*). EF1A is colored as in Fig. 3. EF1B is shown in magenta. Notice that the pseudo twofold symmetry in *A* is vertical in the plane of the figure, whereas in *B* it is perpendicular to the plane of the figure. Although the individual components of both the *E. coli* and the *T. thermophilus* EF1A•EF1B have very similar structures, the overall organization of the pseudo-symmetrical complexes is very different.

from domain 1, thus facilitating the large conformational change of EF1A, which involves creating two completely different interacting areas of the two structural units of EF1A. In both structures, EF1B binds two molecules of EF1A. The structures of EF1B are also very similar, except for the fact that EF1B from *E. coli* has an internal pseudo twofold symmetry and that EF1B from *T. thermophilus* is a dimer. The effect of these differences is that the stoichiometries are $(EF1A)_2$•EF1B in *E. coli* and $(EF1A)_2$•$(EF1B)_2$ in *T. thermophilus*, and that the overall structures are very different.

The switch I region of EF1A in the EF1A•EF1B complex is not well ordered (Kawashima et al. 1996). A curious fact is that a carboxy-terminal extension of the *E. coli* EF1B closely mimics the structure of this region of EF1A•GDP (Polekhina et al. 1996), but with the sequence in the opposite direction (i.e., C → N versus N → C). Whether this structural extension is needed in order to stabilize an otherwise unstable *E. coli* EF1A without nucleotides remains to be investigated.

The solution structure of a carboxy-terminal fragment of human eEF1B has been determined recently (Pérez et al. 1999). Although there are no obvious sequence similarities between prokaryotic EF1B and eukaryotic eEF1B, the fold seen in this structure bears some similarity with that seen in the dimerization domain of *T. thermophilus* EF1B. They both have small β sheets interacting with each other. Curiously, an overlay of the two structures reveals that the similarity is based on sequences running in opposite directions (Pérez et al. 1999).

The Ternary Complex of EF1A

Activated EF1A forms a ternary complex of EF1A•GTP•aa-tRNA (see Fig. 5). The crystal structures of two ternary complexes are now known. One is of yeast Phe-tRNA in complex with *T. aquaticus* EF1A•GDPNP (Nissen et al. 1995) and the other of *E. coli* Cys-tRNA in complex with *T. aquaticus* EF1A•GDPNP (Nissen et al. 1999). It has been shown by small-angle scattering that the structure of the ternary complex in solution is the same as the one found in the crystal structures (Bilgin et al. 1998). Apart from the differences in the tRNA structures (Cys-tRNA is shorter than Phe-tRNA) and the local adaptation of EF1A to these differences, the two structures are very similar. They show that the ternary complex is formed by only minor alterations of the free structures of protein and tRNA. The major contact areas are a broad nonspecific contact between the T-stem helix of tRNA and domain 3 of EF1A, and specific recognitions of the 5′ phosphate and of the CCA-aa end. The 5′ phosphate is recognized in a small binding pocket by well-conserved residues coming from all three domains. The CCA-aa end is bound in a narrow cleft between domains 1 and 2. The terminal A base is bound in a special binding pocket on the surface of domain 2, and the amino ester is specifically recognized by main-chain interactions. The amino acid side chain is found in a binding pocket close to H67, which was earlier found to be cross-linked to εBr-Lys-tRNA (Duffy et al. 1981). The overall structure is surprisingly elongated, with the anticodon of the tRNA pointing away from the protein (Fig. 5).

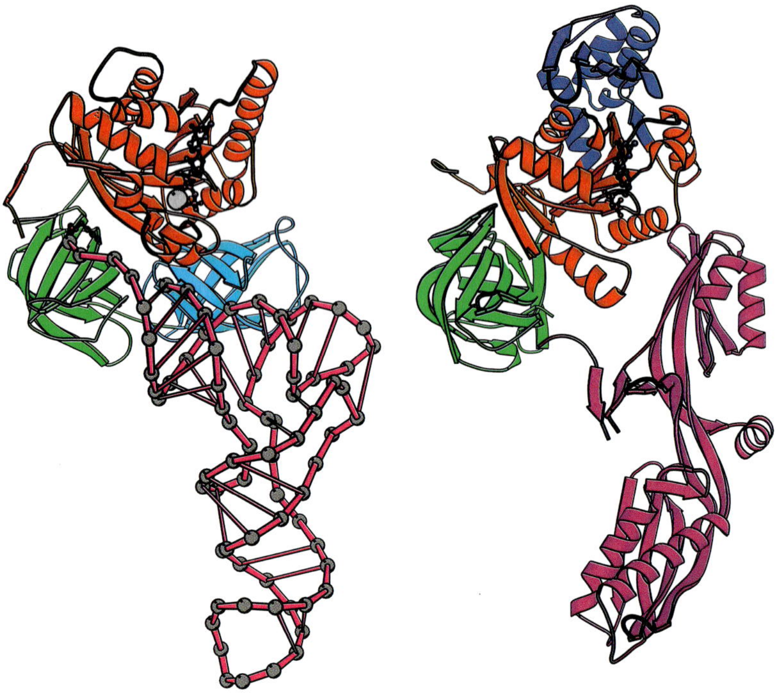

Figure 5 Comparison of the structures of the ternary complex of EF1A and of EF2. In this figure domains 1 of both proteins are red with a dark blue α-helical insert in EF2, domains 2 are green, and domain 3 of EF1A is light blue. The tRNA of the ternary complex of EF1A, and domains 3, 4, and 5 of EF2 are magenta.

EF2

The structure of EF2•GDP from *T. thermophilus* has been determined and is shown in Figure 5 (Czworkowski et al. 1994; Al-Karadaghi et al. 1996). This protein contains five domains. Domains 1 and 2 are very similar to the domains 1 and 2 in EF1A, except for an insertion of a helical domain in domain 1 of EF2. However, the relative conformations of these two domains in the inactive EF2•GDP form are similar to the one found in the active EF1A•GDPNP. Domains 3, 4, and 5 of EF2 have folds similar to some ribosomal proteins (Ævarsson et al. 1994). Domain 3 does not have a very well defined structure in any of the crystals of native EF2•GDP. Domain 4 is very elongated and contains a very unusual left-handed crossover of an α helix on a three-stranded β sheet. The structure of EF2•GTP is not yet known, despite many different attempts to obtain it over the last few years (P.B. Moore and A. Liljas, pers. comm.). At the very tip of domain 4 are three residues, H573, E574, and D576, of which the first is well conserved (see Fig. 1). These residues have been proposed to be of importance for the translocation step (Ævarsson 1995a). This has been verified by mutation experiments (Martemyanov et al. 1998). The third amino acid can be aligned (Cammarano et al. 1992) to the unusual residue diphthamide found in eukaryotes, which can be mono-ADP-ribosylated by diphtheria toxin (Ward 1987). This modification induced by a human pathogen causes inhibition of translocation and cell death. The H573 of *T. thermophilus* EF2 has been mutated into A, without any observable effect on translocation (Martemyanov et al. 1998). However, the crystal structure of this mutant shows EF2•GDP in a "closed" conformation, where domains 3, 4, and 5 are rotated closer to domain 2, such that domain 3 now has a well-defined structure (M. Laurberg and A. Liljas, pers. comm.).

Macromolecular Mimicry

When comparisons were made of the structures of the ternary complex of EF1A and of EF2•GDP, it was obvious that domains 3, 4, and 5 of EF2 are structurally mimicking the tRNA molecule (Nissen et al. 1995). Especially the very elongated domain 4 of EF2 is mimicking the anticodon stem and loop of tRNA (Fig. 5). This gave rise to the concept of "macromolecular mimicry" postulating that some proteins are mimicking the shape of nucleic acids (Nyborg and Kjeldgaard 1996; Nyborg et al. 1996; Clark and Nyborg 1997; Pedersen et al. 1999; Nissen et al. 2000). It was immediately recognized that this astonishing mimicry had to tell something about the similarity in the interactions of the ternary complex of EF1A and of EF2•GDP with the ribosome. One aspect was rather obvious. In the elongation cycle, the two functional forms of these elongation factors are bound to very similar regions of the ribosome in the post-translocational state (see Fig. 2). Thus, it is conceivable that the EF2•GDP, while leaving the ribosome, changes the state of the ribosome such that it leaves behind an imprint of a binding site suitable for the ternary complex (Liljas 1996). It is also very likely that the function of EF2•GTP is to physically force the peptidyl-tRNA in the A/P state of the ribosome out of the A site on the ribosomal 30S subunit, by interaction of the anticodon mimicking domain 4 of EF2•GTP with the site previously occupied by the anticodon stem of tRNA (Nissen et al. 1995; Abel and Jurnak 1996). It is also quite possible that this interaction prevents the rebinding of tRNA to the A site (Nissen et al. 2000). Finally, it can be postulated that all G proteins involved in translation are interacting with the same GTPase activating center of the ribosome. The detailed interaction of the ternary complex and of EF2•GTP with the A-site on the 30S subunit has been shown to transmit a signal through the macromolecules to an activation of the GTPase activity (Rodnina et al. 1997; Pape et al. 1998). It is very likely that this signal in fact results in a very accurate positioning of the G domain of both factors near the GTPase activating center.

It has been suggested from sequence comparisons based on the structural studies that all the G proteins involved in translation (EF1A, EF2, IF2, and RF3) have domains 1 and 2 similar to the ones known now (Ævarsson 1995b). By analogy, it is likely that after the completion of the assembly of the initiation complex, the G domain of IF2 will be in contact with the GTPase center of the 50S subunit of the ribosome. Whether domains of IF2 are mimicking some parts of an A-site-bound tRNA remains to be seen, when structures of IF2 and its ternary complex are determined. Sequence alignments between IF2 and EF2 have been performed with less than convincing results (Brock et al. 1998). For further details of the biochemistry of IF2, see Chapter 2.

Macromolecular mimicry between RF1 or RF2 and tRNA has been proposed (Nissen et al. 1995; Nakamura et al. 1996). Such mimicry must even extend to the recognition of the stop codon in the mRNA. The crystal structure of human eRF1 has been determined (Song et al. 2000), and the shape of eRF1 does indeed mimic that of a tRNA molecule. It contains three domains that resemble the anticodon helix and loop, the acceptor stem with CCA, and the T-stem helix and loop, respectively. The universally conserved GGQ motif (Frolova et al. 1999) is at the tip of the CCA mimic and is proposed to hold a hydrolytic water (Song et al. 2000). Interestingly, domain 3 mimicking the T-stem helix has been shown to be crucial for the interaction with eRF3 (Merkulova et al. 1999), which strongly indicates that the complex eRF1•eRF3•GTP resembles the ternary complex of EF1A•GTP•aa-tRNA. For a more detailed description of the termination factors and their mechanism of action, see Chapter 11.

Finally, the crystal structure of the ribosome recycling factor (RF4) from *Thermotoga maritima* has been determined recently (Selmer et al. 1999). It reveals two domains; one is globular while the other is an extended three-helix bundle. Although it is not a classic "G protein," the shape of RF4 is strikingly similar to that of a tRNA. The implication is that RF4 will bind to the ribosomal A site and, in conjunction with EF2•GTP, induce the dissociation of the subunits (Karimi et al. 1999). For a detailed description of termination see Chapter 11 and for reinitiation in prokaryotes, see Chapter 4.

Structures of the Elongation Factors on the Ribosome

As noted earlier, considerable information on the detailed structure of the bacterial ribosome has been obtained, and recently the cryo-EM reconstruction of a eukaryotic ribosomal particle in vitreous ice has also been published (Dube et al. 1998). Reconstructions of the ribosome at higher and higher resolution have led to very detailed and reliable models of the 23S RNA and of the 16S RNA taking into account the wealth of data on various cross-link and protection experiments (Brimacombe 1995; Mueller and Brimacombe 1997a,b; Mueller et al. 1997). The limit of resolution of the cryo-EM reconstructions has not yet been reached, and can in principle be as high as for crystal diffraction (M. van Heel, pers. comm.). It is likely that reconstructions or crystal structures to a resolution of better than 3 Å of various states of the ribosome will be obtained within the next few years. This structural information will open a completely new era of detailed insight into the function of the ribosome and the details of the elongation cycle of protein biosynthesis.

For the purpose of this review, the structural information available on the interaction of the elongation factors with the ribosome are described and discussed. A cryo-EM reconstruction of the *E. coli* ribosome with a ternary complex of EF1A blocked by the antibiotic kirromycin at a resolution of 18 Å has been published (Stark et al. 1997a). The antibiotic is believed to prevent the dissociation of EF1A•GDP from the ribosome. The ternary complex is thus seen in the testing state with domain 1 of EF1A in contact with the 50S ribosome particle near the stalk region containing the ribosomal protein complex $L10(L7/L12)_4$ and close to the sarcin-ricin loop (SRL) of 23S RNA (Szewczak and Moore 1995; Correll et al. 1998, 1999), to L6 and L14, and to the L11:rRNA (Conn et al. 1999; Wimberly et al. 1999). This area is believed to provide the GTPase activating center of the ribosome. Domain 2 of EF1A is in contact with the 30S subunit. The anticodon of the tRNA is found at the A site of the 30S subunit, such that a possible match with a codon in the mRNA can be formed. The overall structure of the ternary complex is seen to be slightly different from that of the known crystal structure (Nissen et al. 1995), but this can easily be a function of the binding of the antibiotic. It is thus possible that the detailed, functional interaction between the ternary complex in cognate codon/anticodon recognition and the ribosomal GTPase activating center is disturbed in such a complex, preventing the release of tRNA into the A site.

A very similar cryo-EM reconstruction of the *E. coli* ribosome with EF2•GTP blocked by the antibiotic fusidic acid has been published at a resolution of 20 Å (Agrawal et al. 1998). It shows EF2 in a very similar binding mode as the ternary complex, with domain 1 in contact with the 50S subunit and domain 2 in contact with the 30S subunit. This binding mode is thus likely to be similar also for both IF2 and RF3. The tRNA mimicking domain 4 is found in the A site of the 30S subunit. Again, the structure of EF2 as seen in this cryo-EM reconstruction is different from that found in the crystal structures of EF2•GDP (Czworkowski et al. 1994; Al-Karadaghi et al. 1996). Although this, in part, is likely to be due to the binding of the antibiotic as for the ternary complex, there is also the possibility that this structure represents EF2 in a GTP-like form. The tRNA mimicking part of EF2 in this form is rotated into an "open" form resulting in a movement of about 10 Å at the tip of domain 4.

In the crystallographic investigation of the 50S subunit (Ban et al. 1999), models for the binding of the ternary complex of EF1A and of EF2•GDP partly based on the cryo-EM reconstructions have been presented. These show that the sarcin-ricin loop is found very close to the switch I region of EF1A and possibly also of EF2, although this region is not well defined in the crystal structures of EF2•GDP (Czworkowski et al. 1994; Al-Karadaghi et al. 1996). Thus, the sarcin-ricin loop could in principle be involved in the GTPase activation of elongation factors. Structures of heterotrimeric G proteins with the transition-state analog AlF_4^- have shown that two residues are important for stabilization of the GTPase transition state (Coleman et al. 1994; Sondek et al. 1994). The transition-state-stabilizing residues are Q200 from the switch I region and R174 from the switch II region in transducin (Sondek et al. 1994). Similarly, a structure of ras P21 in complex with its GTPase activating protein (GAP), shows the importance of Q61 of ras P21 in the transition-state structure. However, in this case, an R789 from GAP (the "arginine finger") is providing the other stabilizing residue (Scheffzek et al. 1997) as ras P21 does not have an R in its switch II region. Very elegant and comprehensive studies strongly suggest that in ras P21, the general base of the GTPase reaction is the γ phosphate itself (Schweins et al. 1995). In EF1A from *T. aquaticus* the corresponding residues are H85 of the switch I region and R59 of the switch II region. In all known crystal structures these residues are far away from the GTP-binding site. It has been postulated that some ribosomal protein (perhaps L7/L12) could provide an arginine finger, but this has not yet been shown. Another possibility is that the binding of the ternary complex to the ribosome induces a local conformational change of both the switch I and switch II regions, such that H85 and R59 are brought close to the GTP to stabilize the transition state of the GTPase reaction of EF1A.

Future of Structural Studies of Protein Biosynthesis Elongation

Recent years have seen an avalanche of new structures shedding light on the process of the elongation phase of protein biosynthesis. This includes cryo-EM reconstructions of the ternary complex of EF1A (Nissen et al. 1995, 1999) and of the protein EF2 (Czworkowski et al. 1994; Al-Karadaghi et al. 1996) in contact with the prokaryotic ribosome (Stark et al. 1997a; Agrawal et al. 1998). Crystal structures of the 30S and 50S subunits as well as of the whole bacterial ribosome are available at sufficiently high resolution that very detailed models will be built in the near future (Ban et al. 1999; Cate et al. 1999; Clemons et al. 1999). For the first time, these structural determinations will reveal the complicated structures of the ribosomal RNAs, as well as the structures of many of the remaining ribosomal proteins. The first cryo-EM reconstruction of a eukaryotic ribosome particle has also been published (Dube et al. 1998).

If we allow ourselves a daring look into the crystal ball, we will see that within the next 10 years many details about the structure and function of protein biosynthesis elongation will be textbook material. There is no doubt that the detailed interactions of elongation factors with the ribosome will be known. Most likely also structural details of initiation and release factor interactions will be common knowledge. All of this structural information will lead to many new biochemical experiments revealing the detailed intricacies and complexities of the kinetics and thermodynamics of elongation. Furthermore, it is likely that much more will be known about the initiation and elongation phases of the eukaryotic ribosome, supported in large measure by the combination of traditional biochemistry and the more modern techniques available in structural biology and yeast genetics. The very rapid progress in the technique of cryo-EM reconstructions will undoubtedly continue at least for the coming decade. However, most advances from this method will come from the possibility of sorting and averaging many of the various functional states of the ribosome. This perhaps will require handling of millions of micrograph pictures of individual particles, but it will have the enormous benefit of showing the ribosome in action.

Most certainly, all structural details of interactions of antibiotics inhibiting protein biosynthesis will be available. With the structure of the eukaryotic ribosome at hand, this will provide a unique opportunity for studying how small organic molecules can interact with and selectively inhibit the bacterial ribosome. Furthermore, an increased knowledge of the structure and function of translation factors and tRNA-synthetases in both the prokaryotic and eukaryotic systems will provide a large palette of options for fighting bacterial infections.

EUKARYOTIC PROTEIN BIOSYNTHESIS ELONGATION

In general, the eukaryotic elongation cycle is thought to be similar to, if not identical to, the bacterial elongation cycle for which considerably more molecular detail is available (as presented above; see also Merrick 1992; Browning 1996; Merrick and Hershey 1996). As detailed below, the one possible exception may be the reaction catalyzed by fungal eEF3, an elongation factor for which there is no homolog in the bacterial system (Chakraburtty 1999). Table 1 lists the corresponding nomenclature for the factors with previous nomenclatures in parentheses. With the new nomenclature, the function of the corresponding eukaryotic factor should be obvious, given the preceding description of the bacterial elongation cycle. In general terms, the eukaryotic elongation factors are about a third larger than their bacterial counterparts. In evolutionary terms, eEF1A is clearly descended from EF1A, and eEF2 is clearly descended from EF2. What is curious is that none of the subunits of eIF1B appears at all related to EF1B even though the proposed guanine nucleotide exchange mechanism appears to be the same.

eEF1A

As noted earlier, mammalian eEF1A has evolved from EF1A by the insertion of approximately 70 amino acids into 16 different sites, mostly at loop regions based on the crystal structure for EF1A (see Fig. 6, which is adapted from Cavallius and Merrick 1993). As noted previously, amino acid sequence conservation is greatest in domain 1, the GTP-binding domain. eIF1A serves the exact same function as EF1A, namely the formation of a ternary complex (eEF1A•GTP•aa-tRNA) which is then bound to the ribosomal A site in a codon-dependent manner. Hydrolysis of the bound GTP leads to the release of the eEF1A in the form of an

Table 1 Elongation factors

Prokaryotic	Eukaryotic
EF1A (formerly EF-Tu)	eEF1A (formerly eEF-1α or EF-1α)
EF1B (formerly EF-Ts)	eEF1B (formerly eEF-1βγδ)
EF2 (formerly EF-G)	eEF2
—	eEF3

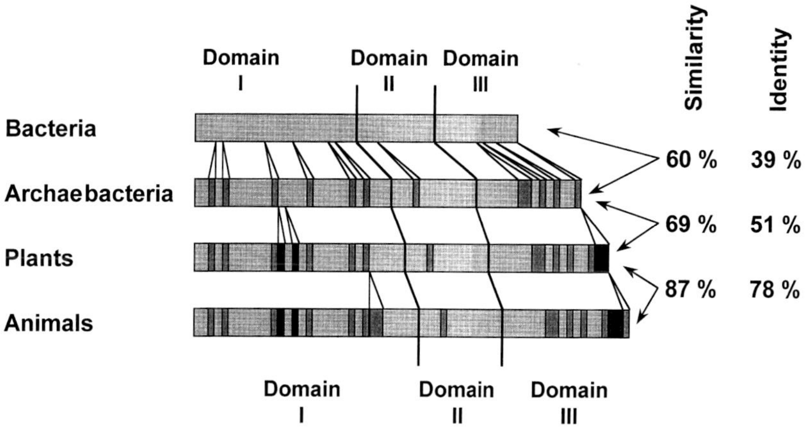

Figure 6 Evolution of eEF1A from EF1A. The apparent inserts into the EF1A sequence are shown which have led to the archeabacterial, plant, and higher eukaryotic forms of eEF1A. As noted in the text, most of the inserts are small and tend to be into regions of EF1A that are loops in the three-dimensional structure. Although the overall structure of eEF1A appears similar to that of EF1A with three distinct structural domains, as noted in Fig. 1, the highest degree of sequence conservation is in the GTP-binding domain, domain 1.

eEF1A•GDP complex. eEF1A has a K_d for GTP and GDP of about 10^{-6} M. Although the off-rate for GDP is still slow enough to make eEF1B essential for growth in yeast, this requirement can be offset by overexpression of eEF1A (Kinzy and Woolford 1995). Thus, the ratio of on and off rates for GTP and GDP is such that a nucleotide exchange factor is required, but just barely.

eEF1A is subject to a number of posttranslational modifications, although these tend to vary depending on the species. Mammalian eEF1A has seven posttranslational modifications, five lysines that are methylated and two glutamic acid residues (301, 374) that form a peptide linkage with glycerylphosphorylethanolamine (Fig. 7) (Dever et al. 1989). This latter modification has only been found on eEF1A and appears to be present in mammalian, plant, and *A. salina* eEF1A, but not in yeast eEF1A. What limited information is available on the lysines that are methylated (both which residue and the number of methyl groups) suggests that these modifications are varied when compared between species. In one study where each of the modified lysines was mutated to an arginine (either singly or in combinations), there was no apparent phenotype (Cavallius et al. 1997). Thus, it is not clear whether the methylated lysines serve a particular function or whether they might block ubiquination sites, thus

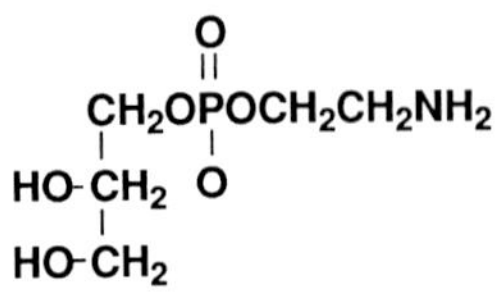

Figure 7 Glycerylphosphorylethanolamine. Shown is the unique posttranslational modification of eEF1A observed in mammalian species.

enhancing the half-life of the protein. An enhanced half-life may, in part, account for the observation that eEF1A is generally 3–5% of the soluble protein in most eukaryotic cells.

The activity of eEF1A (and eEF1B) is also subject to regulation by phosphorylation (Janssen et al. 1988; Venema et al. 1991a,b; Mulner-Lorillon et al. 1994; Kielbassa et al. 1995; Chang and Traugh 1998; Sheu and Traugh 1999), and this information is presented in detail in Chapter 24. The level of eEF1A (and eEF2) mRNA translation is also regulated. Both of these mRNAs are members of the *t*erminal *o*ligo*p*yrimidine (TOP)-containing mRNAs, as are the mRNAs that encode the ribosomal proteins (Loreni et al. 1993; Chapter 22). Translation of these TOP mRNAs is up-regulated upon stimulation for growth, although this regulation may be cell-type-specific.

eEF1B

The guanine nucleotide exchange factor eEF1B serves exactly the same function as bacterial EF1B, to facilitate the off-rate for eEF1A-bound GDP. This protein is made up of three subunits in most species (αβγ), but just two in yeast (αγ). The α and β subunits contain nucleotide exchange activity, whereas the γ subunit has failed to show any particular activity and, in yeast, is not an essential protein (Kinzy et al. 1994). Although there is a slight difference in the sizes of the α and β subunits, their amino acid sequence similarity makes it likely that one arose from the other via a gene duplication event. The 130 or so amino acids at the amino terminus of the β subunit distinguish it from the shorter α subunit. Additionally, by sequence, this region contains a leucine zipper that could potentially be used for dimerization, although the partner has yet to be unequivocally identified. Whereas eEF1B can be purified away from eEF1A, it is likely that most of the eEF1B exists in a complex with eEF1A. At present, two different stoichiometries for the eEF1A•1B complex have been proposed: one of $eEF1A_2$•1B (Janssen et al. 1994) and another of $(eEF1A_2\text{•}1B\text{•}1B\gamma)_2$ (Sheu and Traugh 1999). Studies by many laboratories have shown that the complex of eEF1A•1B can exist in varying degrees of aggregation with the largest complexes approaching 2 million daltons in size.

There are several unexpected features of eEF1B. First and foremost, none of the subunits is related in sequence to EF1B. Second, there has been a report that eEF1A•1B may exist primarily associated with the endoplasmic reticulum (Sanders et al. 1996). Third, there have been numerous reports on the enhanced expression of the mRNAs for one of the subunits under different physiologic states or in tumors (Shepherd et al. 1989; Dje et al. 1990; Sanders et al. 1992; Knudsen et al. 1993; Jefferies et al. 1994; Habben et al. 1995). However, the relevance of this is uncertain, given that only one of the subunit mRNAs was reported to be elevated, not all three.

eEF2

eEF2 catalyzes the exact same reaction as its bacterial counterpart EF2, the GTP-dependent translocation of the peptidyl-tRNA from the A site to the P site (or from the P/A site to the P/P site in the half-site model). However, unlike EF2, eEF2 is subject to two well-known modifications. The first is the multistep conversion of His-715 (for mammalian eEF2) into diphthamide (2-[3-carboxy-amido-3-(trimethylamino)propyl]histidine) that is shown in Figure 8. This modification is the site of the diphtheria-toxin-mediated ADP ribosylation which inactivates eEF2. ADP ribosylation only occurs when His-715 is fully modified to diphthamide; any intermediates in the modification process or histidine itself is not ADP ribosylated. Two-dimensional gel analysis of eEF2 suggests that this modification may be used in the context of the normal regulation of cellular protein synthesis, but to date there has not been a convincing physiologic response identified that might trigger the modification (Fendrick et al. 1992).

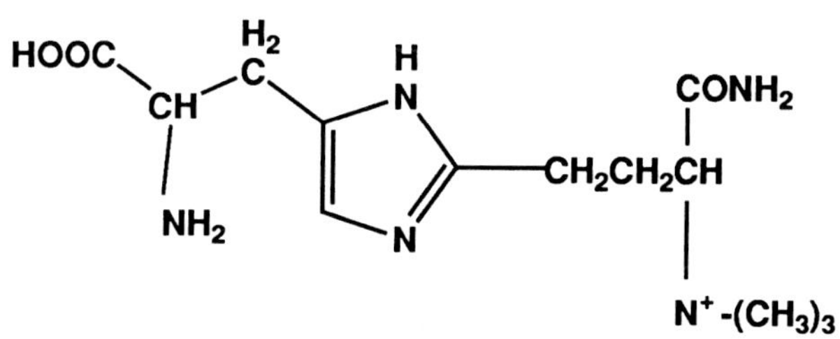

Figure 8 Diphthamide. Shown is the unique posttranslational modification of His-715 of mammalian eEF2.

The second posttranslational modification is phosphorylation (Nygaard et al. 1991; Redpath et al. 1993; Ryazanov and Spirin 1993; Diggle et al. 1998; Hovland et al. 1999) which is discussed in greater detail in Chapter 24. As with the diphthamide modification noted above, the physiologic circumstances that regulate the state of eEF2 phosphorylation in multicelluar organisms are not well defined at this point, although the phosphorylation of eEF2 has been shown to inhibit protein synthesis.

eEF3

eEF3 is a factor unique to fungal protein synthesis. To date there have been no homologs identified in bacteria, archeabacteria, or higher eukaryotes. It is a protein of 1044 amino acids in *Saccharomyces cerevisiae* with a generalized structure as shown in Figure 9. eEF3 is also relatively unique among those proteins with duplicated nucleotide-binding sites in that most of the others are transporter proteins. Biochemical studies indicate that although ATP is most likely the in vivo substrate for eEF3, it is also capable of using UTP, dATP, and TTP with equal efficiency whereas GTP and dGTP are used less well. The energy requirement appears to be associated with the release of the nonacylated tRNA from the E site of the ribosome (Triana-Alonso et al. 1995). The association of eEF3 with the ribosome appears to be through its amino terminus (residues 98–388), which is a sequence with homology to ribosomal protein S5 (Gontarek et al. 1998). Although the experimental evidence is good and eEF3 is an essential protein in yeast, it is surprising that there does not appear to be an equivalent activity in other organisms. Some have suggested that perhaps this activity is present as a part of the ribosome, given that most preparations of either ribosomes or ribosomal subunits display a low level of ATPase or GTPase activity (Rodnina et al. 1994b). However, this intrinsic NTPase activity is greatly reduced relative

Figure 9 eEF3. The numbers below the figure indicate the number of amino acids in each domain. NBS1 and NBS2 are nucleotide-binding sites, and the carboxy-terminal 58 amino acids define the lysine (K)-rich domain (30% lysine residues) (Uritani and Miyazaki 1988; Triana-Alonso et al. 1995; Chakraburtty and Triana-Alonso 1998; Chakraburtty 1999).

to the ribosome-dependent NTPase activities of either eEF1A, eEF2, or eEF3, and thus is unlikely to be kinetically capable of performing the same function as eEF3.

Gene Dosage of Elongation Factors

From bacteria to man, there are at least two functional genes for EF1A/eEF1A. In bacteria and yeast, these genes encode proteins of essentially identical amino acid sequence. However, in higher eukaryotes, the expression of the different eEF1A genes has been shown to be developmentally controlled and the amino acid sequence of the different isoforms shows greater diversity (90–94% identity between different forms). In both *Xenopus laevis* and *Drosophila melanogaster*, the expression of the eEF1A genes is developmentally regulated where one gene appears to be the "housekeeping" gene expressed continuously, and one or two other genes are expressed in a developmentally regulated pattern (Hovemann et al. 1988; Dje et al. 1990). A similar pattern of regulated expression appears to occur in mammals where, during embryogenesis and fetal development, eEF1A1 is the expressed gene. At or near birth, the expression of eEF1A2 (the "muscle-specific isoform") increases in muscle and heart and with time becomes the only form expressed in these tissues (Lee et al. 1992; Knudsen et al. 1993). In mice, the absence of the eEF1A2 gene results in the "wasted" mouse that develops normally through birth (Chambers et al. 1998). By three weeks after birth, the animals show deficiencies in both muscular and neuronal function and usually die a week later. At present, the total number of eEF1A genes that are present in each organism is unknown. Results from the human genome project will soon define the possible limits. That said, in humans and perhaps other mammalian species, it may be necessary to distinguish between authentic genes and pseudogenes (it has been estimated that there are 30–50 pseudogenes for eEF1A in the human genome).

In contrast to the eEF1A multigene family, there is limited evidence that any of the other elongation factors are encoded by a multigene family. This, in part, may be reflected by the greater interest in eEF1A because it may have critical functions unrelated to translation (Chapter 36). It is possible that the existence of other family members will become apparent as the complete sequence of genomes becomes available. There have been reports of two eEF1Bβ (formerly eEF-1δ) genes in *X. laevis* (Minella et al. 1996) and two eEF1Bγ genes in *S. cerevisiae* (Kinzy et al. 1994), which leaves open the possibility that future reports will describe multiple genes for the subunits of eEF1B in other organisms (although not

Caenorhabditis elegans nor *D. melanogaster*). For those organisms where the complete genome is known (yeast, *C. elegans*, and *D. melanogaster*), there are two copies of the eEF2 gene, and thus one expects that two copies of this gene will also be found in higher organisms.

Nontranslational Roles for Elongation Factors

By sheer numbers, the volume of published reports on hints of eEF1A having a nontranslational role in the cell dwarfs all other factors. It should be noted that many reports show an interaction of eEF1A with some other component, but, in general, there has been no genetic evidence to suggest that such an interaction is important. The major concern is that since eEF1A is about 3–5% of the soluble protein in most cells, its association with other cellular components may reflect an artifact of isolation. A complete review of these possibilities is given in Chapter 36.

One question that might be asked is whether there is sufficient eEF1A to provide both protein synthesis and nontranslational roles. A rough calculation based on values obtained from rabbit reticulocyte lysates indicates that the concentration of aminoacyl-tRNA is about 5 μM, whereas the concentration of eEF1A is about 20 μM. If all of the aminoacyl-tRNA were bound to eEF1A, the free concentration of eEF1A would still exceed 10 μM. Thus, it would appear that in these lysates (and in most cell types as well) there is sufficient free eEF1A to participate in a number of nontranslational events.

The Elongation Cycle and the RNA World

In much of the work aimed at elucidation of the exact interactions between factors and the ribosome or between peptidyl- or aminoacyl-tRNAs and the ribosomes, nucleotides within the rRNA have featured prominently (Green and Noller 1997). With the discovery of RNA molecules that could catalyze enzymatic reactions (ribozymes), the suggestion was made that perhaps the first mini- or macromolecules associated with catalysis and life were primarily constituted of RNA (for an extended view, see Gesteland et al. 1999). Perhaps the most convincing evidence for the existence of the RNA world was the demonstration that after exhaustive digestion of the proteins in the 50S subunit, the remaining nucleic acid was capable of effecting the synthesis of peptide bonds (Noller et al. 1992). Although the ionic conditions used in this experiment were nonphysiological, they were the same as those used to assay peptide-bond formation with just 50S subunits. Most importantly, the level of activity observed was significant, 20–40% that of the intact 50S subunit. Thus, the catalytic machinery for early protein synthesis could conceivably have been derived from RNA components exclusively. Although this is an interesting and widely debated topic, it is clear that a number of hurdles would need to be cleared to allow evolution of such an RNA-based form of life (Bartel and Unrau 1999). At the very least, however, it is curious that the biosynthesis of proteins is so extraordinarily dependent on RNAs and on RNA–RNA interactions.

ACKNOWLEDGMENTS

The authors thank Dr. S. Thirup for help with Figure 1, Dr. M. Kjeldgaard for help with Figures 3, 4, and 5, and Dr. J. Cavallius for help with Figure 6. This work has been supported in part by the Programme for Biotechnical Research under the Danish Science Research Council and the EU-project grant BIO-4CT97-2188 (J. N.) and by a grant from the Institute of General Medical Sciences of the National Institutes of Health GM-26796 (W. C. M.).

REFERENCES

Abel K. and Jurnak F. 1996. A complex profile of protein elongation: Translating chemical energy into molecular movement. *Structure* **4:** 229–238.

Abel K., Yoder M.D., Hilgenfeld R., and Jurnak F. 1996. An α to β conformational switch in EF-Tu. *Structure* **4:** 1153–1159.

Ævarsson A. 1995a. On the structure, function and evolution of the translational apparatus: Ribosomal protein L1 and elongation factor G., Ph.D. thesis, University of Lund, Sweden.

———. 1995b. Structure-based sequence alignment of elongation factors Tu and G with related GTPases involved in translation. *J. Mol. Evol.* **41:** 1096–1104.

Ævarsson A., Brazhnikov E., Garber M., Zheltonosova J., Chirgadze Y., Al-Karadaghi S., Svensson L.A., and Liljas A. 1994. Three-dimensional structure of the ribosomal translocase: Elongation factor G from *Thermus thermophilus*. *EMBO J.* **13:** 3669–3677.

Agrawal R.K., Penczek P., Grassucci R.A., and Frank J. 1998. Visualization of the elongation factor G on the *Escherichia coli* 70S ribosome: The mechanism of translocation. *Proc. Natl. Acad. Sci.* **95:** 6134–6138.

Agrawal R.K., Penczek P., Grassucci R.A., Burkhardt N., Nierhaus K.H., and Frank J. 1999. Effect of buffer conditions on the position of tRNA on the 70S ribosome as visualized by cryoelectron microscopy. *J. Biol. Chem.* **274:** 8723–8729.

Agrawal R.K., Penczek P., Grassucci R.A., Li Y., Leith A., Nierhaus K.H., and Frank J. 1996. Direct visualization of A-, P-, and E-site transfer RNAs in the *Escherichia coli* ribosome. *Science* **271:** 1000–1002.

Al-Karadaghi S., Ævarsson A., Garber M., Zheltonosova J., and Liljas A.. 1996. The structure of elongation factor G in complex with GDP: Conformational flexibility and nucleotide exchange. *Structure* **4:** 555–565.

Andersen G.R., Thirup S., Spremulli L.L., and Nyborg J. 2000. High resolution crystal structure of bovine mitochondrial EF-Tu in complex with GDP. *J. Mol. Biol.* **297:** 421–436.

Arnez J.G. and Moras D. 1997. Structural and functional considerations of the aminoacylation reaction. *Trends Biochem. Sci.* **22:** 211–216.

Bairoch A. and Boeckmann B. 1994. "SwissProt database (release 31)." *Nucleic Acids Res.* **22:** 3578–3580.

Ban N., Nissen P., Hansen J., Capel M., Moore P.B., and Steitz T.A. 1999. Placement of protein and RNA structures into a 5 Å-resolution map of the 50S ribosomal subunit. *Nature* **400:** 841–847.

Bartel D.P. and Unrau P.J. 1999. Constructing an RNA world. *Trends Biochem. Sci.* **24:** M9–M13.

Berchtold H., Reshetnikova L., Reiser C.O.A., Schirmer N.K., Sprinzl M., and Hilgenfeld R. 1993. Crystal structure of active elongation factor Tu reveals major domain rearrangements. *Nature* **365:** 126–132.

Bilgin N., Claesens F., Pahverk H., and Ehrenberg M. 1992. Kinetic properties of *Escherichia coli* ribosomes with altered forms of S12. *J. Mol. Biol.* **224:** 1011–1027.

Bilgin N., Ehrenberg M., Ebel C., Zaccai G., Sayers Z., Koch M.H.J., Svergun D., Barberato C., Volkov V., Nissen P., and Nyborg J. 1998. The solution structure of the ternary complex between aminoacyl-tRNA, EF-Tu and GTP. *Biochemistry* **37:** 8163–8172.

Bourne H.R., Sanders D.A., and McCormick F. 1990. The GTPase superfamily: A conserved switch for diverse cell functions. *Nature* **348:** 125–132.

———. 1991. The GTPase superfamily: Conserved structure and molecular mechanism. *Nature* **349:** 117–127.

Brimacombe R. 1995. The structure of ribosomal RNA: A three-dimensional jigsaw puzzle. *Eur. J. Biochem.* **230:** 365–383.

Brock S., Szkaradkiewicz K., and Sprinzl M. 1998. Initiation factors of protein biosynthesis in bacteria and their structural relationship to elongation and termination factors. *Mol. Microbiol.* **29:** 409–417.

Browning K.S. 1996. The plant translational apparatus. *Plant Mol. Biol.* **32:** 107–144.

Cammarano P., Palm P., Creti R., Ceccarelli E., Sanangelantoni A.M., and Tiboni O. 1992. Early evolutionary relationships among known life forms inferred from elongation factor EF-2/EF-G sequences—Phylogenetic coherence and structure of the archaeal domain. *J. Mol. Evol.* **34:** 396–405.

Cate J.H., Yusupov M.M., Yusupov G.Z., Earnest T.N., and Noller H.F. 1999. X-ray crystal structures of 70S ribosome functional complexes. *Science* **285:** 2095–2104.

Cavallius J. and W.C. Merrick. 1993. Eukaryotic translation factors which bind and hydrolyze GTP. In *Hand. Exp. Pharmacol.* **108:** 115–130.

Cavallius J., Popkie A.P., and Merrick W.C. 1997. Site-directed mutants of post-translationally modified sites of yeast eEF1A using a shuttle vector containing a chromogenic switch. *Biochim. Biophys. Acta* **1350:** 345–358.

Chakraburtty K. 1999. Functional interaction of yeast elongation factor 3 with yeast ribosomes. *Inter. J. Biochem. Cell Biol.* **31:** 163–173.

Chakraburtty K. and Triana-Alonso F.J. 1998. Yeast elongation factor 3: Structure and function. *Biol. Chem.* **379:** 831–840.

Chambers D.M., Peters J., and Abbott C.M. 1998. The lethal mutation of the mouse wasted (wst) is a deletion that abolishes expression of a tissue-specific isoform of translation elongation factor 1α, encoded by the Eef1a2 gene. *Proc. Natl. Acad. Sci.* **95:** 4463–4468.

Chang Y.-W.E. and Traugh J.A. 1998. Insulin stimulation of phosphorylation of elongation factor 1 (eEF-1) enhances elongation activity. *Eur. J. Biochem.* **251:** 201–207.

Clark B.F.C. and Nyborg J. 1997. The ternary complex of EF-Tu and its role in protein biosynthesis. *Curr. Opin. Struct. Biol.* **7:** 110–116.

Clark B.F.C., Kjeldgaard M., Barciszewski J., and Sprinzl M. 1995. Recognition of aminoacyl-tRNAs by protein elongation factors. In *tRNA: Structure, biosynthesis and function* (ed. S. Söll and U. RajBhandary), pp. 423–442. ASM Press, Washington, D.C.

Clemons W.M., May J.L.C., Wimberley B.T., McCutcheon J.P., Capel M.S., and Ramakrishnan V. 1999. Structure of a bacterial 30S ribosomal subunit at 5.5 Å resolution. *Nature* **400:** 833–840.

Coleman D.E., Berghuis A.M., Lee E., Linder M.E., Gilman A.G., and Sprang S.R. 1994. Structures of active conformations of $G_{i\alpha 1}$ and the mechanism of GTP hydrolysis. *Science* **265:** 1405–1412.

Conn G.L., Draper D.E., Lattman E.E., and Gittis A.G. 1999. Crystal structure of a conserved ribosomal protein RNA complex. *Science* **284:** 1171–1174.

Correll C.C., Wool I.G., and Munishkin A. 1999. The two faces of the *Escherichia coli* 23S rRNA sarcin/ricin domain: The structure at 1.11 Å resolution. *J. Mol. Biol.* **292:** 275–287.

Correll C.C., Freeborn B., Moore P.B., and Steitz T.A. 1997. Metals, motifs, and recognition in the crystal structure of a 5S rRNA domain. *Cell* **91:** 705–712.

Correll C.C., Munishkin A., Chan Y.L., Ren Z., Wool I.G., and Steitz T.A. 1998. Crystal structure of the ribosomal RNA domain essential for binding elongation factors. *Proc. Natl. Acad. Sci.* **95:** 13436–13441.

Cusack S. 1997. Aminoacyl-tRNA synthetases. *Curr. Opin. Struct. Biol.* **7:** 881–889.

Czworkowski J., Wang J., Steitz T.A., and Moore P.B. 1994. The crystal structure of elongation factor G complexed with GDP, at 2.7Å resolution. *EMBO J.* **13:** 3661–3668.

Dabrowski M., Spahn C.M., Schafer M.A., Patzke S., and Nierhaus K.H. 1998. Protection patterns of tRNAs do not change during ribosomal translocation. *J. Biol. Chem.* **273:** 32793–32800.

Dallas A. and Moore P.B. 1997. The solution structure of the loop E/loop D region of *E.coli* 5S rRNA. *Structure* **5:** 1639–1653.

Dever T.E., Glynias M.J., and Merrick W.C. 1987. The GTP-binding domain: Three consensus sequence elements with distinct spacing. *Proc. Natl. Acad. Sci.* **84:** 1814–1818.

Dever T.E., Costello C.E., Owens C.L., and Merrick W.C. 1989. Location of seven post-translational modifications in rabbit EF-1α including dimethyllysine, trimethyllysine and glycerylphosphorylethanolamine. *J. Biol. Chem.* **264:** 20518–20525.

Diggle T.A., Redpath N.T., Heesom K.J., and Denton R.M. 1998. Regulation of protein-synthesis elongation-factor-2 kinase by cAMP in adipocytes. *Biochem. J.* **336:** 525–529.

Dje M.K., Mazabraud A., Viel A., le Maire M., Denis H., Crawford E., and Brown D.D. 1990. Three genes under different developmental control encode elongation factor 1-α in *Xenopus laevis*. *Nucleic Acids Res.* **18:** 3489–3493.

Dube P., Wieske M., Stark H., Schatz M., Stahk J., Zemlin F., Lutsch G., and van Heel M.

1998. The 80S rat liver ribosome at 25 Å resolution by electron cryomicroscopy and angular reconstitution. *Structure* **6:** 389–399.

Duffy L., Gerber L., Johnson A.E., and Miller D.L. 1981. Identification of a histidine residue near the aminoacyl transfer ribonucleic acid binding site of elongation factor Tu. *Biochemistry* **20:** 4663–4666.

Fendrick J.L., Iglewski W.J., Moehring J.M., and Moehring T.J. 1992. Characterization of the endogenous ADP-ribosylation of wild-type and mutant elongation factor 2 in eukaryotic cells. *Eur. J. Biochem.* **205:** 25–31.

Fourmy D., Recht M.I., Blanchard S.C., and Puglisi J.D. 1996. Structure of the A site of *Escherichia coli* 16S ribosomal RNA complexed with an aminoglycoside antibiotic. *Science* **274:** 1367–1371.

Frank J., Zhu J., Penczek P., Li Y., Srivastava S., Vershoor A., Radermacher M., Grassucci R., Lata R.K., and Agrawal R.K. 1995. A model of protein synthesis based on a new cryo-electron microscopy reconstruction of the *E. coli* ribosome. *Nature* **376:** 441–444.

Frolova L.Y., Tsivkovskii R.Y., Sivolobova G.F., Oparina N.Y., Serpinsky O.I., Blinov V.M., Tatkov S.I., and Kisselev L.L. 1999. Mutations in the highly conserved GGQ motif of class 1 polypeptide release factors abolish ability of human eRF1 to trigger peptidyl-tRNA hydrolysis. *RNA* **5:** 1014–1020.

Gesteland R.F., Cech T.R., and Atkins J.F., Eds. 1999. *The RNA world*, Cold Spring Harbor Laboratory Press, Cold Spring Harbor, New York.

Gontarek R.R., Li H., Nurse K., and Prescott C.D. 1998. The N terminus of eukaryotic translation elongation factor 3 interacts with the 18S rRNA and 80S ribosomes. *J. Biol. Chem.* **273:** 10249–10252.

Green R. and Noller H.F. 1997. Ribosomes and translation. *Annu. Rev. Biochem.* **66:** 679–716.

Habben J.E., Moro G.L., Hunter B.G., Hamaker B.R., and Larkins B.A. 1995. Elongation factor 1α concentration is highly correlated with the lysine content of maize endosperm. *Proc. Natl. Acad. Sci.* **92:** 8640–8644.

Hill W.E., Dahlberg A., Garrett R.A., Moore P.B., Schlessinger D., and Warner J.R., eds. 1990. *The ribosome: Structure, function, and evolution*, ASM Press, Washington, D.C.

Hopfield J.J. 1974. Kinetic proofreading: A new mechanism for reduced errors in biosynthetic processes requiring high specificity. *Proc. Natl. Acad. Sci.* **71:** 261–264.

Hovemann B., Richter S., Walldorf U., and Cziepluch C. 1988. Two genes encode related cytoplasmic elongation factors 1α (EF-1α) in *Drosophila melanogaster* with continuous and stage specific expression. *Nucleic Acids Res.* **16:** 3175–3193.

Hovland R., Eikhom T.S., Proud C.G., Cressey L.I., Lanotte M., Doskeland S.O., and Houge G. 1999. cAMP inhibits translation by inducing Ca^{2+}/calmodulin-independent elongation factor 2 kinase activity in IPC-81 cells. *FEBS Lett.* **444:** 97–101.

Janssen G.M.C., Maessen G.D.F., Amons R., and Möller W. 1988. Phosphorylation of elongation factor 1β by an endogenous kinase affects its catalytic nucleotide exchange activity. *J. Biol. Chem.* **263:** 11063–11066.

Janssen G.M.C., van Damme H.T.F., Kriek J., Amons R., and Möller W. 1994. The subunit structure of elongation factor 1 from *Artemia:* Why two α-chains in this complex? *J. Biol. Chem.* **269:** 31410–31417.

Jefferies H.B.J., Thomas G., and Thomas G.. 1994. Elongation factor-1α mRNA is selectively translated following mitogenic stimulation. *J. Biol. Chem.* **269:** 4367–4372.

Joseph S. and Noller H.F. 1996. Mapping the rRNA neighborhood of the acceptor end of tRNA in the ribosome. *EMBO J.* **15:** 910–916.

Kabsch W. and Sander C. 1983. Dictionary of protein secondary structure: Pattern recognition of hydrogen bonded and geometrical features. *Biopolymers* **22:** 2577–2637.

Karimi R., Pavlov M.Y., Buckingham R.H., and Ehrenberg M. 1999. Novel roles for classical factors at the interface between translation termination and initiation. *Mol. Cell* **3:** 601–609.

Kawashima T., Berthet-Colominas C., Wulff M., Cusack S., and Leberman R. 1996. The structure of the *Escherichia coli* EF-Tu:EF-Ts complex at 2.5 Å resolution. *Nature* **379:** 511–518.

Kielbassa K., Müller H.-J., Meyer H.E., Marks F., and Gschwendt M. 1995. Protein kinase Cδ-specific phosphorylation of the elongation factor eEF-1α and an eEF-1α peptide at threonine 431. *J. Biol. Chem.* **270:** 6156–6162.

Kinzy T.G. and Woolford J.L. 1995. Increased expression of *Saccharomyces cerevisiae* translation elongation factor 1-alpha bypasses the lethality of a TEF5 null allele encoding elongation factor 1-beta. *Genetics* **141:** 481–489.

Kinzy T.G., Ripmaster T.L., and Woolford J.L., Jr. 1994. Multiple genes encode the translation elongation factor EF-1γ in *Saccharomyces cerevisiae. Nucleic Acids Res.* **22:** 2703–2707.

Kjeldgaard M., Nyborg J., and Clark B.F.C. 1996. The GTP-binding motif—Variations on a theme. *FASEB J.* **10:** 1347–1368.

Kjeldgaard M., Nissen P., Thirup S., and Nyborg J. 1993. The crystal structure of elongation factor EF-Tu from *Thermus aquaticus* in the GTP conformation. *Structure* **1:** 35–50.

Knudsen S.M., J. Frydenberg, B.F.C. Clark, and H. Leffers. 1993. Tissue-dependent variation in the expression of elongation factor-1α isoforms: Isolation and characterization of a novel variant of human elongation factor-1α. *Eur. J. Biochem.* **215:** 549–554.

Krab I.M. and Parmeggiani A. 1998. EF-Tu, a GTPase odyssey. *Biochim. Biophys. Acta* **1443:** 1–22.

Krásny L., Mesters J.R., Tieleman L.N., Kraal B., Fucík V., Hilgenfeld R., and Jonák J. 1998. Structure and expression of elongation factor Tu from *Bacillus stearothermophilus*. *J. Mol. Biol.* **283:** 371–381.

Lee S., Francoeur A.-M., Liu S., and Wang E. 1992. Tissue-specific expression in mammalian brain, heart and muscle of S1, a member of the elongation factor-1α gene family. *J. Biol. Chem.* **267:** 24064–24068.

Liljas A. 1996. Protein synthesis: Imprinting through molecular mimicry. *Curr. Biol.* **6:** 247–249.

Liljas A. and Garber M. 1995. Ribosomal proteins and elongation factors. *Curr. Opin. Struct. Biol.* **5:** 721–727.

Loreni F., Francesconi A., and Amaldi F. 1993. Coordinate translational regulation in the syntheses of elongation factor 1α and ribosomal proteins in *Xenopus laevis*. *Nucleic Acids Res.* **21:** 4721–4725.

Louie A. and Jurnak F. 1985. Kinetic studies of *Escherichia coli* elongation factor Tu-guanosine 5′-triphosphate-aminoacyl-tRNA complexes. *Biochemistry* **24:** 6433–6439.

Martemyanov K.A. and Gudkov A.T. 1999. Domain IV of elongation factor G from *Thermus thermophilus* is strictly required for translocation. *FEBS Lett.* **452:** 155–159.

Martemyanov K.A., Yarunin A.S., Liljas A., and Gudkov A.T. 1998. An intact conformation at the tip of elongation factor G domain IV is functionally important. *FEBS Lett.* **434:** 205–208.

Merkulova T.I., Frolova L.Y., Lazar M., Camonis J., and Kisselev L.L. 1999. C-terminal domains of human translation termination factors eFR1 and eRF3 mediate their *in vivo* interaction. *FEBS Lett.* **443:** 41–47.

Merrick W.C. 1992. Mechanism and regulation of eukaryotic protein synthesis. *Microbiol. Rev.* **56:** 291–315.

Merrick W.C. and Hershey J.W.B. (1996). The pathway and mechanism of eukaryotic protein synthesis. In *Translational Control* (ed. J.W.B. Hershey et al.), pp. 31–69. Cold Spring Harbor Laboratory Press, Cold Spring Harbor, New York.

Minella O., Mulner-Lorillon O., Poulhe R., Belle R., and Cormier P. 1996. The guanine-nucleotide-exchange complex (EF-1βγδ) of elongation factor-1 contains two similar leucine-zipper proteins EF-1δ, p34 encoded by EF-1δ^1 and p36 encoded by EF-1δ^2. *Eur. J. Biochem.* **237:** 685–690.

Moazed D. and Noller H.F. 1989. Intermediate states in the movement of transfer RNA in the ribosome. *Nature* **342:** 142–148.

Mueller F. and Brimacombe R. 1997a. A new model for the three-dimensional folding of Escherichia coli 16 S ribosomal RNA. I. Fitting the RNA to a 3D electron microscopic map at 20 Å. *J. Mol. Biol.* **271:** 524–544.

———. 1997b. A new model for the three-dimensional folding of *Escherichia coli* 16 S ribosomal RNA. II. The RNA-protein interaction data. *J. Mol. Biol* **271:** 545–565.

Mueller F., Stark H., van Heel M., Rinke-Appel J., and Brimacombe R. 1997. A new model for the three-dimensional folding of *Escherichia coli* 16 S ribosomal RNA. III. The topography of the functional centre. *J. Mol. Biol.* **271:** 566–587.

Mulner-Lorillon O., Minella O., Cormier P., Capony J.-P., Cavadore J.-C., Morales J., Poulhe R., and Belle R. 1994. Elongation factor EF-1δ, a new target for maturation-promoting factor in *Xenopus* oocytes. *J. Biol. Chem.* **269:** 20201–20207.

Nakamura Y., Ito K., and Isaksson L.A. 1996. Emerging understanding of translation termination. *Cell* **87:** 147–150.

Nierhaus K.H., Franceschi F., Subramanian A.R., Erdmann V.A., and Wittmann-Liebold B., eds. 1993b. *The translational apparatus: Structure, function, regulation, evolution*, Plenum Press, New York.

Nierhaus K.H., Adlung R., Hausner T.-P., Schilling-Bartetzko S., Twardowski T., and Triana F. (1993a). The allosteric three-site model and the mechanism of action of both elongation factors EF-Tu and EF-G. In *The translational apparatus: Structure, function, regulation, evolution.* (ed. K.H. Nierhaus et al.), pp. 263–272. Plenum Press, New York and London.

Nissen P., Kjeldgaard M., and Nyborg J. 2000. Macromolecular mimicry. *EMBO J.* **19:** 489–495.

Nissen P., Thirup S., Kjeldgaard M., and Nyborg J. 1999. The crystal structure of Cys-tRNA^{Cys}:EF-Tu:GDPNP reveals general and specific features in the ternary complex and in tRNA. *Structure* **7:** 143–156.

Nissen P., Kjeldgaard M., Thirup S., Polekhina G., Reshetnikova L., Clark B.F.C., and Nyborg J. 1995. Crystal structure of the ternary complex of Phe-tRNA^{Phe}, EF-Tu, and a GTP analog. *Science* **270:** 1464–1472.

Noller H.F., Hoffarth V., and Zimmniak L. 1992. Unusual resistance of peptidyl transferase to protein extraction procedures. *Science* **256:** 1416–1419.

Nyborg J. and Kjeldgaard M. 1996. Elongation in bacterial protein biosynthesis. *Curr. Opin. Biotechnol.* **7:** 369–375.

Nyborg J., Nissen P., Kjeldgaard M., Thirup S., Polekhina G., Clark B.F.C., and

Reshetnikova L. 1996. Structure of the ternary complex of EF-Tu: Macromolecular mimicry in translation. *Trends Biochem. Sci.* **21:** 81–82.

Nygaard O., Nilsson A., Carlberg U., Nilsson L., and Amons R. 1991. Phosphorylation regulates the activity of the eEF-2-specific Ca^{2+}- and calmodulin-dependent protein kinase III. *J. Biol. Chem.* **266:** 16425–16430.

Pape T., Wintermeyer W., and Rodnina M.V. 1998. Complete kinetic mechanism of elongation factor Tu-dependent binding of aminoacyl-tRNA to the A site of the *E. coli* ribosome. *EMBO J.* **17:** 7490–7497.

———. 1999. Induced fit in initial selection and proofreading of aminoacyl-tRNA on the ribosome. *EMBO J.* **18:** 3800–3807.

Parmeggiani A., Swart G.W.M., Mortensen K.K., Jensen M., Clark B.F.C., Dente L., and Cortese R. 1987. Properties of a genetically engineered G domain of elongation factor Tu. *Proc. Natl. Acad. Sci.* **84:** 3141–3145.

Pedersen G.N., Nyborg J., and Clark B.F.C. 1999. Macromolecular mimicry of nucleic acid and protein. *IUBMB Life* **48:** 13–18.

Pérez J.M.J., Siegal G., Kriek J., Hård K., Dijk J., Canters G.W., and Möller W. 1999. The solution structure of the guanine nucleotide exchange domain of elongation factor 1β reveals a striking resemblance to that of EF-Ts from *Escherichia coli*. *Structure* **7:** 217–226.

Polekhina G., Thirup S., Kjeldgaard M., Nissen P., Lippmann C., and Nyborg J. 1996. Helix unwinding in the effector region of elongation factor EF-Tu: GDP. *Structure* **4:** 1141–1151.

Puglisi E.V., Green R., Noller H.F., and Puglisi J.D. 1997. Structure of a conserved RNA component of the peptidyl transferase centre. *Nat. Struct. Biol.* **4:** 775–778.

Ramakrishnan V. and White S.W. 1998. Ribosomal protein structures: Insights into the architecture, machinery and evolution of the ribosome. *Trends Biochem. Sci.* **23:** 208–212.

Redpath N.T., Price N.T., Severinov K.V., and Proud C.G. 1993. Regulation of elongation factor-2 by multisite phosphorylation. *Eur. J. Biochem.* **213:** 689–699.

Rodnina M.V., Fricke R., and Wintermeyer W. 1994a. Transient conformational states of aminoacyl-tRNA during ribosome binding catalyzed by elongation factor Tu. *Biochemistry* **33:** 12267–12275.

Rodnina M.V., Fricke R., Kuhn L., and Wintermeyer W. 1995. Codon-dependent conformational change of elongation factor Tu preceding GTP hydrolysis on the ribosome. *EMBO J.* **14:** 2613–2619.

Rodnina M.V., Savelsbergh A., Katunin V.I., and Wintermeyer W. 1997. Hydrolysis of GTP by elongation factor G drives tRNA movement on the ribosome. *Nature* **385:** 37–41.

Rodnina M.V., Serebryanik A.I., Ovcharenko G.V., and El'Skaya A.V. 1994b. ATPase strongly bound to higher eukaryotic ribosomes. *Eur. J. Biochem.* **225:** 305–310.

Ryazanov A.G. and Spirin A.S. (1993). Phosphorylation of elongation factor 2. In *Translational regulation of gene expression,* 2nd edition (ed. J. Ilan), pp. 433–455. Plenum Publishers, New York.

Sanders J., Brandsma M., Janssen G.M.C., Dijk J., and Möller W. 1996. Immunofluorescence studies of human fibroblasts demonstrate the presence of the complex of elongation factor-1 βγδ in the endoplasmic reticulum. *J. Cell Sci.* **109:** 1113–1117.

Sanders J., Maassen J.A., and Möller W. 1992. Elongation factor-1 messenger-RNA levels in cultured cells are high compared to tissue and are not drastically affected by oncogenic transformation. *Nucleic Acids Res.* **20:** 5907–5910.

Scheffzek K., Ahmadian M.R., Kabsch W., Wiesmüller L., Lautwein A., Schmitz F., and Wittinghofer A. 1997. The RasGAP complex: Structural basis for the GTPase activation and its loss in oncogenic mutants. *Science* **277:** 333–338.

Schweins T., Geyer M., Scheffzek K., Warshel A., Kalbitzer H.R., and Wittinghofer A. 1995. Substrate-assisted catalysis as a mechanism for GTP hydrolysis of *ras*-p21 and other GTP-binding proteins. *Nat. Struct. Biol.* **2:** 36–44.

Selmer M., Al-Karadaghi S., Hirokawa G., Kaji A., and Liljas A. 1999. Crystal structure of *Thermotoga maritima* ribosome recycling factor: A tRNA mimic. *Science* **286:** 2349–2352.

Semenkov Y.P., Rodnina M.V., and Wintermeyer W. 1996. The "allosteric three-site model" of elongation cannot be confirmed in a well-defined ribosome system from *Escherichia coli*. *Proc. Natl. Acad. Sci.* **93:** 12183–12188.

Shepherd J.C.W., Walldorf U., Hug P., and Gehring W.J. 1989. Fruit flies with additional expression of the elongation factor EF-1α live longer. *Proc. Natl. Acad. Sci.* **86:** 7520–7521.

Sheu G.T. and Traugh J.A. 1999. A structural model for elongation factor 1 (EF-1) and phosphorylation by protein kinase CKII. *Mol. Cell. Biochem.* **191:** 181–186.

Söll D. and RajBhandary U.L., Eds. 1995. *tRNA: Structure, biosynthesis, and function.* ASM Press, Washington, D.C.

Sondek J., Lambright D.G., Noel J.P., Hamm H.E., and Sigler P.B. 1994. GTPase mechanism of G-proteins from the 1.7 Å crystal structure of transducin α•GDP•AlF_4^-. *Nature* **372:** 276–279.

Song H., Parsons M.R., Rowsell S., Leonard G., and Philips S.E.V. 1999. Crystal structure of intact elongation factor EF-Tu from *Escherichia coli* in GDP conformation at 2.05 Å resolution. *J. Mol. Biol.* **285:** 1245–1256.

Song H., Mugnier P., Das A.K., Webb H.M., Evans D.R., Tuite M.F., Hemmings B.A., and Barford D. 2000. The crystal structure of human eukaryotic release factor eRF1: Mechanism of stop codon recognition and peptidyl-tRNA hydrolysis. *Cell* **100:** 311–321.

Spahn C.M. and Nierhaus K.H. 1998. Models of the elongation cycle: An evaluation. *Biol. Chem.* **379:** 753–772.

Stark H., Rodnina M.V., Rinke-Appel J., Brimacombe R., Wintermeyer W., and van Heel M. 1997a. Visualization of elongation factor Tu on the *Escherichia coli* ribosome. *Nature* **389:** 403–406.

Stark H., Müller F., Orlova E.V., Schatz M., Dube P., Erdemir T., Zemlin F., Brimacombe R., and Van Heel M. 1995. The 70S *Escherichia coli* ribosome at 23Å resolution: Fitting the ribosomal RNA. *Structure* **3:** 815–821.

Stark H., Orlova E.V., Rinke-Appel J., Jünke N., Mueller F., Rodnina M., Wintermeyer W., Brimacombe R., and van Heel M. 1997b. Arrangement of tRNAs in pre- and post-translational ribosomes revealed by electron cryomicroscopy. *Cell* **88:** 19–28.

Szewczak A.A. and Moore P.B. 1995. The sarcin/ricin loop, a modular RNA. *J. Mol. Biol.* **247:** 81–98.

Thirup S. and Larsen N.E. 1990. ALMA, an editor for large sequence alignments. *Proteins* **7:** 291–295.

Thompson R.C. 1988. EF-Tu provides an internal kinetic standard for translational accuracy. *Trends Biochem. Sci.* **13:** 91–93.

Triana-Alonso F.J., Chakraburtty K., and Nierhaus K.H. 1995. The elongation factor 3 unique in higher fungi and essential for protein biosynthesis is an E site factor. *J. Biol. Chem.* **270:** 20473–20478.

Uritani M. and Miyazaki M. 1988. Characterization of the ATPase and GTPase activities of elongation factor 3 (EF-3) purified from yeasts. *J. Biochem.* **103:** 522–530.

Venema R.C., Holme H.I., and Traugh J.A. 1991a. Phosphorylation of valyl-tRNA synthetase and elongation factor 1 in response to phorbol esters is associated with stimulation of both activities. *J. Biol. Chem.* **266:** 11993–11998.

Venema R.C., Peters H.I., and Traugh J.A. 1991b. Phosphorylation of elongation factor 1 (EF-1) and valyl-tRNA synthetase by protein kinase C and stimulation of EF-1 activity. *J. Biol. Chem.* **266:** 12574–12580.

Walker J.E., Saraste M., Runswick M.J., and Gay N.J. 1982. Distantly related sequences in the α- and β-subunits of ATP synthase, myosin, kinases and other ATP requiring enzymes and a common nucleotide binding fold. *EMBO J.* **1:** 945–956.

Wang Y., Jiang Y., Meyering-Voss M., Sprinzl M., and Sigler P.B. 1997b. Crystal structure of the EF-Tu:EF-Ts complex from *Thermus thermophilus*. *Nat. Struct. Biol.* **4:** 650–656.

Ward W.H.J. 1987. Diphtheria toxin: A novel cytocidal enzyme. *Trends Biochem. Sci.* **12:** 28–31.

Wierenga R.K., De Mayer M.C.H., and Hol W.G.J. 1985. Interaction of pyrophosphate moieties with alpha-helices in dinucleotide binding proteins. *Biochemistry* **24:** 1346–1357.

Wilson K. and Noller H.F. 1998a. Mapping the position of translational elongation factor EF-G in the ribosome by directed hydrxyl radical probing. *Cell* **92:** 131–139.

———. 1998b. Molecular movement inside the translational engine. *Cell* **92:** 337–349.

Wimberly B.T., Guymon R., McCutcheon J.P., White S.W., and Ramakrishnan V. 1999. A detailed view of a ribosomal active site: The structure of the L11-RNA complex. *Cell* **97:** 491–502.

Woodson S.A. and Leontis N.B. 1998. Structure and dynamics of ribosomal RNA. *Curr. Opin. Struct. Biol.* **8:** 294–300.

Yonath A. and Franceschi F. 1998. Functional universality and evolutionary diversity: Insights from the structure of the ribosome. *Structure* **6:** 679–684.

4

A Comparative View of Initiation Site Selection Mechanisms

Richard J. Jackson
Department of Biochemistry
University of Cambridge
Cambridge CB2 1GA, United Kingdom

Initiation can be defined as the process in which a special initiator tRNA, formyl-Met-tRNA$_f$ or Met-tRNA$_i$, is positioned in the P site of a ribosome located at the correct AUG codon (or in some cases a non-AUG initiation codon). When the initiation stage is complete, the ribosome is capable of dipeptide formation if it is presented with nothing more than a ternary complex of EF1A (formerly EF-Tu), GTP, and the cognate aminoacyl-tRNA appropriate for the A-site codon. A detailed description of the pathway of events involved in initiation in both prokaryotes and eukaryotes is provided in Chapter 2.

What is quite remarkable is that in prokaryotic systems this process of initiation requires just three initiation factor proteins, each of them a single polypeptide chain and with an aggregate mass of 150 kD, in contrast to at least ten separate initiation factors in eukaryotes, totaling over 25 polypeptide chains with an aggregate size approaching 1200 kD. Given this remarkable difference, it doesn't require exceptional imagination to come up with the suggestion that something is fundamentally different between initiation in the two systems! (Note that throughout this chapter, the term "prokaryotic" will be taken to imply eubacterial, specifically *Escherichia coli*, and "eukaryotic" to signify cytoplasmic mRNA translation in eukaryotes. No attempt will be made to cover mRNA translation in *Archaebacteria* or in eukaryotic organelles.)

It is tempting to speculate that this disparity in the complexity of initiation factors is mainly due to differences in the mechanism of initiation site selection, which is obviously so very different between prokaryotes and eukaryotes, whereas other facets of translation initiation seem, at first

Translational Control of Gene Expression
© 2000 Cold Spring Harbor Laboratory Press 0-87969-568-4/00 $5 +. 00

sight, to be very similar in the two systems: the fact that the process starts with separated ribosomal subunits (Guthrie and Nomura 1968; Howard et al. 1970; Blumberg et al. 1979), and the necessity to deliver the initiator tRNA into the ribosomal P site, discriminating against all other aminoacyl-tRNAs including elongator Met-tRNA. Surprisingly, there are quite considerable differences of detail between the two systems with respect to (1) the initiation factors that influence ribosomal subunit association/dissociation dynamics, (2) the features of initiator tRNA that allow it to be discriminated from elongator Met-tRNA, and (3) the structure of the initiation factor (IF2 in prokaryotes and eIF2 in eukaryotes), which forms a ternary complex with GTP and initiator tRNA and delivers the latter to the ribosomal P site (Chapters 2 and 5). Despite these differences between the prokaryotic and eukaryotic initiation factors involved in steps that are at least superficially similar in the two systems, there is little doubt as to the validity of the original premise that the major explanation for the different complexity of initiation factors in the two systems lies in differences in the mechanism of initiation site selection.

It goes without saying that in both systems the initiation pathway involves an obligatory intermediate consisting of the small ribosomal subunit bound to the mRNA and carrying initiator tRNA. Obviously there are two possible routes to this intermediate: Either the small subunit binds mRNA first and then initiator tRNA second, or vice versa. This question is discussed in detail in Chapter 2, and it suffices here to summarize what position will be taken on this issue in this chapter. In the eukaryotic system, there is little doubt that in the usual route the 40S ribosomal subunit binds initiator tRNA (as an eIF2/Met-$tRNA_i$/GTP ternary complex) before binding to mRNA. However, despite the fact that this seems to be the normal route, it is probably not an obligatory sequence of events. Certainly, current models for the regulation of translation of GCN4 mRNA require that a 40S subunit, bereft of bound ternary complex, can scan the mRNA downstream from the first short ORF (Chapter 5), and there are some indications that 40S subunits without associated Met-$tRNA_i$ can bind to mRNA de novo, at or near the 5′ cap, and scan in a 5′→3′ direction, acquiring a ternary complex during the course of scanning (Dasso et al. 1990). Nevertheless, the default position taken in this chapter is that in the normal route the 40S ribosomal subunit binds the ternary complex before it associates with mRNA.

In prokaryotic systems, either route is possible (for review, see Gold et al. 1981), as illustrated by the kinetic analyses of Gualerzi and his colleagues, who concluded that with *E. coli* ribosomes translating poly (A,U,G) and other synthetic mRNAs, the formation of the 30S/f-Met-

$tRNA_f$/mRNA complex is a random order process, with the small ribosomal subunit binding either first to mRNA and subsequently associating with initiator tRNA, or vice versa (Gualerzi et al. 1977; Calogero et al. 1988; Gualerzi and Pon 1990). However, one would imagine that with initiation sites which have strong Shine-Dalgarno (SD) sequences, the pathway in which the 30S/mRNA complex forms before f-Met-$tRNA_f$ binds must be strongly favored, and this is the position that I adopt in this chapter.

SELECTION OF THE CORRECT INITIATION SITE IN PROKARYOTES

In prokaryotes, despite the existence of some mRNAs lacking an SD motif, there is overwhelming evidence that for the vast majority of mRNAs, the SD sequence is the essential, but not necessarily the sole, "identifier" element. The experiments of Steitz and Jakes (1975) demonstrating base-pairing between the mRNA initiation site and the 3′end of 16S rRNA, and the dedicated ribosome experiments of Hui and de Boer (1987), provide irrefutable evidence for recognition of the SD motif by base-pairing with 16S rRNA.

As an obvious consequence of this mechanism of initiation site selection, all cistrons of a polycistronic mRNA are translatable in prokaryotes, in contrast to the case of di- or polycistronic mRNAs introduced into eukaryotic systems. In principle, every cistron of a prokaryotic polycistronic mRNA could be accessed independently via its own SD sequence, and so the ribosomes that translate a downstream cistron need not necessarily have translated the upstream cistrons. Therefore, mutation of the initiation codon of an upstream cistron of a polycistronic mRNA should, in principle, have no influence on the expression of a downstream cistron, although in practice, polarity effects often interfere with such simple predictions, and in addition, there are numerous cases of translational coupling that are discussed later in this chapter. Nevertheless, there are many examples where the prediction of independent and direct access to downstream cistrons is fulfilled. The dicistronic L10 ribosomal protein operon is one such case: The initiation frequency at the downstream L7/L12 cistron is thought to be at least fourfold higher than for the 5′-proximal L10 cistron, implying that ribosomes access the L7/L12 cistron without having previously translated the upstream L10 ORF (Yates et al. 1981).

A particularly important question, from the viewpoint of comparisons with initiation of hepatitis C virus RNA translation in eukaryotic systems, is whether binary 30S/mRNA complexes at the SD motif form as intermediates in the absence of initiation factors or other ligands. Such complexes can, in fact, be detected by toeprinting methods, although they appear to be somewhat less stable than 30S/mRNA complexes formed in the presence of initiation factors and initiator tRNA (Hartz et al. 1991). In the case of binary 30S/mRNA complexes, the reverse transcriptase penetrates to a point 2–5 nucleotides downstream from the 3′ G of the GGAGG SD sequence, whereas in the presence of initiator tRNA the primer extension stops at a point 17–18 nucleotides farther downstream, equivalent to 14–15 nucleotides downstream from the A of the AUG initiation codon (Hartz et al. 1991). Thus, in the 30S/mRNA binary complex, close contacts between the 30S subunit and the mRNA are limited to the SD motif and a few residues on either side, but in the ternary complex, the close contacts extend to 15 residues downstream from the initiation codon. This conclusion is consistent with the fact that site-specific crosslinking by UV irradiation between the 16S rRNA and sites in the mRNA downstream from the initiation codon occurred only in the presence of initiator tRNA, whereas crosslinking to upstream sites was independent of tRNA (McCarthy and Brimacombe 1994). The conversion of the binary complex to a stable ternary complex is accomplished by binding of initiator tRNA and initiation factors. This can be achieved in vitro by IF2 on its own, provided charged f-Met-tRNA$_f$ is present, or by IF3 on its own, in which case the tRNA$_f$ need not be charged, and indeed, the anticodon stem-loop alone suffices to fix the ribosomal subunit at the initiation codon (Hartz et al. 1989).

The widely held premises that initiation efficiency is related to the number of complementary base pairs that can form between the SD motif and the 16S rRNA, and that there is a finite window of allowable spacing between the SD motif and the initiation codon, have been supported in numerous studies of which the most definitive and informative is that of Ringquist et al. (1992), not least because all the mutations were made in a neighboring sequence background designed to eliminate secondary structure. The elegant experiments of de Smit and van Duin (1990a,b) on the effect of mutations around the initiation site of the RNA bacteriophage coat protein cistron demonstrate that the SD sequence and the initiation codon must be in unstructured regions to allow initiation; the frequency of initiation was proportional to the fraction of time during which these elements would be present in an open conformation. A survey of the data for several different initiation sites shows that secondary structure of stability up to –5 to –6 kcal/mole has little influence on initiation efficiency, but with increasing stability beyond this threshold there is a 10-fold decrease in translation initiation rate for every additional –1.4

kcal/mole, exactly as predicted on theoretical grounds (de Smit and van Duin 1994b). The threshold of –5 to –6 kcal/mole is believed to reflect the RNA melting potential of the 30S subunit/SD interaction (de Smit and van Duin 1994a), rather than an influence of any RNA helicases.

Although some components of the prokaryotic translation machinery, notably IF3 and ribosomal protein S1 (Thomas and Szer 1982; Subramanian 1983), have been proposed to have RNA melting properties, they are not ATP-dependent helicases like eukaryotic eIF4A. It is interesting to note that although the prokaryotic translation initiation factors were originally purified mainly through assay of translation of the RNA bacteriophage MS2 (or Qβ, R17, f2) coat protein cistron, in which the initiation site is somewhat occluded in a hairpin structure (de Smit and van Duin 1990a,b), none of the many ATP-dependent helicases present in *E. coli* were isolated as factors that stimulate translation initiation. Perhaps the reason these helicases do not seem to influence initiation efficiency to any significant extent is that there is no mechanism that would focus or direct their action to the particular region of the initiation codon. This has important parallels with initiation of translation of hepatitis C virus and pestivirus RNAs in eukaryotic systems, as discussed below.

It is interesting that although the SD motif and the initiation codon must both be in unstructured regions, hairpin loops between these two elements seem to be allowed. In bacteriophage T4 gene *38* mRNA, the linear spacing between the SD sequence and the AUG initiation codon is so large that efficient initiation would not be expected. However, this intervening region can fold into a hairpin loop with 8 base pairs (Gold 1988), which seems likely to exist in reality given its high GC content and the fact that it is closed by a UUCG tetraloop. This would have the effect of making the spacing between the SD sequence and the AUG codon close to the optimal. If this is indeed the case, it carries the implication that the mRNA lies in a slot or channel in the 70S ribosome, rather than being threaded through anything resembling a tube or hole.

Although the SD motif is the critical sequence defining an initiation site, it is unlikely to be the only feature recognized by the initiating ribosome. A statistical analysis has shown that the sequence of ribosome-binding sites in *E. coli* is not random between positions –20 and +14 (relative to the A of the AUG codon). The most compelling evidence for additional signals is the work of Dreyfus (1988), who used a β-galactosidase construct lacking a ribosome-binding site as a "trap" for sequences able to fulfill the function of an initiation site. In a type of "shot-gun" experiment, he inserted into this site random fragments of 14 *E. coli* genes, which included no fewer than 100 spurious potential initiation sites, i.e., sequences that did not function as initiation sites in vivo but had an acceptable SD/AUG (GUG) tandem. Strikingly, what the screen selected was 12 out of the 14 genuine translation initiation sites (the 2 not selected have very weak SD motifs), but only one out of the 100 internal sites. Clearly then, functional initiation sites must have some additional features besides simply the SD motif, and these peculiarities must be transferable from one gene to another. Moreover, all the active fragments included the –20 to +15 region, but the length of sequences flanking this core was variable (Dreyfus 1988). It is thought that this non-randomness of the ~35-nucleotide segment is not merely due to evolutionary pressure to eliminate secondary structure, but reflects the fact that there is contact between the ribosome and the mRNA throughout the element. Significantly, the mRNA protected by ribosome binding at the initiation site is about 35 nucleotides long and extends to about 15 nucleotides downstream from the initiation codon (Steitz and Jakes 1975; Hartz et al. 1989).

The screen also selected 4 sites (out of a total of 55 potential sites) from random fragments of the SV40 genome, and 5 sites from a part of the renin gene that is largely intronic. Because these mammalian sequences, which fortuitously functioned as translation initiation sites in bacteria, had not been under any selective pressure to perform this function, or to code for the amino terminus of a bacterial protein, Dreyfus (1988) used them to deduce what the additional features of a functional initiation site might be. In agreement with the results of the statistical analyses of Gold et al. (1981) and Stormo et al. (1982), no specific consensus sequence could be found (apart from the SD motif), but the general feature is that they were AT-rich throughout, and especially A-rich downstream from the initiation codon.

Translational Enhancers in Prokaryotic mRNAs

Despite the lack of any obvious consensus other than the SD motif, there are some hints that other sequence motifs in the vicinity of the initiation codon can influence at least the efficiency of initiation, and arguably the actual specificity of initiation site selection. One such indication is provided by those mRNAs that have the AUG initiation codon at the very 5′ end and thus lack any SD motif or upstream sequences; the other is the so-called translational enhancer motifs that influence efficiency rather than specificity of initiation site selection.

The best-known example of a leaderless mRNA is the prophage form of bacteriophage λ cI mRNA (transcribed from the "promoter for repressor maintenance"). Efficient translation of a cI/*lacZ* fusion mRNA required the retention of the 5′-proximal cI coding sequences, mainly the first 4 codons with the sequence 5′-AUGAGCACAAA (Shean and Gottesman 1992). Studies of other “leaderless” mRNAs have shown that there is a much greater stringency for an AUG initiation codon (as opposed to UUG, GUG, etc.) than if there is a complete ribosome-binding site including an SD motif, and that an mRNA completely lacking a 5′UTR is translated more efficiently than if a 5′UTR is present but lacks anything resembling even a vestigial SD motif (van Etten and Janssen 1998).

Translational enhancers that increase the efficiency of initiation have been reviewed recently (Jackson 1996) and so are only summarized here. Some of them have been proposed solely on the basis of database screens of highly expressed genes (Thanaraj and Pandit 1989), whereas others have been proven experimentally: the Epsilon sequence or “Olins box” from bacteriophage T7 gene 10 mRNA (Olins et al. 1988; Olins and Rangwala 1989); an element upstream of the SD motif of the *atpE* gene (McCarthy et al. 1985, 1986); the 5′UTR of tobacco mosaic virus, often known as the Ω sequence (Gallie and Kado 1989); and the so-called downstream box proposed by Sprengart et al. (1990). Some, such as the Epsilon sequence, seem to require an SD motif for maximum effect, whereas others, such as Ω, enhance better in the absence of an SD signal. Surprisingly, some of them lie outside the –20 to +15 window of direct contact between the mRNA and the 30S subunit. Also surprising is the fact that the enhancing effect of some of them, such as the Epsilon sequence, is independent of position relative to the initiation codon, whereas some enhancers that are claimed to occur in many mRNAs, such as the downstream box, are found in different locations in different mRNAs, which could explain why the statistical analysis of initiation site sequences by Gold and his colleagues (Gold et al. 1981; Stormo et al. 1982) failed to detect such motifs. In a great many cases it has been suggested that the enhancer motifs base-pair with the 16S rRNA, but although complementarities can be found on paper, there is no evidence that pairing occurs in reality. In the case of the so-called downstream box, recent evidence demonstrates clearly that the proposed pairing with 16S rRNA does not, in fact, occur (O’Connor et al. 1999), and even the very notion of the downstream box as an enhancer has been questioned (Blasi et al. 1999).

Initiation efficiency is also influenced by ribosomal protein S1, which appears to bind strongly to U-rich sequences found upstream of the SD element in many efficiently translated mRNAs (Boni et al. 1991). Current models envisage an exceedingly elongated structure for S1 (Subramanian 1983; Wallaczek et al. 1990), which therefore might act as a sort of “fish-ing line" to capture and tether the mRNA to the 30S ribosomal subunit, before the initiation site is "landed" by the well-established SD motif/16S rRNA interaction. Significantly, although S1 stimulates initiation on typical prokaryotic mRNAs, there is no stimulation if the 5′UTR is truncated from the 5′ end to just ~20 nucleotides in length, or if it is completely missing so that the initiation codon is at the very 5′ end (Tedin et al. 1997).

TWO DISTINCT MECHANISMS OF INITIATION SITE SELECTION IN EUKARYOTES

It is axiomatic that eukaryotic cellular and viral mRNAs lack a highly conserved motif equivalent to the SD sequences of prokaryotic mRNAs. This correlates with the fact that although the sequence and structure at the 3′ end of the eukaryotic 18S rRNA are very similar to the prokaryotic 16S rRNA, the ..CCUCC.. sequence in 16S rRNA that pairs with the SD element is precisely deleted in all eukaryotic cytoplasmic small ribosomal RNAs (Fig. 1), with the exception of one protozoan, *Giardia lamblia*, which retains it in a slightly corrupted form (Sogin et al. 1989).

It seems highly unlikely that the recognition of initiation sites in eukaryotes involves sequence-specific pairings between the mRNA and the small ribosomal subunit 18S rRNA, at least not base-pairing of comparable stability to the prokaryotic 16S rRNA/SD interactions, since any such pairing would be expected to allow (1) efficient translation of all cistrons of a polycistronic mRNA provided each cistron included the relevant identifier motif and (2) direct binding of (salt-washed) 40S ribosomal subunits to the mRNA in the absence of any initiation factors. It is true that internal ribosome entry sites/segments (IRESs) such as those found in picornavirus 5′UTRs and some cellular mRNAs can potentiate the translation of downstream cistrons of di- or polycistronic mRNAs. However, at least with the one picornavirus IRES that has been rigorously tested, there is no direct binding of salt-washed 40S ribosomal subunits to the mRNA in the absence of canonical initiation factors, and thus such IRESs cannot really be regarded as operationally equivalent to the prokaryotic SD motif.

The only eukaryotic RNAs that can be considered operationally analogous to bacterial mRNAs with respect to initiation mechanism are the IRESs of hepatitis C virus (HCV) and the fairly closely related pestiviruses. These IRESs not only promote translation of downstream cistrons in polycistronic mRNAs, but, more significantly, washed 40S ribosomal subunits bind to the IRES very close to the authentic initiation

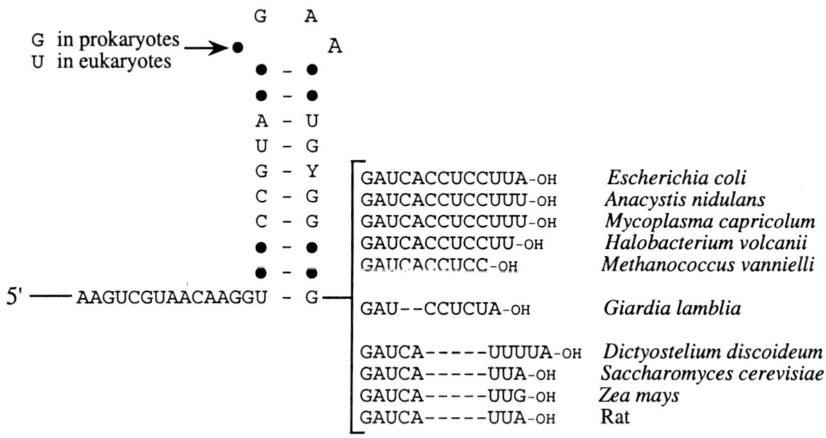

Figure 1 Conservation of the sequence at the 3′ end of the small ribosomal subunit rRNA. The selected sequences shown are from Gutell et al. (1985), except that of *Giardia lamblia* (Sogin et al. 1989). Among the species listed there is complete conservation of (1) the sequence shown upstream of the stem-loop, (2) the length of the stem-loop, and (3) the sequence in three of the positions of the tetraloop. Individual sequences downstream from the stem-loop are given, with gaps (denoted by –) introduced to optimize alignment.

site, even in the absence of initiation factors and Met-tRNA$_i$ (Pestova et al. 1998b). Thus, these IRESs can be considered similar to the prokaryotic SD motif in operational terms, even though it is hard to see why it needs some 300 nucleotides of complex HCV or pestivirus IRES structure (Fig. 2) to perform the same function as the ~5-nucleotide SD sequence. We are also ignorant as to how far the operational similarity extrapolates to a mechanistic similarity. Is the binding of the washed 40S ribosomal subunit driven exclusively by RNA–RNA pairing, possibly pairing involving widely dispersed sites in both the IRES and the 18S rRNA? Or do interactions between the ribosomal proteins and the IRES RNA play a major role?

Despite these uncertainties, the discovery of direct binding of small ribosomal subunits to the HCV and pestivirus IRESs has had a tremendous impact (and is arguably the most significant breakthrough of recent years), because it is correlated with the fact that initiation dependent on these IRESs does not require any of the eIF4 family of initiation factors (Pestova et al. 1998b). In contrast, initiation driven by picornavirus IRESs needs all the canonical translation initiation factors except that eIF4E is completely redundant (apart from one exceptional IRES), and only part of eIF4G is required (Pestova et al. 1996a,b). This contrast between the HCV

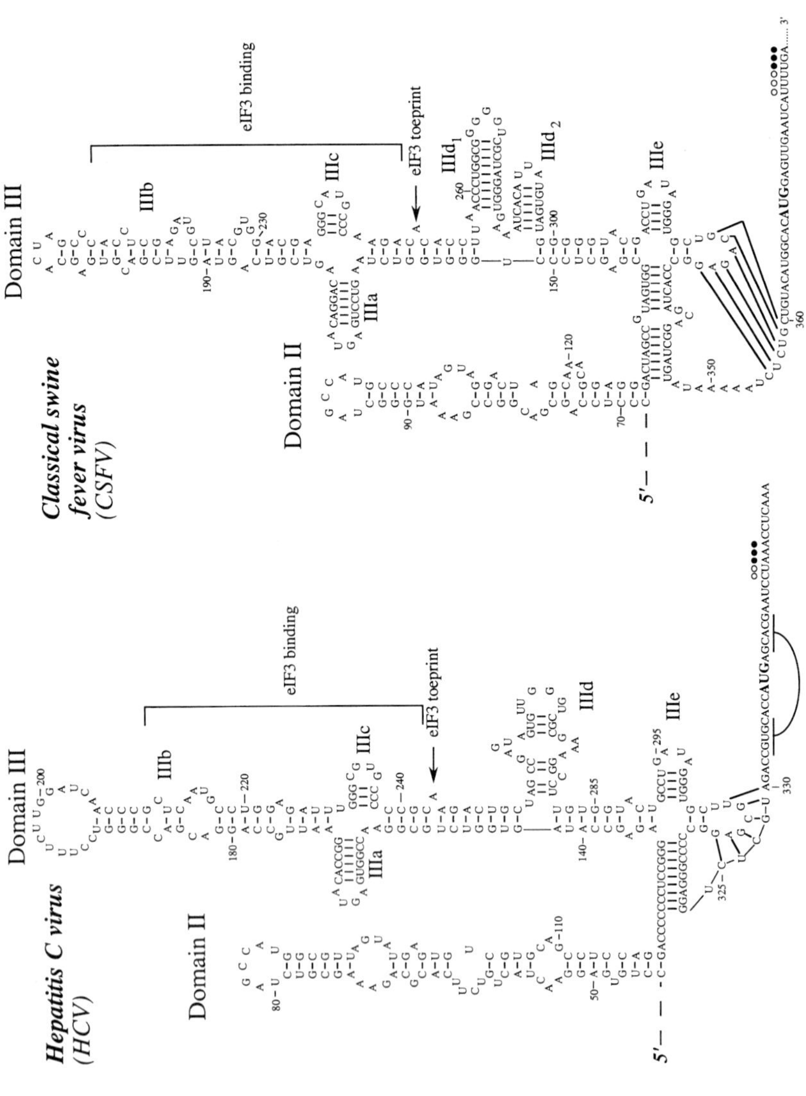

Figure 2 (*See facing page for legend.*)

and picornavirus IRESs has led to the seminal idea that there are just two basic mechanisms for the delivery of the 40S subunit to the correct site on the mRNA: (1) Either the mRNA sequence and structure are such that the 40S subunit binds directly to the correct site in the absence of initiation factors, in which case eIF4A, 4B, 4E, and 4G are not required for initiation; or (2) the 40S subunit cannot bind directly to the mRNA, but has to be delivered by the eIF4 family of factors, in particular by the central domain of eIF4G.

Several variants of the second of these mechanisms appear to exist, as listed in Table 1, and even though these mechanisms may seem very different from one another, what they all have in common is delivery of the 40S ribosomal subunit to the mRNA via eIF4G. Therefore, in the rest of this chapter I discuss these different mechanisms, in the order in which they are listed in Table 1. Although this order may seem somewhat strange in the sense that it starts with unusual or special mechanisms and ends with the most usual pathway (ribosome scanning), it has a certain logic in that it starts with the mechanism that can be regarded as the simplest (in terms of the initiation factor requirements) and ends with the most complex.

A detailed discussion of the structure, function, and interactions of eIF4G is presented in Chapter 2, and so it suffices here to summarize

Figure 2 Secondary structures of HCV and CSFV (a pestivirus) IRESs. The HCV IRES structure is based on a recently revised model (Honda et al. 1999); the CSFV IRES structure is based on direct structure probing and phylogenetic analysis (S.P. Fletcher and R.J. Jackson, in prep.), and is consistent with published models (Rijnbrand et al. 1997). Note the strong similarity between the two structures at the top of Domain II, around the 4-way junction in Domain III, and in the vicinity of the pseudoknot. The extreme 5′-proximal sequences (Domain I) are not shown because they are not part of the functional IRES and show little conservation between HCV and the different pestiviruses. The eIF3-binding site as determined by toeprinting and footprinting is shown (Pestova et al. 1998b; Sizova et al. 1998). Open circles denote the toeprints observed with 40S/IRES binary complexes, and filled circles the toeprints obtained when eIF2, eIF3, Met-tRNA$_i$, and GTP are included (Pestova et al. 1998b). The potential base-pairing between sequences immediately upstream and downstream of the HCV initiation codon (Honda et al. 1996) is shown. Note that ribosome binding at the initiation codon would melt such base-pairing, and that the same pairing does not occur in the CSFV or any other pestivirus IRES.

those features of eIF4G that are pivotal to the topic of this chapter. Any reference to eIF4G should be assumed to relate to mammalian eIF4G, unless otherwise stated, and to relate specifically to what is now known as eIF4GI, since the recently discovered eIF4GII (Gradi et al. 1998) does not appear to differ significantly from eIF4GI with respect to the activities and properties that are relevant to this chapter.

There are two ways of looking at the domain structure of eIF4G. It can either be regarded as divided into an amino-terminal one-third, and carboxy-terminal two-thirds domain, as defined by the site of cleavage of eIF4G by picornavirus proteases (Lamphear et al. 1995). Alternatively, it can be considered as consisting of three domains: the same amino-terminal one-third domain as defined by picornavirus protease action, a central one-third fragment, and a carboxy-terminal one-third domain. In the first perspective, it is the carboxy-terminal two-thirds fragment that interests us here, and in the second it is the central one-third domain. In summary, the critical aspects of this domain organization are as follows (for further details, see Morley et al. 1997; Gingras et al. 1999; Chapter 6):

1. The amino-terminal domain interacts with eIF4E.
2. It has been suggested on the basis of sequence inspection that the central fragment has an RRM-like motif (Goyer et al. 1993), a feature shared by the yeast and plant eIF4Gs (for review, see Morley et al. 1997). However, the putative RNP-1 and RNP-2 motifs are non-canonical (Goyer et al. 1993; Morley et al. 1997), and the spacing between them in the linear amino acid sequence is considerably greater than is typical. In fact, there is no direct evidence as to whether this motif is active as an RNA-binding domain, and, if so, what its specificity is with respect to RNA sequence and/or structure. Indeed, from what we believe we know about eIF4G function, it is hard to imagine that it would exhibit highly sequence-specific binding to RNA, since eIF4G has a generic function that operates on a vast spectrum of RNA sequences.
3. The central fragment, which shows homology with the carboxy-terminal part of yeast and plant eIF4Gs (Morley et al. 1997), interacts with initiation factors eIF3 and eIF4A (Lamphear et al. 1995; Imataka and Sonenberg 1997). The exact site on eIF4G that interacts with eIF3 has not been mapped, nor is it known which polypeptide subunits of eIF3 participate either in this interaction with eIF4G or in the other well-known interaction of eIF3 with the 40S ribosomal subunit. However, it is generally thought that the two interactions of eIF3, (1) with the central domain of eIF4G and (2) with the 40S sub-

Table 1 Classification of eukaryotic initiation site selection mechanisms

A. Salt-washed 40S ribosomal subunits bind directly to the RNA at/near the correct initiation site in the absence of initiation factors. Initiation does not require eIF4A, 4B, 4E, or 4G, or ATP hydrolysis.
   Hepatitis C virus (HCV) and pestivirus IRESs
   Probably the intercistronic IRESs of insect "picornavirus-like" viruses (*Drosophila* C virus, *Plautia stali* intestine virus, *Rhopalosiphum padi* virus)

B. Salt-washed 40S ribosomal subunits do not bind to the RNA in the absence of initiation factors. Initiation requires eIF4A, at least part of eIF4G, and ATP hydrolysis.
   1. Internal initiation; no requirement for eIF4E; eIF4G requirement totally fulfilled by the central domain of eIF4G.
      Picornavirus IRESs, except hepatitis A virus (HAV)
      Laboratory-constructed IRESs dependent on tethered eIF4G
      Probably some cellular mRNA IRESs
   2. Scanning-dependent initiation of uncapped mRNAs; no direct requirement for eIF4E; eIF4G requirement totally fulfilled by the central domain of eIF4G.
      Uncapped versions of normally capped cellular mRNAs
   3. Cap-independent initiation, but requiring eIF4E, probably involving direct binding of eIF4E/4F to an internal site.
      Satellite tobacco necrosis virus (STNV) RNA
      Probably barley yellow dwarf virus RNA
      Possibly hepatitis A virus (HAV) RNA
      Possibly some cellular mRNA IRESs
   4. Scanning-dependent initiation of capped mRNAs; requirement for eIF4E, and for at least part of the amino-terminal domain of eIF4G in addition to the eIF4G central domain.
      Capped cellular and viral mRNAs (includes translation by ribosome shunting, in addition to the more usual mechanism of strictly linear scanning)

unit, are not mutually exclusive, so that in principle a tripartite eIF4G–eIF3–40S subunit interaction relay could occur, which is critical to the models discussed in this chapter.

4. The carboxy-terminal one-third fragment has another site for interaction with eIF4A (Lamphear et al. 1995; Imataka and Sonenberg 1997). The significance of this is not yet clear, since plant and yeast eIF4Gs lack the equivalent of this site (Morley et al. 1997; Gingras et al. 1999), and in many situations the central fragment of mammalian eIF4G seems sufficient to support initiation. Thus, the role of car-

boxy-terminally associated eIF4A is uncertain and cannot be absolutely essential.

5. The three-dimensional structure of eIF4G seems to undergo a radical change on binding of eIF4E (and perhaps also eIF3), as witnessed by the fact that eIF4G cannot be cleaved by picornavirus proteases if eIF4E is not associated with it (Ohlmann et al. 1997), and there are also reports that eIF3 increases the rate of cleavage (Wyckoff et al. 1990).

Singular eIF4A and eIF4F holoenzyme complex both exhibit ATP-dependent RNA helicase activity, provided eIF4B and ATP are also present (Rozen et al. 1990; Jaramillo et al. 1991). Almost unprecedented among RNA and DNA helicases, the activity is reported to be bidirectional. The eIF4F complex appeared to be about fivefold more active on a molar basis than singular eIF4A for unwinding in the 3′→5′ direction, and as much as ~15-fold more active than eIF4A in the 5′→3′ direction, provided the RNA was capped (Rozen et al. 1990). These data suggest that when the cellular complement of initiation factors encounters endogenous mRNAs, the predominant outcome will be cap-dependent unwinding in the 5′→3′ direction (rather than the reverse) carried out by eIF4F complex, rather than singular eIF4A.

Certain dominant negative mutants of eIF4A are extremely potent inhibitors of in vitro translation (Pause et al. 1994a), and this was shown to reflect their potency as inhibitors of the helicase activity of eIF4F. Translation activity can be recovered by addition of either eIF4F or singular eIF4A, but the striking difference is that it requires 6-fold more singular eIF4A than eIF4F (on a molar basis) to effect the same degree of recovery. This has led to the suggestion that the normal function of eIF4A is to recycle or treadmill through the eIF4F holoenzyme complex, and that the dominant negative mutants inhibit by entering the complex but failing to exit and recycle, effectively generating a dead-end eIF4F complex (Pause et al. 1994a). An extrapolation of this interpretation is that the principal role of eIF4A in translation initiation is in association with eIF4G (i.e., as a constituent of the eIF4F complex) and that singular eIF4A has little, if any, influence.

Yeast have an RNA helicase, Ded1p, which, although a member of the same DEAD-box family of helicases, is not closely homologous to eIF4A, yet appears to be involved in translation initiation because it was isolated as a multicopy suppressor of mutations in eIF4E (de la Cruz et al. 1997; Iost et al. 1999). The precise function of this helicase remains unknown, nor is it known whether it has a counterpart in mammalian systems.

INTERNAL INITIATION OF TRANSLATION OF EUKARYOTIC VIRAL AND CELLULAR RNAs

Exceptional Features of Hepatitis C Virus and Pestivirus IRESs

As already mentioned, the truly remarkable feature of these IRESs is that they bind salt-washed 40S ribosomal subunits directly, in the absence of any initiation factors, with high efficiency and accuracy (Pestova et al. 1998b; Pestova and Hellen 1999). Toeprinting shows that in the binary 40S/IRES complex, the ribosome is centered within 3 nucleotides of the initiation codon (Fig. 2). The additional presence of eIF2/GTP/Met-$tRNA_i$ ternary complex is necessary for the ribosomes to precisely lock on to the initiation codon (Pestova et al. 1998b). Initiation factor eIF3 is not strictly necessary for formation of the 40S/mRNA/Met-$tRNA_i$ complex at the correct site, but it is necessary for the subsequent step of subunit joining (Pestova et al. 1998b). Toeprinting and UV-crosslinking assays show that on its own eIF3 binds to the upper part of domain III (Fig. 2), often referred to as domain IIIb (Buratti et al. 1998; Sizova et al. 1998). It seems likely that this is the site of binding of the eIF3 that would normally be expected to be associated with the incoming 40S subunit, rather than the binding site for an additional eIF3 molecule not associated with the 40S subunits. There are also contacts between small ribosomal subunit protein S9 and the IRES, as revealed by UV-crosslinking (Pestova et al. 1998b). The binding site of S9 has not been mapped, but mutations in domain II of the IRES (Fig. 2) can abolish the binding or crosslinking (Fukushi et al. 1997; Pestova et al. 1998b).

Correlated with the unique ability of these IRESs to bind salt-washed 40S subunits in the absence of any initiation factors, another unique feature is that they appear to promote initiation independently of any requirement for, or participation by, eIF4A, 4B, 4E, and 4G, and any requirement for ATP hydrolysis (Pestova et al. 1998b). Not only does the presence or absence of these factors have no influence on initiation complex formation, but these factors do not bind to the IRES, as judged by the lack of any clear toeprint. In addition, translation driven by these IRESs is completely refractory to inhibition by dominant negative eIF4A mutants (Pestova et al. 1998b), yet another property in which these RNAs are unique among all eukaryotic cellular and viral RNAs.

Assay of deletion mutants in the standard bicistronic mRNA assay shows that the IRESs of HCV and the pestiviruses consist of the 300–310 nucleotides immediately preceding the authentic initiation codon, plus, arguably, the first 30–50 nucleotides of viral polyprotein coding sequences (Wang et al. 1993; Reynolds et al. 1995, 1996; Rijnbrand et al.

1995, 1997). The secondary structure of the 5′UTR part of the HCV IRES, as recently revised by Honda et al. (1999), is very similar to the predicted pestivirus IRES structure (Fig. 2). In contrast, the extreme 5′-proximal sequences (domain I), which are not part of the functional IRES according to deletion analyses (Rijnbrand et al. 1995, 1997; Reynolds et al. 1996), show very little conservation between HCV and the different species of pestivirus. There is a complex pseudoknot just upstream of the initiation codon (Fig. 2). Mutational analysis has shown that this pseudoknot structure, but not the primary nucleotide sequence of its paired elements, is essential for IRES activity (Wang et al. 1995; Rijnbrand et al. 1997). Since the pseudoknot is so close to the initiation codon, it would be intriguing to know whether the pseudoknot is unwound as the ribosome enters, or, if not, how the pseudoknot is accommodated within the interior of the ribosome or on its surface.

Given the strong direct interaction of 40S subunits with these IRESs, it is perhaps not surprising that translation driven by the HCV IRES was only slightly reduced when the AUG initiation codon was mutated to CUG or AUU. When it was mutated to something more distantly related to AUG, the system appeared to utilize (at reduced efficiency) an ACG codon two codons farther downstream (Reynolds et al. 1995). However, the initiation "window" of these IRESs is quite narrow (Reynolds et al. 1996; Rijnbrand et al. 1996). An AUG codon present in all pestivirus IRESs just 7 nucleotides upstream of the authentic initiation codon is not functional as an initiation site, and translation is abrogated if the authentic start codon is moved just a short distance downstream (Rijnbrand et al. 1997). Thus, it appears that the 40S subunits do not have the potential to scan much, if at all, following binding to the IRES. It would be interesting to know whether it is the lack of eIF4A, 4B, and 4F in the initiation complex that is the cause of this, or whether the binary 40S/IRES complex is just too tight to allow any ribosome movement.

As mentioned above, there is some controversy as to whether the first 30–50 residues of viral coding sequence are an integral part of these IRESs. In some studies, removal of all viral coding sequences virtually abolished the activity of the HCV IRES, and reduced that of the more potent classical swine fever virus (CSFV) IRES by over 80% (Reynolds et al. 1995; Lu and Wimmer 1996; S.P. Fletcher and R.J. Jackson, in prep.). On the other hand, other workers have questioned these results on the grounds that they could obtain respectable activity using CAT or luciferase reporters fused directly to the initiation codon with no intervening viral coding sequences (Wang et al. 1993; Rijnbrand et al. 1995, 1997; Honda et al. 1996). However, recent publications have shown that the expression of even these CAT and luciferase reporters is enhanced if, rather than being fused directly to the initiation codon, they are expressed as fusion proteins with the amino-terminal segment of the viral polyprotein (Chon et al. 1998; Hahm et al. 1998; Hwang et al. 1998). Alignment of all the sequences that are reasonably permissive to IRES activity when fused directly to the initiation codon shows no motifs highly conserved in both sequence and relative position, but rather just an A-rich character reminiscent of the characteristic of prokaryotic initiation sites identified by Dreyfus (1988), as discussed above.

Structure probing of a set of four CSFV IRES constructs which spanned a 5-fold range of IRES activity revealed that activity was inversely proportional to the apparent degree of secondary structure at and immediately downstream from the initiation codon (S.P. Fletcher and R.J. Jackson, in prep.). Similar results have been reported for the HCV IRES, except that in this case, it was pairing between sequences immediately upstream and downstream of the initiation codon (Fig. 2) that reduced the IRES activity (Honda et al. 1996), whereas in the CSFV constructs, the segment immediately before the initiation codon did not participate in the inhibitory secondary structure. Why should the activity of such IRESs be so sensitive to what appears to be not particularly extensive (Fig. 2), and thus not particularly stable secondary structure? The answer is likely to lie in the fact that these IRESs do not use eIF4F to promote initiation and apparently do not bind eIF4F. Consequently, there is no possibility of focused RNA unwinding by the action of the eIF4A helicase component of eIF4F. If singular eIF4A is capable of RNA unwinding, then presumably its action is normally too dispersed and unfocused to promote local unwinding in the vicinity of the initiation site. Thus, besides the similarity of forming direct binary complexes with small ribosomal subunits, the HCV/pestivirus IRESs show a further resemblance to prokaryotic initiation sites in that initiation efficiency is very sensitive to local secondary structure at or around the initiation codon (de Smit and van Duin 1990a,b, 1994b).

Thus far, HCV and the obviously related pestiviruses are a unique example of this mechanism of initiation in eukaryotic systems. A possible additional (quite unrelated) example is the positive-strand RNA viruses that were once designated as "insect picornaviruses," although this is clearly a misclassification since, instead of the single ORF starting with capsid protein coding sequences characteristic of animal picornaviruses, these insect viruses have two ORFs both encoding polyproteins: The 5′-proximal ORF encodes nonstructural proteins required for RNA replication, and the downstream ORF encodes viral structural proteins (Johnson and Christian 1998; Moon et al. 1998; Sasaki et al. 1998). Assays with bicistronic constructs show that translation of the downstream ORF is unquestionably by internal initiation dependent on an IRES which is active in mammalian systems such as rabbit reticulocyte lysate (Sasaki and Nakashima 1999, 2000). However, a most curious feature is that the downstream ORF does not open with an AUG codon or any codon closely related to AUG. Nevertheless, even though these IRESs differ from the HCV paradigm in that they can promote initiation at codons quite unrelated to AUG, the expectation is that they will turn out to resemble the HCV IRES in not requiring eIF4A, B, E, or G, nor ATP hydrolysis.

Picornavirus IRESs: The Encephalomyocarditis Virus Paradigm

The translation of picornavirus RNAs is discussed in detail in Chapter 31, and therefore in this section I simply summarize those aspects that are pertinent for comparison with the other mechanisms of initiation site selection discussed in this chapter.

Picornavirus IRESs are typically about 450 nucleotides long. At the 3′ end there is invariably an AUG triplet located some 22–25 nucleotides downstream from the start of an oligopyrimidine tract of up to ~10 pyrimidine residues. As with HCV, the extreme 5′-proximal part of the viral genome is not considered to be part of the IRES, although there is some evidence that mutations and protein–RNA interactions in this region may indirectly influence the efficiency of internal initiation (Chapter 31). Likewise, it is generally considered that the viral coding sequences are not an integral part of the IRES. Even though it is true that viral coding sequences can improve the efficiency of internal initiation promoted by the hepatitis A virus IRES (Graff and Ehrenfeld 1998) or alter the dependence of other IRESs on cellular RNA-binding proteins (Kaminski and Jackson 1988), such effects are quantitatively minor compared with the profound effect of the linked coding sequences on HCV and CSFV IRES activity.

By the criteria of primary sequence, and especially secondary structure conservation, the picornavirus IRESs can be divided into one minor and two major groups (for review, see Jackson and Kaminski 1995): (1) hepatitis A virus (the minor group); (2) cardio- and aphthoviruses; and (3) entero- and rhinoviruses. The discussion in this and the following section focuses on the two major groups, and the hepatitis A virus IRES will be mentioned only with respect to those properties in which it differs radically from both of the major groups.

Of all the picornavirus IRESs, it is the cardiovirus, encephalomyocarditis virus (EMCV), that we understand best. Virtually all initiation occurs at the AUG at the 3′ end of the IRES (AUG-11 in the commonly studied EMCV strain R). There is little doubt that the initiating 40S subunit enters at, or very close to, AUG-11 (Fig. 3), since there is almost no initiation at AUG-10, located just 8 nucleotides upstream of AUG-11 (Kaminski et al. 1990, 1994). Thus, internal initiation driven by the EMCV IRES involves direct ribosome entry at the authentic initiation codon, with very little, if any, scanning.

Although this direct entry at the initiation site is a feature shared with the HCV and pestivirus IRESs, the EMCV IRES differs radically from the HCV/pestivirus model in that salt-washed ribosomes do not bind to the EMCV IRES in the absence of initiation factors, as judged by sucrose density gradient centrifugation and toeprinting assays (Pestova et al. 1996a). Binding of the 40S subunit to the correct initiation site on the EMCV IRES absolutely required eIF2, 3, 4A, and either the complete native eIF4F complex or a recombinant fragment of eIF4G that included the central one-third domain (Pestova et al. 1996a,b). There was also a partial requirement for eIF4B, which increased the yield of complexes by about twofold. Thus, internal initiation on the EMCV IRES requires the same set of factors as the scanning mechanism except that eIF4E is completely redundant (Pause et al. 1994b), and the central domain of eIF4G is sufficient to fulfill all eIF4G functions.

Intact eIF4F holoenzyme complex, or just the central domain of eIF4G, binds to the EMCV IRES at a specific site, the J-K domain (Fig. 3), giving a defined toeprint a fairly short distance (~50 nucleotides in terms of primary sequence) upstream of the authentic initiation codon (Pestova et al. 1996b; Kolupaeva et al. 1998). In UV-crosslinking assays there appears to be cooperativity of binding (or of crosslinking) of eIF4A, 4B, and 4G to this region of the IRES (Pestova et al. 1996b). This suggests that one of the critical features of internal initiation is that eIF4G bound at a site toward the 3′ end of the IRES may deliver the 40S ribosomal subunit directly to AUG-11 at the very 3′ terminus of the IRES via the eIF4G–eIF3–40S subunit interaction relay. Obviously this cannot be the whole explanation for the mechanism of internal ribosome entry, for if it were, then the ~150 nucleotides from the start of the J domain up to AUG-11 should be sufficient to function as an IRES, and there would be no requirement for the H and I domains (~300 nucleotides total length) lying upstream of the J domain (Fig. 3). What the functions of these H and I domains might be is a complete mystery. One possibility is that they are sites of direct interactions between the IRES and 40S ribosomal subunits

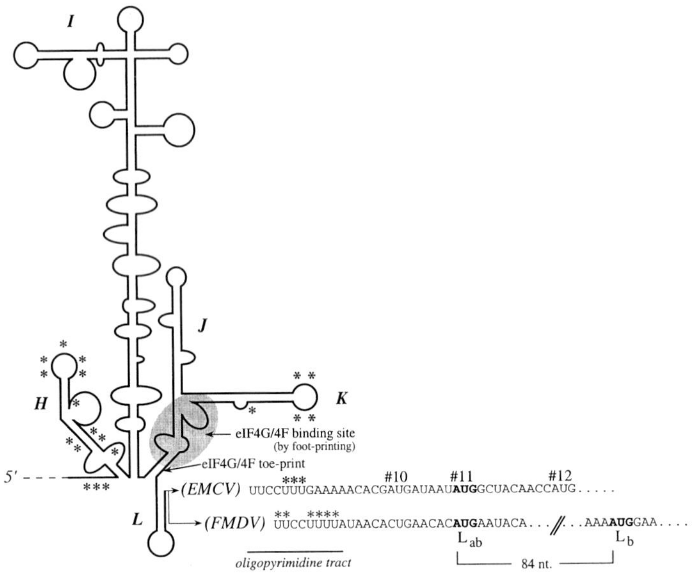

Figure 3 Schematic diagram of the EMCV and foot-and-mouth disease virus (FMDV) IRESs. Initiation sites are shown in bold. The sequences around the single initiation site in EMCV (strain R), and the two initiation sites in FMDV (strain O1K) are shown. The various subdomains of the IRES (H–L) discussed in the text are indicated. The site at which eIF4G (or eIF4F) binds is shown (Pestova et al. 1996b; Kolupaeva et al. 1998), and asterisks denote the regions protected when polypyrimidine tract-binding protein binds to the IRES.

which are too weak to allow stable binding of the (salt-washed) small ribosomal subunit to the IRES, but which might act cooperatively with the 40S subunit–eIF3–eIF4G (central domain)–J-K IRES domain interaction relay.

It must be stressed that even though internal initiation on the EMCV IRES involves direct ribosome entry at the authentic initiation site, probably with no ribosome scanning whatsoever, nevertheless it does require eIF4A and ATP hydrolysis (Pestova et al. 1996a,b). Even when eIF4F holoenzyme complex, which has an eIF4A subunit, is used to drive internal initiation, supplementary singular eIF4A is still required (Pestova et al. 1996a). This is often thought surprising in view of the assumption, prevalent since the mid 1980s, that the function of eIF4A is somehow associated with scanning. However, the more recent evidence provided by dominant negative eIF4A mutants suggests that the main function of eIF4A is in association with eIF4G as a subunit of the eIF4F complex (Pause et al. 1994a). Since the central domain of eIF4G (and associated eIF4A) binds to the J-K domain of the EMCV IRES fairly close to the actual internal ribosome entry site, the likely outcome is not only that the 40S subunit is delivered to the initiation site via the eIF4G–eIF3–40S subunit interaction relay, but also that the RNA-unwinding activity of the associated eIF4A is focused toward the region around the initiation codon. This eIF4G-mediated focused action of the eIF4A helicase may provide an explanation for why it is that although internal initiation on the EMCV and HCV IRESs is similar in the sense that ribosomes are thought to "enter" directly at the authentic initiation codon (rather than scanning from an upstream entry site), there is no evidence that secondary structure around the initiation site of the EMCV IRES has a negative influence, in contrast to the strongly inhibitory effect seen with the HCV and pestivirus IRESs (which function independently of eIF4A or eIF4G).

Other Picornavirus IRESs

Given the close structural similarity, the foot-and-mouth disease virus (FMDV) is assumed to conform closely to the EMCV paradigm, and the limited available evidence, for example, the binding of eIF4B to the J domain (Meyer et al. 1995), supports this presumption. The hepatitis A virus IRES is clearly different from EMCV and all other picornavirus IRESs in that its function is inhibited by cleavage of eIF4G by picornavirus proteases (Borman et al. 1995, 1997; Borman and Kean 1997) or by addition of 4E-BP1 or m^7GpppG cap analog (I.K. Ali and R.J. Jackson, unpubl.): It thus appears to require eIF4E, and perhaps the complete eIF4G, or certainly more than just the central fragment. As for the entero-/rhinovirus IRESs, there is no evidence to suggest that the mechanism is very different from the EMCV paradigm, and thus it is presumed that these too will bind eIF4F or the central domain of eIF4G at a specific site that will allow both delivery of the 40S ribosomal subunit to the appropriate entry site and focused unwinding by the associated eIF4A polypeptide. We have no idea as to where the eIF4G-binding site on the entero-/rhinovirus IRESs might be. These IRESs lack anything resembling the A-rich bulge that is part of the eIF4G-binding site on the EMCV IRES (Pestova et al. 1996b; Kolupaeva et al. 1998).

The different types of picornavirus IRESs differ quite markedly in the requirements for other cellular RNA-binding proteins (Chapter 31). However, unlike the case of eIF4G binding to the IRES, there is no indication that any of these RNA-binding proteins play a direct role in ribosome recruitment or selection of the initiation site. Rather, it is currently believed that their role is to stabilize the appropriate higher-order structure of the IRES. Significantly, all of them are proteins with multiple

RNA-binding motifs, and they could therefore make multipoint contacts with the IRES, as has been shown for the binding of polypyrimidine tract-binding protein (PTB) to cardio-/aphthovirus IRESs (Fig. 3) (Kolupaeva et al. 1996).

Extrapolating from the EMCV IRES to other picornavirus IRESs leads us into the greatest difficulties over the question of where exactly does the ribosome "enter" and how does it access the correct initiation codon. With the EMCV IRES the situation is straightforward: Ribosomes enter directly at the authentic initiation codon (AUG-11), located ~25 nucleotides downstream from the start of the conserved oligopyrimidine tract, and virtually all of them initiate there (Kaminski et al. 1990, 1994). In contrast, in the closely related FMDV IRES, only a minority of the ribosomes (up to ~30% depending on the exact conditions) initiate translation at the equivalent AUG, the L_{ab} initiation site (Fig. 3), and the rest initiate translation at the next AUG downstream, the L_b site (Belsham 1992). In the case of the entero-/rhinovirus IRESs, there is virtually no initiation at the AUG located just downstream from the oligopyrimidine tract, and all initiation is at the next AUG codon farther downstream (Pestova et al. 1994; Ohlmann and Jackson 1999).

In an attempt to make a unified model applicable to both EMCV and FMDV IRESs (and extrapolatable to entero-/rhinovirus IRESs), it was suggested that all ribosomes have to enter at the AUG just downstream from the oligopyrimidine tract, but a slight majority (FMDV) or virtually all of them (entero-/rhinoviruses) then scan on to the next AUG farther downstream. This idea seemed to be fully supported by some rather compelling evidence that an AUG downstream from the oligopyrimidine tract is essential for the translation and infectivity of poliovirus type 1 (Pilipenko et al. 1992). On the other hand, mutation of the L_{ab} AUG in FMDV had very little influence on infectivity, although mutation of the downstream L_b AUG was lethal (Cao et al. 1995). In addition, annealing of antisense oligonucleotides at or just downstream from the L_{ab} AUG had only a small influence on initiation at the L_b site (Lopez de Quinto and Martinez-Salas 1999). These and other observations (Chapter 31) have prompted a certain amount of backtracking toward a position in which internal ribosome entry is still presumed to occur at ~25 nucleotides downstream from the start of the oligopyrimidine tract, but not necessarily precisely at the L_{ab} initiation site, nor necessarily at an AUG codon; and ribosomes are still presumed to access the major initiation site, the L_b site, via entry at an upstream site, although not all of them are necessarily transferred by a strictly linear scanning process, and some may be transferred by shunting.

Given that in the conventional scanning mechanism the 40S subunit enters just downstream of the 5′-cap structure irrespective of the actual nucleotide sequence at this 5′-proximal entry site, if the site of actual internal ribosome entry on picornavirus IRESs must obligatorily be an AUG triplet, even though it may not be used as an initiation site by all (FMDV) or any (entero-/rhinoviruses) of the ribosomes that enter there, we would need to look for a special explanation for this. It would seem to imply that interaction between the primed 40S subunit and the AUG, possibly codon–anticodon interaction involving the 40S subunit-associated Met-tRNA$_i$, is an important stabilizing interaction necessary for internal ribosome entry, and yet some aspect of commitment to initiation, possibly the action of eIF5 and eIF5B, is inefficient and/or delayed for some unknown reason.

A Laboratory-generated IRES Dependent on Tethered eIF4G

The realization that the carboxy-terminal two-thirds fragment, or strictly just the central domain, of eIF4G is the key component in the delivery of initiating 40S subunits to the mRNA led De Gregorio et al. (1999) to a very provocative experiment: the design of a synthetic IRES system in which internal ribosome entry is dependent on tethered eIF4G. The bicistronic construct had one, two, or three copies of the iron response element (IRE) in the intercistronic spacer. The "dedicated" *trans*-acting factor was a fusion between the iron regulatory protein 1 (which binds with high affinity and specificity to the IRE) and the carboxy-terminal two-thirds fragment of eIF4G. In cotransfection experiments, the outcome was a significant level of expression of the downstream cistron that was dependent on both the IRP-eIF4G fusion protein and the *cis*-acting IRE elements, and was independent of translation of the upstream cistron. Deletions of the eIF4G entity showed, not surprisingly, that the central domain of eIF4G, which has the eIF3 interaction site as well as one of the two eIF4A-binding sites, was the essential part, and the activity was abrogated by further deletions, which removed the site on eIF4G to which eIF3 is thought to bind.

Although the efficiency of this synthetic IRES was fairly low by comparison with that of the HCV IRES used as a reference standard, it was very significantly above background (De Gregorio et al. 1999). It is possible that higher efficiencies will be attainable through more fine-tuning of the spatial relationship between the IRE and the downstream cistron.

Alternatively, it may be that the tethered eIF4G is merely providing the same type of ribosome delivery system as the eIF4G/J-K domain interaction in the EMCV IRES, and that the low efficiency is because this synthetic system has nothing equivalent to the undefined and mysterious contributions made by the H and I domains of the EMCV IRES (Fig. 3).

IRESs in Cellular mRNAs

The list of cellular mRNAs claimed to have an IRES grows almost daily, and one sometimes gets the impression (which may well be false because negative results seldom get published) that any 5′UTR of reasonable length scores as an IRES when it is tested as intercistronic spacer in the bicistronic mRNA assay. What usually provokes tests for an IRES is the discovery that a cellular mRNA has a long 5′UTR of a few hundred residues, with some AUG triplets that do not appear to be functional as initiation codons, and/or a high GC content indicative of stable secondary structure—in other words, characteristics that are superficially shared by picornavirus 5′UTRs. However, the presence of upstream AUGs is not in itself a sufficient criterion for postulating the existence of an IRES, as is amply illustrated by the case of yeast GCN4 mRNA (Chapter 5). What is striking is that in the majority of these cellular mRNAs claimed to include an IRES, the frequency of upstream AUGs is much lower than would be expected for random occurrence, and they are generally quite well conserved across all vertebrates. This is in sharp contrast to the picornaviruses, where the number of upstream AUG triplets in the 5′UTR is close to what would be expected for a random occurrence, and most of them are not highly conserved, not even between different clinical isolates of the same virus (Pöyry et al. 1992). This suggests that most of the AUGs in picornavirus 5′UTRs are acquired or lost randomly during evolution and thus have no functional significance, whereas those in the putative IRESs of cellular mRNAs are conserved, presumably for a purpose. This argument is, of course, subject to the big qualification that RNA viral genomes undergo high genetic drift, whereas chromosomal DNA sequences do not.

Some mRNAs with a putative IRES have a 5′UTR sufficiently long that it would certainly be expected to have at least one AUG triplet on the assumption of random probability, yet there are none. The most likely explanation would seem to be that these mRNAs are, and have to be, translated by a scanning mechanism in most circumstances, but an IRES-dependent mechanism operates in special conditions. An example is ornithine decarboxylase mRNA, which, according to recent evidence, is translated by an IRES-dependent mechanism at the G_2/M border, but by a scanning mechanism throughout the rest of the cell cycle (Pyronnet et al. 2000).

What is abundantly clear is that all cellular IRESs described so far are quite different from any known viral IRES (Chapter 11), and thus the all-too-frequent habit of trying to draw parallels between the cellular and viral IRESs may actually be counterproductive. An intriguing but very frustrating feature of many cellular IRESs is that they function only very inefficiently (or even not at all) in cell-free translation systems, or in RNA transfection assays, or in DNA transfections that generate RNA in the cytoplasm via recombinant vaccinia virus encoding bacteriophage RNA polymerases (Iizuka et al. 1995; Stoneley et al. 2000). What seems to be needed for manifestation of IRES activity is a "nuclear experience." However, if we ignore the cynic's suggestion that the nuclear experience is the generation of a minor spliced monocistronic mRNA species that is responsible for the production of all protein encoded by what was the downstream cistron of the original bicistronic construct, there are no indications of what this nuclear event might be: whether it is modification of the RNA, for example, methylation of adenine residues, or association with an RNA-binding protein that is normally located primarily in the nucleus. For further progress in this area of cellular IRESs it would seem important to define the nature of the required nuclear experience.

In the meantime, there is very little that we can say about the mechanism of internal initiation promoted by cellular IRESs. Where exactly is the ribosome entry site? Is the activity of such IRESs dependent on all the canonical initiation factors (as is likely to be the case for hepatitis A virus IRES), or is eIF4E redundant with the consequence that IRES activity can be supported by cleaved eIF4G or just the central domain of eIF4G? In the latter respect, it is thought that the BiP (immunoglobulin heavy-chain-binding protein) IRES functions independently of eIF4E and the amino-terminal part of eIF4G, since BiP mRNA translation persists for a long time following infection of cells by poliovirus (Sarnow 1989; Macejak and Sarnow 1991). The cell-cycle-dependent IRES in ornithine decarboxylase mRNA also functions independently of eIF4E (Pyronnet et al. 2000), but it is not clear whether this is a property shared by all cellular IRESs. It is also worth noting that the criterion of persistence of translation following poliovirus infection would score some mRNAs, for example, adenovirus late mRNAs, as translated by an IRES-dependent mechanism (Castrillo and Carrasco 1987; Dolph et al. 1988), when this is not in fact the case: Adenovirus late mRNAs are translated by a scanning/shunting mechanism, as explained below.

CAP-INDEPENDENT INITIATION VIA THE SCANNING MECHANISM

For almost two decades, dating back long before the discovery of true internal initiation of translation as exemplified by the picornavirus IRESs, the translation characteristics of different eukaryotic cellular and viral mRNAs have been classified as "cap-dependent" or "cap-independent." As we have argued elsewhere (Jackson et al. 1995; Jackson 1996), this is a most inopportune classification, which we believe has hindered the development of ideas in the field, largely because it is seldom clear whether the term is being used as an operational criterion or as a mechanistic explanation; when it is used as a mechanistic interpretation, it is often not at all clear whether what is implied is true internal initiation.

There are only two valid tests for internal initiation: the dicistronic mRNA assay, or even better, the circularized RNA system of Chen and Sarnow (1995). Any mRNA that fails this dicistronic mRNA assay must be presumed to be translated by a 5′-end-dependent scanning mechanism, no matter how cap-independent the translation of the mRNA may appear to be according to the various operational criteria that have been applied: (1) a comparatively high resistance of translation to inhibition by cap analogs (m^7GTP or m^7GDP) or by antibodies against eIF4E; (2) a relatively small difference in translation efficiency between capped and uncapped versions of the same mRNA species; (3) a relatively high translational efficiency in cell-free extracts of poliovirus-infected cells; and (4) persistence of translation in vivo following the general shutoff of host-cell mRNA translation caused by poliovirus infection.

Among the capped mRNAs that best satisfy one or more of these operational criteria, the most closely studied have been alfalfa mosaic virus RNA 4 (AMV RNA 4), the mRNAs coding for the heat shock proteins, and the adenovirus late mRNAs transcribed from the major late promoter. However, as we have argued in detail previously (Jackson et al. 1995; Jackson 1996), a closer scrutiny of the data shows that the translation of these mRNAs is stimulated by capping and is dependent on intact eIF4F holoenzyme complex even though the concentration of eIF4F required is considerably lower than for typical capped mRNAs, reflecting the fact that these mRNAs have an unusually high affinity for eIF4F. Experiments with a fractionated wheat germ system have shown that the concentration of eIF4F required for half-maximal translation efficiency (the apparent K_m) differs widely between different mRNA species, and is particularly low in the case of AMV RNA 4 (Fletcher et al. 1990; Timmer et al. 1993).

Of more interest than hair-splitting arguments about the degree of apparent cap-independence of various capped mRNAs is the question of the mechanism of initiation site selection on uncapped versions of mRNAs that are normally capped. Since decapping mRNA decreases the efficiency of translation initiation but does not usually change the preference for the 5′-proximal AUG codon (Kozak 1989a), the implication is that such decapped mRNAs are also translated by a scanning mechanism, at least in vitro, and the translation characteristics of an uncapped mRNA generated in vivo from an engineered RNA polymerase III promoter are consistent with this supposition (Gunnery et al. 1997). It is also evident that at least part of the eIF4G component of eIF4F is necessary for initiation of translation of such mRNAs. In the fractionated wheat germ system, the apparent K_m for eIF4F was many fold higher for uncapped AMV RNA 4 than for the capped version, but if sufficient eIF4F was added, the uncapped form was translated as efficiently as the capped RNA (Fletcher et al. 1990; Timmer et al. 1993).

Unexpectedly, the efficiency of translation of uncapped mRNAs in mammalian cell-free systems is significantly stimulated either by cleaving the endogenous eIF4G with picornaviral proteases (Ohlmann et al. 1995, 1996) or by supplementing the system (which contains intact eIF4G) with the central domain of eIF4G (De Gregorio et al. 1998). Both of these manipulations also inhibit translation of capped mRNAs, but because it requires a higher concentration of protease or a higher input of eIF4G central domain to inhibit capped mRNA translation than to stimulate translation of uncapped, the stimulation cannot be just a secondary consequence of inhibition of translation of, say, the capped fragments of globin mRNA present in nuclease-treated reticulocyte lysates. The presumption is that uncapped mRNAs must have a higher affinity for the cleavage products of eIF4G than for intact eIF4G or eIF4F. It is therefore rather surprising that the translation of uncapped mRNAs is sensitive to inhibition by added 4E-BP1 (Ohlmann et al. 1996), which would sequester eIF4E (and therefore inhibit any competing translation of capped mRNA fragments). The explanation for this apparent paradox may be that eIF4G undergoes a large conformational change on binding eIF4E, as is witnessed by the fact that it cannot be cleaved by picornavirus proteases if eIF4E is not associated with it (Ohlmann et al. 1997). It appears that the conformation of intact eIF4G in the absence of bound eIF4E is unable to support initiation on uncapped mRNAs.

The central domain of eIF4G thus seems sufficient to promote initiation on uncapped mRNAs, yet such initiation appears to follow a 5′-end-dependent scanning mechanism. It has been suggested that the fidelity of initiation site selection on uncapped mRNAs may be somewhat lower than with a capped version of the same template, but there is only one situation of a serious breakdown in fidelity, namely the translation of uncapped but polyadenylated mRNAs in a yeast cell-free system (Preiss and Hentze 1998; Preiss et al. 1998). If uncapped mRNAs are usually translated by a mechanism of scanning from (near) the 5′ end, and if the 40S ribosomal subunit is delivered to the proximity of the 5′ end by the eIF4G–eIF3–40S subunit interaction relay, the implication is that the eIF4G central domain must interact with the uncapped mRNA at a site close to the 5′ end. We do not know the basis of this apparently specific interaction. The central domain of eIF4G has been suggested to have an RRM-like motif based on sequence inspection (Goyer et al. 1993; Morley et al. 1997), but there is no information as to whether this motif actually binds RNA, let alone whether it shows any preference for a free 5′end.

Cap-independent Initiation of Satellite Tobacco Necrosis Virus RNA Translation

The naturally uncapped satellite tobacco necrosis virus (STNV) RNA appears to be translated by yet another variant of the basic mechanism. The translation of this RNA in a fractionated wheat germ system exhibits a lower apparent K_m for eIF4F than any other mRNA species, and unlike other mRNAs (such as AMV RNA 4) the K_m is uninfluenced by capping (Fletcher et al. 1990; Timmer et al. 1993). This property seems to be conferred largely by a ~100-nucleotide 3′UTR segment located just downstream from the translation termination codon, and predicted to form an irregular stem-loop structure; and to a lesser extent by the short 29-nucleotide 5′UTR. Mutations and deletions in either element reduce the efficiency of translation of uncapped STNV RNA, but this decrease can be reversed by capping (Timmer et al. 1993).

In a yeast three-hybrid assay, the 3′UTR motif was found to bind eIF4E, and in UV-crosslinking assays with eIF4F and the 3′UTR element, only the eIF4E component is crosslinked (K.S. Browning, pers. comm.). This evidence suggests that the 3′UTR "translational enhancer" recruits eIF4F holoenzyme via a specific interaction with eIF4E. It is not clear whether the initiation factor complexes are subsequently transferred from the 3′UTR translational enhancer to the 5′ end, or whether eIF4F remains bound to the 3′UTR element, but the three-dimensional geometry is appropriate for the eIF4G component to deliver the initiating 40S ribosomal subunit at or near the 5′end of the mRNA. That eIF4E should be able to bind to an internal stem-loop structure is very surprising in view of previous evidence that it is highly specific for terminal methylated G residues. However, SELEX experiments have shown us that RNA can adopt structures with high affinity for a bewildering variety of ligands, and so the idea that an internal RNA sequence might be able to mimic an m^7G cap is not out of the question. Another possibility is that the 3′UTR translational enhancer interacts with the eIF4E at some site other than the m^7G binding pocket. What is striking is that the data on the apparent K_m for eIF4F imply that eIF4E (or, strictly speaking, the eIF4F holoenzyme complex) binds to the STNV translational enhancer with higher affinity than it binds to conventional cap structures on typical mRNAs.

It seems likely that a similar mechanism also operates with barley yellow dwarf virus RNA (Wang et al. 1997, 1999), which is an uncapped RNA with a 3′-proximal motif that promotes translation initiation. In addition, there may be some loose parallels with the mechanism of initiation promoted by the hepatitis A virus (HAV) IRES, which, as explained previously, is unique among picornavirus IRESs in requiring eIF4E and more than just the central domain of eIF4G. However, in the case of the HAV IRES, the postulated interaction between eIF4E and an internal segment of the IRES would cause eIF4G to deliver 40S subunits to a site downstream from the putative eIF4E-binding site, rather than upstream of it, as occurs with STNV RNA.

THE SCANNING RIBOSOME MECHANISM

Finally, we come to the scanning ribosome model as the explanation for recognition of the correct initiation site on the overwhelming majority of capped mRNAs (Kozak 1989a, 1999). In terms of initiation factor and other requirements this is more complex than anything discussed so far: It is the only situation that requires a 5′-cap structure, and, except for the HAV IRES, STNV RNA, and perhaps some cellular IRESs, the only one requiring eIF4E and the amino-terminal domain of eIF4G.

In essence, the scanning ribosome model proposes that the primed 40S subunit, with associated initiation factors and Met-$tRNA_i$, first binds to the mRNA close to the 5′-cap structure and then migrates in a 5′→3′ direction, selecting usually the first AUG codon as the initiation site. Mutation of this AUG generally results in initiation at the next AUG codon downstream, whereas insertion of a new AUG upstream of the original 5′-proximal one results in initiation at this new AUG codon. The efficiency of recognition of the 5′-proximal AUG triplet may be influenced by its local sequence context, and, in addition, a hairpin stem-loop motif inserted some 18 nucleotides downstream from the AUG can improve the efficiency with which it is recognized as an initiation codon, probably by slowing the passage of the scanning ribosome at the critical moment when it is centered on the AUG triplet (Kozak 1989a, 1999). Despite these minor caveats, the underlying 5′→3′ directionality in the mechanism of initiation site selection by the scanning mechanism is surely beyond dispute.

Initiation site recognition by the scanning 40S subunit seems to be achieved, at least in part, by base-pairing between the AUG and the anticodon of the Met-tRNA$_i$ in the ternary complex carried on the 40S subunit, since mutation of the anticodon can allow initiation to occur at a complementary non-AUG codon (Cigan et al. 1988). Mutations in any of the three subunits of eIF2 can also result in initiation (with wild-type Met-tRNA$_i$) at non-AUG codons (Donahue et al. 1988; Cigan et al. 1989; Dorris et al. 1995). Although this is sometimes taken as evidence for a direct recognition of the AUG initiation codon by eIF2 itself, this need not necessarily be the case. It could be that the mutations in eIF2 distort the presentation of the anticodon of the Met-tRNA$_i$ sufficiently to allow apparent mismatch pairing in the ribosomal P site. Although it was argued in an early section of this chapter that 40S subunits bereft of an eIF2/Met-tRNA$_i$/GTP complex (but perhaps associated with eIF3) are capable of scanning, there is no reason to suppose that they can recognize the initiation codon unless they have an associated ternary complex. Indeed, according to our current understanding of the regulation of GCN4 mRNA translation, such scanning 40S subunits lacking bound ternary complex do not stop, or even pause for a significant time, on encountering an AUG triplet (Chapter 5).

The underlying 5′→3′ directionality of the initiation site selection process is usually explained in terms of 5′→3′ scanning or migration of primed 40S subunits. However, as Sonenberg (1991, 1993) has pointed out, it could conceivably be due *not* to 40S subunit migration, but to a 5′→3′ directionality of the helicase action of the eIF4A in the eIF4F holoenzyme complex in unwinding the mRNA from the 5′end, coupled with a potential for the 40S subunits to enter at an internal site *provided* the entry site has been unwound by the helicase(s). Although this hypothesis is worth considering, I have argued elsewhere that the balance of the evidence favors a real 5′→3′ movement of the 40S subunits (Jackson 1996).

Nevertheless, direct evidence for such 40S subunit migration is hard to come by. Evidence in favor of actual subunit migration has been claimed from the effects of the antibiotic edeine in promoting the binding of several 40S subunits to the mRNA (Kozak and Shatkin 1978; Kozak 1979), the effect of ATP depletion in blocking a 40S subunit at or near the cap (Kozak 1980), and evidence for 40S subunits queuing on long 5′UTRs (Kozak 1989b, 1991b). However, the trapping of a 40S subunit near the very 5′ end of the mRNA when ATP is depleted is not decisive since the same result would probably be predicted by the model in which the 5′→3′ directionality is a consequence of helicase action directionality. As for queuing of 40S subunits, this has been seen remarkably infrequently, and proof that the additional 40S subunits in the supposed queue are actually located within the 5′UTR is rather meager. To locate a queuing 40S subunit within the 5′UTR it was necessary to have a stable hairpin in the UTR a little way upstream of the initiation codon (Kozak 1989b). It was argued that without this hairpin, the 40S subunit would have scanned off the mRNA fragments during the nuclease treatment step necessary to map the sites at which the queuing 40S subunit is located.

It is frustrating that there appears to be no way of "visualizing" a scanning ribosome in transit through the 5′UTR, for example by stalling it or slowing its movement by omitting specific initiation factors. It was recently reported that if eIF1 and eIF1A were omitted from a fractionated system composed of highly purified initiation factors, 40S subunits failed to reach the authentic initiation codon of globin mRNA but were stalled some distance upstream of it; and that if eIF1 and eIF1A were subsequently added, these complexes of 40S subunits associated with the wrong site on the mRNA disappeared, to be replaced by 40S subunits bound at the initiation codon (Pestova et al. 1998a). However, addition of competitor mRNA at the same time as the delayed addition of eIF1 and eIF1A showed that the stalled subunits were not the elusive intermediates caught in the act of scanning. Rather, the results indicated that the stalled complexes must first dissociate and a de novo attempt must be made to scan to the authentic initiation site. These results suggest two alternative functions, not necessarily mutually exclusive, for eIF1 plus eIF1A: (1) They are necessary for successful scanning to the initiation codon, and as such might be regarded as part of the scanning motor; (2) they dissociate abortive and illegitimate initiation complexes in which the 40S subunit has stalled at a non-AUG triplet that is incapable of acting as an initiation codon.

The second of these two properties echoes results obtained from a yeast genetic screen for mutants that would allow initiation at a non-AUG codon, namely UUG. One such class of mutations was in the *SUI1* gene (Yoon and Donahue 1992), which was subsequently found to encode yeast eIF1 (Kasperaitis et al. 1995; Naranda et al. 1996; Cui et al. 1998). Because mutations in this gene were permissive to initiation at a UUG codon (Yoon and Donahue 1992), the implication is that the wild-type factor would disallow such initiation events, possibly by dissociating 40S subunits that engaged the UUG codon. Curiously, the *SUI1* gene turns out to be the same as *MOF2* (*m*aintenance *o*f *f*rame): Mutations in *SUI1/MOF2* can result in an increased frequency of programmed –1 frameshifting (Cui et al. 1998). This leads to the somewhat surprising conclusion that Sui1p (i.e., eIF1) may monitor anticodon/codon interaction during elongation as well as during scanning and initiation. Even more surprising is that certain specific *mof2* alleles are defective in the nonsense-mediated decay of mRNA (Cui et al. 1998). Admittedly, there is no proof that the mutations in eIF1 have a direct (rather than indirect) effect on nonsense-mediated decay, but it is striking that the human eIF1 gene can substitute for all the functions of *SUI1/MOF2* (Cui et al. 1998).

Another curiosity is that yeast eIF1 exists not only as a singular entity but also as one of the non-core (non-stoichiometric) subunits of eIF3 (Naranda et al. 1996). In contrast, mammalian eIF1 is not considered to be a constituent of eIF3, although interaction with the eIF3 p110 subunit has been reported (Fletcher et al. 1999).

Role of eIF4G/4F in Initiation by the Scanning Mechanism

Translation of capped mRNAs by the scanning mechanism is inhibited by cleavage of eIF4G by picornavirus proteases, by eIF4E-binding proteins (4E-BP) that sequester eIF4E (Pause et al. 1994b), and by dominant negative eIF4A mutants (Pause et al. 1994a). As eIF4E is clearly required, it is not surprising that the central domain of eIF4G (amino acids 613–1090), which was sufficient to fulfill the eIF4G requirement for initiation on uncapped mRNAs (De Gregorio et al. 1998) or the EMCV IRES (Pestova et al. 1996b), cannot support translation of capped mRNAs or 40S initiation complex formation on natural globin mRNA (Morino et al. 2000). However, the whole of eIF4G is not required: Extension of this central fragment just 63 amino acid residues toward the amino terminus (i.e., to amino acid residue 550) to include the eIF4E interaction site was sufficient to give maximum activity in these assays. As for the carboxy-terminal one-third domain of eIF4G, a fragment that included this (amino acids 550–1560) gave 1.5- to 2-fold higher yield in both assays than one which lacked it (residues 550–1090), suggesting that it plays a minor modulatory role. The authors actually claim a more significant 4- to 5-fold effect, but this is on the basis of point mutations in the carboxy-terminal one-third domain, which appear to have a quasi-dominant negative effect and thus show lower activity than if the domain is simply deleted (Morino et al. 2000).

There is some controversy (for review, see Morley et al. 1997; Gingras et al. 1999) as to whether eIF4F enters into the process as a preformed holoenzyme complex or whether it is only assembled from its constituent polypeptides at the actual cap, with eIF4E binding to the cap, and eIF4G being brought in associated with primed 40S subunits (loaded with eIF3 and eIF2/Met-tRNA$_i$/GTP ternary complex). In fact, eIF4E and eIF4G have been shown to be associated with the mRNA under conditions when 40S subunits could not bind to it because of an inhibitory cap-proximal IRP/IRE interaction, an observation that favors the pathway of preformed eIF4F complex (Muckenthaler et al. 1998).

Regardless of this controversy, there is little doubt that what we end up with is eIF4G tethered in the vicinity of the cap by virtue of the eIF4G–eIF4E interaction and the binding of eIF4E to the cap. As a consequence, the helicase action of eIF4A associated with the central domain of eIF4G will be focused to the vicinity of the cap, and the interactions between the central domain of eIF4G and eIF3 (associated with the 40S subunit) could serve to deliver the small subunit to the mRNA, likewise in the vicinity of the cap (Fig. 4). The presumption is that we have a chain of interactions as follows: mRNA 5′-cap–eIF4E–eIF4G–eIF3–40S subunit–mRNA segment just downstream from the cap. Assuming that delivery is followed by scanning, the next question is: When are these interactions disrupted? In the case of the eIF3–40S subunit interaction, there is reasonably firm evidence that this is broken at the stage of subunit joining (Chapter 2), which means that the interaction is maintained until the 40S subunit engages the initiation codon. Of more interest and controversy is the question of when the eIF4G–eIF3 interaction is disrupted. Is there any reason this should be disrupted immediately after scanning starts? If not, does it get disrupted after scanning for 10 residues, or 20, or how many? One possibility is that the eIF4G–eIF3 interaction doesn't break until the 40S subunit engages the initiation codon and the eIF3–40S subunit interaction gets broken at the stage of subunit joining (Fig. 4). This model implies that normally an 80S initiation complex would need to form at the initiation codon before another 40S subunit can be loaded at the 5′ end of the mRNA, and thus it would explain why queuing of 40S subunits on long 5′UTRs has been seen so rarely. The fact that queuing has been reported, albeit not very often, might be construed as demolishing these ideas completely. However, that argument makes the naive assumption that all the interactions between the factors are infinitely tight, whereas in fact we are dealing with interactions of finite affinity that must be expected to break spontaneously from time to time.

Clearly, from a "selfish mRNA" perspective it is important that once a given mRNA has captured an eIF4G molecule it should retain an interaction with it, even if only an indirect interaction. Under the model proposed in Figure 4, there is no reason for the eIF4G to be lost, since even though the eIF4G–eIF3 and eIF3–40S subunit interactions must clearly

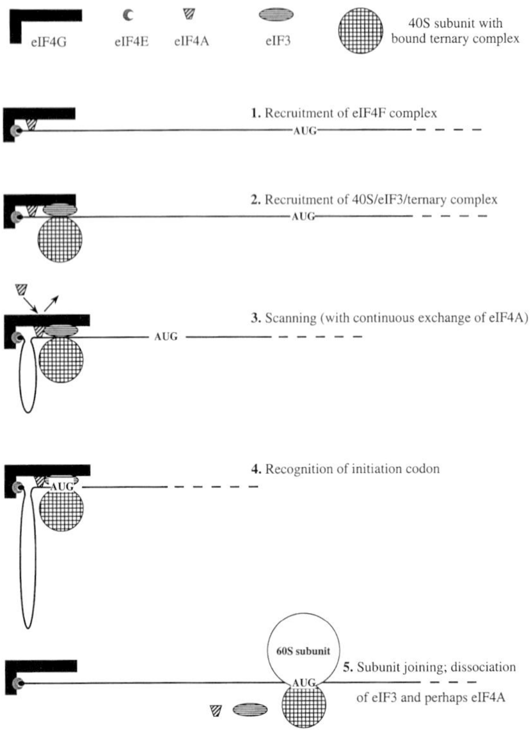

Figure 4 Speculative model for the mechanism of 40S ribosomal subunit scanning on capped mRNAs. The model posits that the eIF4G–eIF3 interaction is maintained at least until the stage of initiation codon recognition and is broken only at the subunit joining step. Although mammalian eIF4G has two sites for binding eIF4A (Imataka and Sonenberg 1997), for simplicity just a single site is shown, which is likely to be the position with yeast and plant eIF4Gs (Morley et al. 1997; Gingras et al. 1999).

be disrupted at some stage, the eIF4E–eIF4G and eIF4E–cap interactions are assumed to be relatively stable and long-lived. However, it may be that these assumptions are wrong, in which case the eIF4G would escape were it not for the interaction between eIF4G and poly(A)-binding protein (PABP) bound to the 3′ poly(A) tail (Imataka et al. 1998). Is this the real explanation for the PABP–eIF4G interaction: not so much to rechannel ribosomes back to the same mRNA via the closed loop, as has been often suggested (Chapter 10), but more to retain eIF4G in proximity to the mRNA? Although it may seem at first sight rather perverse to achieve this by tethering it to the 3′ end, the fact is that tethering it via a protein–RNA interaction in any other region except the 3′UTR would be counterproductive, given that protein–RNA interactions in the 5′UTR or the coding region would either be disrupted by the scanning or the elongating ribosome, or (if more stable) would act as a barrier to scanning or elongation.

Returning to the model shown in Figure 4, an interesting question is: What happens if the first ORF is a short one that is permissive to resumption of scanning, as is the case with the 5′-proximal ORF of yeast GCN4 mRNA (Chapter 5)? If the mRNA 5′-cap–eIF4E–eIF4G–eIF3–40S subunit–mRNA chain of interactions is maintained up until initiation at the 5′-proximal initiation codon, but is then broken, does this mean that the scanning that resumes after translation of the sORF is rather different in its nature and mechanism? Or is it possible that the interaction relay described above can be reformed if the ORF is short? It would be very interesting to know what is the steady-state distribution or packing of ribosomes along the 5′UTR of GCN4 mRNA, and what initiation factors are associated with the 40S subunits that resume scanning after the termination codon of sORF1.

The "Mechanics" of Scanning

Scanning is sometimes viewed as the threading of 40S subunits on the mRNA. However, a threading model, where the thread passes through the 40S subunit itself, implies that after termination of translation, the 40S subunit continues to migrate through the 3′-untranslated region and poly(A) tail (albeit in a state where it is not competent to reinitiate) until it reaches the extreme physical 3′ end of the mRNA. This seems inherently improbable, if only because of the extraordinary length of the 3′UTR of many eukaryotic mRNAs and because it would imply that the 40S subunit would have to displace all the various proteins known to interact with 3′UTR motifs as well as PABP bound to the poly(A) tail. There is in fact no evidence for 40S subunits moving through the 3′-untranslated region. However, this is yet another area where absence of evidence is not evidence of absence, as all attempts to test the idea have to make assumptions about the stability of 40S/3′UTR complexes to sucrose gradient centrifugation or to footprinting techniques, assumptions that may not be warranted.

Perhaps the best evidence against a threading mechanism where the thread-hole is a permanent and integral feature of the 40S subunits is the fact that ribosomes can translate a covalently closed circular RNA, provided that the RNA incorporates an IRES element (Chen and Sarnow 1995). This eliminates threading through a hole that is a permanent feature of the 40S subunit, but it does not eliminate threading through a type of temporary "clasp," composed of the 40S subunit and some initiation factor, perhaps eIF3 and/or eIF4G, particularly the central domain of eIF4G. When initiation occurred and eIF3 and eIF4G dissociated from the ribosome, the thread channel would disappear, perhaps to be replaced by the cleft between the 40S and 60S ribosomal subunits. However, even a model in which scanning involves the threading of the mRNA through a temporary clasp formed of the 40S subunit and possibly eIF3 and/or eIF4G seems incompatible with the phenomenon of ribosome shunting discussed later in this chapter.

Although perhaps not explicitly stated, an implicit feature of the models presented by Kozak would seem to be that ribosome scanning is a processive, systematic, and unidirectional linear migration or search for an appropriate initiation codon, with no significant off-rate. Direct evidence on these issues is still lacking, however. One longstanding problem is that we still do not know whether the ATP hydrolysis, which is unquestionably necessary for initiation on capped mRNAs (Kozak 1980; Jackson 1991), is entirely accounted for by the RNA-dependent ATPase activity of eIF4A, or whether there is a separate ATPase associated directly with 40S subunit movement per se, a type of ATP-dependent scanning "motor" distinct from eIF4A. However, the amino acid sequences of other initiation factors or ribosomal proteins have revealed no canonical ATP-binding site that might help identify this hypothetical ATP-driven scanning motor.

Since dATP can substitute for ATP in supporting the helicase activity of eIF4A and eIF4F (Rozen et al. 1990), one test of this question would be to see whether dATP can substitute completely for ATP in supporting the whole process of initiation (if precharged Met-RNA$_i$ were supplied to bypass any requirement specifically for ATP rather than dATP in the charging reaction). If dATP cannot substitute for ATP, the implication is that there is at least one other specifically ATP-dependent step in addition to the helicase function (although the converse result, where dATP can completely substitute for ATP, has the less satisfactory outcome that no firm conclusion can be drawn). It is known that dATP can substitute for ATP in supporting initiation on the EMCV IRES (Pestova et al. 1996b), but as initiation in this particular case is believed to involve eIF4A action (probably in the form of eIF4A associated with the central domain of eIF4G) but no ribosome scanning, it is hardly a surprising outcome—indeed, it would have been predicted. In addition to this uncertainty as to whether there is an ATP-dependent motor (distinct from eIF4A) driving the scanning process, we also have no idea as to the stoichiometry of ATP hydrolysis and whether this is directly related to the length of the 5′UTR. Clearly, the idea of a systematic unidirectional process would become more credible if there were a defined stoichiometry of distance scanned per ATP hydrolyzed.

An alternative view is that ribosome scanning is merely a random diffusion process, which from time to time may include some movement in the 3′→5′ direction, as well as in the conventional 5′→3′ direction. According to this interpretation, the overall 5′→3′ directionality of the scanning process would merely be a consequence of the fact that the only entry point for the scanning 40S subunit, the point at which the diffusion process starts, is at or very near the 5′ end of the mRNA. In fact, as discussed in the penultimate section of this chapter, there is good evidence that, following translation termination, prokaryotic 30S subunits can undergo limited bidirectional diffusion over a distance of up to ~40 nucleotides from the termination codon. In eukaryotic systems, any scanning following termination is usually considered to be invariably in the 5′→3′ direction, but there are a few, somewhat controversial, reports of limited scanning in the reverse direction, which could be construed as a random, and therefore bidirectional, diffusion.

However, there are some problems with the idea that 40S subunit scanning through the 5′UTR is a random diffusion process. In the first place, the scanning often has to cover distances very much greater than the 40 residues that appear to be the practical limit for scanning by prokaryotic 30S subunits, and so we would need to invoke some special features that allowed the eukaryotic 40S subunit to cover these greater distances. Second, a random diffusion process in which periodic movement in a 3′→5′ direction was as frequent and as extensive as in the 5′→3′ direction would lead to the expectation that the time required for a 40S subunit to traverse the 5′UTR from the cap to the initiation codon would be dependent on the square of the length of the 5′UTR. This implies that translation efficiency would decrease quite sharply with increasing length of 5′UTR, which does not appear to be the case (Kozak 1991b). Our own attempts to measure the lag time between addition of mRNA to a (prewarmed) cell-free translation system and the first initiation events suggested that the lag was more closely related to the length of the 5′UTR than to the square of the length (S. Grünert and R.J. Jackson, unpubl.).

One indication that scanning by eukaryotic ribosomes is not a highly systematic step-wise linear search is the failure of the scanning process to discriminate between two very closely spaced AUG codons. The best example is the NA/NB mRNA of the influenza B viruses, where initiation occurs slightly more frequently at the downstream rather than the upstream of the two AUG codons in the sequence:...AAA**AUG**AA-CA**AUG**CUA...(Williams and Lamb 1989), yet a simple interpretation of the effects of context would suggest that the upstream site should be favored. Of the various mutations and manipulations tested in an attempt to obtain the predicted preference for the 5′-proximal AUG codon, the most effective was, significantly, the introduction of additional sequences to increase the separation of the two AUG codons. Although far from conclusive, this result is more easily explained by a type of diffusion mechanism than by a systematic step-wise linear search. It has since been shown that the upstream of two closely spaced AUG codons will be selected exclusively in vitro provided it is surrounded by *every* known favorable context feature: a 5′UTR of sufficient length, a GCCACC sequence immediately upstream of the 5′-proximal AUG codon, a G at the +4 position, and a hairpin loop at the appropriate distance downstream (Kozak 1995). This result was interpreted as demonstrating that scanning is normally a systematic unidirectional search process (Kozak 1995), but because the dice were loaded so heavily in favor of utilization of the 5′-proximal site, this conclusion may not meet with universal agreement.

Another related and still unsolved, or indeed seldom posed, question, is whether there is a finite off-rate during scanning. Assuming that ribosomes "enter" at the 5′ end of the mRNA at a certain rate, and scan in a 5′→3′ direction, a model in which there is no off-rate requires that all those ribosomes which enter at the 5′-end must initiate translation at some point in the mRNA. Therefore, if the efficiency of the 5′-proximal initiation site is down-regulated, either by changing the context or by mutating it to a non-AUG codon, the decrease in initiation frequency at that site should be exactly matched by increased initiation at downstream sites, and the converse should happen when a weak 5′-proximal site is mutated to a strong one. These predictions certainly seem to be upheld quite well in experiments where the first and second initiation sites are in-frame and quite close together (Kozak 1989c, 1990, 1991a), and have been reported to hold over longer distances (Kozak 1998, 1999), although this seems to be not invariably true (Boeck et al. 1992).

The fact that omission of eIF1 and eIF1A resulted in stalling of 40S subunits near the 5′ cap in complexes that spontaneously dissociate over time shows that scanning 40S subunits can dissociate from the mRNA (Pestova et al. 1998a), albeit under special circumstances. Inasmuch as no interactions are of infinitely high affinity and stability, it seems far from improbable that even if eIF1 and eIF1A are present, they may occasionally dissociate from the scanning ribosome, and this would be expected to give rise to a finite off-rate.

Therefore, the provisional and tentative conclusions concerning the "mechanics" of the scanning process are: (1) It is likely to be predominantly a 5′→3′ process rather than a completely random linear diffusion event, although some limited movement in the 3′→5′ direction is not excluded; (2) it is unlikely to be a highly systematic nucleotide-by-nucleotide inspection of the sequence; and (3) it is probable that there is a finite off-rate so that not every ribosome that is loaded at the 5′-end necessarily accesses an initiation codon on that particular cycle.

Ribosome Shunting or Repositioning

Ribosome shunting is a form of discontinuous scanning. The 40S subunits appear to scan through the 5′-proximal part of the mRNA in the usual way, but then skip the rest of the 5′UTR, landing most probably directly at the initiation codon, or certainly very close to it. At any one time, some ribosomal subunits may be engaged in strictly linear scanning throughout the whole 5′UTR while others may be shunting; the ratio of these two subpopulations of ribosomes may be affected by variables such as initiation factor concentrations. There are three well-documented examples: the 35S mRNAs of plant pararetroviruses such as cauliflower mosaic virus, the late adenovirus RNAs transcribed from the major late promoter, and the P/C mRNA of Sendai virus (and other paramyxoviruses). Certainly in the first two cases, shunting does not seem to be a random process, but rather it is triggered by a specific set of circumstances, although these are not quite the same in the two examples.

The Sendai P/C mRNA encodes many proteins, of which five concern us here, four in one reading frame, and one (the P protein) in another frame. Reading from the capped 5′ end, the order of the initiation sites is: C′ (ACG initiation codon), P (a poor context AUG), C, Y1, and Y2. The relative yield of the C′, P, and C proteins is as would be expected if their initiation sites were accessed by leaky linear scanning, but the Y1 and Y2 initiation sites appear to be accessed by shunting (La Torre et al. 1998). Both the Y1 and Y2 ORFs start with AUG codons, yet the yield of these proteins is unaffected by mutation to ACG. Since the synthesis of Y1 and Y2 is inhibited by cap analogs, or by insertion of a 5′-proximal stem-loop, or by poliovirus infection, the ribosomes that initiate at these sites must start by scanning from the 5′ end. However, it appears that after scanning the first ~30–50 nucleotides from the 5′ cap, they are then shunted or repositioned to the Y1 and Y2 initiation sites. No specific take-off (shunt donor) site could be found, but, on the other hand, there appear to be very specific landing (shunt acceptor) sites; remarkably, if AUG codons with a good context are introduced into the P-protein reading frame at sites around the Y1 and Y2 initiation sites, they are not functional as initiation codons, and their introduction has little effect on the yield of Y1 and Y2 (La Torre et al. 1998).

Plant pararetroviruses, of which cauliflower mosaic virus is the paradigm, express a 35S mRNA with a long (~600 nucleotides) 5′UTR that includes many short ORFs. Structure probing and phylogenetic comparisons suggest that, apart from the 5′-proximal ~80 nucleotides and the region around the authentic initiation codon, most of this 5′UTR folds into a complex irregular stem-loop. The first (5′-proximal) short ORF (sORF-A) lies entirely within the ~80-nucleotide relatively unstructured segment, terminating close to the base of the stem. Study of numerous mutant RNAs in wheat germ or reticulocyte lysate translation systems, and forced evolution of mutants in transfected plant cells, indicate that the critical features for the shunt are sORF-A, the base-paired stem, and the distance between the sORF-A termination codon and the stem (Poogin et al. 1998; Hemmings-Mieszczak and Hohn 1999; Ryabova and Hohn 2000). Shunting efficiency is maintained if the naturally occurring stem-loop is exactly replaced by a shorter (–45 kcal/mole) perfectly base-paired synthetic stem-loop structure (Hemmings-Mieszczak and Hohn 1999). There is evidence that ribosomes translate sORF-A efficiently, and, given the proximity of the termination codon to the stem, this translation will partly melt the base of the stem. This is thought to be the trigger for the shunt, which then bypasses the rest of the stem structure and results in the ribosomes landing close to the authentic initiation codon. Moving the position of sORF-A relative to the base of the stem has some influence on the efficiency of the shunt, but a greater effect on exactly where the shunting ribosome appears to land downstream from the stem (Ryabova and Hohn 2000). The results of mutagenesis suggest that the shunt landing site (shunt acceptor) is determined more by its position than its actual sequence (Hemmings-Mieszczak and Hohn 1999). There are also indications that shunted ribosomes are more prone to initiation at non-AUG codons than is the case with linear scanning (Ryabova and Hohn 2000), and, indeed, in rice tungro bacilliform virus, the authentic initiation codon is actually an AUU codon, although not a very efficiently used one, since mutation to AUG gave ~15-fold higher expression (Fütterer et al. 1996). Thus, in comparison with the Sendai virus P/C mRNA shunt, the plant pararetroviruses seem to have a much more stringently defined shunt donor, but we do not know enough about the shunt acceptors, especially in the Sendai virus system, to be able to make meaningful comparisons.

The late adenovirus RNAs transcribed from the major late promoter all share the same ~220-nucleotide tripartite leader (5′UTR) in common. This does not have any upstream AUG codons, but the 3′ half of the 5′UTR is believed to include a number of stem-loops, whereas the cap-proximal 25 nucleotides is unstructured (Zhang et al. 1989). The 40S ribosomal subunits are believed to scan through this unstructured 5′-proximal region, and then some of them may bypass the stem-loops, landing at or very near the initiation codon (Chapter 32). Introduction of AUGs into the 5′-distal part of the 5′UTR has little influence on initiation at the correct site via shunting, but it does block access by strictly linear scanning, as would be expected. Even more remarkable is the fact that insertion of a very stable (–80 kcal/mole) stem-loop ~20 nucleotides upstream of the correct initiation codon had very little influence on the use of this site by shunting ribosomes, but again blocked access by scanning (Yueh and Schneider 1996). Using this stem-loop insertion experiment as a test for shunting, it was concluded that in uninfected cells, or in the early phase of adenovirus infection, slightly less than half the ribosomes translating an mRNA with the tripartite leader do so by scanning, and the rest access the initiation site by shunting. In contrast, late in adenovirus infection or under heat-shock conditions, when eIF4F activity is reduced, the proportion of shunting as opposed to scanning ribosomes rises to at least 80%, possibly up to 100% (Yueh and Schneider 1996). Since even those ribosomes that subsequently shunt are believed to negotiate the 5′-proximal part of the leader by linear scanning and presumably need eIF4F for this, albeit only in low concentrations, it is not immediately clear why the decision as to whether to continue scanning or to shunt should be influenced by the level of eIF4F activity. It also remains to be seen whether this shunting system too will function efficiently even if a non-AUG codon is substituted at the initiation site.

The tripartite leader has three regions that show complementarity to the base-paired stems of the last helix in 18S rRNA (see Fig. 1). Deletion of each of these regions individually had only a small influence on shunting efficiency, but paired deletion of the second and third regions (or deletion of all three) had a serious negative influence on shunting (but not scanning), as judged by the fact that insertion of the stable stem-loop reduced overall translation efficiency very severely (Yueh and Schneider 2000). Moreover, this mRNA with the deletions was virtually untranslatable in late adenovirus-infected cells.

A database search revealed single elements in the 5′UTRs of human hsp70 mRNA (but not *Drosophila* hsp70 mRNA) and c-fos mRNA with complementarity, albeit less well matched, to the same 18S rRNA element (Yueh and Schneider 2000). By the criterion of the effect of insertion of the stable stem-loop, the translation of human hsp70 mRNA in transfected HeLa cells at 37°C is 40% via shunting, and c-fos mRNA 20%. After heat shock at 44°C for 4 hours, which severely impairs eIF4F activity, the results indicated that strictly linear scanning was completely inhibited, but shunting efficiency was about the same as at 37°C.

It is not at all clear how sequence motifs complementary to the stems of an 18S rRNA helix could promote shunting. The authors are appropriately cautious and point out that there is no evidence for any base-pairing between the 5′UTR motifs and 18S rRNA, a pairing that would require melting of the 18S rRNA stem. It is also true that these complementarities cannot be found in plant pararetrovirus 35S mRNA or Sendai P/C mRNA, but apparently the efficiency of shunting on these mRNAs is considerably lower than is seen with the adenovirus tripartite leader.

THE CONUNDRUM OF REINITIATION AFTER TERMINATION

Although it could be argued that almost every act of initiation is preceded by a termination event (except for newly assembled "virgin" ribosomes), what we are concerned with here are situations where a ribosome that has already translated an upstream cistron of a given mRNA molecule reinitiates translation at another site on the same mRNA molecule.

Reinitiation in Prokaryotes

As discussed in an early section of this chapter, in principle, 30S ribosomal subunits have independent access to each initiation site of a polycistronic mRNA, such that translation of upstream cistrons is not a prerequisite for translation of downstream cistrons. However, there are many examples of translational coupling in bacterial polycistronic mRNAs, where translation of a downstream cistron is absolutely dependent on translation of an upstream cistron (usually, but not necessarily, the immediate upstream cistron). Such coupling "relays" can extend over a very large number of cistrons. There are also cases where a nonsense mutation, a premature termination codon, provokes translation initiation at a nearby site, which would otherwise be silent as an initiation codon.

One explanation for translational coupling in polycistronic mRNAs is that the SD motif or initiation codon of the downstream cistron is buried in secondary structure unless or until translation of the upstream cistron unwinds such secondary structure. This is believed to be the case in at least some ribosomal protein operons (Yates et al. 1981; Nomura et al. 1984) and is also the likely basis whereby translation of the RNA polymerase cistron of the RNA bacteriophages is coupled to translation of the upstream coat protein cistron (Berkhout and van Duin 1985). In other cases, an alternative explanation seems to hold, as exemplified by the coupling of bacteriophage fd gene VII synthesis to translation of the upstream gene V cistron. In this case the gene VII initiation site has only a vestigial and very weak SD sequence, which seems to be recognized at very low efficiency unless ribosomes are "delivered" in close proximity as a result of previously translating the upstream gene V cistron (Ivey-Hoyle and Steege 1992).

In these cases of translational coupling where the termination codon of the upstream cistron lies very close to the initiation codon of the downstream cistron, it is pertinent to ask whether the same individual ribosome translates both cistrons. The results with the fd gene V/gene VII mRNA are most readily explained if there was such "readthrough." Although subunit dissociation is thought to be the ultimate fate of ribosomes following termination, it is not clear whether this occurs instantaneously. Perhaps the 50S subunit is released immediately at termination, but the 30S subunit has the potential to remain bound transiently to the mRNA (Martin and Webster 1975), and to (re)initiate at the gene VII initiation site. Recent experiments studying the cycling of ribosomes between termination and the next initiation event have revealed that after the termination factors have promoted release of the nascent protein chain, ribosome release factor RF4 (RRF) and EF2 (EFG), together with GTP hydrolysis, are necessary to release the 50S ribosomal subunit, and IF3 to release the deacylated tRNA from the P site (or possibly the hybrid P/E site) of the 30S subunit, presumably leaving the 30S subunit associated, however briefly, with the mRNA (Janosi et al. 1996; Karimi et al. 1999).

Another feature of the prokaryotic system, which seems likely to be a close parallel of the translational coupling in polycistronic mRNAs, is the fact that termination codons and nonsense mutations can promote initiation at nearby sites that would otherwise be silent. Reinitiation in these circumstances can occur at AUG or non-AUG codons located up to 40 residues either upstream or downstream of the stop codon, with the small qualification that the efficiency of reinitiation tends to be lower at an upstream rather than a downstream site. This argues that not only do 30S subunits remain associated with the mRNA for a limited time following termination, but they are even capable of limited bidirectional random diffusion from the stop codon, selecting the nearest initiation codon even to the extent of preferring a nearby UUG to a more distant AUG (Adhin and van Duin 1990). However, this reinitiation consequent on random (bidirectional) diffusion may be rather inefficient: In the systems studied by Adhin and van Duin (1990), only ~5% of the terminating ribosomes reinitiated translation at the nearby site.

Reinitiation in Eukaryotes

In eukaryotic systems, reinitiation at downstream AUGs can occur if the 5′-proximal ORF is short, and this is explained by models in which the 40S subunits resume scanning from the termination codon of the upstream ORF and acquire an eIF2/Met-tRNA$_i$/GTP ternary complex in the course of this scanning (Chapter 5). However, it is not generally the case that introduction of a termination codon into the body of a long ORF promotes or allows initiation at a nearby AUG codon that otherwise would be silent, as is the case in prokaryotes. Nevertheless, there are a few reports to this effect, including claims for reinitiation at AUG codons located a short distance (up to 30–40 nucleotides) upstream of the termination codon (Peabody and Berg 1986; Peabody et al. 1986; Thomas and Capecchi 1986). It remains to be seen whether these are really unique exceptions or just the "tip of the iceberg" of a more widespread phenomenon.

Why should what is rather a common event in prokaryotic systems be so rare in eukaryotes? One possible explanation is that in the prokaryotic system interactions between the diffusing 30S subunit and fortuitous SD-like motifs may delay dissociation of the subunit from the mRNA and fix it briefly in a position where it can initiate at an AUG or related codon if it acquires initiator tRNA before it eventually dissociates from the mRNA. In contrast, in the eukaryotic system, even if 40S subunits undergo a similar limited diffusion following termination, there will be no rRNA/mRNA interactions to delay dissociation of the 40S subunit from the mRNA, and nothing to cause the diffusing 40S subunit to pause at an AUG or related codon for a sufficient length of time to acquire an eIF2/Met-tRNA$_i$/GTP ternary complex in order to initiate at that site.

An alternative explanation worth bearing in mind is that any differences between the two systems may lie in differences in the details of the termination process (Chapter 11) rather than initiation itself. For example, although an equivalent of the bacterial RF4 (ribosome release factor) has been found in eukaryotic organelles (Zhang and Spremulli 1998), there seems to be nothing in the yeast genome sequence or in EST databases that might be a cytoplasmic RF4: Either there is no RF4 in the cytoplasmic system, or, if there is, it is so different from the bacterial factor as to be unrecognizable by sequence alignments—which is not impossible given that the mitochondrial protein sequence is quite distant from the eubacterial (Zhang and Spremulli 1998). Thus, the exact mechanics of termination and ribosome release may differ quite considerably between eukaryotes and prokaryotes.

As already mentioned, a special case of reinitiation in eukaryotic systems is when the 5′-proximal ORF is short, as exemplified by yeast

GCN4 mRNA (Chapter 5). This seems to differ significantly from the limited bidirectional diffusion by prokaryotic 30S subunits (and perhaps eukaryotic 40S subunits) in several respects: (1) the 5′-proximal ORF must be short; (2) the subsequent migration of the 40S subunits is unidirectional, at least at the macroscopic level; (3) reinitiation efficiency can be much greater than the ~5% observed by Adhin and van Duin (1990); and (4) it operates over much longer distances than ~40 nucleotides. Thus, the resumed scanning in the translation of mRNAs such as that coding for GCN4 seems to have much more in common with de novo scanning from the 5′end of a capped mRNA than any random limited diffusion process. This suggests that when the 40S subunits resume scanning on GCN4 mRNA, their complement of bound translation initiation factors may be quite similar to when they are scanning through the 5′UTR. According to conventional wisdom, initiation factors that are associated with the 40S subunit while it scans through the 5′UTR will be all released on initiation of translation of the short 5′-proximal ORF, probably before synthesis of the first peptide bond. It may be, however, that this postulate is incorrect. Perhaps some of the factors are released more slowly, or even need to be actively displaced by the growing nascent protein chain. Thus, if the 5′-proximal ORF is short it may be that 40S subunits not only remain associated with the mRNA following termination, but also retain, or easily reacquire, a similar complement of translation initiation factors as they had when scanning from the 5′ end. Whether they retain or reacquire association with eIF3, and indirectly with eIF4G (via the intermediary of eIF3), is a matter of pure speculation: It is hard to envisage how this could happen, and yet this is what might be needed for quasi-unidirectional scanning to resume. Obviously the 40S subunits that resume scanning need to acquire an eIF2/Met-tRNA$_i$/GTP ternary complex in order to be able to (re)initiate at a downstream AUG. This is thought to be a relatively slow process, occurring with some delay after scanning has resumed, which explains why the efficiency of reinitiation increases with increasing distance (up to about 70 nucleotides) between the termination codon of the sORF and the downstream initiation site (Kozak 1987).

Resumption of scanning and reinitiation at a downstream AUG not only requires that the 5′-proximal ORF be short, but also that it can be influenced by the nucleotide sequence around and up to 12 nucleotides downstream from the termination codon of this sORF, as shown by mutagenesis analysis (Grant and Hinnebusch 1994). These studies did not reveal any obvious consensus sequence that is permissive to resumption of scanning; on the contrary, the default outcome seemed to be resumption of scanning, but certain sequences around the termination codon, notably GC-rich sequences, can prevent it. It is not known what actually happens in circumstances that are unfavorable to reinitiation. Do the 40S ribosomal subunits resume scanning the mRNA but remain incompetent to reinitiate, or do they detach from the mRNA at the termination codon of the short ORF? The latter seems intuitively the more plausible of the two alternatives.

CONCLUDING REMARKS

In conclusion, commitment to initiation involves pairing between the initiation codon and the anticodon of the initiator tRNA. Delivery of initiator tRNA to the small ribosomal subunit, and stabilization of its binding, are the responsibility of dedicated initiation factor proteins, while yet other initiation factors monitor the fidelity of the anticodon–codon pairing.

However, the more critical question is how the small ribosomal subunit is delivered to the vicinity of the correct initiation codon against the background of a sea of other AUG triplets present in the mRNA. There appear to be two distinct alternative routes by which this is achieved. Either the small ribosomal subunit can bind directly to the correct site by a mechanism that does not require initiation factor proteins or ATP hydrolysis, or, if the small subunit cannot bind directly, delivery has to be by the initiation factors, in particular by the central domain of eIF4G, via a mechanism that usually involves ATP hydrolysis. The first of these mechanisms is exemplified by the conventional model for initiation site selection in prokaryotic systems, and by initiation on HCV and pestivirus IRESs. Although these differ radically in the sense that one needs a simple linear Shine-Dalgarno primary sequence motif, and the other a complex three-dimensional RNA structure of some 300 nucleotides, in operational terms they are very similar. This type of ribosome entry does not appear to be followed by scanning, perhaps because ribosome/mRNA interactions that are stable enough to allow such direct binding in the absence of factors will prevent any significant ribosome movement. The presence of factors promotes a small accommodation or realignment of the small ribosomal subunit on the mRNA, presumably to place the initiation codon firmly in the P site. An intrinsic feature of this mechanism of initiation site selection is that it is exquisitely sensitive to inhibition by secondary structure (above a certain threshold) at the initiation site, probably because it involves no site-directed action of RNA helicases.

The other main mechanism of initiation site selection is small ribosome subunit delivery by initiation factors, specifically by the central domain of eIF4G and its associated eIF4A helicase. What is required is that the central domain of eIF4G be brought into proximity with the appropriate region of the mRNA, either directly or indirectly. Direct binding of eIF4G to the RNA seems to be the case with the EMCV IRES (and probably all other picornavirus IRESs except HAV) and may also occur in the translation of uncapped mRNAs by the scanning mechanism. In other examples, the association of the eIF4G with the mRNA seems to be mainly indirect: Rather than binding directly to the mRNA it is tethered to it via interaction with another protein, which itself binds directly to the RNA. Such tethering is nearly always achieved via eIF4E as the intermediary, which normally binds to the 5′-cap structure and thereby promotes translation of capped mRNA by the conventional scanning mechanism, or, in the exceptional cases of STNV RNA and perhaps the HAV IRES, the eIF4E may bind to an internal site. Finally, there is the "concept experiment" in which internal ribosome entry is promoted by the central domain of eIF4G tethered, as a fusion with an RNA-binding protein, to an internally located target site of the RNA-binding protein.

As for tethering eIF4G to the mRNA via its interaction with PABP (Chapter 10), this has the advantage of retaining eIF4G in proximity to the mRNA, and so could increase initiation rates on that particular mRNA, but if the eIF4G–PABP–poly(A)tail interaction relay were the only bridge between eIF4G and the mRNA, it is hard to see how a specific AUG could be selected as an initiation site. An additional, direct or indirect, site-specific interaction between the eIF4G and the mRNA would seem to be required in order to ensure initiation at a specific AUG. In support of this supposition, initiation of translation of an uncapped polyadenylated mRNA in a yeast cell-free system occurred at several different sites, whereas a capped polyadenylated mRNA, in which such site-specific interaction between eIF4G and the mRNA would occur via the intermediary of eIF4E, was translated mainly from the 5′-proximal AUG (Preiss and Hentze 1998; Preiss et al. 1998).

A general feature of initiation promoted by eIF4G is that it seems much less sensitive to RNA secondary structure than is the case with the eIF4G-independent mechanism of direct binding of the (salt-washed) small ribosomal subunit to the mRNA in the vicinity of the initiation codon. The likely reason for this is that there is the possibility of site-directed unwinding of RNA structure by the eIF4A associated with the central domain of eIF4G.

Another general feature seems to be that 40S ribosomal subunit delivery to the mRNA by the central domain of eIF4G is coupled with the propensity for scanning (unless the delivery happens to be directly at the AUG initiation codon). It is not clear whether this propensity for scanning is merely a consequence of the fact that the 40S subunit is not usually delivered directly to an AUG codon, or whether it is strictly dependent on some property of the initiation factors that execute the actual delivery.

ACKNOWLEDGMENTS

Although I alone bear responsibility for any outlandish misinterpretations, I thank present and recent past members of my group for their numerous inputs toward the development of these ideas: Ann Kaminski, Sarah Hunt, Iraj Ali, Emma Brown, Tuija Pöyry, Esther Lafuente, Simon Fletcher, Michael Howell, Andrew Borman, Stefan Grünert, Joanna Reynolds, Theo Ohlmann, Catherine Gibbs, Carola Lempke, Paul Crisell, and Annette Lasham. Likewise I acknowledge the valuable contributions of recent collaborators: Matthias Hentze, Bert Semler, Kathie Kean, Sue Milburn, John Hershey, Graham Belsham, and Tim Skern. Work from our own laboratory described herein was supported by grants from the Wellcome Trust.

REFERENCES

Adhin M.R. and van Duin J. 1990. Scanning model for translational reinitiation in eubacteria. *J. Mol. Biol.* **213:** 811–818.

Belsham G.J. 1992. Dual initiation sites of protein synthesis on foot-and-mouth disease virus RNA are selected following internal entry and scanning of ribosomes in vivo. *EMBO. J.* **11:** 1106–1110.

Berkhout B. and van Duin J. 1985. Mechanism of translational coupling between coat protein and replicase genes of RNA bacteriophage MS2. *Nucleic Acids Res.* **13:** 6955–6967.

Blasi U., O'Connor M., Squires C.L., and Dahlberg A.E. 1999. Misled by sequence complementarity: Does the DB - anti-DB interaction withstand scientific scrutiny? *Mol. Microbiol.* **33:** 439–441.

Blumberg B.M., Nakamoto T., and Kezdy F.J. 1979. Kinetics of initiation of bacterial protein synthesis. *Proc. Natl. Acad. Sci.* **76:** 251–255.

Boeck R., Curran J., Matsuoka Y., Compans R., and Kolakofsky D. 1992. The parainfluenza virus type 1 P/C gene uses a very efficient GUG codon to start its C′ protein. *J. Virol.* **66:** 1765–1768.

Boni I., Isaeva D., Musychenko M., and Tzareva N. 1991. Ribosome-messenger recognition: mRNA target sites for ribosomal protein S1. *Nucleic Acids Res.* **19:** 155–162.

Borman A.M. and Kean K.M. 1997. Intact eukaryotic initiation factor 4G is required for hepatitis A virus internal initiation of translation. *Virology* **237:** 129–136.

Borman A.M., Bailly J.-L., Girard M., and Kean K.M. 1995. Picornavirus internal ribosome entry segments: Comparison of translation efficiency and the requirements for optimal internal initiation of translation *in vitro*. *Nucleic Acids Res.* **23:** 3656–3663.

Borman A.M., Le Mercier P., Girard M., and Kean K.M. 1997. Comparison of picornaviral IRES-driven internal initiation in cultured cells of different origins. *Nucleic Acids. Res.* **25:** 925–932.

Buratti E., Tisminetzky S., Zotti M., and Baralle F.E. 1998. Functional analysis of the interaction between HCV 5′ UTR and putative subunits of eukaryotic initiation factor eIF3. *Nucleic Acids Res.* **26:** 3179–3187.

Calogero R.A., Pon C.L., Canonaco M.A., and Gualerzi C.O. 1988. Selection of the mRNA translation initiation region by *Escherichia coli* ribosomes. *Proc. Natl. Acad. Sci.* **85:** 6427–6431.

Cao X.M., Bergmann I.E., Fullkrug R., and Beck E. 1995. Functional analysis of the two alternative translation initiation sites of foot-and-mouth-disease virus. *J. Virol.* **69:** 560–563.

Castrillo J.L. and Carrasco L. 1987. Adenovirus late protein synthesis is resistant to the inhibition of translation induced by poliovirus. *J. Biol. Chem.* **262:** 7328–7334.

Chen C.-Y. and Sarnow P. 1995. Initiation of protein synthesis by the eukaryotic translational apparatus on circular RNAs. *Science* **268:** 415–417.

Chon S.K., Perez D.R., and Donnis R.O. 1998. Genetic analysis of the internal ribosome entry segment of bovine viral diarrhea virus. *Virology* **251:** 370–382.

Cigan A.M., Feng L., and Donahue T.F. 1988. $tRNA_i^{met}$ functions in directing the scanning ribosome to the start site of translation. *Science* **242:** 93–97.

Cigan A.M., Pabich E.K., Feng L., and Donahue T.F. 1989. Yeast translation initiation suppressor *sui2* encodes the α subunit of eukaryotic initiation factor 2 and shares sequence identity with the human α subunit. *Proc. Natl. Acad. Sci.* **86:** 2784–2788.

Cui Y., Dinman J.D., Kinzy T.G., and Peltz S.W. 1998. The Mof2/Sui1 protein is a general monitor of translational accuracy. *Mol. Cell. Biol.* **18:** 1506–1516.

Dasso M.C., Milburn S.C., Hershey J.W.B., and Jackson R.J. 1990. Selection of the 5′-proximal translation initiation site is influenced by mRNA and eIF-2 concentrations. *Eur. J. Biochem.* **187:** 361–371.

De Gregorio E., Preiss T., and Hentze M.W. 1998. Translational activation of uncapped mRNAs by the central part of human eIF4G is 5′ end dependent. *RNA* **4:** 828–836.

———. 1999. Translation driven by an eIF4G core domain *in vivo*. *EMBO J.* **18:** 4865–4874.

de la Cruz J., Iost I., Kressler D., and Linder P. 1997. The p20 and Ded1 proteins have antagonistic roles in eIF4E-dependent translation in *Saccharomyces cerevisiae*. *Proc. Natl. Acad. Sci.* **94:** 5201–5206.

de Smit M. and van Duin J. 1990a. Control of prokaryotic translational initiation by mRNA secondary structure. *Prog. Nucleic Acid Res. Mol. Biol.* **38:** 1–35.

———. 1990b. Secondary structure of ribosome binding site determines translational efficiency. *Proc. Natl. Acad. Sci.* **87:** 7668–7672.

———. 1994a. Translational initiation on structured messengers: Another role for the Shine-Dalgarno interaction. *J. Mol. Biol.* **235:** 173–184.

———. 1994b. Control of translation by messenger RNA secondary structure in *Escherichia coli:* A quantitative analysis of literature data. *J. Mol. Biol.* **244:** 144–150.

Dolph P.J., Racaniello V., Villamarin A., Palladoni F., and Schneider R.J. 1988. The adenovirus tripartite leader eliminates the requirement for cap binding protein during translation initiation. *J. Virol.* **62:** 2059–2066.

Donahue T.F., Cigan A.M., Pabich E.K., and Castilho-Valavicius B. 1988. Mutations at a Zn(II) finger motif in the yeast eIF2β gene alter ribosomal start-site selection during the scanning process. *Cell* **54:** 621–632.

Dorris D.R., Erickson F.L., and Hannig E.M. 1995. Mutations in *GCD11*, the structural gene for eIF2γ in yeast, alter translational regulation of *GCN4* and the selection of the start site for protein synthesis. *EMBO J.* **14:** 2239–2249.

Dreyfus M. 1988. What constitutes the signal for initiation of protein synthesis on *Escherichia coli* mRNAs? *J. Mol. Biol.* **204:** 79–94.

Fletcher C.M., Pestova T.V., Hellen C.U.T., and Wagner G. 1999. Structure and interactions of the translation initiation factor eIF1. *EMBO J.* **18:** 2631–2637.

Fletcher L., Corbin S.D., Browning K.S., and Ravel J.M. 1990. The absence of a m^7G cap on β-globin mRNA and alfalfa mosaic virus RNA 4 increases the amounts of initiation factor 4F required for translation. *J. Biol. Chem.* **265:** 19582–19587.

Fukushi S., Kurihara C., Ishiyama N., Hoshino F.B., Oya A., and Katayama K. 1997. The sequence element of the internal ribosome entry site and a 25-kilodalton cellular protein contribute to efficient internal initiation of hepatitis C virus RNA. *J. Virol.* **71:** 1662–1666.

Fütterer J., Potrykus I., Bao Y., Li L., Burns T.M., Hull R., and Hohn T. 1996. Position-dependent ATT initiation during plant pararetrovirus rice tungro bacilliform virus translation. *J. Virol.* **70:** 2999–3010.

Gallie D.R. and Kado C.I. 1989. A translational enhancer derived from tobacco mosaic virus is functionally equivalent to a Shine-Dalgarno sequence. *Proc. Natl. Acad. Sci.* **86:** 129–132.

Gingras A.-C., Raught B., and Sonenberg N. 1999. eIF4 initiation factors: Effectors of mRNA recruitment to ribosomes and regulators of translation. *Annu. Rev. Biochem.* **68:** 913–963.

Gold L. 1988. Post-transcriptional regulatory mechanisms in *Escherichia coli*. *Annu. Rev. Biochem.* **57:** 199–233.

Gold L., Pribnow D., Schneider T., Shinedling S., Singer B.S., and Stormo G. 1981. Translational initiation in prokaryotes. *Annu. Rev. Microbiol.* **35:** 365–403.

Goyer C., Altmann M., Lee H.S., Blanc A., Deshmukh M., Woolford J.L., Trachsel H., and Sonenberg N. 1993. *TIF4631* and *TIF4632:* Two yeast genes encoding the high molecular weight subunits of cap-binding protein complex (eukaryotic initiation factor 4F) contain an RNA recognition motif-like sequence and carry out an essential function. *Mol. Cell. Biol.* **13:** 4860–4874.

Gradi A., Imataka H., Svitkin Y.V., Rom E., Raught B., Morino S., and Sonenberg N. 1998. A novel functional human eukaryotic translation initiation factor eIF4G. *Mol. Cell. Biol.* **18:** 334–342.

Graff J. and Ehrenfeld E. 1998. Coding sequences enhance internal initiation of translation by hepatitis A virus RNA in vitro. *J. Virol.* **72:** 3571–3577.

Grant C.M. and Hinnebusch A.G. 1994. Effect of sequence context at stop codons on efficiency of reinitiation in GCN4 translational control. *Mol. Cell. Biol.* **14:** 606–618.

Gualerzi C.O. and Pon C.L. 1990. Initiation of mRNA translation in prokaryotes. *Biochemistry* **29:** 5881–5889.

Gualerzi C.O., Risuleo G., and Pon C.L. 1977. Initial rate kinetic analysis of the mechanism of initiation complex formation and the role of initiation factor IF-3. *Biochemistry* **16:** 1684–1689.

Gunnery S., Mälvali Ü., and Mathews M.B. 1997. Translation of an uncapped mRNA involves scanning. *J. Biol. Chem.* **272:** 21642–21646.

Gutell R.R., Weiser B., Woese C.R., and Noller H.F. 1985. Comparative anatomy of 16S-like ribosomal RNA. *Prog. Nucleic Acid Res. Mol. Biol.* **32:** 155–216.

Guthrie C. and Nomura M. 1968. Initiation of protein synthesis; a critical test of the 30S subunit model. *Nature* **219:** 232–235.

Hahm H., Kim Y.K., Kim J.H., Kim T.Y., and Jang S.K. 1998. Heterogeneous nuclear ribonucleoprotein L interacts with the 3′ border of the internal ribosome entry site of hepatitis C virus. *J. Virol.* **72:** 8782–8788.

Hartz D., McPheeters D.S., and Gold L. 1989. Selection of the initiator tRNA by *Escherichia coli* initiation factors. *Genes Dev.* **3:** 1899–1912.

Hartz D., McPheeters D.S., Green L., and Gold L. 1991. Detection of *Escherichia coli* ribosome binding at translation initiation sites in the absence of tRNA. *J. Mol. Biol.* **218:** 99–105.

Hemmings-Mieszczak M. and Hohn T. 1999. A stable hairpin preceded by a short ORF promotes non-linear ribosome migration on a synthetic mRNA leader. *RNA* **5:** 1149–1157.

Honda M., Brown E.A., and Lemon S.M. 1996. Stability of a stem-loop involving the initiator AUG controls the efficiency of internal initiation on hepatitis C virus RNA. *RNA* **2:** 955–968.

Honda M., Beard M.R., Ping L.-H., and Lemon S.M. 1999. A phylogenetically conserved stem-loop structure at the 5′ border of the internal ribosome entry site of hepatitis C virus is required for cap-independent viral translation. *J. Virol.* **73:** 1165–1174.

Hui A. and de Boer H.A. 1987. Specialized ribosome system: Preferential translation of a single mRNA species by a subpopulation of mutated ribosomes in *Escherichia coli*. *Proc. Natl. Acad. Sci.* **84:** 4762–4766.

Howard G.A., Adamson S.D., and Herbert E. 1970. Subunit recycling during translation in a reticulocyte cell-free system. *J. Biol. Chem.* **245:** 6237–6239.

Hwang L.H., Hsieh C.L., Yen A., Chung Y.L., and Chen D.S. 1998. Involvement of the 5′ proximal coding sequences of hepatitis C virus with internal initiation of viral translation. *Biochem. Biophys. Res. Commun.* **252:** 455–460.

Iizuka N., Chen C., Yang Q., Johannes G., and Sarnow P. 1995. Cap-independent translation and internal initiation of translation in eukaryotic cellular mRNA molecules. *Curr. Top. Microbiol. Immunol.* **203:** 155–177.

Imataka H. and Sonenberg N. 1997. Human eukaryotic translation initiation factor 4G (eIF4G) possesses two separate and independent binding sites for eIF4A. *Mol. Cell. Biol.* **17:** 6940–6947.

Imataka H., Gradi A., and Sonenberg N. 1998. A newly identified N-terminal amino acid sequence of human eIF4G binds poly(A)-binding protein and functions in poly(A)-dependent translation. *EMBO J.* **17:** 7480–7489.

Iost I., Dreyfus M., and Linder P. 1999. Ded1p, a DEAD-box protein required for translation initiation in *Saccharomyces cerevisiae*, is an RNA helicase. *J. Biol. Chem.* **274:** 17677–17683.

Ivey-Hoyle M. and Steege D.A. 1992. Mutational analysis of an inherently defective translation initiation site. *J. Mol. Biol.* **224:** 1039–1054.

Jackson R.J. 1991. The ATP requirement for initiation of eukaryotic translation varies according to the mRNA species. *Eur. J. Biochem.* **200:** 285–294.

———. 1996. A comparative view of initiation site selection mechanisms. In *Translational control* (J.W.B. Hershey et al.), pp. 71–112. Cold Spring Harbor Laboratory Press, Cold Spring Harbor, New York.

Jackson R.J. and Kaminski A. 1995. Internal initiation of translation in eukaryotes: The picornavirus paradigm and beyond. *RNA* **1:** 985–1000.

Jackson R.J., Hunt S.L., Reynolds J.E., and Kaminski A. 1995. Cap-dependent and cap-independent translation: Operational distinctions and mechanistic interpretations.

Curr. Top. Microbiol. Immunol. **203:** 1–29.

Janosi L., Hara H., Zhang S.J., and Kaji A. 1996. Ribosome recycling by ribosome recycling factor (RRF): An important but overlooked step of protein biosynthesis. *Adv. Biophys.* **32:** 121–201.

Jaramillo M., Dever T.E., Merrick W.C., and Sonenberg N. 1991. RNA unwinding in translation: Assembly of helicase complex intermediates comprising eukaryotic initiation factors eIF-4F and eIF-4B. *Mol. Cell. Biol.* **11:** 5992–5997.

Johnson K.N. and Christian P.D. 1998. The novel genome organization of the insect picornavirus like *Drosophila* C virus suggests that this virus belongs to a previously undescribed virus family. *J. Gen. Virol.* **79:** 191–203.

Kaminski A. and Jackson R.J. 1988. The polypyrimidine tract binding protein (PTB) requirement for internal initiation of translation of cardiovirus RNAs is conditional rather than absolute. *RNA* **4:** 626–638.

Kaminski A., Belsham G.J., and Jackson R.J. 1994. Translation of encephalomyocarditis virus RNA: Parameters influencing the selection of the internal initiation site. *EMBO. J.* **13:** 1673–1681.

Kaminski A., Howell M.T., and Jackson R.J. 1990. Initiation of encephalomyocarditis virus RNA translation: The authentic initiation site is not selected by a scanning mechanism. *EMBO J.* **9:** 3753–3759.

Karimi R., Pavlov M.Y., Buckingham R.H., and Ehrenberg M. 1999. Novel roles for classical factors at the interface between translation termination and initiation. *Mol. Cell* **3:** 601–609.

Kasperaitis M.A.M., Voorma H.O., and Thomas A.A.M. 1995. The amino acid sequence of eukaryotic initiation factor 1 and its similarity to yeast initiation factor sui-1. *FEBS Lett.* **365:** 49–50.

Kolupaeva V.G., Hellen C.U.T., and Shatsky I.N. 1996. Structural analysis of the interaction of the pyrimidine tract binding protein with the internal ribosome entry segment of encephalomyocarditis virus and foot-and-mouth disease virus. *RNA* **2:** 1100–1212.

Kolupaeva V.G., Pestova T.V., Hellen C.U.T., and Shatsky I.N. 1998. Translation eukaryotic initiation factor 4G recognizes a specific structural element within the internal ribosome entry site of encephalomyocarditis virus RNA. *J. Biol. Chem.* **273:** 18599–18604.

Kozak M. 1979. Migration of 40S ribosomal subunits on messenger RNA when initiation is perturbed by lowering magnesium or adding drugs. *J. Biol. Chem.* **254:** 4731–4735.

———. 1980. Role of ATP in binding and migration of 40S ribosomal subunits. *Cell* **22:** 459–467.

———. 1987. Effects of intercistronic length on the efficiency of reinitiation by eucaryotic ribosomes. *Mol. Cell. Biol.* **7:** 3438–3445.

———. 1989a. The scanning model for translation: An update. *J. Cell Biol.* **108:** 229–241.

———. 1989b. Circumstances and mechanism of inhibition of translation by secondary structure in eukaryotic mRNAs. *Mol. Cell. Biol.* **9:** 5134–5142.

———. 1989c. Context effects and inefficient initiation at non-AUG codons in eukaryotic cell-free translation systems. *Mol. Cell. Biol.* **9:** 5073–5080.

———. 1990. Downstream secondary structure facilitates recognition of initiator codons by eukaryotic ribosomes. *Proc. Natl. Acad. Sci.* **87:** 8301–8305.

———. 1991a. A short leader sequence impairs the fidelity of initiation by eukaryotic ribosomes in vitro. *Gene Expr.* **1:** 111–116.

———. 1991b. Effects of long 5′ leader sequences on initiation by eukaryotic ribosomes *in vitro*. *Gene Expr.* **1:** 117–125.

———. 1995. Adherence to the first-AUG rule when a second AUG codon follows closely upon the first. *Proc. Natl. Acad. Sci.* **92:** 2662–2666.

———. 1998. Primer extension analysis of eukaryotic ribosome-mRNA complexes. *Nucleic Acids Res.* **26:** 4853–4859.

———. 1999. Initiation of translation in prokaryotes and eukaryotes. *Gene* **234:** 187–208.

Kozak M. and Shatkin A.J. 1978. Migration of 40S ribosomal subunits in the presence of edeine. *J. Biol. Chem.* **253:** 6568–6577.

Lamphear B.J., Kirchweger R., Skern T., and Rhoads R.E. 1995. Mapping of functional domains in eukaryotic protein synthesis initiation factor 4G (eIF4G) with picornaviral proteases. *J. Biol. Chem.* **270:** 21975–21983.

La Torre P., Kolakofsky D., and Curran J. 1998. Sendai virus Y proteins are initiated by a ribosomal shunt. *Mol. Cell. Biol.* **18:** 5021–5031.

Lopez de Quinto S.L. and Martinez-Salas E. 1999. Involvement of the aphthovirus RNA region located between two functional AUGs in start codon selection. *Virology* **255:** 324–336.

Lu H.-H. and Wimmer E. 1996. Poliovirus chimeras replicating under the translational control of genetic elements of hepatitis C virus reveal unusual properties of the internal ribosomal entry site of hepatitis C virus. *Proc. Natl. Acad. Sci.* **93:** 1412–1417.

Macejak D.G. and Sarnow P. 1991. Internal initiation of translation mediated by the 5′ leader of a cellular mRNA. *Nature* **353:** 90–94.

Martin J. and Webster R.E. 1975. The in vitro translation of a termination signal by a single *Escherichia coli* ribosome: The fate of the subunits. *J. Biol. Chem.* **250:** 8132–8139.

McCarthy J.E.G. and Brimacombe R. 1994. Prokaryotic translation: The interactive pathway leading to initiation. *Trends Genet.* **10:** 402–407.

McCarthy J.E.G., Schaude H.U., and Sebald W. 1985. Translation initiation frequency of *atp* genes from *Escherichia coli:* Identification of an intercistronic spacer that enhances translation. *EMBO J.* **4:** 519–526.

McCarthy J.E.G., Sebald W., Gross G., and Lammers R. 1986. Enhancement of translational efficiency by the *Escherichia coli atpE* translational initiation region: Its fusion with two human genes. *Gene* **41:** 201–206.

Meyer K., Petersen A., Niepmann M., and Beck E. 1995. Interaction of eukaryotic initiation factor eIF4B with a picornavirus internal translation site. *J. Virol.* **69:** 2819–2824.

Moon J.S., Domier L.L., McCoppin N.K., D'Arcy C.J., and Jin H. 1998. Nucleotide sequence analysis shows that *Rhopalosiphum padi* virus is a member of a novel group of insect-infecting viruses. *Virology* **243:** 54–65.

Morino S., Imataka H., Svitkin Y.V., Pestova T.V., and Sonenberg N. 2000. Eukaryotic translation initiation factor 4E (eIF4E) binding site and the middle one-third of eIF4GI constitute the core domain for cap-dependent translation, and the C-terminal one-third functions as a modulatory region. *Mol. Cell. Biol.* **20:** 468–477.

Morley S.J., Curtis P.S., and Pain V.M. 1997. eIF4G: Translation's mystery factor begins to yield its secrets. *RNA* **3:** 1085–1104.

Muckenthaler M., Gray N.K., and Hentze M.W. 1998. IRP-1 binding to ferritin mRNA prevents the recruitment of the small ribosomal subunit by the cap-binding complex eIF4F. *Mol. Cell* **2:** 383–388.

Naranda T., MacMillan S.E., Donahue T.F., and Hershey J.W.B. 1996. SUI1/p16 is required for the activity of eukaryotic translation initiation factor 3 in *Saccharomyces cerevisiae*. *Mol. Cell. Biol.* **16:** 2307–2313.

Nomura M., Gourse R., and Baughman G. 1984. Regulation of the synthesis of ribosomes and ribosomal components. *Annu. Rev. Biochem.* **53:** 75–117.

O' Connor M., Asai T., Squires C.L., and Dahlberg A.E. 1999. Enhancement of translation by the downstream box does not involve base-pairing of mRNA with the penultimate stem sequence of 16S rRNA. *Proc. Natl. Acad. Sci.* **96:** 8973–8978.

Ohlmann T. and Jackson R.J. 1999. The properties of chimeric picornavirus IRESes show that discrimination between internal translation initiation sites is influenced by the identity of the IRES and not just the context of the AUG codon. *RNA* **5:** 764–778.

Ohlmann T., Rau M., Morley S.J., and Pain V.M. 1995. Proteolytic cleavage of initiation factor eIF-4γ in the reticulocyte lysate inhibits translation of capped mRNAs but enhances that of uncapped mRNAs. *Nucleic Acids Res.* **23:** 334–340.

Ohlmann T., Rau M., Pain V.M., and Morley S.J. 1996. The C-terminal domain of eukaryotic protein synthesis initiation factor (eIF) 4G is sufficient to support cap-independent translation in the absence of eIF4E. *EMBO J.* **15:** 1371–1382.

Ohlmann T., Pain V.M., Wood W., Rau M., and Morley S.J. 1997. The proteolytic cleavage of eukaryotic initiation factor (eIF) 4G is prevented by eIF4E binding protein (PHAS-I; 4E-BP1) in reticulocyte lysate. *EMBO J.* **16:** 844–855.

Olins P.O. and Rangwala S.H. 1989. A novel sequence derived from bacteriophage T7 mRNA acts as an enhancer of translation of the *lacZ* gene of *Escherichia coli*. *J. Biol. Chem.* **264:** 16973–16976.

Olins P.O., Devine C.S., Rangwala S.H., and Kavka K.S. 1988. The T7 phage gene 10 leader RNA, a ribosome-binding site that dramatically enhances the expression of foreign genes in *Escherichia coli*. *Gene* **173:** 227–235.

Pause A.N., Méthot N., Svitkin Y., Merrick W.C., and Sonenberg N. 1994a. Dominant negative mutants of mammalian translation initiation factor eIF4A define a critical role for eIF4F in cap-dependent and cap-independent initiation of translation. *EMBO J.* **13:** 1205–1215.

Pause A., Belsham G.J., Gingras A.-C., Donze O., Lin T.A., Lawrence J.C., and Sonenberg N. 1994b. Insulin-dependent stimulation of protein synthesis by phosphorylation of a regulator of 5′ cap function. *Nature* **371:** 762–767.

Peabody D.S. and Berg P. 1986. Termination-reinitiation occurs in the translation of mammalian cell mRNAs. *Mol. Cell. Biol.* **6:** 2695–2703.

Peabody D.S., Subramani S., and Berg P. 1986. Effect of upstream reading frames on translation efficiency in simian virus 40 recombinants. *Mol. Cell. Biol.* **6:** 2704–2711.

Pestova T.V. and Hellen C.U.T. 1999. Internal initiation of translation of bovine viral diarrhea virus RNA. *Virology* **258:** 249–256.

Pestova T.V., Borukhov S.I., and Hellen C.U.T. 1998a. Eukaryotic ribosomes require initiation factors 1 and 1A to locate initiation codons. *Nature* **394:** 854–859.

Pestova T.V., Hellen C.U.T., and Shatsky I.N. 1996a. Canonical eukaryotic initiation factors determine translation by internal ribosome entry. *Mol. Cell. Biol.* **16:** 6859–6869.

Pestova T.V., Hellen C.U.T., and Wimmer E. 1994. A conserved AUG triplet in the 5′ nontranslated region of poliovirus can function as an initiation codon *in vitro* and *in vivo*. *Virology* **204:** 729–737.

Pestova T.V., Shatsky I.N., and Hellen C.U.T. 1996b. Functional dissection of eukaryotic initiation factor eIF4F: The 4A subunit and the central domain of the 4G subunit are sufficient to mediate internal entry of 43S preinitiation complexes. *Mol. Cell. Biol.* **16:**

6870–6878.

Pestova T.V., Shatsky I.N., Fletcher S.P., Jackson R.J., and Hellen C.U.T. 1998b. A prokaryotic-like mode of cytoplasmic eukaryotic ribosome binding to the initiation codon during internal translation initiation of hepatitis C and classical swine fever virus RNAs. *Genes Dev.* **12:** 67–83.

Pilipenko E.V., Gmyl A.P., Maslova S.V., Svitkin Y.V., Sinyakov A.N., and Agol V.I. 1992. Prokaryotic-like cis elements in the cap-independent internal initiation of translation on picornavirus RNA. *Cell* **68:** 119–131.

Poogin M.M., Hohn T., and Fütterer J. 1998. Forced evolution reveals the importance of short open reading frame A and secondary structure in the cauliflower mosaic virus 35S RNA leader. *J. Virol.* **72:** 4157–4169.

Pöyry T., Kinnunen L., and Hovi T. 1992. Genetic variation in vivo and proposed functional domains of the 5′ noncoding region of poliovirus RNA. *J. Virol.* **66:** 5313–5319.

Preiss T. and Hentze M.W. 1998. Dual function of the messenger RNA cap structure in poly(A)-tail-promoted translation in yeast. *Nature* **392:** 516–520.

Preiss T., Muckenthaler M., and Hentze M.W. 1998. Poly(A) tail promoted translation in yeast: Implications for translational control. *RNA* **4:** 1321–1331.

Pyronnet S., Pradayrol L., and Sonenberg N. 2000. A cell cycle-dependent internal ribosome entry site. *Mol. Cell* **5:** 607–616.

Reynolds J.E., Kaminski A., Carroll A.R., Clarke B.E., Rowlands D.J., and Jackson R.J. 1996. Internal initiation of translation of hepatitis C virus RNA: The ribosome entry site is at the authentic initiation codon. *RNA* **2:** 867–878.

Reynolds J.E., Kaminski A., Kettinen H.J., Grace K., Clarke B.E., Carroll A.R., Rowlands D.J., and Jackson R.J. 1995. Unique features of internal initiation of hepatitis C virus RNA translation. *EMBO J.* **14:** 6010–6020.

Rijnbrand R.C., Abbink T.E.M., Haasnoot P.C.J., Spaan W.J.M., and Bredenbeek P.J. 1996. The influence of AUG codons in the hepatitis C virus 5′ nontranslated region on translation and mapping of the translation initiation window. *Virology* **226:** 47–56.

Rijnbrand R., van der Straaten T., van Rijn P.A., Spaan W.J.M., and Bredenbeek P.J. 1997. Internal entry of ribosomes is directed by the 5′ noncoding region of classical swine fever virus and is dependent on the presence of an RNA pseudoknot upstream of the initiation codon. *J. Virol.* **71:** 451–457.

Rijnbrand R., Bredenbeek P., van der Straaten T., Whetter L., Inachauspe G., Lemon S., and Spaan W. 1995. Almost the entire 5′ non-translated region of hepatitis C virus is required for cap-independent translation. *FEBS Lett.* **365:** 115–119.

Ringquist S., Shinedling S., Barrick D., Green L., Binkley J., Stormo G.D., and Gold L. 1992. Translation initiation in *Escherichia coli*; sequences within the ribosome-binding site. *Mol. Microbiol.* **6:** 1219–1229.

Rozen F., Edery I., Meerovitch K., Dever T.E., Merrick W.C., and Sonenberg N. 1990. Bidirectional RNA helicase activity of eucaryotic initiation factors 4A and 4F. *Mol. Cell. Biol.* **10:** 1134–1144.

Ryabova L.A. and Hohn T. 2000. Ribosome shunting in the cauliflower mosaic virus 35S RNA leader is a special case of reinitiation of translation functioning in plant and animal systems. *Genes Dev.* **14:** 817–829.

Sarnow P. 1989. Translation of glucose regulated protein 78/immunoglobulin heavy chain binding protein mRNA is increased in poliovirus-infected cells at a time when cap-dependent translation of cellular mRNAs is inhibited. *Proc. Natl. Acad. Sci.* **86:** 5795–5799.

Sasaki J. and Nakashima N. 1999. Translation initiation at a CUU codon is mediated by the internal ribosome entry site of an insect picorna-like virus in vitro. *J. Virol.* **73:** 1219–1226.

———. 2000. Methionine-independent initiation of translation in the capsid protein of an insect RNA virus. *Proc. Natl. Acad. Sci.* **97:** 1512–1515.

Sasaki J., Nakashima N., Saito H., and Noda H. 1998. An insect picorna-like virus, *Plautia stali* intestine virus, has genes of capsid proteins in the 3′ part of the genome. *Virology* **244:** 50–58.

Shean C.S. and Gottesman M.E. 1992. Translation of the prophage λ cI transcript. *Cell* **70:** 513–522.

Sizova D.V., Kolupaeva V.G., Pestova T.V., Shatsky I.N., and Hellen C.U.T. 1998. Specific interaction of eukaryotic translation initiation factor 3 with the 5′ nontranslated regions of hepatitis C virus and classical swine fever virus RNAs. *J. Virol.* **72:** 4775–4782.

Sogin M.L., Gunderson J.H., Elwood H.J., Alonso R.A., and Peattie D.A. 1989. Phylogenetic significance of the kingdom concept: An unusual eukaryotic 16S-like ribosomal RNA from *Giardia lamblia*. *Science* **243:** 75–77.

Sonenberg N. 1991. Picornavirus RNA translation continues to surprise. *Trends Genet.* **7:** 105–106.

———. 1993. Remarks on the mechanism of ribosome binding to eukaryotic mRNAs. *Gene Expr.* **3:** 317–323.

Sprengart M.L., Fatscher H.P., and Fuchs E. 1990. The initiation of translation in *Escherichia coli:* Apparent base-pairing between the 16S rRNA and downstream sequences of the mRNA. *Nucleic Acids Res.* **18:** 1719–1723.

Steitz J.A. and Jakes K. 1975. How ribosomes select initiator regions in mRNA: Base-pair formation between the 3′ terminus of 16S rRNA and the mRNA during initiation of protein synthesis in *Escherichia coli*. *Proc. Natl. Acad. Sci.* **72:** 4734–4738.

Stoneley M., Subkhankulova T., Le Quesne J.P.C., Coldwell M.J., Jopling C.L., Belsham G.J., and Willis A.E. 2000. Analysis of the c-*myc* IRES; A potential role for cell-type specific *trans*-acting factors and the nuclear compartment. *Nucleic Acids Res.* **28:** 687–694.

Stormo G.D., Schneider T.D., and Gold L.M. 1982. Characterization of translational initiation sites in *Escherichia coli*. *Nucleic Acids Res.* **10:** 2971–2996.

Subramanian A.R. 1983. Structure and functions of ribosomal protein S1. *Prog. Nucleic Acid Res. Mol. Biol.* **28:** 101–142.

Tedin K., Resch A., and Blasi U. 1997. Requirements for ribosomal protein S1 for translation initiation of mRNAs with and without a 5′ leader sequence. *Mol. Microbiol.* **25:** 189–199.

Thanaraj T.A. and Pandit M.W. 1989. An additional ribosome-binding site on mRNA of highly expressed genes and a bifunctional site on the colicin fragment of 16 S rRNA from *Escherichia coli:* Important determinants of the efficiency of translation initiation. *Nucleic Acids Res.* **17:** 2973–2985.

Thomas J.O. and Szer W. 1982. RNA-helix-destabilising proteins. *Prog. Nucleic Acid Res. Mol. Biol.* **27:** 157–187.

Thomas K.R. and Capecchi M.R. 1986. Introduction of homologous DNA sequences into mammalian cells induces mutations in the cognate gene. *Nature* **324:** 34–38.

Timmer R.T., Benkowski L.A., Schodin D., Lax S.R., Metz A.M., Ravel J.M., and Browning K.S. 1993. The 5′ and 3′ untranslated regions of satellite tobacco necrosis virus RNA affect translational efficiency and dependence on 5′ cap structure. *J. Biol. Chem.* **268:** 9504–9510.

van Etten W.J. and Janssen G.R. 1998. An AUG initiation codon, but not codon-anticodon complementarity, is required for the translation of unleadered mRNA in *Escherichia coli. Mol. Microbiol.* **2:** 987–1001.

Wallaczek J., Albrecht-Ehrlich R., Stöffler G., and Stöffler-Meilicke M. 1990. Three dimensional localization of the NH_2- and carboxyl-terminal domain of ribosomal protein S1 on the surface of the 30S subunit from *Escherichia coli. J. Biol. Chem.* **265:** 11338–11344.

Wang C., Sarnow P., and Siddiqui A. 1993. Translation of human hepatitis C virus RNA in cultured cells is mediated by an internal ribosome-binding mechanism. *J. Virol.* **67:** 3338–3344.

Wang C. , Le S.Y., Ali N., and Siddiqui A. 1995. An RNA pseudoknot is an essential structural element of the internal ribosome entry site located within the hepatitis C virus 5′ noncoding region. *RNA* **1:** 526–537.

Wang S.P., Browning K.S., and Miller W.A. 1997. A viral sequence in the 3′-untranslated region mimics a 5′ cap in facilitating translation of uncapped mRNA. *EMBO J.* **16:** 4107–4116.

Wang S., Guo L., Allen E., and Miller W.A. 1999. A potential mechanism for selective control of cap-independent translation by a viral RNA sequence in *cis* and in *trans. RNA* **5:** 728–738.

Williams M.A. and Lamb R.A. 1989. Effect of mutations and deletions in a bicistronic mRNA on the synthesis of influenza B virus NB and NA glycoproteins. *J. Virol.* **63:** 28–35.

Wyckoff E.E., Hershey J.W.B., and Ehrenfeld E. 1990. Eukaryotic initiation factor III is required for poliovirus 2A protease induced cleavage of the p220 component of eukaryotic initiation factor 4F. *Proc. Natl. Acad. Sci.* **87:** 9529–9533.

Yates J.L., Dean D., Strycharz W.A., and Nomura M. 1981. *E. coli* ribosomal protein L10 inhibits translation of L10 and L7/L12 mRNAs by acting at a single site. *Nature* **294:** 190–192.

Yoon H. and Donahue T.F. 1992. The *sui1* suppressor locus in *Saccharomyces cerevisiae* encodes a translation factor that functions during $tRNA_i^{Met}$ recognition of the start codon. *Mol. Cell. Biol.* **12:** 248–260.

Yueh A. and Schneider R.J. 1996. Selective translation initiation by ribosome jumping in adenovirus-infected and heat-shocked cells. *Genes Dev.* **10:** 1557–1567.

———. 2000. Translation by ribosome shunting on adenovirus and HsP70 mRNAs facilitated by complementarity to 18S rRNA. *Genes Dev.* **14:** 414–421.

Zhang Y.L. and Spremulli L.L. 1998. Identification and cloning of human mitochondrial translational release factor 1 and the ribosome recycling factor. *Biochim. Biophys. Acta* **1443:** 245–250.

Zhang Y., Dolph P.J., and Schneider R.J. 1989. Secondary structure analysis of adenovirus tripartite leader. *J. Biol. Chem.* **264:** 10679–10684.

5

Mechanism and Regulation of Initiator Methionyl-tRNA Binding to Ribosomes

Alan G. Hinnebusch
Laboratory of Eukaryotic Gene Regulation
National Institute of Child Health and Human Development
Bethesda, Maryland 20892

In its simplest terms, the process of translation initiation in eukaryotic organisms consists of the binding of methionyl-initiator tRNA (Met-$tRNA_i^{Met}$) and mRNA to the 40S ribosomal subunit, pairing of the anticodon of Met-$tRNA_i^{Met}$ with the AUG start codon in mRNA, and joining of the 60S ribosomal subunit to form an 80S initiation complex. Each of these steps is stimulated by soluble protein factors known as eukaryotic initiation factors (eIFs). Reconstitution of this process in vitro using purified ribosomes and eIFs indicated that binding of Met-$tRNA_i^{Met}$ to the 40S subunit is a prerequisite for mRNA binding (Benne and Hershey 1978; Trachsel and Staehelin 1979). The Met-$tRNA_i^{Met}$ is transferred to the 40S subunit by a ternary complex consisting of Met-$tRNA_i^{Met}$, the heterotrimeric initiation factor 2 (eIF2), and GTP, and this reaction is stimulated by eIF3, eIF1A, and possibly eIF5B. The resulting 43S preinitiation complex binds mRNA, forming the 48S complex, in a reaction promoted by the mRNA-associated factors (eIF4E, eIF4G, eIF4A, eIF4B, and poly[A]-binding protein) and the eIF3 residing in the 43S complex. The preinitiation complex scans the mRNA, and pairing between the anticodon of Met-$tRNA_i^{Met}$ and the AUG start codon triggers hydrolysis of GTP by eIF2, dependent of the GTPase activating protein (GAP) eIF5. After release of eIF2–GDP and eIF3, the 60S subunit joins the assembly in a reaction stimulated by eIF5B and involves the hydrolysis of a second molecule of GTP (see Chapters 2 and 9).

The eIF2–GDP is inactive for binding Met-$tRNA_i^{Met}$ and must be converted to eIF2–GTP to regenerate the ternary complex. This recycling

© 2000 Cold Spring Harbor Laboratory Press 0-87969-568-4/00 $5 + .00

reaction is stimulated by the heteropentameric guanine nucleotide exchange factor (GEF) eIF2B and is a major target of translational control by a conserved mechanism involving phosphorylation of eIF2. The eIF2 phosphorylated on Ser-51 of its α subunit (eIF2[αP]) is functional for transferring Met-tRNA$_i^{Met}$ to the ribosome; however, the GDP-bound form of the protein is an inhibitor of eIF2B. As eIF2 generally occurs in excess of eIF2B, and phosphorylation of eIF2–GDP increases its affinity for eIF2B, the recycling of eIF2 can be substantially inhibited by phosphorylation of only a fraction of eIF2 (Jackson 1991; Proud 1992).

Four different eIF2α kinases regulated by different signals have been identified in mammalian cells: HRI (heme deprivation), PKR (double-stranded RNA produced in virus-infected cells), PERK (unfolded proteins in the endoplasmic reticulum), and GCN2 (serum starvation) (see Chapters 13, 14, and 15; Berlanga et al. 1999; Sood et al. 2000). GCN2 also exists in *Drosophila melanogaster* (Santoyo et al. 1997; Olsen et al. 1998), *Neurospora crassa* (Sattlegger et al. 1998), and *Saccharomyces cerevisiae* and, at least in the latter two organisms, it is activated by amino acid deprivation (Hinnebusch 1996). All of these kinases phosphorylate Ser-51 of eIF2α and thereby inhibit the recycling of eIF2–GDP to eIF2–GTP and ternary complex formation. Activation of the mammalian kinases leads to a high level of eIF2α phosphorylation that is sufficient to inhibit general translation initiation, which can be viewed as an adaptive response to the stressful conditions that trigger kinase activation. Interestingly, activation of GCN2 in amino acid-starved yeast cells elicits the specific translational induction of GCN4, a transcriptional activator of amino acid biosynthetic genes. Translational control of *GCN4* mRNA and the mechanism of GCN2 activation by uncharged tRNA in *Saccharomyces* are discussed below. Recent findings on the mammalian eIF2α kinases are reviewed separately in Chapters 13, 14, and 15. There are additional modes of regulating eIF2B activity independently of eIF2α phosphorylation, and these are reviewed in Chapter 16.

Many recent advances in our knowledge of the functions of eIF2, its GEF (eIF2B), and its GAP (eIF5) in recruitment of Met-tRNA$_i^{Met}$ and the recognition of AUG start codons have come from genetic analysis of translational control in yeast. Accordingly, these genetic systems are reviewed briefly before we consider the biochemical mechanism of ternary complex formation and its binding to 40S ribosomes. Because these processes seem to be highly conserved between yeast and mammals, findings from the two systems will be integrated to provide a unified picture of these crucial first steps in the translation initiation pathway.

GENETIC DISSECTION OF TRANSLATIONAL CONTROL BY eIF2α PHOSPHORYLATION AND THE MECHANISM OF START SITE SELECTION IN YEAST

Amino acid starvation of yeast cells leads to increased translation of *GCN4* mRNA, encoding a transcriptional activator of many amino acid biosynthetic genes. This response is strongly dependent on the protein kinase GCN2 (Hinnebusch 1992). Substitution of Ser-51 with alanine in yeast eIF2α abolishes its phosphorylation by GCN2 in vivo and in vitro, and impairs *GCN4* translation to the same extent as when *GCN2* is deleted (Dever et al. 1992). These findings provide the key evidence that GCN2 stimulates *GCN4* translation by phosphorylating eIF2α on Ser-51. General translation and cell growth are inhibited in yeast cells only when eIF2α is phosphorylated to higher levels than occurs when GCN2 is activated by amino acid limitation. Such high-level phosphorylation of eIF2α has been achieved by overexpressing mammalian PKR or HRI in *gcn2Δ* mutants (Chong et al. 1992; Dever et al. 1993), or in $GCN2^c$ mutants expressing constitutively activated forms of GCN2 (Wek et al. 1990; Dever et al. 1992; Ramirez et al. 1992; Diallinas and Thireos 1994). Thus, *GCN4* translation is induced by lower levels of eIF2α phosphorylation than are required for general inhibition of protein synthesis in yeast.

The specific induction of *GCN4* translation in response to eIF2α phosphorylation is mediated by four short open reading frames (uORFs) in the leader of *GCN4* mRNA, of which the first (uORF1) and fourth (uORF4) are sufficient for nearly wild-type translational control. According to the current model for *GCN4* translation (Hinnebusch 1996), essentially all ribosomes scanning from the 5′ cap translate uORF1, and half of these resume scanning as 40S subunits. Under nonstarvation conditions, virtually all of these reinitiating ribosomes rebind the ternary complex and reinitiate at uORF4, after which they dissociate from the mRNA. Under starvation conditions, phosphorylation of eIF2α by GCN2 inhibits eIF2B function and lowers the concentration of ternary complexes in the cell. Consequently, ~50% of the ribosomes scanning from uORF1 now reach uORF4 before rebinding the ternary complex and, lacking Met-$tRNA_i^{Met}$, bypass the uORF4 start codon. Most of these ribosomes rebind the ternary complex before reaching *GCN4* and reinitiate translation there. Thus, reducing ternary complex levels by phosphorylating eIF2α allows a fraction of reinitiating ribosomes to bypass the inhibitory uORF4 sequence and reinitiate at *GCN4* instead.

Because *GCN4* translational induction confers resistance to inhibitors of amino acid biosynthesis, growth on media containing these compounds provides a sensitive indicator of eIF2 recycling. This fact has been exploited for genetic dissection of the functions of individual eIF2 or eIF2B subunits in nucleotide exchange and its regulation by phosphorylation of eIF2. Mutations in eIF2γ (Hannig et al. 1992) and the β, γ, δ, and ε subunits of eIF2B (Hinnebusch 1996) were first isolated by their constitutive induction of *GCN4* translation (Gcd⁻ phenotype). These mutations also produce a slow-growth phenotype (Slg⁻) on rich medium, indicating nonlethal impairment of the essential functions of eIF2 or eIF2B in translation initiation. This latter conclusion was confirmed through biochemical analysis of protein synthesis in selected eIF2B or eIF2 mutants (Tzamarias et al. 1989; Cigan et al. 1991; Foiani et al. 1991; Dorris et al. 1995). Mutations in eIF2β and eIF2α elicit the same Gcd⁻ and Slg⁻ phenotypes (Williams et al. 1989) as does deleting two of the four *IMT* genes encoding $tRNA_i^{Met}$ (Dever et al. 1995). The derepression of *GCN4* conferred by the various Gcd⁻ mutations occurs independently of eIF2α phosphorylation by GCN2. The fact that mutations affecting $tRNA_i^{Met}$ or subunits of eIF2 or eIF2B constitutively derepress *GCN4* expression suggested that a reduction in ternary complex levels is sufficient to induce *GCN4* translation and is most likely the consequence of regulated eIF2α phosphorylation in *GCN2*⁺ cells. This conclusion was supported by the demonstration that overexpression of the eIF2 complex interfered with derepression of *GCN4* in starved wild-type cells (Gcn⁻ phenotype) and reversed the Slg⁻ phenotype conferred by high-level eIF2α phosphorylation in *GCN2*^{c} mutants (Dever et al. 1995).

Subsequent genetic analysis provided strong evidence that phosphorylation of eIF2α reduces ternary complex formation in yeast by inhibition of eIF2B. Regulatory mutations were obtained in the α, β, and δ subunits of eIF2B that overcome the derepression of *GCN4* translation in amino-acid-starved *GCN2*⁺ cells and reverse the slow-growth phenotype of *GCN2*^{c} mutants, the same effects observed when the eIF2 complex was overexpressed under these conditions. Accordingly, these mutations appear to make eIF2B insensitive to eIF2(αP) without decreasing its ability to recycle nonphosphorylated eIF2 (Hinnebusch 1996; Pavitt et al. 1997). The results of biochemical analysis of selected eIF2B mutants support this interpretation (Pavitt et al. 1998) (see below). Overexpressing the wild-type eIF2B complex also suppressed the Slg⁻ phenotype of *GCN2*^{c} mutants (Dever et al. 1995), consistent with the idea that eIF2B function is limiting for growth when eIF2α is phosphorylated at high levels. Together, these findings provided in vivo confirmation of the mechanism of translational control by eIF2α phosphorylation derived from biochemical studies of mammalian systems. They further established that *GCN4* translational induction is a unique response to inhibition of eIF2B function and ternary complex formation via eIF2α phosphorylation.

The genetic studies of Donahue and colleagues have provided another valuable entrée into the mechanism of translation initiation, this time at the level of start codon selection by the ternary complex. These workers constructed yeast strains where expression of the histidine biosynthetic gene *HIS4* requires translation initiation at a non-AUG start codon. By characterizing mutations that suppress the histidine requirement of a *his4* start codon mutant (Sui^- phenotype), they demonstrated that the base-pairing between the start codon and Met-$tRNA_i^{Met}$ plays a dominant role in directing the 40S subunit to the initiation site. The Sui^- selection also yielded mutations in all three subunits of eIF2, eIF5 (the GAP for eIF2), and eIF1, thus implicating these factors in stringent selection of the start codon. Biochemical analysis of Sui^- mutants has led to the notion that the intrinsic rate of GTP hydrolysis by eIF2, and its modulation by eIF5 and eIF1, are key determinants of AUG recognition during the scanning process. This work is reviewed more fully in Chapter 12 and is mentioned below where pertinent.

FUNCTIONS OF INITIATOR $tRNA^{Met}$ AND eIF2 SUBUNITS IN TERNARY COMPLEX FORMATION

The ternary complex can be formed in vitro with highly purified eIF2, GTP (or nonhydrolyzable GTP analogs), and charged $tRNA_i^{Met}$ (Proud 1992; Trachsel 1996). As discussed below, a substantial amount of eIF2 in extracts is found associated with eIF5 or eIF2B, raising the question of whether free ternary complex exists in the cell. Nevertheless, the ability to produce ternary complex in vitro from purified components has allowed extensive analysis of the structural features of $tRNA_i^{Met}$ and the subunits of eIF2 that are involved in the formation of this key intermediate, as described next.

Nucleotides in $tRNA_i^{Met}$ That Promote Initiator Function and Restrict Its Activity in Elongation

The eIF2 must discriminate between initiator and elongator forms of $tRNA^{Met}$, and eukaryotic cytoplasmic initiator tRNAs have several unique sequence and structural characteristics that distinguish them from elongator tRNAs. These include the A1:U72 base pair at the end of the acceptor stem and three consecutive G:C base pairs in the anticodon stem (G29:C41, G30:C40, G31:C39) (Fig. 1). Initiators also lack the TψC sequence in loop IV, containing A54 in place of T54 (of the TψC sequence), and contain A60 instead of pyrimidine-60 in this loop. Plant and fungal initiators additionally contain a phosphoribosyl group attached

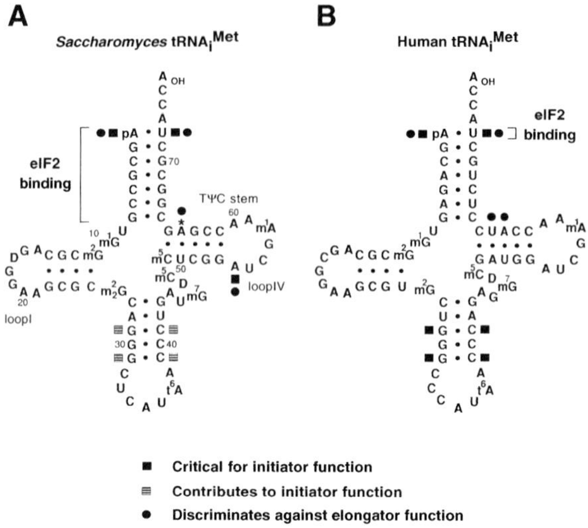

Figure 1 Bases in yeast and human initiator $tRNA^{Met}$ important for initiator function. The sequences of the tRNAs and identities of modified bases are found in Sprinzl et al. (1998). The asterisk at position 64 of yeast initiator designates the phosphoribosyl group attached to the ribose 2′-OH. See text for details. The numbering of bases shown for *Saccharomyces* initiator tRNA also applies to the human initiator tRNA. (Adapted from RajBhandary and Chow 1995.)

to the ribose 2′-OH at position 64 (RajBhandary and Chow 1995). The crystal structure of yeast $tRNA_i^{Met}$ reveals a unique substructure not present in elongator tRNAs formed by tertiary interactions involving residues A54, m^1A58, A59, and A60 in loop IV, and A20 in loop I, which strengthen the connection between these loops (Basavappa and Sigler 1991).

Mutational analysis of yeast $tRNA_i^{Met}$ established the critical importance of the A1:U72 base pair at the end of the acceptor stem for initiator function and cell viability (Fig. 1A) (von Pawel-Rammingen et al. 1992). Given that $tRNA_i^{Met}$ in the fission yeast *Schizosaccharomyces pombe* contains a ψ1:A72 base pair (Sprinzl et al. 1998) and that a U1:A72 substitution of A1:U72 was functional in *S. cerevisiae* yeast $tRNA_i^{Met}$ (Chapman et al. 1992), an A:U base pair seems to be the principal requirement at this position regardless of its orientation. Substitution of A54 in loop IV with U also was lethal, although C and G substitutions were permissible. As the mutant initiators containing G1:C72 and U54 could functionally replace elongator $tRNA^{Met}$ in a strain lacking all elongator $tRNA^{Met}$ genes, these residues seem to be required only for initiator-specific functions (Fig. 1A) (Åström et al. 1993).

Mutations of the conserved G29:C41 and G31:C39 base pairs in the anticodon stem did not detectably impair yeast $tRNA_i^{Met}$ function in vivo (von Pawel-Rammingen et al. 1992). However, the presence of G29:C41 and G31:C39 enhanced the ability of a mutated elongator $tRNA^{Met}$ (also bearing A54 and A1:U72) to substitute for the initiator (Åström et al. 1993) and thus may enhance initiator function (Fig. 1A). Overexpressing the α and β subunits of eIF2 together improved the ability of this multiply mutated elongator to function in initiation, suggesting that its affinity for eIF2 was reduced compared to bona fide $tRNA_i^{Met}$. In contrast, a mutant elongator $tRNA^{Met}$ bearing the entire acceptor stem of initiator tRNA (and no other changes) was a superior functional replacement for wild-type initiator and was insensitive to eIF2 subunit overexpression. Thus, the acceptor stem of $tRNA_i^{Met}$ contains residues important for initiator function beyond A1:U72 and is probably an important binding site for eIF2 (Fig. 1A).

Mutational analysis of human $tRNA_i^{Met}$ showed that the A1:U72 base pair in the acceptor stem was critical and that the G:C base pairs in the anticodon stem had a substantial role in initiator function in a mammalian cell-free translation system; however, a double mutation of A54, A60 to U54, U60 in loop IV had minimal effects in this in vitro assay (Fig. 1B) (Drabkin et al. 1993). Thus, there seem to be some differences between yeast and human initiator regarding sequence requirements in the anticodon stem and loop IV. The purified mutant human initiator bearing G1:C72 had a dissociation constant for eIF2–GTP more than tenfold higher than that of wild-type initiator. However, the ternary complexes formed with mutant or wild-type initiators were indistinguishable in subsequent steps of initiation. Accordingly, the characteristic A1:U72 base pair in the acceptor stem of human initiator $tRNA^{Met}$ is an important determinant for binding eIF2–GTP (Fig. 1B). The corresponding bases in *Escherichia coli* initiator cannot form a Watson-Crick base pair, and this feature contributes to its inability to form a ternary complex with EF1A/GTP (RajBhandary and Chow 1995); perhaps in a related manner the conserved A:U base pair in eukaryotic initiator reflects the need for a weak base pair at this location that can be disrupted in the course of binding eIF2–GTP (Basavappa and Sigler 1991).

In agreement with the findings from yeast (Åström et al. 1993), the A1:U72 base pair is a key feature of human initiator tRNA that discriminates against its activity in elongation, both in vitro and in vivo (Drabkin et al. 1998), in addition to promoting initiator function (Fig. 1A, B). Even more critical discriminating features are the A50:U64 and U51:A63 base pairs in the TψC stem, and mutating these residues together with A1:U72 conferred elongator function in vitro only slightly less than that of wild-type elongator tRNA^{Met}. As these residues vary among different elongator tRNAs, it was proposed that the base pairs in wild-type human initiator tRNA confer a structural perturbation of the TψC helix that blocks eEF1A binding. Interestingly, the initiator tRNAs in fungi and plants contain a unique 2′-O-phosphoribosyl modification of A64 in the TψC helix that prevents elongator function (Kiesewetter et al. 1990; Åström and Byström 1994) and impedes binding to eEF1A–GTP in vitro (Forster et al. 1993). Because the bulky 2′-O-phosphoribosyl group protrudes into the TψC stem (Basavappa and Sigler 1991), it may sterically hinder eEF1A binding. Thus, structural perturbation of the TψC stem may be a common strategy to block initiator binding to eEF1A in all eukaryotes (Drabkin et al. 1998). The only phenotype of inactivating the yeast enzyme responsible for 2′-O-phosphoribosyl modification of A64 (encoded by *RIT1*) is that elongator tRNA^{Met} is not essential for growth; thus, the modification is dispensable for initiator function and serves primarily to block its activity in elongation (Fig. 1A) (Åström and Byström 1994). The absence of *RIT1* is deleterious in strains harboring mutations in eIF2 subunits or lacking a full complement of *IMT* genes encoding $tRNA_i^{Met}$. Presumably, the reductions in ternary complex formation resulting from the latter mutations are intolerable in a *rit1*Δ mutant where unmodified initiator is being diverted into the elongation pathway (Åström et al. 1999).

In addition to the structural features of $tRNA_i^{Met}$, the attached methionyl group may also increase the efficiency of translation initiation. Yeast initiator tRNA charged with isoleucine was inactive because it bound poorly to eIF2 (Wagner et al. 1984). Mutant human initiator tRNA charged with glutamine initiated very poorly from CAG and UAG glutamine codons, whereas a different mutant charged with valine appeared to function well using the GUC valine codon (Drabkin and RajBhandary 1998). It is not known whether glutamine disturbs the structure of human $tRNA_i^{Met}$ or its interactions with eIF2, or instead, whether the particular codon–anticodon pairs formed between glutamine codons and the cognate mutant initiator are incompatible with efficient start-site selection.

eIF2γ Plays a Central Role in Binding Guanine Nucleotides and Initiator tRNA

Yeast eIF2 contains three subunits (α, β, and γ), encoded by the *SUI2, SUI3,* and *GCD11* genes, with molecular masses of 34.7, 31.6, and 57.9 kD, respectively. The molecular masses of mammalian eIF2 subunits are very similar to their yeast counterparts and there is strong sequence similarity between the yeast and mammalian proteins (see Chapter 2). The eIF2γ belongs to the superfamily of GTP-binding proteins and is most closely related to eEF1A and its eubacterial counterpart EF1A, which form ternary complexes with GTP and aminoacylated elongator tRNAs. The sequence similarities between eIF2γ and EF1A extend throughout the G domain, containing three consensus motifs present in GTP-binding proteins, and into domains II and III located carboxy-terminal to the G domain in both proteins (Hannig et al. 1992; Gaspar et al. 1994). The eIF2γ and EF1A proteins are especially related in the regions of EF1A involved in binding GTP and aminoacyl-tRNA, suggesting that eIF2γ can interact directly with GTP and Met-$tRNA_i^{Met}$ (Gaspar et al. 1994). Crosslinking and affinity-labeling experiments indicated that both the β and γ subunits of eIF2 are in close proximity to GTP and Met-$tRNA_i^{Met}$ in the ternary complex (Gaspar et al. 1994; Trachsel 1996); however, eIF2β (particularly from yeast) does not contain a convincing match to the three consensus motifs for GTP binding (Donahue et al. 1988; Pathak et al. 1988). Moreover, a two-subunit form of eIF2 lacking the β subunit could bind GDP but was unable to form a stable ternary complex with Met-$tRNA_i^{Met}$ (Flynn et al. 1993). Thus, it is likely that eIF2γ binds GTP directly, and that the β and γ subunits each make important contributions to binding Met-$tRNA_i^{Met}$ (Fig. 2).

Strong evidence that the γ subunit of yeast eIF2 mediates GTP and Met-$tRNA_i^{Met}$ binding came from biochemical analysis of point mutations in the G domain. The spontaneous Gcd^- allele *gcd11-Y142H* (Harashima and Hinnebusch 1986) contains histidine in place of Tyr-142 (Dorris et al. 1995), corresponding to a histidine residue in *Thermus aquaticus* EF1A that interacts with the phenylalanine moiety of Phe-$tRNA^{Phe}$. The *gcd11-Y142H* mutant has a Slg^- phenotype and decreased polysome content in addition to its Gcd^- phenotype, all suppressible by overproducing $tRNA_i^{Met}$, consistent with diminished ternary complex formation or 40S binding in this mutant. Consistently, purified eIF2 containing the *gcd11-Y142H* subunit had a reduced specific activity for binding Met-$tRNA_i^{Met}$, but normal off-rates for GDP and GTP. The engineered *gcd11-K250R* mutation, which alters the conserved lysine residue in the third consensus motif of the G domain (N<u>K</u>XD), increased the off-rate for GDP and GTP without affecting Met-$tRNA_i^{Met}$ binding by purified eIF2. With wild-type eIF2, the off-rate for GTP greatly exceeds that of GDP, and the eIF2–GTP complex can be stabilized by Met-$tRNA_i^{Met}$. Consistently, addition of Met-$tRNA_i^{Met}$ overcame the GTP-binding defect of the *gcd11-K250R* lesion in vitro and suppressed the Slg^- and Gcd^- phenotypes of this mutation in vivo

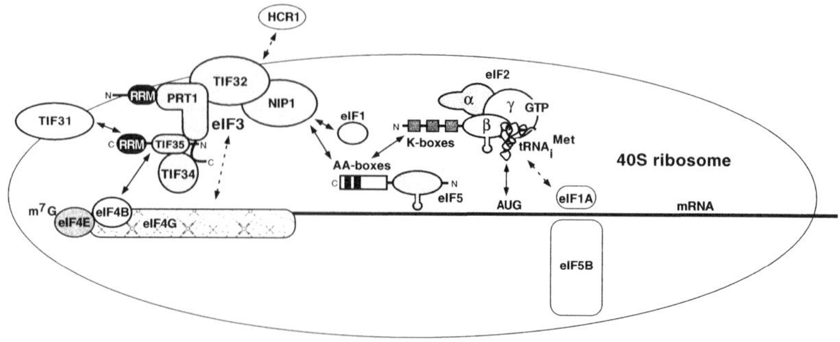

Figure 2. Protein–protein interactions among yeast initiation factors implicated in transferring methionyl-tRNA^{Met} to the 40S ribosomal subunit. Yeast eIF2 contains three subunits (α, β, and γ) with strong sequence similarities to their mammalian counterparts. It is likely that the γ subunit binds GTP directly and that the β and γ subunits each contribute to binding of Met-tRNA$_i^{Met}$.Yeast contains orthologs of the mammalian eIF3 subunits p170, p116, p110, p36, p44, and p35, known as TIF32, PRT1, NIP1, TIF34, TIF35, and HCR1, respectively. Five of these yeast proteins (excluding HCR1) appear to be stoichiometric subunits of a tight complex and were shown to interact directly with one another by two-hybrid or in vitro binding assays (Asano et al. 1998; Phan et al. 1998). These subunit interactions are depicted schematically as points of contact between the shapes representing the five core eIF3 subunits. PRT1 and TIF35 contain RNA recognition motifs (RRM). HCR1 either is a nonstoichiometric eIF3 subunit or is less tightly associated with the complex than are the five core subunits, and it was shown to interact genetically with TIF32 (Valasek et al. 1999). TIF31 has no ortholog in mammalian eIF3 but was physically linked to the eIF3 complex and found to interact directly with TIF35 (Vornlocher et al. 1999); TIF35 also interacted with the yeast eIF4B homolog (encoded by *TIF3*) (Vornlocher et al. 1999). Mammalian eIF4B acts in concert with components of eIF4F, including the cap-binding protein (eIF4E) and eIF4G shown here, to stimulate mRNA binding to the 40S ribosome (see Chapter 2). By analogy with mammalian systems, yeast eIF3 may stimulate binding of mRNA to the 40S subunit through a direct interaction with eIF4G (*dotted arrow*) in addition to its interaction with eIF4B. The amino-terminal portion of eIF5 (the eIF2 GAP) and the carboxy-terminal portion of eIF2β show sequence similarity, including a zinc-finger motif (shown as a prong). The carboxy-terminal domain of eIF5 contains a conserved bipartite motif (AA-boxes) that interacts with the amino-terminal domain of eIF2β containing three lysine-rich segments (K-boxes). The carboxy-terminal domain of eIF5 also interacts with the NIP1 subunit of eIF3, and NIP1 additionally interacts with eIF1. Although yeast eIF3 is required for ternary complex binding to the 40S subunit, no direct interactions between eIF3 and eIF2 have been detected. It is possible that eIF5 contributes to ternary complex binding by bridging an interaction between eIF3 and eIF2; alternatively, the interactions involving eIF5 may be required primarily for accurate start codon recognition. By analogy with mammalian systems, it is expected that yeast eIF1A also stimulates ternary complex binding to the 40S subunit (*dotted arrow*), and it may act in concert with eIF5B in performing this function. (See text for further details.)

(Erickson and Hannig 1996). These data provide strong evidence that eIF2γ is directly involved in binding GTP/GDP and Met-tRNA$_i^{Met}$.

The *N135K* mutation in eIF2γ, isolated for its dominant Sui$^-$ phenotype, also maps in a region of the G domain highly conserved with EF1A. In vitro, this lesion reduced ternary complex formation partly by increasing the rate of spontaneous GTP hydrolysis, as it was partially overcome by a nonhydrolyzable GTP analog, and also by increasing the off-rate of Met-tRNA$_i^{Met}$ from eIF2, without affecting the affinity for GTP. To account for the dominant Sui$^-$ phenotype of this mutation, it was proposed that premature dissociation of Met-tRNA$_i^{Met}$ from the mutant eIF2–GTP complex during the scanning process allows incorrect pairing of the initiator with a UUG codon (Huang et al. 1997; see Chapter 12).

eIF2β: Interactions with Met-tRNA$_i^{Met}$, mRNA, and eIF5

The sequence of eIF2β has several notable features, including the presence of three polylysine stretches in its amino-terminal half and a Cys-4-type zinc-finger motif at its carboxyl terminus (Fig. 2) (Donahue et al. 1988; Pathak et al. 1988; Ye and Cavener 1994). Although no zinc was detected in purified mammalian eIF2 (Pathak et al. 1988), mutational analysis of yeast *SUI3* shows that the cysteine residues are critically required for eIF2β function in vivo (Castilho-Valavicius et al. 1992). Although inviable, a *SUI3* allele lacking the zinc-finger motif had a dominant Gcd$^-$ phenotype, suggesting that the mutant protein can displace wild-type SUI3 and form an eIF2 molecule defective for ternary complex formation or 40S binding. Remarkably, all 13 dominant Sui$^-$ alleles of *SUI3* altered conserved residues in or around the zinc-finger motif (Donahue et al. 1988; Castilho-Valavicius et al. 1992). Biochemical analysis showed that two such Sui$^-$ mutations (*S264Y* and *L254P*) led to increased GTP hydrolysis in the purified ternary complex, independently of the GAP activity of eIF5 (Huang et al. 1997). The *S264Y* mutation also led to increased dissociation of Met-tRNA$_i^{Met}$ from the ternary complex independently of GTP hydrolysis, supporting a role for the β subunit in Met-tRNA$_i^{Met}$ binding. It was proposed that both defects increase the probability that the ternary complex can dissociate during the scanning process and leave Met-tRNA$_i^{Met}$ paired with a UUG codon (Huan et al. 1997; see Chapter 12). Interestingly, the carboxy-terminal two-thirds of eIF2β shows significant similarity to eIF5 (Fig. 2) (Das et al. 1997a), including the putative zinc finger, raising the possibility that the homologous domains in eIF5 and eIF2β interact, or compete, with one another in a way that influences GTP hydrolysis by eIF2.

There are numerous reports that eIF2 can bind mRNA and that this interaction impedes ternary complex formation (for review, see Trachsel 1996). It was also proposed that the binding affinity of mRNA for eIF2 is an important determinant of its translational efficiency (Rosen et al. 1982; Proud 1992). Gonsky et al. (1992) reported that the β subunit has mRNA-binding activity which can survive denaturing conditions that disrupt the eIF2 complex. Flynn et al. (1994) showed that an eIF2 preparation depleted of the β subunit was defective for mRNA binding, and also that 4-thio-UTP-substituted encephalomyocarditis virus (EMCV) RNA could be crosslinked to the carboxy-terminal one-third of eIF2β encompassing the zinc finger. Consistent with these latter findings, recombinant yeast eIF2β bound mRNA in vitro in a manner partially dependent on the segment containing the zinc-finger motif; however, the polylysine repeats in the amino-terminal domain (K-boxes) made an even larger contribution, as deletion of all three K-boxes reduced mRNA binding to ~25% of wild-type. The third K-box was sufficient for nearly wild-type mRNA binding in vitro, and this property was maintained even when altered to a run of arginines. Consistently, deletion of all three K-boxes was lethal, but *SUI3* alleles retaining any single K-box were viable, indicating functional redundancy for the essential function(s) of the K-boxes in vivo (Laurino et al. 1999). Interestingly, removal of K-boxes 1 and 2 abolished the Sui$^-$ phenotype of the *SUI3-S264Y* allele, perhaps by weakening the interaction of ternary complex with the initiation region of mutant *his4* mRNA bearing UUG in place of the AUG start codon (Laurino et al. 1999). Alternatively, this mutation might reduce the interaction between eIF2 and eIF5, which also depends on the K-boxes (as discussed below), and thereby reduce GTP hydrolysis by mutant ternary complexes bearing SUI3-S264Y.

Elimination of all three K-boxes from yeast eIF2β impaired mRNA binding by the purified eIF2 complex, but had no effect on ternary complex formation in vitro. The *SUI3* allele lacking all K-boxes conferred dominant Slg$^-$ and Gcd$^-$ phenotypes in a *SUI3*$^+$ strain, suggesting a defect in ternary complex formation or binding to 40S ribosomes. Ostensibly at odds with this expectation, eIF2 complexes containing the mutant protein were present in 43S or 48S preinitiation complexes in vivo. Thus, the K-boxes are not essential for ternary complex formation or 40S binding (Laurino et al. 1999). However, the dominant Gcd$^-$ phenotype of this *SUI3* allele implies at least a modest reduction in ternary complex levels, and there is evidence that the K-boxes mediate tight interaction between eIF2 and eIF2B in vivo, promoting efficient recycling of eIF2 (Asano et al. 1999). Thus, overexpression of the eIF2β protein lacking K-boxes may decrease eIF2 recycling enough to reduce the rate of ternary complex binding to reinitiating ribosomes on *GCN4* mRNA.

There is considerable evidence that the K-boxes in eIF2β are required for interaction between eIF2 and eIF5 in addition to mRNA binding by eIF2. The eIF5 in rabbit reticulocyte lysates copurified with eIF2 in a complex of M_r=160,000, and a similar complex containing these factors in 1:1 stoichiometry could be formed with purified eIF2 and recombinant eIF5. Radiolabeled recombinant eIF5 seemed to bind eIF2 exclusively in a crude preparation of initiation factors and failed to interact with purified eIF3 or eIF2B (Chaudhuri et al. 1994). Recombinant rat eIF5 bound specifically to the β subunit of purified rabbit eIF2, and in vitro binding assays using rat eIF5 and human eIF2β suggested that the second K-box was necessary and sufficient for their strong interaction. Additionally, yeast eIF5 interacted with rabbit eIF2β, and rat eIF5 interacted with yeast eIF2β, indicating that the eIF5–eIF2 interaction is evolutionarily conserved. Apparently, the eIF2–eIF5 interaction can occur independently of GTP and Met-tRNA$_i^{Met}$ (Das et al. 1997a).

Mutational analysis of yeast eIF2β (SUI3) showed that at least one K-box was required for the interaction in vitro with yeast eIF5 (encoded by *TIF5*) and that each K-box present singly in eIF2β conferred a reduced level of eIF5 binding compared to that seen with all three K-boxes in wild-type eIF2β. Similarly, the presence of K-box 1 or 3 was sufficient for coimmunoprecipitation of eIF5 with epitope-tagged eIF2β from cell extracts, albeit at levels about one-third of that seen with wild-type eIF2β. Thus, as observed for mRNA binding by yeast eIF2β, and also for its essential activity in vivo, the K-boxes have redundant functions in the binding of eIF5. As in the case of mammalian eIF2β, the carboxy-terminal half of yeast eIF2β contributed little to the interaction with eIF5 (Asano et al. 1999).

The binding domain for eIF2β was mapped in vitro to the carboxy-terminal 40% of yeast eIF5 (Fig. 2). This region is highly conserved between yeast and mammalian eIF5 and has at the extreme carboxyl terminus a bipartite motif containing aromatic, hydrophobic, and acidic residues in each of its parts (dubbed AA-boxes for their *a*romatic and *a*cidic constituents). Interestingly, this motif is also found at the carboxyl termini of yeast eIF2Bε (GCD6) and its mammalian homolog (see below, Fig. 3, AA-boxes) (Koonin 1995). Given that eIF2 is a common substrate for eIF5 (the GAP) and eIF2B (the GEF), Koonin suggested that the shared motif could be involved in eIF2 binding. This possibility became more attractive with the discovery that eIF2Bε is the principal catalytic subunit of eIF2B (Fabian et al. 1997; Pavitt et al. 1998). In accordance with this prediction, small deletions of the bipartite motif, or alanine replacements of the conserved residues in the AA-boxes, impaired interaction of eIF5 with recombinant eIF2β and purified eIF2 in vitro. The alanine replacements in the carboxy-terminal AA-box 2 (*tif5-7A* allele) likewise abolished eIF5–eIF2 interaction in cell extracts and conferred Slg⁻ and Ts⁻ phenotypes that could be reduced by overexpressing all three subunits of eIF2 and $tRNA_i^{Met}$. The *tif5-12A* allele bearing alanine replacements in AA-box 1 has a lethal phenotype. These findings suggest that the AA-boxes in eIF5 are required for an important interaction with the ternary complex in vivo (Fig. 2) (Asano et al. 1999).

The binding of eIF5 to the K-box domain of eIF2β may promote the interaction between eIF5 and eIF2 needed to stimulate GTPase activity upon recognition of the start codon. Given that eIF5 is tightly associated with eIF3 in yeast (Phan et al. 1998), and that eIF3 stimulates ternary complex binding to the ribosome (Trachsel 1996), the eIF5–eIF2β interaction might also assist in recruitment of ternary complex to the 40S ribosome (Fig. 2). Seemingly at odds with the latter possibility, *tif5-7A* does not have a Gcd⁻ phenotype (Asano et al. 1999), which would be expected if rebinding of ternary complex to reinitiating ribosomes on *GCN4* mRNA was delayed by the mutation. However, it is not known whether eIF3 is required for ternary complex binding to reinitiating ribosomes. At present, it is unclear whether the essential function of the K-boxes in eIF2β involves an interaction of eIF2 with mRNA or with eIF5, or whether these interactions are mutually exclusive.

eIF2α Contains the Site of Phosphorylation for Inhibition of eIF2B

The α subunit of eIF2 (encoded by *SUI2*) contains the conserved serine residue at position 51 whose phosphorylation converts eIF2–GDP from substrate to inhibitor of eIF2B (Hershey 1991; Dever et al. 1992). The sequence surrounding this residue is highly conserved in eukaryotic eIF2α proteins (Ernst et al. 1987; Cigan et al. 1989; Qu and Cavener 1994), but not in archaea (Bult and al. 1996), consistent with phosphorylation of Ser-51 occurring only in eukaryotes. Interestingly, residues 14–93 in archaeal and eukaryotic eIF2α contain sequence similarities with the S1 RNA-binding domain, a five-stranded antiparallel β barrel originally identified in *E. coli* ribosomal protein S1 (Bycroft et al. 1997) and also present in eubacterial translation initiation factor 1 (IF1) (Sette et al. 1997) and its eukaryotic ortholog eIF1A (Battiste et al. 2000). Sui⁻ mutations in yeast eIF2α alter residues in the amino-terminal region of the protein, one of which (*sui2-2*) harbors a substitution in the β1 strand of the putative S1 motif (Cigan et al. 1989). Thus, this motif might contribute to binding Met-$tRNA_i^{Met}$ or the interaction with mRNA during scanning by eIF2. Other *sui2* mutations in the amino-terminal region reduce the inhibitory effect of phosphorylated eIF2 on its exchange factor eIF2B (Gcn^- phenotype), implicating this domain in the regulatory interaction between eIF2 and eIF2B (Fig. 3 and below).

In addition to the phosphorylation site at Ser-51, yeast eIF2α is phosphorylated both in vitro and in vivo by casein kinase II (CKII) at one or all three serine residues at positions 292, 294, and 301. Although alanine substitutions of these residues did not affect the growth rate or confer a Sui^- phenotype in otherwise wild-type cells, they exacerbated the growth defects of mutants in which recycling of eIF2–GDP to eIF2–GTP by eIF2B was inhibited by high-level phosphorylation of eIF2 (*GCN2*^{c} mutant) or by nonlethal Gcd^- mutations in eIF2Bα (*gcn3*^{c}) or eIF2Bδ (*gcd7*). Thus, lack of CKII phosphorylation probably reduces eIF2 activity by a significant amount only when combined with a defect in eIF2 recycling (Feng et al. 1994); hence, CKII phosphorylation may promote the productive interaction between eIF2–GDP and eIF2B. There is no evidence that this phosphorylation event is regulated in yeast cells. Mammalian eIF2α lacks the CKII sites and is not a substrate for the mammalian kinase in vitro (Proud 1992).

RECRUITMENT OF THE TERNARY COMPLEX TO THE 40S RIBOSOME

eIF3 Promotes Ternary Complex Binding to 40S Ribosomes

In vitro, the mammalian ternary complex can bind to purified 40S subunits in the absence of other factors, and this interaction is stimulated by high, nonphysiological Mg^{++} concentrations (> 2 mM) and the AUG triplet (Peterson et al. 1979). (Use of AUG in place of mRNA obviates the need for the factors required for mRNA binding to the ribosome.) The stimulatory effect of the AUG triplet suggests that base-pairing between the start codon and Met-tRNA$_i^{Met}$ stabilizes ternary complex association with 40S ribosomes. High-level binding of the ternary complex to 40S subunits under more physiological conditions requires initiation factors and is stimulated by eIFs 1, 1A, and 3 (Trachsel et al. 1977; Benne and Hershey 1978). The majority of native free 40S subunits contain eIF3 (Thompson et al. 1977), and ternary complex binding to purified 40S ribosomes can be stimulated by a factor of 2–3 by purified eIF3. The eIF3 can bind to 40S ribosomes in the absence of other factors, although this association may be enhanced by simultaneous binding of the ternary complex (Benne and Hershey 1978; Peterson et al. 1979; Trachsel and Staehelin 1979; Chaudhuri et al. 1997a). Following hydrolysis of the GTP bound to eIF2 in the ternary complex, a reaction stimulated by eIF5, the eIF2 and eIF3 are released from the 48S initiation complex (Benne and Hershey 1978; Peterson et al. 1979; Trachsel and Staehelin 1979). Mammalian eIF3 is a very complicated factor containing 11 subunits: p170, p116, p110, p66, p48, p47, p44, p40, p36, p35, and p28 (see Chapter 2). Yeast contains orthologs of 6 of these proteins (p170, p116, p110, p36, p44, and p35) known as TIF32/RPG1, PRT1, NIP1, TIF34, TIF35, and HCR1, respectively. As discussed below, all of these yeast proteins are associated with the yeast eIF3 complex.

Evidence supporting the critical importance of eIF3 for ternary complex recruitment in vivo has come from studies on yeast. A Ts$^-$ lethal mutation in the *PRT1*-encoded eIF3 subunit (homologous to the 116-kD human eIF3 subunit) leads to dramatic loss of polyribosomes at the restrictive temperature, indicating a severe defect at the initiation step (Hartwell and McLaughlin 1969). Extracts of heat-treated *prt1-1* spheroplasts were defective for ternary complex binding to 40S subunits (Feinberg et al. 1982). This biochemical activity could be heat-inactivated in a *prt1-1* extract prepared from cells grown at the permissive temperature and then rescued with a PRT1-containing complex purified from wild-type cells (Danaie et al. 1995). A similar complex containing polyhistidine-tagged PRT1 was purified by nickel-affinity chromatography and shown to contain the yeast homologs of mammalian eIF3 subunits p170, p110, p36, and p44 (TIF32, NIP1, TIF34, and TIF35, respectively) in addition to PRT1, plus nearly stoichiometric amounts of eIF5 (Fig. 2) (Phan et al. 1998). This purified eIF3–eIF5 complex complemented the defects in translation and ternary complex binding to 40S subunits in heat-treated *prt1-1* extracts. The association of eIF5 with eIF3 first detected in yeast has now been observed in mammalian cells (Bandyopadhyay and Maitra 1999).

Yeast eIF3 was also purified by its ability to substitute for mammalian eIF3 in promoting 80S initiation complex formation and methionyl-puromycin (Met-puromycin) synthesis in the presence of AUG and rabbit reticulocyte factors eIF2, -1A, -5, and -5A (Naranda et al. 1994). The purified complex contained PRT1, TIF34, TIF35, and a 21-kD proteolytic fragment of TIF32, but lacked NIP1 (Naranda et al. 1994, 1997; Hershey et al. 1996). However, it was shown subsequently that yeast extracts depleted of NIP1 were defective for translation and ternary complex binding to 40S subunits, and both defects could be partially rescued with the affinity-purified eIF3–eIF5 complex described above (Phan et al. 1998). Moreover, the translation rate and polysome content were severely reduced following NIP1 depletion in vivo, confirming the importance of this subunit for translation initiation (Greenberg et al. 1998). As expected, significant proportions of both PRT1 (Evans et al. 1995) and NIP1 (Greenberg et al. 1998) were found associated with 40S ribosomes in extracts from wild-type cells.

Analysis of a Ts⁻ lethal mutation in *TIF32* (also known as *RPG1*) showed that this eIF3 subunit is essential for translation initiation in vivo and in vitro (Valasek et al. 1998). Similarly, depletion of wild-type TIF32 (Valasek et al. 1998), TIF34 (Naranda et al. 1997), or TIF35 (Hanachi et al. 1999), or Ts⁻ lethal mutations in *TIF34*, all led to cessation of yeast cell growth and inhibition of translation initiation in vivo (Verlhac et al. 1997; Asano et al. 1998). It is not known whether the TIF32, TIF34, and TIF35 subunits are required for ternary complex binding or for another predicted function of eIF3, such as ribosome- or mRNA-binding. Nor is it known whether the PRT1 and NIP1 subunits are directly involved in ternary complex binding, or needed only to ensure the proper conformation of other eIF3 subunits required for this activity.

PRT1 contains a degenerate RNA recognition motif (RRM) at its amino terminus (Hanic-Joyce et al. 1987; Evans et al. 1995) that is likely required for its essential function in vivo. Expression of a mutant protein lacking the RRM (*prt1-Δ100*) inhibited translation initiation (dominant-negative phenotype), and the mutant protein appeared to reside in eIF3 complexes defective for 40S association (Evans et al. 1995). Although the RRM in PRT1 could be required for binding rRNA, the complexes containing PRT1-Δ100 may have lacked another eIF3 subunit essential for this function. TIF35 (Verlhac et al. 1997) and its mammalian homolog eIF3–p44 (Block et al. 1998) also contain an RRM and were shown to interact directly with fragments of rRNA and globin mRNA in vitro, dependent on the RRM (Block et al. 1998; Hanachi et al. 1999). Surprisingly, removal of the RRM in TIF35 led to a Slg⁻ phenotype but was not lethal in yeast (Hanachi et al. 1999). Nonspecific RNA-binding activity has also been detected for the human homolog of TIF32 (eIF3–p170) (Block et al. 1998). Whether TIF35/eIF3-p44 and eIF3–p170 interact in vivo with rRNA, $tRNA_i^{Met}$, or mRNA remains to be determined.

The five subunits of yeast eIF3 mentioned above are homologous to 5 of the 11 subunits of mammalian eIF3, whereas 5 of the remaining mammalian eIF3 subunits (p66, p48, p47, p40, and p28) have no obvious counterparts encoded in the yeast genome (see Chapter 2). The yeast *HCR1* gene encodes a 30-kD protein with sequence similarity to the remaining mammalian eIF3 subunit, p35, and was recently identified as a dosage suppressor of a Ts⁻ allele of *TIF32* (known as *rpg1-1*). HCR1 is physically associated with TIF32 (and presumably the eIF3 complex) in cell extracts (Fig. 2), although it is not known whether this interaction occurs free of ribosomes. Whereas deleting HCR1 alone had a subtle effect on growth, it exacerbated the Ts$^-$ phenotype of *rpg1-1*, indicating functional interaction between the two proteins (Valasek et al. 1999). It remains to be determined whether HCR1 is a substoichiometric component of eIF3, or simply less tightly associated with the complex than are other eIF3 subunits.

Interactions between the eIF3 Complex and Other Initiation Factors

When purified by their ability to stimulate Met-puromycin synthesis (Naranda et al. 1994), yeast eIF3 preparations contained three polypeptides of 135 kD, 62 kD, and 16 kD in addition to the core eIF3 subunits described above. These additional proteins were subsequently identified as TIF31 (Vornlocher et al. 1999), GCD10 (Garcia-Barrio et al. 1995), and the yeast homolog of eIF1 (encoded by *SUI1*) (Naranda et al. 1996), respectively, and are not related in sequence to any of the known subunits of human eIF3. Affinity purification of eIF3 directed against polyhistidine-tagged TIF35 confirmed the physical association of TIF31 with the complex (Hanachi et al. 1999), and recombinant TIF31 interacted with TIF35 both in vitro and in two-hybrid assays (Fig. 2) (Vornlocher et al. 1999). However, *TIF31* is nonessential, and its deletion has no obvious effect on yeast cell growth or polysome profiles (Vornlocher et al. 1999). As mentioned above, yeast eIF1 was originally identified by the isolation of *sui1* mutations that relax the stringency of AUG recognition on *HIS4* mRNA (Sui$^-$ phenotype) (Yoon and Donahue 1992). The interaction of eIF1 with eIF3 in yeast has been confirmed by its coimmunoprecipitation with other eIF3 subunits (Naranda et al. 1996) and by affinity purification with polyhistidine-tagged PRT1 (Phan et al. 1998). However, its association with eIF3 appears to be very salt-sensitive (Asano et al. 1998; Phan et al. 1998), consistent with the fact that eIF1 does not copurify with eIF3 from mammalian cells (Benne and Hershey 1976; Trachsel et al. 1977). Ribosomal salt washes from a Ts$^-$ *sui1* mutant grown at the nonpermissive temperature lacked both detectable eIF1 protein and eIF3 activity in stimulating Met-puromycin synthesis, suggesting that eIF1 is required for eIF3 activity in this assay (Naranda et al. 1996); however, because the defect was not rescued by purified eIF3, a requirement for eIF1 in expression of eIF3 subunits in vivo cannot be ruled out. Interestingly, recombinant eIF1 interacted specifically with the NIP1 subunit of yeast eIF3 both in vitro and in the yeast two-hybrid assay (Asano et al. 1998), and a similar interaction was demonstrated for the mammalian homologs of these proteins (Fletcher et al. 1999).

As noted above, nearly stoichiometric amounts of yeast eIF5 copurified with eIF3, and this interaction is not salt-labile (Phan et al. 1998). Interestingly, both eIF5 and eIF1 were shown to interact with the NIP1 subunit of eIF3 (Asano et al. 1998; Phan et al. 1998), leading to the suggestion that both factors would be recruited to the 40S subunit through their mutual interactions with the same subunit of eIF3 (Fig. 2). Sui$^-$ mutations have been isolated in eIF5 that increase its GAP activity, and it has been proposed that a hyperactive eIF5 would stimulate the eIF2 GTPase activity inappropriately when the ternary complex is base-paired with non-AUG triplets, allowing their selection as start sites at a higher frequency than occurs in wild-type cells (Huang et al. 1997). The fact that both eIF5 and eIF1(SUI1) modulate AUG recognition and interact with the same subunit of eIF3 suggests that eIF3 may play an important role in juxtaposing these factors with respect to the ternary complex, mRNA, and 40S ribosome for accurate recognition of AUG triplets (Phan et al. 1998). Surprisingly, the interaction between eIF5 and eIF3–NIP1 is dependent on the same bipartite motif at the carboxyl terminus of eIF5 that is required for its binding to eIF2β (Fig. 2) (Asano et al. 1999). If eIF5 can associate simultaneously with eIF3–NIP1 and eIF2β, it might bridge an interaction between eIF3 and eIF2 and thereby promote 40S binding of the ternary complex.

GCD10 was first identified genetically by recessive *gcd10* mutations that led to derepression of *GCN4* translation in cells lacking the eIF2α kinase GCN2 (Gcd$^-$ phenotype) (Harashima and Hinnebusch 1986; Harashima et al. 1987). The association of GCD10 with eIF3 was intriguing because it suggested that eIF3 was required for efficient reinitiation at uORFs 3–4 in the *GCN4* mRNA leader by ribosomes that had previously translated uORF1 and resumed scanning. It was proposed that *gcd10* mutations would reduce the ability of eIF3 to stimulate ternary complex binding to these ribosomes, allowing them to scan past uORFs 3–4 and reinitiate at *GCN4* instead. In this way, *GCN4* translation would be induced in the absence of eIF2α phosphorylation and attendant reduction in levels of the ternary complex (Garcia-Barrio et al. 1995). However, it has not been possible to confirm a direct association of GCD10 with eIF3 by affinity purification or coimmunoprecipitation with tagged eIF3 subunits from cell extracts (Anderson et al. 1998; Phan et al. 1998; Calvo et al. 1999). In addition, no defect in ternary complex binding to 40S subunits was detected in a *gcd10*Δ extract (Anderson et al. 1998).

Subsequent analysis revealed that GCD10 resides in a nuclear complex with the product of *GCD14* (Anderson et al. 1998), another gene first identified as a negative regulator of *GCN4* translation (Cuesta et al. 1998). The GCD10–GCD14 nuclear complex is required for the formation of 1-methyladenosine at position 58 (m^1A58) in all tRNAs containing this modification, including initiator $tRNA^{Met}$ (Anderson et al. 1998). In *gcd10* and *gcd14* mutants, processing of pre-$tRNA_i^{Met}$ is impeded and the steady-state levels of mature $tRNA_i^{Met}$ are reduced as the result of degradation of the initiator (or its precursors). The expression of other tRNAs containing m^1A58 is not impaired by these mutations, and the lethal effects of deleting either *GCD10* or *GCD14* can be suppressed by overexpressing the initiator tRNA (Anderson et al. 1998; Calvo et al. 1999). It was proposed that the strong requirement for m^1A58 in the processing and stability of $tRNA_i^{Met}$ can be explained by its involvement in a tertiary structure unique to the initiator (Basavappa and Sigler 1991). The reduction in ternary complex levels arising from defective $tRNA_i^{Met}$ production in the nucleus can probably account for the Gcd^- phenotypes of *gcd10* and *gcd14* mutants (Anderson et al. 1998) without invoking a cytoplasmic role for GCD10 in eIF3 function.

Other Functions of eIF3 in Assembly of the 48S Preinitiation Complex

In addition to its role in Met-$tRNA_i^{Met}$ recruitment, it was shown that eIF3 could impede association of ribosomal subunits in the absence of ternary complex and that this activity was enhanced by the ternary complex. Moreover, the stimulatory effect of eIF3 on ternary complex binding to 40S subunits was greater when 60S subunits were present under conditions favoring 40S–60S subunit joining (Trachsel and Staehelin 1979). More recently, Chaudhuri et al. (1999) reported that eIF3 does not exhibit ribosome dissociation activity alone, but can prevent 60S subunits from displacing the ternary complex from 40S subunits in the absence of AUG or mRNA, i.e., simultaneous occupancy of the 40S subunit by the ternary complex and eIF3 would form an assembly that resists displacement by a 60S subunit. (It should be noted that the eIF3 preparation of Chaudhuri et al. lacked the p170 subunit, which might be required for the intrinsic ribosome dissociation activity of eIF3.) Electron micrographs show that eIF3 bound to the 40S subunit is oriented away from the interaction site for the 60S subunit (Srivastava et al. 1992), consistent with the idea that the ribosome dissociation activity of eIF3 resides in its ability to stabilize ternary complex binding to the 40S subunit.

The eIF3 also strongly stimulates mRNA binding by the 40S initiation complex (Benne and Hershey 1976, 1978; Trachsel et al. 1977). Because binding of the ternary complex seems to be a prerequisite for binding of mRNA to the 43S initiation complex (Trachsel et al. 1977;

Benne and Hershey 1978), it is expected that eIF3 would stimulate mRNA binding indirectly by promoting the recruitment of ternary complex; however, there appears to be an additional strong requirement for eIF3 in mRNA binding (Trachsel et al. 1977). The latter is generally attributed to interactions between eIF3 and the mRNA-associated factors eIF4G (Lamphear et al. 1995) or eIF4B (Methot et al. 1996). Whereas mammalian eIF4B interacted directly with the p170 eIF3 subunit (homologous to yeast TIF32), the yeast homolog of eIF4B (encoded by *TIF3*) interacted with yeast eIF3 subunit TIF35 (Fig. 2) (Vornlocher et al. 1999). Mammalian eIF3 contains three subunits that bind RNA (p170, p66, and p44) (see Chapter 2) and thus could interact directly with mRNA in the initiation complex. Indeed, eIF3 can bind to the hepatitis C and classic swine fever virus IRES elements, and the p170, p116, p66, and p47 subunits have been UV-crosslinked to these mRNA sequences (Buratti et al. 1998; Sizova et al. 1998). It is noteworthy that prior binding of the ternary complex as a condition for binding of mRNA to the 40S ribosome (Trachsel et al. 1977) is not understood at the molecular level.

eIF1A Functions with eIF3 to Stimulate Ternary Complex Binding to 40S Subunits

Mammalian eIF1A (formerly eIF-4C), a single polypeptide of only ~17 kD, has been implicated in ribosome dissociation, ternary complex binding, and mRNA binding to the ribosome (Merrick and Hershey 1996). In the earlier studies, it was generally less active than eIF3 in promoting ternary complex binding to free 40S subunits (Trachsel et al. 1977; Benne and Hershey 1978), although a greater stimulation could be observed in the presence of 60S subunits and was attributed to its ribosome anti-association activity (Thomas et al. 1980a). More recent work by Maitra's group using purified or recombinant mammalian eIF1A revealed a more pronounced activity in stimulating ternary complex binding to 40S subunits. Under conditions where eIF1A stimulated this reaction almost 20-fold, purified eIF3 conferred only 3-fold stimulation, and the two factors combined showed only slightly greater stimulation than eIF1A alone (Chaudhuri et al. 1997b, 1999). In agreement with earlier work (Benne and Hershey 1978), eIF1A did not cosediment with 40S preinitiation complexes; moreover, it appeared to function catalytically to promote ternary complex binding. Therefore, it was suggested that eIF1A catalyzes ternary complex binding to 40S ribosomes but is not required to stabilize the initiation complex thus formed (Chaudhuri et al. 1997b). In some previous studies, however, eIF1A was associated with native 40S subunits (Goumans et al. 1980) and remained bound to 40S preinitiation complexes formed in vitro during gel filtration (Thomas et al. 1980a). Thus, eIF1A may remain associated with the 40S ribosome and participate in subsequent reactions even if it is not required to stabilize the interaction between ternary complex and the ribosome. The yeast and mammalian proteins are highly conserved in sequence and are functionally interchangeable in the Met-puromycin assay using mammalian factors. Deletion of the yeast gene (*TIF11*) is lethal, supporting an essential role for eIF1A in translation initiation (Wei et al. 1995a).

Although Maitra's group found that eIF1A was more effective than eIF3 in stimulating ternary complex binding to purified 40S subunits, it failed to do so in the presence of 60S subunits under conditions that promote subunit joining (Chaudhuri et al. 1999). This observation seems at odds with the idea that eIF1A has ribosome anti-association activity and, indeed, no such activity was observed by this group (Chaudhuri et al. 1997b). The eIF3, in contrast, could stimulate ternary complex binding in the presence of 60S subunits, and this stimulatory effect disappeared with addition of the AUG triplet. To account for these findings, Chaudhuri et al. proposed that eIF1A and eIF3 are both required to form a stable 40S preinitiation complex under physiological conditions, with eIF1A catalyzing transfer of the ternary complex to 40S ribosomes harboring eIF3. The eIF3, in conjunction with the ternary complex, protects the 43S complex against disruption by a 60S subunit prior to mRNA binding but becomes dispensable for this function after Met-tRNA$_i^{Met}$ is base-paired with the AUG start codon (Chaudhuri et al. 1999).

The eIF1A has an ortholog in archaea and also exhibits significant sequence similarity (21% identity) to prokaryotic initiation factor IF1 (Kyrpides and Woese 1998). The three-dimensional structures of *E. coli* IF1 (Sette et al. 1997) and mammalian eIF1A (Battiste et al. 2000) both contain a 5-stranded antiparallel β barrel known as the oligonucleotide/oligosaccharide-binding (OB) domain (Sette et al. 1997), whereas eIF1A contains an additional α-helical domain not present in bacterial IF1 (Battiste et al. 2000). The eIF1A shows sequence-independent RNA-binding activity in vitro (Wei et al. 1995b), with a K_d of ~15 μM (Battiste et al. 2000). Residues involved in RNA binding were identified by alterations in backbone amide resonances in the nuclear magnetic resonance (NMR) spectrum of eIF1A in the presence of RNA. These residues comprise a belt stretching from the OB domain to the α-helical domain of eIF1A, connected by a groove. Mutations of several such residues reduced RNA binding by eIF1A, and interestingly, the *K67D* mutation of residue Lys-67 (present in the groove) impaired eIF1A-stimulated ternary complex binding to 40S subunits in vitro (Battiste et al. 2000). It was suggested that this last mutation impairs binding of eIF1A to the rRNA in the 40S subunit.

Bacterial IF1 has been implicated in ribosomal subunit dissociation and in stimulating IF2 binding and IF2-dependent recruitment of fMet-tRNA^{Met} to the 30S subunit (Gualerzi and Pon 1990), similar to the varied functions ascribed to eIF1A. IF1 binds to the 30S subunit in 1:1 stoichiometry (Gualerzi and Pon 1990) and, based on footprinting studies (Moazed et al. 1995), makes contact with the A site. Accordingly, it may prevent premature binding of elongator tRNAs to the A site during initiation or may help to position Met-tRNA$_i^{Met}$ in the P site. Based on changes in the NMR spectrum of IF1 produced by 30S subunits, and the results of mutagenesis studies, a large surface of IF1 seems to be involved in ribosome binding (Gualerzi and Pon 1990; Sette et al. 1997). Given the structural similarity between IF1 and eIF1A, it will be interesting to determine whether eIF1A binds to the A site of eukaryotic ribosomes.

Pestova and Hellen showed recently that eIF1A acts in conjunction with eIF1 (in the presence of ternary complex, globin mRNA, eIF3, and the mRNA-associated factors eIF4A, -4B, and -4F) to promote formation of a stable 48S complex with the ribosome positioned at the AUG codon, as judged by toeprint analysis. In the absence of eIF1 and -1A, an unstable complex was formed close to the 5′ end, whereas addition of eIF1, in a manner stimulated by eIF1A, led to dissociation of this complex and formation of the more stable, correctly positioned 48S complex. For EMCV RNA, where ribosome binding to the start codon is directed by an internal ribosome entry site (IRES), eIF1 could direct 40S ribosomes to the correct AUG without eIF1A. Thus, eIF1 seems to possess the critical activity for positioning a 40S ribosome at the start codon (Pestova et al. 1998). Interestingly, mutations in residues on the RNA-binding surface of eIF1A did not impair its ability to disrupt incorrect 48S complexes formed at the cap, but led to the stabilization of incorrect complexes located upstream from the start site (Battiste et al. 2000). These data are consistent with the idea that eIF1A interacts with the A site of a 40S subunit and plays a role in AUG recognition by initiator tRNA during the scanning process.

There are sequence similarities between IF2 or IF1 and different segments of prokaryotic elongation factor EF2 (Brock et al. 1998). The crystal structure of EF2 mimics the shape and surface charge distribution of the aminoacyl-tRNA/EF1A/GTP ternary complex (Nyborg et al. 1996), leading to the hypothesis that EF2 catalyzes translocation by binding to the A site and forcing the peptidyl-tRNA into the P site. It has been suggested that a segment of IF2 and IF1 together mimic the aminoacyl tRNA and bind to the A site where they can interact with fMet-tRNA$_i^{Met}$ in the P site (Brock et al. 1998). Lee et al (1999) further proposed that eIF1A and eIF5B, the eukaryotic orthologs of IF1 and IF2, respectively, may serve a similar function in directing the eIF2/GTP/Met-tRNA$_i^{Met}$ ternary complex into the P site. Deletion of the gene encoding eIF5B in yeast (*FUN12*) is not lethal but leads to a pronounced slow-growth phenotype (Slg$^-$) that can be attributed to a reduced rate of translation initiation. The Slg$^-$ phenotype of the *fun12*Δ mutant was partially suppressed by overexpressing tRNA$_i^{Met}$, suggesting a role for eIF5B in ternary complex binding. Consistently, the *fun12*Δ mutant failed to induce *GCN4* mRNA translation, although the molecular basis for this phenotype is unclear (Choi et al. 1998). In addition to a possible role in Met-tRNA$_i^{Met}$ binding, eIF5B has been implicated in joining of the 60S subunit to the 48S initiation complex following hydrolysis of the GTP bound to eIF2 (Pestova et al. 2000 and Chapter 9).

eIF1 Promotes Correct Interaction between the Ternary Complex and Start Codon

Biochemical studies indicated that eIF1 has a weak stimulatory effect on binding of ternary complex and mRNA to 40S or 80S initiation complexes in the presence of other factors (Trachsel et al. 1977; Benne and Hershey 1978; Thomas et al. 1980b). Its stimulation of Met-puromycin synthesis by 80S initiation complexes was observed only in the absence of AUG, suggesting that it could substitute for AUG in positioning ternary complex in the P site (Thomas et al. 1980b). The eIF1 also appeared to prevent eIF5-catalyzed 60S subunit joining in the absence of mRNA (Trachsel et al. 1977), implying a role in coordinating mRNA and tRNA binding to the initiation complex. These properties are consistent with results from yeast indicating that eIF1(SUI1) modulates eIF5-stimulated GTP hydrolysis by eIF2 during AUG selection (see Chapter 12). They are also in accordance with the results of Pestova and Hellen mentioned above showing that eIF1 acts in concert with eIF1A to promote stable 48S complex formation with the ribosome positioned at the AUG codon (Pestova et al. 1998). It remains to be seen whether eIF1 and eIF1A are required for scanning per se, to destabilize complexes not positioned at an AUG codon or to stabilize the correctly positioned complexes.

The solution structure of eIF1 has been solved by NMR, and the fold resembles that of certain ribosomal proteins and RNA-binding proteins; however, there is no evidence for direct interaction of eIF1 with RNA. The

Sui⁻ alleles of yeast *SUI1* (Yoon and Donahue 1992; Cui et al. 1998) alter residues predicted to be clustered together on the surface of eIF1, along with other residues conserved in eIF1 orthologs from bacteria and archaea. These residues may comprise an important domain for eIF1 function in AUG selection (Fletcher et al. 1999).

Interestingly, a *SUI1* allele known as *mof2-1* increases programed –1 ribosomal frameshifting on yeast L-A virus mRNA. The *mof2-1* allele also has a Sui⁻ phenotype, and the *sui1-1* allele (but not Sui⁻ alleles of *SUI2* or *SUI3* affecting eIF2α or eIF2β, respectively) has a weak Mof⁻ phenotype. The Mof⁻ phenotype was recapitulated in *mof2-1* translation extracts and rescued with recombinant eIF1(SUI1). Thus, eIF1 may have a direct role in accurate decoding during the elongation phase. This unexpected activity seems to be conserved in humans, as human eIF1 cDNA complemented the Mof⁻ phenotype of the *mof2-1* mutant.

eIF2C

It has been reported that mRNA destabilizes the ternary complex at low eIF2 concentrations and that this inhibition can be reversed by a 94-kD polypeptide known as eIF2C (previously Co-eIF-2A) (Gupta et al. 1990). The cDNA encoding the rabbit protein has been isolated and sequenced (Zou et al. 1998) and shows strong homology with proteins in *Caenorhabditis elegans, Drosophila, Arabidopsis,* and fission yeast. One of the *C. elegans* homologs, *rde-1*, is required for double-stranded RNA (dsRNA)-mediated inhibition of gene function (RNAi) (Tabara et al. 1999). It remains to be determined whether eIF2C has a fundamental role in translation initiation in vivo.

REGULATION OF TERNARY COMPLEX FORMATION BY PHOSPHORYLATION OF eIF2α

Mechanism of Guanine Nucleotide Exchange on eIF2 Catalyzed by eIF2B

Following recognition of the AUG codon by Met-tRNA$_i^{Met}$ and hydrolysis of the GTP bound to eIF2 in the ternary complex, it is believed that the resulting eIF2–GDP is released from the ribosome. At physiological Mg^{++} concentrations, the eIF2–GDP complex dissociates slowly, and the affinity of eIF2 is much greater for GDP than for GTP (Proud 1992). Accordingly, the guanine nucleotide exchange factor eIF2B is required to replace the GDP bound to eIF2 with GTP in order to regenerate the ternary complex. The mechanism of the nucleotide exchange reaction catalyzed by eIF2B is uncertain. Rowlands et al. (1988a) presented evidence supporting a substituted enzyme mechanism involving a nucleotide-free eIF2B–eIF2 intermediate; however, this model does not explain why eIF2B is not displaced from eIF2 by GDP (Goss and Parkhurst 1984), nor why a guanine nucleotide is required for displacement of radiolabeled GDP bound to eIF2 by eIF2B (Goss and Parkhurst 1984; Dholakia and Wahba 1989). The latter are both explained, however, by a sequential mechanism involving a GTP–eIF2B–eIF2–GDP quaternary complex, and kinetic data have been obtained consistent with this model (Dholakia and Wahba 1989). It was suggested that under the experimental conditions used by Rowlands et al. (excessively high GDP concentration), it may have been difficult to distinguish between these two mechanisms, and that further work is required before either model can be accepted (Manchester 1997). The sequential mechanism predicts that the eIF2–eIF2B complex should have two guanine-nucleotide-binding sites, one in eIF2 and one in eIF2B. Dholakia and Wahba (1989) reported that eIF2B binds GTP with a K_d of 4 mM, whereas it showed no stable interaction with GDP. Photoaffinity labeling experiments suggested that the β subunit of eIF2B contains a GTP-binding site (Dholakia and Wahba 1989), but this subunit now seems to be dispensable for GEF activity in vitro (Fabian et al. 1997; Pavitt et al. 1998). Although a canonical GTP-binding site (Dever et al. 1987) is not predicted from the amino acid sequences of any eIF2B subunit, a potential nucleotide-binding motif is present in the amino-terminal portions of the γ and ε subunits of eIF2B (Koonin 1995), and it appears that eIF2Bε is sufficient for low-level GEF activity in vitro (Fabian et al. 1997; Pavitt et al. 1998) (see below). Additional work is required to determine whether eIF2Bε can bind GTP and to determine whether this activity is essential for the exchange reaction.

The binary complex eIF2(αP)–GDP (phosphorylated on Ser-51) is a poor substrate for nucleotide exchange catalyzed by eIF2B, both for the mammalian factors (Goss and Parkhurst 1984; Thomas et al. 1984; Rowlands et al. 1988a; Kimball et al. 1998b) and for their yeast counterparts (Pavitt et al. 1998). This does not reflect weak binding of eIF2(αP)–GDP to eIF2B, as phosphorylation of eIF2 increases its affinity for eIF2B (Proud 1992), with estimates ranging from severalfold (Goss and Parkhurst 1984; Pavitt et al. 1997) to more than 100-fold (Rowlands et al. 1988a) for the increase in affinity. It is frequently assumed that eIF2(αP)–GDP forms a highly stable complex with eIF2B that does not dissociate at an appreciable rate, physically sequestering eIF2B in an inactive complex. At odds with this notion, Rowlands et al. reported that the eIF2B–eIF2(αP)–GDP complex dissociated rapidly and that eIF2(αP)–GDP acted as a competitive inhibitor of eIF2B. The increased affinity of eIF2(αP)–GDP versus unphosphorylated eIF2–GDP for eIF2B could reflect an enhanced on-rate, decreased off-rate, or a combination of the two (Rowlands et al. 1988a). Considering that eIF2 is generally present in molar excess over eIF2B, a moderate increase in affinity for eIF2B may account for the strong inhibition of translation that occurs when only a fraction of eIF2 is phosphorylated and rendered ineffective as a substrate, both in mammals (Rowlands et al. 1988b; Jackson 1991) and in yeast (Cigan et al. 1991, 1993; Dever et al. 1992).

Studies in yeast provided in vivo evidence for competitive inhibition of eIF2B by eIF2(αP)–GDP (Dever et al. 1995). Overproduction of eIF2 rescued translation in a strain expressing a hyperactive *GCN2*^{c}-encoded kinase even though the absolute amount of eIF2(αP) increased 3- to 6-fold as a consequence of eIF2 overexpression. Importantly, a smaller proportion of the eIF2α was phosphorylated in the strain overexpressing eIF2 (50%) compared to that seen in a strain containing native eIF2 (80%). Thus, translational inhibition seemed to be determined by the eIF2(αP):eIF2 ratio instead of the absolute amount of eIF2(αP). This finding is consistent with a competitive mode of inhibition by eIF2(αP)–GDP with a relatively high off-rate for the eIF2B–eIF2(αP)–GDP complex. It would be more difficult to explain if the eIF2(αP)–GDP–eIF2B complex had a very low off-rate and dissociated only upon dephosphorylation of eIF2α. The high off-rate is also easier to reconcile with the finding that dephosphorylation of eIF2(αP) by protein phosphatase is impeded by interaction with eIF2B (Crouch and Safer 1984).

Structure of eIF2B Subunits and Interactions with eIF2

Purified eIF2B contains five subunits and, under physiological salt conditions, exists in a 1:1 complex with its substrate eIF2 (Cigan et al. 1991; Proud 1992). The eIF2B subunits have approximate masses of 82, 65, 57, 39, and 30 kD (subunits ε through α, respectively), and their primary structures are well-conserved between yeast and mammals (see Chapter 2). Recessive mutations in the ε , δ, γ, and β subunits (encoded by *GCD6, GCD2, GCD1,* and *GCD7,* respectively) have Ts$^-$ and Gcd$^-$ phenotypes (Hinnebusch 1996), indicative of reduced ternary complex formation and attendant derepression of *GCN4*. Because deletion of each gene is lethal (Hinnebusch 1996), these four subunits are essential for eIF2B function. In contrast, deletion of *GCN3* (encoding eIF2Bα) causes a Gcn$^-$ phenotype (failure to induce *GCN4* in response to eIF2 phosphorylation) and has no effect on cell growth (Hannig and Hinnebusch 1988). Thus, GCN3 seems to be required only for inhibition of eIF2B by eIF2(αP). Special *gcn3*^{c} alleles have been obtained with Gcd^{-} and Slg^{-} phenotypes (Hannig et al. 1990), indicating that GCN3 can be altered in a way that interferes with eIF2B function even though it is dispensable.

The GCD2 and GCD7 subunits show sequence similarity to GCN3 and to one another, which extends over nearly the entire lengths of GCD7 and GCN3 and the carboxy-terminal half of GCD2 (Fig. 3) (Paddon et al. 1989; Bushman et al. 1993). This sequence similarity suggested that GCD7 and GCD2 might cooperate with GCN3 in the regulation of eIF2B by eIF2(αP). This possibility was supported by the fact that overexpressing only these three subunits reduced the inhibitory effect of eIF2(αP) on translation initiation and cell growth in yeast. The excess GCD2, GCD7, and GCN3 formed a stable subcomplex that could be immunoprecipitated from cell extracts. This subcomplex did not compensate for loss of eIF2B function by mutation (Yang and Hinnebusch 1996) and had no GEF activity in cell extracts; however, it could bind to purified eIF2 in a manner stimulated by phosphorylation of eIF2α on Ser-51 (Pavitt et al. 1998). It was proposed that the overexpressed subcomplex binds preferentially to eIF2(αP)–GDP, effectively sequestering this inhibitory complex and allowing native eIF2B to bind and recycle the unphosphorylated eIF2–GDP. None of the individual subunits of this trimeric subcomplex bound specifically to either form of eIF2, suggesting that all three proteins are required for high-affinity binding to eIF2(αP) (Pavitt et al. 1998).

Additional genetic evidence implicating GCD2 and GCD7 in negative regulation of eIF2B came from the isolation of point mutations that reduce the effects of eIF2(αP) on translation in yeast, conferring a Gcn^{-} phenotype and suppressing the lethality of eIF2α hyperphosphorylation by *GCN2*^{c} kinases or human PKR. Some of the *GCD2* and *GCD7* Gcn^{-} alleles suppressed the effects of eIF2 phosphorylation more effectively than a deletion of *GCN3* (Vazquez de Aldana and Hinnebusch 1994; Pavitt et al. 1997). These mutations could decrease the affinity of eIF2B for eIF2(αP), or allow eIF2B to accept eIF2(αP)–GDP as substrate. The latter explanation was favored by the fact that nearly all of the eIF2α was phosphorylated in certain of the mutants. Several lines of evidence ruled out the possibility that these *GCD2* and *GCD7* regulatory mutations simply cause GCN3 to be lost from eIF2B. Thus, it was concluded that GCD2, GCD7, and GCN3 all are directly involved in the negative regulation of eIF2B by eIF2(αP) (Pavitt et al. 1997). Most of the Gcn^{-} mutations fall into two clusters located in regions of strong sequence similarity among GCN3, GCD2, and GCD7 (Fig. 3), and many map within or very close to residues conserved among all three proteins. In several

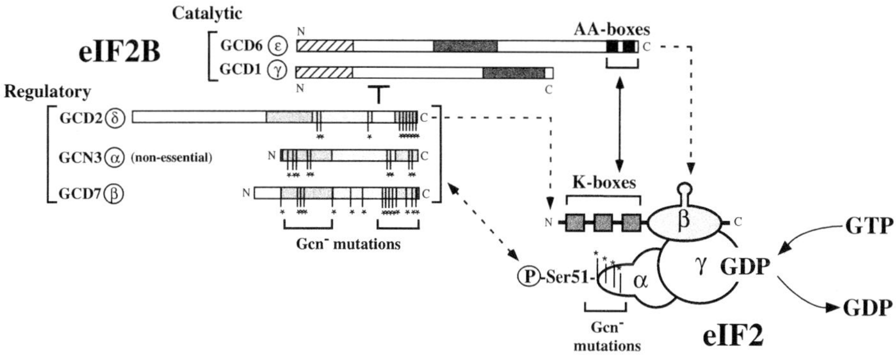

Figure 3 The eIF2B contains two subcomplexes that bind eIF2 independently and mediate catalysis and regulation of GDP–GTP exchange on eIF2. The eIF2 is depicted as in Fig. 2. The five subunits of eIF2B are shown as rectangles depicting their amino acid sequences from amino to carboxyl termini. The eIF2B subunits fall into two classes based on sequence similarities, depicted with identical shading or hatching. Asterisks indicate the locations of single-residue mutations in the three regulatory subunits, or in the α subunit of eIF2, that abrogate the inhibitory effects of eIF2α phosphorylation on GEF activity (Gcn⁻ mutations). The GCD6/GCD1 subcomplex has GEF activity that is insensitive to eIF2α phosphorylation, whereas the GCD2/GCD7/GCN3 subcomplex lacks GEF activity but is required to inhibit the catalytic subcomplex when the substrate is phosphorylated. This interfering effect is symbolized by a bar between the two eIF2B subcomplexes blocking the GCD6/GCD1 subcomplex. The regulatory subcomplex has a binding site for eIF2 with a preference for the phosphorylated protein. Genetic data suggest that this latter interaction involves residues in the eIF2B regulatory subunits and in eIF2α which are altered by Gcn⁻ mutations. By analogy with the mammalian system, interaction between GCD2 and eIF2β (*dotted arrow*) could provide a second contact between eIF2 and the regulatory subcomplex. The AA-boxes at the carboxyl terminus of GCD6 contribute to the stability of the eIF2–eIF2B complex through direct interaction with the amino-terminal half of eIF2β containing the three lysine-rich stretches (K-boxes). Data from the mammalian system also suggest that eIF2Bε contacts eIF2β (*dotted arrow*). (See text for further details.)

instances, Gcn⁻ mutations were obtained at the equivalent residue in two different subunits, e.g., Ser-293 in GCN3 and Ser-359 in GCD7. These genetic data imply that homologous segments in all three proteins have similar regulatory roles, with many residues performing related functions (Pavitt et al. 1997). It was suggested that the homologous regulatory segments in GCN3, GCD2, and GCD7 could be juxtaposed to form a binding pocket for the phosphorylated amino-terminal portion of eIF2α.

Biochemical analysis of yeast eIF2B complexes containing selected Gcn$^-$ mutations in GCD7 or GCN3 provided direct evidence that these subunits mediate the inhibitory effect of eIF2(αP) on GEF activity. In vitro exchange assays using purified eIF2–[^{3}H]GDP phosphorylated in vitro and cell extracts containing overexpressed eIF2B subunits confirmed that eIF2(αP)–GDP is a poor substrate for wild-type yeast eIF2B. In contrast, mutant eIF2B complexes containing the Gcn$^-$ substitutions GCD7–S119P (Pavitt et al. 1997) or GCD7–I118T, D178Y (Vazquez de Aldana and Hinnebusch 1994), and the 4-subunit complex lacking GCN3, all catalyzed nucleotide exchange at nearly identical rates on phosphorylated and unphosphorylated eIF2. The wild-type and 4-subunit eIF2B complexes showed comparable binding to eIF2 and a similar preference for the phosphorylated protein. Thus, deleting GCN3 from eIF2B seemed to overcome the inhibition by allowing eIF2B to accept eIF2(αP) as a substrate rather than by substantially reducing its affinity for this inhibitor (Pavitt et al. 1998).

Remarkably, the 2-subunit complex comprising GCD6 and GCD1 had a specific activity higher than that of native eIF2B and accepted phosphorylated or unphosphorylated eIF2–GDP equally well as substrates. Thus, the GCD6/GCD1 subcomplex has GEF activity but cannot distinguish between the two forms of eIF2–GDP. It was proposed that the GCD2/GCD7/GCN3 regulatory subcomplex is required to impede the catalytic function of the GCD6/GCD1 subcomplex when the substrate is phosphorylated (Pavitt et al. 1998). Because eIF2B contains two independent binding sites for eIF2–GDP, devoted to catalysis or negative regulation, and only the latter is sensitive to the phosphorylation status of eIF2, it was suggested that the interaction of eIF2(αP)–GDP with the regulatory subcomplex impedes its proper binding to the active site of the catalytic subcomplex. In this view, the Gcn$^-$ regulatory mutations would overcome this nonproductive interaction and allow isomerization of the eIF2(αP)–GDP–eIF2B complex to the conformation required for nucleotide exchange (Fig. 4) (Pavitt et al. 1998).

At odds with the findings from yeast, a rabbit eIF2B complex lacking the α subunit (GCN3 homolog) was found to be inactive and did not copurify with eIF2. It was suggested that rabbit eIF2Bα is required for catalytic activity, perhaps by promoting substrate binding (Craddock and Proud 1996). In contrast, a rat eIF2B complex devoid of the α subunit, either overexpressed in insect cell extracts or affinity-purified, showed high-level GEF activity that was relatively insensitive to inhibition by phosphorylated eIF2. The latter results are in keeping with the yeast findings in suggesting a regulatory role for the α subunit of rat eIF2B (Fabian et al. 1997; Kimball et al. 1998b). However, the 4-subunit rat eIF2B

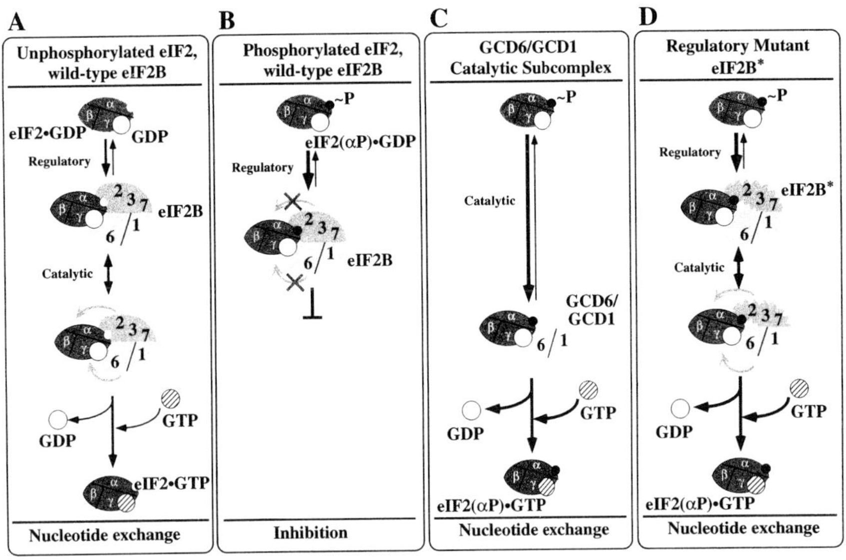

Figure 4 Hypothetical model for eIF2B-catalyzed guanine-nucleotide exchange and its inhibition by eIF2(αP). (*A*) Nucleotide exchange on unphosphorylated eIF2–GDP. The eIF2 (*dark-gray oval* labeled α, β, γ) in a binary complex with GDP (*white ball*) makes initial contact with its binding site in the GCN3/GCD7/GCD2 regulatory subcomplex (*medium-gray shape* labeled 2 3 7). A conformational change in eIF2B (movement indicated by *gray arrows*) allows proper contact between eIF2γ and the GCD6/GCD1 catalytic subcomplex (*light-gray shape* labeled 6 1) required for exchange of GDP with GTP (*hatched ball*). (*B*) Inhibition of nucleotide exchange by phosphorylated eIF2. eIF2(αP)–GDP (as in *A* with added black ball labeled ~P) binds to the eIF2B regulatory subcomplex with higher affinity than does eIF2–GDP (indicated by a broader arrow than in panel *A*), impeding the conformational change required for nucleotide exchange (indicated by "X"s over the arrows). (*C*) In the absence of the eIF2B regulatory subcomplex, direct binding of eIF2γ to the catalytic subcomplex allows nucleotide exchange even with eIF2(αP). (*D*) Regulatory mutant eIF2B can perform nucleotide exchange with eIF2(αP). The eIF2(αP) binds to mutant eIF2B (eIF2B*), making contact with the mutant regulatory subcomplex, as in *A*; however, the mutations permit the conformational change needed for productive interaction between eIF2γ and the catalytic subcomplex with attendant GDP–GTP exchange. (Reprinted, with permission, from Pavitt et al. 1998.)

showed a greater preference for unphosphorylated versus phosphorylated eIF2–GDP as substrate than did the corresponding yeast 4-subunit complex (Kimball et al. 1998b; Pavitt et al. 1998). Two mutations were introduced into the rat δ subunit identical to substitutions in yeast GCD2 that individually rendered yeast eIF2B insensitive to phosphorylated eIF2 in vivo (Gcn⁻ phenotype) (Pavitt et al. 1997). The rat eIF2B bearing the G377K, L381Q double substitution in the δ subunit (eIF2B[δ*]) was only minimally inhibited by preincubation with eIF2(αP), similar to that seen for the 4-subunit eIF2B complex devoid of the α subunit. Unlike the latter, however, the eIF2B(δ*) complex was completely ineffective using eIF2(αP)–GDP as a substrate. Presumably, the eIF2B(δ*) complex escapes inhibition primarily because it binds the phosphorylated inhibitor less tightly than the unphosphorylated substrate (Kimball et al. 1998b). It remains to be seen whether the corresponding mutations in the yeast δ subunit confer reduced binding to eIF2(αP) or increased utilization of eIF2(αP) –GDP as substrate. In any case, evidence from yeast and rat support the notion that the α and δ subunits of eIF2B mediate the inhibitory effects of eIF2(αP) on eIF2B activity.

Point mutations have been isolated in eIF2α near the phosphorylation site at Ser-51 that reduce the inhibitory effect of eIF2(αP) on translation initiation. Several such mutations do not reduce phosphorylation and, therefore, seem to eliminate the inhibitory effect of eIF2(αP)–GDP on eIF2B activity. Alanine substitution of Ser-48 has this effect in mammalian cells when eIF2α is being phosphorylated by PKR or in response to heat shock (Davies et al. 1989; Kaufman et al. 1989; Choi et al. 1992; Murtha-Riel et al. 1993), and the same is true in yeast cells expressing a hyperactive form of GCN2 (Dever et al. 1992). Gcn⁻ mutations at Ile-58, Leu-84, Arg-88, and Val-89 in yeast eIF2α appear to rescue eIF2B activity in the same manner (Fig. 3) (Vazquez de Aldana et al. 1993). It was shown that overexpression of the eIF2α-S48A mutant in mammalian CHO cells rescued eIF2B activity (assayed in cell extracts) from the effects of eIF2α phosphorylation elicited by heat shock in vivo or by addition of exogenous HRI in vitro (Ramaiah et al. 1994). Similarly, when eIF2α-S48A or wild-type eIF2α was expressed in insect cells and added to rabbit reticulocyte lysates, only the mutant protein greatly stimulated eIF2B activity. Presumably, these recombinant eIF2α proteins replaced native eIF2α in the heterotrimeric eIF2 complexes, and the phosphorylated complexes containing eIF2α-S48A inhibited eIF2B less than did those containing wild-type eIF2α (Sudhakar et al. 1999). Interestingly, addition of eIF2α-S48A to the lysates reduced the abundance of 15S complexes containing eIF2, thought to represent inactive eIF2B–eIF2(αP)–GDP complexes stabilized by phosphorylation (Thomas et al. 1985). This last finding supports the idea that Ala-48 reduces the affinity of eIF2(αP)–GDP for eIF2B (Sudhakar et al. 1999).

Although the foregoing results lead to the strong prediction that the eIF2B regulatory subunits directly interact with eIF2α in the region surrounding Ser-51 (Vazquez de Aldana et al. 1993; Hinnebusch 1994), no binding was detected between recombinant mammalian eIF2(αP) and purified eIF2B under conditions where the latter showed significant binding to recombinant eIF2β (Kimball et al. 1998a). Either mammalian eIF2α does not directly interact with eIF2B, or the interaction was too weak to be detected. The carboxy-terminal portion of mammalian eIF2β interacted with both the δ and ε subunits of eIF2B (Kimball et al. 1998a). As discussed below, the amino-terminal portion of yeast eIF2β can interact with eIF2Bε, suggesting that the β subunit of eIF2 provides multiple contacts with eIF2B (Fig. 3).

The rat eIF2Bε (Fabian et al. 1997) and its yeast homolog (GCD6) (Pavitt et al. 1998) can catalyze nucleotide exchange independently of the other subunits, albeit with lower specific activity than native eIF2B. The yeast GCD6/GCD1 subcomplex had much higher GEF activity than GCD6 alone and was comparable in specific activity to the 5-subunit eIF2B. GCD1 alone and its rat homolog, eIF2Bγ, are insufficient for GEF activity (Fabian et al. 1997; Pavitt et al. 1998). The stimulatory effect of GCD1 on the activity of GCD6 was attributed, at least partly, to enhanced binding of eIF2 (Pavitt et al. 1998). GCD6 contains a carboxy-terminal domain harboring the bipartite motif comprising acidic and aromatic residues (AA-boxes) also found at the carboxyl terminus of eIF5 (Figs. 2, 3) (Koonin 1995). As noted above, the AA-boxes in eIF5 are required for a tight complex between eIF5 and the β subunit of yeast eIF2 (SUI3). Similarly, a multiple alanine substitution in the second AA-box of a GST–GCD6 fusion impaired its ability to interact with recombinant eIF2β or purified eIF2 in vitro. The same mutation in native *GCD6 (gcd6-7A)* decreased the stability of the eIF2–eIF2B complex in vivo. Although it did not affect growth rate, *gcd6-7A* conferred a Gcd$^-$ phenotype that could be suppressed by overexpressing eIF2 and initiator tRNA^{Met}, thus implying a reduction in eIF2 recycling. The *gcd6-12A* allele, bearing substitutions in the first AA-box, was lethal, suggesting that the bipartite motif in GCD6 is essential for GEF function (Asano et al. 1999).

The binding of GCD6 to recombinant yeast eIF2β (SUI3), or to native eIF2, is dependent on the three strings of lysine residues (K-boxes) in eIF2β that were implicated in its association with the carboxy-terminal domain of eIF5. All of the single and double K-box mutations in *SUI3* (except for the double mutation in boxes 1 and 3, which was lethal) had a Gcd$^-$ phenotype, consistent with reduced eIF2 recycling. The finding that the carboxy-terminal segments of eIF5 and eIF2Bε both contain AA boxes that interact with the amino-terminal portion of eIF2β bearing the K-boxes suggests that similar molecular interactions are involved in binding of eIF2 to its GAP or GEF. Immunoprecipitation experiments suggested that eIF2B and eIF5 are associated with eIF2 in mutually exclusive complexes in vivo that presumably mediate its recycling and GTPase/AUG-recognition functions, respectively (Asano et al. 1999).

Interestingly, archaea contain all three subunits of eIF2, suggesting that archaeal eIF2 mediates $tRNA_i^{Met}$ binding to ribosomes; however, the β subunit lacks the large amino-terminal domain bearing K-boxes. Moreover, archaea lack recognizable orthologs of eIF5 and eIF2Bε (Bult and al. 1996; Klenk et al. 1997; Smith et al. 1997). Thus, the K-boxes may have been added to eIF2β during evolution to facilitate its interactions with eIF5 and eIF2Bε (Asano et al. 1999). It was reported that archaea contain one or two orthologs of the regulatory subunits of eIF2B (Das et al. 1997b; Dennis 1997); however, these archaeal proteins are more closely related to eukaryotic and eubacterial proteins of unknown function that are distinct from eIF2B subunits (Asano et al. 1999).

A Second Function for eIF2B Late in the Pathway?

The observation that the yeast GCD6/GCD1 eIF2B subcomplex catalyzed nucleotide exchange at the same rate as 5-subunit eIF2B (Pavitt et al. 1998) was surprising, considering that the GCD2 and GCD7 subunits are essential in vivo (Hinnebusch 1996), and it raises the possibility that eIF2B has a second essential function in vivo. This possibility was also suggested previously to explain why temperature-sensitive mutations in *GCD1* (Cigan et al. 1991) or *GCD2* (Foiani et al. 1991) led to accumulation of eIF2 in 43-48S complexes, and why the *gcd2-502* mutant accumulated 40S subunits bound to polysomal mRNAs (halfmer polysomes) (Foiani et al. 1991). These phenotypes imply that initiation is blocked at a step following the association of ternary complex and mRNA with the 40S subunit. This is difficult to explain if eIF2B merely catalyzes nucleotide exchange on nonribosomal eIF2–GDP, because the association of eIF2 and mRNA with 40S ribosomes is dependent on ternary complex formation and this reaction should be impaired in eIF2B mutants. Recently, it was shown that a null mutation in the 40S subunit protein RPS31 suppressed the Gcd^- and Ts^- phenotypes of mutations in *GCD2* and *GCD1*, even though it has a cold-sensitive phenotype in an otherwise wild-type strain (Mueller et al. 1998). This suppression could be explained if elimination of RPS31 from the 40S subunit partially overcomes a requirement for eIF2B in a step carried out on the ribosome. Consistently, eIF2B accumulated in 40S complexes in the *gcd1-101* mutant at the restrictive temperature (Cigan et al. 1991).

Several observations made using rabbit reticulocyte lysates in which translation was inhibited by eIF2α phosphorylation support the idea that eIF2B acts at a late step in the pathway. Baglioni et al. (De Benedetti and Baglioni 1983) found that eIF2(αP) was bound to 48S complexes and that mRNA and initiator $tRNA_i^{Met}$ added to inhibited lysates accumulated in 48S complexes. These workers suggested that 60S joining of the 48S complex was more strongly inhibited than was ternary complex binding in response to eIF2α phosphorylation. Gross et al. (1985, 1987) observed accumulation of 48S complexes and halfmer polysomes containing Met-$tRNA_i^{Met}$ in inhibited lysates that could be reversed by exogenous eIF2B. The 48S complexes lacked eIF2, however, and halfmers did not appear until after protein synthesis was strongly inhibited. Accordingly, Gross et al. proposed that 80S complexes were being formed in the inhibited lysate but could not proceed to the elongation phase, and dissociated into mRNA-bound 40S subunits (halfmers) bearing Met-$tRNA_i^{Met}$ and free 60S subunits.

Another unexpected observation in the studies of Gross et al. was that eIF2–GDP and eIF2(αP)–GDP accumulated on 60S and 80S ribosomes in the inhibited lysate, and only the unphosphorylated species could be displaced from ribosomes by exogenous eIF2B. Because unphosphorylated eIF2–GDP also was found on 60S subunits in uninhibited lysates, they proposed that association of eIF2–GDP with the 60S subunit could have a positive role in subunit joining, and the inability of eIF2B to displace eIF2(αP)–GDP from 60S subunits prevented the formation of stable 80S initiation complexes (Gross et al. 1985, 1987). Levin, London, and colleagues also observed binding of eIF2–GDP and eIF2(αP)–GDP to 60S subunits and polysomes but concluded that phosphorylated eIF2–GDP could be displaced from the 60S subunit by eIF2B. They proposed that eIF2–GDP is transferred from the 48S initiation complex to the 60S subunit following GTP hydrolysis, and that eIF2B releases eIF2–GDP from the 60S subunit for nucleotide exchange and ternary complex formation (Thomas et al. 1985; Ramaiah et al. 1992). Chakrabarti and Maitra (1992) confirmed the transfer of eIF2–GDP from the 40S initiation complex to the 60S subunit in an 80S initiation complex using purified ternary complexes, ribosomes, and eIF5; however, they did not observe displacement of eIF2 from the 80S initiation complex by eIF2B, even though the bound GDP could be exchanged for GTP. In this last study, phosphorylation of eIF2 seemed to have no effect on its interaction with the 60S subunit. In a previous study, Raychaudhuri and Maitra (1986) had concluded that the 40S subunit also contains a binding site for eIF2–GDP.

From the foregoing, it is clear that several groups have observed eIF2–GDP bound to 60S subunits, and it is possible that this complex represents a physiological intermediate in the initiation pathway. It could have a positive role in subunit joining or it could simply represent the product of GTP hydrolysis and release of eIF2–GDP from the 40S to the 60S subunit. If the latter is true, then eIF2B may be required to remove eIF2–GDP from the 60S ribosome in addition to exchanging the GDP for GTP. Moreover, eIF2α phosphorylation may convert 60S-bound eIF2–GDP from a normal intermediate to an inhibitor of the pathway. However, there are discrepancies regarding the ability of eIF2B to dissociate the eIF2–GDP–60S complex and the influence of eIF2α phosphorylation on this reaction. Although it seems likely that eIF2B functions on the ribosome at a late step in the pathway following recruitment of ternary complex and mRNA to the 40S subunit, additional experiments are needed to define the molecular nature of this function and to elucidate the involvement of the ribosome.

There have also been suggestions that eIF2B has an additional function early in the initiation pathway. Gupta et al. (1990) proposed a stimulatory role for eIF2B in ternary complex formation distinct from its GEF activity. Given that the eIF2–GTP binary complex is relatively unstable (Proud 1992), it might not be surprising to find that eIF2B could promote Met-tRNA$_i^{Met}$ binding to eIF2–GTP following GDP–GTP exchange. From their observation of a stable eIF2B–eIF2–GTP–Met-tRNA$_i^{Met}$ quaternary complex, Voorma and coworkers (Salimans et al. 1984) proposed that eIF2B even stimulates the binding of ternary complex to 40S subunits.

REGULATION OF eIF2α KINASE GCN2 BY UNCHARGED tRNA

Evidence That Activation of Kinase Function Requires tRNA Binding, Ribosome Binding, and Dimerization by GCN2

The eIF2α kinase GCN2 is required for the induction of *GCN4* translation in amino-acid-starved yeast cells. GCN2 is expressed constitutively (Wek et al. 1990), and uncharged tRNA appears to be the activating ligand because mutations in aminoacyl-tRNA synthetases stimulate GCN2 function without any limitation for the cognate amino acids (Wek et al. 1995; Hinnebusch 1996). GCN2 contains about 500 residues carboxy-terminal to its kinase domain related to histidyl-tRNA synthetase (HisRS) (Wek et al. 1989), including the conserved "motif 2" sequence that interacts with the acceptor stem of tRNA (Fig. 5) (Ruff et al. 1991). Accordingly, it was proposed that binding of uncharged tRNA to the HisRS-like domain would produce a conformational change in GCN2 that allows the kinase domain to phosphorylate eIF2α (Fig. 6) (Wek et al. 1989). Because starvation for numerous amino acids will activate GCN2 (Wek et al. 1995; Hinnebusch 1996), it was presumed that the HisRS-like domain can bind many, if not all, deacylated tRNAs with similar affinity. In agreement with this model, the HisRS-like domain was shown to bind total uncharged yeast tRNA in vitro dependent on motif-2 residues, and a mutation in motif 2 (*gcn2-Y1119L, R1120L*) abolished GCN2 kinase activity in vivo and in vitro (Wek et al. 1995; Zhu et al. 1996). Furthermore, numerous activating *GCN2ᶜ* mutations alter residues in the HisRS-like region (Fig. 5) (Wek et al. 1990; Ramirez et al. 1992; Diallinas and Thireos 1994). These latter alterations may increase the affinity of GCN2 for uncharged tRNA, allowing activation to occur without amino acid starvation. Other *GCN2ᶜ* mutations mapping in the kinase domain might eliminate an inhibitory conformation that is normally overcome by binding of uncharged tRNA.

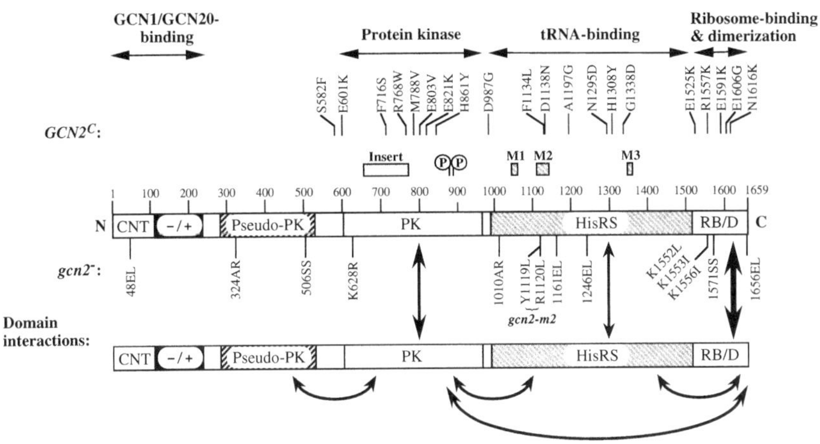

Figure 5 Functional domains of GCN2. The rectangles depict the 1659-amino-acid sequence of GCN2 from amino (N) to carboxyl (C) terminus with the conserved N-terminal domain (CNT), highly charged region (-/+), pseudokinase domain (Pseudo-PK), functional protein kinase domain (PK), histidyl-tRNA synthetase-like region (HisRS), and carboxy-terminal domain (RB/D, for ribosome-binding and dimerization), shown with shading or stippling. Above the upper schematic is shown (from top to bottom) (*1*) the biochemical functions assigned to the various domains; (2) the locations of *GCN2ᶜ* substitutions; (3) the locations of a large insert in the PK domain characteristic of eIF2α kinases (Insert), the autophosphorylation sites at T-882 and T-887 (encircled Ps), and the signature motifs (M1 to M3) of class II aminoacyl-tRNA synthetases; and the amino acid positions in GCN2 (1–1659). Below the top schematic are the locations of *gcn2* substitutions (4). The arrows connecting the identical domains in the top and bottom schematics summarize the known dimerization interactions in GCN2; similarly, the arrows beneath the lower schematic summarize known interdomain interactions. (See text for further details.)

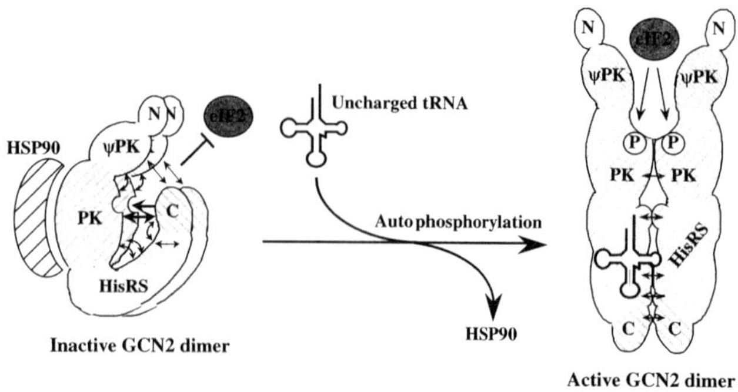

Figure 6 Hypothetical model for the role of domain interactions and dimerization in activation of GCN2 by uncharged tRNA. The domains of GCN2 are depicted schematically and labeled as in Fig. 5 except for the designations of the conserved amino-terminal domain and charged region (N), the pseudokinase domain (ψPK), and carboxy-terminal domain (C). The inactive form of GCN2 present under nonstarvation conditions, when uncharged tRNA is scarce, is depicted as a dimer (*left*) because the tRNA-binding (HisRS) domain is dispensable for dimerization. The C-term domain is shown as an autoinhibitory segment that interacts with the kinase domain (depicted by a broad double-headed arrow) and blocks autophosphorylation by GCN2 and binding of substrates to the active site. The additional interactions between the PK domain and the HisRS or ψPK regions, between the HisRS and C-term domain, and between the ψPK and C-term domains (all double-headed arrows) might contribute to this inactive state. Recent findings suggest that the inactive form of GCN2 is stabilized through complex formation with HSP82, the yeast homolog of mammalian HSP90. Binding of uncharged tRNA to the HisRS region triggers a conformational change in the dimer that disrupts the inhibitory interaction between the PK and C-term segments (*right*) and leads to dissociation of HSP90. *trans*-Autophosphorylation of the GCN2 subunits ensues, altering the structure of the PK domain to permit binding and phosphorylation of the substrate. (See text for further details.)

It was shown that GCN2 can be activated by purine starvation of yeast cells in amino-acid-replete medium, in a manner dependent on the HisRS-like region (Rolfes and Hinnebusch 1993). This observation could be explained by proposing that purine starvation leads to deacylation of one or more tRNAs in vivo, e.g., by interfering with aminoacyl-tRNA synthetase function. Alternatively, tRNA binding to the HisRS-like domain may be a prerequisite for kinase activation, and purine starvation could elicit a modification of the protein that increases its affinity for uncharged tRNA, thereby lowering the concentration of uncharged tRNA required to trigger kinase activation.

There is strong biochemical evidence that GCN2 interacts with translating ribosomes. The GCN2 in cell extracts cosediments with polysomes, monosomes, and free subunits, but interacts primarily with the 60S subunit when 80S ribosomes are dissociated at low [Mg^{++}]. Deletion analysis revealed that 60S-binding was dependent on the carboxy-terminal ~120 residues of GCN2, particularly residues 1536–1570 (Fig. 5) (Ramirez et al. 1991). (Recent results indicate that GCN2 contains 69 amino acids at its amino terminus that were not previously assigned to the protein but are essential for its function in vivo [Garcia-Barrio et al. 2000]; thus, the positions of amino acids cited here may be larger by 69 residues than in the original reports.) The carboxy-terminal segment (C-term) containing residues 1536–1659 expressed in yeast was sufficient for binding to polysomes and 60S subunits in the same manner as full-length GCN2. Substitution of three closely spaced lysines at positions 1552, 1553, and 1556 abolished polysome binding by GCN2 and destroyed its kinase function in vivo. A recombinant form of the C-term fragment bound poly(I)-poly(C) in vitro dependent on these lysine residues. It was proposed that the GCN2 C-term binds to a base-paired segment in 28S rRNA in the 60S subunit and that this association is crucial for GCN2 function (Zhu and Wek 1998). Ribosome binding could be required to localize GCN2 with its substrate on polysomes. Alternatively, it has been suggested that GCN2 is activated by uncharged tRNA base-paired with a cognate codon in the A site of the ribosome (Fig. 7) (Ramirez et al. 1992). This would be akin to activation of the RelA protein of *E. coli* by uncharged tRNA in the stringent response to amino acid starvation (Cashel and Rudd 1987).

A number of activating *GCN2*^{c} mutations map to the C-term, suggesting a more direct function for this segment in regulating kinase activity by uncharged tRNA (Wek et al. 1990; Ramirez et al. 1992; Diallinas and Thireos 1994). The C-term also can dimerize in yeast two-hybrid and in vitro interaction assays. Although the isolated kinase domain also dimerized in these assays, only the C-term was required for coimmunoprecipitation of full-length GCN2 with a lexA–GCN2 fusion expressed in yeast. Because deletions in the HisRS region only slightly reduced GCN2/lexA-GCN2 heterodimers, it appears that tRNA binding is not required for dimerization (Qiu et al. 1998). This conclusion is also consistent with the fact that the yield of GCN2/lexA-GCN2 heterodimers in cell extracts was not increased by amino acid starvation of the cells, although deacylated tRNA may be generated in extracts regardless of the growth conditions (Zhu et al. 1996). If tRNA-binding does not trigger dimerization, then it seems probable that GCN2 exists as a dimer constitutively and that tRNA stabilizes the active conformation of the dimer (Fig. 6).

Two-hybrid analysis of a series of small internal deletions in the C-term suggested the presence of a bipartite dimerization domain spanning positions 1498–1598 (Qiu et al. 1998). All of these internal deletions abolished GCN2 function in vivo, consistent with an important role for dimerization; however, they might also inactivate GCN2 by impairing its ribosome-binding activity. Other evidence supporting an important role for dimerization by GCN2 is that the Slg^- phenotype of *GCN2^c-E1456K* was suppressed by wild-type *GCN2*. To account for this finding, it was suggested that GCN2/GCN2^c heterodimers are less active than GCN2^c/GCN2^c homodimers (Diallinas and Thireos 1994). Similarly, overexpression of GST fusions to the GCN2 C-term or kinase domains suppressed the Slg^- phenotype associated with a *GCN2^c* allele (Qiu et al. 1998). However, these genetic observations could be explained differently by invoking competition between GCN2 or GST–GCN2 fusions and the GCN2^c proteins for binding to the substrate, uncharged tRNA, ribosomes, or the positive effectors GCN1 and GCN20.

GCN2 autophosphorylates in vitro on threonine residues 882 and 887 located in the "activation loop" between kinase subdomains VII and VIII (Fig. 5), and these residues are important (Thr-882) or essential (Thr-887) for GCN2 function in vivo. By analogy with other protein kinases, the subunits of a GCN2 dimer may autophosphorylate these residues in *trans* (Fig. 6). Interestingly, all known eIF2α kinases contain threonine residues at the corresponding positions, and it appears that autophosphorylation of these residues by PKR is important for its function as well (Romano et al. 1997). GCN2 also appears to be a substrate for one or more unknown protein kinases in vivo (Wek et al. 1990), but the sites of phosphorylation remain to be identified.

In vitro protein interaction assays revealed that the GCN2 kinase domain can interact with the C-term, HisRS region, and a degenerate kinase domain located just amino-terminal to the kinase domain. The kinase domain/C-term interaction also was observed using the two-hybrid assay (Fig. 5) (Qiu et al. 1998). Given that many *GCN2^c* alleles map to the C-term, physical interaction between the C-term and kinase domain may be required to repress GCN2 activity under nonstarvation conditions (Fig. 6). Although this proposal seems at odds with the fact that the C-term is required for GCN2 function (Wek et al. 1990), the latter could reflect its positive role in dimerization or ribosome binding. Presumably, the additional domain interactions in GCN2 summarized in Figure 5 are involved in stabilizing the inactive or active conformations of the GCN2 dimer (Figs. 5, 6) (Qiu et al. 1998).

Overexpression of an inactive amino-terminally truncated form of GCN2 impaired the induction of *GCN4* translation in amino-acid-starved cells containing wild-type GCN2. Overexpression of a segment of *GCN4* mRNA leader containing the uORFs had a similar effect, and the negative effects of these inactive components were neutralized by overexpressing them together in the same cells. To explain these results, it was proposed that GCN2 specifically binds to the *GCN4* mRNA leader and this interaction is required for efficient derepression of *GCN4* translation. Presumably, this interaction would promote localized phosphorylation of eIF2 in the vicinity of *GCN4* mRNA, allowing induction of *GCN4* translation with minimal effects on other mRNAs (Tavernarakis and Thireos 1996). Although this mechanism may increase the sensitivity of *GCN4* translation to eIF2 phosphorylation, it does not explain why mutations in eIF2B subunits lead to high-level *GCN4* translation in *gcn2Δ* mutants with little effect on other mRNAs or on *GCN4* derivatives lacking uORFs (Harashima and Hinnebusch 1986; Mueller et al. 1987). It still seems necessary to postulate that reinitiation on *GCN4* mRNA is more sensitive than conventional initiation events to reductions in ternary complex levels.

The GCN1/GCN20 Complex Binds to Ribosomes and Is Required for GCN2 Function In Vivo

The *GCN1*- and *GCN20*-encoded proteins are required for activation of GCN2 in starved cells. Deletions of these genes reduce (*GCN20*) or abolish (*GCN1*) eIF2α phosphorylation by GCN2 in vivo but have no such effect in strains expressing PKR in place of GCN2 (Marton et al. 1993; Vazquez de Aldana et al. 1995). The latter suggests that GCN1 and GCN20 are required to increase GCN2 function, not to inhibit an eIF2α phosphatase. *gcn1* and *gcn20* mutations do not reduce *GCN2* expression, nor do they diminish GCN2 activity in immune-complex kinase assays (Marton et al. 1993; Vazquez de Aldana et al. 1995); accordingly, it was proposed that GCN1 and GCN20 are required to mediate activation of GCN2 by uncharged tRNA under physiological conditions in vivo.

GCN1 and GCN20 exist in a stable complex in vivo (Vazquez de Aldana et al. 1995), and both proteins have sequence similarity to translation elongation factor 3 (eEF3) (Marton et al. 1993; Vazquez de Aldana et al. 1995). eEF3 is an essential protein, unique to fungi, belonging to the ATP-binding cassette (ABC) superfamily, and possesses a ribosome-stimulated ATPase activity (Skogerson and Wakatama 1976). It is thought that eEF3 functions in every round of elongation to stimulate binding of eEF1A/GTP/aminoacyl-tRNA ternary complex to the A site and release of deacylated tRNA from the ribosomal E (exit) site (Triana-Alonso et al. 1995). eEF3 may also promote binding of cognate tRNAs at the expense of noncognate tRNAs to the A site (Uritani and Miyazaki 1988; Kamath and

Chakraburtty 1989; Sandbaken et al. 1990). GCN1 is ~ 297 kD in mass and only its middle one-third is similar to eEF3, showing greatest similarity to the region amino-terminal to the ABCs in eEF3. GCN1 does not contain the invariant residues characteristic of ABC proteins (Marton et al. 1993). With a molecular weight of ~85,000, GCN20 is similar in size to eEF3 and contains two ABCs bearing the conserved signature sequences. GCN20 is more highly related to eEF3 than to all other ABC proteins (except for a single yeast protein of unknown function), but shows little similarity to eEF3 outside of the ABC domains (Vazquez de Aldana et al. 1995). The amino-terminal domain of GCN20 interacts with the eEF3-like region in GCN1, in effect juxtaposing their eEF3-like domains in a heterodimer (Fig. 7). The *gcn1-G1444D* mutation in the eEF3-like region of GCN1 weakens GCN1/GCN20 complex formation in vivo and has a Gcn$^-$ phenotype, consistent with an important role for their association in vivo. Underscoring its functional significance, the eEF3-like region is the most highly conserved segment in a human GCN1 ortholog (Marton et al. 1997).

Substantial amounts of GCN1 and GCN20 cosediment with polysomes and 80S ribosomes in a manner stimulated by ATP. High-level ribosome binding by GCN20 was found to be dependent on complex formation with GCN1. Point mutations in conserved residues of both ABC domains of GCN20, or deletion of the ABC domains, abolished ribosome binding by GCN20 and led to low-level ribosome binding by GCN1 that was relatively insensitive to ATP. Thus, GCN1 seems to have intrinsic ribosome-association activity, whereas GCN20 mediates ATP-enhanced binding to ribosomes by the GCN1/GCN20 complex. The ability of GCN1 and GCN20 to interact with translating ribosomes, combined with the similarity between these proteins and eEF3, led to the suggestion that GCN1/GCN20 function on the ribosome to promote activation of GCN2 by uncharged tRNA bound to the A site (Marton et al. 1997). GCN1 might perform an eEF3-like function to promote A-site binding of uncharged tRNA (that would be independent of eEF1A) or to transfer uncharged tRNA from the A site to the HisRS-like domain of GCN2 (Fig. 7) (Marton et al. 1997). Although the GCN20 ABCs promote ribosome binding by GCN1 and GCN20 in cell extracts, they are dispensable for derepression of *GCN4* in histidine-starved cells (Vazquez de Aldana et al. 1995; Marton et al. 1997). Thus, the high levels of ribosome binding by GCN1/GCN20 observed in vitro in the presence of ATP probably are not required for activation of GCN2 in amino-acid-starved cells. Perhaps high levels of ribosome-bound GCN1/GCN20 permit activation of GCN2 by relatively low levels of uncharged tRNA, providing a mechanism for stimulating GCN2 under stress conditions besides amino acid starvation (Marton et al. 1997).

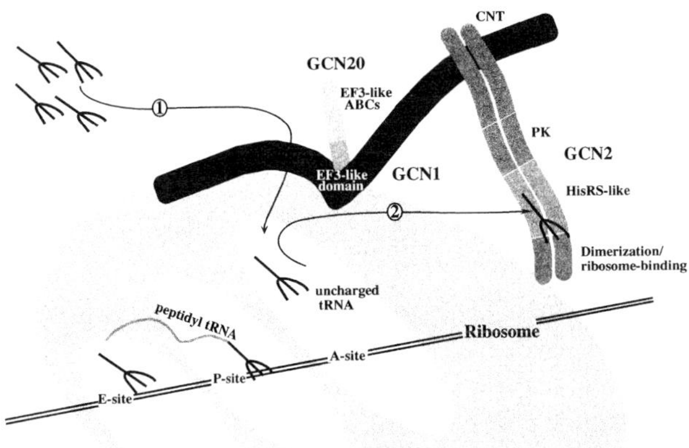

Figure 7 Hypothetical model for the stimulatory role of GCN1/GCN20 complex in the activation of GCN2 by uncharged tRNA in the ribosomal A site. GCN1 is shown as a black ribbon containing near its center the EF-3 like region. GCN20 is shown as a lightly shaded rod, with the EF3-related ATP-binding cassettes (ABCs) located toward the carboxyl terminus and the amino-terminal segment bound to the EF3-like region of GCN1. A GCN2 dimer is shown as a pair of medium gray ribbons with tRNA (a black fork) bound to the HisRS-like domains. Both GCN1/GCN20 and GCN2 have ribosome-binding activities, which for GCN2 is conferred by its carboxy-terminal domain. By analogy with the activation of *E. coli* RelA protein by uncharged tRNA, it was proposed that uncharged tRNA bound to the ribosomal A site and base-paired with the cognate codon in mRNA is the activating ligand for GCN2. On the basis of their similarity to EF3, it was further suggested that GCN1 and GCN20 facilitate binding of uncharged tRNA to the A site (*arrow labeled 1*), or transfer of tRNA from A site to HisRS-like domain in GCN2 (*arrow 2*). The physical association between GCN1/GCN20 and the amino-terminal portion of GCN2 (CNT) is consistent with this second mechanism.

It was shown recently that GCN2 directly interacts with the GCN1/GCN20 complex both in vivo and in vitro. This interaction does not require, but is enhanced by, GCN20, indicating that GCN1 is the principal binding partner for GCN2 in the GCN1/GCN20 complex. The evolutionarily conserved amino-terminal domain of GCN2 and an adjacent highly charged region are necessary and sufficient for complex formation with GCN1. Overexpression of amino-terminal GCN2 segments led to a dominant Gcn⁻ phenotype in a manner correlated with their ability to associate with GCN1/GCN20 and impede interaction between GCN1 and native GCN2. Consistently, this Gcn⁻ phenotype was completely or partially suppressed by overexpressing GCN2, GCN1/GCN20, or $tRNA^{His}$. The requirement for *GCN1* also was lessened by overexpressing $tRNA^{His}$ in a *gcn1Δ* strain. Thus, it appears that binding of GCN1/GCN20 to the amino-terminal region of GCN2 is important for activation of GCN2 by uncharged tRNA in vivo (Garcia-Barrio et al. 2000). If GCN1/GCN20 binds in the vicinity of the ribosomal A site, as suggested above, its interaction with the amino terminus of GCN2 might serve to position the HisRS-like domain for association with uncharged tRNA bound in the A site (Fig. 7).

Evidence for the Involvement of Molecular Chaperone HSP90 in the Expression and Regulation of GCN2

There is recent evidence that the maturation and proper regulation of GCN2 depend on its physical interaction with the protein chaperone HSP90 (known as HSP82 in yeast). Reduced expression of wild-type *HSP82*, or two different nonlethal mutations in *HSP82* (*G313N* and *T525I*), led to derepression of *GCN4* translation under nonstarvation conditions (Gcd^- phenotype), dependent on *GCN2*, and additional derepression occurred in response to a histidine limitation. These findings suggest that the inactivity of GCN2 under nonstarvation conditions is dependent on HSP82. In wild-type yeast strains treated with an inhibitor of HSP82 (macbecin I), or in a temperature-sensitive mutant harboring the *hsp82-G170D* allele at the restrictive temperature, newly synthesized GCN2 induced from a galactose-regulated promoter appeared to be highly unstable. Additionally, mutations in several putative cochaperones of HSP82 (*CDC37*, *SBA1*, and *STI*), conferred sensitivity to 3-aminotriazole, an inhibitor of the histidine biosynthetic enzyme encoded by *HIS3*, a hallmark of Gcn^- mutants. These last findings suggest an additional requirement for HSP82 for the synthesis of GCN2. A fraction of GCN2 was found associated with HSP82 in yeast cells (Donzé and Picard 1999).

To account for these findings, it was proposed that HSP82 binds to nascent GCN2 and plays a critical role in achieving the proper conformation of the mature kinase. Interfering with this function of HSP82 by a severe mutation in *HSP82*, with macbecin I, or by inactivation of cochaperones, would lead to degradation of GCN2 (Gcn^- phenotype). HSP82 would remain bound to GCN2 and help to maintain the latency of the kinase domain under nonstarvation conditions. The nonlethal *G313N* and *T525I* mutations in HSP82 would impair this regulatory activity, leading to inappropriate activation of GCN2 in nonstarved cells (Gcd^- phenotype). Under starvation conditions in wild-type cells, binding of uncharged tRNA would release HSP82 from GCN2, concomitant with, or as the principal means of stimulating, kinase activity (Fig. 6) (Donzé and Picard 1999).

Ostensibly at odds with this model, no detectable HSP82 was coimmunoprecipitated with native GCN2 under conditions where the association of GCN1 with GCN2 was readily detected (Garcia-Barrio et al. 2000). However, if binding uncharged tRNA triggers dissociation of the putative HSP82–GCN2 complex, as proposed, it may be difficult to detect the complex in cell extracts that seem to contain high levels of deacylated tRNA (Zhu et al. 1996). It is important to rule out the possibility that the *G313N* and *T525I* mutations in HSP82 derepress *GCN4* translation indirectly by interfering with amino acid biosynthesis, and to verify that these mutations do not elevate *GCN2* expression (Tzamarias et al. 1989; Wek et al. 1990). Furthermore, it should be demonstrated that the cochaperone mutations reduce GCN2 expression and impair derepression of *GCN4* translation, because there are other possible interpretations of the 3-aminotriazole-sensitive phenotype of these mutants.

Sequence Conservation of GCN2 Found in Fungi, Insects, and Mammals

The domain structure of yeast GCN2 depicted in Figure 5 is conserved in GCN2 orthologs identified in *N. crassa* (Sattlegger et al. 1998), *D. melanogaster* (Santoyo et al. 1997; Olsen et al. 1998), and mouse (Berlanga et al. 1999; Sood et al. 2000). The highest similarity among these proteins occurs in the kinase domains, all of which contain 10 or 11 signature residues conserved among the known eIF2α kinases but not among kinases at large, plus characteristic nonconserved inserts between kinase subdomains IV and V (Ramirez et al. 1992). In addition to these sequence similarities, evidence that the *N. crassa* protein (known as CPC3) is a functional homolog of GCN2 includes the fact that strains disrupted for *cpc-3* phenotypically resemble yeast *gcn2* mutants by failing to induce amino acid biosynthetic genes in response to amino acid starvation. Moreover, *cpc-3* mutants fail to induce CPC1, the structural and functional homolog of GCN4, even though the *cpc-1* transcript is made, and thus appear to be defective for translational induction of CPC1 (Sattlegger et al. 1998). Consistent with this interpretation, the *cpc-1* transcript becomes associated with larger polysomes in response to histidine starvation in wild-type cells (Luo et al. 1995) and contains two short uORFs (Paluh et al. 1988) with limited sequence similarity to *GCN4* uORFs 1 and 4. Therefore, it is very likely that CPC3 regulates *cpc-1* translation by a mechanism similar to that described for GCN2 and *GCN4* mRNA in *S. cerevisiae*. As with yeast GCN2 (Hinnebusch 1996), CPC3 is a nonessential protein required for viability only under starvation conditions (Sattlegger et al. 1998).

Evidence that *Drosophila* GCN2 (dGCN2) functions as an eIF2α kinase includes the fact that it specifically phosphorylates the α subunit of eIF2 in immune-complex assays (Santoyo et al. 1997) and that it restores growth of a yeast *gcn2Δ* mutant on amino-acid-starvation medium dependent on Ser-51 of eIF2α (Olsen et al. 1998). *dGCN2* mRNA is expressed at different levels throughout development and in the adult fly (Santoyo et al. 1997). It appears to be maternally deposited and, thus, may have an important role in early development (Olsen et al. 1998). Interestingly, *dGCN2* mRNA seems to be concentrated in a small subset of cells in the nervous system late in embryogenesis (Santoyo et al. 1997).

Mouse GCN2 (mGCN2) expressed in yeast could substitute for endogenous GCN2 in stimulating amino acid biosynthesis in a manner dependent on eIF2α Ser-51 and the δ subunit of eIF2B (GCD2) (Sood et al. 2000). Moreover, immuno- or affinity-purified mGCN2 phosphorylated eIF2α in vitro dependent on Ser-51, confirming its function as an eIF2α kinase (Berlanga et al. 1999; Sood et al. 2000). The kinase activity of mGCN2 was dependent on conserved residues in motif 2 of its HisRS-like domain, providing strong evidence that this enzyme is activated by uncharged tRNA. Indeed, there is evidence that amino acid deprivation and defects in tRNA aminoacylation lead to increased eIF2α phosphorylation, inhibition of eIF2B function, and decreased protein synthesis in mammalian cells (Clemens et al. 1987; Kimball et al. 1991; Chapter 16). It seems likely that mammalian GCN2 is responsible for the increased eIF2α phosphorylation that occurs under these starvation conditions.

The conserved amino-terminal domain (CNT) of *Drosophila* GCN2 was found to interact with yeast GCN1/GCN20 in yeast cells and its overexpression conferred a dominant Gcn$^-$ phenotype similar to that observed when the corresponding portion of yeast GCN2 was overproduced. Thus, GCN1/GCN20 binding appears to be an evolutionarily conserved function of the GCN2 amino-terminal region (Garcia-Barrio et al. 2000). This observation, together with the identification of GCN1 and GCN20 orthologs in mammals (Vazquez de Aldana et al. 1995; Marton et al. 1997) and *Drosophila* (E. Sattlegger and A.G. Hinnebusch, unpubl.), suggests that the role of GCN1/GCN20 complex in activation of GCN2 by uncharged tRNA is conserved between yeast and mammals. Interestingly, there are different isoforms of mGCN2 that differ at the amino terminus, and only the β-isoform contains the predicted binding domain for GCN1/GCN20. This isoform seems to be the most ubiquitously expressed, whereas the other two exhibit tissue-specific expression (Sood et al. 2000). It was shown that serum starvation of transiently transfected 293T cells expressing recombinant mGCN2 led to increased eIF2α phosphorylation (Berlanga et al. 1999). It is important to investigate whether eIF2α phosphorylation by mGCN2 (and dGCN2) is also stimulated by amino acid limitation, and to investigate the consequences of this regulatory response for cellular metabolism and developmental pathways in these animals.

ACKNOWLEDGMENTS

I thank Tom Dever, and Evelyn Sattlegger, Leos Valasek, and other members of my laboratory for helpful comments on the manuscript; Katsura Asano, Hongfang Qiu, and Evelyn Sattlegger for assistance with graphics; and Bobbie Felix for help in preparation of the manuscript.

REFERENCES

Anderson J., Phan L., Cuesta R., Carlson B.A., Pak M., Asano K., Bjork G.R., Tamame M., and Hinnebusch A.G. 1998. The essential Gcd10p-Gcd14p nuclear complex is required for 1-methyladenosine modification and maturation of initiator methionyl-tRNA. *Genes Dev.* **12:** 3650–3662.

Asano K., Krishnamoorthy T., Phan L., and Hinnebusch A.G. 1999. Conserved bipartite motifs in yeast eIF5 and eIF2Bε, GTPase-activating and GDP-GTP exchange factors in translation initiation, mediate binding to their common substrate eIF2. *EMBO J.* **18:** 1673–1688.

Asano K., Phan L., Anderson J., and Hinnebusch A.G. 1998. Complex formation by all five homologues of mammalian translation initiation factor 3 subunits from yeast *Saccharomyces cerevisiae*. *J. Biol. Chem.* **273:** 18573–18585.

Åström S.U. and Byström A.S. 1994. Rit1, a tRNA backbone-modifying enzyme that mediates initiator and elongator tRNA discrimination. *Cell* **79:** 535–546.

Åström S.U., von Pawel-Rammingen U., and Byström A.S. 1993. The yeast initiator $tRNA^{Met}$ can act as an elongator $tRNA^{Met}$ *in vivo*. *J. Mol. Biol.* **233:** 43–58.

Åström S.U., Nordlund M.E., Erickson F.L., Hannig E.M., and Byström A.S. 1999. Genetic interactions between a null allele of the *RIT1* gene encoding an initiator tRNA-specific modification enzyme and genes encoding translation factors in *Saccharomyces cerevisiae*. *Mol. Gen. Genet.* **261:** 967–976.

Bandyopadhyay A. and Maitra U. 1999. Cloning and characterization of the p42 subunit of mammalian translation initiation factor 3 (eIF3): Demonstration that eIF3 interacts with eIF5 in mammalian cells. *Nucleic Acids Res.* **27:** 1331–1337.

Basavappa R. and Sigler P.B. 1991. The 3 Å crystal structure of yeast initiator tRNA: Functional implications in initiator/elongator discrimination. *EMBO J.* **10:** 3105–3111.

Battiste J.B., Pestova T.V., Hellen C.U.T., and Wagner G. 2000. The eIF1A solution structure reveals a large RNA-binding surface important for scanning function. *Mol. Cell* **5:** 109–119.

Benne R. and Hershey J.W.B. 1976. Purification and characterization of initiation factor IF-E3 from rabbit reticulocytes. *Proc. Natl. Acad. Sci.* **73:** 3005–3009.

—-. 1978. The mechanism of action of protein synthesis initiation factors from rabbit reticulocytes. *J. Biol. Chem.* **253:** 3078–3087.

Berlanga J.J., Santoyo J., and De Haro C. 1999. Characterization of a mammalian homolog of the GCN2 eukaryotic initiation factor 2α kinase. *Eur. J. Biochem.* **265:** 754–762.

Block K.L., Vornlocher H.P., and Hershey J.W.B. 1998. Characterization of cDNAs encoding the p44 and p35 subunits of human translation initiation factor eIF3. *J. Biol. Chem.* **273:** 31901–31908.

Brock S., Szkaradkiewicz K., and Sprinzl M. 1998. Initiation factors of protein biosynthesis in bacteria and their structural relationship to elongation and termination factors. *Mol. Microbiol.* **29:** 409–417.

Bult C.J., White O., Olsen G.J., Zhou L., Fleischmann R.D., Sutton G.G., Blake J.A., FitzGerald L.M., Clayton R.A., Gocayne J.D., Kerlavage A.R., Dougherty B.A., Tomb J.F., Adams M.D., Reich C.I., Overbeek R., Kirkness E.F., Weinstock K.G., Merrick J.M., Glodek A., Scott J.L., Geoghagen N.S.M., and Venter J.C. 1996. Complete genome sequence of the methanogenic archaeon, *Methanococcus jannaschii. Science* **273:** 1058–1073.

Buratti E., Tisminetzky S., Zotti M., and Baralle F.E. 1998. Functional analysis of the interaction between HCV 5'UTR and putative subunits of eukaryotic translation initiation factor eIF3. *Nucleic Acids Res.* **26:** 3179–3187.

Bushman J.L., Asuru A.I., Matts R.L., and Hinnebusch A.G. 1993. Evidence that GCD6 and GCD7, translational regulators of *GCN4* are subunits of the guanine nucleotide exchange factor for eIF-2 in *Saccharomyces cerevisiae. Mol. Cell. Biol.* **13:** 1920–1932.

Bycroft M., Hubbard T.J.P., Proctor M., Freund S.M.V., and Murzin A.G. 1997. The solution structure of the S1 RNA binding domain: A member of an ancient nucleic acid-binding fold. *Cell* **88:** 235–242.

Calvo O., Cuesta R., Anderson J., Gutierrez N., Garcia-Barrio M.T., Hinnebusch A.G., and Tamame M. 1999. GCD14p, a repressor of *GCN4* translation, cooperates with Gcd10p and Lhp1p in the maturation of initiator methionyl-tRNA in *Saccharomyces cerevisiae. Mol. Cell. Biol.* **19:** 4167–4181.

Cashel M. and Rudd K.E. 1987. The stringent response. In *Escherichia coli* and *Salmonella typhimurium: Cellular and molecular biology* (ed. F.C. Neidhardt et al.), pp. 1410–1438. American Society for Microbiology, Washington, D.C.

Castilho-Valavicius B., Thompson G.M., and Donahue T.F. 1992. Mutation analysis of the Cys-X_2-Cys-X_{19}-Cys-X_2-Cys motif in the α subunit of eukaryotic translation factor 2. *Gene Expr.* **2:** 297–309.

Chakrabarti A. and Maitra U. 1992. Release and recycling of eukaryotic initiation factor 2 in the formation of an 80 S ribosomal polypeptide chain initiation complex. *J. Biol. Chem.* **267:** 12964–12972.

Chapman K.B., Byström A.S., and Boeke J.D. 1992. Initiator methionine tRNA is essential for Ty1 transposition in yeast. *Proc. Natl. Acad. Sci.* **89:** 3236–3240.

Chaudhuri J., Chakrabarti A., and Maitra U. 1997a. Biochemical characterization of mammalian translation initiation factor 3 (eIF3). *J. Biol. Chem.* **272:** 30975–30983.

Chaudhuri J., Chowdhury D., and Maitra U. 1999. Distinct functions of eukaryotic translation initiation factors eIF1A and eIF3 in the formation of the 40S ribosomal preinitiation complex. *J. Biol. Chem.* **274:** 17975–17980.

Chaudhuri J., Das K., and Maitra U. 1994. Purification and characterization of bacterially expressed mammalian translation initiation factor 5 (eIF-5): Demonstration that eIF-5 forms a specific complex with eIF-2. *Biochemistry* **33:** 4794–4799.

Chaudhuri J., Si K., and Maitra U. 1997b. Function of eukaryotic translation initiation factor 1A (eIF1A) (formerly called eIF-4C) in initiation of protein synthesis. *J. Biol. Chem.* **272:** 7883–7891.

Choi S.K., Lee J.H., Zoll W.L., Merrick W.C., and Dever T.E. 1998. Promotion of Met-$tRNA_i^{Met}$ binding to ribosomes by yIF2, a bacterial IF2 homolog in yeast. *Science* **280:** 1757–1760.

Choi S.Y., Scherer B.J., Schnier J., Davies M.V., and Kaufman R.J. 1992. Stimulation of protein synthesis in COS cells transfected with variants of the α-subunit of initiation factor eIF-2. *J. Biol. Chem.* **267:** 286–293.

Chong K.L., Feng L., Schappert K., Meurs E., Donahue T.F., Friesen J.D., Hovanessian A.G., and Williams B.R.G. 1992. Human p68 kinase exhibits growth suppression in yeast and homology to the translational regulator GCN2. *EMBO J.* **11:** 1553–1562.

Cigan A.M., Bushman J.L., Boal T.R., and Hinnebusch A.G. 1993. A protein complex of translational regulators of *GCN4* is the guanine nucleotide exchange factor for eIF-2 in yeast. *Proc. Natl. Acad. Sci.* **90:** 5350–5354.

Cigan A.M., Foiani M., Hannig E.M., and Hinnebusch A.G. 1991. Complex formation by positive and negative translational regulators of *GCN4*. *Mol. Cell. Biol.* **11**: 3217-3228.

Cigan A.M., Pabich E.K., Feng L., and Donahue T.F. 1989. Yeast translation initiation suppressor *sui2* encodes the α subunit of eukaryotic initiation factor 2 and shares identity with the human α subunit. *Proc. Natl. Acad. Sci.* **86:** 2784–2788.

Clemens M.J., Galpine A., Austin S.A., Panniers R., Henshaw E.C., Duncan R., Hershey J.W., and Pollard J.W. 1987. Regulation of polypeptide chain initiation in Chinese hamster ovary cells with a temperature-sensitive leucyl-tRNA synthetase. Changes in phosphorylation of initiation factor eIF-2 and in the activity of the guanine nucleotide exchange factor GEF. *J. Biol. Chem.* **262:** 767–771.

Craddock B.L. and Proud C.G. 1996. The α-subunit of the mammalian guanine nucleotide-exchange factor eIF-2B is essential for catalytic activity in vivo. *Biochem. Biophys. Res. Commun.* **220:** 843–847.

Crouch D. and Safer B. 1984. The association of eIF-2 with Met-$tRNA_i$ or eIF-2B alters the specificity of eIF-2 phosphatase. *J. Biol. Chem.* **259:** 10363–10368.

Cuesta R., Hinnebusch A.G., and Tamame M. 1998. Identification of *GCD14* and *GCD15*, novel genes required for translational repression of*GCN4* mRNA in *Saccharomyces cerevisiae*. *Genetics* **148:** 1007–1020.

Cui Y., Dinman J.D., Kinzy T.G., and Peltz S.W. 1998. The Mof2/Sui1 protein is a general monitor of translational accuracy. *Mol. Cell. Biol.* **18:** 1506–1516.

Danaie P., Wittmer B., Altmann M., and Trachsel H. 1995. Isolation of a protein complex containing translation initiation factor Prt1 from *Saccharomyces cerevisiae*. *J. Biol. Chem.* **270:** 4288–4292.

Das S., Maiti T., Das K., and Maitra U. 1997a. Specific interaction of eukaryotic translation initiation factor 5 (eIF5) with the β-subunit of eIF2. *J. Biol. Chem.* **272:** 31712–31718.

Das S., Yu L., Gaitazes C., Rogers R., Freeman J., Bienkowska J., Adams R.M., Smith T.F., and Lindelien J. 1997b. Biology's new Rosetta stone. *Science* **385:** 29–30.

Davies M.V., Furtado M., Hershey J.W.B., Thimmappaya B., and Kaufman R.J. 1989. Complementation of adenovirus-associated RNA I gene deletion by expression of a mutant eukaryotic translation initiation factor. *Proc. Natl. Acad. Sci.* **86:** 9163–9167.

De Benedetti A. and Baglioni C. 1983. Phosphorylation of initiation factor eIF2α, binding of mRNA to 48S complexes, and its reutilization in initiation of protein synthesis. *J. Biol. Chem.* **258:** 14556–14562.

Dennis P.P. 1997. Ancient ciphers: Translation in archaea. *Cell* **89:** 1007–1010.

Dever T.E., Glynias M.J., and Merrick W.C. 1987. GTP-binding domain: Three consensus sequence elements with distinct spacing. *Proc. Natl. Acad. Sci.* **84:** 1814–1818.

Dever T.E., Yang W., Åström S., Byström A.S., and Hinnebusch A.G. 1995. Modulation of $tRNA_i^{Met}$, eIF-2 and eIF-2B expression shows that *GCN4* translation is inversely coupled to the level of eIF-2•GTP•Met-$tRNA_i^{Met}$ ternary complexes. *Mol. Cell. Biol.* **15:** 6351–6363.

Dever T.E., Feng L., Wek R.C., Cigan A.M., Donahue T.D., and Hinnebusch A.G. 1992. Phosphorylation of initiation factor 2α by protein kinase GCN2 mediates gene-specific translational control of *GCN4* in yeast. *Cell* **68:** 585–596.

Dever T.E., Chen J.J., Barber G.N., Cigan A.M., Feng L., Donahue T.F., London I.M., Katze M.G., and Hinnebusch A.G. 1993. Mammalian eukaryotic initiation factor 2α kinases functionally substitute for GCN2 in the *GCN4* translational control mechanism of yeast. *Proc. Natl. Acad. Sci.* **90:** 4616–4620.

Dholakia J.N. and Wahba A.J. 1989. Mechanism of the nucleotide exchange reaction in eukaryotic polypeptide chain initiation: Characterization of the guanine nucleotide exchange factor as a GTP-binding protein. *J. Biol. Chem.* **264:** 546–550.

Diallinas G. and Thireos G. 1994. Genetic and biochemical evidence for yeast GCN2 protein kinase polymerization. *Gene* **143:** 21–27.

Donahue T.F., Cigan A.M., Pabich E.K., and Castilho-Valavicius B. 1988. Mutations at a Zn(II) finger motif in the yeast eIF-2β gene alter ribosomal start-site selection during the scanning process. *Cell* **54:** 621–632.

Donzé O. and Picard D. 1999. Hsp90 binds and regulates the ligand-inducible alpha subunit of eukaryotic translation initiation factor kinase Gcn2. *Mol. Cell. Biol.* **19:** 8422–8432.

Dorris D.R., Erickson F.L., and Hannig E.M. 1995. Mutations in *GCD11*, the structural gene for eIF-2γ in yeast, alter translational regulation of *GCN4* and the selection of the start site for protein synthesis. *EMBO J.* **14:** 2239–2249.

Drabkin H.J. and RajBhandary U.L. 1998. Initiation of protein synthesis in mammalian cells with codons other than AUG and amino acids other than methionine. *Mol. Cell. Biol.* **18:** 5140–5147.

Drabkin H.J., Estrella M., and RajBhandary U.L. 1998. Initiator-elongator discrimination in vertebrate tRNAs for protein synthesis. *Mol. Cell. Biol.* **18:** 1459–1466.

Drabkin H.J., Helk B., and RajBhandary U.L. 1993. The role of nucleotides conserved in eukaryotic initiator methionine tRNAs in initiation of protein synthesis. *J. Biol. Chem.* **268:** 25221–25228.

Erickson F.L. and Hannig E.M. 1996. Ligand interactions with eukaryotic translation initiation factor 2: Role of the γ-subunit. *EMBO J.* **15:** 6311–6320.

Ernst H., Duncan R.F., and Hershey J.W.B. 1987. Cloning and sequencing of complementary DNAs encoding the α-subunit of translational initiation factor eIF-2. *J. Biol. Chem.* **262:** 1206–1212.

Evans D.R.H., Rasmussen C., Hanic-Joyce P.J., Johnston G.C., Singer R.A., and Barnes C.A. 1995. Mutational analysis of the Prt1 protein subunit of yeast translation initiation factor 3. *Mol. Cell. Biol.* **15:** 4525–4535.

Fabian J.R., Kimball S.R., Heinzinger N.K., and Jefferson L.S. 1997. Subunit assembly and guanine nucleotide exchange activity of eukaryotic initiation factor-2B expressed in Sf9 cells. *J. Biol. Chem.* **272:** 12359–12365.

Feinberg B., McLaughlin C.S., and Moldave K. 1982. Analysis of temperature-sensitive mutant *ts*187 of *Saccharomyces cerevisiae* altered in a component required for the initiation of protein synthesis. *J. Biol. Chem.* **257:** 10846–10851.

Feng L., Yoon H., and Donahue T.F. 1994. Casein kinase II mediates multiple phosphorylation of *Saccharomyces cerevisiae* eIF-2α (encoded by *SUI2*), which is required for optimal eIF-2 function in *S. cerevisiae. Mol. Cell. Biol.* **14:** 5139–5153.

Fletcher C.M., Pestova T.V., Hellen C.U.T., and Wagner G. 1999. Structure and interactions of the translation initiation factor eIF1. *EMBO J.* **18:** 2631–2639.

Flynn A., Oldfield S., and Proud C.G. 1993. The role of the β-subunit of initiation factor eIF2 in initiation complex formation. *Biochim. Biophys. Acta* **1174:** 117–121.

Flynn A., Shatsky I.N., Proud C.G., and Kaminski A. 1994. The RNA-binding properties of protein synthesis initiation factor eIF-2. *Biochim. Biophys. Acta* **1219:** 293–301.

Foiani M., Cigan A.M., Paddon C.J., Harashima S., and Hinnebusch A.G. 1991. GCD2, a translational repressor of the *GCN4* gene, has a general function in the initiation of protein synthesis in *Saccharomyces cerevisiae. Mol. Cell. Biol.* **11:** 3203–3216.

Forster C., Chakraburtty K., and Sprinzl M. 1993. Discrimination between initiation and elongation of protein biosynthesis in yeast: Identification assured by a nucleotide modification in the initiator tRNA. *Nucleic Acids Res.* **21:** 5679–5683.

Garcia-Barrio M., Dong J., Ufano S., and Hinnebusch A.G. 2000. Association of GCN1/GCN20 regulatory complex with the conserved N-terminal domain of eIF2α kinase GCN2 is required for GCN2 activation in vivo. *EMBO J.* (in press).

Garcia-Barrio M.T., Naranda T., Cuesta R., Hinnebusch A.G., Hershey J.W.B., and Tamame M. 1995. GCD10, a translational repressor of *GCN4*, is the RNA-binding subunit of eukaryotic translation initiation factor-3. *Genes Dev.* **9:** 1781–1796.

Gaspar N.J., Kinzy T.G., Scherer B.J., Humbelin M., Hershey J.W., and Merrick W.C. 1994. Translation initiation factor eIF-2. Cloning and expression of the human cDNA encoding the γ-subunit. *J. Biol. Chem.* **269:** 3415–3422.

Gonsky R., Itamar D., Harary R., and Kaempfer R. 1992. Binding of ATP and messenger RNA by the β-subunit of eukaryotic initiation factor 2. *Biochimie* **74:** 427–434.

Goss D.J. and Parkhurst L.J. 1984. Studies on the role of eukaryotic nucleotide exchange factor in polypeptide chain initiation. *J. Biol. Chem.* **259:** 7374–7377.

Goumans H., Thomas A., Verhoeven A., Voorma H.O., and Benne R. 1980. The role of eIF-4C in protein synthesis initiation complex formation. *Biochim. Biophys. Acta* **608:** 39–46.

Greenberg J.R., Phan L., Gu Z., deSilva A., Apolito C., Sherman F., Hinnebusch A.G., and Goldfarb D.S. 1998. Nip1p associates with 40S ribosomes and the Prt1p subunit of eIF3 and is required for efficient translation initiation. *J. Biol. Chem.* **273:** 23485–23494.

Gross M., Redman R., and Kaplansky D.A. 1985. Evidence that the primary effect of phosphorylation of eukaryotic initiation factor 2α in rabbit reticulocyte lysate is inhibition of the release of eukaryotic initiation factor-2•GDP from 60S ribosomal subunits. *J. Biol. Chem.* **260:** 9491–9500.

Gross M., Wing M., Rundquist C., and Rubino M.S. 1987. Evidence that phosphorylation of eIF-2(α) prevents the eIF-2B-mediated dissociation of eIF-2•GDP from the 60S subunit of complete initiation complexes. *J. Biol. Chem.* **262:** 6899–6907.

Gualerzi C.O. and Pon C.L. 1990. Initiation of mRNA translation in prokaryotes.

Biochemistry **29:** 5881–5889.

Gupta N.K., Roy A.L., Nag M.K., Kinzy T.G., MacMillan S., Hileman R.F., Dever T.E., Wu S., Merrick W.C., and Hershey J.W.B. 1990. New insights into an old problem: ternary complex (Met-tRNA$_f$• eIF-2•GTP) formation in animal cells. In *Post-transcriptional control of gene expression* (ed. J.E.G. McCarthy and M.F. Tuite), pp. 521–526. Springer-Verlag, Berlin.

Hanachi P., Hershey J.W.B., and Vornlocher H.P. 1999. Characterization of the p33 subunit of eukaryotic translation initiation factor-3 from *Saccharomyces cerevisiae*. *J. Biol. Chem.* **274:** 8546–8553.

Hanic-Joyce P.J., Singer R.A., and Johnston G.C. 1987. Molecular characterization of the yeast *PRT1* gene in which mutations affect translation initiation and regulation of cell proliferation. *J. Biol. Chem.* **262:** 2845–2851.

Hannig E.M. and Hinnebusch A.G. 1988. Molecular analysis of *GCN3*, a translational activator of *GCN4*: Evidence for posttranslational control of *GCN3* regulatory function. *Mol. Cell. Biol.* **8:** 4808–4820.

Hannig E.M., Cigan A.M., Freeman B.A., and Kinzy T.G. 1992. *GCD11,* a negative regulator of *GCN4* expression, encodes the γ subunit of eIF-2 in *Saccharomyces cerevisiae*. *Mol. Cell. Biol.* **13:** 506–520.

Hannig E.M., Williams N.P., Wek R.C., and Hinnebusch A.G. 1990. The translational activator GCN3 functions downstream from GCN1 and GCN2 in the regulatory pathway that couples *GCN4* expression to amino acid availability in *Saccharomyces cerevisiae*. *Genetics* **126:** 549–562.

Harashima S. and Hinnebusch A.G. 1986. Multiple *GCD* genes required for repression of *GCN4*, a transcriptional activator of amino acid biosynthetic genes in *Saccharomyces cerevisiae*. *Mol. Cell. Biol.* **6:** 3990–3998.

Harashima S., Hannig E.M., and Hinnebusch A.G. 1987. Interactions between positive and negative regulators of *GCN4* controlling gene expression and entry into the yeast cell cycle. *Genetics* **117:** 409–419.

Hartwell L.H. and McLaughlin C.S. 1969. A mutant of yeast apparently defective in the initiation of protein synthesis. *Proc. Natl. Acad. Sci.* **62:** 468–474.

Hershey J.W.B. 1991. Translational control in mammalian cells. *Annu. Rev. Biochem.* **60:** 717–755.

Hershey J.W.B., Asano K., Naranda T., Vornlocher H.P., Hanachi P., and Merrick W.C. 1996. Conservation and diversity in the structure of translation initiation factor eIF3 from humans and yeast. *Biochimie* **78:** 903–907.

Hinnebusch A.G. 1992. General and pathway-specific regulatory mechanisms controlling the synthesis of amino acid biosynthetic enzymes in *Saccharomyces cerevisiae*. In *The molecular and cellular biology of the yeast* Saccharomyces: *Gene expression* (ed. J.R. Broach et al.), pp. 319–414. Cold Spring Harbor Laboratory Press, Cold Spring Harbor, New York.

———. 1994. Translational control of *GCN4*: An *in vivo* barometer of initiation factor activity. *Trends Biochem. Sci.* **19:** 409–414.

———. 1996. Translational control of *GCN4*: Gene-specific regulation by phosphorylation of eIF2. In *Translational control* (ed. J.W.B. Hershey et al.), pp. 199–244. Cold Spring Harbor Laboratory Press, Cold Spring Harbor, New York.

Huang H., Yoon H., Hannig E.M., and Donahue T.F. 1997. GTP hydrolysis controls stringent selection of the AUG start codon during translation initiation in *Saccharomyces cerevisiae*. *Genes Dev.* **11:** 2396–2413.

Jackson R.J. 1991. Binding of Met-tRNA. In *Translation in eukaryotes* (ed. H. Trachsel), pp. 193–230. CRC Press, Boca Raton, Florida.

Kamath A. and Chakraburtty K. 1989. Role of yeast elongation factor 3 in the elongation cycle. *J. Biol. Chem.* **264:** 15423–15428.

Kaufman R.J., Davies M.V., Pathak V.K., and Hershey J.W.B. 1989. The phosphorylation state of eucaryotic initiation factor 2 alters translational efficiency of specific mRNAs. *Mol. Cell. Biol.* **9:** 946–958.

Kiesewetter S., Ott G., and Sprinzl M.. 1990. The role of modified purine 64 in initiator/elongator discrimination of $tRNA^{Met}$ from yeast and wheat germ. *Nucleic Acids Res.* **18:** 4677–4682.

Kimball S.R., Antonetti D.A., Brawley R.M., and Jefferson L.S. 1991. Mechanism of inhibition of peptide chain initiation by amino acid deprivation in perfused rat liver. *J. Biol. Chem.* **266:** 1969–1976.

Kimball S.R., Heinzinger N.K., Horetsky R.L., and Jefferson L.S. 1998a. Identification of interprotein interactions between the subunits of eukaryotic initiation factors eIF2 and eIF2B. *J. Biol. Chem.* **273:** 3039–3044.

Kimball S.R., Fabian J.R., Pavitt G.D., Hinnebusch A.G., and Jefferson L.S. 1998b. Regulation of guanine nucleotide exchange through phosphorylation of eukaryotic initiation factor eIF2α. *J. Biol. Chem.* **273:** 12841–12845.

Klenk H.-P., Clayton R.A., Tomb J.F., White O., Nelson K.E., Ketchum K.A., Dodson R.J., Gwinn M., Hickey E.K., Peterson J.D., Richardson D.L., Kerlavage A.R., Graham D.E., Kyrpides N.C., Fleischmann R.D., Quackenbush J., Lee N.H., Sutton G.G., Gill S., Kirkness E.F., Dougherty B.A., McKenney K., Adams M.D., Loftus B., Venter J.C., et al. 1997. The complete genome sequence of the hyperthermophilic sulphate-reducing archaeon *Archaeoglobus fulgidus*. *Nature* **390:** 364–370.

Koonin E.V. 1995. Multidomain organization of eukaryotic guanine nucleotide exchange translation initiation factor eIF-2B subunits revealed by analysis of conserved sequence motifs. *Protein Sci.* **4:** 1608–1617.

Kyrpides N.C. and Woese C.R. 1998. Universally conserved translation initiation factors. *Proc. Natl. Acad. Sci.* **95:** 224–228.

Lamphear B.J., Kirchweger R., Skern T., and Rhoads R.E. 1995. Mapping of functional domains in eukaryotic protein synthesis initiation factor 4G (eIF4G) with picornaviral proteases. *J. Biol. Chem.* **270:** 21975–21983.

Laurino J.P., Thompson G.M., Pacheco E., and Castilho B.A. 1999. The β subunit of eukaryotic translation initiation factor 2 binds mRNA through the lysine repeats and a region comprising the C_2-C_2 motif. *Mol. Cell. Biol.* **19:** 173–181.

Lee J.H., Choi S.K., Roll-Mecak A., Burley S.K., and Dever T.E. 1999. Universal conservation in translation initiation revealed by human and archaeal homologs of bacterial translation initiation factor IF2. *Proc. Natl. Acad. Sci.* **96:** 4342–4347.

Luo Z., Freitag M., and Sachs M.S. 1995. Translational regulation in response to changes in amino acid availability in *Neurospora crassa*. *Mol. Cell. Biol.* **15:** 5235–5245.

Manchester K.L. 1997. Catalysis of guanine nucleotide exchange on eIF-2 by eIF-2B: Is it a sequential or substituted enzyme mechanism? *Biochem. Biophys. Res. Commun.* **239:** 223–227.

Marton M.J., Crouch D., and Hinnebusch A.G. 1993. GCN1, a translational activator of *GCN4* in *S. cerevisiae*, is required for phosphorylation of eukaryotic translation initiation factor 2 by protein kinase GCN2. *Mol. Cell. Biol.* **13:** 3541–3556.

Marton M.J., Vazquez de Aldana C.R., Qiu H., Chakraburtty K., and Hinnebusch A.G.

1997. Evidence that GCN1 and GCN20, translational regulators of *GCN4*, function on elongating ribosomes in activation of the eIF2α kinase GCN2. *Mol. Cell. Biol.* **17:** 4474–4489.

Merrick W.C. and Hershey J.W.B. 1996. The pathway and mechanism of eukaryotic protein synthesis. In *Translational control* (ed. J.W.B. Hershey et al.), pp. 31–69. Cold Spring Harbor Laboratory Press, Cold Spring Harbor, New York.

Methot N., Song M.S., and Sonenberg N. 1996. A region rich in aspartic acid, arginine, tyrosine, and glycine (DRYFG) mediates eukaryotic initiation factor 4B (eIF4B) self-association and interaction with eIF3. *Mol. Cell. Biol.* **16:** 5328–5334.

Moazed D., Samaha R.R., Gualerzi C., and Noller H.F. 1995. Specific protection of 16 S rRNA by translational initiation factors. *J. Mol. Biol.* **248:** 207–210.

Mueller P.P., Harashima S., and Hinnebusch A.G. 1987. A segment of *GCN4* mRNA containing the upstream AUG codons confers translational control upon a heterologous yeast transcript. *Proc. Natl. Acad. Sci.* **84:** 2863–2867.

Mueller P.P., Grueter P., Hinnebusch H.G., and Trachsel H. 1998. A ribosomal protein is required for translational regulation of *GCN4* mRNA. *J. Biol. Chem.* **273:** 32870–32877.

Murtha-Riel P., Davies M.V., Scherer B.J., Choi S.Y., Hershey J.W.B., and Kaufman R.J. 1993. Expression of a phosphorylation-resistant eukaryotic initiation factor 2 α-subunit mitigates heat shock inhibition of protein synthesis. *J. Biol. Chem.* **268:** 12946–12951.

Naranda T., MacMillan S.E., and Hershey J.W.B. 1994. Purified yeast translational initiation factor eIF-3 is an RNA-binding protein complex that contains the PRT1 protein. *J. Biol. Chem.* **269:** 32286–32292.

Naranda T., Kainuma M., McMillan S.E., and Hershey J.W.B. 1997. The 39-kilodalton subunit of eukaryotic translation initiation factor 3 is essential for the complex's integrity and for cell viability in *Saccharomyces cerevisiae*. *Mol. Cell. Biol.* **17:** 145–153.

Naranda T., MacMillan S.E., Donahue T.F., and Hershey J.W. 1996. SUI1/p16 is required for the activity of eukaryotic translation initiation factor 3 in *Saccharomyces cerevisiae*. *Mol. Cell. Biol.* **16:** 2307–2313.

Nyborg J., Nissen P., Kjeldgaard M., Thirup S., Polekhina G., and Clark B.F.C. 1996. Structure of the ternary complex of EF-Tu: Macromolecular mimicry in translation. *Trends Biochem. Sci.* **21:** 81–82.

Olsen D.S., Jordan B., Chen D., Wek R.C., and Cavener D.R. 1998. Isolation of the gene encoding the *Drosophila melanogaster* homolog of the *Saccharomyces cerevisiae GCN2* eIF-2α kinase. *Genetics* **149:** 1495–1509.

Paddon C.J., Hannig E.M., and Hinnebusch A.G. 1989. Amino acid sequence similarity between GCN3 and GCD2, positive and negative translational regulators of *GCN4*: Evidence for antagonism by competition. *Genetics* **122:** 551–559.

Paluh J.L., Orbach M.J., Legerton T.L., and Yanofsky C. 1988. The cross-pathway control gene of *Neurospora crassa, cpc-1,* encodes a protein similar to GCN4 of yeast and the DNA-binding domain of the oncogene v-*jun*-encoded protein. *Proc. Natl. Acad. Sci.* **85:** 3728–3732.

Pathak V.K., Nielsen P.J., Trachsel H., and Hershey J.W.B. 1988. Structure of the β subunit of translational initiation factor eIF-2. *Cell* **54:** 633–639.

Pavitt G.D., Yang W., and Hinnebusch A.G. 1997. Homologous segments in three subunits of the guanine nucleotide exchange factor eIF2B mediate translational regulation by phosphorylation of eIF2. *Mol. Cell. Biol.* **17:** 1298–1313.

Pavitt G.D., Ramaiah K.V.A., Kimball S.R., and Hinnebusch A.G. 1998. eIF2 independently binds two distinct eIF2B subcomplexes that catalyze and regulate guanine-nucleotide exchange. *Genes Dev.* **12:** 514–526.

Pestova T.V., Borukhov S.I., and Hellen C.U.T. 1998. Eukaryotic ribosomes require initiation factors 1 and 1A to locate initiation codons. *Nature* **394:** 854–859.

Pestova T.V., Lomakin I.B., Lee J.H., Choi S.K., Dever T.E., and Hellen C.U.T. 2000. The joining of ribosomal subunits in eukaryotes requires eIF5B. *Nature* **403:** 332–335.

Peterson D.T., Merrick W.C., and Safer B. 1979. Binding and release of radiolabeled eukaroytic initiation factors 2 and 3 during 80 S initiation complex formation. *J. Biol. Chem.* **254:** 2509–2519.

Phan L., Zhang X., Asano K., Anderson J., Vornlocher H.P., Greenberg J.R., Qin J., and Hinnebusch A.G. 1998. Identification of a translation initiation factor 3 (eIF3) core complex, conserved in yeast and mammals, that interacts with eIF5. *Mol. Cell. Biol.* **18:** 4935–4946.

Proud C.G. 1992. Protein phosphorylation in translational control. *Curr. Top. Cell. Regul.* **32:** 243–369.

Qiu H., Garcia-Barrio M.T., and Hinnebusch A.G. 1998. Dimerization by translation initiation factor 2 kinase GCN2 is mediated by interactions in the C-terminal ribosome-binding region and the protein kinase domain. *Mol. Cell. Biol.* **18:** 2697–2711.

Qu S. and Cavener D.R. 1994. Isolation and characterization of the *Drosophila melanogaster* eIF-2α gene encoding the alpha subunit of translation initiation factor eIF-2. *Gene* **140:** 239–242.

RajBhandary U.L. and Chow C.M. 1995. Initiator tRNAs and initiation of protein synthesis. In *tRNA structure, biosynthesis, and function* (ed. D. Soll and U.L. RajBhandary), pp. 511–528. American Society for Microbiology, Washington D.C.

Ramaiah K.V., Davies M.V., Chen J.J., and Kaufman R.J. 1994. Expression of mutant eukaryotic initiation factor 2 α subunit (eIF- 2α) reduces inhibition of guanine nucleotide exchange activity of eIF-2B mediated by eIF-2α phosphorylation. *Mol. Cell. Biol.* **14:** 4546–4553.

Ramaiah K.V.A., Dhindsa R.S., Chen J.J., London I.M., and Levin D. 1992. Recycling and phosphorylation of eukaryotic initiation factor 2 on 60S subunits of 80S initiation complexes and polysomes. *Proc. Natl. Acad. Sci.* **89**: 12063–12067.

Ramirez M., Wek R.C., and Hinnebusch A.G. 1991. Ribosome-association of GCN2 protein kinase, a translational activator of the *GCN4* gene of *Saccharomyces cerevisae*. *Mol. Cell. Biol.* **11:** 3027–3036.

Ramirez M., Wek R.C., Vazquez de Aldana C.R., Jackson B.M., Freeman B., and Hinnebusch A.G. 1992. Mutations activating the yeast eIF-2α kinase GCN2: Isolation of alleles altering the domain related to histidyl-tRNA synthetases. *Mol. Cell. Biol.* **12:** 5801–5815.

Raychaudhuri P. and Maitra U. 1986. Identification of ribosome-bound eukaryotic initiation factor 2•GDP binary complex as an intermediate in polypeptide chain initiation reaction. *J. Biol. Chem.* **261:** 7723–7728.

Rolfes R.J. and Hinnebusch A.G. 1993. Translation of the yeast transcriptional activator *GCN4* is stimulated by purine limitation: Implications for activation of the protein kinase GCN2. *Mol. Cell. Biol.* **13:** 5099–5111.

Romano P.R., Garcia-Barrio M.T., Zhang X., Wang Q., Taylor D.R., Zhang F., Herring C., Mathews M.B., Qin J., and Hinnebusch A.G. 1997. Autophosphorylation in the activation loop is required for full kinase activity in vivo of human and yeast eukaryotic initiation factor 2αkinases PKR and GCN2. *Mol. Cell. Biol.* **18:** 2282–2297.

Rosen H., Di Segni G., and Kaempfer R. 1982. Translational control by messenger RNA competition for eukaryotic initiation factor 2. *J. Biol. Chem.* **257:** 946–952.

Rowlands A.G., Panniers R., and Henshaw E.C. 1988a. The catalytic mechanism of guanine nucleotide exchange factor action and competitive inhibition by phosphorylated eukaryotic initiation factor 2. *J. Biol. Chem.* **263:** 5526–5533.

Rowlands A.G., Montine K.S., Henshaw E.C., and Panniers R. 1988b. Physiological stresses inhibit guanine-nucleotide-exchange factor in Ehrlich cells. *Eur. J. Biochem.* **175:** 93–99.

Ruff M., Krishnaswamy S., Boeglin M., Poterszman A., Mitschler A., Podjarny A., Rees B., Thierry J.C., and Moras D. 1991. Class II aminoacyl transfer RNA synthetases: Crystal structure of yeast aspartyl-tRNA synthetase complexed with $tRNA^{Asp}$. *Science* **252:** 1682–1689.

Salimans M., Goumans H., Amesz H., Beene R., and Voorma H.O. 1984. Regulation of protein synthesis in eukaryotes. Mode of action of eRF, an eIF-2-recycling factor from rabbit reticulocytes in GDP/GTP exchange. *Eur. J. Biochem.* **145:** 91–98.

Sandbaken M., Lupisella J.A., DiDomenico B., and Chakraburtty K. 1990. Isolation and characterization of the structural gene encoding elongation factor 3. *Biochim. Biophys. Acta* **1050:** 230–234.

Santoyo J., Alcalde J., Mendez R., Pulido D., and de Haro C. 1997. Cloning and characterization of a cDNA encoding a protein synthesis initiation factor-2α (eIF-2α) kinase from *Drosophila melanogaster*. *J. Biol. Chem.* **272:** 12544–12550.

Sattlegger E., Hinnebusch A.G., and Barthelmess I.B. 1998. *cpc-3*, the *Neurospora crassa* homologue of yeast *GCN2,* encodes a polypeptide with juxtaposed eIF2α kinase and histidyl-tRNA synthetase-related domains required for general amino acid control. *J. Biol. Chem.* **273:** 20404–20416.

Sette M., van Tilborg P., Spurio R., Kaptein R., Paci M., Gualerzi C.O., and Boelens R. 1997. The structure of the translational initiation factor IF1 from *E. coli* contains an oligomer-binding motif. *EMBO J.* **16:** 1436–1443.

Sizova D.V., Kolupaeva V.G., Pestova T.V., Shatsky I.N., and Hellen C.U. 1998. Specific interaction of eukaryotic translation initiation factor 3 with the 5′ nontranslated regions of hepatitis C virus and classical swine fever RNAs. *J. Virol.* **72:** 4775–4782.

Skogerson L. and Wakatama E. 1976. A ribosome-dependent GTPase from yeast distinct from elongation factor 2. *Proc. Natl. Acad. Sci.* **73:** 73–76.

Smith D.R., Doucette-Stamm L.A., Deloughery C., Lee H., Dubois J., Aldredge T., Bashirzadeh R., Blakely D., Cook R., Gilbert K., Harrison D., Hoang L., Keagle P., Lumm W., Pothier B., Qiu D., Spadafora R., Vicaire R., Wang Y., Wierzbowski J., Gibson R., Jiwani N., Caruso A., Bush D., Reeve J.N., et. al. 1997. Complete genome sequence of *Methanobacterium thermoautotrophicum* deltaH: Functional analysis and comparative genomics. *J. Bacteriol.* **179:** 7135–7155.

Sood R., Porter A.C., Olsen D., Cavener D.R., and Wek R.C. 2000. A mammalian homologue of GCN2 protein kinase important for translational control by phosphorylation of eukaryotic initiation factor-2α. *Genetics* **154:** 787–801.

Sprinzl M., Horn C., Brown M., Loudovitch A., and Steinberg S. 1998. Compilation of tRNA sequences and sequenes of tRNA genes. *Nucleic Acids Res.* **26:** 148–153.

Srivastava S., Verschoor A., and Frank J. 1992. Eukaryotic initiation factor 3 does not prevent association through physical blockage of the ribosomal subunit-subunit interface.

J. Mol. Biol. **220:** 301–304.

Sudhakar A., Krishnamoorthy T., Jain A., Chatterjee U., Hasnain S.E., Kaufman R.J., and Ramaiah K.V. 1999. Serine 48 in initiation factor 2α (eIF2α) is required for high-affinity interaction between eIF2 α(P) and eIF2B. *Biochemistry* **38:** 15398–15405.

Tabara H., Sarkissian M., Kelly W.G., Fleenor J., Grishok A., Timmons L., Fire A., and Mello C.C. 1999. The *rde-1* gene, RNA interference, and transposon silencing in *C. elegans*. *Cell* **99:** 123–132.

Tavernarakis N. and Thireos G. 1996. Genetic evidence for functional specificity of the yeast GCN2 kinase. *Mol. Gen. Genet.* **251:** 613–618.

Thomas A., Goumans H., Voorma H.O., and Benne R. 1980a. The mechanism of action of eukaryotic initiation factor 4C in protein synthesis. *Eur. J. Biochem.* **107:** 39–45.

Thomas A., Spaan W., van Steeg H., Voorma H.O., and Benne R. 1980b. Mode of action of protein synthesis initiation factor eIF-1 from rabbit reticulocytes. *FEBS Lett.* **116:** 67–71.

Thomas N.S.B., Matts R.L., Levin D.H., and London I.M. 1985. The 60S ribosomal subunit as a carrier of eukaryotic initiation factor 2 and the site of reversing factor activity during protein synthesis. *J. Biol. Chem.* **260:** 9860–9866.

Thomas N.S.B., Matts R.L., Petryshyn R., and London I.M. 1984. Distribution of reversing factor in reticulocyte lysates during active protein synthesis and on inhibition by heme deprivation or double-stranded RNA. *Proc. Natl. Acad. Sci.* **81:** 6998–7002.

Thompson H.A., Sadnik I., Scheinbuks J., and Moldave K. 1977. Studies on native ribosomal subunits from rat liver. Purification and characterization of a ribosome dissociation factor. *Biochemistry* **16:** 2221–2230.

Trachsel H. 1996. Binding of initiator methionyl-tRNA to ribosomes. In *Translational control* (ed. J.W.B. Hershey et al.), pp. 113–138. Cold Spring Harbor Laboratory Press, Cold Spring Harbor, New York.

Trachsel H. and Staehelin T. 1979. Initiation of mammalian protein synthesis: The multiple functions of the initiation factor eIF-3. *Biochim. Biophys. Acta* **565:** 305–314.

Trachsel H., Erni B., Schreier M.H., and Staehelin T. 1977. Initiation of mammalian protein synthesis: The assembly of the initiation complex with purified initiation factors. *J. Mol. Biol.* **116:** 755–767.

Triana-Alonso F.J., Chakraburtty K., and Nierhaus K.H. 1995. The elongation factor unique in higher fungi and essential for protein biosynthesis is an E site factor. *J. Biol. Chem.* **270:** 20473–20478.

Tzamarias D., Roussou I., and Thireos G. 1989. Coupling of GCN4 mRNA translational activation with decreased rates of polypeptide chain initiation. *Cell* **57:** 947–954.

Uritani M. and Miyazaki M. 1988. Role of yeast elongation factor 3 (EF-3) at the AA-tRNA binding step. *J. Biochem.* **104:** 118–126.

Valasek L., Trachsel H., Hasek J., and Ruis H. 1998. Rpg1, the *Saccharomyces cerevisiae* homologue of the largest subunit of mammalian translation initiation factor 3, is required for translational activity. *J. Biol. Chem.* **273:** 21253–21260.

Valasek L., Hasek J., Trachsel H., Imre E.M., and Ruis H. 1999. The *Saccharomyces cerevisiae HCRI* gene encoding a homologue of the p35 subunit of human translation initiation factor 3 (eIF3) is a high copy suppressor of a temperature-sensitive mutation in the Rpg1p subunit of yeast eIF3. *J. Biol. Chem.* **274:** 27567–27572.

Vazquez de Aldana C.R. and Hinnebusch A.G. 1994. Mutations in the GCD7 subunit of yeast guanine nucleotide exchange factor eIF-2B overcome the inhibitory effects of phosphorylated eIF-2 on translation initiation. *Mol. Cell. Biol.* **14:** 3208–3222.

Vazquez de Aldana C.R., Dever T.E., and Hinnebusch A.G. 1993. Mutations in the α subunit of eukaryotic translation initiation factor 2 (eIF-2α) that overcome the inhibitory effects of eIF-2α phosphorylation on translation initiation. *Proc. Natl. Acad. Sci.* **90:** 7215–7219.

Vazquez de Aldana C.R., Marton M.J., and Hinnebusch A.G. 1995. GCN20, a novel ATP binding cassette protein, and GCN1 reside in a complex that mediates activation of the eIF-2α kinase GCN2 in amino acid-starved cells. *EMBO J.* **14:** 3184–3199.

Verlhac M.-H., Chen R.-H., Hanachi P., Hershey J.W.B., and Derynck R. 1997. Identification of partners of TIF34, a component of the yeast eIF3 complex, required for cell proliferation and translation initiation. *EMBO J.* **16:** 6812–6822.

von Pawel-Rammingen U., Åström S., and Byström A.S. 1992. Mutational analysis of conserved positions potentially important for initiator tRNA function in *Saccharomyces cerevisiae*. *Mol. Cell. Biol.* **12:** 1432–1442.

Vornlocher H.P., Hanachi P., Ribeiro S., and Hershey J.W.B. 1999. A 110-kilodalton subunit of translation initiation factor eIF3 and an associated 135-kilodalton protein are encoded by the *Saccharomyces cerevisiae TIF32* and *TIF31* genes. *J. Biol. Chem.* **274:** 16802–16812.

Wagner T., Gross M., and Sigler P.B. 1984. Isoleucyl initiator tRNA does not initiate eucaryotic protein synthesis. *J. Biol. Chem.* **259:** 4706–4709.

Wei C.L., Kainuma M., and Hershey J.W.B. 1995a. Characterization of yeast translation initiation factor 1A and cloning of its essential gene. *J. Biol. Chem.* **270:** 22788–22794.

Wei C.L., MacMillan S.E., and Hershey J.W.B. 1995b. Protein synthesis initiation factor eIF-1A is a moderately abundant RNA-binding protein. *J. Biol. Chem.* **270:** 5764–5771.

Wek R.C., Jackson B.M., and Hinnebusch A.G. 1989. Juxtaposition of domains homologous to protein kinases and histidyl-tRNA synthetases in GCN2 protein suggests a mechanism for coupling *GCN4* expression to amino acid availability. *Proc. Natl. Acad. Sci.* **86:** 4579–4583.

Wek R.C., Ramirez M., Jackson B.M., and Hinnebusch A.G. 1990. Identification of positive-acting domains in GCN2 protein kinase required for translational activation of *GCN4* expression. *Mol. Cell. Biol.* **10:** 2820–2831.

Wek S.A., Zhu S., and Wek R.C. 1995. The histidyl-tRNA synthetase-related sequence in the eIF-2α protein kinase GCN2 interacts with tRNA and is required for activation in response to starvation for different amino acids. *Mol. Cell. Biol.* **15:** 4497–4506.

Williams N.P., Hinnebusch A.G., and Donahue T.F. 1989. Mutations in the structural genes for eukaryotic initiation factors 2α and 2β of *Saccharomyces cerevisiae* disrupt translational control of *GCN4* mRNA. *Proc. Natl. Acad. Sci.* **86:** 7515–7519.

Yang W. and Hinnebusch A.G. 1996. Identification of a regulatory subcomplex in the guanine nucleotide exchange factor eIF2B that mediates inhibition by phosphorylated eIF2. *Mol. Cell. Biol.* **16:** 6603–6616.

Ye X. and Cavener D.R. 1994. Isolation and characterization of the *Drosophila melanogaster* gene encoding translation-initiation factor eIF-2β. *Gene* **142:** 271–274.

Yoon H.J. and Donahue T.F. 1992. The *sui1* suppressor locus in *Saccharomyces cerevisiae* encodes a translation factor that functions during $tRNA_i^{Met}$ recognition of the start codon. *Mol. Cell. Biol.* **12:** 248–260.

Zhu S. and Wek R.C. 1998. Ribosome-binding domain of eukaryotic initiation factor-2 kinase GCN2 facilitates translation control. *J. Biol. Chem.* **273:** 1808–1814.

Zhu S., Sobolev A.Y., and Wek R.C. 1996. Histidyl-tRNA synthetase-related sequences in

GCN2 protein kinase regulate *in vitro* phosphorylation of eIF-2. *J. Biol. Chem.* **271:** 24989–24994.

Zou C., Zhang Z., Wu S., and Osterman J.C. 1998. Molecular cloning and characterization of a rabbit eIF2C protein. *Gene* **211:** 187–194.

6

Regulation of Ribosomal Recruitment in Eukaryotes

Brian Raught, Anne-Claude Gingras, and Nahum Sonenberg

Department of Biochemistry and McGill Cancer Research Centre,
University of McGill, Montréal
Québec H3G 1Y6 Canada

The regulation of translation rates, the frequency with which a given mRNA is translated, plays a critical role in many fundamental biological processes, including cell growth (see Chapter 23), development (see Chapters 7 and 27), and the response to biological cues or environmental stresses (many chapters in this book). Dysregulation of translation may also be an important component in the transformation of cells (see Chapter 20). As discussed in other chapters, translation rates are primarily regulated at the initiation phase (see Chapter 2), a multistep process involving the recruitment of the 40S small ribosomal subunit to the 5′ end of an mRNA and the positioning of the ribosome at an initiation codon. This process requires the participation of a large number of initiation factors (see Chapter 2), including the eIF4 group, those proteins that interact directly with the 5′-untranslated region (UTR) of mRNA. Our current understanding of how the activity of the eIF4 initiation factors is regulated by intracellular signaling pathways is the subject of this chapter.

The basic mechanism of ribosomal recruitment to mRNA in eukaryotes is conserved throughout evolution. For example, all eukaryotic organisms studied to date possess an eIF4F-like complex (for review, see Chapter 2), consisting of an mRNA cap-binding protein (eIF4E), a scaffolding protein (eIF4G), and an RNA helicase (eIF4A). An eIF4A cofactor, eIF4B, is also conserved in eukaryotes. However, different organisms appear to have devised very different methods to regulate the activity of these factors. Thus, in this chapter we also summarize our current understanding of how the activity of translation initiation factors is regulated in different types of organisms, as exemplified by mammalian, yeast, and plant cells.

© 2000 Cold Spring Harbor Laboratory Press 0-87969-568-4/00 $5 + .00

REGULATION OF RIBOSOMAL RECRUITMENT IN MAMMALIAN CELLS

eIF4F complex formation and activity are regulated via several different modes in mammalian cells. Interaction with repressor peptides, phosphorylation of its constituent proteins, and proteolysis of the eIF4G subunit (see Chapter 31) may all participate in the modulation of eIF4F activity in different situations. In this way, a network of signal transduction pathways cooperatively regulates translation initiation rates, as described below.

Mammalian eIF4F Formation Is Regulated by a Family of Translation Repressor Proteins (4E-BPs)

Using the Far-Western hybridization technique, Pause et al. (1994) isolated two related human cDNAs encoding small (~12 kD) eIF4E-binding proteins. These proteins, termed 4E-BP1 and 4E-BP2 (eIF*4E*-*b*inding *p*roteins *1* and *2*) share 56% identity and inhibit cap-dependent translation both in a cell-free translation assay and in vivo (Pause et al. 1994). Cap-independent translation is not affected by the 4E-BPs. A third member of the 4E-BP family, 4E-BP3, was subsequently cloned and shares 57% and 59% identity with 4E-BP1 and 4E-BP2, respectively (Poulin et al. 1998). 4E-BP3 possesses a high degree of homology with the other 4E-BPs in the mid-region of the protein (Poulin et al. 1998), which contains the eIF4E-binding site (Mader et al. 1995). 4E-BP3 is also inhibitory to cap-dependent, but not cap-independent, translation (Poulin et al. 1998). The reason for the existence of three mammalian 4E-BPs, all with seemingly identical function, is unknown. However, the 4E-BPs are not expressed to the same levels in all tissues; for example, 4E-BP1 mRNA is expressed to higher levels in skeletal muscle, pancreas, and adipose tissue (Hu et al. 1994; Tsukiyama-Kohara et al. 1996), 4E-BP2 mRNA is ubiquitously expressed to similar levels (Tsukiyama-Kohara et al. 1996), and 4E-BP3 appears to have a limited tissue distribution (F. Poulin and M. Ferraiuolo, unpubl.). The study of "knockout" mice lacking each of the 4E-BPs should yield valuable information regarding the apparent redundancy in function of the three proteins.

How do 4E-BPs inhibit translation? Binding of the 4E-BPs to eIF4E prevents the association between eIF4E and eIF4G and, thus, the assembly of a functional eIF4F complex (Fig. 1) (Haghighat et al. 1995). The interaction with eIF4E is conferred by a conserved amino acid motif (containing the "core" sequence YXXXXLΦ, in which X is any amino acid and Φ is a residue possessing an aliphatic portion, most often L, but

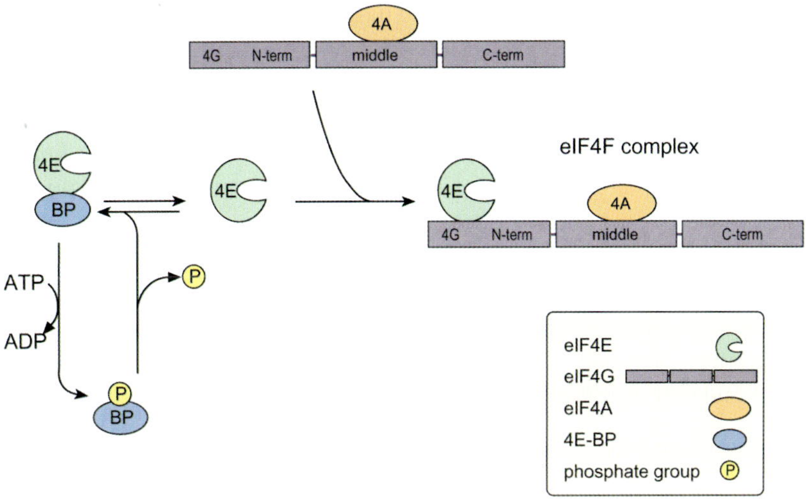

Figure 1 Regulation of eIF4F formation by the 4E-BPs. The 4E-BP and eIF4G proteins compete for a common binding site on the cap-binding protein, eIF4E. 4E-BP binding to eIF4E is modulated by phosphorylation. Various types of extracellular stimuli activate intracellular signaling pathways, leading to hyperphosphorylation of the 4E-BPs. Hyperphosphorylation decreases the affinity of the 4E-BPs for eIF4E, leading to eIF4E release. Free eIF4E can then interact with the eIF4G proteins, forming a functional eIF4F complex. Conversely, a decrease in 4E-BP phosphorylation increases the affinity of the 4E-BPs for eIF4E, which inhibits eIF4F formation.

sometimes M or F) shared by almost every eIF4E-binding protein in all species studied to date (excepting the *Xenopus laevis* maskin protein and the *Drosophilia melanogaster* 4E-BP, which vary somewhat from this consensus; Stebbins-Boaz et al. 1999; M. Miron, unpubl.). Deletion of this sequence or mutation of the tyrosine or leucine residues to alanine(s) abolishes eIF4E binding (Mader et al. 1995; Poulin et al. 1998). In addition, a 20-amino-acid peptide derived from the eIF4E-binding site of the mammalian 4E-BPs or eIF4Gs significantly inhibits cap-dependent translation in an in vitro translation assay (Fletcher et al. 1998; Marcotrigiano et al. 1999). NMR and crystallographic data have provided the structural basis for the importance of this motif in mediating binding to eIF4E (see these structures in Chapter 2). An area of the convex surface of eIF4E exhibits a remarkable evolutionary conservation among all eIF4E proteins (Marcotrigiano et al. 1997). An NMR study in which eIF4E was titrated with 4E-BP protein indicated that this conserved area is the 4E-

BP-binding site (Matsuo et al. 1997). This was later demonstrated directly by crystallographic analysis of eIF4E bound to peptides derived from either 4E-BP1 or eIF4GII (Marcotrigiano et al. 1999). When bound to eIF4E, both peptides exhibit an L-shaped α-helical conformation and bind to the convex dorsal surface of eIF4E. Direct and water-mediated hydrogen bonds, van der Waals, and hydrophobic interactions mediate the binding. As expected, the residues in eIF4G and 4E-BP demonstrated by mutagenesis studies to be crucial for the interaction with eIF4E establish major intermolecular contacts. In addition, consistent with the ability of 4E-BP1 to compete with eIF4G for binding to eIF4E, the eIF4GII and 4E-BP1 peptides establish almost identical contacts with eIF4E. Thus, the 4E-BPs interfere with eIF4G binding to eIF4E by acting as molecular mimics of the eIF4E-binding site in the eIF4G proteins (Marcotrigiano et al. 1999).

The 4E-BPs are largely unstructured in solution (Fletcher et al. 1998; Fletcher and Wagner 1998; Marcotrigiano et al. 1999). How then might this peptide recognize and bind to eIF4E? Specificity appears to be conferred via an "induced fit" mechanism, whereby eIF4E binding fixes a small region of the 4E-BP into the energetically favorable α-helical conformation (Marcotrigiano et al. 1999). This model is bolstered by the observation that an eIF4GII peptide containing the eIF4E-binding site is also unstructured in solution and acquires an α-helical conformation upon eIF4E binding (Marcotrigiano et al. 1999). Structural changes in the eIF4G proteins induced by eIF4E binding may not be limited to the small region cocrystallized with eIF4E: A study performed with yeast eIF4G revealed that following eIF4E binding, a 100-amino-acid peptide acquires secondary structure and becomes resistant to proteolytic cleavage (Hershey et al. 1999).

In sum, a highly efficient mechanism for the regulation of eIF4F formation has evolved in mammals (Fig. 1), whereby the inhibitory 4E-BPs act as molecular mimics of the eIF4E-binding motif present in the eIF4G proteins.

Regulation of 4E-BP Phosphorylation

Phosphorylation of specific serine and threonine residues modulates the affinity of the 4E-BPs for eIF4E (see, e.g., Lin et al. 1994; Pause et al. 1994; Fadden et al. 1997). Although hypophosphorylated 4E-BPs bind efficiently to eIF4E, hyperphosphorylation abrogates this interaction (see, e.g., Lin et al. 1994; Pause et al. 1994; Fadden et al. 1997). 4E-BP phosphorylation levels are modulated by many types of extracellular stimuli (see Table 1 in Gingras et al. 1999a). In this regard, 4E-BP1 was first described ~15 years before its cDNA was cloned as a protein that is highly phosphorylated after insulin or growth factor stimulation of rat adipocytes or murine Swiss 3T3L1 adipocytes (Belsham and Denton 1980; Belsham et al. 1982; Blackshear et al. 1982, 1983). This protein was later biochemically purified and cloned, and termed *p*hosphorylated *h*eat and *a*cid *s*table protein-*i*nsulin responsive, PHAS-I (Hu et al. 1994). The function of PHAS-I was ascertained when it was found to be the rat ortholog of human 4E-BP1 (Lin et al. 1994; Pause et al. 1994). Hormones (insulin, angiotensin II, etc.), growth factors (EGF, PDGF, NGF, IGFI, IGFII, etc.), cytokines (IL-3, GMCSF in combination with steel factor, etc.), mitogens (TPA), G-protein-coupled receptor ligands (gastrin, DAMGO), and adenovirus infection (see Chapter 17) induce hyperphosphorylation of 4E-BP1, accompanied (when assessed) by a resultant decrease in its interaction with eIF4E and an increase in cap-dependent translation rates (for review, see Gingras et al. 1999a). Conversely, serum starvation (see, e.g., von Manteuffel et al. 1996), amino acid deprivation (see below; see also Chapter 16), picornavirus infection (Gingras et al. 1996), and certain environmental stresses such as heat shock (in certain cell types; Vries et al. 1997) or osmotic shock (Parrott and Templeton 1999), lead to a decrease in 4E-BP1 phosphorylation, an increase in its affinity for eIF4E, and an inhibition of cap-dependent translation.

Six Ser/Thr phosphorylation sites have been identified in the mammalian 4E-BP1 protein (Fig. 2) (Fadden et al. 1997; Heesom et al. 1998). No tyrosine phosphorylation has been observed in this protein. Five of the six Ser/Thr sites are followed by a proline residue, and one is followed by glutamine. Two phosphorylated residues, Thr-37 and Thr-46, lie on the

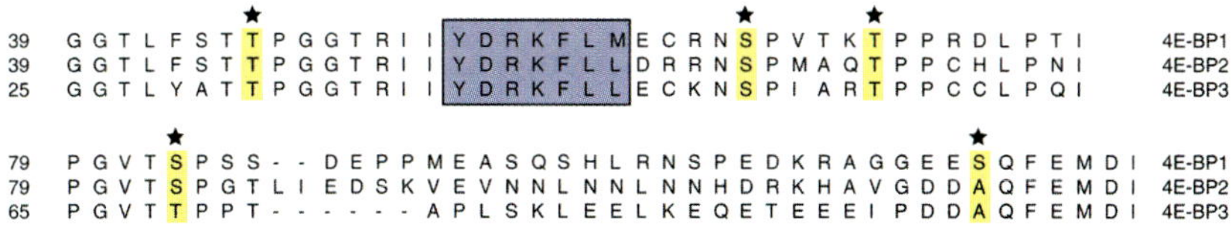

Figure 2 Alignment of the mammalian 4E-BPs through the eIF4E-binding site. The conserved eIF4E-binding motif is boxed in blue. Phosphorylated residues in 4E-BP1 are highlighted in yellow and indicated with a star.

amino-terminal side of the eIF4E-binding motif (located at aa 54–60), and four phosphorylated residues have been identified on the carboxy-terminal side of the eIF4E-binding motif: Ser-65, Thr-70, Ser-83, and Ser-112. Phosphorylated residues may vary somewhat according to cell type and/or species. For example, phosphorylated threonines 37 and 46, Ser-65, Thr-70, and Ser-83 account for the major phosphopeptides observed by two-dimensional phosphopeptide mapping of 4E-BP1 from human embryonic kidney 293 cells (Gingras et al. 1999a,b and in prep.), although some minor phosphopeptides remain unidentified. However, in rat epididymal adipocytes, Ser-112 (Ser-111 in the rat protein) was reported to be a major insulin-stimulated residue (Heesom et al. 1998). In other studies, Ser-112 phosphorylation of rat 4E-BP1 was not detected (Fadden et al. 1997, 1998). The significance of cell-type and/or species-specific differences in 4E-BP1 phosphorylation is not understood.

To determine how each of these phosphorylation events is regulated in vivo, extensive two-dimensional tryptic mapping analyses have been conducted on 4E-BP1 immunoprecipitated from 293 cells. Both Thr-37 and Thr-46 are phosphorylated in 4E-BP1 isolated from serum-deprived cells. Addition of serum to the cell culture media affects the phosphorylation state of Thr-37 and Thr-46 only mildly, increasing it 1.3- to 1.7-fold (Gingras et al. 1999b). Whereas 4E-BP1 from serum-starved cells contains low to undetectable levels of phosphorylated Ser-65 and Thr-70, 4E-BP1 from serum-stimulated cells is highly phosphorylated on these residues (Gingras et al. 1999a,b and in prep.). Whether the phosphorylation state of Ser-83 is modulated by serum remains unclear, as the behavior of the phosphopeptide containing Ser-83 varies from experiment to experiment (Gingras et al. 1998, 1999, and in prep.). A major difference in the sensitivity to the kinase inhibitors wortmannin, LY294002, and rapamycin (discussed below) was also observed for the two sets of phosphorylation sites in 4E-BP1 (von Manteuffel et al. 1997; Gingras et al. 1998, 1999b, and in prep.). In the presence of serum, phosphorylation on Ser-65 and Thr-70 is acutely inhibited by these compounds, whereas the phosphorylation state of Thr-37 and Thr-46 is only mildly affected (Gingras et al. 1999b and in prep.). Thus, 4E-BP1 contains two sets of phosphorylation sites, one on the amino-terminal side of the eIF4E-binding motif that is relatively insensitive to serum-deprivation and kinase inhibitors, and a second set on the carboxy-terminal side of the eIF4E-binding motif that is profoundly sensitive to the presence of serum or kinase inhibitors. Ser-112 may comprise a third class of sites specific to rat adipocytes, which is serum-sensitive but rapamycin-resistant (Heesom et al. 1998).

The same mapping and mutational studies, accompanied by two-dimensional protein gel analysis and the use of phospho-specific antibodies, have demonstrated that 4E-BP1 phosphorylation is a highly ordered, hierarchical process. Phosphorylation on Ser-65 and Thr-70 is only detected in protein species also phosphorylated on Thr-37 and Thr-46 (A.-C. Gingras et al., in prep.). Mutation of Thr-37 or Thr-46 (or both residues) to alanine(s) abolishes phosphorylation of Ser-65 and Thr-70 in serum-replete 293T cells (Gingras et al. 1999b). Furthermore, Ser-65 and Thr-70 are phosphorylated (albeit weakly) when Thr-37 and Thr-46 are mutated to glutamic acid residues, which can mimic phosphate groups in many instances (Gingras et al. 1999b). Mutation of Ser-65 or Thr-70 (or both residues) to alanine has no effect on the phosphorylation state of Thr-37 or Thr-46 (A.-C. Gingras et al., in prep.). Thus, inactivation of 4E-BP1 binding to eIF4E appears to be a two-step process, in which phosphorylation of Thr-37 and Thr-46 acts as a "priming" event for Ser-65 and Thr-70 phosphorylation (Fig. 3). How phosphorylation at Thr-37 and Thr-46 may act as a priming event has not been elucidated; it is possible that the phosphorylated residues recruit a Ser-65/Thr-70 kinase, or phosphorylation could induce a conformational change at the eIF4E/4E-BP1 interface to facilitate access to a kinase. Interestingly, singly phosphorylated 4E-BP1 isoforms are rarely detected, suggesting that the phosphorylation state of Thr-37 and Thr-46 is co-regulated (Gingras et al. 1999b and in prep.). Phosphorylation of the carboxy-terminal residues has been further ordered: Thr-70 phosphorylation precedes that of Ser-65 following serum stimulation, because phosphorylated Ser-65 is detected only in species also phosphorylated on Thr-37, Thr-46, and Thr-70 (A.-C. Gingras et al., in prep.).

Alignment of the human 4E-BPs reveals that all of the phosphorylated residues in 4E-BP1 are conserved in 4E-BP2 and 4E-BP3, except for Ser-112 (Fig. 2). However, the phosphorylation pattern of 4E-BP2 is less complex than that for 4E-BP1: Whereas 4E-BP1 isolated from 293 cells migrates as six species in the two-dimensional gel electrophoresis system (A.-C. Gingras et al., in prep.), 4E-BP2 isolated from 293 cells migrates as only four isoforms (B. Raught et al., unpubl.). Two-dimensional gel analysis and tryptic phosphopeptide mapping indicate that 4E-BP2 is also phosphorylated on both Thr-37 and Thr-46 (B. Raught et al., unpubl.). These data suggest that at least one 4E-BP2 phosphorylation site (or other posttranslational modification) remains to be identified. 4E-BP3 is also a phosphoprotein (Poulin et al. 1998), but the identities of the phosphorylated residues in this 4E-BP have not been established. How these differences in phosphorylation state affect the behavior of the 4E-BPs is an interesting question that remains to be investigated.

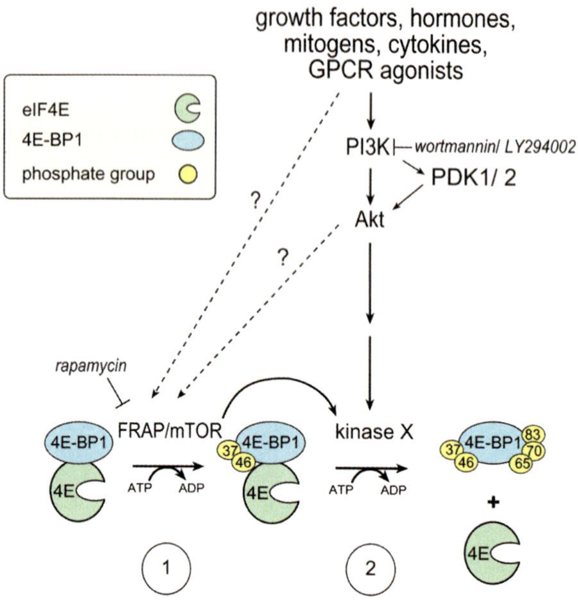

Figure 3 Two-step model for 4E-BP1 inactivation. *Step 1:* FRAP/mTOR phosphorylates 4E-BP1 on two sites, Thr-37 and Thr-46. Phosphorylation of these amino acids appears to be a necessary priming event for subsequent phosphorylation of several residues carboxy-terminal to the eIF4E-binding motif. Whether FRAP/mTOR activity is directly responsive to PI3K or Akt activity is controversial (as indicated by the dashed arrows). *Step 2:* Two inputs are required to activate the 4E-BP1 carboxy-terminal kinase(s) X; one input is required from FRAP/mTOR, and a second input derives from extracellular stimuli (as indicated) that activate the PI3K–PDK–Akt pathway. Kinase(s) X phosphorylates the indicated 4E-BP1 carboxy-terminal residues, effecting eIF4E release. Chemical inhibitors of specific kinases in this pathway are indicated in italics.

Crystallographic studies have suggested a mechanism as to how 4E-BP phosphorylation may disrupt binding to eIF4E: The presence of acidic patches on eIF4E on both sides of the bound 4E-BP peptide suggests that phosphorylation of the 4E-BPs on residues proximal to the eIF4E-binding site could induce electrostatic repulsion between the two proteins (Marcotrigiano et al. 1999). The relative contribution of each of the phosphorylation sites in this process is unclear, however. Even though the ordered phosphorylation of 4E-BP1 culminates in the phosphorylation of Ser-65, phosphorylation of this residue alone is insufficient to induce release from eIF4E, suggesting that phosphorylation of other sites is also required (Fadden et al. 1997; A.-C. Gingras et al., in prep.).

Intracellular Signaling Pathways Modulating 4E-BP Phosphorylation: PI3K, Akt/PKB, and FRAP/mTOR

It is now clear that 4E-BP1 is a downstream target of *p*hospho*i*nositide 3′-OH *k*inase (PI3K; Mendez et al. 1996; von Manteuffel et al. 1996; Gingras et al. 1998) and its downstream effector, the serine/threonine kinase Akt/*p*rotein *k*inase *B* (PKB) (Gingras et al. 1998; Kohn et al. 1998; Dufner et al. 1999; Takata et al. 1999). 4E-BP phosphorylation is also dependent on the kinase *F*KBP12-*r*apamycin *a*ssociated *p*rotein/*m*ammalian *t*arget *o*f *r*apamycin (FRAP/mTOR; Brunn et al. 1997a,b; Hara et al. 1997; Burnett et al. 1998; Gingras et al. 1998). This signaling pathway(s) (for review, see Gingras et al. 1999a) is illustrated in Figure 3 and reviewed briefly below.

PI3 Kinases

The mammalian PI3Ks are a family of enzymes that phosphorylate the hydroxyl group at the D3 position in the inositol ring of phosphatidylinositol. The PI3K family plays a role in the regulation of many critical cellular processes, including proliferation, regulation of cytoskeletal architecture, vesicular trafficking, apoptosis, and protein synthesis (Fruman et al. 1998; Datta et al. 1999). In response to extracellular stimuli, the PI3K regulatory/adapter subunit recruits the catalytic subunit to membranes (including the plasma membrane and internal membranous structures), placing it in close proximity to its lipid substrates (Fruman et al. 1998). Whether particular PI3K isoforms have different effects on translation initiation is unknown at present. Wortmannin irreversibly inhibits the activity of the catalytic subunit of PI3Ks at low concentrations (for review, see Ui et al. 1995), and LY294002 is an unrelated, reversible PI3K inhibitor (Vlahos et al. 1994). It must be noted, however, that at higher concentrations these compounds also inhibit certain PI4Ks and members of the *p*hospho*i*nositide *k*inase (PIK)-related family, including FRAP/mTOR (see below) (Nakanishi et al. 1995; Brunn et al. 1996; Downing et al. 1996; Sarkaria et al. 1998). Wortmannin and LY294002 inhibit phosphorylation of 4E-BP1 at low concentrations, implicating PI3K in 4E-BP1 phosphorylation (Brunn et al. 1996; Lin and Lawrence 1996, 1997; von Manteuffel et al. 1996; Xu et al. 1998b; Gingras et al. 1999b). Platelet-derived growth factor (PDGF) receptor mutants unable to activate the PI3K pathway fail to induce 4E-BP1 phosphorylation, whereas a PDGF receptor possessing only those tyrosine residues required for PI3K activation efficiently effects 4E-BP1 hyperphosphorylation (von Manteuffel et al. 1996). Insulin receptor substrate-1 (IRS-1) mutant proteins possessing only those tyrosine residues necessary for PI3K activation also retain the ability to activate 4E-BP1 (Mendez et al. 1996). Finally, overexpression of a constitutively active p110α PI3K catalytic subunit (targeted to the plasma membrane by a carboxy-terminal farnesylation motif) maintains 4E-BP1 in a constitutively hyperphosphorylated state, even in the absence of serum (Gingras et al. 1998).

Akt/PKB

The Akt/PKB Ser/Thr protein kinases (a family comprising three members in mammals) are activated by PI3K-generated lipid products, which bind to the Akt/PKB amino-terminal pleckstrin homology (PH) domain and target the protein to membranes (for review, see Marte and Downward 1997). This translocation event enables full activation of Akt/PKB via subsequent phosphorylation events catalyzed by the *p*hosphoinositide-*d*ependent *k*inases, PDK1 and PDK2 (Alessi et al. 1998). The introduction of a Src myristoylation signal at the amino terminus of Akt (yielding MyrAkt) creates a constitutively membrane-targeted kinase. Cells expressing MyrAkt survive treatment with apoptosis-inducing agents or growth factor deprivation (Kennedy et al. 1997; Marte and Downward 1997). MyrAkt overexpression induces 4E-BP1 hyperphosphorylation on the same sites phosphorylated in response to serum stimulation (Gingras et al. 1998). A conditionally active MyrAkt protein also enhances the phosphorylation of 4E-BP1 (Kohn et al. 1998), and expression of an activated, but not membrane-targeted, form of Akt/PKB induces 4E-BP1 hyperphosphorylation (Dufner et al. 1999), indicating that the increase in 4E-BP1 phosphorylation is not due to the artificial membrane targeting. Finally, dominant-negative forms of Akt prevent the increase in 4E-BP1 phosphorylation observed following insulin stimulation (Gingras et al. 1998; Dufner et al. 1999; Takata et al. 1999).

Akt/PKB may also be activated by PI3K-independent pathways (Konishi et al. 1996, 1997; Sable et al. 1997) and directly phosphorylated by the Ca^{++}/calmodulin-dependent protein kinase kinase (CaM-KK) both in vitro and in vivo (Yano et al. 1998). Whether these alternative pathways also modulate 4E-BP phosphorylation through Akt/PKB has not been addressed. Importantly, MyrAkt-induced phosphorylation of 4E-BP1 is wortmannin-insensitive, yet remains rapamycin-sensitive, indicating that a rapamycin-sensitive kinase(s) functions downstream from Akt to regulate 4E-BP1 phosphorylation (Gingras et al. 1998).

FRAP/mTOR

FRAP/mTOR (also known as *r*apamycin *a*nd *F*KBP12 *t*arget, RAFT1, or *rap*amycin *t*arget, RAPT1) is the mammalian ortholog of the yeast TOR proteins. The TORs are coded for by two genes isolated in a screen for yeast mutants resistant to the growth-inhibitory effects of the immunosuppressant rapamycin. Rapamycin is an antifungal macrolide that potently represses yeast cell growth and arrests mammalian T-cell proliferation in the G_1 phase of the cell cycle (Abraham and Wiederrecht 1996; Hall 1996). Rapamycin exerts its effects by binding to a highly conserved 12-kD FK506-binding protein, the immunophilin FKBP12. The rapamycin–FKBP "gain-of-function" complex then specifically targets the FRAP/mTOR or yeast TOR proteins to inhibit their kinase activity (for review, see Abraham and Wiederrecht 1996; Hall 1996). FRAP and TOR are members (along with the *a*taxia *t*elangiectasia *m*utated protein, ATM, the *AT*M and *R*ad3-related protein/*F*RAP-*r*elated *p*rotein, ATR/FRP, and others) of a family of kinases termed the PIK-related kinases (Keith and Schreiber 1995; Hoekstra 1997; Thomas and Hall 1997). Although initially identified via their homology with lipid kinases (and especially to PI3Ks), most members of this family appear to function, instead, as protein kinases (Keith and Schreiber 1995; Hoekstra 1997; Thomas and Hall 1997).

In mammalian cells, expression of a rapamycin-resistant FRAP/mTOR mutant protein confers rapamycin resistance to 4E-BP1 phosphorylation (Brunn et al. 1997b; Hara et al. 1997; Gingras et al. 1998), and FRAP/mTOR immunoprecipitates phosphorylate 4E-BP1 and 4E-BP2 in vitro (Brunn et al. 1997a,b; Burnett et al. 1998; Gingras et al. 1999b; B. Raught et al., unpubl.). Whereas an initial report suggested that FRAP/mTOR phosphorylated five Ser/Thr-Pro sites in 4E-BP1 (Brunn et al. 1997a), later studies have demonstrated that FRAP/mTOR phosphorylates only the "priming" sites, Thr-37 and Thr-46 in 4E-BP1 and 4E-BP2 (e.g., Burnett et al. 1998; Gingras et al. 1999b; B. Raught et al., unpubl.). This discrepancy may be reconciled by the observation that the kinase activity in a FRAP immunoprecipitate directed toward Thr-37 and Thr-46 resists stringent washing, whereas a second kinase activity in the immunoprecipitate directed toward Ser-65 and Thr-70 is removed by washing (Heesom and Denton 1999). These data suggest that FRAP/mTOR itself is involved only indirectly in the phosphorylation of Ser-65/Thr-70. It should be emphasized, however, that FRAP/mTOR likely plays a critical regulatory role in the phosphorylation of these amino acids, because Ser-65 and Thr-70 display a higher level of rapamycin sensitivity than Thr-37 and Thr-46 (Fig. 3) (Gingras et al. 1999b and in prep.).

Two hypotheses, which are not mutually exclusive, have been proposed regarding the regulation of FRAP/mTOR activity: (1) FRAP/mTOR is activated directly by growth factors (or other stimuli) and (2) FRAP/mTOR acts as a gatekeeper to "sense," for example, amino acid sufficiency (see below). Reports have provided support for both hypotheses. For example, the kinase activity of immunoprecipitated FRAP/mTOR was reported to increase following treatment with interleukin 3, insulin, or serum, paralleling an increase in ^{32}P incorporation into FRAP/mTOR itself (Scott et al. 1998; Navé et al. 1999; Sekulic et al. 2000). Additionally, in a cell line expressing a conditionally activated Akt/PKB protein, Akt/PKB activation elicits an increase in the kinase activity associated with FRAP/mTOR (Scott et al. 1998). Studies using phosphospecific antibodies revealed that Akt/PKB can phosphorylate FRAP/mTOR directly on Ser-2448 in vitro, and that this phosphorylation event is responsive to insulin and wortmannin treatments in vivo (Navé et al. 1999; Sekulic et al. 2000). However, the role of Ser-2448 phosphorylation is unclear, because phosphorylation at this site is not necessary for signaling to either 4E-BP1 or S6K1 (Sekulic et al. 2000). Support for the gatekeeper hypothesis stems mainly from the results of amino acid deprivation studies suggesting that specific amino acids can modulate 4E-BP1 phosphorylation without the involvement of PI3K or Akt (see below).

A Conserved Rapamycin-sensitive Signaling Pathway in Yeast and Mammals

Although eIF4F formation in *Saccharomyces cerevisiae* and mammals appears to be regulated by different effector proteins (see below), the yeast TOR proteins, like FRAP/mTOR in mammalian cells, play a critical role in the control of yeast translation initiation. Several putative components of a rapamycin-sensitive signaling pathway downstream from the TOR proteins in *S. cerevisiae* have been identified through genetic screening. Various PP2A regulatory subunits (Jiang and Broach 1999), the PP2A-related phosphatase Sit4, and Tap42 (Di Como and Arndt 1996) can all confer partial resistance to rapamycin in this system. *S. cerevisiae* expressing temperature-sensitive mutants of the Tap42 protein exhibit a dramatic defect in translation initiation when grown at the nonpermissive temperature (Di Como and Arndt 1996). Tap42 interacts in a nutrient-dependent manner with the catalytic subunits of the related phosphatases PP2A and Sit4 (Di Como and Arndt 1996). PP2A phosphatases normally function as trimeric heterocomplexes, consisting of a catalytic (C) subunit, an adapter (A) subunit, and a variable regulatory (B) subunit (for review, see Millward et al. 1999). The regulatory subunit interacts with the catalytic subunit through the adapter protein. However, unlike most regulatory subunits, the interaction of Tap42 with PP2A involves direct binding to the catalytic subunit. The association of Tap42 with Sit4 and PP2A is disrupted by rapamycin treatment, and is prevented by expression of a rapamycin-resistant Tor2 mutant protein (Di Como and Arndt 1996). Tap42 is a phosphoprotein, the phosphorylation state of which is sensitive to rapamycin (Jiang and Broach 1999), and the rapamycin-sensitivity of Tap42 phosphorylation is abrogated in a strain expressing a rapamycin-resistant Tor1 protein (Jiang and Broach 1999). In vitro, a Tor1 immunoprecipitate can phosphorylate Tap42 (Jiang and Broach 1999), suggesting that Tor1 (or an associated kinase) directly regulates its binding to PP2A-type phosphatases. Dephosphorylation of Tap42 appears to be mediated by PP2A, as mutations in some PP2A subunits prevent dephosphorylation of Tap42 following rapamycin treatment (Jiang and Broach 1999).

Homologs of Sit4 (the phosphatase PP6) and Tap42 (the B-cell-receptor-binding protein, α4) have been identified in mammalian cells. The interaction between these proteins is also conserved in mammals; α4 binds directly to the catalytic subunits of PP2A (Murata et al. 1997; Inui et al. 1998), PP4, and PP6 (Chen et al. 1998; Nanahoshi et al. 1999). However, how α4 and Tap42 modulate the activity of their binding partners is not well understood; binding has been reported to both increase and decrease phosphatase activity, or to alter substrate specificity (Murata et al. 1997; Nanahoshi et al. 1998). Like Tap42, α4 was demonstrated to be a phosphoprotein in vivo, and the α4-PP2A interaction was reported to be significantly inhibited by rapamycin (Murata et al. 1997; Inui et al. 1998). However, a direct role for the α4/PP2A-like proteins in the regulation of mammalian translation initiation has not been demonstrated. PP2A (or PP2A-like phosphatases) could be involved in 4E-BP1 dephosphorylation induced by rapamycin, since treatment of cells with the phosphatase inhibitor calyculin prevents rapamycin-induced 4E-BP1 dephosphorylation (Peterson et al. 1999). These data suggest that a PP2A-like phosphatase activity directed toward Ser-65 and Thr-70 (the most rapamycin-sensitive sites) may be activated following rapamycin treatment. It has been reported that in vitro both the Tap42 protein and α4 interfere with PP2A-induced 4E-BP1 dephosphorylation (Nanahoshi et al. 1998).

In sum, both the FRAP/mTOR and TOR proteins play critical roles in the regulation of translation. Some of the downstream effectors of these kinases are conserved between yeast and mammalian cells and may share similar functions. Further study is required to evaluate the impact of the PP2A-like phosphatases in regulating phosphorylation of the 4E-BPs.

Other Kinases May Also Modulate 4E-BP1 Activity

The phosphorylation of 4E-BP1 does not appear to be regulated by the ras/raf/MAPK pathway in most cell lines. MAPK ERK activation is not necessary in 293 and Swiss 3T3 cells for 4E-BP1 phosphorylation: In these cells, insulin does not activate MAPK, yet it efficiently induces hyperphosphorylation of 4E-BP1 (von Manteuffel et al. 1996). Furthermore, the MEK (MAPK/ERK kinase, a.k.a. MAPK kinase) inhibitor PD98059 does not prevent 4E-BP1 phosphorylation in response to various stimuli in these cells or in other cell lines (see, e.g., Lin et al. 1995; Azpiazu et al. 1996; Fleurent et al. 1997). However, several recent reports have suggested that 4E-BP1 phosphorylation may be sensitive to PD98059 in certain other cell types. Prostaglandin $F_{2\alpha}$ treatment of growth-arrested rat vascular smooth muscle cells leads to a fourfold increase in 4E-BP1 phosphorylation. The increase in 4E-BP1 ^{32}P-incorporation in this cell type is inhibited by PD98059 (Rao et al. 1999). A similar phenomenon was observed in murine renal epithelial cells, in which pretreatment with PD98059 abrogates an insulin-stimulated increase in 4E-BP1 phosphorylation (B.S. Kasinath, pers. comm.), and in the hematopoietic MO7e cell line, in which GM-CSF/steel factor signaling to 4E-BP1 was inhibited by PD98059 (Aronica et al. 1997). In all of these cases, however, phosphorylation of 4E-BP1 also remained sensitive to rapamycin and PI3K inhibitors. Whether the effect of PD98059 on 4E-BP1 phosphorylation in MO7e, VSMC, and murine renal epithelial cell lines is due to cross-talk between the MAPK and FRAP/mTOR pathways is unknown. In the kidney epithelial cells, PD98059 does not inhibit the activation of Akt/PKB (B.S. Kasinath, pers. comm.), indicating that if any cross-talk occurs from MEK to FRAP/mTOR, it must occur downstream from Akt. It is not likely that ERK itself is responsible for 4E-BP1 phosphorylation in these cell lines: ERK efficiently phosphorylates free 4E-BP1 in vitro (on Ser-65) but cannot phosphorylate 4E-BP1 bound to eIF4E (Fadden et al. 1997; Gingras et al. 1999b). However, it remains possible that pre-phosphorylation of 4E-BP1 on other sites (Thr-37, Thr-46 and Thr-70) could allow phosphorylation of 4E-BP1 by ERK, even in the presence of eIF4E.

An unknown PI3K-activated kinase found in rat fat cells has been reported to phosphorylate 4E-BP1 on Ser-112 (see above; Heesom et al. 1998). The insulin-stimulated Ser-112 kinase activity is inhibited by wortmannin, but not by rapamycin, and was reported to be more active toward a 4E-BP1–eIF4E complex than toward free 4E-BP1 (Heesom et al. 1998). The physiological role for phosphorylation of Ser-112, and the conditions under which this phosphorylation can be detected, remain to be studied.

Finally, a PI3K- and Akt/PKB-independent signaling pathway has also been reported to activate 4E-BP1 phosphorylation in a calcium- and calmodulin-dependent manner (Rybkin et al. 2000). Phenylephrine (PE), an α1 adrenergic receptor agonist, does not activate PI3K and Akt in Rat-1 cells, yet stimulates 4E-BP1 phosphorylation (Ballou et al. 2000; Rybkin et al. 2000). PE-induced phosphorylation is prevented by calcium chelation, or through inhibition of calmodulin, but not by phorbol ester down-regulation of the calcium-dependent PKC isoforms (Rybkin et al. 2000). Phosphorylation of 4E-BP1 in PE-stimulated cells can be fully prevented by treatment with rapamycin or with high concentrations of LY294002, consistent with a direct inhibition of FRAP/mTOR (Rybkin et al. 2000). The possible regulation of FRAP/mTOR activity by calcium remains to be studied.

In sum, a small but growing body of data suggests that, although PI3K, Akt/PKB, and FRAP/mTOR signaling plays a dominant and central role in the modulation of 4E-BP phosphorylation in all cell types, other signaling pathways may function in a co-regulatory capacity or may play a role in the regulation of 4E-BP phosphorylation through "cross-talk" with the FRAP/mTOR signaling pathway in a cell-type-specific manner.

4E-BP1 Phosphorylation Is Regulated by Nutrients

In animal models, a brief period of starvation engenders a potent reduction in protein synthetic rates. Refeeding of the animal rapidly reverses this process. Several in vivo studies aimed at defining the mechanism of this translational up-regulation have suggested that the nutrients themselves (and not, for example, modulation of insulin levels) could be responsible for the observed effects (see Chapter 16). A striking reduction in translation initiation rates also occurs in cultured mammalian cells deprived of amino acids. Re-addition of amino acids to the culture medium readily reverses the translation inhibition, again indicating that the nutrients themselves play a role in translational control (see Chapter 16). In *S. cerevisiae*, amino acid deprivation also induces translational down-regulation, followed by G_1 arrest (Di Como and Arndt 1996; Thomas and Hall 1997). Numerous reports have indicated that amino acid deprivation modulates the phosphorylation state of several translation regulatory factors in mammalian cells, including 4E-BP1 (for review, see Chapter 16).

Amino acid sufficiency is required for insulin-, serum-, or IGF I-mediated 4E-BP1 hyperphosphorylation (see, e.g., Hara et al. 1998; Xu et al. 1998a,b). Incubation of cells in culture medium lacking amino acids results in a reduction of the basal phosphorylation state of 4E-BP1 (Hara et al. 1998). Re-addition of amino acids alone can partially restore 4E-BP1 phosphorylation, and amino acids synergize with insulin or serum to elicit complete phosphorylation (Hara et al. 1998; Xu et al. 1998a,b). Further studies have implicated specific amino acids in the stimulation of 4E-BP1 phosphorylation. In isolated rat adipocytes, increasing the concentration of a complete amino acid mixture to four times the plasma concentration in fasting rats results in an increase in 4E-BP1 phosphorylation, as detected by electrophoretic mobility shift (Fox et al. 1998). However, removal of leucine alone from the amino acid mixture (but not removal of any other amino acid) prevents the shift in electrophoretic mobility (Fox et al. 1998). Leucine addition stimulates phosphorylation of 4E-BP1 in a dose-dependent manner in the presence or absence of other amino acids, with leucine alone being somewhat less potent than the complete amino acid mixture (Fox et al. 1998). Leucine effects are stereospecific, because the D-stereoisomer is much less effective at stimulating 4E-BP1 phosphorylation. These results suggest that the leucine response is mediated through an interaction with a specific receptor, the dose-response curve being compatible with a model involving binding of a ligand at a single site (Fox et al. 1998). Similar results were reported in several other cell systems (Hara et al. 1998; Patti et al. 1998; Wang et al. 1998a; Xu et al. 1998a; Campbell et al. 1999; Kimball et al. 1999; Shigemitsu et al. 1999), with a few exceptions. Xu et al. (1998a) reported that in a pancreatic β cell line (RINm5F), addition of any one of the branched-chain amino acids, leucine, isoleucine, or valine, was able to induce 4E-BP1 phosphorylation. Thus, leucine is a potent inducer of 4E-BP1 phosphorylation in all model systems studied to date. The role of other amino acids (including the other branched amino acids) is not quite as clear, and may be cell-type specific.

How might amino acid signaling be mediated? Several reports have indicated that amino acid deprivation does not prevent PI3K stimulation by insulin, and that amino acid re-addition does not increase PI3K activity (Hara et al. 1998; Patti et al. 1998; Shigemitsu et al. 1999). Treatment of amino-acid-starved cells with exogenous amino acids also does not lead to an increase in Akt activity (Patti et al. 1998; Kimball et al. 1999). Amino acid deprivation does not alter the ability of insulin to fully activate Akt, despite the fact that under the same conditions insulin is unable to induce 4E-BP1 hyperphosphorylation (Hara et al. 1998; Campbell et al. 1999). These results indicate that the signaling pathway to 4E-BP1 phosphorylation modulated by amino acids does not involve PI3K or Akt. It is thus surprising that several reports have indicated that amino acid-induced 4E-BP1 phosphorylation is sensitive to wortmannin treatment (Patti et al. 1998; Wang et al. 1998a; Xu et al. 1998a). *S6 k*inase (S6K) phosphorylation is similarly affected by amino acid starvation (Hara et al. 1998; Wang et al. 1998a; Campbell et al. 1999; Kimball et al. 1999; Shigemitsu et al. 1999). In this regard, a careful analysis revealed that amino acid-stimulated versus insulin-stimulated S6K activities are inhibited to very different degrees by wortmannin (Shigemitsu et al. 1999). Whereas a concentration of 30 nM was sufficient to inhibit 90% of the S6K activity induced by insulin, much higher concentrations (1 μM) were necessary to elicit a similar inhibition of the amino acid-induced activity. Comparable results were observed for 4E-BP1 in H4IIE cells treated with wortmannin: The phosphorylation state of 4E-BP1 is not affected to the same degree following amino acid stimulation as following insulin stimulation (Shigemitsu et al. 1999). A similar experiment performed with rapamycin revealed that a >85% inhibition of both amino acid-stimulated and insulin-stimulated S6K activity occurred at the low concentration of 0.3 ng/ml (Shigemitsu et al. 1999). These data are consistent with the hypothesis that the target of wortmannin following amino acid stimulation is not PI3K, but another kinase with a lower sensitivity to wortmannin. Since the PIK-related kinases (including FRAP/mTOR) are inhibited by wortmannin at higher concentrations than those required for PI3K inhibition (Brunn et al. 1996; Sarkaria et al. 1998), it is possible that the inhibition observed with wortmannin after amino acid stimulation is due to inhibition of FRAP/mTOR itself. The involvement of FRAP/mTOR in amino acid signaling was suggested by the observation that a rapamycin-resistant S6K1 protein was also insensitive to amino acid withdrawal (Hara et al. 1998), and a rapamycin-resistant FRAP/mTOR protein confers resistance to amino acid deprivation to the wild-type S6K1 (Iiboshi et al. 1999). Thus, FRAP/mTOR may play a pivotal checkpoint role, allowing propagation of intracellular signals to the translation apparatus only when sufficient amino acids are present.

4E-BP1 Phosphorylation Is Modulated Via "Translational Homeostasis"

Treatment of cells with translational inhibitors such as anisomycin or cycloheximide leads to a compensatory hyperphosphorylation of both 4E-BP1 and S6K1 (see, e.g., Brown and Schreiber 1996; von Manteuffel et al. 1996). In contrast, in a murine fibroblast cell line transformed by eIF4E overexpression (Lazaris-Karatzas et al. 1990), in which general translation rates are increased (Rosenwald et al. 1999), both 4E-BP1 and S6K1 are maintained in a hypophosphorylated state, as compared to untransformed cells (Khaleghpour et al. 1999). To determine whether the basis for this phenomenon is eIF4E overexpression itself, and not the transformed state it induces, tetracycline-inducible eIF4E-overexpressing fibroblast lines were generated (Khaleghpour et al. 1999). Upon removal of tetracycline from the cell culture medium, these cell lines overexpress eIF4E to varying levels. Strikingly, the degree of eIF4E overexpression was observed to correlate with the degree of 4E-BP1 and S6K1 dephosphorylation (Khaleghpour et al. 1999). These data are consistent with a model in which an unscheduled change in translation initiation rates is sensed by the cell, and the signaling pathways modulating 4E-BP1 and S6K activity respond in a compensatory manner. 4E-BP1 was also observed to be hypophosphorylated in several murine mammary tumor cell lines, in contrast to nontumorigenic parental cell strains (Raught et al. 1996). It is thus also tempting to speculate that transformed cells may acquire the ability to bypass this translational inhibition mechanism.

Mammalian eIF4 Factor Phosphorylation

Although the role of the PI3K-Akt/PKB and FRAP/mTOR signaling pathway(s) in the phosphorylation of the 4E-BPs in mammalian cells is well documented, recent evidence indicates that this pathway also directly modulates the phosphorylation of two eIF4 translation initiation factors, eIF4GI and eIF4B. Phosphorylation of eIF4E is not mediated by this pathway but is modulated by the ERK and p38 MAPK pathways (see below).

eIF4G Phosphorylation

The physical "bridging" of ribosomes to mRNA in mammalian cells is coordinated primarily by two related, modular scaffolding proteins, eIF4GI and eIF4GII (see Chapter 2; also see Hentze 1997; Morley et al. 1997; Gingras et al. 1999a). Through multiple protein–protein and protein–RNA interactions, the eIF4G proteins recruit the small ribosomal subunit to the 5′ ends of mRNAs. The use of viral proteases, which cleave the mammalian eIF4GI protein into three fragments (amino-terminal, middle, and carboxy-terminal; Lamphear et al. 1995), together with biochemical and functional analyses of these segments, has been helpful in clarifying the roles that each of the domains of the eIF4GI protein plays in the formation of a functional translation initiation complex. The amino-terminal fragment interacts directly with eIF4E (Lamphear et al. 1995;

Mader et al. 1995). This region also interacts with the poly (A)-binding protein (PABP; Imataka et al. 1999). Thus, eIF4G is responsible for circularizing the mRNA by binding to proteins that interact with both the 5′ and 3′ UTRs (see Chapter 10). The middle domain of eIF4GI contains binding sites for eIF3 and eIF4A (Lamphear et al. 1995; Imataka and Sonenberg 1997) and possesses RNA-binding activity (Pestova et al. 1996b). The carboxy-terminal fragment of eIF4GI contains a second, independent binding site for eIF4A (Lamphear et al. 1995; Imataka and Sonenberg 1997) and interacts with an eIF4E-kinase termed Mnk1 (see below).

eIF4GII (Gradi et al. 1998) shares 46% overall identity with eIF4GI at the amino acid level. eIF4GI and eIF4GII are functional homologs, such that all of the features described above for eIF4GI are conserved in eIF4GII (Gradi et al. 1998; Imataka et al. 1999; Pyronnet et al. 1999). The mammalian eIF4G family also includes a protein variously referred to as p97, NAT1, or DAP-5 (Imataka et al. 1997; Levy-Strumpf et al. 1997; Yamanaka et al. 1997). p97 is homologous only to the carboxy-terminal two-thirds of the eIF4Gs and does not possess a region corresponding to the amino-terminal one-third of the eIF4Gs. However, like the eIF4Gs, it possesses binding sites for eIF3, eIF4A, and Mnk1. p97 does not interact with eIF4E or PABP and inhibits cap-dependent translation in vivo, presumably by forming nonfunctional initiation complexes (Imataka et al. 1997). Recent evidence suggests that a cleavage product derived from p97 during apoptosis (termed p86) could preferentially enhance translation of IRES-driven mRNAs (see Chapter 19). In addition, an IRES is present in the p97 mRNA itself, suggesting that a positive feedback loop could ensure continuous translation of p97 during apoptosis (Henis-Korenblit et al. 2000).

As discussed previously, the interaction between eIF4E and the eIF4G proteins is regulated by the 4E-BPs, but how the activity of the eIF4G proteins themselves may be regulated has remained obscure. The eIF4Gs have been known for some time to be phosphoproteins (Tuazon et al. 1989; Morley and Traugh 1990, 1993; Donaldson et al. 1991; Bu et al. 1993; Feigenblum and Schneider 1993; Morley and Pain 1995a,b), however, the intracellular signaling pathways regulating their phosphorylation, the location of the phosphorylation sites within the proteins, and the functional consequences of eIF4G phosphorylation have only just begun to be understood. Recently, two distinct sets of phosphorylation sites have been identified in the eIF4GI protein. One set is located in the carboxyl terminus, with the majority of the sites contained in a poorly conserved putative "hinge" region (aa 1035–1190) residing between the middle and carboxy-terminal eIF4A-binding domains (Raught et al. 2000). The second, less characterized set resides in the amino terminus (see below). The phosphorylation status of the carboxy-terminal phosphorylated region changes in response to serum or mitogen treatment, with several major phosphopeptides increasing in intensity, several other phosphopeptides decreasing in intensity, and two phosphopeptides being unaffected (Raught et al. 2000). Mass spectrometric and mutational analyses identified the serum-stimulated phosphorylation sites as serines 1108, 1148, and 1192 (Raught et al. 2000). PI3K and FRAP/mTOR signaling modulates the phosphorylation of these residues, because it is inhibited by wortmannin, LY294002, and rapamycin (Raught et al. 2000). However, these residues are not phosphorylated directly by FRAP/mTOR, S6K1, or S6K2 in an in vitro kinase assay (Raught et al. 2000). Instead, eIF4GI amino-terminal sequences appear to confer serum and mitogen responsiveness (as well as kinase inhibitor sensitivity) to the carboxy-terminal region; truncation mutant proteins lacking the amino terminus are constitutively phosphorylated in the "serum-stimulated" state, even in serum-starved cells, and acquire resistance to kinase inhibitor treatment. Thus, the PI3K–FRAP/mTOR pathway(s) appears to regulate the accessibility of the carboxy-terminal region to other (rapamycin- and wortmannin-insensitive) kinases (Raught et al. 2000). A similar phenomenon has been noted for S6K1, in that autoinhibition and rapamycin sensitivity are conferred by specific amino-terminal sequences (Dennis et al. 1996; Mahalingam and Templeton 1996). Removal of this region results in a rapamycin-insensitive kinase (Dennis et al. 1996; Mahalingam and Templeton 1996).

The amino terminus of eIF4GI also possesses at least one phosphorylation site, as determined by HPLC and mass spectrometric analyses, Ser-274 (S.P. Gygi and B. Raught, unpubl.). Whether phosphorylation of this site is regulated by serum is not known. The amino terminus of eIF4GI can be phosphorylated directly by FRAP/mTOR in vitro (A.-C. Gingras and B. Raught, unpubl.). However, whether Ser-274 phosphorylation is sensitive to rapamycin, and whether this site is phosphorylated by FRAP/mTOR in vivo, remain to be determined.

How then might phosphorylation modulate eIF4GI activity? The carboxy-terminal phosphorylated region does not overlap with the binding site of any known eIF4G-binding partners, and the interactions between eIF4GI and several known binding partners (eIF3, eIF4A and Mnk1) do not appear to be altered by phosphorylation (B. Raught et al., unpubl.). Secondary structure predictions suggest that the phosphorylated region is relatively unstructured (B. Raught and A.-C. Gingras, unpubl.). This region also contains a caspase cleavage site (M. Bushell and S.J. Morley, pers. comm.), suggesting that it is solvent-exposed. Thus, the phosphorylated region may act as a flexible "hinge" between the middle and carboxy-terminal domains. In the absence of evidence for changes in protein–protein interactions, it was suggested that phosphorylation alters intramolecular interactions to cause short- or long-range changes in eIF4GI structure (Raught et al. 2000).

Taking these observations into account, one working model for eIF4GI "stimulation" (Fig. 4) is a two-step activation sequence similar to that proposed for 4E-BP1 inactivation (Gingras et al. 1999b). FRAP/mTOR must first (either directly or indirectly) effect phosphorylation of the eIF4GI amino terminus, or modulate an interaction with an eIF4GI-binding partner. This event leads to a conformational change, such that the carboxy-terminal region becomes accessible to (unknown) rapamycin- and wortmannin-insensitive kinases. Phosphorylation of the carboxy-terminal residues results in a fully "active" protein. How these phosphorylation events affect the ability of eIF4GI to stimulate translation initiation is unknown.

eIF4GII is also a phosphoprotein, but it does not appear to be phosphorylated to a significant extent in the sector corresponding to the eIF4GI carboxy-terminal phosphorylated region (Raught et al. 2000). The eIF4GI carboxy-terminal phosphorylated region shares a very low degree

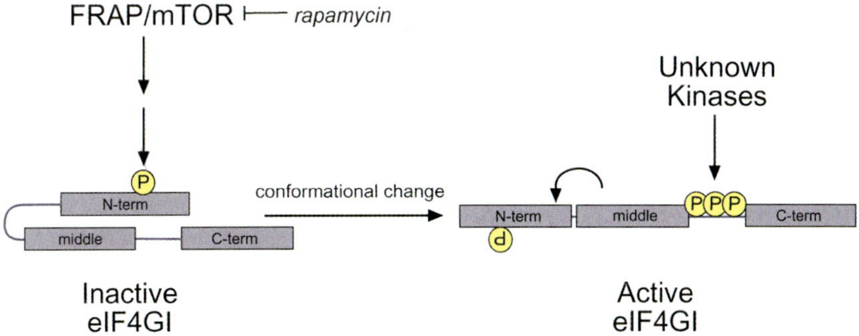

Figure 4 Two-step model for eIF4GI activation. (*Step 1*) FRAP/mTOR directly or indirectly catalyzes the phosphorylation of the eIF4GI amino terminus. This phosphorylation event leads to a conformational change in the protein, which unmasks the carboxy-terminal phosphorylation sites. (*Step 2*) Unknown rapamycin-insensitive kinases phosphorylate these sites, resulting in a fully "active" eIF4GI protein.

of homology with the corresponding area of eIF4GII. In fact, this region is much more homologous to the corresponding fragments of the rabbit and zebrafish eIF4G orthologs than to the same region of the human eIF4GII protein (Raught et al. 2000). Thus, this region of the two human eIF4G proteins appears to have undergone some degree of divergent evolution. Furthermore, eIF4GII phosphorylation does not appear to be responsive to serum or mitogen treatment in 293 cells, suggesting that the two eIF4G proteins may have evolved to respond differently to distinct intracellular signaling pathways. p97 is also a phosphoprotein in 293 cells. Like eIF4GII, the region of p97 corresponding to the eIF4GI carboxy-terminal phosphorylated region is not phosphorylated, and the phosphorylation status of p97 does not appear to be modulated by serum or mitogen treatment (Raught et al. 2000).

eIF4B Phosphorylation

The helicase activity of mammalian eIF4F is significantly stimulated by a ubiquitous cofactor, eIF4B (Lawson et al. 1989; Jaramillo et al. 1990, 1991; Rozen et al. 1990). Although its function is not precisely understood, eIF4B has been proposed to play a multifunctional "matchmaker" role during the initiation process, both by promoting the ATPase and helicase activities of eIF4A (Rozen et al. 1990) and by strengthening the mRNA–rRNA–$tRNA_i^{Met}$ interaction at the initiation codon (Altmann et al. 1995). eIF4B was first purified as an activity capable of stimulating translation and promoting the binding of ribosomes to mRNA (Benne and Hershey 1978; see Chapter 2). More recent studies using a ribosome toe-printing assay have substantiated this function (Pestova et al. 1996a,1998; Morino et al. 1999).

Mammalian eIF4B migrates as multiple isoforms in the two-dimensional isoelectric focusing gel system (Duncan and Hershey 1985), and treatment of cells with serum, insulin, or phorbol esters results in hyperphosphorylation of eIF4B (Duncan and Hershey 1985; Morley and Traugh 1990). eIF4B is phosphorylated in vitro by S6K1, PKC, PKA, CKI, CKII, and PAKI (Morley and Traugh 1989, 1990; F. Peiretti and J.W.B. Hershey, unpubl.). However, how these phosphorylation events may affect eIF4B activity is not understood. Multiple signaling pathways appear to modulate eIF4B phosphorylation. Like eIF4GI and the 4E-BPs, one component of eIF4B phosphorylation appears to be mediated by FRAP/mTOR. Rapamycin treatment of COS-1 cells results in a 60% decrease in eIF4B phosphorylation (F. Peiretti and J.W.B. Hershey, unpubl.). However, unlike the 4E-BPs and eIF4GI, S6K1 was suggested to phosphorylate eIF4B directly in vivo, because expression of a rapamycin-resistant S6K1 protein protects eIF4B phosphorylation from rapamycin treatment. Ser-406-Ala and Ser-422-Ala mutant eIF4B proteins are phosphorylated to about 30% of wild-type levels in vivo, and rapamycin treatment does not further decrease phosphorylation of these proteins, suggesting that these are the sites phosphorylated by S6K1 in vivo (F. Peiretti and J.W.B. Hershey, unpubl.). The relevance of other pathways in eIF4B phosphorylation and the effects of the phosphorylation on its activity remain to be elucidated. The pathways modulating phosphorylation of eIF4B, eIF4GI, and the 4E-BPs are depicted in Figure 5.

eIF4E Phosphorylation

eIF4E is also a phosphoprotein. However, the function of eIF4E phosphorylation is not well understood. Unphosphorylated recombinant eIF4E can stimulate translation in vitro (see, e.g., Svitkin et al. 1996) and can bind to mRNA or mRNA cap analogs (see, e.g., Edery et al. 1988; Carberry et al. 1989). Thus, phosphorylation is not strictly required for eIF4E function. However, as discussed below, the crystal structure of eIF4E suggests that phosphorylation plays a role in the regulation of the eIF4F–mRNA interaction.

The phosphorylation of mammalian eIF4E in response to all stimuli so far examined occurs primarily on a single residue, Ser-209 (numbering

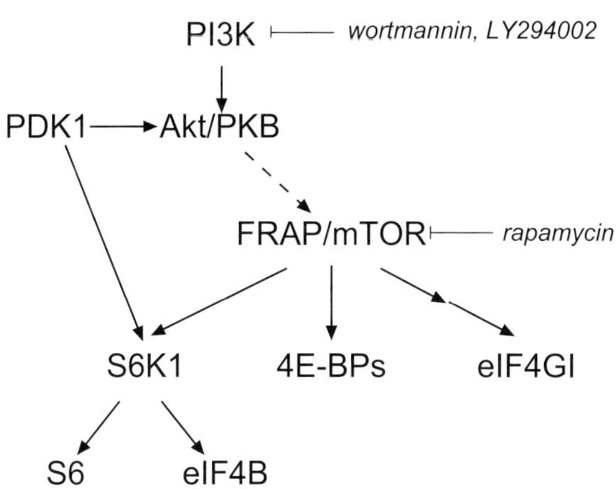

Figure 5 Signaling pathways to the mammalian 4E-BPs, eIF4B, and eIF4GI. A signaling pathway composed of PI3K, Akt/PKB, and FRAP/mTOR modulates 4E-BP, eIF4GI, and eIF4B phosphorylation. Specific kinase inhibitors utilized in the delineation of these pathways are indicated. eIF4B has been suggested to be phosphorylated directly by S6K1, whereas the 4E-BP and eIF4GI proteins are in vitro substrates for FRAP/mTOR.

for the murine protein), with minor phosphorylation detected in certain cases on threonine residues (most likely T210; Bu et al. 1993; Whalen et al. 1996). Although PKC can efficiently phosphorylate eIF4E in vitro on Ser-209 (see, e.g., Whalen et al. 1996; Kleijn et al. 1998), its role in eIF4E phosphorylation in vivo remains unclear. The phosphorylation state of eIF4E, in general, correlates with translation rates and the growth status of the cell (for review, see Sonenberg 1996; Kleijn et al. 1998). eIF4E phosphorylation is modulated in response to a variety of extracellular stimuli: Treatment of cells in culture with hormones, growth factors, cytokines, or mitogens results in a net increase in eIF4E phosphorylation (for review, see Kleijn et al. 1998; Gingras et al. 1999a; Raught and Gingras 1999). eIF4E is hypophosphorylated during mitosis (Bonneau and Sonenberg 1987; Huang and Schneider 1991), a cell cycle phase during which translation rates of most (but not all) mRNAs are low (Fan and Penman 1970; Cornelis et al. 2000; Pyronnet et al. 2000). A putative role for the ras/raf/ERK MAPK pathway (for review, see Waskiewicz and Cooper 1995; Robinson and Cobb 1997) in eIF4E phosphorylation was suggested by the observation that eIF4E phosphorylation is increased in ras- or src-transformed cells (Frederickson et al. 1991; Rinker-Schaeffer et al. 1992). The ERK signaling cascade is activated by extracellular stimuli and is specifically inhibited by PD98059 (for review, see Cobb and Goldsmith 1995; Waskiewicz and Cooper 1995; Robinson and Cobb 1997). Phosphorylation of eIF4E induced by serum or insulin is prevented to a large extent by PD98059 treatment (Flynn and Proud 1996; Morley and McKendrick 1997). However, the ERKs cannot directly phosphorylate eIF4E in vitro, arguing against a direct role for the MAPKs in eIF4E phosphorylation in vivo (Flynn and Proud 1996).

eIF4E Phosphorylation Is Modulated in Response to Environmental Stress. Certain stresses, such as anisomycin or arsenite treatment, increase eIF4E phosphorylation, even though translation rates actually decrease in response to these drugs (Morley and McKendrick 1997). Other types of cellular stress, including heat shock (Duncan et al. 1987) or infection with adenovirus (Huang and Schneider 1991), influenza virus (Feigenblum and Schneider 1993), or encephalomyocarditis virus (Kleijn et al. 1996), are accompanied by a decrease in eIF4E phosphorylation. The p38 subfamily of MAPKs, like the JNK family, is activated in response to many types of environmental stress, including hyperosmolarity, heat shock, UV irradiation, and exposure to lipopolysaccharide, arsenite, or anisomycin (Waskiewicz and Cooper 1995; Robinson and Cobb 1997). p38 MAPK activity (but not JNK activity) is specifically prevented by the pharmacological compound SB203580 (Young et al. 1997). In agreement with a role for p38 MAPK in mediating eIF4E phosphorylation induced by stress, induction of eIF4E phosphorylation by anisomycin is prevented in cells preincubated with SB203580 (Morley and McKendrick 1997; Wang et al. 1998b).

Mnk1 Phosphorylates eIF4E. The mitogen-stimulated pathway acting through the ERKs and the stress-activated pathway acting through the p38 MAPKs converge at a common eIF4E kinase termed Mnk1 (Fukunaga and Hunter 1997; Waskiewicz et al. 1997; MAP kinase interacting kinase 1 or MAP kinase signal integrating kinase 1). Mnk1 was isolated via interaction screening as a substrate for both ERK1 and p38MAPK, and activation of either the ERK or p38 MAPKs (but not the JNK kinases) stimulates Mnk1 kinase activity (Fukunaga and Hunter 1997; Waskiewicz et al. 1997). Mnk1 efficiently phosphorylates eIF4E Ser-209 in vitro (Waskiewicz et al. 1997) and in vivo, following stimulation of either the ERK or p38 MAPK cascades (Pyronnet et al. 1999; Waskiewicz et al. 1999). Mnk1 does not interact directly with eIF4E. Rather, Mnk1 binds to the eIF4G family proteins (Pyronnet et al. 1999; Waskiewicz et al. 1999). Thus, the eIF4Gs recruit Mnk1 to its substrate (Fig. 6). The interaction between eIF4E and eIF4G is required for eIF4E phosphorylation in vivo, because a mutant eIF4E protein that cannot interact with eIF4G is not efficiently phosphorylated in mammalian cells (Pyronnet et al. 1999).

How Does Phosphorylation Affect eIF4E Activity? The three-dimensional structure of the murine and yeast eIF4E proteins, as determined by X-ray crystallography and nuclear magnetic resonance (NMR), respectively (Marcotrigiano et al. 1997; Matsuo et al. 1997), has provided invaluable information as to how eIF4E interacts with the cap, but has also provided important clues as to how phosphorylation of eIF4E may affect this interaction (see Chapter 2). Specific cap binding occurs through a π–π stacking interaction of the cap guanine ring between two eIF4E tryptophan residues (Trp-56 and Trp-102 in the mouse protein; Marcotrigiano et al. 1997). Binding of the cap guanosine is strengthened by hydrogen bonds, notably with a glutamic acid residue (Glu-103), and through a van der Waals contact with Trp-166. The phosphate groups of the cap establish either direct hydrogen bonds or water-mediated hydrogen bonds to basic amino acids located on the concave surface of eIF4E. The crystallographic data also suggest a possible path for bound mRNA, which extends onto a basic area on the concave surface of eIF4E. Bracketing this proposed path, and lying within 7 Å of each other, are a conserved lysine residue

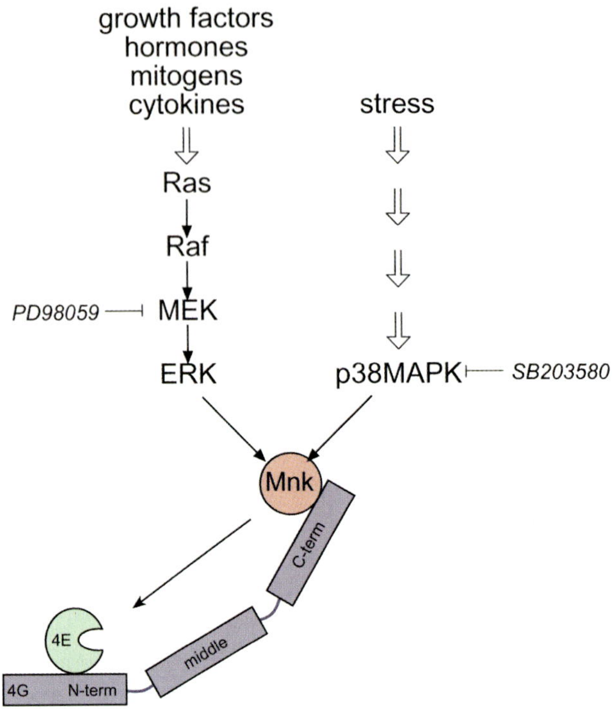

Figure 6 Signaling pathways to eIF4E. Growth factors, hormones, mitogens, cytokines, or other extracellular stimuli activate the Ras/Raf/MEK/ERK kinase cascade. Stress activates the p38MAPK pathway. Both of these pathways can activate Mnk1, an eIF4G-associated kinase. Activated Mnk1 phosphorylates eIF4E on Ser-209. Chemical inhibitors utilized in the delineation of these pathways are indicated in italics.

(Lys-159) and the eIF4E phosphorylation site, Ser-209 (Flynn and Proud 1995; Joshi et al. 1995; Makkinje et al. 1995; Whalen et al. 1996). Ser-209 resides in a flexible region, thus, Lys-159 and a phosphorylated Ser-209 could potentially form a retractable salt bridge to cover and "clamp" bound mRNA. Consistent with this model, phosphorylation of eIF4E has been described to enhance its affinity for mRNA (Minich et al. 1994). Verification of this hypothesis will require crystallization of phosphorylated eIF4E with a capped RNA.

An eIF4E Phosphorylation–mRNA-binding Cycle? The fact that eIF4E is phosphorylated by an eIF4G-bound kinase raises the interesting possibility of an eIF4E phosphorylation–mRNA-binding cycle (Fig. 7). It may be envisioned that upon formation of the eIF4F–mRNA complex,

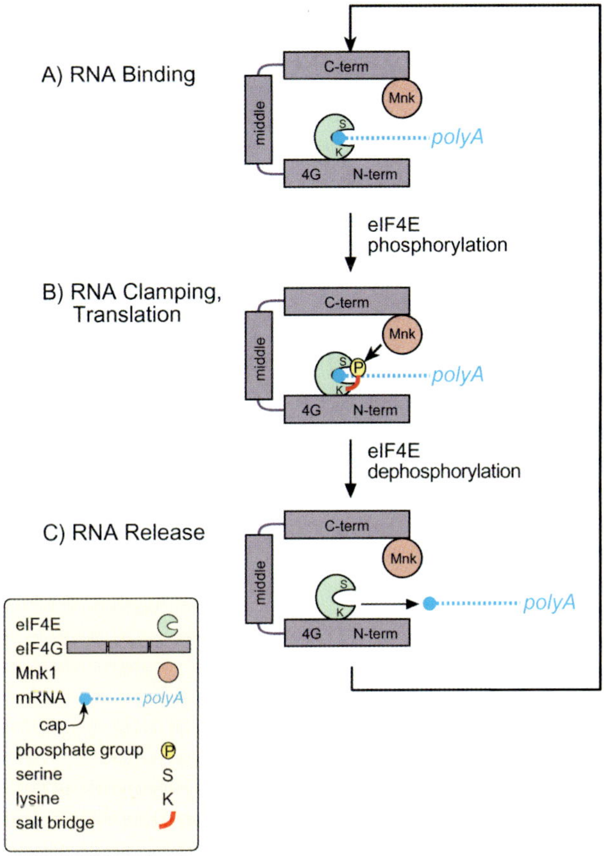

Figure 7 Mnk1/eIF4E clamping cycle model. (*Step A*) eIF4E and Mnk1 interact with the eIF4G proteins. Capped mRNA is bound by unphosphorylated eIF4E. (*Step B*) eIF4E bound to mRNA is phosphorylated by Mnk1. Phosphoserine 209 and a nearby lysine residue (Lys-159) form a salt bridge to cover and "clamp" the mRNA in place. Translation ensues. (*Step C*) eIF4E is dephosphorylated by an unknown phosphatase, effecting mRNA release. Unphosphorylated eIF4E is competent to bind to another mRNA.

eIF4E is phosphorylated by Mnk1 to "clamp" the bound mRNA in place. It is also conceivable that the mRNA must then be "unclamped" to catalyze subsequent rounds of initiation, a task presumably accomplished by an eIF4E phosphatase. eIF4E phosphorylation may thus provide an additional level of control in translation initiation. It could also function to strengthen the eIF4F–mRNA interaction, or to enable more efficient reinitiation (see Chapter 10). In this regard, it is unknown whether eIF4E is released from mRNA after each successive initiation event, or whether all or part of the eIF4F complex remains associated with a given mRNA for more than one round of ribosomal recruitment. There is evidence that eIF4F is assembled before its interaction with the cap structure (Haghighat and Sonenberg 1997; Muckenthaler et al. 1998), arguing against a model in which eIF4E alone remains associated with mRNA, while the rest of the translational machinery is "recycled." Furthermore, UV-induced cross-linking of eIF4E to mRNA cap structures is inefficient (Lee et al. 1985) but is enhanced by the addition of eIF4GI (Haghighat and Sonenberg 1997). This suggests that the eIF4G proteins may stabilize the eIF4E–cap interaction by binding both to eIF4E and to mRNA, and that the eIF4F complex is more likely to bind (initially) to an mRNA than eIF4E alone. It remains possible, however, that phosphorylated eIF4E acquires a more stable interaction with the cap and that the rest of the eIF4F complex can then be recycled. Further study is required to differentiate between these models.

RIBOSOMAL RECRUITMENT TO mRNA IN *S. CEREVISIAE*

S. cerevisiae also possesses an eIF4F complex and, as previously discussed, many of the proteins in a rapamycin-sensitive signaling pathway are conserved between yeast and mammals. However, the factors that regulate eIF4F formation in *S. cerevisiae* differ from those in mammalian cells. Less is known regarding the identity of the intracellular signaling pathways regulating the phosphorylation state of the yeast factors, as well as how phosphorylation affects their activity. No structural homolog of the mammalian 4E-BPs exists in the *S. cerevisiae* genome. However, two yeast-specific eIF4E-binding proteins have been identified, p20 (or Caf20p) and Eap1p (discussed below).

S. cerevisiae eIF4E-binding Proteins

p20

In early studies of *S. cerevisiae* cap-binding proteins, a 20-kD molecule that copurified with eIF4E was suggested to be part of the yeast eIF4F complex (Altmann et al. 1989; Lanker et al. 1992). p20, or Caf20p, binds directly to eIF4E (Altmann et al. 1997), but other than a consensus eIF4E-binding motif (*Y*TIDE*LF*), no significant sequence homology exists between it and the mammalian 4E-BPs. Database searches have failed to reveal any other p20-related proteins in *S. cerevisiae* or in any other organism (A.-C. Gingras, unpubl.). Despite a lack of sequence homology, p20 has been suggested to perform a function similar to the mammalian 4E-BPs, because binding of p20 and the yeast eIF4Gs to eIF4E is mutually exclusive (Altmann et al. 1997), and addition of p20 to a cell-free extract inhibits cap-dependent, but not cap-independent, translation (Altmann et al. 1997). One group reported that disruption of the *CAF20* gene slightly stimulated growth in rich medium (Lanker et al. 1992), but another group failed to observe this effect (de la Cruz et al. 1997). Overexpression of p20 slows cell growth (Altmann et al. 1997; de la Cruz et al. 1997), and the budding index of cells lacking p20 is higher than that of wild-type cells (Altmann et al. 1997). Deletion of *CAF20* partially represses the growth defect of some translation factor mutants (principally eIF4B and eIF4G), and this effect could be reversed by overexpression of p20 (de la Cruz et al. 1997). p20 is a phosphoprotein. The extent of phosphorylated p20 coprecipitating with eIF4E varies following cycloheximide treatment or heat shock (Zanchin and McCarthy 1995). Thus, it has been proposed that binding of p20 to eIF4E may be regulated in a manner similar to that of the mammalian 4E-BPs (Zanchin and McCarthy 1995). However, it is important to note that other groups have observed no change in the amount of p20 bound to eIF4E under many different types of growth conditions, including rapamycin treatment (Altmann et al. 1997; G.P. Cosentino, unpubl.). Thus, there is no clear evidence at this point that a rapamycin-sensitive pathway modulates p20 binding to eIF4E.

Eap1p

Far-Western analysis using a yeast eIF4E probe revealed the presence of an additional 84-kD eIF4E-binding partner, termed *e*IF4E-*a*ssociated *p*rotein 1, Eap1p (Cosentino et al. 2000). Database searches revealed no significant homology with other proteins outside the region containing the eIF4E-binding motif. Binding of Eap1p to eIF4E is disrupted by deletion of this motif, or by mutation of the tyrosine residue in the motif to alanine (Tyr-109-Ala). As with the 4E-BPs and p20, wild-type Eap1p (but not the Tyr-109-Ala mutant) competes with eIF4G for eIF4E binding. Addition of Eap1p to a yeast cell-free translation extract preferentially inhibits cap-dependent translation. Eap1p is not essential for growth and viability; deletion of the *EAP1* gene does not affect growth in rich or defined media, nor does it affect mating or meiosis (Cosentino et al. 2000). However, deletion of *EAP1* does confer partial resistance to growth inhibition by rapamycin, suggesting that Eap1p participates in a rapamycin-sensitive signaling pathway to translation. Further study is required to establish the role of Eap1p in the regulation of translation initiation.

Conservation of *S. cerevisiae* eIF4 Factors

The mammalian eIFs 4A, 4B, 4E, and 4G all have orthologs in *S. cerevisiae* (see Chapter 2). The single gene encoding yeast eIF4E (*CDC33*) is essential for viability (Altmann et al. 1987). Although disruption of either the *TIF4631* (eIF4G1) gene or the *TIF4632* (eIF4G2) gene does not severely affect growth, elimination of both genes is lethal (Goyer et al. 1993). *S. cerevisiae* eIF4A is an essential protein encoded by two genes, *TIF1* and *TIF2* (Linder and Slonimski 1989). Disruption of the yeast eIF4B homolog, *TIF3*, is not lethal, but engenders a slow-growth and temperature-sensitive phenotype (Altmann et al. 1993; Coppolecchia et al. 1993).

The structures of both the murine and *S. cerevisiae* eIF4E proteins reveal a conserved mode of cap binding (Marcotrigiano et al. 1997; Matsuo et al. 1997). The molecular surface responsible for binding to eIF4G and 4E-BP1 is also highly conserved between the yeast and mouse proteins (Marcotrigiano et al. 1997, 1999). Mutation of Trp-75 in the yeast eIF4E protein or Trp-73 in the mouse protein, a residue establishing several contacts with eIF4G, abrogates eIF4G binding in both species (Ptushkina et al. 1998; Marcotrigiano et al. 1999; Pyronnet et al. 1999). The interaction of yeast eIF4E with eIF4G proteins is essential for viability, because the Trp-75 eIF4E mutant is unable to restore yeast viability in *cdc33*-deficient cells (Ptushkina et al. 1998).

Yeast eIF4G proteins share 33% overall similarity with the mammalian eIF4Gs, but lack the carboxy-terminal extension containing eIF4A- and Mnk1-binding sites. The functional consequences of these differences are unknown. As with the mammalian eIF4G proteins, binding of the *S. cerevisiae* eIF4G proteins to eIF4E was delimited to a region harboring the motif YXXXXLL (Mader et al. 1995). Binding of the yeast eIF4G to eIF4A was also recently described (Dominguez et al. 1999; Neff and Sachs 1999) and takes place in a region homologous to the middle domain of the mammalian eIF4Gs. An interaction between eIF3 and eIF4G has not been reported thus far in *S. cerevisiae*. However, the yeast eIF4Gs also interact with the poly(A)-binding protein, Pab1p, through regions homologous to their mammalian counterparts; namely, the amino terminus of eIF4G and RRMs 1 and 2 of Pab1p or PABP. In fact, the interaction between eIF4G proteins and Pab1p was first demonstrated in this system (Tarun and Sachs 1996; Wells et al. 1998). The functional significance of this interaction is reviewed in Chapter 10.

In addition to eIF4A, another member of the DEAD-box family, Ded1p, exhibits ATP-dependent RNA helicase activity (Iost et al. 1999) and is essential for translation initiation in yeast (Chuang et al. 1997; de la Cruz et al. 1997). The functions of Ded1p and the eIF4A homologs Tif1p/2p are not redundant: Deletion of *TIF1* and *TIF2* or *DED1* genes is lethal. Although the function of Ded1p in translation initiation is unknown, a mouse homolog of Ded1p, termed PL10, was demonstrated to substitute for the yeast factor (Chuang et al. 1997), indicating that the role of Ded1p in translation may be conserved in mammals.

Posttranslational Regulation of Yeast eIF4 Factors

eIF4E (CDC33)

Although it is generally believed that eIF4E is a limiting factor for translation in most mammalian cells (Sonenberg 1996), it is not clear that the same is true in *S. cerevisiae*. Overexpression of eIF4E in yeast cells produces no measurable effect on cell growth unless an overexpression of approximately 100-fold is achieved; at this level a slight inhibitory activity is detected (Lang et al. 1994). eIF4E is a phosphoprotein in *S. cerevisiae*. However, the phosphorylation sites and the signaling pathway leading to its phosphorylation differ from those in mammalian cells (McCarthy 1998). The stoichiometry of eIF4E phosphorylation in vivo appears to be low, and the *S. cerevisiae* eIF4E protein does not possess a site equivalent to the human Ser-209. Instead, phosphorylation occurs on serines 2 and 15 (Zanchin and McCarthy 1995). As indicated by the absence of long-range NOEs and by ^{15}N relaxation experiments, the first 35 amino acids of yeast eIF4E, including the two phosphorylation sites, are unstructured (Matsuo et al. 1997). The relevance of phosphorylation to yeast eIF4E activity is thus unclear. The sequence surrounding Ser-2 and Ser-15 fits the casein kinase phosphorylation site consensus, and these residues can be phosphorylated by casein kinase II in vitro (Zanchin and McCarthy 1995). Phosphorylation at these sites is not required for yeast growth under normal conditions (Zanchin and McCarthy 1995), although it is unknown whether it may confer an advantage under other conditions. Although a site corresponding to mammalian Ser-209 is present in the *Schizosaccharomyces pombe* eIF4E, phosphorylation of eIF4E in vivo in logarithmically growing cells is low (J.E.G. McCarthy, pers. comm.). The biological relevance and the site of phosphorylation of the *S. pombe* eIF4E are unknown.

eIF4G

The eIF4Gs are also phosphoproteins in yeast (J.E.G. McCarthy, pers. comm.). However, the location of the phosphorylation sites, the identity of the intracellular signaling pathways mediating phosphorylation of these sites, and whether phosphorylation is modulated under different growth conditions are unknown. Interestingly, unlike their mammalian counterparts, the yeast eIF4G proteins are rapidly degraded following rapamycin treatment (Berset et al. 1998; Powers and Walter 1999). (Other initiation factors are also degraded under these conditions, but with slower kinetics [Powers and Walter 1999].) eIF4G degradation is also observed when yeast cells grow into the diauxic phase (involving a shift from respiration to fermentation), or when they are starved in the stationary phase (Berset et al. 1998; M. Altmann, pers. comm.). eIF4G degradation is not cell-cycle dependent, and is only possible when de novo protein synthesis takes place (Berset et al. 1998; M. Altmann, pers. comm.). Thus, it appears that yeast have evolved a unique mechanism to rapidly remove eIF4G in response to nutrient stress, and that the TOR proteins may function as sensors in this pathway.

REGULATION OF RIBOSOMAL RECRUITMENT IN PLANTS

Translation rates are regulated throughout plant development and, like other organisms, plants have evolved complex regulatory mechanisms to respond to abiotic signals and stresses, such as temperature and light fluctuations and the availability of nutrients, water, and oxygen. One component of this regulation occurs at the translational level. The general mechanism of ribosome binding to mRNA is conserved in plants, with a few notable exceptions. The most striking difference is the presence of two different eIF4F-like complexes, termed eIF4F and eIFiso4F. The plant version of eIF4F, as in mammals, consists of eIF4E, eIF4G, and eIF4A, whereas the eIFiso4F complex consists of eIFiso4E, eIFiso4G, and eIF4A. The significance of having two different eIF4F-like complexes is not well understood, but differences in mRNA binding specificity have been noted (K.S. Browning, pers. comm.). eIF4F and eIFiso4F are the least abundant of the plant initiation factors (Browning et al. 1990), suggesting that these complexes are targets for regulation. eIF4E (p26) and eIFiso4E (p28) were reported to be functionally equivalent in wheat (Browning et al. 1987a; Allen et al. 1992). The *Arabidopsis thaliana* eIF4E and eIFiso4E are, however, not functionally equivalent in a yeast complementation assay, and the two eIF4E species are differentially expressed throughout *Arabidopsis* development (Rodriguez et al. 1998). An eIF4E-related protein, termed nCBP (for *n*ovel *c*ap-*b*inding *p*rotein), is also found in plants (Ruud et al. 1998) and is homologous to the mammalian eIF*4E*-*h*omologous *p*rotein, 4E-HP (Rom et al. 1998), and the *Caenorhabditis elegans* IFE-4 (Keiper et al. 2000). Although the functions of the human and *C. elegans* proteins are unknown, the plant protein was demonstrated to positively function in translation with either eIF4G or eIFiso4G (Ruud et al. 1998). No 4E-BP, Caf20, or Eap1 homologs have been isolated from plant cDNA libraries. The signaling pathways modulating translation initiation factor activity in plants are not well characterized, but phosphorylation of several plant eIFs has been reported, as discussed below.

Plant eIF4 Regulation during Development and in Response to Stress

Translation rates are regulated throughout plant development. For example, during wheat seed maturation, protein synthetic rates are elevated in mid-development to synthesize the bulk of the seed storage proteins, but a rapid, dramatic drop in translation rates occurs at the onset of late development, when the embryo prepares for quiescence. Low translation rates persist until the mature dry stage is reached. As is observed for some mammalian mRNAs, low protein synthesis levels during the maturation phase do not preclude the translation of a subset of mRNAs coding for the "late embryo abundant" proteins, thought to be involved in dessication survival. Mature dry seeds are virtually quiescent, lacking detectable translational activity, but protein synthesis resumes quickly after the onset of germination (Gallie et al. 1998; Le et al. 1998).

In plants, various types of environmental stresses also modulate translation rates. Heat shock leads to a decrease in cap-dependent translation that is in proportion to the severity of the stress (Gallie et al. 1997), suggesting that modulation of eIF4F/eIFiso4F or eIF4B activities may be involved. Oxygen deprivation, which occurs in plant roots following flooding, results in a rapid, global reduction in protein synthesis and polysomal dissociation in maize seedling roots (Bailey-Serres and Freeling 1990). At the same time, a selective enhancement in the translation of some mRNAs (coding for the anaerobic proteins) is observed (see, e.g., Bailey-Serres and Dawe 1996; for review, see Drew 1997). Although it is not yet clear how hypoxia is "sensed" by the root, the response to stress is known to involve a decrease in pH and an increase in cytosolic calcium (for review, see Drew 1997).

plant eIF4A

Phosphorylation of the mammalian and yeast eIF4A proteins has not been observed. However, during wheat seed development, phosphorylated plant eIF4A is observed at early stages, when the endoderm and embryo are provided nutrients by the nucellus, a maternal tissue (Le et al. 1998). In subsequent phases of seed development, and upon induction of germination, only dephosphorylated eIF4A is detected (Le et al. 1998). Forty-eight hours post-germination, when young shoots and roots can be isolated, eIF4A is found only in the unphosphorylated form in shoots, yet both unphosphorylated and phosphorylated isoforms are detected in roots (Le et al. 1998). Thus, the phosphorylation state of eIF4A does not appear to parallel the dramatic changes in translation that occur following seed development in wheat.

eIF4A phosphorylation has also been investigated in the tobacco plant. In tobacco leaves, there are at least 10 expressed eIF4A genes, which can be separated into two divergent families (Owttrim et al. 1994). At least two isoforms of eIF4A are phosphorylated on threonine residue(s) (op den Camp and Kuhlemeier 1998). One of the phosphorylated isoforms, termed NeIF4A8, is specifically expressed in pollen and was proposed to mediate translational control in the developing male gametophyte (Brander and Kuhlemeier 1995). In dry mature pollen, only a small amount of eIF4A is phosphorylated (~ 1–3%). However, 2.5 hours after germination, the amount of phosphorylated eIF4A increases to ~15–20% (op den Camp and Kuhlemeier 1998).

Phosphorylation of eIF4A occurs in the maize root after oxygen deprivation and was proposed to be involved in the hypoxic stress response (Webster et al. 1991). Phosphorylation of wheat leaf eIF4A also occurs following prolonged heat shock (Gallie et al. 1997). However, eIF4A phosphorylation following thermal stress is not likely to be the cause of the immediate reduction in translation efficiency, because translational inhibition occurs much earlier than eIF4A phosphorylation after heat shock (Gallie et al. 1997). Rather, eIF4A phosphorylation was proposed to be part of the plant adaptive response to thermal stress (Gallie et al. 1997). Thus, during the stress response, eIF4A phosphorylation is inversely related to translation rates. This is in contrast to tobacco tube germination, in which phosphorylation of some eIF4A isoforms correlates with an increase in protein synthesis (op den Camp and Kuhlemeier 1998). These observations are not necessarily contradictory: It is possible that phosphorylation occurs on different residues, some of which are inhibitory and some of which are stimulatory. It is also possible that phosphorylation has a positive effect on the function of some eIF4A isoforms and a negative impact on others. Other possibilities may also be envisioned, such as a role in the translation of specific mRNAs, and these remain to be investigated.

Plant eIF4B

Wheat eIF4B can be separated into several isoforms using the two-dimensional isoelectric focusing-SDS polyacrylamide gel electrophoresis system. The most acidic forms can be converted to the most basic forms by phosphatase treatment, indicating that they are phosphorylated species (Gallie et al. 1997). Wheat embryo eIF4B is dephosphorylated, but eIF4B from leaves is predominantly phosphorylated (Gallie et al. 1997). Further studies have indicated that eIF4B is dephosphorylated late in seed development, being almost completely in the hypophosphorylated state in the mature dry seed (Le et al. 1998). Following imbibing, the isoelectric forms of eIF4B were reported to shift to the more acidic species, with kinetics that parallel the increase in protein synthesis observed with maturation (Gallie et al. 1997). Heat shock results in the rapid dephosphorylation of wheat eIF4B (Gallie et al. 1997). Thus, as in mammalian cells, eIF4B phosphorylation correlates with the modulation in protein synthesis following heat shock, and with changes in protein synthesis throughout development.

Plant eIF4E and eIFiso4E

Unphosphorylated eIF4E and eIFiso4E are detected in the maize embryo, whereas only the phosphorylated isoforms are present in leaves (Gallie et al. 1997). In young shoots and roots, even more acidic forms (suggesting a further increase in phosphorylation) of eIF4E and eIFiso4E are present, which disappear in older leaves (Gallie et al. 1997). Thus, the phosphorylation states of both eIF4E and eIFiso4E are regulated during plant development. The role of these phosphorylation events is not understood.

Oxygen deprivation increases the phosphorylation of maize eIF4E (but not of eIFiso4E; Manjunath et al. 1999). This effect is mimicked by treatment with caffeine under aerobic conditions, which also leads to an elevation in cytosolic calcium concentrations (for review, see Drew 1997; Manjunath et al. 1999). Consistent with a role for calcium signaling in the phosphorylation of plant eIF4E, treatment with ruthenium red (which inhibits calcium release from intracellular organelles) prevented hypoxia-induced eIF4E phosphorylation (Manjunath et al. 1999). This situation is analogous to mammalian eIF4E, whose phosphorylation increases following stress (see above). Whether the phosphorylation site(s) and the signaling pathways leading to this increased phosphorylation are conserved between plants and mammals remains to be established.

CONCLUSIONS AND FUTURE PROSPECTS

Since the publication of the first edition of *Translational Control*, significant progress has been made in the characterization of how diverse extracellular stimuli evoke changes in mRNA translation rates in different organisms. In particular, much has been learned regarding the control of ribosomal recruitment to mRNA, a process regulated by the eIF4 group of initiation factors. A primary mode of translational control is the regulation of the formation of a functional eIF4F complex. This process is best understood in mammalian cells, in which a family of translation inhibitors, the 4E-BPs, inhibit eIF4F complex formation through their interaction with eIF4E. In addition to the two yeast eIF4G proteins, two yeast-specific eIF4E-binding proteins have also been characterized, but how their binding to eIF4E is regulated is poorly understood. A second mode of control is via direct phosphorylation of the components of eIF4F. In mammalian cells, eIF4E, eIF4G, and eIF4B phosphorylation is modulated. The eIF4E and eIF4G proteins are also phosphorylated in yeast cells. The plant eIF4E, eIFiso4E, and eIF4B proteins are phosphorylated, and eIF4A phosphorylation has only been observed in plants. How these phosphorylation events affect the activity of these factors is only beginning to be understood.

One of the most interesting topics to emerge in this field is the role of FRAP/mTOR in the regulation of protein synthesis. In fact, FRAP/mTOR (or TOR in *S. cerevisiae*) appears to play a central role in the regulation of metabolism versus catabolism at the cellular level. Along with its regulatory role in the phosphorylation state of the 4E-BPs, eIF4GI, eIF4B, and S6 kinases, it also modulates signaling cascades that activate autophagy (Noda and Ohsumi 1998), nutritionally regulated enzymes (see, e.g., Schmidt et al. 1998), and transcription factors (see, e.g., Beck and Hall 1999). Taken together, these observations suggest that FRAP/TOR activation is an important and highly conserved cellular checkpoint for amino acid sufficiency in both yeast and mammalian cells.

With an improved understanding of how translation initiation is regulated has come the realization that, despite the striking conservation of basic mechanisms through evolution, very interesting variations occur. For instance, both mammals and *S. cerevisiae* have a mechanism in place to regulate the interaction between eIF4E and eIF4G: However, the eIF4E-binding proteins in these species share almost no sequence similarity, except for the very small eIF4E-binding motif. Thus, attesting to its importance, this mode of regulation may have evolved more than once, independently.

What are the future prospects for this field? Much effort will be focused on determining intermediates in the PI3K–Akt/PKB–FRAP/mTOR signaling pathway(s), in the elucidation of how other kinases may invoke "cross-talk" with this pathway, and in ascertaining how nutrient sufficiency regulates FRAP/mTOR activity. Another important problem is the identification of other (possibly FRAP-associated) 4E-BP kinases and phosphatases. We have little understanding as to how phosphorylation modulates the activity of the other eIF4 factors; this will also be an intensive area of research. Finally, a most intriguing problem is how different kinds of signaling inputs may effect specific kinds of changes in translation initiation rates.

ACKNOWLEDGMENTS

We thank Drs. M. Altmann, K.S. Browning, G.P. Cosentino, M.N. Hall, J.W.B. Hershey, H. Imataka, B.S. Kasinath, J.E.G. McCarthy, S. Morino, S.J. Morley, F. Peiretti, S. Pyronnet, and C. Robaglia, as well as M. Ferraiuolo, M. Miron, and F. Poulin for sharing unpublished results. Work in the authors' laboratory was supported by grants from the Howard Hughes Medical Institute, the National Cancer Institute of Canada, the Medical Research Council of Canada, and the Human Frontier Science Program.

REFERENCES

Abraham R.T. and Wiederrecht G.J. 1996. Immunopharmacology of rapamycin. *Annu. Rev. Immunol.* **14:** 483–510.

Alessi D.R., Kozlowski M.T., Weng Q.P., Morrice N., and Avruch J. 1998. 3-Phosphoinositide-dependent protein kinase 1 (PDK1) phosphorylates and activates the p70 S6 kinase in vivo and in vitro. *Curr. Biol.* **8:** 69–81.

Allen M.L., Metz A.M., Timmer R.T., Rhoads R.E., and Browning K.S. 1992. Isolation and sequence of the cDNAs encoding the subunits of the isozyme form of wheat protein synthesis initiation factor 4F. *J. Biol. Chem.* **267:** 23232–23236.

Altmann M., Handschin C., and Trachsel H. 1987. mRNA cap-binding protein: Cloning of the gene encoding protein synthesis initiation factor eIF-4E from *Saccharomyces cerevisiae. Mol. Cell. Biol.* **7:** 998–1003.

Altmann M., Krieger M., and Trachsel H. 1989. Nucleotide sequence of the gene encoding a 20 kDa protein associated with the cap binding protein eIF-4E from *Saccharomyces cerevisiae. Nucleic Acids Res.* **17:** 7520.

Altmann M., Schmitz N., Berset C., and Trachsel H. 1997. A novel inhibitor of cap-dependent translation initiation in yeast: p20 competes with eIF4G for binding to eIF4E. *EMBO J.* **16:** 1114–1121.

Altmann M., Wittmer B., Methot N., Sonenberg N., and Trachsel H. 1995. The *Saccharomyces cerevisiae* translation initiation factor Tif3 and its mammalian homologue, eIF-4B, have RNA annealing activity. *EMBO J.* **14:** 3820–3827.

Altmann M., Muller P.P., Wittmer B., Ruchti F., Lanker S., and Trachsel H. 1993. A *Saccharomyces cerevisiae* homologue of mammalian translation initiation factor 4B contributes to RNA helicase activity. *EMBO J.* **12:** 3997–4003.

Aronica S.M., Gingras A.-C., Sonenberg N., Cooper S., Hague N., and Broxmeyer H.E. 1997. Macrophage inflammatory protein-1 alpha and interferon-inducible protein 10 inhibit synergistically induced growth factor stimulation of MAP kinase activity and suppress phosphorylation of eukaryotic initiation factor 4E and 4E binding protein 1. *Blood* **89:** 3582–3595.

Azpiazu I., Saltiel A.R., DePaoli-Roach A.A., and Lawrence J.C., Jr. 1996. Regulation of both glycogen synthase and PHAS-I by insulin in rat skeletal muscle involves mitogen-activated protein kinase-independent and rapamycin-sensitive pathways. *J. Biol. Chem.* **271:** 5033–5039.

Bailey-Serres J. and Dawe R.K. 1996. Both 5′ and 3′ sequences of maize *adh1* mRNA are requied for enhanced translation under low-oxygen conditions. *Plant Physiol.* **112:** 685–695.

Bailey-Serres J. and Freeling M. 1990. Hypoxic stress-induced changes in ribosomes of maize seedlings. *Plant Physiol.* **94:** 1237–1243.

Ballou L.M., Cross M.E., Huang S., McReynolds E.M., Zhang B.-X., and Lin R.Z. 2000. Differential regulation of the phosphatidylinositol 3-kinase/Akt and p70 S6 kinase pathways by the α1A-adrenergic receptor in Rat1 fibroblasts. *J. Biol. Chem.* **275:** 4803–4809.

Beck T., and Hall M.N. 1999. The TOR signalling pathway controls nuclear localization of nutrient-regulated transcription factors. *Nature* **402:** 689–692.

Belsham G.J. and Denton R.M. 1980. The effect of insulin and adrenaline on the phosphorylation of a 22,000-molecular weight protein within isolated fat cells; possible identification as the inhibitor-1 of the 'general phosphatase.' *Biochem. Soc. Trans.* **8:** 382–383.

Belsham G.J., Brownsey R.W., and Denton R.M. 1982. Reversibility of the insulin-stimulated phosphorylation of ATP citrate lyase and a cytoplasmic protein of M_r 22000 in adipose tissue. *Biochem. J.* **204:** 345–352.

Benne R. and Hershey J.W. 1978. The mechanism of action of protein synthesis initiation factors from rabbit reticulocytes. *J. Biol. Chem.* **253:** 3078–3087.

Berset C., Trachsel H., and Altmann M. 1998. The TOR (target of rapamycin) signal transduction pathway regulates the stability of translation initiation factor eIF4G in the yeast *Saccharomyces cerevisiae*. *Proc. Natl. Acad. Sci.* **95:** 4264–4269.

Blackshear P.J., Nemenoff R.A., and Avruch J. 1982. Preliminary characterization of a heat-stable protein from rat adipose tissue whose phosphorylation is stimulated by insulin. *Biochem. J.* **204:** 817–824.

———. 1983. Insulin and growth factors stimulate the phosphorylation of a M_r-22000 protein in 3T3-L1 adipocytes. *Biochem. J.* **214:** 11–19.

Bonneau A.M. and Sonenberg N. 1987. Involvement of the 24-kDa cap-binding protein in regulation of protein synthesis in mitosis. *J. Biol. Chem.* **262:** 11134–11139.

Brander K.A. and Kuhlemeier C. 1995. A pollen-specific DEAD-box protein related to translation initiation factor eIF-4A from tobacco. *Plant Mol. Biol.* **27:** 637–649.

Brown E.J. and Schreiber S.L. 1996. A signaling pathway to translational control. *Cell* **86:**

517–520.

Browning K.S., Lax S.R., and Ravel J.M. 1987a. Identification of two messenger RNA cap binding proteins in wheat germ. Evidence that the 28-kDa subunit of eIF-4B and the 26-kDa subunit of eIF-4F are antigenically distinct polypeptides. *J. Biol. Chem.* **262:** 11228–11232.

Browning K.S., Maia D.M., Lax S.R., and Ravel J.M. 1987b. Identification of a new protein synthesis initiation factor from wheat germ. *J. Biol. Chem.* **262:** 538–541.

Browning K.S., Humphreys J., Hobbs W., Smith G.B., and Ravel J.M. 1990. Determination of the amounts of the protein synthesis initiation and elongation factors in wheat germ. *J. Biol. Chem.* **265:** 17967–17973.

Brunn G.J., Fadden P., Haystead T.A.J., and Lawrence J.C., Jr. 1997a. The mammalian target of rapamycin phosphorylates sites having a (Ser/Thr)-Pro motif and is activated by antibodies to a region near its COOH terminus. *J. Biol. Chem.* **272:** 32547–32550.

Brunn G.J., Williams J., Sabers C., Wiederrecht G., Lawrence J.C., Jr., and Abraham R.T. 1996. Direct inhibition of the signaling functions of the mammalian target of rapamycin by the phosphoinositide 3-kinase inhibitors, wortmannin and LY294002. *EMBO J.* **15:** 5256–5267.

Brunn G. J., Hudson C.C., Sekulic A., Williams J.M., Hosoi H., Houghton P.J., Lawrence J.C., and Abraham R.T. 1997b. Phosphorylation of the translational repressor PHAS-I by the mammalian target of rapamycin. *Science* **277:** 99–101.

Bu X., Haas D.W., and Hagedorn C.H. 1993. Novel phosphorylation sites of eukaryotic initiation factor-4F and evidence that phosphorylation stabilizes interactions of the p25 and p220 subunits. *J. Biol. Chem.* **268:** 4975-4978.

Burnett P.E., Barrow R.K., Cohen N.A., Snyder S.H., and Sabatini D.M. 1998. RAFT1 phosphorylation of the translational regulators p70 S6 kinase and 4E-BP1. *Proc. Natl. Acad. Sci.* **95:** 1432–1437.

Campbell L.E., Wang X., and Proud C.G. 1999. Nutrients differentially regulate multiple translation factors and their control by insulin. *Biochem. J.* **344:** 433–441.

Carberry S.E., Rhoads R.E., and Goss D.J. 1989. A spectroscopic study of the binding of m7GTP and m7GpppG to human protein synthesis initiation factor 4E. *Biochemistry* **28:** 8078–8083.

Chen J., Peterson R.T., and Schreiber S.L. 1998. α 4 Associates with protein phosphatases 2A, 4, and 6. *Biochem. Biophys. Res. Commun.* **247:** 827–832.

Chuang R.Y., Weaver P.L., Liu Z., and Chang T.H. 1997. Requirement of the DEAD-box protein ded1p for messenger RNA translation. *Science* **275:** 1468–1471.

Cobb M.H. and Goldsmith E.J. 1995. How MAP kinases are regulated. *J. Biol. Chem.* **270:** 14843–14846.

Coppolecchia R., Buser P., Stotz A., and Linder P. 1993. A new yeast translation initiation factor suppresses a mutation in the eIF-4A RNA helicase. *EMBO J.* **12:** 4005–4011.

Cornelis S., Bruynooghe Y., Denecker G., Van Huffel S., and Beyaert R. 2000. Identification and characterisation of a new cell cycle-regulated internal ribosome entry site. *Mol. Cell* **5:** 597–605.

Cosentino G.P., Schmelzle T., Haghighat A., Helliwell S.B., Hall M.N., and Sonenberg N. 2000. Eap1p, a novel eIF4E-associated protein in *Saccharomyces cerevisiae. Mol. Cell. Biol.* (in press).

Datta S.R., Brunet A., and Greenberg M.E. 1999. Cellular survival: A play in three Akts. *Genes Dev.* **13:** 2905–2927.

de la Cruz J., Iost I., Kressler D., and Linder P. 1997. The p20 and Ded1 proteins have antagonistic roles in eIF4E-dependent translation in *Saccharomyces cerevisiae. Proc. Natl. Acad. Sci.* **94:** 5201–5206.

Dennis P.B., Pullen N., Kozma S.C., and Thomas G. 1996. The principal rapamycin-sensitive p70(s6k) phosphorylation sites, T-229 and T-389, are differentially regulated by rapamycin-insensitive kinase kinases. *Mol. Cell. Biol.* **16:** 6242–6251.

Di Como C.J. and Arndt K.T. 1996. Nutrients, via the Tor proteins, stimulate the association of Tap42 with type 2A phosphatases. *Genes Dev.* **10:** 1904–1916.

Dominguez D., Altmann M., Benz J., Baumann U., and Trachsel H. 1999. Interaction of translation initiation factor eIF4G with eIF4A in the yeast *Saccharomyces cerevisiae. J. Biol. Chem.* **274:** 26720–26726.

Donaldson R.W., Hagedorn C.H., and Cohen S. 1991. Epidermal growth factor or okadaic acid stimulates phosphorylation of eukaryotic initiation factor 4F. *J. Biol. Chem.* **266:** 3162–3166.

Downing G..J., Kim S., Nakanishi S., Catt K.J., and Balla T. 1996. Characterization of a soluble adrenal phosphatidylinositol 4-kinase reveals wortmannin sensitivity of type III phosphatidylinositol kinases. *Biochemistry* **35:** 3587–3594.

Drew M.C. 1997. Oxygen deficiency and root metabolism. *Annu. Rev. Plant Physiol. Plant Mol. Biol.* **48:** 223–250.

Dufner A., Andjelkovic M., Burgering B.M., Hemmings B.A., and Thomas G. 1999. Protein kinase B localization and activation differentially affect S6 kinase 1 activity and eukaryotic translation initiation factor 4E-binding protein 1 phosphorylation. *Mol. Cell. Biol.* **19:** 4525–4534.

Duncan R. and Hershey J.W. 1985. Regulation of initiation factors during translational repression caused by serum depletion. Covalent modification. *J. Biol. Chem.* **260:** 5493-5497.

Duncan R., Milburn S.C., and Hershey J.W. 1987. Regulated phosphorylation and low abundance of HeLa cell initiation factor eIF-4F suggest a role in translational control. Heat shock effects on eIF-4F. *J. Biol. Chem.* **262:** 380–388.

Edery I., Altmann M., and Sonenberg N. 1988. High-level synthesis in *Escherichia coli* of functional cap-binding eukaryotic initiation factor eIF-4E and affinity purification using a simplified cap-analog resin. *Gene* **74:** 517–525.

Fadden P., Haystead T.A., and Lawrence J.C., Jr. 1997. Identification of phosphorylation sites in the translational regulator, PHAS-I, that are controlled by insulin and rapamycin in rat adipocytes. *J. Biol. Chem.* **272:** 10240–10247.

———. 1998. Phosphorylation of the translational regulator, PHAS-I, by protein kinase CK2. *FEBS Lett.* **435:** 105–109.

Fan H. and Penman S. 1970. Regulation of protein synthesis in mammalian cells. II. Inhibition of protein synthesis at the level of initiation during mitosis. *J. Mol. Biol.* **50:** 655–670.

Feigenblum D. and Schneider R.J. 1993. Modification of eukaryotic initiation factor 4F during infection by influenza virus. *J. Virol.* **67:** 3027–3035.

Fletcher C.M. and Wagner G. 1998. The interaction of eIF4E with 4E-BP1 is an induced fit to a completely disordered protein. *Protein Sci.* **7:** 1639–1642.

Fletcher C.M., McGuire A.M., Gingras A.-C., Li H., Matsuo H., Sonenberg N., and Wagner G. 1998. 4E binding proteins inhibit the translation factor eIF4E without folded structure. *Biochemistry* **37:** 9–15.

Fleurent M., Gingras A.-C., Sonenberg N., and Meloche S. 1997. Angiotensin II stimulates phosphorylation of the translational repressor 4E-binding protein 1 by a mitogen-activated protein kinase-independent mechanism. *J. Biol. Chem.* **272:** 4006–4012.

Flynn A. and Proud C.G. 1995. Serine 209, not serine 53, is the major site of phosphorylation in initiation factor eIF-4E in serum-treated Chinese hamster ovary cells. *J. Biol. Chem.* **270:** 21684–21688.

———. 1996. Insulin-stimulated phosphorylation of initiation factor 4E is mediated by the MAP kinase pathway. *FEBS Lett.* **389:** 162–166.

Fox H.L., Pham P.T., Kimball S.R., Jefferson L.S., and Lynch C.J. 1998. Amino acid effects on translational repressor 4E-BP1 are mediated primarily by L-leucine in isolated adipocytes. *Am. J. Physiol.* **275:** C1232–C1238.

Frederickson R.M., Montine K.S., and Sonenberg N. 1991. Phosphorylation of eukaryotic translation initiation factor 4E is increased in Src-transformed cell lines. *Mol. Cell. Biol.* **11:** 2896–2900.

Fruman D.A., Meyers R.E., and Cantley L.C. 1998. Phosphoinositide kinases. *Annu. Rev. Biochem.* **67:** 481–507.

Fukunaga R. and Hunter T. 1997. MNK1, a new MAP kinase-activated protein kinase, isolated by a novel expression screening method for identifying protein kinase substrates. *EMBO J.* **16:** 1921–1933.

Gallie D.R., Le H., Tanguay R.L., and Browning K.S. 1998. Translation initiation factors are differentially regulated in cereals during development and following heat shock. *Plant J.* **14:** 715–722.

Gallie D.R., Le H., Caldwell C., Tanguay R.L., Hoang N.X., and Browning K.S. 1997. The phosphorylation state of translation initiation factors is regulated developmentally and following heat shock in wheat. *J. Biol. Chem.* **272:** 1046–1053.

Gingras A.-C., Raught B., and Sonenberg N. 1999a. eIF4 Initiation factors: Effectors of mRNA recruitment to ribosomes and regulators of translation. *Annu. Rev. Biochem.* **68:** 913–963.

Gingras A.-C., Kennedy S.G., O'Leary M.A., Sonenberg N., and Hay N. 1998. 4E-BP1, a repressor of mRNA translation, is phosphorylated and inactivated by the Akt(PKB) signaling pathway. *Genes Dev.* **12:** 502–513.

Gingras A.-C., Svitkin Y., Belsham G.J., Pause A., and Sonenberg N. 1996. Activation of the translational suppressor 4E-BP1 following infection with encephalomyocarditis virus and poliovirus. *Proc. Natl. Acad. Sci.* **93:** 5578–5583.

Gingras A.-C., Gygi S.P., Raught B., Polakiewicz R.D., Abraham R.T., Hoekstra M.F., Aebersold R., and Sonenberg N. 1999b. Regulation of 4E-BP1 phosphorylation: A novel two-step mechanism. *Genes Dev.* **13:** 1422–1437.

Goyer C., Altmann M., Lee H.S., Blanc A., Deshmukh M., Woolford J.L., Jr., Trachsel H., and Sonenberg N. 1993. TIF4631 and TIF4632: Two yeast genes encoding the high-molecular-weight subunits of the cap-binding protein complex (eukaryotic initiation factor 4F) contain an RNA recognition motif-like sequence and carry out an essential function. *Mol. Cell. Biol.* **13:** 4860–4874.

Gradi A., Imataka H., Svitkin Y.V., Rom E., Raught B., Morino S., and Sonenberg N. 1998. A novel functional human eukaryotic translation initiation factor 4G. *Mol. Cell. Biol.* **18:** 334–342.

Haghighat A., and Sonenberg N. 1997. eIF4G dramatically enhances the binding of eIF4E to the mRNA 5′-cap structure. *J. Biol. Chem.* **272:** 21677-21680.

Haghighat A., Mader S., Pause A., and Sonenberg N. 1995. Repression of cap-dependent translation by 4E-binding protein 1: Competition with p220 for binding to eukaryotic initiation factor-4E. *EMBO J.* **14:** 5701–5709.

Hall M.N. 1996. The TOR signalling pathway and growth control in yeast. *Biochem. Soc. Trans.* **24:** 234–239.

Hara K., Yonezawa K., Weng Q.P., Kozlowski M.T., Belham C., and Avruch J. 1998. Amino acid sufficiency and mTOR regulate p70 S6 kinase and eIF-4E BP1 through a common effector mechanism. *J. Biol. Chem.* **273:** 14484–14494.

Hara K., Yonezawa K., Kozlowski M.T., Sugimoto T., Andrabi K., Weng Q.P., Kasuga M., Nishimoto I., and Avruch J. 1997. Regulation of eIF-4E BP1 phosphorylation by mTOR. *J. Biol. Chem.* **272:** 26457–26463.

Heesom, K.J., and Denton R.M. 1999. Dissociation of the eukaryotic initiation factor-4E/4E-BP1 complex involves phosphorylation of 4E-BP1 by an mTOR-associated kinase. *FEBS Lett.* **457:** 489–493.

Heesom K.J., Avison M.B., Diggle T.A., and Denton R.M. 1998. Insulin-stimulated kinase from rat fat cells that phosphorylates initiation factor-4E binding protein 1 on the rapamycin-insensitive site (serine-111). *Biochem. J.* **336:** 39–48.

Henis-Korenblit S., Levy-Strumpf N., Goldstaub D., and Kimchi A. 2000. A novel form of DAP5 protein accumulates in apoptotic cells as a result of caspase cleavage and internal ribosome entry site-mediated translation. *Mol. Cell. Biol.* **20:** 496–506.

Hentze M.W. 1997. eIF4G: A multipurpose ribosome adapter? *Science* **275:** 500–501.

Hershey P.E., McWhirter S.M., Gross J.D., Wagner G., Alber T., and Sachs A.B. 1999. The cap-binding protein eIF4E promotes folding of a functional domain of yeast translation initiation factor eIF4G1. *J. Biol. Chem.* **274:** 21297–21304.

Hoekstra M.F. 1997. Responses to DNA damage and regulation of cell cycle checkpoints by the ATM protein kinase family. *Curr. Opin. Genet. Dev.* **7:** 170–175.

Hu C., Pang S., Kong X., Velleca M., and Lawrence J.C., Jr. 1994. Molecular cloning and tissue distribution of PHAS-I, an intracellular target for insulin and growth factors. *Proc. Natl. Acad. Sci.* **91:** 3730–3734.

Huang J.T. and Schneider R.J. 1991. Adenovirus inhibition of cellular protein synthesis involves inactivation of cap-binding protein. *Cell* **65:** 271–280.

Iiboshi Y., Papst P.J., Kawasome H., Hosoi H., Abraham R.T., Houghton P.J., and Terada N. 1999. Amino-acid-dependent control of p70s6k. *J. Biol. Chem.* **274:** 1092–1099.

Imataka H., Gradi A., and Sonenberg N. 1999. A newly identified N-terminal amino acid stretch of human eIF4G binds poly(A) binding protein and functions in poly(A) dependent translation. *EMBO J.* **17:** 7480–7489.

Imataka, H. and Sonenberg N. 1997. Human eukaryotic translation initiation factor 4G (eIF4G) possesses two separate and independent binding sites for eIF4A. *Mol. Cell. Biol.* **17:** 6940–6947.

Imataka H., Olsen H.S., and Sonenberg N. 1997. A new translational regulator with homology to eukaryotic translation initiation factor 4G. *EMBO J.* **16:** 817–825.

Inui S., Sanjo H., Maeda K., Yamamoto H., Miyamoto E., and Sakaguchi N. 1998. Ig receptor binding protein 1 (α4) is associated with a rapamycin-sensitive signal transduction in lymphocytes through direct binding to the catalytic subunit of protein phosphatase 2A. *Blood* **92:** 539–546.

Iost I., Dreyfus M., and Linder P. 1999. Ded1p, a DEAD-box protein required for translation initiation in *Saccharomyces cerevisiae*, is an RNA helicase. *J. Biol. Chem.* **274:** 17677–17683.

Jaramillo M., Dever T.E., Merrick W.C., and Sonenberg N. 1991. RNA unwinding in translation: Assembly of helicase complex intermediates comprising eukaryotic initiation factors eIF-4F and eIF-4B. *Mol. Cell. Biol.* **11:** 5992–5997.

Jaramillo M., Browning K., Dever T.E., Blum S., Trachsel H., Merrick W.C., Ravel J.M., and Sonenberg N. 1990. Translation initiation factors that function as RNA helicases from mammals, plants and yeast. *Biochim. Biophys. Acta* **1050:** 134–139.

Jiang Y. and Broach J.R. 1999. Tor proteins and protein phosphatase 2A reciprocally regulate Tap42 in controlling cell growth in yeast. *EMBO J.* **18:** 2782–2792.

Joshi B. Cai A.L., Keiper B.D., Minich W.B., Mendez R., Beach C.M., Stepinski J., Stolarski R., Darzynkiewicz E., and Rhoads R.E. 1995. Phosphorylation of eukaryotic protein synthesis initiation factor 4E at Ser-209. *J. Biol. Chem.* **270:** 14597–14603.

Keiper B.D., Lamphear B.J., Deshpande A.M., Jankowska-Anyszka M., Aamodt E.J., Blumenthal T., and Rhoads R.E. 2000. Functional characterization of five eIF4E isoforms in *Caenorhabditis elegans*. *J. Biol. Chem.* **275:** 10590–10596.

Keith C.T. and Schreiber S.L. 1995. PIK-related kinases: DNA repair, recombination, and cell cycle checkpoints. *Science* **270:** 50–51.

Kennedy S.G., Wagner A.J., Conzen S.D., Jordan J., Bellacosa A., Tsichlis P.N., and Hay N. 1997. The PI 3-kinase/Akt signaling pathway delivers an anti-apoptotic signal. *Genes Dev.* **11:** 701–713.

Khaleghpour K., Pyronnet S., Gingras A.-C., and Sonenberg N. 1999. Translational homeostasis: Eukaryotic translation initiation factor 4E control of 4E-binding protein 1 and p70 S6 kinase activities. *Mol. Cell. Biol.* **19:** 4302–4310.

Kimball S.R., Shantz L.M., Horetsky R.L., and Jefferson L.S. 1999. Leucine regulates translation of specific mRNAs in L6 myoblasts through mTOR-mediated changes in availability of eIF4E and phosphorylation of ribosomal protein S6. *J. Biol. Chem.* **274:** 11647–11652.

Kleijn M., Scheper G.C., Voorma H.O., and Thomas A.A.M. 1998. Regulation of translation initiation factors by signal transduction. *Eur. J. Biochem.* **253:** 531–544.

Kleijn, M., Vrins C.L., Voorma H.O., and Thomas A.A. 1996. Phosphorylation state of the cap-binding protein eIF4E during viral infection. *Virology* **217:** 486–494.

Kohn A.D., Barthel A., Kovacina K.S., Boge A., Wallach B., Summers S.A., Birnbaum M.J., Scott P.H., Lawrence J.C., Jr., and Roth R.A. 1998. Construction and characterization of a conditionally active version of the serine/threonine kinase Akt. *J. Biol. Chem.* **273:** 11937–11943.

Konishi H., Matsuzaki H., Tanaka M., Ono Y., Tokunaga C., Kuroda S., and Kikkawa U. 1996. Activation of RAC-protein kinase by heat shock and hyperosmolarity stress through a pathway independent of phosphatidylinositol 3-kinase. *Proc. Natl. Acad. Sci.* **93:** 7639–7643.

Konishi H., Matsuzaki H., Tanaka M., Takemura Y., Kuroda S., Ono Y., and Kikkawa U. 1997. Activation of protein kinase B (Akt/RAC-protein kinase) by cellular stress and its association with heat shock protein Hsp27. *FEBS Lett.* **410:** 493–498.

Lamphear B.J., Kirchweger R., Skern T., and Rhoads R.E. 1995. Mapping of functional domains in eukaryotic protein synthesis initiation factor 4G (eIF4G) with picornaviral proteases. Implications for cap-dependent and cap-independent translational initiation. *J. Biol. Chem.* **270:** 21975–21983.

Lang V., Zanchin N.I. Lunsdorf H., Tuite M., and McCarthy J.E. 1994. Initiation factor eIF-4E of *Saccharomyces cerevisiae*. Distribution within the cell, binding to mRNA, and consequences of its overproduction. *J. Biol. Chem.* **269:** 6117–6123.

Lanker S., Muller P.P., Altmann M., Goyer C., Sonenberg N., and Trachsel H. 1992. Interactions of the eIF-4F subunits in the yeast *Saccharomyces cerevisiae*. *J. Biol. Chem.* **267:** 21167–21171.

Lawson T.G., Lee K.A., Maimone M.M., Abramson R.D., Dever T.E., Merrick W.C., and Thach R.E. 1989. Dissociation of double-stranded polynucleotide helical structures by eukaryotic initiation factors, as revealed by a novel assay. *Biochemistry* **28:** 4729–4734.

Lazaris-Karatzas A., Montine K.S., and Sonenberg N. 1990. Malignant transformation by a eukaryotic initiation factor subunit that binds to mRNA 5′ cap. *Nature* **345:** 544–547.

Le H., Browning K.S., and Gallie D.R. 1998. The phosphorylation state of the wheat translation initiation factors eIF4B, eIF4A, and eIF2 is differentially regulated during seed development and germination. *J. Biol. Chem.* **273:** 20084–20089.

Lee K.A., Edery I., and Sonenberg N. 1985. Isolation and structural characterization of cap-binding proteins from poliovirus-infected HeLa cells. *J. Virol.* **54:** 515–524.

Levy-Strumpf N., Deiss L.P., Berissi H., and Kimchi A. 1997. DAP-5, a novel homolog of eukaryotic translation initiation factor 4G isolated as a putative modulator of gamma interferon-induced programmed cell death. *Mol. Cell. Biol.* **17:** 1615–1625.

Lin T.A. and Lawrence J.C., Jr. 1996. Control of the translational regulators PHAS-I and PHAS-II by insulin and cAMP in 3T3-L1 adipocytes. *J. Biol. Chem.* **271:** 30199–30204.

———. 1997. Control of PHAS-I phosphorylation in 3T3-L1 adipocytes: Effects of inhibiting protein phosphatases and the p70S6K signalling pathway. *Diabetologia* (Suppl 2) **40:** S18–24.

Lin T.A., Kong X., Saltiel A.R., Blackshear P.J., and Lawrence J.C., Jr. 1995. Control of PHAS-I by insulin in 3T3-L1 adipocytes. Synthesis, degradation, and phosphorylation by a rapamycin-sensitive and mitogen-activated protein kinase-independent pathway. *J. Biol. Chem.* **270:** 18531–18538.

Lin T.A., Kong X., Haystead T.A., Pause A., Belsham G., Sonenberg N., and Lawrence J.C., Jr. 1994. PHAS-I as a link between mitogen-activated protein kinase and translation initiation. *Science* **266:** 653–656.

Linder P. and Slonimski P.P. 1989. An essential yeast protein, encoded by duplicated genes TIF1 and TIF2 and homologous to the mammalian translation initiation factor eIF-4A, can suppress a mitochondrial missense mutation. *Proc. Natl. Acad. Sci.* **86:** 2286–2290.

Mader S., Lee H., Pause A., and Sonenberg N. 1995. The translation initiation factor eIF-4E binds to a common motif shared by the translation factor eIF-4 gamma and the translational repressors 4E-binding proteins. *Mol. Cell. Biol.* **15:** 4990–4997.

Mahalingam M. and Templeton D.J. 1996. Constitutive activation of S6 kinase by deletion of amino-terminal autoinhibitory and rapamycin sensitivity domains. *Mol. Cell. Biol.* **16:** 405–413.

Makkinje A., Xiong H., Li M., and Damuni Z. 1995. Phosphorylation of eukaryotic protein synthesis initiation factor 4E by insulin-stimulated protamine kinase. *J. Biol. Chem.* **270:** 14824–14828.

Manjunath S., Williams A.J., and Bailey-Serres J. 1999. Oxygen deprivation stimulates Ca^{++}-mediated phosphorylation of mRNA cap-binding protein eIF4E in maize roots. *Plant J.* **19:** 21–30.

Marcotrigiano J., Gingras A.-C., Sonenberg N., and Burley S.K. 1997. Cocrystal structure of the messenger RNA 5′ cap-binding protein (eIF4E) bound to 7-methyl-GDP. *Cell* **89:** 951–961.

———. 1999. Cap-dependent translation initiation in eukaryotes is regulated by a molecular mimic of eIF4G. *Mol. Cell* **3:** 707–716.

Marte B.M. and Downward J. 1997. PKB/Akt: Connecting phosphoinositide 3-kinase to cell survival and beyond. *Trends Biochem. Sci.* **22:** 355–358.

Matsuo H., Li H., McGuire A.M., Fletcher C.M., Gingras A.-C., Sonenberg N., and Wagner G. 1997. Structure of translation factor eIF4E bound to m7GDP and interaction with 4E-binding protein. *Nat. Struct. Biol.* **4:** 717–724.

McCarthy J.E.G. 1998. Posttranscriptional control of gene expression in yeast. *Microbiol. Mol. Biol. Rev.* **62:** 1492–1553.

Mendez R., Myers M.G., White M., and Rhoads R. 1996. Stimulation of protein synthesis, eukaryotic translation initiation factor 4E phosphorylation, and PHAS-I phosphorylation by insulin requires insulin receptor substrate 1 and phosphatidylinositol 3-kinase. *Mol. Cell. Biol.* **16:** 2857–2864.

Millward T.A., Zolnierowicz S., and Hemmings B.A. 1999. Regulation of protein kinase cascades by protein phosphatase 2A. *Trends in Biochem. Sci.* **24:** 186–191.

Minich W., Balasta M., Goss D., and Rhoads R. 1994. Chromatographic resolution of in vivo phosphorylated and nonphosphorylated eukaryotic translation initiation factor eIF-4E: Increased cap affinity of the phosphorylated form. *Proc. Natl. Acad. Sci.* **91:** 7668–7672.

Morino S., Imataka H., Svitkin Y., Pestova T., and Sonenberg N. 1999. Eukaryotic translation initiation factor (eIF) 4E binding site and the middle one-third of eIF4GI constitute the core domain for cap-dependent translation, and the C-terminal one-third of eIF4GI functions as a modulatory region. *Mol. Cell. Biol.* **20:** 468–477.

Morley S.J. and McKendrick L. 1997. Involvement of stress-activated protein kinase and p38/RK mitogen-activated protein kinase signaling pathways in the enhanced phosphorylation of initiation factor 4E in NIH 3T3 cells. *J. Biol. Chem.* **272:** 17887–17893.

Morley S.J. and Pain V.M. 1995a. Hormone-induced meiotic maturation in *Xenopus* oocytes occurs independently of p70s6k activation and is associated with enhanced initiation factor (eIF)-4F phosphorylation and complex formation. *J. Cell Sci.* **108:** 1751–1760.

———. 1995b. Translational regulation during activation of porcine peripheral blood lymphocytes: Association and phosphorylation of the alpha and gamma subunits of the initiation factor complex eIF-4F. *Biochem. J.* **312:** 627–635.

Morley S.J. and Traugh J.A. 1989. Phorbol esters stimulate phosphorylation of eukaryotic initiation factors 3, 4B, and 4F. *J. Biol. Chem.* **264:** 2401–2404.

———. 1990. Differential stimulation of phosphorylation of initiation factors eIF-4F, eIF-4B eIF-3, and ribosomal protein S6 by insulin and phorbol esters. *J. Biol. Chem.* **265:** 10611–10616.

———. 1993. Stimulation of translation in 3T3-L1 cells in response to insulin and phorbol ester is directly correlated with increased phosphate labelling of initiation factor (eIF-) 4F and ribosomal protein S6. *Biochimie* **75:** 985–989.

Morley S.J., Curtis P.S., and Pain V.M. 1997. eIF4G: Translation's mystery factor begins to yield its secrets. *RNA* **3:** 1085–1104.

Muckenthaler M., Gray N.K., and Hentze M.W. 1998. IRP-1 binding to ferritin mRNA disrupts the bridging between the cap-binding complex and the small ribosomal subunit. *Mol. Cell* **2:** 383–388.

Murata K., Wu J., and Brautigan D.L. 1997. B cell receptor-associated protein α4 displays rapamycin-sensitive binding directly to the catalytic subunit of protein phosphatase 2A. *Proc. Natl. Acad. Sci.* **94:** 10624–10629.

Nakanishi S., Catt K.J., and Balla T. 1995. A wortmannin-sensitive phosphatidylinositol

4-kinase that regulates hormone-sensitive pools of inositolphospholipids. *Proc. Natl. Acad. Sci.* **92:** 5317–5321.

Nanahoshi M., Nishiuma T., Tsujishita Y., Hara K., Inui S., Sakaguchi N., and Yonezawa K. 1998. Regulation of protein phosphatase 2A catalytic activity by α4 protein and its yeast homolog TAP42. *Biochem. Biophys. Res. Commun.* **251:** 520–526.

Nanahoshi M., Tsujishita Y., Tokunaga C., Inui S., Sakaguchi N., Hara K., and Yonezawa K. 1999. α4 Protein as a common regulator of type 2A-related serine/threonine protein phosphatases. *FEBS Lett.* **446:** 108–112.

Navé B.T., Ouwens M., Withers D.J., Alessi D.R., and Shepherd P.R. 1999. Mammalian target of rapamycin is a direct target for protein kinase B: Identification of a convergence point for opposing effects of insulin and amino-acid deficiency on protein translation. *Biochem. J.* **344:** 427–431.

Neff C.L. and Sachs A.B. 1999. Eukaryotic translation initiation factors 4G and 4A from *Saccharomyces cerevisiae* interact physically and functionally. *Mol. Cell. Biol.* **19:** 5557–5564.

Noda T. and Ohsumi Y. 1998. Tor, a phosphatidylinositol kinase homologue, controls autophagy in yeast. *J. Biol. Chem.* **273:** 3963–3966.

op den Camp R.G. and Kuhlemeier C. 1998. Phosphorylation of tobacco eukaryotic translation initiation factor 4A upon pollen tube germination. *Nucleic Acids Res.* **26:** 2058–2062.

Owttrim G.W., Mandel T., Trachsel H., Thomas A.A., and Kuhlemeier C. 1994. Characterization of the tobacco eIF-4A gene family. *Plant Mol. Biol.* **26:** 1747–1757.

Parrott L.A. and Templeton D.J. 1999. Osmotic stress inhibits p70/85 S6 kinase through activation of a protein phosphatase. *J. Biol. Chem.* **274:** 24731–24736.

Patti M.E., Brambilla E., Luzi L., Landaker E.J., and Kahn C.R. 1998. Bidirectional modulation of insulin action by amino acids. *J. Clin. Invest.* **101:** 1519–1529.

Pause A., Belsham G.J., Gingras A.-C., Donzé O., Lin T.A., Lawrence J.C., Jr., and Sonenberg N. 1994. Insulin-dependent stimulation of protein synthesis by phosphorylation of a regulator of 5′-cap function. *Nature* **371:** 762–767.

Pestova T.V., Borukhov S.I., and Hellen C.U.T. 1998. Eukaryotic ribosomes require initiation factors 1 and 1A to locate initiation codons. *Nature* **394:** 854–859.

Pestova T.V., Hellen C.U., and Shatsky I.N. 1996a. Canonical eukaryotic initiation factors determine initiation of translation by internal ribosomal entry. *Mol. Cell. Biol.* **16:** 6859–6869.

Pestova T.V., Shatsky I.N., and Hellen C.U. 1996b. Functional dissection of eukaryotic initiation factor 4F: The 4A subunit and the central domain of the 4G subunit are sufficient to mediate internal entry of 43S preinitiation complexes. *Mol. Cell. Biol.* **16:** 6870–6878.

Peterson R.T., Desai B.N., Hardwick J.S., and Schreiber S.L. 1999. Protein phosphatase 2A interacts with the 70-kDa S6 kinase and is inactivated by inhibition of FKBP12-rapamycin-associated protein. *Proc. Natl. Acad. Sci.* **96:** 4438–4442.

Poulin F., Gingras A.-C., Olsen H., Chevalier S., and Sonenberg N. 1998. 4E-BP3, a new member of the eukaryotic initiation factor 4E-binding protein family. *J. Biol. Chem.* **273:** 14002–14007.

Powers T. and Walter P. 1999. Regulation of ribosome biogenesis by the rapamycin-sensitive TOR-signaling pathway in *Saccharomyces cerevisiae*. *Mol. Biol. Cell* **10:** 987–1000.

Ptushkina M., von der Haar T., Vasilescu S., Frank R., Birkenhäger R., and McCarthy

J.E.G. 1998. Cooperative modulation by eIF4G of eIF4E-binding to the mRNA 5′ cap in yeast involves a site partially shared by p20. *EMBO J.* **17:** 4798–4808.

Pyronnet S., Pradayrol L., and Sonenberg N. 2000. A cell cycle-dependent internal ribosome entry site. *Mol. Cell* **5:** 607–616.

Pyronnet S., Imataka H., Gingras A.-C., Fukunaga R., Hunter T., and Sonenberg N. 1999. Human eukaryotic translation initiation factor 4G (eIF4G) recruits MNK1 to phosphorylate eIF4E. *EMBO J.* **18:** 270–279.

Rao G.N., Madamanchi N.R., Lele M., Gadiparthi L., Gingras A.-C., Eling T.E., and Sonenberg N. 1999. A potential role for extracellular signal-regulated kinases in prostaglandin F2 alpha-induced protein synthesis in smooth muscle cells. *J. Biol. Chem.* **274:** 12925–12932.

Raught B. and Gingras A.-C. 1999. eIF4E activity is regulated at multiple levels. *Int. J. Biochem. Cell Biol.* **31:** 43–57.

Raught B., Gingras A.-C., James A., Medina D., Sonenberg N., and Rosen J.M. 1996. Expression of a translationally regulated, dominant-negative CCAAT/enhancer-binding protein beta isoform and up-regulation of the eukaryotic translation initiation factor 2 alpha are correlated with neoplastic transformation of mammary epithelial cells. *Cancer Res.* **56:** 4382–4386.

Raught B., Gingras A.-C., Gygi S.P., Imataka H., Morino S., Gradi A., Aebersold R., and Sonenberg N. 2000. Serum-induced, rapamycin-sensitive phosphorylation sites in the eukaryotic translation initiation factor 4GI. *EMBO J.* **19:** 434–444.

Rinker-Schaeffer C.W., Austin V., Zimmer S., and Rhoads R.E. 1992. Ras transformation of cloned rat embryo fibroblasts results in increased rates of protein synthesis and phosphorylation of eukaryotic initiation factor 4E. *J. Biol. Chem.* **267:** 10659–10664.

Robinson M.J. and Cobb, M.H. 1997. Mitogen-activated protein kinase pathways. *Curr. Opin. Cell Biol.* **9:** 180–186.

Rodriguez C.M., Freire M.A., Camilleri C., and Robaglia C. 1998. The *Arabidopsis thaliana* cDNAs coding for eIF4E and eIF(iso)4E are not functionally equivalent for yeast complementation and are differentially expressed during plant development. *Plant J.* **13:** 465–473.

Rom E., Kim H.C., Gingras A.-C., Marcotrigiano J., Favre D., Olsen H., Burley S.K., and Sonenberg N. 1998. Cloning and characterization of 4EHP, a novel mammalian eIF4E-related cap-binding protein. *J. Biol. Chem.* **273:** 13104–13109.

Rosenwald I.B., Chen J.-J., Wang S., Savas L., London I.M., and Pullman J. 1999. Upregulation of protein synthesis initiation factor eIF-4E is an early event during colon carcinogenesis. *Oncogene* **18:** 2507–2517.

Rozen F., Edery I., Meerovitch K., Dever T.E., Merrick W.C., and Sonenberg N. 1990. Bidirectional RNA helicase activity of eucaryotic translation initiation factors 4A and 4F. *Mol. Cell. Biol.* **10:** 1134–1144.

Ruud K. A., Kuhlow C., Goss D.J., and Browning K.S. 1998. Identification and characterization of a novel cap-binding protein from *Arabidopsis thaliana*. *J. Biol. Chem.* **273:** 10325–10330.

Rybkin I.I., Cross M.E., McReynolds E.M., Lin R.Z., and Ballou L.M. 2000. α1A Adrenergic receptor induces eukaryotic initiation factor 4E-binding protein 1 phosphorylation via a Ca^{++}-dependent pathway independent of phosphatidylinositol 3-kinase/Akt. *J. Biol. Chem.* **275:** 5460–5465.

Sable C.L., Filippa N., Hemmings B., and Van Obberghen E. 1997. cAMP stimulates protein kinase B in a wortmannin-insensitive manner. *FEBS Lett.* **409:** 253–257.

Sarkaria J.N., Tibbetts R.S., Busby E.C., Kennedy A.P., Hill D.E., and Abraham R.T. 1998. Inhibition of phosphoinositide 3-kinase related kinases by the radiosensitizing agent wortmannin. *Cancer Res.* **58:** 4375–4382.

Schmidt A., Beck T., Koller A., Kunz J., and Hall M.N. 1998. The TOR nutrient signalling pathway phosphorylates NPR1 and inhibits turnover of the tryptophan permease. *EMBO J.* **17:** 6924–6931.

Scott P.H., Brunn G.J., Kohn A.D., Roth R.A., and Lawrence J.C., Jr. 1998. Evidence of insulin-stimulated phosphorylation and activation of the mammalian target of rapamycin mediated by a protein kinase B signaling pathway. *Proc. Natl. Acad. Sci.* **95:** 7772–7777.

Shigemitsu K., Tsujishita Y., Hara K., Nanahoshi M., Avruch J., and Yonezawa K. 1999. Regulation of translational effectors by amino acid and mammalian target of rapamycin signaling pathways. Possible involvement of autophagy in cultured hepatoma cells. *J. Biol. Chem.* **274:** 1058–1065.

Sonenberg N. 1996. mRNA 5′ cap-binding protein eIF4E and control of cell growth. In *Transcriptional Control* (ed. J.W.B. Hershey et al.), pp. 249–269. Cold Spring Harbor Laboratory Press. Cold Spring Harbor, New York.

Stebbins-Boaz B., Cao Q., de Moor C.H., Mendez R., and Richter J.D. 1999. Maskin is a CPEB-associated factor that transiently interacts with eIF-4E. *Mol. Cell* **4:** 1017–1027.

Svitkin Y. V., Ovchinnikov L.P., Dreyfuss G., and Sonenberg N. 1996. General RNA binding proteins render translation cap dependent. *EMBO J.* **15:** 7147–7155.

Takata M., Ogawa W., Kitamura T., Hino Y., Kuroda S., Kotani K., Klip A., Gingras A.-C., Sonenberg N., and Kasuga M. 1999. Requirement for Akt (protein kinase B) in insulin-induced activation of glycogen synthase and phosphorylation of 4E-BP1 (PHAS-1). *J. Biol. Chem.* **274:** 20611–20618.

Tarun S., Jr. and Sachs A.B. 1996. Association of the yeast poly(A) tail binding protein with translation initiation factor eIF-4G. *EMBO J.* **15:** 7168–7177.

Thomas G. and Hall M.N. 1997. TOR signalling and control of cell growth. *Curr. Opin. Cell Biol.* **9:** 782–787.

Tsukiyama-Kohara K., Vidal S.M., Gingras A.-C., Glover T.W., Hanash S.M., Heng H., and Sonenberg N. 1996. Tissue distribution, genomic structure, and chromosome mapping of mouse and human eukaryotic initiation factor 4E-binding proteins 1 and 2. *Genomics* **38:** 353–363.

Tuazon P.T., Merrick W.C., and Traugh J.A. 1989. Comparative analysis of phosphorylation of translational initiation and elongation factors by seven protein kinases. *J. Biol. Chem.* **264:** 2773–2777.

Ui M., Okada T., Hazeki K., and Hazeki O. 1995. Wortmannin as a unique probe for an intracellular signalling protein, phosphoinositide 3-kinase. *Trends Biochem. Sci.* **20:** 303–307.

Vlahos C.J., Matter W.F., Hui K.Y. and Brown R.F. 1994. A specific inhibitor of phosphatidylinositol 3-kinase, 2-(4morpholinyl)-8-phenyl-4H-1-benzopyran-4-one (LY294002). *J. Biol. Chem.* **2169:** 5241–5248.

von Manteuffel S.R., Gingras A.-C., Ming X.F., Sonenberg N., and Thomas G. 1996. 4E-BP1 phosphorylation is mediated by the FRAP-p70^{s6k} pathway and is independent of mitogen-activated protein kinase. *Proc. Natl. Acad. Sci.* **93:** 4076–4080.

von Manteuffel S.R., Dennis P.B., Pullen N., Gingras A.-C., Sonenberg N., and Thomas G. 1997. The insulin-induced signalling pathway leading to S6 and initiation factor 4E binding protein 1 phosphorylation bifurcates at a rapamycin-sensitive point immediately upstream of p70[s6k]. *Mol. Cell. Biol.* **17:** 5426–5436.

Vries R.G., Flynn A., Patel J.C., Wang X., Denton R.M., and Proud C.G. 1997. Heat shock increases the association of binding protein-1 with initiation factor 4E. *J. Biol. Chem.* **272:** 32779–32784.

Wang X., Campbell L.E., Miller C.M., and Proud C.G. 1998a. Amino acid availability regulates p70 S6 kinase and multiple translation factors. *Biochem. J.* **334:** 261–267.

Wang X., Flynn A., Waskiewicz A.J., Webb B.L., Vries R.G., Baines I.A., Cooper J.A., and Proud C.G. 1998b. The phosphorylation of eukaryotic initiation factor eIF4E in response to phorbol esters, cell stresses, and cytokines is mediated by distinct MAP kinase pathways. *J. Biol. Chem.* **273:** 9373–9377.

Waskiewicz A.J. and Cooper J.A. 1995. Mitogen and stress response pathways: MAP kinase cascades and phosphatase regulation in mammals and yeast. *Curr. Opin. Cell Biol.* **7:** 798–805.

Waskiewicz A.J., Flynn A., Proud C.G., and Cooper J.A. 1997. Mitogen-activated protein kinases activate the serine/threonine kinases Mnk1 and Mnk2. *EMBO J.* **16:** 1909–1920.

Waskiewicz A.J., Johnson J.C., Penn B., Mahalingam M., Kimball S.R., and Cooper J.A. 1999. Phosphorylation of the cap-binding protein eukaryotic translation initiation factor 4E by protein kinase Mnk1 in vivo. *Mol. Cell. Biol.* **19:** 1871–1880.

Webster C., Gaut R.L., Browning K.S., Ravel J.M., and Roberts J.K. 1991. Hypoxia enhances phosphorylation of eukaryotic initiation factor 4A in maize root tips. *J. Biol. Chem.* **266:** 23341–23346.

Wells S.E., Hillner P.E., Vale R.D., and Sachs A.B. 1998. Circularization of mRNA by eukaryotic translation initiation factors. *Mol. Cell* **2:** 135–140.

Whalen S.G., Gingras A.-C., Amankwa L., Mader S., Branton P.E., Aebersold R., and Sonenberg N. 1996. Phosphorylation of eIF-4E on serine 209 by protein kinase C is inhibited by the translational repressors, 4E-binding proteins. *J. Biol. Chem.* **271:** 11831–11837.

Xu G., Kwon G., Marshall C.A., Lin T.A., Lawrence J.C., Jr., and McDaniel M.L. 1998a. Branched-chain amino acids are essential in the regulation of PHAS-I and p70 S6 kinase by pancreatic β-cells. A possible role in protein translation and mitogenic signaling. *J. Biol. Chem.* **273:** 28178–28184.

Xu G., Marshall C.A., Lin T.A., Kwon G., Munivenkatappa R.B., Hill J.R., Lawrence J.C., Jr., and McDaniel M.L. 1998b. Insulin mediates glucose-stimulated phosphorylation of PHAS-I by pancreatic beta cells. An insulin-receptor mechanism for autoregulation of protein synthesis by translation. *J. Biol. Chem.* **273:** 4485–4491.

Yamanaka S., Poksay K.S., Arnold K.S., and Innerarity T.L. 1997. A novel translational repressor mRNA is edited extensively in livers containing tumors caused by the transgene expression of the apoB mRNA-editing enzyme. *Genes Dev.* **11:** 321–333.

Yano S., Tokumitsu H., and Soderling T.R. 1998. Calcium promotes cell survival through CaM-K kinase activation of the protein-kinase-B pathway. *Nature* **396:** 584–587.

Young P.R., McLaughlin M.M., Kumar S., Kassis S., Doyle M.L., McNulty D., Gallagher T.F., Fisher S., McDonnell P.C., Carr S.A., Huddleston M.J., Seibel G., Porter T.G., Livi G.P., Adams J.L., and Lee J.C. 1997. Pyridinyl imidazole inhibitors of p38 mitogen-activated protein kinase bind in the ATP site. *J. Biol. Chem.* **272:** 12116–12121.

Zanchin N.I., and McCarthy J.E. 1995. Characterization of the in vivo phosphorylation sites of the mRNA•cap-binding complex proteins eukaryotic initiation factor-4E and p20 in *Saccharomyces cerevisiae*. *J. Biol. Chem.* **270:** 26505–26510.

7

Translational Control of Developmental Decisions

Marvin Wickens

Department of Biochemistry
University of Wisconsin-Madison
Madison, Wisconsin 53706

Elizabeth B. Goodwin

Department of Cell and Molecular Biology
and Lurie Cancer Center
Northwestern University Medical School
Chicago, Illinois 60611

Judith Kimble

Howard Hughes Medical Institute
Departments of Biochemistry and Medical Genetics and
Laboratory of Cell and Molecular Biology
University of Wisconsin-Madison
Madison, Wisconsin 53706

Sidney Strickland

Department of Pharmacology and Program in Genetics
University at Stony Brook
Stony Brook, New York 11794-8651

Matthias W. Hentze

Gene Expression Programme
European Molecular Biology Laboratory
D-69117 Heidelberg, Germany

At fertilization, the calm of oogenesis ends and the egg abruptly begins a flurry of activity. Many crucial steps—decisions concerning when and where to divide, specification of cell fates, and establishment of body axes—rely on materials the egg contains at that moment. In many animals, the first few hours of life proceed with little or no transcription. As a result, developmental regulation at these early stages is dependent on maternal cytoplasm rather than the zygotic nucleus. The regulatory molecules accumulated during oogenesis might, in principle, be of any type, including RNA and protein. It is clear that mRNAs present in the egg

© 2000 Cold Spring Harbor Laboratory Press 0-87969-568-4/00 $5 + .00

before fertilization—so-called maternal mRNAs—play a particularly prominent role in early decisions. Viewed from this perspective, it is not surprising that oocytes and early embryos display an impressive array of posttranscriptional regulatory mechanisms, controlling mRNA stability, localization, and translation.

The mechanisms by which translation of specific maternal mRNAs is controlled, and how those controls contribute to proper development, are the main focus of this chapter. Translational regulation is vital throughout development, in somatic as well as germ cells. The predominant mode of tissue-specific regulation in adult tissues is transcriptional; yet several of the examples we discuss hint that the importance of translational control may be currently underestimated, perhaps dramatically so.

One conclusion emerges exceptionally clearly from studies of translational control during early development: The region between the termination codon and the poly(A) tail—the 3′ untranslated region, or 3′UTR—is a key repository for the regulation of cytoplasmic mRNAs. Other regions of the mRNA will no doubt be found to play critical roles in developmental regulation, but thus far, the 3′UTR is preeminent.

Translational control is defined broadly in this chapter. Ideally, it is demonstrated by comparing the level of a specific, cytoplasmic mRNA to the rate of its translation. However, rates of translation can be difficult to measure directly in vivo. In several cases discussed in this chapter, only steady-state levels of the protein are known; however, translational control is inferred because the regulatory sequences responsible are located outside the protein-coding region. This argument is not airtight, however, and several examples suggest that caution is warranted.

In this chapter, we focus on translational controls that are vital for key developmental decisions. We do not discuss the role of modifications in the level or activity of translation factors, despite their importance in growth and differentiation (for review, see Gingras et al. 1999). Rather, we focus on mRNA-specific regulatory events and the roles of RNA–protein interactions. We first describe examples drawn from a range of biological contexts and organisms, with an emphasis on systems in which genetics has helped reveal biological function. The examples are not intended to be comprehensive, but to provide a reasonably detailed description of a small number of systems, selected to illustrate general points. Drawing on the examples, we consider possible molecular mechanisms and discuss emerging principles about the molecular circuitry of translational control and its regulatory niche.

TRANSLATIONAL CONTROL OF DEVELOPMENTAL EVENTS: SELECTED EXAMPLES

The diversity of developmental decisions in which translational regulation plays a key role is enormous, and the field is expanding explosively. We begin with three examples of regulatory cascades, drawing on the oocytes and embryos of *Caenorhabditis elegans* and *Drosophila melanogaster*.

Cell Fate and Patterning in the *C. elegans* Post-embryonic Germ Line: A Plexus of Controls

As development unfolds, cells assume specific fates and differentiate: For example, one cell becomes a neuron, whereas another becomes a lymphocyte. Although cell-fate regulators often act at the transcriptional level, they can also function at the level of translation. In this section, we describe how a plexus of translational controls regulates cell fates during the growth and differentiation of the *C. elegans* germ line.

Figure 1 summarizes the postembryonic development of the hermaphrodite germ line. *C. elegans* normally develops as either a hermaphrodite or a male, where a hermaphrodite is essentially a female that makes some sperm and then switches to oogenesis. During embryogenesis, two germ-line precursor cells arise from a single germ-line blastomere (Sulston et al. 1983); after the embryo hatches from its eggshell, these two germ-line precursor cells proliferate and differentiate as the animal progresses through four larval stages (L1, L2, L3, and L4) and enters adulthood. During this period of postembryonic development, a cluster of germ-line stem cells resides at the distal end of the growing germ-line tube (Fig. 1, yellow). Cells in meiotic pachytene are first observed during L3 in a "proximal" position (Fig. 1, green). During L4, the most proximal germ-line cells differentiate as sperm (Fig. 1, blue), and later in adulthood germ-line cells switch fates and become oocytes (Fig. 1, pink).

The regulation of germ-line proliferation, survival, and pattern of differentiation all appear to rely on translational controls. Best understood are two 3′UTR-mediated controls that influence the choice between spermatogenesis and oogenesis in the hermaphrodite germ line. The following sections review our current knowledge of these two controls as well as more preliminary studies of the controls governing germ-line proliferation and survival.

The Onset of Hermaphrodite Spermatogenesis: tra-2, *GLD-1, and LAF-1*

The *tra-2* sex-determining gene promotes female cell fates and is predicted to encode a large transmembrane protein, TRA-2A (Hodgkin and

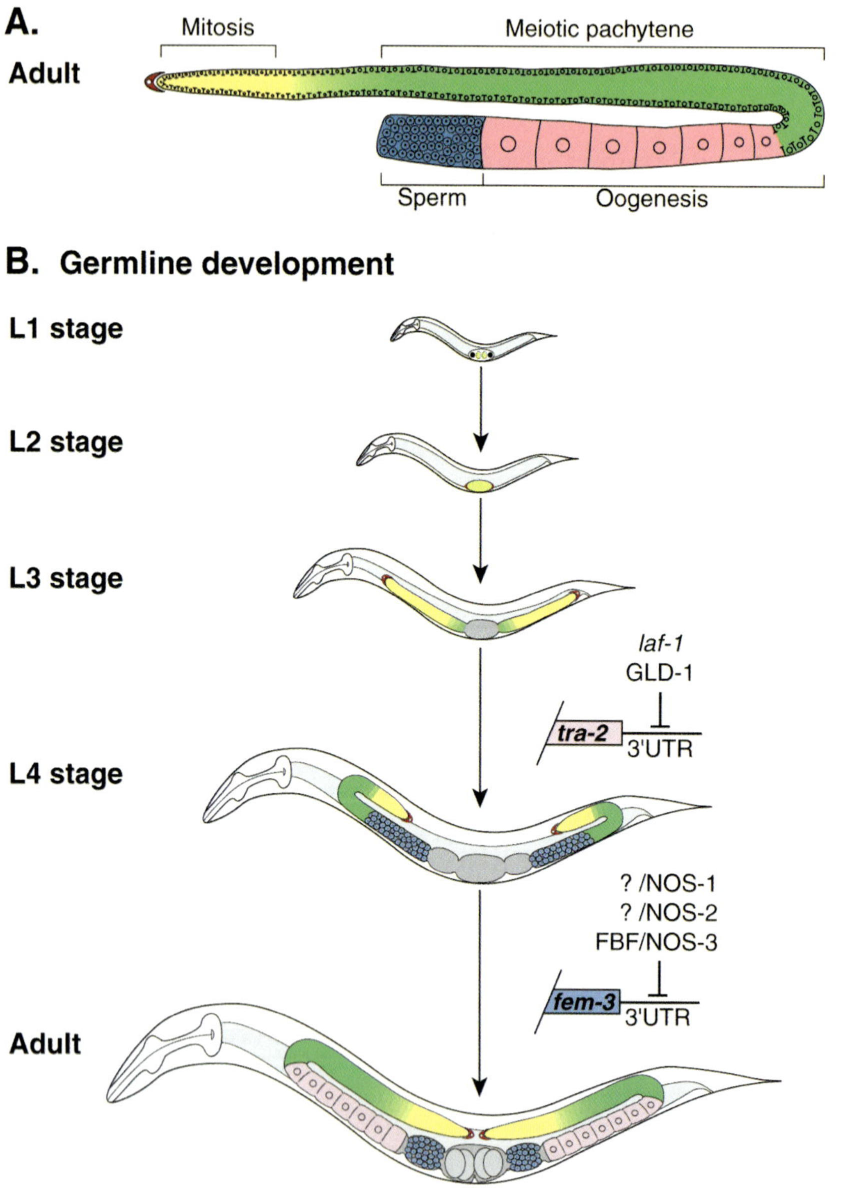

Figure 1. (*See facing page for legend.*)

Brenner 1977; Kuwabara et al. 1992). Male development, including spermatogenesis in hermaphrodites, requires that *tra-2* activity be repressed. Six dominant regulatory mutants, called *tra-2(gf)* (for *g*ain-of-*f*unction), feminize the hermaphrodite germ line so that only oocytes are made (Doniach 1986; Schedl and Kimble 1988). The *tra-2(gf)* mutations therefore identify a site of regulation that is essential for hermaphrodite spermatogenesis. This site is of interest not only for its effect on cell fates, but also for its potential role in the evolution of controls that permit reproduction by hermaphroditism.

The *tra-2(gf)* mutations disrupt two tandemly repeated, *cis*-acting regulatory elements, called TGEs (formerly called DREs) (Fig. 2A). The TGEs are located in the *tra-2* 3′UTR and serve as translational repressor elements (Goodwin et al. 1993). Evidence supporting such a role includes polysome analyses of endogenous *tra-2* mRNAs and a variety of experiments using chimeric reporter mRNAs (Goodwin et al. 1993; Jan et al. 1999). Although the TGEs only partially repress *tra-2* mRNA translation, this is likely to be sufficient because the *tra-2* locus is dosage-sensitive.

The GLD-1 protein appears to be a *trans*-acting repressor of *tra-2* mRNA translation (Fig. 2B). GLD-1 belongs to the STAR family of RNA-binding proteins and is present in the hermaphrodite germ-line cytoplasm (Jones and Schedl 1995; Jones et al. 1996). The phenotype of *gld-1* null mutants suggests that *gld-1* regulates multiple aspects of hermaphrodite germ-line development, including promotion of hermaphrodite spermatogenesis and progression through meiosis during oogenesis (Francis et al. 1995a,b). In addition, *gld-1* controls entry into the meiotic cell cycle (Kadyk and Kimble 1998). The conclusion that GLD-1 controls *tra-2* translation rests on several lines of evidence (Jan et al. 1999). First, GLD-1 binds specifically to TGEs, in both yeast three-hybrid and in vitro

Figure 1 Postembryonic development of the *C. elegans* germ line. (*A*) Pattern of cell fates in the adult hermaphrodite germ line. (*Yellow*) Mitotic germ-line stem cells; (*green*) region of germ line that has entered the meiotic cell cycle and is arrested in the pachytene stage of meiotic prophase I; (*pink*) oogenesis; (*blue*) spermatogenesis. (*B*) Larval development of the germ-line pattern. Color coding same as in *A*. (L1–4) First to fourth larval stages. Repression of the *tra-2* sex-determining mRNA by *laf-1* and GLD-1 is required for the onset of hermaphrodite spermatogenesis; repression of the *fem-3* sex-determining mRNA by FBF and NOS is required for the switch from spermatogenesis to oogenesis. The *mog* genes are also required for the sperm/oocyte switch, but it is not known whether their function is direct or indirect.

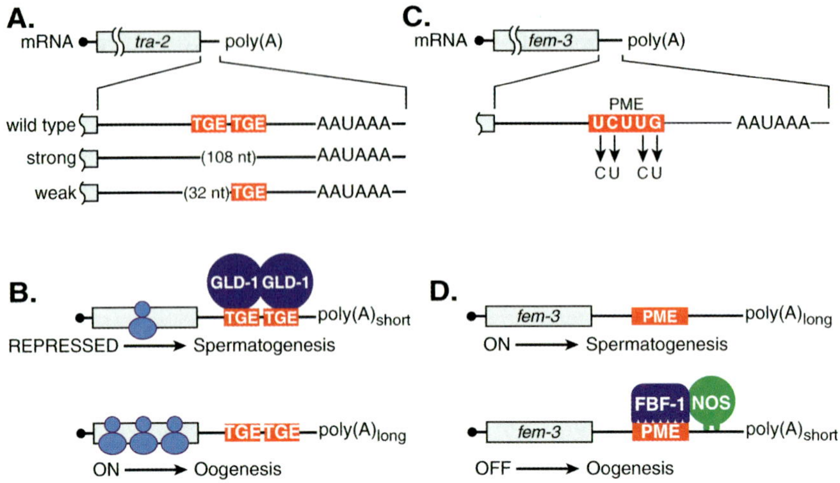

Figure 2 3′UTR regulation of the *C. elegans* sperm/oocyte decision. (*Red*) 3′UTR regulatory element. (*A*) TGE regulatory element is located in the *tra-2* 3′UTR. A strong *tra-2* gain-of-function mutant *(tra-2(gf))* deletes both TGEs; a weak *tra-2(gf)* mutant deletes only one TGE. Two TGEs may serve as a rheostat to regulate *tra-2* translation (Goodwin et al. 1993). (*B*) Model of *tra-2* regulation: The GLD-1 protein binds each copy of the TGE, represses translation activity, and promotes spermatogenesis. (*C*) PME regulatory element is located in the *fem-3* 3′UTR. Point mutations change individual nucleotides within this PME. (*D*) Model of *fem-3* regulation: FBF and NOS interact, and together repress *fem-3* mRNA activity.

binding assays. Second, the level of TRA-2 protein is higher in *gld-1(null)* mutants than in wild-type, without a commensurate increase in the level of *tra-2* mRNA. Third, purified GLD-1 protein specifically represses the translation of TGE-bearing reporter RNAs in vitro. Finally, GLD-1 is a component of DRF, a TGE-specific RNA-binding activity present in crude worm extracts. These findings strongly support the hypothesis that GLD-1 is a translational repressor that acts through TGEs.

The *laf-1* gene also influences TGE activity: loss-of-function mutations in *laf-1* feminize the hermaphrodite germline and disrupt TGE-mediated regulation of reporter transgenes (Goodwin et al. 1997). However, *laf-1* has not been cloned, and its molecular role remains unclear. Interestingly, *laf-1*, like GLD-1, has a complex mutant phenotype, suggesting it too may control multiple mRNAs.

Translational control by TGEs has been broadly conserved in the animal kingdom. TGEs are found in the 3′UTRs of *C. elegans tra-1, C. brig-*

gae tra-2, and the human oncogene GLI mRNAs (Jan et al. 1997). Moreover, TGEs repress translation in nematodes and mammalian cells (Jan et al. 1997), as well as in frog embryos (Thompson et al. 2000).

The mechanism by which translation is repressed by TGEs and GLD-1 is not understood. One clue is that repression is correlated with a change in poly(A) length, such that wild-type mRNAs possess shorter poly(A) tails than their mutant, derepressed counterparts. Similarly, TGEs promote deadenylation in frog embryos, where TGE-mediated repression requires a poly(A) tail (Thompson et al. 2000).

The Hermaphrodite Switch from Spermatogenesis to Oogenesis: fem-3*, FBF, NOS, and MOG*

The *fem-3* sex-determining gene directs male development (Hodgkin 1986; Barton et al. 1987). In a story that is remarkably parallel to that of *tra-2* described in the previous section, genetic selections identified a regulatory element in the *fem-3* 3′UTR that mediates *fem-3* repression and the switch from spermatogenesis to oogenesis. A series of dominant regulatory *fem-3(gf)* mutations masculinize the hermaphrodite germ line: Sperm are made to vast excess and the switch to oogenesis never occurs (Barton et al. 1987). These *fem-3(gf)* mutations therefore identify a site of regulation essential for the sperm/oocyte switch.

The *fem-3(gf)* mutations carry lesions in the *fem-3* 3′UTR: 17 are single nucleotide changes in a 5-bp region (Fig. 2C) (Ahringer 1991; Ahringer and Kimble 1991). The mutated region is presumed to be part of a regulatory element called the point mutation element, or PME. Several lines of evidence support the idea that the PME is a translational control element. First, the *fem-3(gf)* mutations do not detectably affect transcription, splicing, or stability of *fem-3* RNA, and the *fem-3(gf)* mutant RNAs possess a longer poly(A) tail than their wild-type counterparts (Ahringer and Kimble 1991). Second, the FBF and NOS repressors that mediate *fem-3* repression are homologs of Pumilio and Nanos, which are translational repressors in *Drosophila* (see below). Finally, overexpression of the *fem-3* 3′UTR in transgenic animals masculinizes the hermaphrodite germ line, perhaps by titration of the repressor (Ahringer and Kimble 1991). This effect requires the promoter and the PME, suggesting that it relies on the regulatory site in the RNA product.

FBF is a component of the *trans*-acting repressor that acts through the *fem-3* PME (Fig. 2D). *C. elegans* contains two FBF proteins, FBF-1 and FBF-2, that are 91% identical in amino acid sequence; their functions to date are indistinguishable, and so they are often referred to collectively as

FBF. FBF-1 and FBF-2 are both RNA-binding proteins of the Puf family (for *Pu*milio and *F*BF) and are present in the germ-line cytoplasm (Zhang et al. 1997). Animals lacking both *fbf-1* and *fbf-2* make only sperm and fail to switch into oogenesis, consistent with a role for FBF-1 and FBF-2 in *fem-3* repression. Supporting this biological evidence for the role of FBF in *fem-3* repression, both FBF-1 and FBF-2 bind the *fem-3* PME and interact specifically with wild-type, but not mutant, forms of the PME (Zhang et al. 1997). Intriguingly, FBF-deficient germ lines are small, suggesting a broader role for FBF in germ-line proliferation (Zhang et al. 1997).

Three NOS proteins are likely to act together with FBF to repress *fem-3* translation (Fig. 2D) (Kraemer et al. 1999). On the basis of genetic studies, the three *nos* genes appear to be redundant in their regulation of the sperm/oocyte switch. One NOS protein, NOS-3, interacts directly with both FBF-1 and FBF-2, whereas NOS-1 and NOS-2 do not. In one simple model, FBF and NOS-3 function together in a macromolecular complex to repress *fem-3* translation and to regulate the switch from spermatogenesis to oogenesis (Fig. 2D). In this view, recruitment of NOS-3 by FBF either stabilizes a regulatory complex on the *fem-3* 3′UTR or confers repression. However, this model cannot explain involvement of NOS-1 and NOS-2 in the sperm/oocyte switch, since neither protein detectably binds FBF. NOS-1 and NOS-2 may form complexes with FBF and the *fem-3* 3′UTR indirectly, or they may act with other Puf proteins in the *C. elegans* genome to effect *fem-3* repression (Kraemer et al. 1999).

In addition to FBF and NOS, six *mog* genes also are critical for PME-mediated repression and the sperm/oocyte switch (Graham and Kimble 1993; Graham et al. 1993; Gallegos et al. 1998). Hermaphrodites defective in any one of these *mog* genes fail to switch from spermatogenesis to oogenesis. Furthermore, the *mog* genes are required maternally for embryogenesis, suggesting that they may control not only *fem-3*, but other maternal mRNAs as well. Three *mog* genes have now been cloned, and their molecular identity is unexpected and provocative. All three encode members of the DEAH-family of ATP-dependent helicases: *mog-1*, *mog-4,* and *mog-5* encode the *C. elegans* homologs of yeast *PRP16, PRP2,* and *PRP22,* respectively (Puoti and Kimble 1999, 2000). What does this tell us about the molecular function of the MOG proteins? The yeast Prp2p, Prp16p, and Prp22p proteins are integral components of the splicing machinery (Burge et al. 1999). Although a role for the *mog* genes in splicing has not been excluded, no general defect in splicing is observed in *mog-1* null mutants (Puoti and Kimble 1999). The *mog* genes may therefore be evolutionarily related to the *PRP* genes but have acquired a different function. One speculative idea is that the MOG proteins may direct conformational changes in a ribonucleoprotein (RNP) complex involved in PME-mediated repression. MOG-1 is nuclear (Puoti and Kimble 1999), whereas both FBF and NOS-3 are cytoplasmic. Perhaps MOG proteins act in the nucleus to establish an RNP structure that can be accessed by FBF and NOS proteins in the cytoplasm. Such an RNP remodelling function may be analogous to the role of various complexes that remodel chromatin in an ATP-dependent manner (see, e.g., Pazin and Kadonaga 1997). Indeed, one such chromatin remodeler, SWI-2, is a DEAD-box helicase and has been assigned to the same superfamily of helicases as the DEAH-box proteins (Eisen et al. 1995).

TGE- and PME-mediated Repression in Somatic Tissues

TGE- and PME-mediated repression of *tra-2* and *fem-3,* respectively, has crucial roles in regulating germ-line development. However, these controls also occur in somatic tissues. Thus, the strongest *tra-2(gf)* mutation feminizes the intestine of older adult males (Doniach 1986), and the strongest *fem-3(gf)* mutation masculinizes the soma of *tra-1(gf)* XO females (Schedl and Kimble 1988). Although these effects are relatively minor, both demonstrate that TGE- and PME-mediated repression can occur in somatic tissues. In support of this idea, reporter transgenes controlled by TGE- or PME-containing 3′UTRs are translationally controlled in somatic tissues (Goodwin et al. 1997; Gallegos et al. 1998). Certain regulators affect both somatic and germ-line controls: *laf-1* is required for TGE-mediated repression and the *mog* genes for PME-mediated repression in both tissues. In contrast, GLD-1 and FBF are expressed predominantly in the germ line, suggesting that other members of the STAR or Puf gene families may mediate the somatic controls. Therefore, the regulatory machineries for translational controls are found in somatic tissues and are likely to be used there, a theme that is underscored by control of *lin-14* and *lin-41* in hypodermal cells of *C. elegans* (see below).

tra-2 *and* fem-3 *3′UTR Controls and Patterning the Germ Line*

The generation of the hermaphrodite pattern of gametes—first sperm, then oocytes—relies on controls exerted by the *tra-2* and *fem-3* 3′UTRs. How are these controls coordinated to generate the germ-line pattern? Do they act alone or in concert with other modes of regulation? The nature of the TRA-2 and FEM-3 proteins and their regulatory relationship provides some insight into these questions. In particular, TRA-2 protein is itself a *fem-3* repressor (Hodgkin 1986). The intracellular domain of the TRA-2 membrane protein binds FEM-3, suggesting that *fem-3* repression by

TRA-2 may rely on sequestration of FEM-3 (Mehra et al. 1999). By this model, the relative abundance of these two proteins is predicted to be critical for fate specification. Consistent with that idea, the levels of TRA-2 and FEM-3 appear to be poised in a delicate balance in the hermaphrodite germ line: *tra-2(gf)* mutants are predicted to make excess TRA-2 protein, swamp out available FEM-3, and thereby promote oogenesis. Similarly, *fem-3(gf)* mutants are predicted to make excess FEM-3, resulting in free FEM-3 and hence spermatogenesis. Perhaps most important for this discussion, *tra-2(gf); fem-3(gf)* double mutants can possess a self-fertile hermaphrodite germ line with sperm made first and then oocytes (Barton et al. 1987; Schedl and Kimble 1988). In this regard, the strength of the individual *tra-2(gf)* or *fem-3(gf)* allele is critical. Thus, an animal carrying a strong *tra-2(gf)* allele and a weak *fem-3(gf)* allele often makes only oocytes, but an animal carrying both strong *fem-3(gf)* and *tra-2(gf)* alleles is usually self-fertile. It seems likely that when *gf* allelic strengths are matched, the levels of TRA-2 and FEM-3 are comparable, albeit higher than normal, and balance between these two regulatory proteins is restored. The ability of the *tra-2(gf); fem-3(gf)* double mutant to develop a self-fertile hermaphrodite demonstrates that these 3′UTR controls can be bypassed to generate the sperm/oocyte pattern. We suggest, therefore, that this pattern does not rely only on 3′UTR controls, and we speculate that an alternative mechanism acts in parallel to ensure the proper pattern of sperm and then oocytes.

One major unanswered question is how the translational regulators of the *tra-2* and *fem-3* mRNAs are controlled to obtain more FEM-3 early and more TRA-2 later. A simple hypothesis is that all *tra-2* germ-line mRNAs are repressed during larval development, but that *tra-2* mRNAs synthesized in adults are not repressed. This change might rely on a change in activity of the translational repressor or a change in the relative abundance of *tra-2* mRNA to repressor. Similar arguments can be made for *fem-3*.

Pattern Formation in *Drosophila*: A Translational Cascade

In *Drosophila*, asymmetries become evident during oogenesis and early embryogenesis that foreshadow the anterior–posterior and dorsal–ventral axes of the mature organism. Translational controls are critical for establishing body axes (for review, see Wharton 1992; Curtis et al. 1995). Each of the four maternal patterning systems (St Johnston and Nüsslein-Volhard 1992) requires the translational control of one or more mRNAs, representative examples of which are provided in Table 1.

Table 1 Translational control in the four maternal patterning systems of *Drosophila:* Representative examples

Maternal system	Translationally controlled mRNA	Role of protein product
Anterior	*bicoid* (Driever and Nüsslein-Volhard 1988a,b)	Anterior determinant, activates genes required for head and thorax formation (Frohnhöfer and Nüsslein-Volhard 1986; Driever and Nüsslein-Volhard 1988a,b); also required to repress translation of *caudal* mRNA (Struhl 1989), which encodes a homeobox protein (Mlodzik et al. 1985)
Posterior	*nanos* (Gavis and Lehmann 1994)	Posterior determinant (Wang and Lehmann 1991; Wang et al. 1994); collaborates with *pumilio* to suppress translation of posterior maternal *hunchback* mRNA (Hülskamp et al. 1989; Irish et al. 1989; Struhl 1989; Murata and Wharton 1995), which encodes a transcription factor (Hülskamp et al. 1990)
Terminal	*torso* (Casanova and Struhl 1989; Sprenger et al. 1989)	Cell-surface receptor that responds to localized extracellular ligand to generate terminal structures (Stevens et al. 1990; Martin et al. 1994)
Dorso-ventral	*toll* (Gay and Keith 1992)	Cell-surface receptor that responds to localized extracellular ligand to generate ventral structures (Hashimoto et al. 1988; Stein et al. 1991; Morisato and Anderson 1994)

Coordinate Activation

The maternal transcripts of several axis-determining genes are translationally dormant in oocytes but are activated soon after fertilization. This coordinate activation often requires cytoplasmic polyadenylation. mRNAs that encode key regulatory proteins for the anterior, terminal, and dorsal–ventral patterning systems, respectively—*bicoid, torso*, and *toll* (Table 1)—undergo polyadenylation concomitant with their activation.

For *bicoid* mRNA, polyadenylation after egg deposition is critical for translation. Early evidence supporting this idea came from specific (*Bic-D*) mutant embryos that lack anterior structures. These embryos inappropriately express the posterior morphogen *nanos* in the anterior, which blocks production of the anterior determinant *bicoid*. The lack of Bicoid protein production was correlated with a *bicoid* mRNA that has a shortened poly(A) tail (Wharton and Struhl 1991). Subsequent work directly showed that the polyadenylation of *bicoid* mRNA is necessary for its activation (Sallés et al. 1994). Similar experiments have documented that translation of Toll protein, a crucial regulator of dorsal–ventral patterning, is also dependent on poly(A) addition (Schisa and Strickland 1998).

Unlike *bicoid* mRNA, translational activation of mRNAs encoding *nanos* and *oskar*, two crucial posterior determinants, does not involve a detectable change in poly(A) tail length upon fertilization (Sallés et al. 1994; Lie and Macdonald 1999b). However, the ultimate effect of Nanos protein is to control the poly(A) status and translation of maternal *hunchback* mRNA in the posterior (Wreden et al. 1997; considered in detail in the next section). Polyadenylation thus plays a critical role in the anterior, posterior, and dorsal–ventral patterning systems in *Drosophila*.

Translational Cascades: Posterior Patterning

Anterior–posterior patterning hinges in part on a regulatory cascade of translational control. A series of opposing protein gradients help determine the axis, and they are established by regulated mRNA localization and translation, events that are linked in the embryo.

Posterior development of the *Drosophila* embryo is critical both for abdomen formation and for providing the correct environment for germ-cell development (Lehmann and Nüsslein-Volhard 1991). Both of these processes must be restricted to the posterior for normal development to occur; misexpression of the posterior determinant *nanos* in the anterior is lethal to embryos (Wharton and Struhl 1989; Gavis and Lehmann 1992). The difficulties in restricting expression of proteins to the posterior in the absence of transcription illustrates two central features of translational control: regulation in space and in time.

Spatial regulation is necessary since certain mRNAs critical for posterior patterning are found not only in the posterior, but throughout the embryo. Repression of unlocalized mRNAs, coupled with the selective translation of posterior mRNA, ensures that protein production is region-specific. Temporal regulation is also required for these mRNAs. Once they accumulate in the posterior of the oocyte/embryo, their expression must be coordinated with the onset of embryogenesis in the rest of the embryo.

To accomplish posterior regulation, *Drosophila* has evolved a mechanism that coordinates spatial and temporal controls. It is easiest to conceptualize this pathway by starting at the end point. The ultimate goal of the entire system is to repress the translation of maternal *hunchback* mRNA in the posterior: If the posterior determinant *nanos* is lacking, embryos die from a lack of posterior structures, but embryos lacking both *nanos* and maternal *hunchback* are viable (Hülskamp et al. 1989; Irish et al. 1989; Struhl 1989).

Posterior repression of maternal hunchback mRNA requires both Nanos and Pumilio (Fig. 3) (Lehmann and Nüsslein-Volhard 1991; Barker et al. 1992). Pumilio is uniformly distributed (Macdonald 1992) and thus cannot account for the restriction of the process to the posterior (Fig. 3A). Pumilio, a protein structurally related to FBF, binds specifically to *nanos* response elements (NREs) in *hunchback* mRNA's 3′UTR (Murata and Wharton 1995) and likely saturates *hunchback* mRNAs throughout the embryo (Zamore et al. 1999). However, Nanos protein expression is limited to the posterior (Wang and Lehmann 1991; Wang et al. 1994), and this localization underlies the asymmetric repression.

Pumilio recruits Nanos protein to a ternary complex containing the NREs (Fig. 3B) (Sonoda and Wharton 1999). The formation of the ternary complex is critical: Mutant forms of each component that do not regulate in vivo do not form the complex (Sonoda and Wharton 1999). The complex promotes repression and deadenylation of maternal *hunchback* mRNA in the posterior. Shortening of the poly(A) tail is one important factor in repressing its translation (Wreden et al. 1997).

Regions of the ternary complex that may contact the translation or deadenylation machinery have been identified: For example, specific Pumilio mutations permit complex formation but fail to repress (Sonoda and Wharton 1999). Recruitment of Nanos requires specific nucleotides within the NRE as well as specific amino acids in Pumilio, implying that either Pumilio or the RNA undergoes a conformational change upon forming the Pumilio/NRE complex, which then is recognized by Nanos (Fig. 3B) (Sonoda and Wharton 1999).

From the biological standpoint, these results raise the question of how Nanos protein, the localized posterior determinant, is restricted to the posterior. Although *nanos* mRNA is highly concentrated at the posterior pole, there are substantial levels of the transcript throughout the embryo (Bergsten and Gavis 1999). Translation of all the mRNA would be disastrous, since ectopic expression of Nanos protein is lethal (Ephrussi and

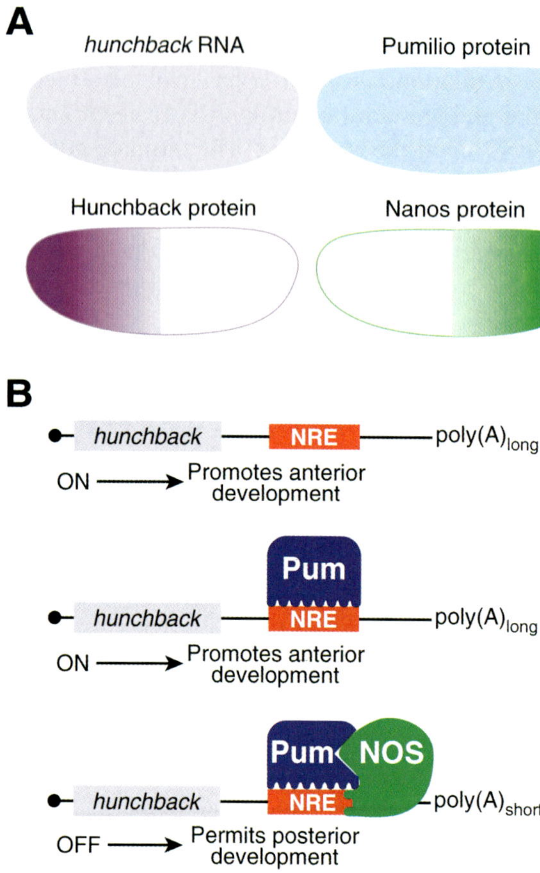

Figure 3 Spatial regulation of *Drosophila hunchback* mRNA by Nanos and Pumilio. (*A*) Distributions of *hunchback* mRNA and of Hunchback, Pumilio, and Nanos proteins in an early syncitial *Drosophila* embryo; anterior to the left, posterior to the right. *hunchback* mRNA is distributed throughout the embryo, but the protein appears only in the anterior portion (*purple*). Pumilio protein (*blue*) is uniformly distributed, while Nanos protein (*green*) is present in a gradient emanating from the posterior pole. (*B*) Pumilio (*blue*), Nanos (*green*), and the NRE (*red*) interact to form a tertiary complex that represses the mRNA. NRE-bound Pumilio is insufficient for repression, leaving *hunchback* mRNA on, and promoting anterior development. Recruitment of Nanos results in the formation of a tertiary complex that represses *hunchback* mRNA and permits posterior development. Formation of the tertiary complex involves an alteration in either the NRE, Pumilio, or both, and is represented by the altered shape of Pumilio in the tertiary complex.

Lehmann 1992; Gavis and Lehmann 1992; Smith et al. 1992). However, translation of the unlocalized mRNA is repressed (Gavis and Lehmann 1994). This repression depends on the uniformly distributed protein, Smaug (Dahanukar et al. 1999), which binds to translational control elements in the 3′UTR of *nanos* mRNA (Smibert et al. 1996; Dahanukar et al. 1999). Once in the posterior, translational repression of *nanos* mRNA by Smaug is overcome by activation of the mRNA by localized Oskar protein (Dahanukar et al. 1999). Thus, *nanos* mRNA is activated in the embryo only in the correct locations at the correct time.

Working backward in the cascade (Fig. 4) prompts the following question: How is the localization of Oskar protein accomplished? The situation here is analogous to regulation of *nanos*: Unlocalized *oskar* mRNA is repressed by repressor proteins that bind to the Bruno response element (BRE) in the 3′UTR (Kim-Ha et al. 1995; Lie and Macdonald 1999a; Castagnetti et al. 2000). However, posterior localization does not automatically trigger *oskar* mRNA's translation. Rather, a 5′ region of *oskar* mRNA is absolutely required to relieve BRE-mediated repression (Gunkel et al. 1998). This activator region is located between the 5′-most AUG and the second AUG of *oskar* mRNA and does not appear to bind Bruno itself. The activator region is not required for the translation of mutant *oskar* mRNAs in which the BREs have been deleted or inactivated. It only functions at the posterior pole, suggesting that at least one limiting component of an active derepressor machinery must be located in this region of the oocyte cytoplasm.

Bruno was the first protein found that acts as a repressor of *oskar* mRNA (Kim-Ha et al. 1995), but it appears to have collaborative partners. Apontic can bind both Bruno protein and to the BREs in the *oskar* mRNA 3′UTR, and there are genetic interactions between the *apontic* and *bruno* genes (the *aret* locus) (Lie and Macdonald 1999a). A 50-kD protein binds both the 5′ end and the 3′ BREs of *oskar* mRNA, and BRE mutants that bind Bruno but not this 50-kD protein have reduced translational repression (Gunkel et al. 1998). Finally, mutants in the *Bic-C* gene, which encodes an RNA-binding protein, prematurely translate *oskar* mRNA (Saffman et al. 1998). Thus, it appears that a multicomponent protein assemblage may regulate translation of unlocalized *oskar* mRNA.

In the posterior, activation of *oskar* mRNA translation also is complex. There are several collaborators: Oskar protein itself, Vasa (Markussen et al. 1997), Orb (a CPEB homolog; Chang et al. 1999), Staufen (St Johnston et al. 1991), and Aubergine (Wilson et al. 1996).

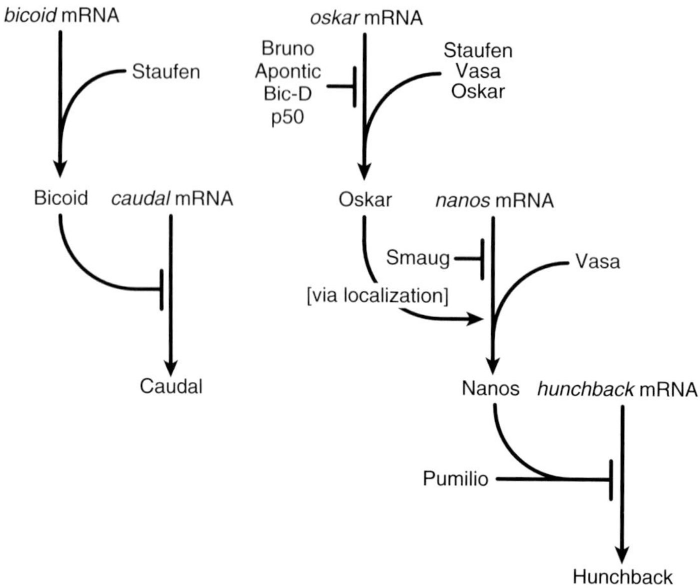

Figure 4 Cascades of translational regulation and localization that control formation of the anterior–posterior axis in *Drosophila*. Arrowheads depict positive events, and blunt ends indicate repressive events. Citations are provided in the text. The events depicted occur either in the growing oocyte or in the syncitial early embryo. mRNAs produced in nurse cells enter the growing oocyte from the presumptive anterior end; some mRNAs must move across the oocyte to the presumptive posterior. Activation of *bicoid* mRNA, which is localized to the anterior and repressed during oogenesis, requires Staufen protein. Bicoid protein then represses the translation in the anterior of uniformly distributed *caudal* mRNA. In the posterior, the initial event is localized expression of Oskar protein. Translation of *oskar* mRNA during its transit from the anterior end of the oocyte is repressed by Bruno in collaboration with Apontic, Bic-D, and p50 (see text). Its localization and activation at the posterior requires Staufen, Vasa, and Oskar protein itself. *nanos* mRNA is also localized to the posterior pole, a process that requires the presence of Oskar protein. Its mis-localized expression is prevented by Smaug, and its activation in the posterior requires Vasa. *hunchback* mRNA is present throughout the embryo, as is Pumilio. Posteriorly localized Nanos acts in concert with Pumilio to repress the *hunchback* mRNA in the posterior. We include in the figure genes and proteins that are discussed in the text; many other genes such as *cappuccino*, *spire*, and *egalitarian* contribute to these processes but have not been included; in particular, proteins that participate in localization but not explicitly in translational regulation are not depicted.

Vasa is an ATP-dependent RNA helicase (Liang et al. 1994), suggesting that its effects may be directly on *oskar* mRNA, altering an RNA structure or promoting RNA–protein transactions. Other proteins required to activate *oskar* mRNA expression act indirectly through their role in mRNA localization. These include gene products that affect cytoskeletal organization and function (see, e.g., Cappuccino and Spire: Ephrussi et al. 1991; Theurkauf 1994; Kim-Ha et al. 1995), as well as proteins that interact with specific mRNAs (e.g., Staufen: St Johnston et al. 1991). Thus, with both *nanos* and *oskar*, the regulatory pathway involves repression in all regions except the posterior, and a separate mechanism to ensure activation in the posterior (Fig. 4).

As a general consideration, repression of unlocalized mRNA translation is the most parsimonious way to achieve posterior specific protein expression. If all mRNA molecules were translated equivalently, a trail of protein would be produced that would have to be either transported posteriorly or destroyed: The spatial control mechanisms provide an intuitively satisfying solution. However, if the oocyte relied on this mechanism alone, with repressor molecules excluded from the posterior, once the mRNA reached this region its translation would commence. A separate activation mechanism in the posterior gives the embryo the temporal control needed to coordinate the patterning systems.

Parallel Cascades in the Anterior and Posterior

As if this complexity were not enough, Bicoid protein, in addition to its role as a transcriptional factor, is required for translational repression of *caudal* mRNA, another mRNA important in axis formation (Fig. 4). In the absence of *bicoid* activity, the normal gradient of Caudal protein—low in the anterior to high in the posterior—is disrupted, with high Caudal now found at the anterior as well (Macdonald and Struhl 1986; Mlodzik and Gehring 1987; Driever and Nüsslein-Volhard 1988b). Bicoid protein binds to *caudal* mRNA, and this interaction appears to be essential for translational repression (Dubnau and Struhl 1996; Rivera-Pomar et al. 1996; Chan and Struhl 1997; Niessing et al. 1999). The key regulatory elements lie in the 3′UTR of *caudal* mRNA. Remarkably, the homeodomain region of Bicoid protein is required to bind both to *caudal* mRNA and to DNA targets in its role as transcriptional activator.

The regulation of the anterior–posterior axis thus involves two parallel cascades of translational control at opposite ends of the embryo (Fig. 5). Many of the key players—*bicoid*, *nanos*, *caudal,* and *hunchback*—initially are translationally dormant and are activated only after fertilization. At the anterior, newly synthesized Bicoid protein represses *caudal* mRNA; at the posterior, Nanos protein represses *hunchback* mRNA. Ultimately, the posteriorly localized *hunchback* mRNA is destroyed (Tautz and Pfeifle 1989). Thus, this web of interactions establishes opposing gradients of Hunchback and Caudal proteins.

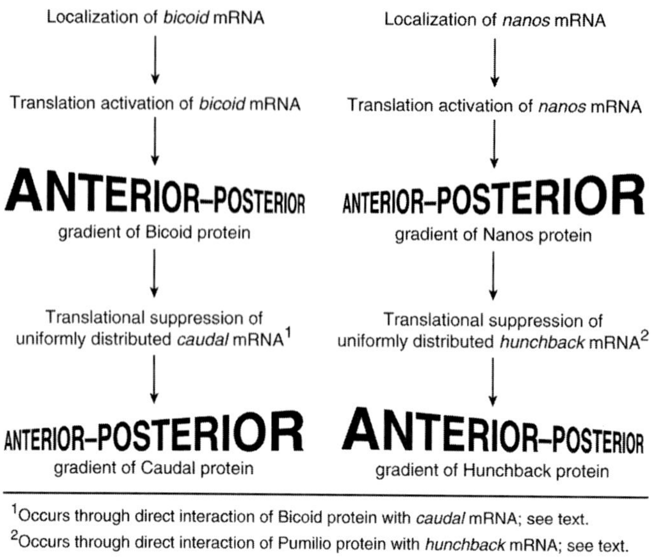

Figure 5 Parallel cascades at opposite ends of the *Drosophila* embryo. See text.

Translational Controls in the *C. elegans* Early Embryo

In *C. elegans*, several key regulators of body axes and blastomere fates are controlled translationally. Some of the controls parallel analogous controls in *Drosophila*; others do not. Perhaps the most important similarity between *Drosophila* and *C. elegans* is the presence of cytoplasmic RNA-enriched granules that are localized to the future posterior of the fertilized zygote, and then segregated into germ-line precursor cells as they are born. These granules, called P granules in *C. elegans*, polar granules in *Drosophila*, and germ plasm in *Xenopus*, appear to be central hubs for translational control and contain at least some related RNAs and proteins.

Figure 6A introduces the *C. elegans* early embryo. The first division establishes the anterior–posterior axis, to a first approximation, and the second division similarly establishes the dorsal–ventral axis. We refer readers to a recent review of *C. elegans* embryogenesis for details (Schnabel and Priess 1997).

Translational Control of glp-1

The *glp-1* gene encodes a Notch-related receptor critical for a cascade of cell–cell interactions specifying dorsal–ventral and left–right axes of the *C. elegans* embryo (for review, see Schnabel and Priess 1997). As shown in Figure 6B, GLP-1 protein first appears at the 2- to 4-cell stage in anterior, but not posterior, blastomeres; in contrast, *glp-1* maternal mRNA is uniformly distributed at this time (Evans et al. 1994). Therefore, *glp-1* mRNA must be subject to at least two distinct translational controls. One is temporal: *glp-1* mRNA is translationally silent in oocytes and the fertilized one-cell embryo, but its translation is activated after the first embryonic division. The second is spatial: *glp-1* is translated only in anterior blastomeres and is kept silent in posterior blastomeres. The elements that mediate both controls reside in the *glp-1* 3′UTR. A U-rich region at the 3′ end of the 3′UTR is required to repress translation in oocytes, and a centrally located stretch of 39 nucleotides is responsible for spatial regulation (Evans et al. 1994). At present, the *trans*-acting factors controlling *glp-1* translation are not known. Such *trans*-acting factors might include posterior repressors, anterior activators, or both. To date, the only genes known to be essential for the asymmetric expression of *glp-1* are the *par* genes, which are critical for asymmetry of the embryo per se (Crittenden et al. 1997).

Translational Control of pal-1 *mRNA by MEX-3*

The PAL-1 homeodomain transcription factor is required for certain posterior fates; it is expressed in posterior blastomeres, largely due to translational regulation conferred by its 3′UTR (Hunter and Kenyon 1996). MEX-3 is a KH-domain RNA-binding protein that may repress *pal-1* mRNA. The location of MEX-3 protein within the embryo complements that of PAL-1 protein (Fig. 6C). MEX-3 is first detected in the cytoplasm of developing oocytes, where it is expressed at high levels, and becomes enriched in AB and its daughters after fertilization (Draper et al. 1996). In contrast, PAL-1 protein is detected for the first time at the 4-cell stage and then only in EMS and P2 (Hunter and Kenyon 1996). Because *pal-1* maternal RNA is evenly distributed in developing oocytes and early embryos, it must be controlled both temporally and spatially, a theme also observed for *glp-1* (see above). In *mex-3* mutants, *pal-1* mRNA is released from those controls: It is expressed early and uniformly, being present throughout oocytes and early embryos (Hunter and Kenyon 1996). A reporter RNA bearing a *pal-1* 3′UTR is expressed in a *pal-1*-like pattern and is similarly derepressed in *mex-3* mutants (Hunter and Kenyon 1996). The simplest interpretation is that MEX-3 acts directly through regulatory elements in the *pal-1* 3′UTR to repress translation of the *pal-1* mRNA.

Translational Control of apx-1

APX-1 is a transmembrane protein that serves as a ligand for the GLP-1 receptor. In the early embryo, APX-1 signals from the P2 blastomere to its neighbor ABp and thereby induces the normal ABp fate (Mello et al. 1994). The *apx-1* maternal mRNA is uniformly distributed in early embryos, but APX-1 protein is found only in specific blastomeres (Fig. 6D) (Mickey et al. 1996). *mex-1* and *pos-1* genes may act in a cascade to control the translation of *apx-1* mRNA (Tabara et al. 1999). MEX-1 and POS-1 are both cytoplasmic proteins that contain two copies of a CCCH "finger" motif (Guedes and Priess 1997; Tabara et al. 1999). A biochemical function for the CCCH motif is unknown, but several proteins with CCCH motifs have been implicated in different aspects of RNA metabolism (Zhang et al. 1992; Barabino et al. 1997; Carballo et al. 1998), sug-

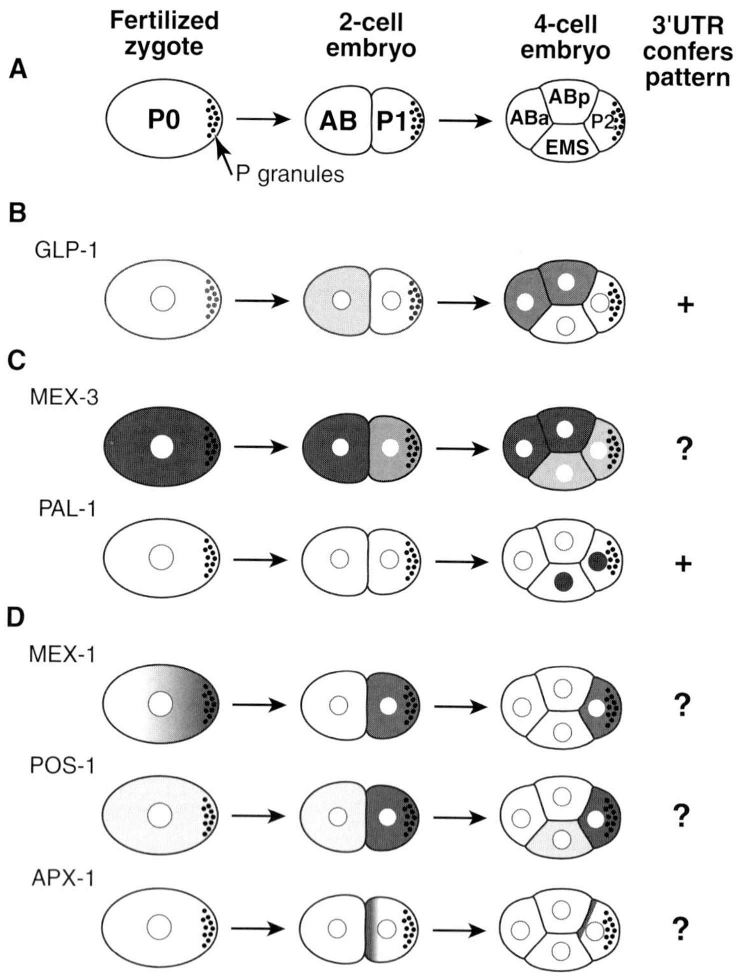

Figure 6 (*See facing page for legend.*)

gesting that it binds RNA. *mex-1* mutant embryos fail to produce both POS-1 and APX-1, and *pos-1* mutant embryos lack APX-1. These findings are consistent with *mex-1* working upstream to control *pos-1* translation, with POS-1 in turn activating translation of *apx-1* mRNA. The temporal expression of the three proteins is also consistent with a regulatory cascade (Fig. 6D). MEX-1 is first detected in oocytes (Guedes and Priess 1997), whereas POS-1 is initially detected at low levels in 1-cell embryos (Tabara et al. 1999). APX-1 is the last protein produced, as it is first detected in P2 of the 2-cell embryo (Mickey et al. 1996). Although these data suggest a linear pathway, it is also possible that MEX-1 and POS-1 act in separate pathways to affect APX-1 expression. Mutations that reduce MEX-1 and POS-1 activities result in different phenotypes, indicating that the two proteins do not only affect APX-1 expression but that they likely have different targets or function at different developmental times. Indeed, MEX-1 is also required for PIE-1 localization, which is essential for germ-line specification (Guedes and Priess 1997; see below).

Translational Control in Germ-line Blastomeres

Specification of the germ-line precursor cells in the early *C. elegans* embryo relies on a combination of transcriptional and posttranscriptional controls. These germ-line precursor cells arise by the segregation of germ-line blastomeres, P1, P2, P3, and P4, in consecutive divisions. The

Figure 6 Translational controls and patterning in the early *C. elegans* embryo. All embryos are oriented with anterior to left and posterior to right. Blastomere names are provided in *A* only; in *B–E*, nuclei are depicted as a circle within the cell. P granules are represented as a cluster of black dots at the posterior end of fertilized zygotes, P1 blastomeres at the 2-cell stage, and P2 blastomeres at the 4-cell stage. (*A*) The fertilized zygote harbors P granules at the posterior end. The 2-cell embryo possesses one larger blastomere, AB, and one smaller one, P1. The 4-cell embryo harbors the daughters of AB, which are called ABa and ABp, and the daughters of P1, which are called EMS and P2. The AB blastomere generates somatic cells that are, for the most part, anterior; the EMS blastomere generates somatic cells, including the intestine, muscle, and hypodermis; P1 and P2 both carry P granules. (*B-D*) Diagrams showing distribution of GLP-1 (*B*), MEX-3 and PAL-1 (*C*), and MEX-1, POS-1, and APX-1 (*D*) proteins at individual stages during early embryogenesis. Maternal mRNAs encoding GLP-1, PAL-1, and APX-1 are uniform in oocytes and early embryos; maternal mRNAs encoding MEX-3, MEX-1, and POS-1 are uniform in oocytes and 1-cell embryos, but become asymmetrically distributed in late-stage 1-cell embryos (MEX-3, POS-1) or later (MEX-1).

RNA-rich P granules are localized to these germ-line blastomeres and are critical for the germ-line fate (Kawasaki et al. 1998; Seydoux and Strome 1999). The germ-line fate relies in part on repression of polymerase II-mediated transcription in germ-line blastomeres by the PIE-1 protein (Seydoux and Fire 1994; Seydoux et al. 1996; Seydoux and Dunn 1997; Batchelder et al. 1999). However, the germ-line fate also appears to rely on posttranscriptional, and likely translational, controls. Specifically, the putative translational regulator *pos-1* is required for specifying the germ-line fate (Tabara et al. 1999). Furthermore, several proteins predicted to control RNA activity or to bind RNA are colocalized with P granules. These include GLH-1 and GLH-2, two homologs of *Drosophila* Vasa that contain DEAD-box helicase motifs (Gruidl et al. 1996); PGL-1, a protein bearing multiple RGG boxes (Kawasaki et al. 1998); the GLD-1 translational regulator (Jones and Schedl 1995; Jan et al. 1999); MEX-1 (Guedes and Priess 1997); MEX-3 (Draper et al. 1996); POS-1 (Tabara et al. 1999); and PIE-1 (Mello et al. 1996). Finally, maternal RNA encoding the translational regulator NOS-2 colocalizes with P granules (Subramaniam and Seydoux 1999). Although the functions of these various proteins and RNAs are not yet fully understood, one idea is that the P granules serve as an RNA control hub in the germ-line blastomere.

The Early Embryonic Cell Cycle and Meiotic Maturation

A dramatic transition from cell cycle arrest to mitotic cleavage occurs upon fertilization. In some species, it is immediately preceded by completion of the meiotic cell cycle, referred to as oocyte maturation. To regulate these transitions, eggs of many species contain mRNAs that encode cell cycle regulators, such as cyclins and cyclin-dependent kinases (CDKs). Control of their translation helps orchestrate the transition from quiescence to meiosis and mitosis, as does their posttranslational modification. For the purposes of this discussion, it is necessary only to know that cyclins and CDKs form complexes that promote the cell cycle. Activation of the complex requires dephosphorylation of the kinase at certain positions by the CDC25 phosphatase, and lack of phosphorylation by the WEE1 kinase.

Translational Regulation of Factors That Contact CDKs: Cyclins, WEE1, and CDC25 Homologs

Translation of cyclin mRNAs appears to be important for proper post-fertilization mitoses in many species, and perhaps for meiotic maturation as well. The analysis of cyclin regulation and function is complicated by the mixed contributions of proteolysis and regulated synthesis to changes in cyclin protein levels, and by the presence of multiple cyclins with overlapping roles. Nevertheless, intensive studies of the translational control of these critical regulators have been informative.

Frog oocytes contain mRNAs encoding several different cyclins. *Xenopus* cyclin A1, B1, and B2 mRNAs are activated at different times during maturation, and to different extents (Kobayashi et al. 1991). Each mRNA receives poly(A) concomitant with its translational stimulation (Sheets et al. 1994). To identify signals involved in these controls, chimeric mRNAs were injected that contained each 3′UTR joined to a translational reporter. The different cyclin 3′UTRs determined when, and how much, translation was stimulated during oocyte maturation. Invariably, translational stimulation required poly(A) addition (Sheets et al. 1994). Thus, 3′UTRs, by controlling polyadenylation, can impose different patterns of translation, stimulating translation at different times and to different extents. Similar results have been obtained with a variety of other mRNAs unrelated to the cell cycle (Chapter 27).

Full translational control of cyclin B1 mRNA appears to be achieved through two separate but related mechanisms: translational repression and polyadenylation. Repression of cyclin B1 mRNA in resting oocytes apparently requires specific sequences in the 3′UTR that overlap with (and may be identical to) those that are required for its subsequent polyadenylation and activation (see below). The role of polyadenylation in derepression of the endogenous mRNA is uncertain, since cyclin B1 protein levels can increase when polyadenylation is blocked by inhibition of the cyclin/CDK complex (Frank-Vaillant et al. 1999), yet injected mRNAs require a poly(A) tail to be derepressed (de Moor and Richter 1999; Barkoff et al. 2000).

Regulation of maternal cyclin mRNAs at the translational level may be common. In *Drosophila* embryos, for example, maternal cyclin B mRNA is localized to pole cells (the presumptive germ line) and is repressed until mitoses resume in the developing gonad, well after fertilization (Dalby and Glover 1993). The regulatory elements responsible for translational control and localization reside in its 3′UTR (Dalby and Glover 1993). *Drosophila* cyclin B1 mRNA is not repressed in *nanos* or *pumilio* mutants: The precocious expression that results may underlie the failure of *nanos* mutant animals to slow the cell cycle and enter mitotic quiescence at the start of germ-cell development (Asaoki-Taguchi et al. 1999; Deshpande et al. 1999).

In surf clams and sea urchins, certain cyclin mRNAs are repressed during oogenesis, then activated dramatically at fertilization, when they receive poly(A) (Rosenthal et al. 1980; Standart 1992). The common regulation of cyclin mRNAs presumably reflects their role after the cell cycle resumes at fertilization, and the deleterious consequences of their premature expression. Other maternal mRNAs that participate in cell-cycle-related events, such as DNA replication and the synthesis of DNA precursors, are also subject to translational control (e.g., histones, ribonucleotide reductase, HGPRT; for review, see Standart 1992).

Proteins that regulate CDK activity by covalent modification are also controlled at the translational level. For example, translation of *Drosophila* CDC25 (*twine*), a phosphatase required to activate CDK2, requires *boule*, an RNA-binding protein of the DAZ family. In the absence of either protein, *Drosophila* oocytes arrest in meiosis. In *Xenopus*, CDC25 levels are constant through maturation and early development, but the level of the inhibitory kinase, WEE1, increases during meiosis (Murakami and Vande Woude 1998). This likely reflects its translational activation.

*c-*mos *mRNA*

The c-*mos* proto-oncogene encodes a protein kinase that is critical in the control of vertebrate meiosis and the early embryonic cell cycle (for review, see Yew et al. 1993; Vande Woude 1994; Gebauer and Richter 1997; Sagata 1997). Consistent with these roles, c-*mos* mRNA is normally found only in the germ line. In frog oocytes, removal of c-*mos* mRNA prevents maturation, whereas its overexpression induces it (Sagata et al. 1988, 1990). Female mice lacking a functional c-*mos* gene display reduced fertility, as well as ovarian cysts and teratomas, consistent with a crucial role in oocyte growth (Colledge et al. 1994; Hashimoto et al. 1994).

In frogs, translation of c-*mos* mRNA apparently increases during oocyte maturation (Sagata et al. 1988). Fox et al. (1989) noted, by sequence inspection, that *Xenopus* c-*mos* mRNA contained signals that could cause cytoplasmic polyadenylation, and proposed that cytoplasmic polyadenylation of c-*mos* mRNA therefore might be a critical control point in meiotic maturation. This hypothesis has since gained substantial support. c-*mos* mRNA receives poly(A) during maturation. Furthermore, the c-*mos* 3′UTR contains signals sufficient for cytoplasmic polyadenylation (Paris and Richter 1990; Sheets et al. 1994), and when linked to a reporter, stimulates translation during maturation (Sheets et al. 1994). Removal of cytoplasmic polyadenylation signals from endogenous c-*mos* mRNA, achieved by targeted RNase H cleavage, prevents maturation

(Sheets et al. 1995). The amputated mRNA, lacking its polyadenylation signals, is stable. Maturation, and the increase in c-*mos* protein levels, can be restored by injection of synthetic c-*mos* mRNA carrying polyadenylation signals, or of a "prosthetic RNA" that brings polyadenylation signals to the amputated endogenous mRNA by base-pairing (Sheets et al. 1995). These experiments strongly argue that polyadenylation, or the presence of a poly(A) tail, is critical in the activation of c-*mos* mRNA.

These studies do not argue that polyadenylation is the only process triggered by progesterone that is critical for c-*mos* activation. The mere presence of a long poly(A) tail, provided by a prosthetic RNA, is sufficient to activate amputated c-*mos* mRNA after addition of progesterone. This rescue by poly(A) is length-dependent: 130 adenosines rescue, whereas 30 do not, corresponding reasonably well with the lengths of poly(A) on c-*mos* mRNA before and after maturation (Barkoff et al. 1998). However, in the absence of progesterone, the presence of a long poly(A) tail does not elevate c-*mos* protein levels, demonstrating that a long tail alone is insufficient to activate. c-*mos* protein levels are controlled not only by changes in translation of c-*mos* mRNA, but also by regulated proteolysis, as is the case with certain cyclins (Nishizawa et al. 1993).

Cytoplasmic polyadenylation of c-*mos* mRNA is also required for the maturation of mouse oocytes (Gebauer et al. 1994). In mouse oocytes, removal of the polyadenylation signals from c-*mos* mRNA does not block completion of first meiosis as in frogs. Rather, these oocytes complete the first meiotic division but fail to progress normally to meiosis II. This phenotype resembles that observed in oocytes derived from females homozygous for a disrupted c-*mos* gene, which undergo parthenogenetic activation after completing first meiosis (Colledge et al. 1994; Hashimoto et al. 1994).

Recent results suggest that cytoplasmic polyadenylation elements (CPEs) are bifunctional, first repressing translation prior to maturation, and later activating. The requirement for polyadenylation may sometimes be simply to prevent removal of the tail due to cytoplasmic deadenylation: For example, it appears that tPA mRNA needs a short poly(A) tail, rather than poly(A) extension per se, to be activated during maturation (Stutz et al. 1998).

In addition to c-*mos*, translational control of at least one other mRNA is likely to be critical in activating maturation in response to progesterone (Nebreda et al. 1995; Barkoff et al. 1998; Frank-Vaillant et al. 1999). Indeed, the translation of cyclin B1 (Frank-Vaillant et al. 1999) and Ringo/Speedy (Ferby et al. 1999; Lenormand et al. 1999) proteins may be critical in inducing maturation.

Perspective

The idiosyncrasies of cell cycle control in the early embryo vary widely among species. c-*mos*, for example, appears to be a vertebrate adaptation (Yew et al. 1993; Gebauer and Richter 1997; Sagata 1997). It is unclear whether there is a widespread and conserved strategy of translational control of a cell cycle component—for example, a common regulator and mRNA target among many species. The apparent conservation of DAZ function in regulating meiosis in both vertebrates and invertebrates suggests this may be such a case, as may control of certain of the cyclins (see above). Regardless, it is clear that many species exploit translational control of specific cell-cycle related mRNAs to help thrust the idling egg through the completion of meiosis and the onset of mitotic cleavage.

Temporal Control of Developmental Events: RNA Regulators

Translational controls are not restricted to maternal mRNAs and early embryos. Indeed, a particularly provocative form of translational control directs progression through the life cycle in the somatic tissues of the nematode *C. elegans*. Normally, *C. elegans* passes through four distinct larval stages, called L1, L2, L3, and L4, to reach maturity as adults (Fig. 1). This progression depends on several "heterochronic" genes, including *lin-14* and *lin-41* (for review, see Ambros and Moss 1994). The key regulators of these two mRNAs appear to be short, repressive RNAs.

lin-14 is required for L1-specific events (Ambros and Horvitz 1984). LIN-14 protein is abundant at the L1 stage, but rare at later stages (Ruvkun and Giusto 1989); in contrast, *lin-14* mRNA is equally abundant throughout larval development (Wightman et al. 1993). Two *lin-14(gf)* mutants, which disrupt the 3′UTR and cause *lin-14* protein levels to remain high throughout larval development (Ambros and Horvitz 1984; Ruvkun et al. 1989; Ruvkun and Giusto 1989; Wightman et al. 1991), reiterate patterns of cell lineage and cell fate normally associated with the L1 larval stage. Temporal repression of *lin-14* at the L1 and later stages requires sequences in its 3′UTR, as well as the *lin-4* gene product (Ambros 1989; Arasu et al. 1991; Wightman et al. 1991, 1993; Lee et al. 1993). Animals lacking *lin-4* activity reiterate L1-specific events (Chalfie et al. 1981), as do *lin-14(gf)* mutants. Remarkably, *lin-4* encodes two short RNAs (22 and 61 nucleotides) with no apparent protein-coding capacity. Instead, both RNAs are complementary to each of seven conserved elements present in *lin-14* mRNA, prompting the proposal that *lin-4*/ *lin-14* RNA duplexes cause translational repression (Fig. 7A) (Lee et al. 1993; Wightman et al. 1993). Indeed, the regions of complementarity are required for repression in vivo, and for base-pairing between the RNAs in vitro (Fig. 7B) (Ha et al. 1996). *lin-28*, another gene that regulates timing of early developmental decisions, is also controlled by *lin-4* and contains only a single sequence complementary to *lin-4* in its 3′UTR that does not form the bulged duplex (Moss et al. 1997).

A later temporal transition in cell fates, from L4 to adult, requires another set of heterochronic genes, including *lin-41* and *let-7*. *lin-41* encodes a RING finger protein of the RBCC subfamily (Slack et al. 2000). Lack of *lin-41* leads to precocious adult fates at the L4 stage without affecting the L1 to L2 transition (Abrahante et al. 1998). *let-7* mutants exhibit a reciprocal phenotype, reiterating L4-stage events in adults. Furthermore, increased *let-7* dosage causes precocious expression of adult events in L4-stage animals (Reinhart et al. 2000). The 3′UTR of *lin-41* causes repression of a reporter gene at the L4/adult transition. These data suggest that *let-7* represses *lin-41* via its 3′UTR (Fig. 7C).

The molecular identity of *let-7* reveals startling parallels with *lin-4*. *let-7* encodes a 21-nucleotide RNA without an open reading frame that is complementary to two segments of the *lin-41* 3′UTR. The structures of the two potential *let-7*/*lin-41* duplexes are similar (Fig. 7D). Although base-pairing has not been demonstrated directly, the complementary sites in the *lin-41* 3′UTR greatly enhance repression of a transgene at the adult stage, and this repression requires *let-7* (Reinhart et al. 2000).

In the simplest view, the early and late developmental transitions are triggered just by the expression of the regulatory RNAs (Fig. 7B). *lin-4* RNA increases in abundance early, as *lin-14* and *lin-28* are repressed (Feinbaum and Ambros 1999). Similarly, abundant *let-7* RNA is first detected at the L4 stage, as *lin-41* is extinguished (Reinhart et al. 2000).

Although regulatory RNAs are critical here, they may not be the whole story. The secondary structures of each potential *lin-4*/*lin-14* hybrid, and the sequence of the "looped-out" regions, are quite similar (Fig. 7B). In particular, they include a bulged C residue whose presence and identity are critical for repression, and which may be part of a protein-binding site (Ha et al. 1996). Similarly, the two putative *let-7*/*lin-41* duplexes are closely related, including bulges with similar sequences (Fig. 7D). Thus the *lin-4*/*lin-14* and *let-7*/*lin-41* interactions may create two distinct RNA structures that are specifically discriminated by proteins. Put another way, the short RNAs create new RNA structures in their targets. Perhaps ATP-dependent RNA helicases implicated in translational regulation (e.g., the Mog and Vasa proteins) act similarly, creating new binding sites for repressor proteins.

The biochemical mechanism by which base-pairing leads to repression is unknown, but appears not to be "simple" interference with initiation: Neither the rate of synthesis of *lin-14* mRNA, its state of polyadenylation, its apparent abundance in the cytoplasm, nor its distribution in a polysome profile changes in response to the accumulation of *lin-4* RNA. These findings suggest that association of *lin-4* RNA with the 3′UTR of *lin-14* mRNA inhibits a step(s) after initiation, such as translational elongation and/or the release of stable LIN-14 protein (Olsen and Ambros 1999).

The identification of the *lin-4* and *let-7* repressors is unambiguous and emphasizes the importance of considering RNA in searching for

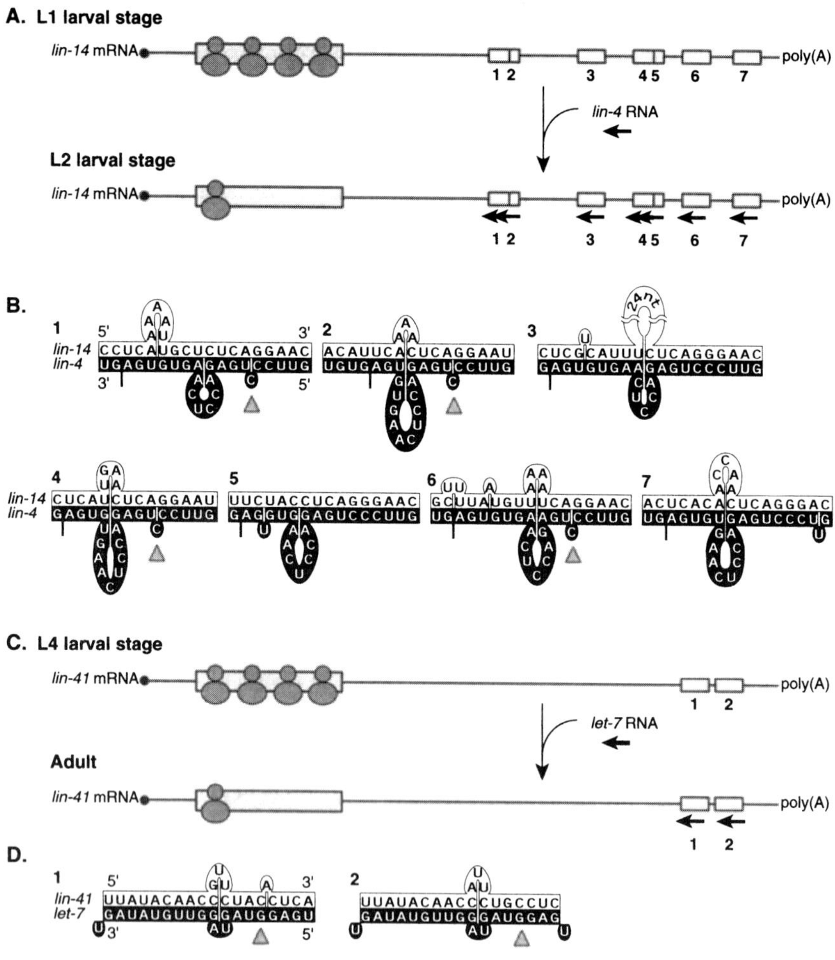

Figure 7 (*See facing page for legend.*)

activities or genes that repress. *lin-4* and *let-7*, and their apparent noninitiation mode of repression, are not likely to be mere deviants, but rather harbingers of other regulatory RNAs and widely used mechanisms (Wickens and Takayama 1994).

Dosage Compensation in *Drosophila*

Dosage compensation balances the transcriptional output of the two female and the single male X chromosomes. In mammals, inactivation of one of the two female X chromosomes implements dosage compensation by adjusting the transcriptional output to that of the male. In *Drosophila*, the transcriptional output from the single male X chromosome is approximately doubled, thus allowing an equal level of expression of X-linked genes in males and females (Baker et al. 1994; Kelley and Kuroda 1995).

The major dosage compensation pathway in *Drosophila* is controlled by a heteromeric complex consisting of the four proteins Maleless (Mle) and Male-Specific Lethal (MSL)-1, -2, and -3. The MSL complex associates with numerous sites along the male X chromosome and probably stimulates transcription by promoting histone acetylation. Although three of the four subunits are expressed in both sexes, MSL complex formation

Figure 7 Repressive RNAs in *C. elegans*: *lin-4* and *let-7*. (*A*) Model for the role of *lin-4* in translational repression of *lin-14* mRNA. (*Shaded circles*) Ribosomes; (*thin lines*) *lin-14* mRNA; (*small open rectangles*) putative regulatory sites to which *lin-4* RNA may bind; (*thick black arrow*) *lin-4* RNA (arrowhead is at the 3′ end of the short [21 nucleotides] *lin-4* RNA). The *lin-14* 3′UTR possesses 7 conserved elements (1–7) that are likely to be translational regulatory elements. During the L1 larval stage, *lin-14* is translated; then the translational repressor, *lin-4*, associates with regulatory elements and *lin-14* becomes translationally repressed. mRNA not drawn to scale. (*B*) Potential hybrids between *lin-14* mRNA and *lin-4* RNA. (*Open rectangles*) Elements in *lin-14* mRNA; (*black rectangles*) *lin-4* RNA. The location of a point mutation in *lin-4* that reduces its activity is indicated by a triangle in hybrids 1, 2, 4, and 6. In addition to the 21-nucleotide *lin-4* RNA, a longer (~60 nucleotides) RNA also is present, with additional sequence beyond the 3′ end of the 21-nucleotide RNA. The 3′ of the short (21 nucleotides) RNA is indicated by a vertical line; any additional complementarity in the longer *lin-4* RNA to the *lin-14* sites is indicated. Note that only a subset of these structures may be needed for repression. (*C*) Model for the role of *let-7* RNA in translational repression of *lin-41* mRNA. See *A* for key. (*D*) Potential hybrids between *lin-41* and *let-7* RNA. The location of a triangle indicates the position of a point mutation that reduces its activity.

is restricted to males by the male-specific expression of the MSL-2 protein. Experimental expression of MSL-2 in females triggers MSL complex assembly, showing that MSL-2 is the limiting subunit (Bashaw and Baker 1995; Kelley et al. 1995).

Expression of MSL-2 is under negative control. In the female, it is inhibited by the female-specific RNA-binding protein Sex-Lethal (SXL). SXL expression is limited to female flies by a combination of transcriptional control and autoregulated splicing (for review, see Gebauer et al. 1997). The SXL protein is composed of two ribonucleoprotein consensus motifs (RRMs) and a glycine/asparagine-rich amino terminus. It binds long oligouridine stretches for high-affinity binding. Affinity may also be modulated by flanking RNA sequences and possible associations with other factors. SXL has been shown to function as a female-specific regulator of splicing that controls the expression of the transformer (*tra*) and its own mRNA.

How does SXL inhibit MSL-2 expression? *msl-2* pre-mRNA harbors two consensus high-affinity SXL-binding sites in its 5′UTR and four in its 3′UTR (egg-shaped symbols in Fig. 8). Interestingly, the two sites in the 5′UTR are both located within an intron that is spliced in a sex-specific fashion. The intron is removed in males but retained in females, due to SXL's effects on splicing (for review, see Gebauer et al. 1997). *msl-2* mRNA is efficiently exported into the cytoplasm in both sexes, but MSL-2 protein is only expressed in males. The retained intron cannot suppress MSL-2 expression per se, because constructs in which the intron is retained due to splice-site mutations are expressed in transfected cells and in transgenic flies if SXL is absent. Several lines of evidence show that SXL acts as a translational repressor in the cytoplasm and that both the 5′ and the 3′UTR-binding sites are important for this (Bashaw and Baker 1997; Kelley et al. 1997; Gebauer et al. 1998). First, reporter constructs bearing the SXL-binding sites only in either the 5′UTR or 3′UTR are not efficiently repressed by SXL in transfected cell lines and transgenic flies. Second, SXL-mediated inhibition of *msl-2* expression does not affect the cytoplasmic levels of *msl-2* mRNA. Third, the regulation of *msl-2* mRNA translation by SXL has recently been recapitulated in a cell-free system from *Drosophila* embryos with recombinant SXL protein and in-vitro-transcribed reporter mRNAs bearing both untranslated regions of *msl-2* mRNA (Gebauer et al. 1999). Mutation of sites in either UTR drastically reduces repression, indicating that the two regions act synergistically (Bashaw and Baker 1997; Kelley et al. 1997). Of the two sites in the 5′UTR, the downstream site is more important, at least in vitro (Gebauer et al. 1999). This indicates that, unlike the IRE/IRP system, cap-proxim-

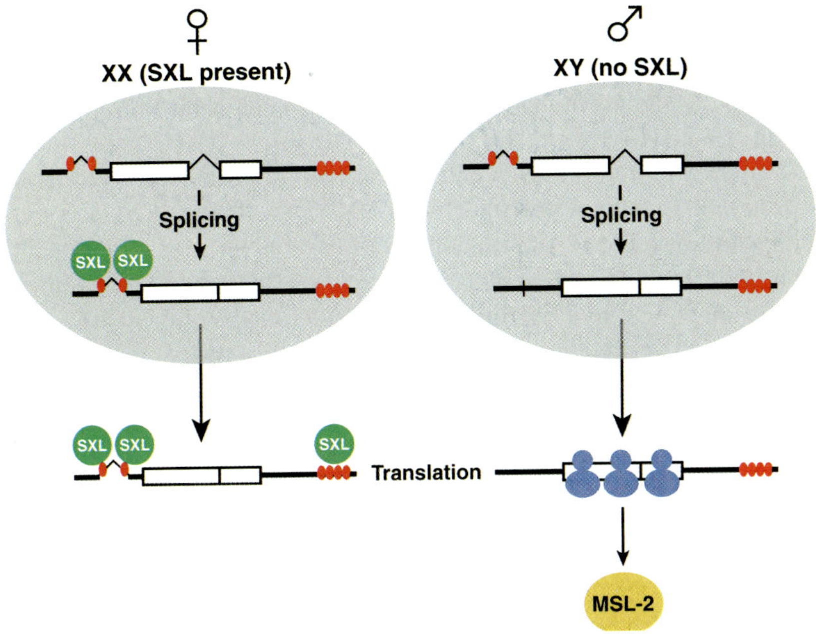

Figure 8 Regulation of *msl-2* expression by sex-lethal (SXL). The nuclear *msl-2* pre-mRNA is depicted with two introns and bearing six SXL binding sites (*red*). SXL is only expressed in female (XX) embryos, where binding to two sites in the intron within the 5′UTR inhibits splicing, enforcing intron retention. Following export into the cytoplasm, SXL binding to the sites within both UTRs represses translation. In male embryos (XY), the MSL-2 protein is expressed following unimpeded splicing and translation.

ity is not important. Numerous other mRNAs from *Drosophila* bearing 3′UTR-binding sites for SXL have been identified (Kelley et al. 1995). Their biological functions remain to be clarified.

In summary, SXL inhibits *msl-2* expression in *Drosophila melanogaster* by an integrated two-step mechanism that involves splicing and translation. Interestingly, the latter but not the former is conserved in evolution: In *Drosophila virilis*, the splice sites are not maintained so that this related organism apparently relies entirely on translational regulation to achieve dosage compensation by the MSL complex.

Mesoderm Specification in *Xenopus*

In frogs, mesoderm arises through a process termed "induction," in which a signal is secreted from endodermal cells at the bottom of the embryo to overlying cells, causing those cells to follow mesodermal fates (for review, see Melton 1994). Members of the fibroblast growth factor (FGF) and transforming growth factor β (TGF-β) families of secreted polypeptides are likely signals in this process, as are the cell-surface receptors to which they bind (for review, see Melton 1994).

Two forms of translational control have been implicated in mesoderm induction. The first involves a maternal mRNA encoding an FGF receptor, FGFR-1 (Robbie et al. 1995). Expression in embryos of a dominant inhibitory form of the FGF receptor interferes with mesoderm induction in vivo, presumably by titrating wild-type receptors into inactive complexes (Amaya et al. 1991, 1993). These and other results strongly suggest that the FGF receptor and its ligand play a key role in mesoderm induction (for review, see Melton 1994). FGFR-1 mRNA is silent in oocytes, but activated during oocyte maturation, prior to fertilization (Musci et al. 1990). The repression is due to a negative regulatory element in the 3′UTR of FGFR-1 mRNA, in the 180 nucleotides immediately downstream from the termination codon. The temporal or spatial control of its de-repression may be important in embryonic induction, although the existence of multiple receptors for FGF-related ligands may complicate the issue.

A second speculative role for translational control in mesoderm induction may be that increased activity of eIF4E in the embryo specifically stimulates the translation of activin, a member of the TGF-β superfamily and a potent inducer of mesoderm (for review, see Melton 1994). This idea is based, in part, on the finding that overexpression of the general translation factor eIF4E in frog embryos induces mesodermal fates in cells that would otherwise form ectoderm (Klein and Melton 1994). Moreover, eIF4E overexpression specifically stimulates translation of injected activin mRNA without affecting either total protein synthesis or other injected mRNAs (Klein and Melton 1994).

Activin and eIF4E may comprise a positive feedback loop. Mesoderm induction by eIF4E is blocked by coexpression of a dominant inhibitory form of the activin receptor (Klein and Melton 1994). Since mRNA injection experiments imply that activin translation may be stimulated by eIF4E, these data suggest a simple autocrine loop: Activin elevates eIF4E levels, which further enhances activin synthesis. A circuit of this type could both amplify the initial inducing signal and explain how one cell that has been induced to form mesoderm can induce mesoderm in an adjacent cell. This model predicts that the level of eIF4E activity is elevated during early development, at least in certain blastomeres, and that specific mRNAs involved in mesoderm induction should be stimulated as a result. Those mRNAs might encode activin or other mesoderm inducers.

Although the activin/eIF4E circuit is speculative, it closely parallels an apparent mechanism of neoplastic transformation of mammalian cells by overexpression of eIF4E (Lazaris-Karatzas et al. 1990; Chapter 6). In that case, as in mesoderm induction, elevation of the levels of a general translation factor has dramatic effects on cell fate.

Terminal Differentiation

Certain genes are expressed late in differentiation, as cells take on their ultimate fates. In the examples of terminal differentiation described below —late spermatogenesis and red blood cell differentiation—the nucleus is effectively silenced: The spermatid pronucleus is highly condensed and inactive, and in mammals, red blood cells lose their nucleus entirely. In these cases, as in the early embryo, the cell must exploit translational control to change the proteins it contains.

Mammalian Spermatogenesis: Protamine mRNAs and DAZ Proteins

Spermatogenesis is a highly conserved process that involves both cell division and cell differentiation. The germ-cell population first expands through mitosis, generating "spermatocytes" that enter meiosis. The haploid products of meiosis ("round spermatids") then differentiate into "elongating spermatids" and "spermatozoa." The entire process takes approximately 3 weeks and occurs throughout adult life.

Regulation of mRNAs appears to play a major role during spermatogenesis. Multiple mRNAs are regulated (for review, see Hecht 1998). Here we focus on two intensely studied examples: protamine mRNAs and their regulators, and the DAZ family of proteins and their likely targets.

Translational Regulation of Protamine Expression. During the terminal stages of spermatogenesis, chromosomes are repackaged with protamines rather than histones to facilitate chromosome condensation. Protamine mRNAs (*Prm-1* and *Prm-2*) that had previously been silent become active. Protamine mRNAs are synthesized in round spermatids, are stored as cytoplasmic ribonucleoprotein (RNP) particles for up to a week, and finally translated in elongated spermatids. Repression of *Prm-1* is imposed by a 3′UTR-mediated mechanism and is essential for normal spermatid differentiation: premature translation of *Prm-1* leads to precocious nuclear condensation and sterility (Braun et al. 1989; Lee et al. 1995).

Several 3′UTR sequences have been implicated in *Prm* mRNA translational control, suggesting redundancy. Sequences at the 5′ and 3′ ends of the *Prm-1* 3′UTR are sufficient to confer *Prm-1*-like translational regulation on a reporter transgene in mice (Fajardo et al. 1997). A protein, called Prbp, binds the 3′ sequence and is present in the cytoplasm of round spermatids, but not in elongated spermatids (Lee et al. 1996). However, Prbp-deficient mice do not prematurely express *Prm-1*, *Prm-2*, or a transgene carrying the 3′ end of the *Prm-1* 3′UTR (Zhong et al. 1999); rather, the activation of these mRNAs in elongated spermatids is defective (Zhong et al. 1999). This suggests a role for Prbp in activation, not repression (Zhong et al. 1999).

In addition, both the *Prm-1* and *Prm-2* 3′UTRs contain two conserved regions, called Y and H boxes. The Y box cross-links to an 18-kD protein present in male germ cells and in testicular RNP particles (Kwon and Hecht 1991). An extract enriched for the 18-kD protein represses translation of reporter RNAs containing the Y and H boxes in vitro (Kwon and Hecht 1993). Interestingly, the 18-kD protein present in round and elongating spermatids binds RNA, whereas the protein found in elongated spermatids does not. Phosphorylation may control the RNA-binding activity, and hence, translation (Kwon and Hecht 1993). The ability of the Y and H boxes to mediate repression in vivo has not been examined.

Efforts to identify proteins responsible for targeting *Prm-1* and related mRNAs to mRNP particles have identified several spermatid mRNP-associated proteins, including poly(A)-binding protein (Gu et al. 1995), spermatid perinuclear RNA-binding protein (Spnr; Schumacher et al. 1995b), testis nuclear RNA-binding protein (Tenr; Schumacher et al. 1995a), and the Y-box proteins (Tafuri et al. 1993). Although their functions are unclear, Y-box proteins nonspecifically bind RNA and may play an important role in forming repressive mRNP particles.

DAZ Proteins and the Regulation of Meiosis. Three regions on the human Y chromosome, called AZFa, AZFb, and AZFc (*Az*oospermia *F*actor) are required for proper spermatogenesis (for review, see Elliot and Cooke 1997). Candidate spermatogenesis genes that encode RNA-binding proteins have been identified in AZFb and AZFc. Deletion of AZFc removes a small family of genes named *DAZ* (*D*eleted in *Az*oospermia; Reijo et al. 1995). Deletion of AZFb region removes another gene family called *RBM* (*R*ibosomal *B*inding *M*otif; Ma et al. 1993; Elliot et al. 1997). Although good correlative evidence suggests that the *DAZ* and *RBM* families are involved in spermatogenesis in humans, mutations that cause spermatogenic defects by affecting only one *DAZ* or *RBM* gene have not been reported.

In mice and flies, genetic evidence demonstrates that *DAZ* family members are required for spermatogenesis. Both mice and flies each have an autosomal *DAZ*-related gene, called Dazla and *boule*, respectively.

Disruption of the mouse Dazla gene results in infertility in both sexes, due to a reduction in the number of germ cells. Thus, Dazla is necessary for development and survival of germ cells in both the ovary and testis (Ruggiu et al. 1997). Loss of *boule* in flies also results in male-specfic infertility, in which spermatogenesis arrests at the G_2/M transition of meiosis I (Castrillon et al. 1993; Eberhart et al. 1996).

twine mRNA is a likely target of Boule protein during male meiosis. *twine* encodes a meiotic, *cdc25*-like phosphatase. *twine* mRNA, but not protein, is present in premeiotic cells, suggesting that *twine* mRNA is repressed at this stage (Alphey et al. 1992; Courtot et al. 1992; White-Cooper et al. 1998). Several lines of evidence suggest that Boule is needed to activate *twine* translation. Spermatocytes in *twine* mutants fail at the G_2/M transition, as do *boule* mutants, and *boule* acts genetically before *twine* in spermatogenesis. Moreover, *boule* is required for translation of a *twine–lacZ* reporter construct (Maines and Wasserman 1999), although a direct interaction between Boule and *twine* mRNA has not been reported. Since *boule* and *twine* have different phenotypes, *boule* probably has other targets (Eberhart et al. 1996).

The function of DAZ family proteins may be conserved. Defects in family members in man, mice, frogs, and flies give similar phenotypes. Moreover, DAZ proteins can function across species: *Xenopus* Xdazl rescues the meiotic defect of *boule* mutant flies (Houston et al. 1998), and human DAZ partially rescues the spermatogenic defect of *Dazl* mutant mice (Slee et al. 1999). Given the similarities in sequence, function, and expression patterns, it seems likely that these proteins commonly control spermatogenesis by regulating translation of specific mRNAs: to date, *Drosophila twine* is the only target mRNA identified.

Red Blood Cell Differentiation: 15-Lipoxygenase mRNA

As mammalian reticulocytes differentiate into erythrocytes, their mitochrondria are destroyed. The enzyme 15-lipoxygenase (LOX) catalyzes deoxygenation of polyenoic fatty acids, even in intact membranes, and is thought to be critical for the destruction of internal membranes and mitochondria (Rapoport and Schewe 1986). Although LOX mRNA apparently is present even at early stages of erythropoiesis, it is not translated until reticulocytes mature into erythrocytes (Thiele et al. 1982). This translational silencing is critical in early erythroid precursor cells and young reticulocytes, which require intact mitochondria for their metabolism.

The 3′UTR of rabbit LOX mRNA contains ten nearly perfect repeats of a 19-nucleotide sequence, whereas the mouse mRNA contains four similar repeats in a comparable location (Hunt 1989; Ostareck-Lederer et al. 1994). These repeats, called differentiation control elements (DICE), mediate translation repression, as demonstrated in vitro (Ostareck-Lederer et al. 1994). Two proteins, hnRNP K and hnRNP E1, can interact with each other, bind to the DICE, and silence LOX mRNA translation both in vitro and in transfected HeLa cells. Importantly, silenced LOX mRNA in early erythroid cells is associated with hnRNP K (Ostareck et al. 1997). Interestingly, repression in vitro appears to be independent of any change in poly(A) length and of the 5′ terminal cap, points to which we later return. Although rabbit LOX mRNA contains ten tandem repeats, two are sufficient for repression (Ostareck et al. 1997).

Masking and CPEB

Masking

The "masking" hypothesis, initially proposed by Spirin more than 30 years ago (Spirin 1966), suggests that specific mRNAs are repressed through the action of proteins that hide them from the translational apparatus. In response to a stimulus, such as fertilization, the masking proteins are removed, the mRNA is revealed, and its translation begins. In its initial formulation, masking was proposed to explain the dramatic increase in protein synthesis observed in sea urchin eggs at fertilization. Classically, masking is defined operationally, using extracts derived from eggs and early embryos. In vivo, a specific mRNA is repressed in the egg but becomes active at fertilization. The patterns of protein synthesis are maintained in extracts of eggs and early embryos; in particular, mRNAs that are repressed in vivo continue to be repressed when translated in vitro, provided they are presented as mRNPs (i.e., with proteins still attached). Removal of the proteins from the mRNPs activates (i.e., "unmasks") the mRNA in vitro. Protein removal can be accomplished crudely, for example, by extraction with organic solvents, or by more subtle means, as described below (for review, see Standart 1992; Standart and Jackson 1994).

Thus, masking is followed by activation, or unmasking. Only some of the mRNAs we have discussed in previous sections—*bicoid*, for example—appear to behave in this way. In contrast, frog cyclin B1 mRNA is already expressed at a low level before oocyte maturation begins, and so, at the least, may not be fully masked; *lin-14* and *lin-41* mRNAs are initially translationally active, and then are shut off, the opposite of the situation in classic masking. It is uncertain whether these different mRNAs are repressed by the masking mechanism used to silence mRNAs from their birth. However, masking is repression, and the differences are likely semantic and historical, not biological.

Clam ribonucleotide reductase mRNA provides a well-studied paradigm for masking (Standart et al. 1990). Unmasking of this mRNA in an extract of surf clam oocytes can be achieved by incubation in 0.5 M KCl and gel filtration, which presumably removes the masking factor. Masking can be restored in the extract by removal of the salt prior to gel filtration, which presumably permits the factor to rebind. Remasking in this fashion requires sequences in the 3′UTR (Standart et al. 1990). Masked ribonucleotide reductase mRNA can be derepressed in oocyte extracts by severing the 3′UTR from the body of the mRNA, using targeted RNase H-cleavage (Standart et al. 1990). The activation appears to be independent of polyadenylation, even though the mRNA receives poly(A) as it is activated in vivo. These data imply that removal of 3′UTR-bound factors is sufficient for derepression, and that derepression in vitro can be uncoupled from poly(A) addition.

CPEs and CPEBs: Going Both Ways

Sequences that control cytoplasmic polyadenylation (CPEs) are located in the 3′UTR; the sequence AAUAAA, located nearby, is also required for the reaction. Most commonly, CPEs have been identified as positive control elements required for polyadenylation and translational activation; injected, mutant mRNAs lacking them are not activated, nor do they receive poly(A) (see Chapter 27).

However, CPEs can also repress, and may mediate masking. This conclusion first emerged in studies of mouse tPA mRNA, in which elements that repress and cause poly(A) removal prior to oocyte maturation overlap with those that activate and cause poly(A) addition once maturation has begun ("ACE" elements; Sallés et al. 1992). More recently, those sequences provided in excess in *trans* have been shown to cause derepression of endogenous mRNAs, presumably by titrating a repressor (Stutz et al. 1998). CPEs of other mRNAs also can mediate repression before being involved in activation (de Moor and Richter 1999; Minshall et al. 1999; Ralle et al. 1999; Barkoff et al. 2000). The duality of CPEs is not invariant, however, as some 3′UTRs that direct polyadenylation do not repress (Barkoff et al. 2000).

The duality of CPEs complicates predictions of the phenotypes of CPE mutations in endogenous genes. For example, suppose that a single CPE first is required to repress an mRNA, and then later to activate it.

Deletion of the CPE in the endogenous gene would result in premature activation of the mRNA, with no subsequent increase. Thus, such CPEs would appear genetically as negative translational control elements, not as positive-acting signals. This raises the possibility that negative control elements described in a variety of systems might also have later, positive-acting functions.

An RNA-binding protein of the RRM family, CPEB, binds to CPEs and is required for cytoplasmic polyadenylation and translational activation of dormant mRNAs (Hake and Richter 1994; Chapter 27). Consistent with this view, mutants lacking a *Drosophila* CPEB homolog, *orb*, fail to activate *oskar* mRNA (Chang et al. 1999), and *Xenopus* oocytes injected with anti-CPEB antibodies fail to activate or polyadenylate c-*mos* mRNA (Stebbins-Boaz et al. 1996). However, CPEB homologs can also cause repression. An 82-kD protein that binds to repressive elements in clam ribonucleotide reductase and cyclin A mRNAs is a CPEB homolog (Minshall et al. 1999; Walker et al. 1999). The duality of this protein's function echoes that of CPEs.

Molecular mechanisms have been proposed for both the repressive and activating activities of CPEBs. CPEB's repressive role involves a second protein, maskin. *Xenopus* CPEB interacts with maskin, which in turn binds the initiation factor, eIF4E: The three proteins are found in a complex in resting oocytes (Stebbins-Boaz et al. 1999). This interaction may preclude binding of eIF4E with eIF4G and thereby cause repression prior to oocyte maturation. In this model, activation is achieved by disrupting the maskin/eIF4E interaction (Stebbins-Boaz et al. 1999; Chapter 27).

The positive-acting properties of CPEB invoke its facilitation of cytoplasmic polyadenylation, ultimately by recruitment of a cytoplasmic poly(A) polymerase (PAP) (Ballantyne et al. 1995; Gebauer and Richter 1995). This event likely requires binding of a cytoplasmic form of cleavage and polyadenylation specificity factor (CPSF) to the AAUAAA sequence of the mRNA, which in turn binds PAP (Bilger et al. 1994; Dickson et al. 1999). CPSF's binding preference for CPE-containing RNAs could facilitate such events (Bilger et al. 1994).

Factors other than, or in addition to, canonical CPEB may also be involved in CPE-mediated events. Two new proteins apparently interact with the CPEs of *Xenopus* lamin mRNA (Ralle et al. 1999); the CPEs of mouse tPA mRNA appear to bind non-CPEB factors as well (Stutz et al. 1998). Moreover, in some organisms, multiple CPEB homologs may have distinct activities. Indeed, the *C. elegans* and zebrafish genomes encode multiple CPEB-related proteins.

MECHANISMS

The examples above illustrate the broad biological range of translational regulation of developmental decisions. Interactions between specific mRNA regulatory sequences and single or multiple proteins are established or broken in response to regulatory cues and control the translation of the respective mRNAs. In this section, we discuss how these events alter translational activity. Most of the examples we discuss hinge on the 3′UTR; nevertheless, we begin by discussing one example of regulation through the 5′UTR. We do so because the relatively detailed mechanistic information sets precedent for how mRNAs can be shut off and activated, and how one regulator can control multiple mRNAs.

In principle, translation can be affected at the levels of initiation, elongation, or termination. Most examples that have been investigated appear to be regulated at the level of initiation (see below), although the number examined in detail is small. In at least one case described (*lin-4* regulation of *lin-14*), regulation occurs after initiation.

Two central questions arise. First, how is translational repression exerted? Second, for those mRNAs that first are repressed and later activated, how is derepression accomplished?

Regulation Via the 5′UTR

Perhaps the most intensively characterized example of translational control via a 5′UTR element is that of ferritin mRNA regulation by iron via iron-responsive elements (IRE) and iron regulatory proteins (IRP) (Chapter 21). We discuss this below, emphasizing that the binding of regulatory proteins to 5′UTR sites can act by two distinct mechanisms—inhibition of 43S recruitment or interference with 43S scanning.

IREs have been identified in the 5′UTR of several different mRNAs that encode proteins involved in iron metabolism (Hentze and Kühn 1996). The IREs are usually located within 40 nucleotides of the cap structure, a feature that is functionally important: Cap-mediated recruitment of the 43S translation preinitiation complex to the mRNA occurs within this region, and is blocked by IRP binding to a cap-proximal IRE (Gray and Hentze 1994). IRP-binding to an IRE still permits assembly of eIF4F on the cap structure in vitro, but the joining of this complex and the small ribosomal subunit is inhibited (Fig. 9) (Muckenthaler et al. 1998). Translation also can be inhibited sterically by high-affinity RNA/protein complexes. Replacement of the IRE by binding sites for RNA-binding proteins that do not play physiological roles in controlling eukaryotic

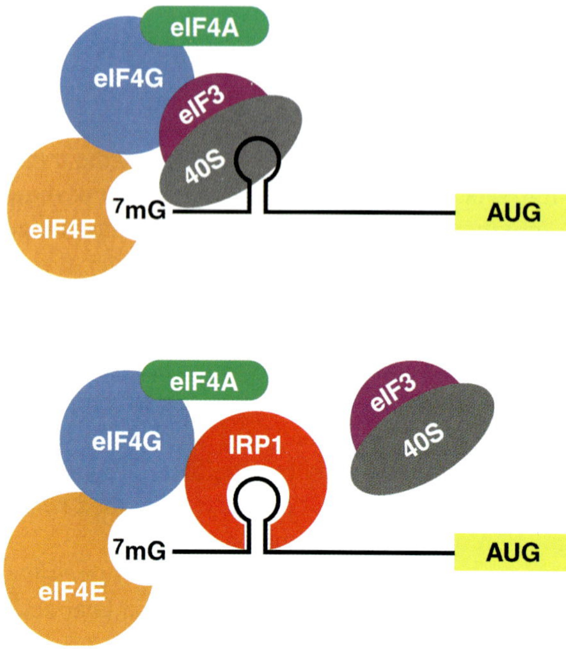

Figure 9 Translational regulation by the IRE/IRP system. The 5′UTR of ferritin mRNA bearing a cap-proximal iron-responsive element (stem-loop structure in black) is depicted. (*Upper panel*) Assembly of a 43S translation initiation complex. (*Lower panel*) Binding of IRP1 (*red*) to the IRE blocks the recruitment of the 40S ribosomal subunit with its associated translation initiation factors to the preassembled cap-binding complex eIF4F.

translation (the spliceosomal protein U1A or the bacteriophage MS2 coat protein) allows specific translational repression by the respective proteins in vitro and in vivo (Stripecke et al. 1994). Other *cis*-acting elements for translational repression that are found within the first 40–50 nucleotides of an mRNA may operate through a similar block of 43S preinitiation complex recruitment.

IRE/IRP complexes in an appropriate position can affect scanning. IRP binding to an IRE cloned farther downstream in the 5′UTR of a reporter mRNA fails to inhibit the recruitment of 43S preinitiation complexes, as expected. Such downstream IRE/IRP complexes still cause some degree of translational inhibition in transfected cells (Goossen and Hentze 1992), although substantially less than cap-proximal IREs. In a cell-free translation system from rabbit reticulocyte lysate this effect was attributed to a kinetic effect on 43S scanning (Paraskeva et al. 1999). While the mammalian initiation machinery is able to eventually overcome cap-distal IRE/IRP complexes, the initiation machinery in wheat germ and yeast translation extracts is not: In these systems, a downstream IRE/IRP complex inhibits efficiently, apparently by stalling the scanning process (Paraskeva et al. 1999). Interestingly, the mRNA encoding one subunit of the *D. melanogaster* succinate dehydrogenase is the only natural example with a cap-distal IRE so far (Kohler et al. 1995; Gray et al. 1996). It should be interesting to explore how the *Drosophila* translation apparatus responds to this IRE/IRP complex.

Links between the 5′ and 3′ Ends

Since many mRNAs are regulated via binding sites in their 3′UTRs or by a combination of 5′UTR and 3′UTR sites, it is important to briefly consider the organization of the two mRNA ends during translation (for a more detailed discussion, see Chapter 10). Although mRNAs are commonly drawn as linear molecules with cap structures on the left and poly(A) tails on the right, cellular mRNAs form local secondary and tertiary structures as well as long-range interactions. Moreover, mRNAs in vivo are not naked nucleic acids, but instead are bound by a multitude of cellular RNA-binding proteins with various specificities and functions. Messenger RNAs hence exist as mRNPs with complex folding patterns, which may or may not juxtapose their two ends.

A wealth of biochemical evidence supports the view that the two ends can be placed in proximity through protein-protein interactions. Poly(A) tail-binding protein (Pab1p/PABP) binds to the amino-terminal region of the translation initiation factor eIF4G, which binds through a neighboring region the cap-binding protein eIF4E. Such interactions have been observed using proteins derived from yeast, plants, and mammals (see Chapter 10). Binding of eIF4E and Pab1p/PABP to eIF4G can occur simultaneously, and hence provides a means to effectively circularize the mRNA (Wells et al. 1998). This end-to-end interaction is likely to be important for translation in cell-free systems (see Chapter 10). In living cells, however, the roles of this complex in translation and its regulation in vivo are unclear, and it is possible that the complex has additional functions.

Juxtaposition of the 5′ and the 3′ end of an mRNA is thought to be important for the synergistic positive effect of the cap structure and the poly(A) tail on translation initiation (Preiss and Hentze 1998; Chapter 10). Regulatory proteins that bind to the untranslated regions and stimulate or inhibit these interactions would be expected to have profound effects on translation (Fig. 10). Furthermore, the effect of these interactions to bring together the 5′ and 3′ ends could also be important for the function of 3′UTR-binding proteins that target a different step in translation initiation but utilize their topographic effects.

Although the interaction between PABP and eIF4G has been demonstrated in cell-free systems, its role in developing gametes and embryos is not clear. In frog oocytes, although the small quantity of PABP apparently is insufficient to occupy the poly(A) tails of all mRNAs (Zelus et al. 1989), endogenous PABP does interact with eIF4G (Keiper and Rhoads 1999). Moreover, the portion of PABP that interacts with eIF4G, bound to a 3′UTR, stimulates translation of that mRNA in these cells (Gray et al. 2000). Cleavage of eIF4G with viral proteases (Keiper and Rhoads 1999) inhibits oocyte maturation and decreases translation of reporter mRNAs.

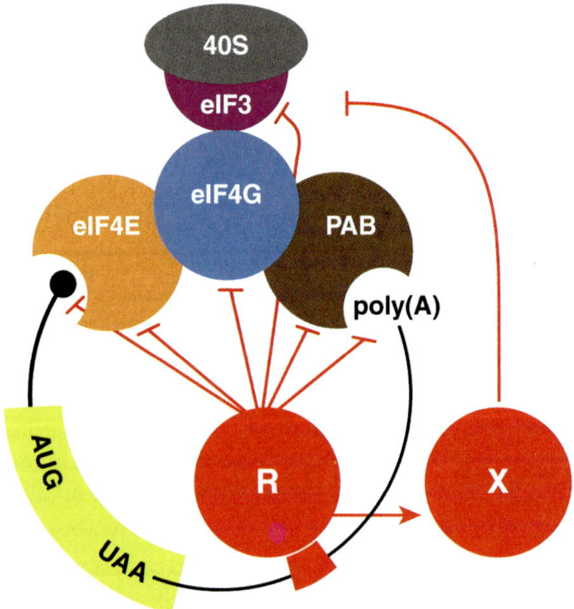

Figure 10 The eIF4E/eIF4G/PABP interaction links the two ends of the mRNA and suggests models for regulation by 3′UTR-bound proteins. Translation initiation factor interactions that contribute to the recruitment of the 40S subunit are depicted. A regulatory element in the 3′UTR (in *red*) is shown to bind a repressor protein (R) and interfere with any of the depicted biochemical interactions, either directly or indirectly by means of a co-repressor (X).

These data argue that the poly(A) tail may mediate its effects in embryos, at least in part, through interactions with PABP and thence eIF4G; however, portions of PABP that do not interact with eIF4G also stimulate translation in oocytes (Gray et al. 2000).

The mammalian protein PAIP-1 (for PABP interacting protein) may also participate in end-to-end communication (Craig et al. 1998). PAIP-1 displays similarity with the central domain of eIF4G and, like eIF4G, interacts with eIF4A. However, it does not appear to interact with eIF4E or eIF3, and hence it is not yet clear how this intriguing player affects translation or effects its control. Another protein with homology to eIF4G is p97/NAT1/DAP-5, which binds eIF4A and eIF3 but does not bind eIF4E and PABP (Imataka et al. 1997; Levy-Strumpf et al. 1997; Yamanaka et al. 1997). Therefore, it is not a prime candidate for being involved in the formation of interactions between the mRNA ends, and its role in translation remains to be more precisely defined.

Role of 5′-end Modifications during Development

Methylation of the 2′ position of the second and third ribose moieties of the mRNA (i.e., 7mGpppG$\underline{m}$G$\underline{m}$) may be linked to polyadenylation and hence to translational control of certain mRNAs. Polyadenylation-dependent ribose methylation has been reported using synthetic B4 mRNA injected into *Xenopus* oocytes (Kuge and Richter 1995). Methylation inhibitors prevent both the modification and translational stimulation (Kuge and Richter 1995), and ribose-methylated mRNAs are translated more efficiently in oocytes (Kuge et al. 1998). However, ribose methylation cannot be the universal cause of the effects of poly(A) on translation in oocytes, since translation of injected reporter RNAs that do not undergo efficient ribose methylation can be dramatically enhanced by polyadenylation (Gillian-Daniel et al. 1998). Nevertheless, a model in which polyadenylation in situ causes cap modification has the merit that it explains repression of mRNAs with respectable tail lengths, simply by their lack of a methyl group prior to polyadenylation.

Deadenylation leads to enzymatic cleavage of the cap structure and hence to mRNA decay in yeast (Chapter 28). A comparable deadenylation-dependent decapping reaction could, in principle, provide a simple mechanism by which poly(A) removal results in translational repression. However, RNAs that are completely deadenylated and repressed retain their caps in a methylated form in *Xenopus* oocytes (Gillian-Daniel et al. 1998).

Mechanisms of Repression Via the 3′UTR

Steric blockage mechanisms are more easily imagined from sites in the 5′UTR than 3′UTR: A priori, one might expect 5′ and 3′UTR-mediated repression mechanisms to differ fundamentally. However, we discuss at least one mechanism that bears strong resemblance with steric repression of translation.

The physical proximity of the 3′UTR and poly(A) tail immediately raises the question of whether translational control is exerted by affecting the length and/or function of the poly(A) tail, or by mechanisms independent of the poly(A). Biology has made use of both possibilities, as discussed below. Clearly, the mechanisms of repression differ among mRNAs and are not mutually exclusive.

Interfering with the Function of the mRNA Ends in 43S Recruitment

The cap structure and the poly(A) tail exert a positive, synergistic effect on translation, involving the eIF4E/eIF4G/PABP interaction. In principle, 3′UTR-binding proteins could regulate translation through interference with this chain of interactions, either by inhibition of eIF4E binding to the cap structure, blocking the eIF4E/eIF4G interaction, the eIF4G/PABP interaction, or the binding of PABP to the poly(A) tail (Fig. 10). The inhibition could be direct, with the repressor touching a translation factor, or could require interaction between the 3′UTR-bound repressor and an intermediary. Furthermore, the function of other translation factors involved in the recruitment of the 43S preinitiation complex could be affected by a 3′UTR-binding protein.

To determine, to a first approximation, whether repression from the 3′UTR requires a cap, one can ask whether it still occurs on an uncapped mRNA or when translation is initiated by a cap-independent, IRES-driven mechanism. In the case of LOX mRNA regulation by hnRNPs K and E1 via a 3′UTR DICE, translational inhibition persists under these conditions, suggesting that the cap structure and eIF4E are not the primary targets (Ostareck et al. 1997).

Analogously, the effect of the poly(A) tail and PABP can be assessed using an assay system that exhibits strong effects of poly(A). This is unfortunately not the case in the popular rabbit reticulocyte and wheat germ systems, but both *Xenopus* oocytes and a newly developed cell-free system from *Drosophila* embryos display this property (Gebauer et al. 1999). Using *Xenopus* embryos, the TGEs of *C. elegans tra-2* mRNA were shown to repress only mRNAs that possessed a poly(A) tail

(Thompson et al. 2000). This is consistent with their either promoting deadenylation or interfering with poly(A)-dependent enhancement of translation (Thompson et al. 2000).

Certain repressors may interfere with cap function: CPEB, through maskin, may bind eIF4E in such a way that it blocks further assembly of a 48S pre-initiation complex (Stebbins-Boaz et al. 1999). This model has the attractive virtue that it can explain why the preexisting poly(A) tail is insufficient to activate, despite its being sufficiently long to bind PABP.

Keeping Poly(A) Tails Short

A related strategy to inhibit translation would be to keep the poly(A) tail short. Although many studies suggest that polyadenylation is an integral part of translational activation, few address how, or whether, poly(A) length is connected to repression. Do negative elements in the 3′UTR act by keeping the poly(A) tail short? Or do they repress through a mechanism that has nothing to do with having a short tail, but which can be relieved by polyadenylation?

mRNA injection experiments support the hypothesis that the repression of a maternal mRNA can be caused by the shortness of its poly(A) tail. In *Drosophila*, injected *bicoid* mRNAs with long poly(A) tails rescue *bicoid* mutant embryos, whereas the same mRNAs with shorter tails do not (Sallés et al. 1994); similarly, injected murine tPA mRNAs are active with long, but not short, poly(A) tails, corresponding to their states before and after oocyte maturation (Huarte et al. 1992).

Whereas short tails lead to less activity than long tails, these differences are clearly not always a sufficient explanation for regulation. For example, poly(A) tails of about 50 nucleotides stimulate translation relative to an mRNA with no tail both in vivo and in vitro, yet repressed mRNAs often have tails longer than this. Furthermore, removal of the poly(A) tail (and 3′UTR) of ribonucleotide reductase turns it on. Translational repression of *msl-2* mRNA by SXL in a poly(A)-responsive extract from *Drosophila* embryos is as efficient when the RNA has a poly(A) tail of 73 nucleotides as when the tail is lacking (Gebauer et al. 1999). Moreover, poly(A) shortening can be a result, rather than a cause, of repression (Muckenthaler et al. 1997), and poly(A) lengthening can occur in the absence of derepression (Culp and Musci 1998).

These considerations suggest that the repression of certain mRNAs which show correlations of translational activity with poly(A) length may be due to a poly(A)-independent mechanism. This does not preclude the possibility that poly(A) addition may play an important role in derepression. Indeed, derepression of mouse tPA (Stutz et al. 1998) and frog cyclin mRNAs (de Moor and Richter 1999) by providing excess CPEs in *trans* requires that a poly(A) tail be present on the mRNA.

Interfering with the Joining of the 60S Subunit or Elongation

Recruitment of the small ribosomal subunit is considered to constitute the rate-limiting step in translation initiation under many conditions (Sachs et al. 1997). This situation predisposes this early step as a target for translational control, but does not preclude subsequent steps from being targeted by inhibitory mechanisms. One example of this is the stalling of scanning by cap-distal IRE/IRP complexes (see above). Another point of interference can be envisaged at the joining step between the small and the large ribosomal subunit at the translation initiation codon. At present, no such example has been reported. However, *lin-14* mRNA appears to remain polysome-associated when repressed by *lin-4* mRNA, implying that repression occurs after initiation (Olsen and Ambros 1999).

Subcellular Localization

Repressors bound to sites in the 3′UTR might move the mRNA into a cellular microenvironment that is translationally compromised or interfere with the movement of an mRNA to a site that is translationally favorable.

Nucleating Assembly of a Repressive Structure

In this model, mRNAs are repressed because they are assembled into a complex that effectively hides them from the translation apparatus. This complex might be an overall structure, that hides the mRNA in much the same way as chromatin condensation hides DNA from the transcription apparatus. As such, this mechanism is an extension of, and quite similar to, a steric interference model.

Y-box proteins, such as FRGY2 (also known as mRNP4), may be important in the formation of structures that cause repression (for review, see Wolffe 1992, 1994). FRGY-2 is expressed in oocytes and not in somatic cells; homologs are present in somatic cells and may have comparable functions. Y-box proteins, including FRGY-2, are bona fide transcription factors (Tafuri and Wolffe 1990, 1992), yet are physically associated with many different maternal mRNAs (see, e.g., Darnbrough and Ford 1981; Dearsly et al. 1985; Murray et al. 1991; Tafuri and Wolffe 1993) and can inhibit their translation (Richter and Smith 1984; Kick et al. 1987; Ranjan et al. 1993; Bouvet and Wolffe 1994). These data suggest several provocative possibilities. For example, Y-box proteins might assemble with the mRNA to form a structure that effectively hides the mRNA. Dephosphorylation of Y-box proteins appears to enhance translational activation of the mRNA with which they are associated (Kick et al. 1987; Murray et al. 1991) yet may have little effect on the binding of Y-box proteins to RNA (for contrary view, see Kick et al. 1987; Tafuri and Wolffe 1993). Thus, phosphorylation and dephosphorylation may influence Y-box protein activity, and hence translation, without modulating their association with RNA. More speculatively, dephosphorylation might conceivably "decondense" a complex structure and reveal the mRNA.

As yet, little sequence specificity has been demonstrated in either the RNA-binding or repressing activities of the Y-box proteins (Marello et al. 1992; Tafuri and Wolffe 1993). Thus, if the Y-box proteins do cause repression of some mRNAs but not others, some other factor must provide the sequence specificity. Proteins bound to negative elements in the 3′UTR could serve such a function, promoting the assembly of Y-box proteins into a repressive form or structure. Y-box proteins can be found associated with active mRNAs, arguing that their binding is insufficient for repression (Tafuri and Wolffe 1993). However, it may be instructive to bear in mind again the analogy with chromatin: Core histones are present on active and inactive genes, but their positions and higher order structures differ and may play a critical role in regulating transcriptional activity. Perhaps sequence-specific regulatory proteins nucleate or disassemble repressive mRNP structures.

Interdependent 5′ and 3′UTRs

In certain instances, translational control requires sites in both UTRs. To repress *msl-2* mRNA in female flies, the protein Sex-Lethal (SXL) must bind to specific sites in the 5′UTR and the 3′UTR of *msl-2* mRNA (see above, Dosage Compensation in *Drosophila*). Localization-dependent translation of *oskar* mRNA involves both its 5′ and 3′UTRs (see above, Pattern Formation in *Drosophila*). In this instance, the 5′UTR element is required for activation rather than repression.

Studies of several mRNAs, particularly plant infectious agents, demonstrate that interactions between 5′ and 3′UTRs can stimulate translation. Some plant viruses harbor positive-acting translational elements in their 3′UTRs: Often, they require the appropriate 5′UTR (Gallie and Walbot 1990). For example, an element in the 3′UTR of the barley yellow dwarf virus genome can act when separated from the stimulated AUG by several ORFs and kilobases of sequence; in this situation, stimulation requires the presence of the natural 5′UTR (Wang and Miller 1995; Wang et al. 1997). When placed at the 5′ end of the mRNA, the element can function on its own (Wang et al. 1997). This implies that base-pairing between the 5′ and 3′UTRs, or protein–protein interactions, are critical in activation. In some cases, the positive elements are often suggested to be the functional equivalents of the cap or poly(A) tail in cognito and may bind basal initiation factors (Gallie and Walbot 1990; Timmer et al. 1993; Wang and Miller 1995). Although these examples involve plant viruses rather than germ cells or embryos, they establish a strong precedent for end-to-end communication in translational regulation.

Derepression/Activation of Translation

Conceptually, the simplest way to activate the translation of a repressed mRNA is to remove the repressor. Although this strategy is frequently followed, there are many informative deviations.

Covalent Modification of the Repressor

In several systems, candidate repressors are phosphorylated as repression is relieved (for review, see Standart and Jackson 1994). The temporal coincidence suggests that phosphorylation could negate the repressor and lead to translational activation. For example, phosphorylation of hnRNPs K and E1 affects their binding activity to RNA in vitro (for review, see Ostareck-Lederer et al. 1998) and may provide a basis for translational derepression of the mRNA. An elegant, phosphorylation-independent mechanism explains the activation of ferritin mRNA in iron-loaded cells: The IRE-binding repressor protein IRP-1 is inactivated posttranslationally by the assembly of an iron–sulfur cluster that prevents access to its RNA-binding sites, whereas IRP-2 is degraded by the proteasome following iron-induced oxidation and ubiquitinylation (Hentze and Kühn 1996; Chapter 21).

Derepression by an Activator Element at the 5′ End

As discussed earlier, relief of Bruno protein's repression of *oskar* mRNA requires an activator element in its 5′UTR. It is possible that the removal of a single component that was initially required to set up a repressed RNP does not suffice to rearrange the active mRNP. The combination of genetic approaches with recently established in vitro assays (Gebauer et al. 1999) may provide the necessary tools to unravel this process.

Derepression and Polyadenylation

For many mRNAs, the transition from silence to activity is accompanied by an increase in poly(A) length. The connection between poly(A) and translation has been discussed elsewhere and is not recapitulated here (Gray and Wickens 1998; Chapters 11 and 27). Instead, we discuss only the connection between polyadenylation and relief of repression by 3′UTR regulatory elements and repressors: How does relief of repression increase poly(A) length, and what does that increase in poly(A) length do to translation?

One obvious consequence of cytoplasmic polyadenylation is to provide more potential binding sites for PABP. However, mRNAs that are silent have sufficiently long poly(A) tails to bind one or more PABP molecules. Thus, longer tails would be expected to enhance translational activity, rather than to flip an off/on switch. It is possible that PABP is not present on the repressed mRNAs, however. PABP attached via a tether to the 3′UTR of a reporter stimulates its translation in a resting oocyte; this implies that, in the absence of other influences, bound PABP would stimulate during early development (Gray et al. 2000). Thus, repressors might interfere with either PABP binding or the interaction of PABP with the translational machinery.

We consider three of many possible connections between repressors, translational activation, and cytoplasmic polyadenylation. In the first, the repressor's primary activity is to keep the tail short; when that activity is lost, the tail gets longer, and that enhances translation. This model accommodates the behavior of certain mRNAs very well, but clearly cannot account for those in which translational activation occurs without polyadenylation. Several instances have been reported of mRNAs that can become active without polyadenylation, even though they normally would undergo it (see above, Keeping Poly(A) Tails Short). In some cases, the presence of a short tail is all that is required to achieve derepression (Stutz et al. 1998); thus, a function of polyadenylation may sometimes be merely to keep a tail there at all, in the face of a competing deadenylation activity (Fox and Wickens 1990; Varnum and Wormington 1990).

In the second pathway, the repressor is inactivated by polyadenylation. For example, bound repressors might be removed or modified by the binding of polyadenylation machinery. This pathway is suggested by experiments in *Xenopus* in which the act of polyadenylation rather than the length of a poly(A) tail appears to be critical for activation (McGrew et al. 1989; Simon et al. 1992; Chapter 27).

In the third pathway, the repressor controls translation and polyadenylation independently. For example, factors bound to the elements might repress by causing formation of an mRNP structure that hides the mRNA from both the translation and polyadenylation machineries. Once the mRNA is exposed, both act. Polyadenylation would then be required to maintain or enhance translational activity. It could do so, for example, by preventing reassociation of the repressor (Standart and Jackson 1994) or by recruiting PABP.

The third pathway accommodates most of the data. It posits that full derepression requires two experimentally separable steps: an initiation step that is independent of polyadenylation, and a second step that is polyadenylation-dependent. Either process individually would yield incomplete, or improperly controlled, translation. The uncoupling of derepression and polyadenylation in vitro would be due to execution of an initiation step without a contribution by a poly(A) tail; the effect of repression in vitro might be substantial, and poly(A)-independent. In vivo, polyadenylation would be required to complete or sustain the derepression. Conversely, the ability of polyadenylation to stimulate translation of an injected mRNA would reflect only the maintenance step; derepression of endogenous mRNAs in vivo would require a separate initiation step.

REGULATORY CIRCUITRY: EMERGING PRINCIPLES AND PROBLEMS

Networks of transcriptional control are commonplace and play crucial roles in development. A single transcription factor can activate some genes and repress others, including those encoding other transcription factors; the intricate interactions of regulatory proteins at a promoter all provide inputs into a single gene's expression. How similar might translational controls be? Are there batteries of mRNAs that are interconnected through common factors? Do differences in the interactions among regulators specify different biological outcomes?

Regulatory Elements: General Features

Although the regulatory elements we have discussed come from many different organisms and control a dramatic array of developmental decisions, they share certain unmistakable similarities. Methods ranging from classic genetics to biochemistry have converged on the 3´UTR as a predominant site of regulation. Indeed, highly conserved sequences in 3´UTRs are likely control elements, although not necessarily ones that act at a translational level (Spicher et al. 1998).

Why the 3´UTR? 3´UTRs are relatively unconstrained in evolution and thus provide fertile ground for the derivation of new regulatory elements (Wickens 1993). In contrast, the 5′UTR must be scanned prior to translation initiation, and alterations in its sequence, structure, or length can affect initiation. The coding region has even more obvious constraints. Although 3′UTRs are in many cases sufficient for regulation, in others, they act in concert with the 5′UTR (see above).

Translational control elements in 3′UTRs may be either on–off switches or adjustable rheostats. Many regulatory elements in 3′UTRs are tandemly repeated. Elimination of some but not all of the regulatory sites in *tra-2* (Goodwin et al. 1993) and *lin-14* (Wightman et al. 1993) yields an intermediate level of translation. Similarly, mRNAs containing a single NRE, rather than two, appear to be repressed less efficiently in vivo (Wharton and Struhl 1991). In wild-type mRNAs, partial occupancy of multiple sites may allow the level of translation to be modulated incrementally. Alternatively, multiple elements might promote cooperative binding of regulatory factors and facilitate concerted repression.

Most of the regulatory elements identified thus far in 3′UTRs of mRNAs critical for development are negative. Some may repress translation as soon as the mRNA enters the cytoplasm, so that the mRNA begins life silently (e.g., *bicoid* and LOX mRNAs). Other negative elements may repress translation only after a period of translational activity (e.g., *lin-14* mRNAs). There are hints that regulatory elements may also be context-dependent. For example, sequence context may influence which mRNAs are stabilized or translationally repressed, as exemplified by globin and LOX regulation in the red blood cell lineage (see below). Certain CPEs are repressive, and others are not (Barkoff et al. 2000).

The preponderance of negative control of translation appears to differ from the predominance of positive control of transcription in mammalian cells (Struhl 1999). This may reflect differences in the basal states of translation and transcription in higher eukaryotes: In the absence of specific information to the contrary, mRNAs are translated, whereas genes are silent.

Translational Regulators with Multiple mRNA Targets

A key emerging principle is that regulators of key developmental decisions often control multiple mRNAs. The importance of this fact is that modulations of a single factor, or regulation of its cofactors, can cause a range of outcomes. The regulation of multiple mRNAs by IRPs modulates cellular iron levels and exemplifies such coordinate control (Hentze and Kühn 1996; Chapter 21).

The existence of multiple targets for single regulators is often inferred from genetic analysis. The logic is straightforward: The phenotype of a mutant that lacks a regulatory site in a single mRNA is a subset of the phenotypes of a mutant that lacks the regulator. For example, consider *fem-3* and its regulator, FBF. *C. elegans* that lack the regulatory element in the 3′UTR of *fem-3* mRNA fail in the sperm/oocyte switch, as do animals that lack FBF. However, animals that lack FBF also exhibit defects in proliferation (Zhang et al. 1997). Similarly, mutants in the regulatory elements of *tra-2* mRNA affect only a single decision in the germ line, whereas mutants that lack its regulator, GLD-1, exhibit a range of germ-line phenotypes (Goodwin et al. 1993; Francis et al. 1995b). *hunchback* mRNA is repressed by Pumilio and Nanos to regulate patterning in the fly embryo, but these proteins regulate germ-line events as well (see below).

Proteins that control poly(A) length during development underlie what appears to be a large network of mRNAs. Many mRNAs undergo polyadenylation as they are activated, or deadenylation as they shut off. A change in the factors responsible (e.g., CPEB, CPSF, PAP, or the deadenylase) could facilitate their coordinate control.

In principle, overexpressing the regulatory signals of a single mRNA might reveal new networks. The feasibility of such an approach has been demonstrated by studies with the negative control element in the 3′UTR of *fem-3* mRNA; overexpression of this element, on its own, masculinizes the germ line (Ahringer and Kimble 1991). The simplest interpretation of this result is that the regulatory factor that binds to the element has been titrated out and can no longer repress the endogenous *fem-3* mRNA. Titration experiments of this type could, in principle, yield unexpected phenotypes that would strongly suggest new targets for the regulatory factor.

Families of Translational Regulators

Many of the translational regulators identified to date are members of much larger families of proteins. In some cases, the similarity is trivial: The regulators merely share the ability to bind RNA. On the other hand, some families appear to have related targets and to share other functions. The importance of this point is twofold. First, such families may share common mechanisms of action: Understanding one regulator may illuminate the whole family. Second, if such proteins often act in complexes, as appears to be the case, interactions among them may be critical for regulation.

ATP-dependent RNA Helicases: Vasa

ATP-dependent RNA helicases can separate RNA duplexes in an ATP-dependent reaction and are characterized by a constellation of conserved amino acids. Here we focus on Vasa, a provocative example of the role of such helicases in translational control.

Drosophila Vasa protein is a member of the DEAD-box protein family of RNA helicases (Hay et al. 1988; Lasko and Ashburner 1988; Liang et al. 1994). It is required for patterning, assembly of the germ plasm, and germ-cell function (Hay et al. 1988; Lasko and Ashburner 1988; Schupbach and Wieschaus 1991; Liang et al. 1994). *vasa* homologs are expressed in the germ cells of many animal species, including planaria (Shibata et al. 1999), *C. elegans* (Gruidl et al. 1996), zebrafish (Olsen et al. 1997; Yoon et al. 1997; Braat et al. 1999), *Xenopus* (Komiya et al. 1994; Ikenishi et al. 1996), mice (Fujiwara et al. 1994), and rats (Komiya and Tanigawa 1995). In *Drosophila* and *C. elegans*, Vasa proteins are components of granules localized to the presumptive germ line (polar granules and P-granules, respectively)—the putative "mRNA control hubs" discussed earlier.

Genetic evidence suggests that Vasa is required to activate a family of germ-line mRNAs, including *oskar, nanos* and *gurken* (Dahanukar and Wharton 1996; Gavis et al. 1996; Styhler et al. 1998; Tinker et al. 1998; Tomancak et al. 1998). Although Vasa binds RNA (Liang et al. 1994), it is unclear that it interacts directly with these putative targets. However, Vasa protein does bind to *Drosophila* IF2 (dIF2; Carrera et al. 2000), a homolog of IF2 of *S. cerevisiae* (yIF2). The dIF2/Vasa complex is likely to be significant in vivo, since dIF2 and *vasa* mutants interact genetically (Carrera et al. 2000). Two functions in translation have been proposed for IF2: to bring initiator tRNAs to the small subunit of the ribosome (Choi et al. 1998; Lee et al. 1999) and to promote 60S subunit joining (Pestova et al. 2000). Thus, Vasa may facilitate activation of specific mRNAs by regulating IF2 activity. The conserved localization and function of Vasa in the germ line suggests that this mode of regulation may be widespread.

Puf and Nanos Families

Drosophila Pumilio and *C. elegans* FBF share eight repeats of approximately 40 amino acids, with distinctive sequences in each repeat; these repeats plus short flanking sequences are necessary and sufficient to bind their specific RNA targets (Zamore et al. 1997; Zhang et al. 1997). These structural features are shared among a large family of proteins, termed Puf proteins. Remarkably, both FBF and Pumilio bind to specific sequences in the 3′UTRs of their targets and cause repression, and both combine with Nanos-related proteins to mediate their effects. Because Pumilio and FBF are distant relatives among Puf proteins, this suggests that other Puf proteins may also be repressors that act through the 3′UTRs of their targets.

The Nanos proteins themselves are weakly conserved, sharing a domain that contains two distinctive CCHC-containing, RNA-binding motifs. This domain is required for all the known functions of *Drosophila nanos* (Arrizabalaga and Lehmann 1999). NOS homologs have been identified in a range of species, including vertebrates, and some are expressed in the germ line (Mosquera et al. 1993; Pilon and Weisblat 1997).

Although *Drosophila nanos* and *pumilio* are best known for their roles in patterning the early embryo, they are also required for various aspects of germ-line development, including the maintenance of germ-line stem cells (Kobayashi et al. 1996; Lin and Spradling 1997; Forbes and Lehmann 1998; Asaoki-Taguchi et al. 1999; Bhat 1999; Deshpande et al. 1999; Parisi and Lin 1999). Similarly, *C. elegans nanos* homologs and several Puf proteins are required redundantly for multiple germ-line functions, including germ-line survival (Kraemer et al. 1999; Subramaniam and Seydoux 1999). In the absence of *nanos*, germ cells in fly embryos ectopically express Sxl as well as the somatic segmentation genes *fts* and *eve* (Deshpande et al. 1999). Both these effects are at the transcriptional level, suggesting that *nanos* may regulate translation of a transcription factor; alternatively, Nanos might control transcription directly, as do other RNA-binding proteins, even including translational repressors (e.g., Bicoid). Thus, the ancestral function(s) of the Puf/Nanos system may have been specific to the germ line. In this view, the specialized roles of the system, such as the sperm/oocyte switch in nematodes and axis formation in *Drosophila*, are later evolutionary accretions (Forbes and Lehmann 1998).

STAR Proteins

C. elegans GLD-1 is a member of the STAR protein family. Members of this family of RNA-binding proteins share a KH-type RNA-binding domain, plus two conserved domains that flank the KH homology region (for review, see Vernet and Artzt 1997). STAR family members are widespread, and include murine Quaking (Ebersole et al. 1996), mammalian SAM-68 (Fumagalli et al. 1994; Taylor and Shalloway 1994) and SF-1 (Kramer 1992; Arning et al. 1996), frog Xqua (Zorn and Krieg 1997), and *Drosophila* HOW proteins (Sidman et al. 1964; Hardy et al. 1996; Baehrecke 1997; Zaffran et al. 1997).

One mouse STAR protein, QKI-6, binds to TGEs and can repress translation of TGE-containing mRNAs in vitro and in *C. elegans* in vivo (Saccomanno et al. 1999). These activities mimic those of *C. elegans*

GLD-1 and suggest that STAR family proteins may commonly mediate translational repression. SF1/BBP, a mammalian STAR protein, is involved in splicing (Abovich and Rosbash 1997; Berglund et al. 1997), suggesting that STAR proteins may have diverse, or multiple, functions. The ability of STAR proteins to hetero- and homodimerize may modulate their activities or the targets they recognize (Chen et al. 1997; Zorn and Krieg 1997).

Multiprotein Complexes

The emerging principle that translational control often involves protein complexes has broad implications. The nature of the complexes may identify which targets are regulated, and what happens to them.

Puf proteins provide an example of the importance of protein–protein interactions among regulators. In both *C. elegans* and *Drosophila*, Puf and Nanos proteins form complexes that regulate target mRNAs. As discussed earlier, FBF and NOS-3 regulate *fem-3* in *C. elegans*, whereas Pumilio and Nanos regulate *hunchback* mRNA in flies. The details of the interactions differ in two respects. First, distinct portions of the *C. elegans* and *Drosophila* Nanos proteins are critical for interaction with their Puf partner (Kraemer et al. 1999; Sonoda and Wharton 1999). Second, the *C. elegans* interaction is RNA-independent, whereas the *Drosophila* interaction requires the RNA and only forms in its presence. Thus, the relative contributions of protein–protein and protein–RNA interactions differ in these two complexes. Nevertheless, the common Puf/Nanos partnership in flies and worms suggests that these protein families may often act in functional pairs. However, this is unlikely to be the only way Puf proteins can function, since *S. cerevisiae*, which possesses five different Puf proteins (Zamore et al. 1997; Zhang et al. 1997), lacks Nanos homologs.

Each member of a regulatory complex on one mRNA may have alternative partners and targets. For example, although FBF and NOS are both required for the sperm/oocyte switch, FBF's role in other *nos*-mediated effects is distinct, and redundant with other Puf proteins (Subramaniam and Seydoux 1999). Moreover, although the other *C. elegans* NOS proteins are required to regulate the sperm/oocyte switch, they do not interact directly with FBF; instead, they may collaborate with other *C. elegans* Puf proteins. The combinatorial nature of regulation by these NOS and Puf proteins prompts an analogy to well-documented principles of transcriptional regulation, in which distinct protein–protein interactions between transcriptional regulators discriminate among various target DNAs and yield specific biological outcomes.

hnRNP E1 (αCP-1) and hnRNP K act together to repress LOX mRNA, but each has additional roles as well. hnRNP E1 is also part of a complex (α-complex) that controls the stability of α-globin mRNA by binding to CU-rich sequences in its 3′UTR (Kiledjian et al. 1995; Wang et al. 1995). The CU-rich sequence of α-globin mRNA cannot substitute for the DICE element of LOX mRNA in mediating translational repression (Ostareck et al. 1997). However, a second protein, E2 (αCP2), which is a close relative of E1, is also involved in globin stability (Kiledjian et al. 1995) and may be able to mediate translational repression via DICE elements (Ostareck et al. 1997). hnRNP K may also function in the transcriptional activation of c-*myc*, which contains a CT-rich promoter (Takimoto et al. 1993; Michelotti et al. 1996). Thus, these proteins seem to be involved in the regulation of transcription, translation, and mRNA stability. This raises the possibility that regulation of one protein, or its partners, could affect a network of genes at several levels.

Linked Processes

As in the film *Rashomon,* a single event—the binding of a protein to a 3′UTR, for example—may be seen quite differently depending on the biological lens through which it is filtered. Translation, stability, and mRNA localization are interconnected.

Translation and Localization

mRNA localization impinges on translational regulation in two modes. In one, the movement of mRNAs to specific but large regions of the cell is critical: *oskar* mRNA is specifically directed to the posterior pole of *Drosophila* oocytes but is not translated until that destination has been reached (Kim-Ha et al. 1995). In the other, more subtle, movements, such as regulated associations with the cytoskeleton, may be targets of regulation. Clearly, these two modes of control may overlap.

As the numbers of examples of localized mRNAs increase, it should become clear whether mRNAs that are mis-localized or still in transit are commonly less active. At this early stage, this seems to be the case. For example, expression of *ASH1* mRNA appears to be more efficient once it is localized in budding yeast (Long et al. 1997). The mechanisms responsible may include the formation of transport particles in which the mRNAs are trafficked but translation does not occur (for review, see Bassell and Singer 1997).

Although many lines of evidence argue that cytoskeletal associations enhance translation, it is unclear that these associations are regulated in a sequence-specific fashion, independent of the large-scale movement of the mRNA. The reconstitution of repression in in vitro systems tends to argue that, at least in those cases, an intact cytoskeleton is not critical.

Translation and Stability

The connections between translation and stability are numerous, and will not be recapitulated here (see, e.g., Jacobson and Peltz 1996; Wickens et al. 1997; Chapters 28 and 29). However, a few comments focused on the early embryo may be useful.

During oogenesis and early embryogenesis, many transcripts are stable, even those that lack a poly(A) tail; presumably this allows mRNAs to accumulate over long periods in the growing oocyte, and to persist until the stage at which their translation is first required. In frog oocytes, mRNAs typically are stable until the so-called mid-blastula transition. Turnover at that stage appears to require deadenylation (see, e.g., Audic et al. 1997; Voeltz and Steitz 1998). It is possible that the same events that would render an mRNA subject to decapping and turnover in yeast cause translational repression in early embryos because the next step (e.g., decapping) simply does not occur. In particular, disruption of the end-to-end complex might cause turnover in yeast, but repression in an oocyte. In this context, the finding that certain sequences that cause instability in mammalian cells cause translational repression in oocytes is provocative, because it suggests that the same event can lead to either outcome. This line of reasoning strongly suggests that understanding modes of decay in yeast may directly shed light on translational control in oocytes and embryos (Wickens et al. 1997).

In some instances, translational repression in an embryo appears to lead to decay, whereas activation avoids that fate. For example, maternal *hunchback* mRNA localized in the anterior is activated and persists, whereas posteriorly localized *hunchback* mRNA never is activated and is rapidly destroyed. Repression places the mRNA in a state that is tolerable until the embryo's decay apparatus has been activated; then the mRNA is destroyed. Translational activation makes the mRNA resistant to that turnover machinery.

Translational Regulators with Other Functions

An mRNA regulator can not only act with different signs—activating or repressing—but can also affect different processes. Sex lethal regulates both splicing in the nucleus and translation in the cytoplasm. *Drosophila*

Bicoid is a transcription factor, but it also represses the translation of *caudal* mRNA by binding to sites within its 3′UTR (Dubnau and Struhl 1996; Rivera-Pomar et al. 1996). hnRNP E1 increases expression of globin, in that it helps stabilize the mRNA, but decreases expression of LOX mRNA in the same cells as a member of a different complex. Thus, a given regulator may act through multiple mechanisms, only one of which is translational, to regulate an mRNA's expression. Modulating its activity, or changing its partners, may have wide and varied repercussions.

WHY TRANSLATIONAL CONTROL?

It is striking that many key decisions in development rely on translational control. Why should this be so? Clearly, during early embryogenesis, when pronuclei or zygotic nuclei are highly condensed and inactive, transcriptional control is not a major option. Controls of protein activities—regulated ligand/receptor interactions, for example—are widely exploited. What are the advantages of controlling maternal mRNA rather than maternal protein? Questions of this type are in one sense futile, as patterns of development evolve and so are restricted by contingency and history. However, within the constraints of a given developmental strategy, translational control can offer unique advantages. For example, activities involved in the earliest stages of pattern formation must be controlled in space and time. The Bicoid protein gradient cannot be established during oogenesis, because diffusion would collapse the gradient before it had a chance to act in the early embryo. For regulatory proteins such as cyclin or *glp-1*, premature translation would clearly disrupt the spatial localization of the regulator, and it also might disrupt the timing of interaction with downstream factors or ligands.

A related and commonly invoked rationale for translational regulation is the quickness and magnitude of the response. Although transcriptional responses can be very rapid, they do not yield high amounts of product as rapidly as translational activation. As a result, one might expect translational regulation in situations requiring a large and instantaneous change in the pattern of protein synthesis. Neuronal plasticity might be such a case, as it appears to require rapid changes at specific, newly stimulated terminals. Indeed, homologs of some of the regulatory proteins discussed here—CPEB and Staufen, for example—are present in mammalian neurons (Wu et al. 1998; Kiebler et al. 1999). Similarly, mouse Quaking protein, the founder of the STAR protein family of which GLD-1 is a member, has neurological functions (Sidman et al. 1964), consistent with the presence of Quaking proteins in oligodendrocytes and Schwann cells

(Hardy et al. 1996). Whether the regulatory circuitry first evolved in the embryo or nervous system is unclear.

Enormous progress recently has been made in identifying regulatory proteins and elements that bind to one another. The stage is now set for delineating how these proteins interact and communicate, and how those events are controlled during development. Who does what to whom, and when? And how do the regulators, wrapped in their own liaisons, communicate with the translational machinery? These events, requiring specific mRNA–protein interactions, must be overlaid on the control of cell growth by modification of the translational apparatus. From the perspective of any one regulator, the end result of these analyses will be a local plexus of interactions and controls; collectively, it will reveal a large, dynamic web of proteins and RNAs, and unanticipated biological richness.

ACKNOWLEDGMENTS

We apologize to those we have not cited due to space limitations. We appreciate helpful and stimulating discussions with the members of our extended labs. We also are grateful to Gary Ruvkun, Robin Wharton, and Ruth Lehmann for helpful comments on the manuscript, and to them and other colleagues for communication of results prior to publication. We appreciate Carol Pfeffer and Anne Helsley-Marchbanks for help with preparation of the manuscript, and Laura Vanderploeg for preparing the figures. Work in the authors' laboratories is supported by National Institutes of Health research grants GM-31892 and GM-50942 to M. W., GM-51836 to E.B.G., and GM-51584, NS-35704, and NS-38472 to S. S. J. K. is an investigator with the Howard Hughes Medical Institute. M.W.H. gratefully acknowledges support from the Deutsche Forschungsgemeinschaft and from Human Frontiers in Science Program grant RG-0038/1999M.

REFERENCES

Abovich N. and M. Rosbash. 1997. Cross-intron bridging interactions in the yeast commitment complex are conserved in mammals. *Cell* **89:** 403–412.

Abrahante J.E., Miller E.A., and Rougvie A.E. 1998. Identification of heterochronic mutants in *Caenorhabditis elegans*: Temporal misexpression of a collagen:green fluorescence protein fusion gene. *Genetics* **149:** 1335–1351.

Ahringer J. 1991. "Post-transcriptional regulation of *fem-3*, a sex-determining gene of *C. elegans*." Ph.D. thesis, University of Wisconsin-Madison.

Ahringer J. and Kimble J. 1991. Control of the sperm-oocyte switch in *Caenorhabditis elegans* hermaphrodites by the *fem-3* 3′ untranslated region. *Nature* **349:** 246–248.

Alphey L., Jimenez J., White-Cooper H., Dawson I., Nurse P., and Glover D.M. 1992. *twine*, a *cdc25* homolog that functions in the male and female germline of *Drosophila*.

Cell **69:** 977–988.

Amaya E., Musci T., and Kirschner M. 1991. Expression of a dominant negative mutant of the FGF receptor disrupts mesoderm formation in *Xenopus* embryos. *Cell* **66:** 257–270.

Amaya E., Stein P., Musci T., and Kirschner M. 1993. FGF signaling in the early specification of mesoderm in *Xenopus*. *Development* **118:** 177–187.

Ambros V. 1989. A hierarchy of regulatory genes controls a larva-to-adult developmental switch in *C. elegans*. *Cell* **57:** 49–57.

Ambros V. and Horvitz H. 1984. Heterochronic mutants of the nematode *Caenorhabditis elegans*. *Science* **226:** 409–416.

Ambros V. and Moss E. 1994. Heterochronic genes and the temporal control of *C. elegans* development. *Trends Genet.* **10:** 123–127.

Arasu P., Wightman B., and Ruvkun G. 1991. Temporal regulation of *lin-14* by the antagonistic action of two other heterochronic genes, *lin-4* and *lin-28*. *Genes Dev.* **5:** 1825–1833.

Arning S., Gruter P., Bilbe G., and Kramer A. 1996. Mammalian splicing factor SF1 is encoded by variant cDNA and binds to RNA. *RNA* **2:** 794–810.

Arrizabalaga G., and Lehmann R. 1999. A selective screen reveals discrete functional domains in *Drosophila nanos*. *Genetics* **153:** 1825–1838.

Asaoki-Taguchi M., Yamada M., Nakamura A., Hanyu K., and Kobayashi S. 1999. Maternal Pumilio acts together with Nanos in germline development in *Drosophila* embryos. *Nat. Cell Biol.* **1:** 431–437.

Audic Y., Omilli F., and Osborne H.B. 1997. Postfertilization deadenylation of mRNA in *Xenopus laevis* embryos is sufficient to cause their degradation at the blastula stage. *Mol. Cell. Biol.* **17:** 209–218.

Baehrecke E. 1997. *who* encodes KH RNA binding protein that functions in muscle development. *Development* **124:** 1323–1332.

Baker B., Gorman M., and Marin I. 1994. Dosage compensation in *Drosophila*. *Annu. Rev. Genet.* **28:** 491–521.

Ballantyne S., Bilger A., Åström J., Virtanen A., and Wickens M. 1995. Poly (A) polymerases in the nucleus and cytoplasm of frog oocytes: Dynamic regulation during oocyte maturation and early development. *RNA* **1:** 64–78.

Barabino S.M., Hubner W., Jenny A., Minvielle-Segastia L., and Keller W. 1997. The 30-kD subunit of mammalian cleavage and polyadenylation specificity factor and its yeast homolog are RNA-binding zinc finger proteins. *Genes Dev.* **11:** 1703–1716.

Barker D., Wang C., Moore J., Dickinson L., and Lehmann R. 1992. *Pumilio* is essential for function but for distribution of the *Drosophila* abdominal determinant nanos. *Genes Dev.* **6:** 2312–2326.

Barkoff A., Ballantyne S., and Wickens M. 1998. Polyadenylation of multiple mRNAs is required for oocyte maturation. *EMBO J.* **17:** 3168–3175.

Barkoff A.F., Dickson K.S., Gray N.K., and Wickens M. 2000. Translational control of cyclin B1 mRNA during meiotic maturation: Coordinated repression and cytoplasmic polyadenylation. *Dev. Biol.* **220:** 97–109.

Barton M., Schedl T., and Kimble J. 1987. Gain-of-function mutations of *fem-3*, a sex-determination gene in *Caenorhabditis elegans*. Genetics **115:** 107–119.

Bashaw G. and Baker B. 1995. The *msl-2* dosage compensation gene of *Drosophila* encodes a putative DNA-binding protein whose expression is sex specifically regulated by *Sex-lethal*. *Development* **121:** 3245–3258.

———. 1997. The regulation of the *Drosophila msl-2* gene reveals a function for *Sex-lethal* in translational control. *Cell* **89:** 789–798.

Bassell G. and Singer R.H. 1997. mRNA and cytoskeletal filaments. *Curr. Opin. Cell Biol.* **9:** 109–115.

Batchelder C., Dunn M.A., Choy B., Suh Y., Cassie C., Yong Shim E., Shin T.H., Mello C., Seydoux G., and Blackwell T.K. 1999. Transcriptional repression by the *Caenorhabditis elegans* germ-line protein PIE-1. *Genes Dev.* **13:** 202–212.

Berglund J.A., Chua K., Abovich N., Reed R., and Rosbash M. 1997. The splicing factor BBP interacts specifically with the pre-mRNA branchpoint sequence UACUAAC. *Cell* **89:** 781–787.

Bergsten S.E. and Gavis E.R. 1999. Role for mRNA localization in translational activation but not spatial restriction of *nanos* RNA. *Development* **126:** 659–669.

Bhat K.M. 1999. The posterior determinant gene *nanos* is required for the maintenance of the adult germline stem cells during *Drosophila* oogenesis. *Genetics* **151:** 1479–1492.

Bilger A., Fox C.A., Wahle E., and Wickens M. 1994. Nuclear polyadenylation factors recognize cytoplasmic polyadenylation elements. *Genes Dev.* **8:** 1106–1116.

Bouvet P. and Wolffe A. 1994. A role for transcription and FRGY2 in masking maternal mRNA within *Xenopus* oocytes. *Cell* **77:** 931–941.

Braat A.K., Zandbergen T., van de Water S., Goos H.J.T., and Zivkovic D. 1999. Characterization of zebrafish primordial germ cells: Morphology and early distribution of vasa RNA. *Dev. Dyn.* **216:** 153–167.

Braun R.E., Peschon J.J., Behringer R.R., Brinster R.L., and Palmiter R.D. 1989. Protamine 3′-untranslated sequences regulate temporal translational control and subcellular localization of growth hormone in spermatids of transgenic mice. *Genes Dev.* **3:** 793–802.

Burge C.B., Tuschl T., and Sharp P.A. 1999. Splicing of precursors to mRNAs by the spliceosomes. *In The RNA world,* 2nd edition (ed. R.F. Gesteland et al.), pp. 525–560. Cold Spring Harbor Laboratory Press, Cold Spring Harbor, New York.

Carballo E., Lai W.S., and Blackshear P.J. 1998. Feedback inhibition of macrophage tumor necrosis factor-α production by tristetraprolin. *Science* **281:** 1001–1005.

Carrera P., Johnstone O., Nakamura A., Casanova J., Jackle H., and Lasko P. 2000. VASA mediates translation through interaction with a *Drosophila* yIF2 homolog. *Mol. Cell* **5:** 181–187.

Casanova J. and Struhl G. 1989. Localized surface activity of torso, a receptor tyrosine kinase, specifies terminal body pattern in *Drosophila*. *Genes Dev.* **3:** 2025–2038.

Castagnetti S., Hentze M.W., Ephrussi A., and Gebauer F. 2000. Control of *oskar* mRNA translation by Bruno in a novel cell-free system from *Drosophila* ovaries. *Development* **127:** 1063–1068.

Castrillon D.H., Bonczy P., Alexander S., Rawson R., Eberhart C.G., Viswanathan S., DiNardo S., and Wasserman S.A. 1993. Toward a molecular genetic analysis of spermatogenesis in *Drosophila melanogaster*: Characterization of a male-sterilemutants generated by single P element mutagenesis. *Genetics* **135:** 489–505.

Chalfie M., Horvits H., and Sulston J. 1981. Mutations that lead to reiterations in the cell lineages of *C. elegans*. Cell **24:** 59–69.

Chan S.K. and Struhl G. 1997. Sequence-specific RNA binding by Bicoid. *Nature* **388:** 634.

Chang J.S., Tan L., and Schedl P. 1999. The *Drosophila* CPEB homolog, orb, is required for oskar protein expression in oocytes. *Dev. Biol.* **215:** 91–106.

Chen T., Damaj B.B., Herrera C., Lasko P., and Richard S. 1997. Self-association of the single-KH-domain family members Sam68, GRP33, GLD-1, and Qk1: Role of the KH domain. *Mol. Cell. Biol.* **17:** 5707–5718.

Choi S.K., Lee J.H., Zoll W.L., Merrick W.C., and Dever T.E. 1998. Promotion of met-tRNA$_i^{met}$ binding to ribosomes by yIF2, a bacterial IF2 homolog in yeast. *Science* **280:** 1757–1760.

Colledge W., Carlton M., Udy G., and Evans M. 1994. Disruption of c-*mos* causes parthenogenetic development of unfertilized mouse eggs. *Nature* **370:** 65–68.

Courtot C., Fankhauser C., Simanis V., and Lehner C.F. 1992. The *Drosophila* cdc-25 homolog *twine* is required for meiosis. *Development* **116:** 405–416.

Craig A.W., Haghighat A., Yu A.T., and Sonenberg N. 1998. Interaction of polyadenylate-binding protein with the eIF4G homologue PAIP enhances translation. *Nature* **392:** 520–523.

Crittenden S.L., Rudel D., Binder J., Evans T.C., and Kimble J. 1997. Genes required for GLP-1 asymmetry in the early *Caenorhabditis elegans* embryo. *Dev. Biol.* **181:** 36–46.

Culp P.A. and Musci T.J. 1998. Translational activation and cytoplasmic polyadenylation of FGF receptor-1 are independently regulated during *Xenopus* oocyte maturation. *Dev. Biol.* **193:** 63–76.

Curtis D., Lehmann R., and Zamore P.D. 1995. Translational regulation in development. *Cell* **81:** 171–178.

Dahanukar A. and Wharton R.P. 1996. The Nanos gradient in *Drosophila* embryos is generated by translational regulation. *Genes Dev.* **10:** 2610–2620.

Dahanukar A., Walker J.A., and Wharton R.P. 1999. Smaug, a novel RNA-binding protein that operates a translational switch in *Drosophila*. *Mol. Cell.* **4:** 209–218.

Dalby G. and Glover D. 1993. Discrete sequence elements control posterior pole accumulation and translational repression of maternal cyclin B RNA in *Drosophila*. *EMBO J.* **12:** 1219–1227.

Darnbrough C. and Ford P. 1981. Identification in *Xenopus laevis* of a class of oocyte-specific proteins bound to messenger RNA. *Eur. J. Biochem.* **113:** 415–424.

Dearsly A., Johnson R., Barrett P., and Sommerville J. 1985. Identification of a 60-kDa phosphoprotein that binds stored messenger RNA of *Xenopus* oocytes. *Eur. J. Biochem.* **150:** 95–103.

de Moor C.H. and Richter J.D. 1999. Cytoplasmic polyadenylation elements mediate masking and unmasking of cyclin B1 mRNA. *EMBO J.* **18:** 2294–2303.

Deshpande G., Calhoun G., Yanowitz J., and Schedl P. 1999. Novel functions of *nanos* in downregulating mitosis and transcription during the development of the *Drosophila* germline. *Cell* **99:** 271–281.

Dickson K.S., Bilger A., Ballantyne S., and Wickens M.P. 1999. The cleavage and polyadenylation specificity factor in *Xenopus laevis* oocytes is a cytoplasmic factor invloved in regulated polyadenylation. *Mol. Cell. Biol.* **19:** 5707–5717.

Doniach T. 1986. Activity of the sex-determining gene *tra-2* is modulated to allow spermatogenesis in the *C. elegans* hermaphrodite. *Genetics* **114:** 53–76.

Draper B.W., Mello C.C., Bowerman B., Hardin J., and Priess J.R. 1996. MEX-3 is a KH domain protein that regulates blastomere identity in early *C. elegans* embryos. *Cell* **87:** 205–216.

Driever W. and Nüsslein-Volhard C. 1988a. The bicoid protein determines position in the *Drosophila* embryo in a concentration-dependent manner. *Cell* **54:** 95–104.

———. 1988b. A gradient of bicoid protein in *Drosophila* embryos. *Cell* **54:** 83–93.

Dubnau J. and Struhl G. 1996. RNA recognition and translational regulation by a homeodomain protein. *Nature* **379:** 694–699.

Eberhart C.G., Maines J.Z., and Wasserman S.A. 1996. Meiotic cell cycle requirement for a fly homologue of human Deleted in Azoospermia. *Nature* **381:** 783–785.

Ebersole T.A., Chen Q., Justice M.J., and Artzt K. 1996. The *quaking* gene product necessary in embryogenesis and myelination combines features of RNA binding and signal transduction proteins. *Nat. Genet.* **12:** 260–265.

Eisen J., Sweder K.S., and Hanawalt P.C. 1995. Evolution of the SNF2 family of proteins: Subfamilies with distinct sequences and functions. *Nucleic Acids Res.* **23:** 2715–2723.

Elliot D.J. and Cooke H.J. 1997. The molecular genetics of male infertility. *BioEssays* **19:** 801–809.

Elliot D.J., Millar M.R., Oghene K., Ross A., Kiesewetter F., Pryor J., McIntyre M., Hargreave T.B., Saunder P.T.K., Vogt P.H., Chandley A.C., and Cooke H. 1997. Expression of RBM in the nuclei of human germ cells is dependent on a critical region of the Y chromosome long arm. *Proc. Natl. Acad. Sci.* **94:** 3848–3853.

Ephrussi A. and Lehmann R. 1992. Induction of germ cell formation by *oskar*. Nature **358:** 387–392.

Ephrussi A., Dickinson L.K., and Lehman R. 1991. *oskar* organizes the germ plasm and directs localization of the posterior determinant *nanos*. Cell **66:** 37–50.

Evans T., Crittenden S., Kodoyianni V., and Kimble J. 1994. Translational control of maternal *glp-1* mRNA establishes an asymmetry in the *C. elegans* embryo. *Cell* **77:** 183–194.

Fajardo M.A., Haugen H.S., Clegg C.H., and Braun R.E. 1997. Separate elements in the 3′ untranslated region of the mouse protamine 1 mRNA regulate translational repression and activation during murine spermatogenesis. *Dev. Biol.* **191:** 42–52.

Feinbaum R. and Ambros V. 1999. The timing of *lin-4* RNA accumulation controls the timing of postembryonic developmental events in *Caenorhabditis elegans*. *Dev. Biol.* **210:** 87–95.

Ferby I., Blazquez M., Palmer A., Eritja R., and Nebreda A.R. 1999. A novel p34(cdc2)-binding and activating protein that is necessary and sufficient to trigger G(2)/M progression in *Xenopus* oocytes. *Genes Dev.* **13:** 2177–2189.

Forbes A. and Lehmann R. 1998. Nanos and Pumilio have critical roles in the development and function of *Drosophila* germline stem cells. *Development* **125:** 679–690.

Fox C.A. and Wickens M. 1990. Poly(A) removal during oocyte maturation: A default reaction selectively prevented by specific sequences in the 3′ UTR of certain maternal mRNAs. *Genes Dev.* **4:** 2287–2298.

Fox C., Sheets M., and Wickens M. 1989. Poly(A) tail addition during maturation of frog oocytes: Distinct nuclear and cytoplasmic activities and regulation by the sequence UUUUUAU. *Genes Dev.* **3:** 2151–2162.

Francis R., Maine E., and Schedl T. 1995a. Analysis of the multiple roles of *gld-1* in germline development: Interactions with the sex determination cascade and the *glp-1* signaling pathway. *Genetics* **139:** 607–630.

Francis R., Barton M.K., Kimble J., and Schedl T. 1995b. *gld-1*, a tumor suppressor gene required for oocyte development in *Caenorhabditis elegans*. *Genetics* **139:** 579–606.

Frank-Vaillant M., Jessus C., Ozon R., Maller J.L., and Haccard O. 1999. Two distinct mechanisms control the accumulation of cyclin B1 and Mos in *Xenopus* oocytes in response to progesterone. *Mol. Biol. Cell* **10:** 3279–3288.

Frohnhoefer H.G. and Nüsslein-Volhard C. 1986. Organization of anterior pattern in the *Drosophila* embryo by the maternal gene *bicoid*. *Nature* **324:** 120–125.

Fujiwara Y., Komiya T., Kawabata H., Sato M., Fujimoto H., Furusawa M., and Noce T.

1994. Isolation of a DEAD-family protein gene that encodes a murine homolog of *Drosophila* vasa and its specific expression in germ cell lineage. *Proc. Natl. Acad. Sci.* **91:** 12258–12262.

Fumagalli S., Totty N.F., Hsuan J.J., and Courtneidge S.A. 1994. A target for Src in mitosis. *Nature* **368:** 871–874.

Gallegos M., Ahringer J., Crittenden S., and Kimble J. 1998. Repression by the 3′UTR of *fem-3*, a sex-determining gene, relies on a ubiquitous *mog*-dependent control in *Caenorhabditis elegans*. *EMBO J.* **17:** 6337–6347.

Gallie D.R. and Walbot V. 1990. RNA pseudoknot domain of tobacco mosaic virus can functionally substitute for a poly(A) tail in plant and animal cells. *Genes Dev.* **4:** 1149–1157.

Gavis E. and Lehmann R. 1992. Localization of *nanos* RNA controls embryonic polarity. Cell **71:** 301–313.

———. 1994. Translational regulation of *nanos* by RNA localization. *Nature* **369:** 315–318.

Gavis E.R., Lunsford L., Bergsten S.E., and Lehmann R. 1996. A conserved 90 nucleotide element mediates translational repression of *nanos* RNA. *Development* **122:** 2791–2800.

Gay N.J. and Keith F.J. 1992. Regulation of translation and proteolysis during the development of embryonic dorso-ventral polarity in *Drosophila*. Homology of easter proteinase with *Limulus* proclotting enzyme and translational activation of Toll receptor synthesis. *Biochim. Biophys. Acta* **1132:** 290–296.

Gebauer F. and Richter J.D. 1995. Cloning and characterization of a *Xenopus* poly(A) polymerase. *Mol. Cell. Biol.* **15:** 1422–1430.

———. 1997. Synthesis and function of Mos: The control switch of vertebrate oocyte meiosis. *BioEssays* **19:** 23–28.

Gebauer F., Merendino L., Hentze M.W., and Valcárcel J. 1997. Novel functions for "nuclear factors" in the cytoplasm: The *Sex-lethal* paradigm. *Semin. Cell Dev. Biol.* **8:** 561–566.

———. 1998. The *Drosophila* splicing regulator *Sex-lethal* directly inhibits translation of *male-specific-lethal 2* mRNA. *RNA* **4:** 142–150.

Gebauer F., Xu W., Cooper G.M., and Richter J.D. 1994. Translational control by cytoplasmic polyadenylation of c-*mos* mRNA is necessary for oocyte maturation in the mouse. *EMBO J.* **13:** 5712–5720.

Gebauer F., Corona D.F.V., Preiss T., Becker P.B., and Hentze M.W. 1999. Translational control of dosage compensation in *Drosophila* by *Sex-lethal*: Cooperative silencing via the 5′ and 3′ UTRs of *msl-2* mRNA is independent of the poly(A) tail. *EMBO J.* **18:** 6146–6154.

Gillian-Daniel D.L., Gray N.K., Astrom J., Barkoff A., and Wickens M. 1998. Modifications of the 5′ cap of mRNAs during *Xenopus* oocyte maturation: Independence from changes in poly(A) length and impact on translation. *Mol. Cell. Biol.* **18:** 6152–6163.

Gingras A.-C., Raught B., and Sonenberg N. 1999. eIF4 initiation factors: Effectors of mRNA recruitment to ribosomes and regulators of translation. *Annu. Rev. Biochem.* **68:** 913–963.

Goodwin E.B., Okkema P.G., Evans T.C., and Kimble J. 1993. Translational regulation of *tra*-2 by its 3′ untranslated region controls sexual identity in *C. elegans*. Cell **75:** 329–339.

Goodwin E.B., Hofstra K., Hurney C.A., Mango S., and Kimble J. 1997. A genetic pathway for regulation of *tra-2* translation. *Development* **124:** 749–758.

Goossen B. and Hentze M.W. 1992. Position is the critical determinant for function of iron-responsive elements as translational regulators. *Mol. Cell. Biol.* **12:** 1959–1966.

Graham P. and Kimble J. 1993. The *mog-1* gene is required for the switch from spermatogenesis to oogenesis in *Caenorhabditis elegans*. *Genetics* **133:** 919–931.

Graham P., Schedl T., and Kimble J. 1993. More *mog* genes that influence the switch from spermatogenesis to oogenesis in the heremaphrodite germ line of *Caenorhabditis elegans*. *Dev. Genet.* **14:** 471–484.

Gray N.K. and Hentze M.W. 1994. Iron regulatory protein prevents binding of the 43S translation pre-initiation complex to ferritin and eALAS mRNAs. *EMBO J.* **13:** 3882–3891.

Gray N.K. and Wickens M. 1998. Control of translation initiation in animals. *Annu. Rev. Cell Dev. Biol.* **14:** 399–458.

Gray N.K., Coller J.M., Dickson K.S., and Wickens M. 2000. Multiple portions of poly(A) binding protein stimulate translation *in vivo*. *EMBO J.* (in press).

Gray N.K., Pantopoulos K., Dandekar T., Ackrell B.A.C., and Hentze M.W. 1996. Translational regulation of mammalian and *Drosophila* citric acid cycle enzymes via iron-responsive elements. *Proc. Natl. Acad. Sci.* **93:** 4925–4930.

Gruidl M., Smith P.A., Kuznicki K.A., McCrone J.S., Kirschner J., Russell D.L., Strome S., and Bennett K.L. 1996. Multiple potential germ-line helicases are components of the germ-line-specific P granules of *Caenorhabditis elegans*. *Proc. Natl. Acad. Sci.* **93:** 13837–13842.

Gu W., Kwon Y., Oko R., Hermo L., and Hecht N.B. 1995. Poly(A) binding protein is bound to both stored and polysomal mRNAs in the mammalian testis. *Mol. Reprod. Dev.* **40:** 273–285.

Guedes S. and Priess J.R. 1997. The *C. elegans* MEX-1 protein is present in germline blastomeres and is a P granule component. *Development* **124:** 731–739.

Gunkel N., Yano T., Markussen F.H., Olsen L.C., and Ephrussi A. 1998. Localization-dependent translation requires a functional interaction between the 5′ and 3′ ends of *oskar* mRNA. *Genes Dev.* **12:** 1652–1664.

Ha I., Wightman B., and Ruvkun G. 1996. A bulged *lin-4/lin-14* RNA duplex is sufficient for *Caenorhabditis elegans lin-14* temporal gradient formation. *Genes Dev.* **10:** 3041–3050.

Hake L.E. and Richter J.D. 1994. CPEB is a specificity factor that mediates cytoplasmic polyadenylation during *Xenopus* oocyte maturation. *Cell* **79:** 617–627.

Hardy R.J., Loushin C.L., Friedrich V.L., Jr., Chen Q., Ebersole T.A., Lazzarini R.A., and Artzt K. 1996. Neural cell type-specific expression of QKI proteins is altered in *quakingviable* mutant mice. *J. Neurosci.* **16:** 7941–7949.

Hashimoto C., Hudson K., and Anderson K. 1988. The *Toll* gene of *Drosophila*, required for dorsal-ventral embryonic polarity, appeares to encode a transmembrane receptor. *Cell* **52:** 269–279.

Hashimoto N., Watanabe N., Furuta Y., Tamemoto H., Sagata N., Yokoyama M., Okazaki K., Nagayoshi M., Takeda N., Ikawa Y., and Aizawa S. 1994. Parthenogenetic activation of oocytes in c-*mos*-deficient mice. *Nature* **370:** 68–71.

Hay B., Jan L.Y., and Jan Y.N. 1988. A protein component of *Drosophila* polar granules is encoded by *vasa* and has extensive sequence similarity to ATP-dependent helicases. *Cell* **55:** 577–587.

Hecht N.B. 1998. Molecular mechanisms of male germ cell differentiation. *BioEssays* **20:** 555–561.

Hentze M.W. and Kühn L.C. 1996. Molecular control of vertebrate iron metabolism: mRNA-based regulatory circuits operated by iron, nitric oxide, and oxidative stress. *Proc. Natl. Acad. Sci.* **93:** 8175–8182.

Hodgkin J. 1986. Sex determination in the nematode *C. elegans:* Analysis of *tra-3* suppressors and characterization of *fem* genes. *Genetics* **114:** 15–52.

Hodgkin J. and Brenner S. 1977. Mutations causing transformation of sexual phenotype in the nematode *Caenorhabditis elegans. Genetics* **86:** 275–287.

Houston D.W., Zhang J., Maines J.Z., Wasserman S.A., and King M.L. 1998. A *Xenopus DAZ*-like gene encodes an RNA component of germ plasm and is a functional homologue of *Drosophila boule. Development* **125:** 171–180.

Huarte J., Stutz A., O'Connell M.L., Gubler P., Belin D., Darrow A.L., Strickland S., and Vassalli J.-D. 1992. Transient translational silencing by reversible mRNA deadenylation. *Cell* **69:** 1021–1030.

Hülskamp M., Pfeifle C., and Tautz D. 1990. A morphogenetic gradient of hunchback protein organizes the expression of the gap genes *Kruppel* and *knirps* in the early *Drosophila* embryo. *Nature* **346:** 577–580.

Hülskamp M., Schroder C., Pfeifle C., Jackle H., and Tautz D. 1989. Posterior segmentation of the *Drosophila* embryo in the absence of a maternal posterior organizer gene. *Nature* **338:** 629–632.

Hunt T. 1989. On the translational control of suicide in red cell development. *Trends Biochem. Sci.* **14:** 393–394.

Hunter C.P. and Kenyon C. 1996. Spatial and temporal controls target *pal-1* blastomere–specification activity to a single blastomere lineage in *C. elegans* embryos. *Cell* **87:** 217–226.

Ikenishi K., Tanaka T.S., and Komiya T. 1996. Spatio-temporal distribution of the protein of *Xenopus vasa* homologue (*Xenopus vasa*-like gene 1, *XVLG1*) in embryos. *Dev. Growth Differ.* **38:** 527–535.

Imataka H., Olsen H.S., and Sonenberg N. 1997. A new translational regulator with homology to eukaryotic translation initiation factor 4G. *EMBO J.* **16:** 817–825.

Irish V., Lehmann R., and Akam M. 1989. The *Drosophila* posterior-group gene *nanos* functions by repressing *hunchback* activity. *Nature* **338:** 646–648.

Jacobson A. and Peltz S.W. 1996. Interrelationships of the pathways of mRNA decay and translation in eukaryotic cells. *Annu. Rev. Biochem.* **65:** 693–739.

Jan E., Motzny C.K., Graves L.E., and Goodwin E.B. 1999. The STAR protein, GLD-1, is a translational regulator of sexual identity in *Caenorhabditis elegans. EMBO J.* **18:** 258–269.

Jan E., Yoon J.W., Walterhouse D., Iannoaccone P., and Goodwin E.B. 1997. Conservation of the *C. elegans tra-2* 3′UTR translational control. *EMBO J.* **16:** 6301–6313.

Jones A.R. and Schedl T. 1995. Mutations in *gld-1*, a female germ cell-specific tumor suppressor gene in *Caebnorhabditis elegans*, affect a conserved domain also found in Src-associated protein Sam68. *Genes Dev.* **9:** 1491–1504.

Jones A.R., Francis R., and Schedl T. 1996. GLD-1, a cytoplasmic protein essential for oocyte differentiation, shows stage-and sex-specific expression during *Caenorhabditis elegans* germline development. *Dev. Biol.* **180:** 165–183.

Kadyk L.C. and Kimble J. 1998. Genetic regulation of entry into meiosis in *Caenorhabditis elegans. Development* **125:** 1803–1813.

Kawasaki I., Shim Y.H., Kirchner J., Kaminker J., Wood W.B., and Strome S. 1998. PGL-1, a predicted RNA-binding component of germ granules, is essential for fertility in *C. elegans. Cell* **94:** 635–645.

Keiper B. and Rhoads R. 1999. Translational recruitment of *Xenopus* maternal mRNAs in response to poly(A) elongation requires initiation factor eIF-4G. *Dev. Biol.* **206:** 1–14.

Kelley R. and Kuroda M. 1995. Equality for X chromosomes. *Science* **270:** 1607–1610.

Kelley R., Wang J., Bell L., and Kuroda M. 1997. *Sex-lethal* controls dosage compensation in *Drosophila* by a non-splicing mechanism. *Nature* **387:** 195–199.

Kelley R., Solovyeva I., Lyman L., Richman R., Solovyev V., and Kuroda M. 1995. Expression of msl-2 causes assembly of dosage compensation regulators on the X chromosomes and female lethality in *Drosophila*. *Cell* **81:** 867–877.

Kick D., Barrett P., Cummings A., and Sommerville J. 1987. Phosphorylation of a 60 kDa polypeptide from *Xenopus* oocytes blocks messenger RNA translation. *Nucleic Acids Res.* **15:** 4099–4109.

Kiebler M., Hemraj I., Verkade P., Köhrmann M., Fortes P., Marión R.M., Ortín J., and Dotti C.G. 1999. The mammalian Staufen protein localizes to the somatodendritic domain of cultured hippocampal neurons: Implications for its involvement in mRNA transport. *J. Neurosci.* **19:** 288–297.

Kiledjian M., Wang X., and Leibhaber A. 1995. Identification of two KH domain proteins in the α-globin mRNP stability complex. *EMBO J.* **14:** 4357–4364.

Kim-Ha J., Kerr K., and Macdonald P.M. 1995. Translational regulation of *oskar* mRNA by bruno, an ovarian RNA-binding protein, is essential. *Cell* **81:** 403–412.

Klein P. and Melton D. 1994. Induction of mesoderm in *Xenopus laevis* embryos by translation initiation factor 4E. *Science* **265:** 803–806.

Kobayashi H., Minshull J., Ford C., Golsteyn R., Poon R., and Hunt T. 1991. On the synthesis and destruction of A- and B-type cyclins during oogenesis and meiotic maturation in *Xenopus laevis*. *J. Cell Biol.* **114:** 755–765.

Kobayashi S., Yamada M., Asaoka M., and Kitamura T. 1996. Essential role of the posterior morphogen nanos for germline development in *Drosophila*. *Nature* **380:** 708–711.

Kohler S.A., Henderson B.R., and Kühn L.C. 1995. Succinate dehydrogenase b mRNA of *Drosophila melanogaster* has a functional iron-responsive element in its 5′-untranslated region. *J. Biol. Chem.* **270:** 30781–30786.

Komiya T. and Tanigawa Y. 1995. Cloning of a gene of the DEAD box protein family which is specifically expressed in germ cells in rats. *Biochem. Biophys. Res. Commun.* **207:** 405–410.

Komiya T., Itoh K., Ikenishi K., and Furusawa M. 1994. Isolation and characterization of a novel gene of the DEAD box protein family which is specifically expressed in germ cells of *Xenopus laevis*. *Dev. Biol.* **162:** 354–363.

Kraemer B., Crittenden S., Gallegos M., Moulder G., Barstead R., Kimble J., and Wickens M. 1999. NANOS-3 and FBF proteins physically interact to control the sperm/oocyte switch in *C. elegans*. *Curr. Biol.* **9:** 1009–1018.

Kramer A. 1992. Purification of splicing factor SF1, a heat-stable protein that functions in the assembly of a presplicing complex. *Mol. Cell. Biol.* **12:** 4545–4552.

Kuge H. and Richter J.D. 1995. Cytoplasmic 3′ poly(A) addition induces 5′ cap ribose methylation: Implications for translational control of maternal mRNA. *EMBO J.* **14:** 6301–6310.

Kuge H., Brownlee G.G., Gershon P.D., and Richter J.D. 1998. Cap ribose methylation of c-*mos* mRNA stimulates translation and oocyte maturation in *Xenopus laevis*. *Nucleic Acids Res.* **26:** 3208–3214.

Kuwabara P.E., Okkema P.G., and Kimble J. 1992. *tra-2* encodes a membrane protein and may mediate cell commuication in the *Caenorhabditis elegans* sex determination pathway. *Mol. Biol. Cell* **3:** 461–473.

Kwon Y.K. and Hecht N.B. 1991. Cytoplasmic protein binding to highly conserved sequences in the 3′ untranslated region of mouse protamine 2 mRNA, a translationally regulated transcript in male germ cells. *Proc. Natl. Acad. Sci.* **88:** 3584–3588.

———. 1993. Binding of a phosphoprotein to the 3′ untranslated region of the mouse protamine 2 mRNA temporally represses its translation. *Mol. Cell. Biol.* **13:** 6547–6557.

Lasko P.F. and Ashburner M. 1988. The product of the *Drosophila* gene *vasa* is very similar to eukaryotic initiation factor-4A. *Nature* **335:** 611–617.

Lazaris-Karatzas A., Montine K., and Sonenberg N. 1990. Malignant transformation by a eukaryotic initiation factor subunit that binds to mRNA 5′ cap. *Nature* **345:** 544–547.

Lee J.H., Choi S.K., Roll-Mecak A., Burley S.K., and Dever T.E. 1999. Universal conservation in translation initiation revealed by human and archaeal homologs of bacterial translation initiation factor IF2. *Proc. Natl. Acad. Sci.* **96:** 4342–4347.

Lee K., Fajardo M.A., and Braun R.E. 1996. A testis cytoplasmic RNA-binding protein that has the properties of a translational repressor. *Mol. Cell. Biol.* **16:** 3023–3034.

Lee K., Haugen H.S., Clegg C.H., and Braun R.E. 1995. Premature translation of *Prm-1* mRNA causes precocious nuclear condensation and arrests spermatid differentiation in mice. *Proc. Natl. Acad. Sci.* **92:** 12451–12455.

Lee R.C., Feinbaum R.L., and Ambros V. 1993. The *C. elegans* heterochronic gene *lin-4* encodes small RNAs with antisense complementary to *lin-14*. *Cell* **75:** 843–854.

Lehmann R., and Nüsslein-Volhard C. 1991. The maternal gene *nanos* has a central role in posterior pattern formation of the *Drosophila* embryo. *Development* **112:** 679–691.

Lenormand J.L., Dellinger R.W., Knudsen K.E., Subramani S., and Donoghue D.G. 1999. Speedy: a novel cell cycle regulator of the G2/M transition. *EMBO J.* **18:** 1869–1877.

Levy-Strumpf N., Deiss L.P., Berissi H., and Kimchi A. 1997. DAP-5, a novel homolog of eukaryotic translation initiation factor 4G isolated as a putative modulator of gamma interferon-induced programmed cell death. *Mol. Cell. Biol.* **17:** 1615–1625.

Liang L., Diehl-Jones W., and Lasko P. 1994. Localisation of vasa protein to the *Drosophila* pole plasm is independent of its RNA-binding and helicase activities. *Development* **120:** 1201–1211.

Lie Y.S. and Macdonald P.M. 1999a. Apontic binds the translational repressor Bruno and is implicated in regulation of *oskar* mRNA translation. *Development* **126:** 1129–1138.

———. 1999b. Translational regulation of *oskar* mRNA occurs independent of the cap and poly(A) tail in *Drosophila* ovarian extracts. *Development* **126:** 4989–4996.

Lin H. and Spradling A. 1997. A novel group of Pumilio mutations affects the asymmetric division of germline stem cells in the *Drosophila* ovary. *Development* **124:** 2463–2476.

Long R.M., Singer R.H., Meng X., Gonzalez I., Nasmyth K., and Jansen R.-P. 1997. Mating type switching in yeast controlled by asymmetric localization of *AHS-1* mRNA. *Science* **277:** 383–387.

Ma K., Inglis J.D., Sharkey A., Bickmore W.A., Hill R.E., Prosser E.J., Speed R.M., Thomson E.J., Jobling M., and Taylor K. 1993. A Y chromosome gene family with RNA-binding protein homology: Candidates for the azoospermia factor AZF controlling human spermatogenesis. *Cell* **75:** 1287–1295.

Macdonald P.M. 1992. The *Drosophila pumilio* gene: An unusually long transcription unit and an unusual protein. *Development* **114:** 221–232.

Macdonald P. and Struhl G. 1986. A molecular gradient in early *Drosophila* embryos and its role in specifying the body pattern. *Nature* **324:** 537–545.

Maines J.Z. and Wasserman S.A. 1999. Post-transcriptional regulation of the meiotic Cdc-

25 protein Twine by the Daxl orthologue Boule. *Nat. Cell Biol.* **1:** 171–174.

Marello K., LaRovere J., and Sommerville J. 1992. Binding of *Xenopus* oocyte masking proteins to mRNA sequences. *Nucleic Acids Res.* **20:** 5593–5600.

Markussen F.H., Breitwieser W., and Ephrussi A. 1997. Efficient translation and phosphorylation of Oskar require Oskar protein and the RNA helicase Vasa. *Cold Spring Harbor Symp. Quant. Biol.* **62:** 13–17.

Martin J., Raibaud A., and Ollo R. 1994. Terminal pattern elements in *Drosophila* embryo induced by the torso-like protein. *Nature* **367:** 741–745.

McGrew L.L., Dworkin R.E., Dworkin M.B., and Richter J.D. 1989. Poly(A) elongation during *Xenopus* oocyte maturation is required for translational recruitment and is mediated by a short sequence element. *Genes Dev.* **3:** 803–815.

Mehra A., Gaudet J., Hick L., Kuwabara P.E., and Spence A.M. 1999. Negative regulation of male development in *Caenorhabditis elegans* by a protein-protein interaction between TRA-2A and FEM-3. *Genes Dev.* **13:** 1453–1463.

Mello C., Draper B., and Priess J. 1994. The maternal genes *apx-1* and *glp-1* and establishment of dorsal-ventral polarity in the early *C. elegans* embryo. *Cell* **77:** 95–106.

Mello C.C., Schubert C., Draper B., Zhang W., Lobel R., and Priess J.R. 1996. The PIE-1 protein and germline specification in *C. elegans* embryos. *Nature* **382:** 710–712.

Melton D. 1994. Vertebrate embryonic induction: Mesodermal and neural patterning. *Science* **266:** 596–604.

Michelotti E.F., Michelotti G.A., Aronsohn A.I., and Leven D. 1996. Heterogeneous nuclear ribonucleoprotein K is a translation factor. *Mol. Cell. Biol.* **16:** 2350–2360.

Mickey K.M., Mello C.C., Montgomery M.K., Fire A., and Preiss J.R. 1996. An inductive interaction in 4-cell stage *C. elegans* embryos involves APX-1 expression in the signaling cell. *Development* **122:** 1791–1798.

Minshall N., Walker J., Dale M., and Standart N. 1999. Dual roles of p82, the clam CPEB homolog, in cytoplasmic polyadenylation and translational masking. *RNA* **5:** 27–38.

Mlodzik M. and Gehring S. 1987. Expression of the caudal gene in the germ line of *Drosophila:* Formation of an RNA and protein gradient during early embryogensis. *Cell* **48:** 465–478.

Mlodzik M., Fjose A., and Gehring W. 1985. Isolation of caudal, a *Drosophila* homeobox-containing gene with maternal expression, whose transcripts form a concentration gradient at the pre-blastoderm stage. *EMBO J.* **4:** 2961–2969.

Morisato D. and Anderson K. 1994. The *spaetzle* gene encodes a component of the extracellular signaling pathway establishing the dorsal-ventral pattern of the *Drosophila* embryo. *Cell* **76:** 677–688.

Mosquera L., Forristall C., Zhou Y., and King M. 1993. A mRNA localized to the vegetal cortex of *Xenopus* oocytes encodes a protein with a *nanos*-like zinc finger domain. *Development* **117:** 377–386.

Moss E.G., Lee R.C., and Ambros V. 1997. The cold shock domain protein LIN-28 controls developmental timing in *C. elegans* and is regulated by the *lin-4* RNA. *Cell* **88:** 637–646.

Muckenthaler M., Gray N.K., and Hentze M.W. 1998. IRP-1 binding to ferritin mRNA prevents the recruitment of the small ribosomal subunit by the cap-binding complex eIF4F. *Mol. Cell* **2:** 383–388.

Muckenthaler M., Gunkel N., Stripecke R., and Hentze M.W. 1997. Regulated poly(A) tail shortening in somatic cells mediated by cap-proximal translational repressor proteins and ribosome association. *RNA* **3:** 983–995.

Murakami M.S. and Vande Woude G.F. 1998. Analysis of the early embryonic cell cycles of *Xenopus*; regulation of cell cycle length by Xe-wee1 and Mos. *Development* **125:** 237–248.

Murata Y. and Wharton R. 1995. Binding of pumilio to *maternal* hunchback mRNA is required for posterior patterning in *Drosophila* embryos. *Cell* **80:** 747–756.

Murray M.T., Krohne G., and Franke W.W. 1991. Different forms of soluble cytoplasmic mRNA binding proteins and particles in *Xenopus laevis* oocytes and embryos. *J. Cell Biol.* **112:** 1–11.

Musci T.J., Amaya E., and Kirschner M.W. 1990. Regulation of the fibroblast growth factor receptor in early *Xenopus* embryos. *Proc. Natl. Acad. Sci.* **87:** 8365–8369.

Nebreda A.R., Gannon J.V., and Hunt T. 1995. Newly synthesized protein(s) must associate with $p34^{cdc2}$ to activate MAP kinase and MPF during progesterone-induced maturation of *Xenopus* oocytes. *EMBO J.* **14:** 5597–5607.

Niessing D., Dostatni N., Jäckle H., and Rivera-Pomar R. 1999. Sequence interval within the PEST motif of Bicoid is imporant for translational repression of *caudal* mRNA in the anterior region of the *Drosophila* embryo. *EMBO J.* **18:** 1966–1973.

Nishizawa M.M., Furuno N., Okazaki K., Tanaka H., Ogawa Y., and Sagata N. 1993. Degradation of Mos by the N-terminal proline (Pro2)-dependent ubiquitin pathway on fertilization of *Xenopus* eggs: Possible significance of natural selection for Pro2 in Mos. *EMBO J.* **12:** 4021–4027.

Olsen L.C., Aasland R., and Fjose A. 1997. A *vasa*-like gene in zebrafish identifies putative primordial germ cells. *Mech. Dev.* **66:** 95–105.

Olsen P.H. and Ambros V. 1999. The *lin-4* regulatory RNA controls developmental timing in *Caenorhabditis elegans* by blocking LIN-14 protein synthesis after the initiation of translation. *Dev. Biol.* **216:** 671–680.

Ostareck D.H., Ostareck-Lederer A., Wilm M., Thiele B.J., Mann M., and Hentze M.W. 1997. mRNA silencing in erythroid differentiation: hnRNP K and hnRNP E1 regulate 15-lipoxygenase translation from the 3′ end. *Cell* **89:** 597–606.

Ostareck-Lederer A., Ostareck D.H., and Hentze M.W. 1998. Cytoplasmic regulatory functions of the KH-domain proteins hnRNPs K and E1/E2. *Trends Biochem. Sci.* **23:** 409–411.

Ostareck-Lederer A., Ostareck D., Standart N., and Thiele B. 1994. Translation of 15-lipoxygenase mRNA is controlled by a protein that binds to a repeated sequence in the 3′ untranslated region. *EMBO J.* **13:** 1476–1481.

Paraskeva E., Gray N.K., Schläger B., Wehr K., and Hentze M.W. 1999. Ribosomal pausing and scanning arrest as mechanisms of translational regulation from cap-distal iron-responsive elements. *Mol. Cell. Biol.* **19:** 807–816.

Paris J. and Richter J.D. 1990. Maturation-specific polyadenylation and translational control: Diversity of cytoplasmic polyadenylation elements, influence of poly(A) tail size, and formation of stable polyadenylation complexes. *Mol. Cell. Biol.* **10:** 5634–5645.

Parisi M. and Lin H. 1999. The *Drosophila pumilio* gene encodes two functional isoforms that play multiple roles in germline development, gonadogenesis, oogenesis and embryogenesis. *Genetics* **153:** 235–250.

Pazin M.J. and Kadonaga J.T. 1997. SWI1/SNF2 and related proteins: ATP-driven motors that disrupt protein-DNA interactions? *Cell* **88:** 737–740.

Pestova T.V., Lomakin I.B., Lee J.H., Choi S.K., Dever T.E., and Hellen C.U. 2000. The joining of ribosomal subunits in eukaryotes requires eIF5B. *Nature* **403:** 332–335.

Pilon M. and Weisblat A. 1997. A *nanos* homolog in leech. *Development* **124:** 1771–1780.

Preiss T. and Hentze M.W. 1998. Dual function of the cap structure in poly(A) tail-promoted translation in yeast. *Nature* **392:** 516–520.

Puoti A. and Kimble J. 1999. The *Caenorhabditis elegans* sex determination gene *mog-1* encodes a member of the DEAH-Box protein family. *Mol. Cell. Biol.* **19:** 2189–2197.

———. 2000. The hermaphrodite sperm/oocyte switch requires the *Caenorhabditis elegans* homologs of PRP2 and PRP22. *Proc. Natl. Acad. Sci.* **97:** 3276–3281.

Ralle T., Gremmels D., and Stick R. 1999. Translational control of nuclear lamin B1 mRNA during oogenesis and early development of *Xenopus*. *Mech. Dev.* **84:** 89–101.

Ranjan M., Tafuri S.R., and Wolffe A.P. 1993. Masking mRNA from translation in somatic cells. *Genes Dev.* **7:** 1725–1736.

Rapoport S.M. and Schewe T. 1986. The maturational breakdown of mitochondria in reticulocytes. *Biochem. Biophys. Acta* **864:** 471–495.

Reijo R., Lee T.Y., Salo P., Alagappan R., Brown L.G., Rosenberg M., Rozen S., Jaffe T., Straus D., and Hovatta O. 1995. Diverse spermatogenic defects in humans caused by Y chromosome deletions encompassing a novel RNA-binding protein gene. *Nat. Genet.* **10:** 383–393.

Reinhart B.J., Slack F.J., Basson M., Pasquinelli A., Bettinger J.C., Rougvie A.E., Horvitz H.R., and Ruvkun G. 2000. The 21-nucleotide *let-7* RNA regulates developmental timing in *Caenorhabditis elegans*. *Nature* **403:** 901–906.

Richter J. and Smith L. 1984. Reversible inhibition of translation by *Xenopus* oocyte-specific proteins. *Nature* **309:** 378–380.

Rivera-Pomar R., Niessing D., Schmidt-Ott U., Gehring W.J., and Jäckle H. 1996. RNA binding and translational suppression by bicoid. *Nature* **379:** 746–749.

Robbie E.P., Peterson M., Amaya E., and Musci T.J. 1995. Temporal regulation of the *Xenopus* FGF receptor in development: A translation inhibitory element in the 3′ untranslated region. *Development* **121:** 1775–1785.

Rosenthal E., Hunt T., and Ruderman J. 1980. Selective translation of mRNA controls the pattern of protein synthesis during early development of the surf clam, *Spisula solidissima*. *Cell* **20:** 487–496.

Ruggiu M., Speed R., Taggart M., McKay S.J., Kilanowski F., Saunders P., Dorin J., and Cooke H.J. 1997. The mouse Dazla gene encodes a cytoplasmic protein essential for gametogenesis. *Nature* **389:** 73–77.

Ruvkun G. and Giusto J. 1989. The *Caenorhabditis elegans* heterochronic gene *lin-14* encodes a nuclear protein that forms a temporal developmental switch. *Nature* **338:** 313–319.

Ruvkun G., Ambros V., Coulson A., Waterston R., Sulston J., and Horvitz H. 1989. Molecular genetics of the *Caenorhabditis elegans* heterochronic gene *lin-14*. *Genetics* **121:** 501–516.

Saccomanno L., Loushin C., Jan E., Punkay E., Artzt K., and Goodwin E.B. 1999. The STAR protein QKI-6 is a translational repressor. *Proc. Natl. Acad. Sci.* **96:** 12605–12610.

Sachs A.B., Sarnow P., and Hentze M.W. 1997. Starting at the beginning, middle, and end: Translation initiation in eukaryotes. *Cell* **89:** 831–838.

Saffman E.E., Styhler S., Rother K., Li W., Richard S., and Lasko P. 1998. Premature translation of *oskar* in oocytes lacking the RNA-binding protein bicaudal-C. *Mol. Cell. Biol.* **18:** 4855–4862.

Sagata N. 1997. What does Mos do in oocytes and somatic cells? *BioEssays* **19:** 13–21.

Sagata N., Daar I., Oskarsson M., Showalter S., and Vande Woude G. 1990. The product of the *mos* proto-oncogene as a candidate "initiator" for oocyte maturation. *Science* **245:** 643–646.

Sagata N., Oskarsson M., Copeland T., Brumbaugh J., and Vande Woude G.F. 1988.

Function of c-*mos* proto-oncogene product in meiotic maturation in *Xenopus* oocytes. *Nature* **335:** 519–525.

Sallés F.J., Darrow A.L., O'Connell M.L., and Strickland S. 1992. Isolation of novel murine maternal messenger RNAs regulated by cytoplasmic polyadenylation. *Genes Dev.* **6:** 1202–1212.

Sallés F.J., Lieberfarb M.E., Wreden C., Gergen J.P., and Strickland S. 1994. Coordinate initiation of *Drosophila* development by regulated polyadenylation of maternal messenger RNAs. *Science* **266:** 1996–1999.

Schedl T. and Kimble J. 1988. *fog-2*, a germ-line-specific sex determination gene required for hermaphrodite spermatogenesis in *Caenorhabditis elegans*. *Genetics* **119:** 43–61.

Schisa J.A. and Strickland S. 1998. Cytoplasmic polyadenylation of Toll mRNA is required for dorsal-ventral patterning in *Drosophila* embryogenesis. *Development* **125:** 2995–3003.

Schnabel R. and Priess J.R. 1997. Specification of cell fates in the early embryo. In *C. elegans* II (ed. D. L. Riddle et al.), pp. 361–382. Cold Spring Harbor Laboratory Press, Cold Spring Harbor, New York.

Schumacher J.M., Lee K., Edelhoff S., and Braun R.E. 1995a. Distribution of Tenr, an RNA-binding protein, in a lattice-like network within the spermatid nucleus in the mouse. *Biol. Reprod.* **52:** 1274–1283.

———. 1995b. Spnr, a murine RNA binding protein that is localized to cytoplasmic microtubules. *J. Cell Biol.* **129:** 1023–1032.

Schupbach T. and Wieschaus R. 1991. Maternal effect mutations altering the anterior-posterior patterning of the *Drosophila* embryo. *Roux's Arch. Dev. Biol.* **195:** 302–317.

Seydoux G. and Dunn M.A. 1997. Transcriptionally repressed germ cells lack a subpopulation of phosphorylated RNA polymerase II in early embryos of *Caenorhabditis elegans* and *Drosophila melanogaster*. *Development* **124:** 2191–2201.

Seydoux G. and Fire A. 1994. Soma-germline asymmetry in the distributions of embryonic RNAs in *Caenorhabditis elegans*. *Development* **120:** 2823–2834.

Seydoux G. and Strome S. 1999. Launching the germline in *Caenorhabditis elegans*: Regulation of gene expression in early germ cells. *Development* **126:** 3275–3283.

Seydoux G., Mellow C.C., Pettitt J., Wood W.B., Priess J.R., and Fire A. 1996. Repression of gene expression in the embryonic germ lineage of *C. elegans*. *Nature* **382:** 713–716.

Sheets M., Wu M., and Wickens M. 1995. Polyadenylation of c-*mos* mRNA as a control point in *Xenopus* meiotic maturation. *Nature* **374:** 511–516.

Sheets M.D., Fox C.A., Hunt T., Vande Woude G.F., and Wickens M. 1994. The 3′-untranslated regions of c-*mos* and cyclin mRNAs stimulate translation by regulating cytoplasmic polyadenylation. *Genes Dev.* **8:** 926–938.

Shibata N., Umesono Y., Oril H., Sajurai T., Watanabe K., and Agata K. 1999. Expression of *vasa (vas)*-related genes in germline cells and totipotent somatic stem cells of planarians. *Dev. Biol.* **206:** 73–87.

Sidman R.L., Dickie M.M., and Appel S.H. 1964. Mutant mice (quaking and jimpy) with deficient myelination in the central nervous system. *Science* **144:** 309–311.

Simon R., Tassan J., and Richter J.D. 1992. Translational control by poly(A) elongation during *Xenopus* development: Differential represion and enhancement by a novel cytoplasmic polyadenylation element. *Genes Dev.* **6:** 2580–2591.

Slack F.J., Basson M., Liu Z., Ambros V., Horvitz H.R., and Ruvkun G. 2000. The *lin-41* RBCC gene acts in the *C. elegans* heterochronic pathway between the *let-7* regulatory RNA and the *lin-29* transcription factor. *Mol. Cell* **5:** 659–669.

Slee R., Grimes B., Speed R.M., Taggart M., Maguire S.M., Ross A., McGill N.I.,

Saunders P.T., and Cooke H.J. 1999. A human DAZ transgene confers partial rescue of the mouse Dazl null phenotype. *Proc. Natl. Acad. Sci.* **96:** 8040–8045.

Smibert C.A., Wilson J.E., Kerr K., and Macdonald P.M. 1996. smaug protein represses translation of unlocalized *nanos* mRNA in the *Drosophila* embryo. *Genes Dev.* **10:** 2600–2609.

Smith J.L., Wilson J.E., and Macdonald P.M. 1992. Overexpression of *oskar* directs ectopic activation of *nanos* and presumptive pole cell formation in *Drosophila* embryos. *Cell* **70:** 849–859.

Sonoda J. and Wharton R.P. 1999. Recruitment of Nanos to *hunchback* mRNA by Pumilio. *Genes Dev.* **13:** 2704–2712.

Spicher A., Guicherit O.M., Duret L., Aslanian A., Sanjines E.M., Denko N.C., Giaccia A.J., and Blau H.M. 1998. Highly conserved RNA sequences that are sensors of environmental stress. *Mol. Cell. Biol.* **18:** 7371–7382.

Spirin A. 1966. On "masked" forms of messenger RNA in early embryogenesis and in other differentiating systems. *Curr. Top. Dev. Biol.* **1:** 1–63.

Sprenger F., Stevens L., and Nüsslein-Volhard C. 1989. The *Drosophila* gene *torso* encodes a putative receptor tyrosine kinase. *Nature* **338:** 478–483.

Standart N. 1992. Masking and unmasking of maternal mRNA. *Semin. Dev. Biol.* **3:** 367–379.

Standart N. and Jackson R.J. 1994. Regulation of translation by specific protein/mRNA interactions. *Biochimie* **76:** 867–879.

Standart N., Dale M., Stewart E., and Hunt T. 1990. Maternal mRNA from clam oocytes can be specifically unmasked in vitro by antisense RNA complementary to the 3′ untranslated region. *Genes Dev.* **4:** 2157–2168.

Stebbins-Boaz B., Hake L.E., and Richter J.D. 1996. CPEB controls the cytoplasmic polyadenlyation of cyclin, Cdk2 and c-*mos* mRNAs and is necessary for oocyte maturation in *Xenopus*. *EMBO J.* **15:** 2582–2592.

Stebbins-Boaz B., Cao Q., de Moor C.H., Mendez R., and Richter J.D. 1999. Maskin is a CPEB-associated factor that transiently interacts with eIF-4E. *Mol. Cell* **4:** 1017–1027.

Stein D., Roth S., Vogelsang E., and Nüsslein-Volhard C. 1991. The polarity of the dorsoventral axis in the *Drosophila* embryo is defined by an extracellular signal. *Cell* **65:** 725–735.

Stevens L., Frohnhoefer H., Klinger M., and Nüsslein-Volhard C. 1990. Localized requirement for torso-like expression in follicle cells for development of terminal anlagen of the *Drosophila* embryo. *Nature* **346:** 660–663.

St Johnston D., and Nüsslein-Volhard C. 1992. The origin of pattern and polarity in the *Drosophila* embryo. *Cell* **68:** 201–219.

St Johnston D., Beuchle D., and Nüsslein-Volhard C. 1991. *staufen*, a gene required to localize maternal RNAs in the *Drosophila* egg. *Cell* **66:** 51–63.

Stripecke R., Oliveira C.C., McCarthy J.E.G., and Hentze M.W. 1994. Proteins binding to 5′ untranslated region sites: A general mechanism for translational regulation of mRNAs in human and yeast cells. *Mol. Cell. Biol.* **14:** 5898–5909.

Struhl G. 1989. Differing strategies for organizing the anterior and posterior body pattern in *Drosophila* embryos. *Nature* **338:** 741–744.

Struhl K. 1999. Fundamentally different logic of gene regulation in eukaryotes and prokaryotes. *Cell* **98:** 1–4.

Stutz A., Conne B., Huarte J., Gubler P., Volkel V., Flandin P., and Vassalli J.-D. 1998. Masking, unmasking, and regulated polyadenylation cooperate in the translational control of a dormant mRNA in mouse oocytes. *Genes Dev.* **12:** 2535–2548.

Styhler S., Nakamura A., Swan A., Suter B., and Lasko P. 1998. *vasa* is required for GURKEN accumulation in the oocyte, and is involved in oocyte differentiation and germline cyst development. *Development* **125:** 1569–1578.

Subramaniam K. and Seydoux G. 1999. *nos-1* and *nos-2*, two genes related to *Drosophila nanos*, regulate primordial germ cell development and survival in *Caenorhabditis elegans*. *Development* **126:** 4861–4871.

Sulston J.E., Schierenberg E., White J.G., and Thomson J.N. 1983. The embryonic cell lineage of the nematode *Caenorhabditis elegans*. *Dev. Biol.* **100:** 64–119.

Tabara H., Hill R.J., Mello C.C., Priess J.R., and Kohara Y. 1999. *pos-1* encodes a cytoplasmic zinc-finger protein essential for germline specification in *C. elegans*. *Development* **126:** 1–11.

Tafuri S., and Wolffe A. 1990. *Xenopus* Y-box transcription factors: Molecular cloning, functional analysis and developmental regulation. *Proc. Natl. Acad. Sci.* **87:** 9028–9032.

———. 1992. DNA binding, multimerization and transcription stimulation by the *Xenopus* Y box proteins in vitro. *New Biol.* **4:** 349–359.

———. 1993. Selective recruitment of masked maternal mRNA from messenger ribonucleoprotein particles containing FRGY2 (mRNNP4). *J. Biol. Chem.* **268:** 24255–24261.

Tafuri S.R., Familari M., and Wolffe A.P. 1993. A mouse Y box protein, MSY1, is associated with paternal mRNA in spermatocytes. *J. Biol. Chem.* **268:** 12213–12220.

Takimoto M., Tomonaga T., Matunis M., Avigan M., Krutzsch H., Dreyfuss G., and Levens D. 1993. Specific binding of heterogeneous ribonucleoprotein particle protein K to the human c-*myc* promoter, in vitro. *J. Biol. Chem.* **268:** 18249–18258.

Tautz D. and Pfeifle C. 1989. A non-radioactive in situ hybridization method for the localization of specific RNAs in *Drosophila* embryos reveals translational control of the segmentation gene hunchback. *Chromosoma* **98:** 81–85.

Taylor S.J. and Shalloway D. 1994. An RNA-binding protein associated with Scr through its SH2 and SH3 domains in mitosis. *Nature* **368:** 867–871.

Theurkauf W.E. 1994. Premature microtubule-dependent cytoplasmic streaming in *cappuccino* and *spire* mutant oocytes. *Science* **265:** 2093–2096.

Thiele B., Andree H., Hohne M., and Rapoport S. 1982. Lipoxygenase mRNA in rabbit reticulocytes. Its isolation, characterization, and translational repression. *Eur. J. Biochem.* **129:** 133–141.

Thompson S.R., Goodwin E.B., and Wickens M. 2000. Rapid deadenylation and poly(A)-dependent translational repression mediated by the *Caenorhabditis elegans tra-2* 3′UTR in *Xenopus* embryos. *Mol. Cell. Biol.* **20:** 2129–2137.

Timmer R.T., Benkowski L.A., Schodin D., Lax S.R., Metz A.M., Ravel J.M., and Browning K.S. 1993. The 5′ and 3′ untranslated regions of satellite tobacco necrosis virus RNA affect translational efficiency and dependence on a 5′ cap structure. *J. Biol. Chem.* **268:** 9504–9510.

Tinker R., Silver D., and Montell D.J. 1998. Requirement for the Vasa RNA helicase in *gurken* mRNA localization. *Dev. Biol.* **199:** 1–10.

Tomancak P., Guichet A., Zavorszky P., and Ephrussi A. 1998. Oocyte polarity depends on regulation of *gurken* by Vasa. *Development* **125:** 1723–1732.

Vande Woude G. 1994. On the loss of *mos*. Nature **370:** 20–21.

Varnum S.M. and Wormington W.M. 1990. Deadenylation of maternal mRNAs during *Xenopus* oocyte maturation does not require specific *cis*-sequences: A default mechanism for translational control. *Genes Dev.* **4:** 2278–2286.

Vernet C. and Artzt K. 1997. STAR, a gene family involved in signal transduction and activation of RNA. *Trends Genet.* **13:** 479–484.

Voeltz G.K., and Steitz J.A. 1998. AUUUA sequence direct mRNA deadenylation uncouled from decay during *Xenopus* early development. *Mol. Cell. Biol.* **18:** 7539–7545.

Walker J., Minshall N., Hake L., Richter J., and Standart N. 1999. The clam 3′UTR masking element-binding protein p82 is a member of the CPEB family. *RNA* **5:** 14–26.

Wang C. and Lehmann R. 1991. nanos is the localized posterior determinant in *Drosophila*. *Cell* **66:** 637–648.

Wang C., Dickinson L., and Lehmann R. 1994. Genetics of *nanos* localization in *Drosophila*. *Dev. Dyn.* **199:** 103–115.

Wang S. and Miller W.A. 1995. A sequence located 4.5 to 5 kilobases from the 5′ end of the barley yellow dwarf virus (PAV) genome strongly stimulates translation of uncapped mRNA. *J. Biol. Chem.* **270:** 13446–13452.

Wang S., Browning K.S., and Miller W.A. 1997. A viral sequence in the 3′-untranslated region mimics a 5′ cap in facilitating translation of uncapped mRNA. *EMBO J.* **16:** 4107–4116.

Wang X., Kiledjian M., Weiss I., and Liebhaber S.A. 1995. Detection and characterization of a 3′ untranslated region ribonucleoprotein complex associated with human α-globin mRNA stability. *Mol. Cell. Biol.* **15:** 1769–1777.

Wells S.E., Hillner P.E., Vale R.D., and Sachs A.B. 1998. Circularization of mRNA by eukaryotic translation initiation factors. *Mol. Cell* **2:** 135–140.

Wharton R.P. 1992. Regulated expression from maternal mRNAs in *Drosophila*. *Semin. Dev. Biol.* **3:** 391–397.

Wharton R.P. and Struhl G. 1989. Structure of the *Drosophila BicaudalD* protein and its role in localizing the posterior determinant nanos. *Cell* **59:** 881–892.

———. 1991. RNA regulatory elements mediate control of *Drosophila* body pattern by the posterior morphogen *nanos*. *Cell* **67:** 955–967.

White-Cooper H., Schafer M.A., Alphey L.S., and Fuller M.T. 1998. Transcriptional and post-transcriptional control mechanisms coordinate the onset of spermatid differentiation with meiosis I in *Drosophila*. *Development* **125:** 125–134.

Wickens M. 1993. Springtime in the desert. *Nature* **363:** 305–306.

Wickens M. and Takayama K. 1994. RNA Deviants – or emissaries. *Nature* **367:** 17–18.

Wickens M., Anderson P., and Jackson R.J. 1997. Life and death in the cytoplasm: Messages from the 3′ end. *Curr. Opin. Genet. Dev.* **7:** 220–232.

Wightman B., Ha I., and Ruvkun G. 1993. Posttranscriptional regulation of the heterochronic gene *lin-14* by *lin-4* mediates temporal pattern formation in *C. elegans*. *Cell* **75:** 855–862.

Wightman B., Burglin T., Gatto J., Arasu P., and Ruvkun G. 1991. Negative regulatory sequences in the *lin-14* 3′-untranslated region are necessary to generate a temporal switch during *Caenorhabditis elegans* development. *Genes Dev.* **5:** 1813–1824.

Wilson J.E., Connell J.E., and Macdonald P.M. 1996. *aubergine* enhances *oskar* translation in the *Drosophila* ovary. *Development* **122:** 1631–1639.

Wolffe A. 1992. The Y-box factors: A family of nucleic acid binding proteins conserved from *Escherichia coli* to man. *New Biol.* **4:** 290–298.

———. 1994. Structural and functional properties of the evolutionary ancient Y-box family of nucleic acid binding proteins. *BioEssays* **16:** 245–251.

Wreden C., Verrotti A.C., Schisa J.A., Lieberfarb M.E., and Strickland S. 1997. *Nanos* and *pumilio* establish embryonic polarity in *Drosophila* by promoting posterior deadenylation of *hunchback* mRNA. *Development* **124:** 3015–3023.

Wu L., Wells D., Tay J., Mendis D., Abbott M.-A., Barnitt A., Quinlan E., Heynen A., Fallon J.R., and Richter J.D. 1998. CPEB-mediated cytoplasmic polyadenylation and the regulation of experience-dependent translation of αCaMKII mRNA at synapses. *Neuron* **21:** 1129–1139.

Yamanaka S., Poksay K.S., Arnold K.S., and Innerarity T.L. 1997. A novel translational repressor mRNA is edited extensively in livers containing tumors caused by the transgene expression of the apoB mRNA-editing enzyme. *Genes Dev.* **11:** 321–333.

Yew N., Strobel M., and Vande Woude G. 1993. *Mos* and the cell cycle: The molecular basis of the transformed phenotype. *Curr. Opin. Genet. Dev.* **3:** 19–25.

Yoon C., Kawakami K., and Hopkins N. 1997. Zebrafish *vasa* homologue RNA is localized to the cleavage planes of 2- and 4-cell-stage embryos and is expressed in the primordial germ cells. *Development* **124:** 3157–3166.

Zaffran S., Astier M., Gratecos D., and Sémériva M. 1997. The *held out wings (how) Drosophila* gene encodes a putative RNA-binding protein involved in the control of muscular and cardiac activity. *Development* **124:** 2087–2098.

Zamore, P.D., Williamson J.R., and Lehmann R. 1997. The Pumilio protein binds RNA through a conserved domain that defines a new class of RNA-binding proteins. *RNA* **3:** 1421–1433.

Zamore P.D., Bartel D.P., Lehmann R., and Williamson J.R. 1999. The PUMILIO-RNA interaction: A single RNA-binding domain monomer recognizes a bipartite target sequence. *Biochemistry* **38:** 596–604.

Zelus B.D., Giebelhaus D.H., Eib D.W., Kenner K.A., and Moon R.T. 1989. Expression of the poly(A)-binding protein during development of *Xenopus laevis. Mol. Cell. Biol.* **9:** 2756–2760.

Zhang B., Gallegos M., Puoti A., Durkin E., Fields S., Kimble J., and Wickens M. 1997. A conserved RNA binding protein that regulates patterning of sexual fates in the *C. elegans* hermaphrodite germ line. *Nature* **390:** 477–484.

Zhang M., Zamore P.D., Carmo-Fonesca M., Lamond A.I., and Green M.R. 1992. Cloning and intracellular localization of the U2 small nuclear ribonucleoprotein auxiliary factor small subunit. *Proc. Natl. Acad. Sci.* **89:** 8769–8773.

Zhong J., Peters A.H., Lee K., and Braun R.E. 1999. A double-stranded RNA binding protein required for activation of repressed messages in mammalian germ cells. *Nat. Genet.* **22:** 171–174.

Zorn A.M. and Krieg P.A. 1997. The KH domain protein encolded by *quaking* functions as a dimer and is essential for notochord development in *Xenopus* embryos. *Genes Dev.* **11:** 2176–2190.

8

Viral Translational Strategies and Host Defense Mechanisms

Tsafi Pe'ery and Michael B. Mathews

Department of Biochemistry and Molecular Biology
New Jersey Medical School
University of Medicine and Dentistry of New Jersey
Newark, New Jersey 07103

Viruses are obligate intracellular parasites or symbionts and depend on cells for their replication. Virus–cell interactions reflect this dependency as well as the efforts of the host to combat infection, and of the virus to counter host defenses. Because of the intimate nature of these relationships between the protagonists, the study of viruses continues to shed light on the detailed workings of host systems at a variety of levels.

Viruses lack the enzymes and associated apparatus for conducting most metabolic and biosynthetic reactions. Instead, they rely on the cells that they infect to supply energy, chemicals, and most of the necessary biosynthetic machinery. Many viruses encode enzymes for nucleic acid biosynthesis, but—with the rare exceptions described below—they do not encode any part of the translational apparatus. They are therefore forced to make use of the cellular apparatus for the synthesis of one of their principal components. As a consequence, and because they can be manipulated with some ease, viral systems have historically provided penetrating insights into the workings of the protein synthetic machinery. Some landmark discoveries made with viruses are listed in Table 1.

In their interactions with the host translation system, viruses do more than simply co-opt the cellular machinery, however. They have adopted regulatory mechanisms from their hosts and perhaps invented some of their own, although several seemingly unorthodox viral mechanisms are now recognized in the cells' own repertoire. Nevertheless, it is difficult to avoid the impression that a greater range and diversity of tactics are employed in the translation of viral than cellular mRNAs. Conspicuously, many viruses impose sweeping changes upon the cellular machinery, modifying the sys-

Translational Control of Gene Expression
© 2000 Cold Spring Harbor Laboratory Press 0-87969-568-4/00 $5 + .00

Table 1 Viruses and milestones in protein synthesis

Concept or discovery	Virus	References
Biochemical evidence for the existence of mRNA	phages T2 and T4	Volkin and Astrachan (1956); Gros et al. (1961); Brenner et al. (1961)
Faithfully initiating cell-free translation systems	phage f2; EMCV	Nathans et al. 1962; Kerr et al. (1966); Mathews and Korner (1970); Smith et al. (1970)
Breaking the genetic code	TMV; phage T4	Wittmann and Wittmann-Liebold (1967); Barnett et al. (1967)
Identification of initiator tRNAs	phages R17 and f2; EMCV	Adams and Capecchi (1966); Webster et al. (1966); Smith and Marcker (1970)
Characterization of ribosome-binding sites	phages Qβ and R17; brome mosaic virus; reovirus	Hindley and Staples (1969); Steitz (1969); Dasgupta et al. (1975);Lazarowitz and Robertson (1977)
Poly(A)	vaccinia virus	Kates (1971)
7-methyl guanosine cap	reovirus; vaccinia virus	Furuichi et al. (1975); Wei and Moss (1975)
Scanning model for initiation site selection	several plant and animal viruses	Kozak (1978)
Frameshifting	RSV	Jacks and Varmus (1985)
Internal ribosome entry site	poliovirus; EMCV	Pelletier and Sonenberg (1988); Jang et al. (1988)
Ribosome hopping	phage T4	Weiss et al. (1990)
Ribosome shunting	CaMV	Fütterer et al. (1993)

Abbreviations: (TMV) Tobacco mosaic virus; (RSV) Rous sarcoma virus; (EMCV) encephalomyocarditis virus; (CaMV) cauliflower mosaic virus.

tem to favor the synthesis of their own proteins at the cells' expense. Furthermore, viruses ingeniously appropriate components of the translation system for other purposes, such as nucleic acid replication. Examples of the redeployment of the protein synthesis apparatus in virus-infected cells are given in Table 2, and the alternative functions of translation factors are fully discussed in Chapter 36.

For their part, cells confronted with a virus do not passively resign themselves to their fate. Infection triggers defensive measures—in prokaryotic as well as eukaryotic cells—that are designed to limit viral multiplication. Host defenses act at all levels, from the translational level (e.g.,

Table 2 Translational components that serve alternative functions in virus-infected cells

Translational components	Alternative function in virus-infected cells	References
EF1A and EF1B ($EF\text{-}T_u$ and $EF\text{-}T_s$) ribosomal protein S1 and HF-I [a]	replication of RNA phages (Qβ, etc.)	Kamen (1970); Kondo et al. (1970); Muffler et al. (1996)
eEF1A, eEF1Bα, and eEF1Bβ [b]	binding to VSV RNA polymerase (replication?)	Das et al. (1998)
Elongation factors, aminoacyl-tRNA synthetases, nucleotidyl- and methyltransferases	binding to tRNA-like sequences at the 3′ ends of plant viral RNAs, e.g., TMV (replication?)	Haenni et al. (1982)
Ribosomal protein L22	binds EBER-1 of EBV (PKR control?)	Toczyski et al. (1994)
tRNA (e.g., $tRNA^{lys3}$)	reverse transcriptase primer for retroviruses (e.g., HIV-1)	Coffin (1996)

Abbreviations: (VSV) Vesicular stomatitis virus; (EBV) Epstein-Barr virus; (TMV) tobacco mosaic virus.

[a] HF-I (host factor I) is a loosely bound ribosomal protein required for translation of σ^s protein in *E. coli* (Muffler et al. 1996).

[b] eEF1 polypeptides engage in several other associations with viral proteins and nucleic acids (see Chapter 36).

through initiation factor modifications) to the level of the cell (as in apoptosis) and the whole organism (via interferon production and mobilization of the immune system). In their turn, these defense mechanisms are blunted by viral countermeasures that aim to sustain viral multiplication.

In this chapter, we discuss the interplay between viruses and the translation system of the cell. We first outline the strategies of viral infection, host defense systems, and viral countermeasures. Focusing on the translation system, we next review viral translational mechanisms and their regulation. Finally, we describe host defenses and viral countermeaures acting at the translational level. Most of the discussion centers on viruses that infect mammals, especially humans, but examples are also drawn from bacteriophages and viruses that infect other eukaryotes.

Throughout this chapter, groups of viruses are usually referred to by their common names or by the name of the best-studied prototype. Formal nomenclature, as well as more comprehensive background information, can be found in standard texts such as Fields et al. (1996) and Flint et al. (1999). For detailed reviews of translational control in cells infected with specific viruses, the reader is referred to Chapters 31–35 on picornaviruses, adenovirus, reovirus, influenza virus, and pox viruses, respectively.

STRATEGIES OF VIRAL INFECTION AND HOST DEFENSE

Viruses as a group are exceptionally heterogeneous and almost certainly polyphyletic in origin. Their strategies for infecting cells and replicating within them are diverse; accordingly, their interactions with their hosts, and with the cellular protein synthesis machinery, are rich and varied. To set these in perspective, we consider features of virus structure and function, describe the impact of infection on host cells, and, last, give an overview of the nature of host defense systems and the viral countermeasures that neutralize them.

Viruses and Viral Infection

Virus Structure

All viruses consist of a nucleic acid genome surrounded by a protective shell of viral protein(s) which forms the capsid. In enveloped viruses (e.g., retroviruses, herpesviruses, and influenza virus) the virion acquires an additional membranous covering, containing viral proteins (often glycoproteins) together with cellular components, as it "buds" through cell membranes. During this process, some viruses also incorporate cytosolic constituents such as tRNA, which serves as a reverse transcription primer for retroviral replication (Coffin 1996; Chapter 36).

Virus particles vary in size over a large range. At one extreme, the virions of picornaviruses and hepatitis delta virus (HDV) are comparable to a ribosome (~30 nm in diameter), whereas the largest approach the dimensions of a mitochondrion (vaccinia particles are a few hundred nm in diameter). Viral genomic complexity varies accordingly: The simplest contain 3–4 kilobases (kb) of nucleic acid (e.g., RNA phages, parvoviruses)—even less (1.7 kb) in the helper-dependent HDV—whereas the most complex are over 200 kb (herpes- and vaccinia viruses). The genetic material may be either DNA or RNA, single- or double-stranded (or partially both) and, if single-stranded, of positive or negative polarity (i.e., equivalent to mRNA or to its complement). Moreover, the genome may be circular or linear, and unipartite or segmented into several "chromosomes." Complete genome sequences are available for a large number of viruses, including prototypical members of most families; these provide a basis for understanding the range of activities of which the viruses are capable, and their reproductive strategies.

Viral Gene Products

Viral genes number from as few as two or three to several hundred, and they may include noncoding RNAs as well as mRNAs. Some noncoding RNAs serve a regulatory role in translation (see Viral Countermeasures against Translation Inhibition), but only the T-even phages are unambiguously proven to carry genes for bona fide translational components. A cluster of tRNA genes, the number and nature of which is a characteristic of the particular T-even phage, assists in host cell shutoff as discussed in the section on tRNA cleavage. Less clear-cut is the role of the eIF2α homolog encoded by iridoviruses (Yu et al. 1999; Chapter 35). These viruses are related to pox viruses, which do not themselves contain such a protein although they encode a shorter protein, K3L, that is homologous to the amino terminus of eIF2α and inhibits the eIF2α kinase PKR (see Viral Countermeasures against Translation Inhibition). It is not known whether the iridovirus eIF2α homolog functions as either an initiation factor subunit or a PKR inhibitor.

Whereas viral genomes must contain all the *information* needed for their own multiplication, they do not necessarily encode all the *enzymes* needed for the replication of their genetic material. Many viruses utilize host-cell DNA and RNA polymerases, and helper-dependent viruses (such as HDV) also require functions provided by larger viruses (hepatitis B virus in this case). Even if they are autonomous for such activities, the nucleic acid genomes of some viruses are not infectious in the absence of virus-coded enzymes that are packaged into the virus particle: Thus, retroviruses need reverse transcriptase (and tRNA) to convert the RNA genome to DNA, some other RNA viruses need RNA-dependent RNA polymerases, and vaccinia virus encodes numerous enzymes including RNA polymerase. The possession of such enzymes has many ramifications, including for the site of viral replication within the cell (discussed in the next section).

Apart from genes for replication functions, viral genomes typically encode one or more structural proteins together with a variable number of regulatory products. These function through interactions with both viral and host components, including components of the translation apparatus. As a general rule, larger and more complex viruses encode more gene products that interact with the translational machinery, but even the simplest viruses engage in sophisticated regulatory interactions.

Sites of Viral Replication

Critical to virus–host interactions is the site of viral genome replication within the cell. Many viruses replicate and are assembled in specific structures termed inclusion bodies, replication compartments, or viral factories. These are not fully defined but comprise both viral and cellular components, including cytoskeletal elements and the viral core in the case of reovirus, and DNA replication enzymes in the case of adeno- and herpesviruses (Knipe 1996).

Whether replication takes place in the nuclear or cytoplasmic compartment of the cell is a fundamental characteristic of each individual virus family that is closely tied to its strategy for mRNA production (Fig. 1A). Viruses that depend on cellular transcription enzymes replicate in the nucleus. This group includes all the DNA viruses except for vaccinia, together with two groups of RNA viruses: the retroviruses (whose genomes go through a chromosomally integrated DNA phase) and influenza virus (which pirates the capped 5′ end of nuclear mRNA precursors as primers for viral transcription). All other known viruses replicate in the cytoplasm, which they are equipped to do because they encode their own replicases or import them in the virion. Vaccinia virions contain viral RNA polymerase as well as enzymes for capping, methylation, and polyadenylation of the products. This virus is not entirely self-sufficient for transcription, however, and cannot replicate in enucleated cells, presumably because of a requirement for host nuclear proteins (Rosales et al. 1994). Among the RNA viruses (retroviruses and influenza viruses excepted), those whose genomes are plus strands (e.g., picornaviruses) generate the requisite enzymes directly by translation; those whose genomes are double-stranded (e.g., reovirus) or of negative polarity (e.g., vesicular stomatitis virus, VSV) package RNA-dependent RNA polymerases in their virions, permitting the generation of plus strands and mRNA.

No virus that is equipped for replication in the cytoplasmic compartment has opted to billet itself in the nucleus, as far as we know, even though it would appear possible for it to do so. Evidently a cytoplasmic location is preferable, but it is unclear whether this is because of proximity to the protein synthesis machinery or to other factors, such as ready entry and exit access or the availability of energy and other supplies.

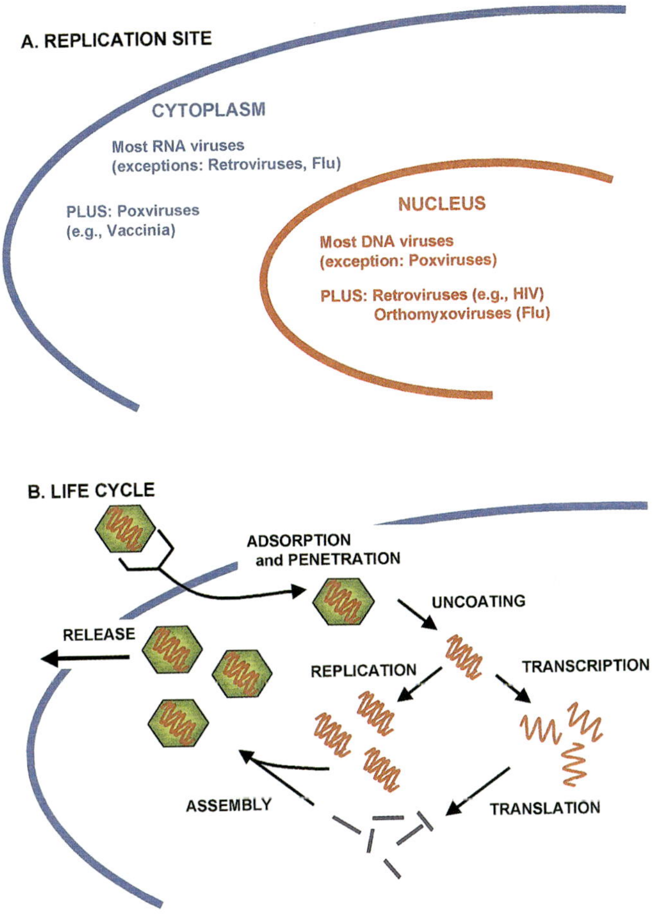

Figure 1 Viral replication. (*A*) The sites of replication for different groups of viruses. (*B*) A generalized viral life cycle.

Life Cycles and Switches

Despite numerous variations, virus life cycles follow the general pattern illustrated in Figure 1B. Infection takes place after the virus has adsorbed to receptors on the cell surface. Susceptible cells carry suitable receptor molecules for the virus in question; other cells are resistant, although they may be infected experimentally by means such as microinjection or transfection. After penetrating the cell, the virus is uncoated, releasing the genome as naked nucleic acid or nucleoprotein, or exposing it in a more limited way as nucleocapsid. Infectious virus then becomes undetectable (the "eclipse" phase) until progeny virions are elaborated.

Many viral genomes, especially those of DNA viruses, are programed to generate products in an orderly fashion and temporal sequence. This is determined largely through transcriptional controls, but also by regulation at subsequent levels. In productive infections with most DNA viruses, the infectious cycle is divided into two phases, early and late, demarcated by the onset of viral DNA replication. During the early phase, cell metabolism is little disturbed and a subset of the viral genes is expressed, generally at a modest level. Early viral products include replication enzymes and regulatory products that set the scene for a more extensive redirection of the host cell activities. These regulatory proteins exert a multitude of functions: Some are transcriptional activators that activate cellular genes needed for viral replication, and others suppress host defenses by interceding in such processes as apoptosis or antigen presentation (see below, Immune Defenses and Viral Countermeasures). In some more complex DNA viruses, such as herpesviruses, the early phase is subdivided into immediate-early and intermediate stages, depending on whether the genes are expressed autonomously or require viral proteins for their expression. During the late phase, template number and transcriptional activity both increase, resulting in the abundant production of viral mRNA. The synthesis of early proteins generally declines, and a new class of late proteins accumulates including the coat protein(s) and other virion components, as well as proteins required for viral morphogenesis and related functions.

Infections with RNA viruses are not generally characterized by well-defined phases of this sort. Nevertheless, their infectious cycles are not wholly undifferentiated, and they may accomplish temporal regulation by mechanisms operating at the level of transcription, mRNA splicing or transport, or translation. Examples of temporal switches operating at the translational level come from the single-stranded RNA phages whose genomes serve directly as mRNAs, as discussed below under Regulation of Viral Gene Expression at the Translational Level.

Infection Outcomes

Infection does not lead inexorably to virus multiplication and cell death, although this is the kind of interaction that underlies most of our discussion to this point and is the most conducive to study in the laboratory. Whether, and to what extent, a virus multiplies depends on its virulence, the permissivity of the cell, and the response of the host, all of which are multifactorial properties operating at many levels. Here we define possible infection outcomes, from the perspective of both the virus and the cell.

Productive and Nonproductive Infection

From the cell's point of view, infection can lead to a wide range of outcomes, including abortive, persistent, and latent infections; oncogenesis; and apoptosis. In a cell that contains all the factors required by the virus, productive infection occurs and viral progeny are assembled. If the virus is endowed with lytic functions, the cell breaks open and virus is released; otherwise, viruses emerge by budding through the cell membrane or are released passively after cell death. Translational control is more easily studied in such infections because the virus often comes to dominate all aspects of cell macromolecular synthesis.

Many cells are nonpermissive or only partially permissive, however. At one extreme, the virus may disappear altogether, its genome may persist in episomal form, or it may integrate some or all of its genome into the cell genome resulting in cellular transformation and possibly in malignancy (Chapter 20). Alternatively, there is slow or intermittent production of infectious virus. Persistently infected cultures continue to produce virus at low levels, either because few cells are productive at any one time or because the infected cells produce virus at a low rate and survive undamaged. Latently infected cells (such as nerve cells infected with herpes virus) contain the viral genome in a quiescent state, but virus production is undetectable until triggered by external stimuli. These more subtle interactions of virus and cell are common but more difficult to study from the perspective of translational control. In such infections, it is assumed that one or more permissivity factors are missing or limiting, and in some cases at least the factors appear to be operating at the translational level. For example, in VSV infections of B lymphocytes, viral mRNA is associated with polysomes but is not translated without cellular activation by mitogens or phorbol esters (Schmidt et al. 1995). The $\gamma_1 34.5$ protein of herpes simplex virus-1 (HSV-1), which is required for growth in neural cells, blocks the shutoff of host protein synthesis by PKR (He et al. 1997). The neuropathogenicity of poliovirus requires a sequence in its internal ribosomal entry site (IRES) that is dispensable for virus growth in nonneuronal cells (Gromeier et al. 1999). Further examples of cell-specific permissivity factors will surely emerge in the future.

Apoptosis

Another possible outcome is the elimination of virus-infected cells by apoptosis. Apoptosis is gene-directed cell suicide and differs from necrotic cell death morphologically and biochemically. During apoptosis, cells undergo DNA fragmentation and form apoptotic bodies that are phagocytosed by macrophages and other cells of the immune system. Apoptosis can be triggered by many different external signals, which are delivered through signal transduction pathways and executed via the activation of caspases, a family of cysteine proteases that are produced in an inactive form (procaspases) and activated through a proteolytic cascade. Along these pathways there are many checkpoints where an anti-apoptotic signal can reverse the process (Wang et al. 1999).

Apoptosis can be triggered from outside the cell by a death signal elicited, for example, by the binding of tumor necrosis factor-α (TNFα) to a death receptor which contains a death domain (DD). Next, adapter molecules (such as FADD), also containing the DD, are recruited by binding to the DD of the receptor. These adapters then bind to the most upstream procaspases via another domain, the death effector domain (DED) common to both proteins, thereby igniting the caspase chain reaction. Caspase activation eventually leads to cell death unless intervention occurs at the appropriate stage (Cryns and Yuan 1998). A major checkpoint in the apoptotic pathway is in the mitochondria. Mitochondria lie on the major pathway leading to apoptosis, and also contain anti-apoptotic members of the Bcl-2 family (e.g., Bcl-x_L) whose prosurvival contribution is to prevent the formation of the "apoptosome." The driving force for apoptosis is sequestration of these proteins by pro-apoptotic members of the same family, such as Bax and Bad. This leads to the release of cytochrome *c* and Apaf-1 (apoptotic protease-activating factor 1) from the mitochondria and the formation of the apoptosome as a complex containing caspase-9. Activation of caspase-9 triggers the apoptotic protease cascade.

Apoptosis can also be triggered from within the cell by inhibition of host macromolecular synthesis resulting from viral infection (discussed further under Apoptosis and Translational Inhibition). One such internal stimulus is the activation of PKR; sensitivity to this stimulus is increased by type I interferons, which induce elevated levels of PKR as discussed below (see Induction of the Antiviral State) and in Chapter 13.

Once the apoptosome is formed, downstream caspases are activated which cleave many cellular proteins that are important for cell survival. Caspase substrates include anti-apoptotic proteins, as well as components of the translation system (Clemens et al. 2000). eIF4GI and eIF4GII are rapidly cleaved; in the case of eIF4GI, caspase 3 cleavage gives rise to a discrete 76-kD fragment (Bushell et al. 1999; see also Apoptosis and Translational Inhibition, below). Additional initiation factors that are cleaved at a later stage of apoptosis are eIF2α, 4E-BP1, and eIF4B (Satoh et al. 1999; Bushell et al. 2000). The signal recognition particle is also cleaved but no functional consequences were observed (Utz et al. 1998).

To ensure their survival, viruses have devised ways to interrupt apoptosis by blocking it at different checkpoints (described in the next section).

Immune Defenses and Viral Countermeasures

With the goal of ridding the organism of viruses and virus-infected cells, or at least of limiting pathogenicity, the host immune system responds to viral invasion in two phases. First is the early inflammatory phase, mediated by the innate branch of the immune system. This phase is regulated by soluble factors, the cytokines (including the interferons), chemokines, and the complement system; its effectors include natural killer cells, macrophages, and neutrophils. Second is the immune-specific phase which leads to acquired immunity and the production of viral antigen-specific cells and antibodies. This stage involves T and B lymphocytes, including cytotoxic T lymphocytes (CTLs), and is also regulated by cytokines and chemokines. Under the pressure of these immune defenses, viruses have developed numerous products acting at several levels to counteract the host immune response and ensure their survival and replication. In the case of lytic viruses, these countermeasures are mostly directed against the innate arm of the immune system since viral replication is usually fast, resulting in rapid destruction of the infected cell (see below, Host Cell Shutoff). Viruses causing persistent and latent infections have developed a larger range of products to defend themselves against both innate and acquired immunity. Here we summarize aspects of the immune system that are related to viral infection and then give examples of viral countermeasures in four broad categories.

Soluble Factors of the Immune System

Cytokines are proteins or polypeptides secreted by numerous cell types in response to appropriate stimuli. Their functions encompass effects on cell proliferation, inflammation, the immune system, and viral infection. Interferons exert powerful antiviral activities, including well-established direct effects on the cellular protein synthetic system (see below, Host Defenses and Viral Countermeasures at the Translational Level). TNF and certain interleukins (especially IL-1) also display antiviral activities, some of which impinge directly on infected cells, whereas others are mediated via the interferons or cells of the immune system (Vilček and Sen 1996). Chemokines are proinflammatory peptides released by multiple cell types in response to stimuli such as injury, antibodies, or invading microorganisms. They mobilize immune system cells toward the site of infection and also induce cytokine production. Chemokines and cytokines bind to specific cell membrane receptors and exert their effects via signal transduction pathways.

The complement system is activated by contact with pathogens, resulting in deposition of complement components on their surfaces. This can prevent viruses from binding to their cellular receptors or can cause aggregation facilitating phagocytosis. Healthy cells are protected from complement attack by the expression of surface proteins that down-regulate complement activity. These surface proteins share a common short consensus repeat sequence (SCR). Complement can also lyse viruses by generating a membrane attack complex (MAC) which forms pores in viral envelopes. Cells protect themselves by sequestering the complement component C9 (Lubinski et al. 1998).

Cells of the Immune System

Two types of immune system cells are relevant to the present discussion. Natural killer (NK) cells are induced rapidly upon viral infection. They have direct cytotoxic activity against virus-infected cells and also produce the antiviral cytokines TNF and interferon-γ. In the later stages of infection, virus-infected cells exhibit on their surface major histocompatibility complex (MHC) class I glycoproteins that present viral immunogenic peptides. The MHC class I antigen presentation pathway has many steps, including translocation of a nascent MHC molecule to the endoplasmic reticulum where the peptide is loaded, and trafficking of the complex via the Golgi compartment to the cell surface.

CTLs recognize high levels of specific MHC class I complexes bound to viral peptides. They exert their cytotoxic effect by lysing the cell, secreting cytokines, or inducing apoptosis. If the level of MHC class I on the surface of an infected cell is down-regulated, the cell can be killed by NK cells, although NK cell killing is inhibited by high levels of MHC class I. Thus, these two types of cytotoxic lymphocytes complement one another in destroying infected cells.

Viroceptors

Viral countermeasures against the immune system include proteins, termed viroceptors, that mimic cellular receptors and trap their natural ligands, the cytokines and chemokines. Poxviruses make viroceptors that sequester ligands of the TNF, interleukin, and interferon families (Krajcsi and Wold 1998). Each viroceptor shares a ligand-binding motif with its corresponding cellular receptor but, lacking a transmembrane domain, the viroceptors are secreted. Chemokines are sequestered by human cytomegalovirus (HCMV) viroceptors, as well as by chemokine-binding proteins of poxviruses which are functional homologs of the viroceptors but do not display sequence homology with known receptors. Other viroceptors of herpes- and pox viruses remain cell-associated but fail to signal appropriately to the immune system (McFadden et al. 1998). Similarly, in the second phase of the immune response, the function of antibodies directed against viral antigens can be blocked. HSV-1 proteins bind to the immunoglobulin G (IgG) Fc domain, preventing an interaction with cellular Fc receptors that triggers phagocytosis and other immune responses (Lubinski et al. 1998).

Inhibitors of Immune Pathways

A second way for viral proteins to counter the immune response to cytokines is by intercepting signals transmitted through the apoptotic pathway. For example, some herpes- and pox viruses encode proteins that contain motifs homologous to the DED domains. They therefore interact with FADD and compete for the binding of procaspase, thereby preventing the activation of caspases. The pox virus CrmA protein interferes with apoptosis at a later stage in the signal transduction pathway, acting as a caspase pseudosubstrate. These inhibitors mimic caspase substrates and form stable complexes with the enzymes (Hardwick 1998). Similarly, vaccinia virus expresses a homolog of complement regulatory proteins, thereby subverting the complement cascade (Lubinski et al. 1998).

Inhibitors of Immune Regulation

A related countermeasure is for viruses to produce proteins that act as mimics or homologs of cellular proteins and interfere with the function of host proteins that regulate the antiviral immune response. In the apoptosis pathway, proteins encoded by herpes- and adenoviruses function in the same way as the anti-apoptotic protein Bcl-2 to reverse the apoptotic signal at the mitochondrial stage. EBV-LMP1 is also involved in shifting the cells into an anti-apoptotic pathway via activation of NF-κB and transactivation of TNF-resistance genes (Krajcsi and Wold 1998). On the other hand, some herpesviruses induce apoptosis selectively in peripheral blood T lymphocytes, thereby delaying removal of the virus from the infected organism (Kieff and Shenk 1998).

In the interferon pathway, adenovirus E1A proteins can interfere with cellular regulators by binding transcriptional activators (CBP/p300) necessary for interferon-stimulated gene transcription, or by repressing the expression of STAT1 proteins that are part of the activator complex (Krajcsi and Wold 1998; see below, Induction of the Antiviral State). Viruses have also developed several strategies to counteract the complement system. For example, vaccinia virus expresses a surface protein that contains SCRs, thereby avoiding detection by the complement system. Other viruses trick this system by incorporating host protective proteins into their envelope during maturation. For example, HIV-1 and HCMV incorporate into their virions the MAC inhibitor CD59 that binds to C9 (Lubinski et al. 1998).

Since viruses that develop strategies to reduce the level of MHC class I complexes on the infected cell are in danger of being eliminated by NK cells, we find examples of viruses such as HCMV that produce decoys which are homologs of MHC class I heavy chain and can even bind peptide and inhibit NK cell action. Acting within the MHC class I pathway are adenovirus, herpesvirus, HCMV, and HIV-1 products that inhibit one or more of its steps—peptide translocation across the endoplasmic reticulum, peptide loading onto the MHC class I complex, and/or trafficking of the complex to the cell membrane (Farrell and Davis-Poynter 1998).

Antigenic Variation

Mutant viral proteins escape recognition by the host immune system, providing a straightforward way to avoid the immune response. Retroviruses (HIV-1 and HTLV-1) take advantage of the high rate of mutation incurred during their replication. Aided by CTL selection pressure, mutations can result either in loss of the immunogenic peptide that binds to MHC class I due to proteolysis of the peptide, or in reduced peptide recognition by the T-cell receptor. The net effect is depletion of the $CD8^+$ CTL population, which characterizes immunodepletion (Gould and Bangham 1998).

VIRUSES AND TRANSLATIONAL CONTROL

Turning to aspects of viral infection that relate closely to protein synthesis, we begin this section with an account of unusual translational mechanisms used by viruses. The section continues with a review of strategies marshalled by viruses to control their own gene expression at the translational level, and concludes with a description of modifications of the translational apparatus that are brought about by viruses in order to usurp the protein synthetic capability of the cell.

Unorthodox Translational Mechanisms Used by Viruses

Despite their reliance on the cellular translational apparatus, viruses have evolved to exploit it in unusual ways. They take advantage of an extensive armory of unconventional translational mechanisms which are depicted in

Figure 2. Although most, if not all, of these are presumably part of the cells' own repertoire (see, for example, Chapter 19), they appear to be employed less frequently or only in special circumstances in the uninfected cell (e.g., during mitosis, heat shock, or apoptosis). For the virus, such mechanisms serve one or more of the following roles: They may be

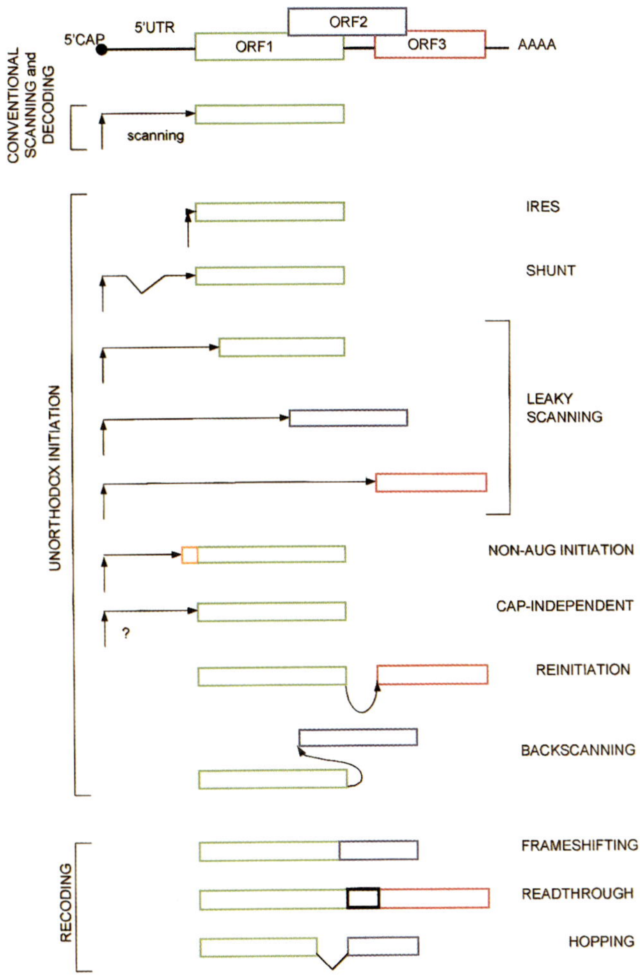

Figure 2 Unconventional translational strategies. The top line represents a capped and polyadenylated mRNA with a 5′ untranslated region (5′UTR) and three open reading frames (ORFs 1–3). Various strategies for initiating protein synthesis on this mRNA and for decoding its information are depicted on the lines below. (?) Although the 5′UTR of an uncapped mRNA is apparently scanned, the site of ribosome entry is not well defined.

a necessary part of the viral translational or replicative strategy (e.g., internal ribosome entry), part of a regulatory mechanism (shunting), or serve to expand the coding capacity of a confined genome (readthrough, frameshifting). In other cases, notably hopping, the raison d'etre is obscure. As described briefly below, these unorthodox tactics represent departures from the orderly and sequential readout of an mRNA by conventional scanning (Kozak 1989) and decoding from the 5′ end: Signals that are ordinarily respected are bypassed or ignored, either routinely or occasionally, in response to overriding or additional signals that specify deviations from the standard mechanism.

Internal Ribosome Entry

The entry of ribosomes at internal sites occurs in response to a signal known as the IRES, found in picornaviruses and a growing number of other viruses (Jang et al. 1988; Pelletier and Sonenberg 1988; see Chapters 4 and 31). An IRES typically spans four or five hundred nucleotides in the 5′-untranslated region (UTR) of the mRNA, and both sequences and higher-order RNA structure are required for its function. No viral proteins have been shown to be essential, although the poliovirus protease 2A plays a stimulatory role (Hambidge and Sarnow 1992), and at least four cellular proteins have been shown to stimulate IRES function in vitro (Chapter 31). Once associated with the mRNA, the ribosome initiates immediately after the IRES in most cases, or migrates to a nearby downstream initiation codon as in poliovirus, for example.

Initiation factor requirements are variable (see Chapters 4 and 31). For many viruses, like poliovirus and encephalomyocarditis virus (EMCV), IRES-mediated initiation dispenses with the requirement for eIF4E and is also insensitive to cleavage of eIF4G caused by viral proteases: Otherwise, all of the standard initiation factors are required. Such relaxed requirements for initiation factors free these viruses from the cap-dependent initiation mechanism and form the basis for host protein synthesis shutoff mechanisms (see below, Modification of the Translation Apparatus by Viruses). Hepatitis C virus (HCV) and the pestiviruses are exceptional in that their IRESs dispense with all the eIF4 family of initiation factors and can direct 40S ribosomal subunits to bind at a site very close to the initiating AUG in the absence of all initiation factors. Hepatitis A virus (HAV), on the other hand, requires a functional eIF4G and is inhibited by 4E-BP1, which can sequester eIF4E (see below, 4E-BP1 Dephosphorylation, and Chapter 6). The IRES also endows viral mRNAs with two other advantages. First, ribosomes can ignore upstream sequences that are irrelevant to translation but play other roles in the viral life cycle—for example, in packaging or replication (Borman et al. 1994). Thus, AUG triplets and secondary structure that would inhibit initiation via the scanning mechanism can be tolerated in the 5′ end of these viral mRNAs. Second, the virus is relieved of the need to replicate in the nucleus where the cellular capping enzyme resides, or to provide its own capping enzyme as most cytoplasmically replicating viruses do.

Ribosome Shunting

Another mechanism for avoiding primary and secondary structure in the 5′UTR is the ribosome shunt (Chapter 4). First characterized in cauliflower mosaic virus (CaMV) 35S RNA (Fütterer et al. 1993) and related plant pararetroviruses, shunting has also been observed in Sendai and papillomaviruses (Latorre et al. 1998; Remm et al. 1999). The mechanism is discussed in detail in Chapter 32 for adenovirus RNAs containing the tripartite leader. Shunting allows 40S ribosomal subunits to bind to the cap, scan for a short distance from the 5′ terminus of the mRNA, then skip to a site up to hundreds of nucleotides downstream without scanning through the intervening sequences (Fig. 2). In CaMV, ribosomes translate a short 5′ ORF in a cap-dependent fashion, traverse an intervening highly structured region, then "land" at or near the initiation site for the major viral ORF. In this case, shunting involves reinitiation after translation of the short upstream ORF (Ryabova and Hohn 2000). No plant or viral protein is essential, but the product of the CaMV ORF VI gene, TAV, is stimulatory. In the Sendai virus P/C mRNA, shunting seems to be triggered by upstream secondary structure or AUG codons, signals that militate against scanning, and ribosomes can initiate at the landing site even if it is mutated away from AUG (Latorre et al. 1998). In adenovirus late mRNAs, ribosomal subunits appear to scan through the 5′-proximal region of the tripartite leader, skip a region of secondary structure, then land and continue scanning for a short distance. Leader sequences complementary to the 3′ end of 18S ribosomal RNA (rRNA) are required, although it is not known whether they actually pair together (Yueh and Schneider 2000). Like the IRES, the shunt permits ribosomes to avoid 5′UTR sequences and structure—even large ORFs in the CaMV case—which are incompatible with conventional scanning. Furthermore, shunting predominates over conventional scanning at late times of adenovirus infection when eIF4E phosphorylation is decreased, consistent with a lower requirement for eIF4F (see below, eIF4E Dephosphorylation, and Chapter 32).

Leaky Scanning

The leaky scanning hypothesis accommodates mRNAs in which a downstream AUG is used for initiation in preference or in addition to the first (i.e., 5′-proximal) AUG. Up to 10% of eukaryotic mRNAs are in this class, and many of these encode two proteins (see Chapter 18). The proteins may be in the same or different reading frames, and dicistronic mRNAs of this type are found in a large number of viruses. One of the principal features defining the initiation site in an mRNA is the context in which the initiation codon is set. The most favorable context, deduced by sequence comparisons and experimentation, is **A**CC<u>AUG</u>**G**, where the A at position –3 and the G at position +4 (relative to the A of the initiator AUG triplet, assigned +1) are the chief determinants of a strong initiation site (Kozak 1989). In dicistronic mRNAs, the first AUG is often in a suboptimal context, suggesting that it functions as a weak initiator. According to the scanning model, this site would be bypassed at a substantial frequency, allowing for initiation at a subsequent site downstream.

Regardless of the detailed mechanism, leaky scanning is widely used in viruses, where it presumably helps to economize on coding space and signals for transcription and RNA processing. In HIV-1, for example, the essential envelope protein (Env) is translated from mRNAs that contain an upstream ORF encoding the accessory protein Vpu in a different reading frame. To permit Env synthesis, the *vpu* initiation site is required to be weak (Schwartz et al. 1992). When the two ORFs are in-frame with each other, the result is a nested pair of proteins with overlapping carboxy-terminal sequences and related functions (e.g., SV40 coat proteins VP2 and VP3). On the other hand, when the two ORFs are in different reading frames, the resultant proteins need not be functionally related (e.g., HIV-1 Vpu and Env); nevertheless, they sometimes are (e.g., adenovirus-5 E1B 19-kD and 55-kD proteins), perhaps as a reflection of their evolutionary origins. In either case, regulatory possibilities abound. For example, there might be interactions between the two ORFs. This has been suggested for the overlapping but out-of-frame ORFs in the reovirus S1 transcript, where ribosomes translating the σ1 ORF may impede those translating the downstream σ1s ORF (Fajardo and Shatkin 1990; Belli and Samuel 1993). Furthermore, subtle adjustments in the translation apparatus might lead to changes in the ratio of the proteins produced by leaky scanning; however, it is not known whether this potential is realized in the infected cell and, if so, under what circumstances.

Initiation at Non-AUG Codons

Initiation can take place at codons other than AUG, either instead of (Fütterer et al. 1996) or in addition to (Corcelette et al. 2000) initiation at the conventional signal. Ribosomes may be predisposed to start at such unorthodox sites when they use mechanisms such as IRESs or shunting that result in internal placement rather than scanning to locate the initiation codon (Latorre et al. 1998; Ryabova and Hohn 2000). In the insect virus PSIV, for example, an IRES confers the ability to initiate with glutamine at a CAA codon (Sasaki and Nakashima 2000). The advantages accruing from this departure from conventional initiation potentially include increased protein-coding capacity and novel regulatory opportunities.

Translation of Uncapped RNA

The RNAs of some plant viruses, such as satellite tobacco necrosis virus (STNV; Wimmer et al. 1968), and of the yeast virus L-A (Nemeroff and Bruenn 1987; Masison et al. 1995) are exceptional in lacking 5′ cap and 3′ poly(A) structures. Unlike picornavirus RNA, which is also not capped, they function without an IRES element. In STNV it appears that a short 5′UTR sequence and a longer, structured 3′UTR sequence collaborate to function somewhat like an IRES (Timmer et al. 1993; see Chapter 4). In the case of L-A, initiation depends on interactions with the cellular SKI gene products (described in the section on The Yeast SKI System, below). Judging from experiments with an artificial uncapped RNA encoding HIV-1 Tat, initiation involves a scanning mechanism despite the absence of a cap (Gunnery et al. 1997).

Reinitiation

Since scanning and 5′ caps are specific to eukaryotes, the foregoing events apply exclusively to viruses that infect eukaryotic cells. Reinitiation provides another means whereby two proteins can be made from a single mRNA. Although rare in eukaryotes, reinitiation is common in prokaryotes where internal ribosome entry is standard. Prokaryotic cellular and viral mRNAs frequently contain nonoverlapping ORFs in tandem, and their ribosomes can generally gain access to all of the initiation sites within a prokaryotic polycistronic mRNA, allowing several proteins to be translated independently. In phage ϕX174, for example, two proteins are translated from the same sequence using separate ribosome-binding sites that give access to different reading frames (Ravetch et al. 1977).

The individual sites are not necessarily utilized with equal efficiencies, however, and translation of a downstream cistron may depend on the translation of an upstream cistron as described below in the section on Translational Coupling.

Evidence that reinitiation can be used by eukaryotes to gain access to downstream ORFs in a dicistronic mRNA has been derived from work with artificially constructed gene fusions (Peabody et al. 1986). In such mRNAs it appears that ribosomes can even "backscan" across a small interval to engage a suitable initiation site (Fig. 2). In natural eukaryotic mRNAs, however, when upstream ORFs occur they are generally short and often serve a regulatory function. Typically, they exert a negative influence over the translation of the downstream ORF, although the effects may be complex (Chapter 18). Yeast GCN4 mRNA contains four short upstream ORFs that are all intrinsically negative in character, yet the first two of them confer inducibility upon the long ORF encoding the GCN4 protein downstream (Chapter 5). In this case, the short encoded peptides do not appear to play a part in the phenomenon; in other cases, however, the effects are sensitive to changes in the upstream coding sequence, implying that the peptide product itself mediates the inhibition. For example, in HCMV the 22-residue peptide produced by translating an upstream ORF represses production of the gpUL4 (gp48) protein encoded downstream by blocking termination and stalling scanning ribosomes (Alderete et al. 1999). The short upstream ORF in CaMV 35S RNA is unusual: It exerts a positive effect on the translation of downstream ORFs and is required for shunting (Ryabova and Hohn 2000). This stimulatory action requires the presence of the viral transactivator protein TAV (see Shunting, above); furthermore, the ORF must encode a peptide of at least 30 amino acids, although its composition does not seem to be critical (Fütterer and Hohn 1992). Evidently, reinitiation is a rarely used strategy in eukaryotic viruses, but when it is used, it is subject to elaborate controls.

Frameshifting

During the decoding of some mRNAs, the advancing ribosome slips forward or back by one nucleotide, resulting in a programmed +1 or –1 change of reading frame (Jacks and Varmus 1985; Chapter 25). Such frameshifting events are common in retroviruses, for many of which a –1 shift is an essential event in reverse transcriptase synthesis. The shift takes place at a "slippery" site, where tRNA can move along the template by one base and reestablish codon:anticodon pairing. Generally, the efficiency of frameshifting is enhanced by a second element, either a pseudoknot or a hairpin positioned downstream, which presumably acts by causing ribosomes to pause at the shift site. The proportion of ribosomes that change reading frames is characteristic of each site, and is usually rather low. Therefore, the translation products consist of two proteins, a majority of the conventionally decoded polypeptide and a minority of the "recoded" form. In retroviruses, the Gag-Pol shifted product is usually about 5% of the unshifted Gag product. Most retroviruses translate *pol* this way, and similar events occur in coronaviruses and yeast L-A virus. There is little evidence that the proportion of frameshifting at a particular site is actively controlled (for review, see Atkins and Gesteland 1996), so the mechanism seems to be a device for increasing the coding capacity of the viral genome and for producing the two products in a fixed ratio. This ratio appears to be of critical importance, since a small increase or decrease in the Gag to Gag-Pol ratio is detrimental to virus assembly and proliferation (for review, see Farabaugh 1995; Dinman et al. 1998). Some progress has been made in developing antiviral drugs acting at this level (Hung et al. 1998).

Readthrough

The suppression or readthrough of stop codons is also used by many viruses to generate a carboxy-terminally extended protein at a fixed ratio to the conventionally translated product (Fig. 2). Examples are found in viruses infecting plants, bacteria, and mammals (e.g., the Moloney murine leukemia virus *gag-pol* protein). Again, there is no firm evidence that the process is controlled during infection.

Hopping

A final instance of recoding is the bypass of 50 nucleotides that occurs during the translation of phage T4 gene *60* mRNA (Fig. 2). This extraordinary event is mediated by the nascent peptide and requires duplications flanking the bypassed sequence, as well as a stop codon and a hairpin structure containing the 5′ end of the gap sequence (Weiss et al. 1990; Chapter 25). The hairpin requirement is reduced by a specific mutation in a 50S ribosomal subunit protein (Herbst et al. 1994). The regulatory significance of hopping is unknown.

Regulation of Viral Gene Expression at the Translational Level

Access of ribosomes to translational initiation sites is both mediated by and controlled through RNA:RNA and RNA:protein interactions.

Accordingly, a major determinant of translational efficiency in bacteria is mRNA secondary structure, particularly in the region of the ribosome-binding site (de Smit and van Duin 1990; Voorma 1996). The degree of secondary structure is subject to regulation by ribosomes, which serve as translational activators, and by *trans*-acting proteins, which generally function as translational repressors. Proteins binding in the vicinity of the initiation site can also influence ribosome binding directly or indirectly (Gold 1988; McCarthy and Gualerzi 1990). These mechanisms are well-known in bacteriophages but less so in eukaryotic viruses.

Translational Coupling

The initiation site is frequently engaged in higher-order structure that obscures an essential feature such as the AUG or the Shine-Dalgarno (SD) sequence, thereby restricting initiation. In some viral and cellular mRNAs, the cistrons are arranged such that this limitation is relieved by ribosomes traversing a different cistron that is usually, but not necessarily, upstream. In essence, translating ribosomes act as derepressors. Examples of such translational coupling have been reported in many phages, although not in viruses of eukaryotes, and have been intensively studied in the RNA phages and phage T4.

In the RNA phages f2, MS2, and Qβ, synthesis of the replicase protein depends on translation of the coat protein cistron that lies upstream. This was first evidenced by the observation that an amber mutation early in the coat protein gene exerts a polar effect on replicase synthesis, whereas an amber mutation later in the gene does not (Lodish and Zinder 1966; Lodish 1975). Similarly, other mutations that prevent the passage of ribosomes across the coat protein cistron also down-regulate replicase synthesis. These effects are due to long-range base-pairing between a sequence in the coat protein cistron and nucleotides immediately upstream of the replicase AUG. This interaction restricts access to the replicase initiation site, a restriction that is lifted by elongating ribosomes which disrupt the base-pairing. Consistent with this model, mutations or chemical treatments that abrogate the base-pairing interaction result in constitutive expression and up-regulation of replicase translation. MS2 phage mutants containing disruptions in the long-range interaction are observed to revert by restoring the same or similar base-pairing, providing strong evidence for the importance of this translational coupling (Licis et al. 1998).

Translation of the gene encoding the assembly or maturation protein, which is located at the 5′ end of the phage RNA, is limited by a cloverleaf-like stem-loop structure that obscures its SD sequence. In this case, access to the initiation site is assured by the delayed formation of the RNA secondary structure (Poot et al. 1997). Disruption of the cloverleaf leads to permanent activation of the gene, which is also selected against during laboratory evolution experiments. These findings emphasize the importance of translational control in the phage life cycle and the dynamic nature of RNA structure (Robertson 1975). Since translation takes place principally on nascent mRNAs in bacteria, and potentially in animal cells infected with viruses that replicate in the cytoplasm, it is necessary to consider temporal effects on protein synthesis related to the growth and folding of the RNA strand.

Most examples of translational coupling involve short-range base-pairing that is broken by terminating, rather than elongating, ribosomes. For example, the lysis peptide ORF of phages f2 and MS2 overlaps the 3′ end of the coat protein cistron. Removal of the coat protein initiation signal or insertion of amber mutations into the coat protein gene eliminates lysis protein production, consistent with translational coupling (Lodish and Zinder 1966; Lodish 1975). Moving the stop codon downstream has the same effect, however, indicating that the mechanism has something to do with the termination event rather than with elongation. Further mutagenesis has led to the proposal that lysis protein initiation makes use of ribosomes that have just finished synthesizing coat protein. Here, the stability of the hairpin containing the lysis protein AUG codon plays an important role (de Smit and van Duin 1990, 1994).

A different kind of translational coupling operates in phage f1 between gene VII and gene V. The gene VII initiation site is intrinsically weak and incapable of directly binding a 30S ribosomal subunit, which is instead supplied to the gene VII initiation site by translation of gene V located upstream (Ivey-Hoyle and Steege 1989, 1992).

Repression and Activation

Translation of the coat protein cistron of the RNA phages is required for initiation at the replicase cistron, as described above, but later in infection the coat protein itself down-regulates replicase synthesis while its own production is amplified. Synthesis of the replicase protein begins shortly after infection but is soon curtailed, well before the synthesis of the other cistrons has peaked (van Duin 1988). It appears that a declining level of replicase synthesis late in infection is advantageous, whereas a sustained high level of coat protein synthesis is plainly required for virion production. The mechanism of replicase repression rests on an RNA–protein interaction (Witherell et al. 1991): coat protein binds to a stem-loop structure that encompasses the replicase initiation site, thereby excluding ribosomes. Both the SD sequence and the initiator AUG codon are blocked. Interestingly, coat protein attached at this site then facilitates virion assembly.

Similar principles apply to several translational repressors found in phage T4 that all interfere with the formation of initiation complexes on their mRNA targets (Gold 1988). The T4 Reg A protein represses the translation of many phage early mRNAs by occluding the AUG codon but not the SD sequence. T4 DNA polymerase, on the other hand, represses its own synthesis by binding to a region of the initiation site that includes the SD sequence but not the AUG itself. Similarly, autogenous regulation of T4 gene *32* protein is due to the occlusion of its SD sequence. In this case, several molecules of the single-stranded DNA-binding protein assemble on the mRNA, beginning at an upstream pseudoknot and extending through cooperative binding interactions until the translation initiation site is covered.

Most such mRNA:protein interactions are negative, but some positively acting effectors are also known. In phage Mu, translation of the Mom protein, a DNA modification enzyme, depends on the Com protein encoded by the same operon. The Mom initiation site is structured such that its AUG and part of its SD sequence are sequestered in a stem; binding of the Com protein upstream of the cryptic Mom initiation site causes a conformational change that makes the site accessible to ribosomes (Hattman et al. 1991; Wulczyn and Kahmann 1991).

Few translational activators or repressors have been reported in eukaryotic viruses, but CaMV TAV may be an example (see sections on Ribosome Shunting and Reinitiation, above).

Modification of the Translation Apparatus by Viruses

Many viruses control the activity of cellular translational components, generally through cleavage or other covalent modifications, thereby influencing the translation of both viral and cellular mRNAs. Such modifications are more pronounced in eukaryotes than prokaryotes, and are often central to the shutoff of host cell protein synthesis. At the translational level, shutoff generally entails two events: The virus places itself at an advantage by compromising some aspect of the translation system, concomitantly engineering a bypass so that it escapes the restriction. We begin with an overview of the shutoff phenomenon, then discuss several translational mechanisms exploited by viruses.

Host Cell Shutoff

During the later stages of a productive infection, many viruses interfere with the production, maturation, or stability of cellular DNA, RNA, and/or proteins. This inhibition of cell-specific macromolecular metabolism is referred to as host cell shutoff. Together with the increasing rate of synthesis of viral products, host shutoff contributes to the viral domination of cellular biosynthetic pathways seen in many infections. Presumably, shutoff allows viruses to usurp the cellular machinery by alleviating competition for cellular resources, thereby accelerating replication and perhaps enhancing virus yield. In addition, shutoff may preempt cellular antiviral defense mechanisms by compromising the synthesis of essential host proteins (see above, Immune Defenses and Viral Countermeasures). Nonetheless, viruses obviously need to keep their hosts alive and functioning to the extent that the viral life cycle can be completed, so shutoff must be delicately balanced.

The severity of host shutoff depends on several factors, including cell type and multiplicity of infection, and its mechanism and phenomenology are both varied. Related viruses may employ different mechanisms to accomplish the same end, or even do without shutoff altogether, as illustrated by members of the picornavirus family (discussed below and in Chapter 31). Moreover, shutoff is not necessarily linked to the abundant synthesis of viral proteins: In poliovirus-infected cells, shutoff precedes the appearance of high levels of viral proteins, whereas the reverse is seen with adenovirus. Indeed, shutoff during adenovirus infection has less to do with the accumulation of viral proteins than with the release of completed virions from degenerating cells (Zhang and Schneider 1994). Shutoff is usually an active process, requiring viral protein synthesis, but protein synthesis may be dispensable if the factor responsible is introduced by infecting virions, as in the cases of vaccinia and herpesviruses (Knipe 1996; see Chapter 35). Mutations that affect protein synthesis shutoff have been recorded in a number of viruses, including VSV, adeno-, herpes-, and polioviruses (Stanners et al. 1977; Read and Frenkel 1983; Bernstein et al. 1985; Hayes et al. 1990). Their analysis, and that of cellular mutants affecting shutoff, should be illuminating.

The preferential synthesis of viral proteins can be achieved simply by mRNA competition: Viral mRNAs may dominate by their overwhelming preponderance at late times of infection (e.g., VSV; Lodish and Porter 1980, 1981) or their greater translational efficiency (e.g., influenza virus; Chapter 34). More often, however, the phenomenon hinges on differences between viral and host cell mRNAs which are accentuated by virus-coded proteins that set them at an advantage (e.g., rotavirus), or are exacerbated by alterations in the translational machinery that place cellular mRNAs at a disadvantage (e.g., polio- and adenoviruses). These modifications of host cell translational components are discussed next. In yet other cases, the picture is more nebulous, and factors including changes in the ionic milieu favoring viral mRNA translation have been implicated (Nuss et al. 1975; Alonso and Carrasco 1981; Lacal and Carrasco 1982).

tRNA Cleavage

The T-even phages bring about many changes in the bacterial protein synthetic apparatus, of which tRNA cleavage reactions are the most conspicuous (Mosig and Eiserling 1988). The destruction of host tRNAs accentuates the dependence of protein synthesis on T4-derived mRNA. For example, the *E. coli* $tRNA^{Leu}$ species that recognizes the codon CUG, rare in T4 mRNAs, is cleaved soon after infection. Phage T4 encodes eight tRNAs (for Arg, Ile, Thr, Ser, Pro, Gly, Leu, and Gln) that are nonessential under standard laboratory conditions; they may be advantageous in a natural setting, however, either because they serve codons that are rare in host mRNAs but frequent in T4 mRNAs, or because they provide a bypass around a virus-induced lesion. Thus, in some *E. coli* strains, the cellular $tRNA^{Ile}$ is cleaved in its anticodon loop and is functionally replaced by the phage-encoded isoaccepting species. Such changes are part of widespread alterations of the cellular machinery that take place after T4 infection to the benefit of the phage: In contrast, the cleavage of $tRNA^{Lys}$, which is not covered by a phage gene, is part of a cellular defense mechanism (described in the section on Bacterial Exclusion Systems).

eIF2 Phosphorylation

Eukaryotic initiation factor 2, eIF2, is composed of three nonidentical subunits. In the form of a ternary complex, eIF2•GTP•Met-$tRNA_i$, it serves to transport the initiator tRNA to the 40S ribosomal particle; it also appears to assist in mRNA binding and initiation site selection. The activity of eIF2 is modulated by phosphorylation on Ser-51 of its α subunit (see Chapters 2 and 5). Eukaryotes possess four kinases that can inhibit initiation by phosphorylating eIF2: PKR, the dsRNA-activated inhibitor (protein kinase, RNA dependent; previously known as DAI, dsI, or P1) is widespread in higher cells; HRI (the heme-regulated inhibitor, also known as HCR) is principally found in red cells and their precursors; GCN2 is present in mammals and best characterized in yeast, where it is responsible for regulating translation of GCN4 and the transcription of amino acid biosynthetic genes; and most recently, PERK (the PKR-like endoplasmic reticulum kinase, also called PEK) which is activated by stress in the endoplasmic reticulum. These kinases are reviewed individually in Chapters 13, 14, 5 and 15, respectively.

PKR is intimately linked with the host response to viral infection, participating in host cell shutoff and apoptosis as well as the interferon-induced antiviral defense mechanism (see below, The Interferon System). Phosphorylation of eIF2 restricts its function by trapping the GTP exchange factor (eIF2B) that is required for eIF2 to recycle, resulting in ternary complex depletion. Since eIF2B is less abundant than eIF2, phosphorylation of a fraction of the eIF2 (~30%) can sequester all of the recycling factor and lead to a complete block to protein synthesis. This outcome contributes to the interferon-induced antiviral response, whereas an intermediate degree of eIF2 phosphorylation occurring in cells infected with many different viruses (for references, see Kozak 1986, 1992) may lead to host shutoff. Support for this idea has been drawn from the observation that host cell shutoff does not take place when PKR-deficient cells are infected with adenovirus (O'Malley et al. 1989; Huang and Schneider 1990), although results with PKR-null cells have not been reported. Three possibilities have been entertained to explain how limited eIF2 phosphorylation could lead to translational selectivity. The first surmises that viral mRNAs are intrinsically more efficient or abundant, and hence less sensitive to a reduction in the effective concentration of initiation factor. The second supposes that the inhibitory effect of eIF2 phosphorylation is compartmentalized in the cell, so that the host mRNAs are preferentially inhibited while the viral mRNAs are spared. Taking note of the observation that the adenovirus PKR inhibitor VA RNA_I (see below, Viral Countermeasures against Translation Inhibition) can interact with the tripartite leader, the third possibility envisions that PKR is not activated in the neighborhood of late adenoviral mRNAs. Similarly, host protein synthesis shutoff in cells infected with different types of reovirus segregates with the viral S4 genome segment (Sharpe and Fields 1982). This segment encodes the outer capsid protein σ3, which is also a PKR inhibitor (see Chapter 33 and Viral Countermeasures against Translation Inhibition, below). As with adenovirus, however, the basis for the selectivity remains ill-defined.

eIF4E Dephosphorylation

The entry of mRNA into the initiation pathway is mediated by factors in the eIF4 group (see Chapters 2 and 6). The cap-binding protein, eIF4E, is the rate-limiting initiation factor in many circumstances (Gingras et al. 1999). It binds to the cap structure at the 5′ end of viral and cellular mRNAs. Together with eIF4A and eIF4G, it comprises the cap-binding complex eIF4F. This complex, with eIF4B, possesses helicase activity and apparently facilitates unwinding of secondary structure in the 5′ end of the mRNA and scanning. The eIF4F complex then catalyzes the binding of the 40S ribosomal subunit to the mRNA. Phosphorylation of eIF4E at Ser-209 correlates with its increased activity in the initiation pathway, possibly because of enhanced mRNA-binding affinity (see Chapter 6), but the degree of eIF4E phosphorylation declines in cells infected with a number of viruses including adeno- and influenza viruses (Huang and Schneider 1991; Feigenblum and Schneider 1993).

eIF4G not only forms the skeleton of the eIF4F assemblage containing eIF4E and eIF4A, but also provides a scaffold for the binding of eIF3, the poly(A)-binding protein (PABP), and the protein kinase Mnk1 that is responsible for eIF4E phosphorylation. At late times of infection with adenovirus, eIF4E is strongly dephosphorylated as a result of the displacement of Mnk1 from its site on eIF4G. Mnk1 is evicted by the adenovirus 100K protein, which accumulates to high levels in the cytoplasm at late times of infection (see Chapter 32). This protein is an RNA-binding protein, although no specificity has yet been demonstrated. A mutation in the 100K protein leads to a defect in the translation of late viral mRNAs without deterring the translation of early viral mRNAs or the shutoff of host protein synthesis in the late phase (Hayes et al. 1990). The milder dephosphorylation of eIF4E that occurs in influenza virus-infected cells, through an unknown mechanism, also correlates with shutoff of host translation (Feigenblum and Schneider 1993).

Dephosphorylation is believed to contribute to host cell shutoff by placing cellular mRNAs at a disadvantage when they are competing against the viral mRNAs. The nature of the competitive advantage enjoyed by the viral mRNAs is beginning to emerge. In the case of adenovirus, late mRNAs carrying the tripartite leader are able to utilize the shunting mechanism of initiation that apparently operates with a lesser demand on eIF4F function (see Chapter 32). In the case of influenza virus mRNAs, sequences in their 5′UTRs seem to confer a selective advantage, possibly by virtue of interactions with cellular and viral proteins (see Chapter 34).

4E-BP1 Dephosphorylation

A similar effect is brought about by different means in cells infected with EMCV and other cardioviruses that initiate via an IRES (see above,

Internal Ribosome Entry). The activity of initiation factor eIF4E is modulated by a family of proteins, the eIF4E-binding proteins, including 4E-BP1. In its under-phosphorylated form, 4E-BP1 sequesters eIF4E, removing it from eIF4G and thereby repressing eIF4F function (see Chapter 6). The phosphorylation of 4E-BP1 decreases in parallel with the inhibition of host protein synthesis in EMCV-infected cells; furthermore, inhibition of 4E-BP1 phosphorylation, by treatment of cells with rapamycin, exacerbates viral replication and host shutoff. Presumably, translation of the viral RNA, which is uncapped and eIF4E-independent, is favored by diminished competition from capped and eIF4E-dependent host mRNAs (Gingras et al. 1996; Svitkin et al. 1998).

eIF4G Cleavage

In cells infected with poliovirus and many other members of the picornavirus family—including the enteroviruses, aphthoviruses, and rhinoviruses, but not the cardioviruses—host protein synthesis shutoff is attributable to proteolysis of initiation factor eIF4G catalyzed by a virus-coded enzyme whose primary role is to cleave the viral polyprotein (for details, see Chapter 31). eIF4G is severed into two fragments. Cleavage is triggered by protease 2A in the enteroviruses (including polio- and coxsackieviruses) and rhinoviruses (the common cold virus), and protease L in foot and mouth disease virus, although it is debatable whether these enzymes cut eIF4G directly or mediate cleavage via cellular proteases. Cleavage of eIF4G effectively separates it into two domains: an amino-terminal part that interacts with eIF4E and PABP, and a carboxy-terminal part that interacts with eIF4A, eIF3, and Mnk1 (see Chapter 6). As a result, cap-dependent initiation is severely inhibited. The IRES-dependent initiation mechanism generally does not need eIF4E (see above, Internal Ribosome Entry), although the central portion of eIF4G is required, so eIF4G cleavage sets these picornavirus mRNAs at an advantage. Shutoff takes hold only after cleavage of eIF4GII, which is completed some time after that of its functional homolog eIF4GI (Gradi et al. 1998). Whereas the proteases of other picornavirus groups—the cardioviruses and hepatoviruses—do not cleave eIF4G, the cardioviruses at least have contrived a different mechanism to achieve protein synthesis shutoff (discussed above in 4E-BP1 Dephosphorylation).

Other Actions on eIF4G

A consequence of the binding of both eIF4E and PABP to eIF4G is that capped and polyadenylated mRNAs can be effectively circularized via a proteinaceous bridge (reviewed in Chapter 10). Rotaviruses, like reoviruses, lack a poly(A) tail and therefore presumably do not bind PABP. However, the viral NSP3 protein seems to fill an analogous role: NSP3 binds to the common 3′-end sequence of rotavirus mRNAs and also interacts with eIF4G (Piron et al. 1998, 1999). At the same time, NSP3 interferes with the eIF4G–PABP interaction, thereby simultaneously facilitating viral mRNA translation and hampering that of cellular mRNAs (see Chapter 33). Conceivably, similar mechanisms operate elsewhere; for example, it has been speculated that the NS1 protein, which can bind poly(A) and eIF4G, may play some such role in influenza virus infection (Chapter 34). The increased eIF4G phosphorylation seen in influenza virus infection could also perhaps contribute (Feigenblum and Schneider 1993), but the significance of this modification is not yet established.

HOST DEFENSES AND VIRAL COUNTERMEASURES AT THE TRANSLATIONAL LEVEL

Interferon was discovered more than 40 years ago as a potent interfering agent produced by influenza-virus-infected chick embryo cells (Isaacs and Lindenmann 1957). Interferons are cytokines (see above, Immune Defenses and Viral Countermeasures) with a particularly direct impact on protein synthesis in the virus-infected cell, inducing enzymes that act at the cellular level as first-line defenses against viral infection. Although the interferons are apparently restricted to vertebrates (Vilček and Sen 1996), similar defensive and evasive systems are widespread in nature. Divergent organisms have developed antiviral defenses that act at the translational level, consistent with the critical role of the protein synthesis system in viral infection.

The Interferon System

Interferon-induced enzymes include the eIF2α kinase PKR (see above, eIF2 Phosphorylation), and an RNA degradation system composed of two enzymes: 2′,5′ oligoadenylate synthetase (2-5A synthetase), which makes an activator of RNase L, and RNase L itself. Together, these comprise a defense system that is present in untreated, uninfected cells, but is neither fully mobilized in the absence of interferon induction nor fully armed without virus infection: dsRNA appears to be central to both phases of this process. Induction and activation of PKR can also lead to host protein synthesis shutoff and trigger apoptosis from inside the cell. In response, a battery of viral products is arrayed to attenuate or neutralize these defensive enzymes, thereby alleviating translation shutoff as well as the onset of apoptosis.

Induction of the Antiviral State

The interferons are multifunctional effectors inhibiting viral replication and cell growth; activating NK cells, CTLs, and monocyte-macrophages; and inducing MHC and Fc receptors. The synthesis of type I interferons (interferons-α and -β) is induced by virus infection, especially by RNA viruses, and by the exposure of cells to other stimuli such as dsRNA (Gilmour and Reich 1995); indeed, intracellular dsRNA of viral origin may be the principal stimulus for interferon induction. Type I interferons are responsible for the primary antiviral response. Type II interferons (interferon-γ) are made by NK cells and activated T cells chiefly in response to antigenic stimulation, and serve to coordinate the immune response and pathogen clearance mechanisms. Almost all cell types display interferon receptors on their cell surfaces. Reflecting the structural differences between them, type I and type II interferons bind distinct receptors. Nevertheless, they activate overlapping signal transduction pathways and, as a result, share a large number of biological functions (Samuel 1991; Vilček and Sen 1996).

Once released from an infected cell, interferon diffuses to adjacent cells in the culture, tissue, or organism, and triggers the transcriptional activation of more than 30 genes via the JAK-STAT signal transduction pathway (Stark et al. 1998). In this chain reaction of tyrosine phosphorylation events, the Janus kinase (JAK) activates the STATs (signal tranducers and activators of transcription), which then translocate to the nucleus. Genes inducible by type I interferons possess a consensus DNA-binding site within their promoter (the interferon-specific response element, ISRE) that binds protein complexes called ISGFs (interferon-stimulated gene factors), including the STAT proteins, that are essential for interferon-mediated gene transcription (Mamane et al. 1999). Some type II interferon-inducible genes contain the ISRE, whereas others contain GAS (γ-activated sequence) or other response elements.

In mice, the type I and type II interferons are essential for antiviral defense (Muller et al. 1994). Interferon-induced gene products establish an antiviral state in recipient cells, during which viral replication may be blocked at several levels from penetration to release of progeny virions. In many cases, infection follows a normal course up to and including the synthesis of viral mRNA, but this mRNA does not become stably associated with polysomes as a result of the actions of PKR and RNase L. Both of these enzymes are found at relatively low concentrations in many uninduced cells; their synthesis is induced by interferon at the transcriptional level, resulting in ~ five- to tenfold and two- to fourfold increased accumulation of PKR and RNase L, respectively (Laurent et al. 1985; Zhou et al. 1993). The 2-5A synthetases, of which there are four isozymes differing in size and subcellular distribution, are also highly inducible by interferon treatment (Kerr et al. 1977). At their uninduced levels, these enzymes confer partial resistance to virus infection; interferon treatment creates a state of heightened readiness—the antiviral state—from which cells can mount a more effective response to infection. On the basis of data from overexpression of the enzymes in tissue culture and knockout mice, as well as circumstantial evidence (e.g., Viral Countermeasures against Translational Inhibition, below), most picornaviruses and possibly vaccinia virus are susceptible to 2-5A synthetase/RNase L, whereas PKR affects a broad range of viruses including adeno-, influenza, reo- and vaccinia viruses and VSV (Vilček and Sen 1996).

Activation of PKR and RNase L

The latent (inactive) forms of these enzymes are both activated by dsRNA, although via different mechanisms (Fig. 3). PKR activation is accompanied by autophosphorylation and occurs as a direct response to dsRNA. The enzyme has two nonidentical copies of a dsRNA-binding motif (dsRBM; Fierro-Monti and Mathews 2000). The binding of dsRNA is thought to cause dimerization and a conformational change in the enzyme that unmasks its kinase activity, thus allowing autophosphorylation, the phosphorylation of eIF2α, and the inhibition of translation initiation (see Chapter 13 and eIF2 Phosphorylation, above). The activation of RNase L occurs by a less direct route. Although lacking a dsRBM (Patel and Sen 1992), the 2-5A synthetases are activated by dsRNA, producing a series of short, 2′-5′ linked oligoadenylates of the form $pppA(pA)_n$, where n is commonly 2 or 3. These oligonucleotides, known collectively as 2-5A, specifically activate RNase L, which degrades RNA chiefly by cutting at the 3′ side of UpUp and UpAp sequences.

The physiological activator of PKR and the 2-5A synthetases is believed to be viral dsRNA generated during the course of viral replication or transcription, or the viral genome itself in the case of dsRNA-containing viruses such as reovirus. This remains a strong likelihood despite doubts as to the existence, free in the cytoplasm, of dsRNA genomes or replicative intermediates, and the presence in cells of nucleases and modifying/unwinding enzymes that seem designed to dispose of potentially toxic, exposed dsRNA. On the other hand, it is possible that such protective mechanisms

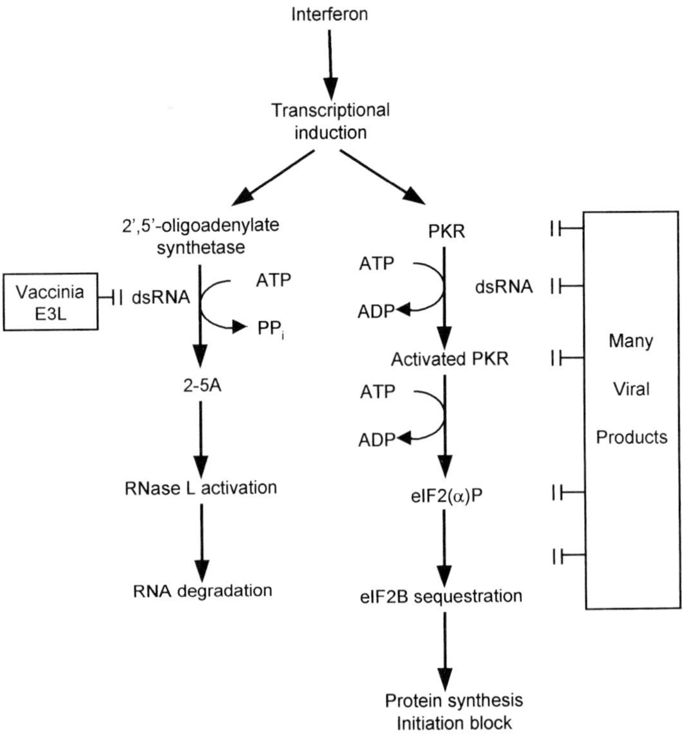

Figure 3 Translational inhibition by interferon-induced pathways. The flowchart summarizes the pathways for induction and activation of PKR and RNase L, and the results of their activation. Positions in the pathways where viral products can intercede to overcome translational inhibition are shown in greater detail in Fig. 4.

are swamped during viral infection. Definitive evidence about the activator is scarce, but dsRNA has been isolated from adenovirus-infected cells (Maran and Mathews 1988), and both strands of EMCV RNA are associated with 2-5A synthetase in immunoprecipitates from infected cells (Gribaudo et al. 1991). Indirect evidence comes from the nature of inhibitors of PKR activation, which include dsRNA-binding proteins and dsRNA analogs (see Viral Countermeasures against Translational Inhibition), and from the action of the anti-poxvirus drug IBT (isatin-β-thiosemicarbazone), which leads to the accumulation of excess dsRNA in infected cells (see Chapter 35). These findings argue that viral dsRNA is indeed the activator. However, it should be borne in mind that PKR can be activated in vitro by ostensibly single-stranded RNAs such as σ3 reovirus mRNA or HDV RNA (Thomis and Samuel 1993; Robertson et al. 1996), by cellular proteins such as PACT and NF90 (Patel and Sen 1998; I. Fierro-

Monti and M.B. Mathews, in prep.), and even by unrelated polyanions such as heparin (Hovanessian and Galabru 1987).

Another enigma stems from the fact that neither PKR nor RNase L is intrinsically specific for viral mRNA. So why are they part of an antiviral strategy? One possible answer is that they act in a localized fashion to affect viral mRNAs while sparing cellular mRNA (in contrast to speculation in eIF2 Phosphorylation, above). This could be achieved if the dsRNA activator were localized near, or actually associated with, the viral mRNA. Although this situation has been modeled in various ways and has received indirect support (Nilsen and Baglioni 1979; DeBenedetti and Baglioni 1984; DeBenedetti et al. 1985; Kaufman and Murtha 1987; Maran et al. 1994), it raises again the question of the nature of the activator. In addition, since the effective inhibitor is not the dsRNA-dependent enzyme itself, but is removed from it by one or two biochemical steps involving presumably diffusible intermediates, it is not obvious how the inhibitor is tethered to the dsRNA activator. An alternative view is that the selective inhibition of viral translation is more apparent than real. Possibly the organism sacrifices infected cells in the interest of restricting virus spread, or perhaps a hiatus in protein synthesis harms the virus more than the cell—either because the viral replicative program is disrupted, or because the cell mobilizes additional defenses in the interim. In this connection, it is worth noting that the two defense pathways have built-in reversibility: 2-5A is labile because of an active 2′,5′ phosphodiesterase, and cellular phosphatases can remove phosphate groups from both PKR and eIF2.

Apoptosis and Translational Inhibition

Translational inhibition, whether caused by viral infection or by chemical inhibitors, can lead to or enhance apoptosis. Both PKR and RNase L have been implicated as mediators of apoptosis (Lee and Esteban 1994; Zhou et al. 1997). PKR probably acts via the phosphorylation of eIF2α and the resultant inhibition of protein synthesis, but activation of NF-κB, IRF1, and p53, which are involved in the multiple pathways leading to apoptosis, may also contribute (Kaufman 1999; Tan and Katze 1999; Chapter 13). In vaccinia virus-infected cells, induction of apoptosis by PKR seems to require both the phosphorylation of eIF2α and activation of NF-κB (Gil et al. 1999). Activation of NF-κB by PKR is supported by several reports (Offermann et al. 1995; Kumar et al. 1997; Gil et al. 1999), but it should be noted that both pro- and anti-apoptotic roles have been attributed to this transcription factor (Sonenshein 1997). Induction of apoptosis by TNFα increases the phosphorylation of eIF2, and overexpression of a nonphosphorylatable form of eIF2α (S51A) partially protects against apoptosis. Conversely, a variant of eIF2α that imitates its phosphorylated form (S51D) can trigger apoptosis (Srivastava et al. 1998). These findings and others (for review, see Tan and Katze 1999) implicate eIF2α phosphorylation directly in apoptosis and suggest that inhibition of translation can signal apoptosis. Further supporting this theme, apoptosis can be suppressed by overexpression of eIF4E, leading to increased translation initiation (Polunovsky et al. 1996). Hence, viruses that can control negative or positive regulators of translation can potentially control apoptosis (see below, Viral Countermeasures against Translation Inhibition).

What follows eIF2α phosphorylation en route to apoptotic cell death is still not clear. Apoptosis is blocked by overexpression of Bcl-2 in cells that are induced to undergo PKR-dependent cell death (Lee et al. 1997), indicating that PKR acts upstream of the Bcl-2 protein and is part of the mitochondria-dependent pathway (see above, Apoptosis). Likewise, RNase L expression can induce apoptosis via general inhibition of protein synthesis, and this action is also blocked by Bcl-2 (Diaz-Guerra et al. 1997). It is possible that apoptosis is a last resort in the event that the antiviral state that shuts down protein synthesis in the infected cell is insufficient to halt viral replication.

Can viruses continue to replicate during apoptosis? The cleavage of eIF4G by caspases during apoptosis is correlated with the inhibition of capped mRNA translation (Marissen and Lloyd 1998), but IRES-mediated translation can continue as demonstrated for c-Myc (Stoneley et al. 2000). Additional proteins involved in apoptosis, Apaf-1 (see above, Apoptosis) and XIAP, a caspase-binding inhibitor of apoptosis, also initiate using an IRES (Holcik et al. 1999; Coldwell et al. 2000). Conceivably, cap-independent translation allows the expression of apoptosis regulators even after the degradation of eIF4G: It will be interesting to learn whether more cellular proteins that are synthesized during apoptosis are translated by IRES-dependent mechanisms. It is not clear whether IRES-mediated translation is supported by the 76-kD degradation product of eIF4GI that is produced during early apoptosis; however, an eIF4G homolog that lacks the eIF4E-binding site, p97/DAP5/NAT1 (see Chapter 6), is cleaved by caspases in apoptotic cells, giving rise to a shorter protein that potentially functions as a translation initiation factor for cellular IRES-containing genes. Not surprisingly, p97 itself contains an IRES and is translated in apoptotic cells (Henis-Korenblit et al. 2000). These findings raise the possibility that IRES-containing viral mRNAs can also be translated during apoptosis by virtue of their reduced requirement for eIF4E and eIF4G.

Viral Countermeasures against Translation Inhibition

Despite considerable variability in the response of cells and viruses to interferon, most viruses induce interferon, and many of them produce dsRNA that can activate PKR and the 2-5A synthetase/RNase L system. PKR in particular poses a serious threat to viral multiplication, judging by the number of mechanisms that viruses have elaborated to counteract its effects on protein synthesis. These countermeasures are diverse in kind, involving viral RNAs and proteins as well as host proteins, which inhibit PKR function at many different levels, as illustrated in Figure 4. The following account is a brief synopsis, since aspects of this topic are covered in Chapters 13 and 32–35.

Reduced PKR Levels. Poliovirus seems to destabilize PKR (Black et al. 1989), which would be expected to dampen the antiviral response. Cells expressing the *tat* gene or infected with HIV-1 virus also contain reduced amounts of PKR, although the mRNA for 2-5A synthetase is unaffected (Roy et al. 1990). In neither case is the mechanism yet known.

Sequestration of dsRNA. Several viruses produce proteins that bind dsRNA, rendering it unavailable to activate PKR and 2-5A synthetase. Both enzymes are inhibited by the vaccinia virus protein, E3L, which has a single copy of the dsRBM that is found as a tandem repeat in PKR (Chang et al. 1992; Rivas et al. 1998). The product of the reovirus S4 gene, σ3, binds dsRNA tightly although it lacks a dsRBM; this protein inhibits PKR activation and can counter the effects of interferon (Beattie et al. 1995b). Similarly, the influenza virus NS1 and HSV-1 U_S11 proteins, neither of which has a dsRBM, can also bind dsRNA and inhibit the activation of PKR (Lu et al. 1995; Cassady et al. 1998; Mulvey et al. 1999).

Activator Analogs. Small, highly structured RNA molecules bind to PKR and antagonize its activation by dsRNA. Adenovirus VA RNA_I is a small (160-nucleotide) RNA that prevents PKR activation at late times of infection. VA RNA_I enhances virus multiplication, confers interferon-resistance, and may be involved in host shutoff (Reichel et al. 1985; Kitajewski et al. 1986; Mathews and Shenk 1991). Other viruses also produce small RNAs that have the capacity to inhibit PKR, but their roles in virus infection are less well established. These include the EBERs of Epstein-Barr virus (Bhat and Thimmappaya 1985; Swaminathan et al. 1992), TAR RNA of HIV-1 (Gunnery et al. 1992; Maitra et al. 1994), and HDV RNA, which is capable of both activating and inhibiting PKR in vitro depending on the circumstances (Robertson et al. 1996). There is

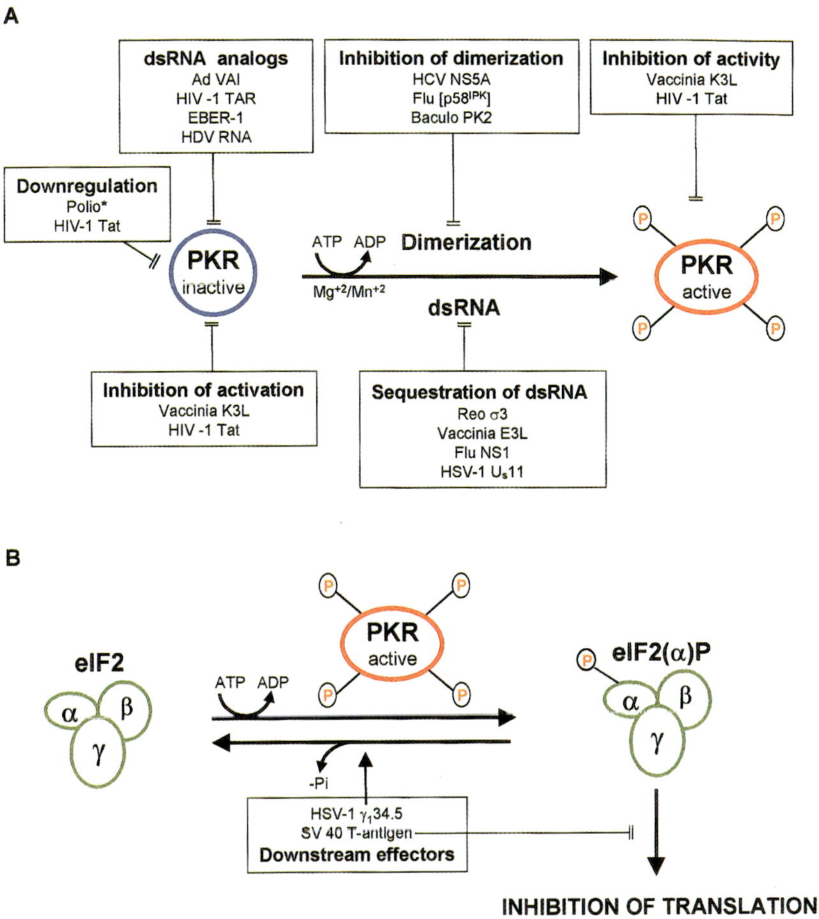

Figure 4 PKR activation and viral countermeasures. (*A*) The activation of PKR by dsRNA-mediated autophosphorylation, and sites of interference with its activation and activity by viral products. (*B*) The PKR-catalyzed phosphorylation of eIF2, and sites of inhibition by viral products. Brackets indicate that the proximal effector is a cellular factor; an asterisk indicates that the putative factor is cellular rather than viral.

also a single report of putative 2-5A derivatives that act as competitive inhibitors of the 2-5A synthetase/RNase L system (Cayley et al. 1984).

Inhibitors of Dimerization. PKR activation entails dimerization that is blocked by several viral proteins. Influenza-virus-infected cells contain an inhibitor of PKR activation and activity called P58^{IPK} (Lee et al. 1994). P58^{IPK} is a cellular protein, a member of the tetratricopeptide family, which is masked in uninfected cells by heat shock protein Hsp40 (Melville et al. 1997). Similarly, HCV NS5A binds PKR and prevents dimerization and activation (Gale et al. 1998). Both influenza virus NS1 and vaccinia E3L may also function in part by binding directly to PKR (Romano et al. 1998; Tan and Katze 1998). Baculovirus PK2 contains a sequence with homology to an eIF2α kinase domain that inhibits PKR phosphorylation by binding to PKR and inhibiting its dimerization or autophosphorylation function (Dever et al. 1998).

Substrate Analogs. Another mechanism of inhibition is by substrate competition. In addition to E3L, described above, vaccinia virus encodes a second protein that interacts with PKR. This protein, K3L, is homologous with the α subunit of eIF2 in the vicinity of its phosphorylation domain and appears to act as a pseudosubstrate for the kinase. It blocks the activation of PKR (and of HRI) as well as the activity of the kinase, once activated. The inhibitor has been shown to counteract the effects of PKR and interferon in vivo (Beattie et al. 1995a). HCV E2, which contains sequences homologous with eIF2α phosphorylation and PKR autophosphorylation sites, also appears to act as an inhibitory pseudosubstrate for PKR (Taylor et al. 1999). HIV Tat is phosphorylated by PKR, leading to inhibition of eIF2α phosphorylation (Brand et al. 1997; Cai et al. 2000), but the physiological import of this observation is uncertain.

Downstream Effectors. Besides viral products that interfere with PKR, HSV-1 and SV40 have developed mechanisms that act subsequent to eIF2 phosphorylation. The HSV-1 $\gamma_1$34.5 protein is a homolog of a phosphatase PP1regulatory subunit, causing dephosphorylation of eIF2α (He et al. 1998; Leib et al. 2000). This gene has an anti-apoptotic function in neuronal cells (Chou and Roizman 1992). Early expression of the U_S11 protein (discussed above) can forestall PKR activation and obviate the need for the $\gamma_1$34.5 protein (Cassady et al. 1998). SV40 large-T antigen induces a bypass of the translation block despite continued eIF2 phosphorylation (Swaminathan et al. 1996): The mechanism in this case remains to be established.

Host–Virus Interactions in Yeast and Bacterial Cells

Systems with interesting parallels to those induced by interferon are known in yeast and bacteria, as diagramed in Figure 5 and described below; coincidentally, one is translational whereas the other concerns mRNA degradation.

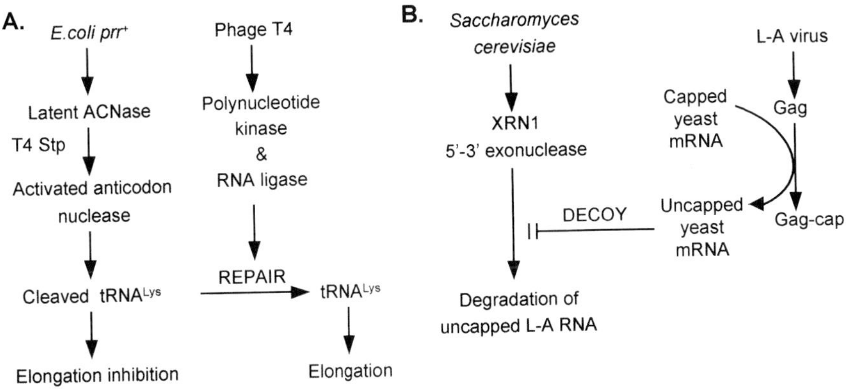

Figure 5 Host/virus interactions in bacteria and yeast. Pathways for translational inhibition in virus-infected bacterial (*A*) and yeast (*B*) cells, and viral mechanisms that restore translational activity.

Bacterial Exclusion Systems

Some *E. coli* strains carry nonessential operons that confer resistance to phage T4 infection by blocking protein synthesis at the elongation stage. Best-studied of these is the *prr* system (Fig. 5A). The *prr* locus contains four ORFs, one of which (*prrC*) encodes a specific endonuclease called anticodon nuclease, ACN. This enzyme inactivates the cellular $tRNA^{Lys}$ by cleaving it at the 5′ border of its anticodon (Penner et al. 1995; Shterman et al. 1995). Parenthetically, ACN also cleaves the mammalian $tRNA^{Lys,3}$ isoacceptor that primes HIV-1 reverse transcriptase. In uninfected bacteria, ACN is present in a latent state, masked by the products of the other three genes of the *prr* locus (*prr A*, *B*, and *D*). Remarkably, unmasking is elicited by a short polypeptide, only 18 amino acids long, that is produced by the phage T4 *stp* gene. Thus, in a scenario that echoes the interferon-induced defense system, the phage Stp polypeptide brings about activation of the cellular anticodon nuclease, resulting in tRNA cleavage, inhibition of late T4 protein synthesis, and aborting the infection.

A homologous system is found in *Neisseria*, and a different anticodon nuclease occurs in colicin E5, implying that such “tRNA toxins” may be widespread (Kaufmann 2000). Furthermore, *E. coli* has another T-even phage exclusion system, *lit*, that targets elongation factor EF1A (formerly EF-Tu). The *lit* system works in a fashion similar to *prr*. A latent protease, Lit, is activated late in infection by the product of the phage major head protein gene; once activated, Lit cleaves a conserved region in EF1A, thereby inhibiting translation (Yu and Snyder 1994; Bingham et al. 2000).

As indicated in Figure 5A, phage T4 is equipped with countermeasures to meet the bacterial *prr* response. The bacteriophage encodes two enzymes that overcome the translational inhibition, polynucleotide kinase and RNA ligase, which act consecutively to repair the severed tRNA. In formal terms, the *prr* system is analogous to the PKR system: In both cases, a viral product activates a latent cellular enzyme, resulting in the modification of a component of the translation apparatus, and in both cases, viruses have elaborated products to neutralize the host defense mechanism.

The Yeast SKI System

SKI1 is one of a group of genes that regulates the “superkiller” phenotype observed in yeast carrying the L-A virus. The mRNA of this virus is uncapped and nonpolyadenylated, resembling an intermediate in the mRNA degradation pathway (see above, Translation of Uncapped RNA, and Chapter 29). Mutations in the *SKI* genes allow enhanced replication of L-A, and of its satellite virus M_1, which produces a secreted toxin called killer toxin. The functions of the *SKI* genes are exerted at least partly at the level of protein synthesis: For example, mutations in several of them (*ski2*, *ski3*, and *ski8*) allow elevated translation of uncapped, nonpolyadenylated RNA (Masison et al. 1995). Ski7p, which displays some similarities to translation factors including eEF1α, also represses the expression of poly(A)$^-$ mRNAs (Benard et al. 1999). Both the translation rate and the stability of such mRNAs are increased in *ski7* mutant cells. Thus, *SKI2*, *3*, *7*, and *8* are functionally related components of the yeast antiviral system.

Whereas these four *SKI* genes are dispensable for normal yeast growth, *SKI1* and *6* are essential genes encoding RNases. *SKI1* is identical with *XRN1* and produces a 5′-3′ exoribonuclease specific for uncapped RNA; mutants in *SKI1/XRN1* permit the accumulation of elevated concentrations of uncapped L-A virus mRNA (as well as uncapped cellular mRNA fragments). Thus, the wild-type *SKI1* nuclease is an antiviral protein, as well as an enzyme in the normal RNA degradation pathway (Masison et al. 1995). Ski6p is part of the 3′-5′ exoribonuclease complex involved in 5.8S rRNA processing, and is a homolog of the bacterial RNase PH that trims the 3′ end of tRNA precursors and 5S RNA during their maturation. Mutations in Ski6p lead to the production of defective 60S ribosomal subunits that permit the elevated expression of poly(A)$^-$ mRNA (Benard et al. 1998).

In the case of L-A virus, the countermeasure consists of an unusual decapping activity. The coat protein of L-A virus, Gag, has the ability to detach the cap structure from mRNA, leaving an uncapped mRNA while the cap remains covalently associated with the Gag protein (Blanc et al. 1992, 1994). The uncapped RNA can then serve as a substrate or "decoy" for the exonuclease, diverting it away from the L-A mRNA, which is therefore spared destruction (Fig. 5B). This cellular defense scheme has some features reminiscent of the 2-5A synthetase/RNase L system, but the viral countermeasures are quite different.

PERSPECTIVES

It is enlightening to consider the changes that have taken place in the half-decade that has elapsed since the first edition of this book was published, and tempting to speculate on developments that are in the offing. What is still true is that viruses are closely engaged with the translation system of the cell and continue to provide insights into the mechanisms of protein synthesis and translational control. Reciprocally, increased knowledge of the mechanism and control of the translation process has deepened our understanding of viral infection strategies and of the interactions between viruses and their hosts. Recent advances include details of translational mechanisms such as IRES-dependent initiation and shunting; a fuller appreciation of host shutoff mechanisms, especially those involving eIF4G, 4E-BP1, and Mnk1; and the recognition of the profound impact of apoptosis on virus–cell interactions at the translational level. Further expansion and refinements can be expected in all these areas. In addition, two promising applications have emerged from research on translational control in viral systems.

The first of these concerns protein expression. Two adenovirus elements have been shown to enhance translation of both viral and nonviral mRNAs. The tripartite leader stimulates the translation of RNAs to which it is attached, perhaps by virtue of shunting (see eIF4E Dephosphorylation and Chapter 32), and VA RNA_I enhances translation, probably by overcoming PKR activation (Kaufman and Murtha 1987; Svensson and Akusjärvi 1990; Chapters 13 and 32). The VA RNA gene has been exploited in commercially developed vectors to boost the level of protein production in transfected tissue-culture cells (Asselbergs and Grand 1993; Linton, 2000). Potentially, such regulators might also find application in vector design for gene therapy.

A second possible application that also impinges on PKR is related to cancer therapy. Recent work established a direct connection between PKR activation and cell transformation (Strong et al. 1998). NIH-3T3 cells are resistant to reovirus infection, but transformed cells in which the Ras signaling pathway is active are permissive for reovirus replication. The resistant cells display a specific translational block that inhibits viral but not cellular mRNAs. This block is apparently due to the phosphorylation and activation of PKR, providing another example of PKR-mediated translational selectivity (discussed above in the sections eIF2 Phosphorylation and Activation of PKR and RNase L). No such block exists in the transformed cells, which are therefore lysed. These findings prompted the use of reovirus to cause tumor regression in mice. Ras-activated tumors were suppressed in both immune-deficient (SCID) and immune-competent mice (Coffey et al. 1998), opening another avenue for using viruses in cancer therapy (for review, see Pennisi 1998). Whether this prospect can be realized in the clinic may depend in part on understanding the Ras-PKR connection in normal cells (Mundschau and Faller 1994, 1995) as well as the involvement of PKR in cancer (Kim et al. 2000; see Chapter 20).

ACKNOWLEDGMENTS

The broad scope of this chapter precluded comprehensive citation of all of the original literature. We apologize to both readers and authors for this exigency. Pertinent original papers can be found by reference to the review articles cited. The authors of this chapter are supported by grants from the National Institute of Allergy and Infectious Diseases, National Institutes of Health.

REFERENCES

Adams J.M. and Capecchi M. 1966. N-formylmethionyl-sRNA as the initiator of protein synthesis. *Proc. Natl. Acad. Sci.* **55:** 147–155.

Alderete J.P., Jarrahian S., and Geballe A.P. 1999. Translational effects of mutations and polymorphisms in a repressive upstream open reading frame of the human cytomegalovirus UL4 gene. *J. Virol.* **73:** 8330–8337.

Alonso M.A. and Carrasco L. 1981. Reversion by hypotonic medium of the shutoff of protein synthesis induced by encephalomyocarditis virus. *J. Virol.* **37:** 535–540.

Asselbergs F.A. and Grand P. 1993. A two-plasmid system for transient expression of cDNAs in primate cells. *Anal. Biochem.* **209:** 327–331.

Atkins J.F. and Gesteland R.F. 1996. Regulatory recoding. In *Translational control* (ed. J.W.B. Hershey et al.), pp. 653–684. Cold Spring Harbor Laboratory Press, Cold Spring Harbor, New York.

Barnett L., Brenner S., Crick F.H.C., Schulman R.G., and Watts-Tobin R.J. 1967. Phase shift and other mutants in the first part of the *rII* cistron of phage T4. *Philos. Trans. R. Soc. Lond. B Biol. Sci.* **252:** 487–560.

Beattie E., Paoletti E., and Tartaglia J. 1995a. Distinct patterns of IFN sensitivity observed in cells infected with vaccinia K3L- and E3L-mutant viruses. *Virology* **210:** 254–263.

Beattie E., Denzler K.L., Tartaglia J., Perkus M.E., Paoletti E., and Jacobs B.L. 1995b. Reversal of the interferon-sensitive phenotype of a vaccinia virus lacking E3L by expression of the reovirus S4 gene. *J. Virol.* **69:** 499–505.

Belli B.A. and Samuel C.E. 1993. Biosynthesis of reovirus-specified polypeptides: Identification of regions of the bicistronic reovirus S1 mRNA that affect the efficiency of translation in animal cells. *Virology* **193:** 16–27.

Benard L., Carroll K., Valle R.C., and Wickner R.B. 1998. Ski6p is a homolog of RNA-processing enzymes that affects translation of non-poly(A) mRNAs and 60S ribosomal subunit biogenesis. *Mol. Cell. Biol.* **18:** 2688–2696.

Benard L., Carroll K., Valle R.C.P., Masison D.C., and Wickner R.B. 1999. The ski7 antiviral protein is an EF1-α homolog that blocks expression of non-poly(A) mRNA in *Saccharomyces cerevisiae. J. Virol.* **73:** 2893–2900.

Bernstein H.D., Sonenberg N., and Baltimore D. 1985. Poliovirus mutant that does not selectively inhibit host protein synthesis. *Mol. Cell. Biol.* **5:** 2913–2923.

Bhat R.A. and Thimmappaya B. 1985. Construction and analysis of additional adenovirus substitution mutants confirm the complementation of VAI RNA function by two small RNAs encoded by Epstein-Barr virus. *J. Virol.* **56:** 750–756.

Bingham R., Ekunwe S.I., Falk S., Snyder L., and Kleanthous C. 2000. The major head protein of bacteriophage T4 binds specifically to elongation factor Tu. *J. Biol. Chem.* (in press).

Black T.L., Safer B., Hovanessian A., and Katze M. 1989. The cellular 68,000-M_r protein kinase is highly autophosphorylated and activated yet significantly degraded during poliovirus infection: Implications for translational regulation. *J. Virol.* **63:** 2244–2251.

Blanc A., Goyer C., and Sonenberg N. 1992. The coat protein of the yeast double-stranded RNA virus L-A attaches covalently to the cap structure of eukaryotic mRNA. *Mol. Cell. Biol.* **12:** 3390–3398.

Blanc A., Ribas J.C., Wickner R.B., and Sonenberg N. 1994. His-154 is involved in the linkage of the *Saccharomyces cerevisiae* L-A double-stranded RNA virus Gag protein to the cap structure of mRNAs and is essential for M_1 satellite virus expression. *Mol. Cell. Biol.* **14:** 2664–2674.

Borman A.M., Deliat F.G., and Kean K.M. 1994. Sequences within the poliovirus internal ribosome entry segment control viral RNA synthesis. *EMBO J.* **13:** 3149–3157.

Brand S.R., Kobayashi R., and Mathews M.B. 1997. The Tat protein of human immunodeficiency virus type 1 is a substrate and inhibitor of the interferon-induced, virally activated protein kinase, PKR. *J. Biol. Chem.* **272:** 8388–8395.

Brenner S., Jacob F., and Meselson M. 1961. An unstable intermediate carrying information from genes to ribosomes for protein synthesis. *Nature* **190:** 576–581.

Bushell M., Wood W., Clemens M.J., and Morley S.J. 2000. Changes in integrity and association of eukaryotic protein synthesis initiation factors during apoptosis. *Eur. J. Biochem.* **267:** 1083–1091.

Bushell M., McKendrick L., Janicke R.U., Clemens M.J., and Morley S.J. 1999. Caspase-3 is necessary and sufficient for cleavage of protein synthesis eukaryotic initiation factor 4G during apoptosis. *FEBS Lett.* **451:** 322–326.

Cai R., Carpick B., Chun R.F., Jeang K.T., and Williams B.R. 2000. HIV-I TAT inhibits PKR activity by both RNA dependent and RNA-independent mechanisms. *Arch. Biochem. Biophys.* **373:** 361–367.

Cassady K.A., Gross M., and Roizman B. 1998. The herpes simplex virus US11 protein effectively compensates for the γ1(34.5) gene if present before activation of protein kinase R by precluding its phosphorylation and that of the α subunit of eukaryotic translation initiation factor 2. *J. Virol.* **72:** 8620–8626.

Cayley P.J., Davies J.A., McCullagh K.G., and Kerr I.M. 1984. Activation of the ppp(A2′p)nA system in interferon-treated, herpes simplex virus-infected cells and evidence for novel inhibitors of the ppp(A2′p)nA-dependent RNase. *Eur. J. Biochem.* **143:** 165–174.

Chang H.-W., Watson J.C., and Jacobs B.L. 1992. The E3L gene of vaccinia virus encodes an inhibitor of the interferon-induced, double-stranded RNA-dependent protein kinase. *Proc. Natl. Acad. Sci.* **89:** 4825–4829.

Chou J. and Roizman B. 1992. The gamma 1(34.5) gene of herpes simplex virus 1 precludes neuroblastoma cells from triggering total shutoff of protein synthesis characteristic of programmed cell death in neuronal cells. *Proc. Natl. Acad. Sci.* **89:** 3266–3270.

Clemens M.J., Bushell M., Jeffrey I.W., Pain V.M., and Morley S.J. 2000. Translation initiation factor modifications and the regulation of protein synthesis in apoptotic cells. *Cell Death Differ.* (in press).

Coffey M.C., Strong J.E., Forsyth P.A., and Lee P.W. 1998. Reovirus therapy of tumors with activated Ras pathway. *Science* **282:** 1332–1334.

Coffin J.M. 1996. Retroviridae: The viruses and their replication. In Field's virology (ed. B.N. Fields et al.), pp. 1767–1830. Lippincott-Raven, Philadelphia.

Coldwell M.J., Mitchell S.A., Stoneley M., MacFarlane M., and Willis A.E. 2000. Initiation of Apaf-1 translation by internal ribosome entry. *Oncogene* **19:** 899–905.

Corcelette S., Masse T., and Madjar J.J. 2000. Initiation of translation by non-AUG codons in human T-cell lymphotropic virus type I mRNA encoding both Rex and Tax regulatory proteins. *Nucleic Acids Res.* **28:** 1625–1634.

Cryns V. and Yuan J. 1998. Proteases to die for (erratum in *Genes Dev.* [1999] **13:** 371). *Genes Dev.* **12:** 1551–1570.

Das T., Mathur M., Gupta A.K., Janssen G.M., and Banerjee A.K. 1998. RNA polymerase of vesicular stomatitis virus specifically associates with translation elongation factor-1 αβγ for its activity. *Proc. Natl. Acad. Sci.* **95:** 1449–1454.

Dasgupta R., Shih D.S., Saris C., and Kaesberg P. 1975. Nucleotide sequence of a viral RNA fragment that binds to eukaryotic ribosomes. *Nature* **256:** 624–628.

DeBenedetti A. and Baglioni C. 1984. Inhibition of mRNA binding to ribosomes by localized activation of dsRNA-dependent protein kinase. *Nature* **311:** 79–81.

DeBenedetti A., Williams G.J., Comeau L., and Baglioni C. 1985. Inhibition of viral mRNA translation in interferon-treated L cells infected with reovirus. *J. Virol.* **55:** 588–593.

de Smit M.H. and van Duin J. 1990. Control of prokaryotic translational initiation by mRNA secondary structure. *Prog. Nucleic Acid Res. Mol. Biol.* **38:** 1–35.

———. 1994. Control of translation by mRNA secondary structure in *Escherichia coli.* A quantitative analysis of literature data. *J. Mol. Biol.* **244:** 144–150.

Dever T.E., Sripriya R., McLachlin J.R., Lu J., Fabian J.R., Kimball S.R., and Miller L.K. 1998. Disruption of cellular translational control by a viral truncated eukaryotic translation initiation factor 2α kinase homolog. *Proc. Natl. Acad. Sci.* **95:** 4164–4169.

Diaz-Guerra M., Rivas C., and Esteban M. 1997. Activation of the IFN-inducible enzyme RNase L causes apoptosis of animal cells. *Virology* **236:** 354–363.

Dinman J.D., Ruiz-Echevarria M.J., and Peltz S.W. 1998. Translating old drugs into new treatments: Ribosomal frameshifting as a target for antiviral agents. *Trends Biotechnol.* **16:** 190–196.

Fajardo J.E. and Shatkin A.J. 1990. Translation of bicistronic viral mRNA in transfected cells: Regulation at the level of elongation. *Proc. Natl. Acad. Sci.* **87:** 328–332.

Farabaugh P.J. 1995. Post-transcriptional regulation of transposition by Ty retrotransposons of *Saccharomyces cerevisiae. J. Biol. Chem.* **270:** 10361–10364.

Farrell H.E. and Davis-Poynter N.J. 1998. From sabotage to camouflage: Viral evasion of cytotoxic T lymphocyte and natural killer cell-mediated immunity. *Semin. Cell Dev. Biol.* **9:** 369–378.

Feigenblum D. and Schneider R.J. 1993. Modification of eukaryotic initiation factor 4F during infection by influenza virus. *J. Virol.* **67:** 3027–3035.

Fields B.N., Knipe D.M., Howley P.M., Chanock R.M., Melnick J.L., Monath T.P., Roizman B., and Straus S.E., Eds. 1996. *Field's virology,* volume 1. Lippincott-Raven, Philadelphia.

Fierro-Monti I. and Mathews M.B. 2000. Proteins binding duplexed RNA: One motif, multiple functions. *Trends. Biochem. Sci.* **25:** 241–246.

Flint S.J., Enquist L.W., Krug R.M., Racaniello V.R., and Skalka A.M. 1999. *Principles of virology: Molecular biology, pathogenesis, and control.* ASM Press, Washington, D.C.

Furuichi Y., Morgan M., Muthukrishnan S., and Shatkin A.J. 1975. Reovirus messenger RNA contains a methylated, blocked 5′-terminal structure: $m^7G(5')ppp(5')G^mpCp$-. *Proc. Natl. Acad. Sci.* **72:** 362–366.

Fütterer J. and Hohn T. 1992. Role of an upstream open reading frame in the translation of polycistronic mRNA in plant cells. *Nucleic Acids Res.* **20:** 3851–3857.

Fütterer J., Kiss-László Z., and Hohn T. 1993. Nonlinear ribosome migration on cauliflower mosaic virus 35S RNA. *Cell* **73:** 789–802.

Fütterer J., Potrykus I., Bao Y., Li L., Burns T.M., Hull R., and Hohn T. 1996. Position-dependent ATT initiation during plant pararetrovirus rice tungro bacilliform virus translation. *J. Virol.* **70:** 2999–3010.

Gale M., Jr., Blakely C.M., Kwieciszewski B., Tan S.L., Dossett M., Tang N.M., Korth M.J., Polyak S.J., Gretch D.R., and Katze M.G. 1998. Control of PKR protein kinase by hepatitis C virus nonstructural 5A protein: Molecular mechanisms of kinase regulation. *Mol. Cell. Biol.* **18:** 5208–5218.

Gil J., Alcami J., and Esteban M. 1999. Induction of apoptosis by double-stranded-RNA-dependent protein kinase (PKR) involves the α subunit of eukaryotic translation initiation factor 2 and NF-κB. *Mol. Cell. Biol.* **19:** 4653–4663.

Gilmour K.C. and Reich N.C. 1995. Signal transduction and activation of gene transcription by interferons. *Gene Expr.* **5:** 1–18.

Gingras A.-C., Raught B., and Sonenberg N. 1999. eIF4 initiation factors: Effectors of mRNA recruitment to ribosomes and regulators of translation. *Annu. Rev. Biochem.* **68:** 913–963.

Gingras A.-C., Svitkin Y., Belsham G.J., Pause A., and Sonenberg N. 1996. Activation of the translational suppressor 4E-BP1 following infection with encephalomyocarditis virus and poliovirus. *Proc. Natl. Acad. Sci.* **93:** 5578–5583.

Gold L. 1988. Posttranscriptional regulatory mechanisms in *Escherichia coli. Annu. Rev. Biochem.* **57:** 199–233.

Gould K.G. and Bangham C.R. 1998. Virus variation, escape from cytotoxic T lymphocytes and human retroviral persistence. *Semin. Cell Dev. Biol.* **9:** 321–328.

Gradi A., Svitkin Y.V., Imataka H., and Sonenberg N. 1998. Proteolysis of human eukaryotic translation initiation factor eIF4GII, but not eIF4GI, coincides with the shutoff of host protein synthesis after poliovirus infection. *Proc. Natl. Acad. Sci.* **95:** 11089–11094.

Gribaudo G., Lembo D., Cavallo G., Landolfo S., and Lengyel P. 1991. Interferon action: Binding of viral RNA to the 40-kilodalton 2′-5′-oligoadenylate synthetase in interferon-treated HeLa cells infected with encephalomyocarditis virus. *J. Virol.* **65:** 1748–1757.

Gromeier M., Bossert B., Arita M., Nomoto A., and Wimmer E. 1999. Dual stem loops within the poliovirus internal ribosomal entry site control neurovirulence. *J. Virol.* **73:** 958–964.

Gros F., Hiatt H., Gilbert W., Kurland G.G., Risebrough R.W., and Watson J.D. 1961. Unstable ribonucleic acid revealed by pulse labelling of *Escherichia coli. Nature* **190:** 581–585.

Gunnery S., Green S.R., and Mathews M.B. 1992. Tat-responsive region RNA of human immunodeficiency virus type 1 stimulates protein synthesis in vivo and in vitro: Relationship between structure and function. *Proc. Natl. Acad. Sci.* **89:** 11557–11561.

Gunnery S., Mäivali Ü., and Mathews M.B. 1997. Translation of an uncapped mRNA involves scanning. *J. Biol. Chem.* **272:** 21642–21646.

Haenni A., Joshi S., and Chapeville F. 1982. tRNA-like structures in the genomes of RNA viruses. *Prog. Nucleic Acid Res. Mol. Biol.* **27:** 101–102.

Hambidge S. and Sarnow P. 1992. Translational enhancement of the poliovirus 5′ noncoding region mediated by virus-encoded polypeptide 2A. *Proc. Natl. Acad. Sci.* **89:** 10272–10276.

Hardwick J.M. 1998. Viral interference with apoptosis. *Semin. Cell Dev. Biol.* **9:** 339–349.

Hattman S., Newman L., Krishna Murthy H.M., and Nagaraja V. 1991. Com, the phage Mu mom translational activator, is a zinc-binding protein that binds specifically to its cognate mRNA. *Proc. Natl. Acad. Sci.* **88:** 10027–10031.

Hayes B.W., Telling G.C., Myat M.M., Williams J.F., and Flint S.J. 1990. The adenovirus L4 100-kilodalton protein is necessary for efficient translation of viral late mRNA species. *J. Virol.* **64:** 2732–2742.

He B., Gross M., and Roizman B. 1997. The γ(1)34.5 protein of herpes simplex virus 1 complexes with protein phosphatase 1alpha to dephosphorylate the alpha subunit of the eukaryotic translation initiation factor 2 and preclude the shutoff of protein synthesis by double-stranded RNA- activated protein kinase. *Proc. Natl. Acad. Sci.* **94:** 843–848.

———. 1998. The γ134.5 protein of herpes simplex virus 1 has the structural and functional attributes of a protein phosphatase 1 regulatory subunit and is present in a high molecular weight complex with the enzyme in infected cells. *J. Biol. Chem.* **273:** 20737–20743.

Henis-Korenblit S., Strumpf N.L., Goldstaub D., and Kimchi A. 2000. A novel form of DAP5 protein accumulates in apoptotic cells as a result of caspase cleavage and internal ribosome entry site-mediated translation. *Mol. Cell. Biol.* **20:** 496–506.

Herbst K.L., Nichols L.M., Gesteland R.F., and Weiss R.B. 1994. A mutation in ribosomal protein L9 affects ribosomal hopping during translation of gene 60 from bacteriophage T4. *Proc. Natl. Acad. Sci.* **91:** 12525–12529.

Hindley J. and Staples D.H. 1969. Sequence of a ribosome binding site in bacteriophage Q-β-RNA. *Nature* **224:** 964–967.

Holcik M., Lefebvre C., Yeh C., Chow T., and Korneluk R.G. 1999. A new internal-ribosome-entry-site motif potentiates XIAP-mediated cytoprotection. *Nat. Cell Biol.* **1:** 190–192.

Hovanessian A.G. and Galabru J. 1987. The double-stranded RNA-dependent protein kinase is also activated by heparin. *Eur. J. Biochem.* **167:** 467–473.

Huang J.T. and Schneider R.J. 1990. Adenovirus inhibition of cellular protein synthesis is prevented by the drug 2-aminopurine. *Proc. Natl. Acad. Sci.* **87:** 7115–7119.

———. 1991. Adenovirus inhibition of cellular protein synthesis involves inactivation of cap binding protein. *Cell* **65:** 271–280.

Hung M., Patel P., Davis S., and Green S.R. 1998. Importance of ribosomal frameshifting for human immunodeficiency virus type 1 particle assembly and replication. *J. Virol.* **72:** 4819–4824.

Isaacs A. and Lindenmann J. 1957. Virus interference. 1. The interferon. *Proc. R. Soc. Lond. B Biol. Sci.* **147:** 258–267.

Ivey-Hoyle M. and Steege D.A. 1989. Translation of phage f1 gene VII occurs from an inherently defective initiation site made functional by coupling. *J. Mol. Biol.* **208:** 233–244.

———. 1992. Mutational analysis of an inherently defective translation initiation site. *J. Mol. Biol.* **224:** 1039–1054.

Jacks T. and Varmus H.E. 1985. Expression of the Rous sarcoma virus *pol* gene by ribosomal frameshifting. *Science* **230:** 1237–1242.

Jang S.K., Krausslich H.G., Nicklin M.J., Duke G.M., Palmenberg A.C., and Wimmer E. 1988. A segment of the 5′ nontranslated region of encephalomyocarditis virus RNA directs internal entry of ribosomes during in vitro translation. *J. Virol.* **62:** 2636–2643.

Kamen R. 1970. Characterization of the subunits of Qβ replicase. *Nature* **228:** 527–533.

Kates J. 1971. Transcription of the vaccinia virus genome and the occurrence of polyriboadenylic acid sequences in messenger RNA. Cold Spring Harbor Symp. Quant. Biol. **35:** 743–752.

Kaufman R.J. 1999. Double-stranded RNA-activated protein kinase mediates virus-induced apoptosis: A new role for an old actor. *Proc. Natl. Acad. Sci.* **96:** 11693–11695.

Kaufman R.J. and Murtha P. 1987. Translational control mediated by eukaryotic initiation factor-2 is restricted to specific mRNAs in transfected cells. *Mol. Cell. Biol.* **7:** 1568–1571.

Kaufmann G. 2000. Anticodon nucleases. *Trends Biochem. Sci.* **25:** 70–74.

Kerr I.M., Brown R.E., and Hovanessian A.G. 1977. Nature of inhibitor of cell-free protein synthesis formed in response to interferon and double-stranded RNA. *Nature* **268:** 540–542.

Kerr I.M., Cohen N., and Work T.S. 1966. Factors controlling amino acid incorporation by ribosomes from krebs II mouse ascites-tumour cells. *Biochem. J.* **98:** 826–835.

Kieff E. and Shenk T. 1998. Modulation of apoptosis by herpesviruses. Semin. Virol. **8:** 471-480.

Kim S.H., Forman A.P., Mathews M.B., and Gunnery S. 2000. Human breast cancer cells contain elevated levels and activity of the protein kinase, PKR. *Oncogene* **19:** 3086–3094.

Kitajewski J., Schneider R.J., Safer B., Munemitsu S.M., Samuel C.E., Thimmappaya B., and Shenk T. 1986. Adenovirus VAI RNA antagonizes the antiviral action of interferon by preventing activation of the interferon-induced eIF-2α kinase. *Cell* **45:** 195–200.

Knipe D.M. 1996. Virus-host cell interactions. In *Field's virology* (ed. B.N. Fields et al.), pp. 273–299. Lippincott-Raven, Philadelphia.

Kondo M., Gallerani R., and Weissmann C. 1970. Subunit structure of Qβ replicase. *Nature* **228:** 525–527.

Kozak M. 1978. How do eucaryotic ribosomes select initiation regions in messenger RNA? *Cell* **15:** 1109–1123.

———. 1986. Regulation of protein synthesis in virus-infected animal cells. *Adv. Virus Res.* **31:** 229–292.

———. 1989. The scanning model for translation: An update. *J. Cell Biol.* **108:** 229–241.

———. 1992. Regulation of translation in eukaryotic systems. *Annu. Rev. Cell Biol.* **8:** 197–225.

Krajcsi P. and Wold W.S. 1998. Inhibition of tumor necrosis factor and interferon triggered responses by DNA viruses. *Semin. Cell Dev. Biol.* **9:** 351–358.

Kumar A., Yang Y.L., Flati V., Der S., Kadereit S., Deb A., Haque J., Reis L., Weissmann C., and Williams B.R. 1997. Deficient cytokine signaling in mouse embryo fibroblasts with a targeted deletion in the PKR gene: Role of IRF-1 and NF-κB. *EMBO J.* **16:** 406–416.

Lacal J.C. and Carrasco L. 1982. Relationship between membrane integrity and the inhibition of host translation in virus-infected mammalian cells. Comparative studies between encephalomyocarditis virus and poliovirus. *Eur. J. Biochem.* **127:** 359–366.

Latorre P., Kolakofsky D., and Curran J. 1998. Sendai virus Y proteins are initiated by a ribosomal shunt. *Mol. Cell. Biol.* **18:** 5021–5031.

Laurent A.G., Krust B., Galabru J., Svab J., and Hovanessian A.G. 1985. Monoclonal antibodies to an interferon-induced M_r 68,000 protein and their use for the detection of double-stranded RNA-dependent protein kinase in human cells. *Proc. Natl. Acad. Sci.* **82:** 4341–4345.

Lazarowitz S.G. and Robertson H.D. 1977. Initiator regions from the small size class of reo-virus mRNA protected by rabbit reticulocyte ribosomes. *J. Biol. Chem.* **252:** 7842–7849.

Lee S.B. and Esteban M. 1994. The interferon-induced double-stranded RNA-activated protein kinase induces apoptosis. *Virology* **199:** 491–496.

Lee S.B., Rodriguez D., Rodriguez J.R., and Esteban M. 1997. The apoptosis pathway triggered by the interferon-induced protein kinase PKR requires the third basic domain, initiates upstream of Bcl- 2, and involves ICE-like proteases. *Virology* **231:** 81–88.

Lee T.G., Tang N., Thompson S., Miller J., and Katze M.G. 1994. The 58,000-dalton cellular inhibitor of the interferon-induced double-stranded RNA-activated protein kinase (PKR) is a member of the tetratricopeptide repeat family of proteins. *Mol. Cell. Biol.* **14:** 2331–2342.

Leib D.A., Machalek M.A., Williams B.R.G., Silverman R.H., and Virgin H.W. 2000. Specific phenotype restoration of an unattenuated virus by knockout of a host resistance gene. *Proc. Natl. Acad. Sci.* **97:** 6097–6101.

Licis N., van Duin J., Balklava Z., and Berzins V. 1998. Long-range translational coupling in single-stranded RNA bacteriophages: An evolutionary analysis. *Nucleic Acids Res.* **26:** 3242–3246.

Linton B. 2000. Promega Life Science Catalog, 11.2. Promega Corporation, Madison, Wisconsin.

Lodish H.F. 1975. Regulation of in vitro protein synthesis by bacteriophage RNA by RNA tertiary structure. In RNA phages (ed. N.D. Zinder), pp. 301–318. Cold Spring Harbor

Laboratory, Cold Spring Harbor, New York.
Lodish H.F. and Porter M. 1980. Translational control of protein synthesis after infection by VSV. *J. Virol.* **36:** 719–733.
———. 1981. Vesicular stomatitis virus mRNA and inhibition of translation of cellular mRNA—Is there a P function in vesicular stomatitis virus? *J. Virol.* **38:** 504–517.
Lodish H.F. and Zinder N.D. 1966. Mutants of the bacteriophage f2. VIII. Control mechanisms for phage-specific syntheses. *J. Mol. Biol.* **19:** 333–348.
Lu Y., Wambach M., Katze M.G., and Krug R.M. 1995. Binding of the influenza virus NS1 protein to double-stranded RNA inhibits the activation of the protein kinase that phosphorylates the elF-2 translation initiation factor. *Virology* **214:** 222–228.
Lubinski J., Nagashunmugam T., and Friedman H.M. 1998. Viral interference with antibody and complement. *Semin. Cell Dev. Biol.* **9:** 329–337.
Maitra R.K., McMillan N.A.J., Desai S., McSwiggen J., Hovanessian A.G., Sen G., Williams B.R.G., and Silverman R.H. 1994. HIV-1 TAR RNA has an intrinsic ability to activate interferon-inducible enzymes. *Virology* **204:** 823–827.
Mamane Y., Heylbroeck C., Genin P., Algarte M., Servant M.J., LePage C., DeLuca C., Kwon H., Lin R., and Hiscott J. 1999. Interferon regulatory factors: The next generation. *Gene* **237:** 1–14.
Maran A. and Mathews M.B. 1988. Characterization of the double-stranded RNA implicated in the inhibition of protein synthesis in cells infected with a mutant adenovirus defective for VA RNA_I. *Virology* **164:** 106–113.
Maran A., Maitra R.K., Kumar A., Dong B., Xiao W., Li G., Williams B.R.G., Torrence P.F., and Silverman R.H. 1994. Blockage of NF-κB signaling by selective ablation of an mRNA target by 2-5A antisense chimeras. *Science* **265:** 789–792.
Marissen W.E. and Lloyd R.E. 1998. Eukaryotic translation initiation factor 4G is targeted for proteolytic cleavage by caspase 3 during inhibition of translation in apoptotic cells. *Mol. Cell. Biol.* **18:** 7565–7574.
Masison D.C., Blanc A., Ribas J.C., Carroll K., Sonenberg N., and Wickner R.B. 1995. Decoying the cap mRNA degradation system by a double-stranded RNA virus and poly(A)$^-$ mRNA surveillance by a yeast antiviral system. *Mol. Cell. Biol.* **15:** 2763–2771.
Mathews M.B. and Korner A. 1970. The inhibitory action of a mammalian viral RNA on the initiation of protein synthesis in a reticulocyte cell-free system. *Nature* **228:** 661–663.
Mathews M.B. and Shenk T. 1991. Adenovirus virus-associated RNA and translational control. *J. Virol.* **65:** 5657–5662.
McCarthy J.E.G. and Gualerzi C. 1990. Translational control of prokaryotic gene expression. *Trends Genet.* **6:** 78–85.
McFadden G., Lalani A., Everett H., Nash P., and Xu X. 1998. Virus-encoded receptors for cytokines and chemokines. *Semin. Cell Dev. Biol.* **9:** 359–368.
Melville M.W., Hansen W.J., Freeman B.C., Welch W.J., and Katze M.G. 1997. The molecular chaperone hsp40 regulates the activity of P58IPK, the cellular inhibitor of PKR. *Proc. Natl. Acad. Sci.* **94:** 97–102.
Mosig G. and Eiserling F. 1988. Phage T4 structure and metabolism. In The bacteriophages (ed. R. Calendar), vol. 2, pp. 521–606. Plenum Press, New York.
Muffler A., Fischer D., and Henegge-Aronis R. 1996. The RNA-binding protein HF-I, known as a host factor for phage Qβ RNA replication, is essential for rpoS translation in *Escherichia coli.* *Genes Dev.* **10:** 1143–1151.

Muller U., Steinhoff U., Reis L.F., Hemmi S., Pavlovic J., Zinkernagel R.M., and Aguet M. 1994. Functional role of type I and type II interferons in antiviral defense. *Science* **264:** 1918–1921.

Mulvey M., Poppers J., Ladd A., and Mohr I. 1999. A herpesvirus ribosome-associated, RNA-binding protein confers a growth advantage upon mutants deficient in a GADD34-related function. *J. Virol.* **73:** 3375–3385.

Mundschau L.J. and Faller D.V. 1994. Endogenous inhibitors of the dsRNA-dependent eIF-2α protein kinase PKR in normal and *ras*-transformed cells. *Biochimie* **76:** 792–800.

———. 1995. Platelet-derived growth factor signal transduction through the interferon-inducible kinase PKR. Immediate early gene induction. *J. Biol. Chem.* **270:** 3100–3106.

Nathans D., Notani G., Schwartz J.H., and Zinder N.D. 1962. Biosynthesis of the coat protein of coliphage f2 by *E. coli* extracts. *Proc. Natl. Acad. Sci.* **48:** 1424–1431.

Nemeroff M.E. and Bruenn J.A. 1987. Initiation by the yeast viral transcriptase in vitro. *J. Biol. Chem.* **262:** 6785–6787.

Nilsen T.W. and Baglioni C. 1979. Mechanism for discrimination between viral and host mRNA in interferon-treated cells. *Proc. Natl. Acad. Sci.* **76:** 2600–2604.

Nuss D.L., Oppermann H., and Koch G. 1975. Selective blockage of initiation of host protein synthesis in RNA virus-infected cells. *Proc. Natl. Acad. Sci.* **72:** 1258–1262.

O'Malley R.P., Duncan R.F., Hershey J.W.B., and Mathews M.B. 1989. Modification of protein synthesis initiation factors and shut-off of host protein synthesis in adenovirus-infected cells. *Virology* **168:** 112–118.

Offermann M.K., Zimring J., Mellits K.H., Hagan M.K., Shaw R., Medford R.M., Mathews M.B., Goodbourn S., and Jagus R. 1995. Activation of the double-stranded-RNA-activated protein kinase and induction of vascular cell adhesion molecule-1 by poly (I).poly (C) in endothelial cells. *Eur. J. Biochem.* **232:** 28–36.

Patel R.C. and Sen G.C. 1992. Identification of the double-stranded RNA-binding domain of the human interferon-inducible protein kinase. *J. Biol. Chem.* **267:** 7671–7676.

———. 1998. PACT, a protein activator of the interferon-induced protein kinase, PKR. *EMBO J.* **17:** 4379–4390.

Peabody D.S., Subramani S., and Berg P. 1986. Effect of upstream reading frames on translational efficiency in simian virus 40 recombinants. *Mol. Cell. Biol.* **6:** 2704–2711.

Pelletier J. and Sonenberg N. 1988. Internal initiation of translation of eukaryotic mRNA directed by a sequence derived from poliovirus RNA. *Nature* **334:** 320–325.

Penner M., Morad I., Snyder L., and Kaufmann G. 1995. Phage T4-coded stp: Double-edged effector of coupled DNA and tRNA-restriction systems. *J. Mol. Biol.* **249:** 857–868.

Pennisi E. 1998. Training viruses to attack cancers (comments). *Science* **282:** 1244–1246.

Piron M., Delaunay T., Grosclaude J., and Poncet D. 1999. Identification of the RNA-binding, dimerization, and eIF4GI-binding domains of rotavirus nonstructural protein NSP3. *J. Virol.* **73:** 5411–5421.

Piron M., Vende P., Cohen J., and Poncet D. 1998. Rotavirus RNA-binding protein NSP3 interacts with eIF4GI and evicts the poly(A) binding protein from eIF4F. *EMBO J.* **17:** 5811–5821.

Polunovsky V.A., Rosenwald I.B., Tan A.T., White J., Chiang L., Sonenberg N., and Bitterman P.B. 1996. Translational control of programmed cell death: Eukaryotic translation initiation factor 4E blocks apoptosis in growth-factor-restricted fibroblasts with physiologically expressed or deregulated Myc. *Mol. Cell. Biol.* **16:** 6573–6581.

Poot R.A., Tsareva N.V., Boni I.V., and van Duin J. 1997. RNA folding kinetics regulates translation of phage MS2 maturation gene. *Proc. Natl. Acad. Sci.* **94:** 10110–10115.

Ravetch J.V., Model P., and Robertson H.D. 1977. Isolation and characterization of ϕX174 ribosome binding sites. *Nature* **265:** 698–702.

Read G.S. and Frenkel N. 1983. Herpes simplex virus mutants defective in the virion-associated shutoff of host polypeptide synthesis and exhibiting abnormal synthesis of alpha (immediate early) polypeptides. *J. Virol.* **46:** 498–512.

Reichel P.A., Merrick W.C., Siekierka J., and Mathews M.B. 1985. Regulation of a protein synthesis initiation factor by adenovirus virus-associated RNAI. *Nature* **313:** 196–200.

Remm M., Remm A., and Ustav M. 1999. Human papillomavirus type 18 E1 protein is translated from polycistronic mRNA by a discontinuous scanning mechanism. *J. Virol.* **73:** 3062–3070.

Rivas C., Gil J., Melkova Z., Esteban M., and Diaz-Guerra M. 1998. Vaccinia virus E3L protein is an inhibitor of the interferon (i.f.n.)-induced 2-5A synthetase enzyme. *Virology* **243:** 406–414.

Robertson H.D. 1975. Functions of replicating RNA in cells infected by RNA bacteriophages. In RNA phages (ed. N.D. Zinder), pp. 113–145. Cold Spring Harbor Laboratory, Cold Spring Harbor, New York.

Robertson H.D., Manche L., and Mathews M.B. 1996. Paradoxical interactions between human delta hepatitis agent RNA and the cellular protein kinase PKR. *J. Virol.* **70:** 5611–5617.

Romano P.R., Zhang F., Tan S.L., Garcia-Barrio M.T., Katze M.G., Dever T.E., and Hinnebusch A.G. 1998. Inhibition of double-stranded RNA-dependent protein kinase PKR by vaccinia virus E3: Role of complex formation and the E3 N-terminal domain. *Mol. Cell. Biol.* **18:** 7304–7316.

Rosales R., Sutter G., and Moss B. 1994. A cellular factor is required for transcription of vaccinia viral intermediate-stage genes. *Proc. Natl. Acad. Sci.* **91:** 3794–3798.

Roy S., Katze M.G., Parkin N.T., Edery I., Hovanessian A.G., and Sonenberg N. 1990. Control of the interferon-induced 68-kilodalton protein kinase by the HIV-1 *tat* gene product. *Science* **247:** 1216–1219.

Ryabova L.A. and Hohn T. 2000. Ribosome shunting in the cauliflower mosaic virus 35S RNA leader is a special case of reinitiation of translation functioning in plant and animal systems. *Genes Dev.* **14:** 817–829.

Samuel C.E. 1991. Antiviral actions of interferon: Interferon-regulated cellular proteins and their surprisingly selective antiviral activities. *Virology* **183:** 1–11.

Sasaki J. and Nakashima N. 2000. From the cover: Methionine-independent initiation of translation in the capsid protein of an insect RNA virus. *Proc. Natl. Acad. Sci.* **97:** 1512–1515.

Satoh S., Hijikata M., Handa H., and Shimotohno K. 1999. Caspase-mediated cleavage of eukaryotic translation initiation factor subunit 2α. *Biochem. J.* **342:** 65–70.

Schmidt M.R., Gravel K.A., and Woodland R.T. 1995. Progression of a vesicular stomatitis virus infection in primary lymphocytes is restricted at multiple levels during B cell activation. *J. Immunol.* **155:** 2533–2544.

Schwartz S., Felber B.K., and Pavlakis G.N. 1992. Mechanism of translation of monocistronic and multicistronic human immunodeficiency virus type 1 mRNAs. *Mol. Cell.*

Biol. **12:** 207–219.

Sharpe A.H. and Fields B.N. 1982. Reovirus inhibition of cellular RNA and protein synthesis: Role of the S4 gene. *Virology* **122:** 381–391.

Shterman N., Elroy-Stein O., Morad I., Amitsur M., and Kaufmann G. 1995. Cleavage of the HIV replication primer $tRNA^{Lys,3}$ in human cells expressing bacterial anticodon nuclease. *Nucleic Acids Res.* **23:** 1744–1749.

Smith A.E. and Marcker K.A. 1970. Cytoplasmic methionine transfer RNAs from eukaryotes. *Nature* **226:** 607–610.

Smith A.E., Marcker K.A., and Mathews M.B. 1970. Translation of RNA from encephalomyocarditis virus in a mammalian cell-free system. *Nature* **225:** 184–187.

Sonenshein G.E. 1997. Rel/NF-κB transcription factors and the control of apoptosis. *Semin. Cancer Biol.* **8:** 113–119.

Srivastava S.P., Kumar K.U., and Kaufman R.J. 1998. Phosphorylation of eukaryotic translation initiation factor 2 mediates apoptosis in response to activation of the double-stranded RNA-dependent protein kinase. *J. Biol. Chem.* **273:** 2416–2423.

Stanners C.P., Francoeur A.M., and Lam T. 1977. Analysis of VSV mutant with attenuated cyto-pathogenicity in viral function, P, for inhibition of protein synthesis. *Cell* **11:** 273–281.

Stark G.R., Kerr I.M., Williams B.R., Silverman R.H., and Schreiber R.D. 1998. How cells respond to interferons. *Annu. Rev. Biochem.* **67:** 227–264.

Steitz J.A. 1969. Polypeptide chain initiation: Nucleotide sequences of the three ribosomal binding sites in bacteriophage R17 RNA. *Nature* **224:** 957–964.

Stoneley M., Chappell S.A., Jopling C.L., Dickens M., MacFarlane M., and Willis A.E. 2000. c-Myc protein synthesis is initiated from the internal ribosome entry segment during apoptosis. *Mol. Cell. Biol.* **20:** 1162–1169.

Strong J.E., Coffey M.C., Tang D., Sabinin P., and Lee P.W. 1998. The molecular basis of viral oncolysis: Usurpation of the Ras signaling pathway by reovirus. *EMBO J.* **17:** 3351–3362.

Svensson C. and Akusjärvi G. 1990. A novel effect of adenovirus VA RNA_I on cytoplasmic mRNA abundance. *Virology* **174:** 613–617.

Svitkin Y.V., Hahn H., Gingras A.-C., Palmenberg A.C., and Sonenberg N. 1998. Rapamycin and wortmannin enhance replication of a defective encephalomyocarditis virus. *J. Virol.* **72:** 5811–5819.

Swaminathan S., Huneycutt B.S., Reiss C.S., and Kieff E. 1992. Epstein-Barr virus-encoded small RNAs (EBERs) do not modulate interferon effects in infected lymphocytes. *J. Virol.* **66:** 5133–5136.

Swaminathan S., Rajan P., Savinova O., Jagus R., and Thimmapaya B. 1996. Simian virus 40 large-T bypasses the translational block imposed by the phosphorylation of eIF-2α. *Virology* **219:** 321–323.

Tan S.L. and Katze M.G. 1998. Biochemical and genetic evidence for complex formation between the influenza A virus NS1 protein and the interferon-induced PKR protein kinase. *J. Interferon Cytokine Res.* **18:** 757–766.

———. 1999. The emerging role of the interferon-induced PKR protein kinase as an apoptotic effector: A new face of death? *J. Interferon Cytokine Res.* **19:** 543–554.

Taylor D.R., Shi S.T., Romano P.R., Barber G.N., and Lai M.M.C. 1999. Inhibition of the interferon-inducible protein kinase PKR by HCV E2. *Science* **285:** 107–110.

Thomis D.C. and Samuel C.E. 1993. Mechanism of interferon action: Evidence of intermolecular autophosphorylation and autoactivation of the interferon-induced, RNA-dependent protein kinase PKR. *J. Virol.* **67:** 7695–7700.

Timmer R.T., Benkowski L.A., Schodin D., Lax S.R., Metz A.M., Ravel J.M., and Browning K.S. 1993. The 5′ and 3′ untranslated regions of satellite tobacco necrosis virus RNA affect translational efficiency and dependence on a 5′ cap structure. *J. Biol. Chem.* **268:** 9504–9510.

Toczyski D.P., Matera A.G., Ward D.C., and Steitz J.A. 1994. The Epstein-Barr virus (EBV) small RNA EBER1 binds and relocalizes ribosomal protein L22 in EBV-infected human B lymphocytes. *Proc. Natl. Acad. Sci.* **91:** 3463–3467.

Utz P.J., Hottelet M., Le T.M., Kim S.J., Geiger M.E., van Venrooij W.J., and Anderson P. 1998. The 72-kDa component of signal recognition particle is cleaved during apoptosis. *J. Biol. Chem.* **273:** 35362–35370.

van Duin J. 1988. Single-stranded RNA bacteriophages. In *The bacteriophages* (ed. R. Calendar), vol. 1, pp. 117–167. Plenum Press, New York.

Vilček J. and Sen G.C. 1996. Interferons and other cytokines. In *Field's virology* (ed. B.N. Fields et al.), pp. 375–399. Lippincott-Raven, Philadelphia.

Volkin E. and Astrachan L. 1956. Phosphorus incorporation in *E. coli* ribonucleic acid after infection with bacteriophage T2. *Virology* **2:** 146–161.

Voorma H.O. 1996. Control of translation initiation in prokaryotes. In *Translational control* (ed. J.W.B. Hershey et al.), pp. 759–777. Cold Spring Harbor Laboratory Press, Cold Spring Harbor, New York.

Wang E., Marcotte R., and Petroulakis E. 1999. Signaling pathway for apoptosis: A racetrack for life or death. J. Cell. Biochem. (suppl. 32/33): 95–102.

Webster R.E., Engelhardt D., and Zinder N. 1966. In vitro protein synthesis: Chain initiation. *Proc. Natl. Acad. Sci.* **55:** 155–161.

Wei C.M. and Moss B. 1975. Methylated nucleotides block 5′-terminus of vaccinia virus messenger RNA. *Proc. Natl. Acad. Sci.* **72:** 318–322.

Weiss R.B., Huang W.M., and Dunn D.M. 1990. A nascent peptide is required for ribosomal bypass of the coding gap in bacteriophage T4 gene 60. *Cell* **62:** 117–126.

Wimmer E., Chang A.Y., Clark J.M., Jr., and Reichmann M.E. 1968. Sequence studies of satellite tobacco necrosis virus RNA. Isolation and characterization of a 5′-terminal trinucleotide. *J. Mol. Biol.* **38:** 59–73.

Witherell G.W., Gott J.M., and Uhlenbeck O.C. 1991. Specific interaction between RNA phage coat proteins and RNA. *Prog. Nucleic Acid Res. Mol. Biol.* **40:** 185–220.

Wittmann H.G. and Wittmann-Liebold B. 1967. Protein chemical studies of two RNA viruses and their mutants. *Cold Spring Harbor Symp. Quant. Biol.* **31:** 163–172.

Wulczyn F.G. and Kahmann R. 1991. Translational stimulation: RNA sequence and structure requirements for binding of Com protein. *Cell* **65:** 259–269.

Yu Y.T. and Snyder L. 1994. Translation elongation factor Tu cleaved by a phage-exclusion system. *Proc. Natl. Acad. Sci.* **91:** 802–806.

Yu Y.X., Bearzotti M., Vende P., Ahne W., and Bremont M. 1999. Partial mapping and sequencing of a fish iridovirus genome reveals genes homologous to the frog virus 3 p31, p40 and human eIF2α. *Virus Res.* **63:** 53–63.

Yueh A. and Schneider R.J. 2000. Translation by ribosome shunting on adenovirus and hsp70 mRNAs facilitated by complementarity to 18S rRNA. *Genes Dev.* **14:** 414–421.

Zhang Y. and Schneider R.J. 1994. Adenovirus inhibition of cell translation facilitates release of virus particles and enhances degradation of the cytokeratin network. *J. Virol.* **68:** 2544–2555.

Zhou A., Hassel B.A., and Silverman R.H. 1993. Expression cloning of 2-5A-dependent

RNAase: A uniquely regulated mediator of interferon action. *Cell* **72:** 753–765.

Zhou A., Paranjape J., Brown T.L., Nie H., Naik S., Dong B., Chang A., Trapp B., Fairchild R., Colmenares C., and Silverman R.H. 1997. Interferon action and apoptosis are defective in mice devoid of 2′,5′-oligoadenylate-dependent RNase L. *EMBO J.* **16:** 6355–6363.

9

Ribosomal Subunit Joining

Tatyana V. Pestova[1,2] and Christopher U.T. Hellen[1]

[1]Department of Microbiology and Immunology
Morse Institute for Molecular Genetics
State University of New York Health Science Center
Brooklyn, New York 11203
[2]A.N. Belozersky Institute of Physico-Chemical Biology
Moscow State University, 119899 Moscow, Russia

Thomas E. Dever

Laboratory of Eukaryotic Gene Regulation
National Institute of Child Health and Human Development
National Institutes of Health
Bethesda, Maryland 20892-2716

Ribosomal subunit joining is the final stage in translation initiation in eukaryotes and follows formation of a 48S initiation complex at the initiation codon of an mRNA. Biochemical studies have identified individual steps leading to 48S complex formation on the majority of mRNAs (Chapter 2). First, a 43S complex is assembled from eukaryotic initiation factor (eIF) 1A, eIF3, an [eIF2/GTP/initiator Met $tRNA_i^{Met}$] ternary complex, and a 40S ribosomal subunit. The 5′-terminal cap is bound by eIF4F, which with eIFs 4A and 4B unwinds RNA structure in the 5′-nontranslated region (5′NTR) and facilitates binding of the 43S complex to the 5′ end of the mRNA. eIF1 and eIF1A are required for 43S complexes to locate the initiation codon (Pestova et al. 1998), presumably by scanning until base-pairing is established between the first AUG triplet and the anticodon of $tRNA_i^{Met}$. The resulting 48S complex arrested at the initiation codon is stable and after fractionation by sucrose density gradient centrifugation or gel filtration remains associated with factors that include eIFs 1, 1A, 3 and the [eIF2/ GTP/Met-$tRNA_i^{Met}$] complex (Benne and Hershey 1978; Trachsel and Staehelin 1978; Peterson et al. 1979a; Thomas et al. 1980; Pestova et al. 2000; Chapter 2).

Joining of a 60S ribosomal subunit to the 48S complex is associated with two linked events: hydrolysis of GTP bound to eIF2 in the 48S com-

© 2000 Cold Spring Harbor Laboratory Press 0-87969-568-4/00 $5 + . 00

plex and displacement of factors. Replacement of GTP by GMPPNP (a nonhydrolyzable analog) does not impair ternary or 48S complex formation but prevents the 48S complex from joining a 60S subunit to form an 80S ribosome that is competent to begin translation (Trachsel et al. 1977; Benne and Hershey 1978; Peterson et al. 1979a, b). eIF2 exists either in an active, GTP-bound form that can bind initiator tRNA and associate with the 40S subunit or in an inactive GDP-bound form that cannot bind initiator tRNA. GTP hydrolysis therefore results in release of eIF2 in an inactive GDP-bound form and is thought to leave Met-tRNA$_i^{Met}$ in the ribosomal P site. 80S ribosomes are not associated with eIF1, eIF1A, eIF2, or eIF3, which are removed or displaced prior to or as a result of subunit joining (Benne and Hershey 1978; Trachsel and Staehelin 1978; Peterson et al. 1979a; Thomas et al. 1980; Pestova et al. 2000).

IDENTIFICATION OF A HIGH-MOLECULAR-WEIGHT SUBUNIT-JOINING FACTOR

Simple addition of 60S subunits to 48S complexes does not yield 80S ribosomes, so additional factors are required for subunit joining. Studies on this process began in the early 1970s when investigators started to use in vitro reconstituted systems. In reconstituted systems mRNA can be replaced by an AUG triplet. 48S and 80S complex formation on AUG triplets is more efficient and much simpler than on native mRNAs. For example, it does not require the whole set of initiation factors and can occur in the absence of ATP, eIF1, eIF4F, eIF4A, eIF4B, and even eIF3 (Benne and Hershey 1978). Although the use of AUG triplets instead of mRNA in reconstituted systems is very attractive for reasons of simplicity and efficiency, 48S complexes assembled on AUG triplets lack some RNA–protein interactions present in 48S complexes assembled on native mRNA. These interactions are likely to contribute to the process of 80S complex formation, and thus initiation on native mRNAs is likely more complicated than on AUG triplets. Studies done using both AUG triplets and native mRNAs resulted in the identification of a single polypeptide factor of molecular weight 125,000–168,000, variously named IF-L2 (Levin et al. 1973), IF-II (Cashion and Stanley 1974), F-0.25 (Suzuki and Goldberg 1974), IF-M2A (Merrick et al. 1975), IF-E5 (Benne et al. 1976), IF-E2 (Grummt 1974), and eIF5 (Schreier et al. 1977; Benne et al. 1978; Peterson et al. 1979a,b; Meyer et al. 1981). The term eIF5 was used to describe this factor for a number of years, but for reasons described below, was eventually dropped. We shall refer to this factor as eIF5B. It may be a phosphoprotein, but this modification has not been linked with any regulatory function (Traugh et al. 1976; Duncan and Hershey 1983). eIF5B was reported to mediate joining of 60S subunits to 48S complexes to form active 80S ribosomes as assayed by a model reaction (synthesis of methionylpuromycin [MP]) that mimics formation of the first peptide bond (Leder and Bursztyn 1966). The structure of puromycin resembles the 3′ end of aminoacylated tRNA closely enough so that it can bind to the ribosomal A site and react with methionyl-tRNA in the P site to form MP (Yarmolinsky and de la Hara 1959; Bretscher and Marcker 1966). eIF5B catalytically induces the hydrolysis of eIF2-bound GTP on 48S complexes and can do so on an AUG template in the absence of 60S subunits (Odom et al. 1978; Peterson et al. 1979b). The translation component with which eIF5B interacts to induce hydrolysis of eIF2-bound GTP is not known. Assembly of active 80S ribosomes on AUG triplets can occur without additional factors if 48S complexes are first pre-incubated with eIF5B to induce hydrolysis of eIF2-bound GTP and are then purified by sucrose density gradient centrifugation before incubation with 60S subunits (Peterson et al. 1979b). Assembly of active 80S ribosomes can therefore be separated into at least two distinct stages: hydrolysis of eIF2-bound GTP, which in this case is sufficient to prepare 48S complexes for joining, and a subsequent subunit-joining event. eIF1A enhances the assembly and stability of 48S complexes on AUG triplets and their joining with 60S subunits to form 80S ribosomes (Peterson et al. 1979a).

Incubation of 48S complexes assembled on globin mRNA or AUG triplets with eIF5B and 60S subunits and subsequent fractionation on sucrose density gradients leads to loss of GTP, eIF2, and eIF3 from assembled 80S complexes in a step that depends on GTP hydrolysis (Benne and Hershey 1978; Trachsel and Staehelin 1978; Peterson et al. 1979a). It is not clear whether factor release from the 40S subunit in both instances precedes, or is the result of, 60S subunit joining. Release of factors occurred when 60S subunits were omitted from assays done using eIF5B and 48S complexes assembled on AUG triplets with or without eIF3 and in the absence of eIFs 1, 4A, 4B, and 4F (Peterson et al. 1979b). Nevertheless, it is important to note that sucrose density gradient centrifugation of these 48S complexes may itself lead to the loss of initiation factors which might normally still be bound to 48S complexes after hydrolysis of eIF2-bound GTP. Moreover, the presence of additional factors (eIFs 1, 4A, 4B, and 4F) and natural mRNA instead of AUG triplets in 48S complexes may further stabilize the binding of factors with 40S subunits in intermediate preinitiation complexes after GTP hydrolysis. For these reasons, a firm conclusion concerning the timing of initiation factor release is not yet possible.

Puromycin-reactive 80S complexes were formed when 48S complexes assembled on AUG and purified by gel filtration were incubated with eIF5B and 60S subunits but without additional GTP (Merrick 1979). This observation led to the suggestion that subunit joining requires hydrolysis of only one GTP molecule, which is bound to eIF2 in 48S complexes (Merrick 1979). However, as reported earlier, eIF5B alone mediates GTP hydrolysis in a reaction that required just 40S and 60S subunits and no other initiation factors with estimated K_m for GTP 10 μM and a V_{max} for hydrolysis of 1.2 pmole/μg eIF5B/min (Merrick et al. 1975).

A SECOND SUBUNIT-JOINING FACTOR IS A GTPASE ACTIVATING PROTEIN

More recently, a second subunit-joining factor was identified by Maitra and colleagues that mediated 60S subunit joining to "minimal" 48S complexes assembled on AUG triplets using only 40S subunits and the [eIF2/GTP/Met-$tRNA_i^{Met}$] ternary complex (Raychaudhuri et al. 1985a,b; Chakravarti et al. 1993). This new factor was a homogeneous monomeric phosphoprotein of apparent M_r = 58,000–62,000, which had many (but not all) of the activities of the 125,000-168,000 factor. Immunological characterization indicated that this novel 58-kD factor is not generated from the larger protein by proteolysis during purification (Ghosh et al. 1989; Chevesich et al. 1993). This conclusion has been confirmed by sequence analysis of rat, human, yeast, and maize cDNAs (Chakravarti and Maitra 1993; Das et al. 1993; Si et al. 1996; Ribera et al. 1997). Recombinant forms of this novel eIF5 were as active as the native factor in all assays (Chakravarti and Maitra 1993; Das et al. 1993). Because of the reported high specific enzymatic activity of the recombinant 58-kD protein in this subunit-joining assay, it was suggested that this protein was the active component of the previously purified subunit-joining activity and that the larger protein was merely an inactive contaminant. At this point, the term "eIF5" was taken from the ~150-kD protein and given to the new 58-kD factor. To avoid further confusion, we shall use the revised nomenclature: the 58-kD factor will be referred to as eIF5, and the larger factor will be referred to as eIF5B.

Incubation of eIF5 and "minimal" 48S complexes with or without 60S subunits led to rapid, quantitative hydrolysis of eIF2-bound GTP and release of eIF2–GDP, and in the presence of 60S subunits, yielded puromycin-reactive 80S ribosomes (Raychaudhuri et al. 1985a,b). In the absence of 60S subunits, the eIF2–GDP complex was detected on 40S subunits purified by gel filtration but not by sucrose density gradient centrifugation (Raychaudhuri and Maitra 1986). eIF5 itself was not detected on these gel-filtered complexes (Chakrabarti and Maitra 1991), consistent with its highly catalytic role in inducing hydrolysis of eIF2-bound GTP on 48S complexes (Ghosh et al. 1989; Chakravarti et al. 1993). In simplified assay systems, eIF5 therefore resembles eIF5B in that it can induce hydrolysis by eIF2 of bound GTP but differs from eIF5B in that it is not a ribosome-dependent GTPase (Raychaudhuri et al. 1985a; Chakrabarti and Maitra 1991).

eIF5 is an essential single copy gene in *Saccharomyces cerevisiae* (Chakravarti and Maitra 1993). Sequence homology between eIF5 from different organisms exceeds 50% (Ribera et al. 1997), and mammalian eIF5 can substitute for yeast eIF5 in vitro and in vivo (Maiti and Maitra 1997). Sequence homology is greatest in an amino-terminal ~160-amino-acid residue domain that contains a putative C_2-C_2 zinc finger and is strongly homologous to the zinc-finger domain of eIF2β (Das et al.1997; Ribera et al.1997). The less-conserved carboxy-terminal domain of eIF5 mediates interaction with eIF2β (Chaudhuri et al. 1994; Koonin 1995; Das et al. 1997; Asano et al. 1999). eIF5 also associates with yeast and mammalian eIF3, and in yeast the eIF5-binding site has been localized to the Nip1p subunit of eIF3 (Phan et al. 1998; Bandyopadhyay and Maitra 1999). This mammalian and yeast subunit of eIF3 also binds eIF1 (Naranda et al. 1996; Asano et al. 1998; Fletcher et al. 1999). The same carboxy-terminal region of eIF5 is required for the interaction with eIF2β and the Nip1p subunit of eIF3 (Asano et al. 1999). These interactions with eIF2 and eIF3 are thought to recruit eIF5 to the 43S complex and to enable eIF5 to activate GTP hydrolysis by eIF2.

Several recent reports have begun to reveal how the interaction of eIF5 with the 48S complex causes the hydrolysis of bound GTP. eIF5 alone does not hydrolyze free GTP or GTP bound to ternary complexes in the absence of 40S subunits, so the context of the ribosomal initiation complex is needed for the interaction of eIF5 with eIF2 to cause a conformational change that activates its GTPase activity. Substitution of a conserved residue (G31R) near the zinc finger in yeast eIF5 that increases utilization of a UUG triplet as the initiation codon (*Sui*$^-$ phenotype) does so as a result of an increase in the specific activity of eIF5 in stimulating GTP hydrolysis by ribosome-bound eIF2 (Huang et al. 1997; Chapter 12). *Sui*$^-$ mutations in a putative zinc finger in the carboxyl end of eIF2β conferred an intrinsic GTPase activity on eIF2 independent of eIF5 (Huang et al. 1997). *Sui*$^-$ alleles have also been identified in yeast eIF1 (Yoon and Donahue 1992), which binds to the same eIF3 subunit as eIF5. Accurate recognition of AUG as the start codon is therefore dependent on the functions of the eIF1/eIF2/eIF3/eIF5 complex and may be coupled directly to the rate of hydrolysis of GTP bound to eIF2 in 48S complexes. Donahue and colleagues have suggested that the signal for eIF5-dependent hydrolysis of GTP bound to eIF2 is establishment of base-pairing between an AUG initiation codon and the CAU anticodon of Met-tRNA$_i^{Met}$ and that mutations that uncouple this linkage can lead to recognition of non-AUG triplets as the initiation codon (Huang et al. 1997).

A HOMOLOG OF PROKARYOTIC INITIATION FACTOR IF2

eIF5 was identified on the basis of its activity in joining 60S subunits to 48S complexes that had been assembled on AUG triplets using only 40S subunits and the [eIF2/GTP/Met-tRNA$_i^{Met}$] complex. Although it is necessary, eIF5 is not sufficient to promote joining of 60S subunits to 48S complexes that had been assembled on a natural (globin) mRNA using a complete set of factors (eIFs 1, 1A, 2, 3, 4A, 4B, and 4F), and a 175-kD protein is also required (Pestova et al. 2000). This factor is termed eIF5B and is a eukaryotic homolog of the prokaryotic initiation factor IF2 (Pestova et al. 2000). eIF5B has been identified in various eukaryotes (Sutrave et al. 1994; Choi et al. 1998; Nagase et al. 1998; Lee et al. 1999; Wilson et al. 1999; Carrera et al. 2000). A role for eukaryotic IF2 in translation was first revealed by studies in yeast. Deletion of the *FUN12* gene encoding the yeast IF2 homolog resulted in a severe slow-growth phenotype, and polyribosome profile analyses and in vitro translation assays revealed a translation initiation defect in the *fun12Δ* strain (Choi et al. 1998). Importantly, the translation defect in extracts from fun12Δ strains could be fully restored by addition of recombinant yeast IF2 or native and recombinant human eIF5B (Choi et al. 1998; Lee et al. 1999; Pestova et al. 2000). Several observations indicated that yeast "IF2" is a general translation factor and is not required simply for a subset of cellular mRNAs. For example, polysome profiles revealed that translation of the majority of cellular mRNAs was impaired in the *fun12Δ* strain, and second, the expression pattern of human eIF5B in various tissues is similar to that of other general translation factors such as eIF2 and eIF4G (Lee et al. 1999). A *trans*-dominant mutant form of human eIF5B had an inhibitory effect on translation in human cells (Wilson et al. 1999). Common physical and biochemical properties suggest that eIF5B is the ~150-kD protein previously implicated in subunit joining (see above), and this has been confirmed by Western blotting, using native human eIF5B (from J.W.B. Hershey), native mouse eIF5B, and affinity-purified antibodies raised against recombinant human eIF5B (T. Pestova, unpubl.).

There is an apparent contradiction between a requirement for both eIF5 and eIF5B in subunit joining (Pestova et al. 2000) and earlier indications that these factors have overlapping and even redundant functions in this process. In fact, the overlap in functions is apparent only when "minimal" 48S complexes are assembled on AUG triplets using only eIF2, and a requirement for both factors becomes apparent in the presence of the full set of factors required for 48S complex formation on natural mRNA (Pestova et al. 2000). In a complete assay system lacking only eIF5, eIF5B loses the ability to stimulate the hydrolysis of GTP bound to eIF2, and without eIF5B, GTP hydrolysis induced by eIF5 is not sufficient for subunit joining. The influence of individual factors on the activities of eIF5 and eIF5B in promoting hydrolysis of eIF2-bound GTP and assembly of puromycin-reactive 80S ribosomes was assayed by systematically adding factors to "minimal" 48S complexes assembled on AUG triplets from eIF1A, the [eIF2/GTP/Met-tRNA$_i^{Met}$] ternary complex and a 40S subunit. eIF1 enhanced the activity of eIF5 but not eIF5B, whereas eIF3 enhanced the activity of eIF5B and reduced the activity of eIF5. Together, eIF1 and eIF3 reduced the individual activity of eIF5B and to a greater extent that of eIF5. MP synthesis was strongly increased by the synergistic activities of eIF5 and eIF5B in the presence of eIF1 and eIF3. The ability of eIF5 and eIF5B to stimulate hydrolysis of eIF2-bound GTP also depended on the presence of different initiation factors. eIF5 and eIF5B stimulated GTP hydrolysis equally when 48S complexes were assembled only with eIF2 and eIF1A. Inclusion of eIF3 and eIF1 into such 48S complexes reduced the activity of eIF5B but not eIF5. The reduced ability of eIF5B to stimulate hydrolysis of eIF2-bound GTP in the presence of eIF1 and eIF3 can account for the reduced ability of eIF5B to promote MP synthesis in the presence of eIF1 and eIF3. Amino-terminally truncated human eIF5B(587-1220) had a much lower ability to induce GTP hydrolysis than the full-length native protein (which probably reflects the difference in affinities of the two forms of eIF5B for components of the 48S complex) but was equally active in MP synthesis in the presence of eIF5. This observation indicates that the hydrolysis of eIF2-bound GTP in normal circumstances is induced by eIF5. Taken together, these results suggest that if 48S complexes are assembled only with eIF2 and eIF1A, hydrolysis of eIF2-bound GTP induced either by eIF5 or by eIF5B is sufficient to prepare the 48S complex for joining. This preparatory step may involve ejection of eIF2–GDP from the 48S complex. If 48S complexes are associated with the complete set of factors, then eIF5-induced GTP hydrolysis is not sufficient for joining, and eIF5B is also required. The inability of 48S complexes assembled with all factors to join 60S subunits after eIF5-induced GTP hydrolysis in the absence of eIF5B could be because these factors are not displaced from the 40S subunit by eIF5 and still block the subunit interface. They may be displaced by eIF5B directly or by the 60S subunit as a consequence of eIF5B-mediated joining. However, incubation of 48S complexes assembled on encephalomyocarditis virus (EMCV) mRNA with eIF5 and eIF5B and subsequent purification by sucrose density centrifugation did not yield ribosomal complexes that are competent for subunit joining. The addition of 60S subunits to these purified complexes did not lead to formation of 80S ribosomes, and the addition of eIF5B, but not eIF5, together with 60S subunits was necessary and sufficient for the formation of 80S complexes (T. Pestova, unpubl.). This observation suggests that eIF5B participates directly in the subunit-joining step rather than in just preparing 48S complexes for subsequent joining. Unfortunately, in the conditions of this experiment it was not possible to determine whether incubation with eIF5 and eIF5B of 48S complexes assembled on the EMCV internal ribosome entry site (IRES) led to the loss of eIFs 2, 3, 1, and 1A because 48S complexes migrated very close to 43S complexes, which contained these four factors.

STRUCTURE OF eIF5B

One surprise encountered in the genome sequences of *S. cerevisiae* and the archaeon *Methanococcus jannaschii* was the identification of homologs for prokaryotic IF2 in both. Subsequent studies have indicated that IF2/eIF5B is present in all organisms and that the sequences of IF2/eIF5B homologs are highly conserved (Fig. 1). The structures of these proteins can be subdivided into three regions, corresponding approximately to amino acids 1–628, 629–850, and 851–1220 of human eIF5B. The first of these regions is highly variable, whereas the second and third are strongly conserved in all Archaea, Bacteria, and Eucarya. The role of IF2 in initiation has been studied in the greatest detail in prokaryotes, and each of these three domains has been assigned a specific function in this process. Prokaryotic IF2 therefore serves as a benchmark against which related proteins can be compared. IF2 binds four major ligands (GTP, initiator tRNA, 30S and 50 ribosomal subunits) and is involved in two essential steps in translation initiation. First, it stimulates binding of initiator tRNA to the P site of the 30S subunit (Petersen et al. 1979; Pon et al. 1985). IF1 prevents premature dissociation of IF2 from the 30S preinitiation complex (Benne et al. 1973). Second, IF2 promotes subunit association (Godefroy-Colburn et al. 1975). The GTPase activity of IF2 is activated after subunit joining (Kolakofsky et al.1968).

Alternative hypotheses suggest that this activity is necessary either (1) to correctly adjust the position of initiator tRNA in the P site (La Teana et al. 1996) or (2) for (accelerated) release of IF2 from the 70S ribosome after assembly (Lockwood et al. 1972; Luchin et al. 1999).

The amino-terminal region of IF2/eIF5B proteins is variable in length and shows little sequence conservation (Bremaud et al. 1997; Lee et al. 1999; Carrera et al. 2000). However, it does contain an unusually high proportion of charged residues that are responsible for the aberrant electrophoretic mobility of these proteins. This region is absent in archaea and some prokaryotes; deletion of it from *Escherichia coli* IF2 caused a severe slow-growth phenotype (Laalami et al. 1991). A similar deletion made in the *S. cerevisiae* factor had no effect on cell growth, and the truncated protein is able to fully restore translation in extracts from *fun12*Δ strains (Choi et al. 1998). Deletion of the amino-terminal third of *E. coli* IF2 abrogated its binding to the small (30S) ribosomal subunit and impaired its binding to the large (50S) subunit and to the 70S ribosome; binding to 30S subunits and 70S ribosomes but not 50S subunits was restored by addition of IF1, a homolog of eIF1A (Moreno et al. 1999). These observations suggested that the amino-terminal domain of IF2 interacts with the 30S subunit. An analogous function for this region of human eIF5B would be consistent with the effects of deleting it on the activity of eIF5B in inducing hydrolysis of eIF2-bound GTP in 48S complexes.

The functionally critical and most strongly conserved portions of IF2/eIF5B comprise a GTP-binding domain of ~200 amino acid residues and a larger carboxy-terminal region. The sequence of the G domain is homologous to the corresponding domains of the elongation factors EF1A (EF-Tu) and EF2 (EF-G) (Laalami et al. 1996; Brock et al. 1998). Free IF2 binds GTP with a tenfold lower affinity than GDP, but both nucleotides are bound less efficiently than by elongation factors (Pon et al. 1985). These binding affinities may reflect higher rates of dissociation, which could in turn enable IF2 to cycle from the GDP state to the GTP state without a requirement for a guanine nucleotide exchange factor (GEF). Free IF2 does not show GTPase activity under physiological conditions; it is induced by binding of the 50S subunit during subunit joining (Kolakofsky et al. 1968). Significantly, the isolated G domain retains the ability to bind to the 50S subunit and displays ribosome-dependent GTPase activity (Gualerzi et al. 1991). The importance of the G domain for IF2 function and thus of GTP hydrolysis in translation initiation has been assessed by mutational analysis of highly conserved residues likely to be involved in GTP binding or hydrolysis (Laalami et al. 1994). These residues are present at identical positions in human eIF5B (Fig. 1). Six of

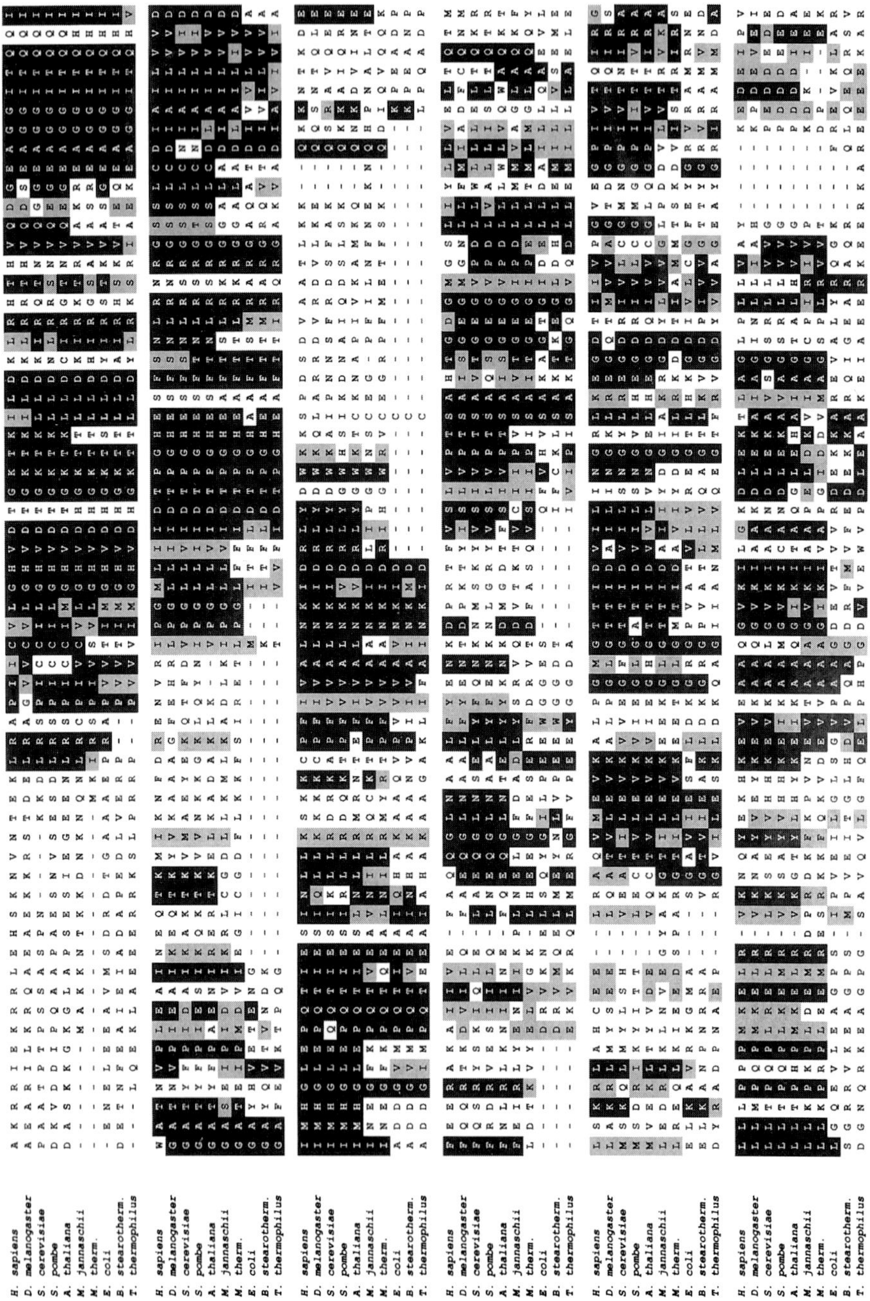
H. sapiens
D. melanogaster
S. cerevisiae
S. pombe
A. thaliana
M. jannaschii
M. therm.
E. coli
B. stearotherm.
T. thermophilus

Figure 1 Alignment of amino acid sequences of the GTP-binding domain and carboxy-terminal regions of IF2 homologs from eukaryotes, archaea, and prokaryotes. The SwissProt (SP) or GenBank (GB) accession numbers and the number of the first residue of each sequence in the alignment are: Eukaryotes: *Homo sapiens* (SP:O60841; first residue 609), *Drosophila melanogaster* (GB:AF143207, first residue 669), *S. cerevisiae* (SP:P39730, first residue 387), *Schizosaccharomyces pombe* (SP:Q10251, first residue 462), *Arabadopsis thaliana* (GB:AC010718.5, first residue 669); archaea: *M. jannaschii* (GB:Q57710, first residue 1), *Methanobacterium thermoautotrophicum* (GB:AE000812.1, first residue 1); prokaryotes: *E. coli* (SP:P02995, first residue 370), *Bacillus stearothermophilus* (SP:P04766, first residue 224), *Thermus thermophilus* (SP:P48515, first residue 56). The sequences were aligned using the program Clustal W (version 1.74) (Thompson et al. 1994), and the figure was generated using the program Boxshade (version 3.31) utilizing a web-based interface established by Peter FitzGerald (NIH, Bethesda, MD). Identical residues are printed in reverse type and conserved residues are shaded.

seven mutations were lethal, and of these, two were dominant negative inhibitors of translation that had an unchanged affinity for GTP but had defects in GTP hydrolysis and in release from 70S ribosomes after subunit joining (Luchin et al. 1999). The importance of the G domain for eIF5B's function in initiation has been confirmed in a similar manner (see below).

Prokaryotic IF2 stimulates binding of fMet-tRNA^{fMet} to the P site of the ribosomal 30S subunit (Petersen et al. 1979; Pon et al. 1985). IF2 interacts with fMet-tRNA^{fMet} through its carboxy-terminal domain (Spurio et al. 2000). There appear to be two sites of interaction between IF2 and fMet-tRNA^{fMet} (Petersen et al. 1979; Yusupova et al. 1996). Structural analyses of binary complexes by X-ray diffraction and nuclear magnetic resonance (NMR) (see, e.g., Förster et al. 1999) may soon reveal the structural basis of this interaction. In contrast, human eIF5B(598-1220) has a carboxy-terminal domain that is ~25% identical to the corresponding domain of *E. coli* IF2 (Lee et al. 1999) but does not form stable complexes with Met-tRNA$_i^{Met}$ in the presence or absence of GTP (T. Dever and T. Pestova, unpubl.). The molecular basis for this difference is not yet known. However, the slow-growth phenotype of the *fun12*Δ *S. cerevisiae* strain could be partially suppressed by overexpression of tRNA$_i^{Met}$ (Choi et al. 1998). This observation suggests that even though eIF5B does not bind stably to Met-tRNA$_i^{Met}$ to form binary or ternary complexes (with GTP), yeast eIF5B may nevertheless stabilize or enhance binding of Met-tRNA$_i^{Met}$ to ribosomal complexes. This proposed function for eIF5B would show some similarity with the function of prokaryotic IF2 in promoting binding of fMet-tRNA^{fMet} to the 30S subunit. Sequence comparisons reveal that archaeal IF2 is more similar to the eukaryotic than the prokaryotic factor. The eukaryotic and archaeal sequences contain insertions within and immediately carboxy-terminal to the G domain that are absent in the prokaryotic IF2 sequences (Fig. 1). Consistent with its structural similarity to the eukaryotic factor, IF2 from *M. jannaschii* was able to substitute partially for the yeast factor in vivo and to stimulate translation in vitro in extracts from *fun12*Δ strains (Lee et al. 1999).

In addition to the sequence similarity between eIF5B and prokaryotic IF2, the eukaryotic factor eIF1A is similar to prokaryotic IF1 (Kyrpides and Woese 1998; Battiste et al. 2000). As mentioned above, IF1 stabilizes binding of IF2 to the 30S preinitiation complex (Benne et al. 1973), and crosslinking studies have revealed that IF1 and IF2 are in close proximity when bound to ribosomes (Boileau et al. 1983). These observations raise the possibility that eIF5B and eIF1A interact directly. Yeast two-hybrid assays have confirmed that *S. cerevisiae* eIF1A and eIF5B do interact, and they have mapped a binding site for eIF1A to the carboxyl terminus of eIF5B (aa 850–1002 of yeast eIF5B). Interestingly, the eIF1A-binding site in eIF5B maps to the region that corresponds to the Met-tRNA-binding site in IF2 (Spurio et al. 2000). Co-immunoprecipitation and GST pull-down experiments confirmed the interaction between eIF5B and eIF1A, and binding studies with purified, recombinant forms of eIF5B and eIF1A demonstrated that the factors interacted directly (T. Dever et al., unpubl.). Genetic studies in *S. cerevisiae* have indicated that the carboxyl terminus of eIF5B is critical for growth in vivo, and overexpression of eIF1A exacerbates the growth defect of yeast strains expressing carboxy-terminally truncated eIF5B. These findings indicate that eIF5B and eIF1A interact functionally in vivo. As eIF1A is thought to facilitate ternary complex binding to ribosomes (Benne and Hershey 1978), it is tempting to propose that the eIF5B–eIF1A interaction helps to direct or stabilize Met-tRNA$_i^{Met}$ binding to the ribosomal P site.

FUNCTION OF THE G DOMAIN OF eIF5B IN THE SUBUNIT-JOINING PROCESS

Sequence analyses suggested that the amino acid sequence of eIF5B contains four of the sequence motifs G1–G5 characteristic of GTP-binding proteins (Fig. 2) (Bourne et al. 1991; Laalami et al. 1996; Brock et al. 1998; Lee et al. 1999) and that, as a result, eIF5B would have GTP-binding and hydrolysis activities, and that it may not need a GEF to be recycled (Keeling and Doolittle 1995; Lee et al. 1999). These hypotheses have been confirmed experimentally (Pestova et al. 2000). eIF5B binds GTP but has no detectable intrinsic GTPase activity; its interaction with GTP is not dependent on either 40S or 60S subunits, but its hydrolysis of GTP is strongly activated by them together and considerably less so by 60S subunits alone. These data are in agreement with the previous results obtained using the 150-kD subunit-joining factor (Merrick et al. 1975). GTP hydrolysis by eIF5B is therefore coupled to subunit joining. GTP bound to eIF5B exchanges readily, and it also appears probable that eIF5B–GDP does not require a recycling GEF because eIF5B can catalyze multiple rounds of GTP hydrolysis and of 80S complex assembly in a fully reconstituted assay system (Odom et al. 1978; Peterson et al. 1979b; Pestova et al. 2000). The degree of dependence of eIF5B's activity in assembly of puromycin-reactive 80S complexes on binding to GTP was determined by the integrity of the protein. The activity of eIF5B(587-1220) was completely GTP-dependent, whereas full-length native eIF5B retained low activity in the absence of GTP but was nevertheless stimulated 3–4 times by binding to GTP (Pestova et al. 2000; T. Pestova,

	G1	G2	G3	G4	G5
Cons.	GxxxxGKT/S	DxxxxxxxxxxT	DxxGh	NKxD	SAK
IF2	GhvdhGKT	tkvasgeaggiT	DtpGh	NKiD	SAK
eIF5B	GhvdtGKT	thvqdgeaggiT	DtpGh	NKiD	
	638 645	654 665	702-6	756-9	

Figure 2 Sequence alignment of GTP-binding protein consensus sequence motifs G1–G5 (Laalami et al. 1996) and sequence motifs G1–G4 of *E. coli* IF2 (Sacerdot et al. 1984) and human eIF5B (Nagase et al. 1998; Lee et al. 1999; Wilson et al. 1999).

unpubl.). These observations indicate that eIF5B has to bind GTP to adopt the active conformation required for subunit joining. In addition, they indicate that the amino-terminal part of eIF5B may influence the structure of the protein and its affinity for components of the translational apparatus, which may reduce the dependence of eIF5B's activity on binding to GTP. Hydrolysis of GTP by eIF5B is not necessary for subunit joining or for removal of other factors from assembled 80S ribosomes: Both processes can occur in the presence of nonhydrolyzable analogs such as GMPPNP, but in this case eIF5B is no longer able to act catalytically. Hydrolysis of GTP by eIF5B is necessary for eIF5B to dissociate from the 80S ribosome after its assembly. If GTP is substituted by GMPPNP, eIF5B-GMPPNP remains associated with the 80S ribosome and blocks its ability to react with puromycin (Pestova et al. 2000). The ability of bound eIF5B to block MP synthesis suggests that eIF5B may bind directly to the ribosomal A site.

Since the model for eIF5B function suggests that it adopts an active conformation when bound to GTP and that GTP hydrolysis is required for release of eIF5B from assembled 80S ribosomes, the phenotype of eIF5B G-domain mutants would be expected to depend on their ability to bind and hydrolyze GTP. Overexpression in 293 T cells of human G643R mutant eIF5B impaired reporter gene translation in vivo (Wilson et al. 1999). This mutation is in motif G1 (which interacts with the phosphates of GTP). Another mutation (V640G) in the same G1 motif yielded human eIF5B that was able to restore translation in vitro in extracts from fun12Δ cells (Lee et al. 1999) and to promote 80S subunit joining and MP synthesis in vitro in a reconstituted assembly reaction (T. Dever and T. Pestova, unpubl.). An H706Q mutant retained partial activity in the in vitro translation extract and in promoting MP synthesis in a reconstituted reaction, whereas H706E and D759N substitutions in eIF5B almost abrogated its activity (Lee et al. 1999; T. Dever and T. Pestova, unpubl.). The activity of these mutants correlated with their ability to hydrolyze GTP: H706E and D759N mutants did not hydrolyze GTP in the presence of ribosomal subunits, whereas the V640G mutant hydrolyzed GTP at half the rate of *wt* eIF5B (T. Dever and T. Pestova, unpubl.). However, careful analysis of a D759N eIF5B mutant demonstrated clearly that two GTP hydrolysis steps are required during translation initiation in eukaryotes. The substitution of Asp-759 by asparagine in this mutant switches eIF5B's nucleotide specificity from GTP to xanthosine triphosphate (XTP). This D759N substitution abolishes the ability of human eIF5B to hydrogen-bond with the 2-amino group on GTP, but hydrogen bonding is restored when (as in XTP) a keto group is placed at the 2 position on the purine ring. Utilizing this nucleotide specificity switch mutant, it was found that MP synthesis required both GTP (for eIF2) and XTP (for eIF5B), and thus demonstrated that two nucleotide hydrolysis events are required during translation initiation (T. Dever and T. Pestova, unpubl.). A recent kinetic analysis of MP synthesis by 80S complexes in rabbit reticulocyte lysate indicated that there is an additional step in initiation after 80S complex formation that can be inhibited by GMPPNP (Lorsch and Herschlag 1999). This step is required to convert the 80S complex to an 80S* complex that is 30-fold more active in MP synthesis; after conversion, MP synthesis by these activated complexes is insensitive to inhibition by GMPPNP. These results suggest that energy released during this additional GTP hydrolysis step may be required for conformational reorganization of the 80S complex. The relationship between these observations and the requirement for eIF5B-catalyzed GTP hydrolysis in formation of 80S ribosomes that are active in MP synthesis has still to be investigated.

CONCLUDING REMARKS

The process of ribosomal subunit joining in eukaryotes requires two initiation factors, eIF5 and eIF5B, which together promote the release of other initiation factors from the 48S complex and the subsequent joining of a 60S subunit to form an active 80S ribosome. The function of eIF5 in activating hydrolysis of eIF2-bound GTP is clearly defined. This is the first and was until recently thought to be the only GTP hydrolysis step in subunit joining. A distinct requirement for eIF5B has only recently been recognized, and the mechanism of its action has not been characterized in such detail. eIF5B is a ribosome-activated GTPase, but the role of this activity in steps other than its own release is not yet known. Nevertheless, it is apparent that ribosomal subunit joining involves two successive GTP hydrolysis events.

The close homology of eIF5B and the prokaryotic initiation factor IF2 was unexpected. It provides further evidence that the initiation processes in prokaryotes and eukaryotes are more closely related than had previously been recognized. However, the most fundamental difference between them lies in the mechanism of initiation codon selection. Instead of direct binding of the small ribosomal subunit at the initiation codon, as in prokaryotes, eukaryotes utilize a scanning mechanism for initiation codon selection, for which a sophisticated apparatus of initiation factors is used to promote binding of the small ribosomal subunit at the capped 5′ end of the mRNA, followed by scanning to the initiation codon. In prokaryotes, IF2 plays dual roles of promoting recruitment of initiator tRNA to the small ribosomal subunit and in mediating subunit joining. In eukaryotes, initiator tRNA recruitment is mediated by eIF2, a factor that has no homolog in prokaryotes, and eIF5B, the homolog of IF2, plays a role in subunit joining.

ACKNOWLEDGMENTS

We are grateful to Bill Merrick for many helpful discussions, and to Jon Lorsch, Dan Herschlag, and Paul Lasko for communicating results prior to publication. The authors' work is supported by grants from the National Institutes of Health and the National Science Foundation.

REFERENCES

Asano K., Phan L., Anderson J., and Hinnebusch A.G. 1998. Complex formation by all five homologues of mammalian translation initiation factor 3 subunits from yeast *Saccharomyces cerevisiae*. *J. Biol. Chem.* **273:** 18573–18585.

Asano K., Krishnamoorthy T., Phan L., Pavitt G.D., and Hinnebusch A. 1999. Conserved bipartite motifs in yeast eIF5 and eIF2Bε, GTPase-activating and GDP-GTP exchange factors in translation initiation, mediate binding to their common substrate, eIF2. *EMBO J.* **18:** 1673–1688.

Bandyopadhyay A. and Maitra U. 1999. Cloning and characterization of the p42 subunit of mammalian translation initiation factor 3 (eIF3): Demonstration that eIF3 interacts with eIF5 in mammalian cells. *Nucleic Acids Res.* **27:** 1331–1337.

Battiste J.L., Pestova T.V., Hellen C.U.T., and Wagner G. 2000. The eIF1A solution structure reveals a large RNA-binding surface important for scanning function. *Mol. Cell* **5:** 109–119.

Benne R. and Hershey J.W.B. 1978. The mechanism of action of protein synthesis initiation factors from rabbit reticulocytes. *J. Biol. Chem.* **253:** 3078–3087.

Benne R., Brown-Luedi M.L., and Hershey J.W.B. 1978. Purification and characterization of protein synthesis initiation factors eIF-1, eIF-4C, eIF-4D, and eIF-5 from rabbit reticulocytes. *J. Biol. Chem.* **253:** 3070–3077.

Benne R., Naaktgeboren N., Gubbens J., and Vooma H.O. 1973. Recycling of initiation factors IF-1, IF-2 and IF-3. *Eur. J. Biochem.* **32:** 372–380.

Benne R., Wong C., Luedi M., and Hershey J.W.B. 1976. Purification and characterization of initiation factor IF-E2 from rabbit reticulocytes. *J. Biol. Chem.* **251:** 7675–7681.

Boileau G., Butler P., Hershey J.W.B., and Traut R.R. 1983. Direct cross-links between initiation factors 1, 2, and 3 and ribosomal proteins promoted by 2-iminothiolane. *Biochemistry* **22:** 3162–3170.

Bourne H.R., Saunders D.A., and McCormick F. 1991. The GTPase superfamily: Conserved structure and molecular mechanism. *Nature* **349:** 117–127.

Bremaud L., Laalami S., Derijard B., and Cenatiempo Y. 1997. Translation initiation factor IF2 of the myxobacterium *Stigmatella aurantiaca:* Presence of a single species with an unusual N-terminal sequence. *J. Bacteriol.* **179:** 2348–2355.

Bretscher M.S. and Marcker K.A. 1966. Polypeptidyl-sRibonucleic acid and amino-acyl-sRibonucleic acid bind sites on ribosomes. *Nature* **211:** 380–384.

Brock S., Szkaradiewicz K., and Sprinzl M. 1998. Initiation factors of protein biosynthesis in bacteria and their structural relationship to elongation and termination factors. *Mol. Microbiol.* **29:** 409–417.

Carrera P., Johnstone O., Nakamura A., Casanova J., Jackle H., and Lasko P. 2000. *Vasa* mediates translation through interaction with a *Drosophila* yIF2 homolog. *Mol. Cell:* **5:** 181–187.

Cashion L.M. and Stanley W.M. 1974. Two eukaryotic initiation factors (IF-I and IF-II) of protein synthesis are required to form an initiation complex with rabbit reticulocyte ribosomes. *Proc. Natl. Acad. Sci.* **71:** 436–440.

Chakrabarti A. and Maitra U. 1991. Function of eukaryotic initiation factor 5 in the formation of an 80S ribosomal polypeptide chain initiation complex. *J. Biol. Chem.* **266:** 14039–14045.

Chakravarti D. and Maitra U. 1993. Eukaryotic translation initiation factor 5 from *Saccharomyces cerevisiae.* Cloning, characterization and expression of the gene encoding the 45,346-Da protein. *J. Biol. Chem.* **268:** 10524–10533.

Chakravarti D., Maita T., and Maitra U. 1993. Isolation and immunochemical characterization of eukaryotic translation initiation factor 5 from *Saccharomyces cerevisiae. J. Biol. Chem.* **268:** 5754–5762.

Chaudhuri J., Das K., and Maitra U. 1994. Purification and characterization of bacterially expressed mammalian translation initiation factor 5 (eIF5): Demonstration that eIF-5 forms a specific complex with eIF-2. *Biochemistry* **33:** 4794–4799.

Chevesich J., Chaudhuri J., and Maitra U. 1993. Characterization of mammalian translation initiation factor 5 (eIF-5). Demonstration that eIF-5 is a phosphoprotein and is present in cells as a single molecular form of apparent M_r 58,000. *J. Biol. Chem.* **268:** 20659–20667.

Choi S.K., Lee J.H., Zoll W.L., Merrick W.C., and Dever T.E. 1998. Promotion of binding Met-$tRNA_i^{Met}$ to ribosomes by yIF2, a bacterial IF2 homolog in yeast. *Science* **280:** 1757–1760.

Das K., Chevesich J., and Maitra U. 1993. Molecular cloning and expression of cDNA for mammalian translation initiation factor 5. *Proc. Natl. Acad. Sci.* **90:** 3058–3062.

Das S., Maiti T., Das K., and Maitra U. 1997. Specific interaction of eukaryotic translation initiation factor 5 (eIF5) with the β-subunit of eIF2. *J. Biol. Chem.* **272:** 31712–31718.

Duncan R. and Hershey J.W.B. 1983. Identification and quantitation of levels of protein synthesis initiation factors in crude HeLa cell lysates by two-dimensional polyacrylamide gel electrophoresis. *J. Biol. Chem.* **258:** 7228–7235.

Fletcher C.M., Pestova T.V., Hellen C.U.T., and Wagner G. 1999. Structure and interactions of the translation initiation factor eIF1. *EMBO J.* **18:** 2631–2637.

Förster C., Krafft C., Welfle H., Gualerzi C.O., and Heinemann U.. 1999. Preliminary characterization by X-ray diffraction and Raman spectroscopy of a crystalline complex of *Bacillus stearothermophilus* initiation factor C-domain and fMet-tRNA^{fMet}. *Acta Crystallogr.* **D55:** 712–716.

Ghosh S., Chesevich J., and Maitra U. 1989. Further characterization of eukaryotic initiation factor 5 from rabbit reticulocytes. Immunochemical characterization and phosphorylation by casein kinase II. *J. Biol. Chem.* **264:** 5134–5140.

Godefroy-Colburn T., Wolfe A.D., Dondon J., and Grunberg-Manago M. 1975. Light scattering studies showing the effect of initiation factors on the reversible dissociation of *Escherichia coli* ribosomes. *J. Mol. Biol.* **94:** 461–478.

Grummt F. 1974. Studies on two initiation factors of protein synthesis from rat-liver cytoplasm. *Eur. J. Biochem.* **43:** 337–342.

Gualerzi C.O., Severin M., Spurio R., La Teana A., and Pon C.L. 1991. Molecular dissection of translation initiation factor IF2. Evidence for two structural and functional domains. *J. Biol. Chem.* **266:** 16356–16362.

Huang H.-K., Yoon H., Hannig E.M., and Donahue T.F. 1997. GTP hydrolysis controls stringent selection of the AUG start codon during translation initiation in *Saccharomyces cerevisiae. Genes. Dev.* **11:** 2396–2413.

Keeling P.J. and Doolittle W.F. 1995. Archaea: Narrowing the gap between prokaryotes and eukaryotes. *Proc. Natl. Acad. Sci.* **92:** 5761–5764.

Kolakofsky D., Ohta T., and Thach R.E. 1968. Junction of the 50S ribosomal subunit with the 30S initiation complex. *Nature* **220:** 244–247.

Koonin E.V. 1995. Multidomain organization of eukaryotic guanine nucleotide exchange translation initiation factor eIF-2B subunits revealed by analysis of conserved sequence motifs. *Protein Sci.* **4:** 1608–1617.

Kyrpides N.C. and Woese C.R. 1998. Universally conserved translation initiation factors. *Proc. Natl. Acad. Sci.* **95:** 224–228.

La Teana, A., Pon C.L., and Gualerzi C.O. 1996. Late events in translation initiation. Adjustment of fMet-tRNA in the ribosomal P site. *J. Mol. Biol.* **256:** 667–675.

Laalami S., Grentzmann G., Bremaud L., and Cenatiempo Y. 1996. Messenger RNA translation in prokaryotes: GTPase centers associated with translational factors. *Biochimie* **78:** 577–589.

Laalami S., Putzer H., Plumbridge J.A., and Grunberg-Manago M. 1991. A severely truncated form of translation initiation factor 2 supports growth of *Escherichia coli. J. Mol. Biol.* **220:** 335–349.

Laalami S., Timofeev A.V. , Putzer H., Leautey J., and Grunberg-Manago M. 1994. In vivo study of engineered G-domain mutants of *Escherichia coli* translation initiation factor IF2. *Mol. Microbiol.* **11:** 293–302.

Leder P. and Bursztyn H. 1966. Initiation of protein synthesis II. A convenient assay for the ribosome-dependent synthesis of N-formyl-C^{14}-methionylpuromycin. *Biochem. Biophys. Res. Commun.* **25:** 233–238.

Lee J.H., Choi S.K., Roll-Mecak A., Burley S.K., and Dever T.E. 1999. Universal conservation in translation initiation revealed by human and archaeal homologs of bacterial translation factor IF2. *Proc. Natl. Acad. Sci.* **96:** 1066–1070.

Levin D.H., Kyner D., and Acs G. 1973. Protein synthesis in eukaryotes. Characterization of ribosomal factors from mouse fibroblasts. *J. Biol. Chem.* **248:** 6416–6425.

Lockwood A.H., Sarkar P., and Maitra U. 1972. Release of polypeptide chain initiation factor IF-2 during initiation complex formation. *Proc. Natl. Acad. Sci.* **69:** 3602–3605.

Lorsch J.R. and Herschlag D. 1999. Kinetic dissection of fundamental processes of eukaryotic translation initiation in vitro. *EMBO J.* **18:** 6705–6717.

Luchin S., Putzer H., Hershey J.W.B., Cenatiempo Y., Grunberg-Manago M., and Laalami S. 1999. In vitro study of two dominant inhibitory GTPase mutants of *Escherichia coli* translation initiation factor IF2. Direct evidence that GTP hydrolysis is necessary for factor recycling. *J. Biol. Chem.* **274:** 6074–6079.

Maiti T. and Maitra U. 1997. Characterization of translation initiation factor 5 (eIF5) from *Saccharomyces cerevisiae*. Functional homology with mammalian eIF5 and the effect of depletion of eIF5 on protein synthesis in vivo and in vitro. *J. Biol. Chem.* **272:** 18333–18340.

Merrick W.C. 1979. Evidence that a single GTP is used in the formation of 80S initiation complexes. *J. Biol. Chem.* **254:** 3708–3711.

Merrick W.C., Kemper W.M., and Anderson W.F. 1975. Purification and characterization of homogenous initiation factor M2A from rabbit reticulocytes. *J. Biol. Chem.* **250:** 5556–5562.

Meyer L.J., Brown-Luedi M.L., Corbett S., Tolan D.R., and Hershey J.W.B. 1981. The purification and characterization of multiple forms of protein synthesis eukaryotic initiation factors 2, 3 and 5 from rabbit reticulocytes. *J. Biol. Chem.* **256:** 351–356.

Moreno J. M.P., Dyrskjøtersen L., Kristensen J.E., Mortensen K.K., and Sperling-Petersen H.U. 1999. Characterization of the domains of *E. coli* initiation factor IF2 responsible for recognition of the ribosome. *FEBS Lett.* **455:** 130–134.

Nagase T., Ishikawa K., Suyama M., Kikuno R., Miyajima N., Tanaka A., Kotani H., Nomura N., and Ohara O. 1998. Prediction of the coding sequences of unidentified human genes. XI. The complete sequences of 100 new cDNA clones from brain which code for large proteins in vitro. *DNA Res.* **5:** 277–286.

Naranda T., McMillan S.E., Donahue T.F., and Hershey J.W.B. 1996. SUI1/p16 is required for the activity of eukaryotic initiation factor 3 in *Saccharomyces cerevisiae*. *Mol. Cell. Biol.* **17:** 145–153.

Odom O.W., Kramer G., Henderson A.B., Pinphanichakarn P., and Hardesty B. 1978. GTP hydrolysis during methionyl-$tRNA_f$ binding to 40S ribosomal subunits and the site of edeine inhibition. *J. Biol. Chem.* **253:** 1807–1816.

Pestova T.V., Borukhov S.I., and Hellen C.U.T. 1998. Eukaryotic ribosomes require initiation factors 1 and 1A to locate initiation codons. *Nature* **394:** 854–859.

Pestova T.V., Lomakin I.B., Lee J.H., Choi S.K., Dever T.E., and Hellen C.U.T. 2000. The ribosomal subunit joining reaction in eukaryotes requires eIF5B. *Nature* **403:** 332–335.

Petersen H.U., Roll T., Grunberg-Manago M., and Clark B.F.C. 1979. Specific interaction of initiation factor IF2 of *E. coli* with formylmethionyl-$tRNA_f^{met}$. *Biochim. Biophys. Res. Commun.* **91:** 1068–1074.

Peterson D.T., Safer B., and Merrick W.C. 1979a. Binding and release of radiolabeled eukaryotic initiation factors 2 and 3 during 80S initiation complex formation. *J. Biol. Chem.* **254:** 2509–2516.

———. 1979b. Role of eukaryotic initiation factor 5 in the formation of 80S initiation complexes. *J. Biol. Chem.* **254:** 7730–7735.

Phan L., Zhang X., Asano K., Anderson J., Vornlocher H.-P., Greenberg J.R., Qin J., and Hinnebusch A.G. 1998. Identification of a translation initiation factor 3 (eIF3) core complex, conserved in yeast and mammals, that interacts with eIF5. *Mol. Cell. Biol.* **18:** 4935–4946.

Pon C.L., Paci M., Pawlik R.T., and Gualerzi C.O. 1985. Stucture-function relationship in *Escherichia coli* initiation factors. Biochemical and biophysical characterization of the interaction between IF2 and guanosine nucleotides. *J. Biol. Chem.* **260:** 8918–8924.

Raychaudhuri P. and Maitra U. 1986. Identification of ribosome-bound eukaryotic initiation factor 2.GDP binary complex as an intermediate in polypeptide chain initiation reaction. *J. Biol. Chem.* **261:** 7723–7728.

Rayhaudhuri P., Chaudhuri A., and Maitra U. 1985a. Eukaryotic initiation factor 5 from calf liver is a single polypeptide chain protein of M_r = 62, 000. *J. Biol. Chem.* **260:** 2132–2139.

———. 1985b. Formation and release of eukaryotic initiation factor 2-GDP complex during eukaryotic ribosomal polypeptide chain initiation complex formation. *J. Biol. Chem.* **260:** 2140–2145.

Ribera I.L., Ruiz-Avila L., and Puigdomenech P. 1997. The eukaryotic translation initiation factor 5, eIF-5, a protein from *Zea mays*, containing a zinc-finger structure, binds nucleic acids in a zinc-dependent manner. *Biochem. Biophys. Res. Commun.* **236:** 510–516.

Sacerdot C., Dessen P., Hershey J.W.B., Plumbridge J.A., and Grunberg-Manago M. 1984. Sequence of initiation factor IF2 gene; unusual protein features and homologies with elongation factors. *Proc. Natl. Acad. Sci.* **81:** 7787–7791.

Schreier M.H., Erni B., and Staehelin T. 1977. Initiation of protein synthesis. I. Purification and characterization of seven initiation factors. *J. Mol. Biol.* **116:** 727–753.

Si K., Das K., and Maitra U. 1996. Characterization of multiple mRNAs that encode mammalian translation initiation factor 5 (eIF-5). *J. Biol. Chem.* **271:** 16934–16938.

Spurio R., Brandi L., Caserta E., Pon C.L., Gualerzi C.O., Misselwitz R., Krafft C., Welfle K., and Welfle H. 2000. The C-terminal subdomain (IF-C2) contains the entire fMet-tRNA binding site of initiation factor IF2. *J. Biol. Chem.* **275:** 2447–2454.

Sutrave P., Shafer B.K., Strathern J.N., and Hughes S.H. 1994. Isolation, identification and characterization of the FUN12 gene of *Saccharomyces cerevisiae*. *Gene* **146:** 209–213.

Suzuki H. and Goldberg I.H. 1974. Reversal of pactamycin inhibition of methionyl-puromycin synthesis and 80S initiation complex formation by a ribosomal joining factor. *Proc. Natl. Acad. Sci.* **71:** 4259–4263.

Thomas A., Goumans H., Voorma H.O., and Benne R. 1980. The mechanism of action of eukaryotic initiation factor 4C in protein synthesis. *Eur. J. Biochem.* **107:** 39–45.

Thompson J. D., Higgins D.G., and Gibson T.J. 1994. CLUSTAL W: Improving the sensitivity of progressive multiple sequence alignment through sequence weighting, positions-specific gap penalties and weight matrix choice. *Nucleic Acids Res.* **22:** 4673–4680.

Trachsel H. and Staehelin T. 1978. Binding and release of eukaryotic initiation factor eIF-2 and GTP during protein synthesis initiation. *Proc. Natl. Acad. Sci.* **75:** 204–208.

Trachsel H., Erni B., Schreier M.H., and Staehelin T. 1977. Initiation of protein synthesis. II. The assembly of the initiation complex with purified initiation factors. *J. Mol. Biol.* **116:** 755–767.

Traugh J.A., Tahara S.M., Sharp S.B., Safer B., and Merrick W.C. 1976. Factors involved in initiation of haemoglobin synthesis can be phosphorylated in vitro. *Nature* **263:** 163–165.

Wilson S.A., Sieiro-Vazquez C., Edwards N.J., Iourin O., Byles E.D., Kotsopoulou E., Adamson C.S., Kingsman S.M., Kingsman A.J., and Martin-Rendon E. 1999. Cloning and characterization of hIF2, a human homologue of bacterial translation initiation factor 2, and its interaction with HIV-1 matrix. *Biochem. J.* **342:** 97–103.

Yoon H.J. and Donahue T.F. 1992. The *sui1* supressor locus in *Saccharomyces cerevisiae* encodes a translation factor that functions during $tRNA_i^{Met}$ recognition of the start codon. *Mol. Cell. Biol.* **12:** 248–260.

Yarmolinsky M.B. and de la Hara G.L. 1959. Inhibition by puromycin of amino acid incorporation into protein. *Proc. Natl. Acad. Sci.* **45:** 1721–1729.

Yusupova G., Reinbolt J., Wakao H., Laalami S., Grunberg-Manago M., Romby P., Ehresmann B., and Ehresmann C. 1996. Topography of the *Escherichia coli* initiation factor 2/ fMet-$tRNA^{fMet}$ complex as studied by cross-linking. *Biochemistry* **35:** 2978–2984.

10

Physical and Functional Interactions between the mRNA Cap Structure and the Poly(A) Tail

Alan Sachs
Department of Molecular and Cell Biology
University of California,
Berkeley, California 94720-3202

The 5′-end of all nuclear-encoded messenger RNAs in eukaryotes contains a cap structure (m^7GpppX). With few exceptions, the 3′-end also contains a poly(A) tail. These structures are added by specific enzymes during or just after mRNA transcription. Not surprisingly, they play important roles in many aspects of mRNA metabolism, including mRNA splicing, transport, stability, localization, and translation. In this chapter, I focus primarily on the role of the poly(A) tail in translation initiation and how it interacts with the 5′-cap structure. The reader is referred to other chapters in this volume for more detailed information on some of these other areas.

THE CAP STRUCTURE AND THE eIF4F COMPLEX

The involvement of the cap structure in mRNA translation was suggested soon after its discovery in the early 1970s (Shatkin 1976). Using an in vitro translation system, it was found that the cap structure stimulated expression of mRNA (Both et al. 1975; Muthukrishnan et al. 1975). Soon after this discovery, the 24-kD protein eIF4E was shown to be associated with the cap structure (Sonenberg et al. 1978). eIF4E could be easily purified from cell extracts by affinity chromatography with m^7GDP-agarose (Sonenberg et al. 1979). Two other major proteins were found to copurify with eIF4E: The 220-kD protein was named eIF4G, and the 46-kD protein was the already-characterized initiation factor eIF4A. In subsequent years it was shown in both mammalian and yeast systems that eIF4G pro-

Translational Control of Gene Expression
© 2000 Cold Spring Harbor Laboratory Press 0-87969-568-4/00 $5 + .00

vided the scaffold to which eIF4E and eIF4A became associated. The complex of eIF4E with eIF4G and eIF4A is often referred to as the eIF4F complex (for review, see Chapter 6).

A current working model for the way in which eIF4F mediates translational stimulation envisions that the association of eIF4E with the cap structure brings the eIF4G and eIF4A proteins into the vicinity of the mRNA. In conjunction with the eIF4B protein, eIF4A unwinds secondary structure within the leader of the mRNA as it hydrolyzes ATP. The small ribosomal subunit, in association with the multisubunit eIF3 complex, binds to the mRNA via an interaction with eIF4G and, potentially, eIF4B. After disengaging from these factors, the ribosome scans in a 5′–3′ direction along the unstructured leader in search of an initiator codon. (See Chapter 2 for a more detailed description of these early steps in the initiation process.)

THE POLY(A) TAIL AND THE POLY(A)-BINDING PROTEIN Pab1p

The discovery of the poly(A) tail in the early 1970s (Lim and Canellakis 1970; Kates 1971) was followed by many years in which its function in mRNA metabolism was debated. The role of poly(A) in mRNA stability is discussed in Chapter 28. With respect to translation and the poly(A) tail, a detailed account of literature prior to 1990 can be found in Jacobson (1996). In brief, as of 1990 it was known that the addition of a poly(A) tail to mRNA had modest (2- to 3-fold) stimulatory effects on the translation of mRNA added to mammalian translation extracts. In reticulocyte lysates this stimulation was due to enhanced rates of 60S ribosomal subunit binding (Munroe and Jacobson 1990). It had been shown that the addition of exogenous poly(A) to translation extracts could stimulate the translation of capped, poly(A)-deficient mRNA and inhibit the translation of polyadenylated mRNA (Jacobson and Favreau 1983; Munroe and Jacobson 1990). Finally, a high degree of correlation between the cytoplasmic polyadenylation of maternal mRNAs and their recruitment onto polyribosomes was observed in vertebrate oocytes and developing embryos, as was the inhibition of this recruitment when polyadenylation was prevented (for review, see Wickens 1990a,b). These observations led to the working hypothesis that the mRNA poly(A) tail was in some way involved in the translation process.

The discovery and subsequent characterization of the eukaryotic poly(A)-binding protein (Pab1p) during the 1980s (Baer and Kornberg 1983; Sachs and Kornberg 1985; Adam et al. 1986; Sachs et al. 1986, 1987) also supported the hypothesis that the poly(A) tail was involved in translation. Mutations within the yeast *PAB1* gene resulted in an inhibition of translation and cell growth (Sachs and Davis 1989). Suppressors of a deletion of the essential *PAB1* gene also altered the level of the 60S ribosomal subunit (Sachs and Davis 1989, 1990). These suppressors included a ribosomal protein and a DEAD-box protein involved in rRNA biogenesis. Together with the above experiments studying poly(A), these data suggested that the poly(A)/Pab1p complex played an important role in the translation process.

SIGNIFICANT RECENT ADVANCES FOR STUDIES ON CAP AND POLY(A) TAIL FUNCTION IN TRANSLATION

Two major advances in understanding the role of the poly(A) tail and its interactions with the cap structure during translation were reported in the early 1990s. Each of these opened new avenues for exploration of these areas. One introduced the concept of translational synergy into the field, and the other introduced a method to prepare translation extracts that exhibited significant stimulation when mRNA was capped or polyadenylated.

In the first of these important advances, Daniel Gallie and coworkers (Gallie 1991) showed that animal, plant, and yeast cells exhibited significant translation of mRNA that was delivered by electroporation. Using this system, it was found that the cap structure and the poly(A) tail provided a translational advantage to the mRNA. When both of these structures were added to the mRNA, their stimulatory effect was greater than additive. This synergy of translational stimulation by the cap and the poly(A) tail became a hallmark by which the validity of in vitro translation extracts was evaluated.

The second important advance was provided by Peter Sarnow and coworkers (Iizuka et al. 1994). In a series of experiments designed to investigate internal ribosome entry site function in vitro, these scientists developed a translation extract derived from the yeast *Saccharomyces cerevisiae* that barely translated an mRNA lacking a cap structure or a poly(A) tail. In the presence of either of these structures, the mRNA's translation was stimulated between 5- and 50-fold. When both the cap and the poly(A) tail were on the mRNA, a further 5- to 10-fold synergistic stimulation was observed. More recent work indicates that the amount of synergism in these extracts is in part determined by the degree of competition from endogenous mRNA, since synergism is most readily observed in non-nuclease-treated extracts or in nuclease-treated extracts that have been supplemented with excess competitor RNA (Tarun and Sachs 1997; Preiss and Hentze 1998).

ELUCIDATION OF A MECHANISM OF TRANSLATIONAL STIMULATION BY THE CAP AND POLY(A) TAIL IN YEAST

These new yeast extracts were used to study how poly(A) tails stimulated translation. The potential involvement of Pab1p in poly(A)-dependent translation was suggested by the previous genetic analysis of this protein. Direct evidence for this involvement was provided by the observation that immunoneutralization or depletion of Pab1p from the yeast extract inhibited poly(A)-dependent but not cap-dependent translation (Tarun and Sachs 1995). Importantly, the synergistic stimulation of translation by the cap and the poly(A) tail was also lost in the absence of Pab1p. An analysis of translation intermediates in extracts lacking Pab1p revealed that Pab1p stimulated 40S ribosomal subunit binding to the mRNA. Because 40S subunit binding was also stimulated by the cap structure, this indicated that the poly(A) tail and the cap structure performed similar functions in the translation initiation pathway. Whether Pab1p and the poly(A) tail also stimulate the 60S joining reaction in the yeast system, as they do in the reticulocyte system (Munroe and Jacobson 1990), remains to be explored.

That the cap structure and the poly(A) tail performed similar functions implied that they might have a common target which would mediate this function. Previous work had shown that eIF4G was required for eIF4E to exert its stimulatory effect on translation (see, e.g., Etchison et al. 1982). The possible association of yeast Pab1p with eIF4G was therefore investigated (Tarun and Sachs 1996). First, it was found that purification of yeast eIF4E by m^{7}GDP-agarose chromatography yielded not only eIF4E, eIF4G, and CAF20 (a negative regulator of eIF4E), but also Pab1p. The association of Pab1p with eIF4E was shown to depend on eIF4G, thereby implicating eIF4G as the scaffold to which other factors bound. Evidence for the specificity of this interaction was that recombinant Pab1p bound to recombinant eIF4G in a poly(A)-dependent manner (Tarun and Sachs 1996).

Further dissection of the eIF4G–Pab1p interaction revealed that the first and second RNA-recognition motifs (RRMs) of Pab1p are needed for eIF4G binding, and that substitutions of as few as two amino acids within RRM2 of Pab1p can prevent this binding (Table 1) (Kessler and Sachs 1998; Otero et al. 1999). Furthermore, a fragment as small as 110 amino acids from eIF4G1 has been shown to bind to yeast Pab1p specifically, and amino acid substitutions within this region of the full-length eIF4G1 protein have also been found to diminish Pab1p binding (Table 1) (Tarun and Sachs 1996; Tarun et al. 1997).

A clear advantage of the yeast system is that the effects of mutations within an individual factor on translation can be investigated in the context of a crude extract simply by preparing the extract from the mutant strain. Such an approach was used to confirm that the association of eIF4E or Pab1p with eIF4G was needed for the extracts to exhibit cap- or poly(A)-dependent translation, respectively, and that translational synergism required both proteins to interact with eIF4G (see Table 1) (Tarun and Sachs 1997; Tarun et al. 1997; Kessler and Sachs 1998). It also became clear that the mechanism by which poly(A) tails stimulate translation is likely to be more complex than a simple interaction between Pab1p and eIF4G, since mutations within Pab1p (Pab1-16p) that inhibited poly(A)-dependent translation did not inhibit eIF4G1 binding (Kessler and Sachs 1998), and mutations that disrupted the Pab1p–eIF4G interaction in vitro (Pab1-ΔRRM2 or eIF4G1-ΔN300) did not result in cell inviability (Tarun et al. 1997). Future studies should clarify what other contacts Pab1p makes with the translation initiation apparatus that might be redundant with its contacts with eIF4G.

One expected consequence of the simultaneous association of eIF4E and Pab1p with eIF4G is that mRNA would become circularized. The hypothesis that mRNA would be circularized as a result of an interaction between the two ends of mRNA had been proposed by many investigators to explain several of the key observations being made in the translation and mRNA stability fields (for review, see Jacobson 1996). This hypothesis was supported by the direct demonstration of circular polyribosomes in electron micrographs of serial sections through rough ER membranes (Christensen et al. 1987). The availability of recombinant eIF4E, eIF4G, and Pab1 proteins, as well as mutations within eIF4G that prevented its association with either eIF4E or Pab1p, permitted a test of whether RNA could be circularized by them (Fig. 1) (Wells et al. 1998). After assembling these recombinant proteins on a double-stranded RNA molecule that had single-stranded capped 5′ and polyadenylated 3′ ends, it was shown through the use of atomic force microscopy that circles were readily visible. Thus, not only were eIF4E and Pab1p capable of stimulating translation in vitro, they were also capable of stabilizing a unique conformation of mRNA. The implications of this latter result are discussed below.

IDENTIFICATION OF THE Pab1p–eIF4G INTERACTION IN MAMMALIAN CELLS

The high degree of conservation between the yeast and mammalian translation initiation factors suggested that the interaction between eIF4G and Pab1p would also be conserved in mammalian cells. However, Pab1p had never been reported to copurify with eIF4G via eIF4E using m^7GDP-

Table 1 Summary of Pab1 and eIF4G mutagenesis studies

Protein[a]	eIF4G–Pab1 interaction[b]	Poly(A)-dependent translation[c]	Translational synergy[d]	References
yPab1	yes	yes	yes	Tarun and Sachs (1995); Kessler and Sachs (1998)
yPab1-101 (deletes RRM2)	no	no	no	Kessler and Sachs (1998)
yPab-105 (contains only RRM1,2)	yes	yes	N.D.	Kessler and Sachs (1998)
yPab1-16 (point mutations in RRM1,2	yes	no	yes	Kessler and Sachs (1998)
yPab1-180 (point mutations in RRM2)	no	no	yes	Otero et al. (1999)
yPab1-184 (point mutations in RRM2)	no	no	yes	Otero et al. (1999)
hPab1	yes	N.D.	N.D.	Imataka et al. (1998); Piron et al. (1998)
hPab1-RRM1,2 (contains only RRM1 and 2)	yes	N.D.	N.D	Imataka et al. (1998)
yeIF4G1	yes	yes	yes	Tarun and Sachs (1996); Tarun et al. (1997)

yeIF4G1-188/299 (contains only residues 188–299)	yes	N.D.	N.D.	Tarun et al. (1997)
yeIF4G1-213 (changes residues 213–216 to alanine)	no	<5%[e]	yes	Tarun et al. (1997)
yeIF4G1-ΔN300 (deletes amino-terminal 300 residues)	no	no	no	Tarun et al. (1997)
heIF4G1	yes	N.D.	N.D.	Imataka et al. (1998)
heIF4G1-132/160 (contains only residues 132-160)	yes	N.D.	N.D.	Imataka et al. (1998)
heIF4G1–134 (changes residues 134–138 to alanine)	no	N.D.	N.D.	Imataka et al. (1998)

N.D., not determined.

[a] The yeast (y) or human (h) protein, including the nature of its mutations, is shown.

[b] The ability of the recombinant protein to bind to eIF4G or Pab1p is indicated. These assays are less sensitive than the translation assays, and thus a failure to detect binding does not rule out the possibility of weak binding.

[c] The ability of the recombinant protein to allow translation of uncapped, polyadenylated mRNA in extracts is indicated.

[d] The ability of extracts containing the indicated protein to exhibit translational synergy is shown.

[e] This protein exhibits low but detectable activity.

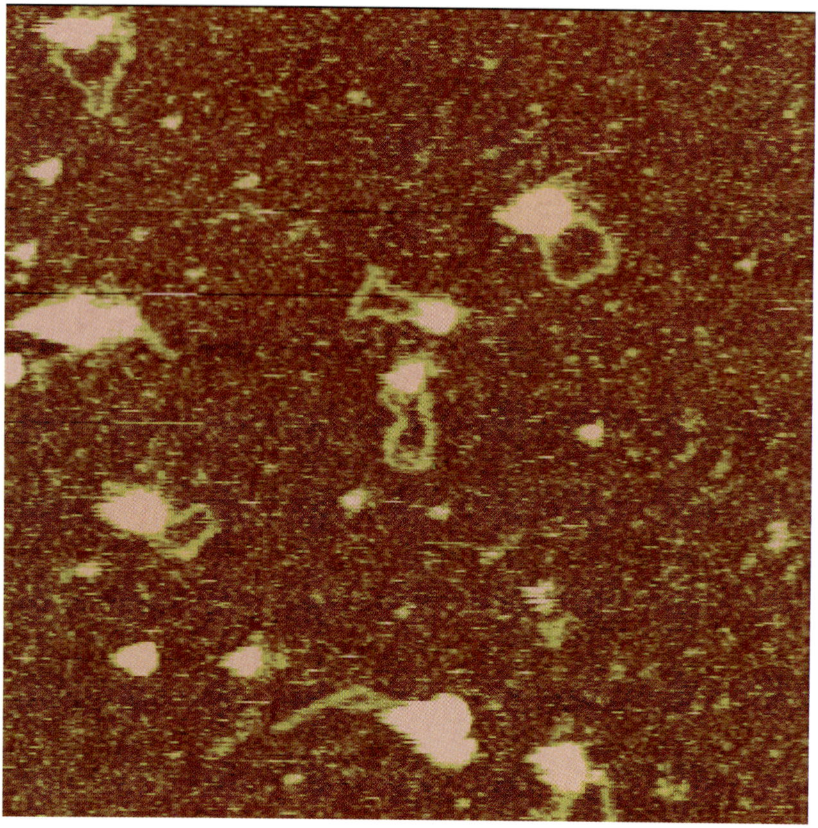

Figure 1 Circularization of RNA by the eIF4E, eIF4G, and Pab1 proteins. 1.2-kb double-stranded RNA molecules containing a single-stranded, capped 5′ end and a 3′ poly(A) tail were incubated with recombinant Pab1p, eIF4E, and a 315-amino acid fragment of eIF4G1 that is capable of binding to both Pab1p and eIF4E. Following purification of nucleic acid–protein complexes by gel exclusion chromatography, samples were absorbed onto a mica surface and visualized in liquid by atomic force microscopy. The large white particles connecting the two ends of the dsRNA contain each of the three proteins. (For details, see Wells et al. 1998.)

agarose chromatography, and the yeast and mammalian eIF4G proteins did not exhibit any significant degree of sequence homology within their amino termini. As a result, it was unclear whether the findings within the yeast system would be generally applicable to the mammalian system (Craig et al. 1998).

The discovery of the mammalian Pab1p–eIF4G interaction occurred within the same year in two different labs. Poncet and coworkers discovered that the original cDNA sequence of the mammalian eIF4G1 protein was incorrect, and that in fact it had an extended amino terminus (Piron et al. 1998). This region of the protein was shown to enable eIF4F to bind to Pab1p.

Sonenberg and coworkers also discovered that the original cDNA sequence of mammalian eIF4G1 was incorrect, and that the newly identified amino terminus interacted directly with Pab1p (Imataka et al. 1998). By utilizing a series of deletion constructs within both eIF4G1 and Pab1p, these workers were able to narrow the Pab1p-binding site on eIF4G to a 29-amino fragment and to show that conserved charged residues within this fragment were important for Pab1p binding (Table 1). Likewise, they were able to show that RRM1 and RRM2 of human Pab1p were required for its binding to eIF4G1, as they are for yeast Pab1p binding to yeast eIF4G (Table 1). Interestingly, although the interaction between eIF4G and Pab1p appears to have been conserved throughout evolution, no obvious sequence homology exists between human and yeast eIF4G within their Pab1p-binding domains. Similarly, although human and yeast Pab1p share extensive sequence homology throughout their RRM1 and RRM2 regions, the human Pab1p cannot bind to yeast eIF4G (Otero et al. 1999). These data suggest that the human eIF4G and Pab1 proteins have diverged in sequence to a point whereby they can still bind to each other but fail to bind to their more primordial partners.

Another indication of such divergence came from the discovery of an additional Pab1-binding partner in human cells, the PAIP protein (Craig et al. 1998). PAIP was discovered through its ability to be bound by human Pab1p in a Far Western assay. PAIP appears to play a role in translation, since its overexpression leads to enhanced translation of reporter mRNAs following their transfection into cells. Interestingly, PAIP shares extensive sequence homology with the central region of eIF4G that binds to eIF4A, and PAIP was shown to bind to eIF4A. Thus, it appears that in higher cells Pab1p can interact with at least two translation factors, eIF4G and PAIP, each of which can bind to eIF4A. Future work on PAIP should reveal in more detail what role it plays in conjunction with Pab1p in the translation process.

IDENTIFICATION OF THE Pab1p–eIF4G INTERACTION IN PLANT CELLS

Gallie and coworkers have examined the effects of poly(A) addition to translation extracts and have found that it has a greater inhibitory effect on the expression of uncapped than capped mRNA (Gallie and Tanguay 1994). Supplementation of poly(A)-inhibited extracts with various purified translation factors revealed that either eIF4F or eIF4B was able to rescue the inhibitory effects of the poly(A). From these data, it was concluded that the inhibitory effects of poly(A) were due to sequestration of these initiation factors. As a corollary, it was noted that the simultaneous association of these factors to both the cap structure and the poly(A) tail could in some way account for their synergistic stimulation of translation.

In more recent work, it was discovered by Far Western analysis that Pab1p from wheat germ could interact with immobilized eIF4B and either of the two isoforms of wheat eIF4F (eIF4F or iso-eIF4F) (Le et al. 1997). More specifically, it was found that Pab1p interacted with iso-eIF4G, and through fluorescence titration studies, that the association of Pab1p with this and the other factors was in the 10- to 50-nM range. Finally, the functional significance of this interaction was implied by the finding that the affinity of Pab1p for poly(A) was increased when it was bound by these factors. It was also discovered that the amino terminus of iso-eIF4G was needed for this affinity-enhancing effect.

One major caveat to the interpretation of these recent experiments is the absence of a clear demonstration of the association between plant Pab1p and these other factors. For instance, the reconstitution of the complex using recombinant proteins in solution and measured through standard techniques such as immunoprecipitation, affinity chromatography, or even exclusion chromatography has not been demonstrated. Instead, these experiments have relied heavily on indirect measures of association between two proteins, and thus must be considered more preliminary than the data with the yeast or human proteins. Nonetheless, in general, these results on the wheat germ system do support the conclusion that Pab1p and eIF4G interact in plant cells. They also leave open the possibility that eIF4B is involved in the interaction between the two ends of mRNA.

MECHANISTIC ASPECTS OF TRANSLATIONAL SYNERGY

The available data from both the in vitro and in vivo systems indicate that the simultaneous presence of a cap structure and a poly(A) tail on mRNA stimulates translation initiation in a cooperative manner. Recent work has begun to address the mechanisms by which this synergy occurs. This area of study is still quite new, and therefore only preliminary conclusions can be drawn from the data.

A significant contribution to understanding synergy and the separate functions of the cap and poly(A) tail in translation has been provided by Thomas Preiss and Matthias Hentze (Preiss and Hentze 1998; Preiss et al. 1998). They have shown using yeast extracts and electroporation of mRNA into cells that the poly(A) tail is capable of promoting internal translation initiation on mRNA, and that the cap structure prevents this by directing the recruited ribosome to the 5′end of the mRNA. The consequence of this tethering effect is that the amount of translation of full-length versus truncated (i.e., internally initiated) proteins is increased.

Since the cap structure can redirect the initiation machinery to the 5′ end, it is possible that translational synergy is solely due to the proper placement of the poly(A)-tail-recruited ribosome to the correct position on the mRNA, and not some additional activity gained when the two ends of the mRNA interact. However, several observations argue against this model. For instance, mutations in Pab1p (Pab1-180, -184p) or eIF4G (eIF4G1-213) that significantly decrease poly(A)-tail-dependent translation do not have the same deleterious effect on synergy (Table 1) (Tarun et al. 1997; Otero et al. 1999). This suggests the processes are mechanistically different. Additionally, nuclease-treated yeast translation extracts show high levels of cap and poly(A)-dependent translation but very little synergy (Tarun and Sachs 1997; Preiss and Hentze 1998; Preiss et al. 1998). Since the cap structure is still tethering the ribosome and Pab1p is still interacting with eIF4G in nuclease-treated extracts, these data do not support the above model. Finally, by themselves the cap and poly(A) tail only poorly activate translation of mRNA under competitive conditions, but together they still mediate synergy (Preiss and Hentze 1998; Preiss et al. 1998). This observation is not consistent with synergy being the result of ribosomes recruited to the mRNA via the poly(A) tail being placed at the 5′ end due to the interaction of eIF4E with the cap structure.

One possible mechanism of synergy is that eIF4G binding to eIF4E and Pab1p increases their affinity for mRNA. Data that support this mechanism have been obtained. The association of eIF4G with eIF4E increases its affinity for mRNA cap structures (Haghighat and Sonenberg 1997; Ptushkina et al. 1998, 1999). The association of wheat germ Pab1p with eIF4F increases the affinity of eIF4F for a cap analog (Wei et al. 1998) and the affinity of Pab1p for poly(A) (Le et al. 1997).

An alternative (although not exclusive) mechanism underlying synergy is that the simultaneous binding of eIF4E and Pab1p to eIF4G could alter their conformation so as to allow one or more of them to bind more efficiently to a target in the translational apparatus. Evidence for allosteric transitions within eIF4G upon binding to eIF4E has been reported (Hershey et al. 1999). Furthermore, the observation that the yeast Pab1-16 protein can bind to eIF4G but not activate translation (Kessler and Sachs 1998) suggests that conformational changes resulting from the eIF4G/Pab1p association lead to translational stimulation.

Finally, the finding that the association of plant eIF4F with Pab1p enhances their affinities for the cap and poly(A) tail, respectively, also supports the hypothesis that each of these proteins may undergo structural transitions when associated with the other.

Besides providing optimal binding surfaces for other proteins, allostery could also result in the activation of proteins already bound to eIF4G. For example, Pab1p can activate the translation of capped mRNA *in trans* (Otero et al. 1999). Although Pab1p binding to eIF4G is not required for this effect, the target of Pab1p depends on the presence of the amino terminus of eIF4G. This target could be a translation factor that becomes fully activated when both eIF4E and Pab1p associate with it. The future development of a reconstituted translation initiation system that reflects the cap- and poly(A)-dependence of ribosome binding to mRNA should give greater insights into mechanisms underlying translational synergy.

REGULATION OF THE eIF4E–eIF4G–Pab1p INTERACTION

The importance of the eIF4E–eIF4G interaction was originally underscored by the observation that cleavage of eIF4G during enterovirus infection separates eIF4E from the majority of eIF4G (for review, see Belsham and Sonenberg 1996). This was proposed to lead to the shutoff of host cell translation, since recognition of the cap structure by eIF4E would not result in the concomitant recruitment of eIF4G and its associated proteins to the mRNA. More recently, it has become appreciated that Pab1p is also removed from eIF4G during enterovirus infection as a result of its binding site being located amino-terminal to the eIF4E-binding site, and that Pab1p is itself cleaved by the enterovirus protease (Joachims et al. 1999; Kerekatte et al. 1999). These data suggest that modulation of Pab1p and its interaction with eIF4G are additional mechanisms employed by enteroviruses to shut down host cell translation.

Another example of how Pab1p binding to eIF4G could be disrupted during viral infection was presented by Poncet and coworkers (Piron et al. 1998). It was discovered that the rotavirus NSP3 protein competes with Pab1p binding to eIF4G. These authors proposed that late in viral infection, the shutoff of host cell protein synthesis occurs in part because eIF4G is bound by NSP3 and not Pab1p. As a result, eIF4G is redirected to the NSP3-binding sites on the viral 3′ end and away from the poly(A) tails on the host cell mRNA.

More examples of regulation of Pab1p binding to eIF4G may be forthcoming from studies on mRNA expression during early development. As described above, the polyadenylation of mRNA during early development resulted in the recruitment of the mRNA into polyribosomes. The maternal mRNAs that have their translation controlled by regulated polyadenylation are also translationally repressed prior to their translational activation. It has recently been shown that the cytoplasmic polyadenylation element in the 3′ ends of these maternal mRNAs can be responsible for both the repression and activation of the mRNA (Stutz et al. 1998; de Moor and Richter 1999; Minshall et al. 1999). One mechanism of silencing may involve masking of the cap structure by proteins associated with the CPE through protein bridging interactions (see Chapter 27). In this way, the circle formed by an interaction with the short poly(A) tail on dormant maternal mRNA could be disrupted by competition with the CPE-associated protein. As a result, the enhanced translatability of the mRNA that results from the simultaneous association of the two ends of the mRNA would be lost, and the mRNA would be translationally silenced. Because the Pab1p found in adult tissue appears to be underrepresented in the oocyte (Zelus et al. 1989; Paynton 1998), one new avenue for investigation will be the identification of Pab1p-like proteins bound to stored and newly polyadenylated mRNA that mediate the effects of the poly(A) tail during early development.

The above discussion has assumed that the translational machinery can be activated by the poly(A)–Pab1p complex and that in the absence of poly(A)–Pab1p this machinery is in a "neutral" state. However, several recent reports by Reed Wickner and colleagues suggest that poly(A) and Pab1p could stimulate translation due to derepression of an inhibitory complex consisting of several yeast Ski proteins (Widner and Wickner 1993; Masison et al. 1995; Benard et al. 1998, 1999). The data in support of this hypothesis are that capped, poly(A)-deficient mRNAs, derived either from the endogenous RNA viruses in yeast or from mRNA introduced into yeast by electroporation, are poorly expressed in wild-type strains but highly expressed in the *ski* mutants. Although there is some question about whether this Ski phenotype is a direct result of changes to the translational apparatus or an indirect result of changes in the RNA degradation and/or ribosome biogenesis pathways (Mitchell et al. 1997; Anderson and Parker 1998), it is still the case that there are yeast mutants which express to high levels poly(A)-deficient mRNA. It will be very exciting over the next few years to understand in more detail how these Ski proteins lead to activation of translation of poly(A)-deficient mRNA.

WHY FORM A CIRCLE?

Although it is clear that the introduction of Pab1p into the eIF4E–eIF4G complex can stimulate the translation initiation process, it is much less clear what function, if any, mRNA circularization plays in mRNA metabolism. For instance, it could be argued from a teleological point of view that the poly(A)/Pab1p complex is situated on the 3′ end of mRNA to ensure that only full-length (i.e., non-degraded) mRNAs are translated efficiently. Furthermore, the observation that capped mRNA can have its translation activated *in trans* by Pab1p in vitro suggests that the initiation process does not require the two ends of the mRNA to be linked in order for the mRNA's expression to be stimulated by the poly(A) tail (Otero et al. 1999). Such viewpoints suggest that circles themselves provide no functional advantage to mRNA.

Nonetheless, it is interesting to consider the possible ramifications of mRNA circularization on mRNA metabolism. Over the past several years, it has become clear that removal of the mRNA cap structure can be the rate-limiting step for mRNA degradation in yeast (see Chapter 28). It has also been shown that deadenylation of yeast mRNA precedes its decapping (see, e.g., Decker and Parker 1993), although the contribution of deadenylation to the control of decapping rates is still debated (Morrissey et al. 1999). Given these findings, it seems logical to conclude that the maintenance of a poly(A) tail and a cap structure will lead to mRNA stabilization, and that these structures will be more stable if they are associated with initiation factors. Recent work from the Parker lab supports this contention (Schwartz and Parker 1999). Thus, mRNA circularization may stabilize mRNA by maintaining the association of translation initiation factors with its two ends.

Although mRNA may be stabilized by circularization, this still does not address the positive (rather than protective) functions the circle could have on mRNA metabolism. In this regard, it seems logical that translation reinitiation could be stimulated by circularization. For instance, once the ribosome terminates translation on mRNA, it might not dissociate back into the cytoplasm. Instead, it may slide off its 3′ end and rebind rapidly to the 5′ end. The proximity of the two mRNA ends resulting from circularization could enhance this rebinding process. Very little is known about what happens to the ribosome after it terminates translation on mRNA, and about the requirements for initiation factors during the reinitiaton process (Karimi et al. 1999). Perhaps the future development of translation extracts that are both poly(A)-dependent and capable of exhibiting reinitiation will reveal that mRNA circularization is an essential component for the reinitiation process.

FUTURE DIRECTIONS

The recent progress in understanding the mechanisms by which the poly(A) tail and the cap structure work together to stimulate translation has helped to settle the debate about whether poly(A) plays an integral role in the translation process. However, this progress has also created new questions to be answered.

Among the most important of these is how poly(A) stimulates translation in higher cells. Until now the yeast in vitro translation system has been the only means by which to study this phenomenon. It is anticipated that the development of extracts from higher cells that exhibit poly(A)-dependent translation and synergy, in conjunction with the development of reagents that will replace the use of mutant proteins in crude yeast extracts, will permit a more detailed investigation into the degree of conservation of the poly(A) tail's role and the eIF4E–eIF4G–Pab1p interaction in the initiation process. Recent results from the Hentze lab using *Drosophila* embryo extracts suggest that such studies will be forthcoming in the near future (Gebauer et al. 1999). Furthermore, given the wealth of *trans*-acting factors that are known to regulate the translation of an mRNA via binding to its 3′ end, these systems should allow a dissection of the processes by which the interaction between the 5′ and 3′ ends is regulated.

An important question that remains in the yeast system is why mutations that disrupt the interaction between Pab1p and eIF4G do not lead to cell death. It is difficult to imagine that the conservation of the interaction between Pab1p and the translation initiation machinery does not reflect conservation of an underlying essential process. It is therefore likely that Pab1p is making other contacts within the translational apparatus that also allow translational stimulation and possibly circularization. In this scenario, disruption of Pab1p's single contact with eIF4G may not be a lethal event in vivo as a result of these other redundant interactions. Future work aimed at discovering these interactions should explain in more detail how the mRNA 3′ end interacts with the 5′ end to stimulate translation.

Finally, and most generally, the role of mRNA circularization in mRNA metabolism remains unknown. Although translational reinitiation and mRNA stabilization are the obvious possible events that will be affected by circularization, it is also possible that there are undiscovered roles for circularization. In a similar vein, understanding the regulation of the poly(A) tail's contribution to mRNA expression within the cell may provide key insight into how gene expression is controlled. It is anticipated that progress in many of these areas will be achieved during the next decade.

REFERENCES

Adam S.A., Nakagawa T., Swanson M.S., Woodruff T.K., and Dreyfuss G. 1986. mRNA polyadenylate-binding protein: Gene isolation and sequencing and identification of a ribonucleoprotein consensus sequence. *Mol. Cell. Biol.* **6:** 2932–2943.

Anderson J.S. and Parker R. 1998. The 3′ to 5′ degradation of yeast mRNAs is a general mechanism for mRNA turnover that requires the SKI2 DEVH box protein and 3′ to 5′ exonucleases of the exosome complex. *EMBO J.* **17:** 1497–1506.

Baer B.W. and Kornberg R.D. 1983. The protein responsible for the repeating structure of cytoplasmic poly(A)-ribonucleoprotein. *J. Cell Biol.* **96:** 717–721.

Belsham G.J. and Sonenberg N. 1996. RNA-protein interactions in regulation of picornavirus RNA translation. *Microbiol. Rev.* **60:** 499–511.

Benard L., Carroll K., Valle R.C.P., and Wickner R.B. 1998. Ski6p is a homolog of RNA-processing enzymes that affects translation on non-poly(A) mRNAs and 60S ribosomal subunit biogenesis. *Mol. Cell. Biol.* **18:** 2688–2696.

Benard L., Carroll K., Valle R.C.P., Masison D.C., and Wickner R.B. 1999. The Ski7 antiviral protein is an EF1-α homolog that blocks expression in non-poly(A) mRNA in *Saccharomyces cerevisiae*. *J. Virol.* **73:** 2893–2900.

Both G.W., Banerjee A.K., and Shatkin A.J. 1975. Methylation-dependent translation of viral messenger RNAs in vitro. *Proc. Natl. Acad. Sci.* **72:** 1189–1193.

Christensen A.K., Kahn L.E., and Bourne C.M. 1987. Circular polysomes predominate on the rough endoplasmic reticulum of somatotropes and mammotropes in the rat anterior pituitary. *Am. J. Anat.* **178:** 1–10.

Craig A.W.B., Haghighat A., Yu A.T.K., and Sonenberg N. 1998. Interaction of polyadenylate-binding protein with the eIF4G homologue PAIP enhances translation. *Nature* **392:** 520–523.

Decker C.J. and Parker R. 1993. A turnover pathway for both stable and unstable mRNAs in yeast: Evidence for a requirement for deadenylation. *Genes Dev.* **7:** 1632–1643.

de Moor C.H. and Richter J.D. 1999. Cytoplasmic polyadenylation elements mediate masking and unmasking of cyclin B1 mRNA. *EMBO J.* **18:** 2294–2303.

Etchison D., Milburn S.C., Edery I., Sonenberg N., and Hershey J.W. 1982. Inhibition of HeLa cell protein synthesis following poliovirus infection correlates with the proteolysis of a 220,000-dalton polypeptide associated with eucaryotic initiation factor 3 and a cap binding protein complex. *J. Biol. Chem.* **257:** 14806–14810.

Gallie D.R. 1991. The cap and the poly(A) tail function synergistically to regulate mRNA translational efficiency. *Genes Dev.* **5:** 2108–2116.

Gallie D.R. and Tanguay R. 1994. Poly(A) binds to initiation factors and increases cap-dependent translation in vitro. *J. Biol. Chem.* **269:** 17166–17173.

Gebauer M., Corona D.F., Preiss T., Becker P.B., and Hentze M. 1999. Translational control of dosage compensation in *Drosophilla* by Sex-lethal: Cooperative silencing via the 5′ and 3′ UTRs of msl-2 mRNA is independent of the poly(A) tail. *EMBO J.* **18:** 6146–6154.

Haghighat A. and Sonenberg N. 1997. eIF4G dramatically enhances the binding of eIF4E to the mRNA 5′-cap structure. *J. Biol. Chem.* **272:** 21677–21680.

Hershey P.E., McWhirter S.M., Gross J.D., Wagner G., Alber T., and Sachs A.B. 1999. The Cap-binding protein eIF4E promotes folding of a functional domain of yeast translation initiation factor eIF4G1. *J. Biol. Chem.* **274:** 21297–21304.

Iizuka N., Najita L., Franzusoff A., and Sarnow P. 1994. Cap-dependent and cap-independent translation by internal initiation of mRNAs in cell extracts prepared from *Saccharomyces cerevisiae*. *Mol. Cell. Biol.* **14:** 7322–7330.

Imataka H., Gradi A., and Sonenberg N. 1998. A newly identified N-terminal amino acid sequence of human eIF4G binds poly(A)-binding protein and functions in poly(A)-dependent translation. *EMBO J.* **17:** 7480–7489.

Jacobson A. 1996. Poly(A) metabolism and translation: The closed-loop model. In *Translational control* (ed. J.W.B. Hershey et al.), pp. 451–480. Cold Spring Harbor Laboratory Press, Cold Spring Harbor, New York.

Jacobson A. and Favreau M. 1983. Possible involvement of poly(A) in protein synthesis. *Nucleic Acids Res.* **11:** 6353–6368.

Joachims M., Van Breugel P.C., and Lloyd R.E. 1999. Cleavage of poly(A)-binding protein by enterovirus proteases concurrent with inhibition of translation in vitro. *J. Virol.* **73:** 718–727.

Karimi R., Pavlov M.Y., Buckingham R.H., and Ehrenberg M. 1999. Novel roles for classical factors at the interface between translation termination and initiation. *Mol. Cell* **3:** 601–609.

Kates J. 1971. Transcription of the vaccinia virus genome and the occurrence of polyriboadenylic acid sequences in messenger RNA. *Cold Spring Harbor Symp. Quant. Biol.* **35:** 743–752.

Kerekatte V., Keiper B.D., Badorff C., Cai A., Knowlton K.U., and Rhoads R.E. 1999. Cleavage of poly(A)-binding protein by coxsackievirus 2A protease in vitro and in vivo: Another mechanism for host protein synthesis shutoff? *J. Virol.* **73:** 709–717.

Kessler S.H. and Sachs A.B. 1998. RNA recognition motif 2 of yeast Pab1p is required for its functional interaction with eukaryotic translation initiation factor 4G. *Mol. Cell. Biol.* **18:** 51–57.

Le H., Tanguay R.L., Balasta M.L., Wei C.C., Browning K.S., Metz A.M., Goss D.J., and Gallie D.R. 1997. Translation initiation factor eIF-iso4G and eIF4B interact with the poly(A)-binding protein and increase its RNA binding activity. *J. Biol. Chem.* **272:** 16247–16255.

Lim L. and Canellakis E.D. 1970. Adenine-rich polymer associated with rabbit reticulocyte messenger RNA. *Nature* **270:** 710–712.

Masison D.C., Blanc A., Ribas J.C., Carroll K., Sonenberg N., and Wickner R.B. 1995. Decoying the cap-mRNA degradation system by a double-stranded RNA virus and poly(A)-mRNA surveillance by a yeast antiviral system. *Mol. Cell. Biol.* **15:** 2763–2771.

Minshall N., Walker J., Dale M., and Standart N. 1999. Dual roles of p82 the clam CPEB homolog, in cytoplasmic polyadenylation and translational masking. *RNA* **5:** 27–38.

Mitchell P., Petfalski E., Shevchenko A., Mann M., and Tollervey D. 1997. The exosome: A conserved eukaryotic RNA processing complex containing multiple 3′→5′ exoribonucleases. *Cell* **91:** 457–466.

Morrissey J.P., Deardorff J.A., Hebron C., and Sachs A.B. 1999. Decapping of stabilized, polyadenylated mRNA in a yeast *pab1* mutant. *Yeast* **15:** 687–702.

Munroe D. and Jacobson A. 1990. mRNA poly(A) tail, a 3′ enhancer of translation initiation. *Mol. Cell. Biol.* **10:** 3441–3455.

Muthukrishnan S., Both G.W., Furuichi Y., and Shatkin A.J. 1975. 5′-Terminal 7-methylguanosine in eukaryotic mRNA is required for translation. *Nature* **255:** 33–37.

Otero L., Ashe M., and Sachs A. 1999. The yeast poly(A)-binding protein Pab1p stimulates *in vitro* poly(A)-dependent and cap-dependent translation by distinct mechanisms. *EMBO J.* **18:** 3153–3163.

Paynton B.V. 1998. RNA-binding proteins in mouse oocytes and embryos: Expression of genes encoding Y box, DEAD box RNA helicase, and polyA binding proteins. *Dev. Genet.* **23:** 285–298.

Piron M., Vende P., Cohen J., and Poncet D. 1998. Rotavirus RNA-binding protein NSP3 interacts with eIF4G1 and evicts poly(A) binding protein from eIF4F. *EMBO J.* **17:** 5811–5821.

Preiss T. and Hentze M.W. 1998. Dual function of the messenger RNA cap structure in poly(A)-tail-promoted translation in yeast. *Nature* **392:** 516–520.

Preiss T., Muckenthaler M., and Hentze M.W. 1998. Poly(A)-tail-promoted translation in yeast: Implications for translational control. *RNA* **4:** 1321–1331.

Ptushkina M., von der Haar T., Karim M.M., Hughes J.M., and McCarthy J.E. 1999. Repressor binding to a dorsal regulatory site traps human eIF4E in a high cap-affinity state. *EMBO J.* **18:** 4068–4075.

Ptushkina M., von der Haar T., Vasilescu S., Frank R., Birkenhager R., and McCarthy J.E. 1998. Cooperative modulation by eIF4G of eIF4E-binding to the mRNA 5′ cap in yeast involves a site partially shared by p20. *EMBO J.* **17:** 4798–4808.

Sachs A.B. and Davis R.W. 1989. The poly(A) binding protein is required for poly(A) shortening and 60S ribosomal subunit-dependent translation initiation. *Cell* **58:** 857–867.

———. 1990. Translation initiation and ribosomal biogenesis: Involvement of a putative rRNA helicase and RPL46. *Science* **247:** 1077–1079.

Sachs A.B. and Kornberg R.D. 1985. Nuclear polyadenylate-binding protein. *Mol. Cell. Biol.* **5:** 1993–1996.

Sachs A.B., Bond M.W., and Kornberg R.D. 1986. A single gene from yeast for both nuclear and cytoplasmic polyadenylate-binding proteins: Domain structure and expression. *Cell* **45:** 827–835.

Sachs A.B., Davis R.W., and Kornberg R.D. 1987. A single domain of yeast poly(A)-binding protein is necessary and sufficient for RNA binding and cell viability. *Mol. Cell. Biol.* **7:** 3268–3276.

Schwartz D.C. and Parker R. 1999. Mutations in translation initiation factors lead to increased rates of deadenylation and decapping of mRNAs in *Saccharomyces cerevisiae*. *Mol. Cell. Biol.* **19:** 5247–5256.

Shatkin A.J. 1976. Capping of eucaryotic mRNAs. *Cell* **9:** 645–653.

Sonenberg N., Morgan M.A., Merrick W.C., and Shatkin A.J. 1978. A polypeptide in eukaryotic initiation factors that crosslinks specifically to the 5′-terminal cap in mRNA. *Proc. Natl. Acad. Sci.* **75:** 4843–4847.

Sonenberg N., Rupprecht K.M., Hecht S.M., and Shatkin A.J. 1979. Eukaryotic mRNA cap binding protein: Purification by affinity chromatography on Sepharose-coupled m^7GDP. *Proc. Natl. Acad. Sci.* **76:** 4345–4349.

Stutz A., Conne B., Huarte J., Gubler P., Volkel V., Flandin P., and Vassalli J.D. 1998. Masking, unmasking, and regulated polyadenylation cooperate in the translational control of a dormant mRNA in mouse oocytes. *Genes Dev.* **12:** 2535–2548.

Tarun S.Z. and Sachs A.B. 1995. A common function for mRNA 5′ and 3′ ends in translation initiation in yeast. *Genes Dev.* **9:** 2997-3007.

———. 1996. Association of the yeast poly(A) tail binding protein with translation initiation factor eIF-4G. *EMBO J.* **15:** 7168–7177.

———. 1997. Binding of eIF4E to eIF4G represses translation of uncapped mRNA. *Mol. Cell. Biol.* **17:** 6876–6886.

Tarun S.Z., Wells S.E., Deardorff J.A., and Sachs A.B. 1997. Translation initiation factor eIF-4G mediates in vitro poly(A) tail-dependent translation. *Proc. Natl. Acad. Sci.* **94:** 9046–9051.

Wei C.C., Balasta M.L., Ren J., and Goss D.J. 1998. Wheat germ poly(A) binding protein enhances the binding affinity of euckaryotic initiation factor 4G and (iso)4F for cap analogues. *Biochemistry* **37:** 1910–1916.

Wells S.E., Hillner P., Vale R., and Sachs A.B. 1998. Circularization of mRNA by eucaryotic translation initiation factors. *Mol. Cell* **2:** 135–140.

Wickens M. 1990a. How the messenger got its tail: Addition of poly(A) in the nucleus. *Trends Biochem. Sci.* **15:** 277–281.

———. 1990b. In the beginning is the end: Regulation of poly(A) addition and removal during early development. *Trends Biochem. Sci.* **15:** 320–324.

Widner W.R. and Wickner R.B. 1993. Evidence that the *SKI* antiviral system of *Saccharomyces cerevisiae* acts by blocking expression of viral mRNA. *Mol. Cell. Biol.* **13:** 4331–4341.

Zelus B.D., Giebelhaus D.H., Eib D.W., Kenner K.A., and Moon R.T. 1989. Expression of the poly(A)-binding protein during development of *Xenopus laevis*. *Mol. Cell. Biol.* **9:** 2756–2760.

11

Translation Termination: It's Not the End of the Story

Ellen M. Welch, Weirong Wang, and Stuart W. Peltz

Department of Molecular Genetics and Microbiology
University of Medicine and Dentistry of New Jersey
Robert Wood Johnson Medical School
and the Cancer Institute of New Jersey
Piscataway, New Jersey 08854

The protein synthesis apparatus must carry out translation initiation, elongation, and termination to synthesize a complete protein product. Each step of this process has evolved to occur with great accuracy and can be utilized to regulate gene expression. Until relatively recently, however, translation termination was the least investigated and understood aspect of the translation process. Recent results, however, suggest that translation termination also can be used to modulate gene expression. The fact that sequences in the 3′-untranslated regions (3′UTRs) of mRNAs bind to RNA-binding proteins and can dramatically affect the translation initiation process suggests that steps which occur at translation termination may strongly affect subsequent events in the translation process. It is conceivable that translation termination may be involved in determining whether ribosomes reinitiate at the translation start site of the same transcript for subsequent rounds of translation.

In recent years, there has been a significant advancement in our knowledge of the translation termination process. The goal of this short review is to describe the state of knowledge concerning the mechanism of translation termination. Results from studies using both prokaryotic and eukaryotic systems are discussed. In particular, we attempt to address the potential role of the translation termination process in the regulation of gene expression. In addition, we discuss the modulation of the termination process as a potential target to treat diseases that arise as a consequence of mutations that prematurely terminate translation.

Translational Control of Gene Expression
© 2000 Cold Spring Harbor Laboratory Press 0-87969-568-4/00 $5 + .00

GENERAL PRINCIPLES OF TRANSLATION TERMINATION

Nearly 40 years ago, Crick and colleagues deduced the "general nature of the genetic code" from the results of crosses between mutants in the T4 rIIB cistron (Crick et al. 1961). They recognized that the code must be nonoverlapping, degenerate, and read from a fixed point; moreover, they predicted that some triplets must be nonsense, i.e., noncoding, and that the possible function of nonsense triplets may be to signal the end of synthesis of the polypeptide chain. Subsequent work from numerous laboratories has substantiated these predictions and demonstrated that, in most genetic systems, 61 of the 64 possible triplets encode amino acids and that the triplets UAA, UAG, and UGA are noncoding and promote translational termination (Osawa et al. 1992). The polypeptide chain-terminating effects of UAA, UAG, and UGA triplets have been amply documented and characterized (Craigen et al. 1990).

The general framework of the translation termination process has been elucidated (Fig.1) (for review, see Craigen et al. 1990; Nakamura et al. 1996). The presence of one of the three termination codons in the A site of the ribosome signals the polypeptide chain release factors to bind and recognize the termination signal. A subsequent hydrolysis of the ester bond between the 3′ nucleotide of the tRNA located in the P site and the nascent polypeptide chain ensues. Ultimately, the polypeptide is released as a completed protein from the tRNA. The steps following peptide hydrolysis have been studied in *Escherichia coli*. The 50S ribosomal subunit is dissociated from the mRNA/30S subunit/tRNA complex to participate in additional rounds of translation by the GTP-bound EF2 and RF4. Subsequently, the action of IF3 clears the deacylated tRNA from the P site of the 30S subunit, allowing the 30S subunit to participate in additional rounds of translation (Fig. 1) (Karimi et al. 1999). The 30S subunit bound to the mRNA either dissociates spontaneously or scans the mRNA to initiate at a nearby initiation site.

GENERAL APPROACHES TO STUDY THE TRANSLATION TERMINATION PROCESS

Both genetic and biochemical approaches have been used to investigate the translation termination process (Caskey et al. 1970; Cox et al. 1980; Donahue et al. 1981; Tuite et al. 1981; Wakem and Sherman 1990; Freistroffer et al. 1997; Grentzmann and Kelly, 1997). A biochemical assay that monitors release of radiolabeled fMet-tRNA from purified ribosomes has been used to identify and characterize the functions of translation termination factors in *E. coli* (Caskey et al. 1970; Grentzmann

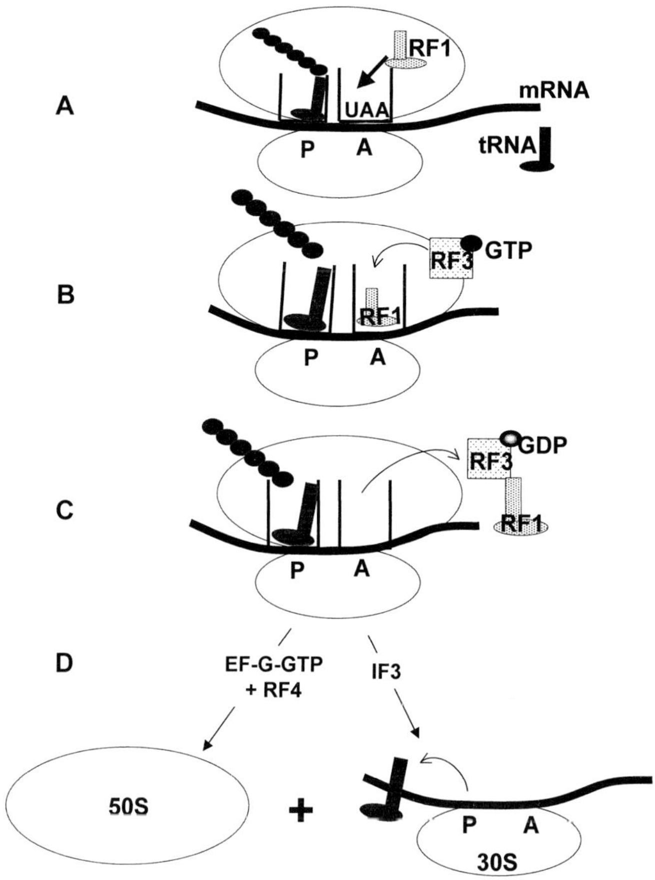

Figure 1 Proposed mechanism for translation termination. (*A*) Translation termination is initiated when one of three stop codons (indicated by UAA) is present in the ribosomal A site, resulting in binding of the RF to the A site. (*B*) Hydrolysis of the peptide bond results in a deacylated tRNA in the ribosomal P site. (*C*) RF1 is removed from the ribosome in a GTP-dependent reaction involving RF3. (*D*) Dissociation of the 70S/mRNA complex by GTP-bound EF2 and RF4 (ribosome recycling factor) into the 50S ribosomal subunit and the 30S/mRNA/deacylated tRNA complex. IF3 is necessary for recycling of the 30S ribosomal subunit by removing the deacylated tRNA from the P site. See text for details.

and Kelly 1997; Pavlov et al. 1997b; Karimi et al. 1999). Recently, more elegant and sophisticated biochemical assays have been developed to characterize the steps that occur during translation termination, as well as the subsequent events that follow peptidyl hydrolysis (see Freistroffer et al. 1997; Grentzmann and Kelly 1997). For example, small transcripts, or mini-messengers, have been recently used in translation termination assays (Grentzmann and Kelly 1997). Release of the radiolabeled tRNA can be efficiently induced by an RNA containing only AUGUAA in which the stop codon was localized to the ribosomal A site and the fMet-tRNA was bound to the AUG codon in the P site. A second in vitro translation termination assay has been developed to characterize how factors promote translation termination of ribosome-bound RNAs that have paused at a termination codon (see Freistroffer et al. 1997). In this assay, termination complexes are prepared by translation of a small mRNA in the absence of release factors. The substrate for the termination reaction contains ribosomes harboring a tetrapeptidyl-tRNA bound to the ribosomal P site and a stop codon bound to the A site. The complexes containing the paused ribosomes at the stop codon were purified from the rest of the translation components by gel filtration. The different steps in the translation termination process are subsequently characterized by addition of the purified release factors, as well as other factors putatively thought to be involved in translation termination, and subsequently monitoring the release of the radiolabeled tetrapeptide. As described below, this assay has been invaluable in characterizing the mechanism of translation termination.

Genetic studies have also been used to identify and characterize factors involved in the translation termination process. An in vivo assay for the measurement of translation termination efficiency is nonsense suppression, which results when a termination codon is suppressed (for review, see Stansfield and Tuite 1994; Czaplinski et al. 1999). Suppression results when a near cognate tRNA or a mutant suppressor tRNA successfully competes with the release factors at a termination codon. In contrast, antisuppression results when the efficiency of tRNAs to translate termination codons is reduced (for review, see Stansfield and Tuite 1994).

FACTORS THAT CONTROL THE TRANSLATION TERMINATION PROCESS

Both prokaryotes and eukaryotes have two classes of specialized factors, called release factors, that are required for the termination reaction. The roles of class 1 release factors are to recognize a stop codon and promote the hydrolysis of the ester bond linking the polypeptide chain with the peptidyl (P) site tRNA by the peptidyl transferase center of the ribosome. Class 2 release factors are GTPases and stimulate the class 1 release factor activity (Mikuni et al. 1994; Stansfield et al. 1995; Zhouravleva et al. 1995; Frolova et al. 1996). Translation termination in prokaryotes is catalyzed by two codon-specific class 1 release factors: RF1, which recognizes the UAG and UAA stop codons, and RF2, which recognizes the UGA and UAA termination codons. Interestingly, a single amino acid substitution in RF2 allows recognition of all three termination codons (Ito et al. 1998b, 2000), a situation that is more similar to the characteristics of the eukaryotic eRF1 (see below). In vitro, RF1 and RF2 do not require additional factors for activity (Caskey et al. 1970); however, a class 2 release factor, RF3, stimulates the termination reaction in a GTP-dependent manner (Mikuni et al. 1994; Freistroffer et al. 1997; Grentzmann et al. 1998).

In eukaryotes, translation termination is catalyzed by eRF1, which recognizes all three termination codons. eRF1 homologs have been identified in humans, hamsters, *Saccharomyces cerevisiae*, *Xenopus laevis,* and *Arabidopsis thaliana* and show a high degree of sequence identity (Urbero et al. 1997). Sequence analysis of class 1 release factors had revealed the existence of a GGQ sequence motif that is universal to all species. Substitutions in the GGQ motif abolished the ability of human eRF1 to trigger peptidyl-tRNA hydrolysis (Frolova et al. 1999; Song et al. 2000). The crystal structure of human eRF1 protein has recently been determined (Fig. 2) (Song et al. 2000). The polypeptide chain of eRF1 is organized into three domains in a structure reminiscent of the letter Y. The overall shape and dimensions of eRF1 resemble a tRNA molecule with domains 1, 2, and 3 of eRF1 corresponding to the anticodon loop, aminoacyl acceptor stem, and T stem of a tRNA molecule, respectively. A second release factor, eRF3, stimulates eRF1 activity in a GTP-dependent fashion (Frolova et al. 1994; Stansfield et al. 1995; Zhouravleva et al. 1995). The GTPase activity of eRF3 requires eRF1 and the ribosome (Frolova et al. 1996). Since no significant similarity exists between the eukaryotic/Archaea RF sequences and their prokaryotic/mitochondrial counterparts, the prokaryotic and eukaryotic release factors may belong to two distinct protein families (Frolova et al. 1994).

In the yeast *S. cerevisiae*, eRF1 and eRF3 are encoded by the *SUP45* and *SUP35* genes, respectively. eRF1 and eRF3 interact in vitro and in vivo (Stansfield et al. 1995; Zhouravleva et al. 1995; Paushkin et al. 1997; Frolova et al. 1998; Ito et al. 1998a; Ebihara and Nakamura 1999; Merkulova et al. 1999). The Sup35p is divided into three domains—N (amino-terminal), M (middle), and C (carboxy-terminal)—based on amino acid features and functions. The carboxy-terminal sequence (domain C) of eRF3 protein is essential for viability. This domain is similar to translation elongation factor EF1A and harbors a GTPase domain (for review, see Kushnirov and Ter-Avanesyan 1998). Although the

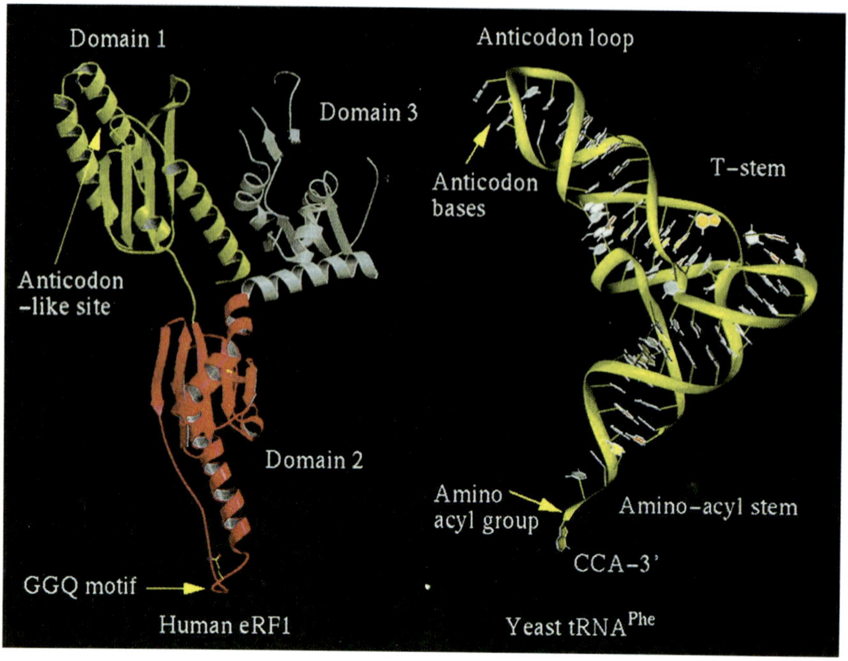

Figure 2 Ribbon diagrams of the human eRF1 and yeast tRNA-Phe crystal structures as described by Song and colleagues (2000). The overall shape and dimensions of eRF1 resemble a tRNA molecule with domains 1, 2, and 3 of eRF1 corresponding to the anticodon loop, aminoacyl acceptor stem, and T stem of the tRNA molecule, respectively. The site of attachment of an aminoacyl group at the CCA stem of the tRNA and the position of the GGQ motif in domain 2 of eRF1 are indicated. See text for details (Song et al. 2000).

amino-terminal domain (N domain) of 123 amino acids and the middle region of 253 amino acids (M domain) are not essential, the N domain is responsible for the prion-like properties associated with the protein (Paushkin et al. 1997; for review, see Liebman and Derkatch 1999; Serio and Lindquist 1999; Wickner et al. 1999). Sup35p contains two binding sites for Sup45p: One is formed by the N and M domains, and the other is located within the C domain (Paushkin et al. 1997; for review, see Serio and Lindquist 1999; Wickner et al. 1999). In *Schizosaccharomyces pombe*, a single interaction domain between the carboxyl terminus of eRF1 and eRF3 was mapped (Ebihara and Nakamura 1999).

Recently, other genes whose protein products interact with eRF3 have been identified and include *UPF1* (Czaplinski et al. 1998); the carboxy-terminal portion (411 residues) of the cytoskeletal assembly protein *SLA1* (Bailleul et al. 1999); two genes involved in the control of glucose metabolism *REG1* and *ENO2*; the translation elongation factor eEF2 (Bailleul et al. 1999); the heat shock protein HSP104 (Schirmer and Lindquist 1997; Czaplinski et al. 2000); and the poly(A)-binding protein (PABP; Hoshino et al. 1999; see below).

THE PROCESS OF TRANSLATION TERMINATION

The molecular mimicry model was first proposed on the basis of the observation that the structure of the GDP-bound form of EF2 resembles that of the GTP-bound form of EF1A in association with tRNA, which suggests that protein elongation factors function in translation by mimicking the structure of tRNA molecules (Chapter 3). Based on additional comparisons between the conserved protein motifs of the release factors RF1 and RF2 and EF2, the molecular mimicry model was extended to include the release factors and RF4 (RRF) (Nissen et al. 1995; Ito et al. 1996, 2000; Nakamura et al. 1996; Buckingham et al. 1997; Selmer et al. 1999).

The eukaryotic eRF1 protein functionally and structurally supports the tRNA mimicry model. The eRF1 protein functionally mimics tRNA molecules not only by interacting at the peptidyl transferase center of the large subunit, but also by recognizing mRNA codons within the small subunit and by association with a GTPase (Song et al. 2000). The eRF1 structure suggests that there is a shallow groove present on the surface of domain 1 that has been proposed to function as the stop codon decoding site of eRF1 (Fig. 2) (Song et al. 2000). This groove is 80 Å from the conserved GGQ motif, a distance quite close to the 75 Å observed between the anticodon bases and the amino group of the aminoacylated tRNA molecule. The residues within the putative groove are highly conserved among eukaryotic and Archaea class 1 release factors. Interestingly, mutations in eRF1 that result in nonsense suppression have been mapped to this region (Song et al. 2000).

Based on the structure of eRF1, the molecular mechanisms by which eRF1 decodes stop codons and promotes ribosome-catalyzed peptidyl-tRNA hydrolysis have been proposed (Song et al. 2000). The universally conserved GGQ motif is located in the region of eRF1 protein that appears equivalent to the aminoacyl group attached to the CCA-3′ sequence found in the aminoacyl stem of a tRNA molecule (Song et al. 2000). Thus, the eRF1 protein is thought to bind to the A site of the ribosome so that the Gln-185 residue of the GGQ motif is in coordination with a water molecule and can mediate a nucleophilic attack on the ester bond of the peptidyl-tRNA molecule in the P site, resulting in hydrolysis of the nascent polypeptide chain (Song et al. 2000). The occurrence of the GGQ motif in all prokaryotic and eukaryotic release factors despite the overall low sequence similarity may reflect a similar mechanism in all translation termination reactions.

EVENTS THAT OCCUR AFTER PEPTIDE HYDROLYSIS

The steps following peptide hydrolysis have been most intensively investigated in *E. coli* and may be different between prokaryotes and eukaryotes. (Fig. 1) (Karimi et al. 1999; for review, see Craigen et al. 1990; Buckingham et al. 1997). Peptide hydrolysis is followed by steps that involve removing the release factors from the A site of the ribosome, separating the ribosomal subunits, and dissociating the tRNA. A number of factors, including RF3, elongation factor EF2, the "ribosome recycling factor" RF4, initiation factor IF3, and GTP, are required to prepare the ribosomal subunits for recycling (Hirashima and Kaji 1972; Janosi et al. 1996, 1998; MacDougall et al. 1997; Pavlov et al. 1997a,b; Heurgué-Hamard et al. 1998; Karimi et al. 1999). RF3 is necessary to promote dissociation of the RF1/RF2, clearing the A site for the subsequent ribosomal recycling step (Fig. 1) (Freistroffer et al. 1997). The RF4 and EF2 then promote dissociation of terminated 70S ribosomes into free 50S subunits and 30S subunit–mRNA complexes (Fig. 1). IF3 functions to remove the deacylated tRNA from the P site of the 30S subunit. In the presence of IF3 and RF4, both the 50S and 30S subunits are able to recycle, but IF3 is only necessary for the recycling of 30S subunits (Fig. 1) (Karimi et al. 1999).

The events that occur after peptide hydrolysis in eukaryotes are not so well understood as in *E. coli*. No gene with similarity to RF4 has been identified. Furthermore, the role of the eukaryotic translation initiation factors in post-peptide hydrolysis events has not been investigated. However, the GTPase function of eRF3 is thought to be involved in dissociation of the eRF1/eRF3 complex from the ribosome (see Frolova et al. 1996).

MODULATORS OF TRANSLATION TERMINATION EFFICIENCY

Sequence Context Surrounding the Termination Codon Can Modulate Translation Termination Efficiency

Although the three nonsense codons encode the major determinants that signal the translation apparatus to initiate translation termination, it is clear that sequences surrounding the termination codon can affect translation termination efficiency. Results from a large number of studies indicate that the nucleotide immediately downstream from the termination codon is nonrandom and can influence termination efficiency (Angenon et al. 1990; Brown et al. 1990, 1993; Bonetti et al. 1995; McCaughan et al. 1995). In *E. coli*, translation termination is most efficient when UAAU is the termination codon and least efficient at UGAC (Poole et al. 1995; Pavlov et al. 1998). Nucleotides 3′ of the termination codon in yeast also influence termination efficiency (Bonetti et al. 1995). The preferred termination signals in eukaryotes are UAA(A/G) and UGA(A/G) (Brown et al. 1990, 1993).

trans-Acting Factors Can Modulate Translation Termination Efficiency

A number of *trans*-acting factors that influence translation termination efficiency also have been identified (Leeds et al. 1992; Cui et al. 1995; He and Jacobson 1995; Lee and Culbertson 1995; Weng et al. 1996a,b). The nonessential *UPF* genes (*UPF1–3*) encode a set of factors that have been shown to be involved in the nonsense-mediated mRNA decay pathway (NMD; for review, see Czaplinski et al. 1999; Chapter 29). In addition, *upf1*Δ, *upf2*Δ, or *upf3*Δ strains demonstrate a nonsense suppression phenotype, indicating that the Upf1p, Upf2p, and Upf3p are all involved in the regulation of translation termination (Leeds et al. 1992; Weng et al. 1996a,b; Czaplinski et al. 1998). The Upf2p has been shown to interact with Upf1p and Upf3p, indicating that the Upfps form a complex (Weng et al. 1996a; He et al. 1997; see below).

The Upf1p has been investigated most intensively. It is an RNA helicase that has been shown to interact directly with the peptidyl release factors eRF1 and eRF3 (Czaplinski et al. 1995, 1998). Recent evidence suggests that the Upf2p and Upf3p can also interact with the release factors eRF1 and eRF3 (K. Czaplinski and S.W. Peltz, unpubl.). These results suggest that Upf/release factor complexes are important for efficient translation termination efficiency and that aberrant forms of the complex result in reduced translation termination efficiency.

Mutations in several genes in yeast lead to reduced termination efficiency at nonsense codons, and hence increase nonsense suppression. Mutations in several genes were identified that enhance the efficiency of sup35 and sup45 suppressor mutations (Cox et al. 1980). These genes include *SAL1*, *SAL2 (upf1)*, *SAL5*, and *SAL6*. The *SAL1* and *SAL5* genes have not yet been identified, but the *SAL6* gene encodes a protein having homology with type I serine/threonine phosphatases (Vincent et al. 1994).

Consistent with its role in modulating translation termination, double mutants of *sal6* and *sup45* (eRF1) demonstrate a synergistic nonsense suppression phenotype and show a cold-sensitive growth phenotype (Song and Liebman 1987). Recent results demonstrate that a *sal6*Δ strain increases TY1 frameshifting in yeast (Burck et al. 1999). In addition, the Sal6p interacts with both Upf1p and the release factors (X. Han and S.W. Peltz, unpubl.). Consistent with Sal6p functioning in termination, a regulatory subunit of the phosphatase PP2A interacts with eRF1 in mammalian cells (Andjelkovic et al. 1996). Although the target of Sal6p has not been identified, potential and interesting candidates include the Upf proteins and/or the RFs. Characterization of the Sal6p will undoubtedly provide additional insight into the regulation of translation termination. Although studies on the phosphorylation status of the factors known to be involved in translation termination have not been undertaken, the results described above suggest that the phosphorylation/dephosphorylation of factors involved in translation termination may play important roles in regulating this process.

TRANSLATION TERMINATION AS A FORM OF GENE REGULATION

It has now been recognized that the efficiency of translation termination can act as a control point to regulate gene expression (Grant and Hinnebusch 1994; Moriarty et al. 1998; Steneberg et al. 1998; Eaglestone et al. 1999; Hoshino et al. 1999; Vilela et al. 1999; Chapter 18). For example, the headcase (hdc) gene in *Drosophila* is required for tracheal development. The hdc cDNA sequence has an ORF which is interrupted by an internal UAA stop codon. Translational readthrough of the internal stop codon due to inefficient termination is necessary because both the shorter and the longer forms of the protein are required for hdc function. The rescue of the tracheal mutant phenotype requires full-length hdc mRNA (Steneberg et al. 1998).

Regulation of translation termination is also important for the production of selenocysteine-containing proteins (for review, see Böck et al. 1991; Low and Berry 1996; Chapter 26).The mammalian mRNA for selenium-dependent glutathione peroxidase I (Se-GPx1) contains a UGA codon that is recognized by the nonstandard amino acid selenocysteine (Sec). Low levels of selenium (Se) result in a decreased abundance of Se-GPx1 mRNA due to the inability of the cell to suppress the nonsense codon, and the subsequent degradation of the mRNA by the NMD pathway. Hepatocytes isolated from rats whose diets were depleted of selenium resulted in a 50-fold reduction in Se-GPx1 activity and a 20-fold reduction in Se-GPx1 mRNA abundance compared to hepatocytes isolated from rats whose diets were supplemented with selenium (Moriarty et al. 1998).

Regulated translation termination can also control gene expression in certain viruses. A 22-codon open reading frame (uORF2) in the human cytomegalovirus UL4 mRNA leader is a *cis*-acting element that inhibits downstream translation (Alderete et al. 1999; Chapter 18). The peptide product of the uORF2 has been shown to inhibit subsequent translation of the UL4 protein by reducing the translation termination efficiency at its own stop codon, thus stalling the ribosome on the mRNA. In the presence of the peptide product of the uORF2, peptidyl-tRNA hydrolysis is negligible even after 30 minutes, whereas in the presence of a mutant form of the peptide, peptidyl-tRNA hydrolysis occurs in less than 1 minute (Alderete et al. 1999). The peptides synthesized from the uORFs of the *CPA1* and *CPC1* genes in fungi have similar properties in modulating translation termination (see Chapter 18).

The factors that modulate translation termination may also influence other translational processes. A specific mutation in the *UPF1* gene *(mof4-1)*, as well as a *upf3*Δ strain, led to enhanced –1 programed ribosomal frameshifting as well as nonsense suppression (Cui et al. 1996; Ruiz-Echevarria et al. 1998). The *sal6-1* allele increased +1 frameshifting in Ty1 and promotes nonsense suppression (Burck et al. 1999). Interestingly, the *suf12* allele of the eRF3 has also been shown to increase nonprogramed frameshift events and nonsense suppression (Doel et al. 1994). In *E. coli*, the RF2 coding region contains an in-frame UGA stop codon at position 26. The RF2 protein can only be successfully synthesized when a +1 frameshift occurs at the UGA codon. The amount of RF2 protein synthesized is controlled by an autoregulatory mechanism involved in modulation of readthrough of the UGA termination codon and the +1 frameshift (Craigen and Caskey 1986). Clearly, events at the UGA termination codon must allow the ribosome to shift frames to produce the RF2 protein. Interestingly, these observations suggest that there is a cross-talk between the factors that modulate translation termination and translation elongation.

TERMINATION AS A GENERAL REGULATOR OF TRANSLATION EFFICIENCY

Recent results from a number of groups have suggested that factors associated with the 5′ and 3′ ends of the mRNA interact in order to regulate both the stability and translation efficiency of the transcript (Jacobson 1996; see Chapter 10). Key players in these interactions include the poly(A)-binding protein (PABP) associated with the poly(A) tail at the 3′ end of the transcript and the translation initiation factor eIF4G, one of the proteins that is part of the eIF4F complex and interacts with the 5′ cap structure (see Chapters 2 and 10). Undoubtedly, additional factors are also involved in the interactions between the 5′ and 3′ ends of the mRNA. These results have led to the model that the mRNA forms a "closed loop" in which the interactions at the ends of the RNA facilitate terminating ribosomes to migrate and reinitiate at the protein-coding region of the mRNA (for review, see Jacobson 1996). Furthermore, the closed loop is thought to prevent rapid degradation of the transcript. Loss of the poly(A) tail will reduce the interactions between the 5′ and 3′ ends of the RNA, reducing translation efficiency and making the RNA more susceptible to RNA degradation (Chapter 29).

An intriguing observation was recently reported demonstrating an interaction of the eRF3p with PABP (Hoshino et al. 1999). The interaction between eRF3 and PABP appears to also have some effect on the cooperative binding of PABP to the poly(A) tail. This suggests that the terminating ribosomes may influence the RNP structure of the proposed "loop" structure, perhaps forming a "pretzel-like" mRNP structure (Fig. 3). Such a structure would allow terminating ribosomes to recycle onto the same transcript through the interactions between the eRF3p, PABP, and factors interacting at the 5′ cap structure (Fig. 3). The interesting aspect of this model is that it suggests that factors involved in translation termination are also critical for positioning ribosomes for subsequent rounds of translation, thus affecting the translation efficiency. Such processes may be modulated by regulatory factors, such as phosphatases (i.e., Sal6p or PP2A) that might regulate both translation and termination efficiencies (Vincent et al. 1994; Andjelkovic et al. 1996). Recent results have shown that these phosphatases interact with the release factors (Andjelkovic et al. 1996; S.W. Peltz, unpubl.). Modification of these proteins may lead to altered translation termination efficiencies as well as modulate their affinity for proteins either involved in translation termination or involved in events following termination.

It is becoming increasingly clear that translation termination is an important and interesting process that is now just beginning to be elucidated. Future experiments should help to clarify the role of termination in regulating gene expression. Nonetheless, recent results tantalizingly suggest that the termination process may not be simply the end of synthesizing a protein. Rather, like most cellular processes, translation termination is most likely an integrated event that is involved in modulating both the translation efficiency and stability of a given transcript. Elucidation of the detailed mechanisms involved remains a major challenge of the field.

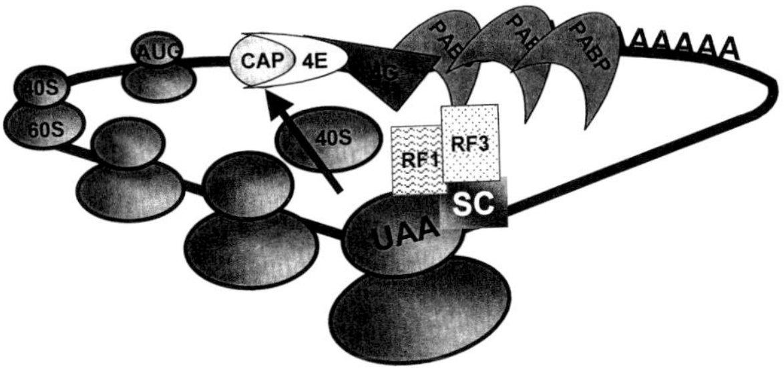

Figure 3 A model for how translation termination may be a general regulator of translation efficiency. The putative "pretzel-like" RNP structure of a wild-type mRNA with ribosomes and associated factors is schematically represented. This structure is proposed to allow terminating ribosomes to recycle onto the same transcript through interactions between the eRF(s), poly(A)-binding protein, and factors that interact with the 5′ cap. The arrow between the terminating ribosome and eIF4E indicates recycling of the 40S ribosomal subunit from the termination codon (UAA) to the initiation codon (AUG) at the 5′ end of the transcript. The 5′ cap structure is indicated by the sphere labeled "CAP." Abbreviations: (4E) eIF4E; (4G) eIF4G; (PABP) poly(A)-binding protein; (RF1) eRF1; (RF3) eRF3; (SC) surveillance complex, which consists of, at a minimum, the Upf proteins. The SC complex is thought to monitor the fidelity of multiple processes in translation and mRNA decay (Chapter 29).

MODULATION OF TERMINATION EFFICIENCY AS A THERAPEUTIC APPROACH TO TREAT DISEASES

Nonsense mutations cause approximately 20–40% of the individual cases of more than 240 different inherited diseases, including cystic fibrosis, hemophilia, Duchenne muscular dystrophy, and Marfan syndrome (McKusick 1994; see Chapter 30). For many diseases in which only 1% of the functional protein is produced, patients suffer serious disease symptoms, whereas boosting expression to only 5% of normal levels can greatly reduce the severity or eliminate the disease. In addition, a remarkably large number of the most common forms of colon, breast, esophageal, lung, head and neck, and bladder cancers result from frameshift and nonsense mutations in regulatory genes (e.g., p53, *BRCA1*, *BRCA2*; McKusick 1994). Correcting nonsense mutations in the regulatory genes to permit synthesis of the respective proteins should cause death of the cancer cells.

Clinical approaches that target the termination event to promote nonsense suppression have recently been described for model system studies of cystic fibrosis and muscular dystrophy (DMD; Bedwell et al. 1997; Clancy et al. 1998; Barton-Davis et al. 1999). The antibiotic gentamicin stabilizes the nonsense-containing CFTR mRNA and allows sufficient synthesis of CFTR and dystrophin proteins to either restore biochemical function of CFTR or produce high enough levels of the dystrophin protein to protect the tissue from muscle injury (Howard et al. 1996; Bedwell et al. 1997; Clancy et al. 1998; Barton-Davis et al. 1999). Gentamicin is an aminoglycoside antibiotic that causes translational misreading and allows the insertion of amino acids at the site of the nonsense codon (Palmer et al. 1979; Recht et al. 1999). Although gentamicin can influence both the stability of a nonsense-containing transcript and translational readthrough, modulating the translation termination machinery is probably the most important aspect of obtaining a sufficient amount of the readthrough product to treat diseases due to nonsense mutations. Results that support this hypothesis have shown that increased stability of nonsense-containing transcripts does not always result in nonsense suppression (Weng et al. 1996a,b). Thus, these results suggest that drugs which promote nonsense suppression by reducing translation termination efficiency at a premature termination codon would be the best class of drugs to treat diseases due to nonsense mutations. Taken together, these exciting results suggest that small-molecular-weight molecules which promote readthrough of nonsense codons may be effective therapies to treat patients who have diseases due to nonsense mutations.

REFERENCES

Alderete J.P., Jarrahian S., and Geballe A.P. 1999. Translational effects of mutations and polymorphisms in a repressive upstream open reading frame of the human cytomegalovirus UL4 gene. *J. Virol.* **73:** 8330–8337.

Andjelkovic N., Zolnierowicz S., Van Hoof C., Goris J., and Hemmings B.A. 1996. The catalytic subunit of protein phosphatase 2A associates with the translation termination factor eRF1. *EMBO J.* **15:** 7156–7167.

Angenon G., Van Montagu M., and Depicker A. 1990. Analysis of the stop codon context in plant nuclear genes. *FEBS Lett.* **271:** 144–146.

Bailleul P.A., Newnam G.P., Steenbergen J.N., and Chernoff Y.O. 1999. Genetic study of interactions between the cytoskeletal assembly protein sla1 and prion-forming domain of the release factor sup35 (eRF3) in *Saccharomyces cerevisiae*. *Genetics* **153:** 81–94.

Barton-Davis E.R., Cordier L., Shoturma D.I., Leland S.E., and Sweeney H.L. 1999. Aminoglycoside antibiotics restore dystrophin function to skeletal muscles of mdx mice. *J. Clin. Invest.* **104:** 375–381.

Bedwell D.M., Kaenjak A., Benos D.J., Bebok Z., Bubien J.K., Hong J., Tousson A., Clancy J.P., and Sorscher E.J. 1997. Suppression of a CFTR premature stop mutation in a bronchial epithelial cell line. *Nat. Med.* **3:** 1280–1284.

Böck A., Forchhammer K., Heider J., and Baron C. 1991. Selenoprotein synthesis: An expansion of the genetic code. *Trends Biochem. Sci.* **16:** 463–467.

Bonetti B., Fu L., Moon J., and Bedwell D.M. 1995. The efficiency of translation termination is determined by a synergistic interplay between upstream and downstream sequences in *Saccharomyces cerevisiae*. *J. Mol. Biol.* **251:** 334–345.

Brown C.M., Dalphin M.E., Stockwell P.A., and Tate W.P. 1993. The translational termination signal database. *Nucleic Acids Res.* **21:** 3119–3123.

Brown C.M., Stockwell P.A., Trotman C.N., and Tate W.P. 1990. The signal for the termination of protein synthesis in procaryotes. *Nucleic Acids Res.* **18:** 2079–2086.

Buckingham R., Grentzmann G., and Kisselev L. 1997. Polypeptide chain release factors. *Mol. Microbiol.* **24:** 449–456.

Burck C.L., Chernoff Y.O., Liu R., Farabaugh P.J., and Liebman S.W. 1999. Translational suppressors and antisuppressors alter the efficiency of the Ty1 programmed translational frameshift. *RNA* **5**: 1451–1457.

Caskey T., Scolnick E., Tompkins R., Goldstein J., and Milman G. 1970. Peptide chain termination, codon, protein factor, and ribosomal requirements. *Cold Spring Harbor Symp. Quant. Biol.* **34:** 479–488.

Clancy J.P., Hong J.S., Bebok Z., King S.A., Demolombe S., Bedwell D.M., and Sorscher E.J. 1998. Cystic fibrosis transmembrane conductance regulator (CFTR) nucleotide-binding domain 1 (NBD-1) and CFTR truncated within NBD-1 target to the epithelial plasma membrane and increase anion permeability. *Biochemistry* **37:** 15222–15230.

Cox B.S., Tuite M.F., and Mundy C.J. 1980. Reversion from suppression to nonsuppression in SUQ5 [psi+] strains of yeast: The classification of mutations. *Genetics* **95:** 589–609.

Craigen W.J. and Caskey C.T. 1986. Expression of peptide chain release factor 2 requires high-efficiency frameshift. *Nature* **322:** 273–275.

Craigen W.J., Lee C.C., and Caskey C.T. 1990. Recent advances in peptide chain termination. *Mol. Microbiol.* **4:** 861–865.

Crick F.H.C., Barnett L., Brenner S., and Watts-Tobin R.J. 1961. General nature of the genetic code for proteins. *Nature* **192:** 1227–1232.

Cui Y., Dinman J.D., and Peltz S.W. 1996. *Mof4-1* is an allele of the *PF1/IFS2* gene which affects both mRNA turnover and -1 ribosomal frameshifting efficiency. *EMBO J.* **15:** 5726–5736.

Cui Y., Hagan K.W., Zhang S., and Peltz S.W. 1995. Identification and characterization of genes that are required for the accelerated degradation of mRNAs containing a premature translational termination codon. *Genes Dev.* **9:** 423–436.

Czaplinski K., Majlesi N., Banerjee T., and Peltz S.W. 2000. Mtt1 is a Upf1-like helicase that interacts with the translation termination factors and whose overexpression can modulate termination efficiency. *RNA* **6:** 730–743..

Czaplinski K., Ruiz-Echevarria M.J., Gonzales C.I., and Peltz S.W. 1999. Should we kill the messenger? The role of the surveillance complex in translation termination and mRNA turnover. *BioEssays* **21:** 685–696.

Czaplinski K., Weng Y., Hagan K.H., and Peltz S.W. 1995. Purification and characterization of the Upf1p protein: A factor involved in translation and mRNA degradation. *RNA* **1:** 610–623.

Czaplinski K., Ruiz-Echevarria M.J., Paushkin S.V., Han X., Weng Y., Perlick H.A., Dietz H.C., Ter-Avanesyan M.D., and Peltz S.W. 1998. The surveillance complex interacts with the translation release factors to enhance termination and degrade aberrant mRNAs. *Genes Dev.* **12:** 1665–1677.

Doel S.M., McCready S.J., Nierras C.R., and Cox B.S. 1994. The dominant PNM2-mutation which eliminates the psi factor of *Saccharomyces cerevisiae* is the result of a missense mutation in the SUP35 gene. *Genetics* **137:** 659–670.

Donahue T.F., Farabaugh P.J., and Fink G.R. 1981. Suppressible glycine and proline four base codons. *Science* **212:** 455–457.

Eaglestone S.S., Cox B.S., and Tuite M.F. 1999.Translation termination efficiency can be regulated in *Saccharomyces cerevisiae* by environmental stress through a prion-mediated mechanism. *EMBO J.* **18:** 1974–1981.

Ebihara K. and Nakamura Y. 1999. C-terminal interaction of translational release factors eRF1 and eRF3 of fission yeast: G-domain uncoupled binding and the role of conserved amino acids. *RNA* **5:** 739–750.

Freistroffer D.V., Pavlov M.Y., MacDougall J., Buckingham R.H., and Ehrenberg M. 1997. Release factor RF3 in *E. coli* accelerates the dissociation of release factors RF1 and RF2 from the ribosome in a GTP-dependent manner. *EMBO J.* **16:** 4126–4133.

Frolova L., Le Goff X., Zhouravleva G., Davydova E., Philippe M., and Kisselev L. 1996. Eukaryotic polypeptide chain release factor eRF3 is an eRF1- and ribosome-dependent guanosine triphosphatase. *RNA* **4:** 334–341.

Frolova L.Y., Tsivkovskii R.Y., Sivolobova G.F., Oparina N.Y., Serpinsky O.I., Blinov V.M., Tatkov S.I., and Kisselev L.L. 1999. Mutations in the highly conserved GGQ motif of class 1 polypeptide release factors abolish ability of human eRF1 to trigger peptidyl-tRNA hydrolysis. *RNA* **5:** 1014–1020.

Frolova L.Y., Simonsen J.L., Merkulova T.I., Litvinov D.Y., Martensen P.M., Rechinsky V.O., Camonis J.H., Kisselev L.L., and Justesen. J. 1998. Functional expression of eukaryotic polypeptide chain release factors 1 and 3 by means of baculovirus/insect cells and complex formation between the factors. *Eur. J. Biochem.* **256:** 36–44.

Frolova L., Le Goff X., Rasmussen H.H., Cheperegin S., Drugeon G., Kress M., Arman I., Haenni A.L., Celis L.E., Phillippe M., Justesen J., and Kisselev L. 1994. A highly conserved eukaryotic protein family possessing properties of a polypeptide chain release factor. *Nature* **372:** 701–703.

Grant C.M. and Hinnebusch A.G. 1994. Effect of sequence context at stop codons on efficiency of reinitiation in GCN4 translational control. *Mol. Cell. Biol.* **14:** 606–618.

Grentzmann G. and Kelly P.J. 1997. Ribosomal binding site of release factors RF1 and RF2. A new translational termination assay in vitro. *J. Biol. Chem.* **272:** 12300–12304.

Grentzmann G., Kelly P.J., Laalami S., Shuda M., Firpo M.A., Cenatiempo Y., and Kaji A. 1998. Release factor RF-3 GTPase activity acts in disassembly of the ribosome termination complex. *RNA* **4:** 973–983.

He F. and Jacobson A. 1995. Identification of a novel component of the nonsense-mediated mRNA decay pathway using an interacting protein screen. *Genes Dev.* **9:** 437–454.

He F., Brown A., and Jacobson A. 1997. Upf1p, Nmd2p, and Upf3p are interacting components of the yeast nonsense-mediated mRNA decay pathway. *Mol. Cell. Biol.* **17:** 1580–1594.

Heurgué-Hamard V., Karimi R., Mora L., MacDougall J., Leboeuf C., Grentzmann G., Ehrenberg M., and Buckingham R.H. 1998. Ribosome release factor RF4 and termination factor RF3 are involved in dissociation of peptidyl-tRNA from the ribosome. *EMBO J.* **17:** 808–816.

Hirashima A. and Kaji A. 1972. Purification and properties of ribosome-releasing factor. *Biochemistry* **11:** 4037–4044.

Hoshino S., Imai M., Kobayashi T., Uchida N., and Katada T. 1999.The eukaryotic polypeptide chain releasing factor (eRF3/GSPT) carrying the translation termination signal to the 3′-poly(A) tail of mRNA. Direct association of erf3/GSPT with polyadenylate-binding protein. *J. Biol. Chem.* **274:** 16677–16680.

Howard M., Frizzell R.A., and Bedwell D.M. 1996. Aminoglycoside antibiotics restore CFTR function by overcoming premature stop mutations. *Nat. Med.* **2:** 467–469.

Ito K., Ebihara K., and Nakamura Y. 1998a. The stretch of C-terminal acidic amino acids of translational release factor eRF1 is a primary binding site for eRF3 of fission yeast. *RNA* **4:** 958–972.

Ito K., Uno M., and Nakamura Y. 1998b. Single amino acid substitution in prokaryote polypeptide release factor 2 permits it to terminate translation at all three stop codons. *Proc. Natl. Acad. Sci.* **95:** 8165–8169.

———. 2000. A tripeptide anticodon deciphers stop codons in messenger RNA. *Nature* **403:** 680–684.

Ito K., Ebihara K., Uno M., and Nakamura Y. 1996. Conserved motifs in prokaryotic and eukaryotic polypeptide release factors: tRNA-protein mimicry hypothesis. *Proc. Natl. Acad. Sci.* **93:** 5443–5448.

Jacobson A. 1996. Poly(A) metabolism and translation: The closed-loop model. In *Translational control* (ed. J.W.B. Hershey et al.), pp. 451–480. Cold Spring Harbor Laboratory Press, Cold Spring Harbor, New York.

Janosi L., Hara H., Zhang S., and Kaji A. 1996. Ribosome recycling by ribosome recycling factor (RRF) — An important but overlooked step of protein biosynthesis. *Adv. Biophys.* **32:** 121–201.

Janosi L., Mottagui-Tabar S., Isaksson L.A., Sekine Y., Ohtsubo E., Zhang S., Goon S., Nelken S., Shuda M., and Kaji A. 1998. Evidence for in vivo ribosome recycling, the fourth step in protein biosynthesis. *EMBO J.* **17:** 1141–1151.

Karimi R., Pavlov M.Y., Buckingham R.H., and Ehrenberg M. 1999. Novel roles for classical factors at the interface between translation termination and initiation. *Mol. Cell* **3:** 601–609.

Kushnirov V.V. and Ter-Avanesyan M.D. 1998. Structure and replication of yeast prions. *Cell* **94:** 13–16.

Lee B.S. and Culbertson M.R. 1995. Identification of an additional gene required for eukaryotic nonsense mRNA turnover. *Proc. Natl. Acad. Sci.* **92:** 10354–10358.

Leeds P., Wood J.M., Lee B.S., and Culbertson M.R. 1992. Gene products that promote mRNA turnover in *Saccharomyces cerevisiae. Mol. Cell. Biol.* **12:** 2165–2177.

Liebman S.W. and Derkatch I.L. 1999. The yeast [PSI+] prion: Making sense of nonsense. *J. Biol. Chem.* **274:** 1181–1184.

Low S.C. and Berry M.J. 1996. Knowing when not to stop: Selenocysteine incorporation in eukaryotes. *Trends Biol. Sci.* **21:** 203–208.

MacDougall J., Holst-Hansen P., Mortensen K.K., Freistroffer D.V., Pavlov M.Y., Ehrenberg M., and Buckingham R.H. 1997. Purification of active *Escherichia coli* ribosome recycling factor (RRF) from an osmo-regulated expression system. *Biochimie* **79:** 243–246.

McCaughan K.K., Brown C.M., Dalphin M.E., Berry M.J., and Tate W.P. 1995. Translational termination efficiency in mammals is influenced by the base following the stop codon. *Proc. Natl. Acad. Sci.* **92:** 5431–5435.

McKusick V.A. (with the assistance of Francomano C.A., Antonarakis S.E., and Pearson P.L.). 1994. *Mendelian inheritance in man: A catalog of human genes and genetic disorders.* Johns Hopkins University Press, Baltimore, Maryland (web site: http://www.ncbi.nim.nih.gov./Omim/).

Merkulova T.I., Frolova L.Y., Lazar M., Camonis J., and Kisselev L.L. 1999. C-terminal domains of human translation termination factors eRF1 and eRF3 mediate their in vivo interaction. *FEBS Lett.* **443:** 41–47.

Mikuni O., Ito K., Moffat J., Matsumura K., McCaughan K., Nobukuni T., Tate W., and Nakamura Y. 1994. Identification of the prfC gene, which encodes peptide chain release factor 3 of *Escherichia coli*. *Proc. Natl. Acad. Sci.* **91:** 5798–5802.

Moriarty P.M., Reddy C.C., and Maquat L.E. 1998. Selenium deficiency reduces the abundance of mRNA for Se-dependent glutathione peroxidase 1 by a UGA-dependent mechanism likely to be nonsense codon-mediated decay of cytoplasmic mRNA. *Mol. Cell. Biol.* **18:** 2932–2939.

Nakamura Y., Ito K., and Isaksson L. 1996. Emerging understanding of translation termination. *Cell* **87:** 147–150.

Nissen P., Kjeldgaard M., Thirup S., Polekhina G., Reshetnikova L., Clark B.F., and Nyborg J. 1995. Crystal structure of the ternary complex of Phe-tRNAPhe, EF-Tu, and a GTP analog. *Science* **270:** 1464–1472.

Osawa S., Jukes T.H., Watanabe K., and Muto A. 1992. Recent evidence for evolution of the genetic code. *Microbiol. Rev.* **56:** 229–264.

Palmer E., Wilhelm J.M., and Sherman F. 1979. Phenotypic suppression of nonsense mutants in yeast by aminoglycoside antibiotics. *Nature* **277:** 148–150.

Paushkin S.V., Kushnirov V.V., Smirnov V.N., and Ter-Avanesyan M.D. 1997. Interaction between yeast Sup35p (eRF3) and Sup45p (eRF1) polypeptide chain release factors: Implications for prion-dependent regulation. *Mol. Cell. Biol.* **17:** 2798–2805.

Pavlov M.Y., Freistroffer D.V., Heurgué-Hamard V., Buckingham R.H., and Ehrenberg M. 1997a. Release factor RF3 abolishes competition between release factor RF1 and ribosome recycling factor (RRF) for a ribosome binding site. *J. Mol. Biol.* **273:** 389–401.

Pavlov M.Y., Freistroffer D.V., MacDougall J., Buckingham R.H., and Ehrenberg M. 1997b. Fast recycling of *Escherichia coli* ribosomes requires both ribosome recycling factor (RRF) and release factor RF3. *EMBO J.* **16:** 4134–4141.

Pavlov M.Y., Freistroffer D.V., Dincbas V., MacDougall J., Buckingham R.H., and Ehrenberg M. 1998. A direct estimation of the context effect on the efficiency of termination. *J. Mol. Biol.* **284:** 579–590.

Poole E.S., Brown C.M., and Tate W.P. 1995. The identity of the base following the stop codon determines the efficiency of in vivo translational termination in *Escherichia coli*. *EMBO J.* **14:** 151–158.

Recht M.I., Douthwaite S., and Puglisi J.D. 1999. Basis for prokaryotic specificity of action of aminoglycoside antibiotics. *EMBO J.* **18:** 3133–3138.

Ruiz-Echevarria M.J., Gonzalez C., and Peltz S.W. 1998. Identifying the right stop: Determining how the surveillance complex recognizes and degrades an abberant mRNA. *EMBO J.* **17:** 575–589.

Schirmer E.C. and Lindquist S. 1997. Interactions of the chaperone Hsp104 with yeast Sup35 and mammalian PrP. *Proc. Natl. Acad. Sci.* **94:** 13932–13937.

Selmer M., Al-Karadaghi S., Hirokawa G., Kaji A., and Liljas A. 1999. Crystal structure of *Thermotoga maritima* ribosome recycling factor: A tRNA mimic. *Science* **286:** 2349–2352.

Serio T.R. and Lindquist S.L. 1999. [PSI+]: An epigenetic modulator of translation termination efficiency. *Annu. Rev. Cell. Dev. Biol.* **15:** 661–703.

Song J.M. and Liebman S.W. 1987. Allosuppressors that enhance the efficiency of omnipotent suppressors in *Saccharomyces cerevisiae*. *Genetics* **115:** 451–460.

Song H., Mugnier P., Das A.K., Webb H.M., Evans D.R., Tuite M.F., Hemmings B.A., and

Barford D. 2000. The crystal structure of human eukaryotic release factor eRF1—Mechanism of stop codon recognition and peptidyl-tRNA hydrolysis. *Cell* **100:** 311–321.

Stansfield I. and Tuite M.F. 1994. Polypeptide chain termination in *Saccharomyces cerevisiae*. *Curr. Genet.* **25:** 385–395.

Stansfield I., Jones K.M., Kushnirov V.V., Dagakesamanskaya A.R., Poznyakov A.I., Paushkin S.V., Nierras C.R., Cox B.S., Ter-Avanesyan M.D., and Tuite M.F. 1995. The products of the *SUP45* (eRF1) and *SUP35* genes interact to mediate translation termination in *Saccharomyces cerevisiae*. *EMBO J.* **14:** 4365–4373.

Steneberg P., Englund C., Kronhamn J., Weaver T.A., and Samakovlis C. 1998. Translational readthrough in the hdc mRNA generates a novel branching inhibitor in the *Drosophila* trachea. *Genes Dev.* **12:** 956–967.

Tuite M.F., Cox B.S., and McLaughlin C.S. 1981. An homologous *in vitro* assay for yeast nonsense suppressors. *J. Biol. Chem.* **256:** 7298–7304.

Urbero B., Eurwilaichitr L., Stansfield I., Tassan J.P., Le Goff X., Kress M., and Tuite M.F. 1997. Expression of the release factor eRF1 (Sup45p) gene of higher eukaryotes in yeast and mammalian tissues. *Biochimie* **79:** 27–36.

Vilela C., Ramirez C.V., Linz B., Rodrigues-Pousada C., and McCarthy J.E. 1999. Post-termination ribosome interactions with the 5′UTR modulate yeast RNA stability. *EMBO J.* **18:** 3139–3152.

Vincent A., Newnam G., and Liebman S.W. 1994. The yeast translational allosuppressor, SAL6: A new member of the PP1-like phosphatase family with a long serine-rich N-terminal extension. *Genetics* **138:** 597–608.

Wakem L.P. and Sherman F. 1990. Isolation and characterization of omnipotent suppressors in the yeast *Saccharomyces cerevisiae*. *Genetics* **124:** 515–522.

Weng Y., Czaplinski K., and Peltz S.W. 1996a. Genetic and biochemical characterization of mutations in the ATPase and helicase regions of the Upf1 protein. *Mol. Cell. Biol.* **16:** 5477–5490.

———. 1996b. Identification and characterization of mutations in the *UPF1* gene that affect nonsense suppression and the formation of the Upf protein complex but not mRNA turnover. *Mol. Cell. Biol.* **16:** 5491–5506.

Wickner R.B., Taylor K.L., Edskes H.K., Maddelein M.L., Moriyama H., and Roberts B.T. 1999. Prions in *Saccharomyces* and *Podospora spp.*: Protein-based inheritance. *Microbiol. Mol. Biol. Rev.* **63:** 844–861.

Zhouravleva G., Frolova L., LeGoff X., LeGuellec R., Inge-Vechtomov S., Kisselev L., and Phillippe M. 1995. Termination of translation in eukaryotes is governed by two interacting polypeptide chain release factors, eRF1 and eRF3. *EMBO J.* **14:** 4065–4072.

12

Genetic Approaches to Translation Initiation in *Saccharomyces cerevisiae*

Thomas F. Donahue
Department of Biology
Indiana University
Bloomington, Indiana 47405

Historically, biochemistry has been the main approach for studying the eukaryotic translation initiation process, with most effort directed toward mammalian systems from which the factorology of translation initiation became well developed. Less than 15 years ago, little was known of the factorology of the yeast system, let alone whether a similar mechanism for initiating protein synthesis existed. At that time, the genetics of yeast began to prove itself as an effective approach to study the mechanism of eukaryotic translation. The similarities between the mammalian and yeast initiation processes, now well established, enable us to assess the mechanism of initiation by combining biochemical and genetic data, a very powerful means to dissect the eukaryotic translation machinery. The ability to do genetics provides additional dimensions to the analysis of translation initiation. As discussed below, it affords direct selection for suppressor mutations in genes that encode factors. These suppressor genes help to define those factors that function at particular steps in the initiation pathway and to show how individual subunits contribute to the activity of multisubunit initiation factors. This has always been one of the rate-limiting problems associated with the biochemical analysis of translation initiation, because the complexity of the process was not compatible, in many cases, with developing specific assays for factors or subunits of factors at defined steps in the pathway. Genetic suppressor analysis can also be a very sensitive method for detecting factor function. It has the potential to define new factors, which may not be realized by biochemical approaches, especially if the factors are not abundant proteins. Aside from suppressor analysis, genetics has also allowed the ability to establish the biological significance of factors for translation initiation, to

Translational Control of Gene Expression
© 2000 Cold Spring Harbor Laboratory Press 0-87969-568-4/00 $5 +. 00

define translational control mechanisms that operate in response to different physiological conditions, and to help define important functional and physical interactions between different initiation factors. The purpose of this chapter is to provide an overview of these genetic studies in yeast with special emphasis on genetic studies that relate to factors involved in the recognition of the AUG start codon.

SACCHAROMYCES CEREVISIAE UTILIZES A SCANNING MECHANISM FOR TRANSLATION INITIATION

The first genetic attempt to determine the signals for directing translation initiation at a yeast mRNA was performed by Sherman and colleagues nearly 30 years ago (Stewart et al. 1971). Using reversion analysis of *CYC1* initiator codon mutants, they found that the most 5′-proximal AUG serves as the start codon for translation initiation when present at any site within a 37-nucleotide region (Stewart et al. 1971; Sherman et al. 1980). These genetic data provided the initial evidence that AUG was the primary signal for translation initiation in yeast mRNA and also contributed to the establishment of the scanning model (Kozak 1978). The findings were corroborated by other genetic studies in yeast. A genetic selection scheme was designed to identify mutations that have a negative effect on translation initiation at the AUG start codon in the *HIS4* mRNA (Donahue and Cigan 1988). The scheme was designed in such a way that yeast cells which failed to initiate translation at the first AUG initiator codon could be detected by their ability to initiate at a downstream AUG codon. All mutations identified that altered initiation at the first AUG were mutations within the AUG itself. No mutations in other regions of the leader were identified, suggesting that the first AUG codon nearest the 5′ end of mRNA served as the signal for translation.

The sequence context neighboring the AUG initiator codon of a mammalian gene has been shown by mutational analysis to contribute significantly to the efficiency of initiation (Kozak 1986). The consensus derived from this and comparative studies, 5′-CACCAUGG-3′, is believed to be optimal for initiation. Comparative analysis of yeast mRNAs indicates that yeast leader regions are A-rich (Cigan and Donahue 1987). The sequence context of yeast initiator regions is biased toward having an A nucleotide at the –3 position and a U nucleotide at the +4 position (the A of the AUG as +1). However, mutational analysis of the sequence context at both the *CYC1* and *HIS4* genes showed that changing the sequence surrounding the AUG start codon had no more than a twofold effect on the efficiency of initiation (Baim et al. 1988; Cigan et al. 1988b). In addition, introduction of the mammalian consensus sequence at *HIS4* did not alter the efficiency of initiation at the AUG start codon. Thus, in yeast, AUG appears to be the primary signal for translation initiation.

The studies described above provide a simple mechanism whereby the preinitiation complex scans the leader and recognizes the first AUG codon nearest the 5′ end of yeast mRNA. However, perhaps the best evidence for yeast using a scanning mechanism comes from studies of the mechanism of *GCN4* translational control. Here, Hinnebusch and colleagues have shown that ribosomes are capable of traversing ~600 nucleotides of leader length to arrive at the *GCN4* start codon during the general amino acid control response (Abastado et al. 1991; Chapter 5).

RECOGNITION OF THE AUG START CODON BY $tRNA_i^{Met}$

In light of the fact that in yeast the AUG initiator codon solely serves as the start signal for translation, it was logical to assume that a base-pair interaction between the anticodon of $tRNA_i^{Met}$ and the AUG was an important component for ribosomal recognition of a start site. To test this idea, genetic studies were initiated in which the anticodon of one of the $tRNA_i^{Met}$ genes from yeast was mutated from 3′-UAC-5′ to 3′-UCC-5′ and the initiator codon at the *HIS4* gene was changed to AGG to be complementary with UCC-$tRNA_i^{Met}$ (Cigan et al. 1988a). Both genes when present in single copy in yeast cells would not allow translation initiation to occur at *HIS4* as noted by a His$^-$ phenotype. This result was anticipated, as the anticodon mutation was expected to result in the inability of mutant tRNA to be charged by the methionyl-tRNA synthetase (Schulman and Pelka 1984). However, the UCC-$tRNA_i^{Met}$ could restore translation at *his4* that was specific for the AGG codon in either of two ways: first, by mutation in the methionyl-tRNA synthetase, which now had the ability to charge this UCC-$tRNA_i^{Met}$, and second, by overexpression either of the wild-type synthetase gene (in the presence of single copy, UCC-initiator tRNA) or of the UCC initiator-tRNA. Presumably, overexpression of either gene led to increased levels of charged mutant tRNA that could now support an initiation event at the AGG codon at *his4*.

The ability to put this genetic system together enabled testing of the importance of the anticodon:codon interaction in directing the ribosome to the site of initiation (Cigan et al. 1988a). A *his4* allele was constructed to contain an AGG codon upstream and out of frame with the AGG codon at the initiator site. As observed for the effects of an AUG codon that is upstream and out of frame, the UCC-initiator tRNA directed the ribosome to the first AGG codon in the *his4* message, which precluded translation initiation at the downstream AGG. These data indicated that at least one element which is important to the ability of the ribosome to recognize the initiator region in mRNA is a cognate anticodon:codon interaction between the $tRNA_i^{Met}$ and the first AUG codon in mRNA. It is important to note that these studies indicated that the translation initiation factors can work with this aberrant initiator-tRNA, suggesting that the factors themselves are not directly "recognizing" a translational start site. Rather, the anticodon of initiator-tRNA appears to recognize the proper translational start site through a 3-bp anticodon:codon interaction.

CONTROL OF AUG RECOGNITION

One fundamental difference between yeast and bacterial translation initiation is that the former is quite stringent in only selecting an AUG start codon. For example, none of the nine possible point mutations of the *HIS4*, AUG start codon can serve as a signal for translation initiation (Donahue and Cigan 1988). In contrast, prokaryotic translation initiation at some genes uses alternative codons, such as UUG and GUG (Gualerzi and Pon 1990). Nevertheless, there is nothing fundamentally important about an AUG per se for yeast translation initiation. As described above, an AGG codon can serve as a start site provided a 3-bp codon:anticodon interaction can be established between the AGG codon and the initiator-tRNA.

To gain further insight into the mechanism of ribosome recognition of an AUG start codon, a genetic approach was used whereby extragenic suppressors were sought that can restore translation initiation at *HIS4* despite the absence of an AUG start codon. This involved a genetic reversion analysis whereby initiator codon mutants at *HIS4* that confer a His^- phenotype were reverted to His^+ and screened genetically for extragenic suppression events (Donahue et al. 1988). The rationale was that a mutation in a component of the preinitiation complex which functions in ribosomal recognition of a start codon would now allow the ribosome to initiate at a non-AUG codon and restore *HIS4* expression. The characterization of these encoded suppressor gene products would then point toward the components of the preinitiation complex that might function in mediating the start-site selection process.

Five unlinked genes, *sui1*, *sui2*, *SUI3*, *SUI4* (*GCD11*), *SUI5* (*TIF5*), were identified by this genetic selection scheme using either haploid or diploid yeast strains (Castilho-Valavicius et al. 1990, 1992; Huang et al. 1997). Cloning and sequencing of these suppressor loci provided important insight into the nature of translation initiation factors that participate in the AUG recognition process. The *sui2* and *SUI3* and *SUI4/GCD11* suppressor genes encode mutated forms of the α, β, and γ subunits of eIF2, respectively (Donahue et al. 1988; Cigan et al. 1989; Hannig et al. 1993; Huang et al. 1997). The early observations that the *sui2* and *SUI3* genes showed strong conservation in their amino acid sequence to the mammalian eIF2 subunits strongly supported a conserved mechanism for translation initiation (Pathak et al. 1988; Cigan et al. 1989). The *SUI5* suppressor locus (Huang et al. 1997) corresponded to a mutated form of the *TIF5* gene that encodes the translational initiation factor eIF5 (Chakravarti and Maitra 1993). The cloning of the *sui1* gene product did not give immediate insight into its initiation factor assignment (Yoon and Donahue 1992). This came later after the amino acid sequence of the mammalian translation factor eIF1 was identified and shown to possess significant identity to the predicted amino acid sequence of the Sui1 gene product (Kasperaitis et al. 1995).

Each of these suppressor mutants shares in common the same in vivo mechanism of suppression at *his4*; that is, to allow a mismatched base-pair interaction between a UUG codon (the second codon downstream from the *HIS4* start site; referred to as +3 UUG codon) and the anticodon (3′-UAC-5′) of the wild-type initiator-tRNA. This was established either by direct sequencing of the His4 protein produced by *sui1*, *sui2,* or *SUI3* suppression events (Donahue et al. 1988; Yoon and Donahue 1992 and unpubl.), or indirectly for the *SUI4* and *SUI5* suppressors based on similar suppression specificity to that of *sui1*, *sui2*, and *SUI3* (Huang et al. 1997). The suppressor strains appear to prefer to use a UUG codon as the start site for suppression, using a GUG codon less efficiently (Huang et al. 1997). It is not clear what leads to this preference, but it does not underlie the original selection scheme. For example, selection for extragenic suppressors when GUG is replaced for the +3 UUG codon in the *HIS4* coding region identifies either *sui1* or *SUI3,* which confer a weak His^+ suppressor phenotype. However, when either of these suppressor alleles is combined through genetic crosses with the *his4* allele that has the +3 UUG codon, the His^+ suppressor phenotype is more robust (H.J. Yoon and T.F. Donahue, unpubl.).

The ability of the suppressor mutants to allow initiation at a non-AUG codon represents a breakdown in the fidelity of the eukaryotic translation initiation process. Primarily, the mutation characterized in the *SUI4* allele, eIF2γ^{N135K}, which is located in the G_2 domain of the G protein, and the identification of the *SUI5* suppressor allele corresponding to eIF5, known biochemically to stimulate the GTPase activity associated with the γ subunit of eIF2, strongly suggested defects related to GTP binding/GTP hydrolysis (Huang et al. 1997). By using purified components and in vitro assays, insight into the nature of suppression was obtained. eIF2 containing the eIF2γ^{N135K} suppressor gene product was initially found to be defective in ternary complex formation, with eIF2 binding to initiator-tRNA in a GTP-dependent fashion. However, further assays indicated that the major defect of the mutant eIF2 complex was that it dissociated from the initiator-tRNA, presumably in the absence of GTP hydrolysis (Huang et al. 1997). A more complex assay was required to measure GTPase activation function of eIF5 (Chakravarti et al. 1993; Huang et al. 1997). In this assay, a preinitiation complex was formed using purified components (eIF2:initiator-tRNA:[γ-^{32}P]GTP, 40S ribosomes, and an AUG triplet). Subsequent addition of the mutant *SUI5* suppressor gene product, eIF5^{G31R}, to this complex stimulated GTP hydrolysis twofold faster than wild-type eIF5 (Huang et al. 1997).

All *SUI3* suppressor alleles so far characterized contain mutations in amino acids that map near or within a CysX$_2$Cys X$_{19}$CysX$_2$Cys domain in the carboxyl terminus of the eIF2β subunit (Donahue et al. 1988; Castilho-Valavicius et al. 1992). This domain and the amino acids affected in *SUI3* suppressor alleles are conserved in the human eIF2β subunit (Pathak et al. 1988; Castilho-Valavicius et al. 1992). Two of the mutants were characterized biochemically in in vitro reactions. One of the mutants assayed, eIF2β^{S264Y}, originated from the selection scheme in haploid cells (Donahue et al. 1988). The second mutant that was assayed, eIF2β^{L254P}, originated from the selection scheme in diploid cells and is lethal when present in a haploid cell (Castilho-Valavicius et al. 1992). eIF2 complexes containing either eIF2β^{S264Y} or eIF2β^{L254P} are defective in the ability to bind initiator-tRNA in the presence of GTP, and this defect stems from the mutations conferring intrinsic GTP hydrolysis activity on the eIF2 complex in the absence of eIF5 (Huang et al. 1997).

On the basis of these data and the biochemistry of eIF2 and eIF5, initiation can be viewed as a process that involves a molecular switch. The active state of eIF2 is the GTP-bound form that can bind initiator-tRNA and associate with the 40S ribosome to scan mRNA. The inactive state of eIF2 is the GDP-bound form that cannot bind initiator-tRNA. GTP hydrolysis, stimulated by eIF5 at the time of translation initiation, serves as a switch to convert eIF2 from the active to the inactive form that allows it to dissociate from the 48S preinitiation complex. This leaves the initiator-tRNA in the P site of the ribosome that signals the transition from initiation to elongation. The signal for the switch would appear to be the establishment of a 3-bp codon:anticodon interaction between the AUG start codon and the initiator-tRNA. The inability to establish a 3-bp codon:anticodon interaction in wild-type cells, such as encountering a UUG codon, inhibits or does not lead to a productive "conformation" to signal or stimulate the GTP hydrolysis step, and the ribosome continues to scan.

For initiation to occur at a non-AUG codon, the main obstacle to overcome is to dissociate initiation factors and leave the initiator-RNA in the P site of the ribosome, mismatched base-paired with a non-AUG codon. The biochemical characterizations of the *SUI3*, *SUI4*, and *SUI5* suppressor mutants are consistent with such a mechanism for aberrant initiation events (Huang et al. 1997). When a 48S preinitiation complex encounters a UUG codon, either GTP hydrolysis occurs inappropriately or too quickly (eIF2β and eIF5 suppressor mutants), or eIF2 dissociates from the initiator-tRNA prematurely (eIF2γ suppressor mutant). Each of these events would lead to leaving the initiator-tRNA in the P site, mismatched base-paired with a UUG codon, and the preinitiation complex is now committed to make the transition from initiation to elongation. These data suggest that the initiation factors eIF2 and eIF5 maintain the fidelity of translation initiation and that the GTP hydrolysis step is an important link in signaling or responding to ribosomal recognition of an AUG start codon.

The *sui1* suppressor mutant was isolated in the same genetic selection scheme and allows a mismatched base-pair interaction between the initiator-tRNA at the +3 UUG codon at *his4* (Yoon and Donahue 1992). It therefore stands to reason that in light of *sui1*'s similar in vivo properties with *SUI3*, *SUI4*, and *SUI5* suppressor mutants, mutations in *sui1* would confer similar biochemical defects either in eIF2 dissociation or GTP hydrolysis to account for its suppression phenotype. However, addition of purified Sui1 protein to in vitro ternary complex formation assays or GTP hydrolysis assays, as described above, has no effect on these reactions (H.-K. Huang and T.F. Donahue, unpubl.). It has recently been shown that Sui1 protein copurifies with eIF3, although not in stoichiometric amounts, and interacts with the Nip1 subunit of yeast eIF3 (Naranda et al. 1996; Asano et al. 1998; Phan et al. 1998). Thus, its association with Nip1/eIF3 may be required for its functional activity, and the assays used to characterize *SUI3*, *SUI4,* and *SUI5* do not require eIF3. Sui1 protein promotes Met-puromycin formation in in vitro translation initiation assays (Naranda et al. 1996) but does not appear to be required for eIF3 core complex activity (Phan et al. 1998). The mammalian counterpart of Sui1, eIF1, has been characterized in toeprinting assays to dissociate 43S complexes stalled on mRNA near the 5′-terminal cap and to promote formation of 48S complexes at the correct translation initiation site (Pestova et al. 1998). These activities also required eIF1A (Pestova et al. 1998). On the basis of these toeprinting studies, the authors proposed that eIF1/Sui1 might need to associate with the preinitiation complex prior to or at the time of binding to mRNA and serve as a clamp to stabilize the 48S complex until it locates the initiation complex. However, the specific biochemical function of Sui1/eIF1 remains unknown.

An additional genetic approach to gain insight into the biochemical function of Sui1/eIF1 has been undertaken. Starting with *sui1* mutants that are temperature-sensitive for growth at 37°C, revertants were selected that grow better at the restrictive temperature, and mutations in unlinked genes were identified (S.I. Reed and T.F. Donahue, unpubl.). The rationale for this approach is the assumption that the poor growth properties observed with *sui1* mutants are a result of a functional defect during translation initiation. Thus, a mutation in an unlinked gene that counteracts this functional defect will allow *sui1* mutants to grow better. More importantly, the gene product mutated to produce this effect may be functionally related to Sui1/eIF1, and if this gene product is more tractable biochemically, it may provide indirect insight into the biochemical function of Sui1/eIF1.

One suppressor locus identified by this scheme, *ssu2* (*s*uppressor of *sui1*), corresponds to the *TIF5* gene that encodes eIF5. Biochemical characterization of eIF5 purified from *ssu2-1* or *ssu2-2* strains showed that each of the mutant proteins is ~25% as active as wild-type eIF5 in stimulating hydrolysis of GTP (H.-K. Huang and T.F. Donahue, unpubl.). Thus, these mutations in eIF5 lower the GTPase activation function that must lower the rate of GTP hydrolysis in vivo. Thus, by lowering the rate of GTP hydrolysis, *sui1* mutants can grow better. It therefore seems logical that *sui1* mutants are defective in some aspect of the GTP hydrolysis step during translation initiation, perhaps resulting in an increased rate in GTP hydrolysis similar to *SUI3* and *SUI5* mutants, and this confers the poor growth properties in a *sui1* strain.

There are additional reasons to suspect that at least one function of Sui1/eIF1 during translation initiation is tightly linked to the GTP hydrolysis reaction at the time of ribosomal recognition of a start codon. Both Sui1/eIF1 and eIF5 have been shown to bind the Nip1 subunit of eIF3 that at least places Sui1 and eIF5 in a similar location in the preinitiation complex. eIF5 and eIF2β have also been shown to interact physically (Das et al. 1997; Asano et al. 1999). This interaction takes place between the carboxy-terminal end of eIF5 and the amino-terminal end of eIF2β (Asano et al. 1999). This is interesting, as all mutations in eIF5 identified to alter its GTPase activation function map in the amino-terminal end of eIF5, and this suggests that it is the amino-terminal end of eIF5 that catalyzes the GTPase activation function (Huang et al. 1997 and unpubl.). In addition, all *SUI3* suppressor alleles characterized at the molecular level map near or within the carboxy-terminal end of the eIF2β subunit. Thus, functional interactions as well as physical interactions or proximity between Sui1/eIF1, eIF2β, and eIF5 make an interesting case for proposing Sui1 to have a specific function in controlling AUG recognition through the coupled GTP hydrolysis step.

However, other observations about *sui1* suggest that the protein product is more complex functionally. *mof2/sui1* mutants have been characterized that lead to –1 ribosomal frameshifting and loss of killer virus, and alter nonsense-mediated decay properties of a mini-PGK1 mRNA (Cui et al. 1999). This has resulted in the proposal that Sui1 may be a general regulator of both translation and mRNA turnover. It has been proposed that the 70% of Sui1 identified as a free subunit (not associated with eIF3) might function independently of its role in translation initiation (Cui et al. 1999).

Sui1/eIF1 has also been shown to have bacterial and archeal homologs (Kyrpides and Woese 1998). IF1 and IF2 also appear to be conserved translation initiation factors (Choi et al. 1998; Kyrpides and Woese 1998). It seems clear that to fully appreciate the function of eukaryotic Sui1/eIF1 we'll need to understand how its bacterial homolog functions.

THE GENERAL AMINO ACID CONTROL RESPONSE

In yeast, reinitiation of translation is used to regulate the expression of the *GCN4* gene, which encodes a positive transcriptional activator of amino acid biosynthetic genes subject to the general amino acid control response (Chapter 5). The general amino acid control response was studied extensively using genetic approaches even before it was known that *GCN4* was subject to regulation at the level of initiation. The basic approach was similar to that performed with bacteria that attempted to determine transcriptional regulation of amino acid biosynthetic genes. One aspect of the genetic analysis was designed to select for up-mutants (*gcd* mutants; general control depressed) that are resistant to an amino acid analog, such as 1,2,4-triazole-3-alanine, a histidine analog that is toxic to wild-type yeast cells (Wolfner et al. 1975; Hinnebusch and Fink 1983). The complementary genetic approach was to screen for down mutants (*gcn* mutants; general control nonderepressed) that could not grow in the presence of an amino acid analog such as the competitive inhibitor 3-amino-1,2,4,-triazole (Wolfner et al. 1975; Hinnebusch and Fink 1983).

Using this approach, a number of translation initiation factors and regulators were identified, including the *GCD1*, *GCD2*, *GCD6*, *GCD7,* and *GCN3* genes that have been shown to encode the five subunits of eIF2B, the guanine nucleotide exchange factor for eIF2 (Bushman et al. 1993a,b; Cigan et al. 1993); the *GCD11* gene that encodes the γ subunit of eIF2 (Hannig et al. 1993); and *GCN2,* the eIF2α kinase (Wek et al. 1989). The genetics of this analysis is quite consistent with the regulation of *GCN4* (Dever et al. 1992; Wek et al. 1995; Chapter 5). Mutations in subunits of eIF2, including *sui2* and *SUI3* (Williams et al. 1989), artificially reduce the level of ternary complex by impairment of eIF2 function. Mutations in subunits of eIF2B artificially reduce the level of ternary complex by impairment of the nucleotide exchange reaction. This mimics the regulatory response. Activation of Gcn2 by amino acid starvation leads to phosphorylation of eIF2α and results in inhibition of eIF2B. This inhibition of eIF2B results in lower levels of the active form of eIF2 in the GTP-bound state. In either case, it is the lower levels of eIF2 that change the rate at which ternary complex rebinds to ribosomes after initiation at uORF1 that results in bypass of uORF2-4 and reinitiation at the *GCN4* start position.

GENETIC IDENTIFICATION OF OTHER TRANSLATION INITIATION FACTORS

The genetic analyses of AUG utilization at *GCN4* and non-AUG utilization at *HIS4* have clearly proved most successful in the genetic identification of translation initiation factors. Primarily, eIF1, eIF2, eIF2B, and eIF5 have stemmed from these two analyses and represent the genetic identification of 10 proteins out of an estimated total of 25 proteins to participate in translation initiation, excluding ribosomal proteins. Limited success has been obtained in the genetic identification of the remaining translation initiation factors/subunits, perhaps because relevant genetic selection schemes or screens have not yet been realized. Below is a brief summary of additional genetic approaches that have identified translation initiation factors.

eIF4-related Factors

The eIF4E gene encoding the cap-binding protein was identified genetically as the cell cycle mutant, *cdc33* (Pringle and Hartwell 1982). The *cdc33* mutant was originally isolated as a temperature-sensitive mutant that arrested at G_1 upon shifting cells from 25°C to 35°C (Reed 1980). eIF4E isolated from this strain was shown to have a defect in cap-binding activity (Altmann and Trachsel 1989). It has been speculated that the G_1 arrest of the *cdc33* mutant may represent a specific involvement of eIF4E in controlling the cell cycle, as cells with a conditional defect in protein synthesis would be expected to arrest at random points in the cell cycle (Brenner et al. 1988).

There are two genes in the yeast haploid genome, *TIF1* and *TIF2*, that encode eIF4A, an RNA helicase. The *TIF1* gene was isolated by a genetic suppression scheme but not in such a way that might have been predicted by the proposed function of this protein in translation initiation. *TIF1* was isolated as a high-copy suppressor of a missense mutation in the mitochondrially encoded *OXI2* gene (Linder and Slonimski 1989). The *oxi2* mutation causes respiratory insufficiency due to the inability to make the mitochondrial cytochrome oxidase III subunit, and this mutant phenotype can be detected by the inability of the strain to grow on a nonfermentable carbon source such as glycerol. High-copy suppression by *TIF1* enabled the *oxi2* mutant to grow on glycerol. A likely explanation for the suppression of the *oxi2* mutant phenotype is that increased levels of eIF4A may lead to an increase in translation of a nuclear-encoded protein that functions in *oxi2* expression in the mitochondria. When this hypothetical protein is abundant in the mitochondria it can suppress the defect conferred by the *oxi2* mutation.

The yeast gene, *TIF3*, encodes eIF4B, which is believed to work in conjunction with eIF4A by providing RNA-binding specificity. *TIF3* has been identified by independent approaches (Altmann et al. 1993; Coppolccchia ct al. 1993). One of the approaches that led to its identification was a genetic screen for high-copy suppressors of a temperature-sensitive mutation in eIF4A (Coppolecchia et al. 1993). The observation that eIF4B (*STM1*) can suppress a conditional mutant of eIF4A provides in vivo evidence that both proteins may function together.

eIF3

eIF3 is a multisubunit complex that has been implicated in various aspects of the initiation process. It promotes 80S dissociation that is needed to recycle ribosome subunits, associates with 40S ribosomal subunits, and stabilizes binding of the ternary complex (eIF2/GTP/Met-$tRNA_i^{Met}$) to the 40S ribosomal subunit (Chapter 2). In *S. cerevisiae*, eIF3 has been identified as a core complex of five different proteins, encoded by *NIP1*, *PRT1*, *TIF32*, *TIF34*, and *TIF35*. *PRT1* was described genetically over 30 years ago as a temperature-sensitive mutant, *prt1-1*, that was not able to efficiently incorporate amino acids into proteins at the restrictive temperature (Hartwell and McLaughlin 1968, 1969). Also, the *prt1* mutant polysomes disintegrated immediately following the temperature shift, indicative of a defect in translation initiation. Cell extracts were also defective in converting 40S to 45S ribosomal particles that represent a 40S–eIF3 complex, and were defective in ternary complex association with the 40S, two activities associated with eIF3. *PRT1* was also identified genetically as the temperature-sensitive cell cycle mutant, *cdc63*, that arrests at the G_1 phase of the cell cycle when shifted to the restrictive temperature, and similar observations have been made with the temperature-sensitive *prt1* mutant strain (Bedard et al. 1981; Hanic-Joyce et al. 1987). Perhaps eIF4E and Prt1 both play a role in cell proliferation in addition to their roles in translation initiation.

The only other subunit of eIF3 that has been identified genetically is Nip1. The *NIP1* locus was identified originally in a genetic screen of temperature-sensitive mutants that were defective in nuclear import (Gu et al. 1992). However, *nip1* mutants also show defects expected of a protein that functions in translation initiation (Greenberg et al. 1998), and Nip1 interacts with Sui1/eIF1 and with eIF5, aside from being a subunit of eIF3 (Phan et al. 1998). These observations suggest that Nip1 may have dual functions or that nuclear import and translation initiation may be coupled in yeast.

Poly(A)-binding Protein

Although the *PAB1* gene encoding poly(A)-binding protein (PABP) has never been implicated directly by genetic analysis to be involved in translation initiation, subsequent genetic analyses of a constructed mutant at least implicated a role for it in translation (see Chapter 10). A mutation in the *SPB4* gene that encodes one of the 60S ribosomal proteins can suppress the lethality resulting from a deletion of the *pab1* gene (Sachs and Davis 1990). Because the PABP–poly(A) complex also enhanced translational efficiency in vitro by recruiting 40S subunits to mRNAs, and this enhancement was cap-binding-protein (eIF4E)-independent (Tarun and Sachs 1995), this suggested that an *spb4* mutant might suppress a *pab1* deletion by increasing the concentration of free 40S ribosomes. This implicated PABP to play some role in enhancing recruitment of 40S to mRNA. The significance of this genetic observation most likely relates to a number of more recent studies which have suggested that the translation initiation complex is built on mRNA through a series of interactions involving the 5′ cap structure, eIF4F, eIF3, eIF2, and PABP (Chapter 2). Thus, mRNA may be translated maximally when the 3′ end of mRNA is in direct contact with the 5′ end of mRNA. This contact could localize dissociating 40S and 60S ribosomes near the 5′ end of mRNA after the

80S completed elongation. The major mRNA decay pathway that was determined for yeast also implicates a communication between the 5′ end and the 3′ end of mRNA (Chapter 28). Genetic observations have also implied an important role for PABP in inhibiting removal of the 5′ cap structure of mRNA that leads to rapid 5′ to 3′ exonucleolytic degradation of mRNA (1995). Thus, the major function of PABP may be to couple mRNA decay with efficient translation through its interaction with eIF4G at the 5′ end of mRNA and its binding to poly(A) at the 3′ end of mRNA.

REFERENCES

Abastado J.P., Miller P.F., Jackson B.M., and Hinnebusch A.G. 1991. Suppression of ribosomal reinitiation at upstream open reading frames of amino acid-starved cells forms the basis for *GCN4* translational control. *Mol. Cell. Biol.* **11:** 486–496.

Altmann M. and Trachsel H. 1989. Altered mRNA cap recognition activity of initiation factor 4E in the yeast cell cycle division mutant *cdc33*. *Nucleic Acids Res.* **17:** 5923–5931.

Altmann M., Muller P.P., Wittmer B., Ruchti F., Lanker S., and Trachsel H. 1993. A *Saccharomyces cerevisiae* homologue of mammalian translation initiation factor 4B contributes to RNA helicase activity. *EMBO J.* **12:** 3997–4003.

Asano K., Phan L., Anderson J., and Hinnebusch A.G. 1998. Complex formation by all five homologues of mammalian translation initiation factor 3 subunits from yeast *Saccharomyces cerevisiae*. *J. Biol. Chem.* **273:** 18573–18585.

Asano K., Krishnamoorthy T., Phan L., Pavitt G.D., and Hinnebusch A.G. 1999. Conserved bipartite motifs in yeast eIF5 and eIF2Bε, GTPase-activating and GDP-GTP exchange factors in translation initiation, mediate binding to their common substrate eIF2. *EMBO J.* **18:** 1673–1688.

Baim S.B. and Sherman F. 1988. mRNA structures influencing translation in the yeast *Saccharomyces cerevisiae*. *Mol. Cell. Biol.* **8:** 1591–1601.

Bedard D.P., Johnston G.C., and Singer R.A. 1981. New mutations in the yeast *Saccharomyces cerevisiae* affecting completion of "start." *Curr. Genet.* **4:** 205–214.

Brenner C., Nakayama N., Goebl M., Tanaka K., Toh-e A., and Matsumoto K. 1988. *CDC33* encodes mRNA cap-binding protein eIF-4E of *Saccharomyces cerevisiae*. *Mol. Cell. Biol.* **8:** 3556–3559.

Bushman J.L., Asuru A.I., Matts R.L., and Hinnebusch A.G. 1993a. Evidence that *GCD6* and *GCD7*, translational regulators of *GCN4*, are subunits of the guanine nucleotide exchange factor for eIF-2 in *Saccharomyces cerevisiae*. *Mol. Cell. Biol.* **13:** 1920–1932.

Bushman J.L., Foiani M.. Cigan A.M., Paddon C.J., and Hinnebusch A.G. 1993b. Guanine nucleotide exchange factor for eukaryotic translation initiation factor 2 in *Saccharomyces cerevisiae:* Interactions between the essential subunits *GCD2*, *GCD6*, and *GCD7* and the regulatory subunit *GCN3*. *Mol. Cell. Biol.* **13:** 4618–4631.

Caponigro G. and Parker R. 1995. Multiple functions for the poly(A)-binding protein in mRNA decapping and deadenylation in yeast. *Genes Dev.* **9:** 2421–2432.

Castilho-Valavicius B., Thompson G.M., and Donahue T.F. 1992. Mutation analysis of the Cys-X_2-Cys-X_{19}-Cys-X_2-Cys motif in the β subunit of eukaryotic translation initiation factor 2. *Gene Expr.* **2:** 297–309.

Castilho-Valavicius B., Yoon H., and Donahue T.F. 1990. Genetic characterization of the *Saccharomyces cerevisiae* translational initiation suppressors *sui1*, *sui2* and *SUI3* and their effects on *HIS4* expression. *Genetics* **124:** 483–495.

Chakravarti D. and Maitra U. 1993. Eukaryotic translation initiation factor 5 from *Saccharomyces cerevisiae.* Cloning, characterization, and expression of the gene encoding the 45,346-Da protein. *J. Biol. Chem.* **268:** 10524–10533.

Chakravarti D., Maiti T., and Maitra U. 1993. Isolation and immunochemical characterization of eukaryotic translation initiation factor 5 from *Saccharomyces cerevisiae. J. Biol. Chem.* **268:** 5754–5762.

Choi S.K., Lee J.H., Zoll W.L., Merrick W.C., and Dever T.E. 1998. Promotion of Met-$tRNA_i^{Met}$ binding to ribosomes by yIF2, a bacterial IF2 homolog in yeast. *Science* **280:** 1757–1760.

Cigan A.M. and Donahue T.F. 1987. Sequence and structural features associated with translational initiator regions in yeast — Review. *Gene* **59:** 1–18.

Cigan A.M., Feng L., and Donahue T.F. 1988a. $tRNA_i^{met}$ functions in directing the scanning ribosome to the start site of translation. *Science* **242:** 93–97.

Cigan A.M., Pabich E.K., and Donahue T.F. 1988b. Mutational analysis of the *HIS4* translational initiator region in *Saccharomyces cerevisiae. Mol. Cell. Biol.* **8:** 2964–2975.

Cigan A.M., Bushman J.L., Boal T.R., and Hinnebusch A.G. 1993. A protein complex of translational regulators of *GCN4* mRNA is the guanine nucleotide-exchange factor for translation initiation factor 2 in yeast. *Proc. Natl. Acad. Sci.* **90:** 5350–5354.

Cigan A.M., Pabich E.K., Feng L., and Donahue T.F. 1989. Yeast translation initiation suppressor *sui2* encodes the α subunit of eukaryotic initiation factor 2 and shares identity with the human α subunit. *Proc. Natl. Acad. Sci.* **86:** 2784–2788.

Coppolecchia R., Buser P., Stotz A., and Linder P. 1993. A new yeast translation initiation factor suppresses a mutation in the eIF-4A RNA helicase. *EMBO J.* **12:** 4005–4011.

Cui Y., Gonzalez C.I., Kinzy T.G., Dinman J.D., and Peltz S.W. 1999. Mutations in the *MOF2/SUI1* gene affect both translation and nonsense-mediated mRNA decay. *RNA* **5:** 794–804.

Das S., Maiti T., Das K., and Maitra U. 1997. Specific interaction of eukaryotic translation initiation factor 5 (eIF5) with the β-subunit of eIF2. *J. Biol. Chem.* **272:** 31712–31718.

Dever T.E., Feng L., Cigan A.M., Wek R., Donahue T.F., and Hinnebusch A.G. 1992. Phosphorylation of initiation factor 2α by protein kinase GCN2 mediates gene specific translational control of *GCN4* in yeast. *Cell* **68:** 585–596.

Donahue T.F. and Cigan A.M. 1988. Genetic selection for mutations that reduce or abolish ribosomal recognition of the *HIS4* translational initiator region. *Mol. Cell. Biol.* **8:** 2955–2963.

Donahue T.F., Cigan A.M., Pabich E.K., and Castilho-Valavicius B. 1988. Mutations at a Zn(II) finger motif in the yeast eIF-2β gene alter ribosomal start-site selection during the scanning process. *Cell* **54:** 621–632.

Greenberg, J.R., Phan L., Gu Z., deSilva A., Apolito C., Sherman F., Hinnebusch A.G., and Goldfarb D.S. 1998. Nip1p associates with 40S ribosomes and the Prt1p subunit of eukaryotic initiation factor 3 and is required for efficient translation initiation. *J. Biol. Chem.* **273:** 23485–23494.

Gu Z., Moerschell R.P., Sherman F., and Goldfarb D.S. 1992. *NIP1*, a gene required for nuclear transport in yeast. *Proc. Natl. Acad. Sci.* **89:** 10355–10359.

Gualerzi C.O. and Pon C.L. 1990. Initiation of mRNA translation in prokaryotes. *Biochemistry* **29:** 5881–5889.

Hanic-Joyce P.J., Singer R.A., and Johnson G.C. 1987. Molecular characterization of the yeast *PRT1* gene in which mutations affect translation initiation and regulation of cell proliferation. *J. Biol. Chem.* **262:** 2845–2851.

Hannig E.M., Cigan A.M., Freeman B.A., and Kinzy T.G. 1993. *GCD11*, a negative regulator of *GCN4* expression, encodes the γ subunit of eIF-2 in *Saccharomyces cerevisiae*. *Mol. Cell. Biol.* **13:** 506–520.

Hartwell L.H. and McLaughlin C.S. 1968. Temperature-sensitive mutants of yeast exhibiting a rapid inhibition of protein synthesis *J. Bacteriol.* **96:**1664–1671.

———. 1969. A mutant of yeast apparently defective in the initiation of protein synthesis. *Proc. Natl. Acad. Sci.* **62:** 468–474.

Hinnebusch A.G. and Fink G.R. 1983. Positive regulation in the general amino acid control of *Saccharomyces cerevisiae*. *Proc. Natl. Acad. Sci.* **80:** 5374–5378.

Huang H.-K., Yoon H., Hannig E.M., and Donahue T.F. 1997. GTP hydrolysis controls stringent selection of the AUG start codon during translation initiation in *Saccharomyces cerevisiae*. *Genes Dev.* **11:** 2396–2413.

Kasperaitis M.A., Voorma H.O., and Thomas A.A. 1995. The amino acid sequence of eukaryotic translation initiation factor 1 and its similarity to yeast initiation factor SUI1. *FEBS. Lett.* **365:** 47–50.

Kozak M. 1978. How do eukaryotic ribosomes select initiation regions in messenger RNA? *Cell* **15:** 1109–1123.

———. 1986. Point mutations define a sequence flanking the AUG initiator codon that modulates translation by eukaryotic ribosomes. *Cell* **44:** 283–292.

Kyrpides N.C. and Woese C.R. 1998. Universally conserved translation initiation factors. *Proc. Natl. Acad. Sci.* **95:** 224–228.

Linder P. and Slonimski P.P. 1989. An essential yeast protein, encoded by duplicated genes *TIF1* and *TIF2* and homologous to the mammalian translation initiation factor eIF-4A, can suppress a mitochondrial missense mutation. *Proc. Natl. Acad. Sci.* **86:** 2286–2290.

Naranda T., MacMillan S.E., Donahue T.F., and Hershey J.W.B. 1996. SUI1/p16 is required for the activity of eukaryotic translation initiation factor 3 in *Saccharomyces cerevisiae*. *Mol. Cell. Biol.* **16:** 2307–2313.

Pathak V.K., Nielson P., Trachsel H., and Hershey J.W.B. 1988. Structure of the β subunit of translational initiation factor eIF-2. *Cell* **54:** 633–639.

Pestova T.V., Borukhov S.I., and Hellen C.U.T. 1998. Eukaryotic ribosomes require initiation factors 1 and 1A to locate initiation codons. *Nature* **394:** 854–859.

Phan L., Zhang X., Asano K., Anderson J., Vornlocher H.-P., Greenberg J.R., Qin J., and Hinnebusch A.G. 1998. Identification of a translation initiation factor 3 (eIF3) core complex, conserved in yeast and mammals, that interact with eIF5. *Mol. Cell. Biol.* **18:** 4935–4946.

Pringle J.R. and Hartwell L.H. 1982. The *Saccharomyces cerevisiae* cell cycle. In *The molecular biology of the yeast* Saccharomyces: *Life cycle and inheritance* (ed. J.N. Strathern et al.), pp. 97–142. Cold Spring Harbor Laboratory, Cold Spring Harbor, New York.

Reed S.I. 1980. The selection of *S. cerevisiae* mutants defective in the start event of cell division. *Genetics* **95:** 561–577.

Sachs A.B. and Davis R.W. 1990. Translation initiation and ribosomal biogenesis:

Involvement of a putative RNA helicase and *RPL46*. *Science* **247:** 1077–1079.

Schulman L.H. and Pelka H. 1984. Recognition of tRNAs by aminoacyl-tRNA synthetases: *Escherichia coli* $tRNA^{Met}$ and *E. coli* methionyl-tRNA synthetase. *Fed. Proc.* **43:** 2977–2980.

Sherman F., Stewart J.W., and Schweingruber A.M. 1980. Mutants of yeast initiating translation of iso-1-cytochrome *c* within a region spanning 37 nucleotides. *Cell* **20:** 215–222.

Stewart J.W., Sherman F., Shipman N.A., and Schweingruber A.M. 1971. Identification and mutational relocation of the AUG codon initiating translation of iso-1-cytochrome *c* in yeast. *J. Biol. Chem.* **246:** 7429–7445.

Tarun S.Z., Jr., and Sachs A.B. 1995. A common function for mRNA 5′ and 3′ ends in translation initiation in yeast. *Genes Dev.* **9:** 2997–3007.

Wek R.C., Jackson B.M., and Hinnebusch A.G. 1989. Juxtaposition of domains homologous to protein kinases and histidyl-tRNA synthetases in GCN2 protein suggests a mechanism for coupling *GCN4* expression to amino acid availability. *Proc. Natl. Acad. Sci.* **86**: 4579–4583.

Wek S.A., Zhu S., and Wek R.C. 1995. The histidyl-tRNA synthetase-related sequence in the eIF-2α protein kinase *GCN2* interacts with tRNA and is required for activation in response to starvation for different amino acids. *Mol. Cell. Biol.* **15**: 4497–4506.

Williams N.P., Hinnebusch A.G., and Donahue T.F. 1989. Mutations in the structural genes for eukaryotic translation initiation factors 2α and 2β of *Saccharomyces cerevisiae* disrupt translational control of GCN4 mRNA. *Proc. Natl. Acad. Sci.* **86:** 7515–7519.

Wolfner M., Yep D., Messenguy F., and Fink G.R. 1975. Integration of amino acid biosynthesis into the cell cycle of *Saccharomyces cerevisiae*. *J. Mol. Biol.* **96:** 273–290.

Yoon H.J. and Donahue T.F. 1992. The *sui1* suppressor locus in *Saccharomyces cerevisiae* encodes a translation factor that functions during tRNA(iMet) recognition of the start codon. *Mol. Cell. Biol.* **12:** 248–260.

22388 32B

13

The Double-stranded RNA-activated Protein Kinase PKR

Randal J. Kaufman

Department of Biological Chemistry
Howard Hughes Medical Institute
University of Michigan Medical Center
Ann Arbor, Michigan 48109-0650

Cells respond to environmental stress by immediate changes in their capacity and specificity to select mRNAs for translation. Most translational control occurs at the level of initiation. The rate of translation initiation depends on mRNA abundance, the number of ribosomes, the availability of initiator $tRNA_i^{Met}$, and the amount and activity of eukaryotic translation initiation factors (eIFs). Reversible covalent modification of these eIFs is a fundamental mechanism that determines the rate of initiation. One modification used for translational control in response to environmental stress is the reversible phosphorylation of eIF2. Phosphorylation of heterotrimeric eIF2 on the α subunit (eIF2α) rapidly reduces the level of functional eIF2 and limits initiation events on all cellular mRNAs within the cell. In higher eukaryotes, it is not known whether some mRNAs are selectively translated when eIF2 is inactivated by phosphorylation, as demonstrated for the *Saccharomyces cerevisiae* mRNA encoding GCN4 (see Chapter 5). Reversible phosphorylation of eIF2α provides the cell with an efficient, rapid, and reversible means to respond to a variety of different stimuli.

Initiation of polypeptide chain synthesis requires the recycling of eIF2 between GTP- and GDP-bound states in a reaction catalyzed by the guanine nucleotide exchange factor, eIF2B (for review, see Pain 1996; Chapters 2 and 5). Phosphorylation of eIF2α at residue Ser-51 stabilizes the eIF2/GDP/eIF2B complex and prevents the GDP–GTP exchange reaction. Because the cellular levels of eIF2B are lower than the levels of eIF2 (Oldfield et al. 1994), the exchange process is inhibited when only a fraction (i.e., 20–30%) of eIF2α is phosphorylated and complexed with the majority of eIF2B in the cell.

© 2000 Cold Spring Harbor Laboratory Press 0-87969-568-4/00 $5 + .00

In higher eukaryotic cells, the status of eIF2α phosphorylation is controlled by at least four known Ser/Thr protein kinases that primarily utilize eIF2α as a substrate. These are the hemin-regulated inhibitor kinase (HRI) (see Chapter 14), the general control of amino amino acid biosynthesis kinase (GCN2) (see Chapter 5), the double-stranded (ds) RNA-activated protein kinase (PKR), and the pancreatic eIF2α kinase or the PKR-endoplasmic reticulum (ER)-related kinase (PEK/PERK) (see Chapter 15). PKR (previously known as p68, DAI, and P1-eIF2 kinase) signals the interferon antiviral and antiproliferative responses (Stark et al. 1998). In addition, PKR activation leads to induction of the cell death program, apoptosis (Kaufman 1999; Tan and Katze 1999). Increasing evidence supports additional roles for PKR in cell differentiation, cytokine signaling, and transcriptional regulation of dsRNA-activated genes, such as interferon β (Zinn et al. 1988; Kumar et al. 1994; Kumar et al. 1997; Wong et al. 1997; Cuddihy et al. 1999b; Gil et al. 1999; Williams 1999).

The PKR gene is expressed in the majority of cells in higher vertebrates and plants. Its transcription is induced by type 1 interferons (α and β) (Samuel et al. 1997; Stark et al. 1998). The PKR promoter contains a number of potential regulatory elements that include the interferon-stimulated response element (ISRE) (Kuhen and Samuel 1999). It appears that interferon activates PKR transcription through interferon regulatory factor 1 (IRF-1) binding to the ISRE (Kirchhoff et al. 1995; Beretta et al. 1996).

PKR STRUCTURE AND MECHANISM OF ACTIVATION

PKR mRNA is translated to yield a latent protein kinase of apparent mass 68 kD in humans. The PKR protein contains a dsRNA-binding domain (dsRBD) in its amino terminus followed by a short linker with the Ser/Thr protein kinase domain in its carboxyl terminus (Fig. 1) (Green and Mathews 1992; St Johnston et al. 1992). The dsRBD comprises two dsRNA-binding motifs (dsRBMs) of approximately 70 amino acid residues. dsRBMs are present in at least 9 families of functionally distinct dsRNA-binding proteins (Fierro-Monti and Mathews 2000). The carboxy-terminal residues represent the most conserved region of the dsRBM and are rich in basic amino acids. The first dsRBM (dsRBM1) has a better match to the consensus dsRBM, and this correlates with its higher affinity for dsRNA demonstrated by mutagenesis studies (Green and Mathews 1992; Schmedt et al. 1995). However, both dsRNA-binding motifs are required for the specific and high-affinity binding that is required for activation of PKR kinase activity (Patel and Sen 1992;

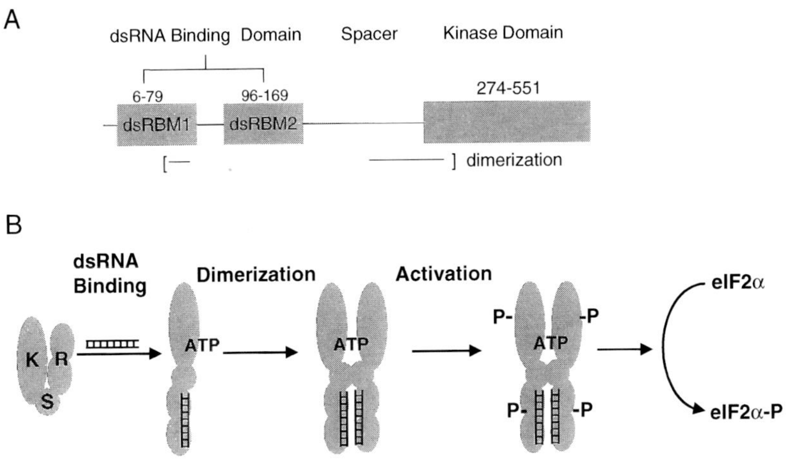

Figure 1 PKR structure and mechanism of activation. (*A*) dsRNA-binding motifs and the regions proposed for dimerization. (*B*) Mechanism of PKR activation. (R), the spacer (S), and the protein kinase domain (K). See text for details.

McCormack et al. 1994; Green et al. 1995; Romano et al. 1995; Bevilacqua and Cech 1996; Patel et al. 1996).

A variety of dsRNAs or RNA molecules with a high degree of secondary structure activate PKR (Minks et al. 1979; Bischoff and Samuel 1989; Manche et al. 1992). In addition to dsRNA and highly structured RNAs, polyanions such as heparin can activate PKR, although this activation does not require the dsRBD (Galabru and Hovanessian 1987; Patel et al. 1994). Binding of dsRNA to PKR exposes the ATP-binding site (Galabru and Hovanessian 1987; Bischoff and Samuel 1989) and induces dimerization. Dimerization subsequently stimulates *trans*-autophosphorylation (Thomis and Samuel 1993; Ortega et al. 1996). PKR autophosphorylation is essential to convert the kinase into a catalytically active form that can bind and phosphorylate eIF2α in the absence of dsRNA (Fig. 1) (Galabru and Hovanessian 1987; Galabru et al. 1989; Kostura and Mathews 1989; Wu and Kaufman 1997).

dsRNA binds to PKR in an RNA-sequence-independent manner. Viral dsRNA genomes (e.g., reovirus), replication intermediates, and mRNA transcripts with extensive secondary structure resembling dsRNA are potent activators of PKR. The activation curve for dsRNA is bimodal: Low concentrations of dsRNA activate, and high concentrations of dsRNA inhibit, activation of PKR kinase activity (Hunter et al. 1975; Kostura and Mathews 1989). Short RNA duplexes of 16 bp are capable of binding PKR, although the dsRBD occupies an 11-bp segment (Schmedt et al. 1995).

However, the binding affinity and activation increase with length up to 85 bp, and longer dsRNA have no greater effect (Manche et al. 1992; Schmedt et al. 1995; Bevilacqua and Cech 1996). Selection of dsRBD-binding RNA sequences from a large library suggested that the dsRBD can bind to RNAs with particular secondary structure, such as tandem mismatches:

$$\frac{5'\text{-AG-}3'}{3'\text{-GA-}5'},$$

as long as the RNA has an overall A-form geometry (Bevilacqua et al. 1998).

Biochemical analyses demonstrated that the dsRNA-binding domain of PKR binds dsRNA through electrostatic interactions and does not bind dsDNA or RNA–DNA (Hunter et al. 1975). This specificity is primarily due to interaction with 2′-OHs and possibly some phosphate groups within the minor groove of the dsRNA (Bevilacqua and Cech 1996). A solution structure of the 20-kD amino-terminal dsRBD was determined by multidimensional nuclear magnetic resonance (NMR) spectroscopy (Nanduri et al. 1998). The dsRBD adopts a dumbbell shape in which the two dsRBMs are separated by a flexible linker of 22 residues. Each dsRBM contains an α–β–β–β–α fold in which the two α helices are stacked against a three-stranded antiparallel β sheet. Mutations in dsRBM1 that are known to disrupt dsRNA binding (Arg-39, Phe-41, Ser-59, Lys-60, Lys-61, and Lys-64; Green et al. 1995; McMillan et al. 1995a; Patel et al. 1996) form a positively charged surface for electrostatic interaction with negatively charged phosphate backbone oxygen atoms within the RNA. The interaction with the phosphate backbone can explain the dsRNA sequence-independent interaction. The NMR results suggest that the two dsRBMs wrap around the RNA duplex, which may account for the bimodal dependence on dsRNA for activation. At low concentrations of dsRNA, dsRBM1, with its higher dsRNA-binding affinity, could facilitate dsRNA binding to the lower-affinity dsRBM2. At higher concentrations of dsRNA, the binding may not be cooperative, so that the two dsRBMs in PKR would bind to two different dsRNA molecules, thereby preventing a conformational change required for kinase activation.

Some studies suggest that dsRNA is required for efficient dimerization (Cosentino et al. 1995). However, the consensus from several laboratories using different approaches supports the view that dsRNA binding is required for PKR activation but is not required for dimerization (Patel et al. 1995; Ortega et al. 1996; Wu and Kaufman 1996; Carpick et al. 1997). PKR exists in an equilibrium between monomeric and dimeric forms (Langland and Jacobs 1992; Carpick et al. 1997). Dimer formation increases upon addition of activator dsRNA, such as the HIV-1 *trans*-activating region (TAR) RNA element (Carpick et al. 1997). At saturating concentrations of TAR RNA, an activation complex forms consisting of two molecules of PKR and one RNA molecule. Mutations that disrupt dsRNA binding did not affect dimerization (Wu and Kaufman 1996). Two dimerization motifs have been mapped. One is localized to hydrophobic residues in an amphipathic α helix between residues 60 and 75 in the first dsRBM (Patel and Sen 1998b), a region that may be influenced by dsRNA binding. The other is localized between residues 244 and 296 (Tan et al. 1998), outside the dsRBD.

Since all three tryptophan residues in PKR are located in the carboxy-terminal catalytic domain, it was possible to measure conformational changes in the catalytic domain by fluorescence quenching. Upon addition of HIV TAR RNA to intact PKR, the tryptophan fluorescence quenching increased, suggesting a conformational change in this region of the molecule (Carpick et al. 1997). Small-angle neutron scattering demonstrated that a PKR dimer undergoes an extensive conformational change to an elongated form upon binding to one molecule of HIV TAR RNA. Mutagenesis studies demonstrated that deletion of the dsRBD (residues 1–227) yields a constitutively active, apparently monomeric, eIF2α kinase (Wu and Kaufman 1997). On the basis of these observations, it was proposed that the dsRBD folds back to block the ATP-binding site in the PKR catalytic domain. Upon binding to dsRNA, the dsRBD extends out to elongate the molecule and make accessible the ATP-binding site within the catalytic domain (Fig. 1). Support for this model is also derived from the observation that the amino-terminal region of PKR can interact with the carboxy-terminal kinase domain in a yeast two-hybrid assay (Sharp et al. 1998). This type of kinase regulation is reminiscent of pp60 Src, in which the regulatory domain acts as an inhibitor of the kinase activity. The conformational change induced by dsRNA binding may also promote dimerization and subsequent *trans*-activation (Wu and Kaufman 1997).

The PKR autophosphorylation sites that are required for PKR functional activity have been characterized (Taylor et al. 1996; Romano et al. 1998b). Numerous protein kinases are activated by phosphorylation of residues between kinase subdomains VII and VIII (Johnson et al. 1996). These activating phosphorylation sites are located at the amino-terminal side of the conserved AlaProGlu (APE) motif in kinase subdomain VIII. Phosphorylation within this so-called activation loop may stimulate either (1) the binding of protein substrate or (2) the binding of ATP and the rate of phosphoryl transfer. PKR contains two threonine residues at positions 446 and 451 within this activation loop. Mass spectrometry analysis confirmed phosphorylation at Thr-446, but not at Thr-451, in PKR prepared from yeast (Romano et al. 1998b). Mutation of Thr-446 to alanine substantially reduced PKR activity, supporting a role for phosphorylation at this site for activation of PKR (Romano et al. 1998b). However, mutation at Thr-551 completely abolished activity. Therefore, it is not known whether Thr-551 is phosphorylated or whether the threonine residue is functionally important for PKR activity.

A cluster of phosphorylation sites was identified between the dsRBD and the protein kinase domain. Mutation at Thr-258 reduced the efficiency of autophosphorylation and substrate phosphorylation in vitro and partially inhibited kinase function in vivo (Taylor et al. 1996). However, mutations at two neighboring sites of autophosphorylation, Ser-242 and Thr-255, had little effect on kinase activity. The triple mutant retained significant activity, suggesting that other sites within PKR must be required for full activation. Mass spectrometry also detected phosphorylation at multiple residues from 88 and 97 between the two dsRBMs, although mutation of these residues had no effect (Romano et al. 1998b). Phosphorylation between residues 80 and 90 in the dsRBD was also detected (D. Taylor and M.B. Mathews, in prep.)

PKR was originally characterized as an inhibitory activity present in reticulocyte lysate translation reactions (Hunter et al. 1975). It is a ribosome-associated protein that is removed by washing with high-salt buffers (Levin and London 1978; Samuel et al. 1986; Langland and Jacobs 1992). Whereas PKR associated with the ribosome is monomeric, free cytosolic PKR is dimeric (Langland and Jacobs 1992). Interestingly, mammalian ribosomes inhibit PKR activation (Raine et al. 1998). At least one 60S ribosomal protein, subunit L18, was shown to bind dsRBM1 in PKR and inhibit PKR activation by dsRNA in a competitive manner (Kumar et al. 1999). However, human PKR expressed in yeast bound to the small ribosomal subunit (Zhu et al. 1997). In contrast, analysis of PKR overexpressed in COS-1 monkey cells demonstrated association with the 60S ribosomal subunit (Kumar et al. 1999). Two ribosome-binding sites were identified, one in each dsRBM, and a higher-affinity site was localized to the PKR kinase domain (Wu et al. 1998). Immunofluorescence demonstrated that PKR is localized on the surface of the rough endoplasmic reticulum (ER) as well as in the nucleolus, consistent with its ribosomal association (Schwemmle et al. 1992; Jimenez-Garcia et al. 1993; Jeffrey et al. 1995).

PKR IN THE CELLULAR ANTIVIRAL RESPONSE

Eukaryotic viruses and their hosts have coevolved complex interrelationships that permit virus reproduction without destruction of the host.

Viruses have evolved gene products that act through different mechanisms to inhibit virus-induced apoptosis and are vital for viruses to replicate and/or establish latent states. These anti-death mechanisms include modulation of Bcl-2, inactivation of the tumor suppressor p53, and inhibition of caspase (Shen and Shenk 1995; Teodoro and Branton 1997). It is recognized that PKR is also a cellular target that viruses inactivate to prevent apoptosis. dsRNA that is produced as a viral replication intermediate activates PKR to mount an antiviral response through several different mechanisms that include (1) induction of interferon gene transcription through IRF-1 and NF-κB (Chu et al. 1999), (2) inhibition of viral protein synthesis at the level of initiation, and (3) induction of apoptosis with subsequent autodigestion of the cell. These processes are discussed below. To prevent cell death and establish a productive infection, viruses must overcome the interferon-induced blockade on viral replication imposed by PKR. For those viruses that do not have effective means for inactivating PKR, pathogenesis may be attributed to virus-induced host-cell apoptosis. For example, overexpression of PKR renders cells resistant to infection with encephalomyocarditis virus (EMCV), an interferon-sensitive virus, because cells undergo rapid apoptosis (Meurs et al. 1992). In addition, increased PKR levels sensitize cells to apoptosis induced by influenza virus infection (Balachandran et al. 2000). In contrast, reduction in PKR levels converts a lytic EMCV infection into a persistent infection (Yeung et al. 1999). Therefore, the level of PKR and its activation state in the cell may predict the outcome of a viral infection. The role of PKR in viral infection is underscored by the observation that viruses, such as hepatitis C virus (HCV) (Gale et al. 1998a, 1999; Tan and Katze 1998; Taylor et al. 1999), vaccinia virus (Davies et al. 1993; Sharp et al. 1997, 1998), and possibly HIV (Gunnery et al. 1990, 1992; Brand et al. 1997), have individually evolved multiple mechanisms to inactivate the PKR pathway. These observations suggest that pharmacological intervention directed at the PKR pathway may be a productive strategy to control viral pathogenesis.

Eukaryotic viruses display diverse mechanisms to inhibit PKR function (Fig. 2) (for review, see Gale and Katze 1998; Chapters 8 and 32–35). Adenovirus (see Chapter 32) and Epstein-Barr virus encode small RNA polymerase III transcripts that bind and prevent PKR activation by dsRNA (Mathews 1993; Jacobs and Langland 1996). Hepatitis δ virus genomic RNA also interferes with PKR activation in response to dsRNA (Robertson et al. 1996). Numerous viruses encode proteins that inhibit PKR activation. The baculovirus PK2 gene product is a truncated eIF2α kinase homolog that binds and prevents PKR autophosphorylation (Dever et al. 1998). The influenza NS1 gene product interacts with the PKR

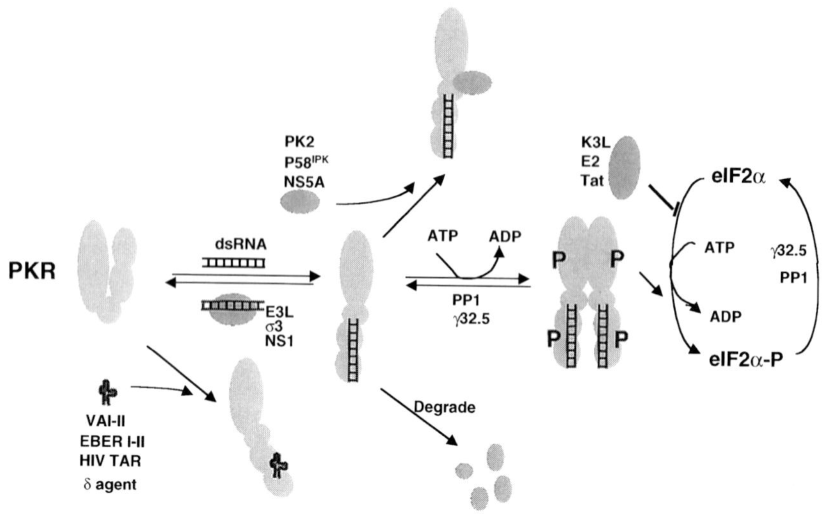

Figure 2 Viral inhibitors of PKR. The mechanism of PKR activation is depicted showing the different sites for regulation by viral gene products. See text for details.

dsRBD and also binds dsRNA to inhibit PKR activation (Tan and Katze 1998; Chapter 34). The HCV NS5A protein binds to PKR and prevents dimerization (Gale et al. 1998b). Vaccinia virus E3L (Romano et al. 1998a; Sharp et al. 1998; Chapter 35) and reovirus σ3 (Yue and Shatkin 1997; Chapter 33) are dsRNA-binding proteins that bind and sequester dsRNA to prevent PKR activation. Vaccinia virus K3L (Davies et al. 1993; Kawagishi-Kobayashi et al. 1998; Chapter 35) and HCV E2 (Taylor et al. 1999) have homology with the eIF2α phosphorylation site and act as pseudosubstrates that bind and prevent PKR interaction with eIF2α. In contrast, HIV Tat acts as a substrate that is phosphorylated by PKR and functions as a competitive inhibitor of eIF2α phosphorylation (McMillan et al. 1995b; Brand et al. 1997). Alternatively, some viruses activate cellular gene products to prevent PKR signaling. Influenza virus activates the cellular PKR inhibitor $P58^{IPK}$ to bind to PKR and prevent PKR dimerization (Tan et al. 1998). The herpes simplex virus (HSV1) protein γ34.5 activates a cellular protein phosphatase 1 (PP1) that dephosphorylates PKR and eIF2α (He et al. 1998). Poliovirus activates a protease that induces PKR degradation (Black et al. 1989). Finally, the large-T antigen of simian virus 40 (SV40) induces a bypass of the translational block imposed by eIF2α phosphorylation, although the mechanism is unknown (Swaminathan et al. 1996).

PKR IN GROWTH CONTROL AND CELL STRESS

In addition to its role in preventing translation initiation, PKR regulates cell growth and apoptosis. Overexpression of wild-type PKR is toxic to yeast, insect, and mouse cells (Chong et al. 1992; Thomis and Samuel 1992; Dever et al. 1998). Expression of a catalytically inactive mutant PKR transforms NIH-3T3 cells to form tumors in nude mice and promotes growth in HeLa cells (Koromilas et al. 1992; Meurs et al. 1993; Barber et al. 1995). This observation was ascribed to a *trans*-dominant inhibitory effect of the mutant enzyme on the endogenous wild-type PKR and implicated *PKR* as a tumor suppressor gene. In addition, overexpression of a nonphosphorylatable Ser-51Ala mutant eIF2α (Donze et al. 1995), as well as the cellular PKR inhibitors $p58^{IPK}$ (Barber et al. 1994) and HIV TAR element binding protein (TRBP) (Benkirane et al. 1997), transformed NIH-3T3 cells, similar to dominant-negative mutants of PKR. A tumor suppressor activity for PKR is consistent with a rearrangement detected in the *PKR* gene in a lymphocytic leukemia cell line (Abraham et al. 1998). However, the levels of PKR activity observed in human breast cancer cells are a source of controversy. Savinova et al. (1999) reported elevated expression of PKR in human breast cancer cells, although the actual level of PKR activity was reduced. In contrast, Kim et al. (2000) observed elevated PKR protein, PKR activity, and eIF2α phosphorylation in human breast cancer cells, suggesting a positive regulatory role for PKR in growth control. At present, there is no compelling evidence that supports a tumor suppressor function for PKR in humans.

Subsequent studies demonstrated that the growth inhibitory activity of PKR could be attributed to apoptosis when overexpressed and/or activated in mammalian cells (Lee and Esteban 1994; Takizawa et al. 1996; Der et al. 1997; Kibler et al. 1997). Transient or inducible overexpression of PKR activates apoptosis in several cell systems (Lee and Esteban 1994; Balachandran et al. 1998; Srivastava et al. 1998; Donze et al. 1999). Overexpression of Bcl-2 protects cells from PKR-induced apoptosis, suggesting that Bcl-2 is downstream from PKR (Lee et al. 1997).

PKR activity is regulated during the cell cycle (Zamanian-Daryoush et al. 1999). PKR-mediated phosphorylation of eIF2α is required for the control of cyclin D1 translation and G_1 cell cycle arrest that occur in response to PKR activation (Aktas et al. 1998). When PKR expression was induced, cells accumulated in the G_1 phase of the cell cycle and increased expression of effector molecules forming death-induced signaling complexes (DISC) such as Fas, the TNFα receptor (TNFR-1), Fas-associated protein with death domain (FADD), and caspase 8. In contrast, expression of a dominant-negative PKR mutant reduced the transcription of genes encoding molecules in DISC (Balachandran et al. 1998; Donze et al. 1999). Activated PKR may directly activate transcription factors for genes encoding pro-apoptotic functions. Alternatively, activated PKR may affect the levels of specific transcription factors through an eIF2α phosphorylation-dependent pathway, as exemplified by the *GCN2*-mediated translational control of *GCN4* mRNA (Chapter 5).

Activation of PKR is associated with differentiation in some cell culture systems. PKR activation occurs with growth arrest of murine 3T3-F442A fibroblasts subsequent to their differentiation into adipocytes (Judware and Petryshyn 1991, 1992; Petryshyn et al. 1997). In the myogenic cell line L8 there is an association between the extent of PKR activation and the level of muscle-specific protein expression (Salzberg et al. 1995). In this system, expression of a dominant-negative PKR mutant interfered with myogenesis (Salzberg et al. 2000).

The localization of PKR to the rough ER membrane suggests that it may signal in response to ER stress. Indeed, activation of PKR occurs in response to calcium depletion from the ER (Srivastava et al. 1995). In addition, expression of a dominant-negative PKR mutant inhibited eIF2α phosphorylation and protected from translational inhibition in response to ER stress induced by calcium depletion or reduction of disulfide bonds mediated by dithiothreitol (Srivastava et al. 1995; Brostrom et al. 1996). In contrast, translation and eIF2α phosphorylation were not affected upon induction of ER stress in PKR-deficient mouse embryonic fibroblasts (MEFs) (Harding et al. 1999). These results support the view that PKR does not mediate eIF2α phosphorylation in response to ER stress, and suggest that the dominant-negative mutant PKR may act to inhibit PKR alone. More recently, it was shown that either deletion of PERK/PEK or expression of a PERK/PEK dominant-negative mutant protected cells from translational inhibition induced by calcium depletion from the ER and from accumulation of unfolded protein in the ER (Harding et al. 1999; see Chapter 15), directly implicating a role for PERK/PEK in ER stress-induced phosphorylation of eIF2α.

Because of the effect of PKR on fundamental cellular processes of cell growth and apoptosis, it was surprising that deletion of the PKR gene did not have a significant phenotype or an increase in tumor incidence (Yang et al. 1995; Abraham et al. 1999). The PKR-null mouse derived by Yang et al. deleted only the amino-terminal first two exons of the PKR gene. As a consequence, this deletion has the potential to produce a protein that, on the basis of site-directed deletion mutagenesis experiments, may be expected to have residual PKR activity (Barber et al. 1995). However, Abraham et al. (1999) deleted the carboxy-terminal region in the kinase domain of PKR and obtained a similar normal growth phenotype for the mouse. In the mouse derived by Yang et al. the induction of type 1 interferon gene expression by dsRNA or by virus was not affected. However, the antiviral response induced by interferon γ and dsRNA toward EMCV was diminished (Yang et al. 1995). In contrast, there was no effect on vaccinia virus (an interferon-resistant virus) replication or on influenza virus (an interferon-sensitive virus) lethality in the presence or absence of interferon-α/β treatment (Abraham et al. 1999). These studies support the view that another gene product may be functioning in a parallel pathway to PKR.

MEFs from the two strains of PKR-deleted mice displayed a low constitutive level of eIF2α phosphorylation, indicating that at least one additional eIF2α kinase is functioning (Abraham et al. 1999). MEFs homozygous for the PKR catalytic domain deletion displayed wild-type apoptotic responses to TNFα and influenza virus infection (Abraham et al. 1999). In contrast, MEFs prepared from the homozygous amino-terminal PKR-null MEFs had an impaired apoptotic response to TNFα and dsRNA-mediated induction of type I interferon and activation of NF-κB (Yang et al. 1995). However, pretreatment with either type I or type II interferon restored the dsRNA responsiveness to one-half the level of that observed for wild-type MEFs (Yang et al. 1995). Under these conditions, interferon treatment may activate transcription of the targeted PKR gene to increase the level of the amino-terminal deleted protein. Alternatively, these results suggest that another interferon-inducible gene exists which encodes a function similar to PKR. At present, it is not known what is responsible for the different phenotypes observed between the two PKR-null mouse lines. Because these PKR-null mice were backcrossed into different strains, this may contribute to the differences observed.

eIF2α PHOSPHORYLATION IN PKR-INDUCED APOPTOSIS

To elucidate the role of eIF2α phosphorylation in PKR-induced apoptosis, studies were performed with a Ser-51Ala mutant eIF2α that is resistant to phosphorylation. Expression of this Ser-51Ala mutant eIF2α protected cells from serum-deprivation-induced, TNFα-induced, and vaccinia virus-induced apoptosis (Srivastava et al. 1998; Gil et al. 1999). Importantly, expression of a Ser-51Asp mutant of eIF2α that mimics phosphorylated serine inhibits protein synthesis and induces apoptosis in COS-1 monkey cells (Srivastava et al. 1998). Apoptosis in these cells correlates with activation of caspase 3. These studies point to a direct role for eIF2α phosphorylation in promoting apoptosis under certain conditions.

More direct evidence was recently obtained through generation of a Ser-51Ala mutation within the germ line of the mouse by knock-in technology using Cre-Lox “in and out” homologous recombination. This approach was used to introduce the Ser-51Ala mutation into the endogenous murine eIF2α gene without disrupting expression of the eIF2α gene. This mutation prevents eIF2α phosphorylation in response to all known eIF2α kinases. Homozygous Ala/Ala mutant MEFs derived from these mice were resistant to a variety of apoptotic stimuli, including treatment with type I interferon in the presence of poly rI:rC, treatment with TNFα, and serum deprivation (D. Scheuner and R.J. Kaufman, in prep.). No resistance was observed upon analysis of the heterozygous Ser/Ala MEFs. These results demonstrate that these treatments require eIF2α phosphorylation to effectively induce apoptosis.

Although it is possible that phosphorylation of eIF2α influences some process other than translation initiation to activate apoptosis, it is most reasonable to expect that inhibition of translation initiation is primarily responsible. Further studies are required to elucidate how a reduction in protein synthesis initiation can induce an apoptotic cascade. Possible mechanisms include (1) the inhibition of synthesis of an anti-apoptotic gene product, such as an inhibitor of apoptosis (IAP; Deveraux and Reed 1999); (2) the paradoxical preferential translation of mRNA(s) encoding an effector(s) of apoptosis, similar to translational control of *GCN4* (see Chapter 5); or (3) an unidentified feedback mechanism from the translational apparatus into the proapoptotic pathway.

Mice that are heterozygous for the Ser-51 Ala mutation displayed no significant phenotype. However, upon mating of the heterozygotes, homozygous Ser51Ala progeny were not detected. Close analysis revealed that the homozygous mice did develop, but died within 24 hours after birth (D. Scheuner and R.J. Kaufman, in prep.). The mouse lethality in the homozygous Ser-51 Ala mice was due to severe hypoglycemia. Surprisingly, eIF2α phosphorylation, a ubiquitous regulatory step in protein synthesis, is not required for mammalian development in utero. Further studies will elucidate how altered regulation of protein synthesis initiation may interfere with glucose metabolism.

CELLULAR REGULATORS OF PKR ACTIVATION

Although numerous cellular proteins can modulate PKR activation in vitro, there are few examples where these regulators have been unambiguously demonstrated to have physiological significance in vivo. The identity and functional significance of PKR activator(s) in the absence of a viral infection are unknown. A number of proteins that contain dsRBMs are able to interact with PKR and alter its activation (Fig. 3). PACT was isolated as a PKR-interacting protein that has three dsRBMs (Patel and Sen 1998a). RAX is the murine homolog of human PACT (Ito et al. 1999). PACT/RAX and NF90 (I. Fierro-Monti and M.B. Mathews, unpubl.) are the only known protein activators of PKR. PACT/RAX binds the dsRBD of PKR in a manner that does not require dsRNA and elicits its activation. In contrast to PACT, RAX did not directly activate PKR in vivo. However, upon treatment of cells with sodium arsenite, heat shock, or peroxide, or upon IL-3 deprivation from IL-3-dependent cell lines, RAX was rapidly phosphorylated, associated with PKR, and induced PKR autophosphorylation and activation (Ito et al. 1999). It was proposed that PACT/RAX may couple membrane stress with regulation of protein synthesis through PKR.

PKR may also be regulated through numerous cellular inhibitors (Fig. 3). P58^{IPK} is activated by influenza virus infection that binds PKR, and prevents its dimerization and activation (Tan et al. 1998; Chapter 34). It is proposed that P58^{IPK} may activate PKR in response to protein misfolding.

All known dsRNA-binding proteins have the potential to regulate PKR activation. TRBP is cellular protein that binds the HIV TAR ele-

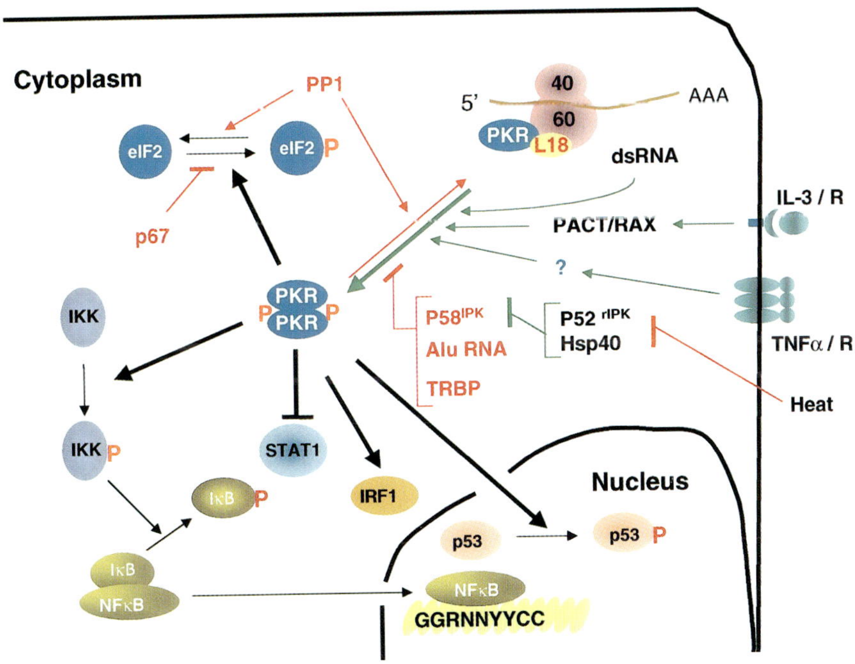

Figure 3 Cellular regulators of PKR signaling and downstream effectors. The mechanism of action for PKR cellular inhibitors is depicted in red. Pathways for PKR activation are shown in green. The major downstream effectors of PKR function are shown as bold arrows. Heat stress may signal through PACT/RAX or through p58^{IPK}.

ment, has a dsRNA-binding motif similar to PKR, and can inhibit PKR through preventing PKR dimerization (Cosentino et al. 1995). Transformation induced by overexpression of either p58^{IPK} or TRBP in NIH-3T3 cells correlated with inhibition of apoptosis, similar to dominant-negative mutants of PKR (Barber et al. 1994; Benkirane et al. 1997; Cassady et al. 1998).

p67 is an eIF2-associated 67-kD glycoprotein that protects eIF2α from phosphorylation by binding to eIF2 and providing steric interference from PKR (Wu et al. 1996). p67 transcription correlates with cell growth, and it was proposed that p67 may couple cell growth to translation via its inhibitory effect on eIF2α (Gupta et al. 1997).

As mentioned earlier, L18 is a 60S ribosomal protein subunit that was shown to inhibit PKR activity in vitro and in vivo. Although L18 protein did not directly bind dsRNA, it did compete with dsRNA for binding to the first dsRBM of PKR (Kumar et al. 1999). L18 is frequently overexpressed in colorectal cancer tissue, and this may potentiate protein synthesis in the transformed cells (Barnard et al. 1993; Kumar et al. 1999).

Alu RNAs are RNA polymerase III transcripts that are induced upon stress conditions and, at high concentrations, can inhibit PKR through binding to its dsRBD in vitro and in vivo (Chu et al. 1998). At low concentrations, Alu RNA can activate PKR (Williams 1999). Alu RNAs were proposed to facilitate tolerance to cell stress through reversing translational inhibition mediated by PKR activation (Chu et al. 1998).

Less well characterized cellular modulators of PKR activity have been identified. A 15-kD protein inhibitor of PKR (dRF) was identified that is induced upon differentiation of 3T3-F442A cells into adipocytes (Judware and Petryshyn 1991, 1992). An inhibitor was induced upon Ras transformation of BALB/C 3T3 fibroblasts (Mundschau and Faller 1992) that may also play a role in Ras-dependent inhibition of PKR which is required to permit viral protein synthesis in Ras-transformed cells (Strong et al. 1998). Future studies are required to identify the physiological significance of these inhibitors.

PKR IN SIGNAL TRANSDUCTION AND TRANSCRIPTIONAL ACTIVATION

PKR is implicated in signaling from PDGF and IL-3 (Fig. 3) (Ito et al. 1994; Mundschau and Faller 1995). IL-3 deprivation from IL-3-dependent cells results in PKR activation and a decreased rate of protein synthesis (Ito et al. 1994). Interfering with PKR expression disrupted PDGF signaling (Mundschau and Faller 1995). However, the mechanism(s) by which PKR mediates these responses is not known.

PKR can lead to activation of the transcription factors NF-κB (Kumar et al. 1994, 1997; Offermann et al. 1995; Gil et al. 1999), interferon regulatory factor 1 (IRF1) (Kumar et al. 1997), and p53 (Cuddihy et al. 1999b). These factors promote pro-inflammatory and pro-apoptotic responses. The NF-κB heterodimer is held in an inactive complex with its inhibitor IκB (for review, see Karin 1999). In response to activators, IκB is phosphorylated on serine residues 32 and 36 to induce its ubiquitin-dependent proteasome-mediated degradation, allowing NF-κB to translocate to the nucleus. Activation of PKR in cells leads to phosphorylation of IκB and activation of NF-κB (Kumar et al. 1994), possibly by directly phosphorylating the IκB kinase complex (IKK) (Zandi et al. 1997; Chu et al. 1999; Zamanian-Daryoush et al. 2000). Cells defective in PKR cannot activate NF-κB in response to dsRNA (Yang et al. 1995; Kumar et al. 1997). Cellular responses to TNFα and macrophage responses to lipopolysaccharide (LPS) are defective in PKR-null cells, suggesting these molecules signal through PKR (Kumar et al. 1994; Der et al. 1997). Although NF-κB provides primarily an anti-apoptotic function, numerous reports have correlated NF-κB activation with apoptosis (Gil et al. 1999 and references therein). NF-κB induces transcription of several death-promoting transcription factors, including p53 and cMyc, as well as other death-promoting genes, e.g., Fas, Fas ligand, IRF-1, and caspase-1.

The tumor suppressor p53 is a transcription factor that regulates the cell cycle in response to DNA damage induced by genotoxic stress. TNFα-induced apoptosis in U937 cells correlates with induction of p53 (Yeung and Lau 1998). PKR can bind and phosphorylate the carboxyl terminus of p53 at residue Ser-392, although the significance of this phosphorylation is not known (Cuddihy et al. 1999a). Both the transcriptional activity and cell cycle arrest functions of p53 are impaired in PKR-null cells (Cuddihy et al. 1999b).

PKR mediates interferon and dsRNA-signaling pathways by modulating the function of the signal transducer and activator of transcription STAT1. PKR binds to STAT1 and inhibits its transcription-activating function (Wong et al. 1997). In addition, STAT1 binding to PKR inhibits its PKR activation and ability to phosphorylate eIF2α (A.H. Wong and A.E. Koromlov, in prep.). Therefore, it appears that PKR regulates STAT1 activity and STAT1 also regulates PKR and translational control.

Although it is apparent that PKR may signal into multiple transcription pathways, the significance of PKR in regulation of these pathways needs to be characterized in vivo. To date, there have not been any physiological consequences of PKR gene deletion in the mouse other than responses to cytokines and viral infection. It will be important to carefully evaluate different responses in the PKR-deficient mice.

FUTURE DIRECTIONS

Extensive information supports a fundamental role for PKR in the interferon-induced antiviral response. The structure of PKR and its mechanism of activation by dsRNA with subsequent inhibition of protein synthesis through phosphorylation of eIF2α are well characterized. Recent results support the conclusion that PKR signaling is fundamental in responses to cytokine and cellular stress, transcriptional activation, cell growth, regulation of cell differentiation, and induction of apoptosis. Analysis of cells that express a non-phosphorylatable mutant eIF2α indicates that reduction in translation initiation may be a primary mechanism to activate apoptosis in response to a variety of stress conditions. It is not known how reduced translation initiation may activate apoptosis. Studies are required to identify whether there is preferential translation of mRNAs encoding pro-apoptotic functions or whether there is reduced translation of inhibitors of apoptosis that have short half-lives. It is puzzling that studies of PKR-null mice indicate that these cellular processes are not significantly perturbed. It is possible that PKR provides primarily a stress-response signaling pathway that is only activated upon appropriate signals, such as viral infection or heat shock. Therefore, the PKR-null mice should be carefully evaluated under a variety of different stress conditions. The importance and mechanism of PKR activation and action in cellular processes other than translation initiation need to be elucidated. Although many mechanisms concerning the role of PKR in apoptosis have been proposed, future studies are required to identify how these mechanisms are integrated together, as well as with the known mediators and executioners of cell death. Finally, the potential for pharmacological intervention in these processes makes PKR an attractive therapeutic target for drug discovery to reduce viral pathogenesis.

REFERENCES

Abraham N., Jaramillo M.L., Duncan P.I., Methot N., Icely P.L., Stojdl D.F., Barber G.N., and Bell J.C. 1998. The murine PKR tumor suppressor gene is rearranged in a lymphocytic leukemia. *Exp. Cell Res.* **244:** 394-404.

Abraham N., Stojdl D.F., Duncan P.I., Methot N., Ishii T., Dube M., Vanderhyden B.C., Atkins H.L., Gray D.A., McBurney M.W., Koromilas A.E., Brown E.G., Sonenberg N., and Bell J.C. 1999. Characterization of transgenic mice with targeted disruption of the catalytic domain of the double-stranded RNA-dependent protein kinase, PKR. *J. Biol.*

Chem. **274:** 5953–5962.

Aktas H., Fluckiger R., Acosta J.A., Savage J.M., Palakurthi S.S., and Halperin J.A. 1998. Depletion of intracellular Ca^{++} stores, phosphorylation of eIF2α, and sustained inhibition of translation initiation mediate the anticancer effects of clotrimazole. *Proc. Natl. Acad. Sci.* **95:** 8280–8285.

Balachandran S., Kim C.N., Yeh W.C., Mak T.W., Bhalla K., and Barber G.N. 1998. Activation of the dsRNA-dependent protein kinase, PKR, induces apoptosis through FADD-mediated death signaling. *EMBO J.* **17:** 6888–6902.

Balachandran S., Roberts P.C., Kipperman T., Bhalla K.N., Compans R.W., Archer D.R., and Barber G.N. 2000. Alpha/beta interferons potentiate virus-induced apoptosis through activation of the FADD/caspase-8 death signaling pathway. *J. Virol.* **74:** 1513–1523.

Barber G.N., Wambach M., Thompson S., Jagus R., and Katze M.G. 1995. Mutants of the RNA-dependent protein kinase (PKR) lacking double-stranded RNA binding domain I can act as transdominant inhibitors and induce malignant transformation. *Mol. Cell. Biol.* **15:** 3138–3146.

Barber G.N., Thompson S., Lee T.G., Strom T., Jagus R., Daveau A., and Katze M.G. 1994. The 58-kilodalton inhibitor of the interferon induced ds-RNA activated protein kinase is a tetratricopeptide repeat protein with oncogenic properties. *Proc. Natl. Acad. Sci.* **91:** 4278–4282.

Barnard G.F., Staniunas R.J., Mori M., Puder M., Jessup M.J., Steele G.D., Jr., and Chen L.B. 1993. Gastric and hepatocellular carcinomas do not overexpress the same ribosomal protein messenger RNAs as colonic carcinoma. *Cancer Res.* **53:** 4048–4052.

Benkirane M., Neuveut C., Chun R.F., Smith S.M., Samuel C.E., Gatignol A., and Jeang K.T. 1997. Oncogenic potential of TAR RNA binding protein TRBP and its regulatory interaction with RNA-dependent protein kinase PKR. *EMBO J.* **16:** 611–624.

Beretta L., Gabbay M., Berger R., Hanash S.M., and Sonenberg N. 1996. Expression of the protein kinase PKR in modulated by IRF-1 and is reduced in 5q-associated leukemias. *Oncogene* **12:** 1593–1596.

Bevilacqua P.C. and Cech T.R. 1996. Minor-groove recognition of double-stranded RNA by the double-stranded RNA-binding domain from the RNA-activated protein kinase PKR. *Biochemistry* **35:** 9983–9994.

Bevilacqua P.C., George C.X., Samuel C.E., and Cech T.R. 1998. Binding of the protein kinase PKR to RNAs with secondary structure defects: Role of the tandem A-G mismatch and noncontiguous helixes. *Biochemistry* **37:** 6303–6316.

Bischoff J.R. and Samuel C.E. 1989. Mechanism of interferon action. Activation of the human P1/eIF-2α protein kinase by individual reovirus s-class mRNAs: s1 mRNA is a potent activator relative to s4 mRNA. *Virology* **172:** 106–115.

Black T.L., Safer B., Hovanessian A., and Katze M.G. 1989. The cellular 68,000-M_r protein kinase is highly autophosphorylated and activated yet significantly degraded during poliovirus infection: Implications for translational regulation. *J. Virol.* **63:** 2244–2251.

Brand S.R., Kobayashi R., and Mathews M.B. 1997. The Tat protein of human immunodeficiency virus type 1 is a substrate and inhibitor of the interferon-induced, virally activated protein kinase, PKR. *J. Biol. Chem.* **272:** 8388–8395.

Brostrom C.O., Prostko C.R., Kaufman R.J., and Brostrom M.A. 1996. Inhibition of translational initiation by activators of the glucose-regulated stress protein and heat shock protein stress response systems. Role of the interferon-inducible double-stranded

RNA-activated eukaryotic initiation factor 2α kinase. *J. Biol. Chem.* **271:** 24995–25002.

Carpick B.W., Graziano V., Schneider D., Maitra R.K., Lee X., and Williams B.R.G. 1997. Characterization of the solution complex between the interferon-induced, double-stranded RNA-activated protein kinase and HIV-I *trans*-activating region RNA. *J. Biol. Chem.* **272:** 9510–9516.

Cassady K.A., Gross M., and Roizman B. 1998. The herpes simplex virus US11 protein effectively compensates for the γ1(34.5) gene if present before activation of protein kinase R by precluding its phosphorylation and that of the α subunit of eukaryotic translation initiation factor 2. *J. Virol.* **72:** 8620–8626.

Chong K.L., Feng L., Schappert K., Meurs E., Donahue T.F., Friesen J.D., Hovanessian A.G., and Williams B.R.G. 1992. Human p68 kinase exhibits growth suppression in yeast and homology to the translational regulator GCN2. *EMBO J.* **11:** 1553–1562.

Chu W.M., Ballard R., Carpick B.W., Williams B.R., and Schmid C.W. 1998. Potential Alu function: Regulation of the activity of double-stranded RNA-activated kinase PKR. *Mol. Cell. Biol.* **18:** 58–68.

Chu W.M., Ostertag D., Li Z.W., Chang L., Chen Y., Hu Y., Williams B., Perrault J., and Karin M. 1999. JNK2 and IKKβ are required for activating the response to viral infection. *Immunity* **11:** 721–731.

Cosentino G.P., Venkatesan S., Serluca F.C., Green S.R., Mathews M.B., and Sonenberg N. 1995. Double-stranded-RNA-dependent protein kinase and TAR RNA-binding protein form homo- and heterodimers in vivo. *Proc. Natl. Acad. Sci.* **92:** 9445–9449.

Cuddihy A.R., Wong A.H., Tam N.W., Li S., and Koromilas A.E. 1999b. The double-stranded RNA activated protein kinase PKR physically associates with the tumor suppressor p53 protein and phosphorylates human p53 on serine 392 in vitro. *Oncogene* **18:** 2690–2702.

Cuddihy A.R., Li S., Tam N.W., Wong A.H., Taya Y., Abraham N., Bell J.C., and Koromilas A.E. 1999b. Double-stranded-RNA-activated protein kinase PKR enhances transcriptional activation by tumor suppressor p53. *Mol. Cell. Biol.* **19:** 2475–2484.

Davies M.V., Chang H.-W., Jacobs B.L., and Kaufman R.J. 1993. The E3L and K3L vaccinia virus gene products stimulate translation through inhibition of the double-stranded RNA-activated protein kinase by different mechanisms. *J. Virol.* **67:** 1688–1692.

Der S.D., Yang Y.L., Weissmann C., and Williams B.R. 1997. A double-stranded RNA-activated protein kinase-dependent pathway mediating stress-induced apoptosis. *Proc. Natl. Acad. Sci.* **94:** 3279–3283.

Dever T.E., Sripriya R., McLachlin J.R., Lu J., Fabian J.R., Kimball S.R., and Miller L.K. 1998. Disruption of cellular translational control by a viral truncated eukaryotic translation initiation factor 2α kinase homolog. *Proc. Natl. Acad. Sci.* **95:** 4164–4169.

Deveraux Q.L. and Reed J.C. 1999. IAP family proteins-suppressors of apoptosis. *Genes Dev.* **13:** 239–252.

Donze O., Dostie J., and Sonenberg N. 1999. Regulatable expression of the interferon-induced double-stranded RNA dependent protein kinase PKR induces apoptosis and fas receptor expression. *Virology* **256:** 322–329.

Donze O., Jagus R., Koromilas A.E., Hershey J.W., and Sonenberg N. 1995. Abrogation of translation initiation factor eIF-2 phosphorylation causes malignant transformation of NIH 3T3 cells. *EMBO J.* **14:** 3828–3834.

Fierro-Monti I. and Mathews M.B. 2000. Proteins binding to duplexed RNA: One motif, multiple functions. *Trends Biochem. Sci.* **25:** 241–246.

Galabru J. and Hovanessian A. 1987. Autophosphorylation of the protein kinase dependent on double-stranded RNA. *J. Biol. Chem.* **262:** 15538–15544.

Galabru J., Katze M.G., Robert N., and Hovanessian A.G. 1989. The binding of double-stranded RNA and adenovirus VAI RNA to the interferon-induced protein kinase. *Eur. J. Biochem.* **178:** 581–589.

Gale M., Jr. and Katze M.G. 1998. Molecular mechanisms of interferon resistance mediated by viral-directed inhibition of PKR, the interferon-induced protein kinase. *Pharmacol. Ther.* **78:** 29–46.

Gale M., Jr., Kwieciszewski B., Dossett M., Nakao H., and Katze M.G. 1999. Antiapoptotic and oncogenic potentials of hepatitis C virus are linked to interferon resistance by viral repression of the PKR protein kinase. *J. Virol.* **73:** 6506–6516.

Gale M.J., Jr., Blakely C.M., Hopkins D.A., Melville M.W., Wambach M., Romano P.R., and Katze M.G. 1998a. Regulation of the interferon-induced protein kinase, PKR: Modulation of P58^{IPK}, the cellular inhibitor of PKR. *Mol. Cell. Biol.* **18:** 859–871.

Gale M., Jr., Blakely C.M., Kwieciszewski B., Tan S.L., Dossett M., Tang N.M., Korth M.J., Polyak S.J., Gretch D.R., and Katze M.G. 1998b. Control of PKR protein kinase by hepatitis C virus nonstructural 5A protein: Molecular mechanisms of kinase regulation. *Mol. Cell. Biol.* **18:** 5208–5218.

Gil J., Alcami J., and Esteban M. 1999. Induction of apoptosis by double-stranded-RNA-dependent protein kinase (PKR) involves the α subunit of eukaryotic translation initiation factor 2 and NF-κB. *Mol. Cell. Biol.* **19:** 4653–4663.

Green S.R. and Mathews M.B. 1992. Two RNA binding mofits in the double-stranded RNA activated protein kinase, DAI. *Genes Dev.* **6:** 2478–2490.

Green S.R., Manche L., and Mathews M.B. 1995. Two functionally distinct RNA-binding motifs in the regulatory domain of the protein kinase DAI. *Mol. Cell. Biol.* **15:** 358–364.

Gunnery S., Green S.R., and Mathews M.B. 1992. Tat-responsive region RNA of human immunodeficiency virus type 1 stimulates protein synthesis *in vivo* and *in vitro:* Relationship between structure and function. *Proc. Natl. Acad. Sci.* **89:** 11557–11561.

Gunnery S., Rice A.P., Robertson H.D., and Mathews M.B. 1990. Tat-responsive region RNA of human immunodeficiency virus 1 can prevent activation of the double-stranded-RNA-activated protein kinase. *Proc. Natl. Acad. Sci.* **87:** 8687–8691.

Gupta S., Bose A., Chatterjee N., Saha D., Wu S., and Gupta N.K. 1997. p67 transcription regulates translation in serum-starved and mitogen-activated KRC-7 cells. *J. Biol. Chem.* **272:** 12699–12704.

Harding H.P., Zhang Y., and Ron D. 1999. Protein translation and folding are coupled by an endoplasmic-reticulum-resident kinase. *Nature* **397:** 271–274.

He B., Gross M., and Roizman B. 1998. The γ 134.5 protein of herpes simplex virus 1 has the structural and functional attributes of a protein phosphatase 1 regulatory subunit and is present in a high molecular weight complex with the enzyme in infected cells. *J. Biol. Chem.* **273:** 20737–20743.

Hunter T., Hunt T., Jackson R.J., and Robertson H.D. 1975. The characteristics of inhibition of protein synthesis by double-stranded ribonucleic acid in reticulocyte lysates. *J. Biol. Chem.* **250:** 409–417.

Ito T., Jagus R.M., and May W.S. 1994. Interleukin-3 stimulates protein synthesis by regulating ds RNA-dependent protein kinase (PKR). *Proc. Natl. Acad. Sci.* **91:** 7455–7459.

Ito T., Yang M., and May W.S. 1999. RAX, a cellular activator for double-stranded RNA-dependent protein kinase during stress signaling. *J. Biol. Chem.* **274:** 15427–15432.

Jacobs B.L. and Langland J.O. 1996. When two strands are better than one: The mediators and modulators of the cellular responses to double-stranded RNA. *Virology* **219:** 339–349.

Jeffrey I.W., Kadereit S., Meurs E.F., Metzger T., Bachmann M., Schwemmle M., Hovanessian A.G., and Clemens M.J. 1995. Nuclear localization of the interferon-inducible protein kinase PKR in human cells and transfected mouse cells. *Exp. Cell Res.* **218:** 17–27.

Jimenez-Garcia L.F., Green S.R., Mathews M.B., and Spector D.L. 1993. Organization of the double-stranded RNA-activated protein kinase DAI and virus-associated VA RNA1 in adenovirus-2-infected HeLa cells. *J. Cell Sci.* **106:** 11–22.

Johnson L.N., Noble M.E., and Owen D.J. 1996. Active and inactive protein kinases: Structural basis for regulation. *Cell* **85:** 149–158.

Judware R. and Petryshyn R. 1991. Partial characterization of a cellular factor that regulates the double-stranded RNA-dependent eIF-2 alpha kinase in 3T3-F442A fibroblasts. *Mol. Cell. Biol.* **11:** 3259–3267.

———. 1992. Mechanism of action of a cellular inhibitor of the dsRNA-dependent protein kinase from 3T3-F442A cells. *J. Biol. Chem.* **267:** 21685–21690.

Karin M. 1999. How NF-κB is activated: The role of IκB kinase (IKK). *Oncogene* **18:** 6867–6874.

Kaufman R.J. 1999. Double-stranded RNA-activated protein kinase mediates virus-induced apoptosis: A new role for an old actor. *Proc. Natl. Acad. Sci.* **96:** 11693–11695.

Kawagishi-Kobayashi M., Silverman J.B., Ung T.L., and Dever T.E. 1997. Regulation of the protein kinase PKR by the vaccinia virus pseudosubstrate inhibitor K3L is dependent on residues conserved between the K3L protein and the PKR substrate eIF2α. *Mol. Cell. Biol.* **17:** 4146–4158.

Kibler K.V., Shors T., Perkins K.B., Zeman C.C., Banaszak M.P., Biesterfeldt J., Langland J.O., and Jacobs B.L. 1997. Double-stranded RNA is a trigger for apoptosis in vaccinia virus-infected cells. *J. Virol.* **71:** 1992–2003.

Kim S.H., Forman A.P., Mathews M.B., and Gunnery S. 2000. Human breast cancer cells contain elevated levels and activity of the protein kinase, PKR. *Oncogene* (in press).

Kirchhoff S., Koromilas A.E., Schaper F., Grashoff M., Sonenberg N., and Hauser H. 1995. IRF-1 induced cell growth inhibition and interferon induction requires the activity of the protein kinase PKR. *Oncogene* **11:** 439–445.

Koromilas A.E., Roy S., Barber G.N., Katze M.G., and Sonenberg N. 1992. Malignant transformation by a mutant of the interferon-inducible double-stranded RNA dependent protein-kinase. *Science* **257:** 1685–1689.

Kostura M. and Mathews M.B. 1989. Purification and activation of the double-stranded RNA-dependent eIF-2 kinase DAI. *Mol. Cell. Biol.* **9:** 1576–1586.

Kuhen K.L. and Samuel C.E. 1999. Mechanism of interferon action: Functional characterization of positive and negative regulatory domains that modulate transcriptional activation of the human RNA-dependent protein kinase PKR promoter. *Virology* **254:** 182–195.

Kumar A., Hague J., Lacoste J., Hiscott J., and Williams B.R.G. 1994. Double-stranded RNA-dependent protein kinase activates transcription factor NF-κB by phosphorylating I-κB. *Proc. Natl. Acad. Sci.* **91:** 6288–6292.

Kumar A., Yang Y.L., Flati V., Der S., Kadereit S., Deb A., Haque J., Reis L., Weissmann

C., and Williams B.R. 1997. Deficient cytokine signaling in mouse embryo fibroblasts with a targeted deletion in the PKR gene: Role of IRF-1 and NF-κB. *EMBO J.* **16:** 406–416.

Kumar K.U., Srivastava S.P., and Kaufman R.J. 1999. Double-stranded RNA-activated protein kinase (PKR) is negatively regulated by 60S ribosomal subunit protein L18. *Mol. Cell. Biol.* **19:** 1116–1125.

Langland J.O. and Jacobs B.L. 1992. Cytosolic double-stranded RNA-dependent protein kinase is likely a dimer of partially phosphorylated M_r = 66,000 subunits. *J. Biol. Chem.* **267:** 10729–10736.

Lee S.B. and Esteban M. 1994. The interferon-induced double-stranded RNA-activated protein kinase induces apoptosis. *Virology* **199:** 491–496.

Lee S.B., Rodriguez D., Rodriguez J.R., and Esteban M. 1997. The apoptosis pathway triggered by the interferon-induced protein kinase PKR requires the third basic domain, initiates upstream of Bcl-2, and involves ICE-like proteases. *Virology* **231:** 81–88.

Levin D. and London I.M. 1978. Regulation of protein synthesis: Activation by double-stranded RNA of a protein kinase that phosphorylates eukaryotic initiation factor 2. *Proc. Natl. Acad. Sci.* **75:** 1121–1125.

Manche L., Green S.R., Schmedt C., and Mathews M.B. 1992. Interactions between double-stranded RNA regulators and the protein kinase DAI. *Mol. Cell. Biol.* **12:** 5238–5248.

Mathews M.B. 1993. Viral evasion of cellular defense mechanisms: Regulation of the protein kinase DAI by RNA effectors. *Semin. Virol.* **4:** 247–257.

McCormack S.J., Ortega L.G., Doohan J.P., and Samuel C.E. 1994. Mechanism of interferon action motif I of the interferon-induced, RNA-dependent protein kinase (PKR) is sufficient to mediate RNA-binding activity. *Virology* **198:** 92–99.

McMillan N.A., Carpick B.W., Hollis B., Toone W.M., Zamanian-Daryoush M., and Williams B.R. 1995a. Mutational analysis of the double-stranded RNA (dsRNA) binding domain of the dsRNA-activated protein kinase, PKR. *J. Biol. Chem.* **270:** 2601–2606.

McMillan N.A.J., Chun R.F., Sideovski D.P., Galabru J., Toone W.M., Samuel C.E., Mak T.W., Hovanessian A.G., Jeang K.T., and Williams B.R.G. 1995b. HIV-1 Tat directly interacts with the inteferon-induced double-stranded RNA-dependent kinase PKR. *Virology* **213:** 413–424.

Meurs E.F., Galabru J., Barber G.N., Katze M.G., and Hovanessian A.G. 1993. Tumor suppressor function of the interferon-induced double-stranded RNA-activated protein kinase. *Proc. Natl. Acad. Sci.* **90:** 232–236.

Meurs E.F., Watanabe Y., Kadereit S., Barber G.N., Katze M.G., Chong K., Williams B.R., and Hovanessian A.G. 1992. Constitutive expression of human double-stranded RNA-activated p68 kinase in murine cells mediates phosphorylation of eukaryotic initiation factor 2 and partial resistance to encephalomyocarditis virus growth. *J. Virol.* **66:** 5804–5814.

Minks M.A., Benvin S., and Baglioni C. 1979. Structural requirements of double-stranded RNA for the activation of 2′,5′-oligo(A) polymerase and protein kinase of interferon-treated HeLa cells. *J. Biol. Chem.* **254:** 10180–10183.

Mundschau L.J. and Faller D.V. 1992. Oncogenic Ras induces an inhibitor of ds-RNA dependent eIF2α kinase activation. *J. Biol. Chem.* **267:** 23092–23098.

———. 1995. Platelet-derived growth factor signal transduction through the interferon-inducible kinase PKR. *J. Biol. Chem.* **270:** 3100–3106.

Nanduri S., Carpick B.W., Yang Y., Williams B.R., and Qin J. 1998. Structure of the double-stranded RNA-binding domain of the protein kinase PKR reveals the molecular basis of its dsRNA-mediated activation. *EMBO J.* **17:** 5458–5465.

Oldfield S., Jones B.L., Tanton D., and Proud C.G. 1994. Use of monoclonal antibodies to study the structure and function of eukaryotic protein synthesis initiation factor eIF2B. *Eur. J. Biochem.* **221:** 399–410.

Offermann M.K., Zimring J., Mellits K.H., Hagan M.K., Shaw R., Medford R.M., Mathews M.B., Goodbourn S., and Jagus R. 1995. Activation of the double-stranded-RNA-activated protein kinase and induction of vascular cell adhesion molecule-1 by poly (I).poly (C) in endothelial cells. *Eur. J. Biochem.* **23:** 28–36.

Ortega L.G., McCotter M.D., Henry G.L., McCormack S.J., Thomis D.C., and Samuel C.E. 1996. Mechanisms of interferon action: Biochemical and genetic evidence for the intermolecular association of the RNA-dependent protein kinase PKR from human cells. *Virology* **215:** 31–39.

Pain V.M. 1996. Initiation of protein synthesis in eukaryotic cells. *Eur. J. Biochem.* **236:** 747–771.

Patel R.C. and Sen G.C. 1992. Identification of the double-stranded RNA-binding domain of the human interferon-inducible protein kinase. *J. Biol. Chem.* **267:** 7671–7676.

———. 1998a. PACT, a protein activator of the interferon-induced protein kinase, PKR. *EMBO J.* **17:** 4379–4390.

———. 1998b. Requirement of PKR dimerization mediated by specific hydrophobic residues for its activation by double-stranded RNA and its antigrowth effects in yeast. *Mol. Cell. Biol.* **18:** 7009–7019.

Patel R.C., Stanton P., and Sen G.C. 1994. Role of the amino-terminal residues of the interferon-induced protein kinase in its activation by double-stranded RNA and heparin. *J. Biol. Chem.* **269:** 18593–18598.

———. 1996. Specific mutations near the amino terminus of double-stranded RNA-dependent protein kinase (PKR) differentially affect its double-stranded RNA binding and dimerization properties. *J. Biol. Chem.* **271:** 25657–25663.

Patel R.C., Stanton P., McMillan N.M., Williams B.R., and Sen G.C. 1995. The interferon-inducible double-stranded RNA-activated protein kinase self-associates in vitro and in vivo. *Proc. Natl. Acad. Sci.* **92:** 8283–8287.

Petryshyn R.A., Ferrenz A.G., and Li J. 1997. Characterization and mapping of the double-stranded regions involved in activation of PKR within a cellular RNA from 3T3-F442A cells. *Nucleic Acids Res.* **25:** 2672–2678.

Raine D.A., Jeffrey I.W., and Clemens M.J. 1998. Inhibition of the double-stranded RNA-dependent protein kinase PKR by mammalian ribosomes. *FEBS Lett.* **436:** 343–348.

Robertson H.D., Manche L., and Mathews M.B. 1996. Paradoxical interactions between human delta hepatitis agent RNA and the cellular protein kinase PKR. *J. Virol.* **70:** 5611–5617.

Romano P.R., Green S.R., Barber G.N., Mathews M.B., and Hinnebusch A.G. 1995. Structural requirements for double-stranded RNA binding, dimerization, and activation of the human eIF-2 alpha kinase DAI in *Saccharomyces cerevisiae*. *Mol. Cell. Biol.* **15:** 365–378.

Romano P.R., Zhang F., Tan S.L., Garcia-Barrio M.T., Katze M.G., Dever T.E., and Hinnebusch A.G. 1998a. Inhibition of double-stranded RNA-dependent protein kinase PKR by vaccinia virus E3: Role of complex formation and the E3 N-terminal domain. *Mol. Cell. Biol.* **18:** 7304–7316.

Romano P.R., Garcia-Barrio M.T., Zhang X., Wang Q., Taylor D.R., Zhang F., Herring C., Mathews M.B., Qin J., and Hinnebusch A.G. 1998b. Autophosphorylation in the activation loop is required for full kinase activity in vivo of human and yeast eukaryotic initiation factor 2alpha kinases PKR and GCN2. *Mol. Cell. Biol.* **18:** 2282–2297.

Salzberg S., Mandelboim M., Zalcberg M., Shainberg A., and Mandelbaum M.M. 1995. Interruption of myogenesis by transforming growth factor beta 1 or EGTA inhibits expression and activity of the myogenic-associated (2′-5′) oligoadenylate synthetase and PKR. *Exp. Cell Res.* **219:** 223–232.

Salzberg S., Vilchik S., Cohen S., Heller A., and Kronfeld-Kinar Y. 2000. Expression of a PKR dominant-negative mutant in myogenic cells interferes with the myogenic process. *Exp. Cell Res.* **254:** 45–54.

Samuel C.E., Knutson G.S., Berry M.J., Atwater J.A., and Lasky S.R. 1986. Purification of double-stranded RNA-dependent protein kinase from mouse fibroblasts. *Methods Enzymol.* **119:** 499–516.

Samuel C.E., Kuhen K.L., George C.X., Ortega L.G., Rende-Fournier R., and Tanaka H. 1997. The PKR protein kinase — An interferon-inducible regulator of cell growth and differentiation. *Int. J. Hematol.* **65:** 227–237.

Savinova O., Joshi B., and Jagus R. 1999. Abnormal levels and minimal activity of the dsRNA-activated protein kinase, PKR, in breast carcinoma cells. *Int. J. Biochem. Cell Biol.* **31:** 175–189.

Schmedt C., Green S.R., Manche L., Taylor D.R., Ma Y., and Mathews M.B. 1995. Functional characterization of the RNA-binding domain and motif of the double-stranded RNA-dependent protein kinase DAI (PKR). *J. Mol. Biol.* **249:** 29–44.

Schwemmle M., Clemens M.J., Hilse K., Pfeifer K., Troster H., Muller W.E., and Bachmann M. 1992. Localization of Epstein-Barr virus-encoded RNAs EBER-1 and EBER-2 in interphase and mitotic Burkitt lymphoma cells. *Proc. Natl. Acad. Sci.* **89:** 10292–10296.

Sharp T.V., Witzel J.E., and Jagus R. 1997. Homologous regions of the alpha subunit of eukaryotic translational initiation factor 2 (eIF2α) and the vaccinia virus K3L gene product interact with the same domain within the dsRNA-activated protein kinase (PKR). *Eur. J. Biochem.* **250:** 85–91.

Sharp T.V., Moonan F., Romashko A., Joshi B., Barber G.N., and Jagus R. 1998. The vaccinia virus E3L gene product interacts with both the regulatory and the substrate binding regions of PKR: Implications for PKR autoregulation. *Virology* **250:** 302–315.

Shen Y. and Shenk T.E. 1995. Viruses and apoptosis. *Curr. Opin. Genet. Dev.* **5:** 105–111.

Srivastava S.P., Davies M.V., and Kaufman R.J. 1995. Calcium depletion from the endoplasmic reticulum activates the double-stranded RNA-dependent protein kinase (PKR) to inhibit protein synthesis. *J. Biol. Chem.* **270:** 16619–16624.

Srivastava S.P., Kumar K.U., and Kaufman R.J. 1998. Phosphorylation of eukaryotic translation initiation factor 2 mediates apoptosis in response to activation of the double-stranded RNA-dependent protein kinase. *J. Biol. Chem.* **273:** 2416–2423.

Stark G.R., Kerr I.M., Williams B.R., Silverman R.H., and Schreiber R.D. 1998. How cells respond to interferons. *Annu. Rev. Biochem.* **67:** 227–264.

St Johnston D., Brown N.H., Gall J.G., and Jantsch M. 1992. A conserved double-stranded RNA-binding domain. *Proc. Natl. Acad. Sci.* **89:** 10979–10983.

Strong J.E., Coffey M.C., Tang D., Sabinin P., and Lee P.W. 1998. The molecular basis of viral oncolysis: Usurpation of the Ras signaling pathway by reovirus. *EMBO J.* **17:** 3351–3362.

Swaminathan S., Rajan P., Savinova O., Jagus R., and Thimmapaya B. 1996. Simian virus 40 large-T bypasses the translational block imposed by the phosphorylation of eIF-2α. *Virology* **219:** 321–323.

Takizawa T., Ohashi K., and Nakanishi Y. 1996. Possible involvement of double-stranded RNA-activated protein kinase in cell death by influenza virus infection. *J. Virol.* **70:** 8128–8132.

Tan S.L. and Katze M.G. 1998. Biochemical and genetic evidence for complex formation between the influenza A virus NS1 protein and the interferon-induced PKR protein kinase. *J. Interferon Cytokine Res.* **18:** 757–766.

———. 1999. The emerging role of the interferon-induced PKR protein kinase as an apoptotic effector: A new face of death? *J. Interferon Cytokine Res.* **19:** 543–554.

Tan S.L., Gale M.J., Jr., and Katze M.G. 1998. Double-stranded RNA-independent dimerization of interferon-induced protein kinase PKR and inhibition of dimerization by the cellular P58IPK inhibitor. *Mol. Cell. Biol.* **18:** 2431–2443.

Taylor D.R., Shi S.T., Romano P.R., Barber G.N., and Lai M.M. 1999. Inhibition of the interferon-inducible protein kinase PKR by HCV E2 protein. *Science* **285:** 107–110.

Taylor D.R., Lee S.B., Romano P.R., Marshak D.R., Hinnebusch A.G., Esteban M., and Mathews M.B. 1996. Autophosphorylation sites participate in the activation of the double-stranded-RNA-activated protein kinase PKR. *Mol. Cell. Biol.* **16:** 6295–6302.

Teodoro J.G. and Branton P.E. 1997. Regulation of apoptosis by viral gene products. *J. Virol.* **71:** 1739–1746.

Thomis D.C. and Samuel C.E. 1992. Mechanism of interferon action: Autoregulation of RNA-dependent P1/eIF-2 alpha protein kinase (PKR) expression in transfected mammalian cells. *Proc. Natl. Acad. Sci.* **89:** 10837–10841.

———. 1993. Mechanism of interferon action: Evidence for intermolecular autophosphorylation and autoactivation of the interferon-induced, RNA-dependent protein kinase PKR. *J. Virol.* **67:** 7695–7700.

Williams B.R. 1999. PKR; a sentinel kinase for cellular stress. *Oncogene* **18:** 6112–6120.

Wong A.H., Tam N.W., Yang Y.L., Cuddihy A.R., Li S., Kirchhoff S., Hauser H., Decker T., and Koromilas A.E. 1997. Physical association between STAT1 and the interferon-inducible protein kinase PKR and implications for interferon and double-stranded RNA signaling pathways. *EMBO J.* **16:** 1291–1304.

Wu S. and Kaufman R.J. 1996. Double-stranded RNA binding is required and dimerization is not sufficient for the activation of the dsRNA-dependent protein kinase (PKR). *J. Biol. Chem.* **271:** 1756–1763.

———. 1997. A model for the double-stranded RNA (dsRNA)-dependent dimerization and activation of the dsRNA-activated protein kinase PKR. *J. Biol. Chem.* **272:** 1291–1296.

Wu S., Kumar K.U., and Kaufman R.J. 1998. Identification and requirement of three ribosome binding domains in dsRNA-dependent protein kinase (PKR). *Biochemistry* **37:** 13816–13826.

Wu S., Rehemtulla A., Gupta N.K., and Kaufman R.J. 1996. An eIF2-associated 67 kDa glycoprotein partially reverses protein synthesis inhibition by activated PKR in intact cells. *Biochemistry* **35:** 8275–8280.

Yang Y.-L., Reis L.F., Pavlovic J., Aguzzi A., Schäfer R., Kumar A., Williams B.R.G., Aguet M., and Weissmann C. 1995. Deficient signaling in mice devoid of double-stranded RNA-dependent protein kinase. *EMBO J.* **14:** 6095–6106.

Yeung M.C. and Lau A.S. 1998. Tumor suppressor p53 as a component of the tumor necrosis factor-induced, protein kinase PKR-mediated apoptotic pathway in human promonocytic U937 cells. *J. Biol. Chem.* **273:** 25198–25202.

Yeung M.C., Chang D.L., Camantigue R.E., and Lau A.S. 1999. Inhibitory role of the host apoptogenic gene PKR in the establishment of persistent infection by encephalomyocarditis virus in U937 cells. *Proc. Natl. Acad. Sci.* **96:** 11860–11865

Yue Z. and Shatkin A.J. 1997. Double-stranded RNA-dependent protein kinase (PKR) is regulated by reovirus structural proteins. *Virology* **234:** 364–371.

Zamanian-Daryoush M., Der S.D., and Williams B.R. 1999. Cell cycle regulation of the double stranded RNA activated protein kinase, PKR. *Oncogene* **18:** 315–326.

Zamanian-Daryoush M., Mogensen T.H., DiDonato J.A., Williams B.R. 2000. NF-κB activation by double-stranded-RNA-activated protein kinase (PKR) is mediated through NF-κB-inducing kinase and IκB kinase. *Mol. Cell. Biol.* **20:** 1278–1290.

Zandi E., Rothwarf D.M., Delhase M., Hayakawa M., and Karin M. 1997. The IκB kinase complex (IKK) contains two kinase subunits, IKKα and IKKβ, necessary for IκB phosphorylation and NF-κB activation. *Cell* **91:** 243–252.

Zhu S., Romano P.R., and Wek R.C. 1997. Ribosome targeting of PKR is mediated by two double-stranded RNA-binding domains and facilitates in vivo phosphorylation of eukaryotic initiation factor-2. *J. Biol. Chem.* **272:** 14434–14441.

Zinn K., Keller A., Whittemore L.A., and Maniatis T. 1988. 2-Aminopurine selectively inhibits the induction of beta-interferon, c-*fos*, and c-*myc* gene expression. *Science* **240:** 210–213.

14
Heme-regulated eIF2α Kinase

Jane-Jane Chen
Harvard-MIT Division of Health Sciences and Technology
Massachusetts Institute of Technology
Cambridge, Massachusettes 02139

Iron-deficiency anemia is one of the most prevalent human diseases and is characterized by hypochromic microcytic erythrocyctes (for review, see Fairbanks and Beutler 1983). This observation led to studies in the 1950s, which demonstrated that inorganic iron stimulates the synthesis of hemoglobin in immature erythroid cells (Kruh and Borsook 1956; Kassenaar et al. 1957; Morell et al. 1958). Further studies showed that heme, not iron per se, is required for protein synthesis in reticulocyte lysates, since the iron-chelating agent, desferrioxamine, does not block the stimulatory effect of heme (Bruns and London 1965; Grayzel et al. 1966). During the last step of heme biosynthesis, iron is incorporated into protoporphyrin by ferrochelatase to form heme. Consequently, iron deficiency leads to heme deficiency. Heme-iron accounts for the majority of the iron in the human body, and hemoglobin (the most abundant hemoprotein) contains as much as 70% of the total iron content of a normal adult. Thus, iron and heme play very important roles in hemoglobin synthesis and erythroid cell differentiation. In reticulocytes, iron deficiency causes inhibition of protein synthesis with disaggregation of polysomes, which can be prevented by the addition of iron or hemin (Waxman and Rabinovitz 1965, 1966; Grayzel et al. 1966). This shutoff of protein synthesis is a result of the activation of the heme-regulated inhibitor (HRI), which is a heme-regulated eIF2α kinase (for review, see Chen 1993; Chen and London 1995).

HRI belongs to the family of eIF2α kinases. This family also includes the double-stranded RNA-dependent eIF2α kinase (PKR) (Chapter 13), the GCN2 protein kinase (Chapter 5), and the recently identified mammalian ER resident kinase, PERK, which is identical to the enzyme PEK that is highly expressed in pancreas (Chapter 15). The eIF2α kinases share extensive homology in the kinase catalytic domains (Meurs et al. 1990; Chen et al. 1991a; Ramirez et al. 1991; Chong et al. 1992; Berlanga

Translational Control of Gene Expression
© 2000 Cold Spring Harbor Laboratory Press 0-87969-568-4/00 $5 +. 00

et al. 1998; Shi et al. 1998; Harding et al. 1999) and phosphorylate eIF2α at residue Ser-51 (Colthurst et al. 1987; Dever et al. 1993; Shi et al. 1998; Harding et al. 1999). Phosphorylation of eIF2α results in the inactivation of eIF2B, the factor that is required for the recycling of eIF2, and thus inhibits protein synthesis (for review, see Chapter 5). However, the regulatory domains and the regulatory mechanisms of each of these eIF2α kinases are very different. HRI activity is down-regulated by heme. Regulation of other eIF2α kinases is discussed in Chapters 5, 13, and 15.

REGULATION OF THE ACTIVITY AND BIOGENESIS OF HRI BY HEME AND CHAPERONES

Although it has been known for 30 years that HRI is activated in heme deficiency (Maxwell and Rabinovitz 1969), the underlying molecular mechanism of this activation of HRI is still not well understood. Multiple stages of the activation of HRI, multiple phosphorylation of HRI and consequently multiple forms of HRI ranging from inactive proinhibitor, heme-reversible HRI to heme-irreversible HRI, have been described (for review, see Chen 1993). These earlier studies were based mainly on the inhibition of protein synthesis and the kinase activity of HRI. The molecular cloning of HRI cDNA, the expression of recombinant HRI, and the production of high-titer HRI antibody have facilitated progress toward the characterization of different forms of HRI.

Inhibition of HRI Autophosphorylation and eIF2α Phosphorylation by Heme

Activation of HRI in heme-deficient reticulocyte lysates is accompanied by its phosphorylation and the phosphorylation of eIF2α (for review, see Chen 1993; Chen and London 1995; Clemens 1996 and references therein). Hemin (the oxidized form of heme with Fe^{+++}) binds to highly purified HRI, and this binding results in the inhibition of both autophosphorylation and HRI-mediated eIF2α phosphorylation (Trachsel et al. 1978; Fagard and London 1981; Chefalo et al. 1998). Rabbit reticulocyte HRI and recombinant HRI expressed from baculovirus-infected insect Sf9 cells (BV-HRI) have been purified to near homogeneity (Chefalo et al. 1998). Purified HRI is a stable homodimer, undergoes autophosphorylation, and phosphorylates eIF2α. Both autokinase and eIF2α kinase activities are inhibited by submicromolar concentrations of hemin with an apparent K_i of 0.5 μM. Although it had been assumed that HRI was a hemoprotein, direct proof has been generated only recently. One important line of evidence is that active BV-HRI purifies as a hemoprotein with the three characteristic absorption peaks in the visible wavelength, the γ band at 424 nm and the α and β bands around 550 nm (Chefalo et al. 1998). Sf9 cells are low in heme content, and expression of functional hemoproteins such as cytochrome P450 and nitric oxide synthase (Asseffa et al. 1989; Richards and Marletta 1994) requires the addition of hemin to the culture medium. Thus, the finding that HRI purifies as a hemoprotein from Sf9 cells without added exogenous heme supports the idea that HRI has a high affinity for heme in vivo.

Purified BV-HRI contains two major species that can be separated by native polyacrylamide gel electrophoresis: a predominant fast-migrating species I and a slower-migrating species II. Assays of the heme-binding activity of these two species of HRI indicate that there are two distinct types of heme-binding sites per HRI homodimer (Chefalo et al. 1998). One type of binding site is nearly saturated with stably bound endogenous heme in purified HRI, whereas the other is available to bind [^{14}C]hemin reversibly. HRI with heme bound on the first site (species I) migrates faster than HRI with heme bound on both sites (species II) in native polyacrylamide gels. Purified BV-HRI containing mainly species I is active both as an autokinase and eIF2α kinase, and its kinase activities are regulated by exogenous hemin even though it contains one heme. Thus, it seems that the second reversible heme-binding site is responsible for the down-regulation of HRI activity by heme (Chefalo et al. 1998). The finding that one type of heme-binding site is stable and of high affinity suggests that heme may also be important in the proper folding of HRI into an active kinase and in maintaining its stability.

Hemin promotes intersubunit disulfide-bond formation in HRI homodimers (Chen et al. 1989; Yang et al. 1992; Chefalo et al. 1998). There is a positive correlation between intersubunit disulfide formation of HRI and the inhibition of HRI kinase activities in vitro by various porphyrin compounds that are structurally similar to hemin. Furthermore, there is a positive correlation between the abilities of various porphyrin compounds to promote intersubunit disulfide formation of HRI and the maintenance of protein synthesis, the reversal of the inhibition of protein synthesis, and the phosphorylation of eIF2α in rabbit reticulocyte lysates (Yang et al. 1992). These findings underscore the importance of intersubunit disulfide-bond formation in the regulation of HRI by heme.

Purified HRI binds ATP, but the ATP binding is inhibited by prior treatment of HRI with hemin in a concentration-dependent manner (Chen et al. 1991b). This result suggests that the heme-regulated inhibition of autophosphorylation and eIF2α phosphorylation is likely to be the result of blocking ATP binding to HRI. Since hemin also promotes disulfide-bond formation, one may infer that free sulfhydryl groups are important for the binding of ATP to HRI. This is consistent with an earlier finding that the presence of ATP prevents the inactivation of lysate HRI by high concentrations of *N*-ethylmaleimide (NEM) (Hunt 1979). Hemin may inhibit the binding of ATP to HRI by inducing a conformational change in HRI that brings the essential sulfhydryl groups into close proximity, thereby facilitating disulfide-bond formation. Since hemin does not promote complete disulfide-bond formation of all HRI, in vitro or in reticulocyte lysates, it is likely that the conformational changes involving these essential sulfhydryl groups may be sufficient to inhibit HRI activity (Chen et al. 1989; Yang et al. 1992). Metal-free protoporphyrin IX (PP), SnPP, NiPP, and ZnPP, which do not promote disulfide-bond formation, probably do not effect equivalent conformational changes and therefore do not inhibit HRI activity.

These biochemical results, a model of the regulation of HRI by heme, are proposed and presented in Figure 1. The formation of species I HRI from newly synthesized HRI is discussed below. The purified stable active HRI is a homodimer whose subunits are joined by noncovalent interactions (species I). Species I HRI has one stable heme bound per subunit to its amino terminus and is already autophosphorylated at both serine and threonine, since it is recognized by antibodies to phosphoserine and phosphothreonine (M. Rafie-Kolpin and J-J. Chen, unpubl.). In the presence of heme, the activity of species I HRI is down-regulated by heme binding to the kinase insert (KI). Heme promotes intersubunit disulfide-bond formation in HRI (species II) and inhibits its autokinase and eIF2α kinase activity. In heme deficiency, species I HRI undergoes further autophosphorylation on serine, threonine, and, to a lesser extent, tyrosine residues (M. Rafie-Kolpin and J.-J. Chen, unpubl.), resulting in the slower electrophoretic mobility observed in SDS-polyacrylamide gels (species III). Autophosphorylation is always observed in active HRI. Although the function of the additional phosphorylation observed in species III is yet to be determined, it is likely that the autophosphorylation of HRI is required for its eIF2α kinase activity. Autophosphorylation of many protein kinases including PKR and GCN2 (Romano et al. 1998) at the activation loop is important for substrate phosphorylation (for review, see Johnson et al. 1996). At present nothing is known about the reversal of species II or species III to species I, although dephosphorylation and reduction of disulfide bridges are required.

The Amino-terminal Region Is Required for Stable Heme Binding

Alignment of HRI with the other eIF2α kinases indicates the presence of three unique regions in HRI: the amino terminus (amino acids 1–160), the kinase insert (KI) region (amino acids 236–380), and the carboxyl termi-

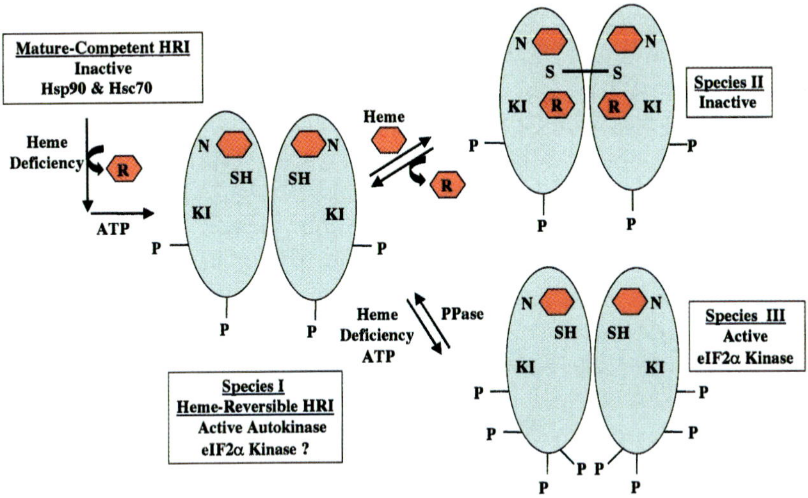

Figure 1 Schematic model of the regulation of HRI by heme and autophosphorylation. In this model, heme and autophosphorylation are proposed to regulate HRI in a two-stage event. During the biosynthesis of HRI in hemin-supplemented reticulocyte lysates, Hsp90 and Hsc70 are associated cotranslationally with HRI. These chaperones are necessary for the proper folding and the maintenance of HRI in an inactive mature-competent form. It is most likely that both of the heme-binding sites of the mature-competent HRI are loaded. In heme deficiency, mature-competent HRI is converted to stable and active species I HRI by dissociation of the heme bound to KI and subsequent autophosphorylation. Purified stable HRI is a homodimer held by noncovalent interactions. This HRI (species I) is autophosphorylated and has one heme stably bound per subunit to its amino-terminal domain. Heme molecules are represented by red hexagon symbols and the amino-terminal domain is labeled with N. Species I is multiply phosphorylated at both serine and threonine residues. For simplicity, two phosphorylation sites are marked with the P symbols in species I. The exact number of phosphorylated sites is yet to be determined. Species I HRI is an active autokinase regulated by heme. When heme concentration is high, it binds to the KI of species I HRI, promotes intersubunit disulfide formation in the HRI homodimer (species II), and inhibits its activity. This reversible heme binding is marked by R in the heme molecules. In low heme concentrations, species I HRI undergoes further multiple autophosphorylation (species III) and phosphorylates eIF2α. Whether the additional autophosphorylation seen in species III is required for its eIF2α kinase activity is yet to be determined. More detailed discussion of the model can be found in the text.

nus (amino acids 507–626) (Wek et al. 1989; Meurs et al. 1990, Chen et al. 1991a; Shi et al. 1998). These unique regions may be involved in the regulation of HRI by heme.

To determine the role of the amino terminus in the heme regulation of HRI, two amino-terminal deletion mutants of rabbit HRI were prepared as illustrated schematically in Figure 2. Met2 HRI begins at the second methionine and lacks the first 103 amino acids. Met3 HRI begins at the third methionine and lacks the first 130 amino acids. Wild-type, Met2, and Met3 HRI are expressed at comparable levels in baculovirus-infected Sf9 cells. Both Met2 and Met3 HRI are active eIF2α kinases and autokinases; their specific eIF2α kinase activities are about 50% of that of the wild-type HRI. These results suggest that the amino terminus may be important for achieving the high specific eIF2α kinase activity but is not essential for the kinase activity of HRI (Rafie-Kolpin et al. 2000). Although the eIF2α kinase activities of the wild-type, Met2, and Met3 HRI are heme-regulated, the eIF2α kinase activity of wild-type HRI is more sensitive to inhibition by heme than that of Met2 or Met3 HRI. The concentration of hemin required for 50% inhibition of eIF2α kinase activity is 0.2 μM for wild type, 2 μM for Met2, and 3 μM for Met3 HRI in S10 extracts. Furthermore, the extent of the inhibition of Met2 and Met3 HRI by heme is less than that of the wild-type HRI at all hemin concentrations tested.

In contrast to wild-type and K199R HRI (an inactive mutant with the conserved Lys-199 mutated to arginine in kinase domain II; Chefalo et al. 1994), Met2 and Met3 HRI display very little absorption at 424 nm, less than 5% of the wild-type and K199 HRI. Thus, the amino-terminal 103 amino acids of HRI are necessary for stable high-affinity heme-binding to HRI. Nevertheless, when exogenous hemin is added to purified Met2 and Met3 HRI, these mutants, like wild-type HRI, can bind heme at the low-affinity site and exhibit the characteristic hemoprotein peak at 420 nm (Rafie-Kolpin et al. 2000). These results show that regions of HRI other than the amino terminus are also responsible for heme-binding.

Two Heme-binding Domains in HRI

HRI comprises five domains: the amino terminus, the first kinase lobe, the kinase insert, the second kinase lobe, and the carboxyl terminus (Fig. 2). All five domains have been expressed as His-tagged fusion proteins in *Escherichia coli,* purified by Ni^{++} column chromatography, and analyzed for heme-binding by difference UV-visible spectroscopy (Rafie-Kolpin et al. 2000). In some of the expressed proteins, there is a slight overlap in the amino-terminal portion. This is done purposely to maintain the predicted secondary structure. The first domain contains the amino-terminal 154 amino acids and exhibits the characteristic Soret band upon addition

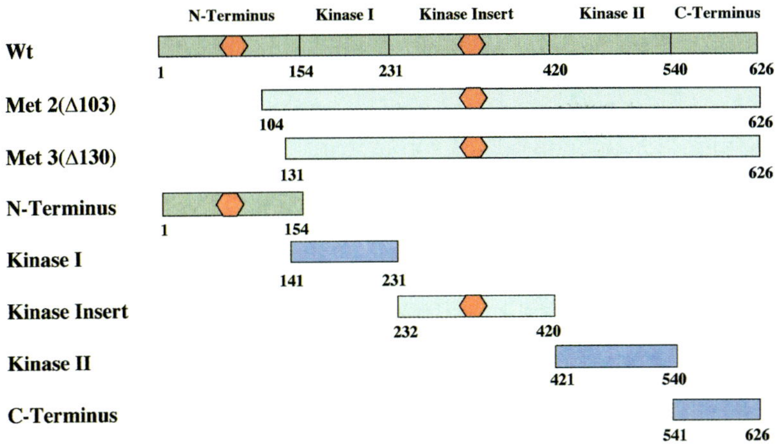

Figure 2 Schematic illustration of the domains of HRI. HRI comprises five domains as indicated. The two heme-binding domains are marked with red heme molecules. The numbers underneath the proteins are the positions of the amino acids in the rabbit HRI sequence. Truncated forms of HRI (Met2 and Met3) described in the text are also shown.

of exogenous heme. Thus, the amino terminus by itself can bind heme. When expressed in the presence of 5 μM hemin, this domain purifies as a hemoprotein with visible pink color. A similar observation has also been made with the amino terminus expressed as a β-galactosidase fusion protein (Uma et al. 2000). Excepting the first 5–15 amino acids, the amino-terminal domain of HRI is highly conserved among mammalian HRIs. The mouse amino-terminal domain (amino acids 1–138) also binds heme (Rafie-Kolpin et al. 2000).

The predicted secondary structure of the amino-terminal 11–118 amino acids of HRI has a significant similarity to the known structure of mammalian α-globin. Furthermore, the amino terminus of HRI adopts mainly helical structure and can be modeled to fit the helical structure of α-globin with His-83 in the HRI amino terminus as the predicted proximal heme ligand (Uma et al. 1999). In hemoglobin, heme is held by hydrophobic interactions of amino acid residues brought together by the α helices of the protein. Thus, the amino-terminal stable heme-binding domain of HRI may resemble that of hemoglobin.

The second domain consists of amino acids 141–231, which corresponds mostly to the first lobe of the kinase domain (kinase I). The fourth domain consists of amino acids 421–540, corresponding to the second kinase lobe (kinase II). The fifth domain contains the carboxy-terminal amino acids 541–626. This domain is unique to HRI and is well conserved among HRI proteins from different species. No heme binding is observed in any of these three domains.

The third domain consists of amino acids 219–420 and contains the entire rabbit KI sequence unique to HRI and includes one putative heme regulatory motif (Zhang and Guarente 1995). Although rabbit KI expressed as a His-tagged fusion protein is insoluble, and its UV-visible spectrum cannot be obtained, a subfragment of the rabbit KI domain (301–420) is expressed as a soluble protein and binds heme. The entire KI domain of mouse HRI (241–406) is also expressed as soluble protein which can also bind heme. In contrast to the amino-terminal domain, mouse KI does not purify as a hemoprotein when it is expressed in the presence of 5 μM hemin. This observation suggests that KI may be the heme-binding site responsible for the reversible heme regulation of HRI (Rafie-Kolpin et al. 2000).

In addition to hemoglobin, myoglobin, and the cytochromes, the structures of several hemoproteins such as FixL (Gong et al. 1998; Miyatake et al. 1999) and nitric oxide synthase (Crane et al. 1997; Raman et al. 1998) have recently been solved. Of particular interest is HasA of *Serratia marcescens,* a pathogenic bacterium. HasA takes up heme from hemoglobin and transfers it to a receptor called HasR, which in turn releases heme into the bacterium (Arnoux et al. 1999). HRI and HasA are similar in that both proteins bind heme reversibly and heme does not serve as an oxygen carrier in either protein. In HasA, heme is held by hydrophobic and stacking interactions of amino acid residues present in the two loops at the interface of α and β structures (Arnoux et al. 1999). Since the secondary structure prediction of the KI indicates that it may exist mainly as a loop structure, the reversible heme binding in the KI of HRI may resemble that of HasA.

The detection of two heme-binding domains in HRI is consistent with the stoichiometry of two heme-binding sites per HRI monomer (Chefalo et al. 1998). The heme binding to the amino-terminal domain is stable and remains with the purified protein, whereas the heme binding to the KI region is reversible and does not remain with purified HRI or KI of HRI. Taken together, the data indicate that heme binding to KI is responsible for rapid regulation of HRI by heme. In addition, it appears that the two heme-binding sites cooperate for optimal eIF2α kinase activity and heme regulation of HRI, since Met2 and Met3 HRI are less active and less heme-responsive.

Although HRI contains two putative heme regulatory motifs (HRMs), Cys 406/Pro 407 and Cys 548/Pro 549, that are not present in other eIF2α kinases (Lathrop and Timko 1993; Chen and London 1995; Zhang and Guarente 1995), the significance of this motif in heme regulation is not clear. The HRM contains a basic amino acid followed by a conserved Cys-Pro dipeptide often flanked downstream by a hydrophobic residue (Lathrop and Timko 1993; Zhang and Guarente 1995). Cys-406 and Cys-548 have been mutated individually to serine with no apparent effect on the heme-responsiveness of HRI (M. Rafie-Kolpin and J.-J. Chen, unpubl.). Since the carboxyl terminus (541–626) that contains the second HRM of HRI does not bind heme, this HRM by itself may not be sufficient to bind heme.

Hsp90, Hsc70, and Heme in the Biogenesis and Activation of HRI

HRI can also be activated in hemin-supplemented lysates by heat shock or oxidative stress (for review, see Chen 1993; Chen and London 1995; Clemens 1996). Interactions of HRI with heat-shock proteins (Hsp90, Hsc70, p59, and p23) in reticulocyte lysates have been demonstrated by coimmunoadsorptions (Matts et al. 1992; Xu et al. 1997). Copurification of Hsp90 in their preparation of HRI led Mendez et al. (1992) to propose that hemin regulates HRI by promoting a disulfide-linked HRI–Hsp90 heterodimer and that dissociation of Hsp90 from HRI is required for the activation of HRI by NEM. However, the following recent data are not consistent with their model: (1) HRI activity is inhibited by hemin in Hsp90-depleted reticulocyte lysates (Xu et al. 1997) and in pure HRI preparations free of Hsp90 (Chen et al. 1991b; Chefalo et al. 1998). (2) The association of Hsp90 with HRI is maintained in heat-shocked, NEM- or Hg^{++}-treated hemin-supplemented reticulocyte lysates. (3) No monomeric HRI is generated under these stress conditions. Thus, Hsp90 is not required for the heme regulation of HRI activity. Hsp90 acts as a molecular chaperone to protect mature competent HRI from denaturation during stress conditions (Uma et al. 1997; Xu et al. 1997).

Heme and autophosphorylation appear to regulate HRI in a two-step event. The first step is the conversion of newly synthesized (mature-competent) HRI to stable species I HRI by autophosphorylation. The second step is the down-regulation of HRI activity by reversible heme binding and suppression of species I HRI activity (Fig. 1). Recently, [^{35}S]Met-labeled HRI, His-tagged and newly synthesized in hemin-supplemented reticulocyte lysates, has been used as a homogeneous species called mature-competent HRI to study its maturation and activation. Both Hsp90 and Hsc70 are associated with nascent HRI cotranslationally on the polyribosomes and are required for the folding of the newly synthesized

HRI into the mature-competent HRI (Uma et al. 1997, 1999). The two heme-binding sites of the mature-competent HRI are likely to be filled. In heme-deficiency, a portion of the mature-competent HRI can be converted into an active kinase (transformed HRI) by autophosphorylation with concomitant upshift in migration in SDS-polyacrylamide gels (Uma et al. 1997). The biochemical properties of the transformed HRI (Uma et al. 1997) are the same as those of the species I HRI purified from rabbit reticulocytes or recombinant BV-HRI described in Figure 1, and transformed HRI is considered as species I in this review. Although Hsp90 and Hsc70 are required for the conversion, once HRI is converted to species I, it no longer interacts with Hsp90. Thus, Hsp90 and Hsc70 serve as molecular chaperones that act to fold HRI into an active autokinase conformation capable of undergoing autophosphorylation to form species I.

These studies demonstrate that autophosphorylation of HRI is important in maintaining not only the activity, but also the proper conformation of the protein. In this respect, HRI is very similar to the catalytic subunit of cAMP-dependent protein kinase (PKA). Phosphorylation of the catalytic subunit of PKA is necessary for its activity and stability (Steinberg et al. 1993; Adams et al. 1995; Yonemoto et al. 1997). Species I HRI is more resistant than the mature-competent HRI to aggregation induced by depletion of chaperones in reticulocyte lysates (Uma et al. 1997). Similarly, when expressed in *E. coli*, the phosphorylated form of HRI is found in the soluble fraction and is active, whereas the unphosphorylated HRI is found in insoluble inclusion bodies and is inactive (Bauer et al. 1999).

In addition to assisting in the maturation and the conversion of newly synthesized HRI to species I HRI (Fig. 1), Hsc70 also suppresses the kinase activity of species I HRI. The level of Hsc70 in lysates appears to be inversely related to the degree of translational inhibition in hemin-supplemented lysates under conditions of heat and oxidative stress (Matts and Hurst 1992). These results suggest that Hsc70 may be required to maintain HRI in an inactive form. Indeed, the addition of denatured proteins that sequester Hsc70 activates HRI in hemin-supplemented lysates (Matts et al. 1993; Uma et al. 1999). Purified Hsc70 prevents the activation of HRI by suppressing its autophosphorylation in heme deficiency and in hemin-supplemented reticulocyte lysates treated with heat, Hg^{++}, oxidized glutathione, or denatured proteins (Gross et al. 1994; Thulasiraman et al. 1998; Uma et al. 1999). In addition, Dna K (the *E. coli* homolog of Hsp70), but not Hsp90, copurifies with the phosphorylated soluble HRI (B.N. Bauer and J.-J. Chen, unpubl.). Thus, Hsc70 appears to play a dual role in the regulation of HRI.

PHYSIOLOGICAL FUNCTION OF HRI

Phylogenetic Conservation of HRI

HRI cDNAs from rabbit (Chen et al. 1991a), rat (Mellor et al. 1994), and mouse (Berlanga et al. 1998; L. Lu and J.-J. Chen, unpubl.) have been cloned. The entire human HRI cDNA sequence is present in several overlapping expressed sequence tag (EST) clones and can be composed from these sequences. Alignments of the amino acids of HRI from human, mouse, rat, and rabbit show that HRI is very well conserved among these mammals with approximately 77% identity and 83% homology. The only regions that are divergent are the KI and the amino-terminal 15 amino acids. Evidence for the presence of HRI in Sugano Kawakami zebrafish has been provided by two EST clones of zebrafish that have high homology with HRI from amino acids 427 to the carboxy-terminal end and amino acids 65–169. There is also one EST clone from *Xenopus laevis* that has sequence homology with HRI amino acids 29–175. Although the malarial parasite eIF2α kinase, PfPk4, has been suggested to be heme-regulated, no significant homology is found in a domain that is well-conserved in other HRI proteins, i.e., the amino-terminal stable heme-binding domain (amino acids 16–140) of mammalian HRI (Mohrle et al. 1997). *Schizosaccharomyces pombe* also has two putative eIF2α kinases that have higher homology with HRI than with other eIF2α kinases. The heme regulation of these two eIF2α kinases remains to be established.

There is no HRI in *Saccharomyces cerevisiae, Caenorhabditis elegans,* or *Drosophila melanogaster*. In keeping with a function of HRI for the balanced synthesis of heme and globins, it is tempting to speculate that HRI evolved when blood circulation was established wherein hemoglobin is carried in red blood cells. This reasoning would predict the presence of HRI in vertebrates in which diffusion alone can no longer supply the increasing demands for oxygen.

Tissue Distribution of HRI

In adult rabbits, HRI protein and mRNA are expressed predominantly in immature erythroid cells. In addition, HRI mRNA levels are increased during erythroid differentiation of mouse erythroleukemic (MEL) cells, and this increase is dependent on the presence of heme (Crosby et al. 1994). Two reports have described the presence of small amounts of HRI mRNA in nonerythroid rat and mouse tissues, and HRI-like activity in mouse liver and NIH-3T3 cells (Mellor et al. 1994; Berlanga et al. 1998). These reports suggested a possible role of HRI in translational regulation in nonerythroid cells in addition to its role in erythroid cells. However, the presence of HRI protein was not established by Western blot analysis in these studies. The criterion used to demonstrate HRI in 3T3 cells and mouse liver was the inhibition of eIF2α kinase activity by rather high hemin concentrations (apparent K_i of about 5 μM). It may also be relevant that NIH-3T3 cells are embryonic in origin, and erythropoiesis in mice persists in the liver of young adults.

The question of whether HRI plays a role in the regulation of protein synthesis in nonerythroid cells is currently being studied using HRI $^{-/-}$ mice (for phenotype, see below; A. Han and J.-J. Chen, unpubl.). Western blot analyses of several tissues of wild-type adult mice using polyclonal antibodies raised against the amino-terminal 138 amino acids and the KI (amino acids 241–406) of mouse HRI indicated that there is no detectable HRI protein in the liver, spleen, heart, kidney, pancreas, or lung, whereas HRI is readily detected in reticulocytes, blood, and bone marrow. As expected, no HRI was detected in HRI null mice. The level of phosphorylated eIF2α was also examined in these tissues by Western blot analysis using anti-eIF2 (αP) antibody (DeGracia et al. 1997). The only tissue extracts analyzed that have different levels of eIF2α phosphorylation between wild-type and HRI null mice are bone marrow, reticulocytes, and blood, which have higher levels of eIF2α (P) in the wild-type than in null mice. No difference was observed in brain, liver, spleen, pancreas, kidney, muscle, lung, or heart, consistent with the previous report (Crosby et al. 1994) that HRI protein is expressed predominantly in erythroid cells.

Regulation of Hemoglobin Synthesis and Proliferation of Differentiating Erythroid Cells

To examine directly the physiological function of HRI, wild-type and inactive mutant proteins have been expressed in both nonerythroid 3T3 cells and erythroid MEL cells (Crosby et al. 2000). Forced expression of wild-type HRI in 3T3 cells inhibits cell growth and ultimately causes cell death. The addition of hemin to the culture medium maintains the growth of wild-type HRI-expressing 3T3 cells. Upon removal of hemin, wild-type HRI becomes active and shuts off protein synthesis, resulting in cell death. This lethal effect of overexpression of wild-type HRI in 3T3 cells is similar to that of PKR overexpression (Lee and Esteban 1994; for review, see Chapter 13). Coexpression of eIF2α Ser51Ala with wild-type HRI protects the cells from the inhibition of protein synthesis and the antiproliferative and lethal effects of HRI, indicating that this effect of HRI is mediated through eIF2α phosphorylation. Although coexpression of Bcl-2 also blocks HRI-induced cell death, Bcl-2 failed to restore proliferation of wild-type HRI-expressing cells in contrast to hemin or eIF2α Ser51Ala. This observation is similar to the finding that Bcl-2 expression also inhibits PKR-induced apoptosis in HeLa cells (Lee et al. 1997). In addition, PKR-induced inhibition of protein synthesis is not prevented by Bcl-2. Therefore, Bcl-2 acts downstream from eIF2α phosphorylation.

Overexpression of PKR mutant proteins such as PKRΔ6 in NIH-3T3 cells results in malignant transformation (Koromilas et al. 1992; Barber et al. 1995a,b; Romano et al. 1998). Because PKR is a dimer, it is believed that PKRΔ6 protein acts in a transdominant negative manner to form an inactive wild-type PKR-PKRΔ6 heterodimer with the endogenous PKR (Koromilas et al. 1992). In soft agar growth assays, expression of wild-type HRI prevents the transformation of NIH-3T3 cells by PKRΔ6 or eIF4E (Crosby et al. 2000). However, overexpression of the inactive HRI mutant K199R, Δ10 (deletion of amino acids 375–384), or Δ20 (deletion of amino acids 375–394) does not result in any morphological change that would be indicative of malignant transformation. These cells do not grow in soft agar or form tumors in nude mice (Crosby et al. 2000). Coexpression of wild-type HRI with K199R HRI in baculovirus-infected insect cells leads to diminished kinase activity of the wild-type HRI (Chefalo et al. 1994). This reduced activity is thought to be the result of the formation of an inactive HRI heterodimer. Similar results are obtained using the inactive dominant-negative HRI deletion mutants Δ10 and Δ20. The observation that forced expression of K199R, Δ10, or Δ20 HRI fails to confer upon NIH-3T3 cells a transformed phenotype (J.S. Crosby et al., in prep.) suggests that if there is HRI protein present in NIH-3T3 cells, it is probably not active under our experimental conditions.

In contrast to the absence of a phenotypic change in 3T3 cells, forced expression of the HRI mutants in differentiating MEL cells results in increased hemoglobin synthesis and proliferative capacity of these cells. These phenotypic changes not only provide direct evidence for the function of HRI in the regulation of hemoglobin synthesis in erythroid cells and for its dimerization in vivo, but also suggest an additional role of HRI in the regulation of the proliferation of the differentiating erythroid cells, most likely through the inhibition of protein synthesis (Crosby et al. 2000).

Targeted Disruption of the HRI Gene in Mice

HRI-null mice appear to be normal, without any detectable effect on their hematology, and are fertile under normal iron-sufficient dietary conditions (A. Han and J.-J. Chen, unpubl.). The phenotype of $HRI^{-/-}$ mice has been characterized in vitro using reticulocytes obtained from phenylhydrazine-induced anemic mice. There is much less eIF2α phosphorylation in the intact reticulocytes of the HRI$^{-/-}$ mice than in those of HRI$^{+/+}$ mice. Thus, it seems that HRI is the major eIF2α kinase in reticulocytes under the experimental conditions. In addition, the rate of protein synthesis is severalfold higher in the reticulocytes of HRI$^{-/-}$ mice as compared to the +/+ mice. Protein synthesis in the intact reticulocytes is increased by the addition of hemin to the media in the wild-type but not in HRI $^{-/-}$ mice. Inhibition of protein synthesis by succinylacetone, an inhibitor of heme biosynthesis, was also observed in the +/+ but not in –/– mice. Therefore, isolated reticulocytes display a phenotype that is as expected from the known enzymatic activity of HRI. Future work should elicit phenotypic changes in HRI$^{-/-}$ mice by making mice heme-deficient, thereby illuminating the physiological function of HRI in vivo.

PERSPECTIVES

There are four known eIF2α kinases, PKR, PERK, GCN2, and HRI, that regulate initiation of protein synthesis in eukaryotic cells under various stress conditions. Significant progress has been made in understanding the structure of HRI in relation to heme binding and heme regulation. The crystal structures of HRI and its heme-binding domains will further our appreciation of the molecular mechanism of heme regulation and the role of autophosphorylation in the stabilization and the activation of HRI. Analysis of the role of chaperones in the regulation of eIF2α kinases has revealed the existence of distinct species of HRI during its biogenesis and activation. The importance of chaperones in regulating GCN2 and PKR is also emerging. Last, the availability of the HRI knockout mice provides a valuable tool to further examine the physiological function of HRI. Studies using knockout mice lacking the individual eIF2α kinases singly and in combination will help us understand the contribution of each member of this kinase family in the regulation of protein synthesis under different stress conditions.

ACKNOWLEDGMENTS

I thank Drs. Lee Gehrke, Irving M. London, and members of my laboratory for their critical readings of the manuscript. I also acknowledge Drs. Ronald Wek and Thomas Dever for communicating their recent work currently in press. The research in my laboratory is supported by grants from the National Institutes of Health (DK-16272 and DK-53223).

REFERENCES

Adams J.A., McGlone M.L., Gibson R., and Taylor S.S. 1995. Phosphorylation modulates catalytic function and regulation in the cAMP-dependent protein kinase. *Biochemistry* **34:** 2447–2454.

Arnoux P., Haser R., Izadie N., Lecroisey A., Delepierre M., Wandersman C., and Czjzek M. 1999. The crystal structure of HasA, a hemophore secreted by *Serratia marcescens*. *Nat. Struct. Biol.* **6:** 516–520.

Asseffa A., Smith S.J., Nagata K., Gillette J., Gelboin H.V., and Gonzalez F.J. 1989. Novel exogenous heme-dependent expression of mammalian cytochrome P450 using baculovirus. *Arch. Biochem. Biophys.* **274:** 481–490.

Barber G.N., Jagus R., Meurs E.F., Hovanessian A.G., and Katze M.G. 1995a. Molecular mechanisms responsible for malignant transformation by regulatory and catalytic domain variants of the interferon-induced enzyme RNA-dependent protein kinase. *J. Biol. Chem.* **270:** 17423–17428.

Barber G.N., Wambach M., Thompson S., Jagus R., and Katze M.G. 1995b. Mutants of the RNA-dependent protein kinase (PKR) lacking double-stranded RNA binding domain I can act as transdominant inhibitors and induce malignant transformation. *Mol. Cell. Biol.* **15:** 3138–3146.

Bauer B.N., Lu L., Chuu Y., and Chen J.-J. 1999. Autophosphorylations of heme-regulated eIF-2α kinase at multiple sites are essential for its stability, activity and heme-response. *FASEB J.* **13:** A1365.

Berlanga J.J., Herrero S., and de Haro C. 1998. Characterization of the hemin-sensitive eukaryotic initiation factor 2α kinase from mouse nonerythroid cells. *J. Biol. Chem.* **273:** 32340–32346.

Bruns G.P. and London I.M. 1965. The effect of hemin on the synthesis of globin. *Biochem. Biophys. Res. Commun.* **18:** 236–242.

Chefalo P., Oh J., Rafie-Kolpin M., and Chen J.-J. 1998. Heme-regulated eIF-2α kinase purifies as a hemoprotein. *Eur. J. Biochem.* **258:** 820–830.

Chefalo P.J., Yang J.M., Ramaiah K.V.A., Gehrke L., and Chen J.-J. 1994. Inhibition of protein synthesis in insect cells by baculovirus-expressed heme-regulated eIF-2α kinase. *J. Biol. Chem.* **269:** 25788–25794.

Chen J.-J. 1993. Translational regulation in reticulocytes: The role of heme-regulated eIF-2α kinase. In *Translational control of gene expression,* 2nd edition (ed. J. Ilan), pp. 349–372. Plenum Press, New York.

Chen J.-J. and London I.M. 1995. Regulation of protein synthesis by heme-regulated eIF-2α kinase. *Trends Biochem. Sci.* **20:** 105–108.

Chen J.-J., Yang J.M., Petryshyn R., Kosower N., and London I.M. 1989. Disulfide bond formation in the regulation of eIF-2α kinase by heme. *J. Biol. Chem.* **264:** 9559–9564.

Chen J.-J., Throop M.S., Gehrke L., Kuo I., Pal J.K., Brodsky M., and London I.M. 1991a. Cloning of the cDNA of the heme-regulated eukaryotic initiation factor 2α (eIF-2α) kinase of rabbit reticulocytes: Homology to yeast GCN2 protein kinase and human double-stranded-RNA-dependent eIF-2α kinase. *Proc. Natl. Acad. Sci.* **88:** 7729–7733.

Chen J.-J., Pal J.K., Petryshyn R., Kuo I., Yang J.M., Throop M.S., Gehrke L., and London I.M. 1991. Amino acid microsequencing of the internal tryptic peptides of heme- regulated eukaryotic initiation factor 2α subunit kinase: Homology to protein kinases. *Proc. Natl. Acad. Sci.* **88:** 315–319.

Chong K.L., Schappert K., Meurs E., Feng F., Donahue T.F., Friesen J.D., Hovanessian A.G., and Williams B.R.G. 1992. Human p68 kinase exhibits growth suppression in yeast and homology to the translational regulator GCN2. *EMBO J.* **11:** 1553–1562.

Clemens M.J. 1996. Protein kinases that phosphorylate eIF2 and eIF2B, and their role in eukaryotic cell translational control. In *Translational control* (ed. J.W.B. Hershey et al.), pp. 139–172. Cold Spring Harbor Laboratory Press, Cold Spring Harbor, New York.

Colthurst D.R., Campbell D.G., and Proud C.G. 1987. Structure and regulation of eukaryotic initiation factor eIF-2. Sequence of the site in the alpha subunit phosphorylated by the haem-controlled repressor and by the double-stranded RNA-activated inhibitor. *Eur. J. Biochem.* **166:** 357–363.

Crane B.R., Arvai A.S., Gachhui R., Wu C., Ghosh D.K., Getzoff E.D., Stuehr D.J., and Tainer J.A. 1997. The structure of nitric oxide synthase oxygenase domain and inhibitor complexes. *Science* **278:** 425–431.

Crosby J.S., Lee K., London I.M., and Chen J.-J. 1994. Erythroid expression of the heme-regulated eIF-2α kinase. *Mol. Cell. Biol.* **14:** 3906–3914.

Crosby J.S., Chefalo P.J., Yeh I., Ying S., London I.M., Leboulch P., and Chen J.-J. 2000. Regulation of hemoglobin synthesis and proliferation of differentiating erythroid cells by heme-regulated eIF-2α kinase. *Blood.* (in press).

DeGracia D.J., Sullivan J.M., Neumar R.W., Alousi S.S., Hikade K.R., Pittman J.E., White B.C., Rafols J.A., and Krause G.S. 1997. Effect of brain ischemia and reperfusion on the localization of phosphorylated eukaryotic initiation factor 2 alpha. *J. Cereb. Blood Flow Metab.* **17:** 1291–1302.

Dever T.E., Chen J.-J., Barber G.N., Cigan A.M., Feng L., Donahue T.F., London I.M., Katze M.G., and Hinnebusch A.G. 1993. Mammalian eIF-2α kinases functionally substitute for GCN2 in the GCN4 translational control mechanism of yeast. *Proc. Natl. Acad. Sci.* **90:** 4616–4620.

Fagard R. and London I.M. 1981. Relationship between the phosphorylation and activity of the heme-regulated eIF-2α kinase. *Proc. Natl. Acad. Sci.* **78:** 866–870.

Fairbanks V.F. and Beutler E. 1983. Iron deficiency. In *Hematology*, 3rd edition (ed. W.J. Williams et al.), pp. 466–488. McGraw-Hill, New York,

Gong W., Hao B., Mansy S.S., Gonzalez G., Gilles-Gonzalez M.A., and Chan M.K. 1998. Structure of a biological oxygen sensor: A new mechanism for heme-driven signal transduction. *Proc. Natl. Acad. Sci.* **95:** 15177–15182.

Grayzel A.I.P., Horchner P., and London I.M. 1966. The stimulation of globin synthesis by heme. *Proc. Natl. Acad. Sci.* **55:** 650–655.

Gross M., Olin A., Hessefort S., and Bender S. 1994. Control of protein synthesis by hemin. *J. Biol. Chem.* **269:** 22738–22748.

Harding H.P., Zhang Y., and Ron D. 1999. Protein translation and folding are coupled by an endoplasmic-reticulum-resident kinase. *Nature* **397:** 271–274.

Hunt T. 1979. The control of protein synthesis in rabbit reticulocyte lysates. In *Miami Winter Symp.* **16:** 321–345.

Johnson L.N., Noble M.E.M., and Owen D.J. 1996. Active and inactive protein kinases: Structural basis for regulation. *Cell* **85:** 149–158.

Kassenaar A., Morell H., and London I.M. 1957. The incorporation of glycine into globin and the synthesis of heme in vitro in duck erythrocytes. *J. Biol. Chem.* **229:** 423–435.

Koromilas A.E., Roy S., Barber G.N., Katze M.G., and Sonenberg N. 1992. Malignant transformation by a mutant of the IFN-inducible dsRNA-dependent protein kinase. *Science* **257:** 1685–1689.

Kruh J. and Borsook G. 1956. Hemoglobin synthesis in rabbit reticulocytes in vitro. *J. Biol. Chem.* **220:** 905–915.

Lathrop J.T. and Timko M.P. 1993. Regulation by heme of mitochondrial protein transport through a conserved amino acid motif. *Science* **259:** 522–525.

Lee S.B. and Esteban M. 1994. The interferon-induced double-stranded RNA-activated protein kinase induces apoptosis. *Virology* **199:** 491–496.

Lee S.B., Rodriguez D., Rodriguez J.R., and Esteban M. 1997. The apoptosis pathway triggered by the interpheron-induced protein kinase PKR requires the third basic domain, initiates upstream of Bcl-2, and involves ICE-like proteases. *Virology* **231:** 81–88.

Matts R.L. and Hurst R. 1992. The relationship between protein synthesis and heat shock protein levels in rabbit reticulocytes. *J. Biol. Chem.* **267:** 18168–18174.

Matts R.L., Hurst R., and Xu Z. 1993. Denatured proteins inhibit translation in hemin-supplemented rabbit reticulocyte lysate by inducing the activation of the heme-regulated eIF-2α kinase. *Biochemistry* **32:** 7323–7328.

Matts R.L., Xu Z., Pal J.K., and Chen J.-J. 1992. Interactions of the heme-regulated eIF-2α kinase with heat shock proteins in rabbit reticulocyte lysates. *J. Biol. Chem.* **267:** 18160– 18167.

Maxwell C.R. and Rabinovitz M. 1969. Evidence for an inhibitor in the control of globin synthesis by hemin in a reticulocyte lysate. *Biochem. Biophys. Res. Commun.* **35:** 79–85.

Mellor H., Flowers K.M., Kimball S.R., and Jefferson L.S. 1994. Cloning and characterization of cDNA encoding rat hemin-sensitive initiation factor-2α (eIF-2α) kinase. *J. Biol. Chem.* **269:** 10201–10204.

Mendez R., Moreno A., and de Haro C. 1992. Regulation of heme-controlled eukaryotic polypeptide chain initiation factor 2 α-subunit kinase of reticulocyte lysates. *J. Biol. Chem.* **267:** 11500–11507.

Meurs E., Chong K., Galabru J., Thomas N.S.B., Kerr I.M., Williams B.R.G., and Hovanessian A.G. 1990. Molecular cloning and characterization of human double-stranded RNA activated protein kinase induced by interferon. *Cell* **62:** 379–390.

Miyatake H., Kanai M., Adachi S., Nakamura H., Tamura K., Tanida H., Tsuchiya T., Iizuka T., and Shiro Y. 1999. Dynamic light-scattering and prelimininary crystallographic studies of the sensor domain of the haem-based oxygen sensor FixL from *Rhizobium meliloti. Acta Crystallogr. D Biol. Crystallogr.* **55:** 1215–1218.

Mohrle J.J., Zhao Y., Wernli B., Franklin R.M., and Kappes B. 1997. Molecular cloning, characterization and localization of PfPK4, an eIF-2α kinase-related enzyme from the malarial parasite *Plasmodium falciparum. Biochem. J.* **328:** 677–687.

Morell H., Savoie J.C., and London I.M. 1958. The biosynthesis of heme and the incorporation of glycine into globin in rabbit bone marrow in vitro. *J. Biol. Chem.* **233:** 923–929.

Rafie-Kolpin M., Chefalo P.J., Hussain Z., Hahn J., Uma S., Matts R.L., and Chen J.-J. 2000. Two heme-binding domains of heme-regulated eIF-2α kinase: N-terminus and kinase insertion. *J. Biol. Chem.* **275:** 5171–5178.

Raman C.S., Li H., Martasek P., Kral V., Masters B.S.S., and Poulos T.L. 1998. Crystal structure of constitutive endothelial nitric oxide synthase: A paradigm for pterin function involving a novel metal center. *Cell* **95:** 939–950.

Ramirez M., Wek R.C., and Hinnebusch A.G. 1991. Ribosome association of GCN2 protein kinase, a translational activation of the GCN4 gene of *Saccharomyces cerevisiae. Mol. Cell. Biol.* **11:** 3027–3036.

Richards M.K. and Marletta M.A. 1994. Characterization of neuronal nitric oxide synthase and a C415H mutant, purified from a baculovirus overexpression system. *Biochemistry* **33:** 14723– 14732.

Romano P.R., Garcia-Barrio M.T., Zhang X., Wang Q., Taylor D.R., Zhang F., Herring C.,

Mathews, M.B., Qin J., and Hinnebusch, A.G. 1998. Autophosphorylation in the activation loop is required for full kinase activity in vivo of human and yeast eukaryotic initiation factor 2α kinases PKR and GCN2. *Mol. Cell. Biol.* **18:** 2282–2297.

Shi Y., Vattem K.M., Sood R., An J., Liang J., Stramm L., and Wek R.C. 1998. Identification and characterization of pancreatic eukaryotic initiation factor 2 α-subunit kinase, PEK, involved in translational control. *Mol. Cell. Biol.* **18:** 7499–7509.

Steinberg R.A., Cauthron R.D., Symcox M.M., and Shuntoh H. 1993. Autoactivation of catalytic (C alpha) subunit of cyclic AMP-dependent protein kinase by phosphorylation of threonine 197. *Mol. Cell. Biol.* **13:** 2332–2341.

Thulasiraman V., Xu Z., Uma S., Gu Y., Chen J.-J., and Matts R.L. 1998. Evidence that Hsc70 negatively modulates the activation of the heme-regulated eIF-2α kinase in rabbit reticulocyte lysate. *Eur. J. Biochem.* **255:** 552–562.

Trachsel H., Ranu R.S., and London I.M. 1978. Regulation of protein synthesis in rabbit reticulocyte lysates: Purification and characterization of heme-reversible translational inhibitor. *Proc. Natl. Acad. Sci.* **75:** 3654–3658.

Uma S., Thulasiraman V., and Matts R.L. 1999. Dual role for Hsc70 in the biogenesis and regulation of the heme-regulated kinase of the α subunit of eukaryotic translation initiation factor 2. *Mol. Cell. Biol.* **19:** 5861–5871.

Uma S., Hartson S.D., Chen J.-J., and Matts R.L. 1997. Hsp90 is obligatory for the heme-regulated eIF-2α kinase to acquire and maintain an activatable conformation. *J. Biol. Chem.* **272:** 11648–11656.

Uma S., Matts R.L., Guo Y., White S., and Chen J.-J. 2000. The N-terminal region of the heme-regulated eIF2α kinase is an autonomous heme binding domain. *Eur. J. Biochem.* **267:** 1–10.

Waxman H.S. and Rabinovitz M. 1965. Iron supplementation in vitro and the state of aggregation and function of reticulocyte ribosomes in hemoglobin synthesis. *Biochem. Biophys. Res. Commun.* **19:** 538–545.

———. 1966. Control of reticulocyte polyribosome content and hemoglobin synthesis by heme. *Biochim. Biophys. Acta* **129:** 369–379.

Wek R.C., Jackson B.M., and Hinnebusch A.G. 1989. Juxtaposition of domains homologous to protein kinases and histidyl-tRNA synthetases in GCN2 protein suggests a mechanism for coupling GCN4 expression to amino acid availability. *Proc. Natl. Acad. Sci.* **86:** 4579–4583.

Xu Z., Pal J.K., Thulasiraman V., Hahn H.P., Chen J.-J., and Matts R.L. 1997. The role of the 90-kDa heat-shock protein and its associated cohorts in stabilizing the heme-regulated eIF-2α kinase in reticulocyte lysates during heat stress. *Eur. J. Biochem.* **246:** 461–470.

Yang J.M., London I.M., and Chen J.-J. 1992. Effects of hemin and porphyrin compounds on intersubunit disulfide formation of heme-regulated eIF-2α kinase and the regulation of protein synthesis in reticulocyte lysates. *J. Biol. Chem.* **267:** 20519–20524.

Yonemoto W., McGlone M.L., Grant B., and Taylor S.S. 1997. Autophosphorylation of the catalytic subunit of cAMP-dependent protein kinase in *Escherichia coli*. *Protein Eng.* **10:** 915–925.

Zhang L. and Guarente L. 1995. Heme binds to a short sequence that serves a regulatory function in diverse proteins. *EMBO J.* **14:** 313–320.

15

PERK and Translational Control by Stress in the Endoplasmic Reticulum

David Ron and Heather P. Harding

Skirball Institute of Biomolecular Medicine
Departments of Medicine, Cell Biology and the Kaplan Cancer Center
New York University School of Medicine
New York, New York 10016

TRANSLATIONAL CONTROL IN THE UNFOLDED PROTEIN RESPONSE: HISTORICAL BACKGROUND

Protein folding is the last step in gene expression. The information required for folding is encoded in the genome and is an intrinsic property of the translated polypeptide. However, in the crowded environment of the cell this process is rarely spontaneous; specific chaperone systems have evolved to assist in folding (Anfinsen 1973). The work they perform is energetically very costly and their function is dependent on the environment in the different cellular compartments. Consequently, physical, biochemical, and genetic perturbations can unfavorably affect the process of folding, resulting in the accumulation of malfolded proteins. The biological consequence of production and accumulation of malfolded proteins is not completely understood; however, it is a highly disadvantageous situation for the cell. For this reason, cells from all phyla have evolved means of dealing with the stress imposed by malfolded and abnormal proteins. One of the best-characterized responses consists of a global attenuation in translation initiation rates. This response is thought to be adaptive insomuch as it reduces the burden on the folding machinery during times of excessive demand. In mammalian cells it appears that this conceptually simple feedback loop is compartmentalized and utilizes sensing and effecting modalities that have considerable specificity to both compartment and challenge. Other chapters in this book deal with translational control in the context of heat shock (Chapter 17), viral infection (Chapter 9) and nutrient deprivation (Chapter 5). This chapter focus-

© 2000 Cold Spring Harbor Laboratory Press 0-87969-568-4/00 $5 + .00

es on a signaling pathway that operates in the endoplasmic reticulum (ER) and serves to match protein synthesis rates with rates of protein folding in that compartment.

Proteins destined for the exocytic compartment are synthesized on ER-bound ribosomes that are intimately associated with a pore structure that allows the cotranslational insertion of polypeptides into the lumen of the ER. Posttranslational modification and folding takes place in the lumen of the ER and is guided by chaperones that function in an environment that is unique to that organelle. The ER lumen has a high concentration of calcium and ATP and is an oxidizing environment. These features are essential for folding and disulfide-bond formation that affect most proteins which fold in the ER. In addition, proper folding is influenced by protein glycosylation, a posttranslational modification that is unique to the ER. The folding environment in the ER is thus very different from that of the cytoplasm and other organelles and is therefore susceptible to perturbation by specific physiological and pharmacological interventions.

The folding environment in the ER is continuously monitored by a stress-response pathway that regulates the transcription rate of genes encoding ER chaperones. Conditions associated with protein malfolding in the ER trigger an increase in chaperone production. This signaling pathway is known as the unfolded protein response (UPR) and is conserved in all eukaryotes (Gething and Sambrook 1992). Chaperone gene mRNA levels thus serve as convenient markers for the state of the folding environment in the ER (Kozutsumi et al. 1988). The UPR can be triggered by adverse physiological states, such as hypoxia and glucose deprivation, genetic defects that alter protein structure, and specific toxins that affect the unique folding environment in the ER (Lee 1992). The latter include agents such as thapsigargin, an inhibitor of the ER calcium pump; reducing agents such as dithiothreitol (DTT); and tunicamycin, a toxin that blocks protein glycosylation. Historically, study of the UPR has concentrated on the analysis of signaling pathways that culminate in transcriptional regulation and has relied heavily on the aforementioned pharmacological reagents.

One of the earliest suggestions of the existence of a translational regulatory component to the UPR was presented by work from the Brostrom lab in which it was demonstrated that adding calcium back to the media of calcium-depleted cells had a profound stimulatory effect on global translation rates (Brostrom et al. 1983). Initially this was interpreted as a requirement for cytosolic calcium in translation, but subsequent work shifted the emphasis to the ER. Pharmacological experiments provided the first clues to this ER link: Thapsigargin, an agent that depletes ER calcium but elevates cytosolic calcium, was found to have the same inhibitory effect on translation as manipulations that deplete calcium from both compartments; depletion from both compartments was effected by exposing cells to a calcium ionophore in an environment of low extracellular calcium (Prostko et al. 1992). Agents that perturb ER function without having major effects on calcium homeostasis, such as DTT and tunicamycin, were also found to inhibit translation, suggesting the existence of an ER nexus for translational control (Prostko et al. 1993; Wong et al. 1993).

These studies also identified a correlation between translation inhibition by ER stress, phosphorylation of eIF2α, and inhibition of the nucleotide exchange factor eIF2B. Comparison of polysome profiles of stressed and unstressed cells and measurements of elongation rates suggested that translation was inhibited at the initiation stage (for review, see Brostrom and Brostrom 1998). These results were thus consistent with the well-characterized role of eIF2α phosphorylation in blocking the formation of a 43S initiation complex (see Chapters 2 and 5). The unfolded protein response was thus demonstrated to consist of two components: A transcriptional regulatory component that matched chaperone expression to the needs of the organelle and a component that attenuated translation in response to ER stress through the phosphorylation of eIF2α. The nature of the signaling pathway that links ER stress to eIF2α phosphorylation and the effector mechanisms involved were not known.

IDENTIFICATION OF PERK AND CHARACTERIZATION OF ITS BIOCHEMICAL PROPERTIES

At the time, three eIF2α kinases were candidates for linking ER stress to inhibition of translation initiation: GCN2, a yeast eIF2α kinase known to respond to amino acid deprivation (see Chapter 5); HRI, a mammalian kinase known to couple heme availability to translation rates in reticulocytes (see Chapter 14); and PKR, an interferon-inducible kinase that is activated by double-stranded RNA (see Chapter 13). Because of its physiological regulation by the stress of viral infection, PKR seemed a good candidate for mediating the effects of ER stress on translation. Indeed, treatment of cells with agents that deplete ER calcium stores was found to activate PKR (Prostko et al. 1995) and expression of a dominant negative form of PKR interfered with the ability of ER calcium depletion to attenuate translation rates (Srivastava et al. 1995). The nature of the link between ER stress and PKR activation remained a mystery.

Meanwhile, yeast genetics was being used to unravel the transcriptional regulatory components of the UPR. This arm of the signaling pathway was shown to depend on the function of a type-I transmembrane ER-resident protein encoded by the *IRE1* gene that serves as an ER stress receptor. Recognition of the ER stress signal by the lumenal domain of Ire1p is thought to lead to oligomerization of the protein and results in a transphosphorylation event that takes place in its carboxy-terminal cytoplasmic kinase domain. This phosphorylation event serves to activate Ire1p effector function and culminates in the productive splicing of an mRNA that encodes a transcription factor required for the activation of chaperone gene expression (for review, see Shamu et al. 1994; Shamu 1997; Kaufman 1999). Recently, homologs of IRE1p have been identified in other species (Sidrauski and Walter 1997; Tirasophon et al. 1998; Wang et al. 1998). This provided an opportunity to examine the conserved features of the amino acid sequence of the amino-terminal ER-stress-sensing domain of IRE1 and led to a search for related proteins that might be implicated in responding to ER stress. Such a search, carried out in our laboratory, led to the identification of a *Caenorhabditis elegans* gene that was predicted to encode a type-I transmembrane protein with a lumenal domain related to IRE1 and a carboxy-terminal cytoplasmic domain that had all the features of a protein kinase. However, the novel *C. elegans* gene lacked the endonuclease domain found in IRE1 proteins (the latter is required for the downstream splicing event that is an essential feature of IRE1 signaling), instead its carboxy-terminal kinase domain was most closely related to known eIF2α kinases PKR, GCN2, and HRI. The predicted membrane topology of the protein, and the juxtaposition of a lumenal domain related to the sensing domain of IRE1 with a kinase domain related to known eIF2α kinases, suggested that the protein might be implicated in transducing ER stress to eIF2α phosphorylation and inhibition of translation initiation (Fig. 1). The existence of such an entity was predicted by the observation that *Pkr* knockout cells are unimpaired in their ability to respond to ER stress by attenuating translation rates (Harding et al. 1999).

A mouse homolog of the *C. elegans* protein was cloned, and an expression vector for a tagged version of the protein was prepared. The encoded protein was localized to the endoplasmic reticulum and tentatively named *P*KR-like *ER K*inase, or PERK. When expressed in bacteria and purified as a fusion protein with glutathione-*S*-transferase, the kinase domain of PERK is highly active, both in self-phosphorylating and in phosphorylating its substrate, eIF2α. When overexpressed in mammalian cells, PERK kinase is active and exerts a profound inhibitory effect on protein translation rates (Harding et al. 1999). Similar results were obtained with the human homolog of PERK, a protein named *p*ancreatic

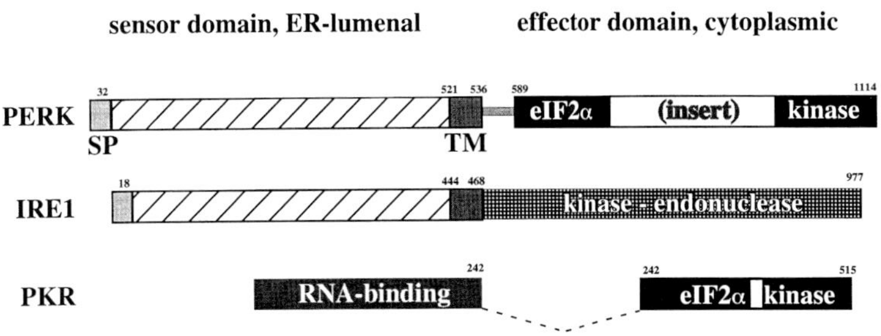

Figure 1 Cartoon diagram of the domain organization of PERK and a comparison with that of IRE1 and PKR. The residues corresponding to the boundaries of the different domains are all based on the murine sequences, with IRE1 being the β isoform. (SP), Signal peptide; (TM), transmembrane domain. The lumenal domain of PERK and IRE1 is cross-hatched and the predicted two lobes of the eIF2α kinase domain, conserved between PERK and PKR (*black rectangles*), are separated by a large insert in PERK and by a much smaller insert in PKR (*white rectangle*). Also shown is the kinase-endonuclease domain of IRE1 and the RNA-binding domain of PKR. (Reprinted, with permission, from Harding et al 1999. [copyright Macmillan Magazines].)

*e*IF2α *k*inase, or PEK, that was identified as a novel kinase expressed at high levels in the pancreas (Shi et al. 1998). The link between PERK/PEK and the phosphorylation of eIF2α and inhibition of translation initiation is thus rather firm.

PERK's potential for autophosphorylation is reflected by the fact that the bacterially expressed active form of the kinase exhibits considerably slower mobility on SDS-PAGE than do mutant forms of the protein that are devoid of kinase activity. The slower-migrating forms of active PERK can be converted to more rapidly migrating forms by treatment with alkaline phosphatase, suggesting that phosphorylation of the protein retards its mobility. In cells, agents that cause ER stress lead to a shift in PERK mobility to the slower-migrating form, and the immune-precipitated slower-migrating form of PERK exhibits increased kinase activity when assayed in vitro. Mutant PERK that lacks kinase activity does not undergo this shift in mobility in response to ER stress, suggesting that the shift is due to an autophosphorylation event rather than to phosphorylation of PERK by an upstream activating kinase. This correlation between PERK activation status and its mobility on SDS-PAGE allowed us to examine the specificity of the relationship between PERK activation and ER stress. To date, all manipulations known to induce ER stress (as reflected in their ability to activate chaperone gene expression) also activate PERK phosphorylation, whereas the stress of heatshock, arsenite treatment, or amino acid deprivation, events that are thought to be initiated outside the ER, fail to activate the kinase (Harding et al. 1999, 2000). Consistent with its structure and homology with IRE1, PERK thus appears to have evolved to respond specifically to ER stress.

More recently, our lab has obtained further insight into the role of PERK in the attenuation of protein translation in response to ER stress: Embryonic stem cells (ES cells) and mouse embryonic fibroblasts (MEFs) that are nullizygous for *Perk* were created and, when exposed to agents such as thapsigargin, tunicamycin, or DTT, the *Perk*$^{-/-}$ cells exhibited no attenuation in translation rates. In wild-type cells, ER stress leads to a profound inhibition in translation rates serving as a positive control. The refractoriness of *Perk*$^{-/-}$ cells to translational inhibition in response to ER stress was mirrored by the absence of increased eIF2α phosphorylation (Harding et al. 2000). These results suggest that in ES cells and MEFs, PERK plays a nonredundant role in coupling ER stress to eIF2α phosphorylation and translational inhibition. Given the evidence that PKR may be activated by calcium release (Prostko et al. 1995), it remains possible that in cells other than ES cells, with a different complement of eIF2α kinases, PERK's role might prove less essential.

The biochemical details of PERK activation by ER stress are not well understood. IRE1 is capable of undergoing oligomerization, and this appears to be a property of the amino-terminal stress-sensing lumenal domain (although a contribution of the transmembrane domain has not been excluded). Oligomerization was demonstrated biochemically by the observation that full-length and carboxy-terminally truncated forms of IRE1p associate in a complex when overexpressed in yeast (Shamu and Walter 1996). Unfortunately, these overexpression experiments were not able to reveal stress-dependence of the oligomerization event. Nonetheless, the prevailing hypothesis for IRE1 activation accommodates a stress-inducible oligomerization event that results in transphosphorylation and "unmasking" of an effector function (for review, see Shamu et al. 1994; Kaufman 1999). In this sense, IRE1 is thought to function like a classic transmembrane receptor protein kinase.

PKR, the kinase most closely related to PERK, undergoes oligomerization upon binding of double-stranded RNA to its amino-terminal regulatory domain (Kostura and Mathews 1989; Langland and Jacobs 1992; Wu and Kaufman 1997). This triggers a transphosphorylation event that enhances the kinase activity of PKR (Thomis and Samuel 1993). Two lines of evidence suggest that PERK activation may be similarly regulated: We have expressed a recombinant form of PERK in which its cytosolic kinase domain is fused to the extracellular and transmembrane domain of the T-cell coreceptor CD4. This chimeric protein is expressed on the cell surface and is inactive. When the cells are treated with an antibody that cross-links the extracellular portion of the CD4–PERK chimera, its kinase is activated, and eIF2α is phosphorylated (Bertolotti et al. 2000). This experiment reveals a potential role for oligomerization in regulating PERK kinase activity. In other experiments, we have analyzed the size of the endogenous PERK-containing complexes extracted by detergent from the ER membrane of stressed and unstressed cells. In cells exposed to ER stress, PERK shifts from a complex of relatively low molecular weight (mean size of ~240 kD) to one of a very high molecular weight. PERK in the high-molecular-weight complex is entirely phosphorylated, suggesting that it is activated (Bertolotti et al. 2000). One interpretation of this finding is that in response to ER stress, PERK oligomerizes to form a high-molecular-weight complex in which transphosphorylation and kinase activation take place.

The trigger for activation of PERK has not yet been clearly identified; however, it is likely to be shared by IRE1. Evidence for this comes from domain swap experiments in which the lumenal and transmembrane domain of IRE1β has been found to impart ER-stress regulation on the kinase domain of PERK (Bertolotti et al. 2000). We have recently found that in non-stressed cells, PERK (and IRE1) are associated in a stable complex with the soluble ER chaperone BiP (or GRP78). This association is rapidly disrupted by exposing cells to ER stress. BiP binding is rapidly restored upon reversal of the stressful condition. Furthermore, overexpression of BiP attenuates the activation of PERK. This is reflected in delayed and reduced phosphorylation of PERK by agents that cause ER stress in the BiP overexpressing cells. One model for PERK regulation during ER stress would assign an inhibitory role to BiP binding, perhaps by steric hindrance of PERK oligomerization. According to this model, accumulation of malfolded proteins in the ER would lead to dissociation of the PERK–BiP complex. PERK molecules that have lost their inhibitory BiP partner would undergo oligomerization, transphosphorylation, and activation. The ER stress signal is thus postulated to be inversely related to the availability of BiP to bind to the lumenal domain of PERK. The major features of this model have been presented in cartoon form in Figure 2. Competing hypotheses are also consistent with the data. For example, the existence of specific activating ligands that are elaborated during ER stress cannot be excluded. BiP overexpression might attenuate signaling by these ligands through its ability to bind them directly or indirectly by interfering with their production.

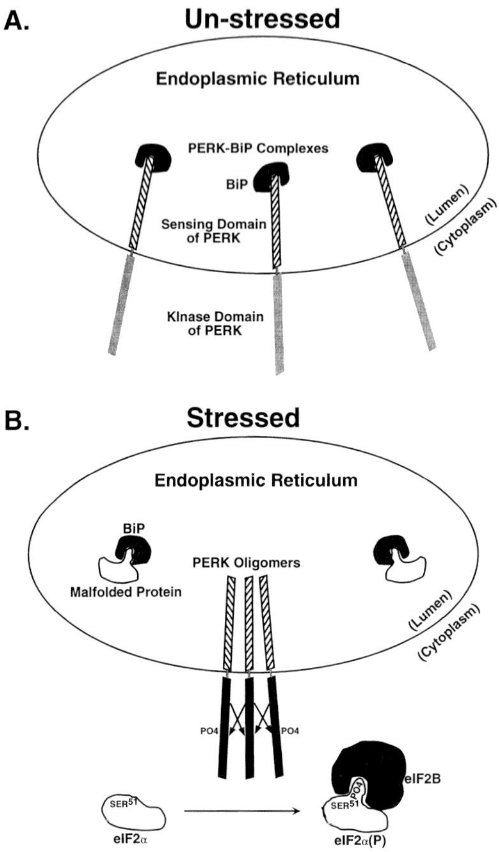

Figure 2 Cartoon describing features of PERK activation by ER stress. (*A*) In unstressed cells, PERK exists in a complex with the ER chaperone BiP. PERK's cytoplasmic kinase domain is inactive (the inactive state of the cytoplasmic kinase domain is colored light gray). (*B*) Upon induction of ER stress, the PERK–BiP interaction is disrupted. PERK oligomerizes (arbitrarily depicted here as a trimer; the true composition of the oligomers is not known), transphosphorylation of the cytoplasmic domain takes place (depicted here by the crossed arrows and the appearance of PO4 moieties), and PERK kinase activity markedly increases (the activated cytoplasmic kinase domain is colored solid black). This results in the phosphorylation of eIF2α on serine 51 (SER^{51}), sequestration of eIF2B in an inactive complex, and inhibition of the rate of formation of a 43S translation initiation complex.

SPECULATION ON THE PHYSIOLOGICAL SIGNIFICANCE OF *PERK* SIGNALING

Perk$^{-/-}$ cells exhibit markedly decreased survival compared to *Perk*$^{+/+}$ and *Perk*$^{+/-}$ cells when exposed to conditions causing ER stress. This defect can be partially reversed by treating the knockout cells with low-dose cycloheximide during application of the ER stress (Harding et al. 2000). In wild-type ES cells, treatment with agents that cause ER stress (thapsigargin, tunicamycin, and DTT) leads to a global *decrease* in the quantity of newly synthesized proteins that associate with BiP. This presumably reflects the fact that PERK-mediated translational attenuation is very effective at reducing the total load on the ER. In *Perk*$^{-/-}$ cells, in contrast, ER stress leads to an increase in the amount of newly synthesized proteins associated with BiP (H. Harding et al., unpublished observations). These results suggest that translation inhibition during the UPR plays an important role in mitigating the stress on the ER and that *Perk*$^{-/-}$ cells that are deficient in this regulatory pathway exhibit increased levels of ER stress. Additional support for this point comes from the observation that IRE1 activation is enhanced in tunicamycin-treated *Perk*$^{-/-}$ cells, compared to their wild-type counterparts.

It is thus reasonable to speculate that PERK is part of a signaling pathway that matches rates of translation to rates of protein folding. The studies reviewed here rely exclusively on pharmacological agents to induce ER stress and activate PERK signaling. Very little is known about the physiological and pathophysiological pathways that result in PERK activation and attenuation of translation. Numerous studies have documented eIF2α phosphorylation and translational inhibition in response to hormones that activate calcium signaling and deplete ER calcium stores (Takuma et al. 1984; Kimball and Jefferson 1990, 1992; Perkins and Pandol 1992). The physiological significance of this regulatory pathway is not understood, but it appears likely that PERK plays a role in this response. Neuronal ischemia is also associated with the striking coincidence of profound ER calcium depletion and eIF2α phosphorylation (for review, see Paschen and Doutheil 1999), and one might imagine a role for PERK in mediating the translational inhibition in this setting as well. In this latter system, it is not known whether the translational inhibition is salutary from the point of view of neuronal survival or whether it plays a role in promoting neuronal death.

There are numerous examples of disease states associated with the production of proteins that malfold in the endoplasmic reticulum (for review, see Carrell and Lomas 1997). In some of these, the genetics suggest a gain-of-function phenotype related to the accumulation of the abnormal protein, rather than a deficiency in the synthesis of the normal protein (e.g., see Hudson and Nadon 1992; Gow and Lazzarini 1996; Ito and Jameson 1997; Wang et al. 1999). We do not know whether production of malfolded proteins encoded by these mutant alleles of highly expressed genes is sufficient to trigger PERK activation. However, given the homology between the transcriptional and translational arms of the UPR, and knowing that many of the aforementioned disease states are associated with elevated levels of BiP, we speculate that PERK might be activated under such circumstances. One informative line of experimentation would be to evaluate the genetic interactions between these putative ER-stress-causing mutations and PERK deficiency. If PERK deficiency was found to enhance the phenotype of the mutations in secreted proteins, this would provide evidence for a role for ER stress in causing cellular dysfunction and would support the hypothesis that overloading the folding machinery of the ER can promote cellular dysfunction. The relevance of PERK activity may project beyond the rare genetic disorders cited above. It is possible that some of the common aging-related diseases prevalent in the human population have a component of cellular dysfunction related to chronic ER stress. Type II diabetes, for example, is a disorder associated with peripheral insulin resistance and a long presymptomatic phase of insulin hyperproduction and hypersecretion. It ultimately results in the "exhaustion" of the insulin-producing pancreatic islet β-cell. The cellular pathways that mediate this exhaustion phenomenon are not known, but chronic ER stress may have a role in its development. It is interesting to note that PERK is expressed at dramatically high levels in the endocrine pancreas (Shi et al. 1998), suggesting that PERK activity may affect this important disease state.

Interpretation of the results of the "synthetic" interactions between PERK deficiency and malfolding protein disorders may not be so simple, particularly if PERK signaling itself leads, under some circumstances, to cellular dysfunction. One reason for considering this possibility is the evidence that PKR-mediated eIF2α phosphorylation may participate in promoting programmed cell death during viral infection (Der et al. 1997; Srivastava et al. 1998). Given the homology between PKR and PERK and the ability of both proteins to phosphorylate eIF2α, it is possible that under some circumstances PERK may play a similar role in response to ER stress. One situation in which this type of response might be advantageous to the cell is viral infection. It is possible to imagine that ER stress triggered by a massive influx of viral glycoproteins might activate PERK, contributing to eIF2α phosphorylation that is already induced by activated PKR. It will be interesting to determine whether the two kinases function synergistically in providing an antiviral state and in causing virally infected cells to undergo programmed cell death. It is interesting to note that at least one virally encoded PKR antagonist, the K3L gene of vaccinia, is able to inactivate PERK (Sood et al. 2000).

Further complicating the issue of PERK's role in the adaptation to stress is the possibility that in addition to regulating translation rates, PERK plays a role in the induction of gene expression in response to ER stress. We have found that *Perk*$^{-/-}$ cells are severely impaired in activation of *Chop* gene transcription in response to tunicamycin, thapsigargin, and DTT (H.P. Harding et al., unpubl.). Recent findings from the Kaufman lab suggest that this defect is due to failure to phosphorylate eIF2α. These investigators found that cells in which the wild-type eIF2α has been replaced by a mutant form of the protein lacking the regulatory serine 51 phosphorylation site are unable to activate *Chop* transcription during ER stress. Together, our findings suggest that the phenotype of PERK deficiency may be a composite of the direct translational effects — leading to more ER stress and an impairment of a component of the transcriptional response to ER stress that is normally mediated by eIF2α phosphorylation. It is perhaps relevant to note that *Chop* encodes an ER stress-inducible transcription factor that may play a role in promoting programmed cell death under some conditions (Zinszner et al. 1998).

The relationship between PERK, eIF2α, and *Chop* revealed by these studies also raises the interesting possibility of the existence of a mammalian gene expression pathway related to that mediating regulation of GCN4 by amino acid starvation in the yeast (see Chapter 5). It is notable in this regard that *Chop* is also positively regulated by amino acid starvation (Bruhat et al. 1997; Jousse et al. 1999), but this response is preserved in *Perk*$^{-/-}$ cells (H.P. Harding et al., unpubl.), suggesting that a different eIF2α kinase may be mediating *Chop* induction through an identical downstream pathway.

Beyond the intuitively appealing notion that malfolded proteins in the ER predispose to cellular dysfunction, direct evidence that this is the case has been hard to come by and next to nothing is known about the pathways that might mediate the "toxic" effects of malfolded proteins. Defective function of the signaling pathway mediated by PERK would be expected to lead to cellular dysfunction due to the enhanced propensity to synthesize malfolded proteins. Availability of the *Perk* knockout mouse may also make it possible to study the physiological implications of chronic ER stress under conditions in which the mitigating effect of translational inhibition is disabled.

ACKNOWLEDGMENTS

We are grateful to our colleagues Anne Bertolotti, Isabel Novoa, and Huiqing Zeng for contributing unpublished results and for their critical comments. Work in the author's lab is supported by National Institutes of Health grants DK-47119 and ES-08681. H.P.H. is the recipient of NRSA award, CA-73181. D.R. is a Stephen Birnbaum Scholar of the Leukemia Society of America.

REFERENCES

Anfinsen C.B. 1973. Principles that govern the folding of protein chains. *Science* **181:** 223–230.

Bertolotti A., Zhang Y., Hendershot L., Harding H., and Ron D. 2000. Dynamic interaction of BiP and the ER stress transducers in the unfolded protein response. *Nature Cell Biology* **6:** 326–332.

Brostrom C.O. and Brostrom M.A. 1998. Regulation of translational initiation during cellular responses to stress. *Prog. Nucleic. Acid. Res. Mol. Biol.* **58:** 79–125.

Brostrom C.O., Bocckino S.B., and Brostrom M.A. 1983. Identification of a $Ca2^{+}$ requirement for protein synthesis in eukaryotic cells. *J. Biol. Chem.* **258:** 14390–14399.

Bruhat A., Jousse C., Wang X.-Z., Ron D., Ferrara M., and Fafournoux F. 1997. Amino acid limitation induces expression of *CHOP*, a CCAAT/enhancer binding protein-related gene, at both transcriptional and post-transcriptional levels. *J. Biol. Chem.* **272:** 17588–17593.

Carrell R.W. and Lomas D.A. 1997. Conformational disease. *Lancet* **350:** 134–138.

Der S.D., Yang Y.L., Weissmann C., and Williams B.R. 1997. A double-stranded RNA-activated protein kinase-dependent pathway mediating stress-induced apoptosis. *Proc. Natl. Acad. Sci.* **94:** 3279–3283.

Gething M.J. and Sambrook J. 1992. Protein folding in the cell. *Nature* **355:** 33–45.

Gow A. and Lazzarini R.A. 1996. A cellular mechanism governing the severity of Pelizaeus-Merzbacher disease. *Nat. Genet.* **13:** 422–428.

Harding H., Zhang Y., and Ron D. 1999. Translation and protein folding are coupled by an endoplasmic reticulum resident kinase. *Nature* **397:** 271–274.

Harding H., Zhang Y., Bertolotti A., Zeng H., and Ron D. 2000. *Perk* is essential for translational regulation and cell survival during the unfolded protein response. *Mol. Cell* **5:** 897–904.

Hudson L. and Nadon N. 1992. Amino acid substitutions in proteolipid protein that cause dysmyelination. In *Myelin: Biology and chemistry* (ed. R. Martenson), pp. 677–703. CRC Press, Boca Raton, Florida.

Ito M. and Jameson J.L. 1997. Molecular basis of autosomal dominant neurohypophyseal diabetes insipidus. Cellular toxicity caused by the accumulation of mutant vasopressin precursors within the endoplasmic reticulum. *J. Clin. Invest.* **99:** 1897–1905.

Jousse C., Bruhat A., Harding H.P., Ferrara M., Ron D., and Fafournoux P. 1999. Amino acid limitation regulates CHOP expression through a specific pathway independent of the unfolded protein response. *FEBS Lett.* **448:** 211–216.

Kaufman R.J. 1999. Stress signaling from the lumen of the endoplasmic reticulum: Coordination of gene transcriptional and translational controls. *Genes Dev.* **13:** 1211–1233.

Kimball S.R. and Jefferson L.S. 1990. Mechanism of the inhibition of protein synthesis by vasopressin in rat liver. *J. Biol. Chem.* **265:** 16794–16798.

———. 1992. Regulation of protein synthesis by modulation of intracellular calcium in rat liver. *Am. J. Physiol.* **263:** E958–E964.

Kostura M. and Mathews M.B. 1989. Purification and activation of the double-stranded RNA-dependent eIF-2 kinase DAI. *Mol. Cell. Biol.* **9:** 1576–1586.

Kozutsumi Y., Segal M., Normington K., Gething M.J., and Sambrook J. 1988. The presence of malfolded proteins in the endoplasmic reticulum signals the induction of glucose-regulated proteins. *Nature* **332:** 462–464.

Langland J.O. and Jacobs B.L. 1992. Cytosolic double-stranded RNA-dependent protein kinase is likely a dimer of partially phosphorylated M_r = 66,000 subunits. *J. Biol. Chem.* **267:** 10729–10736.

Lee A. 1992. Mammalian stress response: Induction of the glucose-regulated protein family. *Curr. Biol.* **4:** 267–273.

Paschen W. and Doutheil J. 1999. Disturbances of the functioning of endoplasmic reticulum: A key mechanism underlying neuronal cell injury? *J. Cereb. Blood Flow Metab.* **19:** 1–18.

Perkins P.S. and Pandol S.J. 1992. Cholecystokinin-induced changes in polysome structure regulate protein synthesis in pancreas. *Biochim. Biophys. Acta.* **1136:** 265–271.

Prostko C.R., Brostrom M.A., and Brostrom C.O. 1993. Reversible phosphorylation of eukaryotic initiation factor 2 alpha in response to endoplasmic reticular signaling. *Mol. Cell. Biochem.* **127–128:** 255–265.

Prostko C.R., Brostrom M.A., Malara E.M., and Brostrom C.O. 1992. Phosphorylation of eukaryotic initiation factor (eIF) 2 alpha and inhibition of eIF-2B in GH3 pituitary cells by perturbants of early protein processing that induce GRP78. *J. Biol. Chem.* **267:** 16751–16754.

Prostko C.R., Dholakia J.N., Brostrom M.A., and Brostrom C.O. 1995. Activation of the double-stranded RNA-regulated protein kinase by depletion of endoplasmic reticular calcium stores. *J. Biol. Chem.* **270:** 6211–6215.

Shamu C.E. 1997. Splicing together the unfolded-protein response. *Curr. Biol.* **7:** R67–R70.

Shamu C.E. and Walter P. 1996. Oligomerization and phosphorylation of the Ire1p kinase during intracellular signaling from the endoplasmic reticulum to the nucleus. *EMBO J.* **15:** 3028–3039.

Shamu C., Cox J., and Walter P. 1994. The unfolded-protein-response pathway in yeast. *Trends Cell. Biol.* **4:** 56–60.

Shi Y., Vattem K.M., Sood R., An J., Liang J., Stramm L., and Wek R.C. 1998. Identification and characterization of pancreatic eukaryotic initiation factor 2 alpha-subunit kinase, PEK, involved in translational control. *Mol. Cell. Biol.* **18:** 7499–7509.

Sidrauski C. and Walter P. 1997. The transmembrane kinase Ire1p is a site-specific endonuclease that initiates mRNA splicing in the unfolded protein response. *Cell* **90:** 1031–1039.

Sood R., Porter A.C., Ma K., Quilliam L.A., and Wek R.C. 2000. Pancreatic eukaryotic initiation factor-2α kinase (PEK) homologues in humans, *Drosophila melanogaster* and *Caenorhabditis elegans* that mediate translational control in response to endoplasmic reticulum stress. *Biochem J.* **346:** 281–293.

Srivastava S.P., Davies M.V., and Kaufman R.J. 1995. Calcium depletion from the endoplasmic reticulum activates the double-stranded RNA-dependent protein kinase (PKR) to inhibit protein synthesis. *J. Biol. Chem.* **270:** 16619–16624.

Srivastava S.P., Kumar K.U., and Kaufman R.J. 1998. Phosphorylation of eukaryotic translation initiation factor 2 mediates apoptosis in response to activation of the double-stranded RNA- dependent protein kinase. *J. Biol. Chem.* **273:** 2416–2423.

Takuma T., Kuyatt B.L., and Baum B.J. 1984. Alpha 1-adrenergic inhibition of protein synthesis in rat submandibular cells. *Am. J. Physiol.* **247:** G284–G289.

Thomis D.C. and Samuel C.E. 1993. Mechanism of interferon action: Evidence for intermolecular autophosphorylation and autoactivation of the interferon-induced, RNA-dependent protein kinase PKR. *J. Virol.* **67:** 7695–7700.

Tirasophon W., Welihinda A.A., and Kaufman R.J. 1998. A stress response pathway from the endoplasmic reticulum to the nucleus requires a novel bifunctional protein kinase/endoribonuclease (Ire1p) in mammalian cells. *Genes. Dev.* **12:** 1812–1824.

Wang J., Takeuchi T., Tanaka S., Kubo S.K., Kayo T., Lu D., Takata K., Koizumi A., and Izumi T. 1999. A mutation in the insulin 2 gene induces diabetes with severe pancreatic beta-cell dysfunction in the Mody mouse. *J. Clin. Invest.* **103:** 27–37.

Wang X.Z., Harding H.P., Zhang Y., Jolicoeur E.M., Kuroda M., and Ron D. 1998. Cloning of mammalian Ire1 reveals diversity in the ER stress responses. *EMBO. J.* **17:** 5708–5717.

Wong W.L., Brostrom M.A., Kuznetsov G., Gmitter-Yellen D., and Brostrom C.O. 1993. Inhibition of protein synthesis and early protein processing by thapsigargin in cultured cells. *Biochem. J.* **289:** 71–79.

Wu S. and Kaufman R.J. 1997. A model for the double-stranded RNA (dsRNA)-dependent dimerization and activation of the dsRNA-activated protein kinase PKR. *J. Biol. Chem.* **272:** 1291–1296.

Zinszner H., Kuroda M., Wang X., Batchvarova N., Lightfoot R.T., Remotti H., Stevens J.L., and Ron D. 1998. CHOP is implicated in programmed cell death in response to impaired function of the endoplasmic reticulum. *Genes. Dev.* **12:** 982–995.

16

Regulation of Translation Initiation in Mammalian Cells by Amino Acids

Scot R. Kimball and Leonard S. Jefferson
Department of Cellular and Molecular Physiology
The Pennsylvania State University
College of Medicine
Hershey, Pennsylvania 17033

Translation initiation in mammalian cells is exquisitely sensitive to modulation by variations in amino acid supply. Modulation involves the eIF2/eIF2B, eIF4F, and p70^{S6k} regulatory steps in translation initiation and, dependent on the prevailing experimental conditions, can manifest as a change in the global rate of protein synthesis and/or a change in translation of specific mRNAs. Limitation of essential amino acids down-regulates global rates of protein synthesis through uncharacterized protein kinases; one modulates phosphorylation of Ser-51 on the α subunit of eIF2 and another modulates phosphorylation of an unidentified site on the ε subunit of eIF2B. In contrast, the amino acid leucine, at concentrations within the physiological range, plays a unique role involving modulation of eIF4F and p70^{S6k} regulatory steps through a mechanism involving the protein kinase mTOR (the *m*ammalian *t*arget *o*f *r*apamycin) (Chapter 6). Leucine may also modulate eIF2B through an mTOR-insensitive mechanism. These effects of leucine appear to involve a ligand-specific recognition site that has some overlap of specificity with isoleucine but not other amino acids. In this chapter, we summarize recent evidence that supports this view of the role of amino acids in the modulation of translation initiation in mammalian cells.

PROTEIN SYNTHESIS IN MAMMALIAN TISSUES IS REGULATED BY VARIATIONS IN AMINO ACID SUPPLY

An abundance of evidence generated from in vivo studies in both humans and animals has established a predominant role for amino acids in the regulation of protein synthesis. Of the nutrients and hormones whose

Translational Control of Gene Expression
© 2000 Cold Spring Harbor Laboratory Press 0-87969-568-4/00 $5 +. 00

plasma concentrations change in response to food intake in humans, amino acids and insulin play the primary roles in promoting whole-body protein anabolism (Castellino et al. 1987; Tessari et al. 1987; Luzi et al. 1990; Inchiostro et al. 1992). Amino acid regulation accounts for as much as 90% of postprandial protein anabolism in humans, whereas insulin produces only a modest additional effect (Volpi et al. 1996). Modulation of protein synthesis due to variations in amino acid supply is observed in a variety of tissues including liver and skeletal muscle. For example, changes in protein synthesis in liver contribute importantly to the whole-body response in humans, with the rate of albumin secretion from the liver being increased approximately 50% by amino acid intake and additionally increased approximately 20% by postprandial hyperinsulinemia (Volpi et al. 1996). Indeed, the increase in albumin synthesis accounts for 28% of the increase in whole-body protein synthesis associated with feeding (De Feo et al. 1992). Similarly, amino acids have an overall anabolic effect on protein metabolism in skeletal muscle. In adult humans, oral administration of a complete mixture of amino acids (Volpi et al. 1996; Biolo et al. 1997) or a mixture of only the essential amino acids (Tipton et al. 1999) results in a stimulation of muscle protein synthesis.

Modulation of protein synthesis in liver and skeletal muscle in response to food intake has been studied in detail in animals. In rat liver, the protein synthetic rate undergoes circadian variations with more than a twofold change in amplitude that correlates with the rhythmic feeding behavior of the animal (Lardeux et al. 1978). The variations in protein synthesis are correlated with the protein and RNA contents of the liver, which are reduced by as much as 50% by food deprivation and are restored rapidly to peak values by food intake (Ramsey and Steele 1976; Lardeux et al. 1978). The protein synthetic response of the liver to food intake is mediated by the protein component of the meal and is independent of increases in plasma insulin concentrations (Yoshizawa et al. 1998). A similar conclusion can be drawn from studies of skeletal muscle. In rats or mice subjected to an 18-hour fast, plasma insulin concentrations are reduced to approximately one-half of the value observed in fed animals (Svanberg et al. 1996; Yoshizawa et al. 1998). Feeding the fasted animals either a complete diet or an isocaloric diet consisting of carbohydrate alone returns insulin to control, freely fed, values in both cases. However, only the protein-containing diet stimulates protein synthesis in skeletal muscle (Gautsch et al. 1998; Yoshizawa et al. 1998).

The stimulation of protein synthesis in response to food intake associated with activation of translation initiation is evidenced by rapid reaggregation of polysomes (Lardeux et al. 1978; Rannels et al. 1978;

Yokogoshi and Yoshida 1986). Translation initiation, as judged by polysome aggregation, varies in relation to the nutritional quality of dietary proteins (Yokogoshi et al. 1980) and with the amount of single essential amino acids (Fleck et al. 1965; Sidransky et al. 1968, 1971; Yokogoshi and Yoshida 1980). Similar responses in translation initiation are observed in vitro in preparations of perfused rat liver (Flaim et al. 1982) and skeletal muscle (Vary et al. 1999) exposed to variations in physiologically balanced mixtures of amino acids, unbalanced mixtures, or limiting amounts of single essential amino acids.

THE eIF2/eIF2B REGULATORY STEP IN TRANSLATION INITIATION IS INVOLVED IN MEDIATING THE EFFECTS OF AMINO ACID DEPRIVATION ON GLOBAL RATES OF PROTEIN SYNTHESIS

Numerous studies have shown that deprivation of essential amino acids causes a reduction in global rates of protein synthesis in cells in culture. The possibility that an accumulation of uncharged tRNA caused by limitation of amino acid supply might be involved in the inhibition had been suggested by early studies showing that amino acid analogs, such as histidinol, dramatically attenuate protein synthesis (Hansen et al. 1972; Vaughan and Hansen 1973; Grummt and Grummt 1976; Warrington et al. 1977; Thomas and Mathews 1984). These compounds act as competitive inhibitors of aminoacyl-tRNA synthetases and result in accumulation of deacylated tRNA. The reduction in protein synthesis associated with deacylation of tRNA is a result of a fall in the rate of translation initiation relative to elongation as evidenced by the disaggregation of polysomes that occurs in cells exposed to amino acid analogs (Vaughan and Hansen 1973; Warrington et al. 1977; Thomas and Mathews 1984). A more recent study (Kimball and Jefferson 1991) confirms the results of the earlier studies and extends them to show that the inhibition of protein synthesis caused by amino acid limitation in perfused rat liver is associated with an increase in phosphorylation of the α subunit of eukaryotic initiation factor (eIF)2 and an inhibition of the guanine nucleotide exchange activity of eIF2B. Phosphorylation of eIF2α on Ser-51 converts the protein from a substrate into a competitive inhibitor of eIF2B and results in a decrease in the amount of the eIF2•GTP binary complex available to bind Met-$tRNA_i$ (for review, see Chapters 2 and 5). Addition of histidinol to the perfusion medium magnifies the effect of histidine deprivation on protein synthesis, eIF2α phosphorylation, and eIF2B activity (Kimball and Jefferson 1991), suggesting that deacylation of histidyl-tRNA is causative in the effects.

The guanine nucleotide exchange activity of eIF2B is also regulated by amino acid deprivation through mechanisms distinct from eIF2α phosphorylation. In cultures of L6 myoblasts, deprivation of either leucine or histidine results in an inhibition of protein synthesis that is associated with enhanced phosphorylation of eIF2α as well as a reduction in eIF2B activity (Kimball et al. 1998a). However, the proportion of eIF2 in the phosphorylated form is never more than 5% of the total, even in amino-acid-deprived L6 myoblasts (Kimball et al. 1998a). Because the ratio of eIF2 to eIF2B in L6 myoblasts is approximately 2:1, it is unlikely that phosphorylation of 5% of total eIF2 accounts completely for the inhibition of protein synthesis or eIF2B activity. Likewise, in perfused rat livers, provision of an unbalanced mixture of amino acids containing an excess of leucine in relation to other amino acids results in an inhibition of protein synthesis in association with a 50% reduction in the nucleotide exchange activity of eIF2B (Shah et al. 1999). Of interest is the finding that in *Saccharomyces cerevisiae*, addition of leucine to normal growth medium in excess of the normal amount results in activation of the general control response (Niederberger et al. 1981). If valine and isoleucine are added together with leucine, no induction of the general control response is observed, suggesting that it is an imbalance in branched-chain amino acids that is responsible for the effect. In yeast, the general control response is normally induced as a result of amino acid deprivation and involves increased phosphorylation of eIF2α as well as inhibition of eIF2B (for review, see Hinnebusch 1994, 1997). However, in perfused rat liver, no change in the phosphorylation state of eIF2α was observed in response to leucine imbalance (Shah et al. 1999). A possible explanation for the inhibition of nucleotide exchange activity under these conditions is modulation of the phosphorylation state of the ε subunit of eIF2B, which is a substrate in vitro for three different protein kinases, i.e., glycogen synthase kinase (GSK)-3, casein kinase (CK)-I, and CK-II (Dholakia and Wahba 1988; Oldfield and Proud 1992; Welsh and Proud 1993; Aroor et al. 1994; Singh et al. 1994, 1996; Welsh et al. 1997a, 1998; Jefferson et al. 1999). Phosphorylation by either CK-I (Singh et al. 1996) or CK-II (Dholakia and Wahba 1988; Singh et al. 1994) reportedly stimulates the guanine nucleotide exchange activity of eIF2B. In contrast, phosphorylation of eIF2B partially purified from CHO.T cells by GSK-3 is inhibitory (Welsh et al. 1998). A precedent for this mode of regulation has been established in CHO.T cells where phosphorylation of eIF2Bε by GSK-3 is an important mechanism for mediating the action of insulin on the guanine nucleotide exchange activity of eIF2B (Welsh et al. 1997b, 1998). In CHO.T cells, insulin regulates GSK-3 through a PI3-kinase-dependent pathway; however, this pathway does not appear to be modulated by amino acid supply in FAO (Patti et al. 1998) or Jurkat T-lymphoblastoid (Iiboshi et al. 1999) cells. In L6 myoblasts, deprivation of either leucine or histidine results in a decrease in the activity of an uncharacterized eIF2Bε kinase but has no effect on GSK-3 activity (Kimball et al. 1998a). Furthermore, the direction of change in eIF2Bε kinase activity observed in leucine- or histidine-deprived L6 myoblasts is the opposite of what would be expected if the eIF2Bε kinase were GSK-3. Thus, because phosphorylation by GSK-3 is inhibitory, a decrease in eIF2B activity should be associated with an increase in GSK-3 activity if GSK-3 is causative in the effect. Overall, it is unlikely that GSK-3 mediates the regulation of eIF2B activity by amino acid supply. However, because eIF2Bε kinase activity decreases in response to amino acid deprivation, a role for other kinases in the regulation of eIF2B in L6 myoblasts must be considered.

THE eIF4F REGULATORY STEP IN TRANSLATION INITIATION IS INVOLVED IN MEDIATING THE EFFECTS OF AMINO ACID SUPPLY ON BOTH GLOBAL RATES OF PROTEIN SYNTHESIS AND TRANSLATION OF SPECIFIC mRNAS

The stimulation of protein synthesis observed in intact rats and mice in response to ingestion of a protein-containing meal is accompanied by a decrease in association of the mRNA cap-binding protein, eIF4E, with the translational repressor, 4E-BP1, in both liver and skeletal muscle (Yoshizawa et al. 1998). As reviewed in Chapters 2 and 6, binding of mRNA to the 40S ribosomal subunit is mediated by the eIF4F complex, which consists of eIF4E, eIF4A (an RNA helicase), and eIF4G (a scaffolding protein to which eIF4E and eIF4A bind). Binding of 4E-BP1 to eIF4E prevents formation of the active eIF4F complex and thereby interferes with mRNA binding to the 40S ribosomal subunit. Association of eIF4E with 4E-BP1 is modulated by changes in phosphorylation of 4E-BP1 where the hyperphosphorylated form of 4E-BP1 does not bind to eIF4E (Chapter 6). Thus, in both liver and skeletal muscle, a protein-containing meal causes an increase in phosphorylation of 4E-BP1 (Yoshizawa et al. 1997, 1998). Moreover, the reduced binding of 4E-BP1 to eIF4E is inversely proportional to alterations in association of eIF4G with eIF4E. A recent study using diabetic mice shows that changes in protein synthesis, 4E-BP1 phosphorylation, and 4E-BP1 and eIF4G bound to eIF4E in response to feeding are the same in skeletal muscle of diabetic animals as in controls, even though plasma insulin concentrations are increased in control animals, but not in diabetic animals (Svanberg et al. 1997). Taken together, the findings of these studies strongly implicate amino acid supply, perhaps in combination with a permissive concentration of insulin, as a mediator of the response to feeding of protein synthesis and the eIF4F regulatory step in liver and skeletal muscle.

Of all the amino acids, leucine appears to be most effective in recapitulating the responses in protein synthesis and the eIF4F regulatory step produced by feeding a protein-containing meal. For example, in skeletal muscle, oral administration of leucine, or a combination of leucine and carbohydrate, restores protein synthesis to the rate observed in freely fed animals, whereas carbohydrate alone does not (Anthony et al. 2000). In addition, leucine administration stimulates 4E-BP1 phosphorylation 3-fold, as measured by reduced migration on SDS-polyacrylamide gels, compared to the value observed in fasted animals. Serum insulin is elevated 2.6- and 3.7-fold in animals administered carbohydrate alone or a combination of leucine and carbohydrate, respectively, but is unchanged in animals given leucine alone. These findings suggest that leucine administration stimulates protein synthesis in skeletal muscle in part by enhancing eIF4F formation, and this effect is independent of increases in serum insulin.

The effect of leucine to stimulate phosphorylation of 4E-BP1 and cause dissociation of the 4E-BP1•eIF4E complex is highly specific. For example, in skeletal muscle of rats fasted for 18 hours, oral administration of leucine causes a 7-fold increase in the amount of 4E-BP1 in the most highly phosphorylated form (Anthony et al. 2000). In comparison, isoleucine administration stimulates 4E-BP1 phosphorylation only 3-fold, and valine administration is without effect. Similar results have been observed in perfused rat livers exposed to an unbalanced mixture of amino acids. Thus, perfusion with medium containing amino acids at concentrations found in the plasma of fasted rats, except that leucine, glutamine, and tyrosine are at four times those values, leads to increased phosphorylation of 4E-BP1, dissociation of the 4E-BP1•eIF4E complex, and enhanced binding of eIF4E to eIF4G (Shah et al. 1999). The combination of leucine, glutamine, and tyrosine was used because earlier work had shown that those three amino acids play a key role in regulating autophagy in liver (Mortimore et al. 1987). Modulation of eIF4F complex formation under these conditions is leucine-specific because substitution of leucine with isoleucine is ineffective in stimulating 4E-BP1 phosphorylation or binding of eIF4E to eIF4G.

A number of recent studies have utilized cells in culture to examine the mechanism(s) by which amino acids modulate the eIF4F regulatory step in translation initiation. In a variety of cell types, amino acid deprivation causes a rapid dephosphorylation of 4E-BP1 (Fox et al. 1998; Hara et al. 1998; Kimball et al. 1998a; Wang et al. 1998; Xu et al. 1998a). Dephosphorylation of 4E-BP1 is associated with increased association of the protein with eIF4E (Fox et al. 1998; Kimball et al. 1998a; Wang et al. 1998) and decreased binding of eIF4E to eIF4G (Kimball et al. 1998a). In all of the studies, leucine accounts for the majority of the effect of amino acids on 4E-BP1 phosphorylation. The transamination product of leucine, α-ketoisocaproic acid (α-KIC), reverses the changes in 4E-BP1 phosphorylation caused by amino acid deprivation (Fox et al. 1998; Patti et al. 1998; Xu et al. 1998a). However, the effect of α-KIC, but not leucine, on 4E-BP1 phosphorylation is attenuated by the transaminase inhibitor, (aminooxy)-acetate, suggesting that conversion of α-KIC to leucine is required for the effect (Fox et al. 1998). Thus, leucine, and not α-KIC, is the effector molecule in enhancing 4E-BP1 phosphorylation.

In addition to modulating global rates of protein synthesis, alterations in amino acid supply also cause changes in translation of specific mRNAs. As noted above, deprivation of either leucine or histidine in L6 myoblasts results in inhibition of global rates of protein synthesis due to a reduction in eIF2B activity (Kimball et al. 1998a). Leucine deprivation, but not histidine deprivation, additionally modulates the eIF4F regulatory step under these conditions. Thus, restoration of the leucine concentration in the culture medium of the amino-acid-deprived cells causes a preferential increase in the synthesis of ornithine decarboxylase (ODC), over and above the observed stimulation of global rates of protein synthesis. The 5′-untranslated region of the mRNA encoding ODC is long and predicted to form extensive secondary structure, and the translation of such mRNAs is regulated through eIF4F (for review, see Klein et al. 1998). The effect of leucine resupplementation is prevented by rapamycin, a specific inhibitor of the protein kinase known as the mammalian target of rapamycin (mTOR). mTOR serves as a bifurcation point in the signal transduction pathways to 4E-BP1 and $p70^{S6k}$ that are activated by growth-promoting hormones such as insulin and insulin-like growth factor I as well as cytokines and G-coupled receptor agonists (Chapter 6). Thus, leucine stimulates 4E-BP1 phosphorylation and promotes formation of the active eIF4F complex through an mTOR-dependent process.

The preponderance of studies examining regulation of the eIF4F regulatory step in translation initiation by amino acids shows that changes in binding of eIF4E to eIF4G inversely correlate with alterations in association of eIF4E with 4E-BP1 (for review, see Gingras et al. 1999). Such findings support a model whereby amino acids stimulate 4E-BP1 phosphorylation, resulting in dissociation of the 4E-BP1•eIF4E complex, freeing eIF4E to bind to eIF4G. In contrast, a recent study in perfused rat skeletal muscle shows that increasing amino acids in the perfusate from concentrations found in plasma of fasted rats to 10 times that amount stimulates protein synthesis 2-fold and increases the amount of eIF4G bound to eIF4E 8-fold (Vary et al. 1999), with no change in the amount of 4E-BP1 bound to eIF4E. This result suggests that binding of 4E-BP1 to eIF4E is not the sole determinant regulating the interaction of eIF4E with eIF4G. This idea is supported by other studies demonstrating alterations in association of eIF4E with eIF4G that occur independent of changes in binding of eIF4E to 4E-BP1 (Morley and McKendrick 1997; Kimball et al. 1998b; Fraser et al. 1999). In such cases, 4E-BP2 or 4E-BP3 may be involved in modulating binding of eIF4E to eIF4G. However, few studies have examined such a possibility.

Another potential mechanism for regulating the mRNA-binding step in translation initiation involves phosphorylation of eIF4E. In cells in culture, increased incorporation of $^{32}P_i$ into eIF4E correlates with enhanced protein synthesis under a variety of conditions (for review, see Flynn et al. 1997; Morley 1997). In contrast, in rats fasted for 18 hours and then either refed a complete diet (Yoshizawa et al. 1998) or administered oral leucine alone (Anthony et al. 2000) the phosphorylation state of eIF4E, as quantitated by Western blot analysis of phosphorylated and unphosphorylated forms of the protein separated by isoelectric focusing gel electrophoresis, is reduced in both liver and skeletal muscle. As discussed in greater detail in Chapter 6, one possible explanation for these apparent discrepancies is that the turnover of phosphate on eIF4E may be more important in controlling protein synthesis than is the steady-state phosphorylation of the protein.

PHOSPHORYLATION OF RIBOSOMAL PROTEIN S6 IS REGULATED BY AMINO ACID SUPPLY

The role of ribosomal protein S6 phosphorylation in the regulation of mRNA translation is reviewed in detail in Chapters 22 and 23. Briefly, S6 phosphorylation by the 70-kD S6 kinase ($p70^{S6k}$) is unlikely to play an important role in the global control of mRNA translation, but instead may regulate the synthesis of a particular subset of proteins encoded by mRNAs containing a 5′-terminal oligopyrimidine tract (5′TOP). 5′TOP mRNAs encode primarily proteins involved in translation, such as ribosomal proteins (Meyuhas et al. 1996) and elongation factors eEF1A and eEF2 (Jefferies et al. 1994; Terada et al. 1994). Activation of $p70^{S6k}$ promotes the binding of 40S ribosomal subunits to TOP mRNAs, resulting in increased translation of these messages. This effect is also observed in vivo. In rats subjected to an 18-hour fast, $p70^{S6k}$ is predominantly in the hypophosphorylated, least active form (Anthony et al. 2000). Oral administration of leucine, or a combination of leucine and carbohydrate, stimulates phosphorylation of $p70^{S6k}$ to levels observed in freely fed animals. In contrast, feeding carbohydrate alone has little effect on $p70^{S6k}$ phosphorylation. In cells in culture, leucine also promotes complete activation of $p70^{S6k}$, although other amino acids, such as methionine, can partially activate the kinase (Shigemitsu et al. 1999b).

Phosphorylation and activation of $p70^{S6k}$ in response to amino acids is also observed in a variety of cells in culture. In L6 myoblasts, phosphorylation of $p70^{S6k}$ correlates with increased phosphorylation of its substrate, S6 (Kimball et al. 1999). Furthermore, activation of $p70^{S6k}$ caused by leucine addition to leucine-deprived cells has functional consequences in that synthesis of eEF1A is preferentially increased compared to the synthesis of the majority of cellular proteins. The leucine-induced phosphorylation of $p70^{S6k}$ is completely prevented by treatment with rapamycin (Kimball et al. 1999; Shigemitsu et al. 1999b), indicating that the effect is mediated through mTOR.

WHICH STEP IN TRANSLATION INITIATION IS RATE-CONTROLLING FOR PROTEIN SYNTHESIS?

As discussed above, the essential amino acids leucine and histidine stimulate translation initiation in L6 myoblasts at both the Met-$tRNA_i$ and mRNA-binding steps (Kimball et al. 1998a). In particular, both amino acids activate eIF2B. In contrast, leucine, but not histidine, additionally causes a redistribution of eIF4E from the inactive eIF4E•4E-BP1 complex to the active eIF4E•eIF4G complex concomitant with increased phosphorylation of 4E-BP1. Leucine, but not histidine, also stimulates phosphorylation of both $p70^{S6k}$ and ribosomal protein S6 (Kimball et al. 1998a, 1999). Moreover, the changes in 4E-BP1 phosphorylation, eIF4E redistribution, and $p70^{S6k}$ activation promoted by leucine result in a specific increase in the synthesis of translationally regulated mRNAs such as those encoding ODC and eEF1A (Kimball et al. 1999). The mTOR inhibitor, rapamycin, prevents all of the leucine-induced effects involving eIF4E, activation of $p70^{S6k}$, and preferential stimulation of ODC and eEF1A synthesis, without affecting the rate of global protein synthesis or eIF2B activity. Overall, the results suggest that both leucine and histidine regulate global rates of protein synthesis through modulation of eIF2B activity. Leucine, acting through an mTOR-dependent pathway, also stimulates the translation of specific mRNAs both by increasing the availability of eIF4E and by stimulating phosphorylation of S6. Importantly, in L6 myoblasts deprived of leucine, alterations in eIF4E availability are not rate-controlling for global protein synthesis, but are necessary for regulation of translation of specific mRNAs.

The possibility that modulation of eIF4E availability following amino acid treatment might not regulate the global rate of protein synthesis is also supported by a recent study showing that in skeletal muscle of 18-hour fasted rats administered a solution containing only isoleucine, binding of eIF4E to eIF4G is significantly increased, eIF4E binding to 4E-BP1 is reduced, and $p70^{S6k}$ is activated (J.C. Anthony et al., in prep.); however, global rates of protein synthesis are unchanged, suggesting that the observed changes in eIF4E availability and $p70^{S6k}$ phosphorylation are not sufficient to activate synthesis of the majority of muscle proteins. In contrast, in cells in culture that overexpress eIF4E, both global rates of protein synthesis and the synthesis of specific proteins are increased (Rosenwald et al. 1999). Thus, when present in excess of the normal physiological amount, eIF4E can stimulate global protein synthesis.

HOW DO CELLS RECOGNIZE AND RESPOND TO VARIATIONS IN AMINO ACID SUPPLY?

Presently, little is known about how cells recognize and respond to variations in amino acid supply. The best-characterized response to amino acid limitation is in *S. cerevisiae* and involves GCN2, an eIF2α kinase containing a domain with sequence similarity to histidyl-tRNA synthetase (for review, see Chapter 5). This domain is thought to recognize deacylated tRNAs and mediate activation of GCN2. Increased phosphorylation of eIF2α also occurs in mammalian cells that are depleted of amino acids, or that are treated with histidinol (Kimball and Jefferson 1991), a competitive inhibitor of His-tRNA synthetase, or that contain a temperature-sensitive mutant of Leu-tRNA synthetase (Austin et al. 1986; Clemens et al. 1987). These observations are consistent with a mechanism whereby accumulation of deacylated-tRNA would activate an eIF2α kinase in mammalian cells. However, despite the recent discovery of a human GCN2 homolog (Berlanga et al. 1999; Sood et al. 2000), identification of an eIF2α kinase activated in response to amino acid depletion in mammalian cells is still lacking. Furthermore, this potential mechanism, although possibly explaining the effects of severe amino acid limitation, would not account for modulation of protein synthesis by physiological variations in amino acid supply (Vary et al. 1999), where tRNA would be expected to be in a fully charged state because the amino acid concentration would be above the K_m for the tRNA synthetase.

In addition to putatively initiating the signal for eIF2α phosphorylation, tRNA aminoacylation has also been implicated in amino-acid-mediated control of $p70^{S6k}$ (Iiboshi et al. 1999). That study shows that a variety of amino acid derivatives, including leucinol and histidinol, rapidly inhibit $p70^{S6k}$ activity in Jurkat cells. Furthermore, $p70^{S6k}$ activity is attenuated in tsBN250 cells containing a temperature-sensitive variant of histidinyl-tRNA synthetase when the cells are incubated at the nonpermissive temperature (Iiboshi et al. 1999). These results provide support for a model whereby deacylated tRNA generates a signal that inhibits both the Met-$tRNA_i$- and mRNA-binding steps in translation initiation. However, a recent study suggests that this model should be viewed with caution; in isolated adipocytes, leucinol is without effect on 4E-BP1 phosphorylation, although leucine deprivation causes dramatic changes in phosphorylation of the protein (Lynch et al. 2000). Thus, there may be tissue-specific modulation of the mRNA-binding step. Alternatively, $p70^{S6k}$ and 4E-BP1 phosphorylation may be independently modulated by leucine.

An alternative explanation for how cells recognize variations in amino acid supply is provided by studies investigating leucine suppression of deprivation-induced macroautophagy in rat liver (Mortimore et al. 1994). In these studies, a nontransportable multiple antigen derivative constructed by attaching eight leucine residues to a lysine core (Leu_8-MAP) is as effective as leucine in suppressing macroautophagy. Furthermore, the Leu_8-MAP substrate photoaffinity-labels a membrane protein of approximately 340,000 M_r. The possibility of a signal being generated from an amino acid recognition molecule located in the plasma membrane is consistent with recent studies in *S. cerevisiae* involving Ssy1p, a protein in yeast having homology with amino acid permeases and displaying structural features similar to those that distinguish the Snf3p and Rgt2p glucose sensors from sugar transport proteins (Didion et al. 1998; Iraqui et al. 1999; Klasson et al. 1999). Ssy1p has no detectable amino acid transport activity but is required for transcriptional induction by leucine of several genes encoding amino acid permeases and peptide transporters (Didion et al. 1998). Furthermore, Ssy1p regulates protein degradation by signaling through the F-box protein Grr1p, a component of the SCF^{Grr1} ubiquitin–protein ligase complex (Iraqui et al. 1999). However, a mammalian homolog of Ssy1p is still to be discovered, and a role for Ssy1p in regulating translation has not yet been defined.

MEDIATION OF THE RESPONSE TO VARIATIONS IN AMINO ACID SUPPLY INVOLVES MULTIPLE SIGNAL TRANSDUCTION PATHWAYS

As discussed above, amino acids mediate their effects on 4E-BP1 phosphorylation and p70^{S6k} activation through signal transduction pathways that involve mTOR. In addition, pretreating cells with wortmannin, an inhibitor of PI3-kinase, which is upstream of mTOR in the insulin signaling pathway, also prevents 4E-BP1 phosphorylation in response to amino acids (Wang et al. 1998; Xu et al. 1998a). However, amino acids do not stimulate either PI3-kinase (Patti et al. 1998; Iiboshi et al. 1999) or PKB (Hara et al. 1998; Patti et al. 1998; Wang et al. 1998; Iiboshi et al. 1999), suggesting that, in contrast to insulin or IGF-I, amino acids are acting downstream from these two kinases to regulate 4E-BP1 and p70^{S6k} phosphorylation. Thus, the inhibitory effect of wortmannin on amino-acid-stimulated phosphorylation of 4E-BP1 may reflect either a nonspecific effect of the inhibitor on a kinase other than PI3-kinase, or a requirement for a basal amount of PI3-kinase activity in order for amino acids to stimulate 4E-BP1 phosphorylation. Overall, the results suggest that leucine functions through at least three distinct signal transduction pathways to modulate translation initiation. The first pathway is mTOR-dependent

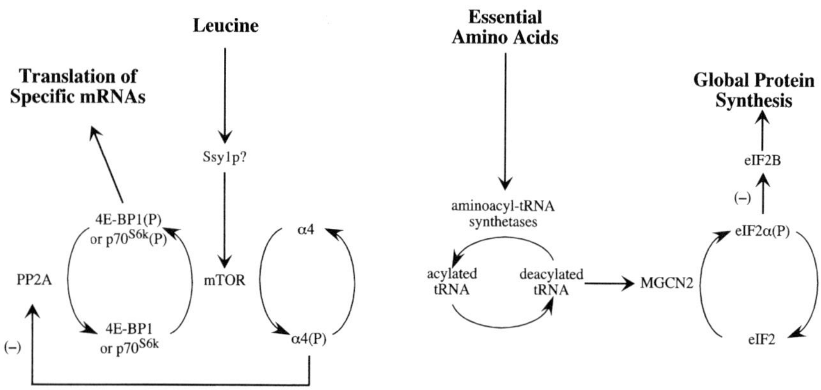

Figure 1 Signal transduction pathways from amino acids to translation initiation. The figure depicts two pathways through which amino acids potentially signal to translation initiation. On the left, increased translation of specific mRNAs is mediated through activation of p70^{S6k} and phosphorylation of 4E-BP1. Although leucine is depicted as the activator of this pathway, structurally related amino acids such as isoleucine and methionine additionally act as regulators, although to a lesser extent than leucine. On the right, all essential amino acids regulate global rates of protein synthesis through modulation of eIF2B activity.

and results in preferential translation of specific mRNAs through stimulation of eIF4F function and activation of p70^{S6k} (Fig. 1). The second is independent of mTOR and causes an increase in global rates of protein synthesis through changes in eIF2B activity. The third is independent of mTOR and is apparently mediated by severe limitation of leucine or other essential amino acids, resulting in activation of an eIF2α kinase.

COORDINATED REGULATION OF TRANSLATION INITIATION INVOLVING AMINO ACIDS AND HORMONES

In addition to a direct effect to stimulate eIF4F complex formation, amino acids also play an important permissive role in the insulin-mediated stimulation of protein synthesis (Fig. 2). This fact is underscored by the observation that in most cell types, neither insulin (Patti et al. 1998; Xu et al. 1998b; Shigemitsu et al. 1999b) nor IGF-I (Xu et al. 1998a) stimulates 4E-BP1 or p70^{S6k} phosphorylation in the absence of amino acids. The lack of effect is not due to a defect in activation of components of the insulin signal transduction pathway. For example, in the absence of amino acids, insulin still causes phosphorylation of the insulin receptor, IRS-1 and -2, PKB, and ERK1 and 2, promotes binding of the 85-kD subunit of PI3-kinase to IRS-1, and stimulates the activity of PI3-kinase (Hara et al. 1998; Patti et al. 1998). Thus, in the absence of amino acids, the ability of

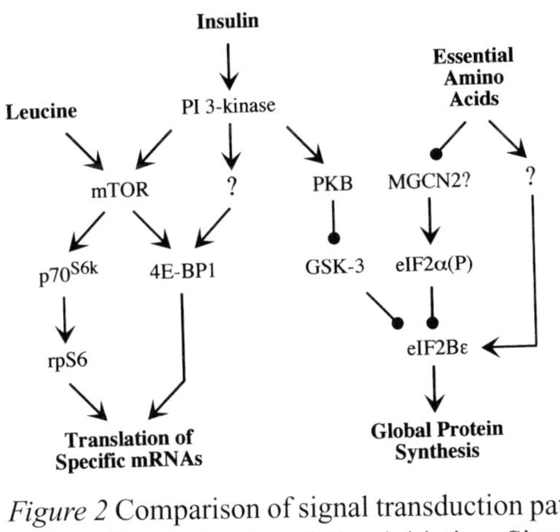

Figure 2 Comparison of signal transduction pathways through which amino acids and insulin regulate translation initiation. Signal transduction pathways activated by insulin and amino acids converge on eIF2B, 4E-BP1, and ribosomal protein S6 (rpS6), resulting in enhanced protein synthesis on a global (eIF2B) or specific (4E-BP1 and rpS6) basis.

insulin to activate kinases upstream of 4E-BP1 and p70^{S6k} is maintained. Interestingly, in HEK293 cells, amino acid deprivation prevents insulin-stimulated phosphorylation of Ser-2448 on mTOR, the site phosphorylated by PKB (Nave et al. 1999). Furthermore, purified PKB phosphorylates mTOR immunoprecipitated from cells maintained in amino-acid-replete medium, but not from amino-acid-deprived cells, suggesting that lack of amino acids results in modification of mTOR, or a protein associated with mTOR, that prevents phosphorylation by PKB. An apparent exception to these observations is cells of hepatic origin. In either FAO (Patti et al. 1998) or H4IIE (Shigemitsu et al. 1999a) hepatoma cells, insulin stimulates phosphorylation of both 4E-BP1 and p70^{S6k} in the absence of exogenous amino acids. The apparent contradiction between results obtained with hepatoma cells compared to other cell lines derives from the high rate of proteolysis in hepatocytes. In hepatoma cells, amino acids liberated as a result of protein degradation are sufficient for insulin-stimulated phosphorylation of 4E-BP1 and p70^{S6k}. Thus, in the presence of the inhibitor of protein degradation, 3-methyladenine, insulin no longer stimulates p70^{S6k} activity or 4E-BP1 phosphorylation in amino-acid-deprived H4IIE cells (Shigemitsu et al. 1999a). The simplest interpretation of the data is that phosphorylation of 4E-BP1 and p70^{S6k} requires input from two independent signal transduction pathways, one stimulated by amino acids and the other by insulin or IGF-I. One model that would account for these findings is that insulin or IGF-I stimulates a 4E-BP1 and p70^{S6k} kinase(s), whereas amino acids inhibit an mTOR-regulated phosphatase.

ACKNOWLEDGMENTS

The work in this chapter that was performed in the laboratory of Dr. Leonard S. Jefferson was supported in part by grant DK-15658 from the National Institutes of Diabetes and Digestive and Kidney Disease.

REFERENCES

Anthony J.C., Anthony T.G., Kimball S.R., Vary T.C., and Jefferson L.S. 2000. Orally administered leucine stimulates protein synthesis in skeletal muscle of post-absorptive rats in association with increased eIF4F formation. *J. Nutr.* **130:** 139–145.

Aroor A.R., Denslow N.D., Singh L.P., O'Brien T.W., and Wahba A.J. 1994. Phosphorylation of rabbit reticulocyte guanine nucleotide exchange factor in vivo. Identification of putative casein kinase II phosphorylation sites. *Biochemistry* **33:** 3350–3357.

Austin S.A., Pollard J.W., Jagus R., and Clemens M.J. 1986. Regulation of polypeptide chain initiation and activity of initiation factor eIF-2 in Chinese-hamster-ovary cell mutants containing temperature-sensitive aminoacyl-tRNA synthetases. *Eur. J. Biochem.* **157:** 39–47.

Berlanga J.J., Santoyo J., and de Haro C. 1999. Characterization of a mammalian homolog of GCN2 eukaryotic initiation factor 2α kinase. *Eur. J. Biochem.* **265:** 754–762.

Biolo G., Tipton K.D., Klein S., and Wolfe R.R. 1997. An abundant supply of amino acids enhances the metabolic effect of exercise on muscle protein. *Am. J. Physiol.* **273:** E122–E129.

Castellino P., Luzi L., Simonson D.C., Haymond M., and DeFronzo R.A. 1987. Effect of insulin and plasma amino acid concentrations on leucine metabolism in man. Role of substrate availability on estimates of whole body protein synthesis. *J. Clin. Invest.* **80:** 1784–1793.

Clemens M.J., Galpine A., Austin S.A., Panniers R., Henshaw E.C., Duncan R., Hershey J.W.B., and Pollard J.W. 1987. Regulation of polypeptide chain initiation in Chinese hamster ovary cells with a temperature-sensitive leucyl-tRNA synthetase. Changes in phosphorylation of initiation factor eIF-2 and in the activity of the guanine nucleotide exchange factor GEF. *J. Biol. Chem.* **262:** 767–771.

De Feo P., Horber F.F., and Haymond M.W. 1992. Meal stimulation of albumin synthesis: A significant contributor to whole body protein synthesis in humans. *Am. J. Physiol.* **263:** E794–E799.

Dholakia J.N. and Wahba A.J. 1988. Phosphorylation of the guanine nucleotide exchange factor from rabbit reticulocytes regulates its activity in polypeptide chain initiation. *Proc. Natl. Acad. Sci.* **85:** 51–54.

Didion T., Regenberg B., Jorgensen M.U., Kielland-Brandt M.C., and Andersen H.A. 1998. The permease homologue Ssy1p controls the expression of amino acid and peptide transporter genes in *Saccharomyces cerevisiae*. *Mol. Microbiol.* **27:** 643–650.

Flaim K.E., Liao W.S.L., Peavy D.E., Taylor J.M., and Jefferson L.S. 1982. The role of amino acids in the regulation of protein synthesis in perfused rat liver. II. Effects of amino acid deficiency on peptide chain initiation, polysomal aggregation, and distribution of albumin mRNA. *J. Biol. Chem.* **257:** 2939–2946.

Fleck A., Shepherd J., and Munro H.N. 1965. Protein synthesis in rat liver: Influence of amino acids in diet on microsomes and polysomes. *Science* **150:** 628–629.

Flynn A., Vries R.G.J., and Proud C.G. 1997. Signalling pathways which regulate eIF4E. *Biochem. Soc. Trans.* **25:** 192S.

Fox H.L., Pham P.T., Kimball S.R., Jefferson L.S., and Lynch C.J. 1998. Amino acid effects on translational repressor 4E-BP1 are mediated primarily by L-leucine in isolated adipocytes. *Am. J. Physiol.* **275:** C1232–C1238.

Fraser C.S., Pain V.M., and Morley S.J. 1999. The association of initiation factor 4F with poly(A)-binding protein is enhanced in serum-stimulated *Xenopus* kidney cells. *J. Biol. Chem.* **274:** 196–204.

Gautsch T.A., Anthony J.C., Kimball S.R., Paul G.L., Layman D.K., and Jefferson L.S. 1998. Availability of eIF4E regulates skeletal muscle protein synthesis during recovery from exercise. *Am. J. Physiol.* **274:** C406–C414.

Gingras A.-C., Raught B., and Sonenberg N. 1999. eIF4 initiation factors: Effectors of mRNA recruitment to ribosomes and regulators of translation. *Annu. Rev. Biochem.* **68:** 913–963.

Grummt F. and Grummt I. 1976. Studies on the role of uncharged tRNA in pleiotypic response of animal cells. *Eur. J. Biochem.* **64:** 307–312.

Hansen B.S., Vaughan M.H., and Wang L.-J. 1972. Reversible inhibition by histidinol of protein synthesis in human cells at the activation of histidine. *J. Biol. Chem.* **247:** 3854–3857.

Hara K., Yonezawa K., Weng Q.-P., Kozlowski M.T., Belham C., and Avruch J. 1998. Amino acid sufficiency and mTOR regulate p70 S6 kinase and eIF-4E BP1 through a common effector mechanism. *J. Biol. Chem.* **273:** 14484–14494.

Hinnebusch A.G. 1994. Translational control of *GCN4*: An in vivo barometer of initiation factor activity. *Trends Biochem. Sci.* **19:** 409–414.

———. Translational regulation of yeast GCN4 — A window on factors that control initiator–tRNA binding to the ribosome. *J. Biol. Chem.* **272:** 21661–21664.

Iiboshi Y., Papst P.J., Kawasome H., Hosoi H., Abraham R.T., Houghton P.J., and Terada N. 1999. Amino acid-dependent control of $p70^{S6k}$. Involvement of tRNA aminoacylation in the regulation. *J. Biol. Chem.* **274:** 1092–1099.

Inchiostro S., Biolo G., Bruttomesso D., Fongher C., Sabadin L., Carlini M., Duner E., Tiengo A., and Tessari P. 1992. Effects of insulin and amino acid infusion on leucine and phenylalanine kinetics in type 1 diabetes. *Am. J. Physiol.* **262:** E203–E210.

Iraqui I., Vissers S., Bernard F., de Craene J.O., Boles E., Urrestarazu A., and André B. 1999. Amino acid signaling in *Saccharomyces cerevisiae*: A permease-like sensor of external amino acids and F-Box protein Grr1p are required for transcriptional induction of the AGP1 gene, which encodes a broad-specificity amino acid permease. *Mol. Cell. Biol.* **19:** 989–1001.

Jefferies H.B.J., Reinhard C., Kozma S.C., and Thomas G. 1994. Rapamycin selectively represses translation of the "polypyrimidine tract" mRNA family. *Proc. Natl. Acad. Sci.* **91:** 4441–4445.

Jefferson L.S., Fabian J.R., and Kimball S.R. 1999. Glycogen synthase kinase-3 is the predominant insulin-regulated eukaryotic initiation factor 2B kinase in skeletal muscle. *Int. J. Biochem. Cell Biol.* **31:** 191–200.

Kimball S.R. and Jefferson, L.S. 1991. Mechanism of inhibition of peptide chain initiation by amino acid deprivation in perfused rat liver. Regulation involving inhibition of eukaryotic initiation factor 2α phosphatase activity. *J. Biol. Chem.* **266:** 1969–1976.

Kimball S.R., Horetsky R.L., and Jefferson L.S. 1998a. Implication of eIF2B rather than eIF4E in the regulation of global protein synthesis by amino acids in L6 myoblasts. *J. Biol. Chem.* **273:** 30945–30953.

———. 1998b. Signal transduction pathways involved in the regulation of protein synthesis by insulin in L6 myoblasts. *Am. J. Physiol.* **274:** C221–C228.

Kimball S.R., Shantz L.M., Horetsky R.L., and Jefferson L.S. 1999. Leucine regulates translation of specific mRNAs in L6 myoblasts through mTOR-mediated changes in availability of eIF4E and phosphorylation of ribosomal protein S6. *J. Biol. Chem.* **274:** 11647–11652.

Klasson H., Fink G.R., and Ljungdahl P.O. 1999. Ssy1p and Ptr3p are plasma membrane components of a yeast system that senses extracellular amino acids. *Mol. Cell. Biol.* **19:** 5405–5416.

Kleijn M., Scheper G.C., Voorma H.O., and Thomas A.A.M. 1998. Regulation of translation initiation factors by signal transduction. *Eur. J. Biochem.* **253:** 531–544.

Lardeux B., Bourdel G., and Girard-Globa A. 1978. Regulation of hepatic synthesis of proteins by the chronology of protein ingestion. *Biochim. Biophys. Acta* **518:** 113–124.

Luzi L., Castellino P., Simonson D.C., Petrides A.S., and DeFronzo R.A. 1990. Leucine metabolism in IDDM: Role of insulin and substrate availability. *Diabetes* **39:** 38–48.

Lynch C.G., Fox H.L., Vary T.C., Jefferson L.S., and Kimball S.R. 2000. Regulation of amino acid-sensitive TOR signaling by leucine analogs in adipocytes. *J. Cell. Biochem.* **77:** 234–251.

Meyuhas O., Avni D., and Shama S. 1996. Translational control of ribosomal protein mRNAs in eukaryotes. In *Translational control* (ed. J.W.B. Hershey et al.), pp. 363–388. Cold Spring Harbor Laboratory Press, Cold Spring Harbor, New York.

Morley S.J. 1997. Intracellular signalling pathways regulating initiation factor eIF4E phosphorylation during the activation of cell growth. *Biochem. Soc. Trans.* **25:** 503–509.

Morley S.J. and McKendrick L. 1997. Involvement of stress-activated protein kinase and p38/RK mitogen-activated protein kinase signaling pathways in the enhanced phosphorylation of initiation factor 4E in NIH 3T9 cells. *J. Biol. Chem.* **272:** 17887–17893.

Mortimore G.E., Pösö A.R., Kadowaki M., and Wert J.J.J. 1987. Multiphasic control of hepatic protein degradation by regulatory amino acids. General features and hormonal modulation. *J. Biol. Chem.* **262:** 16322–16327.

Mortimore G.E., Wert J.J., Miotto G., Venerando R., and Kadowaki M. 1994. Leucine-specific binding of photoreactive Leu7-MAP to a high molecular weight protein on the plasma membrane of the isolated rat hepatocyte. *Biochem. Biophys. Res. Commun.* **203:** 200–208.

Nave B.T., Ouwens D.M., Withers D.J., Alessi D.R., and Shepherd P.R. 1999. Mammalian target of rapamycin is a direct target for protein kinase B: Identification of a convergence point for opposing effects of insulin and amino acid deficiency on protein translation. *Biochem. J.* **344:** 427–431.

Niederberger P., Miozzari G., and Hütter R. 1981. Biological role of the general control of amino acid biosynthesis in *Saccharomyces cerevisiae. Mol. Cell. Biol.* **1:** 584–593.

Oldfield S. and Proud C.G. 1992. Purification, phosphorylation and control of the guanine-nucleotide-exchange factor from rabbit reticulocyte lysates. *Eur. J. Biochem.* **208:** 73–81.

Patti M.-E., Brambilla E., Luzi L., Landaker E.J., and Kahn C.R. 1998. Bidirectional modulation of insulin action by amino acids. *J. Clin. Invest.* **101:** 1519–1529.

Ramsey J.C. and Steele W.J. 1976. Effect of starvation of the distribution of free and membrane-bound ribosomes in rat liver and on the content of phospholipid and glycogen in purified ribosomes. *Biochim. Biophys. Acta* **447:** 312–318.

Rannels D.E., Pegg A.E., Rannels S.R., and Jefferson L.S. 1978. Effect of starvation on initiation of protein synthesis in skeletal muscle and heart. *Am. J. Physiol.* **235:** E126–E133.

Rosenwald I.B., Chen J.J., Wang S.T., Savas L., London I.M., and Pullman J. 1999. Upregulation of protein synthesis initiation factor eIF-4E is an early event during colon carcinogenesis. *Oncogene* **18:** 2507–2517.

Shah O.J., Antonetti D.A., Kimball S.R., and Jefferson L.S. 1999. Leucine, glutamine, and tyrosine reciprocally modulate the translation initiation factors eIF4F and eIF2B in rat liver. *J. Biol. Chem.* **274:** 36168–36175.

Shigemitsu K., Tsujishita Y., Hara K., Nanahoshi M., Avruch J., and Yonezawa K. 1999a. Regulation of translational effectors by amino acids and mammalian target of rapamycin signaling pathways. Possible involvement of autophagy in cultured hepatoma cells. *J. Biol. Chem.* **274:** 1058–1065.

Shigemitsu K., Tsujishita Y., Miyake H., Hidayat S., Tanaka N., Hara K., and Yonezawa K. 1999b. Structural requirement of leucine for activation of p70 S6 kinase. *FEBS Lett.* **447:** 303–306.

Sidransky H., Verney E., and Sarma D.S. 1971. Effect of tryptophan on polyribosomes and protein synthesis in liver. *Am. J. Clin. Nutr.* **24:** 779–785.

Sidransky H., Sarma D.S., Bongiorno M., and Verney E. 1968. Effect of dietary tryptophan on hepatic polyribosomes and protein synthesis in fasted mice. *J. Biol. Chem.* **243:** 1123–1132.

Singh L.P., Aroor A.R., and Wahba A.J. 1994. Phosphorylation of the guanine nucleotide exchange factor and eukaryotic initiation factor 2 by casein kinase II regulates guanine nucleotide binding and GDP/GTP exchange. *Biochemistry* **33:** 9152–9157.

Singh L.P., Denslow N.D., and Wahba A.J. 1996. Modulation of rabbit reticulocyte guanine nucleotide exchange factor activity by casein kinases 1 and 2 and glycogen synthase kinase 3. *Biochemistry* **35:** 3206–3212.

Sood R., Porter A.C., Olsen D., Cavener D.R., and Wek R.C. 2000. A mammalian homologue of GCN2 protein kinase important for translational control by phosphorylation of eukaryotic initiation factor-2α. *Genetics* **154:** 787–801.

Svanberg E., Jefferson L.S., Lundholm K., and Kimball S.R. 1997. Postprandial stimulation of muscle protein synthesis is independent of changes in insulin. *Am. J. Physiol.* **272:** E841–E847.

Svanberg E., Zachrisson H., Ohlsson C., Iresjo B.-M., and Lundholm K.G. 1996. Role of insulin and IFG-I in activation of muscle protein synthesis after oral feeding. *Am. J. Physiol.* **270:** E614–E620.

Terada N., Patel H.R., Takase K., Kohno K., Nairn A.C., and Gelfand E.W. 1994. Rapamycin selectively inhibits translation of mRNAs encoding elongation factors and ribosomal proteins. *Proc. Natl. Acad. Sci.* **91:** 11477–11481.

Tessari P., Inchiostro S., Biolo G., Trevisan R., Fantin G., Marescotti M.C., Iori E., Tiengo A., and Crepaldi G. 1987. Differential effects of hyperinsulinemia and hyperaminoacidemia on leucine carbon metabolism in vivo. Evidence for distinct mechanisms in regulation of net amino acid deposition. *J. Clin. Invest.* **79:** 1062–1069.

Thomas G.P. and Mathews M.B. 1984. Alterations of transcription and translation in HeLa cells exposed to amino acid analogs. *Mol. Cell. Biol.* **4:** 1063–1072.

Tipton K.D., Gurkin B.E., Matin S., and Wolfe R.R. 1999. Nonessential amino acids are not necessary to stimulate net muscle protein synthesis in healthy volunteers. *J. Nutr. Biochem.* **10:** 89–95.

Vary T.C., Jefferson L.S., and Kimball S.R. 1999. Amino acid-induced stimulation of translation initiation in rat skeletal muscle. *Am. J. Physiol.* **277:** E1077–E1086.

Vaughan M.H. and Hansen B.S. 1973. Control of initiation of protein synthesis in human cells. Evidence for a role of uncharged transfer ribonucleic acid. *J. Biol. Chem.* **248:** 7087–7096.

Volpi E., Lucidi P., Cruciani G., Monacchia F., Reboldi G., Brunetti P., Bolli G.B., and De Feo P. 1996. Contribution of amino acids and insulin to protein anabolism during meal absorption. *Diabetes* **45:** 1245–1252.

Wang X., Campbell L.E., Miller C.M., and Proud C.G. 1998. Amino acid availability regulates p70 S6 kinase and multiple translation factors. *Biochem. J.* **334:** 261–267.

Warrington R.C., Wratten N., and Hechtman R. 1977. L-histidinol inhibits specifically and reversibly protein and ribosomal RNA synthesis in mouse L cells. *J. Biol. Chem.* **252:** 5251–5257.

Welsh G.I. and Proud C.G. 1993. Glycogen synthase kinase-3 is rapidly inactivated in response to insulin and phosphorylates eukaryotic initiation factor eIF-2B. *Biochem. J.* **294:** 625–629.

Welsh G.I., Patel J.C., and Proud C.G. 1997a. Peptide substrates suitable for assaying glycogen synthase kinase-3 in crude cell extracts. *Anal. Biochem.* **244:** 16–21.

Welsh G.I., Miller C.M., Loughlin A.J., Price N.T., and Proud C.G. 1998. Regulation of eukaryotic initiation factor eIF2B: Glycogen synthase kinase-3 phosphorylates a conserved serine which undergoes dephosphorylation in response to insulin. *FEBS Lett.* **421:** 125–130.

Welsh G.I., Stokes C.M., Wang X., Sakaue H., Ogawa W., Kasuga M., and Proud C.G. 1997b. Activation of translation initiation factor eIF2B by insulin requires phosphatidyl inositol 3-kinase. *FEBS Lett.* **410:** 418–422.

Xu G., Kwon G., Marshall C.A., Lin T.-A., Lawrence J.C., and McDaniel M.L. 1998a. Branched-chain amino acids are essential in the regulation of PHAS-I and p70 S6 kinase by pancreatic β-cells. A possible role in protein translation and mitogenic signaling. *J. Biol. Chem.* **273:** 28178–28184.

Xu G., Marshall C.A., Lin T.-A., Kwon G., Munivenketappa R.B., Hill J.R., Lawrence J.C., and McDaniel M.L. 1998b. Insulin mediates glucose-stimulated phosphorylation of PHAS-I by pancreatic beta cells. An insulin-receptor mechanism for autoregulation of protein synthesis by translation. *J. Biol. Chem.* **273:** 4485–4491.

Yokogoshi H. and Yoshida A. 1980. Effects of supplementation and depletion of a single essential amino acid on hepatic polyribosome profile in rats. *J. Nutr.* **110:** 375–382.

———. 1986. Time-dependent changes in aggregation of hepatic ribosomes after meal feeding of rats. *J. Nutr.* **116:** 472–476.

Yokogoshi H., Sakuma Y., and Yoshida A. 1980. Relationships between nutritional quality of dietary proteins and hepatic polyribosome profiles in rats. *J. Nutr.* **110:** 383–387.

Yoshizawa F., Kimball S.R., and Jefferson L.S. 1997. Modulation of translation initiation in rat skeletal muscle and liver in response to food intake. *Biochem. Biophys. Res. Commun.* **240:** 825–831.

Yoshizawa F., Kimball S.R., Vary T.C., and Jefferson L.S. 1998. Effect of dietary protein on translation initiation in rat skeletal muscle and liver. *Am. J. Physiol.* **275:** E814–E820.

17

Translational Control during Heat Shock

Robert J. Schneider

Department of Microbiology
New York University School of Medicine
New York, New York 10016

HEAT SHOCK RESPONSE INVOLVES TRANSLATIONAL CONTROL

Heat shock is a thermally induced stress response in cells or organisms at 5–10°C above their normal physiological temperature (for review, see Lindquist 1986). Cells respond to heat in a conserved manner to protect against hyperthermic lethality and to accelerate recovery following heat stress. The heat shock response in most organisms and cells includes inhibition of protein synthesis above a characteristic threshold temperature, with the exclusive translation of heat shock protein mRNAs (for review, see Panniers 1994). The extent of inhibition of non-heat shock ("normal") mRNA translation increases with the severity of heat shock. The higher the temperature and the longer the exposure of cells, the greater the inhibition of normal mRNA translation and the longer the interval for recovery of protein synthesis following heat shock (for review, see Duncan 1996). Following heat shock, cells can acquire translational thermotolerance, in which they possess a more rapid ability to recover normal protein synthesis following heat stress, and sometimes a greater ability to maintain translation during subsequent rounds of heat shock (Mizzen and Welch 1988; Beck and De Maio 1994). In most cells (except yeast), mRNAs are not degraded or modified during heat shock, and they can be translated in vitro (Kruger and Benecke 1981). The translational apparatus is therefore altered to prevent translation of normal mRNAs. The block in normal mRNA translation during heat shock involves disassembly of polysomes and inhibition of initiation (for review, see Panniers 1994; Duncan 1996). In *Drosophila* cells there is also some evidence for alteration of translation elongation rates (Ballinger and Pardue 1983).

Translational Control of Gene Expression
© 2000 Cold Spring Harbor Laboratory Press 0-87969-568-4/00 $5 + .00

The inhibition of normal mRNA translation during heat shock is thought to enhance cell survival by limiting the accumulation of misfolded or denatured proteins that result from heat stress. In this chapter, I review the understanding of translational control during the heat shock response, primarily in mammalian and *Drosophila* cells where it has been intensively studied. The heat shock response is complex and multifaceted, involving alteration of many different aspects of cellular physiology. The reader is guided to reviews on other aspects of the heat shock response, including effects on cellular structure, DNA replication, transcriptional control, mRNA processing and transport, calcium flux, and endoplasmic reticulum function (Roti Roti et al. 1997; Brostrom and Brostrom 1998; Morimoto 1998; Chapter 15).

ROLE OF HEAT SHOCK PROTEINS IN THE HEAT SHOCK RESPONSE

A variety of cell stresses, in addition to heat, can trigger a heat shock response, including exposure to heavy metals, hypoxia, and deprivation of glucose (for review, see Duncan 1996). All of these responses have in common the abundant expression of heat shock proteins (Hsps) (for review, see Morimoto 1998), which is vital for protection of cells against lethality (Mizzen and Welch 1988). Most members of the Hsp family are molecular chaperones, normally assisting in protein folding, transport and assembly of multiprotein complexes, and, during heat shock, in protecting proteins, repairing damaged proteins, and promoting degradation of severely damaged proteins (Morimoto 1998). Hsps 100, 90, 70, 60, and 27 trap and bind unfolding protein intermediates, maintaining them in partially unfolded states of denaturation. Hsps therefore prevent irreversible denaturation of proteins during heat shock and promote refolding of proteins to their native states following thermal stress (Morimoto 1998). In mammalian cells, Hsp27 functions as an ATP-independent chaperone to limit protein denaturation and aggregation and to enhance thermotolerance of cells (Landry et al. 1989; Ehrnsperger et al. 1997). During recovery from heat shock, abundant expression of ATP-dependent chaperones Hsp100, 90, 70, and 60 facilitates the native refolding of proteins trapped by Hsp27, promoting increased cell survival (Ehrnsperger et al. 1997; Glover and Lindquist 1998). Proteins that are excessively aggregated (insolubilized) and/or misfolded are degraded through the ubiquitin-proteasome pathway. As described later, Hsps play important roles during heat shock in the control of translation. The signals that elicit activation of the heat shock response during thermal stress are reviewed elsewhere (Morimoto 1998).

SUPPRESSION OF mRNA TRANSLATION DURING HEAT SHOCK

Overview

Impaired initiation is the primary translation block to normal cellular mRNAs during heat shock, involving modification of at least three factors, eIF2, eIF2B, and eIF4F (covered below). Calcium mobilization from the endoplasmic reticulum (ER), which is enhanced by heat shock, can also be associated with inhibition of normal cell mRNA translation in some systems (for review, see Brostrom and Brostrom 1998; Chapter 15). Treatment of cells with calcium chelators prior to heat shock reduces inhibition of translation, whereas it is increased by agents that release free calcium. PKR does not seem to play a major role in inhibition of protein synthesis by ER stress and calcium release, which activates the ER-activated PKR-related kinase (PERK), which can phosphorylate eIF2α (see Harding et al. 1999; Chapter 15).

Role of eIF2 and eIF2B in Repression of Protein Synthesis during Heat Shock

Regulation of eIF2 activity is a major mechanism for control of protein synthesis in eukaryotic cells (see Chapters 2, 5, and 13). In mammalian cells, phosphorylation of eIF2 on its α subunit (eIF2αP), by the dsRNA activated protein kinase (PKR), the heme-controlled inhibitor (HRI), or PERK, results in inhibition of protein synthesis. eIF2αP binds tightly to the limiting amounts of eIF2–GTP exchange factor (eIF2B), blocking recycling of GDP for GTP on eIF2, thereby preventing subsequent formation of initiating ribosomes. Because the level of eIF2B is typically much lower than the level of eIF2, a small increase in eIF2αP is usually sufficient to block translation initiation.

The phosphorylation of eIF2α is increased, and the activities of both eIF2 and eIF2B are reduced, in some cell types during heat shock. The inactivation of eIF2/eIF2B is enhanced in a temperature- and cell-type-dependent fashion (Duncan and Hershey 1984, 1989; Scorsone et al. 1987; Rowlands et al. 1988; Scheper et al. 1997). For example, only a small percentage of eIF2α is phosphorylated in heat shocked HeLa cells, despite significant inhibition of protein synthesis (Duncan and Hershey 1984), whereas in Ehrlich cells there is a large increase in eIF2α phosphorylation (Scorsone et al. 1987). It is possible that low levels of eIF2B make HeLa cells more sensitive to translational inhibition upon small increases in eIF2α phosphorylation. On the other hand, mild or moderate heat shock of many cell types, including HeLa cells, results in little if any enhanced phosphorylation of eIF2α or inactivation of eIF2B activity (Mariano and Siekierka 1986; Duncan and Hershey 1989), although protein synthesis is largely impaired. Severe heat shock induces much stronger eIF2α phosphorylation in many cells that have been examined (Duncan 1996). However, even severe heat shock of many cell types strongly inhibits protein synthesis without any detectable phosphorylation of eIF2α (Duncan and Hershey 1984, 1989; Duncan et al. 1995). Moreover, normal mRNA translation in heat shocked extracts from HeLa or *Drosophila* cells could be strongly recovered by addition of purified eIF4F but not eIF2 (Panniers et al. 1985; Zapata et al. 1991). In addition, overexpression of a mutant form of eIF2α protein containing a serine to Ala-51 mutation, the site of inactivating phosphorylation, reduced repression of normal protein synthesis during heat shock, but by only 30–50% (Murtha-Riel et al. 1993). Thus, inactivation of eIF2 can play a role in inhibition of protein synthesis during heat shock, but it is dependent on cell type and severity of heat shock treatment. Since PKR is not sufficiently activated during heat shock to block protein synthesis (Dubois et al. 1991), it is more likely that increased eIF2αP results from ER-associated PERK activation in most mammalian cells (Harding et al. 1999), HRI in red blood cells and reticulocytes, or possibly other related protein kinases such as the pancreatic-enriched kinase PEK, which is also found at low levels in other tissues (Shi et al. 1998).

Role of HRI in Reticulocyte Translation Inhibition during Heat Shock

HRI is the primary regulator of protein synthesis in reticulocytes and red blood cells, where it is primarily but possibly not exclusively located (see Chapter 14). HRI is strongly activated and phosphorylates eIF2α during heat shock of reticulocytes. The mechanism for activation of HRI by heat shock is complex. Newly synthesized HRI requires the chaperone activity of Hsp90 and Hsc70, the constitutively produced form of Hsp70, for mature folding into a native and active conformation (Uma et al. 1997, 1999). However, Hsc70 also blocks HRI kinase activity in the presence of heme (Matts et al. 1993; Uma et al. 1999), accounting for kinase repression by heme. Binding of Hsc70 to HRI probably induces a conformational change that prevents kinase activity (Uma et al. 1999). Heme also appears to displace Hsp90 but not Hsc70 from HRI, thereby promoting the formation of a kinase-inactive complex. Heat stress disrupts the negative regulatory HRI-Hsc70 interaction, despite the presence of heme, presumably by promoting a conformational change in HRI that releases

Hsc70, thereby activating HRI kinase activity (Uma et al. 1999). It is not presently known whether Hsp90 and Hsc70 are similarly involved in regulation of PKR or PERK activities, in reticulocytes or other cells.

Role of eIF4F in Repression of Protein Synthesis during Heat Shock

eIF4F is a cap-binding complex of proteins containing cap-binding protein eIF4E, the ATP-dependent RNA helicase eIF4A, and the adapter protein eIF4G. eIF4F is also associated with poly(A)-binding protein (PABP), the eIF4E kinase (Mnk1), and eIF3 (see Chapters 2 and 6). The major function of eIF4F/eIF3 is to promote 40S ribosome subunit association with capped and polyadenylated mRNA, probably by facilitating the 5′ unwinding of mRNA secondary structure. eIF4F is a major target for regulation of protein synthesis in mammalian cells. For the most part, increased phosphorylation of eIF4E correlates with enhanced translation of most capped mRNAs, and decreased phosphorylation with reduced translation (see Chapter 6). Exceptions include increased phosphorylation of eIF4E in arsenite-treated cells, despite reduced mRNA translation (Morley and McKendrick 1997; Wang et al. 1998). In this case, other mechanisms likely block protein synthesis, and the increased phosphorylation of eIF4E is a compensatory response to maintain some level of translation.

Heat shock was shown by Duncan and Hershey (1984) to strongly decrease the phosphorylation of eIF4E, consistent with inhibition of normal cell mRNA translation. Subsequent studies in a wide variety of cell types and species confirmed this effect (Panniers and Henshaw 1984; Panniers et al. 1985; Duncan et al. 1987; Duncan and Hershey 1987, 1989; Lamphear and Panniers 1990, 1991; Zapata et al. 1991; Joshi-Barve et al. 1992; Yueh and Schneider 1996). Since phosphorylation of eIF4E at Ser-209 is thought to clamp eIF4E/4F to capped mRNAs (Marcotrigiano et al. 1997), Hsp mRNAs likely possess either a reduced dependence on eIF4F or an enhanced ability to utilize the low levels that remain. Heat shock also substantially reduces the abundance of eIF4F complexes (Lamphear and Panniers 1991), which would not be predicted merely by dephosphorylation of eIF4E, implicating other effects of heat shock on the formation of eIF4F (see below). Thus, suppression of normal mRNA translation during heat shock correlates well with eIF4E dephosphorylation, inactivation of eIF4F activity, and decreased levels of eIF4F complexes.

A family of eIF4E-binding proteins, known as 4E-BP1,-2,-3, were shown to displace eIF4E from eIF4G and sequester it in a manner controlled by 4E-BP phosphorylation, thereby blocking formation of eIF4F (Lin et al. 1994, 1995; Haghighat et al. 1995; Mader et al. 1995; Poulin et al. 1998). 4E-BPs become hyperphosphorylated in response to growth-stimulating factors, which blocks their binding to eIF4E. In growth-arrested cells, 4E-BPs are hypophosphorylated and strongly bind eIF4E (Lin et al. 1994). Studies have shown that 4E-BPs become hypophosphorylated during heat shock, concomitant with increased binding to eIF4E and inhibition of normal mRNA translation (see Fig. 1) (Feigenblum and Schneider 1996; Vries et al. 1997). The mechanism by which heat shock reduces 4E-BP phosphorylation involves control of the rapamycin-sensitive FRAP/TOR signaling pathway (Feigenblum and Schneider 1996), as

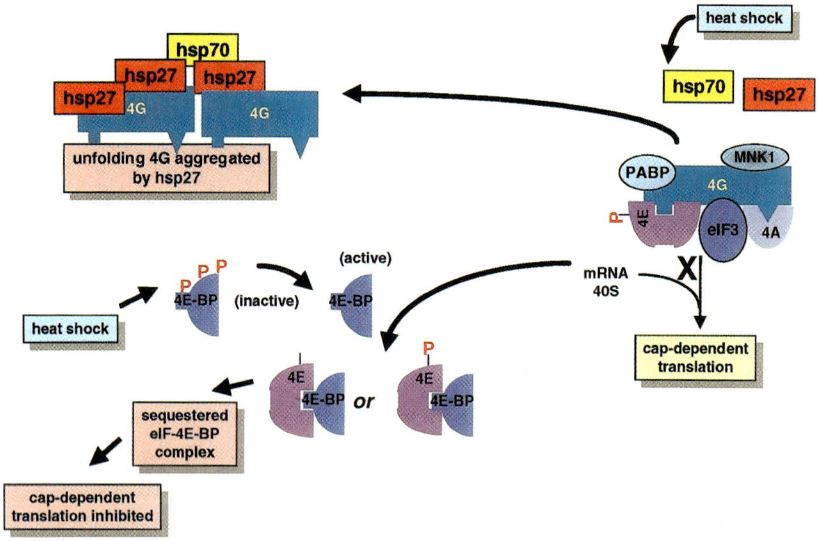

Figure 1 Model for inhibition of eIF4F-dependent mRNA translation during heat shock. Shown is the eIF4F cap-binding complex with associated eIF3, poly(A)-binding protein (PABP), and the eIF4E kinase, Mnk1. Heat shock induces abundant accumulation of Hsp70 and Hsp27. Hsp27 (and possibly Hsp70) is involved in aggregation of unfolding eIF4G concomitant with displacement of PABP, Mnk1, eIF4A, and eIF3. eIF4E-binding proteins (4E-BPs) are activated for binding to eIF4E early in the heat shock response by inhibition of the FRAP/TOR signaling pathway leading to dephosphorylation. 4E-BPs bind to eIF4E and aid in its removal from eIF4G.The dissociation of the eIF4F complex by Hsp27/Hsp70, and the sequestration of eIF4E by 4E-BPs, both contribute to inhibition of eIF4F-dependent (normal) mRNA translation.

expected for 4E-BP phosphorylation (see Chapter 6). However, in rat hepatoma cells no decrease in 4E-BP phosphorylation, nor increased association with eIF4E, was observed during heat shock (Scheper et al. 1997). These results suggest that although 4E-BP inhibition of eIF4F complex formation is an important and widespread mechanism for inhibition of eIF4F-dependent mRNA translation during heat shock, it is not universal.

Role of Heat Shock Proteins in Regulation of Protein Synthesis during Heat Shock

Considerable evidence links expression of Hsps to control of translation during heat shock. For instance, a mild heat shock, which induces expression of Hsps, programs cells to survive a subsequent severe heat shock that would ordinarily be lethal. Survival is linked to a more rapid and robust recovery of protein synthesis following heat shock, and to some extent, to a greater ability to sustain normal protein synthesis during heat shock (Mizzen and Welch 1988; Liu et al. 1992; Beck and De Maio 1994). Hsp synthesis is essential for protection of the translational apparatus, since cycloheximide treatment of cells during heat shock blocks translational thermotolerance (Laszlo and Li 1985; Carper et al. 1997). Hsp70 and Hsp27 are primarily responsible for protection of the protein synthetic machinery during heat shock. Cells stably transfected with Hsp70 expression vectors, or cell lines selected for enhanced translational thermotolerance, have demonstrated the importance of abundant expression of Hsp70 protein for protection and recovery of normal translation following heat shock (Li et al. 1991; Liu et al. 1992; De Maio et al. 1993; Beck and De Maio 1994). It is presumed, but not proven, that the ATP-dependent chaperone activity of Hsp70 is involved in protecting the translational apparatus, possibly by stabilizing the native conformation of translation factors during heat shock, or refolding key factors to a native state during recovery.

Hsp27 (in humans) and Hsp25 (in mouse) are also strongly linked to translational thermotolerance during heat shock. Hsp27/25 levels rapidly increase during heat shock, first by phosphorylation and disaggregation of Hsp27/25 from large complexes, then by induction of new Hsp27/25 synthesis, followed by aggregation into perinuclear heat shock granules (Landry et al. 1989, 1991). Hsp27/25 mediates aggregation of proteins and, therefore, blocks their irreversible denaturation, enhancing cell thermotolerance and survival (Landry et al. 1989; Ehrnsperger et al. 1997). Following heat shock, Hsp70 is thought to assist Hsp27/25 in facilitating native refolding of partially denatured proteins (Ehrnsperger et al. 1997;

Lee et al. 1997). Enhanced expression of Hsp27 in cell lines protects the translation apparatus and enhances the rate and extent of recovery of normal protein synthesis following heat shock (Mizzen and Welch 1988; Liu et al. 1992; Carper et al. 1997).

Recent studies have begun to elucidate a molecular mechanism for the role of Hsp27 and Hsp70 in maintaining and recovering eIF4F-dependent (normal) mRNA translation in heat-shocked mammalian cells. In heat-shocked HeLa, 293, and 293T cells, eIF4G is specifically aggregated with Hsp27 into heat shock granules with kinetics that parallel the disassembly of eIF4F complexes and the inhibition of normal cell mRNA translation (see Fig. 1) (Cuesta et al. 2000). Approximately 90–95% of eIF4G is aggregated with Hsp27, which is highly selective, in that eIF4E, eIF4A, eIF3, PABP, and Mnk1 are all displaced from eIF4G, remain unaggregated, and do not bind Hsp27. Small amounts of Hsp70 are in eIF4G/Hsp27 complexes, but no other Hsp is specifically associated. It is not known whether *hsp* mRNAs preferentially utilize the small amount of eIF4F that remains intact, or possibly translate with little if any requirement for eIF4G. Recovery of normal mRNA translation following heat shock correlates with release of eIF4G from aggregated Hsp27 complexes and reformation of eIF4F. It is currently not known how Hsp27 interacts with and traps eIF4G, which presumably occurs due to thermally induced local unfolding of eIF4G, and how this leads to release of eIF4G bound factors and disassembly of eIF4F. It is also not known whether displacement of eIF4E from eIF4G by its binding to 4E-BPs is required to promote aggregation of eIF4G with Hsp27.

SELECTIVE TRANSLATION OF HSP mRNAS DURING HEAT SHOCK

Translation efficiency of *hsp* mRNAs increases with the severity (temperature and/or duration) of heat shock, despite the inhibition of normal protein synthesis. The features of Hsp mRNAs responsible for translation during heat shock remain poorly characterized. Most Hsp 5′UTRs are long (150–250 nucleotides) and rich in A residues (~50%), which reduces secondary structure (Lindquist and Petersen 1990). The ability to translate during heat shock resides in the Hsp 5′UTR. For example, deletion of the *Drosophila hsp70* or *hsp22* 5′UTR blocks translation during heat shock, whereas the cap-proximal 30 nucleotides partially preserve it (Hultmark et al. 1986; Lindquist and Petersen 1990). Although reduced secondary structure is important for translation of *hsp* mRNAs during heat shock, an artificial A-rich *hsp70* 5′UTR does not overcome the translation block, implicating a requirement for specific sequences, structural elements, or both (Lindquist and Petersen 1990). A detailed mutational analysis of the *Drosophila hsp70* 5′UTR found that reduced secondary structure and uncharacterized segments of the 3′ region of the leader were both required for efficient translation of the mRNA during heat shock (Hess and Duncan 1996).

The Grp78/BiP mRNA, which encodes an ER-associated variant of Hsp70, translates in an eIF4F-independent manner by utilizing an internal ribosome entry site (IRES) element (Macejak and Sarnow 1991), whereas the *Drosophila* and human *hsp*70 5′UTRs do not contain an IRES (McGarry and Lindquist 1985; Hess and Duncan 1996; Yueh and Schneider 1996, 2000). Instead, the human *hsp70* mRNA 5′UTR was shown to translate by ribosome shunting (Yueh and Schneider 2000), which reduces or eliminates a requirement for eIF4F (see Chapter 32). The *Drosophila hsp70* 5′UTR does not promote ribosome shunting, consistent with earlier results showing that a stable hairpin blocks its ability to direct translation during heat shock (Hess and Duncan 1996). The human, but not the *Drosophila, hsp70* 5′UTR contains two copies of an unusual complementarity to the 3′ hairpin of 18S rRNA, present as well in the adenovirus late tripartite leader 5′UTR. The complementarity is important for both 5′UTRs to direct a high level of translation by ribosome shunting (Yueh and Schneider 2000). Nevertheless, other unidentified elements must also be involved in promoting translation by the human *hsp70* 5′UTR, since deletion of the 18S rRNA complementarities reduces but does not eliminate translation.

CONCLUDING REMARKS

The mechanism by which heat shock inhibits cellular protein synthesis is complex, and current evidence indicates that it likely involves multiple mechanisms. Future studies need to determine the respective roles for eIF4E sequestration by 4E-BP, and eIF4G aggregation by the small heat shock protein family (Hsp25/27), in translation inhibition in different cell types during heat shock. In particular, it is not clear how the duration and temperature of the heat shock response influences the involvement of eIF4E sequestration by 4E-BP, and eIF4G aggregation by Hsp25/27. Whereas the role of eIF2α phosphorylation by activated PKR and PERK has now been well studied, the involvement of other potential protein kinase family members might be more fully examined. With regard to the selective synthesis of heat shock proteins, studies should be directed to elucidating the mechanisms that permit selective translation of the whole family of heat shock mRNAs. Both internal initiation of translation and ribosome shunting mechanisms have been described for two heat shock protein mRNAs. The mechansims that permit selective translation of the remaining members of this large family of mRNAs during heat shock have yet to be investigated in depth. It is likely that ever more interesting modes of translation will be discovered, leading to new insights in the process of protein synthesis.

ACKNOWLEDGMENTS

I thank R. Cuesta for thoughtful discussions. The author's work described in this review was supported by a grant from the National Institutes of Health (CA-42357).

REFERENCES

Ballinger D.G. and Pardue M.L. 1983. The control of protein synthesis during heat shock in *Drosophila* cells involves altered polypeptide elongation rates. *Cell* **33:** 103–113.

Beck S.C. and De Maio A. 1994. Stabilization of protein synthesis in thermotolerant cells during heat shock: Association of heat shock protein-72 with ribosomal subunits of polysomes. *J. Biol. Chem.* **269:** 21803–21811.

Brostrom C.O. and Brostrom M.A. 1998. Regulation of translational initiation during cellular responses to stress. *Prog. Nucleic Acid Res. Mol. Biol.* **58:** 79–125.

Carper S.W., Rocheleau T., Cimino D., and Storm F.K. 1997. Heat shock protein 27 stimulates recovery of RNA and protein synthesis following a heat shock. *J. Cell. Biochem.* **66:** 153–164.

Cuesta R., Laroia G., and Schneider R.J. 2000. Chaperone Hsp27 inhibits translation during heat shock by binding eIF4G and facilitating dissociation of cap-initiation complexes. *Genes Dev.* **14:** 1460–1470.

De Maio A., Beck S.C., and Buchman T.G. 1993. Heat shock gene expression and development of translational thermotolerance in human hepatoblastoma cells. *Circ. Shock* **40:** 177–186.

Dubois M.F., Hovanessian A.G., and Bensaude O. 1991. Heat-shock-induced denaturation of proteins. Characterization of the insolubilization of the interferon-induced p68 kinase. *J. Biol. Chem.* **266:** 9707–9711.

Duncan R.F. 1996. Translational control during heat shock. In *Translational control* (ed. J.W.B. Hershey et al.), pp. 271–294. Cold Spring Harbor Laboratory Press, Cold Spring Harbor, New York.

Duncan R. and Hershey J.W.B. 1984. Heat shock-induced translational alterations in HeLa cells. Initiation factor modifications and the inhibition of translation. *J. Biol. Chem.* **259:** 11882–11889.

———. 1987. Initiation factor protein modifications and inhibition of protein synthesis. *Mol. Cell. Biol.* **7:** 1293–1295.

———. 1989. Protein synthesis and protein phosphorylation during heat stress, recovery, and adaptation. *J. Cell Biol.* **109:** 1467–1481.

Duncan R.F., Cavener D.R., and Qu S. 1995. Heat shock effects on phosphorylation of protein synthesis initiation factor proteins eIF-4E and eIF-2α in *Drosophila*. *Biochemistry* **34:** 2985–2997.

Duncan R., Milburn S.C., and Hershey J.W.B. 1987. Regulated phosphorylation and low abundance of Hela cell initiation factor eIF-4F suggest a role in translational control. Heat shock effects on eIF-4F. *J. Biol. Chem.* **262:** 380–388.

Ehrnsperger M., Graber S., Gaestel M., and Buchner J. 1997. Binding of non-native protein to hsp25 during heat shock creates a reservoir of folding intermediates for reactivation. *EMBO J.* **16:** 221–229.

Feigenblum D. and Schneider R.J. 1996. Cap-binding protein (eukaryotic initiation factor 4E) and 4E-inactivating protein BP-1 independently regulate cap-dependent translation. *Mol. Cell. Biol.* **16:** 5450–5457.

Glover J.R. and Lindquist S. 1998. Hsp104, Hsp70, and Hsp40: A novel chaperone system that rescues previously aggregated proteins. *Cell* **94:** 73–82.

Haghighat A., Mader S., Pause A., and Sonenberg N. 1995. Repression of cap-dependent translation by 4E-binding protein 1: Competition with p220 for binding to eukaryotic initiation factor-4E. *EMBO J.* **14:** 5701–5709.

Harding H.P., Zhang Y., and Ron D. 1999. Protein translation and folding are coupled by an endoplasmic-reticulum-resident kinase. *Nature* **397:** 271–274.

Hess M.A. and Duncan R.F. 1996. Sequence and structure determinants of *Drosophila* Hsp70 mRNA translation: 5′UTR secondary structure specifically inhibits heat shock protein mRNA translation. *Nucleic Acids Res.* **24:** 2441–2449.

Hultmark D., Klemenz R., and Gehring W.J. 1986. Translational and transcriptional control elements in the untranslated leader of the heat shock gene hsp22. *Cell* **44:** 429–438.

Joshi-Barve S., DeBenedetti A., and Rhoads R.E. 1992. Preferential translation of heat shock mRNAs in Hela cells deficient in protein synthesis initiation factors eIF-4E and eIF-gamma. *J. Biol. Chem.* **267:** 21038–21043.

Kruger C. and Benecke B.J. 1981. In vitro translation of *Drosophila* heat-shock and non-heat-shock mRNAs in heterologous and homologous cell-free systems. *Cell* **23:** 595–603.

Lamphear B.J. and Panniers R. 1990. Cap binding protein complex that restores protein synthesis in heat shocked Ehrlich cell lysates contains highly phosphorylated eIF-4E. *J. Biol. Chem.* **265:** 5333–5336.

———. 1991. Heat shock impairs the interaction of cap binding protein complex with 5′ mRNA cap. *J. Biol. Chem.* **266:** 2789–2794.

Landry J., Chretien P., Laszlo A., and Lambert H. 1991. Phosphorylation of HSP27 during development and decay of thermotolerance in Chinese hamster cells. *J. Cell. Physiol.* **147:** 93–101.

Landry J., Chretien P., Lambert H., Hickey E., and Weber L.A. 1989. Heat shock resistance conferred by expression of the human hsp27 gene in rodent cells. *J. Cell Biol.* **109:** 7–15.

Laszlo A. and Li G.C. 1985. Heat-resistant variants of Chinese hamster fibroblasts altered in expression of heat shock protein. *Proc. Natl. Acad. Sci.* **82:** 8029–8033.

Lee G.J., Roseman A.M., Saibil H.R., and Vierling E. 1997. A small heat shock protein stably binds heat-denatured model substrates and can maintain a substrate in a folding-competent state. *EMBO J.* **16:** 659–671.

Li G.C., Li. L., Liu Y.-K., Mak J.K., Chen L., and Lee W.M.F. 1991. Thermal response of rat fibroblasts stably transfected with the human 70-kDa heat shock protein-encoding genes. *Proc. Natl. Acad. Sci.* **88:** 1681–1685.

Lin T.A., Kong X., Saltiel A.R., Blackshear P.J., and Lawrence J.C. 1995. Control of PHAS-I by insulin in 3T3-L1 adipocytes. *J. Biol. Chem.* **270:** 18531–18538.

Lin T.-A., Kong X., Haystead T.A.J., Pause A., Belsham G., Sonenberg N., and Lawrence J.C. 1994. PHAS-1 as a link between mitogen-activated protein kinase and translation initiation. *Science* **266:** 653–656.

Lindquist S. 1986. The heat-shock response. *Annu. Rev. Biochem.* **55:** 1151–1191.

Lindquist S. and Petersen R. 1990. Selective translation and degradation of heat shock messenger RNAs in *Drosophila*. *Enzyme* **44:** 147–166.

Liu R.Y., Li X., Li L., and Li G.C. 1992. Expression of human hsp70 in rat fibroblasts enhances cell survival and facilitates recovery from translational and transcriptional inhibition following heat shock. *Cancer Res.* **52:** 3667–3673.

Macejak D.G. and Sarnow P. 1991. Internal initiation of translation mediated by the 5′ leader of a cellular mRNA. *Nature* **353:** 90–94.

Mader S., Lee H., Pause A., and Sonenberg N. 1995. The translation initiation factor eIF-4E binds to a common motif shared by the translation factor eIF-4γ and the translational repressors 4E-binding proteins. *Mol. Cell. Biol.* **15:** 4990–4997.

Marcotrigiano J., Gingras, A.C., Sonenberg, N., and Burley S.K. 1997. Cocrystal structure of the messenger RNA 5′ cap-binding protein (eIF4E) bound to 7-methyl-GDP. *Cell* **89:** 951–961.

Mariano T.M. and Siekierka J. 1986. Inhibition of HeLa cell protein synthesis under heat shock conditions in the absence of initiation factor eIF-2α phosphorylation. *Biochem. Biophys. Res. Commun.* **138:** 519–525.

Matts R.L., Hurst R., and Xu Z. 1993. Denatured proteins inhibit translation in hemin-supplemented rabbit reticulocyte lysate by inducing the activation of the heme-regulated eIF-2α kinase. *Biochemistry* **32:** 7323–7328.

McGarry T.J. and Lindquist S. 1985. The preferential translation of *Drosophila* hsp70 mRNA requires sequences in the untranslated leader. *Cell* **42:** 903–911.

Mizzen L.A. and Welch W.J. 1988. Characterization of the thermotolerant cell. I. Effects on protein synthesis activity and the regulation of heat-shock protein 70 expression. *J. Cell Biol.* **106:** 1105–1116.

Morimoto R.I. 1998. Regulation of the heat shock transcriptional response: Cross talk between a family of heat shock factors, molecular chaperones, and negative regulators. *Genes Dev.* **12:** 3788–3796.

Morley, S. J. and McKendrick, L. 1997. Involvement of stress-activated protein kinase and p38/RK mitogen-activated protein kinase signaling pathways in the enhanced phosphorylation of initiation factor 4E in NIH 3T3 cells. *J. Biol. Chem.* **262:** 17887–17893.

Murtha-Riel P., Davies M.V., Scherer B.J., Choi S.Y., Hershey J.W.B., and Kaufman R.J. 1993. Expression of a phosphorylation-resistant eukaryotic initiation factor 2 alpha-subunit mitigates heat shock inhibition of protein synthesis. *J. Biol. Chem.* **268:** 12946–12951.

Panniers R. 1994. Translational control during heat shock. *Biochimie* **76:** 737–747.

Panniers R. and Henshaw E.C. 1984. Mechanism of inhibition of polypeptide chain initiation in heat-shocked Ehrlich ascites tumour cells. *Eur. J. Biochem.* **140:** 209–214.

Panniers R., Stewart E.B., Merrick W.C., and Henshaw E.C. 1985. Mechanism of inhibition of polypeptide chain initiation in heat shocked Ehrlich cells involves reduction of eukaryotic initiation factor 4F activity. *J. Biol. Chem.* **260:** 9648–9653.

Poulin F., Gingras A.C., Olsen H., Chevalier S., and Sonenberg N. 1998. 4E-BP3, a new member of the eukaryotic initiation factor 4E-binding protein family. *J. Biol. Chem.* **273:** 14002–14007.

Roti Roti J.L., Wright W.D., and VanderWaal R. 1997. The nuclear matrix: A target for heat shock effects and a determinant of stress response. *Crit. Rev. Eukaryot. Gene Expr.* **7:** 343–360.

Rowlands A.G., Montine K.S., Henshaw E.C., and Panniers R. 1988. Physiological stresses inhibit guanine-nucleotide-exchange factor in Ehrlich cells. *Eur. J. Biochem.* **175:** 93–99.

Scheper G.C., Mulder J., Kleijn M., Voorma H.O., Thomas A.A., and van Wijk R. 1997. Inactivation of eIF2B and phosphorylation of PHAS-I in heat-shocked rat hepatoma cells. *J. Biol. Chem.* **272:** 26850–26856.

Scorsone K.A., Panniers R., Rowlands A.G., and Henshaw E.C. 1987. Phosphorylation of eukaryotic initiation factor 2 during physiological stresses which affect protein synthesis. *J. Biol. Chem.* **262:** 14538–14543.

Shi Y., Vattem K.M., Sood R., An J., Liang J., Stramm L., and Wek R.C. 1998. Identification and characterization of pancreatic eukaryotic initiation factor 2 alpha-subunit kinase, PEK, involved in translational control. *Mol. Cell. Biol.* **18:** 7499–7509.

Uma S., Thulasiraman V., and Matts R.L. 1999. Dual role for Hsc70 in the biogenesis and regulation of the heme-regulated kinase of the alpha subunit of eukaryotic translation initiation factor 2. *Mol. Cell. Biol.* **19:** 5861–5871.

Uma S., Hartson S.D., Chen J.J., and Matts R.L. 1997. Hsp90 is obligatory for the heme-regulated eIF-2α kinase to acquire and maintain an activable conformation. *J. Biol. Chem.* **272:** 11648–11656.

Vries R.G., Flynn A., Patel J.C., Wang X., Denton R.M., and Proud C.G. 1997. Heat shock increases the association of binding protein-1 with initiation factor 4E. *J. Biol. Chem.* **272:** 32779–32784.

Wang X., Flynn A., Waskiewicz A.J., Webb B.L.J., Vries R.G., Baines I.A., Cooper J.A., and Proud C.G. 1998. The phosphorylation of eukaryotic initiation factor eIF-4E in response to phorbol esters, cell stresses and cytokines is mediated by distinct MAP kinase pathways. *J. Biol. Chem.* **273:** 9373–9377.

Yueh A. and Schneider R.J. 1996. Selective translation by ribosome jumping in adenovirus infected and heat shocked cells. *Genes Dev.* **10:** 1557–1567.

———. 2000. Translation by ribosome shunting on adenovirus and Hsp70 mRNAs facilitated by complementarity to 18S rRNA. *Genes Dev.* **14:** 414–421.

Zapata J.M., Maroto F.G., and Sierra J.M. 1991. Inactivation of mRNA cap-binding protein complex in *Drosophila melanogaster* embryos under heat shock. *J. Biol. Chem.* **266:** 16007–16014.

18

Translational Control by Upstream Open Reading Frames

Adam P. Geballe
Divisions of Human Biology and Clinical Research
Fred Hutchinson Cancer Research Center
Seattle, Washington 98109-1024
Departments of Medicine and Microbiology
University of Washington, Seattle, Washington 98115

Matthew S. Sachs
Department of Biochemistry and Molecular Biology
Oregon Graduate Institute of Science & Technology
Beaverton, Oregon 97006-8921
Department of Molecular Microbiology and Immunology
Oregon Health Sciences University
Portland, Oregon 97201-3098

A commonly held but not always correct conception of eukaryotic messenger RNA structure envisions one major open reading frame (ORF) encoding a single polypeptide product. Flanking the ORF are sequences that may regulate properties such as the translational efficiency, stability, and subcellular distribution of the mRNA. The RNA sequence between the methyl guanosine cap and the initiation codon of the ORF is conventionally known as the 5′-untranslated region (5′UTR), even though this term is a misnomer when applied to transcript leader sequences containing upstream AUG (uAUG) codons that function as translation initiation sites. A large number (though a small fraction, overall) of eukaryotic mRNAs contain uAUGs in their 5′UTRs.

Three potential configurations of ORFs that follow uAUG codons based on the position of their termination codons are shown in Figure 1. A nonoverlapping upstream open reading frame (uORF) results from a termination codon being in the 5′UTR. An overlapping uORF has its termination codons within the major coding ORF of the mRNA. Finally, a reading frame initiated by the uAUG codon may be in-frame with the downstream ORF and end at the same termination site. In this latter case,

© 2000 Cold Spring Harbor Laboratory Press 0-87969-586-4/00 $5 +. 00

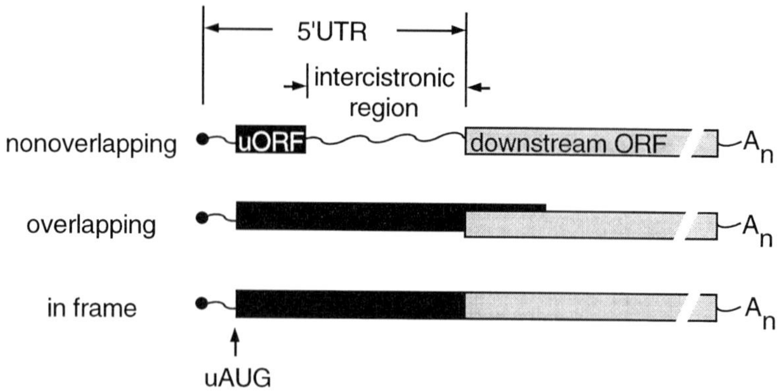

Figure 1 Three possible configurations of ORFs following upstream AUG codons.

the resulting amino-terminal extension to the major protein product may provide additional domains that confer important auxiliary functions to the protein. The control of expression of such alternative proteins appears determined by initiation codon selection (see Chapters 5 and 12). As described in this chapter, additional control mechanisms operate to regulate expression downstream from nonoverlapping and overlapping uORFs. Other reviews that discuss the functions of uORFs can be found in Chapters 5 and 29 and elsewhere (Lovett and Rogers 1996; Morris 1997; Gray and Wickens 1998; McCarthy 1998).

EUKARYOTIC GENES WITH UPSTREAM INITIATION CODONS

Despite the accumulation of nucleotide sequence data and the compilation of 5′UTR sequence databases (Dalphin et al. 1999), identifying transcripts that have uAUG codons remains problematic. The precise 5′UTR sequences of the mRNA are often unclear because mRNA end-mapping data may be unavailable or the cDNA sequence may contain cloning artifacts (see Kozak 1996). In some genes, alternative splicing or transcription initiation site usage may result in expression of multiple transcripts having different 5′UTRs. Furthermore, although the vast majority of known initiation sites are AUG codons, non-AUG codons sometimes function as initiators (Kozak 1999). Finally, database annotations are often ambiguous or inaccurate in specifying 5′UTR boundaries. Thus, there is as yet no reliable method to exploit large-scale sequencing efforts to identify genes that have uAUG codons in their transcripts.

Several surveys have estimated the prevalence of uAUG codons in eukaryotic mRNAs. In 1989, the frequency of uAUG codons in known vertebrate mRNAs was observed to be approximately 10% (Kozak 1987a), although subsequent analysis indicated that this is likely an overestimate for this set of genes (Kozak 1992). uAUG codons appear to be present in at least 3% of *Saccharomyces cerevisiae* mRNAs (Vilela et al. 1998) and are also found in mRNAs of filamentous fungi (Sachs 1998). They appear to be substantially more frequent in some subsets of transcripts, including *Drosophila* mRNAs (Cavener and Cavener 1993) and oncogene mRNAs (Kozak 1991).

Although sequencing has revealed that many eukaryotic transcripts contain uORFs, the impact of uORFs on translation has been investigated for only a small subset of these. Table 1 lists examples in which the expression of proteins from mRNAs containing a uORF is altered by point mutations that eliminate the uORF start codon. Although these genes have diverse functions, the occurrence of functional uORFs in genes such as *CLN3* (Polymenis and Schmidt 1997), *bcl-2* (Harigai et al. 1996), *mdm-2* (Brown et al. 1999), and c-*mos* (Steel et al. 1996) suggests that functional uORFs frequently contribute to the control of cellular growth. For the uORF in the *S. cerevisiae* mRNA encoding the G_1-cyclin Cln3p, a specific role in growth control is strongly implicated because elimination of the uORF accelerates progression through the cell cycle (Polymenis and Schmidt 1997); additional experiments should clarify the role of *trans*-acting factors in the translational regulation of *CLN3* (Danaie et al. 1999). Other genes not listed in Table 1 also contain uORFs that appear to affect gene expression, but the effects of eliminating the uORF initiation codons have not been reported. For example, a mutation predicted to cause premature termination of a uORF in the mRNA of the *Arabidopsis thaliana ATR1* transcription factor increases translation (Bender and Fink 1998). In other cases, mRNAs that contain uORFs are found on smaller polysomes than alternatively spliced variant mRNAs that lack the uORFs (Foyt et al. 1991; Mariottini et al. 1999).

In addition to in vivo analyses of protein expression, RNA expression data are also necessary to evaluate the level(s) at which uORFs function. In most of the examples shown in Table 1, the mRNAs containing the wild-type uORF and the uAUG mutation express similar levels of mRNA (presumably with similar structures), supporting the conclusion that these uORFs act through translational mechanisms. In a few cases, translation of the uORF regulates mRNA stability; this phenomenon is emerging as a potentially important regulatory effect of uORFs, as discussed below and elsewhere (see Chapters 29 and 30).

There are also instances in which mutations of uORF initiation codons do not appear to affect gene expression (see, e.g., Biegalke and Geballe 1990; Krummeck et al. 1991; di Blasi et al. 1993; Vivier et al.

Table 1 uORFs that affect downstream translation in vivo

mRNA	References
Viruses	
baculovirus gp64	Chang and Blissard (1997)
cauliflower mosaic virus	Fütterer and Hohn (1992); Pooggin et al. (1998)
cytomegalovirus *gpUL4*[a]	Schleiss et al. (1991); Degnin et al. (1993); Alderete et al. (1999)
influenza NB/NA	Williams and Lamb (1989)
reovirus S1	Fajardo and Shatkin (1990); Belli and Samuel (1993)
Rous sarcoma virus	Donzé and Spahr (1992); Moustakas et al. (1993b)
SV40 16S and 19S RNAs	Grass and Manley (1987); Perez et al. (1987); Sedman and Mertz (1988); Sedman et al. (1989)
Fungi	
arg-2[a]	Freitag et al. (1996); Luo and Sachs (1996)
CLN3	Polymenis and Schmidt (1997)
CPA1[a]	Werner et al. (1987)
cyc1-362	Pinto et al. (1992)
GCN4	Hinnebusch (1997)
HOL1	Wright et al. (1996)
stuA	Wu and Miller (1997)
YAP2	Vilela et al. (1998, 1999)
Plants	
Lc	Wang and Wessler (1998)
Opaque-2	Lohmer et al. (1993)
pma1	Michelet et al. (1994)
pma3	Lukaszewicz et al. (1998)
Mammals	
A_{2A} adenosine receptor	Lee et al. (1999)
S-adenosylmethionine decarboxylase[a]	Hill and Morris (1992, 1993); Mize et al. (1998)
angiotensin II type 1A receptor	Mori et al. (1996)
β2 adrenergic receptor[a]	Parola and Kobilka (1994)
bcl-2	Harigai et al. (1996)
BTEB	Imataka et al. (1994)
C/EBPα	Lincoln et al. (1998)
C/EBPβ	Lincoln et al. (1998)

Table 1 (*continued*)

mRNA	References
erythrocyte carbonic anhydrase inhibitor	Bergenhem et al. (1992)
fibroblast growth factor 5	Bates et al. (1991)
Fli-1	Sarrazin et al. (2000)
her-2/neu	Child et al. (1999a,b)
lck	Marth et al. (1988)
mdm-2	Brown et al. (1999)
c-*mos*	Steel et al. (1996)
muscle acylphosphatase	Fiaschi et al. (1997)
ornithine decarboxylase	Manzella and Blackshear (1990); Shantz and Pegg (1999)
placenta growth factor	Maglione et al. (1993)
PR65 (PP2A regulatory subunit)	Wera et al. (1995)
retinoic acid receptor β2[a]	Zimmer et al. (1994); Reynolds et al. (1996)
serine hydroxymethyltransferase	Byrne et al. (1995)
thrombpoietin	Ghilardi et al. (1998)
transforming growth factor β3	Arrick et al. (1991)

[a] uORFs whose peptide coding sequence is critical for controlling gene expression.

1999). In such cases, the uORF start codon may be very poorly recognized, and/or ribosomes that have terminated translation of the uORF may reinitiate efficiently. Alternatively, ribosomes might bypass the uAUG codons by ribosome shunting or by the use of internal ribosomal entry sites to initiate translation downstream (see Chapters 4, 19, and 32).

MECHANISMS OF TRANSLATIONAL CONTROL BY uORFS

Determinants of Initiation at uAUG Codons

Although eukaryotic ribosomes appear to be able to choose translation initiation sites in several ways, the preponderance of the evidence indicates that scanning is the most common mechanism for determining start sites on capped and adenylated mRNAs (Kozak 1999; Chapters 2 and 4). The efficiency with which a scanning ribosome chooses an AUG codon to serve as a translation initiation site is modulated by a variety of parameters that are independent of whether that codon is a uAUG or the AUG initiating the major coding ORF. The context of nucleotides flanking an AUG codon is a well-established determinant of initiation efficiency (Kozak 1999). Flanking sequences are also important when non-AUG codons function as initiators (Boeck and Kolakofsky 1994; Grunert and Jackson 1994; Kozak 1999). Close proximity of the AUG codon to the cap site, binding of regulatory factors to RNA, or extensive secondary structure preceding the AUG can each reduce the efficiency of initiation at an AUG codon (McCarthy 1998; Kozak 1999). One noteworthy consequence of a ribosome bypassing the first AUG codon in an mRNA (when that AUG initiates translation of the major gene product) is that a short ORF whose initiation codon is downstream from the first AUG codon may function like the conventional uORFs shown in Figure 1, for example, by affecting RNA stability (Welch and Jacobson 1999) or by producing a peptide (Bullock and Eisenlohr 1996). In contrast to mechanisms that interfere with AUG codon recognition, secondary structure in the mRNA downstream from an AUG codon may slow scanning and facilitate initiation at an otherwise inefficient start site (Kozak 1999).

Ribosomal recognition of a uAUG codon is necessary but not sufficient for that codon to have a regulatory effect on downstream translation. In addition, the efficiencies of translation elongation and termination of the uORF, the properties of the uORF-encoded peptide, and post-termination events affecting reinitiation must each be considered to understand the impact of the uORF on downstream translation. Because of such additional considerations, even relatively inefficient initiation at a uAUG codon can result in substantial inhibition of downstream translation (see, e.g., Cao and Geballe 1995).

Regulation by Eukaryotic Sequence-dependent uORFs

Repression of downstream translation has been shown to depend on the uORF-encoded peptide sequences in a variety of eukaryotic systems including vertebrates, viruses, and fungi (Table 1), indicating that such uORFs have widespread functions in biology. The relative frequencies of occurrence of uORFs whose functions are peptide sequence-dependent or sequence-independent are not yet clear. It is also possible that a given uORF might have sequence-dependent and/or -independent effects on translation, depending on the environment in which it is translated.

The determination that a uORF acts by a sequence-dependent mechanism depends on finding that at least some missense mutations in the uORF alter expression of the downstream ORF. Independent selection schemes were used to identify *cis*-acting mutations affecting the arginine-specific regulation of an arginine biosynthetic gene in two different fungi (*S. cerevisiae CPA1* and *Neurospora crassa arg-2*). In both fungi, negative regulation by arginine was lost by changing an Asp residue in a highly conserved uORF-encoded peptide to Asn (Werner et al. 1987; Freitag et al. 1996). The amino acids encoded by the fourth and fifth codons of the six-codon uORF in the mammalian *S*-adenosyl-methionine decarboxylase (AdoMetDC) transcript are necessary for inhibition of downstream translation in both mammalian cells and *S. cerevisiae,* whereas mutants having changes in the second and third codons, and some but not all changes in the sixth codon, retain the inhibitory properties of the wild-type uORF (Hill and Morris 1993; Mize et al. 1998). In the second uORF in the human cytomegalovirus (HCMV) *gpUL4* (*gp48*) transcript, specific codons dispersed throughout the uORF are necessary for the uORF to inhibit translation (Degnin et al. 1993; Alderete et al. 1999). Thus far, no consensus sequences that distinguish between the classes of sequence-dependent and sequence-independent uORFs have been recognized.

Expression downstream from sequence-dependent uORFs is conditional in some cases. As mentioned above, synthesis of CPA1 and ARG-2 proteins is strongly repressed by uORFs in the presence of high levels of arginine. Similarly, polyamines act in concert with a uORF to repress AdoMetDC translation (Ruan et al. 1996). One of the retinoic acid receptor β2 uORFs represses translation in heart and brain, but not in other tissues of transgenic mice (Reynolds et al. 1996).

Mechanisms of Inhibition by Sequence-dependent uORFs

The requirement for specific codons within a uORF for inhibitory effects on downstream translation implicates the uORF-encoded peptide product as the mediator of the repressive activity. Although ribosomes may stall at codons simply because the cognate aminoacylated-tRNA is scarce, as occurs in attenuation control of *Escherichia coli* amino acid biosynthetic operons (Landick et al. 1996), naturally occurring examples of rare codon-dependent mechanisms have not been observed for eukaryotic sequence-dependent uORFs. Yet, these uORFs generally act only in *cis,* suggesting a requirement for a particular orientation of the nascent peptide that might occur only transiently and only in or near the ribosome that synthesized the peptide. The ability of *cis*-acting nascent peptide products of short ORFs to alter ribosomal function and thereby impact downstream translation also appears to be important for regulation of some prokaryotic genes (Lovett and Rogers 1996; Konan and Yanofsky 1999).

Several lines of evidence suggest that the nascent peptides of sequence-dependent uORFs mediate inhibition of downstream translation by interfering with a step during translation termination (Fig. 2). First, primer extension inhibition ("toeprint") assays reveal that ribosomes stall on the HCMV *UL4, N. crassa arg-2,* and *S. cerevisiae CPA1* uORFs when the termination codon of the uORF reaches the ribosomal A site (Cao and Geballe 1996a; Wang and Sachs 1997; Wang et al. 1999). Second, extending the *UL4* and *AdoMetDC* uORFs by mutating their termination codons eliminates the inhibition of downstream translation, revealing that these uORFs do not function without a properly positioned termination codon. Finally, the nascent uORF peptide in the *UL4* mRNA appears to interfere with the translation termination reaction prior to the peptidyl-tRNA hydrolysis step, as judged by the sequence-dependent persistence of the covalent linkage of the uORF-encoded peptide to the tRNA that decodes the final uORF codon (Cao and Geballe 1996b; 1998).

Despite the striking similarity of the ribosomal stalling at the end of the fungal *arg-2/CPA1* uORFs and the *UL4* uORF, the precise mechanisms used by these uORFs are not identical. Ribosomes stall after trans-

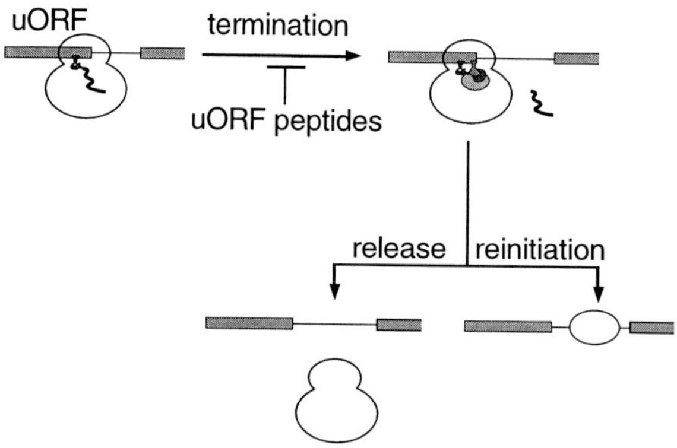

Figure 2 Model of uORF translation termination and downstream reinitiation. When a ribosome reaches the end of a uORF, eukaryotic release factors 1 and 3, GTP, the ribosomal peptidyl transferase center, and possibly other factors catalyze the hydrolysis of the peptidyl-tRNA ester linkage (Nakamura et al. 1996; Czaplinski et al. 1999). Several uORF peptides interfere with the termination reaction, as discussed in the text. After termination, the ribosome may dissociate from the mRNA or may resume scanning and possibly reinitiate translation downstream. Whether the form of the ribosome that resumes scanning after termination is the 40S subunit alone (as depicted) or the entire 80S ribosome is not known.

lation of the *arg-2* and *CPA1* uORFs even when the termination codon has been eliminated (Wang et al. 1998, 1999). Thus, these uORFs can induce ribosome stalling during elongation as well as termination, in contrast to the *UL4* uORF. These observations suggest that interfering with ribosomal progression, whether by an effect on ribosomal elongation or on termination, may be a common mechanism by which sequence-dependent uORFs block downstream translation.

The nascent peptide moieties of these uORFs may interact with—and disrupt—the function of a ribosomal component or a translation factor that is required for efficient termination of translation. As a result, hydrolysis of the nascent peptidyl-tRNA bond is impaired, ribosomes fail to vacate the termination site of the uORF for a prolonged time period, and access to the downstream cistron is obstructed. Although the factors with which these uORF peptides interact have not yet been identified, ribosomal RNA(s), ribosomal protein(s), or translation factors are likely candidates. In *E. coli,* a uORF-encoded peptide of the *cat* gene distorts the structure and inhibits the activity of the ribosomal peptidyl transferase center (Lovett and Rogers 1996). Since the peptidyl transferase center is required for both elongation and termination, it should be considered a candidate target of the *arg-2* and *CPA1* uORFs. In cases such as the *AdoMetDC* and *UL4* uORFs that inhibit terminating but not elongating ribosomes, the target of the uORF could possibly be a termination-specific factor. In prokaryotes, another factor known as ribosome recycling factor (RF4) catalyzes release of the ribosome from the mRNA (Janosi et al. 1996; Chapter 11). Although no corresponding factor has been found in eukaryotes (except in prokaryote-derived organelles), such an activity would be another potential target of the uORF peptides. Differences among the sequences of these uORFs suggest that there may be various interactions capable of interfering with efficient translation termination.

The full inhibitory effects of the *AdoMetDC* and *arg-2*/*CPA1* uORFs require elevated levels of polyamines and arginine, respectively. In principle, these coregulators could act in a number of ways. They might stabilize key interactions between the uORF-encoded peptides and the ribosomes or ribosome-associated translation factors, thereby interfering with subsequent steps in translation. Other data consistent with the capacity of polyamines to regulate translation are the examples of programmed ribosomal frameshifting mediated by spermidine (Matsufuji et al. 1995, 1996). Alternatively, as with any uORF, whether sequence-dependent or -independent, a change in the efficiency of initiation codon recognition has the potential to modulate the uORF's inhibitory activity. In the case of *AdoMetDC*, alteration in the recognition of the uAUG codon accounts for the inactivity of the uORF in some cell types (Ruan et al. 1994), but does not explain the regulation mediated by polyamine levels (Ruan et al. 1996).

Control of Reinitiation

Regardless of considerations of the role of the uORF peptide sequence, a ribosome that has terminated translation of a uORF may either dissociate from the mRNA or resume scanning on the same mRNA and potentially reinitiate at a downstream AUG codon (Fig. 2). Gene expression can be regulated by modulating the efficiency of reinitiation, as is clearly illustrated by studies of the mechanism of *GCN4* translational control (see Chapter 5). As mentioned above, a uORF might appear to have no impact on downstream translation if reinitiation (at the start codon of the ORF specifying the gene product) is very efficient. In such cases, there may be as-yet-undefined conditions under which reinitiation efficiency is reduced, and thus, uORFs that appear to be translationally inactive may ultimately be found to be regulatory.

Although the signals that determine the fate of the ribosome after it has translated a uORF are not fully understood, it is clear that RNA sequences between the termination codon of the uORF and the downstream ORF influence reinitiation efficiency (Chapter 5). Lengthening short intercistronic regions tends to augment downstream reinitiation (see, e.g., Kozak 1987b; Child et al. 1999b). In addition to length, the sequence of the intercistronic region has a major impact on reinitiation frequency. In the case of the C/EBPα mRNA, translation downstream from the 7-nucleotides intercistronic spacer can be considerably increased by substitution mutations (Lincoln et al. 1998). Similarly, alterations of sequences downstream from a uORF in the plant *Lc* mRNA greatly augment reinitiation (Wang and Wessler 1998).

The mechanisms by which the intercistronic sequences affect reinitiation efficiency are unknown. Following translation of a uORF, the ribosome may encounter RNA sequences or structures that promote dissociation of the ribosome from the mRNA and thus eliminate the potential for reinitiation. Even if the ribosome remains attached to the mRNA, it presumably needs to reacquire initiation factors (such as eIF2-GTP) and Met-$tRNA_i^{Met}$ before it can reinitiate downstream. Longer intercistronic spacing may augment reinitiation by allowing more time for ribosomes to acquire the necessary factors, as appears to be the case for *GCN4* (Chapter 5).

In addition to the intercistronic sequences, other elements in the mRNA affect reinitiation. In one case, lengthening a uORF reduces the efficiency of downstream reinitiation (Luukkonen et al. 1995), for reasons that remain to be determined. Sequences upstream from the uAUG and those of the downstream cistron also affect reinitiation (Grant et al. 1994, 1995). Elucidating the controls that influence the fate of the ribosome and its reinitiation potential after translation of a uORF will be critical for understanding the regulatory mechanisms used by many uORFs.

ADDITIONAL ROLES FOR uORFS

Stimulatory uORFs

In contrast to most eukaryotic uORFs that are either neutral or repress translation of the downstream ORF, a few act positively to augment downstream translation. In prokaryotes, 5′-proximal ORFs that increase translation of downstream ORFs are relatively common because translational coupling is used to modulate expression of different polypeptide coding regions within an operon (McCarthy 1998 and references therein). In eukaryotes, uORFs that appear to stimulate translation may not have inherent stimulatory activity but may enable ribosomes to bypass other *cis*-acting inhibitory elements in the mRNA. For example, the first uORF in the *GCN4* 5′UTR allows ribosomes to bypass the strongly repressive fourth uORF under derepressing conditions (see Chapter 5). The *Aspergillus nidulans* gene *stunted* expresses two mRNAs; one (α) contains a uORF that augments expression of the downstream protein by an as-yet-unknown mechanism (Wu and Miller 1997).

A fascinating example of a uORF acting as a positive effector of downstream translation occurs in plant viruses. The 5′UTR of the polycistronic 35S cauliflower mosaic virus (CaMV) RNA is long, highly structured, and contains up to nine uAUGs, the first of which initiates a four-codon uORF. Expression of the downstream cistrons in CaMV appears to occur by a ribosome shunting mechanism; shunting is enhanced by the first uORF and requires a viral *trans*-activator protein and secondary structural elements in the 5′UTR (Dominguez et al. 1998). Although mutants in which the first uAUG codon is absent are infectious, upon passage in plants, revertants consistently emerge in which the capacity to translate the first uORF is restored. These observations suggest that the uORF confers a selective advantage for viral growth (Pooggin et al. 1998). One model of the underlying mechanism proposes that, following translation of the first uORF, an RNA secondary structure forms that enables the ribosome to shunt to a far downstream ORF. Further elucidation of this mechanism should be facilitated by the reconstruction of shunting in cell-free translation extracts and in yeast (Sha et al. 1995; Dominguez et al. 1998).

Effects on RNA Stability

The stability and translation of eukaryotic mRNAs are coupled processes in some cases. Pertinent to gene regulation by uORFs is the phenomenon of nonsense-mediated decay (see Chapters 29 and 30), in which mRNAs are destabilized by nonsense mutations in the 5′-proximal portion of an ORF. A downstream sequence element and *trans*-acting factors (such as are encoded by *UPF* genes in *S. cerevisiae*) are also important for nonsense-mediated decay. uORFs mimic the configuration of mRNAs that have premature stop codons, and they have the potential to trigger mRNA decay. For example, addition of a functional downstream element to a transcript containing one or more *GCN4* uORFs activates nonsense-mediated decay of the mRNA (Ruiz-Echevarria and Peltz 1996). An aberrant initiation codon in the transcript leader of the *cyc1-362* allele creates a uORF and reduces transcript levels by 100-fold (Pinto et al. 1992). The *YAP2* mRNA contains two uORFs that each contribute to mRNA instability but by a mechanism that seems not to depend on the *UPF1*, suggesting that there may be multiple mechanisms by which uORFs destabilize mRNAs (Vilela et al. 1999). Thus, although the translation of many of the uORFs listed in Table 1 does not appear to alter the stability of their mRNAs, it is clear that the translation of at least some uORFs can affect mRNA stability.

Role in Viral Packaging

For some viruses, transcripts serve both as genomes and as mRNAs. The mechanisms that determine whether a particular RNA molecule is packaged into new virions or is translated are largely unknown, nor are these necessarily mutually incompatible fates. Because the 5′ ends of viral RNAs are often important in virus replication as well as in translation and packaging, experimental dissection of the various functions of these sequences is complex.

In a few viruses, uAUG codons affect encapsidation of the viral genome (Nassal et al. 1990; Donzé and Spahr 1992; Moustakas et al. 1993a; Donzé et al. 1995). The 5′UTRs of avian sarcoma-leukosis retroviruses contain three uORFs that are conserved in length and position within the 5′UTR, but not in amino-acid-coding content. Ribosomes appear to initiate translation at the first and third uAUG codons and, at least in cell-free translation assays, they synthesize the seven-amino-acid peptide product of the first uORF (Hackett et al. 1986). Mutations of the first and third uAUG codons interfere with production of infectious virus, suggesting that translation of the associated uORFs might be required for packaging. However, recent studies suggest that mutations of the third uAUG codon diminish viral packaging by perturbing an RNA secondary structure (Banks and Linial 2000). Thus, the role of translation of the uORFs in packaging of these viruses remains unresolved.

Peptide Products Resulting from Initiation at uAUG Codons

Bicistronic transcripts in which the polypeptide encoded in the upstream region has been demonstrated to play a functional role outside of translational control are very rare in eukaryotes (Gray et al. 1999). In the case of short uORFs, the encoded peptide products have been detected in a few cases, typically in cell-free translation assays (Hackett et al. 1986; Khalili et al. 1987; Cao and Geballe 1996b; Wang and Wessler 1998). However, fusions of uORFs to downstream reporter gene ORFs often yield detectable fusion proteins in vivo, suggesting that these uORFs are in fact translated. Several uORFs are highly conserved in peptide sequence (see, e.g., Maglione et al. 1993; Parola and Kobilka 1994; Lincoln et al. 1998; Martinez-Garcia et al. 1998; Child et al. 1999a). For many it remains to be determined whether this conservation is important for translational control or for other functions.

CONCLUSIONS

The relatively common occurrence of uORFs in diverse genes from multiple systems indicates that uORFs are involved in a broad range of biological processes. uORFs appear to be unusually prevalent in mRNAs of genes involved in controlling cellular growth, such as oncogenes, growth factors, and cellular receptors, and in several of these cases the uORFs have been shown to control gene expression (Table 1). These observations suggest that one general role of uORFs may be to participate in the control of cellular growth and division by monitoring cellular translational capacity. uORFs may also serve more specialized functions, such as their apparent role in discriminating genomic RNAs from mRNA during viral encapsidation. In other cases, uORF may produce small peptides with roles beyond translation. As additional examples of genes with functional uORFs are identified, a more complete understanding of the significance of these elements should emerge.

Experimental data illustrate that uORFs control gene expression by a variety of mechanisms. Although the efficiency of initiation at uAUG codons modulates the impact of a uORF, the model that inhibition of downstream translation is explained solely by uAUG codons preemptively capturing scanning ribosomes is clearly an oversimplification. Subsequent steps during and after translation of the uORF also influence uORF function. Translation of a uORF in some cases produces a *cis*-acting peptide that, as a result of its inhibitory effect on translation elongation or termination, causes ribosomes to stall at the end of the uORF and to block access to the downstream ORF. In other cases, the uORF peptide sequence appears unimportant, at least under the experimental conditions used, but the length or position of the uORF or other features of the mRNA determine the efficiency with which ribosomes reinitiate translation at the downstream ORF. Translation of a uORF can also control gene expression by affecting the stability of the mRNA.

For both peptide sequence-dependent and -independent uORFs, *trans*-acting factors may participate in the regulatory mechanism. In some examples where reinitiation or mRNA stability is regulated by a uORF, the availability of general factors such as eIF2-GTP or factors involved in the nonsense-mediated decay pathway are critical for the uORF function. Translation of the downstream ORF may be regulated by the concentrations of the specific coregulators acting in conjunction with the uORF peptide. Future studies will undoubtedly reveal unexpected functions and mechanisms of regulation by uORFs.

ACKNOWLEDGMENTS

Work in the authors' laboratories is supported by National Institutes of Health grants AI-26672 to A.P.G. and GM-47498 to M.S.S. We thank David Morris for critical reading of the manuscript.

REFERENCES

Alderete J.P., Jarrahian S., and Geballe A.P. 1999. Translational effects of mutations and polymorphisms in a repressive upstream open reading frame of the human cytomegalovirus UL4 gene. *J. Virol.* **73:** 8330–8337.

Arrick B.A., Lee A.L., Grendell R.L., and Derynck R. 1991. Inhibition of translation of transforming growth factor-beta 3 mRNA by its 5′ untranslated region. *Mol. Cell. Biol.* **11:** 4306–4313.

Banks J.D. and Linial M.L. 2000. Secondary structure analysis of a minimal avian leukosis-sarcoma virus packaging signal. *J. Virol.* **74:** 456–464.

Bates B., Hardin J., Zhan X., Drickamer K., and Goldfarb M. 1991. Biosynthesis of human fibroblast growth factor-5. *Mol. Cell. Biol.* **11:** 1840–1845.

Belli B.A. and Samuel C.E. 1993. Biosynthesis of reovirus-specified polypeptides: Identification of regions of the bicistronic reovirus S1 mRNA that affect the efficiency of translation in animal cells. *Virology* **193:** 16–27.

Bender J. and Fink G.R. 1998. A Myb homologue, *ATR1*, activates tryptophan gene expression in *Arabidopsis*. *Proc. Natl. Acad. Sci.* **95:** 5655–5660.

Bergenhem N.C., Venta P.J., Hopkins P.J., Kim H.J., and Tashian R.E. 1992. Mutation creates an open reading frame within the 5′ untranslated region of macaque erythrocyte carbonic anhydrase (CA) I mRNA that suppresses CA I expression and supports the scanning model for translation. *Proc. Natl. Acad. Sci.* **89:** 8798–8802.

Biegalke B.J. and Geballe A.P. 1990. Translational inhibition by cytomegalovirus transcript leaders. *Virology* **177:** 657–667.

Boeck R. and Kolakofsky D. 1994. Positions +5 and +6 can be major determinants of the efficiency of non-AUG initiation codons for protein synthesis. *EMBO J.* **13:** 3608–3617.

Brown C.Y., Mize G.J., Pineda M., George D.L., and Morris D.L. 1999. Role of two upstream open reading frames in the translational control of oncogene *mdm2*. *Oncogene* **18:** 5631–5637.

Bullock T.N. and Eisenlohr L.C. 1996. Ribosomal scanning past the primary initiation codon as a mechanism for expression of CTL epitopes encoded in alternative reading frames. *J. Exp. Med.* **184:** 1319–1329.

Byrne P.C., Sanders P.G., and Snell K. 1995. Translational control of mammalian serine hydroxymethyltransferase expression. *Biochem. Biophys. Res. Commun.* **214:** 496–502.

Cao J. and Geballe A.P. 1995. Translational inhibition by a human cytomegalovirus upstream open reading frame despite inefficient utilization of its AUG codon. *J. Virol.* **69:** 1030–1036.

———. 1996a. Coding sequence-dependent ribosomal arrest at termination of translation. *Mol. Cell. Biol.* **16:** 603–608.

———. 1996b. Inhibition of nascent-peptide release at translation termination. *Mol. Cell. Biol.* **16:** 7109–7114.

———. 1998. Ribosomal release without peptidyl tRNA hydrolysis at translation termination in a eukaryotic system. *RNA* **4:** 181–188.

Cavener D.R. and Cavener B.A. 1993. Translation start sites and mRNA leaders. In *An atlas of* Drosophila *genes* (ed. G. Maroni), pp. 359-377. Oxford University Press, New York.

Chang M.J. and Blissard G.W. 1997. Baculovirus *gp64* gene expression: Negative regulation by a minicistron. *J. Virol.* **71:** 7448–7460.

Child S.J., Miller M.K., and Geballe A.P. 1999a. Cell type-dependent and -independent control of HER-2/neu translation. *Int. J. Biochem. Cell. Biol.* **31:** 201–213.

———. 1999b. Translational control by an upstream open reading frame in the HER-2/*neu* transcript. *J. Biol. Chem.* **274:** 24335–24341.

Czaplinski K., Ruiz-Echevarria M.J., Gonzalez C.I., and Peltz S.W. 1999. Should we kill the messenger? The role of the surveillance complex in translation termination and mRNA turnover. *Bioessays* **21:** 685–696.

Dalphin M.E., Stockwell P.A., Tate W.P., and Brown C.M. 1999. TransTerm, the translational signal database, extended to include full coding sequences and untranslated regions. *Nucleic Acids Res.* **27:** 293–294.

Danaie P., Altmann M., Hall M.N., Trachsel H., and Helliwell S.B. 1999. *CLN3* expression is sufficient to restore G1-to-S-phase progression in *Saccharomyces cerevisiae* mutants defective in translation initiation factor eIF4E. *Biochem. J.* **340:** 135–141.

Degnin C.R., Schleiss M.R., Cao J., and Geballe A.P. 1993. Translational inhibition mediated by a short upstream open reading frame in the human cytomegalovirus gpUL4 (gp48) transcript. *J. Virol.* **67:** 5514–5521.

di Blasi F., Carra E., de Vendittis E., Masturzo P., Burderi E., Lambrinoudaki I., Mirisola M.G., Seidita G., and Fasano O. 1993. The SCH9 protein kinase mRNA contains a long 5′ leader with a small open reading frame. *Yeast* **9:** 21–32.

Dominguez D.I., Ryabova L.A., Pooggin M.M., Schmidt-Puchta W., Fütterer J., and Hohn T. 1998. Ribosome shunting in cauliflower mosaic virus. Identification of an essential and sufficient structural element. *J. Biol. Chem.* **273:** 3669–3678.

Donzé O. and Spahr P.-F. 1992. Role of the open reading frames of Rous sarcoma virus leader RNA in translation and genome packaging. *EMBO J.* **11:** 3747–3757.

Donzé O., Damay P., and Spahr P.-F. 1995. The first and third uORFs in RSV leader RNA are efficiently translated: Implications for translational regulation and viral RNA packaging. *Nucleic Acids Res.* **23:** 861–868.

Fajardo J.E. and Shatkin A.J. 1990. Translation of bicistronic viral mRNA in transfected cells: Regulation at the level of elongation. *Proc. Natl. Acad. Sci.* **87:** 328–332.

Fiaschi T., Marzocchini R., Raugei G., Veggi D., Chiarugi P., and Ramponi G. 1997. The 5′-untranslated region of the human muscle acylphosphatase mRNA has an inhibitory effect on protein expression. *FEBS Lett.* **417:** 130–134.

Foyt H.L., LeRoith D., and Roberts C.T., Jr. 1991. Differential association of insulin-like growth factor I mRNA variants with polysomes *in vivo*. *J. Biol. Chem.* **266:** 7300–7305.

Freitag M., Dighde N., and Sachs M.S. 1996. A UV-induced mutation that affects translational regulation of the *Neurospora arg-2* gene. *Genetics* **142:** 117–127.

Fütterer J. and Hohn T. 1992. Role of an upstream open reading frame in the translation of polycistronic mRNAs in plant cells. *Nucleic Acids Res.* **20:** 3851–3857.

Ghilardi N., Wiestner A., and Skoda R.C. 1998. Thrombopoietin production is inhibited by a translational mechanism. *Blood* **92:** 4023–4030.

Grant C.M., Miller P.F., and Hinnebusch A.G. 1994. Requirements for intercistronic distance and level of eukaryotic initiation factor 2 activity in reinitiation on *GCN4* mRNA vary with the downstream cistron. *Mol. Cell. Biol.* **14:** 2616–2628.

———. 1995. Sequences 5′ of the first upstream open reading frame in *GCN4* mRNA are required for efficient translational reinitiation. *Nucleic Acids Res.* **23:** 3980–3988.

Grass D.S. and Manley J.L. 1987. Selective translation initiation on bicistronic simian virus 40 late mRNA. *J. Virol.* **61:** 2331–2335.

Gray N.K. and Wickens M. 1998. Control of translation initiation in animals. *Annu. Rev. Dev. Biol.* **14:** 399–458.

Gray T.A., Saitoh S., and Nicholls R.D. 1999. An imprinted, mammalian bicistronic transcript encodes two independent proteins. *Proc. Natl. Acad. Sci.* **96:** 5616–5621.

Grunert S. and Jackson R.J. 1994. The immediate downstream codon strongly influences the efficiency of utilization of eukaryotic translation initiation codons. *EMBO J.* **13:** 3618–3630.

Hackett P.B., Petersen R.B., Hensel C.H., Albericio F., Gunderson S.I., Palmenberg A.C., and Barany G. 1986. Synthesis in vitro of a seven amino acid peptide encoded in the leader RNA of Rous sarcoma virus. *J. Mol. Biol.* **190:** 45–57.

Harigai M., Miyashita T., Hanada M., and Reed J.C. 1996. A *cis*-acting element in the *BCL-2* gene controls expression through translational mechanisms. *Oncogene* **12:** 1369–1374.

Hill J.R. and Morris D.R. 1992. Cell-specific translation of S-adenosylmethionine decarboxylase mRNA. Regulation by the 5′ transcript leader. *J. Biol. Chem.* **267:** 21886–21893.

———. 1993. Cell-specific translational regulation of S-adenosylmethionine decarboxylase mRNA: Dependence on translation and coding capacity of the *cis*-acting upstream open reading frame. *J. Biol. Chem.* **269:** 726–731.

Hinnebusch A.G. 1997. Translational regulation of yeast *GCN4:* A window on factors that control initiator-tRNA binding to the ribosome. *J. Biol. Chem.* **272:** 21661–21664.

Imataka H., Nakayama K., Yasumoto K., Mizuno A., Fujii-Kuriyama Y., and Hayami M. 1994. Cell-specific translational control of transcription factor BTEB expression. The role of an upstream AUG in the 5′-untranslated region. *J. Biol. Chem.* **269:** 20668–20673.

Janosi L., Hara H., Zhang S., and Kaji A. 1996. Ribosome recycling by ribosome recycling factor (RRF) - An important but overlooked step of protein biosynthesis. *Adv. Biophys.* **32:** 121–201.

Khalili K., Brady J., and Khoury G. 1987. Translational regulation of SV40 early mRNA defines a new viral protein. *Cell* **48:** 639–645.

Konan K.V. and Yanofsky C. 1999. Role of ribosome release in regulation of *tna* operon expression in *Escherichia coli*. *J. Bacteriol.* **181:** 1530–1536.

Kozak M. 1987a. An analysis of 5′-noncoding sequences from 699 vertebrate messenger RNAs. *Nucleic Acids Res.* **15:** 8125–8148.

———. 1987b. Effects of intercistronic length on the efficiency of reinitiation by eucaryotic ribosomes. *Mol. Cell. Biol.* **7:** 3438–3445.

———. 1991. An analysis of vertebrate mRNA sequences: Intimations of translational control. *J. Cell Biol.* **115:** 887–903.

———. 1992. Regulation of translation in eukaryotic systems. *Annu. Rev. Cell Biol.* **8:** 197–225.

———. 1996. Interpreting cDNA sequences: Some insights from studies on translation. *Mamm. Genome* **7:** 563–574.

———. 1999. Initiation of translation in prokaryotes and eukaryotes. *Gene* **234:** 187–208.

Krummeck G., Gottenöf T., and Rödel G. 1991. AUG codons in the RNA leader sequences of the yeast *PET* genes *CBS1* and *SCO1* have no influence on translation efficiency. *Curr. Genet.* **20:** 465–469.

Landick R., Turnbough C.L.J., and Yanofsky C. 1996. Transcription attenuation. In Escherichia coli *and* Salmonella: *Cellular and molecular biology* (ed. F.C. Neidhardt et al.), pp. 1263–1286. ASM Press, Washington, D.C.

Lee Y.C., Chang C.W., Su C.W., Lin T.N., Sun S.H., Lai H.L., and Chern Y. 1999. The 5′ untranslated regions of the rat A_{2A} adenosine receptor gene function as negative translational regulators. *J. Neurochem.* **73:** 1790–1798.

Lincoln A.J., Monczak Y., Williams S.C., and Johnson P.F. 1998. Inhibition of CCAAT/enhancer-binding protein alpha and beta translation by upstream open reading frames. *J. Biol. Chem.* **273:** 9552–9560.

Lohmer S., Maddaloni M., Motto M., Salamini F., and Thompson R.D. 1993. Translation of the mRNA of the maize transcriptional activator *Opaque-2* is inhibited by upstream open reading frames present in the leader sequence. *Plant Cell* **5:** 65–73.

Lovett P.S. and Rogers E.J. 1996. Ribosome regulation by the nascent peptide. *Microbiol. Rev.* **60:** 366–385.

Lukaszewicz M., Jerouville B., and Boutry M. 1998. Signs of translational regulation within the transcript leader of a plant plasma membrane H(+)-ATPase gene. *Plant J.* **14:** 413–423.

Luo Z. and Sachs M.S. 1996. Role of an upstream open reading frame in mediating arginine-specific translational control in *Neurospora crassa*. *J. Bacteriol.* **178:** 2172–2177.

Luukkonen B.G., Tan W., and Schwartz S. 1995. Efficiency of reinitiation of translation on human immunodeficiency virus type 1 mRNAs is determined by the length of the upstream open reading frame and by intercistronic distance. *J. Virol.* **69:** 4086–4094.

Maglione D., Guerriero V., Rambaldi M., Russo G., and Persico M.G. 1993. Translation of the placenta growth factor mRNA is severely affected by a small open reading frame localized in the 5′ untranslated region. *Growth Factors* **8:** 141–152.

Manzella J.M. and Blackshear P.J. 1990. Regulation of rat ornithine decarboxylase mRNA translation by its 5′-untranslated region. *J. Biol. Chem.* **265:** 11817–11822.

Mariottini P., Shah Z.H., Toivonen J.M., Bagni C., Spelbrink J.N., Amaldi F., and Jacobs H.T. 1999. Expression of the gene for mitoribosomal protein S12 is controlled in human cells at the levels of transcription, RNA splicing, and translation. *J. Biol. Chem.* **274:** 31853–31862.

Marth J.D., Overell R.W., Meier K.E., Krebs E.G., and Perlmutter R.M. 1988. Translational activation of the *lck* proto-oncogene. *Nature* **332:** 171–173.

Martinez-Garcia J.F., Moyano E., Alcocer M.J., and Martin C. 1998. Two bZIP proteins from *Antirrhinum* flowers preferentially bind a hybrid C-box/G-box motif and help to define a new sub-family of bZIP transcription factors. *Plant J.* **13:** 489–505.

Matsufuji S., Matsufuji T., Wills N.M., Gesteland R.F., and Atkins J.F. 1996. Reading two bases twice: Mammalian antizyme frameshifting in yeast. *EMBO J.* **15:** 1360–1370.

Matsufuji S., Matsufuji T., Miyazaki Y., Murakami Y., Atkins J.F., Gesteland R.F., and Hayashi S. 1995. Autoregulatory frameshifting in decoding mammalian ornithine decarboxylase antizyme. *Cell* **80:** 51–60.

McCarthy J.E.G. 1998. Posttranscriptional control of gene expression in yeast. *Microbiol. Mol. Biol. Rev.* **62:** 1492–1553.

Michelet B., Lukaszewicz M., Dupriez V., and Boutry M. 1994. A plant plasma membrane proton-ATPase gene is regulated by development and environment and shows signs of a translational regulation. *Plant Cell* **6:** 1375–1389.

Mize G.J., Ruan H., Low J.J., and Morris D.R. 1998. The inhibitory upstream open reading frame from mammalian S-adenosylmethionine decarboxylase mRNA has a strict sequence specificity in critical positions. *J. Biol. Chem.* **273:** 32500–32505.

Mori Y., Matsubara H., Murasawa S., Kijima K., Maruyama K., Tsukaguchi H., Okubo N., Hamakubo T., Inagami T., Iwasaka T., and Inada M. 1996. Translational regulation of angiotensin II type 1A receptor: Role of upstream AUG triplets. *Hypertension* **28:** 810–817.

Morris D.R. 1997. *Cis*-acting mRNA structures in gene-specific translational control. In *mRNA metabolism and post-transcriptional gene regulation* (ed. J.B. Harford and D.R. Morris), pp. 165–180. Wiley-Liss, New York.

Moustakas A., Sonstegard T.S., and Hackett P.B. 1993a. Alterations of the three short open reading frames in the Rous sarcoma virus leader RNA modulate viral replication and gene expression. *J. Virol.* **67:** 4337–4349.

———. 1993b. Effects of the open reading frames in the Rous sarcoma virus leader RNA on translation. *J. Virol.* **67:** 4350–4357.

Nakamura Y., Ito K., and Isaksson L.A. 1996. Emerging understanding of translation termination. *Cell* **87:** 147–150.

Nassal M., Junker-Niepmann M., and Schaller H. 1990. Translational inactivation of RNA function: Discrimination against a subset of genomic transcripts during HBV nucleocapsid assembly. *Cell* **63:** 1357–1363.

Parola A.L. and Kobilka B.K. 1994. The peptide product of a 5′ leader cistron in the β_2 adrenergic receptor mRNA inhibits receptor synthesis. *J. Biol. Chem.* **269:** 4497–4505.

Perez L., Wills J.W., and Hunter E. 1987. Expression of the Rous sarcoma virus *env* gene from a simian virus 40 late-region replacement vector: Effects of upstream initiation codons. *J. Virol.* **61:** 1276–1281.

Pinto I., Na J.G., Sherman F., and Hampsey M. 1992. *Cis*- and *trans*-acting suppressors of a translation initiation defect at the *cyc1* locus of *Saccharomyces cerevisiae*. *Genetics* **132:** 97–112.

Polymenis M. and Schmidt E.V. 1997. Coupling of cell division to cell growth by translational control of the G_1 cyclin *CLN3* in yeast. *Genes Dev.* **11:** 2522–2531.

Pooggin M.M., Hohn T., and Fütterer J. 1998. Forced evolution reveals the importance of short open reading frame A and secondary structure in the cauliflower mosaic virus 35S RNA leader. *J. Virol.* **72:** 4157–4169.

Reynolds K., Zimmer A.M., and Zimmer A. 1996. Regulation of RARβ2 mRNA expression: Evidence for an inhibitory peptide encoded in the 5′-untranslated region. *J. Cell Biol.* **134:** 827–835.

Ruan H., Hill J.R., Fatemie-Nainie S., and Morris D.R. 1994. Cell-specific translational regulation of S-adenosylmethionine decarboxylase mRNA: Influence of the structure of the 5′ transcript leader on regulation by the upstream open reading frame. *J. Biol. Chem.* **269:** 17905–17910.

Ruan H., Shantz L.M., Pegg A.E., and Morris D.R. 1996. The upstream open reading frame of the mRNA encoding S-adenosylmethionine decarboxylase is a polyamine-responsive translational control element. *J. Biol. Chem.* **271:** 29576–29582.

Ruiz-Echevarria M.J. and Peltz S.W. 1996. Utilizing the GCN4 leader region to investigate the role of the sequence determinants in nonsense-mediated mRNA decay. *EMBO J.* **15:** 2810–2819.

Sachs M.S. 1998. Posttranscriptional control of gene expression in filamentous fungi. *Fungal Genet. Biol.* **23:** 117–124.

Sarrazin S., Starck J., Gonnet C., Donbeikovski A., Melet F., and Mork F. 2000. Negative and translation termination-dependent positive control of *Fli-1* protein synthesis by conserved overlapping 5′ upstream open reading frames in Fli-1 mRNA. *Mol. Cell. Biol.* **20:** 2959–2969.

Schleiss M.R., Degnin C.R., and Geballe A.P. 1991. Translational control of human cytomegalovirus gp48 expression. *J. Virol.* **65:** 6782–6789.

Sedman S.A. and Mertz J.E. 1988. Mechanisms of synthesis of virion proteins from the functionally bigenic late mRNAs of simian virus 40. *J. Virol.* **62:** 954–961.

Sedman S.A., Good P.J., and Mertz J.E. 1989. Leader-encoded open reading frames modulate both the absolute and relative rates of synthesis of the virion proteins of simian virus 40. *J. Virol.* **63:** 3884–3893.

Sha Y., Broglio E.P., Cannon J.F., and Schoelz J.E. 1995. Expression of a plant viral polycistronic mRNA in yeast, *Saccharomyces cerevisiae*, mediated by a plant virus translational transactivator. *Proc. Natl. Acad. Sci.* **92:** 8911–8915.

Shantz L.M. and Pegg A.E. 1999. Translational regulation of ornithine decarboxylase and other enzymes of the polyamine pathway. *Int. J. Biochem. Cell Biol.* **31:** 107–122.

Steel L.F., Telly D.L., Leonard J., Rice B.A., Monks B., and Sawicki J.A. 1996. Elements in the murine c-*mos* messenger RNA 5′-untranslated region repress translation of downstream coding sequences. *Cell Growth Differ.* **7:** 1415–1424.

Vilela C., Linz B., Rodrigues-Pousada C., and McCarthy J.E. 1998. The yeast transcription factor genes *YAP1* and *YAP2* are subject to differential control at the levels of both translation and mRNA stability. *Nucleic Acids Res.* **26:** 1150–1159.

Vilela C., Ramirez C.V., Linz B., Rodrigues-Pousada C., and McCarthy J.E. 1999. Post-termination ribosome interactions with the 5′UTR modulate yeast mRNA stability. *EMBO J.* **18:** 3139–3152.

Vivier M.A., Sollitti P., and Pretorius I.S. 1999. Functional analysis of multiple AUG codons in the transcripts of the STA2 glucoamylase gene from *Saccharomyces cerevisiae*. *Mol. Gen. Genet.* **261:** 11–20.

Wang L. and Wessler S.R. 1998. Inefficient reinitiation is responsible for upstream open reading frame-mediated translational repression of the maize R gene. *Plant Cell* **10:** 1733–1746.

Wang P., Gaba A., and Sachs M.S. 1999. A highly conserved mechanism of regulated ribosome stalling mediated by fungal arginine attenuator peptides that appears independent of the charging status of arginyl-tRNAs. *J. Biol. Chem.* **274:** 37565–37574.

Wang Z. and Sachs M.S. 1997. Ribosome stalling is responsible for arginine-specific translational attenuation in *Neurospora crassa*. *Mol. Cell. Biol.* **17:** 4904–4913.

Wang Z., Fang P., and Sachs M.S. 1998. The evolutionarily conserved eukaryotic arginine attenuator peptide regulates the movement of ribosomes that have translated it. *Mol. Cell. Biol.* **18:** 7528–7536.

Welch E.M. and Jacobson A. 1999. An internal open reading frame triggers nonsense-mediated decay of the yeast SPT10 mRNA. *EMBO J.* **18:** 6134–6145.

Wera S., Fernandez A., Lamb N.J., Turowski P., Hemmings-Mieszczak M., Mayer-Jaekel R.E., and Hemmings B.A. 1995. Deregulation of translational control of the 65-kDa regulatory subunit (PR65 alpha) of protein phosphatase 2A leads to multinucleated cells. *J. Biol. Chem.* **270:** 21374–21381.

Werner M., Feller A., Messenguy F., and Piérard A. 1987. The leader peptide of yeast *CPA1* is essential for the translational repression of its expression. *Cell* **49:** 805–813.

Williams M.A. and Lamb R.A. 1989. Effect of mutations and deletions in a bicistronic mRNA on the synthesis of influenza B virus NB and NA glycoproteins. *J. Virol.* **63:** 28–35.

Wright M.B., Howell E.A., and Gaber R.F. 1996. Amino acid substitutions in membrane-spanning domains of Hol1, a member of the major facilitator superfamily of transporters, confer nonselective cation uptake in *Saccharomyces cerevisiae*. *J. Bacteriol.* **178:** 7197–7205.

Wu J. and Miller B.L. 1997. *Aspergillus* asexual reproduction and sexual reproduction are differentially affected by transcriptional and translational mechanisms regulating *stunted* gene expression. *Mol. Cell. Biol.* **17:** 6191–6201.

Zimmer A., Zimmer A.M., and Reynolds K. 1994. Tissue specific expression of the retinoic acid receptor-beta 2: Regulation by short open reading frames in the 5′-noncoding region. *J. Cell Biol.* **127:** 1111–1119.

19

Cellular Internal Ribosome Entry Site Elements and the Use of cDNA Microarrays in Their Investigation

Mark S. Carter, Kenneth M. Kuhn, and Peter Sarnow

Department of Microbiology and Immunology
Stanford University School of Medicine
Stanford, California 94305

The discovery that picornaviral RNA genomes are translated by an internal ribosome entry mechanism (see Chapter 4) raised the question whether host cell mRNAs could use this initiation mechanism as well. This idea was supported by the finding that internal initiation could operate on viral internal ribosome entry sites (IRES) both in cultured cells (Pelletier and Sonenberg 1988; Jang et al. 1989) and in cell-free systems (Jang et al. 1988; Pelletier and Sonenberg 1989) in the absence of any viral factors, suggesting that the eukaryotic translation apparatus can perform internal initiation.

As discussed in Chapter 4, IRES elements can promote translation in the absence of a functional 5′ cap-binding protein complex eIF4F, composed of factors eIF4E, eIF4A, and eIF4G. Thus, it was not too surprising when the first cellular IRES-containing mRNA, encoding the immunoglobulin binding protein BiP (Ting and Lee 1988), was identified by its ability to be translated in poliovirus-infected cells when overall translation of host cell mRNAs was inhibited as a result of eIF4G degradation (Sarnow 1989). Translation assays with dicistronic mRNAs that contained the 5′ noncoding region (5′NCR) of BiP inserted between two reporter cistrons revealed that the BiP 5′NCR could mediate translation of the second cistron without a requirement for ribosomes to traverse the first cistron (Macejak and Sarnow 1991). This result strongly suggested that the BiP 5′NCR functions as an IRES which can bind ribosomes directly and mediate translation of open reading frames, located downstream.

Translational Control of Gene Expression
© 2000 Cold Spring Harbor Laboratory Press 0-87969-568-4/00 $5 + .00

The identification of further cellular mRNAs with IRES elements has been cumbersome, because it has been difficult to identify putative IRES elements in mRNAs. Traditionally, mRNAs which are burdened with long 5′NCRs that contain numerous AUG codons upstream of the start-site AUG codon, yet are efficiently translated, have been suspected of containing IRES elements. To test for IRES activity, suspected 5′NCRs are inserted between two cistrons in a dicistronic mRNA, and translational assays are performed. Even if translation of the second cistron is observed, the suspected RNA element has to pass numerous tests before it can be classified as an IRES. First, controls need to be performed to exclude the possibility that any accumulation of the second cistron product resulted from the translation of functionally monocistronic mRNAs generated by cryptic promoter elements, fortuitously spliced mRNA species, or partially degraded mRNAs. To address this concern, examination of the expressed RNAs using Northern analysis or RNase protection assays are usually performed. If expression of the second cistron is low compared to the first cistron, further analyses using in-vitro-synthesized dicistronic RNAs are warranted to exclude the possibility that minor species of functionally monocistronic RNAs mediated second cistron expression. The assumption in these experiments is that RNAs synthesized in vitro by T7 RNA polymerase, which is then used in RNA transfection of cultured cells, cell-free translation assays, or ribosome-binding experiments, will contain fewer breaks, or at least different cleavages than RNA transcribed by cellular polymerase II. Second, experiments are needed to examine whether ribosomes that translated the second cistron entered the mRNA at the putative IRES, located between the two cistrons, or at the 5′ end of the mRNA. In the latter case, ribosomes could have translated the first cistron, traversed the intergenic spacer, and reinitiated translation at the second cistron (Peabody and Berg 1986). To distinguish between these possibilities, stable RNA structures can be inserted at the 5′ end of the mRNA. It is predicted that such structures should impede entry and 5′-end-dependent scanning of ribosomes, resulting in decreased translation of the first cistron; yet, the translational efficiency of the second cistron should not be affected in such dicistronic mRNAs if the intercistronic region contains an IRES. Of course, failure to detect second cistron translation by a putative IRES element could be due to inactivity of the IRES in the constraint of a dicistronic mRNA. Clearly, additional functional assays for IRES activities are needed.

In this chapter, we first summarize what is known about the handful of known cellular IRES elements and then describe a cDNA microarray technology that can be used to identify IRES-containing mRNAs in virally infected cells or other conditions which might enhance the translation of mRNAs that require only low concentrations of intact cap-binding protein complex eIF4F.

CELLULAR mRNAS THAT CONTAIN IRES ELEMENTS IN THEIR 5′ NONCODING REGIONS

In Table 1, we list the cellular mRNAs whose 5′NCRs have been shown to contain IRES activities in dicistronic assays. As the names for most of these genes imply, their gene products are involved in a variety of cellular processes, including development (*Antennapedia* and *Ultrabithorax*), differentiation (PDGF2/c-*sis*), stress (VEGF), growth/cancer (c-*myc*, FGF2), and programmed cell death (DAP5, XIAP). Although the current list is small, it is interesting to point out that several gene products of

Table 1 Cellular mRNAs that harbor IRES elements

Gene	References
Mammalian	
immunoglobulin heavy-chain binding protein (BiP)	Macejak and Sarnow (1991)
fibroblast growth factor 2 (FGF2)	Vagner et al. (1995)
insulin-like growth factor II (IGF2)	Teerink et al. (1995)
platelet-derived growth factor B (PDGF/c-*sis*)	Bernstein et al. (1997)
c-*myc*	Nanbu et al. (1997); Stoneley et al. (1998)
DNA binding protein MYT2	Kim et al. (1998)
vascular endothelial growth factor (VEGF)	Akiri et al. (1998); Huez et al. (1998); Stein et al. (1998)
cardiac voltage-gated potassium channel	Negulescu et al. (1998)
eukaryotic initiation factor 4G (eIF4G)[a]	Gan and Rhoads (1996)
eukaryotic initiation factor 4GI (eIF4GI)[a]	Johannes and Sarnow (1998)
X-linked inhibitor-of-apoptosis (XIAP)	Holcik et al. (1999)
death-associated protein 5 (DAP5)	Henis-Korenblit et al. (2000)
ornithine decarboxylase	Pyronnet et al. (2000)
protein kinase p58^{PITSLRE}	Cornelis et al. (2000)
Gtx homeodomain protein	Chappell et al. (2000)
Drosophila	
Antennapedia	Oh et al. (1992); Ye et al. (1997)
Ultrabithorax	Ye et al. (1997)

[a]Both mRNAs are identical except their 5′ noncoding regions. See text for details.

IRES-containing mRNAs, such as VEGF, FGF2, PDGF, and Cyr61 (see below) have been shown to play roles in angiogenesis, suggesting that IRES-mediated translation may be used during angiogenesis, which is the growth of new blood vessels in situations such as wound healing, cardiovascular and cerebral ischemia, and many types of cancer. Recently, two IRES elements have been identified that are specifically active in the G_2/M phase of the cell cycle when cap-dependent translation is reduced (Cornelis et al. 2000; Pyronnet et al. 2000): An IRES located in ornithine decarboxylase (ODC) mRNA enhances the synthesis of ODC, resulting in increased amounts polyamines which might aid in the formation of the mitotic spindle (Pyronnet et al. 2000); similarly, an IRES residing in protein kinase $p58^{PITSLRE}$ mRNA functions only during G_2/M, indicating a role of the $p58^{PITSLRE}$ protein in this particular phase of the cell cycle (Cornelis et al. 2000). Many of these genes listed in Table 1 are under the regulation of complex transcriptional and posttranscriptional control, and several of the genes express mRNAs that encode different protein isoforms. For example, expression of VEGF transcripts is influenced by hypoxic conditions at the level of transcription and RNA stability, and the four isoforms of angiogenic VEGF are generated by alternative splicing (Stein et al. 1995; Warren et al. 1996; Ferrara and Keyt 1997).

Interestingly, the mRNA encoding the eIF4G subunit of eIF4F was shown to contain an IRES in its 5′NCR (Gan and Rhoads 1996; Gan et al. 1998). More recently, an additional eIF4G transcript was identified, termed eIF4GI, which differed from eIF4G in its 5′NCR (Gradi et al. 1998b). Because the 5′NCRs of eIF4G and eIF4GI diverged precisely at a predicted consensus 3′ splice site, it was suggested that the eIF4G transcript represented an intron-containing pre-mRNA (Gradi et al. 1998b). Curiously, both 5′NCRs were shown to contain IRES elements in dicistronic assays. Consistent with the idea that at least one of these IRES elements functions intracellularly, natural eIF4GI mRNA was found to be associated with polysomes in poliovirus-infected human HeLa cells at a time when most cap-dependent translation was inhibited, arguing that eIF4GI mRNA can be translated in the presence of low amounts of intact eIF4F (Johannes and Sarnow 1998). Furthermore, the mRNA encoding the death-associated protein DAP5 (p97), a member of the eIF4G family that lacks an eIF4E binding site (Imataka et al. 1997; Levy-Strumpf et al. 1997; Shaughnessy et al. 1997; Yamanaka et al. 1997), has recently been shown to contain an IRES whose activity remains unaffected when cells enter FAS-induced apoptosis, a scenario that results in the inhibition of cap-dependent translation (Henis-Korenblit et al. 2000). Overall, these findings suggest that various forms of eIF4G, an important adapter molecule involved in the recruitment of ribosomes (Hentze 1997; Gingras et al. 1999); see Chapter 2), can be synthesized in cells when overall cap-dependent translation is reduced, emphasizing the important role of eIF4G in cellular mRNA translation.

Features in Cellular IRES Elements

A common feature of IRES-containing 5′NCRs is that they are longer than 5′NCRs which lack IRES elements. Although an average eukaryotic 5′NCR tends to be fewer than 100 nucleotides in length (Cigan and Donahue 1987), the IRES-containing 5′NCRs, for example, of BiP and *Antennapedia* are 220 and 1727 nucleotides in length, respectively (Iizuka et al. 1995). Furthermore, the 5′NCRs of IRES-containing transcripts contain relatively high GC nucleotide contents, often harbor multiple AUG triplets, and are predicted to fold into higher-order structures with greater stability than most RNAs (Iizuka et al. 1995). Thus, features in mRNAs that have been shown to impede the 5′-end-dependent scanning of 43S ribosomal subunits (Pelletier and Sonenberg 1985; Kozak 1986) are often found in IRES elements. There is ample evidence that even small deletions or nucleotide insertions in IRES elements affect IRES activity in viral and cellular mRNAs (Meerovitch et al. 1991; Martinez Salas et al. 1993; Wang and Siddiqui 1995; Paulin et al. 1996; Willis et al. 1997), indicating that the overall RNA structure needs to be maintained for ribosome recruitment.

Deletion analyses have indicated that IRES elements tend to be located toward the 3′ end of the 5′NCRs of cellular mRNAs. Juxtaposition of an IRES with an open reading frame will position the 43S ribosomal subunit close to the AUG start codon, allowing 80S formation without the need to scan long distances from the initial binding site to the initiation codon. In addition, the presence of nucleotide elements preceding the IRES may have important regulatory functions. For example, small open reading frames near the 5′ end of the mRNA may force ribosomal subunits that entered the mRNA at the capped 5′ end to initiate translation. This "capture" of ribosomes may be important to prevent the unwinding of IRES structures by scanning ribosomes. In addition, 5′-proximal sequence elements may affect functional half-life, subcellular localization, or IRES activity of the mRNA. Evidence for the latter was provided in polioviral mRNAs, in which a 5′-proximal cloverleaf-type structure is known to communicate with the downstream-located IRES by binding of the poly(rC) binding protein PCBP (Blyn et al. 1996, 1997; Gamarnik and Andino 1997, 1998; Parsley et al. 1997). The hypothesis that IRESs in cellular mRNAs could be modulated by regulatory elements received support from the finding that IRES activity of both PDGF2/c-*sis* and VEGF was highest with the full-length 5′NCRs compared to smaller subfragments (Bernstein et al. 1997; Huez et al. 1998). Furthermore, the IRES in PDGF2/c-*sis* was found to be induced during megakaryocytic differentiation, and the "differentiation-sensing" element in the PDGF2/*c-sis* IRES was recently mapped to the central portion of the 5′NCR (Sella et al. 1999). In addition, transgenic *Drosophila* that express dicistronic constructs harboring either the 5′ NCR of *Antennapedia* or *Ultrabithorax* in the intercistronic region showed accumulation of the second cistron product in a spatially and temporally regulated manner that partially correlated with the normal expression of Antennapedia and Ultrabithorax proteins during development (Ye et al. 1997).

Despite the identification of several cellular and many viral IRES elements, there is no consensus sequence that identifies an IRES. Most viral IRES elements contain an oligopyrimidine-rich sequence near the 3′ border of the IRES (see Chapter 4) that is present in some (Bernstein et al. 1997; Negulescu et al. 1998) but not all cellular IRES elements. Computer-aided analysis has predicted a secondary structure motif that may be common to several cellular IRES elements: a Y-shaped double-hairpin structure followed by a smaller single-hairpin structure upstream of the start AUG codon (Le and Maizel 1997). This motif is found conserved between the diverse BiP 5′NCRs in human, hamster, rat, *Saccharomyces cerevisiae, Caenorhabditis elegans, Trypanosoma brucei,* and *Giardia lamblia* (Le and Maizel 1997). A similar motif can also be predicted in FGF2, *Antennapedia*, VEGF, and PDGF2/c-*sis* IRES elements (Le and Maizel 1997; Huez et al. 1998; Sella et al. 1999). However, it remains to be established whether the presence of this motif plays any role in cellular IRES function. Recently, Chappell et al. (2000) have noted that various segments of the 5′NCR of the Gtx homeodomain protein mRNA (Komuro et al. 1993) can function as an IRES in the dicistronic assay. Curiously, one of those segments consisted of a nine-nucleotide sequence element with perfect base-complementarity to 18S rRNA. Insertion of multiple copies of this sequence element in the intercistronic spacer region in dicistronic mRNAs cooperatively increased the translation of the second cistron (Chappell et al. 2000). Surprisingly, the same sequence element decreased translational efficiency when placed in the 5′NCR of a monocistronic mRNA (Hu et al. 1999). Thus, this 18S rRNA-complementary sequence element can function as a translational enhancer or repressor, dependent on its location in the mRNA. Although the mechanisms underlying these observations are unclear, it is possible that the inhibitory effect of the element in monocistronic mRNAs reflects the kinetics of binding and subsequent scanning of 40S subunits that were simultaneously recruited to the 5′ end of the mRNA and to an internal region of the 5′NCR via the 18S rRNA-complementary sequence element. These findings suggest that structural features, or sequences with complementarity to ribosomal RNAs, are important elements by which IRESs recruit ribosomal subunits.

Regulation of IRES Elements by *trans*-Acting Factors

Studies using viral IRES elements have shown that with the exception of the cap-binding protein eIF4E, all known canonical initiation factors are involved in internal initiation (see Chapter 4); the hepatitis C virus IRES presents an exception because it does not need eIF3, eIF4A, eIF4B, or eIF4F for binding of ribosomal subunits (Pestova et al. 1998). The polypyrimidine tract binding (PTB) protein (Hellen et al. 1993), La autoantigen (Meerovitch et al. 1993), poly(rC) binding protein PCBP (Blyn et al. 1997; Gamarnik and Andino 1998) and the "upstream of n-ras" factor unr (Hunt et al. 1999) have been described as noncanonical factors that can function as modulators of viral IRES activity (see Chapters 4 and 31). Curiously, none of these noncanonical factors has been shown to play any functional role in the regulation of cellular IRES activities. Instead, crosslinking studies revealed the interaction of an array of other proteins with various cellular IRESs (Vagner et al. 1996; Yang and Sarnow 1997; Huez et al. 1998; Paulin et al. 1998). However, it has remained unclear whether the specific interactions of these proteins with the IRES elements have functional consequences.

Recently, Elroy-Stein and colleagues provided evidence that hnRNP C regulates the differentiation-induced IRES in PDGF2/c-*sis* (Sella et al. 1999). It was shown that hnRNP C binds specifically to a region in the 5′NCR of PDGF2/c-*sis* that is required for the induction of IRES function during megakaryocytic differentiation (Bernstein et al. 1995, 1997). Intriguingly, while most hnRNP C remained in the nucleus upon megakaryocytic differentiation, a phosphorylated form of hnRNP C, that was found only during differentiation, was shown to interact with the PDGF2/c-*sis* IRES. The phosphorylated form of hnRNP C could also be detected in a fraction isolated from salt-washed ribosomes (Sella et al. 1999). Although it is not yet known whether the phosphorylated hnRNP C-IRES complex is required for IRES activity, it is curious that hnRNP C, like the viral IRES-binding protein PTB, plays a role in pre-mRNA splicing (Nakielny and Dreyfuss 1996; Nanbu et al. 1997). Further characterization of other cellular IRES-binding proteins should test the hypothesis that IRES-containing mRNAs, expressed by RNA polymerase II in the nucleus, recruit nuclear RNA-binding proteins that aid in their journey to the cytoplasmic translation apparatus. For example, cellular IRES-binding proteins could aid in the stability, transport, or proper folding of the mRNA molecules.

USE OF cDNA MICROARRAYS TO STUDY TRANSLATIONAL CONTROL

The identification of IRES-containing mRNAs or mRNAs that are translationally regulated during cell growth, differentiation, or environmental changes has been difficult. With the advent of cDNA microarrays, such as those developed in Pat Brown's laboratory at Stanford University that contain 10,000 cDNA probes (DeRisi and Iyer 1999; Eisen and Brown 1999; Iyer et al. 1999), systematic searches for translationally controlled mRNAs can be launched. This is an exciting and promising research venue because even smaller cDNA arrays, containing 1200 gene probes, have already proven valuable in detecting mRNAs that are translationally regulated during mitogenic activation of resting fibroblasts (Zong et al. 1999). This section describes some approaches in which translationally regulated mRNAs can be identified using cDNA microarray display; the next section then summarizes results obtained from cDNA microarray studies that have identified cellular mRNAs that can be translated when intracellular eIF4F concentrations are low.

Comparison of the Distribution of mRNAs in mRNP/80S and Polysomal Fractions

The position of a given mRNA in a polysome gradient can be attributed to the number of associated ribosomes. Barring events such as arrested ribosomes or slowed elongation rates, the number of ribosomes per mRNA reflects the translational efficiency of the mRNA. An mRNA that is poorly translated or translationally inactive will typically accumulate in the messenger ribonucleoprotein (mRNP) or "monosome" (80S) fraction of the gradient; translationally active mRNAs will sediment in the polysome fraction of the gradient (Fig. 1). This feature allows the physical separation of mRNAs that are translationally repressed from those that are activated under certain cellular conditions. For example, mRNP/80S ($mRNP_r$) isolated from a reference condition can be fluorescently labeled with Cy3-dUTP (green), and polysomal ($Poly_r$) mRNAs isolated from the same gradient can be labeled with Cy5-dUTP (red), during cDNA synthesis (Fig. 1). When these cDNA probes are mixed and hybridized to a cDNA microarray (Eisen and Brown 1999), spots of cDNAs encoding mRNAs that are translationally inactive will be labeled green, those that are translationally active will be red, and those that are equally distributed in the two fractions will be yellow (Fig. 1). Similarly labeled cDNA

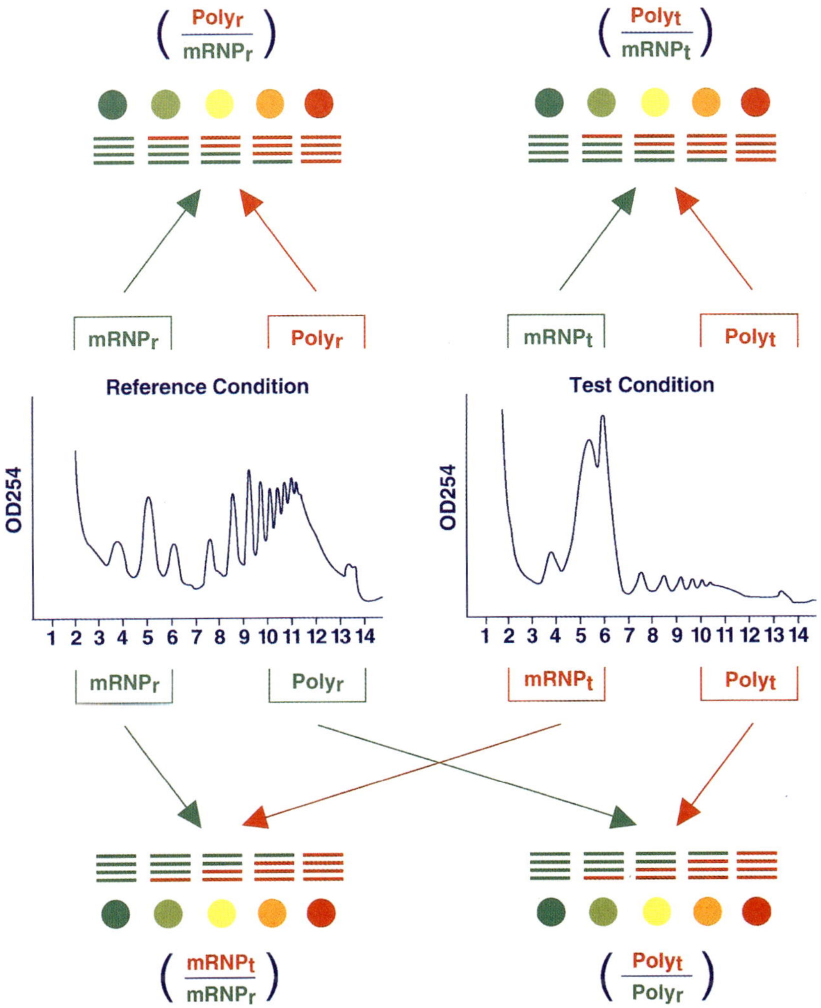

Figure 1 Experimental outline for combining polysomal fractionation and cDNA microarray analysis to identify cellular mRNAs that are regulated at the translational level. Details of the Poly vs. Poly and mRNP vs. Poly methods are explained in the text. Red and green arrows/brackets indicate cDNA labeled with Cy5-dUTP and Cy3-dUTP, respectively. Colored lines represent labeled cDNAs, and colored circles represent the cDNA microarray spots.

copies of mRNA samples from a different test condition ($mRNP_t$ and $Poly_t$) can be hybridized to a second, identical cDNA microarray (Fig.1). As a control, unfractionated mRNAs obtained from the reference and test condition should be analyzed. In this way, the polysomal association of thousands of mRNAs species can be evaluated simultaneously.

Several advantages are inherent in this experimental approach (Fig. 1, top). First, mRNA for each microarray is isolated from a single gradient; therefore, the hybridization results are relatively independent of fluctuations in steady-state levels of mRNAs. Second, translational activation of an mRNA should be reflected in a redistribution of mRNA from the mRNP/80S fraction to the polysomal fraction, a shift that is easily detectable. For example, if one assumes that an mRNA is translationally repressed under the reference condition, so that 75% of the mRNA is found in the mRNP/80S fraction and 25% is found in the polysomal fraction, the $Poly_r/mRNP_r$ ratio is 1/3. If the mRNA is translationally activated during the test condition, the mRNA becomes associated with polysomes and the ratio can reverse. For example, if 75% of the activated mRNA is found in the polysomal fraction and 25% is found in the mRNP/80S fraction during the test condition, the $Poly_t/mRNP_t$ ratio becomes 3. Therefore, comparison of the two data points will yield a ninefold difference in signal.

Comparison of the Distribution of mRNAs in Two Different Polysomal Fractions

When mRNAs from polysomal fractions of reference and test conditions are compared directly (Fig. 1, bottom), different limitations and advantages come into play. Because the collected $Poly_r$ and $Poly_t$ fractions are likely to contain different amounts of mRNA, control experiments are required to determine whether a change in relative abundance of a given mRNA species reflected either redistribution of the mRNA within the polysome gradient or change in rate of transcription or stability. Thus, the total intracellular abundance of the mRNA (Tot_t/Tot_r) needs to be determined. Furthermore, an $mRNP_r$ vs. $mRNP_t$ cDNA microarray should show for an mRNA that is mobilized into polysomes during the test condition a simultaneous loss from the mRNP/80S fraction. Again, such a control increases the sensitivity of the experiment. For the example given above, $(Poly_t/Poly_r)/(mRNP_t/mRNP_r)$ again equals nine. A direct comparison of polysomal mRNA isolated from reference and test conditions has proven to be an effective strategy in the identification of mRNAs that remain associated with polysomes in poliovirus-infected of human HeLa cells (Johannes et al. 1999) and after glucose depletion in yeast (K.M. Kuhn and P. Sarnow, in prep.).

One issue deserves consideration prior to performing such a cDNA microarray experiment. After pooling the appropriate gradient fractions from reference and test conditions, should equal amounts of RNA or equal volumes of the fractions be used for first-strand cDNA synthesis? Each method has advantages and disadvantages, and both have been used successfully to identify translationally regulated mRNA molecules. If equal amounts of RNA are compared, more material needs to be collected from fractions containing fewer polysomes. As a result, cDNA copies of polysome-associated mRNA species become preferentially enriched. This approach is useful for the identification of mRNAs that are translated at a time when overall mRNA translation is reduced. In performing a $Poly_t$ vs. $Poly_i$ experiment under conditions that led to a global translational repression by infection of human cells with poliovirus, or glucose starvation of yeast, we have used equal amounts of RNA for first-strand cDNA synthesis. However, due to this enrichment step, the cDNA microarray expression ratio (i.e., $Poly_t/Poly_r$) will not accurately reflect the relative abundance of a given mRNA within the gradient fraction.

Alternatively, mRNA from equal volumes of pooled fractions can be used as a source for first-strand cDNA synthesis. Although it is possible to lose mRNA during purification from a sample with less starting material, this does not seem to be problematic since heparin, present in the sucrose gradient, acts as a carrier during the initial precipitation steps. In addition, the final LiCl precipitation step provides a high percentage of recovery over a wide range of RNA concentrations. Thus, using equal volumes from gradient fractions as starting material, the measured expression ratio is likely to be an accurate indicator of the abundance of individual mRNA species in the gradient fractions. This is the preferential method if associations of mRNAs in mRNP fractions and polysomal fractions are compared.

Comparison of the Distribution of mRNAs Isolated from Total and Polysomal Fractions

By comparing the ratio of polysomal mRNA to total cytoplasmic mRNA ($Poly_r/Tot_r$) for any condition, it should be possible to estimate the percentage of a given mRNA that is associated with polysomes. A change in this percentage from one condition to another could indicate that the mRNA is translationally controlled. Although we have not carried out an extensive analysis of this approach, preliminary experiments have indicated that it is effective.

Additional Comments about the Use of cDNA Microarrays in Studying Translational Control

Prior to initiating cDNA microarray experiments using gradient-fractionated mRNAs, several issues deserve consideration. First, the reproducibility of the assay will depend on the purity of the RNA template. Especially RNA present in the top fractions of the gradient (i.e., mRNP fractions), where many soluble proteins sediment, needs to be purified carefully. Gradient fractions can be collected directly into a solution of guanidine hydrochloride to immediately denature proteins, especially ribonucleases. Following precipitation by alcohol, RNA is reprecipitated in the presence of 1.5M LiCl to remove heparin, which can inhibit first-strand cDNA synthesis (Johannes et al. 1999).

As a rule, a two- to threefold change in relative abundance of polysome-associated mRNA is an indicator of a translational control event. Ideally, each reference and test sample should be hybridized to at least two cDNA arrays, and the results from each experiment should be verified by two independent methods, for example, an mRNP vs. Poly and a Poly vs. Poly experiment (see above). The cDNA microarray assay is very sensitive to small changes in relative abundance of a target mRNA. In general, a twofold change in relative abundance can be reliably detected for an mRNA present at 0.1 to 1 copy per cell.

By combining polysome fractionation with cDNA microarray display, a variety of translational regulatory events can now be studied. Novel targets of translational control are likely to be discovered and will increase our understanding of the regulatory pathways that occur when the cell is confronted with changing environmental conditions. For example, cDNA arrays were used to show that mitogenic stimulation of resting fibroblasts resulted in the translational regulation of less than 1% of the identified mRNAs, indicating that this process is very selective (Zong et al. 1999). In another example, cDNA microarrays identified mRNAs that are associated with polysomes in poliovirus-infected cells, as described below.

IDENTIFICATION BY cDNA MICROARRAY OF NATURAL mRNAS THAT ARE TRANSLATED DURING LOW LEVELS OF INTACT CAP-BINDING PROTEIN COMPLEX eIF4F

Polysomal Association of Cellular mRNAs in Poliovirus-infected Cells

Infection of susceptible human cells with poliovirus, a positive-stranded RNA virus, results in the inhibition of translation of capped host cell mRNAs (see Chapter 31). Translational inhibition is caused by the proteolytic degradation of both forms of eIF4G, eIF4GI and eIF4GII, resulting in inactivation of the cap-binding protein complex eIF4F (Gradi et al. 1998a); however, translation of the IRES-containing viral mRNA is not inhibited. We reasoned that any cellular mRNA that was associated with polysomes during poliovirus infection must not require significant amounts of eIF4F for its translation. Possible reasons for a low eIF4F requirement include (1) the presence of an IRES element, (2) the presence of unstructured 5′NCRs that have a low requirement for eIF4F, and (3) a block in translation elongation or termination.

Combining polysomal fractionation and cDNA microarray technology (see above), we found that 200 of 7000 analyzed mRNAs were associated with polysomes in poliovirus-infected cells under conditions in which most or all eIF4G was proteolyzed (Johannes et al. 1999). Pulse-labeling experiments showed that polysomal-associated mRNAs were translationally active (Johannes and Sarnow 1998), suggesting that many of the 200 mRNAs are being translated during poliovirus infection. Many of these mRNAs encoded immediate-early transcription factors, kinases, and phosphatases of the mitogen-activated protein kinase pathways, receptors, and several proto-oncogenes (Johannes et al. 1999). Using dicistronic assays (see above), it was subsequently shown that the 5′NCRs of several polysomal-associated mRNAs, encoding BiP, c-myc, Pim-1, Map kinase phosphatase MKP-1, and transcription factor Cyr61, contained IRES elements (Johannes et al. 1999). These findings indicated that the cDNA microarray approach could serve as an assay to identify natural IRES-containing mRNAs.

Properties of Identified mRNAs

Among the newly identified IRES-containing mRNAs were two genes known to participate in angiogenesis: Cyr61 and MKP-1. The Cyr61 gene encodes a serum-induced secreted protein that induces movement of vascular endothelial cells in tissue culture and induces tumor growth and angiogenesis in mice (Babic et al. 1998). Its 5′NCR is 209 nucleotides in length and is predicted to fold into a secondary structure with a calculated free energy of –83 kcal/mole (Martinerie et al. 1997). MKP-1 is an immediate-early gene whose transcription is induced within one hour after exposure of cells to growth factors or hypoxic conditions (Kwak et al. 1994). The 5′NCR of the MKP-1 mRNA is 248 nucleotides in length and is predicted to fold into a secondary structure with a calculated free energy of –99 kcal/mole. Substrates for the MKP-1 phosphatase include the kinases involved in the stress-activated MAP kinase pathways (Groom et al. 1996).

MKK3b mRNA, encoding a kinase that operates within the stress-activated MAP kinase pathway, was also shown to be associated with polysomes in poliovirus-infected cells (Johannes et al. 1999), although it has not yet been determined whether this is due to IRES function. Among the "downstream" substrates for MKK3b are the stress-activated kinase p38 and kinase Mnk1 (Fukunaga and Hunter 1997; Waskiewicz et al. 1997), which can phosphorylate the cap-binding protein eIF4E (Waskiewicz et al. 1999). It has been hypothesized that phosphorylation of eIF4E by Mnk1 could stabilize the interaction of eIF4F with the cap structure (Pyronnet et al. 1999; Waskiewicz et al. 1999). Therefore, cap-independent translation of MKK3b mRNA could produce "active" eIF4F, which could be used as soon as a cell recovered from the stress response that initially had resulted in the inhibition of cap-dependent translation.

Several mRNAs that encode proto-oncogenes, such as TGFβ3, B94, and Pim-1, were found to be associated with polysomes when the cellular concentration of eIF4F was low (Johannes et al. 1999). The Pim-1 gene encodes a serine-threonine protein kinase that cooperates with c-myc in cellular transformation in hematopoietic cells (Domen et al. 1987; Leverson et al. 1998). The Pim-1 mRNA has a long 5′NCR containing 354 nucleotides and is predicted to fold into a secondary structure with a calculated free energy of –153 kcal/mole (Domen et al. 1987). The possibility of coordinate regulation of the translation of these mRNAs is currently under investigation.

Certain receptor-encoding mRNAs, including the mRNA for glucocorticoid receptor (GR), androgen receptor (AR), transforming growth factor β III receptor (TGFβIIIR), and the mitogen-induced orphan receptor (MINOR), were identified in this screen as likely to be translated cap-independently (Johannes et al. 1999). GR and AR are members of the steroid/thyroid/retinoic acid receptor superfamily (Evans 1988). The 5′NCR for the GR mRNA is 493 nucleotides in length and is predicted to fold into a secondary structure with a calculated free energy of –233 kcal/mole (Encio and Detera-Wadleigh 1991). The GR gene expresses two mRNAs produced by alternative splicing that encode two proteins which differ in their carboxyl termini (Hollenberg et al. 1985). Because the alternative splicing event occurs in the 3′ end of the coding region, both GR mRNAs should have the same 5′NCR and, therefore, are both predicted to be translated when the concentration of eIF4F is low. The 5′NCR of AR mRNA is 1124 nucleotides in length but contains only one AUG codon that is not used as a start site (Mizokami et al. 1994). The AR 5′NCR is predicted to fold into a secondary structure with a calculated free energy of –407 kcal/mole. Similar to the GR gene, expression of the AR gene is complex, because translation of the single mRNA results in the synthesis of two proteins; the smaller protein initiates at an AUG codon internal to the coding sequence of the larger one (Zoppi et al. 1993). The 5′NCR of the TGFβIIIR mRNA is 348 nucleotides in length and has a predicted secondary structure with a calculated free energy of –91 kcal/mole (Moren et al. 1992). In addition to the strong predicted secondary structure, the 5′NCR contains five upstream AUG triplets with five predicted open reading frames preceding the start AUG codon. The 5′NCR for the MINOR mRNA is 209 nucleotides in length and contains four upstream AUG triplets (Hedvat and Irving 1995), although the primary sequence for its 5′NCR folds into a predicted secondary structure with the calculated free energy of only –33 kcal/mole. Thus, this predicted unstructured 5′NCR might require only low amounts of eIF4F for ribosome recruitment.

CONCLUDING REMARKS

The growing list of cellular mRNAs that contain IRES elements and the data obtained from cDNA microarray analyses suggest that internal initiation might be more prevalent than has previously been anticipated. On the basis of the functions encoded by IRES-containing mRNAs, IRES elements seem to operate during a variety of stress responses including angiogenesis, inflammation, and response to serum. Important questions that need to be addressed include: What is the mechanism of IRES-mediated ribosome recruitment? What is the role, if any, of IRES-binding proteins such as hnRNP C in regulating IRES activity? One of the largest tasks will be to determine the locations of IRES elements in natural mRNAs identified in assays such as cDNA microarray. It is quite possible that IRES elements may be identified in coding regions as well as in putatively assigned noncoding regions. For example, a functionally dicistronic mRNA that is regulated by two independent IRES elements has been recently discovered in the cricket paralysis virus (CrPV) genome (Wilson et al. 2000). We have found that one of these IRES elements can assemble 80S monosomes without eIF2, Met-tRNA$_i$ or GTP hydrolysis, using a CCU triplet positioned in the ribosomal P site and a GCU triplet in the ribosomal A site (Wilson et al. 2000). Monitoring the positioning of ribosomal subunits on wild-type and mutated IRES elements by toeprinting assays, and protein sequencing analysis of translation products, has revealed that the amino-terminal amino acid was encoded by the triplet in the ribosomal A site, arguing that the P site was not decoded (J.E. Wilson et al., in prep.). Similar results were obtained with another insect virus: plantia stali intestinal virus (PSIV) (Sasaki and Nakashima 2000). In this case, translation from the second ORF initiated with glutamine. Translational initiation without Met-tRNA$_i$ suggests that the repertoire of translated open reading frames in eukaryotic mRNAs may be greater than anticipated.

ACKNOWLEDGMENTS

We thank Karla Kirkegaard for critical reading of this chapter. Work in the authors' laboratory was supported by National Institutes of Health grant R01 GM-55979 (P.S.) and by American Cancer Society grant PF-98-16-01-MBC (M.S.C.).

REFERENCES

Akiri G., Nahari D., Finkelstein Y., Le S.Y., Elroy-Stein O., and Levi B.Z. 1998. Regulation of vascular endothelial growth factor (VEGF) expression is mediated by internal initiation of translation and alternative initiation of transcription. *Oncogene* **17:** 227–236.

Babic A.M., Kireeva M.L., Kolesnikova T.V., and Lau L.F. 1998. CYR61, a product of a growth factor-inducible immediate early gene, promotes angiogenesis and tumor growth. *Proc. Natl. Acad. Sci.* **95:** 6355–6360.

Bernstein J., Shefler I., and Elroy-Stein O. 1995. The translational repression mediated by the platelet-derived growth factor 2/c-*sis* mRNA leader is relieved during megakaryocytic differentiation. *J. Biol. Chem.* **270:** 10559–10565.

Bernstein J., Sella O., Le S.Y., and Elroy-Stein O. 1997. PDGF2/c-*sis* mRNA leader contains a differentiation-linked internal ribosomal entry site (D-IRES). *J. Biol. Chem.* **272:** 9356–9362.

Blyn L.B., Towner J.S., Semler B.L., and Ehrenfeld E. 1997. Requirement of poly(rC) binding protein 2 for translation of poliovirus RNA. *J. Virol.* **71:** 6243–6246.

Blyn L.B., Swiderek K.M., Richards O., Stahl D.C., Semler B.L., and Ehrenfeld E. 1996. Poly(rC) binding protein 2 binds to stem-loop IV of the poliovirus RNA 5′ noncoding region: Identification by automated liquid chromatography-tandem mass spectrometry. *Proc. Natl. Acad. Sci.* **93:** 11115–11120.

Chappell S.A., Edelman G.M., and Mauro V.P. 2000. A 9-nt segment of a cellular mRNA can function as an internal ribosome entry site (IRES) and when present in linked multiple copies greatly enhances IRES activity. *Proc. Natl. Acad. Sci.* **97:** 1536–1541.

Cigan A.M. and Donahue T.F. 1987. Sequence and structural features associated with translational initiator regions in yeast — A review. *Gene* **59:** 1–18.

Cornelis S., Bruynooghe Y., Denecker G., van Huffel S., Tinton S., and Beyaert F. 2000. Identification and characterization of a new cell cycle regulated internal ribosome entry site. *Mol. Cell* **5:** 597–605.

DeRisi J.L. and Iyer V.R. 1999. Genomics and array technology. *Curr. Opin. Oncol.* **11:** 76–79.

Domen J., Von Lindern M., Hermans A., Breuer M., Grosveld G., and Berns A. 1987. Comparison of the human and mouse PIM-1 cDNAs: Nucleotide sequence and immunological identification of the in vitro synthesized PIM-1 protein. *Oncogene Res.* **1:** 103–112.

Eisen M.B. and Brown P.O. 1999. DNA arrays for analysis of gene expression. *Methods Enzymol.* **303:** 179–205.

Encio I.J. and Detera-Wadleigh S.D. 1991. The genomic structure of the human glucocorticoid receptor. *J. Biol. Chem.* **266:** 7182–7188.

Evans R.M. 1988. The steroid and thyroid hormone receptor superfamily. *Science* **240:** 889–895.

Ferrara N. and Keyt B. 1997. Vascular endothelial growth factor: Basic biology and clinical implications. *EXS* **79:** 209–232.

Fukunaga R. and Hunter T. 1997. MNK1, a new MAP kinase-activated protein kinase, isolated by a novel expression screening method for identifying protein kinase substrates. *EMBO J.* **16:** 1921–1933.

Gamarnik A.V. and Andino R. 1997. Two functional complexes formed by KH domain containing proteins with the 5′ noncoding region of poliovirus RNA. *RNA* **3:** 882–892.

———. 1998. Switch from translation to RNA replication in a positive-stranded RNA virus. *Genes Dev.* **12:** 2293–2304.

Gan W. and Rhoads R.E. 1996. Internal initiation of translation directed by the 5′-untranslated region of the mRNA for eIF4G, a factor involved in the picornavirus-induced switch from cap-dependent to internal initiation. *J. Biol. Chem.* **271:** 623–626.

Gan W., Celle M.L., and Rhoads R.E. 1998. Functional characterization of the internal ribosome entry site of eIF4G mRNA. *J. Biol. Chem.* **273:** 5006–5012.

Gingras A.-C., Raught B., and Sonenberg N. 1999. eIF4 initiation factors: Effectors of mRNA recruitment to ribosomes and regulators of translation. *Annu. Rev. Biochem.* **68:** 913–963.

Gradi A., Svitkin Y.V., Imataka H., and Sonenberg N. 1998a. Proteolysis of human eukaryotic translation initiation factor eIF4GII, but not eIF4GI, coincides with the shutoff of host protein synthesis after poliovirus infection. *Proc. Natl. Acad. Sci.* **95:** 11089–11094.

Gradi A., Imataka H., Svitkin Y.V., Rom E., Raught B., Morino S., and Sonenberg N. 1998b. A novel functional human eukaryotic translation initiation factor 4G. *Mol. Cell. Biol.* **18:** 334–342.

Groom L.A., Sneddon A.A., Alessi D.R., Dowd S., and Keyse S.M. 1996. Differential regulation of the MAP, SAP and RK/p38 kinases by Pyst1, a novel cytosolic dual-specificity phosphatase. *EMBO J.* **15:** 3621–3632.

Hedvat C.V. and Irving S.G. 1995. The isolation and characterization of MINOR, a novel mitogen-inducible nuclear orphan receptor. *Mol. Endocrinol.* **9:** 1692–1700.

Hellen C.U., Witherell G.W., Schmid M., Shin S.H., Pestova T.V., Gil A., and Wimmer E. 1993. A cytoplasmic 57-kDa protein that is required for translation of picornavirus RNA by internal ribosomal entry is identical to the nuclear pyrimidine tract-binding protein. *Proc. Natl. Acad. Sci.* **90:** 7642–7646.

Henis-Korenblit S., Strumpf N.L., Goldstaub D., and Kimchi A. 2000. A novel form of DAP5 protein accumulates in apoptotic cells as a result of caspase cleavage and internal ribosome entry site-mediated translation. *Mol. Cell. Biol.* **20:** 496–506.

Hentze M.W. 1997. eIF4G: A multipurpose ribosome adapter? *Science* **275:** 500–501.

Holcik M., Lefebvre C., Yeh C., Chow T., and Korneluk R.G. 1999. A new internal-ribosome-entry-site motif potentiates XIAP-mediated cytoprotection. *Nat. Cell Biol.* **1:** 190–192.

Hollenberg S.M., Weinberger C., Ong E.S., Cerelli G., Oro A., Lebo R., Thompson E.B., Rosenfeld M.G., and Evans R.M. 1985. Primary structure and expression of a functional human glucocorticoid receptor cDNA. *Nature* **318:** 635–641.

Hu M.C., Tranque P., Edelman G.M., and Mauro V.P. 1999. rRNA-complementarity in the 5′ untranslated region of mRNA specifying the Gtx homeodomain protein: Evidence that base- pairing to 18S rRNA affects translational efficiency. *Proc. Natl. Acad. Sci.* **96:** 1339–1344.

Huez I., Creancier L., Audigier S., Gensac M.C., Prats A.C., and Prats H. 1998. Two independent internal ribosome entry sites are involved in translation initiation of vascular endothelial growth factor mRNA. *Mol. Cell. Biol.* **18:** 6178–6190.

Hunt S.L., Hsuan J.J., Totty N., and Jackson R.J. 1999. unr, a cellular cytoplasmic RNA-binding protein with five cold-shock domains, is required for internal initiation of translation of human rhinovirus RNA. *Genes Dev.* **13:** 437–448.

Iizuka N., Chen C., Yang Q., Johannes G., and Sarnow P. 1995. Cap-independent translation and internal initiation of translation in eukaryotic cellular mRNA molecules. *Curr. Top. Microbiol. Immunol.* **203:** 155–177.

Imataka H., Olsen H.S., and Sonenberg N. 1997. A new translational regulator with homology to eukaryotic translation initiation factor 4G. *EMBO J.* **16:** 817–825.

Iyer V.R., Eisen M.B., Ross D.T., Schuler G., Moore T., Lee J.C.F., Trent J.M., Staudt L.M., Hudson J., Jr., Boguski M.S., Lashkari D., Shalon D., Botstein D., and Brown P.O. 1999. The transcriptional program in the response of human fibroblasts to serum. *Science* **283:** 83–87.

Jang S.K., Davies M.V., Kaufman R.J., and Wimmer E. 1989. Initiation of protein synthesis by internal entry of ribosomes into the 5′ nontranslated region of encephalomyocarditis virus RNA in vivo. *J. Virol.* **63:** 1651–1660.

Jang S.K., Krausslich H.G., Nicklin M.J., Duke G.M., Palmenberg A.C., and Wimmer E. 1988. A segment of the 5′ nontranslated region of encephalomyocarditis virus RNA directs internal entry of ribosomes during in vitro translation. *J. Virol.* **62:** 2636–2643.

Johannes G. and Sarnow P. 1998. Cap-independent polysomal association of natural mRNAs encoding c-*myc*, BiP, and eIF4G conferred by internal ribosome entry sites. *RNA* **4:** 1500–1513.

Johannes G., Carter M.S., Eisen M.B., Brown P.O., and Sarnow P. 1999. Identification of eukaryotic mRNAs that are translated at reduced cap binding complex eIF4F concentrations using a cDNA microarray. *Proc. Natl. Acad. Sci.* **96:** 13118–13123.

Kim J.G., Armstrong R.C., Berndt J.A., Kim N.W., and Hudson L.D. 1998. A secreted DNA-binding protein that is translated through an internal ribosome entry site (IRES) and distributed in a discrete pattern in the central nervous system. *Mol. Cell. Neurosci.* **12:** 119–140.

Komuro I., Schalling M., Jahn L., Bodmer R., Jenkins N.A., Copeland N.G., and Izumo S. 1993. Gtx: A novel murine homeobox-containing gene, expressed specifically in glial cells of the brain and germ cells of testis, has a transcriptional repressor activity in vitro for a serum-inducible promoter. *EMBO J.* **12:** 1387–1401.

Kozak M. 1986. Influences of mRNA secondary structure on initiation by eukaryotic ribosomes. *Proc. Natl. Acad. Sci.* **83:** 2850–2854.

Kwak S.P., Hakes D.J., Martell K.J., and Dixon J.E. 1994. Isolation and characterization of a human dual specificity protein-tyrosine phosphatase gene. *J. Biol. Chem.* **269:** 3596–3604.

Le S.Y. and Maizel J.V., Jr. 1997. A common RNA structural motif involved in the internal initiation of translation of cellular mRNAs. *Nucleic Acids Res.* **25:** 362–369.

Leverson J.D., Koskinen P.J., Orrico F.C., Rainio E.M., Jalkanen K.J., Dash A.B., Eisenman R.N., and Ness S.A. 1998. Pim-1 kinase and p100 cooperate to enhance c-Myb activity. *Mol. Cell* **2:** 417–425.

Levy-Strumpf N., Deiss L.P., Berissi H., and Kimchi A. 1997. DAP-5, a novel homolog of eukaryotic translation initiation factor 4G isolated as a putative modulator of gamma interferon-induced programmed cell death. *Mol. Cell. Biol.* **17:** 1615–1625.

Macejak D.G. and Sarnow P. 1991. Internal initiation of translation mediated by the 5′ leader of a cellular mRNA (see comments). *Nature* **353:** 90–94.

Martinerie C., Viegas-Pequignot E., Nguyen V.C., and Perbal B. 1997. Chromosomal mapping and expression of the human cyr61 gene in tumour cells from the nervous system. *Mol. Pathol.* **50:** 310–316.

Martinez-Salas E., Saiz J.C., Davilla M., Belsham G.J., and Domingo E. 1993. A single nucleotide substitution in the internal ribosome entry site of foot-and-mouth disease virus leads to enhanced cap-independent translation in vivo. *J. Virol.* **67:** 3748–3755.

Meerovitch K., Nicholson R., and Sonenberg N. 1991. In vitro mutational analysis of *cis*-acting RNA translational elements within the poliovirus type 2 5′ untranslated region. *J. Virol.* **65:** 5895–5901.

Meerovitch K., Svitkin Y.V., Lee H.S., Lejbkowicz F., Kenan D.J., Chan E.K.L., Agol V.I., Keene J.D., and Sonenberg N. 1993. La autoantigen enhances and corrects aberrant translation of poliovirus RNA in reticulocyte lysate. *J. Virol.* **67:** 3798–3807.

Mizokami A., Yeh S.Y., and Chang C. 1994. Identification of 3′,5′-cyclic adenosine monophosphate response element and other *cis*-acting elements in the human androgen receptor gene promoter. *Mol. Endocrinol.* **8:** 77–88.

Moren A., Ichijo H., and Miyazono K. 1992. Molecular cloning and characterization of the human and porcine transforming growth factor-beta type III receptors. *Biochem. Biophys. Res. Commun.* **189:** 356–362.

Nakielny S. and Dreyfuss G. 1996. The hnRNP C proteins contain a nuclear retention sequence that can override nuclear export signals. *J. Cell Biol.* **134:** 1365–1373.

Nanbu R., Montero L., D′Orazio D., and Nagamine Y. 1997. Enhanced stability of urokinase-type plasminogen activator mRNA in metastatic breast cancer MDA-MB-231 cells and LLC-PK1 cells down-regulated for protein kinase C — Correlation with cytoplasmic heterogeneous nuclear ribonucleoprotein C. *Eur. J. Biochem.* **247:** 169–174.

Negulescu D., Leong L.E., Chandy K.G., Semler B.L., and Gutman G.A. 1998. Translation initiation of a cardiac voltage-gated potassium channel by internal ribosome entry. *J. Biol. Chem.* **273:** 20109–20113.

Oh S.K., Scott M.P., and Sarnow P. 1992. Homeotic gene *Antennapedia* mRNA contains 5′-noncoding sequences that confer translational initiation by internal ribosome binding. *Genes Dev.* **6:** 1643–1653.

Parsley T.B., Towner J.S., Blyn L.B., Ehrenfeld E., and Semler B.L. 1997. Poly (rC) binding protein 2 forms a ternary complex with the 5′-terminal sequences of poliovirus RNA and the viral 3CD proteinase. *RNA* **3:** 1124–1134.

Paulin F.E.M., Chappell S.A., and Willis A.E. 1998. A single nucleotide change in the c-*myc* internal ribosome entry segment leads to enhanced binding of a group of protein factors. *Nucleic Acids Res.* **26:** 3097–3103.

Paulin F.E., West M.J., Sullivan N.F., Whitney R.L., Lyne L., and Willis A.E. 1996. Aberrant translational control of the c-*myc* gene in multiple myeloma. *Oncogene* **13:** 505–513.

Peabody D.S. and Berg P. 1986. Termination-reinitiation occurs in the translation of mammalian cell mRNAs. *Mol. Cell. Biol.* **6:** 2695–2703.

Pelletier J. and Sonenberg N. 1985. Insertion mutagenesis to increase secondary structure within the 5′ noncoding region of a eukaryotic mRNA reduces translational efficiency. *Cell* **40:** 515–526.

———. 1988. Internal initiation of translation of eukaryotic mRNA directed by a sequence derived from poliovirus RNA. *Nature* **334:** 320–325.

———. 1989. Internal binding of eucaryotic ribosomes on poliovirus RNA: Translation in HeLa cell extracts. *J. Virol.* **63:** 441–444.

Pestova T.V., Shatsky I.N., Fletcher S.P., Jackson R.J., and Hellen C.U. 1998. A prokaryotic-like mode of cytoplasmic eukaryotic ribosome binding to the initiation codon during internal translation initiation of hepatitis C and classical swine fever virus RNAs. *Genes Dev.* **12:** 67–83.

Pyronnet S., Pradayrol L., and Sonenberg N. 2000. A cell cycle-dependent internal ribosome entry site. *Mol. Cell* **5:** 607–616.

Pyronnet S., Imataka H., Gingras A.C., Fukunaga R., Hunter T., and Sonenberg N. 1999. Human eukaryotic translation initiation factor 4G (eIF4G) recruits Mnk1 to phosphorylate eIF4E. *EMBO J.* **18:** 270–279.

Sasaki J. and Nakashima N. 2000. Methionine-independent initiation of translation in the capsid protein of an insect RNA virus. *Proc. Natl. Acad. Sci.* **97:** 1512–1515.

Sarnow P. 1989. Translation of glucose-regulated protein 78/immunoglobulin heavy-chain binding protein mRNA is increased in poliovirus-infected cells at a time when cap-dependent translation of cellular mRNA is inhibited. *Proc. Natl. Acad. Sci.* **86:** 5795–5799.

Sella O., Gerlitz G., Le S.Y., and Elroy-Stein O. 1999. Differentiation-induced internal translation of c-*sis* mRNA: Analysis of the *cis* elements and their differentiation-linked binding to the hnRNP C protein. *Mol. Cell. Biol.* **19:** 5429–5440.

Shaughnessy J.D., Jr., Jenkins N.A., and Copeland N.G. 1997. cDNA cloning, expression analysis, and chromosomal localization of a gene with high homology to wheat eIF-(iso)4F and mammalian eIF-4G. *Genomics* **39:** 192–197.

Stein I., Neeman M., Shweiki D., Itin A., and Keshet E. 1995. Stabilization of vascular endothelial growth factor mRNA by hypoxia and hypoglycemia and coregulation with other ischemia-induced genes. *Mol. Cell. Biol.* **15:** 5363–5368.

Stein I., Itin A., Einat P., Skaliter R., Grossman Z., and Keshet E. 1998. Translation of vascular endothelial growth factor mRNA by internal ribosome entry: Implications for translation under hypoxia. *Mol. Cell. Biol.* **18:** 3112–3119.

Stoneley M., Paulin F.E., Le Quesne J.P., Chappell S.A., and Willis A.E. 1998. C-Myc 5′ untranslated region contains an internal ribosome entry segment. *Oncogene* **16:** 423–428.

Teerink H., Voorma H.O., and Thomas A.A. 1995. The human insulin-like growth factor II leader 1 contains an internal ribosomal entry site. *Biochim. Biophys. Acta* **1264:** 403–408.

Ting J. and Lee A.S. 1988. Human gene encoding the 78,000-dalton glucose-regulated protein and its pseudogene: Structure, conservation, and regulation. *DNA* **7:** 275–286.

Vagner S., Gensac M.C., Maret A., Bayard F., Amalric F., Prats H., and Prats A.C. 1995. Alternative translation of human fibroblast growth factor 2 mRNA occurs by internal entry of ribosomes. *Mol. Cell. Biol.* **15:** 35–44.

Vagner S., Touriol C., Galy B., Audigier S., Gensac M.C., Amalric F., Bayard F., Prats H., and Prats A.C. 1996. Translation of CUG- but not AUG-initiated forms of human fibroblast growth factor 2 is activated in transformed and stressed cells. *J. Cell Biol.* **135:** 1391–402.

Wang C. and Siddiqui A. 1995. Structure and function of the hepatitis C virus internal ribosome entry site. *Curr. Top. Microbiol. Immunol.* **203:** 99–115.

Warren R.S., Yuan H., Matli M.R., Ferrara N., and Donner D.B. 1996. Induction of vascular endothelial growth factor by insulin-like growth factor 1 in colorectal carcinoma. *J. Biol. Chem.* **271:** 29483–29488.

Waskiewicz A.J., Flynn A., Proud C.G., and Cooper J.A. 1997. Mitogen-activated protein kinases activate the serine/threonine kinases Mnk1 and Mnk2. *EMBO J.* **16:** 1909–1920.

Waskiewicz A.J., Johnson J.C., Penn B., Mahalingam M., Kimball S.R., and Cooper J.A. 1999. Phosphorylation of the cap-binding protein eukaryotic translation initiation factor 4E by protein kinase Mnk1 in vivo. *Mol. Cell. Biol.* **19:** 1871–1880.

Willis A.E., Paulin F.E., West M.J., and Whitney R.L. 1997. Investigation of aberrant translational control of c-*myc* in cell lines derived from patients with multiple myeloma. *Curr. Top. Microbiol. Immunol.* **224:** 269–276.

Wilson J.E., Pestova T.V., Hellen C.U.T., and Sarnow P. 2000. Initiation of protein synthesis from the A-site of the ribosome. *Cell.* **120:** (in press).

Wilson J.E., Powell M.J., Hoover S.E., and Sarnow P. 2000. Naturally occurring dicistronic cricket paralysis virus RNA is regulated by two internal ribosome entry sites. *Mol. Cell. Biol.* **20:** 4990–4999.

Yamanaka S., Poksay K.S., Arnold K.S., and Innerarity T.L. 1997. A novel translational repressor mRNA is edited extensively in livers containing tumors caused by the transgene expression of the apoB mRNA-editing enzyme. *Genes Dev.* **11:** 321–333.

Yang Q. and Sarnow P. 1997. Location of the internal ribosome entry site in the 5′ noncoding region of the immunoglobulin heavy-chain binding protein (BiP) mRNA: Evidence for specific RNA-protein interactions. *Nucleic Acids Res.* **25:** 2800–2807.

Ye X., Fong P., Iizuka N., Choate D., and Cavener D.R. 1997. *Ultrabithorax* and *Antennapedia* 5′ untranslated regions promote developmentally regulated internal translation initiation. *Mol. Cell. Biol.* **17:** 1714–1721.

Zong Q., Schummer M., Hood L., and Morris D.R. 1999. Messenger RNA translation state: The second dimension of high-throughput expression screening. *Proc. Natl. Acad. Sci.* **96:** 10632–10636.

Zoppi S., Wilson C.M., Harbison M.D., Griffin J.E., Wilson J.D., McPhaul M.J., and Marcelli M. 1993. Complete testicular feminization caused by an amino-terminal truncation of the androgen receptor with downstream initiation. *J. Clin. Invest.* **91:** 1105–1112.

20

Translational Control and Cancer

John W. B. Hershey and Suzanne Miyamoto
Department of Biological Chemistry
School of Medicine
University of California
Davis, California 95616

Cell growth is regulated by the carefully balanced expression of genes that either promote or inhibit cell proliferation. Cancerous cells arise when a number of these genes are mutated, thereby disturbing the balanced regulatory network. The stepwise accumulation of mutations that cause malignancy is thought to occur primarily in genes encoding growth factors, growth factor receptors, protein kinases involved in signal transduction cascades, and transcription factors (McCormick 1999). The genetic mutations, either inherited or generated by environmental factors such as irradiation, chemical carcinogens, and viral infection, cause dysregulation of cell growth and/or proliferation. In this chapter, we review the evidence supporting the notion that protein synthesis plays an important role in regulating cell proliferation and that aberrations in its regulation may contribute to loss of cell cycle control.

The cell cycle consists of a precisely regulated series of reactions or processes leading to cell growth and division. Protein synthesis and translational control of specific gene expression is required for the orderly transition from one phase of the cell cycle to the next. In yeast, for example, a conditional mutation in *cdc33* encoding eIF4E causes arrest of cells at the G_1/S boundary, possibly due to inefficient synthesis of the cyclin CLN3 (Polymenis and Schmidt 1997; Danaie et al. 1999). Therefore, it is plausible that mutations in the genes encoding translational components that affect translational control could be oncogenic. The translational apparatus also is a target of signal transduction pathways that are activated by mitogens. Therefore, perturbations in signaling pathways that lead to altered levels or activities of translational components such as initiation factors might contribute to the malignant state of cells.

Translational Control of Gene Expression
© 2000 Cold Spring Harbor Laboratory Press 0-87969-568-4/00 $5 + . 00

It is useful in considering the role of protein synthesis in the cell cycle to distinguish between cell growth (resulting in an increase in cell mass) and cell proliferation (resulting in an increase in cell number). Protein synthesis is required for cell growth, and cell growth is thought to be dominant over cell proliferation (Conlon and Raff 1999). In this regard, a *Drosophila* mutant lacking the p70 S6 kinase is only half the size of the wild-type fly, due to smaller cells, but not fewer cells (Montagne et al. 1999). Therefore, cell growth, not cell proliferation, is affected by the lack of the S6 kinase, an activator of protein synthesis. However, protein synthesis also affects cell proliferation, as a substantial body of evidence links translation and progression through the cell cycle. It is generally recognized that proto-oncogene and tumor-suppressor gene expression is controlled at multiple levels, among them the regulation of protein synthesis. Transformed cells generally show higher rates of protein synthesis than normal cells (Heys et al. 1991). The key question addressed in this chapter is: Do enhanced rates of protein synthesis contribute to the *cause* of cancer, or is the high translation rate in malignant cells a necessary *consequence* of the increased proliferation rate? A definitive answer to this question is not yet available. Many studies have asked: Is the abundance or activity of one or more components of the translational machinery altered in tumors? Does a specific change in a translational component cause malignant transformation of immortal cells? Altered protein synthesis resulting from the dysregulation of key signaling pathways also may result in aberrant translation, thereby contributing to higher growth rates. The view is developed that translational control is intimately connected to signal transduction pathways and to the expression of a large variety of transcription factor and signaling component genes (proto-oncogenes and tumor suppressor genes).

OVEREXPRESSION OF eIF4E CAUSES CELL MALIGNANCY

Sonenberg and coworkers made the startling discovery that overexpression of eIF4E induces the malignant transformation of immortal NIH-3T3 cells (Lazaris-Karatzas et al. 1990). The transformed cells exhibit anchorage-independent growth, focus formation in soft agar, and tumor formation in nude mice. CHO cells expressing Max (a partner of Myc) also are transformed by overexpression of eIF4E (De Benedetti et al. 1994). eIF4E is the m^7G-cap-binding protein that is thought to be rate-limiting for initiation (Chapter 6). eIF4E overexpression activates the *ras* oncogene, whereas overexpression of GAP, a negative regulator of Ras, reverses the transformed phenotype (Lazaris-Karatzas et al. 1992). In addition, eIF4E has been shown to cooperate with two immortalizing oncogenes, v-myc and adenovirus E1A, to transform rat embryo fibroblasts (Lazaris-Karatzas et al. 1992), thereby satisfying the definition of an oncogene in the classic two-oncogene transformation assay.

The importance of eIF4E for growth control was further demonstrated by suppressing its activity in transformed cells either by overexpressing the translational inhibitors 4E-BP1 or 4E-BP2 (Rousseau et al. 1996a) or by expression of eIF4E antisense RNA (De Benedetti et al. 1991). The reduced eIF4E level or activity results in partial reversal of the transformed phenotype. Interpretation of the antisense results is complicated by the fact that the level of eIF4G also was observed to decrease, so an effect of eIF4E alone is uncertain.

eIF4E also is implicated in apoptosis. Elevated eIF4E levels prevent apoptosis of NIH-3T3 cells subjected to serum deprivation (Polunovsky et al. 1996). *ras*-transformed fibroblasts are resistant to apoptosis, but overexpression of 4E-BP1 or treatment with rapamycin, both leading to diminished eIF4E activity, increases their susceptibility to apoptosis (Tan et al. 2000). An anti-apoptotic role for eIF4E also was shown in rat embryo fibroblasts expressing c-Myc, in that high eIF4E levels rescue the cells from genotoxic and cytostatic drugs (Tan et al. 2000). Thus, eIF4E-dependent protein synthesis suppresses apoptosis and contributes to cell survival. In effect, all of the results cited above strongly implicate translational control in growth regulation and suggest that the failure to down-regulate eIF4E activity and translation initiation may lead to loss of growth control. Possible explanations are discussed below for how eIF4E may affect pathways leading to cell proliferation.

INVOLVEMENT OF eIF2α PHOSPHORYLATION AND PKR IN GROWTH CONTROL

Another important mechanism for regulating global rates of protein synthesis is the phosphorylation of eIF2α, which leads to reduced levels of ternary complex (Met-tRNA$_i$•eIF2•GTP) and an inhibition of initiation (Chapters 5 and 16). Expression of a dominant negative mutant form of the eIF2α kinase, PKR, causes the malignant transformation of immortal NIH-3T3 cells (Koromilas et al. 1992b), suggesting that PKR acts as a tumor suppressor (Meurs et al. 1993). Overexpression of TRBP (Benkirane et al. 1997) or p58 (Barber et al. 1994), cellular inhibitors of PKR, also results in cell malignancy. However, because PKR has phosphorylation targets other than eIF2α (e.g., IκB, the inhibitor of the transcription factor NF-κB), its function in growth control may not be through translation. This issue was addressed by overproducing an eIF2α mutant protein with alanine substituting for Ser-51, the phosphorylation target of eIF2α kinases (Donze et al. 1995). Cells transfected with the Ala-51 mutant form become malignant, whereas cells transfected with wild-type eIF2α do not. The results indicate that inhibition of protein synthesis by phosphorylation of eIF2 is important for growth control.

However, the evidence supporting a role for PKR in cell proliferation is complex and sometimes contradictory (for review, see Jagus et al. 1999; Chapter 13). PKR levels tend to increase with cell differentiation, whereas overexpression of the wild-type kinase is toxic to cells. Indeed, induced synthesis of PKR in HeLa cells activates apoptosis (Lee and Esteban 1994), and murine fibroblasts overexpressing wild-type PKR are more sensitive to double-stranded RNA-promoted apoptosis (Balachandran et al. 1998). Consistent with a role as a tumor suppressor, PKR levels are lower in murine lymphoblastic leukemic cells (Jaramillo et al. 1995) and in some human leukemic and preleukemic myeloid cells (Beretta et al. 1996). In addition, low PKR levels are found in head and neck squamous carcinomas (Haines et al. 1998). In this case, PKR levels are inversely proportional to the levels of the proliferating cell nuclear antigen (PCNA), and high levels of PKR are associated with the disease-free state and overall survival (Haines et al. 1993). In contrast, PKR levels are higher in breast cancer (Haines et al. 1996) and in hepatic carcinomas (Shimada et al. 1998) than in normal cells. Furthermore, lymphocytes derived from leukemia patients have normal levels of PKR (Basu et al. 1997). Thus, the data implicating PKR as a tumor suppressor are not consistent and conclusive.

An important limitation to the experiments above is that only PKR *levels* were measured, whereas the relevant parameter is PKR *activity*. PKR is normally latent and is activated by polyanions such as dsRNA, highly structured single-stranded RNAs, heparin and poly(L-glutamine), and by a cellular activator, PACT (see Chapter 13). In a recent study, PKR activity in human breast carcinoma cell lines was found to be high relative to a nontransformed cell line (Kim et al. 2000). This resulted in elevated eIF2α phosphorylation, although protein synthesis was not inhibited, perhaps due to an unusually high level of eIF2B (see Chapter 2). Taken together, PKR appears not to function as a tumor suppressor in these breast cancer cell lines.

The strongest evidence against a necessary role for PKR in controlling cell proliferation comes from experiments with transgenic mice. The PKR knock-out mouse ($Pkr^{-/-}$) is normal and shows no tendency toward neoplasia (Yang et al. 1995). However, primary fibroblast cell lines derived from it grow more slowly because passage through the G_1/S phase of the cell cycle is retarded (Zamanian-Daryoush et al. 1999). The results suggest that PKR functions as a positive growth regulator rather than as an inhibitor of cell proliferation. The difficulty in demonstrating a convincing role for PKR in growth control or apoptosis may be due to the presence of other eIF2α kinases such as HRI and PERK, which may substitute for PKR (Chapters 13–15).

OVEREXPRESSION OF PROTEINS OF THE TRANSLATIONAL APPARATUS IN TUMOR CELLS

The finding that overexpression of eIF4E causes malignant transformation of immortal cells led investigators to examine eIF4E protein or mRNA levels in naturally occurring tumor cells. Elevated levels of eIF4E were found in a broad spectrum of transformed cell lines and solid tumors (Miyagi et al. 1995). eIF4E is particularly abundant in breast carcinoma (Miyagi et al. 1995; Anthony et al. 1996) and head and neck squamous cell carcinomas (Nathan et al. 1997). High levels of eIF4E also have been observed in non-Hodgkin's lymphoma (Wang et al. 1999) and in superficial and muscle invasive primary bladder cancers (Crew et al. 2000). The difference in levels between normal and breast cancer cells is so large that eIF4E abundance might be suitable as a prognostic tumor marker (Li et al. 1997; Nathan et al. 1999).

There are a number of other initiation factors whose protein or mRNA levels appear higher in tumors than in nonmalignant cells. High levels of eIF4AI mRNA are detected in human melanoma cells, whereas mRNAs encoding other initiation factors such as eIF4AII, eIF2γ, eIF4B, and eIF4E are not elevated (Eberle et al. 1997). Increased levels of eIF2α are found in non-Hodgkin's lymphoma (Wang et al. 1999) and mouse mammary tumors (Raught et al. 1996), whereas eIF2B is elevated in some human breast cancer cell lines (Kim et al. 2000). In addition, high levels of eIF4G have been observed in 30% of squamous lung carcinomas (Brass et al. 1997). In this regard, overexpression of eIF4G in NIH-3T3 cells promotes increased focus formation, anchorage-independent growth, and tumor formation in nude mice (Fukuchi-Shimogori et al. 1997). Since the concentration of eIF4E did not change in these cells, it appears that eIF4G is able to transform cells without altering eIF4E expression.

A number of eIF3 subunits are overexpressed in various cancer cells. High levels of the p170 subunit occur in human mammary carcinomas and esophageal squamous cell carcinomas (Bachmann et al. 1997), and overexpression of the p40 subunit has been reported in breast and prostate cancer (Nupponen et al. 1999). The p48 subunit of eIF3 is encoded by the *int-6* gene, which is a frequent site of integration of the mouse mammary tumor virus (MMTV) genome in mouse mammary tumors (Marchetti et al. 1995; Asano et al. 1997). Human p48 interacts with the Tax viral oncoprotein of human T-cell leukemia virus type I (Desbois et al. 1996) and is found in promyelocytic leukemia nuclear bodies, perhaps through its binding to the Ret finger protein (Rfp) (Morris-Desbois et al. 1999). Since other subunits of eIF3 are not located in the nucleus, the results with p48 suggest that it may serve functions outside of eIF3.

Further evidence that eIF3 may be important for the regulation of cell proliferation comes from studies of yeast. A strain of *Saccharomyces cerevisiae* with a mutated p110 subunit (human p170 homolog) is defective in passage through the G_1 phase of the cell cycle (Kovarik et al. 1998). The *Schizosaccharomyces pombe* homolog (moe1) of the human p66 subunit interacts with the Ras effector, Scd1 (Chen et al. 1999), suggesting involvement in signal transduction pathways. Another *S. pombe* eIF3 subunit, homologous to human p36, is a high-copy number suppressor of a cell cycle checkpoint mutant (*cdc2-3w*) that is defective in receiving the checkpoint signal for progression from S to G_2/M phase (Humphrey and Enoch 1998). Although precise mechanisms are not known for how eIF3 causes the observed phenotypes, the sum of the evidence strongly suggests that eIF3 plays a role in cell cycle control or that its subunits possess alternate functions.

Other translational components also are found at abnormally high levels in certain cancer cells. eEF1A (formerly eEF-1α) is elevated in tumors of the pancreas, colon, breast, lung, and stomach (Grant et al. 1992). High levels of eEF1Bβ (formerly eEF-1γ) mRNA have been reported in pancreatic cancers (Lew et al. 1992) and in colorectal (Mathur et al. 1998) and gastric (Mimori et al. 1995) carcinomas. Finally, several ribosomal protein genes are overexpressed in prostate cancer cell lines and tissues (Vaarala et al. 1998) and in other malignant cells (Ou et al. 1987; Pogue-Geile et al. 1991).

The impressive list above strongly suggests a link between the translational apparatus and cancer. Unfortunately, most of the studies cited examined only one or a small number of proteins comprising the translational apparatus. What is lacking is a systematic examination of a much larger set of proteins. For example, do malignant tissues possess high levels of just a few initiation factors, or do they possess high levels of most of them? If only one or a small number are expressed at a high frequency in a specific cancer, this suggests that the altered level (or activity) might be a contributing cause of the malignancy. On the other hand, if essentially all translational components are elevated or activated, this would indicate that the increased translation rate is a consequence of rapid cell proliferation.

A recent study that addresses this question employed oligonucleotide microarray technology to assess differences in mRNA levels between tumor and normal colon tissues (Alon et al. 1999). Unfortunately, only a few of the initiation factor mRNAs were represented in the arrays. Another study utilizing cDNA microarrays analyzed the ratios of 5000 gene transcripts in a number of different cancer cells (13 breast tumor tissues and 3 malignant cell lines) compared to a control human mammary epithelial cell line growing in culture (Perou et al. 1999). The arrays included one or more probes for 12 of the 28 initiation factor mRNAs. A single cDNA probe for eIF2Bε, eIF3-p170, eIF3-p40, and eIF4B showed greater than fivefold elevations in some, but not all, cancers examined. However, a duplicate probe for a different region of the same transcript often gave the opposite result, i.e., detecting over fivefold underexpression in the same cancer sample. Such technological problems make firm conclusions about factor mRNA levels in these tissue preparations difficult, if not impossible. Further efforts are required to resolve whether or not a small subset of specific initiation factor mRNAs are elevated, or whether there is a general increase in the level of the translational apparatus. Nevertheless, measuring mRNA levels alone is not sufficient, as initiation factor activity is the relevant parameter that needs to be assessed, along with phosphorylation states of the proteins.

HOW ARE THE LEVELS AND ACTIVITIES OF TRANSLATIONAL COMPONENTS REGULATED?

Little is known about the normal regulation of initiation factor genes in mammalian cells, much less about their dysregulation. It appears that at least some genes encoding parts of the translational machinery are susceptible to regulation by oncogenes such as c-Myc, Ras, and viral oncogenes. For example, eIF4E and eIF2α are overexpressed in immortal cells transformed with c-Myc (Rosenwald et al. 1993b). The promoter for the eIF4E gene contains a region (the E boxes) which is a target of c-Myc (Jones et al. 1999). A different mechanism for overexpression, gene amplification, accounts for the high levels of eIF3-p40 in breast and prostate cancers (Nupponen et al. 1999), and eIF4GI in 30% of squamous lung carcinoma tumors (Brass et al. 1997). A challenge for the future will be to elucidate how genes encoding components of the translational apparatus are regulated and coordinately expressed. Such knowledge is essential for understanding aberrant regulation of genes that lead to overexpression.

More is known about how individual initiation factor activities are regulated by phosphorylation/dephosphorylation. Phosphorylation of eIF2α (and eEF2) causes inhibition (Chapter 5), whereas phosphorylation of the 4E-BPs stimulates initiation (Chapter 6). In addition, increased phosphorylation of eIF3, eIF4B, eIF4E, eIF4G, and ribosomal protein S6 correlates with activation of protein synthesis (Chapters 2 and 23). Thus phosphorylation/dephosphorylation of the translational apparatus is a major mechanism for modulating the overall initiation potential.

The protein kinases and phosphatases involved in determining initiation factor phosphorylation states are regulated by signal transduction pathways. Most descriptions of signal transduction pathways point to regulatory targets in the nucleus (McCormick 1999; Gibbs 2000), but it has become abundantly clear that cytoplasmic regulation of the translational apparatus also occurs. A detailed description of signal transduction pathways is beyond the scope of this chapter, but several reviews of how these pathways affect protein synthesis are available (Brown and Schreiber 1996; Morley 1996; Rhoads 1999; Chapters 6 and 23). Briefly, two major pathways signal to activate protein synthesis upon treatment of cells with mitogens, growth factors, hormones, cytokines, or mitogens (Fig. 1). The Ras pathway, through MEK and ERK, activates Mnk1, which phosphorylates eIF4E. The PI3-K pathway, operating through PDK, PKB, and mTOR, leads to the phosphorylation of ribosomal protein S6, eIF4B, eIF4GI, and the 4E-BPs, whereas another branch affects the phosphorylation of eIF2B. The PI3-K pathway is especially important for activating translational components (for detailed discussion, see Chapters 6 and 23). Still other signaling pathways alter the levels of IP3 and Ca^{++}, thereby affecting PKC and PKR (Chapter 13). These pathways also target the nucleus, affecting transcriptional regulation of gene expression. Although distinct pathways are drawn in Figure 1, it is recognized that the pathways actually are interconnected to form a weblike network. The complexity is further demonstrated by the fact that different pathways can affect the same target. The major idea to be stressed is that not only transcription, but also translation, is regulated by the signal transduction network.

Many proto-oncogenes or tumor suppressor genes encode components of the signal transduction network involved in regulating protein synthesis. In particular, the Ras pathway drives uncontrolled proliferation and is dysregulated in many cancers (McCormick 1999). Components of the PI3-K pathway also are implicated in tumorigenesis and apoptosis. For example, PTEN, which is mutated in many advanced cancers, is a

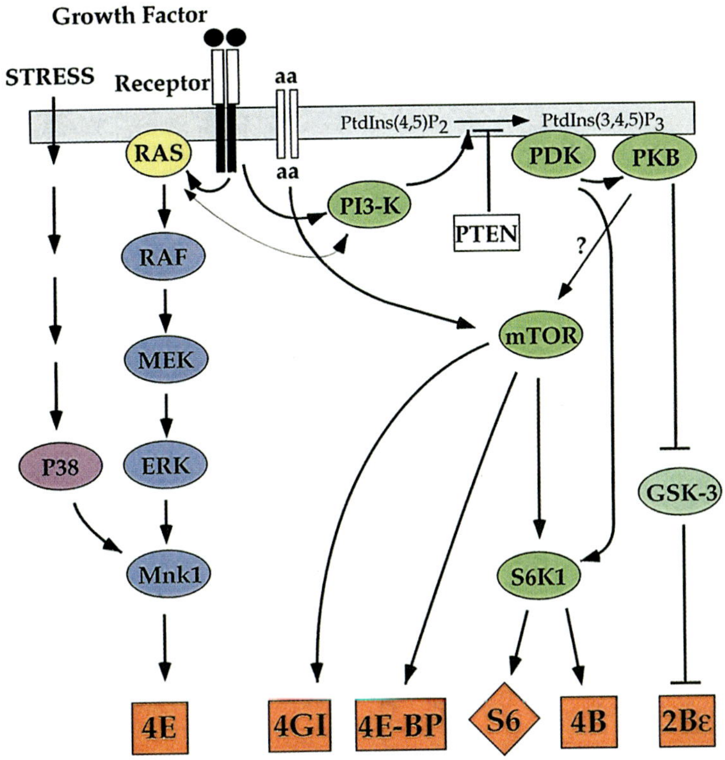

Figure 1 Signal transduction pathways affecting the translational apparatus. The model of the signaling pathways shows the MAPK pathway in blue and the mTOR pathway in green. The plasma membrane is shown in light gray. Depicted in the membrane are a growth factor receptor, a channel for amino acid (aa) transport, and phosphatidylinositol(4,5) diphosphate (PtdIns[4,5]P_2) and phosphatidylinositol(3,4,5) triphosphate (PtdIns[3,4,5]P_3). Protein kinases are shown as colored ovals; the protein phosphatase PTEN is shown as a white rectangle. The translation targets are shown in red as rectangles (for initiation factors) and as a diamond (for ribosomal protein S6).

phosphoinositide 3-phosphatase, dephosphorylating the third position of the inositol ring of phosphatidyl-inositide(3,4,5)P_3 needed for activation of PKB (Besson et al. 1999). PKB overexpression is seen in breast cancer cells and in pancreatic and ovarian carcinomas, and generally has anti-apoptotic effects (for review, see Vanhaesebroeck and Alessi 2000). PKB has many downstream targets that are involved in apoptosis (e.g., BAD, caspase-9, IκB kinase) and cell proliferation (e.g., BRCA1, eNOS, Raf; Vanhaesebroeck and Alessi 2000).

One of the consequences of dysregulation of signaling pathways may be the activation of protein synthesis through enhanced transcription of translation factor genes in the nucleus or through the altered phosphorylation states of their protein products in the cytoplasm. For example, cells transformed with Ras (or Src) exhibit elevated eIF4E phosphorylation, which is thought to lead to increased eIF4E activity (Frederickson et al. 1991; Rinker-Schaeffer et al. 1992). It was suggested that a positive feedback autocrine loop may be established in which eIF4E stimulates the translation of growth factor mRNAs whose protein products then activate Ras and subsequently eIF4E phosphorylation (Sonenberg 1996).

TRANSLATIONAL CONTROL OF GROWTH-REGULATING GENES

An attractive hypothesis to explain how protein synthesis might contribute to the regulation of cell proliferation states that the overall potential for translation initiation determines the identity and relative amounts of the proteins synthesized. When the initiation potential is increased, the translation rates of poorly translated mRNAs are stimulated to a much higher degree than those of strongly translated mRNAs. Conversely, when the initiation potential is decreased, poorly translated mRNAs are repressed more than other mRNAs. A poorly translated mRNA is one that competes ineffectively for the translational apparatus, most importantly for the m^7G-cap-binding complex, eIF4F (see Chapters 2 and 6). Such inefficiently translated mRNAs often have long, highly structured 5′UTRs that may interfere with the binding of eIF4F and/or inhibit preinitiation complex scanning (for a listing of such mRNAs that encode transcription factors, growth factors, receptors, and tyrosine kinases, see Kozak 1991). It is therefore plausible that activation of translation initiation leads to more efficient translation of this class of mRNAs, resulting in overproduction of growth-regulating proteins. In normal cells the initiation potential is held in check by several mechanisms, e.g., eIF2 phosphorylation or the action of the 4E-BP family of proteins on eIF4E, and the mRNAs for growth-regulating proteins are much less efficiently translated.

Attempts to demonstrate a relative increase in translational efficiency of poorly translated mRNAs due to stimulation of initiation most often have employed cells overproducing eIF4E. As described above, high levels of eIF4E cause malignant transformation of immortal cells. One of the best-studied poorly translated mRNAs encodes ornithine decarboxylase (ODC), the rate-limiting enzyme in polyamine synthesis (polyamines are required for entry into S phase). ODC mRNA is translated inefficiently in normal cells, but when the highly structured 5′UTR is removed, ODC synthesis is greatly stimulated (Manzella and Blackshear 1990). ODC mRNA is translated up to 30 times more efficiently in cells overexpressing eIF4E (Shantz and Pegg 1994; Rousseau et al. 1996b). The ODC mRNA is representative of many other mRNAs whose long 5′UTRs are G/C-rich and presumably highly structured. Examples of overproduction in a high eIF4E background include cyclin D1, involved in regulating a protein kinase required for the G_1/S transition (Rosenwald et al. 1993a); c-Myc, a transcription factor (De Benedetti et al. 1994); human fibroblast growth factor 2 (FGF-2); and vascular endothelial growth factor (VEGF), a cytokine (Kevil et al. 1996). Consistent with this, the translation of mRNAs with secondary structures inserted into their 5′UTR is increased in eIF4E-overproducing cells (Koromilas et al. 1992a). A comprehensive review of genes whose expression is altered by eIF4E activity has been published recently (De Benedetti and Harris 1999). Clearly, overproduction of eIF4E, as observed in many tumor cells, contributes to the malignant state by enhancing the synthesis of growth-promoting proteins. It can be anticipated that the application of high throughput functional genomic methods to assess polysome size and translational efficiency will soon enable identification of essentially all of the mRNAs whose translation is stimulated by overexpressed eIF4E.

Initiation by cap-dependent scanning or by internal binding at an IRES (see Chapter 4) provides an mRNA the means to evade certain regulatory pathways. A number of the mRNAs described above can initiate at an IRES. Examples are the fibroblast growth factor FGF-2 (Vagner et al. 1995), vascular epithelial growth factor VEGF (Stein et al. 1998), the X-linked inhibitor of apoptosis XIAP (Holcik et al. 1999), the proto-oncogene c-Myc (Nanbru et al. 1997; Stoneley et al. 1997), and ornithine decarboxylase (ODC) (Pyronnet et al. 2000). The ODC mRNA again provides a good example, as its translation during the cell cycle is biphasic. At the G_1/S transition, ODC synthesis is stimulated by overproduction of eIF4E as described above, and is inhibited by rapamycin that acts on the 4E-BPs, indicating a cap-dependent mechanism. At the G_2/M transition when global rates of protein synthesis are reduced, ODC synthesis is no longer inhibited by rapamycin, but rather appears to occur by internal initiation (Pyronnet et al. 2000), a pathway that does not require eIF4E. c-Myc synthesis also occurs during mitosis (Pyronnet et al. 2000), presumably by a similar mechanism.

Many of the IRES-containing mRNAs described above are translated during cell stress. Treatment of cells with γ-irradiation induces XIAP and

ODC synthesis (Song et al. 1998; Holcik et al. 1999), whereas hypoxia induces both ODC activity (Longo et al. 1993) and VEGF synthesis (Stein et al. 1998). Sonenberg and coworkers (Pyronnet et al. 2000) point out that these genes encode growth-promoting proteins that play a role in anti-apoptotic signaling, and therefore may be important for cell survival during therapeutic radiation.

Another example of cell-cycle-regulated gene expression dependent on an IRES concerns the PITSLRE protein kinases (Cornelis et al. 2000). Members of this family of cyclin-dependent kinases are tumor suppressors that affect cell cycle progression, and the PITSLRE locus is deleted in many cancers. Two isoforms, $p110^{PITSLRE}$ and $p58^{PITSLRE}$ are translated from the same mRNA, the latter by internal initiation. Thus $p58^{PITSLRE}$ is translated during G_2/M, whereas $p110^{PITSLRE}$ is not. It is tempting to speculate that initiation at IRESs requires a special *trans*-acting protein that is activated during M phase and is involved in stimulating cell proliferation.

In addition, there are a number of other growth-regulated genes whose mRNAs contain upstream open reading frames (uORFs); e.g., CLN3, bcl2, c-mos; Her-2 (Kozak 1991; Chapter 18). The uORFs act to reduce the translational efficiency of these mRNAs. It seems possible, but not yet demonstrated, that such mRNAs may partially evade the translational repression of the uORF, resulting in more efficient translation in tumors. This is the case for CLN3 in yeast, where failure of the *cdc33* (eIF4E) mutant strain to progress from G_1 into the S phase of the cell cycle can be overcome by expressing an altered CLN3 mRNA with a 5′UTR lacking a uORF (Danaie et al. 1999). Instead of arrest at the G_1/S boundary, cells arrest throughout the cell cycle.

EVALUATION OF THE TRANSLATIONAL LINK TO CANCER

The experimental results reviewed in the sections above make a strong case for the involvement of translation in the regulation of cell proliferation. The demonstration that the eIF4E gene satisfies the definition of an oncogene represents the strongest argument, which is further supported by the finding that eIF4E levels are elevated in many cancers. Other correlations of unregulated activation of the initiation potential and malignant transformation of cells have been established. Furthermore, there is a plausible mechanistic explanation for how such activation leads to cell proliferation. Many of the mRNAs encoding proteins in the growth-promoting pathway are inefficiently translated. Since their translation is stimulated at high initiation potential, a resulting imbalance in the levels of their protein products versus growth-retarding proteins is readily envisioned. This attractive hypothesis remains to be demonstrated by rigorous experimentation. This is not possible simply by examining a small number of mRNAs, but requires a functional genomic approach where essentially all of the cell's transcripts are monitored, not only for cellular level, but also for translational efficiency. Given the overall complexity of growth control and the network of proteins involved, a clear interpretation of data generated by functional genomic methods will be a major challenge in the future. Nevertheless, there are sufficient indications of an important role for translation in the regulation of the cell cycle that the translational apparatus already can be considered an attractive target for therapeutic intervention.

ACKNOWLEDGMENTS

The authors thank Shoba Gunnery for communicating results prior to publication, and Anne-Claude Gingras, Brian Raught, and members of the Davis laboratory for helpful comments during the preparation of the text and figure.

REFERENCES

Alon U., Barkai N., Notterman D.A., Gish K., Ybarra S., Mack D., and Levine A.J. 1999. Broad patterns of gene expression revealed by clustering analysis of tumor and normal colon tissues probed by oligonucleotide arrays. *Proc. Natl. Acad. Sci.* **96:** 6745–6750.

Anthony B., Carter P., and De Benedetti A. 1996. Overexpression of the protooncogene-translation factor eIF-4E in breast carcinoma cell lines. *Int. J. Cancer* **65:** 858–863.

Asano K., Merrick W.C., and Hershey J.W.B. 1997. The translation initiation factor eIF3-p48 subunit is encoded by int-6, a site of frequent integration by the mouse mammary tumor virus genome. *J. Biol. Chem.* **272:** 23477–23480.

Bachmann F., Bänziger R., and Burger M.M. 1997. Cloning of a novel protein overexpressed in human mammary carcinoma. *Cancer Res.* **57:** 988–994.

Balachandran S., Kim C.N., Yeh W.C., Mak T.W., Bhalla K., and Barber G.N. 1998. Activation of the dsRNA-dependent protein kinase, PKR, induces apoptosis and employs the death signal transducer FADD. *EMBO J.* **17:** 6888–6902.

Barber G.N., Thompson S., Lee T.G., Strom T., Jagus R., Darveau A., and Katze M.G. 1994. The 58-kilodalton inhibitor of the interferon-induced double-stranded RNA-activated protein kinase is a tetratricopeptide repeat protein with oncogenic properties. *Proc. Natl. Acad. Sci.* **91:** 4278–4282.

Basu S., Panayiotidis P., Hart S.M., He L.Z., Man A., Hoffbrand A.V., and Ganeshaguru K. 1997. Role of double-stranded RNA-activated protein kinase in human hematological malignancies. *Cancer Res.* **57:** 943–947.

Benkirane M., Neuveut C., Chun R.F., Smith S.M., Samuel C.E., Gatignol A., and Jeang K.T. 1997. Oncogenic potential of TAR RNA binding protein TRBP and its regulatory interaction with RNA-dependent protein kinase PKR. *EMBO J.* **16:** 611–624.

Beretta L., Gabbay M., Berger G., Hanash S.M., and Sonenberg N. 1996. Expression of the protein kinase PKR is modulated by IRF-1 and is reduced in 5q-associated leukemias. *Oncogene* **12:** 1593–1596.

Besson A., Robbins S.M., and Yong V.W. 1999. PTEN/MMAC1/TEP1 in signal transduction and tumorigenesis. *Eur. J. Biochem.* **263:** 605–611.

Brass N., Heckel D., Sabin U., Pfreundschuh M., Sybrecht G.W., and Meese E. 1997. Translation initiation factor eIF-4γ is encoded by an amplified gene and induces an immune response in squamous cell lung carcinoma. *Hum. Mol. Genet.* **6:** 33–39.

Brown E.J. and Schreiber S.L. 1996. A signaling pathway to translational control. *Cell* **86:** 517–520.

Chen C.-R., Li Y.-C., Chen J., Hou M.-C., Papadaki P., and Chang E.C. 1999. Moe1, a conserved protein in *Schizosaccaromyces pombe*, interacts with a Ras effector, Scd1, to affect proper spindle formation. *Proc. Natl. Acad. Sci.* **96:** 517–522.

Conlon I. and Raff M. 1999. Size control in animal development. *Cell* **96:** 235–244.

Cornelis S., Bruynooghe Y., Denecker G., Van Huffle S., Tinton S., and Beyaert R. 2000. Identification and characterization of a novel cell cycle-regulated internal ribosome entry site. *Mol. Cell* **5:** 597–605.

Crew J.P., Fuggle S., Bicknell R., Cranston D.W., De Benedetti A., and Harris A.L. 2000. Eukaryotic initiation factor-4E in superficial and muscle invasive bladder cancer and its correlation with vascular endothelial growth factor expression and tumor progression. *Br. J. Cancer* **82:** 161–166.

Danaie P., Altmann M., Hall M.N., Trachsel H., and Helliwell S.B. 1999. CLN3 expression is sufficient to restore G1-to-S-phase progression in *Saccharomyces cerevisiae* mutants defective in translation initiation factor eIF4E. *Biochem. J.* **340:** 135–141.

De Benedetti A. and Harris A.L. 1999. eIF4E expression in tumors: Its possible role in progression of malignancies. *Int. J. Biochem. Cell Biol.* **31:** 59–72.

De Benedetti A., Joshi B., Graff J.R., and Zimmer S.G. 1994. CHO cells transformed by the initiation factor eIF-4E display increased c-*myc* expression but require overexpression of Max for tumorigenicity. *Mol. Cell. Differ.* **2:** 347–371.

De Benedetti A., Joshi-Barve S., Rinker-Schaeffer C., and Rhoads R.E. 1991. Expression of antisense RNA against initiation factor eIF-4E mRNA in HeLa cells results in lengthened cell division times, diminished translation rates, and reduced levels of both eIF-4E and the p220 component of eIF-4F. *Mol. Cell. Biol.* **11:** 5435–5445.

Desbois C., Rousset R., Bantignies F., and Jalinot P. 1996. Exclusion of Int-6 from PML nuclear bodies by binding to the HTLV-I Tax oncoprotein. *Science* **273:** 951–953.

Donze O., Jagus R., Koromilas A.E., Hershey J.W.B., and Sonenberg N. 1995. Abrogation of translation initiation factor eIF-2 phosphorylation causes malignant transformation of NIH 3T3 cells. *EMBO J.* **14:** 3828–3834.

Eberle J., Krasagakis K., and Orfanos C.E. 1997. Translation initiation factor eIF-4AI mRNA is consistently overexpressed in human melanoma cells in vitro. *Int. J. Cancer* **71:** 396–401.

Frederickson R.M., Montine K.S., and Sonenberg N. 1991. Phosphorylation of eukaryotic translation initiation factor 4E is increased in Src-transformed cell lines. *Mol. Cell. Biol.* **11:** 2896–2900.

Fukuchi-Shimogori T., Ishii I., Kashiwagi K., Mashiba H., Ekimoto H., and Igarashi K. 1997. Malignant transformation by overproduction of translation initiation factor eIF4G. *Cancer Res.* **57:** 5041–5044.

Gibbs J.B. 2000. Mechanism-based target identification and drug discovery in cancer research. *Science* **287:** 1969–1973.

Grant A.G., Flomen R.M., Tizard M.L., and Grant D.A. 1992. Differential screening of a human pancreatic adenocarcinoma lambda gt11 expression library has identified increased transcription of elongation factor EF-1 alpha in tumour cells. *Int. J. Cancer* **50:** 740–745.

Haines G.K., Becker S., Ghadge G., Kies M., Pelzer H., and Radosevich J.A. 1993. Expression of PKR(p68) in squamous cell carcinoma of the head and neck region. *Arch. Otolaryngol. Head Neck Surg.* **119:** 1142–1147.

Haines G.K., Cajulis R., Hayden R., Duda R., Talamonti M., and Radosevich J.A. 1996. Expression of the double-stranded RNA-dependent protein kinase (p68) in human breast cancer. *Tumor Biol.* **17:** 5–12.

Haines G.K., Panos R.J., Bak P.M., Brown T., Zielinski M., Leyland J., and Radosevich J.A. 1998. Interferon-responsive protein kinase (p68) and proliferating cell nuclear antigen are inversely distributed in head and neck squamous cell carcinoma. *Tumor Biol.* **19:** 52–59.

Heys S.D., Park K.G., McNurlan M.A., Calder A.G., Buchan V., Blessing K., Eremin O., and Garlick P.J. 1991. Measurement of tumor protein synthesis in vivo in human colorectal and breast cancer. *Clin. Sci.* **80:** 587–593.

Holcik M., Lefebvre C., Yeh C., Chow T., and Korneluk R.G. 1999. A novel internal ribosome entry site (IRES) motif potentiates XIAP mediated cytoprotection. *Nat. Cell Biol.* **3:** 190–193.

Humphrey T. and Enoch T. 1998. Sum1, a highly conserved WD-repeat protein, suppresses S-M checkpoint mutants and inhibits the osmotic stress cell cycle response in fission yeast. *Genetics* **148:** 1731–1742.

Jagus R., Joshi B., and Barber G.N. 1999. PKR, apoptosis and cancer. *Int. J. Biochem. Cell Biol.* **31:** 123–138.

Jaramillo M.L., Abraham N., and Bell J.C. 1995. The interferon system: A review with emphasis on the role of PKR in growth control. *Cancer Invest.* **13:** 327–338.

Jones R.M., Branda J., Kohnston K.A., Polymenis M., Gadd M., Rustgi A., Callanan L., and Schmidt E.V. 1999. An essential E box in the promoter of the gene encoding the mRNA cap-binding protein (eukaryotic initiation factor 4E) is a target for activation by c-*myc*. *Mol. Cell. Biol.* **16:** 4754–4764.

Kevil C., De Benedetti A., Payne K.D., Coe L.L., Laroux S., and Alexander S. 1996. Translational regulation of vascular permeability factor by eukaryotic initiation factor 4E: Implications for tumor angiogenesis. *Int. J. Cancer* **65:** 785–790.

Kim S.H., Forman A.P., Mathews M.B., and Gunnery S. 2000. Human breast cancer cells contain elevated levels and activity of the protein kinase, PKR. *Oncogene* (in press).

Koromilas A.E., Lazaris-Karatzas A., and Sonenberg N. 1992a. mRNAs containing extensive secondary structure in their 5′ non-coding region translate efficiently in cells overexpressing initiation factor eIF-4E. *EMBO J.* **11:** 4153–4158.

Koromilas A.E., Roy S., Barber G.N., Katze M.G., and Sonenberg N. 1992b. Malignant transformation by a mutant of the IFN-inducible dsRNA-dependent protein kinase. *Science* **257:** 1685–1689.

Kovarik P., Hasek J., Valasek L., and Ruis H. 1998. *RPG1:* An essential gene of *Saccharomyces cerevisiae* encoding a 110 kDa protein required for passage through the G_1 phase. *Curr. Genet.* **33:** 100–109.

Kozak M. 1991. An analysis of vertebrate mRNA sequences: Intimations of translational control. *J. Cell Biol.* **115:** 887–903

Lazaris-Karatzas A., Montine K.S., and Sonenberg N. 1990. Malignant transformation by a eukaryotic initiation factor subunit that binds to mRNA 5' cap. *Nature* **345:** 544–547.

Lazaris-Karatzas A., Smith M.R., Frederickson R.M., Jaramillo M.L., Liu Y.-L., Kung H.-F., and Sonenberg N. 1992. Ras mediates translation initiation factor 4E-induced malignant transformation. *Genes Dev.* **6:** 1631–1642.

Lee S.B. and Esteban M. 1994. The interferon-induced double-stranded RNA-activated protein kinase induces apoptosis. *Virology* **199:** 491–496.

Lew Y., Jones D.V., Mars W.M., Evans D., Byrd D., and Frazier M.L. 1992. Expression of elongation factor-1 gamma-related sequence in human pancreatic cancer. *Pancreas* **7:** 144–152.

Li B.D., Liu L., Dawson M., and De Benedetti A. 1997. Overexpression of eukaryotic initiation factor 4E (eIF4E) in breast carcinoma. *Cancer* **79:** 2385–2390.

Longo L.D., Packianathan S., McQueary J.A., Stagg R.B., Byus C.V., and Cain C.D. 1993. Acute hypoxia increases ornithine decarboxylase activity and polyamine concentrations in fetal rat brain. *Proc. Natl. Acad. Sci.* **90:** 692–696.

Manzella J.M. and Blackshear P.J. 1990. Regulation of rat ornithine decarboxylase mRNA translation by its 5′-untranslated region. *J. Biol. Chem.* **265:** 11817–11822.

Marchetti A., Buttitta F., Miyazaki S., Gallahan D., Smith G.H., and Callahan R. 1995. *Int-6*, a highly conserved, widely expressed gene, is mutated by mouse mammary tumor virus in mammary preneoplasia. *J. Virol.* **69:** 1932–1938.

Mathur S., Cleary K.R., Inamdar N., Kim Y.H., Steck P., and Frazier M.L. 1998. Overexpression of elongation factor-1γ protein in colorectal carcinoma. *Cancer* **82:** 816–821.

McCormick F. 1999. Signalling networks that cause cancer. *Trends Cell Biol.* **9:** 53–59.

Meurs E.F., Galabru J., Barber G.N., Katze M.G., and Hovanessian A.G. 1993. Tumor suppressor function of the interferon-induced double-stranded RNA-activated protein kinase. *Proc. Natl. Acad. Sci.* **90:** 232–236.

Mimori K., Mori M., Tanaka S., Akiyoshi T., and Sugimachi K. 1995. The overexpression of elongation factor 1γ mRNA in gastric carcinoma. *Cancer* **75:** 1446–1449.

Miyagi Y., Sugiyama A., Asai A., Okazuki T., Kuchino Y., and Kerr S. 1995. Elevated levels of eukaryotic initiation factor eIF-4E mRNA in a broad spectrum of transformed cell lines. *Cancer Lett.* **91:** 247–252.

Montagne J., Stewart M.J., Stocker H., Hafen E., Kozma S.C., and Thomas G. 1999. *Drosophila* S6 kinase: A regulator of cell size. *Science* **285:** 2126–2129.

Morley S.J. 1996. Regulation of components of the translational machinery by protein phosphorylation. In *Protein phosphorylation in cell growth regulation* (ed. M.J. Clemens), pp. 197–224. Harwood Academic Publishers, Amsterdam, The Netherlands.

Morris-Desbois C., Bochard V., Reynaud C., and Jalinot P. 1999. Interaction between the Ret finger protein and the *int-6* gene product and co-localization into nuclear bodies. *J. Cell Sci.* **112:** 3331–3342.

Nanbru C., Lafon I., Audigier S., Gensac M.C., Vagner S., Huez G., and Prats A.C. 1997. Alternative translation of the proto-oncogene c-*myc* by an internal ribosome entry site. *J. Biol. Chem.* **272:** 32061–32066.

Nathan C.A., Liu L., Li B., Nandy I., Abreco F., and De Benedetti A. 1997. Detection of the proto-oncogene eIF4E in surgical margins may predict recurrence to head and neck cancer. *Oncogene* **15:** 579–584.

Nathan C.A., Franklin S., Abreo F.W., Nassar R., De Benedetti A., Williams J., and Stucker F.J. 1999. Expression of eIF4E during head and neck tumorigenesis: Possible role in angiogenesis. *Laryngoscope* **109:** 1253–1258.

Nupponen N.N., Porkka K., Kakkola L., Tanner M., Persson K., Borg A., Isola J., and Visakorpi T. 1999. Amplification and overexpression of p40 subunit of eukaryotic translation initiation factor 3 in breast and prostate cancer. *Am. J. Pathol.* **154:** 1777–1783.

Ou J.H., Yen T.S., Wang Y.F., Kam W.K., and Rutter W.J. 1987. Cloning and characterization of a human ribosomal protein gene with enhanced expression in fetal and neoplastic cells. *Nucleic Acids Res.* **11:** 8919–8934.

Perou C.M., Jeffrey S.G., van der Rijn M., Rees C.A., Eisen M.B., Ross D.T., Pergamenschikov A., Williams C.F., Zhu S.X., Lee J.C., Lashkari D., Shalon D., Brown P.O., and Botstein D. 1999. Distinctive gene expression patterns in human mammary epithelial cells and breast cancers. *Proc. Natl. Acad. Sci.* **96:** 9212–9217.

Pogue-Geile K., Geiser J.R., Shu M., Miller C., Wool I.G., Meisler A.I., and Pipas J.M. 1991. Ribosomal protein genes are overexpressed in colorectal cancer: Isolation of a cDNA clone encoding the human S3 ribosomal protein. *Mol. Cell. Biol.* **11:** 3842–3849.

Polunovsky V.A., Rosenwald I.B., Tan A.T., White J., Chiang L., Sonenberg N., and Bitterman P.B. 1996. Translational control of programmed cell death: Eukaryotic translation initiation factor 4E blocks apoptosis in growth-factor-restricted fibroblasts with physiologically expressed or deregulated Myc. *Mol. Cell. Biol.* **16:** 6573–6581.

Polymenis M. and Schmidt E.V. 1997. Coupling of cell division to cell growth by translational control of the G1 cyclin CLN3 in yeast. *Genes Dev.* **19:** 2522–2531.

Pyronnet S., Pradayrol L., and Sonenberg N. 2000. A cell cycle-dependent internal ribosome entry site. *Mol. Cell* **5:** 607–616.

Raught B., Gingras A.-C., James A., Medina D., Sonenberg N., and Rosen J.M. 1996. Expression of a translationally regulated, dominant negative CCAAT/enhancer-binding protein β isoform and up-regulation of the eukaryotic translation initiation factor 2α are correlated with neoplastic transformation of mammary epithelial cells. *Cancer Res.* **56:** 4382–4386.

Rhoads R.E. 1999. Signal transduction pathways that regulate eukaryotic protein synthesis. *J. Biol. Chem.* **274:** 30337–30340.

Rinker-Schaeffer C.W., Austin V., Zimmer S., and Rhoads R.E. 1992. *ras* transformation of cloned rat embryo fibroblasts results in increased rates of protein synthesis and phosphorylation of eucaryotic initiation factor 4E. *J. Biol. Chem.* **267:** 10659–10664.

Rosenwald I.B., Lazaris-Karatzas A., Sonenberg N., and Schmidt E.V. 1993a. Elevated levels of cyclin D1 protein in response to increased expression of eukaryotic initiation factor 4E. *Mol. Cell. Biol.* **13:** 7358–7363.

Rosenwald I.B., Rhoads D.B., Callanan L.D., Isselbacher K.J., and Schmidt E.V. 1993b. Increased expression of eukaryotic translation initiation factors eIF-4E and eIF2α in response to growth induction by c-*myc*. *Proc. Natl. Acad. Sci.* **90:** 6175–6178.

Rousseau D., Gingras A.-C., Pause A., and Sonenberg N. 1996a. The eIF4E-binding proteins 1 and 2 are negative regulators of cell growth. *Oncogene* **13:** 2415–2420.

Rousseau D., Kaspar R., Rosenwald I., Gehrke L., and Sonenberg N. 1996b. Translation initiation of ornithine decarboxylase and nucleocytoplasmic transport of cyclin D1 mRNA are increased in cells overexpressing eukaryotic initiation factor 4E. *Proc. Natl. Acad. Sci.* **93:** 1065–1070.

Shantz L.M. and Pegg A.E. 1994. Overproduction of ornithine decarboxylase caused by relief of translational repression is associated with neoplastic transformation. *Cancer Res.* **54:** 3213–3216.

Shimada A., Shiota G., Miyata H., Kamahora T., Kawasaki H., Shiraki K., Hino S., and Terada T. 1998. Aberrant expression of double-stranded RNA-dependent protein kinase in hepatocytes of chronic hepatitis and differentiated hepatocellular carcinoma. *Cancer Res.* **58:** 4434–4438.

Sonenberg N. 1996. mRNA 5′ cap-binding protein eIF4E and control of cell growth. In *Translational control* (ed. J.W.B. Hershey et al.), pp. 245–269. Cold Spring Harbor Laboratory Press, Cold Spring Harbor, New York.

Song H.J., Kim T.H., Cho C.K., Yoo S.Y., Park K.S., and Lee Y.S. 1998. Increased expression of ornithine decarboxylase by gamma-ray in mouse epidermal cells: Relationship with protein kinase C signaling pathway. *J. Radiat. Res.* **39:** 175–184.

Stein I., Itin A., Einat P., Skaliter R., Grossman Z., and Keshet E. 1998. Translation of vascular endothelial growth factor mRNA by internal ribosome entry: Implication for translation under hypoxia. *Mol. Cell. Biol.* **18:** 3112–3119.

Stoneley M., Paulin F.E.M., Le Quesne J.P.C., Chappell S.A., and Willis A.E. 1997. C-Myc 5′ untranslated region contains an internal ribosome entry segment. *Oncogene* **16:** 423–428.

Tan A., Bitterman P., Sonenberg N., Peterson M., and Polunovsky V. 2000. Inhibition of Myc-dependent apoptosis by eukaryotic translation initiation factor 4E requires cyclin D1. *Oncogene* **19:** 1437–1447.

Vaarala M.H., Porvari K.S., Kyllonen A.P., Mustonen M.V., Lukkarinen O., and Vihko P.T. 1998. Several genes encoding ribosomal proteins are over-expressed in prostate-cancer cell lines: Confirmation of L7a and L37 over-expression in prostate-cancer tissue samples. *Int. J. Cancer* **78:** 27–32.

Vagner S., Gensac M.-C., Maret A., Bayard F., Amalric F., Prats H., and Prats A.-C. 1995. Alternative translation of human fibroblast growth factor 2 mRNA occurs by internal entry of ribosomes. *Mol. Cell. Biol.* **15:** 35–44.

Vanhaesebroeck B. and Alessi D.R. 2000. The PI3K-PDK1 connection: More than just a road to PKB. *Biochem. J.* **346:** 561–576.

Wang S., Rosenwald I.B., Hutzler M.J., Pihan G.A., Savas L., Chen J.-J., and Woda B.A. 1999. Expression of the eukaryotic translation initiation factors 4E and 2α in non-Hodgkin's lymphomas. *Am. J. Pathol.* **155:** 247–255.

Yang Y.L., Reis L.F., Pavlovic J., Aguzzi A., Schafer R., Kumar A., Williams B.R., Aguet M., and Weissmann C. 1995. Deficient signaling in mice devoid of double-stranded RNA-dependent protein kinase. *EMBO J.* **14:** 6095–6106.

Zamanian-Daryoush M., Der S.D., and Williams B.R. 1999. Cell cycle regulation of the double stranded RNA activated protein kinase, PKR. *Oncogene* **18:** 315–326.

21
Translational Control of Ferritin Synthesis

Tracey A. Rouault
Cell Biology and Metabolism Branch
National Institute of Child Health and Human Disease
Bethesda, Maryland 20892

Joe B. Harford
National Cancer Institute
Bethesda, Maryland 20892

REGULATION OF IRON HOMEOSTASIS

Iron is an important nutrient in mammalian cells, and the processes of uptake and sequestration of iron are highly regulated to ensure that supplies are adequate and toxicity is controlled. A major mediator of iron uptake is the transferrin receptor, a plasma membrane protein that binds serum diferric transferrin at the cell surface. Upon transferrin binding, the transferrin receptor– diferric transferrin complex is internalized in endosomes; within endosomes, iron is released from transferrin and transported into the cytosol and other compartments of the cell such as the mitochondria. When iron levels in the cell are high, toxic products generated by the reaction of iron with oxygen species can oxidize and damage cellular proteins, lipids, and nucleic acids (Halliwell and Gutteridge 1992). To protect against iron toxicity, cells synthesize ferritin, a 24-subunit spherical protein that sequesters as many as 4,500 iron atoms within precipitates in its interior (Theil 1987). Although ferritin serves a useful purpose in cells with high cytosolic iron levels, excess production of ferritin can be deleterious in iron-depleted cells because it may compete with other proteins for the limited amounts of available iron (Picard et al. 1998). Translational regulation of ferritin expression is the mechanism utilized by mammalian cells to ensure that ferritin is expressed primarily when it is needed for sequestration of excess iron (Fig. 1). Binding of cytosolic iron-sensing proteins known as iron regula-

© 2000 Cold Spring Harbor Laboratory Press 0-87969-568-4/00 $5 +. 00

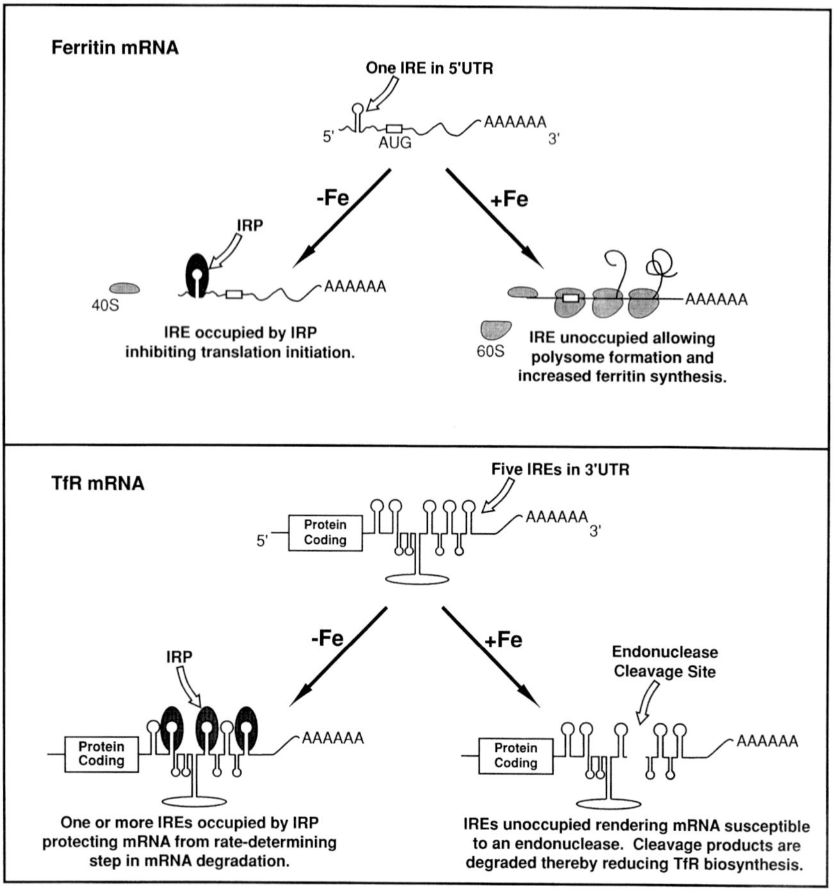

Figure 1 Coordinate regulation of the expression of ferritin and the transferrin receptor. The 5′ untranslated region of ferritin mRNA contains a single iron-responsive element (IRE) (*upper panel*). The 3′ untranslated region of the transferrin receptor (TfR) mRNA contains five similar elements (*lower panel*). The consensus structure of the IRE stem-loop is shown in the inset of the upper panel. When iron is scarce, cytoplasmic iron regulatory proteins (IRPs) interact with the IREs of both mRNAs. There are two related IRPs termed IRP1 and IRP2 in mammalian cells (see text for details). In the case of the ferritin mRNA, the IRE–IRP interaction represses the translation of the mRNA. In the case of the TfR mRNA, the interaction of one or more IRPs with the IREs appears to protect the mRNA from cleavage by an endonuclease. Since the cleavage products are rapidly degraded, the endonucleolytic scission of the transcript appears to be the rate-determining step in TfR mRNA decay. Thus, comparable interactions between RNA elements and RNA-binding proteins mediate the coordinate but opposite regulation of ferritin and TfR levels in response to changes in cellular iron levels.

tory proteins (IRPs) to a conserved RNA stem-loop element in the ferritin transcript interferes with ferritin translation in iron-depleted cells (Klausner et al. 1993; Hentze and Kuhn 1996). Translation remains repressed until cytosolic iron levels increase and the iron-sensing regulatory proteins are released from the transcript. In this chapter, we focus primarily on the mechanisms by which these regulatory proteins sense iron levels and accordingly modify their affinity for the RNA stem-loop element transcripts known as iron-responsive elements (IREs). Because this subject area has been extensively reviewed in the past (Munro 1990; Klausner et al. 1993; Hentze and Kuhn 1996; Aisen et al. 1999), emphasis is placed on reviewing more recent work on the IRE structure and the mechanisms of iron-sensing utilized by each IRP. The molecular basis for translational repression is not reviewed in detail here, because the effect of IRP binding on translational initiation has been discussed previously (Rouault et al. 1996; Chapter 7). In addition, a recently described human disease that results from mutations in the ferritin L-chain IRE is discussed, and future research directions are considered.

IRON-RESPONSIVE ELEMENTS

Ferritin was one of the first proteins recognized as being translationally regulated. Years before direct measurements of ferritin transcript levels were possible, it was known that rates of ferritin biosynthesis could change dramatically without changes in estimated mRNA levels (Zahringer et al. 1976). When the two distinct ferritin gene products known as ferritin H and L chains were cloned and expressed, iron-dependent translation of each transcript was observed (Aziz and Munro 1986). Although the sequences of ferritin H- and L-chain transcripts are quite dissimilar, a common regulatory element required for iron-dependent translational regulation was identified within the 5′UTR of both ferritin genes and in several different species. The sequence element, known as an iron-responsive element or IRE, was also capable of conferring iron-dependent translational regulation upon unrelated reporter genes when placed in the 5′UTR of reporter gene transcripts (Hentze et al. 1987; Munro 1990). Phylogenetic comparisons, mutational analyses, and comparisons among functional IREs in different transcripts were used to identify functionally important features. IREs are stem-loop structures that consist of A-form RNA helical stem regions separated by an unpaired cytosine 5 base pairs removed from a 6-nucleotide loop. The sequence of the loop is almost always CAGUGN, where N can be any base but G (for review, see Klausner et al. 1993). Sequences in the upper and lower stems of the IRE may vary as long as base-pairing is maintained (Fig. 2) (Leibold et al. 1990; Bettany et al. 1992).

IRE sequences have been found near the 5′ end of several transcripts including ferritin, the erythrocyte-specific form of the heme biosynthetic enzyme 5-aminolevulinate synthase (Bhasker et al. 1993; Melefors et al. 1993), mitochondrial aconitase (Gray et al. 1996; Kim et al. 1996), succinate dehydrogenase b of *Drosophila melanogaster* (Kohler et al. 1995; Gray et al. 1996), and more recently in an intestinal iron exporter known as ferriportin (Donovan et al. 2000) or IREG1 (McKie et al. 2000). In these examples, the IRE sequence confers the capacity for iron-dependent translational regulation upon its transcript. IREs are also found in the 3′UTR of several genes. In the mRNA that encodes the transferrin receptor, a regulatory region that contains five IREs and an endonucleolytic cleavage site determines how rapidly the mRNA is degraded (for summa-

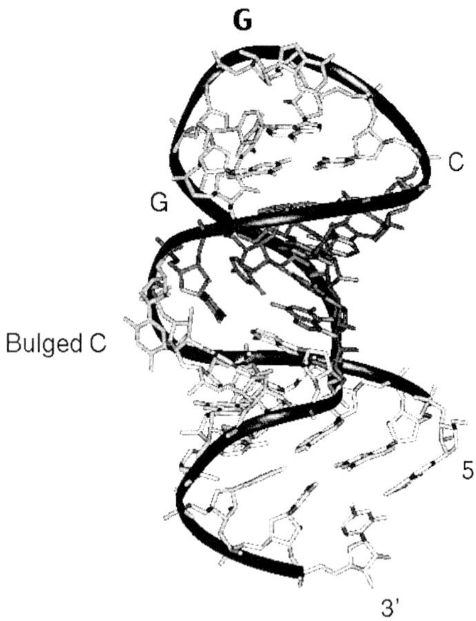

Figure 2 Structure of the IRE. The IRE is composed of A-form helical stem regions, an internal bulge, and a Watson-Crick C-G base pair between the first and fifth nucleotides of the loop. There is stacking between the second position of the loop and the G in position 5. Several residues, including the conserved C in the internal bulge and residues 3, 4, and 6 of the loop, are dynamic in this structure (for further details, see text). (Reprinted in modified form, by permission of Academic Press, from Addess et al. 1997.)

ry, see Binder et al. 1994). A single IRE is found in the 3′UTR of a plasma membrane iron transporter known as the divalent metal transporter (DMT1) (Andrews et al. 1999) or as DCT1 (Gunshin et al. 1997) or Nramp2 (Fleming et al. 1997). The IRE in DMT1 differs from other IREs in that the upper stem contains an unpaired residue (Gunshin et al. 1997). An alternatively spliced form of DMT1 contains the IRE, and it remains unclear how this IRE affects DMT1 expression. Interestingly, levels of the IRE containing a form of DMT1 are increased in patients with genetic hemochromatosis (Fleming et al. 1999).

IRON REGULATORY PROTEINS

IREs are binding sites for cytosolic IRPs (for review, see Rouault and Klausner 1997). There are two known functional IRPs in mammalian cells, and these genes belong to a larger gene family that includes IRP1 and IRP2, bacterial aconitases, and mammalian mitochondrial aconitases. The sequence of IRP1 is approximately 60% identical to *Escherichia coli* aconitase, and IRP1 is a fully functional aconitase in the cytosol of iron-replete mammalian cells (Beinert et al. 1996; Gruer et al. 1997). IRP1 was initially purified and cloned on the basis of its ability to bind to IRE sequences and shift the electrophoretic mobility of radiolabeled IRE-containing RNA in a nondenaturing gel (Rouault et al. 1990; Hirling et al. 1992; Patino and Walden 1992; Yu et al. 1992). Cloning revealed that the amino acid sequence of IRP1 was close to 30% identical to that of porcine mitochondrial aconitase (Rouault et al. 1991). Because of the observed sequence similarity, insights obtained from biochemical (Beinert and Kennedy 1989; Beinert 1990) and structural analyses of mitochondrial aconitase (Robbins and Stout 1989; Lauble et al. 1992) have been used extensively to guide studies of IRP1 function.

Subsequent studies including cloning, overexpression, and biochemical characterizations have shown that IRP1 is a bifunctional protein which exists in two distinct physical forms distinguished by the presence or absence of an iron–sulfur cluster (for review, see Klausner and Rouault 1993). The form of IRP1 that binds RNA is an apoprotein that lacks an iron–sulfur cluster, whereas the form of IRP1 that contains an iron–sulfur cluster has aconitase function and interconverts citrate and isocitrate in the cytosol. In mitochondria, interconversion of citrate and isocitrate is known to be a critical reaction in the citric acid cycle, whereas it is not yet clear why another aconitase is present in cytosol. Iron–sulfur clusters are prosthetic groups most commonly found in proteins that participate in oxidation–reduction reactions (Beinert et al. 1997). Many iron–sulfur proteins facilitate electron transfer reactions in the mitochondrial respiratory chain, but IRP1 is the only known cytosolic iron–sulfur protein in mammalian cells. In iron–replete cells, a cubane iron–sulfur cluster composed of four iron atoms and four sulfur atoms is ligated to cysteines of IRP1 (Kennedy et al. 1992). Iron–sulfur clusters that are accessible to solvent and oxidants are notoriously unstable, and although no formal turnover studies have been done, it is likely that the iron–sulfur cluster of IRP1 is recycled much more rapidly than the protein itself (Tang et al. 1992). The overall conversion of cellular IRP1 from the RNA-binding form to the active enzyme and vice versa requires several hours from the time when the iron status of cells in tissue culture is changed by adding iron salts or iron chelators to the culture medium. It is likely that this period of time is required either for the iron salts or chelators to change available iron pools, or for cluster disassembly and reassembly to take place within the population of preexisting IRP1 molecules (Beinert and Kiley 1996; Rouault and Klausner 1996). Interestingly, sensing of iron levels by IRP2 also requires several hours, even though sensing is accomplished through an entirely different mechanism (see below).

The process of iron–sulfur cluster reassembly is facilitated by a group of enzymes initially described in bacteria that provide sulfur and iron in the appropriate oxidation forms for cluster synthesis (Zheng et al. 1998). One such protein found in mammalian cells is homologous to a bacterial enzyme known as NifS, a cysteine desulfurase that provides inorganic sulfur for cluster synthesis (Land and Rouault 1998). When cells are iron-replete, there is sufficient iron to rebuild the cluster, whereas when they are depleted of iron, it is likely that iron released to the cytosol by disassembly of the iron–sulfur cluster is incorporated into other iron proteins. An increase in oxidative stress can increase the amount of IRP1 that is in the RNA-binding form (Hanson and Leibold 1999), and treatment of cells with hydrogen peroxide has the same effect, although cluster disassembly after hydrogen peroxide exposure may require activation of a signal transduction pathway (Pantopoulos and Hentze 1998). Nitric oxide can also cause disassembly of iron–sulfur clusters and attendant changes in IRE binding activity (Pantopoulos and Hentze 1995; Kennedy et al. 1997). Phosphorylation of IRP1 has been demonstrated, and the phosphorylation state of the protein may affect synthesis or stability of the cluster (Schalinske et al. 1997b; Brown et al. 1998). As a result of the relative accessibility of the iron in its iron–sulfur cluster, a possible role of IRP1 may be to function as a short-term reservoir of iron. Iron released from turnover of the iron–sulfur cluster could be utilized in synthesis of other iron proteins, leaving insufficient iron for resynthesis of the iron–sulfur cluster. The accumulated apoprotein could then implement a program of regulation in which iron uptake is enhanced through the transferrin receptor and iron sequestration by ferritin is repressed.

A second iron regulatory protein known as IRP2 is 57% identical to IRP1 and also binds IREs in the cytosol of mammalian cells. Like IRP1, IRP2 regulates the expression of iron proteins, and it appears to have arisen as a result of a gene duplication event. An important aspect of gene duplication is that it allows functional divergence of members of a gene family such that increased specialization is possible (Nadeau and Sankoff 1997). IRP2 differs from IRP1 in several important respects. Unlike IRP1, IRP2 does not appear to utilize an iron–sulfur cluster for iron sensing. IRP2 lacks aconitase function, and although it has retained the cysteines required for cluster ligation within its primary sequence, there is no evidence that an iron–sulfur cluster is incorporated. Instead, IRP2 is degraded in iron-replete cells, and the protein is stable only in cells that are iron-depleted (Guo et al. 1994; Samaniego et al. 1994). Deletional analysis has demonstrated that an IRP2-specific peptide sequence known as the iron-dependent degradation domain is responsible for the iron-dependent turnover of IRP2. When this domain is ligated into the corresponding position of IRP1, these 73 amino acids can mediate iron-dependent degradation of the IRP1/IRP2 chimeric protein (Iwai et al. 1995). Degradation of IRP2 is mediated by proteasomes (Guo et al. 1995), and proteasomal inhibitors have been used to characterize forms of IRP2 that are marked for degradation. These studies have revealed that like many other proteins destined for proteasomal degradation, IRP2 undergoes ubiquitination in iron-replete cells (Iwai et al. 1998). When ubiquitin multimers are ligated to proteins, the proteins are tethered to a ubiquitin receptor on the 26S proteasome, and they are then unfolded and degraded by the proteasome (Hochstrasser 1996; Hershko and Ciechanover 1998). Another modification that may be important in the degradation pathway has also been identified in IRP2; in the presence of iron, IRP2 undergoes oxidation, revealed by formation of carbonyl groups (Levine et al. 1994), both in vitro and in vivo (Iwai et al. 1998). Carbonyl formation may occur when amino acid side chains in the vicinity of a proposed iron-binding site in the degradation domain are oxidized by reactive hydroxyl groups generated by the bound iron. In vitro ubiquitination assays indicate that an oxidized form of protein may be the preferred ubiquitination target (Iwai et al. 1998). Thus, a likely pathway for iron-dependent degradation of IRP2 involves direct binding of iron by the degradation domain, iron-dependent oxidation of IRP2, ubiquitination of oxidized IRP2, and subsequent proteasomal degradation (Iwai et al. 1998).

MOLECULAR BASIS OF IRP EFFECTS ON TRANSLATION AND mRNA TURNOVER

IRPs appear to function passively by sterically hindering assembly of the translation initiation complex. IRP binding leads to translational regulation of transcripts in which the IRE is found near the cap structure of the mRNA (Goossen and Hentze 1992). Studies have shown that IRP binding does not interfere with binding of cap-binding factors, including eIF4F and eIF4B, to the mRNA. However, IRP binding interferes with the ability of the cap-binding complex to recruit and bind the small ribosomal subunit, most likely because of steric hindrance, and translation is therefore inhibited (Muckenthaler et al. 1998). IRP binding to IREs in the 3′UTR of TfR mRNA is proposed to prolong the half-life of TfR mRNA by sterically interfering with endonuclease access to a vulnerable cleavage site (Binder et al. 1994). The endonuclease responsible for the cleavage event has not yet been characterized. If steric hindrance is the mechanism by which IRPs function, then IRPs may one day be revealed to be important in regulation of other processes that affect mature transcripts and depend on RNA–protein interactions.

FUNCTIONAL DIVERSIFICATION OF IRPs

Even though the mechanisms of iron-sensing differ markedly between IRP1 and IRP2, the two proteins appear to be functionally redundant with respect to regulation of ferritin and transferrin receptor expression (Schalinske et al. 1997a; T. Rouault, unpubl.). Most naturally occurring IREs are equally well bound by each of the two iron regulatory proteins, IRP1 and IRP2. Measurements of binding and Scatchard analysis estimate a K_d of 5–50 pM for binding of the consensus IRE to IRP1 (Haile et al. 1989), and equal K_d values have been measured for binding of the consensus IRE to IRP2 (Butt et al. 1996). In addition, each of the two IRPs functions equally well as a translational repressor in in vitro translation assays (Kim et al. 1995). There are as yet no examples of a tissue in which only one of the two IRPs is expressed. A CHO cell line that does not express IRP1 has been demonstrated to regulate expression of the transferrin receptor and ferritin (Schalinske et al. 1997a). Targeted deletion of IRP1 in mice does not interfere with viability or normal iron metabolism (T. Rouault, unpubl.).

Although endogenous transcripts that are regulated by only one of the two IRPs have not yet been identified, it is quite possible that such sequences will be identified in the future, because artificial IRE-like sequences that are recognized by only one of the two IRPs have been identified (Butt et al. 1996; Henderson et al. 1996). These sequences were generated in SELEX procedures (Tuerk and Gold 1990) in which ligation of random IRE-like sequences to IRPs was followed by selection, amplification, and sequencing. Incorporation of artificial IRE sequences that are bound by only one of the two IRPs into the 5′UTR of reporter transcripts has demonstrated that IRP-specific regulation of such transcripts containing either type of ligand can occur (Menotti et al. 1998). An alternative ligand for IRP1 discovered in SELEX experiments revealed the importance of structure in the loop. When positions 1 and 5 of the loop, which are normally C and G residues, respectively, were replaced by a U at position 1 and an A at position 5, substitutions that allow base-pair formation, the IRE was bound normally by IRP1 although IRP2 did not bind to this ligand (Henderson et al. 1994; Butt et al. 1996). Subsequently, nuclear magnetic resonance (NMR) solution structures have confirmed that a Watson-Crick C-G base pair exists between positions 1 and 5 of the loop, and furthermore, that the adenine at position 2 stacks on position 5, leading to a high degree of structure within the loop (Addess et al. 1997; Gdaniec et al. 1998). Three residues in the loop are dynamic in solution (Fig. 2). SELEX studies reveal that whereas substitutions of other residues at positions 4 and 6 do not interfere with IRE function, the guanosine at position 3 cannot be replaced without loss of high-affinity binding (Butt et al. 1996). In NMR solution structure, this guanosine adopts a somewhat unusual *syn* conformation that may be important in sequence-specific recognition of the RNA by the protein (Addess et al. 1997).

IREs AND IRPs IN HUMAN DISEASE

The importance of the IRE in translational repression of ferritin synthesis has been confirmed by observations made in humans with hereditary hyperferritinemia-cataract syndrome (HHCS), a novel genetic disorder characterized by elevated serum ferritin and early-onset cataract formation (Beaumont et al. 1995; Girelli et al. 1995). Initially, patients with this disorder were identified when hyperferritinemia was detected in patients that were not clinically iron-overloaded. When such patients were phlebotomized, they developed iron deficiency anemia quite readily, implying that hyperferritinemia was present in the absence of iron overload. An autosomal dominant mode of inheritance together with the absence of iron overload in HHCS patients led to analysis of ferritin sequences in HHCS patients and identification of mutations occurring within the IRE of L-ferritin. Sequencing of the ferritin L-chain IRE in unrelated kindreds has revealed that numerous different mutations, most of which occur as single point mutations, interfere with high-affinity binding by IRPs. Unrepressed synthesis of ferritin L-chain leads to significant accumulations of ferritin in serum, thus establishing that ferritin L-chain can be a source of serum ferritin, although the mechanism for ferritin release to the serum is not known. Cataract development is the only clinically significant manifestation of the disease, and the molecular basis for development of these cataracts remains unclear. The clinical severity of HHCS is remarkably varied; serum ferritin levels can range anywhere from two- to tenfold higher than normal, and lens opacities can be severe enough to require surgical intervention, or mild enough to have no detectable effect on visual acuity. The severity of the syndrome correlates strongly with the degree of impairment of IRP binding as measured in RNA gel retardation studies (Allerson et al. 1999). The mutations discovered in the ferritin L-chain IRE are the only naturally occurring mutations known to disrupt IRP recognition.

There is potential that IRP–IRE interactions will prove to be significant in other human diseases. Targeted deletion of IRP2 causes a progressive neurodegenerative disease in mice, and it is possible that IRP2 mutations cause neurodegenerative disease in humans (T. Rouault et al., in prep.).

IRPs AND TOTAL BODY IRON HOMEOSTASIS

An important area of research in the future will be to determine how regulation of iron metabolism is performed in specific tissue compartments of mammals. In mammalian iron homeostasis, there are two sites that are particularly important in regulating total body iron metabolism—the gut and macrophages. Cells of the gut are responsible for adjusting uptake of elemental iron and heme from the intestinal lumen to meet the needs of the organism. Similarly, macrophages are responsible for metabolizing senescent red cells and returning iron obtained from heme metabolism to the circulation. It is likely that IRPs play an important role in adjusting the overall iron metabolism program by sensing iron levels and regulating iron metabolism proteins in these critical cells. In genetic hemochromatosis, IRE-binding activity is inappropriately high in monocytes (Cairo et al. 1997) and in intestinal mucosal cells of affected patients (Pietrangelo et al. 1995). These inappropriate IRP activities are likely responsible for causing gut cells to take up excessive amounts of iron and macrophages to inappropriately release iron under the same circumstances (Bothwell et al. 1979). Increased IRP activity at these sites results in expression of proteins that coordinate an influx of iron into the serum, thereby correcting iron depletion under normal circumstances, or contributing to pathology in genetic hemochromatosis. In a mouse model of hemochromatosis, increases in an IRE-containing mRNA that encodes an iron importer, the divalent metal transporter (DMT1), were detected in intestinal mucosal cells (Fleming et al. 1999). DMT1 is responsible for uptake of ferrous iron across the apical mucosa of intestinal epithelial cells (Fleming et al. 1997; Gunshin et al. 1997), and an increase in its expression would be expected to lead to increased mucosal iron uptake. Recently, an iron exporter expressed at the basolateral membrane of the intestinal mucosa, known as IREG1 (McKie et al. 2000), or ferroportin (Donovan et al. 2000), has been cloned. Interestingly, the transcript has an IRE in its 5′UTR and is therefore likely to be translationally regulated by IRPs, although this would not explain the increase in IREG1 transcript levels observed in patients with genetic hemochromatosis (McKie et al. 2000). Thus, it is likely that IRPs regulate iron metabolism transcripts not only to meet the needs of individual cells, but also to regulate the amount of iron that is transported into compartments that can be reached only by transport across cell membranes. The roles of IRPs and IREs in iron uptake into the serum, the central nervous system, the testes, and placenta are likely to be important future research areas. The effect of targeted deletions of IRP1 or IRP2 on iron homeostasis in mice will reveal much about this complex regulatory system. The recent discoveries of IREs in an intestinal iron transport protein (Gunshin et al. 1997) as well as in a candidate intestinal iron exporter (Donovan et al. 2000; McKie et al. 2000) underscore the pivotal role of IRPs in the regulation of mammalian iron metabolism.

REFERENCES

Addess K.J., Basilion J.P., Klausner R.D., Rouault T.A., and Pardi A. J. 1997. Structure and dynamics of the iron responsive element RNA: Implications for binding of the RNA by iron regulatory proteins. *J. Mol. Biol.* **274:** 72–83.

Aisen P., Wessling-Resnick M., and Leibold E.A. 1999. Iron metabolism. *Curr. Opin. Chem. Biol.* **3:** 200–206.

Allerson C.R., Cazzola M., and Rouault T.A. 1999. Clinical severity and thermodynamic effects of iron-responsive element mutations in hereditary hyperferritinemia-cataract syndrome. *J. Biol. Chem.* **274:** 26439–26447.

Andrews N.C., Fleming M.D., and Gunshin H. 1999. Iron transport across biologic membranes. *Nutr. Rev.* **57:** 114–123.

Aziz N. and Munro H.N. 1986. Both subunits of rat liver ferritin are regulated at a translational level by iron induction. *Nucleic Acids Res.* **14:** 915–927.

Beaumont C., Leneuve P., Devaux I., Scoazec J.Y., Berthier M., Loiseau M.N., Grandchamp B., and Bonneau D. 1995. Mutation in the iron responsive element of the

L ferritin mRNA in a family with dominant hyperferritinaemia and cataract. *Nat. Genet.* **11:** 444–446.

Beinert H. 1990. Recent developments in the field of iron–sulfur proteins. *FASEB J.* **4:** 2483–2491.

Beinert H. and Kennedy M.C. 1989. 19th Sir Hans Krebs lecture. Engineering of protein bound iron–sulfur clusters. A tool for the study of protein and cluster chemistry and mechanism of iron–sulfur enzymes. *Eur. J. Biochem.* **186:** 5–15.

Beinert H. and Kiley P. 1996. Redox control of gene expression involving iron–sulfur proteins. Change of oxidation-state or assembly/disassembly of Fe-S clusters? *FEBS Lett.* **382:** 218-221.

Beinert H., Holm R.H., and Munck E. 1997. Iron–sulfur clusters: Nature's modular, multipurpose structures. *Science* **277:** 653–659.

Beinert H., Kennedy M.C., and Stout D.C. 1996. Aconitase as iron–sulfur protein, enzyme, and iron-regulatory protein. *Chem. Rev.* **96:** 2335–2373.

Bettany A.J., Eisenstein R.S., and Munro H.N. 1992. Mutagenesis of the iron-regulatory element further defines a role for RNA secondary structure in the regulation of ferritin and transferrin receptor expression. *J. Biol. Chem.* **267:** 16531–16537.

Bhasker C.R., Burgiel G., Neupert B., Emery-Goodman A., Kuhn L.C., and May B.K. 1993. The putative iron-responsive element in the human erythroid 5-aminolevulinate synthase mRNA mediates translational control. *J. Biol. Chem.* **268:** 12699–12705.

Binder R., Horowitz J.A., Basilion J.P., Koeller D.M., Klausner R.D., and Harford J.B. 1994. Evidence that the pathway of transferrin receptor mRNA degradation involves an endonucleolytic cleavage within the 3'UTR and does not involve poly(A) tail shortening. *EMBO J.* **13:** 1969–1980.

Bothwell T.H., Charlton R.W., Cook J.D., and Finch C.A. 1979. *Iron metabolism in man.* Blackwell Scientific, Oxford, United Kingdom.

Brown N.M., Anderson S.A., Steffen D.W., Carpenter T.B., Kennedy M.C., Walden W.E., and Eisenstein R.S. 1998. Novel role of phosphorylation in Fe-S cluster stability revealed by phosphomimetic mutations at Ser-138 of iron regulatory protein 1. *Proc. Natl. Acad. Sci.* **95:** 15235–15240.

Butt J., Kim H.Y., Basilion J.P., Cohen S., Iwai K., Philpott C.C., Altschul S., Klausner R.D., and Rouault T.A. 1996. Differences in the RNA binding sites of iron regulatory proteins and potential target diversity. *Proc. Natl. Acad. Sci.* **93:** 4345–4349.

Cairo G., Recalcati S., Montosi G., Castrusini E., Conte D., and Pietrangelo A. 1997. Inappropriately high iron regulatory protein activity in monocytes of patients with genetic hemochromatosis. *Blood* **89:** 2546–2553.

Donovan A., Brownlie A., Zhou Y., Shepard J., Pratt S.J., Moynihan J., Paw B.H., Drejer A., Barut B., Zapata A., Law T.C., Brugnara C., Lux S.E., Pinkus G.S., Pinkus J.L., Kingsley P.D., Palis J., Fleming M.D., Andrews N.C., and Zon L.I. 2000. Positional cloning of zebrafish ferroportin 1 identifies a conserved vertebrate iron exporter. *Nature* **403:** 776–781.

Fleming M.D., Trenor C.C., III, Su M.A., Foernzler D., Beier D.R., Dietrich W.F., and Andrews N.C. 1997. Microcytic anaemia mice have a mutation in Nramp2, a candidate iron transporter gene. *Nat. Genet.* **16:** 383–386.

Fleming R.E., Migas M.C., Zhou X., Jiang J., Britton R.S., Brunt E.M., Tomatsu S., Waheed A., Bacon B.R., and Sly W.S. 1999. Mechanism of increased iron absorption in murine model of hereditary hemochromatosis: Increased duodenal expression of the iron transporter DMT1. *Proc. Natl. Acad. Sci.* **96:** 3143–3148.

Gdaniec Z., Sierzputowska-Gracz H., and Theil E.C. 1998. Iron regulatory element and internal loop/bulge structure for ferritin mRNA studied by cobalt (III) hexammine binding, molecular modeling, and NMR spectroscopy (erratum in *Biochemistry* [1999] **38:** 5676). *Biochemistry* **37:** 1505–1512.

Girelli D., Corrocher R., Bisceglia L., Olivieri O., De Franceschi L., Zelante L., and Gasparini P. 1995. Molecular basis for the recently described hereditary hyperferritinemia-cataract syndrome: A mutation in the iron-responsive element of ferritin L-subunit gene. *Blood* **86:** 4050–4053.

Goossen B. and Hentze M.W. 1992. Position is the critical determinant for function of iron-responsive elements as translational regulators. *Mol. Cell. Biol.* **12:** 1959–1966.

Gray N.K., Pantopoulos K., Dandekar T., Ackrell B.A.C., and Hentze M.W. 1996. Translational regulation of mammalian and *Drosophila* citric-acid cycle enzymes via iron-responsive elements. *Proc. Natl. Acad. Sci.* **93:** 4925–4930.

Gruer M.J., Artymiuk P.J., and Guest J.R. 1997. The aconitase family: Three structural variations on a common theme. *Trends Biochem. Sci.* **22:** 3–6.

Gunshin H., Mackenzie B., Berger U.V., Gunshin Y., Romero M.F., Boron W.F., Nussberger S., Gollan J.L., and Hediger M.A. 1997. Cloning and characterization of a mammalian proton-coupled metal-ion transporter. *Nature* **388:** 482–488.

Guo B., Yu Y., and Leibold E.A. 1994. Iron regulates cytoplasmic levels of a novel iron-responsive element-binding protein without aconitase activity. *J. Biol. Chem.* **269:** 24252–24260.

Guo B., Phillips J.D., Yu Y., and Leibold E.A. 1995. Iron regulates the intracellular degradation of iron regulatory protein 2 by the proteasome. *J. Biol. Chem.* **270:** 21645–21651.

Haile D.J., Hentze M.W., Rouault T.A., Harford J.B., and Klausner R.D. 1989. Regulation of interaction of the iron-responsive element binding protein with iron-responsive RNA elements. *Mol. Cell. Biol.* **9:** 5055–5061.

Halliwell B. and Gutteridge J.M. 1992. Biologically relevant metal ion-dependent hydroxyl radical generation. An update. *FEBS Lett.* **307:** 108–112.

Hanson E.S. and Leibold E.A. 1999. Regulation of the iron regulatory proteins by reactive nitrogen and oxygen species. *Gene Expr.* **7:** 367–376.

Henderson B.R., Menotti E., and Kuhn L.C. 1996. Iron regulatory proteins 1 and 2 bind distinct sets of RNA target sequences. *J. Biol. Chem.* **271:** 4900–4908.

Henderson B.R., Menotti E., Bonnard C., and Kuhn L.C. 1994. Optimal sequence and structure of iron-responsive elements. Selection of RNA stem-loops with high-affinity for iron regulatory factor. *J. Biol. Chem.* **269:** 17481–17489.

Hentze M.W. and Kuhn L.C. 1996. Molecular control of vertebrate iron metabolism: mRNA-based regulatory circuits operated by iron, nitric oxide, and oxidative stress. *Proc. Natl. Acad. Sci.* **93:** 8175–8182.

Hentze M.W., Caughman S.W., Rouault T.A., Barriocanal J.G., Dancis A., Harford J.B., and Klausner R.D. 1987. Identification of the iron-responsive element for the translational regulation of human ferritin mRNA. *Science* **238:** 1570–1573.

Hershko A. and Ciechanover A. 1998. The ubiquitin system. *Annu. Rev. Biochem.* **67:** 425–479.

Hirling H., Emery-Goodman A., Thompson N., Neupert B., Seiser C., and Kuhn L.C. 1992. Expression of active iron regulatory factor from a full-length human cDNA by in vitro transcription/translation. *Nucleic Acids Res.* **20:** 33–39.

Hochstrasser M. 1996. Ubiquitin-dependent protein degradation. *Annu. Rev. Genet.* **30:** 405–440.

Iwai K., Klausner R.D., and Rouault T.A. 1995. Requirements for iron-regulated degradation of the RNA binding protein, iron regulatory protein 2. *EMBO J.* **14:** 5350–5357.

Iwai K., Drake S.K., Wehr N.B., Weissman A.M., LaVaute T., Minato N., Klausner R.D., Levine R.L., and Rouault T.A. 1998. Iron-dependent oxidation, ubiquitination, and degradation of iron regulatory protein 2: Implications for degradation of oxidized proteins. *Proc. Natl. Acad. Sci.* **95:** 4924–4928.

Kennedy M.C., Antholine W.E., and Beinert H. 1997. An EPR investigation of the products of the reaction of cytosolic and mitochondrial aconitases with nitric oxide. *J. Biol. Chem.* **272:** 20340–20347.

Kennedy M.C., Mende-Mueller L., Blondin G.A., and Beinert H. 1992. Purification and characterization of cytosolic aconitase from beef liver and its relationship to the iron-responsive element binding protein (IRE-BP). *Proc. Natl. Acad. Sci.* **89:** 11730–11734.

Kim H.Y., Klausner R.D., and Rouault T.A. 1995. Translational repressor activity is equivalent and is quantitatively predicted by in vitro RNA binding for two iron-responsive element binding proteins, IRP1 and IRP2. *J. Biol. Chem.* **270:** 4983–4986.

Kim H.Y., LaVaute T., Iwai K., Klausner R.D., and Rouault T.A. 1996. Identification of a conserved and functional iron-responsive element in the 5'UTR of mammalian mitochondrial aconitase. *J. Biol. Chem.* **271:** 24226–24230.

Klausner R.D. and Rouault T.A. 1993. A double life: Cytosolic aconitase as a regulatory RNA binding protein. *Mol. Biol. Cell* **4:** 1–5.

Klausner R.D., Rouault T.A., and Harford J.B. 1993. Regulating the fate of mRNA: The control of cellular iron metabolism. *Cell* **72:** 19–28.

Kohler S.A., Henderson B.R., and Kuhn L.C. 1995. Succinate dehydrogenase b mRNA of *Drosophila melanogaster* has a functional iron-responsive element in its 5'-untranslated region. *J. Biol. Chem.* **270:** 30781–30786.

Land T. and Rouault T.A. 1998. Targeting of a human iron–sulfur cluster assembly enzyme, nifs, to different subcellular compartments is regulated through alternative AUG utilization. *Mol. Cell* **2:** 807–815.

Lauble H., Kennedy M.C., Beinert H., and Stout C.D. 1992. Crystal structures of aconitase with isocitrate and nitroisocitrate bound. *Biochemistry* **31:** 2735–2748.

Leibold E.A., Laudano A., and Yu Y. 1990. Structural requirements of iron-responsive elements for binding of the protein involved in both transferrin receptor and ferritin mRNA post-transcriptional regulation. *Nucleic Acids Res.* **18:** 1819–1824.

Levine R.L., Williams J.A., Stadtman E.R., and Shacter E. 1994. Carbonyl assays for determination of oxidatively modified proteins. *Methods Enzymol.* **233:** 346–357.

McKie A.T., Marciani P., Rolfs A., Brennan K., Wehr K., Barrow D., Miret S., Bomford A., Peters T.J., Farzaneh F., Hediger M.A., Hentze M.W., and Simpson R.J. 2000. A novel duodenal iron-regulated transporter, IREG1, implicated in the basolateral transfer of iron to the circulation. *Mol. Cell* **5:** 299–309.

Melefors O., Goossen B., Johansson H.E., Stripecke R., Gray N.K., and Hentze M.W. 1993. Translational control of 5-aminolevulinate synthase mRNA by iron-responsive elements in erythroid cells. *J. Biol. Chem.* **268:** 5974–5978.

Menotti E., Henderson B.R., and Kuhn L.C. 1998. Translational regulation of mRNAs with distinct IRE sequences by iron regulatory proteins 1 and 2. *J. Biol. Chem.* **273:** 1821–1824.

Muckenthaler M., Gray N.K., and Hentze M.W. 1998. IRP-1 binding to ferritin mRNA prevents the recruitment of the small ribosomal subunit by the cap-binding complex eIF4F. *Mol. Cell* **2:** 383–388.

Munro H.N. 1990. Iron regulation of ferritin gene expression. *J. Cell. Biochem.* **44:** 107–115.

Nadeau J.H. and Sankoff D. 1997. Comparable rates of gene loss and functional divergence after genome duplications early in vertebrate evolution. *Genetics* **147:** 1259–1266.

Pantopoulos K. and Hentze M.W. 1995. Nitric oxide signaling to iron-regulatory protein: Direct control of ferritin mRNA translation and transferrin receptor mRNA stability in transfected fibroblasts. *Proc. Natl. Acad. Sci.* **92:** 1267–1271.

———. 1998. Activation of iron regulatory protein-1 by oxidative stress in vitro. *Proc. Natl. Acad. Sci.* **95:** 10559–10563.

Patino M.M. and Walden W.E. 1992. Cloning of a functional cDNA for the rabbit ferritin mRNA repressor protein: Demonstration of a tissue specific pattern of expression. *J. Biol. Chem.* **267:** 19011–19016.

Picard V., Epsztejn S., Santambrogio P., Cabantchik Z.I., and Beaumont C. 1998. Role of ferritin in the control of the labile iron pool in murine erythroleukemia cells. *J. Biol. Chem.* **273:** 15382–15386.

Pietrangelo A., Casalgrandi G., Quaglino D., Gualdi R., Conte D., Milani S., Montosi G., Cesarini L., Ventura E., and Cairo G. 1995. Duodenal ferritin synthesis in genetic hemochromatosis (erratum in *Gastroenterology* [1995] **108:** 1963). *Gastroenterology*. **108:** 208–217.

Robbins A.H. and Stout C.D. 1989. The structure of aconitase. *Proteins* **5:** 289–312.

Rouault T.A. and Klausner R.D. 1996. Iron–sulfur clusters as biosensors of oxidants and iron. *Trends Biochem. Sci.* **21:** 174–177.

———. 1997. Regulation of iron metabolism in eukaryotes. *Curr. Top. Cell. Regul.* **35:** 1–19.

Rouault T.A., Klausner R.D., and Harford J.B. 1996. Translational control of ferritin. In *Translational control* (ed. J.W.B. Hershey et al.), pp. 335-362. Cold Spring Harbor Laboratory Press, Cold Spring Harbor, New York.

Rouault T.A., Stout C.D., Kaptain S., Harford J.B., and Klausner R.D. 1991. Structural relationship between an iron-regulated RNA-binding protein (IRE-BP) and aconitase: Functional implications. *Cell* **64:** 881–883.

Rouault T.A., Tang C.K., Kaptain S., Burgess W.H., Haile D.J., Samaniego F., McBride O.W., Harford J.B., and Klausner R.D. 1990. Cloning of the cDNA encoding an RNA regulatory protein — The human iron-responsive element-binding protein. *Proc. Natl. Acad. Sci.* **87:** 7958–7962.

Samaniego F., Chin J., Iwai K., Rouault T.A., and Klausner R.D. 1994. Molecular characterization of a second iron responsive element binding protein, iron regulatory protein 2 (IRP2): Structure, function and post-translational regulation. *J. Biol. Chem.* **269:** 30904–30910.

Schalinske K.L., Blemings K.P., Steffen D.W., Chen O.S., and Eisenstein R.S. 1997a. Iron regulatory protein 1 is not required for the modulation of ferritin and transferrin receptor expression by iron in a murine pro-B lymphocyte cell line. *Proc. Natl. Acad. Sci.* **94:** 10681–10686.

Schalinske K.L., Anderson S.A., Tuazon P.T., Chen O.S., Kennedy M.C., and Eisenstein R.S. 1997b. The iron–sulfur cluster of iron regulatory protein 1 modulates the accessibility of RNA binding and phosphorylation sites. *Biochemistry* **36:** 3950–3958.

Tang C.K., Chin J., Harford J.B., Klausner R.D., and Rouault T.A. 1992. Iron regulates the activity of the iron-responsive element binding protein without changing its rate of synthesis or degradation. *J. Biol. Chem.* **267:** 24466–24470.

Theil E.C. 1987. Ferritin: Structure, gene regulation, and cellular function in animals, plants, and microorganisms. *Annu. Rev. Biochem.* **56:** 289–315.

Tuerk C. and Gold L. 1990. Systematic evolution of ligands by exponential enrichment: RNA ligands to bacteriophage T4 DNA polymerase. *Science* **249:** 505–510.
Yu Y., Radisky E., and Leibold E.A. 1992. The iron-responsive element binding protein: Purification, cloning and regulation in rat liver. *J. Biol. Chem.* **267:** 19005–19010.
Zahringer J., Baliga B.S., and Munro H.N. 1976. Novel mechanism for translational control in regulation of ferritin synthesis by iron. *Proc. Natl. Acad. Sci.* **73:** 857–861.
Zheng L., Cash V.L., Flint D.H., and Dean D.R. 1998. Assembly of iron–sulfur clusters. Identification of an iscSUA-hscBA-fdx gene cluster from *Azotobacter vinelandii*. *J. Biol. Chem.* **273:** 13264–13272.

22

Translational Control of TOP mRNAs

Oded Meyuhas and Eran Hornstein

Department of Biochemistry
The Hebrew University-Hadassah Medical School
Jerusalem 91120, Israel

Fundamental biological processes like cell proliferation, differentiation, development, and malignant transformation in eukaryotes involve changing requirements for the translational machinery. Observations made in recent years have disclosed that synthesis of many components of the translational apparatus is regulated at the translational level (Table 1). These include ribosomal proteins (rps) (Meyuhas et al. 1996); elongation factors 1A and 2; poly(A)-binding protein (PABP), which has been implicated in both translation initiation and ribosome assembly (see Sachs Chapter 10); P40, a protein that was isolated from rat 40S ribosomal subunit (Tohgo et al. 1994), whose yeast counterpart has been implicated in processing of the 20S rRNA precursor and maturation of the 40S ribosomal subunit (Ford et al. 1999); B23, a major phosphoprotein in the interphasic nucleolus (Zatsepina et al. 1999), which is a putative ribosome assembly factor; hnRNP A1, a protein that functions in nuclear splicing and export of mRNAs (Dreyfuss et al. 1993; Izaurralde et al. 1997); as well as a P23 that has been shown to be associated with microtubules, but whose function has not yet been identified (Gachet et al. 1999). It seems that resources and energy are wasted by resting cells for maintaining a large excess of inefficiently translated mRNAs, but this actually enables them to respond rapidly to mitogenic stimulations by having immediately available a large protein synthesis capacity.

A structural hallmark common to all the mRNAs above is the presence of a 5′ *t*erminal *o*ligo*p*yrimidine tract (5′TOP), and therefore they are termed TOP mRNAs (Table 1). This structural motif comprises the core of the translational *cis*-regulatory element (TLRE) of these mRNAs. 5′TOP also has been identified in QM mRNA, whose translation has so far not been shown to be regulated, yet it encodes a highly conserved

Translational Control of Gene Expression
© 2000 Cold Spring Harbor Laboratory Press 0-87969-568-4/00 $5 + .00

Table 1 Sequences around the transcription start site of translationally regulated non-rp genes

Function of gene product	Sequence (+1 at start of underlined 5′TOP)	5′TOP (nucl.)	C/T ratio	5′UTR (nucl.)	Accession number	Translational control
Ribosome biogensis and assembly						
h PABP	GGGTGCAGGAGGCGGTGCTTCCCCTTCTCCCCGGCGGTTA	12	9/3	505	U68093	yes[a]
m P40	CTTTTTTCGCGCTACCCGGG	8	2/6	79	J02870, X06406	yes[b]
h QM	ATGCGCAGGCGGAGGAGCGCCTCTTTCCCTTCGGTGTGGT	12	6/6	143	U37218, D28410	n.d.
h B23	TAAGCGCGGGAGCCTGCGTCCTTTCCCTGGTGTGATTCCG	8	4/4	100	U89309, D28343	yes[c]
Elongation factors						
h eEF1A	CAGTAGTCGCCGTGAACGTTCTTTTTCGCAACGGGTTTGC	7	2/5	63	J04617	yes[d]
h eEF1B	CTTTTTCCTCTCTTCAGCGT	15	6/9	84	D28350	n.d.
x EF1A	TGGTTAAGGCCCGGTTCGCTCTCTTCCTCACCGGGTCTGC	9	5/4	60	M25697	yes[e]
c.h. eEF2	CGGGCGTTGACGTCAGCGGTCTCTTCCGCCGCAGCCGCCG	7	4/3	83	J03200	yes[f]
Small nucleolar RNA host transcripts						
h *UHG*	TCTCCCCGGGGAGATTCGTTCTCATTTTTCTACTGCTCGT	3	2/1	—	U40580	n.a.
m*UHG*	CACACGTTGTTCTTGTCGGCCTCATTTTTCCTTGTTCGGG	3	2/1	—	U40654	n.a.

h *U17HG*	ATGCTTCGCGTTTTCTCTCTCCTTTTTGGAGACAGATTCG	7	2/5	—	D00591	n.a.
m*U17HG*	CTCTCTAGGCGTCGCTCTCT	6	3/3	—	AJ006836	n.a.
m *gas5*	CGACTCAGCCGCCATCTCAGCCTTTCGGAGCTGTGCGGCA	6	3/3	—	X67267	yes[g]
Pre-mRNA splicing and RNA export						
h hnRNP A1	CAGATACGTTCGTCAGCTTGCTCCTTTCTGCCCGTGGACG	9	4/5	63	X12671, X73096	yes[h]
Proteins of unknown function						
m P23	CTTTTTTCCGCCCGCTCCCC	9	3/6	101	X06407	yes[i]
rb P23	TGGGGAGCGGCTGAGTCGGCCTTTTCCGCCCGCTCCCCCC	7	3/4	116	AJ225898	n.d.
h P23	CTTTTCCGCCCGCTCCCCCC	7	4/3	90	D28408, NM_003295	n.d.
Translationally invariant						
h β1-tubulin	GCCCTCCCTCTCCTTTCTCCCTCTCAGAACCTTCCTGCCG	5	3/2	155	J00314	no[j]
h β2-tubulin	GCGGACGGTCGGTTGTAGCACTCTGCGCGCCCGCTCTTCT	4	2/2	73	X02344	no[j]

(h) human; (x) *Xenopus laevis*; (d) *Drosophila melanogaster*; (c.h.) Chinese hamster; (m) mouse; (rb) rabbit; (n.d.) not determined; (n.a.) not applicable.
[a] Hornstein et al. (1999); [b]Yenofsky et al. (1983); [c]Zong et al. (1999); [d]Thomas et al. (1981); Thomas and Thomas (1986); Rao and Slobin (1987); Avni et al. (1994); Jefferies et al. (1994a); [e]Wormington (1989); Loreni et al. (1993); [f]Terada et al. (1994); Avni et al. (1997); [g]Smith and Steitz (1998); [h]Camacho-Vanegas et al. (1997); [i] Yenofsky et al. (1983); [j]Avni et al. (1997).

basic protein that is required for joining the 40S and 60S subunits (Dick et al. 1997; Eisinger et al. 1997; Nguyen et al. 1998). Moreover, 5′TOP motifs have been shown in a few mRNAs whose translation efficiency has been demonstrated to be invariant, or which are even non-protein-coding transcripts (for details see below).

In this review, we focus on the mechanism underlying the translational control of vertebrate TOP mRNAs, with special emphasis on the characterization of the TLRE and the possible *trans*-acting factors. It should be noted, however, that these mRNAs and their unique mode of regulation are not confined to vertebrates, as translationally controlled TOP mRNAs have been demonstrated previously in insects (Meyuhas et al. 1996 and references therein). In contrast to higher eukaryotes, none of the yeast proteins associated with the translational apparatus has been shown to be encoded by a TOP mRNA. Finally, several aspects of the translational control of TOP mRNAs, which are only briefly mentioned here, have been discussed thoroughly elsewhere (Meyuhas et al. 1996; Amaldi and Pierandrei-Amaldi 1997).

GENERAL FEATURES OF THE GROWTH-DEPENDENT TRANSLATIONAL CONTROL OF TOP mRNAS

TOP mRNAs are selectively shifted from polysomes in growing cells into mRNP particles (subpolysomal fraction) in quiescent cells. This behavior has been demonstrated during the transition of cells, in culture and in whole animals, between growing and nongrowing states, and in response to a wide variety of physiological and pharmacological stimuli (Table 2). Collectively, TOP mRNAs display the following translational features: (1) The proportion of TOP mRNAs actively engaged in protein synthesis, i.e., associated with polysomes in growing cells, is significantly lower than that of other housekeeping non-TOP mRNAs. Plausible explanations for the underutilization of TOP mRNAs, even during rapid proliferation, have been discussed previously (Meyuhas et al. 1996). (2) Translation of TOP mRNAs is selectively repressed upon growth arrest at different phases (G_0, S, and M) of the cell cycle (Table 2). (3) Close inspection of the polysome profiles of TOP mRNAs from growing or quiescent cells discloses a distinct bimodal distribution, suggesting two discrete populations of mRNA molecules: translationally active (polysome-associated) and translationally inactive (masked in mRNP particles). Interestingly, the translationally active population, even in nongrowing cells, is fully loaded with ribosomes (Agrawal and Bowman 1987; Meyuhas et al. 1987; Loreni and Amaldi 1992; Shima et al. 1998; Hornstein et al. 1999). Thus,

TOP mRNAs alternate between repressed and active states, and, when in the active state, they are translated at near maximum efficiency (an "all-or-none" phenomenon). This typical apportionment of TOP mRNAs between mRNP and polysomes clearly indicates that the translational repression results from a blockage at the translational initiation step. It should be mentioned, however, that in a few cases TOP mRNAs were shown to shift upon growth arrest from large polysomes into disomes and monosomes, rather than into mRNPs (Jefferies et al. 1994a, b, 1997). The reason for this exceptional behavior of TOP mRNAs is presently unknown. (4) TOP mRNAs are not only sequestered in mRNP in a translatable form (Baum and Wormington 1985; Thomas and Thomas 1986; Hammond et al. 1991), but also display a similar translational activity in vitro as do their counterparts extracted from translationally active polysomes (Shama and Meyuhas 1996). (5) The selective increase in the proportion of TOP mRNAs associated with polysomes, observed upon mitogenic stimulations, seems to reflect recruitment of mRNA, which is sequestered in mRNP. Two lines of circumstantial evidence support this notion. First, the appearance of TOP mRNAs in the polysomal fractions coincides with their disappearance from the subpolysomal fractions without a parallel alteration in the abundance or rate of accumulation of the respective mRNAs (Geyer et al. 1982; Meyuhas et al. 1987; Loreni and Amaldi 1992). Second, growth stimulation of resting Swiss 3T3 cells by serum induced the synthesis of several proteins encoded by TOP mRNAs even in the presence of actinomycin D (Thomas et al. 1981; Thomas and Thomas 1986; Jefferies et al. 1994a). This latter argument, however, should be cautiously referred to, as several inhibitors of transcription, including actinomycin D, have recently been shown to induce recruitment of TOP mRNAs into polysomes (F. Loreni et al., unpubl.).

TRANSLATIONAL CONTROL OF TOP mRNAS DURING DEVELOPMENT AND DIFFERENTIATION

The appearance of newly synthesized TOP mRNAs during embryogenesis of *Xenopus laevis* seems to occur at gastrulation and precedes the onset of synthesis of the corresponding proteins, which occurs in tailbud embryos. This selective translational repression is evidenced by the fact that mRNAs encoding rps and eEF1A initially appear in mRNP particles (stage 15) and start to be mobilized into polysomes around stage 26 (Pierandrei-Amaldi et al. 1982; Baum and Wormington 1985; Loreni et al. 1993). Likewise, accumulation of PABP mRNA precedes the synthesis of the respective protein (Zelus et al. 1989). It is worth noting, how-

Table 2 Relationships between growth status and translational efficiency of TOP mRNAs

TOP mRNAs	Cell line/tissue	Stimulation	Cell cycle phase	Translational efficiency	Reference
rps	*Xenopus* embryo	early development	growing	low	Pierandrei-Amaldi et al. (1982)
rps	mouse MM24DZ myoblasts	insulin	growing	high	Hammond and Bowman (1988)
rps	mouse MC-9 mast cells	IL-3	growing	high	Zhang et al. 1998)
various	rat liver	regeneration	growing	high	Aloni et al. (1992); Hornstein et al. (1999)
rps	bovine primary T-lymphocyte	concanavalin a	growing	high	Kaspar et al. (1992)
eEF1A	human primary T cells	PDBu+ ionomycin	growing	high	Terada et al. (1995)
rps, eEF2	mouse P1798 lymphosarcoma	dexamethasone	resting (G_0)	low	Meyuhas et al. (1987, 1990); Avni et al. (1997)
rps	mouse MM24DZ myoblasts	differentiation	resting (G_0)	low	Agrawal and Bowman (1987)
various	mouse erythroleukemia cells	differentiation	resting (G_0)	low	Yenofsky et al. 1983; Avni et al. (1994)
rps	human HL-60 promyelocytic	differentiation	resting (G_0)	low	Krichevsky et al. (1999)

rps, eEF1A	*Xenopus* B3.2 kidney cells	serum starvation	resting (G_0)	low	Loreni et al. (1993)
rps	mouse 3T6 fibroblasts	serum starvation	resting (G_0)	low	Geyer et al. (1982)
rps	mouse Swiss 3T3 fibroblasts	serum starvation	resting (G_0)	low	Kaspar et al. (1990)
eEF1A	mouse embryo fibroblasts	serum starvation	resting (G_0)	low	Shima et al. (1998)
eEF1A	mouse ES cells	serum starvation	resting (G_0)	low	Kawasome et al. (1998)
hnRNP A1	human HeLa cells	serum starvation	resting (G_0)	low	Camacho-Vanegas et al. (1997)
various	human HEK 293 cells	serum starvation	resting (G_0)	low	Jefferies et al. (1997)
rps	mouse NIH-3T3 fibroblasts	contact inhibition	resting (G_0)	low	Shama et al. (1995)
eEF1A	mouse erythroleukemia cells	confluence	resting (G_0)	low	Rao and Slobin (1987)
various	mouse NIH-3T3 fibroblasts	aphidicolin	resting (S)	low	Avni et al. (1994)
various	human HEK 293 cells	aphidicolin	resting (S)	low	Avni et al. (1997)
rps	mouse NIH-3T3 fibroblasts	hydroxyurea	resting (S)	low	Shama et al. (1995)
various	Chinese hamster ovary cells	hydroxyurea	resting (S)	low	Avni et al. (1997)
various	human HeLa cells	nocodazole	resting (M)	low	Avni et al. (1994)

ever, that this is a unique physiological condition, in which translational repression of TOP mRNAs occurs in rapidly proliferating cells.

Translational repression of TOP mRNAs has been observed upon induced differentiation of various mammalian cell lines (Table 2). Possibly, the apparent unloading of TOP mRNAs from polysomes in terminally differentiating cells reflects, as in most other cases, the cessation of cell proliferation, rather than differential gene expression associated with the new phenotype.

THE TRANSLATIONAL *CIS*-REGULATORY ELEMENT OF TOP mRNAS

Comparative analysis of sequences in the vicinity of the cap site of numerous vertebrate TOP mRNAs indicates the following structural features: (1) The hallmark of members of this family is a C residue at the cap site, followed by an uninterrupted stretch of 4–14 pyrimidines (Table 1) (Meyuhas et al. 1996); (2) the composition of the 5′TOP, although varying among TOP mRNAs even within a species, generally maintains a similar proportion of C and U residues; and finally (3) these mRNAs possess 5′-untranslated regions (5′UTRs), which vary in length from 21 to 505 nucleotides, but are devoid of upstream AUGs (Table 1) (Meyuhas et al. 1996). It should be emphasized that initiation of transcription at a C residue is rare among eukaryotic genes, which normally start at an A residue (Bucher 1990), whereas the percentage of mammalian transcripts with a C residue at the cap site is only 17% (Schibler et al. 1977).

Initially it was shown that the 35-nucleotide-long 5′UTR of frog rpS19 mRNA is sufficient to confer translational repression on a reporter mRNA in a developmentally dependent manner (Mariottini and Amaldi 1990). Subsequently, it has been shown that the first 27–34 nucleotides of mammalian mRNAs encoding for rpS16, rpL32, rpL13a, eEF2, and PABP are sufficient to confer the typical behavior of TOP mRNAs on human growth hormone mRNA (Levy et al. 1991; Avni et al. 1994, 1997; Hornstein et al. 1999).

The large majority of vertebrate TOP mRNAs for which complete sequence information is available have a 5′TOP of eight or more residues (Table 1) (Meyuhas et al. 1996). However, translational control of rat rpP2 mRNA, which contains a 5-pyrimidine-long 5′TOP, is indistinguishable from that of rp mRNAs with a 5′TOP of eight or more residues (Aloni et al. 1992; Avni et al. 1994).

The activity of the mammalian TLRE appears to be position-dependent. Thus, it fails to exert its effect when located internally, even when preceded by a single A residue (Avni et al. 1994). Replacement of five out of the eight pyrimidines by purines, or just the C residue at the cap site by an A residue, completely abolished the translational regulatory properties of the TLRE of mouse rpS16 mRNA (Levy et al. 1991). The first 28 nucleotides of rat β-actin mRNA bears remarkable sequence similarity with the respective sequence of mouse rpS16 mRNA, yet its translation is refractory to growth arrest. Gain-of-function experiments showed that two purine-to-pyrimidine substitutions at the cap site and at position +5 of rat β-actin mRNA generated a 7-nucleotide-long 5′TOP. Consequently, the resulting transcript became translationally controlled in a growth-dependent manner (Biberman and Meyuhas 1997). Collectively these results indicate the critical role played by the 5′TOP in the translational control mechanism.

Heterologous transfection experiments revealed that the similar structural properties of the TLREs in mammalian and amphibian rp mRNAs reflect a complete evolutionary conservation of all the elements required for the growth-dependent translational control mechanism in these two distantly related vertebrate classes (Avni et al. 1994).

Cell-type-specific Translational Control of TOP mRNAs

Studies of the translational behavior of eEF2 mRNA in cells of lymphatic origin established this mRNA as translationally regulated in a growth-dependent manner (Terada et al. 1994, 1995). However, unlike the translation of mRNAs encoding rps and eEF1A, which are ubiquitously regulated, the translation of eEF2 seems to be confined to hematopoietic cells (Avni et al. 1997). Similarly, PABP mRNA, although translationally regulated in some cell types (Hornstein et al. 1999), displays resistance to growth arrest in HeLa cells and is constitutively repressed in lymphosarcoma cells (D. Avni et al., unpubl.). Transfection experiments with chimeric mRNAs demonstrated that the first 29 and 34 nucleotides of the mRNAs encoding eEF2 and PABP, respectively, can confer growth-dependent translational control on a reporter mRNA, even in cells where the endogenous mRNAs are refractory to this mode of regulation (Avni et al. 1997 and unpubl.). These results suggest that downstream sequences in some TOP mRNAs can override the regulatory features of the 5′TOP in a cell-type-specific manner. This notion is further supported by the fact that the first 10 nucleotides of rpS16 mRNA are sufficient to confer on a heterologous mRNA translational control in lymphosarcoma cells, but not in fibroblasts. The latter cells require, in addition, a CG motif spanning positions 12–15 of rpS16 mRNA (Avni et al. 1997; Biberman and Meyuhas 1997).

mRNAs Containing a 5′ TOP Are Not Necessarily Subject to Translational Control, and Vice Versa

Human mRNAs encoding β1-tubulin and β2-tubulin contain all the hallmarks of a typical TOP mRNA (Table 1), yet their translation is refractory to growth arrest of any of the examined cell lines. Transfection experiments indicate that the first 53 nucleotides of β1-tubulin mRNAs suffice for conferring growth-dependent translational control on a reporter mRNA in any of the examined cell lines (Avni et al. 1997).

Whatever the mechanism involved in the lack of translational regulation of the endogenous β1- or β2-tubulin, our observations clearly indicate that the presence of a 5′TOP, per se, cannot be used as an ultimate diagnostic tool to seek out mRNAs regulated at the translational level. Similarly, some mRNAs, like the 6-kb mRNA encoding insulin-like growth factor II (IGF-II leader 3 mRNA) and mitochondrial rpS12 mRNA, are translationally repressed upon growth arrest, yet they lack a 5′TOP motif (Raizis et al. 1993; Nielsen et al. 1995; Mariottini et al. 1999).

MANY TOP TRANSCRIPTS CONTAIN SMALL NUCLEOLAR RNAs WITHIN THEIR INTRONS

Small nucleolar RNAs (snoRNAs), most of which are encoded within introns of protein-encoding genes (Maxwell and Fournier 1995; Lafontaine and Tollervey 1998), are involved in maturation of eukaryotic rRNA. Interestingly, a large proportion of vertebrate transcripts, which include snoRNAs within their introns, start with a 5′TOP motif (Pelczar and Filipowicz 1998; Smith and Steitz 1998). In addition, three of the bona fide TOP transcripts (Table 1), which contain within their introns snoRNAs, do not appear to specify a protein product. Nevertheless, the spliced forms of two of these transcripts, *UHG* and *gas5* RNAs, have been shown to be associated with monosomes (Tycowski et al. 1996; Pelczar and Filipowicz 1998; Smith and Steitz 1998). Perhaps association of noncoding exonic sequences with monosomes enables their rapid cytoplasmic degradation by a nonsense-mediated decay mechanism (see also Chapter 28).

One plausible explanation for the relatively high proportion of snoRNAs within TOP transcripts is that the 5′TOP sequences mediate the coordinate expression not only of the various protein components and factors involved in ribosome biogenesis, but also of the RNA counterparts. Perhaps the evolutionary conservation of 5′TOP sequences also among non-protein-coding transcripts implies that this motif plays an additional regulatory role within the nucleus as well. Thus, the 5′TOP sequence might render the respective transcripts accessible for specialized splicing factor(s), enabling the retrieval of the snoRNAs from within the intronic sequences, which are otherwise doomed to complete hydrolysis. Disruption of the 5′TOP motif in a snoRNA-host gene or insertion of a snoRNA sequence into an intron of a chimeric TOP gene might enable us to verify the possible role played by the 5′TOP motif in intronic sequence-preserving splicing.

THE TRANSLATION OF RP mRNAS CONTINUES IN VIRUS-INFECTED CELLS

Infection of cultured cells by poliovirus is characterized by a marked and rapid decrease of host-cell protein synthesis (for review, see Belsham and Jackson, Chapter 31). However, viral as well as cellular mRNAs, which contain an internal ribosome entry site (IRES) and therefore are subject to cap-independent translation, are efficiently translated in these cells (Johannes and Sarnow 1998). Surprisingly, it has recently been shown that rp mRNAs, even though they lack a bona fide IRES, also escape the host-cell shutoff following poliovirus infection of HE-p2 cells (Cardinali et al. 1999). Likewise, selective resistance to host-cell shutoff has been demonstrated for rp mRNAs in HeLa cells infected with herpes simplex virus type 1 (Greco et al. 1997). It should be noted, however, that persistence of translation has been shown only for rp mRNAs, and therefore, the translational behavior of other TOP mRNAs has yet to be monitored.

It seems unlikely that continuous synthesis of rp mRNA is maintained by the poliovirus to ensure an adequate supply of ribosomes needed for its replication, as the synthesis of the rRNA is inhibited shortly after infection (Rubinstein and Dasgupta 1989). In vitro experiments have previously disclosed that TOP mRNAs are sensitive to the inhibitory effect of a cap analog, to the same extent as are non-TOP mRNAs (Shama et al. 1995). However, we cannot formally exclude the possibility that preferential translation of rp mRNAs reflects a cap-independent mechanism operating in vivo. Moreover, it has been shown that infection by various viruses evokes phosphorylation of ribosomal protein S6 (Dahl et al. 1996; Greco et al. 1997; Cardinali et al. 1999 and references therein), which has been implicated in translational control of TOP mRNAs (for further details, see below and Chapter 23).

POSSIBLE INVOLVEMENT OF GENERAL COMPONENTS OF THE TRANSLATIONAL MACHINERY IN THE TRANSLATIONAL CONTROL OF TOP mRNAS

A simple model to account for the selective growth-dependent translational control of rp mRNAs assumes the participation of a factor that is a component of the general protein synthesis machinery. If such a factor had a particularly low affinity for rp mRNAs due to their unique sequence at the 5′ terminus, a decrease in its activity or amount might lead to disproportionate discrimination against these mRNAs (Lodish 1974; Alton and Lodish 1977). A prime candidate for such a factor is the limiting initiation factor eIF4E (for review, see Chapter 2). However, experiments with an NIH-3T3-derived eIF4E-overexpressing cell line showed that surplus amounts of phosphorylated eIF4E did not prevent the translational repression of TOP mRNAs upon growth arrest (Shama et al. 1995). It appears, therefore, that the decrease in the phosphorylation of eIF4E upon growth arrest does not play a role in the repression of translation of TOP mRNAs.

Currently, the possibility that the translational efficiency of TOP mRNAs is determined by the activity of a single component or a combination of general components of the translational machinery cannot be formally excluded. However, two lines of circumstantial evidence argue against this notion: (1) The majority of TOP mRNAs are totally excluded from polysomes upon growth arrest (Agrawal and Bowman 1987; Meyuhas et al. 1987; Loreni and Amaldi 1992; Hornstein et al. 1999) rather than being gradually shifted into lighter polysomes, as observed upon partial inhibition of translation initiation by pactamycin (Loreni and Amaldi 1997). Hence, the translational repression of TOP mRNAs appears to involve a specific factor that sequesters these mRNAs rather than diminishing competitiveness for a limiting initiation factor. (2) The translation of *Xenopus* TOP mRNAs is repressed, not only upon growth arrest, but also during early embryogenesis under conditions of extremely rapid proliferation and extensive protein synthesis (Amaldi and Pierandrei-Amaldi 1990). It is highly unlikely, therefore, that the selective translational repression of TOP mRNAs in the developing *Xenopus* embryo can be ascribed simply to a temporary deficiency of one or more general translational factors.

THE PHOSPHORYLATION STATE OF rpS6 TIGHTLY CORRELATES WITH THE TRANSLATIONAL EFFICIENCY OF TOP mRNAS

Numerous studies have shown that mitogenic stimulation of quiescent cells induces phosphorylation of rpS6 through activation of at least one of two respective rpS6 kinases, S6K1 (also known as $p70^{S6K}$) and S6K2 (for review, see Chapter 23). The concomitant translational activation of TOP mRNAs under these circumstances led Thomas and his colleagues to propose that rpS6 phosphorylation increases the affinity of ribosomes for TOP mRNAs and thus facilitates their initiation (Thomas and Thomas

1986; Jefferies et al. 1994a,b). This notion is supported by additional correlative evidence: (1) Regenerating rat liver shows a large increase in S6K1 activity (Nemenoff et al. 1988), phosphorylation of rpS6 (Gressner and Wool 1974b), and translational efficiency of TOP mRNAs (Aloni et al. 1992; Hornstein et al. 1999); (2) changes in S6K1 activity following induction of meiotic maturation by progesterone of stage VI *X. laevis* oocytes appear to coincide with increased translational efficiency of rp mRNAs (Hyman and Wormington 1988; Schwab et al. 1999); (3) treatment of eukaryotic cells with a low concentration of cycloheximide, which slows elongation of the nascent peptide, activates S6K1 (Price et al. 1989; Iiboshi et al. 1999) and leads to phosphorylation of rpS6 (Gressner and Wool 1974a), polysomal association of rp mRNAs (Agrawal and Bowman 1987; Pierandrei-Amaldi et al. 1991), and synthesis of rps (Ignotz et al. 1981); (4) viral infections increase both the relative translational efficiency of TOP mRNAs and the phosphorylation of rpS6 (Greco et al. 1997; Cardinali et al. 1999).

Attempts have been made to examine the cause-and-effect relationship between rpS6 phosphorylation and translational control of TOP mRNAs, using the immunosuppressant rapamycin. This compound indirectly blocks the activation of S6K1 and S6K2 as well as rpS6 phosphorylation (Chung et al. 1992; Price et al. 1992), by a mechanism that is not fully understood (see Chapter 23), and concomitantly represses the translation of TOP mRNAs in a selective manner (Terada et al. 1995; Kawasome et al. 1998; Shima et al. 1998). Rapamycin has also been shown to block cell cycle progression and to inhibit the proliferation of a variety of cell types (for review, see Sehgal 1998). However, it has been shown that rapamycin inhibits the proliferation of embryonic stem (ES) cells and represses the translation of eEF1A mRNA in these cells without affecting the phosphorylation status of their rpS6, which is permanently unphosphorylated, due to knockout of the S6K1 gene and most probably due to the absence of S6K2 activity in these cells (Kawasome et al. 1998). Hence, results derived by using this drug should be interpreted cautiously, as its effect on the translation of TOP mRNAs may reflect a response to growth arrest, similar to that observed with other antiproliferative agents (see Table 2).

An alternative approach is based on transfection of cells with expression vectors encoding a mutant version of the kinase, which appears to mistarget upstream signals. Indeed, the translational activation of a chimeric TOP mRNA following mitogenic stimulation was partially, yet selectively, inhibited by an excess of a kinase-dead mutant, $p70^{S6K}$A229, lacking a critical phosphorylatable threonine at position 229 of S6K1 (Jefferies et al. 1997).

It is tempting to assume, therefore, that activation of S6K1 and/or S6 phosphorylation might be determinants in the control of the translational efficiency of TOP mRNAs. Nevertheless, several observations suggest that these alterations are not sufficient to fully account for this mode of regulation: (1) Three temporal peaks of S6K1 activity were observed before stage 13 of early embryogenesis of *X. laevis* (Schwab et al. 1999). However, these waves of activity do not lead to translational activation of TOP mRNAs, whose onset of recruitment into polysomes occurs around stage 26 (Pierandrei-Amaldi et al. 1982; Baum and Wormington 1985; Loreni et al. 1993). (2) If initiation of translation on TOP mRNAs becomes less efficient upon rpS6 dephosphorylation, one should expect a shift of TOP mRNAs into smaller-sized polysomes, rather than the apparent all-or-none phenomenon reported for many of these mRNAs. (3) Rapamycin treatment, like serum starvation, almost eliminates the activity of S6K1 in both NIH-3T3 and 293 cells. However, whereas the latter leads to substantial unloading of endogenous or chimeric TOP mRNAs from polysomes (about a third remained in polysomes), rapamycin leads to only a marginal shift into mRNPs (about two-thirds of these mRNAs remained in polysomes) (Jefferies et al. 1994a, 1997). (4) A mutant possessing a high basal S6K1 activity, $p70^{S6K}$E389, failed to relieve the translational repression of eEF1A mRNA in serum-starved 293 cells (Jefferies et al. 1997). Finally, (5) disruption of the S6K1 gene in mouse ES cells not only did not lead to constitutive repression of the examined TOP mRNA, but this mRNA also remained mostly associated with polysomes even in resting cells (Kawasome et al. 1998). A quite different picture was observed in mouse embryonic fibroblasts derived from S6K1 knockout mice (Shima et al. 1998). In this case, a TOP mRNA displayed a similar growth-dependent translational control in the S6K1-deficient cells to that in wild-type cells. It appears, therefore, that the activity of the missing S6K1 is neither sufficient nor necessary for this mode of regulation. Noteworthy, however, is the fact that S6K1-deficient mouse embryonic fibroblasts still express S6K2 (Shima et al. 1998).

POSSIBLE INVOLVEMENT OF SPECIFIC *TRANS*-ACTING FACTORS IN THE TRANSLATIONAL CONTROL OF TOP mRNAS

One plausible explanation for the all-or-none translational behavior of the rp mRNAs is that the selective packaging of rp mRNAs into RNP particles renders them unavailable to the translational machinery. For the packaging to be selective, the rp mRNAs would have to interact directly or indirectly with a specific repressor. Alternatively, recruitment of these mRNAs into polysomes might require a specific activator. Two lines of evidence suggest that the putative *trans*-acting factor functions as a selective repressor: (1) A high-salt wash of ribonucleoprotein particles from growth-arrested erythroleukemia cells selectively represses in vitro the translation of several mRNAs. One of these was identified as a TOP mRNA encoding eEF1A (Slobin and Rao 1993). (2) Both reticulocyte lysate and wheat germ extract, which derive from nongrowing cells, discriminate against TOP mRNAs (Shama and Meyuhas 1996). Preincubation of both these cell-free translation systems with a short synthetic RNA oligonucleotide containing the first 16 nucleotides of rpS16 mRNA completely relieves the translational repression of chimeric TOP mRNAs. In contrast, this oligonucleotide failed to stimulate the translation of non-TOP mRNAs. Moreover, mutations in the pyrimidine tract, or in the immediate downstream sequence within this oligonucleotide, eliminated its ability to relieve the translational repression. It appears, therefore, that translational repression of TOP mRNAs is achieved in vitro by the accumulation of a titratable repressor rather than the loss of an activator, and that this repressor recognizes multiple TOP mRNAs with a diverse set of 5′TOP motifs (Biberman and Meyuhas 1999).

Establishing the role of the 5′TOP and the circumstantial evidence concerning the involvement of a specific repressor in the translational control of TOP mRNAs has led to the hypothesis that this motif may bind a specific *trans*-acting factor. A cytoplasmic protein of about 56 kD from mouse T lymphocytes, $p56^{L32}$, was shown to specifically bind DNA, or RNA sequences containing the 34-nucleotide-long TLRE of mouse rpL32 mRNA. Moreover, this binding was dependent on the presence of an intact oligopyrimidine tract as well as of the CG-rich region immediately downstream (Kaspar et al. 1992; Severson et al. 1995).

A sequence containing the first 52 nucleotides of *Xenopus* rpL1 mRNA was demonstrated to bind proteins both to the oligopyrimidine tract and to a sequence immediately downstream from this motif (Cardinali et al. 1993). Later studies have established that the *X. laevis* homologs of La autoantigen and cellular nucleic-acid-binding protein (CNBP) bind the 5′TOP and the CG-rich sequence immediately downstream within the 5′UTR of rp mRNAs, respectively (Pellizzoni et al. 1996, 1997). Accordingly, mutations in the pyrimidine stretch, known to disturb the translational control of rp mRNA in vivo or in vitro, impair the binding of both these proteins to their cognate sequences (Pellizzoni et al. 1996, 1998). The binding of La and CNBP requires the assistance of a third protein, Ro60 autoantigen (Pellizzoni et al. 1998). Interestingly, La and CNBP seem to compete for interaction with Ro60, and their binding to the 5′UTR is mutually exclusive. It seems, however, that if indeed CNBP is a component of the *trans*-acting complex, it is not universally essential, as translational control of TOP mRNAs in mouse lymphosarcoma cells requires just the 5′TOP motif (Avni et al. 1994, 1997; Biberman and Meyuhas 1997).

The apparent correlation between the structural requirements for a functional translational *cis*-regulatory element and the binding specificity of mammalian $p56^{L32}$ or amphibian La and CNBP makes these proteins attractive candidates for the translational control of TOP mRNAs. However, no experimental data are currently available to show that any of these proteins individually or in combination can affect the translation of TOP mRNA in vitro or in vivo. Possibly, an additional, yet-unidentified protein(s), which might operate through protein–protein interaction, is required for demonstrating the repression in functional assays. Furthermore, unlike the translational control of TOP mRNAs, the binding activity of $p56^{L32}$ is not regulated in a growth-dependent manner in mouse cells (Kaspar et al. 1992), and that of La and CNBP is not developmentally regulated in *Xenopus* (Cardinali et al. 1993). It should be pointed out, however, that binding activity does not have to directly or inversely correlate with the translation efficiency of the target mRNAs. A protein might constitutively bind to the recognition sequence and serve as a docking site for yet another protein, which functions as a translational modulator. Alternatively, the binding protein might undergo a modification, like phosphorylation or dephosphorylation, that alters its repressor or activator potential in a growth- or development-dependent manner, without affecting its binding activity. An example of this latter possibility is the cAMP response element-binding protein (CREB), which is constitutively bound to its cognate sequence, yet acquires its transcription *trans*-activation potential upon phosphorylation (Karin 1994).

CONCLUDING REMARKS

Significant progress has been made in recent years toward understanding the mechanism of the translational control of vertebrate TOP mRNAs. Nonetheless, two fundamental entities remain ununknown (1) the number and the nature of the *trans*-acting factors involved in the regulatory apparatus and (2) the pathway(s) through which growth signals are transduced to the translational machinery.

Clues concerning putative specific *trans*-acting factors have been obtained from RNA-protein-binding experiments. Nevertheless, the relevance of the various binding proteins has been questioned, due to the inconsistency between the binding activity and the translational behavior of rp mRNAs. Clearly, establishing the role of these proteins in the translational control of TOP mRNA can only be demonstrated unambiguously by a functional assay or by genetic experiments. Likewise, despite circumstantial and experimental evidence supporting the involvement of S6K1 and/or S6K2 in the translational control of TOP mRNAs, it is clear that they cannot fully account for the fluctuations in the translational efficiency of these mRNAs. Hence, verifying the relative contribution of these kinases to the translational control of TOP mRNAs should await the establishment of S6K1/S6K2 double knockout mice. If indeed these kinases are shown to be essential for this mode of regulation, disruption of the phosphorylation sites in rpS6 will still be necessary to elucidate whether the effect of S6K1/S6K2 is exerted solely through phosphorylation of rpS6 or through other, yet-unknown, substrates.

One plausible explanation, which might reconcile most of these contradictory observations, is based on the assumption that the translational control of 5′TOP-containing mRNAs is carried out by an interaction between specific and general factors. According to this model, the translation of these mRNAs is completely repressed through the attachment of a specific protein(s) to the 5′TOP. Translation can initiate, however, upon displacement of this repressor by a general component of the translational machinery. The binding activity of the repressor to its cognate sequence would remain constant regardless of the growth status, yet the general factor would acquire its displacement potential upon growth or development. A prime candidate for the general factor is the small ribosomal subunit with the associated initiation factors, which could detach the repressor during scanning of the 5′UTR. Conceivably, the displacement potential of the 40S subunit could be increased when it undergoes growth-dependent phosphorylation on ribosomal protein S6. The involvement of a putative specific repressor is consistent with the all-or-none phenomenon, and the presumptive engagement of the 40S subunit accords with the correlation between the phosphorylation status of rpS6 and the translation efficiency of TOP mRNAs.

ACKNOWLEDGMENTS

We are indebted to Francesco Amaldi and his colleagues for sharing data prior to their publication. We are grateful to Robert P. Perry for critically reviewing the manuscript and for helpful comments. Research in the authors' laboratory is supported by grants to O. M. from the Israel Science Foundation founded by the Israel Academy of Sciences and

Humanities and from the United States–Israel Binational Science Foundation (BSF-97-00055).

REFERENCES

Agrawal A.G. and Bowman L.H. 1987. Transcriptional and translational regulation of ribosomal protein formation during mouse myoblast differentiation. *J. Biol. Chem.* **262:** 4868–4875.

Aloni R., Peleg D., and Meyuhas O. 1992. Selective translational control and nonspecific posttranscriptional regulation of ribosomal protein gene expression during development and regeneration of rat liver. *Mol. Cell. Biol.* **12:** 2203–2212.

Alton T.H. and Lodish H.F. 1977. Translational control of protein synthesis during early stages of differentiation of the slime mold *Dictyostelium discoideum*. *Cell* **12:** 301–310.

Amaldi F. and Pierandrei-Amaldi P. 1990. Translational regulation of the expression of ribosomal protein genes in *Xenopus laevis*. *Enzyme* **44:** 93–105.

———. 1997. TOP genes: A translationally controlled class of genes including those coding for ribosomal proteins. *Prog. Mol. Subcell. Biol.* **18:** 1–17.

Avni D., Biberman Y., and Meyuhas O. 1997. The 5′ terminal oligopyrimidine tract confers translational control on TOP mRNAs in a cell type- and sequence context-dependent manner. *Nucleic Acids Res.* **25:** 995–1001.

Avni D., Shama S., Loreni F., and Meyuhas O. 1994. Vertebrate mRNAs with a 5′-terminal pyrimidine tract are candidates for translational repression in quiescent cells: Characterization of the translational *cis*-regulatory element. *Mol. Cell. Biol.* **14:** 3822–3833.

Baum E.Z. and Wormington W.M. 1985. Coordinate expression of ribosomal protein genes during *Xenopus* development. *Dev. Biol.* **111:** 488–489.

Biberman Y. and Meyuhas O. 1997. Substitution of just five nucleotides at and around the transcription start site of rat β-actin promoter is sufficient to render the resulting transcript a subject for translational control. *FEBS Lett.* **405:** 333–336.

———. 1999. TOP mRNAs are translationally inhibited by a titratable repressor in both wheat germ extract and reticulocyte lysate. *FEBS Lett.* **456:** 357–360.

Bucher P. 1990. Weight matrix descriptions of four eukaryotic RNA polymerase II promoter elements derived from 502 unrelated promoter sequences. *J. Mol. Biol.* **212:** 563–578.

Camacho-Vanegas O., Weighardt F., Ghigna C., Amaldi F., Riva S., and Biamonti G. 1997. Growth-dependent and growth-independent translation of messengers for heterogeneous nuclear ribonucleoproteins. *Nucleic Acids Res.* **25:** 3950–3954.

Cardinali B., Di Cristiana M., and Pierandrei-Amaldi P. 1993. Interaction of proteins with the mRNA for ribosomal protein L1 in *Xenopus*: Structural characterization of *in vivo* complexes and identification of proteins that bind in vitro to its 5′ UTR. *Nucleic Acids Res.* **21:** 2301–2308.

Cardinali B., Fiore L., Campioni N., De Dominicis A., and Pierandrei-Amaldi P. 1999. Resistance of ribosomal protein mRNA translation to protein synthesis shutoff induced by poliovirus. *J. Virol.* **73:** 7070–7076.

Chung J., Kuo C.J., Crabtree G.R., and Blenis J. 1992. Rapamycin-FKBP specifically blocks growth-dependent activation of and signaling by the 70 kd S6 kinases. *Cell* **69:** 1227–1236.

Dahl J., Freund R., Blenis J., and Benjamin T. 1996. Studies of partially transforming polyomavirus mutants establish a role for phosphatidylinositol 3-kinase in activation of pp70 S6 kinase. *Mol. Cell. Biol.* **16:** 2728–2735.

Dick F., Karamanou S., and Trumpower B. 1997. QSR1, an essential yeast gene with a genetic relationship to a subunit of the mitochondrial cytochrome bc1 complex, codes for a 60S ribosomal subunit protein. *J. Biol. Chem.* **272:** 13372–13379.

Dreyfuss G., Matunis M.J., Pinol-Roma S., and Burd C.G. 1993. hnRNP proteins and the biogenesis of mRNA. *Annu. Rev. Biochem.* **62:** 289–321.

Eisinger D., Dick F., and Trumpower B. 1997. Qsr1p, a 60S ribosomal subunit protein, is required for joining of 40S and 60S subunits. *Mol. Cell. Biol.* **17:** 5136–5145.

Ford C., Randal-Whitis L., and Ellis S. 1999. Yeast proteins related to the p40/laminin receptor precursor are required for 20S ribosomal RNA processing and the maturation of 40S ribosomal subunits. *Cancer Res.* **59:** 704–710.

Gachet Y., Tournier S., Lee M., Lazaris-Karatzas A., Poulton T., and Bommer U. 1999. The growth-related, translationally controlled protein P23 has properties of a tubulin binding protein and associates transiently with microtubules during the cell cycle. *J. Cell Sci.* **112:** 1257–1271.

Geyer P.K., Meyuhas O., Perry R.P., and Johnson L.F. 1982. Regulation of ribosomal protein mRNA content and translation in growth-stimulated mouse fibroblasts. *Mol. Cell. Biol.* **2:** 685–693.

Greco A., Laurent A., and Madjar J. 1997. Repression of beta-actin synthesis and persistence of ribosomal protein synthesis after infection of HeLa cells by herpes simplex virus type 1 infection are under translational control. *Mol. Gen. Genet.* **256:** 320–327.

Gressner A. and Wool I. 1974a. The stimulation of the phosphorylation of ribosomal protein S6 by cycloheximide and puromycin. *Biochem. Biophys. Res. Commun.* **60:** 1482–1490.

———. 1974b. The phosphorylation of liver ribosomal proteins *in vivo*. Evidence that only a single small subunit protein (S6) is phosphorylated. *J. Biol. Chem.* **249:** 6917–6925.

Hammond M.L. and Bowman L.H. 1988. Insulin stimulates the translation of ribosomal proteins and the transcription of rDNA in mouse myoblasts. *J. Biol. Chem.* **263:** 17785–17791.

Hammond M.L., Merrick W., and Bowman L.H. 1991. Sequences mediating the translation of mouse S16 ribosomal protein mRNA during myoblast differentiation and *in vitro* and possible control points for the in vitro translation. *Genes Dev.* **5:** 1723–1736.

Hornstein E., Git A., Braunstein I., Avni D., and Meyuhas O. 1999. The expression of poly (A)-binding protein gene is translationally regulated in a growth dependent fashion through a 5′-terminal oligopyrimidine tract motif. *J. Biol. Chem.* **274:** 1708–1714.

Hyman L.E. and Wormington W.M. 1988. Translational inactivation of ribosomal protein mRNAs during *Xenopus* oocyte maturation. *Genes Dev.* **2:** 598–605.

Ignotz G.G., Hokari S., DePhilip R.M., Tsukada K., and Lieberman I. 1981. Lodish model and regulation of ribosomal protein synthesis by insulin-deficient chick embryo fibroblasts. *Biochemistry* **20:** 2550–2558.

Iiboshi Y., Papst P., Kawasome H., Hosoi H., Abraham R., Houghton P., and Terada N. 1999. Amino acid-dependent control of p70^{s6k}. Involvement of tRNA aminoacylation in the regulation. *J. Biol. Chem.* **274:** 1092–1099.

Izaurralde E., Jarmolowski A., Beisel C., Mattaj I., Dreyfuss G., and Fischer U. 1997. A role for the M9 transport signal of hnRNP A1 in mRNA nuclear export. *J. Cell Biol.* **137:** 27–35.

Jefferies H.B., Thomas G., and Thomas G. 1994a. Elongation factor-1α mRNA is selectively translated following mitogenic stimulation. *J. Biol. Chem.* **269:** 4367–4372.

Jefferies H.B., Reinhard C., Kozma S.C., and Thomas G. 1994b. Rapamycin selectively represses translation of the 'polypyrimidine tract' mRNA family. *Proc. Natl. Acad. Sci.* **91:** 4441–4445.

Jefferies H., Fumagalli S., Dennis P., Reinhard C., Pearson R., and Thomas G. 1997. Rapamycin suppresses 5′TOP mRNA translation through inhibition of p70^{s6k}. *EMBO J.* **16:** 3693–3704.

Johannes G. and Sarnow P. 1998. Cap-independent polysomal association of natural mRNAs encoding c-*myc*, BiP, and eIF4G conferred by internal ribosome entry sites. *RNA* **4:** 1500–1513.

Karin M. 1994. Signal transduction from the cell surface to the nucleus through the phosphorylation of transcription factors. *Curr. Opin. Cell Biol.* **6:** 415–424.

Kaspar R.L., Kakegawa T., Cranston H., Morris D.R., and White M.W. 1992. A regulatory *cis* element and a specific binding factor involved in the mitogenic control of murine ribosomal protein L32 translation. *J. Biol. Chem.* **267:** 508–514.

Kaspar R.L., Rychlik W., White M.W., Rhoads R.E., and Morris D.R. 1990. Simultaneous cytoplasmic redistribution of ribosomal protein L32 mRNA and phosphorylation of eukaryotic initiation factor 4E after mitogenic stimulation of Swiss 3T3 cells. *J. Biol. Chem.* **265:** 3619–3622.

Kawasome H., Papst P., Webb S., Keller G., Johnson G., Gelfand E., and Terada N. 1998. Targeted disruption of p70$^{(s6k)}$ defines its role in protein synthesis and rapamycin sensitivity. *Proc. Natl. Acad. Sci.* **95:** 5033–5038.

Krichevsky A., Metzer E., and Rosen H. 1999. Translational control of specific genes during differentiation of HL-60 cells. *J. Biol. Chem.* **274:** 14295–14305.

Lafontaine D. and Tollervey D. 1998. Birth of the snoRNPs: The evolution of the modification-guide snoRNAs. *Trends Biochem. Sci.* **23:** 383–388.

Levy S., Avni D., Hariharan N., Perry R.P., and Meyuhas O. 1991. Oligopyrimidine tract at the 5′ end of mammalian ribosomal protein mRNAs is required for their translational control. *Proc. Natl. Acad. Sci.* **88:** 3319–3323.

Lodish H.F. 1974. Model for the regulation of mRNA translation applied to haemoglobin synthesis. *Nature* **251:** 385–388.

Loreni F. and Amaldi F. 1992. Translational regulation of ribosomal protein synthesis in *Xenopus* cultured cells: mRNA relocation between polysomes and RNP during nutritional shifts. *Eur. J. Biochem.* **205:** 1027–1032.

———. 1997. Translational control of terminal oligopyrimidine mRNAs requires a specific regulator. *FEBS Lett.* **416:** 239–242.

Loreni F., Francesconi A., and Amaldi F. 1993. Coordinate translational regulation in the synthesis of elongation factor 1α and ribosomal proteins in *Xenopus laevis*. *Nucleic Acids Res.* **21:** 4721–4725.

Mariottini P. and Amaldi F. 1990. The 5′ untranslated region of mRNA for ribosomal protein S19 is involved in its translational regulation during *Xenopus* development. *Mol. Cell. Biol.* **10:** 816–822.

Mariottini P., Shah Z.H., Toivonen J.M., Bagni C., Spelbrink J.N., Amaldi F., and Jacobs H.T. 1999. Expression of the gene for mitoribosomal protein S12 is controlled in human cells at the levels of transcription, RNA splicing, and translation. *J. Biol. Chem.* **274:** 31853–31862.

Maxwell E. and Fournier M. 1995. The small nucleolar RNAs. *Annu. Rev. Biochem.* **64:** 897–893.

Meyuhas O., Avni D., and Shama S. 1996. Translational control of ribosomal protein mRNAs in eukaryotes. In *Translational control* (ed. J.W.B. Hershey et al.), pp. 363–384. Cold Spring Harbor Laboratory Press, Cold Spring Harbor, New York.

Meyuhas O., Thompson A.E., and Perry R.P. 1987. Glucocorticoids selectively inhibit the translation of ribosomal protein mRNAs in P1798. lymphosarcoma cells. *Mol. Cell. Biol.* **7:** 2691–2699.

Meyuhas O., Baldin V., Bouche G., and Amalric F. 1990. Glucocorticoids repress ribosome biosynthesis in lymphosarcoma cells by affecting gene expression at the level of transcription, posttranscription and translation. *Biochim. Biophys. Acta* **1049:** 38–44.

Nemenoff R., Price D., Mendelsohn M., Carter E., and Avruch J. 1988. An S6 kinase activated during liver regeneration is related to the insulin-stimulated S6 kinase in H4 hepatoma cells. *J. Biol. Chem.* **263:** 19455–19460.

Nguyen Y., Mills A., and Stanbridge E. 1998. Assembly of the QM protein onto the 60S ribosomal subunit occurs in the cytoplasm. *J. Cell. Biochem.* **68:** 281–285.

Nielsen F., Ostergaard L., Nielsen J., and Christiansen J. 1995. Growth-dependent translation of IGF-II mRNA by a rapamycin-sensitive pathway. *Nature* **377:** 358–362.

Pelczar P. and Filipowicz W. 1998. The host gene for intronic U17 small nucleolar RNAs in mammals has no protein-coding potential and is a member of the 5′-terminal oligopyrimidine gene family. *Mol. Cell. Biol.* **18:** 4509–4518.

Pellizzoni L., Lotti F., Maras B., and Pierandrei-Amaldi P. 1997. Cellular nucleic acid binding protein binds a conserved region of the 5′ UTR of *Xenopus laevis* ribosomal protein mRNAs. *J. Mol. Biol.* **267:** 264–275.

Pellizzoni L., Lotti F., Rutjes S., and Pierandrei-Amaldi P. 1998. Involvement of the *Xenopus laevis* Ro60 autoantigen in the alternative interaction of La and CNBP proteins with the 5′UTR of L4 ribosomal protein mRNA. *J. Mol. Biol.* **281:** 593–608.

Pellizzoni L., Cardinali B., Lin-Marq N., Mercanti D., and Pierandrei-Amaldi P. 1996. A *Xenopus laevis* homologue of the La autoantigen binds the pyrimidine tract of the 5′UTR of ribosomal protein mRNAs in vitro: Implication of a protein factor in complex formation. *J. Mol. Biol.* **259:** 904–915.

Pierandrei-Amaldi P., Campioni N., and Cardinali B. 1991. Experimental changes in the amount of maternally stored ribosomes affect the translation efficiency of ribosomal protein mRNA in *Xenopus* embryo. *Cell Mol. Biol.* **37:** 227–238.

Pierandrei-Amaldi P., Campioni N., Beccari E., Bozzoni I., and Amaldi F. 1982. Expression of ribosomal protein genes in *Xenopus laevis* development. *Cell* **30:** 163–171.

Price D., Nemenoff R., and Avruch J. 1989. Purification of a hepatic S6 kinase from cycloheximide-treated rats. *J. Biol. Chem.* **26:** 13825–13833.

Price D.J., Grove J.R., Calvo V., Avruch J., and Bierer B.E. 1992. Rapamycin-induced inhibition of the 70-kilodalton protein kinase. *Science* **257:** 973–977.

Raizis A., Eccles M., and Reeve A. 1993. Structural analysis of the human insulin-like growth factor-II P3 promoter. *Biochem. J.* **289:** 133–139.

Rao T.R. and Slobin L.I. 1987. Regulation of the utilization of mRNA for eucaryotic elongation factor Tu in Friend erythroleukemia cells. *Mol. Cell. Biol.* **7:** 687–697.

Rubinstein S. and Dasgupta A. 1989. Inhibition of rRNA synthesis by poliovirus: Specific inactivation of transcription factors. *J. Virol.* **63:** 4689–4696.

Schibler U., Kelley D.E., and Perry R.P. 1977. Comparison of methylated sequences in messenger RNA and heterogeneous nuclear RNA from mouse L cells. *J. Mol. Biol.* **115:** 695–714.

Schwab M., Kim S., Terada N., Edfjäll C., Kozma S., Thomas G., and Maller J. 1999.

p70^{S6K} controls selective mRNA translation during oocyte maturation and early embryogenesis in *Xenopus laevis*. *Mol. Cell. Biol.* **19:** 2485–2494.

Sehgal S.N. 1998. Rapamune (RAPA, rapamycin, sirolimus): Mechanism of action immunosuppressive effect results from blockade of signal transduction and inhibition of cell cycle progression. *Clin. Biochem.* **31:** 335–340.

Severson W.E., Mascolo P.L., and White M.W. 1995. Lymphocyte p56^{L32} is a RNA/DNA-binding protein which interacts with conserved elements of the murine L32 ribosomal protein mRNA. *Eur. J. Biochem.* **229:** 426–432.

Shama S. and Meyuhas O. 1996. The translational *cis*-regulatory element of mammalian ribosomal protein mRNAs is recognized by the plant translational apparatus. *Eur. J. Biochem.* **236:** 383–388.

Shama S., Avni D., Frederickson R.M., Sonenberg N., and Meyuhas O. 1995. Overexpression of initiation factor eIF-4E does not relieve the translational repression of ribosomal protein mRNAs in quiescent cells. *Gene Expr.* **4:** 241–252.

Shima H., Pende M., Chen Y., Fumagalli S., Thomas G., and Kozma S. 1998. Disruption of the p70^{s6k}/p85^{s6k} gene reveals a small mouse phenotype and a new functional S6 kinase. *EMBO J.* **17:** 6649–6659.

Slobin L.I. and Rao M.N. 1993. Translational repression of EF-1α mRNA *in vitro*. *Eur. J. Biochem.* **213:** 919–926.

Smith C. and Steitz J. 1998. Classification of *gas5* as a multi-small-nucleolar-RNA (snoRNA) host gene and a member of the 5′-terminal oligopyrimidine gene family reveals common features of snoRNA host genes. *Mol. Cell. Biol.* **18:** 6897–6909.

Terada N., Takase K., Papst P., Nairn A.C., and Gelfand E.W. 1995. Rapamycin inhibits ribosomal protein synthesis and induces G1 prolongation in mitogen-activated T lymphocytes. *J. Immunol.* **155:** 3418–3426.

Terada N., Patel H.R., Takase K., Kohno K., Nairn A.C., and Gelfand E.W. 1994. Rapamycin selectively inhibits translation of mRNAs encoding elongation factors and ribosomal proteins. *Proc. Natl. Acad. Sci.* **91:** 11477–11481.

Thomas G. and Thomas G. 1986. Translational control of mRNA expression during the early mitogenic response in Swiss mouse 3T3 cells: Identification of specific proteins. *J. Cell Biol.* **103:** 2137–2144.

Thomas G., Thomas G., and Luther H. 1981. Transcriptional and translational control of cytoplasmic proteins after serum stimulation of quiescent Swiss 3T3 cells. *Proc. Natl. Acad. Sci.* **78:** 5712–5716.

Tohgo A., Takasawa S., Munakata H., Yonekura H., Hayashi N., and Okamoto H. 1994. Structural determination and characterization of a 40 kDa protein isolated from rat 40S ribosomal subunit. *FEBS Lett.* **340:** 133–138.

Tycowski K.T., Shu M.D., and Steitz J.A. 1996. A mammalian gene with introns instead of exons generating stable RNA products. *Nature* **379:** 464–466.

Wormington W.M. 1989. Developmental expression and 5S rRNA-binding activity of *Xenopus laevis* ribosomal protein L5. *Mol. Cell. Biol.* **9:** 5281–5288.

Yenofsky R., Careghini S., Krowczynska A., and Brawerman G. 1983. Regulation of mRNA utilization in mouse erythroleukemia cells induced to differentiate by exposure to dimethyl sulfoxide. *Mol. Cell. Biol.* **3:** 1197–1203.

Zatsepina O., Rousselet A., Chan P., Olson M., Jordan E., and Bornens M. 1999. The nucleolar phosphoprotein B23 redistributes in part to the spindle poles during mitosis. *J. Cell Sci.* **112:** 455–466.

Zelus B.D., Giebelhaus D.H., Eib D.W., Kenner K.A., and Moon R.T. 1989. Expression of the poly(A)-binding protein during development of *Xenopus laevis*. *Mol. Cell. Biol.* **9:** 2756–2760.

Zhang C., Nechushtan H., Jacob-Hirsch Y., Avni D., Meyuhas O., and Razin E. 1998. Growth-dependent and PKC-mediated translational control of the upstream stimulating factor 2 (USF2) mRNA in hematopoietic cell. *Oncogene* **16:** 763–769.

Zong Q., Schummer M., Hood L., and Morris D. 1999. Messenger RNA translation state: The second dimension of high-throughput expression screening. *Proc. Natl. Acad. Sci.* **96:** 10632–10636.

23

S6 Phosphorylation and Signal Transduction

Stefano Fumagalli and George Thomas

Friedrich Miescher Institute
Basel, Switzerland

The process of growth plays a fundamental role in the life cycle of a cell, serving two distinct purposes depending on the cell type and the developmental stage of the organism (Thomas and Hall 1997; Su and O'Farrell 1998; Conlon and Raff 1999). Two scenarios of growth are readily distinguished. In the first, in response to mitogens, cells enter the cell cycle, grow during G_1 phase to a critical mass permissive for the transition to S phase, then continue to increase in size during G_2 before initiating mitosis and giving rise to two daughter cells (Pardee 1989). In the second, in response to an agonist, terminally differentiated cells increase in size to accommodate a specific demand without subsequent cell division (Laser et al. 1998). In addition, besides playing an important role in the normal life cycle of an organism, the process of cell growth is subverted in several pathological conditions, as in hyperplasia and cancer or diseases where cells grow abnormally (hypertrophy) or where they lose their capacity to grow (atrophy). An understanding at the molecular level of the processes that control cell growth is a fundamental biological issue and will be important for identifying targets for therapies aimed at fighting specific disease states. An obligatory requirement for cell growth is the activation of protein synthesis (Pardee 1989). The major products being generated by the protein synthetic apparatus are components of the translational machinery, notably ribosomal proteins. Indeed, it has been calculated that this process may consume as much as 80% of a growing cell's energy (Schmidt 1999). Recent studies have shown that ribosomal protein production in metazoans is largely controlled at the translational level (see Meyuhas and Hornstein, Chapter 22), in part through the activation of the S6 Ser/Thr kinase (Jefferies et al. 1997), initially termed $p70^{s6k}$ and

© 2000 Cold Spring Harbor Laboratory Press 0-87969-568-4/00 $5 + .00

more recently S6K1 (Shima et al. 1998), which phosphorylates 40S ribosomal protein S6. These findings suggest that S6K1 plays a pivotal role in the process of cell growth, possibly through increased S6 phosphorylation. Recent years have seen an increase in our understanding of the mechanisms involved in S6K1 activation, and genetic approaches have been useful in revealing the in vivo role of S6K1 in cell growth (Shima et al. 1998; Montagne et al. 1999). It is now important to test whether the effects of S6K1 are mediated through phosphorylation of S6 and/or through other potential substrates. The purpose of this review is to summarize the state of our knowledge of the S6K1 signaling pathway and to discuss potential strategies that will further increase our understanding of the role of this pathway in downstream signaling and cell growth.

RIBOSOMAL PROTEIN S6

S6 is one of 33 40S ribosomal proteins, which, with a single strand of 18S rRNA, is present in one copy per 40S ribosome (Wool et al. 1996). Earlier studies had shown that it is assembled into the 40S subunit at an early stage in ribosome biogenesis (Todorov et al. 1983). Its importance in the maturation of 40S subunits is demonstrated by the fact that in the livers of mice where the S6 gene has been conditionally deleted, ribosome biogenesis is abolished with an accumulation of a 34S rRNA precursor and a block in 18S rRNA processing (Volarevic et al. 2000). As determined by cross-linking and protection studies, S6 is located near the mRNA/tRNA binding site, at the interface between the small and the large subunit where it has been shown to make direct contact with the 28S rRNA of the larger 60S subunit (Nygärd and Nika 1982; Nygärd and Nilsson 1990). Consistent with these findings, S6 has been cross-linked with a number of initiation factors of the translational machinery, suggesting that it exerts its effects at the level of initiation of protein synthesis (Nygärd and Nilsson 1990).

Increased S6 phosphorylation is observed when quiescent cells reenter the cell cycle in response to mitogens (Thomas et al. 1979). This phenomenon has been reported both in vivo, in the liver during regeneration (Gressner and Wool 1974) or following fasting and refeeding (Kozma et al. 1989), and in a large number of model cell-culture systems (Kozma and Thomas 1994). Thus, it appears to be a general response to signals that induce cell growth. Phosphorylation of S6 is concomitant with an increase in protein synthesis, and therefore, it had been postulated to be involved in enhancing the rate of translation, in particular at the level of initiation (Thomas et al. 1982). Earlier studies also showed that increased

S6 phosphorylation correlated with synthesis of a number of proteins whose levels were not affected by the transcription inhibitor actinomycin D, and were therefore argued to be controlled at the translational level (Thomas et al. 1980). One of these proteins is elongation factor eEF1α (Thomas and Thomas 1986). eEF1α is encoded by an mRNA found later to belong to the 5′TOP mRNA family (Jefferies et al. 1994a), whose translation is up-regulated by the S6K1 pathway in response to mitogenic stimuli (see below). These mRNAs are characterized by an oligopyrimidine tract at their 5′ transcriptional start site and encode for components of the translational apparatus (see Meyuhas and Hornstein Chapter 22). At this point, direct proof for a role for S6 phosphorylation in this process is lacking, but both biochemical and genetic approaches will be useful in addressing this issue (see below). The S6 phosphorylation sites have been mapped to five clustered residues, including Ser-235, Ser-236, Ser-240, Ser-244, and Ser-247, which are located at the carboxyl terminus of the protein (Fig.1) (Krieg et al. 1988). Phosphorylation occurs in an ordered fashion, with Ser-236 being the first residue phosphorylated, followed in order by Ser-235, Ser-240, Ser-244 and Ser-247. S6 phosphorylation is also observed in *Drosophila melanogaster*. Insulin stimulates phosphorylation of S6 in two *D. melanogaster* cell lines. Again, the position of the phosphorylation sites has been mapped to five serines present at the carboxyl terminus of the protein, namely Ser-233, Ser-235, Ser-239, Ser-242, and Ser-245 (Fig. 1) (Radimerski et al. 2000). In analogy with the findings in vertebrates, it suggests that S6 phosphorylation plays an important role in the initiation of protein synthesis, which has been conserved in metazoans.

Species:	**Sequence:**
Homo sapiens	RRRL**SS**LRA**S**TSK**S**ES**S**QK^{249}
Mus musculus	RRRL**SS**LRA**S**TSK**S**ES**S**QK^{249}
Rattus norvegicus	RRRL**SS**LRA**S**TSK**S**ES**S**QK^{249}
Xenopus laevis	RRRL**SS**LRA**S**TSK**S**ES**S**QK^{249}
Drosophila melanogaster	RRR**S**A**S**IRE**S**KS**S**VS**S**DKK^{248}

Figure 1 Sequences of the carboxy-terminal region of S6 from different vertebrate species and from *D. melanogaster*. The identified phosphorylation sites are in bold.

IDENTIFICATION OF S6 KINASE 1 ($p70^{s6k}/p85^{s6k}$)

S6K1 was purified to homogeneity as a 70-kD polypeptide from extracts of cells stimulated with mitogens (Jenö et al. 1988; Kozma et al. 1989; Price et al. 1989). Subsequent peptide sequencing and screening of a cDNA library revealed the existence of two isoforms of S6K1, $p70^{s6k}$ and $p85^{s6k}$, which utilize different translational start codons on the same mRNA (Banjerjee et al. 1990; Kozma et al. 1990; Grove et al. 1991; Reinhard et al. 1992). The two isoforms are therefore identical in their sequence except for a 23-amino-acid extension at the amino terminus of $p85^{s6k}$, which contains a nuclear localization signal that targets $p85^{s6k}$ to the nucleus, whereas $p70^{s6k}$ is largely localized in the cytoplasm (Fig. 2) (Reinhard et al. 1994). This form of S6K1 can be divided into five functional domains (Fig. 2). The amino terminus, rich in acidic residues, is important in conferring rapamycin sensitivity to the enzyme (Cheatham et al. 1995; Weng et al. 1995; Dennis et al. 1996). This is followed by the catalytic domain, which belongs to the protein A, protein G, and protein C (AGC) family of protein kinases (Hanks and Hunter 1995). In this domain, the most significant similarity is observed in the residues flanking the phosphorylation site in the activation loop, T229. The similarities between S6K1 and the kinases of the AGC family extend beyond the kinase domain, in what has been called the linker region in S6K1. Two critical mitogen-induced phosphorylation sites, S371 and T389, reside in this region (Pearson et al. 1995; Moser et al. 1997). S371 is flanked in +1 position by a proline, whereas T389 is flanked both in +1 and –1 positions by large bulky amino acids, as are T229 and the corresponding sites in the AGC kinases (Pearson et al. 1995). Carboxy-terminal to the linker domain is a region that shares homology with the sequence surrounding the phosphorylation sites in S6. This region is thought to function as an autoinhibitory domain (Banerjee et al. 1990; Ferrari et al. 1992) and is absent in the other AGC family members. Five phosphorylation sites are present in this domain, including S404, S411, S418, T421, and S424 (Ferrari et al. 1992; Pearson et al. 1995). With the exception of S404, which is in a context similar to those observed for T229 and T389, these sites are followed in the +1 position by a proline, similar to S371. Finally, the extreme carboxyl terminus of S6K1 has been shown to contain a distinct domain that interacts with a neuron-specific F-actin-binding protein, neurabin (Burnett et al. 1998b). This interaction has been speculated to specifically target S6K1 to nerve terminals and may be exploited by other targeting molecules in a similar manner.

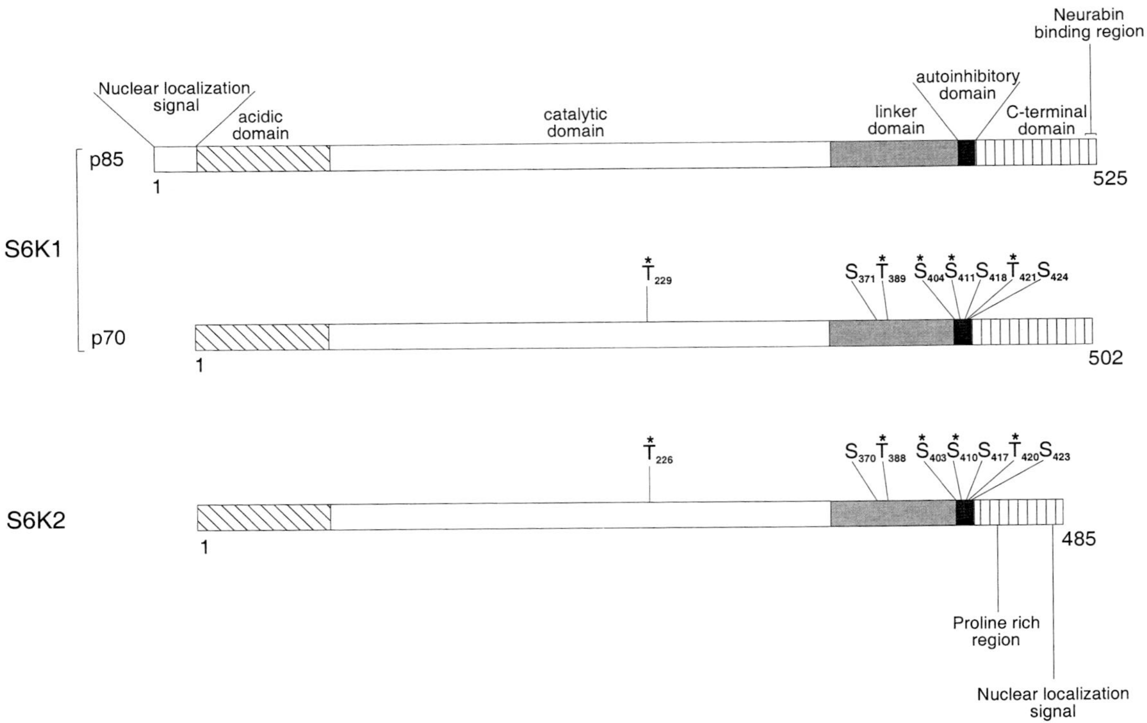

Figure 2 Schematic representations of S6K1 and of S6K2. The known phosphorylation sites in p70 S6K1, and the homologous sites in S6K2, are shown. Rapamycin-sensitive sites are marked with an asterisk. Rapamycin-sensitive amino terminus (*hatched*), catalytic domain (*white*), linker domain (*gray*), autoinhibitory domain (*black*), and carboxyl terminus (*vertical bars*).

MECHANISMS OF ACTIVATION

Activation of S6K1 requires an interplay between multiple phosphorylation sites described above and specific regulatory domains of the molecule (Pullen and Thomas 1997). The first step is thought to involve the disruption of an interaction between the amino- and carboxy-terminal parts of the molecule. This event is mediated by the phosphorylation of the S/TP sites in the autoinhibitory domain by an as-yet-unidentified kinase, followed by phosphorylation of T389 in the linker domain (Dennis et al. 1998). The mammalian target of rapamycin, mTOR, has been proposed as the T389 kinase (Burnett et al. 1998a) consistent with recent in vitro reconstitution studies (Isotani et al. 1999), however, not all the data are consistent with this hypothesis (see below). Phosphorylation of T389 and the S/TP sites promotes phosphorylation of T229 in the activation loop of the catalytic domain, resulting in kinase activation (Dennis et al. 1998). Phosphorylation of T229 is mediated by the phosphoinositide-dependent kinase PDK1 (Alessi et al. 1998; Pullen et al. 1998). PDK1 was identified originally as the kinase for the activation loop site in PKB (Alessi et al. 1997; Stephens et al. 1998) which, as in the case of T229, is flanked, at both +1 and –1 position, by large hydrophobic residues. PDK1 can phosphorylate and activate both in vitro and in vivo a form of S6K1 where the residues in the autoinhibitory domain and the T389 site have been mutated to acidic amino acids to mimic the phosphorylation state (Dennis et al. 1998; Pullen et al. 1998). Interestingly, PDK1 can also phosphorylate T389 (Balendran et al. 1999), however, phosphorylation of T389 is sensitive to wortmannin (Dennis et al. 1996), whereas PDK1 activity is resistant to the inhibitor (Pullen et al. 1998). S371 phosphorylation also plays an important role in S6K1 activation. When S371 is mutated to either alanine or aspartic acid, the kinase is catalytically inactive even when T389 is mutated to an aspartate (Moser et al. 1997). In addition, the same mutations in S371 impair T389 phosphorylation. These findings suggest that phosphorylation of S371 is important, not only in maintaining the kinase activity, but also in regulating phosphorylation of T389 (Moser et al. 1997).

THE SIGNALING PATHWAY

Although S6K1 is phosphorylated in vitro by MAP kinase (Mukhopadhayay et al. 1992), early studies indicated that S6K1 resides on a signaling pathway distinct from the ras/MAP kinase (Ballou et al. 1991). Subsequent experiments with the inhibitors wortmannin and LY 294002, and with receptors mutated in binding sites for signaling molecules, indicated that the signal to S6K1 activation was initiated at the receptor level by the phosphoinositide 3OH-dependent kinase, PI3K (Chung et al. 1994; Ming et al. 1994). Although the specificity of inhibitors is questionable (Brunn et al. 1996) and the data with receptor mutants may be controversial (Ming et al. 1994), the role of PI3K as an important upstream regulator of S6K1 is generally recognized (Fig. 3). Nevertheless, the effects of PI3K on S6K1 do not appear to be direct. Several signaling molecules have been shown to lie downstream from PI3K and for some of them a role in regulation of S6K1 has been proposed. Among these, a constitutively membrane-targeted form of protein kinase B, PKB, which has high basal kinase activity, was shown to activate S6K1 in cotransfection experiments (Burgering and Coffer 1995). However, subsequent studies showed that dominant-negative forms of PKB, although effective in inhibiting insulin-induced initiation factor 4E binding protein (4EBP1) phosphorylation and glycogen synthase 3 kinase (GSK3) inactivation, did not affect S6K1 activation (Dufner et al. 1999). Consistent with these findings, an activated allele of PKB led to S6K1 activation only when targeted to the membrane, which is not a requirement for either 4EBP1 or GSK3 inactivation (Dufner et al. 1999). These results suggest that S6K1 and PKB lie on parallel pathways downstream from PI3K.

The atypical protein kinase Cs, PKCλ and PKCζ, have been recently implicated in signaling to S6K1. Both are known to be downstream effectors of PI3K (Fig. 3) (Nakanishi et al. 1993; Akimoto et al. 1996). In one study, overexpression of the inactive catalytic domain or the regulatory domain of PKCλ was shown to partially inhibit S6K1 activation. Similar results were observed when a kinase dead variant of PKCζ was coexpressed with S6K1 (Romanelli et al. 1999). In addition, overexpression of constitutively active PKCζ led to a small activation of S6K1 but was very effective when coexpressed with PDK1 (Romanelli et al. 1999). Neither study addressed whether the observed effects were mediated through direct phosphorylation of S6K1. In addition, it should be noted that PDK1 phosphorylates the activation loop sites of all the PKCs (Le Good et al. 1998), suggesting that PDK1 may control S6K1 activity both directly, through T229 phosphorylation, and indirectly, through activation of the atypical PKCs (Fig. 3).

A pivotal role in the regulation of S6K1 is played by the mammalian target of rapamycin, mTOR. mTOR is a 290-kD protein harboring a carboxy-terminal protein kinase domain related to phosphatidylinositide-3OH kinases (Dennis et al. 1999). It was initially discovered as the target of the immunosuppressant rapamycin, which is known to inhibit in vivo

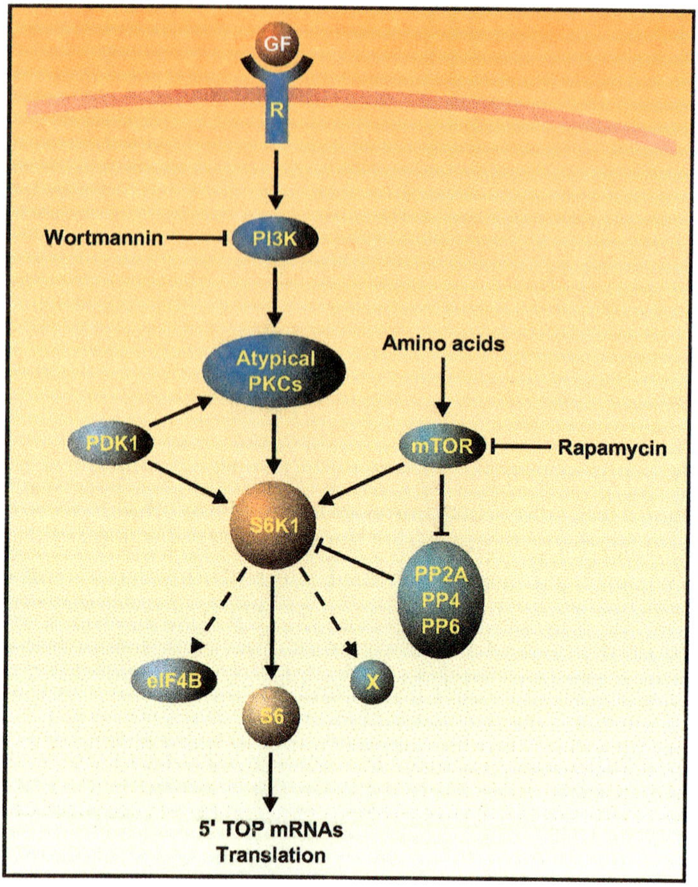

Figure 3 Present model of the S6K1 signaling pathway. (GF) Growth factor, (R) receptor. See the text for details.

mitogen-mediated S6K1 activation. Rapamycin, in a gain-of-function inhibitory complex with the prolyl-isomerase FKBP12, binds mTOR in the FRB (*F*KBP12-*r*apamycin *b*inding) domain, inhibiting mTOR autophosphorylation and phosphorylation of its putative substrate 4EBP1 (Thomas and Hall 1997; Dennis et al. 1999). In the case of S6K1, it is known that a rapamycin-resistant form of mTOR protects S6K1 from the inhibitory effects of rapamycin. Conversely, expression of a kinase dead form of mTOR inhibits mitogen-induced S6K1 activation (Burnett et al. 1998a). Rapamycin induces the dephosphorylation of a subset of sites of S6K1 (Pearson et al. 1995) and, as mentioned above, it has been reported that in vitro mTOR can phosphorylate the principal rapamycin-sensitive site, T389 (Burnett et al. 1998a). This is surprising in light of two observations: First, unlike the sites in 4EBP1 phosphorylated by mTOR, which are flanked in +1 position by prolines (Burnett et al. 1998a; Gingras et al. 1999), T389 is flanked at both the +1 and –1 positions by large bulky amino acids. Second, an S6K1 variant lacking the carboxyl and amino termini is responsive to mitogens, resistant to rapamycin, and still phosphorylated at T389 (Dennis et al. 1996). Indeed, it has recently been shown that mTOR can phosphorylate both T389 and two of the S/TP sites in the autoinhibitory domain; however, given the caveats above, the authors suggest that mTOR probably plays a minor role in mitogen-induced phosphorylation of T389 (Isotani et al. 1999). The effect of mTOR on S6K1 activity might rather be exerted at the level of a phosphatase (Fig. 3). In yeast, mTOR is known to control protein phosphatase 2A (PP2A) and suppressor of initiation of transcription 4 (Sit4) phosphatase activity by modulating their association with Tap42 (Di Como and Arndt 1996). In mammalian cells, PP2A is thought to be the phosphatase involved in regulating S6K1 dephosphorylation (Ballou et al. 1988). Consistent with the findings above, it has been reported that the decrease in PP2A activity in response to mitogens was abolished in cells that were pretreated with either rapamycin or wortmannin (Begum and Ragolia 1996). In addition, in vivo association between S6K1 and PP2A, which was dependent on the amino terminus of S6K1, has been reported (Peterson et al. 1999; Westphal et al. 1999). A Tap-42-related protein, termed α4, has been shown in mammals to associate with PP2A, protein phosphatase 4 (PP4), and protein phosphatase 6 (PP6), the Sit4 homolog (Murata et al. 1997; Chen et al. 1998; Inui et al. 1998; Nanahoshi et al. 1998). However, it is not clear whether mTOR is regulating the formation of those complexes.

Data from yeast suggest that TOR plays an important role by modulating the opposing effects of protein synthesis (Barbet et al. 1996) and protein degradation (Noda and Ohsumi 1998) depending on the availability of nutrients (Dennis et al. 1999); (see also Kimball and Jefferson Chapter 16 for the effects of amino acid deprivation on protein synthesis). TOR-induced protein degradation is mediated by autophagy, a process of bulk degradation of proteins in which cytoplasmic components, such as ribosomes, are enclosed by double-membrane structures known as autophagosomes for delivery to lysosomes or vacuoles for degradation (Dunn 1994). In the case of ribosomes, their degradation is crucial for the generation of amino acids and nucleotides that can be used for energy and survival during periods of fasting. In yeast, a number of the components involved in regulating this response have been identified through genetic screens (Mizushima et al. 1998). For some of the yeast genes, mammalian homologs have also been identified (Mizushima et al. 1998; Liang et al.

1999). In mammals, the role of S6 in this process was initially suggested from studies in rat hepatocytes, where inhibition of autophagy by amino acids was directly paralleled by increased S6 phosphorylation and S6 kinase activity (Blommaart et al. 1995). These studies were subsequently extended to show that amino acids are potent activators not only of S6K1 (Fig. 3), but also of 4EBP1 phosphorylation (Hara et al. 1998; Patti et al. 1998; Wang et al. 1998; Xu et al. 1998). In addition, it was shown that insulin failed to activate S6K1 in the absence of amino acids (Hara et al. 1998). S6K1 activity was rapidly abolished if amino acids were omitted from the medium, in a way reminiscent of the effect of rapamycin. Interestingly, the activity of a rapamycin-resistant form of S6K1 was not affected by amino acid withdrawal, suggesting that mTOR modulates the activity of a regulator of S6K1 according to amino acid availability (Hara et al. 1998). In agreement with this model, a rapamycin-resistant form of mTOR was shown to protect amino-acid-induced S6K1 activation from the inhibitory effects of rapamycin (Iiboshi et al. 1999). It can be speculated that the activation of S6K1 by amino acids responds to a demand for cell growth in the absence of cell proliferation; for example, in the growth of liver cells in response to feeding following fasting. In contrast, amino acid deprivation induces autophagy, which is reversed upon re-addition of amino acids. It can be speculated that S6K1 acts as positive regulator of the protein content of a cell, through stimulation of protein synthesis and possibly through inhibition of protein degradation (for the effects of amino acids deprivation on protein synthesis, see Kimball and Jefferson Chapter 16).

S6K1-DEFICIENT MICE AND IDENTIFICATION OF S6 KINASE 2

To gain insight into the physiological function of S6K1 in vivo, mice were generated in which one or both S6K1 alleles were deleted. The mice were viable and fertile (Shima et al. 1998). When compared to wild-type mice, S6K1-deficient mice were smaller in size, an effect that was more pronounced during embryonic development. Indeed, during embryonic development, heterozygous mice showed haploid insufficiency with respect to size (Shima et al. 1998). Surprisingly, induction of S6 phosphorylation, measured both in the liver after re-feeding and in embryonic fibroblasts after stimulation with serum, was unaffected in cells derived from the S6K1-deficient animals. Similarly, translation of 5′TOP mRNAs was up-regulated normally in fibroblasts from S6K1-deficient animals treated with mitogens, as was reentry into the cell cycle, as measured by BrdU incorporation. Just as importantly, all three responses were inhibited by rapamycin (Shima et al. 1998). Together these findings suggested that an unknown rapamycin-sensitive kinase was involved in these responses. An explanation came from the identification in public databases of a gene that encodes a kinase highly homologous to S6K1, termed S6K2 or S6Kβ (Fig. 2) (Gout et al. 1998; Saitoh et al. 1998; Shima et al. 1998), which might compensate, at least in part, for the lack of S6K1 expression. In agreement with this hypothesis, the S6K2 mRNA was found to be up-regulated in all tissues examined in the S6K1-deficient animals (Shima et al. 1998). Mice harboring deletion of the S6K2 gene, as well as mice harboring deletion of both S6K genes, will be valuable in establishing their relative importance in regulating S6 phosphorylation.

The highest degree of identity between S6K1 and S6K2, more than 80%, is observed in the catalytic, linker, and autoinhibitory domains, where all the essential known phosphorylation sites are conserved (Shima et al. 1998). The kinases are 40% identical in the amino-terminal domain, which in S6K2, like S6K1, is highly acidic, whereas strong divergence is observed in the carboxyl terminal domain, which exhibits only 20% identity (Fig. 2). The carboxyl terminus of S6K2 is unique in that it is proline-rich and may serve to mediate interactions with signaling molecules through specialized domains, such as SH3 domains. In addition, a stretch of basic amino acids has been shown to fulfill the requirements of a nuclear localization signal (Fig. 2) (Koh et al. 1999), although a potential S6K1-like variant having a basic amino terminus has been reported (Gout et al. 1998). Antibodies raised against a unique peptide sequence in the amino-terminal domain of S6K2 recognize the kinase on Western blots migrating with an anomalously high molecular mass of 68 kD. As in the case of S6K1, the activation of S6K2 is inhibited in cells treated with rapamycin.

The major phenotype of the S6K1-deficient animals is their size defect (Shima et al. 1998). This abnormality is more evident in embryos and, in part, is due to a delay in development. How does the lack of S6K1 account for such a phenotype? On the basis of tissue-culture studies, we assumed that S6K1, due to its role in ribosome biogenesis, is important in driving cell growth and proliferation. Thus, a defect in ribosome production at a very early stage of development might either block or delay the growth of the S6K1-deficient embryos, until a stage where S6K2 would begin to compensate. Alternatively, it is possible that S6K1 controls the production of a humoral factor involved in the early growth of the embryo, such as the insulin-like growth factor II, IGFII. Interestingly, it has been shown that one of the mRNAs encoding IGF-II is translationally regulated and that its translation is inhibited by rapamycin (Nielsen et al. 1995). Moreover, an oligopyrimidine tract is present in the 5′UTR of the IGF-II mRNA (Nielsen et al. 1995). It will be of interest to measure IGF-II levels in S6K1-deficient versus wild-type animals.

S6K1 AND THE REGULATION OF 5′TOP mRNA TRANSLATION

The strict correlation between activation of the S6 kinase and S6 phosphorylation and increases in the rates of protein synthesis has led investigators to postulate a causal relationship between the two events (Jefferies and Thomas 1996). However, direct evidence for the involvement of the S6 kinase pathway in the regulation of translation has been lacking until recently. The observation that rapamycin inhibits mitogen-induced S6K1 activation and S6 phosphorylation (Chung et al. 1992; Kuo et al. 1992; Price et al. 1992) provided scientists with a potent pharmacological tool to obtain insight into this issue. Rapamycin was found to inhibit protein synthesis to varying degrees, depending on the cell type analyzed and the concentrations used, raising the possibility that rapamycin targets a component in the mTOR pathway which controls selectively the translation of a subset of the total mRNAs present in a cell (Jefferies et al. 1994b; Terada et al. 1994; Beretta et al. 1996; Mendez et al. 1996). In agreement with this model, it was shown that rapamycin selectively inhibits the mitogen-induced up-regulation of 5′TOP mRNAs (Jefferies et al. 1994b; Terada et al. 1994). In contrast, the translation of other mRNAs lacking this motif were not affected by rapamycin treatment. The effect of rapamycin on these transcripts was not fully inhibitory, arguing that, at least in this cell system, translation of 5′TOP mRNAs is under the control of at least two pathways, only one of which is sensitive to the action of rapamycin (Jefferies et al. 1994b). Using chimeric transcripts containing the 5′TOP of ribosomal protein S16, either wild type or mutated, it was shown that the inhibitory effect of rapamycin requires an intact wild-type 5′TOP (Jefferies et al. 1997). From parallel studies, it was evident that inhibition of mTOR by rapamycin not only inhibits S6K1 and S6 phosphorylation, but also blocks 4EBP phosphorylation (Beretta et al. 1996; Von Manteuffel et al. 1996). Furthermore, it has been shown that these two pathways bifurcate upstream of S6K1 (Von Manteuffel et al. 1997). To address the issue of whether the rapamycin effects on 5′TOP mRNA translation are mediated through inhibition of S6 kinase, mutants of the kinase were employed in transfection experiments. It was shown that overexpression of a kinase dead form of S6K1, which acts as a dominant-negative allele, suppressed mitogen-induced translation of a chimeric construct with an intact 5′TOP. Conversely, overexpression of a rapamycin-resistant, but not the wild-type, form of the kinase was able to protect the translational up-regulation of eEF1α mRNA (Jefferies et al. 1997). These results demonstrate that the S6 kinase is a necessary component in the regulation of 5′TOP mRNA translation and, as a consequence, ribosome biogenesis (Fig. 3).

S6: THE TARGET OF S6K1 FUNCTION

Several lines of evidence support the concept that S6 is the substrate by which S6K1 and S6K2 mediate their effects on cell growth. It is clear that, despite considerable efforts from different investigators, no other proteins have been found that fulfill the requirements for an S6K1 substrate, with the possible exception of eIF4B (Fig. 3) (J. Hershey, pers. comm.). These investigators have shown that eIF4B is multiply phosphorylated by S6K1 in vitro, likely at S406 and S422. Mitogenic stimulation of cells results in rapamycin-sensitive eIF4B phosphorylation in vivo at the same sites. However, eIF4B phosphorylation levels are not affected by rapamycin when cells express a rapamycin-insensitive form of S6K1, indicating that this kinase is involved in vivo. Oddly, excess eIF4B is inhibitory when added to the reticulocyte lysate. Interestingly, conversion of S406 and S422 to acidic residues relieves this inhibition and modestly increases eIF4B binding to mRNA. It may be that this change in mRNA affinity reflects a preference for specific mRNAs and that eIF4B acts in unison with S6 to regulate 5′TOP mRNA translation. In the case of S6, it is reasonable to postulate that the effects of S6K1 on 5′TOP mRNAs translation are mediated, at least in part, through its phosphorylation. As mentioned above, the position of S6 in the ribosome suggests that it has an active role in protein synthesis. Phosphorylated subunits are selectively found associated with actively translating polysomes (Duncan and McConkey 1982; Thomas et al. 1982), pointing to a positive effect of S6 phosphorylation. At this stage, it is important to test whether S6 phosphorylation directly affects the translation of 5′TOP mRNAs. Until now, this issue has been difficult to assess due to the lack of in vitro assay systems to examine the effects of modified translational components on protein synthesis. Recently, a very sensitive in vitro assay has been developed, which makes use of purified translational components to study the initiation of protein synthesis in eukaryotes (Pestova et al. 1998). Since translation of 5′TOP mRNAs is thought to be controlled at the level of initiation (Meyuhas et al. 1996), it will be important to apply such assay systems to test the importance of phosphorylated subunits, as well as eIF4B, on formation of initiation complex with 5′TOP mRNAs. This approach also would be very useful in gaining insights into the role of other factors, which are postulated as well to have either positive (La protein) or negative effects (CNBP) on 5′TOP mRNA translation (Pellizzoni et al. 1998). Alternatively, a genetic approach might also be helpful. Such a genetic system has been developed, in which deletion of the S6 gene can be induced in the liver, leading to a gradual loss of 40S subunits (Volarevic et al. 2000). In theory it should be possible, using viral vectors, to drive in the liver of these animals ectopic expression of phosphorylation site mutants of S6 and to test the ability of 40S subunits bearing the mutated S6 to promote translation of 5′TOP mRNAs. Such an approach might also be undertaken using the genetics of *D. melanogaster* and mutated alleles of the S6 gene.

The case for S6 phosphorylation regulating translation of 5′TOP mRNAs is not as strong in *Xenopus laevis* oocytes. Progesterone-induced maturation of *X. laevis* oocytes leads to transient activation of S6K1 and phosphorylation of S6 on three of five serines (Martin-Perez et al. 1986). Treatment with rapamycin blocks progesterone-induced S6K1 activation, suppresses 5′TOP mRNA translation but, surprisingly, has little effect on S6 phosphorylation (Schwab et al. 1999). These results suggest (1) that another enzyme, possibly the p90 ribosomal protein S6 kinase, p90^{rsk}, and not S6K1, phosphorylates S6 during oocyte maturation and (2) that at least in this system, the effect of S6K1 on 5′TOP mRNA translation is exerted through a substrate other than S6. However, less than 1% of the ribosomes are translationally active in oocytes. Thus, S6K1 may mediate its effect on 5′TOP mRNA translation through phosphorylation of a small subset of actively translating polysomes. Under such conditions, it would be very difficult to detect changes in S6 phosphorylation on a small subset of active polysomes.

Finally, it has been reported that the translation of 5′TOP mRNAs is not sensitive to the inhibitory effects of poliovirus infection. In addition, poliovirus stimulates S6K1 kinase activity and S6 phosphorylation (Cardinali et al. 1999). Inhibition of host protein synthesis by poliovirus is due to cleavage of eIF4G and, as a consequence, leads to a shutoff of cap-dependent translation. This may indicate that the translation of 5′TOP mRNAs is not cap-dependent and may instead utilize a distinct mechanism for initiation, such as those employed by viruses (for review, see Chapter 4).

S6K IN *DROSOPHILA*

D. melanogaster is an ideal metazoan system for studying signal transduction pathways because, unlike mammals, most genes are present in a single copy and many of the signal transduction pathways appear to be conserved (Miklos and Rubin 1996). A cDNA encoding a *Drosophila* homolog of S6K1 has been previously cloned and is 78% identical in the catalytic domain, with an overall identity of 54% to the mammalian enzyme (Stewart et al. 1996; Watson et al. 1996). Furthermore, the essential phosphorylation sites and autoinhibitory domain are conserved, as is the sensitivity of the kinase to rapamycin and wortmannin (Stewart et al. 1996; Watson et al. 1996). The cytological location of the *dS6K* gene was determined, and a mutant line was found bearing a P-element insertion 48 bp upstream of the translation start site (Montagne et al. 1999). This insertion resulted in female sterility, a developmental delay of 3 days and semi-lethality, as only about 25% of the expected homozygous mutant flies emerged as adults. The homozygous mutants were slightly smaller in size, with the P-element insertion resulting in decreased expression of *dS6K* mRNA (Montagne et al. 1999). The wild-type phenotype was recovered by excising the P-element, whereas imprecise deletion of the gene led to the isolation of a line that was null for *dS6K*. The phenotype of these flies is more severe than the homozygous P-element-bearing flies. The few adults who survive have a 5-day delay in development and are 50% smaller than wild-type flies (Montagne et al. 1999). Since cell growth is assumed to be dominant over cell proliferation (Conlon and Raff 1999), it was predicted that the size defect in the mutant animals was due to fewer cells rather than smaller cells. Surprisingly, the number of cells was found to be the same in the wing and the eye of mutant versus wild-type flies, but all cells were uniformly smaller. Further studies showed that cells from mutant flies progress through the cell cycle at a slower rate, with all the phases of the cell cycle being affected equally (Montagne et al. 1999). Since S6K1 has been implicated in the synthesis of hormones involved in the growth of the organism (Nielsen et al. 1995; Leibiger et al. 1998), it is possible that the phenotype of the mutant flies was due to a humoral effect and was not cell-autonomous. To test this possibility, somatic clones that lacked the *dS6K* gene were generated in a heterozygous background. Only the size of the cells of the mutant clones was affected, whereas the surrounding tissue was normal, arguing that the effect of dS6K on cell size is cell-autonomous (Montagne et al. 1999). Given the involvement of S6 kinase in ribosome biogenesis, it is surprising that the phenotype of *dS6K* mutant flies is distinct from ribosomal protein gene mutants termed *Minutes* (Lambertsson 1998). In fact, although delayed in development, *Minutes* have a normal cell size and number. Furthermore, *dS6K* mutants do not show a defect in bristle length (Montagne et al. 1999), which is one of the hallmarks of *Minutes*. These differences argue that, besides controlling ribosome biogenesis, *dS6K* has additional functions. One possibility is that *dS6K* directly regulates the synthesis of a cell-size regulator. Alternatively, in the absence of *dS6K*, the translation of 5´TOP mRNAs encoding ribosomal proteins would be inhibited. Since these mRNAs constitute up to 30% of the mRNA transcripts present in a cell, the translation of mRNAs lacking a 5´TOP would be favored. If such mRNAs encoded cell cycle regulators, these regulators could reach a concentration critical to promote transition through the cell cycle at a smaller cell size (Thomas 2000). In support of this model, it has been shown that inhibition of S6 kinase and of 5´TOP mRNA translation by rapamycin during early *Xenopus* mitotic cell cycles accelerates the translation of the cell cycle regulator cdc25A (Schwab et al. 1999).

PERSPECTIVES

In recent years, our understanding of the mechanism of S6K1 regulation has increased. Some of the upstream components of the pathway that have been identified lead to phosphorylation of residues crucial for activation of the kinase (Fig. 3). In the future, both biochemical and genetic approaches will be useful in identifying new regulators and providing a more complete picture of the pathway that leads to the activation of the S6 kinases. A major breakthrough in this field has been the discovery that S6K1 is a positive effector of translation of 5´TOP mRNAs. It will now be important to see whether the effect of S6K1 is mediated through phosphorylation of S6. In this respect, the use of genetic approaches will help. In particular, the expression of mutant S6 alleles in S6-deficient mice or in *Drosophila* will provide information on the importance of S6 phosphorylation in the translation of 5´TOP mRNAs. Nevertheless, results obtained with these approaches will have to be evaluated in light of data which suggest that S6 may have extra-ribosomal functions. Such a possibility emerges from studies in *Drosophila,* where *DS6* mutants develop melanotic tumors in lymph glands, where the hemocytes, the cells devoted to the immune response, are greatly enlarged (Watson et al. 1992; Stewart and Denell 1993). This phenotype is quite distinct from *Minutes* (Lambertsson 1998), suggesting that *DS6* may function as a tumor suppressor gene. In addition, S6 is found in a free form in the nucleus of mammalian cells, where it can be phosphorylated in response to mitogens (Franco and Rosenfeld 1990). Biochemical studies employing in vitro systems will therefore be required in order to understand the mechanism by which S6 phosphorylation specifically affects translation. It is also possible that the effects of S6K1 on translation are mediated through phosphorylation of substrates other than S6. In this respect, genetic screens in *Drosophila* will be useful in identifying novel effectors of S6K1. Thus, through a combination of genetic and biochemical approaches, it should be possible to resolve whether the effects of this pathway are confined to S6 and translation, or whether they are also exerted by distinct molecules through regulation of other anabolic processes.

ACKNOWLEDGMENTS

S.F. is a recipient of a long-term fellowship from the European Molecular Biology Organization.

REFERENCES

Akimoto K., Takahashi R., Moriya S., Nishioka N., Takayanagi J., Kimura K., Fukui Y., Osada S., Mizuno K., Hirai S., Kazlauskas A., and Ohno S. 1996. EGF or PDGF receptors activate atypical PKCλ through phosphatidylinositol 3-kinase. *EMBO J.* **15:** 788–798.

Alessi D.R., Kozlowski M.T., Weng Q.P., Morrice N., and Avruch J. 1998. 3-Phosphoinositide-dependent protein kinase 1 (PDK1) phosphorylates and activates the p70 S6 kinase in vivo and in vitro. *Curr. Biol.* **8:** 69–81.

Alessi D.R., James S.R., Downes C.P., Holmes A.B., Gaffney P.R., Reese C.B., and Cohen P. 1997. Characterization of a 3-phosphoinositide-dependent protein kinase which phosphorylates and activates protein kinase Bα. *Curr. Biol.* **7:** 261–269.

Balendran A., Currie R., Armstrong C.G., Avruch J., and Alessi D.R. 1999. Evidence that 3-phosphoinositide-dependent protein kinase-1 mediates phosphorylation of p70 S6 kinase in vivo at Thr-412 as well as Thr-252. *J. Biol. Chem.* **274:** 37400–37406.

Ballou L.M., Jenö P., and Thomas G. 1988. Protein phosphatase 2A inactivates the mitogen-stimulated S6 kinase from Swiss mouse 3T3 cells. *J. Biol. Chem.* **263:** 1188–1194.

Ballou L.M., Luther H., and Thomas G. 1991. MAP2 kinase and 70k S6 kinase lie on distinct signalling pathways. *Nature* **349:** 348–350.

Banerjee P., Ahamad M.F., Grove J.R., Kozlosky C., Price D.J., and Avruch J. 1990. Molecular structure of a major insulin/mitogen-activated 70kDa S6 protein kinase. *Proc. Natl. Acad. Sci.* **87:** 8550–8554.

Barbet N.C., Schneider U., Helliwell S.B., Stansfield I., Tuite M.F., and Hall M.N. 1996. TOR controls translation initiation and early G1 progression in yeast. *Mol. Biol. Cell* **7:** 25–42.

Begum N. and Ragolia L. 1996. cAMP counter-regulates insulin-mediated protein phosphatase-2A inactivation in rat skeletal muscle cells. *J. Biol. Chem.* **271:** 31166–31171.

Beretta L., Gingras A.-C., Svitkin Y.V., Hall M.N., and Sonenberg N. 1996. Rapamycin blocks the phosphorylation of 4E-BP1 and inhibits cap-dependent initiation of translation. *EMBO J.* **15:** 658–664.

Blommaart E.F., Luiken J.J., Blommaart P.J., van Woerkom G.M., and Meijer A.J. 1995. Phosphorylation of ribosomal protein S6 is inhibitory for autophagy in isolated rat hepatocytes. *J. Biol. Chem.* **270:** 2320–2326.

Brunn G.J., Williams J., Sabers C., Wiederrecht G., Lawrence J.C., Jr., and Abraham R.T.

1996. Direct inhibition of the signaling functions of the mammalian target of rapamycin by the phosphoinositide 3-kinase inhibitors, wortmannin and LY294002. *EMBO J.* **15:** 5256–5267.

Burgering B.M. and Coffer P.J. 1995. Protein kinase B (c-*Akt*) in phosphatidylinositol-3-OH kinase signal transduction. *Nature* **376:** 599–602.

Burnett P.E., Barrow R.K., Cohen N., Snyder S.H., and Sabatini D.M. 1998a. RAFT1 Phosphorylation of the translational regulators p70 S6 kinase and 4E-BP1. *Proc. Natl. Acad. Sci.* **95:** 1432–1437.

Burnett P.E., Blackshaw S., Lai M.M., Qureshi I.A., Burnett A.F., Sabatini D.M., and Snyder S.H. 1998b. Neurabin is a synaptic protein linking p70 S6 kinase and the neuronal cytoskeleton. *Proc. Natl. Acad. Sci.* **95:** 8351–8356.

Cardinali B., Fiore L., Campioni N., De Dominicis A., and Pierandrei-Amaldi P. 1999. Resistance of ribosomal protein mRNA translation to protein synthesis shutoff induced by poliovirus. *J. Virol.* **73:** 7070–7076.

Cheatham L., Monfar M., Chou M.M., and Blenis J. 1995. Structural and functional analysis of $pp70^{s6k}$. *Proc. Natl. Acad. Sci.* **92:** 11696–11700.

Chen J., Peterson R.T., and Schreiber S.L. 1998. α4 Associates with protein phosphatase 2A, 4 and 6. *Biochem. Biophys. Res. Commun.* **247:** 827–832.

Chung J., Kuo C.J., Crabtree G.R., and Blenis J. 1992. Rapamycin-FKBP specifically blocks growth-dependent activation of and signaling by the 70 kd S6 protein kinases. *Cell* **69:** 1227–1236.

Chung J., Grammer T.C., Lemon K.P., Kazlauskas A., and Blenis J. 1994. PDGF- and insulin-dependent $pp70^{s6k}$ activation mediated by phosphatidylinositol-3-OH kinase. *Nature* **370:** 71–75.

Conlon I. and Raff M. 1999. Size control in animal development. *Cell* **96:** 235-244.

Dennis P.B., Fumagalli S., and Thomas G. 1999. Target of rapamycin (TOR): Balancing the opposing forces of protein synthesis and degradation. *Curr. Opin. Genet. Dev.* **9:** 49–54.

Dennis P.B., Pullen N., Kozma S.C., and Thomas G. 1996. The principal rapamycin-sensitive $p70^{s6k}$ phosphorylation sites T_{229} and T_{389} are differentially regulated by rapamycin-insensitive kinase-kinases. *Mol. Cell. Biol.* **16:** 6242–6251.

Dennis P.B., Pullen N., Pearson R.B., Kozma S.C., and Thomas G. 1998. Phosphorylation sites in the autoinhibitory domain participate in $p70^{s6k}$ activation loop phosphorylation. *J. Biol. Chem.* **273:** 14845–14852.

Di Como C.J. and Arndt K.T. 1996. Nutrients, via the Tor proteins, stimulate the association of Tap42 with type 2A phosphatases. *Genes Dev.* **10:** 1904–1916.

Dufner A., Andjelkovic M., Burgering B.M.T., Hemmings B.A., and Thomas G. 1999. Protein kinase B localization and activation differentially affect S6 kinase1 activity and eukaryotic translation initiation factor 4E-binding protein1 phosphorylation. *Mol. Cell. Biol.* **19:** 4525–4534.

Duncan R. and McConkey E.H. 1982. Rapid alterations in initiation rate and recruitment of inactive RNA are temporally correlated with S6 phosphorylation. *Eur. J. Biochem.* **123:** 539–544.

Dunn W.A.J. 1994. Autophagy and related mechanisms of lysosome-mediated protein degradation. Trends *Cell Biol.* **4:** 139–143.

Ferrari S., Bannwarth W., Morley S.J., Totty N.F., and Thomas G. 1992. Activation of $p70^{s6k}$ is associated with phosphorylation of four clustered sites displaying Ser/Thr-Pro motifs. *Proc. Natl. Acad. Sci.* **89:** 7282–7285.

Franco R. and Rosenfeld M.G. 1990. Hormonally inducible phosphorylation of a nuclear pool of ribosomal protein S6. *J. Biol. Chem.* **265:** 4321–4325.

Gingras A.-C., Gygi S.P., Raught B., Polakiewicz R.D., Abraham R.T., Hoekstra M.F., Aebersold R., and Sonenberg N. 1999. Regulation of 4E-BP1 phosphorylation: A novel two-step mechanism. *Genes Dev.* **13:** 1422–1437.

Gout I., Minami T., Hara K., Tsujishita Y., Filonenko V., Waterfield M.D., and Yonezawa K. 1998. Molecular cloning and characterization of a novel p70 S6 kinase, p70 S6 kinase b containing a proline-rich region. *J. Biol. Chem.* **273:** 30061–30064.

Gressner A.M. and Wool I.G. 1974. The phosphorylation of liver ribosomal proteins in vivo. Evidence that only a single small subunit (S6) is phosphorylated. *J. Biol. Chem.* **249:** 6917–6925.

Grove J.R., Banerjee P., Balasubramanyam A., Coffer P.J., Price D.J., Avruch J., and Woodgett J.R. 1991. Cloning and expression of two human p70 S6 kinase polypeptides differing only at their amino termini. *Mol. Cell. Biol.* **11:** 5541–5550.

Hanks S.K. and Hunter T. 1995. The eukaryotic protein kinase superfamily. In *The protein kinase facts book. Protein-serine kinases* (ed. G. Hardie and S.K. Hanks), pp. 7–47. Academic Press, London.

Hara K., Yonezawa K., Weng Q.-P., Kozlowski M.T., Belham C., and Avruch J. 1998. Amino acid sufficiency and mTOR regulate p70 S6 kinase and eIF-4E BP1 through a common effector mechanism. *J. Biol. Chem.* **273:** 14484–14494.

Iiboshi Y., Papst P.J., Kawasome H., Hosoi H., Abraham R.T., Houghton P.J., and Terada N. 1999. Amino acid-dependent control of p70(s6k). Involvement of tRNA aminoacylation in the regulation. *J. Biol. Chem.* **274:** 1092–1099.

Inui S., Sanjo H., Maeda K., Yamamoto H., Miyamoto E., and Sakaguchi N. 1998. Ig receptor binding protein 1 (α4) is associated with a rapamycin-sensitive signal transduction in lymphocytes through direct binding to the catalytic subunit of protein phosphatase 2A. *Blood* **92:** 539–546.

Isotani S., Hara K., Tokunaga C., Inoue H., Avruch J., and Yonezawa K. 1999. Immunopurified mammalian target of rapamycin phosphorylates and activates p70 S6 kinase alpha in vitro. *J. Biol. Chem.* **274:** 34493–34498.

Jefferies H.B.J. and Thomas G. 1996. Ribosomal protein S6 phosphorylation and signal transduction. In *Translational control* (ed. J.W.B. Hershey et al.), pp. 389–409. Cold Spring Harbor Laboratory Press, Cold Spring Harbor, New York.

Jefferies H.B.J., Thomas G., and Thomas G. 1994a. Elongation factor-1α mRNA is selectively translated following mitogenic stimulation. *J. Biol. Chem.* **269:** 4367–4372.

Jefferies H.B.J., Reinhard C., Kozma S.C., and Thomas G. 1994b. Rapamycin selectively represses translation of the "polypyrimidine tract" mRNA family. *Proc. Natl. Acad. Sci.* **91:** 4441–4445.

Jefferies H.B.J., Fumagalli S., Dennis P.B., Reinhard C., Pearson R.B., and Thomas G. 1997. Rapamycin suppresses 5′TOP mRNA translation through inhibition of $p70^{s6k}$. *EMBO J.* **12:** 3693–3704.

Jenö P., Ballou L.M., Novak-Hofer I., and Thomas G. 1988. Identification and characterization of a mitogenic-activated S6 kinase. *Proc. Natl. Acad. Sci.* **85:** 406–410.

Koh H., Jee K., Lee B., Kim J., Kim D., Yun Y.-H., Kim J.W., Choi H.-K., and Chung J. 1999. Cloning and characterization of a nuclear S6 kinase, S6 kinase-related kinase (SRK); a novel nuclear target of Akt. *Oncogene.* **18:** 5115–5119.

Kozma S.C. and Thomas G. 1994. p70s6k/p85s6k: Mechanism of activation and role in mitogenesis. *Cancer Biol.* **5:** 255–266.

Kozma S.C., Ferrari S., Bassand P., Siegmann M., Totty N., and Thomas G. 1990. Cloning of the mitogen-activated S6 kinase from rat liver reveals an enzyme of the second messenger subfamily. *Proc. Natl. Acad. Sci.* **87:** 7365–7369.

Kozma S.C., Lane H.A., Ferrari S., Luther H., Siegmann M., and Thomas G. 1989. A stimulated S6 kinase from rat liver: Identity with the mitogen-activated S6 kinase from 3T3 cells. *EMBO J.* **8:** 4125–4132.

Krieg J., Hofsteenge J., and Thomas G. 1988. Identification of the 40 S ribosomal protein S6 phosphorylation sites induced by cycloheximide. *J. Biol. Chem.* **263:** 11473–11477.

Kuo C.J., Chung J., Fiorentino D.F., Flanagan W.M., Blenis J., and Crabtree G.R. 1992. Rapamycin selectively inhibits interleukin-2 activation of p70 S6 kinase. *Nature* **358:** 70–73.

Lambertsson A. 1998. The minute genes in *Drosophila* and their molecular functions. *Adv. Genet.* **38:** 69–134.

Laser M., Kasi V.S., Hamawaki M., Cooper G.T., Kerr C.M., and Kuppuswamy D. 1998. Differential activation of p70 and p85 S6 kinase isoforms during cardiac hypertrophy in the adult mammal. *J. Biol. Chem.* **273:** 24610–24619.

Le Good J.A., Ziegler W.H., Parekh D.B., Alessi D.R., Cohen P., and Parker P.J. 1998. Protein kinase C isotypes controlled by phosphoinositide 3-kinase through the protein kinase PDK1. *Science* **281:** 2042–2045.

Leibiger I.B., Leibiger B., Moede T., and Berggren P.O. 1998. Exocytosis of insulin promotes insulin gene transcription via the insulin receptor/PI-3 kinase/p70 s6 kinase and CaM kinase pathways. *Mol. Cell* **1:** 933–938.

Liang X.H., Jackson S., Seaman M., Brown K., Kempkes B., Hibshoosh H., and Levine B. 1999. Induction of autophagy and inhibition of tumorigenesis by beclin 1. *Nature* **402:** 672–676.

Martin-Perez J., Rudkin B.B., Siegmann M., and Thomas G. 1986. Activation of ribosomal protein S6 phosphorylation during meiotic maturation of *Xenopus laevis* oocytes: In vitro ordered appearance of S6 phosphopeptides. *EMBO J.* **5:** 725–731.

Mendez R., Myers M.G., Jr., White M.F., and Rhoads R.E. 1996. Stimulation of protein synthesis, eukaryotic translation initiation factor 4E phosphorylation, and PHAS-I phosphorylation by insulin requires insulin receptor substrate 1 and phosphatidylinositol 3-kinase. *Mol. Cell. Biol.* **16:** 2857–2864.

Meyuhas O., Avni D., and Shama S. 1996. Translational control of ribosomal protein mRNAs in eukaryotes. In *Translational control* (J.W.B. Hershey et al.), pp. 363–388. Cold Spring Harbor Laboratory Press, Cold Spring Harbor, New York.

Miklos G.L.G. and Rubin G.M. 1996. The role of the genome project in determining gene function: Insights from model organisms. *Cell* **86:** 521–529.

Ming X.F., Burgering B.M.T., Wennström S., Claesson-Welsh L., Heldin C.H., Bos J.L., Kozma S.C., and Thomas G. 1994. Activation of p70/p85 S6 kinase by a pathway independent of $p21^{ras}$. *Nature* **371:** 426–429.

Mizushima N., Noda T., Yoshimori T., Tanaka Y., Ishii T., George M.D., Klionsky D.J., Ohsumi M., and Ohsumi Y. 1998. A protein conjugation system essential for autophagy (comments). *Nature* **395:** 395–398.

Montagne J., Stewart M.J., Stocker H., Hafen E., Kozma S.C., and Thomas G. 1999. *Drosophila* S6 kinase: A regulator of cell size. *Science* **285:** 2126–2129.

Moser B.A., Dennis P.B., Pullen N., Pearson R.B., Williamson N.A., Wettenhall E.H., Kozma S.C., and Thomas G. 1997. Dual requirement for a newly identified phosphorylation site in $p70^{s6k}$. *Mol. Cell. Biol.* **17:** 5648–5655.

Mukhopadhayay N.K., Price D.J., Kyriakis J.M., Pelech S.L., Sanghera J., and Avruch J. 1992. An array of insulin-activated, proline-directed serine/threonine protein kinases phosphorylate the p70 S6 kinase. *J. Biol. Chem.* **267:** 3325–3335.

Murata K., Wu J., and Brautigan D.L. 1997. B cell receptor-associated protein α4 displays rapamycin-sensitive binding to the catalytic subunit of protein phosphatase 2A. *Proc. Natl. Acad. Sci.* **94:** 10624–10629.

Nakanishi H., Brewer K.A., and Exton J.H. 1993. Activation of the zeta isozyme of protein kinase C by phosphatidylinositol 3,4,5-trisphosphate. *J. Biol. Chem.* **268:** 13–16.

Nanahoshi M., Nishiuma T., Tsujishita Y., Hara K., Inui S., Sakaguchi N., and Yonezawa K. 1998. Regulation of protein phosphatase 2A catalytic activity by α4 and its yeast homolog Tap42. *Biochem. Biophys. Res. Commun.* **251:** 520–526.

Nielsen F.C., Ostergaard L., Nielsen J., and Christiansen J. 1995. Growth-dependent translation of IGF-II mRNA by a rapamycin-sensitive pathway. *Nature* **377:** 358–362.

Noda T. and Ohsumi Y. 1998. Tor, a phosphatidylinositol kinase homologue, controls autophagy in yeast. *J. Biol. Chem.* **273:** 3963–3966.

Nygärd O. and Nika H. 1982. Identification by RNA-protein cross-linking of ribosomal proteins located at the interface between the small and the large subunits of mammalian ribosomes. *EMBO J.* **1:** 357–362.

Nygärd O. and Nilsson L. 1990. Translational dynamics. Interactions between the translational factors, tRNA and ribosomes during eukaryotic protein synthesis. *Eur. J. Biochem.* **191:** 1–17.

Pardee A.B. 1989. G_1 events and regulation of cell proliferation. *Science* **246:** 603–608.

Patti M.E., Brambilla E., Luzi L., Landaker E.J., and Kahn C.R. 1998. Bidirectional modulation of insulin action by amino acids. *J. Clin. Invest.* **101:** 1519–1529.

Pearson R.B., Dennis P.B., Han J.W., Williamson N.A., Kozma S.C., Wettenhall R.E.H., and Thomas G. 1995. The principal target of rapamycin-induced $p70^{s6k}$ inactivation is a novel phosphorylation site within a conserved hydrophobic domain. *EMBO J.* **21:** 5279–5287.

Pellizzoni L., Lotti F., Rutjes S.A., and Pierandrei-Amaldi P. 1998. Involvement of the *Xenopus laevis* Ro60 autoantigen in the alternative interaction of La and CNBP proteins with the 5′UTR of L4 ribosomal protein mRNA. *J. Mol. Biol.* **281:** 593–608.

Pestova T.V., Borukhov S.I., and Hellen C.U. 1998. Eukaryotic ribosomes require initiation factors 1 and 1A to locate initiation codons (comments). *Nature* **394:** 854–859.

Peterson R.T., Desai B.N., Hardwick J.S., and Schreiber S.L. 1999. Protein phosphatase 2A interacts with the 70-kDa S6 kinase and is activated by inhibition of FKBP12-rapamycin-associated protein. *Proc. Natl. Acad. Sci.* **96:** 4438–4442.

Price D.J., Nemenoff R.A., and Avruch J. 1989. Purification of a hepatic S6 kinase from cycloheximide-treated rats. *J. Biol. Chem.* **264:** 13825–13833.

Price D.J., Grove J.R., Calvo V., Avruch J., and Bierer B.E. 1992. Rapamycin-induced inhibition of the 70-kilodalton S6 protein kinase. *Science* **257:** 973–977.

Pullen N. and Thomas G. 1997. The modular phosphorylation and activation of $p70^{s6k}$. *FEBS Lett.* **410:** 78–82.

Pullen N., Dennis P.B., Andjelkovic M., Dufner A., Kozma S., Hemmings B.A., and Thomas G. 1998. Phosphorylation and activation of $p70^{s6k}$ by PDK1. *Science* **279:** 707–710.

Radimerski T., Mini T., Schneider U., Wettenhall R.E.H., Thomas G., and Jenö P. 2000. Identification of insulin-induced sites of ribosomal protein S6 phosphorylation in *Drosophila melanogaster*. *Biochemistry* (in press).

Reinhard C., Thomas G., and Kozma S.C. 1992. A single gene encodes two isoforms of the p70 S6 kinase: Activation upon mitogenic stimulation. *Proc. Natl. Acad. Sci.* **89:** 4052–4056.

Reinhard C., Fernandez A., Lamb N.J.C., and Thomas G. 1994. Nuclear localization of $p85^{s6k}$: Functional requirement for entry into S phase. *EMBO J.* **13:** 1557–1565.

Romanelli A., Martin K.A., Toker A., and Blenis J. 1999. p70 S6 kinase is regulated by protein kinase C zeta and participates in a phosphoinositide 3-kinase-regulated signalling complex. *Mol. Cell. Biol.* **19:** 2921–2928.

Saitoh M., ten Dijke P., Miyazono K., and Ichijo H. 1998. Cloning and characterization of $p70^{s6k}\beta$ defines a novel family of p70 S6 kinases. *Biochem. Biophys. Res. Commun.* **253:** 470–476.

Schmidt E.V. 1999. The role of c-*myc* in cellular growth control. Oncogene **18:** 2988–2996.

Schwab M.S., Kim S.H., Terada N., Edfjall C., Kozma S.C., Thomas G., and Maller J.L. 1999. p70(S6K) controls selective mRNA translation during oocyte maturation and early embryogenesis in *Xenopus laevis*. *Mol. Cell. Biol.* **19:** 2485–2494.

Shima H., Pende M., Chen Y., Fumagalli S., Thomas G., and Kozma S.C. 1998. Disruption of the $p70^{s6k}/p85^{s6k}$ gene reveals a small mouse phenotype and a new functional S6 kinase. *EMBO J.* **17:** 6649–6659.

Stephens L., Anderson K., Stokoe D., Erdjument-Bromage H., Painter G.F., Holmes A.B., Gaffney P.R., Reese C.B., McCormick F., Tempst P., Coadwell J., and Hawkins P.T. 1998. Protein kinase B kinases that mediate phosphatidylinositol 3,4,5-trisphosphate-dependent activation of protein kinase B. *Science* **279:** 710–714.

Stewart M.J. and Denell R. 1993. Mutations in the *Drosophila* gene encoding ribosomal protein S6 cause tissue overgrowth. *Mol. Cell. Biol.* **13:** 2524–2535.

Stewart M.J., Berry C.O.A., Zilberman F., Thomas G., and Kozma S.C. 1996. The *Drosophila* $p70^{s6k}$ homolog exhibits conserved regulatory elements and rapamycin sensitivity. *Proc. Natl. Acad. Sci.* **93:** 10791–10796.

Su T.T. and O'Farrell P.H. 1998. Size control: Cell proliferation does not equal growth. *Curr. Biol.* **8:** R687–R689.

Terada N., Patel H.R., Takase K., Kohno K., Narin A.C., and Gelfand E.W. 1994. Rapamycin selectively inhibits translation of mRNAs encoding elongation factors and ribosomal proteins. *Proc. Natl. Acad. Sci.* **91:** 11477–11481.

Thomas G. 2000. An encore for ribosome biogenesis in the control of cell proliferation. *Nat. Cell Biol.* **2:** E71–E72.

Thomas G. and Hall M.N. 1997. TOR signalling and control of cell growth. *Curr. Opin. Cell Biol.* **9:** 782–787.

Thomas G. and Thomas G. 1986. Translational control of mRNA expression during the early mitogenic response in Swiss mouse 3T3 cells: Identification of specific proteins. *J. Cell Biol.* **103:** 2137–2144.

Thomas G., Siegmann M., and Gordon J. 1979. Multiple phosphorylation of ribosomal protein S6 during transition of quiescent 3T3 cells into early G_1, and cellular compartmentalisation of the phosphate donor. *Proc. Natl. Acad. Sci.* **76:** 3952–3956.

Thomas G., Martin-Pèrez J., Siegmann M., and Otto A.M. 1982. The effect of serum, EGF, PGF_{2a} and insulin on S6 phosphorylation and the initiation of protein and DNA synthesis. *Cell* **30:** 235–242.

Thomas G., Siegmann M., Kubler A., Gordon J., and Jimenez de Asua L. 1980. Regulation of 40S ribosomal protein S6 phosphorylation in Swiss mouse 3T3 cells. *Cell* **19:**

1015–1023.

Todorov I.T., Noll F., and Hadjiolov A.A. 1983. The sequential addition of ribosomal proteins during the formation of the small ribosomal subunit in Friend erythroleukemia cells. *Eur. J. Biochem.* **131:** 271–275.

Volarevic S., Stewart M.J., Ledermann B., Zilberman F., Terracciano L., Montini E., Grompe M., Kozma S.C., and Thomas G. 2000. Proliferation but not growth, blocked by conditional deletion of 40S ribosomal protein S6. *Science* **288:** 2045–2042.

Von Manteuffel S.R., Gingras A.-C., Ming X.-F., Sonenberg N., and Thomas G. 1996. 4E-BP1 phosphorylation is mediated by the FRAP-p70^{s6k} pathway and is independent of mitogen-activated protein kinase. *Proc. Natl. Acad. Sci.* **93:** 4076–4080.

Von Manteuffel S.R., Dennis P.B., Pullen N., Gingras A.-C., Sonenberg N., and Thomas G. 1997. The insulin-induced signalling pathway leading to S6 and initiation factor 4E binding protein 1 phosphorylation bifurcates at a rapamycin-sensitive point immediately upstream of p70^{s6k}. *Mol. Cell. Biol.* **17:** 5426–5436.

Wang X., Campbell L.E., Miller C.M., and Proud C.G. 1998. Amino acid availability regulates p70 S6 kinase and multiple translation factors. *Biochem. J.* **334:** 261–267.

Watson K.L., Konrad K.D., Woods D.F., and Bryant P.J. 1992. *Drosophila* homolog of the human S6 ribosomal protein is required for tumor suppression in the hematopoietic system. *Proc. Natl. Acad. Sci.* **89:** 11302–11306.

Watson K.L., Chou M.M., Blenis J., Gelbart W.M., and Erickson R.L. 1996. A *Drosophila* gene structurally and functionally homologous to the mammalian 70-kDa S6 kinase gene. *Proc. Natl. Acad. Sci.* **93:** 13694–13698.

Weng Q.-P., Andrabi K., Kozlowski M.T., Grove J.R., and Avruch J. 1995. Multiple independent inputs are required for activation of the p70 S6 kinase. *Mol. Cell. Biol.* **15:** 2333–2340.

Westphal R.S., Coffee R.L., Jr., Marotta A., Pelech S.L., and Wadzinski B.E. 1999. Identification of kinase-phosphatase signaling modules composed of p70 S6 kinase-protein phosphatase 2A (PP2A) and p21-activated kinase-PP2A. *J. Biol. Chem.* **274:** 687–692.

Wool I.G., Chan Y.-L., and Glück A. 1996. Mammalian ribosomes: The structure and the evolution of the proteins. In *Translational control* (ed. J.W.B. Hershey et al.), pp. 685–732. Cold Spring Harbor Laboratory Press, Cold Spring Harbor, New York

Xu G., Kwon G., Marshall C.A., Lin T.A., Lawrence J.C., Jr., and McDaniel M.L. 1998. Branched-chain amino acids are essential in the regulation of PHAS-I and p70 S6 kinase by pancreatic beta-cells. A possible role in protein translation and mitogenic signaling. *J. Biol. Chem.* **273:** 28178–28184.

24

Control of the Elongation Phase of Protein Synthesis

Christopher Proud

Department of Anatomy and Physiology
University of Dundee
Dundee DD1 5EH
United Kingdom

EUKARYOTIC ELONGATION FACTORS

Peptide-chain elongation is the programmed assembly of amino acids into a polypeptide as dictated by the sequence of bases in the mRNA. In all organisms it requires ancillary proteins termed elongation factors (see Chapter 3). These factors fall into two groups: (1) those required for the recruitment of aminoacyl-tRNAs to the ribosome and (2) those involved in the subsequent translocation event in which the ribosome moves along the mRNA. The proteins mediating these steps in the cytoplasm of eukaryotic cells are termed eEF1A[1] and eEF2, respectively (Table 1). Both are GTP-binding proteins and the GTP is hydrolyzed during the events associated with the function of these proteins. Each protein thus ends up bound to GDP, in which state it is inactive and must be recycled to its GTP-bound form. For eEF2 this appears to occur spontaneously (due to rapid dissociation of the GDP), whereas for eEF1A/EF-Tu a guanine nucleotide-exchange factor (GEF) termed eEF1B is needed. Mammalian eEF1B consists of three subunits (α–γ), two of which (α and γ) have nucleotide-exchange activity, at least in vitro.

It is the regulation of these proteins, primarily the higher eukaryotic factors, with which this chapter is concerned. The structural organization of eEF1A, eEF1B, and eEF2 is summarized in Figure 1, which also indicates the location of known sites of phosphorylation in these proteins.

[1]eEF1 was formerly regarded as a complex of four subunits. More recently, to be more consistent, e.g., with the initiation factors eIF2 and eIF2B, they have been redefined as a GTP- and amino acyl-tRNA binding factor (eEF1A, formerly termed eEF1α) and a trimeric guanine nucleotide-exchange factor (eEF1B, with subunits α-γ, formerly called eEF1$\beta\gamma\delta$).

© 2000 Cold Spring Harbor Laboratory Press 0-87969-568-4/00 $5 +. 00

Table 1 Eukaryotic elongation factors

Factor (earlier name)	Molecular mass (kD)	Function	Regulation?	Other comments
eEF1A (EF-1; eEF1α)	50.1	binds GTP and aminoacyl-tRNA		undergoes multiple posttranslational modifications (Dever et al. 1989; Cavallius et al. 1993); tissue-specific isoform identified associated with wasted phenotype (Chambers et al. 1998; Kahns et al. 1998)
eEF1B (eEF1βγδ)	24.8 (α); 50.0 (γ); 31.1 (β)	catalyzes guanine nucleotide exchange on eEF1A		
eEF2	95.2	translocation; binds GTP	phosphorylation at Thr-56 inhibits activity	also inhibited by ADP-ribosylation catalyzed by *Diphtheria* and *Pseudomonas* A toxins

REGULATION OF ELONGATION

Why Regulate Elongation?

Altering the rate of elongation will alter the rate at which peptide chains are completed and thus is expected to contribute to the overall regulation of protein synthesis, e.g., to its activation by stimuli such as insulin (see below). Under conditions where the cell needs to conserve energy or divert it into other processes, inhibition of elongation would serve temporarily to reduce the amount of energy being consumed by protein synthesis, thus allowing, e.g., nucleoside triphosphates to be utilized for other processes (e.g., muscle contraction, see below). Regulating elongation under such circumstances can be viewed as preferable to affecting initiation, which is generally regarded as the major locus for control of translation: Inhibiting elongation will not cause disaggregation of polysomes, thus allowing the effect to be reversed quickly, and would be unlikely to have differential effects on the translation of different mRNAs, which *are* likely to arise if initiation were inhibited.

Another reason for regulating elongation rates when overall protein synthesis is being activated or inhibited relates to the fidelity of transla-

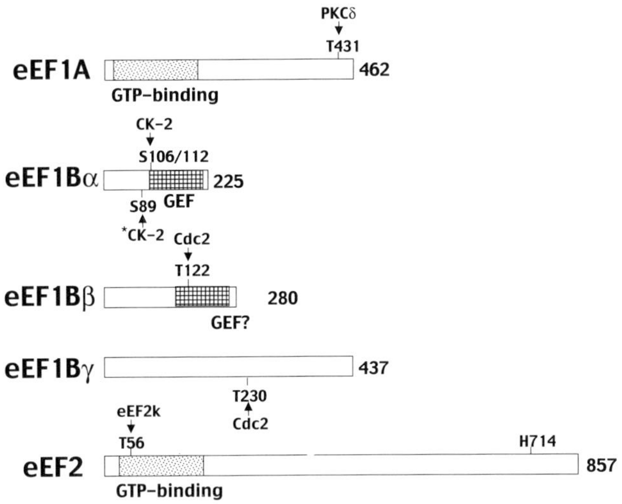

Figure 1 Organization of eukaryotic elongation factors. The figure depicts the structures of the eukaryotic elongation factors eEF1A, eEF1B, and eEF2. To the right of each is indicated the number of residues in the mammalian proteins. The stippled regions in eEF1A and eEF2 show their GTP-binding domains, and for eEF1Bα and β the cross-hatched areas indicate their guanine nucleotide-exchange factor (GEF) domains. Information above or below the box indicates phosphorylation sites, using the single-letter code for amino acids. In the case of eEF2, the site of ADP-ribosylation at diphthamide (a modified form of His-714) is also indicated. The kinase(s) responsible are indicated by abbreviations as follows: (cdc2) $p34^{cdc2}$; (CK-2) casein kinase-2; (eEF2k) eEF2 kinase; (PKCδ) protein kinase Cδ. The figure is not intended to imply that these phosphorylation events occur in all species: Unless otherwise indicated, the information was obtained in studies on the mammalian factors. See the text for further information. *Phosphorylation site in *Asteria salina* factor.

tion. The activity of elongation factors modulates the accuracy of the process; decreased activity of eEF1A (or eEF1B), for example, results in increased fidelity (see, e.g., Carr-Schmid et al. 1999 and references therein). Thus, it may be important to match elongation factor activity closely to the actual requirements of the cell instead of allowing a high basal activity of elongation, which might compromise translational accuracy and potentially lead to increased missense errors as well as to termination readthrough. Thus, where protein synthesis is activated, for example, it would be important to increase translation factor activity to avoid the possibility that the rate of elongation would limit overall protein synthesis. Conversely, when translation initiation is inhibited, it may be important to slow the rate of elongation, thus helping to maintain polysome integrity. This could help protect mRNAs against degradation and/or allow active translation to resume quickly once initiation was switched on again.

Conditions under Which the Rate of Elongation Has Been Shown to Be Altered

Early studies (Nielsen and McConkey 1980) showed that stimulation of HeLa with serum resulted in a twofold reduction in the ribosomal transit time, indicative of a two-fold acceleration of elongation. Insulin has been shown to increase rapidly (within times < 1 hour) the rate of elongation, again measured by ribosomal transit times, in several cell-types. These include 3T3-L1 cells (Chang and Traugh 1997), Chinese hamster ovary (CHO) cells expressing the insulin receptor (Redpath et al. 1996a), and mouse soleus muscle (Yoshizawa et al. 1995). Conversely, in reticulocytes, anoxic or hypertonic stress caused a decrease in the overall rate of elongation (Binninger and Weber 1984). Glucagon has been shown to inhibit protein synthesis and cause accumulation of polysomes in rat hepatocytes, suggesting that it also inhibits elongation (Ayuso-Parrilla et al. 1976). However, the control of elongation remains a relatively neglected area of translation control, and the growing evidence that the activities of elongation factors are subject to acute control, e.g., by phosphorylation, suggests that it merits more detailed study.

Regulation of the Cellular Levels of Elongation Factors

The effects of insulin on elongation rates are rapid, occurring within about 1 hour after treatment of the cells, and are likely to involve changes in the intrinsic activity of components involved in elongation. At longer times, the cellular levels of elongation factors are also subject to control. In higher eukaryotes, the mRNA encoding eEF1A possesses a 5′-terminal tract of pyrimidines (5′TOP), which confers on it regulation through the rapamycin-sensitive mTOR (target of rapamycin) pathway (see Chapter 22). The eEF1A mRNA is poorly translated in unstimulated cells but shifts into polysomes upon serum-treatment, leading to increased synthesis of eEF1A (Jefferies and Thomas 1994; Jefferies et al. 1994). Amino acids also modulate mTOR-linked signalling events in mammalian cells and have been shown to regulate, positively, the synthesis of eEF1A (see Chapter 16). Thus, treatment with cells with certain hormones and changes in amino acid supply each modulate the synthesis of eEF1A: It clearly makes physiological "sense" for the cellular level of eEF1A to be controlled by agents that enhance (e.g., hormones) or are required for (amino acids) mRNA translation. Furthermore, insulin rapidly (< 20 minutes) induces the synthesis of eEF2 by activating the translation of its mRNA (Levenson et al. 1989), the 5′ terminus of which also appears to contain a TOP sequence, suggesting it is regulated in a similar way to eEF1A and other 5′-TOP mRNAs such as those for ribosomal proteins (Meyuhas et al. 1996).

REGULATION OF ELONGATION FACTORS BY PHOSPHORYLATION

eEF1A, all three components of eEF1B, and eEF2 have all been shown to undergo phosphorylation, suggesting that this may offer mechanisms for modulating their activities in vivo in response to particular conditions or stimuli. Understanding of the regulation of eEF2 is more advanced than is the case for eEF1A/B, and it is discussed first.

Regulation of eEF2

Phosphorylation Inhibits eEF2 Activity

eEF2 is a monomer of approximately 100 kD. eEF2 from mammals and insects undergoes phosphorylation in its amino terminus, i.e., within the region that binds guanine nucleotides (Nairn and Palfrey 1987; Ovchinnikov et al. 1990; Price et al. 1991; Oldfield and Proud 1993). Most studies have focused on mammalian eEF2, and it has been shown that phosphorylation inhibits its activity (Nairn and Palfrey 1987; Ryazanov et al. 1988; Carlberg et al. 1990; Redpath et al. 1993), probably by impairing its interaction with ribosomes (Carlberg et al. 1990). Early experiments studying the effect of eEF2 phosphorylation employed the poly(U)-directed synthesis of polyphenylalanine and suggested that phosphorylated eEF2 exerted a dominant inhibitory effect (Ryazanov et al. 1988; Redpath et al. 1993). However, this was not borne out by studies in the more "physiological" reticulocyte lysate system, where even very high levels of eEF2 phosphorylation did not completely inhibit translation (Redpath et al. 1993). This is consistent with the idea that phosphorylation causes a loss of activity, where phosphorylated eEF2 fails to bind to ribosomes (Carlberg et al. 1990) and is thus absent from polyribosomes (Redpath and Proud 1989). The principal site of phosphorylation is Thr-56, but Thr-53 and Thr-58 can also be phosphorylated in vitro, albeit more slowly than Thr-56 (Nairn and Palfrey 1987; Ovchinnikov et al. 1990; Price et al. 1991; Redpath et al. 1993). In vivo, only the species phosphorylated at Thr-56 is detected (Palfrey et al. 1987b; Mackie et al. 1989). Antisera are available that specifically detect this form of eEF2 (Marin et al. 1997).

eEF2 Kinase Is an Unusual Protein Kinase

The protein kinase responsible for the phosphorylation of eEF2 was initially identified as a Ca^{++}/calmodulin (CaM)-dependent enzyme termed CaM-kinase III (Nairn et al. 1985; Palfrey et al. 1987a), prior to the discovery that its 100-kD substrate was eEF2 (Nairn and Palfrey 1987). eEF2 is the only known substrate for this enzyme, which has since been renamed eEF2 kinase. cDNAs encoding eEF2 kinase have been cloned from several species (Redpath et al. 1996b; Ryazanov et al. 1997), and homologs have been identified in others mainly from searching databases (Ryazanov et al. 1997, 1999). eEF2 kinase is not closely related to the main Ser/Thr/Tyr kinase superfamily or to the histidine kinase/mitochondrial kinase family either (Redpath et al. 1996b; Ryazanov et al. 1997). Instead, it appears to belong to a small group of enzymes that also includes myosin heavy-chain kinases in *Dictyostelium discoideum* and several mammalian proteins, which were identified only from searching databases and are therefore of unknown function (Ryazanov et al. 1999). The phosphorylation sites on the myosin heavy chain lie in regions whose structures are α-helical. Given that in EF1A (which is related to eEF2), the region corresponding to that containing Thr-56 of eEF2 is also α-helical, Ryazanov has suggested that these kinases may target phosphorylation sites in α-helical regions rather than the β-turns in which many sites for other, orthodox, kinases lie. He has therefore termed these enzymes α-kinases (Ryazanov et al. 1999). Studies on the structural organization of eEF2 kinase are therefore of considerable interest, as it is a paradigm for this novel group of kinases. There is evidence for the existence of multiple forms of eEF2 kinase (or additional eEF2 kinases) in mammalian cells based on the lack of immunoreactivity to anti-eEF2 kinase antisera in some cell types which nevertheless possess substantial eEF2 kinase activity (Hait et al. 1996).

Redpath's group (Diggle et al. 1999) has recently begun to dissect the functional domains in eEF2 kinase. Their data further highlight the fact that eEF2 kinase is an unorthodox enzyme. First, its calmodulin-binding domain is amino-terminal to the catalytic domain, whereas it is carboxy-terminal in all other Ca^{++}/CaM-dependent kinases (Hardie and Hanks 1995). Second, members of the main serine/threonine kinase superfamily contain a GXGXXG sequence followed (after a short spacing) by an essential lysyl residue (Hardie and Hanks 1995). Although eEF2 kinase does contain a GXGXXG motif, neither of the two lysyl residues that follow it is required for activity. The data also suggest that the extreme carboxyl terminus of eEF2 kinase is required for its activity against eEF2, perhaps because this region is required for substrate binding.

Caenorhabditis elegans, the only metazoan whose entire genome has to date (late 1999) been sequenced, possesses homologs of mammalian eEF2 kinase (Ryazanov et al. 1999). In contrast, the yeast (*Saccharomyces cerevisiae*) genome does not contain one. Yeast eEF2 is a substrate for mammalian eEF2 kinase in vitro (Donovan and Bodley 1991), and some evidence has been presented that it is phosphorylated by an endogenous kinase present in yeast cells. However, little further work has been carried out to study the phosphorylation of yeast eEF2 and there is no information on its possible role in regulating mRNA translation in vivo.

The activity of eEF2 kinase is normally completely dependent on Ca^{++}ions and calmodulin at neutral pH (Nairn et al. 1987; Mitsui et al. 1993; Redpath and Proud 1993b; Hovland et al. 1999), but following treatment with cAMP-dependent protein kinase (PKA), which phosphorylates eEF2 kinase at multiple sites (Redpath and Proud 1993a), it acquires Ca^{++}/CaM-independent activity. In this respect, it resembles certain other Ca^{++}/CaM-dependent protein kinases, e.g., phosphorylase kinase (Heilmeyer and Kilimann 1995) (which acquires Ca-independent activity after phosphorylation by PKA) and the type II Ca^{++}/CaM-kinase (Hanson and Schulman 1992) (where Ca-independent activity arises after autophosphorylation). This property of eEF2 kinase should allow it to be activated in response to treatments that elevate cytoplasmic cAMP without increasing Ca^{++}ion levels, and thus allow agonists that raise cAMP, as well as those that increase $[Ca^{++}]_i$, to activate eEF2 kinase, bring about the phosphorylation of eEF2, and thus inhibit peptide-chain elongation (Fig. 2).

Regulation of Elongation by Agents That Raise Cytoplasmic Ca^{++} Levels

A variety of agents and treatments that raise cytoplasmic Ca^{++} levels have been shown to result in increased phosphorylation of eEF2. These include serum (Palfrey et al. 1987b; Redpath et al. 1996a) (but see also next section), bradykinin (Palfrey et al. 1987b) in fibroblasts, thrombin and histamine in endothelial cells (Mackie et al. 1989; Marino et al. 1996). Several different secretagogues increased eEF2 phosphorylation in acinar or chromaffin cells (Haycock et al. 1987; Hincke and Nairn 1992). Phosphorylation of eEF2 was transient, like the change in cytoplasmic Ca^{++} levels, but reached high levels, e.g., up to 90% in response to thrombin (Mackie et al. 1989). Peptide mapping indicated that the main site of phosphorylation was Thr-56 (Mackie et al. 1989).

The neurotransmitter glutamate increases Ca^{++} levels in neurons and plays a key role in long-term synaptic plasticity. Brain ischemia causes a massive release of glutamate and this is implicated in the damaging

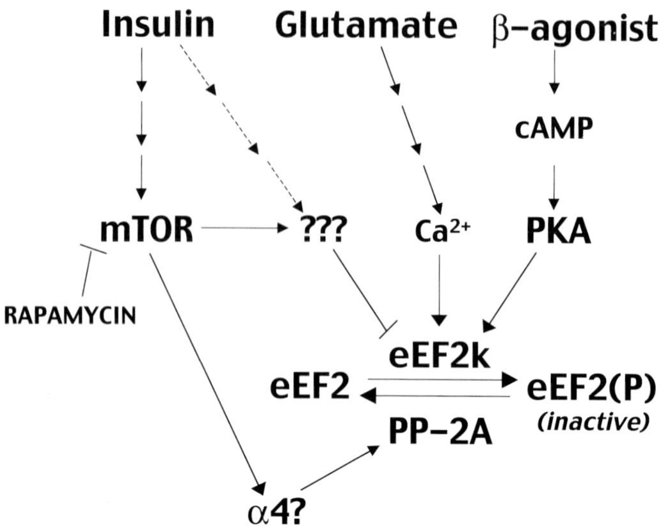

Figure 2 Signaling pathways impinging on eEF2. Elongation factor eEF2 can be converted from its active nonphosphorylated form (eEF2) to the inactive phosphorylated species (eEF2[P]) by phosphorylation by eEF2 kinase, which occurs at Thr-56. The phosphatase that catalyzes the reverse reaction and activates eEF2 is probably a form of PP-2A (see text). Insulin induces the rapid dephosphorylation of eEF2 through signaling events that are blocked by rapamycin and thus involve mTOR, although it is not clear whether these effects involve modulation of the activity of eEF2 kinase or PP-2A or both. Regulation of PP-2A might involve the phosphatase binding protein α4, which shows structural similarity to Tap42, which in turn is implicated in TOR signaling in yeast (see text). It is not clear how mTOR might regulate eEF2 kinase (indicated by ??? and dotted arrows). Glutamate is an example of a stimulus which by raising cytoplasmic Ca^{++} could activate eEF2 kinase (via calmodulin) and has been shown to cause phosphorylation of eEF2 (Marin et al. 1997) in neuronal cells. Other agents that raise cytoplasmic Ca^{++} concentrations are likely to have the same effect. β-Adrenergic agonists activate adenyl cyclase and increase cellular cAMP levels, thus activating PKA. Since PKA can phosphorylate eEF2 kinase and render its activity partially independent of Ca^{++} -ions, this would also lead to increased phosphorylation of eEF2, as has now been shown in several cell types (Diggle et al. 1998; Hovland et al. 1999).

effects of this condition (Coyle and Puttfarcken 1993), e.g., neuronal cell death. Neuronal death is preceded by inhibition of protein synthesis, and glutamate appears to play an important role in this (see Marin et al. 1997 and references therein). Recent studies (Marin et al. 1997) have shown that glutamate treatment of primary neuronal cultures causes a rapid and pronounced increase in eEF2 phosphorylation (Fig. 2), which correlates with changes in intracellular Ca^{++} levels and with inhibition of protein synthesis. The increased phosphorylation of eEF2 was localized to the cell bodies and proximal, but not distal, processes. On the basis of these and other data, the authors suggested that phosphorylation of eEF2 and inhibition of translation might actually constitute a protective mechanism against glutamate-induced cytotoxicity.

Activation of the *N*-methyl-D-aspartate (NMDA) receptor also regulates Ca^{++} levels and has been shown to bring about phosphorylation of eEF2 in tadpole tecta. In adult tecta, visual activity also causes phosphorylation of eEF2 (Scheetz et al. 1997). It has been suggested that these effects and those of glutamate referred to above may play a role in reprogramming the translational machinery as part of the process of synaptic plasticity. Synaptic plasticity involves changes in protein expression, raising the question of how newly synthesized proteins would be targeted to specific synapses: It is possible, instead, that localized changes in mRNA translation are involved, allowing new proteins to be made where they are required. However, it is not obvious how a general block in elongation, the predicted consequence of elevated eEF2 phosphorylation, would accomplish this — one might expect that this would actually tend to bring about the accumulation of ribosomes into existing polysomes, which would not have the proposed effect. However, studies from Thach and colleagues provided evidence that the elongation inhibitor cycloheximide actually enhanced the translation of certain mRNAs in mouse fibroblasts by driving them from untranslated particles or small polysomes into larger polysomes (Ray et al. 1985; Walden and Thach 1986). They proposed a model to explain these effects and suggested that inhibiting elongation would only reduce the translation of efficient or "strong" mRNAs, for which elongation is rate-limiting, whereas translation of "weak" mRNAs, for which initiation rather than elongation is limiting, would actually be increased when elongation rates are reduced (see Brendler et al. 1981 and references therein).

Regulation of Elongation by Agents That Raise Cytoplasmic cAMP Levels

As described above, eEF2 kinase can be phosphorylated by PKA in vitro, whereupon it acquires partial Ca^{++}/CaM-independent activity (Redpath and Proud 1993a). Consistent with this, hormonal or pharmacological treatments that activate PKA in intact cells do indeed render eEF2 kinase independent of Ca^{++}/CaM and increase the level of phosphorylation of eEF2 (Fig. 2) (Diggle et al. 1998; Hovland et al. 1999). For example, in both primary fat cells and 3T3-L1 adipocytes, isoproterenol or a cell-permeant cAMP analog (chlorophenylthio-cAMP, CPT-cAMP) induced Ca^{++}-independent eEF2 kinase activity. In 3T3-L1 adipocytes, these and other treatments designed to elevate cAMP caused increased phosphorylation of eEF2 (Diggle et al. 1998), and this was associated with inhibition of total protein synthesis and reduced rates of elongation (as indicated by increased transit times, a measurement of the rate of elongation [Diggle et al. 1998]). In IPC-81 fibroblasts, CPT-cAMP also caused increased eEF2 phosphorylation and retardation of elongation, as well as accumulation of polyribosomes (Hovland et al. 1999). These data show that the rate of elongation can be regulated by signals transduced via cAMP (Fig. 2). Early work showed that glucagon inhibited elongation in rat hepatocytes (Ayuso-Parrilla et al. 1976). Because glucagon increases cAMP levels in these cells, the effect may also have been exerted through phosphorylation and inhibition of eEF2.

What could be the physiological significance of the inhibition of elongation in response to cAMP? As cAMP levels are generally increased in response to stress- or starvation-related conditions (e.g., by adrenaline or glucagon), the most obvious explanation is that it would serve to inhibit protein synthesis, acutely and reversibly, during periods when cellular energy is required for other processes. For example, following adrenalin treatment of muscle, it might be important to maximize the availability of ATP for contraction. In the case of glucagon, it may also serve to help divert amino acids from protein synthesis into gluconeogenesis for glucose production.

Regulation of Elongation by Insulin Links to mTOR Signaling

Insulin activates protein synthesis in a wide range of cell types (for review, see Proud and Denton 1997). Although it is well-established that this effect involves the activation of the process of peptide-chain initiation, several studies have now shown that insulin also rapidly brings about the dephosphorylation of eEF2 (in CHO cells [Redpath et al. 1996a], adipocytes [Diggle et al. 1998], and adult cardiomyocytes [Wang et al. 2000]). In the first case, this was shown to be associated, as expected, with an acceleration of elongation (Redpath et al. 1996a), and the effects of insulin on eEF2 phosphorylation and transit time were both blocked by the immunosuppressant drug rapamycin. Rapamycin acts by inhibiting the “mammalian target of rapamycin,” a protein that has similarity to lipid kinases of the phosphoinositide kinase family but has not so far been shown to have lipid kinase activity itself (see Chapter 6). A number of reports have indicated that it has protein kinase activity, at least in vitro (Brunn et al. 1997; Burnett et al. 1998), against other translational regulators that are controlled through this rapamycin-sensitive pathway; i.e., the initiation factor eIF4E binding protein, 4E-BP1, and the ribosomal protein S6 kinase, p70 S6k (Proud 1996; Lawrence and Abraham 1997;

Chapters 6 and 23). It could be that mTOR phosphorylates and regulates eEF2 kinase: If so, it would be expected that phosphorylation would inhibit eEF2 kinase. However, treatment of cells with either insulin or rapamycin has only very modest effects on the activity of eEF2 kinase when its activity is subsequently measured in vitro. Perhaps the effects are more marked under in vivo conditions. Alternatively, these agents may also, or mainly, affect the activity of the protein phosphatase acting on eEF2, most likely a form of protein phosphatase-2A (PP-2A). A number of connections between rapamycin and PP-2A activity have been reported in mammalian cells as well as in yeast (Di Como and Arndt 1996; Murata et al. 1997; Peterson and Schreiber 1998; Jiang and Broach 1999), and PP-2A has been found associated with p70 S6 kinase (Peterson et al. 1999). In budding yeast, the protein Tap42p interacts with the catalytic subunits of PP-2A and plays a key role in TOR-dependent signaling (Di Como and Arndt 1996). In mammalian cells, α4 is homologous to Tap42p and also associates with the catalytic subunit of PP-2A. Association of PP-2A with α4 has been reported to affect its activity against several substrates (Nanahoshi et al. 1998, 1999). Forced overexpression of α4 in COS-7 cells has recently been shown to cause dephosphorylation of eEF2 (Chung et al. 1999), perhaps because it affects the activity of PP-2A against eEF2 or against eEF2 kinase. mTOR signaling is an active area of research (see Chapter 6), and further work will be required to establish the role played by mTOR in the control of eEF2 phosphorylation. Current views are summarized in Figure 2.

Role of Nutrients in Modulating the Phosphorylation of eEF2

Early work by van Venrooij et al. (1970) showed that the availability of glucose and amino acids regulated both the initiation and elongation stages of translation. Subsequent studies have shed light on the molecular mechanisms that may underlie these observations. In several types of mammalian cells, amino acids have been shown to modulate mRNA translation through the mTOR pathway (see Chapter 16 Kimball and Jefferson): In the absence of amino acids, p70 S6 kinase undergoes dephosphorylation and inactivation, while the eIF4E-binding protein 4E-BP1 also becomes dephosphorylated and is then able to bind to and inhibit eIF4E (Fox et al. 1998a,b; Hara et al. 1998; Patti et al. 1998; Wang et al. 1998; Shigemitsu et al. 1999). eEF2 also responds to amino acid withdrawal, undergoing a marked increase in phosphorylation (Wang et al. 1998). All three changes contribute to a reduction in the rate of translation: Such inhibition clearly makes physiological sense in the absence of the precursors for protein synthesis. Amino acids are required for the regulation of p70 S6k and 4E-BP1 by insulin and other stimuli (Hara et al. 1998; Wang et al. 1998). Surprisingly, they are not sufficient to facilitate the insulin-induced dephosphorylation of eEF2, at least in CHO cells, because glucose is also required for this effect (Campbell et al. 1999), suggesting that the control of eEF2 involves other signals in addition to those emanating from mTOR. Such glucose-dependent signals might be allosteric regulators of eEF2 or eEF2 kinase, which are generated from glucose (e.g., sugar phosphates), or they could be connected with a mammalian glucose signaling pathway similar to that which operates in budding yeast in response to this sugar (Liang and Gaber 1996; Ozcan et al. 1998). It would make physiological sense for cells to assess the availability of glucose (energy source) or amino acids (precursors) when responding to stimuli that stimulate translation. In this context, it is interesting to note that the ability of insulin to activate another translation factor required for overall protein synthesis, the initiation factor eIF2B, also requires glucose and amino acids (Campbell et al. 1999).

REGULATION OF eEF1A AND eEF1B

eEF1A and eEF1B are Phosphoproteins

A number of reports have described the phosphorylation of eEF1A and of subunits of eEF1B from a range of eukaryotic species. Rabbit eEF1A and the α and γ subunits of eEF1B are substrates in vitro for protein kinase C (PKC) (Venema et al. 1991a), and all three undergo increased phosphorylation when reticulocytes are treated with phorbol esters, which activate classic and atypical isoforms of PKC (Venema et al. 1991b). Data for eEF1A indicated that PKC phosphorylated the same sites in vitro as are phosphorylated in response to phorbol esters in vivo (Venema et al. 1991a). When the eEF1A/B complex was isolated (as a complex with valyl-tRNA synthetase) from phorbol-ester-treated cells and assayed for activity, it was found to support a twofold higher rate of elongation than the complex from untreated cells (Venema et al. 1991b). Similarly, treatment of the purified complex with PKC increased its activity in guanine nucleotide exchange and thus in the formation of complexes with charged tRNA (Peters et al. 1995). Mammalian eEF1A is phosphorylated by PKCδ at Thr-431 (Kielbassa et al. 1995) (which is conserved in almost all known eEF1A sequences), but whether this affected its activity was not tested. To date, no data have been presented to show whether phorbol esters affect the rate of elongation in intact cells.

Insulin brings about increased phosphorylation of eEF1A and the α and γ subunits of eEF1B (Chang and Traugh 1999). The complex isolated from insulin-treated cells showed twofold higher activity than that from serum-starved cells, although it is not clear how the in vivo phosphorylation affects the activity of the complex — i.e., whether it alters the activity of eEF1A or of the GEF complex eEF1B. The data also suggest that PKC is not directly responsible for the phosphorylation of eEF1A/B in response to insulin and point instead to a role here for an enzyme termed the multipotential protein kinase. This kinase phosphorylates eEF1A and the α/γ-subunits of eEF1B in vitro and also enhances the activity of the complex as assayed in vitro (Chang and Traugh 1997).

It therefore appears that, in mammalian cells, the activity of the eEF1A/B complex may be enhanced by insulin or phorbol esters through phosphorylation by distinct kinases that target three polypeptide components of this complex (eEF1A, eEF1Bα and β). However, it still remains to be established whether the phosphorylation of eEF1A/B in vivo actually modulates the rate of elongation in the cell. This is clearly a key issue. The data of Peters et al. (1995) indicate that the phosphorylation of eEF1 by PKC in vitro leads to the stimulation of guanine nucleotide exchange on eEF1A, and there is evidence that this process may be a rate-controlling step in elongation (Janssen and Moller 1988). Increasing the rate of nucleotide exchange would accelerate the regeneration of active eEF1A•GTP complexes and potentially increase the formation of the eEF1A•GTP•aminoacyl-tRNA complexes that recruit aminoacyl-tRNAs to the ribosomal A-site during elongation (but it remains to be shown that phosphorylation of eEF1A/B in vivo, e.g., in response to insulin, actually affects nucleotide exchange). There is potentially a clear parallel here with the initiation of translation, and specifically with eIF2B (Webb and Proud 1998), the GEF for eIF2, the factor that recruits the initiator methionyl-tRNA to the 40S initiation complex. The activity of eIF2B regulates the regeneration of active eIF2, and it is well-established that eIF2B activity is regulated under a wide range of conditions in eukaryotic cells, both through phosphorylation of its substrate eIF2 (see Chapter 5) and of its own ε subunit (Welsh et al. 1998). As discussed above, a second potential mechanism for activating elongation in response to insulin has been identified (Redpath et al. 1996a) involving dephosphorylation of eEF2. Both this and activation of elongation rates by insulin were inhibited by the immunosuppressant rapamycin, whereas rapamycin had no effect on the insulin-induced phosphorylation of subunits of eEF1A/B (Chang and Traugh 1999). These data are consistent with the notion that regulation of eEF2 plays a major role in modulating elongation rates in response to insulin, and phosphorylation of eEF1A/B is not involved, at least in the cell type studied (Redpath et al. 1996a).

Phosphorylation of eEF1B by CK-2

There is also some evidence for regulation of the nucleotide exchange process in other animal species: Janssen et al. (1988) demonstrated that eEF1Bα, from *Artemia salina* can be phosphorylated by a kinase which copurifies with it. The phosphorylation site was identified as Ser-89, and the kinase responsible was identified (based largely on indirect evidence) as casein kinase-2 (CK-2). Phosphorylation decreased the protein's guanine nucleotide exchange activity. Sequence conservation is poor in this region of the protein, but many species possess potential CK-2 phosphorylation sites in roughly this part of their structure. Indeed, the eEF1Bα polypeptides from rabbit and wheat can be phosphorylated by CK-2 (Matsumoto et al. 1993; Chen and Traugh 1995).

Traugh's group has extensively studied the phosphorylation of subunits of mammalian eEF1B by CK-2. They find that phosphorylation by CK-2 of eEF1Bα occurs at seryl residues 106 and 112 (Chen and Traugh 1995): These sites are present in the sequences of eEF1Bα from other mammalian species, chicken and *Xenopus laevis*. CK-2 also phosphorylates eEF1Bβ (Sheu and Traugh 1999), but the authors were unable to demonstrate any effect of phosphorylation of eEF1Bα or β, alone or when associated with other subunits, upon the rate of elongation, measured using the poly(U)-directed synthesis of polyphenylalanine (Sheu and Traugh 1997). Possible effects of phosphorylation do not appear to have been tested in a more physiologically relevant assay system, and it is thus unclear whether phosphorylation of these proteins by CK-2 might play a role in controlling translation. In this context, it should be noted that phosphorylation of eEF2 affects its activity in elongation with different characteristics when assayed in poly(U)-directed synthesis of polyphenylalanine or reticulocyte lysates (Redpath et al. 1993; see above). Although earlier studies suggested that CK-2 was activated by insulin or growth factors, more recent work has questioned this (Litchfield et al. 1994), adding further doubt to the significance of CK-2-mediated phosphorylation in the acute control of eEF1A/B.

Cell-cycle Regulation of the Phosphorylation of eEF1B

Belle and colleagues have presented a substantial amount of data showing that subunits of eEF1B can be phosphorylated by the cell-cycle-regulated kinase, p34[cdc2]. These studies stemmed from the initial observation that a 47-kD protein from *X. laevis*, which is phosphorylated by the meiotic maturation-promoting factor that contains p34[cdc2], was eEF1Bγ (Belle et al. 1989; Janssen et al. 1991). eEF1Bγ becomes phosphorylated during meiosis. Subsequent work identified the phosphorylation site as Thr-230 (Mulner-Lorillon et al. 1992), which is present in the sequences of mammalian eEF1B proteins but not in those from *Artemia*, fission yeast, or several other nonvertebrate species.

eEF1Bβ in *Xenopus* is also a substrate for p34[cdc2] (Mulner-Lorillon et al. 1994), with phosphorylation occurring on seryl and threonyl residues. One site of phosphorylation was identified as Thr-122, which lies in a motif suitable for phosphorylation by p34[cdc2], but is absent from the rabbit or human proteins (Mulner-Lorillon et al. 1994). During oocyte maturation, this site underwent phosphorylation at metaphase, i.e., when p34[cdc2] is active, and, as assessed from changes in the electrophoretic migration of eEF1Bβ, persisted into early stages of development (Minella et al. 1994). To date, no studies on the phosphorylation of mammalian eEF1A/B by cell-cycle kinases or during the cell-cycle have appeared, and there appear to be no data at all on the functional effect of these phosphorylation events. It is therefore not yet possible to make any conclusions about their physiological rolc(s).

Other Conditions under Which Elongation Is Known to Be Regulated

A number of such conditions have already been discussed, such as insulin, agents that raise cytoplasmic cAMP or Ca^{++} levels, and thereby modulate the phosphorylation of eEF2. Early work from Ballinger and Purdue (1983) indicated that, in *Drosophila* S2 Schneider cells, heat shock resulted in decreased rates of peptide-chain elongation on non-heat shock mRNAs, since material sedimenting in a similar fashion to polysomes accumulated although protein synthesis was inhibited. From our understanding of the control of elongation factor activity, it is not clear how heat shock could inhibit elongation or how this could be specific for certain classes of mRNA. It may be that the mRNAs are actually packaged into large non-polysomal complexes rather than actually building up as true polysomes.

CONCLUSIONS AND PERSPECTIVES

The regulation of peptide-chain elongation remains a relatively neglected area of translational control, and further work is required to extend our rather sketchy knowledge of the conditions under which elongation rates are modulated. This comment applies both to the regulation of eEF2, which has been studied only under a small set of conditions in a few cell types, and even more to eEF1A/B, as our understanding of physiological control of these proteins is particularly poor. In the case of eEF2, the very topical area of the links to mTOR and to nutrient availability merits further detailed investigation. As a relatively well understood member of a "new" enzyme family, the structure and regulation of eEF2 kinase are also of particular interest.

REFERENCES

Ayuso-Parrilla M.S., Martin-Requero A., Perez-Diaz J., and Parrilla R. 1976. Role of glucagon in the control of hepatic protein synthesis in the rat *in vivo*. *J. Biol. Chem.* **251:** 7785–7790.

Ballinger D.G. and Pardue M.L. 1983. The control of protein synthesis during heat shock in *Drosophila* cells involves altered polypeptide elongation rates. *Cell* **33:** 103–114.

Belle R., Derancourt J., Poulhe R., Capony J.P., Ozon R., and Mulner-Lorillon O. 1989. A purified complex from *Xenopus* oocytes contains a p47 protein, an in vivo substrate of MPF, and a p30 protein respectively homologous to elongation factors EF-1 gamma and EF-1 beta. *FEBS Lett.* **255:** 101–104.

Binninger D. and Weber L.A. 1984. Coordinate regulation of polypeptide chain initiation and elongation in rabbit reticulocytes. *J. Cell. Physiol.* **118:** 267–276.

Brendler T., Godefroy-Colburn T., Yu S., and Thach R.E. 1981. The role of mRNA competition in regulating translation. III. Comparison of in vitro and in vivo results. *J. Biol. Chem.* **256:** 11755–11761.

Brunn G.J., Hudson C.C., Sekulic A., Williams J.M., Hosoi H., Houghton P.J., Lawrence J.C., and Abraham R.T. 1997. Phosphorylation of the translational repressor PHAS-I by the mammalian target of rapamycin. *Science* **277:** 99–101.

Burnett P.E., Barrow R.K., Cohen N.A., Snyder S.H., and Sabatini D.M. 1998. RAFT1 phosphorylation of the translational regulators p70 S6 kinase and 4E-BP1. *Proc. Natl. Acad. Sci.* **95:** 1432–1437.

Campbell L.E., Wang X., and Proud C.G. 1999. Nutrients differentially modulate multiple translation factors and their control by insulin. *Biochem. J.* **344:** 433–441.

Carlberg U., Nilsson A., and Nygård O. 1990. Functional properties of phosphorylated elongation factor 2. *Eur. J. Biochem.* **191:** 639–645.

Carr-Schmid A., Valente L., Loik V.I., Williams T., Starita L.M., and Kinzy T.G. 1999. Mutations in elongation factor 1β, a guanine nucleotide exchange factor, enhance translational fidelity. *Mol. Cell. Biol.* **19:** 5257–5266.

Cavallius J., Zoll W., Chakraburtty K., and Merrick W.C. 1993. Characterization of yeast EF-1 alpha: Non-conservation of post- translational modifications. *Biochim. Biophys. Acta* **1163:** 75–80.

Chambers D.M., Peters J., and Abbott C.M. 1998. The lethal mutation of the mouse wasted (wst) is a deletion that abolishes expression of a tissue-specific isoform of translation elongation factor 1alpha, encoded by the Eef1a2 gene. *Proc. Natl. Acad. Sci.* **95:** 4463–4468.

Chang Y.W. and Traugh J.A. 1997. Phosphorylation of elongation factor-1 and ribosomal protein S6 by multipotential S6 kinase and insulin stimulation of translational elongation. *J. Biol. Chem.* **272:** 28252–28257.

———. 1999. Insulin stimulation of phosphorylation of elongation factor 1 (eEF-1) enhances elongation activity. *Eur. J. Biochem.* **251:** 201–207.

Chen C.J. and Traugh J.A. 1995. Expression of recombinant elongation factor 1 beta from rabbit in *Escherichia coli*. Phosphorylation by casein kinase II. *Biochim. Biophys. Acta* **1264:** 303–311.

Chung H., Nairn A.C., Murata K., and Brautigan D.L. 1999. Mutation of Tyr307 and Leu309 in the protein phosphatase 2A catalytic subunit favours association with the alpha4 subunit which promotes dephosphorylation of elongation factor-2. *Biochemistry* **10:** 10371–10376.

Coyle J.T. and Puttfarcken P. 1993. Oxidative stress, glutamate and neurodegenerative disorders. *Science* **262:** 689–695.

Dever T.E., Costello C.E., Owens C.L., Rosenberry T.L., and Merrick W.C. 1989. Location of seven post-translational modifications in rabbit elongation factor 1 alpha including dimethyllysine, trimethyllysine, and glycerylphosphorylethanolamine. *J. Biol. Chem.* **264:** 20518–20525.

Di Como C.J. and Arndt K.T. 1996. Nutrients, via the TOR proteins, stimulate the association of Tap42 with Type 2A phosphatases. *Genes Dev.* **10:** 1904–1916.

Diggle T.A., Redpath N.T., Heesom K.J., and Denton R.M. 1998. Regulation of protein synthesis elongation factor-2 kinase by cAMP in adipocytes. *Biochem. J.* **336:** 525–529.

Diggle T.A., Seehra C.K., Hase S., and Redpath N.T. 1999. Analysis of the domain structure of elongation factor-2 kinase by mutagenesis. *FEBS Lett.* **457:** 189–192.

Donovan M.G. and Bodley J.W. 1991. *Saccharomyces cerevisiae* elongation factor 2 is phosphorylated by an endogenous kinase. *FEBS Lett.* **291:** 303–306.

Fox H.L., Kimball S.R., Jefferson L.S., and Lynch C.J. 1998a. Amino acids stimulate phosphorylation of p70(S6k) and organisation of rat adipocytes into multicellular clusters. *Am. J. Physiol.* **43:** C206–C213.

Fox H.L., Pham P.T., Kimball S.R., Jefferson L.S., and Lynch C.J. 1998b. Amino acid effects on translational repressor 4E-BP1 are mediated primarily by L-leucine in isolated adipocytes. *Am. J. Physiol.* **44:** C1232–C1238.

Hait W.N., Ward M.D., Trakht I.N., and Ryazanov A.G. 1996. Elongation factor-2 kinase — Immunological evidence for the existence of tissue-specific isoforms. *FEBS Lett.* **397:** 55–60.

Hanson P.I. and Schulman H. 1992. Neuronal Ca^{2+}/calmodulin-dependent protein kinases. *Annu. Rev. Biochem.* **61:** 559–601.

Hara K., Yonezawa K., Weng Q.-P., Kozlowski M.T., Belham C., and Avruch J. 1998. Amino acid sufficiency and mTOR regulate p70 S6 kinase and eIF4E BP1 through a common effector mechanism. *J. Biol. Chem.* **273:** 14484–14494.

Hardie G. and Hanks S.K., Eds. 1995. *The protein kinase facts book.* Academic Press, London.

Haycock J.W., Browning M.D., and Greengard P. 1987. Cholinergic regulation of protein phosphorylation in bovine adrenal chromaffin cells. *Proc. Natl. Acad. Sci.* **85:** 1677–1681.

Heilmeyer L.M.G. and Kilimann M.W. 1995. Phosphorylase kinase (vertebrates). In *The protein kinase facts book,* part I: *Protein-serine kinases* (ed. G. Hardie and S.K. Hanks), pp. 146–152. Academic Press, London.

Hincke M.T. and Nairn A.C. 1992. Phosphorylation of elongation factor 2 during $Ca^{(2+)}$-

mediated secretion from rat parotid acini. *Biochem. J.* **282:** 877–882.

Hovland R., Eikhom T.S., Proud C.G., Cressey L.I., Lanotte M., Doskeland S.O., and Houge G. 1999. cAMP inhibits translation by inducing Ca^{2+}/calmodulin-independent elongation factor 2 kinase activity in IPC-81 cells. *FEBS Lett.* **444:** 97–101.

Janssen G.M. and Moller W. 1988. Kinetic studies on the role of elongation factors 1 beta and 1 gamma in protein synthesis. *J. Biol. Chem.* **263:** 1773–1778.

Janssen G.M., Maessen G.D., Amons R., and Moller W. 1988. Phosphorylation of elongation factor 1 beta by an endogenous kinase affects its catalytic nucleotide exchange activity. *J. Biol. Chem.* **263:** 11063–11066.

Janssen G.M., Morales J., Schipper A., Labbe J.C., Mulner-Lorillon O., Belle R., and Moller W. 1991. A major substrate of maturation promoting factor identified as elongation factor 1 beta gamma delta in *Xenopus laevis*. *J. Biol. Chem.* **266:** 14885–14888.

Jefferies H.B.J. and Thomas G. 1994. Elongation factor-1alpha mRNA is selectively translated following mitogenic stimulation. *J. Biol. Chem.* **269:** 4367–4372.

Jefferies H.B.J., Reinhard G., Kozma S.C., and Thomas G. 1994. Rapamycin selectively represses translation of the 'polypyrimidine tract' mRNA family. *Proc. Natl. Acad. Sci.* **91:** 4441–4445.

Jiang Y. and Broach J.R. 1999. Tor proteins and protein phosphatase 2A reciprocally regulate Tap42p in controlling cell growth in yeast. *EMBO J.* **18:** 2782–2792.

Kahns S., Lund A., Kristensen P., Knudsen C.R., Clark B.F., Cavallius J., and Merrick W.C. 1998. The elongation factor 1 A-2 isoform from rabbit: Cloning of the cDNA and characterization of the protein. *Nucleic Acids Res.* **26:** 1884–1890.

Kielbassa K., Muller H.-J., Meyer H.E., Marks F., and Gschwendt M. 1995. Protein kinase C delta-specific phosphorylation of the elongation factor eEF-1 alpha and an eEF-1 alpha peptide at threonine 431. *J. Biol. Chem.* **270:** 6156–6162.

Lawrence J.C. and Abraham R.T. 1997. PHAS/4E-BPs as regulators of mRNA translation and cell proliferation. *Trends Biochem. Sci.* **22:** 345–349.

Levenson R.M., Nairn A.C., and Blackshear P.J. 1989. Insulin rapidly induces the biosynthesis of elongation factor 2. *J. Biol. Chem.* **264:** 11904–11911.

Liang H. and Gaber R.F. 1996. A novel signal transduction pathway in *Saccharomyces cerevisiae* defined by Snf3-regulated expression of HXT6. *Mol. Cell. Biol.* **7:** 1953–1966.

Litchfield D.W., Dobrowolska G., and Krebs E.G. 1994. Regulation of casein kinase-II by growth factors: A re-evaluation. *Cell. Mol. Biol. Res.* **40:** 373–381.

Mackie K.P., Nairn A.C., Hampel G., Lam G., and Jaffe E.A. 1989. Thrombin and histamine stimulate the phosphorylation of elongation factor 2 in human umbilical vein endothelial cells. *J. Biol. Chem.* **264:** 1748–1753.

Marin P., Nastiuk K.L., Daniel N., Girault J., Czernik A.J., Glowinski J., Nairn A.C., and Premont J. 1997. Glutamate dependent phosphorylation of elongation factor 2 and inhibition of protein synthesis in neurons. *J. Neurosci.* **17:** 3445–3454.

Marino M.W., Dunbar J.D., Wu L.W., Ngaiza J.R., Han H.M., Guo D., Matsushita M., Nairn A.C., Zhang Y., Kolesnick R., Jaffe E.A., and Donner D.B. 1996. Inhibition of tumor necrosis factor signal transduction in endothelial cells by dimethylaminopurine. *J. Biol. Chem.* **271:** 28624–28629.

Matsumoto S., Mizoguchi M., Oizumi N., Tsuruga S., Shinozaki K., Taira H., and Ejiri S. 1993. Analysis of phosphorylation of wheat elongation factor-1-β and factor-β′ by casein kinase-II. *Biosci. Biotechnol. Biochem.* **57:** 1740–1742.

Meyuhas O., Avni D., and Shama S. 1996. Translational control of ribosomal protein mRNAs in eukaryotes. In *Translational control* (ed. J.W.B. Hershey et al.), pp.

363–388. Cold Spring Harbor Laboratory Press, Cold Spring Harbor, New York.

Minella O., Cormier P., Morales J., Poulhe R., Belle R., and Mulner-Lorillon O. 1994. cdc2 kinase sets a memory phosphorylation signal on elongation factor EF-1 delta during meiotic cell division, which perdures in early development. *Cell. Mol. Biol.* **40:** 521–525.

Mitsui K., Brady M., Palfrey H.C., and Nairn A.C. 1993. Purification and characterization of calmodulin-dependent protein kinase III from rabbit reticulocytes and rat pancreas. *J. Biol. Chem.* **268:** 13422–13433.

Mulner-Lorillon O., Cormier P., Cavadore J.C., Morales J., Poulhe R., and Belle R. 1992. Phosphorylation of *Xenopus* elongation factor-1 gamma by cdc2 protein kinase: Identification of the phosphorylation site. *Exp. Cell Res.* **202:** 549–551.

Mulner-Lorillon O., Minella O., Cormier P., Capony J.P., Cavadore J., Morales J., Poulhe R., and Belle R. 1994. Elongation factor EF-1delta, a new target for maturation-promoting factor in *Xenopus* oocytes. *J. Biol. Chem.* **269:** 20201–20207.

Murata K., Wu J., and Brautigan D.L. 1997. B cell receptor-associated protein α4 displays rapamycin-sensitive binding directly to the catalytic subunit of protein phosphatase 2A. *Proc. Natl. Acad. Sci.* **94:** 10624–10629.

Nairn A.C. and Palfrey H.C. 1987. Identification of the major M_r 100,000 substrate for calmodulin-dependent protein kinase III in mammalian cells as elongation factor-2. *J. Biol. Chem.* **262:** 17299–17303.

Nairn A.C., Bhagat B., and Palfrey H.C. 1985. Identification of calmodulin-dependent protein kinase III and its major M_r 100,000 substrate in mammalian tissues. *Proc. Natl. Acad. Sci.* **82:** 7939–7943.

Nairn A.C., Nichols R.A., Brady M.J., and Palfrey H.C. 1987. Nerve growth factor treatment or cAMP elevation reduces Ca^{2+}/calmodulin-dependent protein kinase III activity in PC12 cells. *J. Biol. Chem.* **262:** 14265–14272.

Nanahoshi M., Nishiuma T., Tsujishita Y., Hara K., Inui S., Sakaguchi N., and Yonezawa K. 1998. Regulation of protein phosphatase 2A catalytic activity by α4 protein and its yeast homolog Tap42. *Biochem. Biophys. Res. Commun.* **251:** 520–526.

Nanahoshi M., Tsujishita Y., Tokunaga C., Inui S., Sakaguchi N., Hara K., and Yonezawa K. 1999. α4 Protein as a common regulator of type 2A-related serine/threonine protein phosphatases. *FEBS Lett.* **446:** 108–112.

Nielsen P.J. and McConkey E.H. 1980. Evidence for control of protein synthesis in HeLa cells via the elongation rate. *J. Cell. Physiol.* **104:** 269–281.

Oldfield S. and Proud C.G. 1993. Phosphorylation of elongation factor-2 from the lepidopteran insect, *Spodoptera frugiperda*. *FEBS Lett.* **327:** 71–74.

Ovchinnikov L.P., Motuz L.P., Natapov P.G., Averbuch L.J., Wettenhall R.E., Szyszka R., Kramer G., and Hardesty B. 1990. Three phosphorylation sites in elongation factor 2. *FEBS Lett.* **275:** 209–212.

Ozcan S., Dover J., and Johnston M. 1998. Glucose sensing and signaling by two glucose receptors in the yeast *Saccharomyces cerevisiae*. *EMBO J.* **17:** 2566–2573.

Palfrey H.C., Nairn A.C., Muldoon L.L., and Villereal M.L. 1987a. Rapid activation of calmodulin-dependent protein kinase III in mitogen-stimulated human fibroblasts. *J. Biol. Chem.* **262:** 9785–9792.

———. 1987b. Rapid activation of calmodulin-dependent protein kinase III in mitogen-stimulated human fibroblasts. *J. Biol. Chem.* **262:** 9785–9792.

Patti M.-E., Brambilla E., Luzi L., Landaker E.J., and Kahn C.R. 1998. Bidirectional modulation of insulin action by amino acids. *J. Clin. Invest.* **101:** 1519–1529.

Peters H.I., Chang Y.W.E., and Traugh, J.A. 1995. Phosphorylation of elongation factor-1

(EF-1) by protein kinase C stimulates GDP/GTP-exchange activity. *Eur. J. Biochem.* **234:** 550–556.

Peterson R.T. and Schreiber S.L. 1998. Translation control: Connecting mitogens and the ribosome. *Curr. Biol.* **8:** R248–R250.

Peterson R.T., Desai B.N., Hardwick J.S., and Schreiber S.L. 1999. Protein phosphatase 2A interacts with the 70 kDa S6 kinase and is activated by inhibition of FKBP-12-rapamycin-associated protein. *Proc. Natl. Acad. Sci.* **96:** 4438–4442.

Price N.T., Redpath N.T., Severinov K.V., Campbell D.G., Russell J.M., and Proud C.G. 1991. Identification of the phosphorylation sites in elongation factor-2 from rabbit reticulocytes. *FEBS Lett.* **282:** 253–258.

Proud C.G. 1996. p70 S6 kinase: An enigma with variations. *Trends Biochem. Sci.* **21:** 181–185.

Proud C.G. and Denton R.M. 1997. Molecular mechanisms for the activation of protein synthesis by insulin. *Biochem. J.* **328:** 329–341.

Ray A., Walden W.E., Brendler T., Zenger V.E., and Thach R.E. 1985. Effect of medium hypotonicity on reovirus translation rates. An application of kinetic modeling *in vivo*. *Biochemistry* **24:** 7525–7532.

Redpath N.T. and Proud C.G. 1989. The tumour promoter okadaic acid inhibits reticulocyte-lysate protein synthesis by increasing the net phosphorylation of elongation factor 2. *Biochem. J.* **262:** 69–75.

———. 1993a. Cyclic AMP-dependent protein kinase phosphorylates rabbit reticulocyte elongation factor-2 kinase and induces calcium-independent activity. *Biochem. J.* **293:** 31–34.

———. 1993b. Purification and phosphorylation of elongation factor-2 kinase from rabbit reticulocytes. *Eur. J. Biochem.* **212:** 511–520.

Redpath N.T., Foulstone E.J., and Proud C.G. 1996a. Regulation of translation elongation factor-2 by insulin via a rapamycin-sensitive signalling pathway. *EMBO J.* **15:** 2291–2297.

Redpath N.T., Price N.T., and Proud C.G. 1996b. Cloning and expression of cDNA encoding protein synthesis elongation factor-2 kinase. *J. Biol. Chem.* **271:** 17547–17554.

Redpath N.T., Price N.T., Severinov K.V., and Proud C.G. 1993. Regulation of elongation factor-2 by multisite phosphorylation. *Eur. J. Biochem.* **213:** 689–699.

Ryazanov A.G., Pavur K.S., and Dorovkov M.V. 1999. Alpha kinases: A new class of protein kinases with a novel catalytic domain. *Curr. Biol.* **9:** R43–R45.

Ryazanov A.G., Shestakova E.A., and Natapov P.G. 1988. Phosphorylation of elongation factor 2 by EF-2 kinase affects rate of translation. *Nature* **334:** 170–173.

Ryazanov A.G., Ward M.D., Mendola C.E., Pavur K.S., Dorovkov M.V., Wiedmann M., Erdjument-Bromage H., Tempst P., Parmer T.G., Prostko C.R., Germino F.J., and Hait W.N. 1997. Identification of a new class of protein kinases represented by eukaryotic elongation factor-2 kinase. *Proc. Natl. Acad. Sci.* **94:** 4884–4889.

Scheetz A.J., Nairn A.C., and Constantine-Paton M. 1997. N-methyl-D-aspartate receptor activation and visual activity induce elongation factor-2 phosphorylation in amphibian tecta: A role for N-methyl-D-aspartate receptors in controlling protein synthesis. *Proc. Natl. Acad. Sci.* **94:** 14770–14775.

Sheu G.T. and Traugh J.A. 1997. Recombinant subunits of mammalian elongation factor 1 expressed in *Escherichia coli:* Subunit interactions, elongation activity, and phosphorylation by protein kinase CKII. *J. Biol. Chem.* **272:** 33290–33297.

———. 1999. A structural model for elongation factor (EF-1) and phosphorylation by protein kinase CKII. *Mol. Cell. Biochem.* **191:** 181–186.

Shigemitsu K., Tsujishita Y., Hara K., Nanahoshi M., Avruch J., and Yonezawa K. 1999. Regulation of translational effectors by amino acid and mammalian target of rapamycin signaling pathways: Possible involvement of autophagy in cultured hepatoma cells. *J. Biol. Chem.* **274:** 1058–1065.

van Venrooij W.J.W., Henshaw E.C., and Hirsch C.A. 1970. Nutritional effects on the polyribosome distribution and rate of protein synthesis in Ehrlich ascites tumor cells in culture. *J. Biol. Chem.* **245:** 5947–5953.

Venema R.C., Peters H.I., and Traugh J.A. 1991a. Phosphorylation of elongation factor 1 (EF-1) and valyl-tRNA synthetase by protein kinase C and stimulation of EF-1 activity. *J. Biol. Chem.* **266:** 12574–12580.

———. 1991b. Phosphorylation of valyl-tRNA synthetase and elongation factor 1 in response to phorbol esters is associated with stimulation of both activities. *J. Biol. Chem.* **266:** 11993–11998.

Walden W.E. and Thach R.E. 1986. Translational control of gene expression in a normal fibroblast. Characterisation of a subclass of mRNAs with unusual kinetic properties. *Biochemistry* **22:** 2033–2041.

Wang L., Wang X., and Proud C.G. 2000. Activation of mRNA translation by insulin in rat cardiomyocytes involves multiple rapamycin-sensitive steps. *Am. J. Physiol.* **278:** H1056–H1068.

Wang X., Campbell L.E., Miller C.M., and Proud C.G. 1998. Amino acid availability regulates p70 S6 kinase and multiple translation factors. *Biochem. J.* **334:** 261–267.

Webb B.L.J. and Proud C.G. 1998. Eukaryotic initiation factor 2B (eIF2B). *Int. J. Biochem. Cell Biol.* **29:** 1127–1131.

Welsh G.I., Miller C.M., Loughlin A.J., Price N.T., and Proud C.G. 1998. Regulation of eukaryotic initiation factor eIF2B: Glycogen synthase kinase-3 phosphorylates a conserved serine which undergoes dephosphorylation in response to insulin. *FEBS Lett.* **421:** 125–130.

Yoshizawa F., Tonouchi A., Miura Y., Yagasaki K., and Funabiki R. 1995. Insulin-stimulated polypeptide chain elongation in the soleus muscle of mice. *Biosci. Biotechnol. Biochem.* **59:** 348–349.

25

Programmed Translational Frameshifting, Hopping, and Readthrough of Termination Codons

Philip J. Farabaugh, Qiang Qian, and Guillaume Stahl

Department of Biological Sciences and
Program in Molecular and Cell Biology
University of Maryland Baltimore County
Baltimore, Maryland 21250

The conventional understanding of the term translational control is that it refers to an effect occurring during translation which causes a change in the amount of completed product protein per unit amount of mRNA. This chapter concerns a different type of translational control, one in which the effect of the control is to regulate the structure of the encoded protein. Genes that use programmed translational frameshifting or programmed readthrough of termination codons produce two primary translation products. The first is a protein decoded from the gene using the canonical rules of translation, corresponding one-to-one to the sequence of codons in the mRNA. The second product has a structure not predicted by the mRNA sequence, because at a discrete point in the mRNA the rules of translation are changed to allow a shift in reading frame or bypassing of a termination codon. For this type of alternative decoding, also called recoding, genes can use any of a variety of mechanisms to subvert normal translational elongation. The mechanisms employed interfere with translational fidelity so they may illuminate how ribosomes normally maintain accuracy. The purpose of recoding can be to create a second form of a protein used in a morphogenetic process, to autogenously control gene expression, or to provide alternative enzymatic activities.

The phenomena of suppression and recoding appear on their face to be unrelated. Suppression, either of nonsense or frameshift mutations, results from a permanent change to a translational component that results in occasional misapplication of the rules of translation: Ribosomes errantly translocate more than 3 nucleotides, thereby changing reading

Translational Control of Gene Expression
© 2000 Cold Spring Harbor Laboratory Press 0-87969-568-4/00 $5 + .00

frame, or they misinterpret a nonsense codon as sense. Recoding results when a sequence in an mRNA interferes with the orderly process of translation, causing a local deviation from its rules. Yet each of these phenomena through different means imposes a type of fuzzy logic in translation. They create a situation where either of two events can occur—frameshifting or in-frame decoding, termination or readthrough. The suppressors can be thought of as creating an addendum to the rules of translation allowing a different outcome when a particular sequence occurs in any mRNA. Some suppressors alter tRNA molecules. These mutations literally change the meaning of nucleic acid sequences. The change is not absolute, since in many cases a mutant tRNA is the only isoacceptor that can be used to decode conventionally one or more codons. Some suppressors affect other elements of the translational apparatus: elongation factors, rRNA, or ribosomal proteins. These must affect the accuracy with which the ribosome interprets the mRNA sequence, altering the stringency of the required match between tRNAs and mRNAs.

Recoding sites also create fuzzy logic, allowing a proportion of ribosomes to deviate from canonical decoding, but the change is imposed in this case by the structure of the mRNA itself. A recoding site acts to interfere with continued normal decoding and promotes an alternative noncanonical event. A common feature of all recoding sites is the imposition of a translational pause. Ribosomes pause at these sites either because a stable mRNA structure (e.g., a pseudoknot) blocks their progress down the mRNA, or because the sense or nonsense codon in the A site is recognized slowly. Although noncanonical decoding is presumed to be kinetically very slow, pausing per se is not sufficient to stimulate the recoding event; other sequences may induce similar pauses, yet they do not interfere with normal decoding. Recoding sites require other sequences that increase the efficiency of noncanonical decoding. One of the contested points about recoding is whether these sequences simply stimulate the efficiency of nonprogrammed translational errors or whether they cause novel phenomena unrelated to random errors. The phenomenology of recoding sites is sufficiently diverse that there is no single resolution of this issue. Rather, recoding sites appear to exist on a continuum from the simplest, which resemble simple translational error sites, to the most complex, which may promote a novel mechanism of translation.

PROGRAMMED FRAMESHIFTING

Sequences in mRNAs that predispose ribosomes to shift reading frames are termed programmed frameshift sites. These sites shift reading one base in either the 5′ direction (–1 frameshifting) or in the 3′ direction (+1 frameshifting). The mechanisms underlying the two types of events appear to be quite different from each other, suggesting that they interfere with different aspects of proper frame maintenance.

Programmed –1 Frameshifting

Retrovirus, other viruses, and retrotransposons commonly use –1 frameshifting to express a defined ratio of a long fusion protein (e.g., a *gag–pol* fusion) to a shorter protein (*gag*) (for review, see Brierley 1995; Miller et al. 1995; Farabaugh 1996; Fütterer and Hohn 1996; Maia et al. 1996). Some bacterial insertion sequences, *IS* elements (for review, see Chandler and Fayet 1993), and cellular genes (Tsuchihashi and Brown 1992) also use this type of translational control.

All –1 frameshift sites appeared to use a single mechanism, although some exceptions are now recognized. In canonical –1 frameshifting, first described by Jacks and collaborators studying Rous sarcoma virus (RSV) (Jacks and Varmus 1985; Jacks et al. 1988a), frameshifting happens at a slippery heptamer with the sequence X-XXY-YYZ, where X can be any nucleotide, Y is a weakly base-pairing nucleotide A or U, and Z is species-specific. Mutagenesis and protein sequencing demonstrate that two tRNAs decoding XXY-YYZ slide one nucleotide backward on the messenger to XXX-YYY, with each retaining at least 2 bp. This mechanism is termed simultaneous slippage. The types of heptamers used differ among organisms. For example, the heptamer A-AAA-AAG is very slippery in prokaryotes (Weiss et al. 1989; Tsuchihashi and Brown 1992) but allows little or no frameshifting in eukaryotes (Tsuchihashi and Brown 1992; Garcia et al. 1993). Some frameshift sites do not fit the X-XXY-YYZ canonical description, for example, G-UUA-AAC for the equine arthritis virus (Brierley et al. 1992) or a G-GAU-UUA at the *pro-pol* junction in mouse mammary tumor virus (Jacks et al. 1987). In these cases, it is unclear how frameshifting occurs, since pairing in the –1 shifted frame would be limited.

In several cases frameshifting appears to involve slippage by a single peptidyl-tRNA. In potato virus M frameshifting occurs at A-AAA-UGA, stimulated by the UGA stop codon (Gramstat et al. 1994). Peptide sequencing provided evidence for 30% single slippage in human immunodeficiency virus 1 (HIV–1) (Jacks et al. 1988b) and in the *Escherichia coli IS1* element (Sekine et al. 1992; Sekine and Ohtsubo 1992).

Efficient –1 frameshifting usually requires a secondary structure 3′ to the slippery heptamer. The structures can be as simple as a hairpin, as in

HIV-1 (Parkin et al. 1992), or as complicated as a pseudoknot with branched secondary stem-loops, as in RSV (Marczinke et al. 1998), or a three-way hairpin junction, as in *IS911* (Rettberg et al. 1999). (Pseudoknots are hairpin structures in which loop nucleotides base-pair with a region farther downstream, forming a quasi-continuous double helix joined by single-stranded loops.) The structures are located a precise distance from the slippery heptamer (usually 4–9 nucleotides), and one-nucleotide changes in spacing can abolish their enhancing effect (Brierley et al. 1992). However, there are also frameshift sites without any apparent stimulating structure (ten Dam et al. 1990).

The secondary structures have been divided into four classes (for review, see Brierley 1995). These include short, simple stem-loops; longer stem-loops containing bulges, internal loops, and mismatches; short H-type (hairpin-like) pseudoknots; and longer, more complex pseudoknots. Figure 1 cartoons the structure of an H-type pseudoknot, which consists of two short stems connected by two loops. Nuclear magnetic resonance (NMR) studies have determined structures of H-type pseudoknots that include two unusual features. First, an A residue at the junction of the two stems acts as a wedge to provoke a bend in the pseudoknot (Chen et al. 1995; Shen and Tinoco 1995). In Figure 1 the single A between stem 1 and stem 2 is shown inducing a kink in the pseudoknot. Second, tertiary interactions between loop nucleotides and stem 1 also occur in frameshift-inducing pseudoknots (Liphardt et al. 1999; Su et al. 1999). In the structure cartooned in Figure 1, the A-rich region of loop 2 interacts in the major groove with stem 1 (Su et al. 1999). The unusual structure of these pseudoknots induces frameshifting but is not explicitly necessary since the stimulatory effect of more complex pseudoknots depends on the length of stem 1 rather than any of these structural motifs (Napthine et al. 1999). The effect of simple hairpin loops also appears to relate to their stability, as shown in bacteria for *dnaX* (Larsen et al. 1997) or in eukaryotes for HIV-1 (Bidou et al. 1997). Simple hairpin loops cannot replace equivalently stable pseudoknots, so they may enhance frameshifting only on especially shifty heptamers.

The fact that the ability of certain pseudoknots to stimulate frameshifting depends on structural features suggests that they may do so by specifically interacting with the ribosome. This interaction could be analogous to the effect of Shine-Dalgarno (SD) mRNA•rRNA interactions on both +1 and –1 frameshifting (Larsen et al. 1995). The pseudoknot pauses the ribosome on the mRNA, but it must have some other function as well, since hairpins that also pause the ribosome cannot induce frameshifting (Tu et al. 1992; Somogyi et al. 1993). Recent in vitro

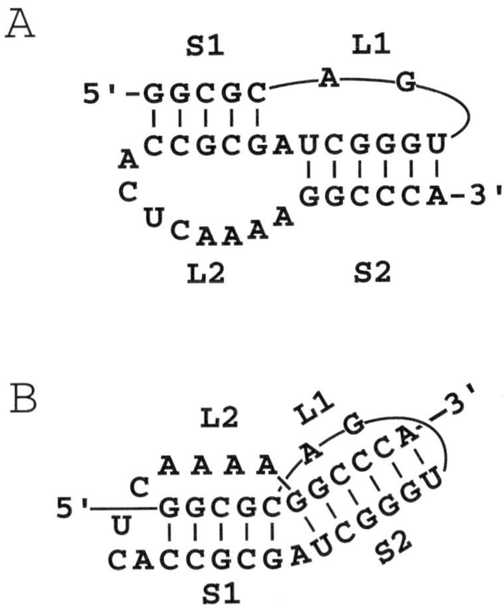

Figure 1 Structure of a frameshift-inducing, hairpin-type pseudoknot VPK. (*A*) This exploded view clearly shows the continuity of the RNA chain through the two base-paired stems (S1 and S2) and the intervening loops (L1 and L2), but poorly represents the three-dimensional structure. (*B*) Since S1 and S2 stack coaxially, this view gives a better representation of the structure. The single adenosine between stem 1 and stem 2 creates a wedge that causes a > 30° kink in the structure. The two loops lie down the major grooves of the two stems, and the adenosine-rich region of L2 forms a set of triplex interactions with S1 (Su et al. 1999). Structures of H-type pseudoknots that induce frameshifting can be found at the Protein Data Bank (at http://www.rcsb.org/pdb), using PDB codes 1kaj, 1kpd, 1rnk, and 437d.

data show that when HIV-1 frameshifting is assayed in an *E. coli* extract, addition of error-inducing antibiotics stimulates frameshifting, but only when the downstream hairpin loop is present (Brunelle et al. 1999). These data suggest a possible functional interaction between the hairpin and the "fidelity center" of the ribosome. Several mechanistic models have been proposed to account for the –1 frameshifting-specific requirement for secondary structures. A simple model in which pausing allows stochastic reframing on monotonous sequences (Jacks et al. 1988a) now seems unlikely, although a torsional resistance model may better explain the different effects of pseudoknots and hairpins (Dinman 1995). Other models posit a direct effect of the pseudoknot on ribosomal function, either interfering with peptidyl-tRNA translocation (Farabaugh 1996) or a secondary structure unwinding/rewinding process occurring after translocation (Farabaugh 1996). These models have not yet been specifically tested. Attempts to identify proteins that bind sequence-specifically to these structures have been unsuccessful (Dinman and Wickner 1994; ten Dam et al. 1994; Chen et al. 1995).

Some viruses that use a frameshift mechanism of expression can provoke hypomodification of tRNAs (Hatfield et al. 1989). Although tRNA hypomodification can increase frameshifting in vitro (Hatfield et al. 1992; Carlson et al. 1999), surprisingly there appears to be no similar effect in vivo (see, e.g., Brierley et al. 1997), nor is frameshifting altered in chronically virally infected cells in culture (Cassan et al. 1994; Reil et al. 1994). These data argue that viral infection does not induce the production of specially altered tRNAs that could act as suppressor tRNAs to induce frameshifting.

The fact that altering frameshift efficiency can reduce viral propagation (for review, see Farabaugh 1995; Dinman et al. 1998) suggests a possible therapeutic importance of frameshifting. According to this view, certain antibiotics that appear to alter the efficiency of frameshifting in vitro, like the peptidyltransferase inhibitors anisomycin or sparsomycin, could be used as antiviral drugs (Honigman et al. 1995; Dinman et al. 1998). More specific drugs are currently being investigated in vivo for their capabilities of blocking HIV-1 replication (Hung et al. 1998).

Programmed +1 Frameshifting

Programmed +1 frameshifting occurs ubiquitously, but much less commonly than does –1 frameshifting. The best-studied examples of this type of frameshift occur in the *prfB* gene of *E. coli* (Craigen and Caskey 1986) encoding peptide release factor 2 (RF2), in a region between the *GAG* and *POL* genes of the *Saccharomyces cerevisiae* retrotransposons of the Ty family (Belcourt and Farabaugh 1990; Farabaugh et al. 1993), and in genes encoding ornithine decarboxylase antizyme from metazoans (Matsufuji et al. 1995; Ivanov et al. 1998a,b). Two chromosomal genes in *S. cerevisiae* use programmed +1 frameshifting: *EST3* (Morris and Lundblad 1997), encoding a subunit of telomerase, and *APB140* (Asakura et al. 1998), encoding an actin-binding protein. Those viruses that use programmed frameshifting almost invariably use the –1 simultaneous slippage mechanism, but plant viruses of the closterovirus group appear to use +1 frameshifting (Agranovsky et al. 1994; Karasev et al. 1995; ten Dam 1995).

Programmed +1 frameshifting occurs at a different step in the translation elongation cycle than does simultaneous slippage –1 frameshifting. The clearest evidence for this is the fact that the rate of occupancy of the codon immediately following the last zero-frame codon regulates the efficiency of frameshifting. Slow recognition of this codon is presumed to stimulate a translational pause that allows the slow frameshift event to occur. The codon can be a poorly recognized termination codon (one with an inappropriate 3′ sequence context; for review, see Tate et al. 1996) as in the *prfB* gene of *E. coli* (Curran and Yarus 1989; Donly et al. 1990) and the ornithine decarboxylase antizyme genes of metazoans (Matsufuji et al. 1995). Alternatively, it can be a sense codon recognized by a low-abundance tRNA, as in the Ty family of retrotransposons in *S. cerevisiae* (Belcourt and Farabaugh 1990; Farabaugh et al. 1993). These data imply that the ribosomal A site is not occupied by aminoacyl-tRNA prior to the frameshift event. Later data showed that the slow recognition of the first +1 frame codon reduces frameshifting (Pande et al. 1995) suggesting that the aminoacyl-tRNA for that codon must enter the A site to promote frameshifting.

The heptamer representing the three codons relevant to +1 frameshifting—the last decoded zero-frame codon, the next zero-frame codon, and the overlapping +1 frame codon—is essential to +1 frameshifting. In nearly all of the programmed sites, one or more sequences outside the essential heptamer stimulate frameshifting. In the prokaryotic system, *prfB*, there is an interaction between the SD interaction site in 16S rRNA (Shine and Dalgarno 1974) and a sequence immediately preceding the heptamer (Weiss et al. 1988). The SD site is placed slightly closer than the optimum spacing used during initiation; slippage of the mRNA at the decoding site in the 5′ direction to reduce the strain of the interaction would result in +1 slippage of the ribosome-bound peptidyl-tRNA (Larsen et al. 1995). Other stimulatory sequences may function indirectly to stimulate frameshifting. One such site is a 14-nucleotide sequence that lies immediately downstream from the heptamer in the Ty3 retrotransposon frameshift site from *S. cerevisiae*. Recent data suggest that this sequence stimulates frameshifting about 7.5-fold by interfering with the ribosomal accuracy center (Z. Li and P.J. Farabaugh, unpubl.). The sequence is complementary to a sequence in the 530 loop of the 16S rRNA, a region of the ribosome that must undergo a structural rearrangement in order to select cognate aminoacyl-tRNAs in the ribosome (for review, see Powers and Noller 1994). The 530 loop cross-links to the mRNA 6 nucleotides downstream from the A-site (Dontsova et al. 1992) in register with the putative mRNA•rRNA context pairing. This suggests a model in which pairing between the mRNA and rRNA interferes with the rearrangement of the 530 loop, leading to loss of selectivity in the A site and consequently increased frameshifting. Genetic data support this conclusion (Z. Li and P. Farabaugh, unpubl.).

The developing model of programmed +1 frameshifting proposes that frameshifting occurs in competition with continued zero-frame decoding. The stimulatory sequences function either by forcing slippage of the mRNA to align the +1 frame codon correctly in the A site, as in *prfB*, or by reducing the efficiency of cognate decoding in the zero frame, as in Ty3. Effects that increase in-frame recognition tend to reduce frameshifting. In the *prfB* gene, for example, changing the availability of peptide release factor 2 (RF2), the product of the gene, regulates frameshifting: Overexpressing RF2 reduces frameshifting (Craigen and Caskey 1986; Donly et al. 1990), whereas expressing partially nonfunctional forms of RF2 increases frameshifting (Donly et al. 1990). This effect allows *prfB* to be autogenously regulated.

There now appear to be two types of programmed +1 frameshifts: those that occur by peptidyl-tRNA slippage, and those that involve out-of-frame recognition in the A site without peptidyl-tRNA slippage. The sequences of the +1 frameshift sites from the *E. coli prfB* gene (Craigen and Caskey 1986) and the *S. cerevisiae* Ty1 element (Belcourt and Farabaugh 1990) clearly imply that the peptidyl-tRNA may slip +1. In fact, among mutants of the *prfB* site, frameshifting efficiency correlates directly with the stability of +1 frame-pairing by the peptidyl-tRNA (Curran 1993). For both the Ty3 element in *S. cerevisiae* (Farabaugh et al. 1993) and the metazoan ornithine decarboxylase antizyme gene (Matsufuji et al. 1995) it was strange that frameshifting appeared to occur without the possibility of tRNA slippage; i.e., that in each case the peptidyl-tRNA is unable to base-pair with the overlapping codon in the +1 reading frame. In yeast, in contrast to *E. coli*, there is no correlation between frameshifting and slippage ability by the peptidyl-tRNA (Vimaladithan and Farabaugh 1994). It was unclear how certain peptidyl-tRNAs could provoke a change in reading frames without themselves slipping on the mRNA.

Recent data reveal that frameshifting in yeast occurs on codons that are read by near-cognate tRNAs, those that make a suboptimal interaction with the codon (Sundararajan et al. 1999). Slippage-dependent frameshifting occurs when the peptidyl-tRNA that occupies the P site makes a very weak pyrimidine•pyrimidine interaction in the wobble position. This weak pairing presumably encourages transient dissociation of peptidyl-tRNA from the mRNA, allowing the tRNA to slip +1 spontaneously during the translational pause caused by slow reading of the next codon. Nonslippage frameshifting is also caused by a wobble position mismatch by the peptidyl-tRNA, a purine•purine clash, or a weak U•G pair. Since these tRNAs do not appear to dissociate from the mRNA and pair to the +1 shifted codon, how do they provoke frameshifting? A clue comes from the observation that even normal wobble interactions in the P site reduce the efficiency of A-site cognate decoding (Kato et al. 1990). The more severe disruptions that occur at nonslippage frameshift sites may reduce cognate decoding of the next in-frame codon sufficiently that out-of-frame cognate decoding can effectively compete for the A site. Since the ribosome must recognize a correct in-frame tRNA by the structure it forms with the mRNA, disturbing the structure of the P site may interfere with cognate recognition in the A site by disrupting that structure.

Various groups have reported *trans*-acting factors that increase the efficiency of programmed frameshifting. Altering the availability of tRNAs affects +1 frameshifting in *S. cerevisiae* by either reducing translational pausing (Belcourt and Farabaugh 1990; Farabaugh et al. 1993), altering +1 frame reading (Pande et al. 1995) or affecting near-cognate decoding (Sundararajan et al. 1999). Mutations affecting EF1A (formerly EF-Tu) in bacteria (Hughes et al. 1987; Vijgenboom and Bosch 1989) or eEF1A (formerly EF-1α) in *S. cerevisiae* (Sandbaken and Culbertson 1988) suppress nonsense and frameshift mutations. In *S. cerevisiae* these mutants increase programmed +1 and –1 frameshifting (Dinman and Kinzy 1997), probably by reducing ribosomal accuracy (Farabaugh and Vimaladithan 1998). Both Dinman and Wickner (1994) and Lee et al. (1995) isolated suppressors that stimulate –1 programmed frameshifting, but some also increase +1 frameshifting. In particular, mutant forms of 5S rRNA stimulate both types of frameshifting, although the reason for this effect is unclear. Others of these suppressors affect elements of the surveillance complex, which is involved in degradation of mRNAs containing premature termination codons. Their effect on frameshifting implies that the surveillance complex monitors general translational fidelity (Ruiz-Echevarria et al. 1998). Recent data have suggested that this conclusion should be reevaluated. In particular, Muhlrad and Parker (1999) have demonstrated that the surveillance complex reduces the specific translational efficiency (initiation per mRNA) of nonsense-containing mRNAs, which could account for the increase in expression of frameshift reporters in mutants affecting the complex. Furthermore, Bidou et al. (2000) have shown that surveillance complex mutations actually do not affect the rate of frameshifting, arguing that the suppression effect seen results from increased translational initiation. These data have demonstrated the need for extreme care when characterizing *trans*-acting effectors of programmed translational frameshifting.

A Connection between Recoding and Frameshift Suppression

The fact that the anticodon of the tRNA consists of three nucleotides complementary to the codon (Holley 1965) suggested the tRNA may measure out the codon using the anticodon as a yardstick. This idea was reinforced by the finding that almost all the +1 frameshift suppressors contain an expanded anticodon, although the structures of most have never been confirmed directly at the RNA level. For example, *sufD42* is a mutant of $tRNA^{Pro}_{GGG}$ with a C insertion in its CCC anticodon that suppresses at GGG-G sites (Riddle and Carbon 1973). The structure of these tRNAs suggested a simple mechanism, termed the quadruplet translocation model, in which a 4-base anticodon base-pairs with a 4-base codon at the A site followed by quadruplet translocation to shift the reading frame +1 (Roth 1981). Pairing in the wobble position appeared unnecessary (Bossi and Roth 1981; Gaber and Culbertson 1984), implying that the fourth base may only sterically block access by the next in frame tRNA. The model appeared inconsistent with the fact that some frameshift suppressors actually have normal-sized anticodon loops (Hüttenhofer et al. 1990; Sroga et al. 1992; Qian and Björk 1997). Recent data have further eroded support for the model. Suppressor forms of $tRNA^{Pro}$ isoacceptors are modified so that, despite having an expanded anticodon loop, they are precluded from reading a 4-nucleotide codon (Qian et al. 1998). Suppression by these mutant tRNAs occurs in competition with normal decoding in the A site, implying that access to the next in-frame codon is not restricted by the suppressor, as the quadruplet decoding model suggests (Qian et al. 1998).

These data are inconsistent with the quadruplet translocation model and suggest an alternative peptidyl-tRNA slippage model (Qian et al. 1998; Farabaugh and Björk 1999). A suppressing tRNA decodes the last in-frame codon (e.g., the GGG of a GGG-N suppression site), often as a near-cognate by base-pairing only to the first 2 nucleotides of the codon. This tRNA moves to the ribosomal P site where it can slip +1 (e.g., from GGG to GGN), with translation then continuing in the +1 frame. Slow recognition of the next in-frame codon would stimulate suppression, as observed previously (Qian et al. 1998). In principle, slippage by the suppressing tRNA should not be restricted to the +1 direction, and in fact the *sufD* suppressor can stimulate both −1 and +1 frameshifting, presumably by slipping in either direction (Weiss et al. 1990b). The directionality of suppressors may therefore be imposed by the sequence context of the suppression site: +1 suppressors may usually only be able to successfully rebind after dissociation to a codon in the +1 direction. A model of this type has been proposed to explain how certain mutant tRNAs can promote short translational hops by dissociation and repairing to the mRNA (O'Connor et al. 1989). A more general form of this model could explain all frameshift suppression.

The proposed slippage mechanism strongly resembles the process of programmed +1 frameshifting in *S. cerevisiae* (Sundararajan et al. 1999). In both cases, frameshifting occurs because of weak pairing in the normal frame. Weak pairing also stimulates −1 frameshifting at the *dnaX* programmed frameshift site (Tsuchihashi and Brown 1992), suggesting that near-cognate decoding may generally stimulate programmed frameshifting.

TRANSLATIONAL HOPPING

Arguably the most bizarre form of translational recoding occurs in gene *60* of bacteriophage T4 (Huang et al. 1988). Ribosomes that initiate translation of gene *60* mRNA encounter a UAG terminator at codon 47. About 50% of the time, peptidyl-$tRNA^{Gly}$ dissociates from the 46th codon of the mRNA (GGA), while remaining bound to the ribosome. The tRNA rebinds to an identical codon 48 nucleotides downstream and translation then resumes, having bypassed entirely a 50-nucleotide region of the gene (Fig. 2). Three *cis*-acting elements stimulate this event: the in-frame UAG terminator at codon 47, an upstream primary sequence of the nascent protein, and a downstream hairpin structure (Weiss et al. 1990a). Slow recognition of the terminator probably provides a translational pause to allow peptidyl-tRNA dissociation. The functions of the nascent protein sequence and the downstream hairpin are less clear. The structural complexity of gene *60* hopping implied that the ribosome must be forced to allow dissociation and rebinding of the peptidyl-tRNA to the mRNA, implying that hopping likely occurs much less frequently than frameshifting, which requires much less elaborate mRNA structures. It was surprising, then, when Gallant and Lindsley (1998) showed that starving ribosomes for amino acids induces translational hopping at certain codons, previously shown to induce +1 frameshifting. In fact, the hopping occurred nearly as efficiently as did frameshifting in this system (the absolute efficiency of the event is hard to gauge from the published data but is likely to be rather low). The efficiency is inversely related to the

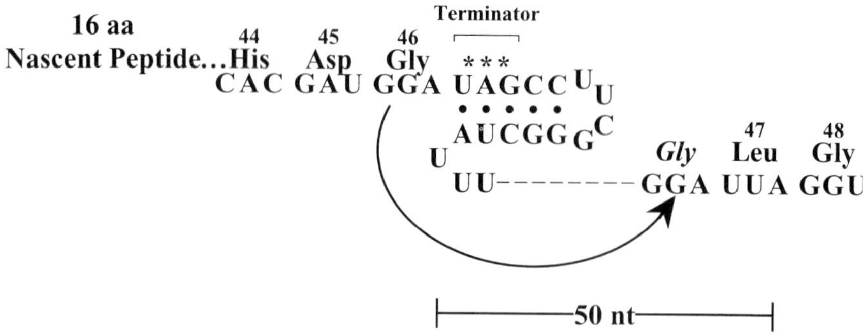

Figure 2 Structure of the T4 gene *60* translational hop site. The hop occurs between the incorporation of amino acids 46 and 47 of topoisomerase. The hop is stimulated by the upstream 16-amino-acid nascent protein, the in-frame UAG terminator adjacent to codon 46, the stem loop shown, and the matching take-off and landing GGA codons. The landing codon is labeled in italics to indicate that the peptidyl-tRNA hops to that codon before the ribosome resumes elongation. The amino acids of topoisomerase are numbered.

length of the hop, probably because the peptidyl-tRNA can dissociate from the ribosome before rebinding to the mRNA, analogous to ribosomal editing as described by Menninger (1977). Very long bypass events, as in gene *60*, should therefore be very inefficient. The stimulatory elements in gene *60* may increase efficiency by preventing peptidyl-tRNA from dissociating from the ribosome during the hop.

The importance of the data on starvation-induced hopping is that it questions the idea that hopping is an illegal event on the ribosome, requiring quite specific sequences and structures to reprogram the translation apparatus. Instead, it appears that hopping is a normal, though very infrequent, translational error. The elaborate gene *60* stimulatory sequences may merely raise hopping efficiency to levels sufficient to provide the kind of output required by the gene.

NONSENSE SUPPRESSION AND PROGRAMMED READTHROUGH OF TERMINATION CODONS

Nonsense suppressors are mutations that allow efficient reading of termination codons as sense. As with frameshift suppressors, they include mutations that alter the structure of tRNAs and other components of the translational machinery. Most mutations creating nonsense suppressor tRNAs alter the anticodon so as to allow base-pairing with a termination codon; for example, introducing a CUA anticodon that can pair with a

UAG codon. Others alter nucleotides outside the anticodon, presumably to allow noncognate recognition of the termination codon, e.g., a mutant of tRNA $^{\text{Trp}}_{\text{CCA}}$ that decodes UGA (Hirsh 1971). Non-tRNA suppressors include mutants affecting subunits of peptide release factors, in bacteria including *prfA* (Ryden et al. 1986), *prfB* (Kawakami et al. 1988; Karow et al. 1998), and *prfC* (Grentzmann et al. 1994; Mikuni et al. 1994), and in *S. cerevisiae SUP35* and *SUP45* (for review, see Stansfield and Tuite 1994).

The importance of having a sufficient supply of active release factor to promote efficient termination, and to ensure that termination codons are not mistakenly read as sense, is underscored by the existence of a prion form of Sup35 protein, [PSI$^+$], (for review, see Liebman and Derkatch 1999), which forms intracellular aggregates that restrict its availability during translation. This restriction leads to readthrough of termination codons and frameshifting at slippery stops, i.e., sites using a terminator as pause codon (Liebman and Sherman 1979; Bidou et al. 2000). Programmed nonsense readthrough sites also facilitate readthrough of termination codons by ensuring inefficient recognition by release factor. They do this by providing a sequence context that is inappropriate for termination. The simplest sites, for example in Sindbis (Li and Rice 1993) or tobacco mosaic virus (Skuzeski et al. 1991), consist of a termination codon followed by an short context sequence, as little as UGA-C in Sindbis. Since the immediate 3′ context of a termination codon modulates the efficiency of release factor recognition (Bonetti et al. 1995; McCaughan et al. 1995; Poole et al. 1995), these sites may simply reduce the efficiency of termination sufficiently to allow noncognate decoding.

Some programmed readthrough sites are more complex; the paradigmatic site comes from the retrovirus Moloney murine leukemia virus (MoMLV) (Yoshinaka et al. 1985). The site consists of a UAG termination codon and a downstream stimulatory region (Panganiban 1988), a pseudoknot that begins 8 nucleotides downstream from the UAG (Wills et al. 1991; Feng et al. 1992). The primary sequence of the pseudoknot is important to its stimulatory effect (Feng et al. 1992; Wills et al. 1994), implying that the ribosome or a ribosome-associated factor might have to interact specifically with the pseudoknot to induce frameshifting. Candidates for an interacting factor include rRNA, ribosomal proteins, and release factor.

Incorporation of selenocysteine into nascent peptides occurs by suppression of specific UGA codons (for review, see Chapter 26). The system is an example of programmed readthrough of an extraordinary kind, since it involves a dedicated tRNA$^{\text{[Ser]Sec}}$ that binds only to a dedicated analog of

EF1A, the SelB protein. An RNA secondary structure, the selenocysteine insertion sequence (SECIS), is the target of SelB, directing recruitment of a ternary complex of selenocysteyl-tRNA$^{[Ser]Sec}$•SelB•GTP to the ribosome at the suppressed UGA codon. Like a programmed readthrough site, the efficiency of incorporation of selenocysteine is probably improved because release factor inefficiently recognizes the UGA termination codon.

CONCLUDING REMARKS

What is most striking about recoding events is their diversity of mechanism. There are some commonalties among the various events, the need for some type of translational pause to allow a slow unconventional event to occur being the most obvious. However, recoding events subvert the rules of translation in distinctive ways. We still have little understanding of how programmed –1 frameshifting occurs. Evidence is growing for specific secondary structural motifs in frameshift-inducing pseudoknots, but how they manipulate the ribosome to cause the shift remains elusive. In +1 frameshifting we know that in yeast it is an unusual codon•anticodon interaction in the P site that stimulates errant decoding in the A site. The fact that this same mechanism seems not to operate in *E. coli* suggests that there are other ways to encourage a +1 shift of frame. Thus, even within a well-defined class of recoding events, the mechanisms underlying the events can differ substantially. Since recoding sites must manipulate the error-correcting ability of ribosomes, emerging structural information about ribosomes may provide key insights that will lead to a more fundamental understanding of how recoding sites work, perhaps ultimately leading to a more unified theory of recoding events as we discover that the seeming mechanistic diversity masks a more fundamental similarity in function.

Little mention has been made in this chapter of the regulatory potential of recoding sites. There are clear examples of genes being regulated directly by a recoding event. The frameshift in *prfB* of *E. coli* competes directly with termination and forms an autogenous feedback loop for the product of the gene, peptide release factor 2. Ornithine decarboxylase antizyme controls the activity of ornithine decarboxylase (ODCase), and expression of antizyme is regulated by frameshifting in response to polyamines, the ultimate product of ODCase. There are few other examples of clear regulatory functions of recoding sites. Most of the others allow the expression of alternative proteins, either for a morphogenetic purpose (as in retroviruses and transposons) or perhaps to allow expression of enzymes with distinct catalytic properties (as with the *dnaX* gene encoding a subunit of DNA polymerase in *E. coli*). There are hints of possible regulatory effects in the observation that +1 frameshifting controls expression of the Est3 gene, a subunit of telomerase. Regulating expression of this protein by frameshifting would have profound effects on cellular physiology, including senescence, but we have no evidence of such a regulatory effect. Perhaps clear examples of other cases of regulation of recoding events will emerge from genome sequence efforts.

REFERENCES

Agranovsky A.A., Koonin E.V., Boyko V.P., Maiss E., Frotschl R., Lunina N.A., and Atabekov J.G. 1994. Beet yellows closterovirus: Complete genome structure and identification of a leader papain-like thiol protease. *Virology* **198:** 311–324.

Asakura T., Sasaki T., Nagano F., Satoh A., Obaishi H., Nishioka H., Imamura H., Hotta K., Tanaka K., Nakanishi H., and Takai Y. 1998. Isolation and characterization of a novel actin filament-binding protein from *Saccharomyces cerevisiae. Oncogene* **16:** 121–130.

Belcourt M.F. and Farabaugh P.J. 1990. Ribosomal frameshifting in the yeast retrotransposon Ty: tRNAs induce slippage on a 7 nucleotide minimal site. *Cell* **62:**339–352.

Bidou L., Stahl G., Grima B., Liu H., Cassan M., and Rousset J.P. 1997. In vivo HIV–1 frameshifting efficiency is directly related to the stability of the stem–loop stimulatory signal. *RNA* **3:** 1153–1158.

Bidou L., Stahl G., Hatin I., Namy O., Rousset J.-P., and Farabaugh P.J. 2000. Nonsense mediated decay mutants do not affect programmed –1 frameshifting. RNA (in press).

Bonetti B., Fu L., Moon J., and Bedwell D.M. 1995. The efficiency of translation termination is determined by a synergistic interplay between upstream and downstream sequences in *Saccharomyces cerevisiae. J. Mol. Biol.* **251:** 334–345.

Bossi L. and Roth J.R. 1981. Four-base codons ACCA, ACCU, and ACCC are recognized by frameshift suppressor *sufJ*. *Cell* **25:** 489–496.

Brierley I. 1995. Ribosomal frameshifting on viral RNAs. *J. Gen. Virol.* **76:** 1885–1892.

Brierley I., Jenner A.J., and Inglis S.C. 1992. Mutational analysis of the "slippery-sequence" component of a coronavirus ribosomal frameshift signal. *J. Mol. Biol.* **227:** 463–479.

Brierley I., Meredith M.R., Bloys A.J., and Hagervall T.G. 1997. Expression of a coronavirus ribosomal frameshift signal in *Escherichia coli:* Influence of tRNA anticodon modification on frameshifting. *J. Mol. Biol.* **270:** 360–373.

Brunelle M.N., Payant C., Lemay G., and Brakier-Gingras L. 1999. Expression of the human immunodeficiency virus frameshift signal in a bacterial cell-free system: Influence of an interaction between the ribosome and a stem-loop structure downstream from the slippery site. *Nucleic Acids Res.* **27:** 4783–4791.

Carlson B.A., Kwon S.Y., Chamorro M., Oroszlan S., Hatfield D.L., and Lee B.J. 1999. Transfer RNA modification status influences retroviral ribosomal frameshifting. *Virology* **255:** 2–8.

Cassan M., Delaunay N., Vaquero C., and Rousset J.P. 1994. Translational frameshifting at the *gag-pol* junction of human immunodeficiency virus type 1 is not increased in infected T-lymphoid cells. *J. Virol.* **68:** 1501–1508.

Chandler M. and Fayet O. 1993. Translational frameshifting in the control of transposition in bacteria. *Mol. Microbiol.* **7:** 497–503.

Chen X., Chamorro M., Lee S.I., Shen L.X., Hines J.V., Tinoco I.J., and Varmus H.E. 1995. Structural and functional studies of retroviral RNA pseudoknots involved in ribosomal frameshifting: Nucleotides at the junction of the two stems are important for efficient ribosomal frameshifting. *EMBO J.* **14:** 842–852.

Craigen W.J. and Caskey C.T. 1986. Expression of peptide chain release factor 2 requires high-efficiency frameshift. *Nature* **322:** 273–275.

Curran J.F. 1993. Analysis of effects of tRNA:message stability on frameshift frequency at the *Escherichia coli* RF2 programmed frameshift site. *Nucleic Acids Res.* **21:** 1837–1843.

Curran J.F. and Yarus M. 1989. Rates of aminoacyl-tRNA selection at 29 sense codons *in vivo*. *J. Mol. Biol.* **209:** 65–77.

Dinman J.D. 1995. Ribosomal frameshifting in yeast viruses. *Yeast* **11:** 1115–1127.

Dinman J. and Kinzy T. 1997. Translational misreading: Mutations in translation elongation factor 1α differentially affect programmed ribosomal frameshifting and drug sensitivity. *RNA* **3:** 870–881.

Dinman J.D. and Wickner R.B. 1994. Translational maintenance of frame: Mutants of *Saccharomyces cerevisiae* with altered –1 ribosomal frameshifting efficiencies. *Genetics* **136:** 75–86.

Dinman J.D., Ruiz-Echevarria M.J., and Peltz S.W. 1998. Translating old drugs into new treatments: Ribosomal frameshifting as a target for antiviral agents. *Trends Biotechnol.* **16:** 190–196.

Donly B.C., Edgar C.D., Adamski F.M., and Tate W.P. 1990. Frameshift autoregulation in the gene for *Escherichia coli* release factor 2: Partly functional mutants result in frameshift enhancement. *Nucleic Acids Res.* **18:** 6517–6522.

Dontsova O., Dokudovskaya S., Kopylov A., Bogdanov A., Rinke-Appel J., Junke N., and Brimacombe R. 1992. Three widely separated positions in the 16S RNA lie in or close to the ribosomal decoding region; a site-directed cross-linking study with mRNA analogues. *EMBO J.* **11:** 3105–3116.

Farabaugh P. 1995. Post-transcriptional regulation of transposition by Ty retrotransposons of *Saccharomyces cerevisiae*. *J. Biol. Chem.* **270:** 10361–10364.

———. 1996. Programmed translational frameshifting. *Microbiol. Rev.* **60:** 103–134.

Farabaugh P.J. and Björk G.R. 1999. How translational accuracy influences reading frame maintenance. *EMBO J.* **18:** 1427–1434.

Farabaugh P.J. and Vimaladithan A. 1998. Effect of frameshift-inducing mutants of elongation factor 1α on programmed +1 frameshifting in yeast. *RNA* **4:** 38–46.

Farabaugh P.J., Zhao H., and Vimaladithan A. 1993. A novel programed frameshift expresses the *POL3* gene of retrotransposon Ty3 of yeast: Frameshifting without tRNA slippage. *Cell* **74:** 93–103.

Feng Y.-X., Yuan H., Rein A., and Levin J.G. 1992. Bipartite signal for read-through suppression in murine leukemia virus mRNA: An eight-nucleotide purine-rich sequence immediately downstream of the *gag* termination codon followed by an RNA pseudoknot. *J. Virol.* **66:** 5127–5132.

Fütterer J. and Hohn T. 1996. Translation in plants—Rules and exceptions. *Plant Mol. Biol.* **32:** 159–189.

Gaber R.F. and Culbertson M.R. 1984. Codon recognition during frameshift suppression in *Saccharomyces cerevisiae*. *Mol. Cell. Biol.* **4:** 2052–2061.

Gallant J.A. and Lindsley D. 1998. Ribosomes can slide over and beyond "hungry" codons, resuming protein chain elongation many nucleotides downstream. *Proc. Natl. Acad. Sci.* **95:** 13771–13776.

Garcia A., van Duin J., and Pleij C.W. 1993. Differential response to frameshift signals in eukaryotic and prokaryotic translational systems. *Nucleic Acids Res.* **21:** 401–406.

Gramstat A., Prufer D., and Rohde W. 1994. The nucleic acid-binding zinc finger protein of potato virus M is translated by internal initiation as well as by ribosomal frameshifting involving a shifty stop codon and a novel mechanism of P-site slippage. *Nucleic Acids Res.* **22:** 3911–3917.

Grentzmann G., Brechemier-Baey D., Heurgue V., Mora L., and Buckingham R.H. 1994. Localization and characterization of the gene encoding release factor RF3 in *Escherichia coli*. *Proc. Natl. Acad. Sci.* **91:** 5848–5852.

Hatfield D.L., Levin J.G., Rein A., and Oroszlan S. 1992. Translational suppression in retroviral gene expression. *Adv. Virus Res.* **41:** 193–239.

Hatfield D., Feng Y.X., Lee B.J., Rein A., Levin J.G., and Oroszlan S. 1989. Chromatographic analysis of the aminoacyl-tRNAs which are required for translation of codons at and around the ribosomal frameshift sites of HIV, HTLV-1 and BLV. *Virology* **173:** 736–742.

Hirsh D. 1971. Tryptophan transfer RNA as the UGA suppressor. *J. Mol. Biol.* **58:** 439–458.

Holley R.W. 1965. Structure of an alanine transfer ribonucleic acid. *J. Am. Med. Assoc.* **194:** 868–871.

Honigman A., Falk H., Mador N., Rosental T., and Panet A. 1995. Translation efficiency of the human T-cell leukemia virus (HTLV-2) *gag* gene modulates the frequency of ribosomal frameshifting. *Virology* **208:** 312–318.

Huang W.M., Ao S.-Z., Casjens S., Orlandi R., Zeikus R., Weiss R., Winge D., and Fang M. 1988. A persistent untranslated sequence within bacteriophage T4 DNA topoisomerase gene 60. *Science* **239:** 1005–1012.

Hughes D., Atkins J.F., and Thompson S. 1987. Mutants of elongation factor Tu promote ribosomal frameshifting and nonsense readthrough. *EMBO J.* **6:** 4235–4239.

Hung M., Patel P., Davis S., and Green S.R. 1998. Importance of ribosomal frameshifting for human immunodeficiency virus type 1 particle assembly and replication. *J. Virol.* **72:** 4819–4824.

Hüttenhofer A., Weiss-Brummer B., Dirheimer G., and Martin R.P. 1990. A novel type of + 1 frameshift suppressor: A base substitution in the anticodon stem of a yeast mitochondrial serine-tRNA causes frameshift suppression. *EMBO J.* **9:** 551–558.

Ivanov I.P., Gesteland R.F., and Atkins J.F. 1998a. A second mammalian antizyme: Conservation of programmed ribosomal frameshifting. *Genomics* **52:** 119–129.

Ivanov I.P., Simin K., Letsou A., Atkins J.F., and Gesteland R.F. 1998b. The *Drosophila* gene for antizyme requires ribosomal frameshifting for expression and contains an intronic gene for snRNP Sm D3 on the opposite strand. *Mol. Cell. Biol.* **18:** 1553–1561.

Jacks T. and Varmus H.E. 1985. Expression of the Rous sarcoma virus *pol* gene by ribosomal frameshifting. *Science* **230:** 1237–1242.

Jacks T., Madhani H.D., Masiarz F.R., and Varmus H.E. 1988a. Signals for ribosomal frameshifting in the Rous sarcoma virus *gag-pol* region. *Cell* **55:** 447–458.

Jacks T., Townsley K., Varmus H.E., and Majors J. 1987. Two efficient ribosomal frameshifting events are required for synthesis of mouse mammary tumor virus *gag*-related polyproteins. *Proc. Natl. Acad. Sci.* **84:** 4298–4302.

Jacks T., Power M.D., Masiarz F.R., Luciw P.A., Barr P.J., and Varmus H.E. 1988b. Characterization of ribosomal frameshifting in HIV-1 *gag-pol* expression. *Nature* **331:** 280–283.

Karasev A.V., Boyko V.P., Gowda S., Nikolaeva O.V., Hilf M.E., Koonin E.V., Niblett C.L., Cline K., Gumpf D.J., Lee R.F., et al. 1995. Complete sequence of the citrus tristeza virus RNA genome. *Virology* **208:** 511–520.

Karow M.L., Rogers E.J., Lovett P.S., and Piggot P.J. 1998. Suppression of TGA mutations in the *Bacillus subtilis* spoIIR gene by prfB mutations. *J. Bacteriol.* **180:** 4166–4170.

Kato M., Nishikawa K., Uritani M., Miyazaki M., and Takemura S. 1990. The difference in the type of codon-anticodon base pairing at the ribosomal P-site is one of the determinants of the translational rate. *J. Biochem.* **107:** 242–247.

Kawakami K., Inada T., and Nakamura Y. 1988. Conditionally lethal and recessive UGA-suppressor mutations in the prfB gene encoding peptide chain release factor 2 of *Escherichia coli*. *J. Bacteriol.* **170:** 5378–5381.

Larsen B., Gesteland R.F., and Atkins J.F. 1997. Structural probing and mutagenic analysis of the stem-loop required for *Escherichia coli* dnaX ribosomal frameshifting: programmed efficiency of 50%. *J. Mol. Biol.* **271:** 47–60.

Larsen B., Peden J., Matsufuji S., Matsufuji T., Brady K., Maldonado R., Wills N.M., Fayet O., Atkins J.F., and Gesteland R.F. 1995. Upstream stimulators for recoding. *Biochem. Cell Biol.* **73:** 1123–1129.

Lee S., Umen J., and Varmus H. 1995. A genetic screen identifies cellular factors involved in retroviral –1 frameshifting. *Proc. Natl. Acad. Sci.* **92:** 6587–6591.

Li G. and Rice C.M. 1993. The signal for translational readthrough of a UGA codon in Sindbis virus RNA involves a single cytidine residue immediately downstream of the termination codon. *J. Virol.* **67:** 5062–5067.

Liebman S.W. and Derkatch I.L. 1999. The yeast [PSI+] prion: Making sense of nonsense. *J. Biol. Chem.* **274:** 1181–1184.

Liebman S.W. and Sherman F. 1979. Extrachromosomal psi+ determinant suppresses nonsense mutations in yeast. *J. Bacteriol.* **139:** 1068–1071.

Liphardt J., Napthine S., Kontos H., and Brierley I. 1999. Evidence for an RNA pseudoknot loop-helix interaction essential for efficient -1 ribosomal frameshifting. *J. Mol. Biol.* **288:** 321–335.

Maia I.G., Seron K., Haenni A.L., and Bernardi F. 1996. Gene expression from viral RNA genomes. *Plant Mol. Biol.* **32:** 367–391.

Marczinke B., Fisher R., Vidakovic M., Bloys A.J., and Brierley I. 1998. Secondary structure and mutational analysis of the ribosomal frameshift signal of rous sarcoma virus. *J. Mol. Biol.* **284:** 205–225.

Matsufuji S., Matsufuji T., Miyazaki Y., Murakami Y., Atkins J.F., Gesteland R.F., and Hayashi S. 1995. Autoregulatory frameshifting in decoding mammalian ornithine decarboxylase antizyme. *Cell* **80:** 51–60.

McCaughan K.K., Brown C.M., Dalphin M.E., Berry M.J., and Tate W.P. 1995. Translational termination efficiency in mammals is influenced by the base following the stop codon. *Proc. Natl. Acad. Sci.* **92:** 5431–5435.

Menninger J. 1977. Ribosome editing and the error catastrophe hypothesis of cellular aging. *Mech. Ageing Dev.* **6:** 131–142.

Mikuni O., Ito K., Moffat J., Matsumura K., McCaughan K., Nobukuni T., Tate W., and

Nakamura Y. 1994. Identification of the *prfC* gene, which encodes peptide-chain-release factor 3 of *Escherichia coli. Proc. Natl. Acad. Sci.* **91:** 5798–5802.

Miller W.A., Dinesh-Kumar S.P., and Paul C.P. 1995. Luteovirus gene expression. *Crit. Rev. Plant Sci.* **14:** 179–211.

Morris D.K. and Lundblad V. 1997. Programmed translational frameshifting in a gene required for yeast telomere replication. *Curr. Biol.* **7:** 969–976.

Muhlrad D. and Parker R. 1999. Recognition of yeast mRNAs as "nonsense containing" leads to both inhibition of mRNA translation and mRNA degradation: Implications for the control of mRNA decapping. *Mol. Biol. Cell* **11:** 3971–3978.

Napthine S., Liphardt J., Bloys A., Routledge S., and Brierley I. 1999. The role of RNA pseudoknot stem 1 length in the promotion of efficient –1 ribosomal frameshifting. *J. Mol. Biol.* **288:** 305–320.

O'Connor M., Gesteland R.F., and Atkins J.F.. 1989. tRNA hopping: Enhancement by an expanded anticodon. *EMBO J.* **8:** 4315–4323.

Pande S., Vimaladithan A., Zhao H., and Farabaugh P.J. 1995. Pulling the ribosome out of frame +1 at a programmed frameshift site by cognate binding of aminoacyl-tRNA. *Mol. Cell. Biol.* **15:** 298–304.

Panganiban A.T. 1988. Retroviral *gag* gene amber codon suppression is caused by an intrinsic *cis*-acting component of the viral mRNA. *J. Virol.* **62:** 3574–6580.

Parkin N.T., Chamorro M., and Varmus H.E. 1992. Human immunodeficiency virus type 1 *gag-pol* frameshifting is dependent on downstream mRNA secondary structure: Demonstration by expression *in vivo*. *J. Virol.* **66:** 5147–5151.

Poole E.S., Brown C.M., and Tate W.P. 1995. The identity of the base following the stop codon determines the efficiency of in vivo translational termination in *Escherichia coli. EMBO J.* **14:** 151–158.

Powers T. and Noller H.F. 1994. The 530 loop of 16S rRNA: A signal to EF-Tu? *Trends Genet.* **10:** 27–31.

Qian Q. and Björk G.R. 1997. Structural alterations far from the anticodon of the $tRNA^{Pro}{}_{GGG}$ of *Salmonella typhimurium* induce +1 frameshifting at the peptidyl-site. *J. Mol. Biol.* **273:** 978–992.

Qian Q., Li J.N., Zhao H., Hagervall T.G., Farabaugh P.J., and Björk G.R. 1998. A new model for phenotypic suppression of frameshift mutations by mutant tRNAs. *Mol. Cell* **1:** 471–482.

Reil H., Hoxter M., Moosmayer D., Pauli G., and Hauser H. 1994. CD4 expressing human 293 cells as a tool for studies in HIV-1 replication: The efficiency of translational frameshifting is not altered by HIV-1 infection. *Virology* **205:** 371–375.

Rettberg C.C., Prere M.F., Gesteland R.F., Atkins J.F., and Fayet O. 1999. A three-way junction and constituent stem-loops as the stimulator for programmed –1 frameshifting in bacterial insertion sequence IS911. *J. Mol. Biol.* **286:** 1365–1378.

Riddle D. and Carbon J. 1973. Frameshift suppression: A nucleotide addition in the anticodon of a glycine tRNA. *Nat. New Biol.* **242:** 230–234.

Roth J.R. 1981. Frameshift suppression. *Cell* **24:** 601–602.

Ruiz-Echevarria M.J., Yasenchak J.M., Han X., Dinman J.D., and Peltz S.W. 1998. The upf3 protein is a component of the surveillance complex that monitors both translation and mRNA turnover and affects viral propagation. *Proc. Natl. Acad. Sci.* **95:** 8721–8726.

Ryden M., Murphy J., Martin R., Isaksson L., and Gallant J. 1986. Mapping and complementation studies of the gene for release factor 1. *J. Bacteriol.* **168:** 1066–1069.

Sandbaken M.G. and Culbertson M.R. 1988. Mutations in elongation factor EF-1α affect the frequency of frameshifting and amino acid misincorporation in *Saccharomyces cerevisiae*. *Genetics* **120:** 923–934.

Sekine Y. and Ohtsubo E. 1992. DNA sequences required for translational frameshifting in production of the transposase encoded by IS1. *Mol. Gen. Genet.* **235:** 325–332.

Sekine Y., Nagasawa H., and Ohtsubo E. 1992. Identification of the site of translational frameshifting required for production of the transposase encoded by insertion sequence IS 1. *Mol. Gen. Genet.* **235:** 317–324.

Shen L.X. and Tinoco I. 1995. The structure of an RNA pseudoknot that causes efficient frameshifting in mouse mammary tumor virus. *J. Mol. Biol.* **247:** 963–978.

Shine J. and Dalgarno L. 1974. The 3′-terminal sequence of *Escherichia coli* 16S ribosomal RNA: Complementarity to nonsense triplets and ribosome binding sites. *Proc. Natl. Acad. Sci.* **71:** 1342–1346.

Skuzeski J.M., Nichols L.M., Gesteland R.F., and Atkins J.F. 1991. The signal for a leaky UAG stop codon in several plant viruses includes the two downstream codons. *J. Mol. Biol.* **218:** 365–373.

Somogyi P., Jenner A.J., Brierley I., and Inglis S.C. 1993. Ribosomal pausing during translation of an RNA pseudoknot. *Mol. Cell. Biol.* **13:** 6931–6940.

Sroga G.E., Nemoto F., Kuchino Y., and Björk G.R. 1992. Insertion (*sufB*) in the anticodon loop or base substitution (*sufC*) in the anticodon stem of tRNA(Pro)2 from *Salmonella typhimurium* induces suppression of frameshift mutations. *Nucleic Acids Res.* **20:** 3463–3469.

Stansfield I. and Tuite M.F. 1994. Polypeptide chain termination in *Saccharomyces cerevisiae*. *Curr. Genet.* **25:** 385–395.

Su L., Chen L., Egli M., Berger J.M., and Rich A. 1999. Minor groove RNA triplex in the crystal structure of a ribosomal frameshifting viral pseudoknot. *Nat. Struct. Biol.* **6:** 285–292.

Sundararajan A., Michaud W.A., Qian Q., Stahl G., and Farabaugh P.J. 1999. Near-cognate peptidyl-tRNAs promote +1 programmed translational frameshifting in yeast. *Mol. Cell.* **4:** 1005–1015.

Tate W.P., Poole E.S., Dalphin M.E., Major L.L., Crawford D.J., and Mannering S.A. 1996. The translational stop signal: Codon with a context, or extended factor recognition element? *Biochimie* **78:** 945–952.

ten Dam E. 1995. "Pseudoknot-dependent ribosomal frameshifting." Ph.D. thesis, Leiden University, Leiden, the Netherlands.

ten Dam E.B., Pleij C.W., and Bosch L. 1990. RNA pseudoknots: Translational frameshifting and readthrough on viral RNAs. *Virus Genes* **4:** 121–136.

ten Dam E.B., Brierley I., Inglis S.C., and Pleij C.W. 1994. Identification and analysis of the pseudoknot-containing gag-pro ribosomal frameshift signal of simian retrovirus –1. *Nucleic Acids Res.* **22:** 2304–2310.

Tsuchihashi Z. and Brown P.O. 1992. Sequence requirements for efficient translational frameshifting in the *Escherichia coli dnaX* gene and the role of an unstable interaction between $tRNA^{Lys}$ and an AAG lysine codon. *Genes Dev.* **6:** 511–519.

Tu C., Tzeng T.H., and Bruenn J.A. 1992. Ribosomal movement impeded at a pseudoknot required for frameshifting. *Proc. Natl. Acad. Sci.* **89:** 8636–8640.

Vijgenboom E. and Bosch L. 1989. Translational frameshifts induced by mutant species of the polypeptide chain elongation factor Tu of *Escherichia coli*. *J. Biol. Chem.* **264:** 13012–13017.

Vimaladithan A. and Farabaugh P.J. 1994. Special peptidyl-tRNA molecules promote translational frameshifting without slippage. *Mol. Cell. Biol.* **14:** 8107–8116.

Weiss R., Huang W., and Dunn D. 1990a. A nascent peptide is required for ribosomal bypass of the coding gap in bacteriophage T4 gene *60*. *Cell* **62:** 117–126.

Weiss R., Dunn D., Atkins J., and Gesteland R. 1990b. Ribosomal frameshifting from –2 to +50 nucleotides. *Prog. Nucleic Acid Res. Mol. Biol.* **39:** 159–183.

Weiss R., Dunn D., Dahlberg A., Atkins J., and Gesteland R. 1988. Reading frame switch caused by base-pair formation between the 3′ end of 16S rRNA and the mRNA during elongation of protein synthesis in *Escherichia coli*. *EMBO J.* **7:** 1503–1507.

Weiss R.B., Dunn D.M., Shuh M., Atkins J.F., and Gesteland R.F. 1989. *E. coli* ribosomes re-phase on retroviral frameshift signals at rates ranging from 2 to 50 percent. *New Biol.* **1:** 159–169.

Wills N.M., Gesteland R.F., and Atkins J.F. 1991. Evidence that a downstream pseudoknot is required for translational read-through of the Moloney murine leukemia virus *gag* stop codon. *Proc. Natl. Acad. Sci.* **88:** 6991–6995.

———. 1994. Pseudoknot-dependent read-through of retroviral *gag* termination codons: Importance of sequences in the spacer and loop 2. *EMBO J.* **13:** 4137–4144.

Yoshinaka Y., Katoh I., Copeland T.D., and Oroszlan S. 1985. Murine leukemia virus protease is encoded by the *gag-pol* gene and is synthesized through suppression of an amber termination codon. *Proc. Natl. Acad. Sci.* **82:** 1618–1622.

26

Recoding UGA as Selenocysteine

Marla J. Berry
Thyroid Division, Department of Medicine
Brigham and Women's Hospital
Harvard Medical School
Boston, Massachusetts 02115

THE VERSATILE UGA CODON

The UGA codon is more versatile than any other code word in the genetic alphabet. Best known as a termination codon, its primary function in most organisms, UGA also serves in several capacities as a sense codon. These include being decoded as tryptophan in mitochondria and mycoplasma, and as cysteine in some ciliates, due to divergence in the release factor and tRNA specificities in these organisms or organelle (for review, see Watanabe and Osawa 1995). At a small number of UGA codons in organisms whose primary use of UGA is to dictate cessation of protein synthesis, the default termination mode is apparently overridden by signals in the mRNAs, redirecting the translation machinery to incorporate an amino acid instead. UGA codons are the sites of +1 frameshifting in bacterial release factor 2 and eukaryotic antizyme, in both cases resulting in asparagine incorporation (for review, see Gesteland and Atkins 1996; Atkins et al. 1999). Examples of readthrough of UGA codons in the absence of frameshifting are known in phage Qβ, *Sindbis* virus, plant viruses, insects, and mammals, and may occur in all organisms. Of the numerous functions that UGA codons can assume, perhaps the most intriguing is "recoding" as selenocysteine. This usage is implemented in a wide variety of species in the eubacteria, archaea, and eukarya kingdoms.

Considerable knowledge has been gained about the process of selenocysteine incorporation in *Escherichia coli*, predominantly through elegant genetic and biochemical studies carried out in the Böck laboratory. The mechanism of selenocysteine biosynthesis and incorporation in bacteria and speculation on the evolution of selenocysteine recoding have been given excellent coverage in recent reviews (Böck 1994; Huttenhofer

Translational Control of Gene Expression
© 2000 Cold Spring Harbor Laboratory Press 0-87969-568-4/00 $5 + . 00

and Böck 1998; Atkins et al. 1999; Commans and Böck 1999). Much less is known about the incorporation pathway in eukaryotes and archaea. In this chapter, I will attempt to provide an overview of the current state of the field, incorporating some of the essential background information from prokaryotes but focusing primarily on eukaryotes. Distinguishing features between the two will be highlighted, and some of the more pressing questions that remain will be raised—the "where, how, why, and how much" of selenoprotein synthesis. Where does selenocysteine incorporation occur: at which UGA codons, in what proteins, in what tissues and organisms? How is selenocysteine specified instead of stop at specific UGA codons? How efficient is selenocysteine recoding? Does termination sometimes occur instead? If so, under what circumstances? Why go to the trouble to evolve a specific elongation factor and mRNA recoding signals dedicated solely to this amino acid—what's so special about it?

WHERE DOES SELENOCYSTEINE INCORPORATION OCCUR?

Although the process of selenocysteine incorporation is likely ancient, it is clearly not universal. Selenoproteins have been identified in only 5 out of 20 bacterial species examined (Atkins et al. 1999). In the eukaryotic kingdom, genes encoding selenoproteins or components of the selenocysteine incorporation machinery are found in species ranging from protozoa to vertebrates. Numerous mammalian, amphibian, avian, and fish selenoproteins have been identified and characterized, and selenoproteins have recently been identified in *Caenorhabditis elegans* (Buettner et al. 1999; Gladyshev et al. 1999) and *Dictyostelium discoideum* (C. Buettner and R. Kay, pers. comm). The list continues to grow through biochemical studies and searches of genome sequence databases. However, among the eukarya, there are also exceptions. Upon completion of the *Saccharomyces cerevisiae* genome sequence, components of the selenocysteine incorporation pathway were noted to be conspicuously absent, eliminating hopes of employing yeast genetics to study the process in eukaryotes. The presence of tRNAs decoding UGA as tryptophan or cysteine in mycoplasma and ciliates, respectively, likely precludes its recoding as selenocysteine in these species.

HOW IS SELENOCYSTEINE SPECIFIED INSTEAD OF STOP?

Selenocysteine incorporation in *E. coli* is highly context-dependent. Specific mRNA stem-loop structures residing immediately downstream of each of the selenocysteine codons in this species are required for recoding each UGA as selenocysteine instead of stop. The sequence and position of these stem-loops are tightly constrained — single point mutations in the loop or movement of the stem upstream or downstream by as little as three nucleotides reduces or abolishes incorporation (Heider et al. 1992).

In eukaryotic selenoprotein mRNAs, the required signals, also stem-loops but differing in sequence and structure from their prokaryotic counterparts, are located in the 3′ untranslated regions (UTR) of the messages, at considerable distances from the selenocysteine codons. These signals, termed selenocysteine insertion sequences, or SECIS elements, serve to recode the entire message, functioning for any upstream in-frame UGA codon (Berry et al. 1993; Hill et al. 1993), provided a minimal spacing requirement of ~60 nucleotides is met (Martin et al. 1996). In most eukaryotic selenoprotein mRNAs, the SECIS element is located several hundred nucleotides to several kilobases downstream of the UGA codon(s). Hence, the end of the open reading frame is typically marked by either UAA or UAG. However, in two known exceptions, mammalian selenoprotein W and *Schistosoma mansoni* glutathione peroxidase, the first in-frame UGA encodes selenocysteine and the second specifies termination. In these mRNAs, the spacing between the second UGA and the SECIS element is less than the experimentally determined functional minimum (Martin et al. 1996).

Identification of the signals thought to direct selenocysteine incorporation in archaea brought perhaps the biggest surprise of all. Analysis of the genome sequence of the archaeon, *Methanococcus jannaschii*, revealed seven putative selenoprotein sequences, with conserved predicted stem-loops in the 3′UTRs in six of these, but in the 5′UTR in the seventh (Wilting et al. 1997). Thus, in this organism, recoding apparently can occur from either end of the mRNA. Further investigation of this process in archaea by genetic means has been hindered by the inability to transform *M. jannaschii*, but considerable mechanistic insights have been gained through biochemical analyses (see below).

Although the stem-loops directing selenocysteine incorporation in the three kingdoms differ in sequence and predicted structure (Fig. 1), they appear to be functionally, if not mechanistically, analogous. The term selenocysteine incorporation sequence, or SECIS, was originally proposed for the eukaryotic structures but has since been adopted for the prokaryotic and archaebacterial elements as well, and will be used in this chapter when referring to these structures.

In addition to the *cis*-acting SECIS elements signaling recoding, cotranslational incorporation of selenocysteine requires several special-

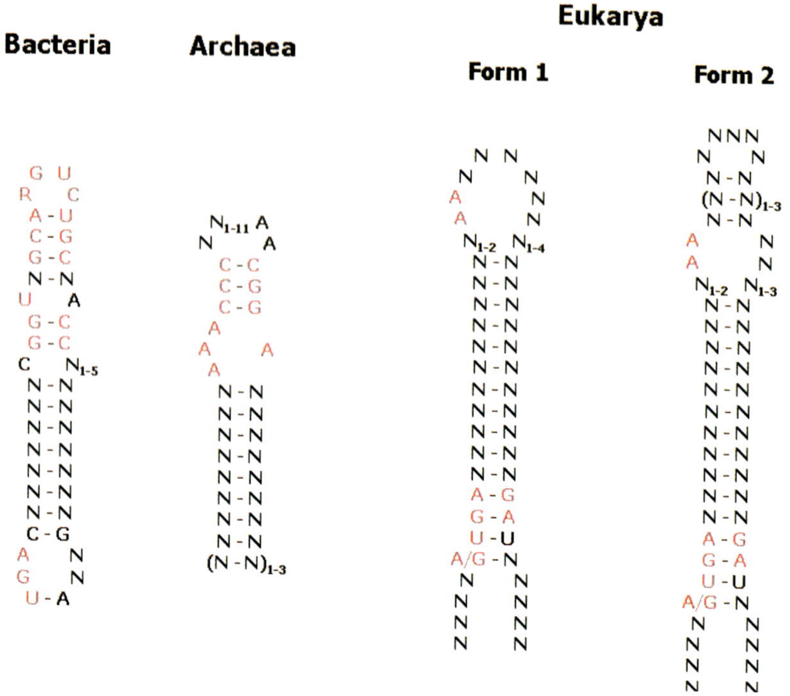

Figure 1 Structures of bacterial, archaeal, and two forms of eukaryotic SECIS elements. Conserved nucleotides are indicated in red.

ized components of the translational machinery. A minor seryl-tRNA species that decoded UGA had been reported in mammals as early as 1970 (Hatfield and Portugal 1970; Maenpaa and Bernfield 1970); nearly two decades later it was shown that serine provided the carbon backbone for selenocysteine in the first known mammalian selenoprotein, glutathione peroxidase (Sunde and Evenson 1987). At about the same time as the latter discovery, genetic studies of formate metabolism in *E. coli* led to the identification and characterization of four genes required for biosynthesis of the formate dehydrogenase selenoenzymes. Designated *selA, B, C,* and *D*, these genes encode a unique tRNA, a unique selenocysteine-specific elongation factor, and enzymes involved in the biosynthesis of the amino acid (Leinfelder et al. 1988; for review, see Böck 1994; Huttenhofer and Böck 1998; Atkins et al. 1999). As suggested from the mammalian studies, the bacterial selenocysteine-specific tRNA (the *selC* gene product) proved to be a serine isoacceptor with a UGA-complementary anticodon: It is recognized by seryl-tRNA synthetase and charged with serine. Conversion of serine to selenocysteine occurs on the tRNA, through the action of the *selA* gene product, selenocysteine synthase. Selenophosphate, synthesized from an unknown physiological selenium donor and ATP by the *selD* gene product, serves as the selenium donor for this reaction.

This pathway for selenocysteine biosynthesis earned the designation of the substrate as $tRNA^{[Ser]Sec}$. The minor seryl-tRNA species identified early in mammals was subsequently also shown to be a $tRNA^{[Ser]Sec}$ (Lee et al. 1989), and homologs were rapidly identified in a number of other eukaryotes (Lee et al. 1990). The $tRNA^{[Ser]Sec}$ gene is present in single copy in every species examined, with the exception of the zebrafish, *Danio rerio*, which carries two copies (Xu et al. 1999). *SelA* gene homologs have so far only been reported in bacteria and archaea. However, like the tRNA gene, homologs of the bacterial *selD* gene, designated selenophosphate synthetase, or SPS, have been found in many species through database searches. Two SPS genes are present in mammals, the SPS-2 gene containing an in-frame TGA encoding a selenocysteine residue in the putative active site and possessing a higher catalytic activity than the non-selenoenzyme. This suggests the presence of an autoregulatory positive feedback loop at an early step in the selenoprotein synthesis pathway. When selenium levels are limiting, synthesis of selenophosphate would be catalyzed predominantly by the low-specific-activity non-selenoenzyme. When selenium is abundant, the highly active selenoenzyme form of SPS would be made, resulting in increased selenophosphate, and hence selenocysteine, biosynthesis.

Following generation of selenocysteyl-tRNA via the *selA*, *C*, and *D* gene products, the amino acid is then incorporated into protein at specifically designated UGA codons. Unique structural features of $tRNA^{[Ser]Sec}$ result in its being poorly recognized by the standard elongation factors, EF1A (also called EF-Tu) or eEF1A (Baron and Böck 1991; Jung et al. 1994), which in turn circumvents its functioning as a nonspecific UGA suppressor. Instead, selenocysteyl-$tRNA^{[Ser]Sec}$ is bound by the selenocysteine-specific elongation factor, SELB. SELB exhibits specificity not only for $tRNA^{[Ser]Sec}$, but also for the presence of selenocysteine on the tRNA: It does not appreciably bind the seryl-tRNA precursor (Forchhammer et al. 1989). This discrimination on the part of the elongation factor prevents misincorporation of serine at UGA selenocysteine codons. SELB in prokaryotes possesses an amino-terminal domain with high homology to EF1A and a carboxy-terminal extension that binds the bacterial SECIS element (Fig. 2), recruiting the factor to UGA selenocysteine codons (Baron et al. 1993; Kromayer et al. 1996; Huttenhofer and Böck 1998). In striking contrast to the relative ease of identifying eukaryotic homologs of the tRNA and *selD* genes, the selenocysteine-specific elongation factor, SELB, has proved much more difficult in this respect. The factor has been identified and characterized in an archaebacterial species, but only very recently have homologs been found in any eukaryotic species (see below).

A MODEL FOR SELENOCYSTEINE INCORPORATION IN EUKARYOTES

The position of eukaryotic SECIS elements in the 3´UTR and the ability of these elements to function for multiple UGA codons (in nature in selenoprotein P and shown experimentally for type 1 deiodinase) suggested that a mechanism distinct from that in prokaryotes would be employed to elicit recoding from a distance in eukaryotes. This distinction led to the proposal of a model whereby the eukaryotic SECIS elements recruit one or more factors, possibly including a selenocysteyl-tRNA-specific elongation factor analogous to prokaryotic SELB (Berry et al. 1993). Assembly of a complex containing selenocysteyl-tRNA and the putative elongation factor on the SECIS element in the 3´UTR, followed by delivery of this complex to a UGA codon occupying the ribosomal A

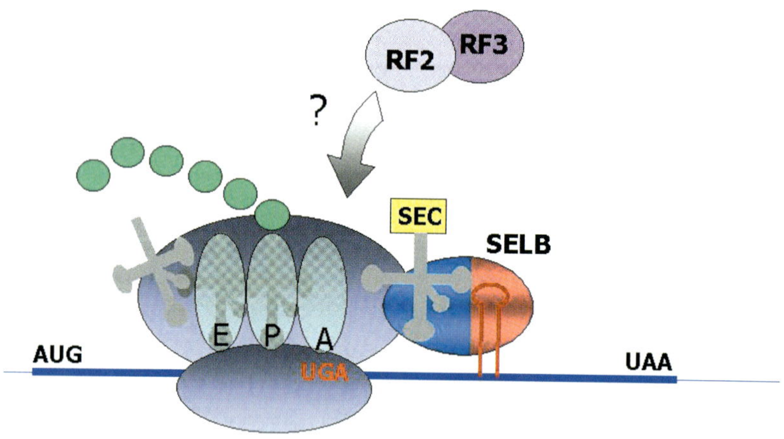

Figure 2 Prokaryotic selenoprotein synthesis. Translation of a prokaryotic selenoprotein mRNA (*blue line*) involves binding of selenocysteyl-tRNA (*grey*, with selenocysteine shown in *yellow*) by the SELB elongation factor domain (*blue half of oval*). The SELB SECIS-binding domain is indicated by the red half of the oval, bound to the SECIS element (*red*). Green circles represent amino acids of the nascent protein. Large grey ovals indicate ribosomal subunits, with A, P, and E sites designated by light green ovals. Release factors are indicated in purple and pink.

site, would allow translation of UGA codons without strict context requirements. The intervening mRNA would be looped out, accommodating the variable spacing between UGA and SECIS element—from a few hundred base pairs to nearly 5 kb—found in eukaryotic selenoprotein mRNAs. The minimal spacing requirement discussed above may reflect the spacing needed to loop back and properly orient a SECIS RNA–protein complex at the ribosome.

Following identification of eukaryotic SECIS elements, considerable efforts were concentrated on delineating the specific sequence and structural features of these elements to gain insight into the mechanism by which they functioned. These studies would also yield the information needed to design reagents for identification of SECIS-specific binding proteins. Whereas *E. coli* SECIS elements comprise a conserved hexanucleotide loop bounded by a less-conserved stem, eukaryotic SECIS elements are characterized by a small number of conserved nucleotides at specific positions in both the stem and the loop (Fig. 1) (Berry et al. 1991b, 1993). These are foremost, the sequence motifs A/GUGA and GA opposite each other at the 5′ and 3′ bases of the stem, respectively. On the basis of chemical and enzymatic probing and computer modeling analysis of SECIS sequences, Walczak et al. (1996) proposed that this region forms a quartet of non Watson-Crick base pairs, with a G-A, A-G tandem pair at the center. This region is termed the SECIS core (Martin et al. 1998). The other conserved sequence is a stretch of two or three adenosines in a loop or bulge at the top of the stem. Finally, the stem separating the SECIS core from the conserved adenosines is fixed at 9–11 base pairs, approximately one helical turn of an A-form RNA helix.

All eukaryotic SECIS elements identified to date possess these features, and mutagenesis studies have confirmed the importance of the conserved nucleotides, the requirement for base-pairing in the stem, and the conserved stem length for SECIS function (Berry et al. 1993; Shen et al. 1995; Kollmus et al. 1996; Martin et al. 1996, 1998; Grundner-Culemann et al. 1999). Furthermore, the 5′ and 3′ borders defining the minimal functional SECIS element were shown to correspond precisely with the conserved nucleotides (A/GUGA and GA) on the 5′ and 3′ sides of the stem (Martin et al. 1996). These boundaries were initially defined for type 1 deiodinase and have subsequently been confirmed for all eukaryotic SECIS elements examined (Martin et al. 1998). Using this information, the search for binding proteins that discriminate between functional wild-type and nonfunctional mutant SECIS elements ensued. Several laboratories reported identification of putative SECIS-binding proteins (Hubert et al. 1996; Lesoon et al. 1997; Shen et al. 1998), but until recently, none of these was shown to play a role in selenoprotein synthesis (see below).

In parallel with efforts to identify SECIS-binding proteins, a second prong of attack on the problem was to conduct searches of eukaryotic sequence databases to identify potential homologs of *E. coli* SELB. Homology searches proved difficult, due in part to the presence of multiple EF-related proteins and pseudogenes in many genomes. Furthermore, while producing high numbers of hits, these searches failed to yield a protein with a carboxy-terminal extension analogous to that in SELB in any eukaryotic database (M. Berry, unpubl.).

Determination of the complete sequence of the genome of the archaeon, *M. jannaschii*, revealed the presence of three EF genes, two bearing homology with eEF1A and eEF2, and the third unassigned (for review, see Dennis 1997). Selenoproteins had previously been identified in this organism and a $tRNA^{[Ser]Sec}$ gene was also present in the genome sequence. This suggested that the unassigned EF might function as a selenocysteyl-tRNA-specific elongation factor, but the sequence contained no carboxy-terminal extension analogous to that found in *E. coli* SELB. Nonetheless, it appeared to be the only potential candidate in the genome sequence. Conserved secondary structures proposed to function as SECIS elements were identified in the 3′ and 5′ UTRs of the archaea selenoprotein genes, suggesting that archaea and eukaryotes might employ similar recoding mechanisms. Searches of eukaryotic sequence databases with the putative archaea SELB sequence led to identification of homologs in the *C. elegans* and *D. melanogaster* genomes, and subsequently, in the mammalian est databases (Tujebajeva et al. 2000).

Recent and ongoing studies are beginning to shed light on the mechanism of selenocysteine incorporation directed by sequences in the 3′UTR. The unassigned archaeal EF gene was shown to exhibit specificity for selenocysteyl-tRNA, but did not bind the putative SECIS elements. Electrophoretic mobility shift assays performed with proteins from *M. jannaschii* extracts revealed a separate SECIS-binding activity (M. Rother and A. Böck, pers. comm.). Despite the lack of either a carboxy-terminal extension or SECIS-binding activity in the EF, use of the term SELB homolog will be employed herein in discussing this factor in order to distinguish it from the SECIS-binding activity.

At about the same time, a mammalian SECIS-binding protein was purified and cloned by Copeland et al. (2000). This protein, designated SECIS-binding protein 2 (SBP-2), exhibits specificity for wild-type but not mutant mammalian SECIS elements, does not contain an EF-homologous domain, but functions to enhance selenocysteine incorporation in an in vitro assay. Taken together, these data suggest that the functional equivalent of bacterial SELB in eukaryotes and archaea consists of at least two distinct proteins. Strikingly, although the SECIS core region is required for binding of SBP-2, the conserved adenosines in the loop are not, suggesting that the requirement for this conserved region may be conferred by an additional factor or factors. Purified recombinant SBP-2 binds SECIS elements in vitro with no requirement for additional proteins or other factors. Thus, any additional putative factor(s) may be recruited independent of SBP-2, or may require SBP-2 for binding. Interestingly, gel filtration analysis of SBP-2 indicates an aggregate molecular mass of ~500 kD (Copeland and Driscoll 1999).

A model that emerges from these results involves the SECIS element recruiting SBP-2, which in turn recruits the selenocysteine-specific elongation factor directly or through additional factors (Fig. 3). Recent studies demonstrate that the murine SELB homolog interacts with SBP-2 and that the SECIS element is required for this interaction (Tujebajeva et al. 2000). An alternative, though not mutually exclusive, function of the SECIS element has been suggested to explain the ability of a single element to serve multiple UGAs. In this model, the SECIS element delivers a signal to, or confers a modification upon, upstream ribosomes, presumably through a SECIS-bound factor or factors, such that the ribosomes acquire the ability to read UGA as selenocysteine throughout the transit of that message (Atkins et al. 1999). Intriguingly, SBP-2 contains a

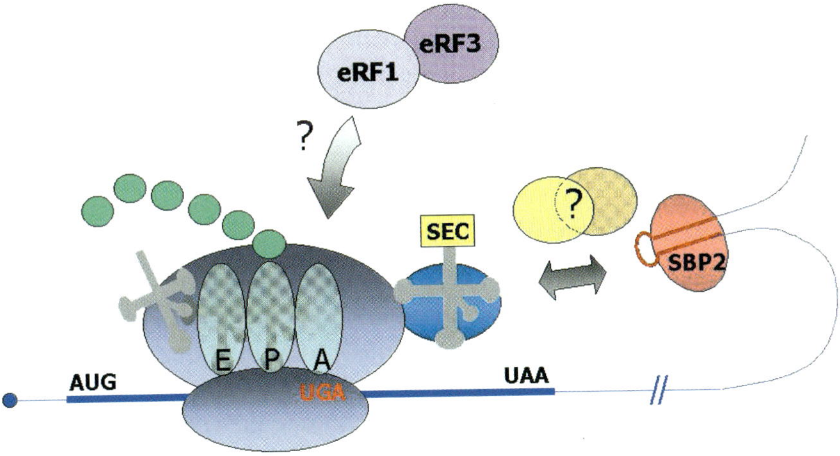

Figure 3 Eukaryotic selenoprotein synthesis. The eukaryotic selenocysteine-specific elongation factor is indicated by the blue oval, and SECIS-binding protein 2 (SBP-2) by the red oval. Orange and yellow ovals represent other possible cofactors. Components are as in Fig. 2.

domain homologous to the yeast SUP1 omnipotent suppressor of translation termination, suggesting it may serve an anti-termination function. This function may act in concert with a SELB-recruiting activity, residing in either SBP-2 or additional SECIS-binding proteins, to confer both selenocysteine incorporation and suppression of termination.

HOW EFFICIENT IS SELENOCYSTEINE INCORPORATION?

The efficiency of selenocysteine incorporation has recently been investigated in *E. coli* and found to be very low, ~4–5 % (Suppmann et al. 1999), with the majority of the translation products terminating. Overexpression of $tRNA^{[Ser]Sec}$ and the *selA* and *selB* genes increased this level by only ~2-fold. Following selenocysteine incorporation, SELB would be required to dissociate from the stem-loop to allow its translation, and a ribosome occupying this region would likely prevent refolding of the hairpin and SELB reassociation. If the next ribosome arrived at the UGA codon before the stem could refold and recruit SELB, termination would likely ensue. The effects of translation initiation rates and ribosome spacing on efficiency have not been reported to date. However, unlike the situation in prokaryotes, there is no need for eukaryotic SECIS-binding factor(s) or complexes to ever dissociate from the time they first bind mRNA until either the mRNA or proteins are degraded.

Nonetheless, studies in transiently transfected mammalian cells showed that selenocysteine incorporation into type 1 deiodinase was inefficient, with significant levels of termination at the UGA codon (Berry et al. 1994). Substitution of a cysteine codon for the selenocysteine codon increased expression of full-length protein by up to two orders of magnitude (Berry et al. 1992). Termination levels at the UGA codon in the wild-type mRNA were subsequently found to be greatly underestimated due to rapid turnover of the truncated peptide (Warner et al. 2000). Low efficiency of incorporation may be due to titration of the endogenous components of the selenoprotein synthesis machinery by excess selenoprotein mRNA, an effect that would be particularly pronounced when expressing from plasmids optimized for efficient transcription. This inefficiency was partially overcome by cotransfecting the $tRNA^{[Ser]Sec}$ and/or *selD* genes, or supplementing media with selenium, but not to the levels of the corresponding cysteine mutant. However, other components of the incorporation machinery had not been identified in these earlier studies and thus could not be tested. Recent studies indicate that SBP-2 is limiting in reticulocyte lysates, even in the presence of very low levels of selenoprotein mRNAs. Similarly, cotransfection of SBP-2 and selenoprotein mRNAs in mammalian cells increases selenoprotein expression by ~2 to 3-fold (P. Copeland and M. Berry, unpubl.).

It is unlikely that the inefficiency observed in transfected cells reflects the situation in the intact organism. As an example of this, whereas $tRNA^{[Ser]Sec}$ is limiting in transient transfection systems, this does not appear to be the case in studies of endogenous selenoproteins in cultured cells. Neither an ~10–fold increase in expression of the tRNA via introduction of multiple stable transgenes, nor an ~50% reduction due to targeted disruption of one copy of the endogenous tRNA gene had any detectable effect on selenoprotein expression patterns (Chittum et al. 1997b; Moustafa et al. 1998). Furthermore, if selenocysteine incorporation were inefficient in a protein with a single selenocysteine residue per polypeptide, it would be difficult to envision how full-length selenoprotein P, containing 10–12 selenocysteine residues, would ever be synthesized. In fact, most of the selenoprotein P in plasma from rats maintained on a selenium-sufficient diet is full-length, whereas a product resulting from termination at the second UGA codon is detected when the animals are selenium-deprived (Himeno et al. 1996). Termination at the first UGA would not have been detected in these studies, due to the use of radiolabeled selenium to detect the protein.

Early studies showed that selenium status affects not only selenoprotein synthesis, but also the levels of some selenoprotein mRNAs (Saedi et al. 1988; Sunde et al. 1989; Christensen and Burgener 1992). More recently, it has been demonstrated that under conditions of limiting selenium, degradation of cytoplasmic glutathione peroxidase mRNA occurs via nonsense mediated decay (NMD) (Moriarty et al. 1998; Weiss and Sunde 1998). The NMD pathway is thought to serve as an mRNA surveillance mechanism that functions to eliminate mRNAs containing premature termination codons (Maquat 1995; Hentze and Kulozik 1999; Maquat, Chapter 30). Truncated polypeptides encoded by such mRNAs could be nonfunctional, or might function in a dominant-negative fashion, proving deleterious to the cell. A critical feature in discrimination between physiological and premature termination codons in mammalian cells appears to be the position of the last intron in the pre-mRNA relative to the termination codon. An intron greater than ~50 nucleotides downstream of a termination codon will define this stop codon as premature (Nagy and Maquat 1998; Thermann et al. 1998).

The subcellular localization of NMD is a point of considerable intrigue. NMD of most mammalian mRNAs studied is apparent in RNA associated with the nuclear fraction (Maquat et al. 1981; Daar and Maquat 1988; Belgrader et al. 1994), leading some to speculate on the existence of a nuclear mechanism for scanning open reading frames. However, such a mechanism may not be required to explain this observation; cytoplasmic ribosomes can begin translating mRNAs during the process of nuclear export, as they emerge from, but remain tethered to, the nuclear pore complex.

The NMD pathway may play an important physiological role in maintaining the appropriate balance between mRNAs and the factors required for selenocysteine incorporation. How? If these factors are present in sufficient levels such that they are recruited by the SECIS element prior to or early in the first round of translation, selenocysteine will be incorporated and the message will not be targeted for NMD. If, on the other hand, selenoprotein mRNA levels exceed the levels of any of the required factors, some selenocysteine codons will be recognized as stop, and those mRNAs will be targeted to the NMD pathway. NMD may thus serve to maintain the appropriate stoichiometry between selenoprotein mRNAs and the necessary factors, to ensure efficient selenocysteine incorporation on the surviving mRNAs. Interestingly, most selenoprotein genes contain an intron meeting the minimal spacing requirement to define the selenocysteine codon as premature. Because the NMD pathway would not be operative on mRNAs expressed from transfected cDNAs due to the requirement for the prior presence of at least one intron, caution must be used, as suggested above, in interpreting efficiency studies employing cDNA-based expression systems. Finally, it is of considerable interest that the deduced protein sequences of both SBP-2 and the murine selenocysteine-specific elongation factor contain putative nuclear localization signals. It is tempting to speculate that nuclear import of these proteins could result in their binding SECIS elements in the nucleus immediately after transcription of the selenoprotein mRNAs, beginning the process of complex assembly prior to RNA export and susceptibility to NMD.

REGULATION OF SELENIUM INCORPORATION

Selenoprotein synthesis in mammals exhibits a hierarchy of tissue preferences, with brain and endocrine organs at the top of this pyramid. Under conditions of selenium depletion, these tissues preferentially retain available body selenium stores and the ability to synthesize selenoproteins. Virtually nothing is known about the mechanism of preferential selenium storage, but two lines of study have implications for mechanisms of preferential selenoprotein synthesis. First, Northern blot analysis for SBP-2 mRNA expression reveals two larger RNA species present in most tissues, and a smaller species detected only in testes, but at very high levels. This mRNA may be required for the high levels of expression of phospholipid hydroperoxide glutathione peroxidase in mammalian sperm, where it accumulates and serves a structural role (Ursini et al. 1999; see below). Second, selenium supplementation induces a specific methylation in $tRNA^{[Ser]Sec}$. This modification has been shown to decrease the turnover rate of $tRNA^{[Ser]Sec}$, increasing tRNA pool size when selenium is plentiful (Chittum et al. 1997a). Strikingly, no increase in the modified species is detected in either testes or brain following selenium supplementation.

WHY IS SELENOCYSTEINE PRESENT?

Although selenoproteins are required only for anaerobic metabolism in *E. coli*, they appear to be essential for life in mammals and possibly in lower eukaryotes. The most compelling evidence for this stems from targeted disruption of the $tRNA^{[Ser]Sec}$ gene in mice, which resulted in an embryonic lethal phenotype apparent from day 3 in utero (Bosl et al. 1997). Disruption of the *selD* gene in *D. melanogaster* resulted in arrested larval development and death, indicating a requirement for selenoprotein synthesis in this organism (Alsina et al. 1999). However, the specific selenoprotein(s) conferring this requirement has not been determined in any eukaryote. Of the selenoenzymes whose functions are known, most catalyze oxido-reduction reactions. The critical role selenocysteine plays in the activities of several of these enzymes has been illustrated by mutagenesis studies. Replacement of the active site selenocysteines by cysteines decreased catalytic activities of these enzymes by two to three orders of magnitude (Axley et al. 1991; Berry et al. 1991a; Rocher et al. 1992; Gasdaska et al. 1999). The cysteine-mutant enzymes are catalytically active, but selenocysteine likely confers strong adaptive advantages.

Information on the biological functions and properties of eukaryotic selenoproteins is accumulating at an unprecedented pace. The characteristics and functions of a few of the best-studied examples are given in Table 1 and discussed briefly herein; for more detailed information the reader is referred to recent reviews and the references cited therein (Stadtman 1996; Gladyshev and Hatfield 1999). The glutathione peroxidases, the most extensively studied family of selenoenzymes, function in protection of cells and membranes from oxidative damage caused by the free-radical by-products of hydroperoxides and lipid peroxides (Sunde 1994; Sunde et al. 1997; Ursini et al. 1997). One of the members of this multigene family, phospholipid hydroperoxide glutathione peroxidase, has recently been shown to have a dual function. It serves as a soluble peroxidase in spermatids, then aggregates through oxidative cross-links to form a structural protein in mature spermatozoa (Ursini et al. 1999). This

Table 1 Eukaryotic selenoproteins

Gene family	Functions
Glutathione peroxidases	ROOH → ROH + H_2O; protection from peroxide damage
cytoplasmic	
plasma	
phospholipid hydroperoxide	dual role, first as a peroxidase, later as an aggregate functioning as a sperm midpiece structural protein, preventing male sterility
gastrointestinal	
Iodothyroinine deiodinases	activation and inactivation of thyroid hormones
types 1 and 2	T4 → T3; thyroid hormone activation
type 3	T3 → T2,T1,T0; thyroid hormone inactivation
Thioredoxin reductases types 1, 2, 3	maintain cellular redox status, provide reducing equivalents for ribonucleotide reductase, function in folding of NFκB and other transcription factors
Selenophosphate synthetase	synthesis of selenophosphate from selenide and ATP
Selenoprotein P	function unknown, plasma protein, binds heavy metals, may promote neuronal survival, mRNAs contain 10–12 UGAs
Selenoprotein W	function unknown, muscle protein implicated in white muscle disease
15-kD prostate selenoprotein	function unknown, polymorphism may correlate with prostate cancer

latter function may explain the association between selenium deficiency and impaired sperm motility and hence, male infertility. A glutathione peroxidase gene was identified in the sequence of the human poxvirus, *Molluscum contagiosum* virus, representing the first known "viral" selenoenzyme (Senkevich et al. 1996). The sequence bears high homology with the human cellular enzyme, from whence it likely originated. Evidence has been presented for a role in protecting host cells from oxidative damage by UV or peroxides (Shisler et al. 1998).

Another important family of redox enzymes recently identified as selenoenzymes are the thioredoxin reductases (Bjornstedt et al. 1997), whose functions include maintaining cellular redox balance through their substrate, thioredoxin. Selenoprotein W is a muscle protein whose function may be associated with white muscle disease in selenium-deficient sheep and cattle (Vendeland et al. 1995). Demonstration of glutathione binding suggests a possible redox function for this protein (Gu et al. 1999). The functions of the iodothyronine deiodinase selenoenzymes are the activation (D1 and D2) or inactivation (D3) of thyroid hormones (Larsen and Berry 1995; St. Germain and Galton 1997), which in turn coordinate the expression of thyroid-hormone-responsive genes. Intriguingly, the dechlorinase selenoenzyme recently identified in *D. discoideum* bears sequence homology in the region of the active site selenocysteine with the deiodinases, suggesting a common ancestral dehalogenase (C. Buettner and R. Kay, unpubl.).

Selenoprotein P is a plasma selenoprotein containing 10–12 selenocysteine residues; the function(s) of this protein are under intensive investigation (Burk and Hill 1999). Coadministration of selenium has long been known to play a role in reducing the toxic effects of mercury (Koelman et al. 1973; Kosta et al. 1975), and recent studies have implicated selenoprotein P as a possible scavenger of toxic heavy metals (Yoneda and Suzuki 1997). Another recent study identified selenoprotein P as the component in serum that promotes neuronal survival in culture (Yan and Barrett 1998). The list of new members of the selenoprotein family continues to grow, with some of the most recent additions being identified through a novel strategy employing a software program designed to identify putative SECIS elements in sequence databases (Kryukov et al. 1999).

OUTLOOK AND FUTURE DIRECTIONS

Evidence has been presented by several groups (Barbarese et al. 1995; Stapulionis and Deutscher 1995; Negrutskii et al. 1999) that translation occurs on supramolecular complexes including not only the ribosomes, elongation factors, and tRNAs, but also the tRNA synthetases. According to this model, tRNAs are directly transferred from aminoacyl-tRNA synthetases to elongation factor to ribosomes without dissociation into the cellular fluid. Upon leaving the ribosome, deacylated tRNAs are directly transferred to their cognate synthetases. Extending this model to selenocysteine incorporation, the SECIS element may serve as the nucleating center of a "selenosome," recruiting not only SBP-2 and the elongation factor-tRNA complex, but perhaps also one or more of the enzymes involved in tRNA aminoacylation. This could serve to increase the efficiency of recharging, keeping the elongation factor primed and ready for the next ribosome or in the case of a multi-UGA mRNA, the next UGA codon.

Recent studies support a relatively long half-life for the putative selenosome complex, circumventing the need for repeated disassembly and reassembly, thus further increasing the efficiency of incorporation.

Overexpression of either the wild-type tRNA$^{[Ser]Sec}$ gene or a nonfunctional mutant tRNA$^{[Ser]Sec}$ gene significantly increased or decreased, respectively, selenocysteine incorporation directed by cotransfected selenoprotein genes (Berry et al. 1994; G. Warner and J. Faust, pers. comm.). Endogenous selenoprotein expression, however, was not discernibly affected. Similar specific enhancement was observed with cotransfected mammalian SPS and SBP-2 genes (Low et al. 1995; M. Berry, unpubl.). These results may reflect the presence of the majority of the endogenous selenoprotein mRNAs and other factors in translation complexes in the cytoplasm at the time of transfection. Selenoprotein mRNAs, tRNAs, and factors expressed from transfected plasmids would undergo concomitant synthesis and assembly into new complexes. Thus, the complexes assembled following transfection would undergo minimal exchange with the complexes previously assembled from endogenously expressed components.

The rapid pace of recent progress in the study of eukaryotic selenoprotein synthesis has left the field primed for future studies focusing on identification of additional factors involved in this process and their assembly into complexes competent for translation of selenoproteins. These studies will undoubtedly be greatly facilitated by genomics, as well as genetic studies in lower eukaryotes such as *C. elegans* and *D. discoideum*.

Study of this small family of proteins has not, to date, resulted in significant revision of textbook discussions of protein synthesis or the genetic code. Perhaps some of the publicity generated by elucidation of the many critical roles of selenium, as well as recent progress in unravelling the mechanism of selenocysteine incorporation, will result in a remedy to this.

ACKNOWLEDGMENTS

August Böck, Paul Copeland, Donna Driscoll, Vadim Gladyshev, Dolph Hatfield, and Greg Warner are gratefully acknowledged for communicating results prior to publication. I am indebted to John Atkins, Dolph Hatfield, Matthias Hentze, and Glover Martin III for many helpful and constructive suggestions on this chapter. The author is supported by National Institutes of Health grants RO1-DK-47320 and DK-52963.

REFERENCES

Alsina B., Corominas M., Berry M.J., Baguna J., and Serras F. 1999. Disruption of selenoprotein biosynthesis affects cell proliferation in the imaginal discs and brain of *Drosophila melanogaster. J. Cell Sci.* **112:** 2875–2884.

Atkins J.F., Böck A., Matsufuji S., and Gesteland R.F. 1999. Dynamics of the genetic code. In *The RNA world*, 2nd edition (ed. R.F. Gesteland et al.), pp. 637–673. Cold Spring Harbor Laboratory Press, Cold Spring Harbor, New York.

Axley M.J., Böck A., and Stadtman T.C. 1991. Catalytic properties of an *Escherichia coli* formate dehydrogenase mutant in which sulfur replaces selenium. *Proc. Natl. Acad. Sci.* **88:** 8450–8454.

Barbarese E., Koppel D.E., Deutscher M.P., Smith C.L., Ainger K., Morgan F., and Carson J.H. 1995. Protein translocation components are colocalized in granules in oligodendrocytes. *J. Cell Sci.* **108:** 2781–2790.

Baron C. and Böck A. 1991. The length of the aminoacyl-acceptor stem of the selenocysteine-specific $tRNA^{Sec}$ of *Escherichia coli* is the determinant for binding to elongation factors SELB or Tu. *J. Biol. Chem.* **266:** 20375–20379.

Baron C., Heider J., and Böck A. 1993. Interaction of translation factor SELB with the formate dehydrogenase H selenopolypeptide mRNA. *Proc. Natl. Acad. Sci.* **90:** 4181–4185.

Belgrader P., Cheng J., Zhou X., Stephenson L.S., and Maquat L.E. 1994. Mammalian nonsense codons can be *cis*-effectors of nuclear mRNA half-life. *Mol. Cell. Biol.* **14:** 8219–8228.

Berry M.J., Banu L., Harney J.W., and Larsen P.R. 1993. Functional characterization of the eukaryotic SECIS elements which direct selenocysteine insertion at UGA codons. *EMBO J.* **12:** 3315–3322.

Berry M.J., Harney J.W., Ohama T., and Hatfield D.L. 1994. Selenocysteine insertion or termination: Factors affecting UGA codon fate and complementary anticodon:codon mutations. *Nucleic Acids Res.* **22:** 3753–3759.

Berry M.J., Kieffer J.D., Harney J.W., and Larsen P.R. 1991a. Selenocysteine confers the biochemical properties of the type I iodothyronine deiodinase. *J. Biol. Chem.* **266:** 14155–14158.

Berry M.J., Maia A.L., Kieffer J.D., Harney J.W., and Larsen P.R. 1992. Substitution of cysteine for selenocysteine in type I iodothyronine deiodinase reduces the catalytic efficiency of the protein but enhances its translation. *Endocrinology* **131:** 1848–1852.

Berry M.J., Banu L., Chen Y., Mandel S.J., Kieffer J.D., Harney J.W., and Larsen P.R. 1991b. Recognition of UGA as a selenocysteine codon in type I deiodinase requires sequences in the 3′ untranslated region. *Nature* **353:** 273–276.

Bjornstedt M., Kumar S., Bjorkhem L., Spyrou G., and Holmgren A. 1997. Selenium and the thioredoxin and glutaredoxin systems. *Biomed. Environ. Sci.* **10:** 271–279.

Böck A. 1994. Incorporation of selenium into bacterial selenoproteins. In *Selenium in biology and human health* (ed. R.F. Burk), pp. 9–24. Springer-Verlag, New York.

Bosl M.R., Takaku K., Oshima M., Nishimura S., and Taketo M.M. 1997. Early embryonic lethality caused by targeted disruption of the mouse selenocysteine tRNA gene (Trsp). *Proc. Natl. Acad. Sci.* **94:** 5531–5534.

Buettner C., Harney J.W., and Berry M.J. 1999. The *Caenorhabditis elegans* homologue of thioredoxin reductase contains a selenocysteine insertion sequence (SECIS) element that differs from mammalian SECIS elements but directs selenocysteine incorporation. *J. Biol. Chem.* **274:** 21598–21602.

Burk R.F. and Hill K.E. 1999. Orphan selenoproteins. *Bioessays* **21:** 231–237.

Chittum H.S., Hill K.E., Carlson B.A., Lee B.J., Burk R.F., and Hatfield D.L. 1997a. Replenishment of selenium deficient rats with selenium results in redistribution of the selenocysteyl tRNA population in a tissue specific manner. *Biochim. Biophys. Acta* **1359:** 25–34.

Chittum H.S., Baek, H.J., Diamond A.M., Fernandez-Salguero P., Gonzalez F., Ohama T.,

Hatfield D.L., Kuehn M., and Lee B.J. 1997b. Selenocysteine $tRNA^{[Ser]Sec}$ levels and selenium-dependent glutathione peroxidase activity in mouse embryonic cells heterozygous for a targeted mutation in $tRNA^{[Ser]Sec}$ gene. *Biochemistry* **36:** 8634–8639.

Christensen M.J. and Burgener K.W. 1992. Dietary selenium stabilizes glutathione peroxidase mRNA in rat liver. *J. Nutr.* **122:** 1620–1626.

Commans S. and Böck A. 1999. Selenocysteine inserting tRNAs: An overview. *FEMS Microbiol. Rev.* **23:** 335–351.

Copeland P.R. and Driscoll D.M. 1999. Purification, redox sensitivity, and RNA binding properties of SECIS-binding protein 2, a protein involved in selenoprotein biosynthesis. *J. Biol. Chem.* **274:** 25447–25454.

Copeland P.R., Fletcher J.E., Carlson B.A., Hatfield D.L., and Driscoll D.M. 2000. A novel RNA binding protein, SBP2, is required for the translation of mammalian selenoprotein mRNAs. *EMBO J.* **19:** 306–314.

Daar I.O. and Maquat L.E. 1988. Premature translation termination mediates triosephosphate isomerase mRNA degradation. *Mol. Cell. Biol.* **8:** 802–813.

Dennis P.P. 1997. Ancient ciphers: Translation in Archaea. *Cell* **89:** 1007–1010.

Forchhammer K., Leinfelder W., and Böck A. 1989. Identification of a novel translation factor necessary for the incorporation of selenocysteine into protein. *Nature* **342:** 453–456.

Gasdaska J.R., Harney J.W., Gasdaska P.Y., Powis G., and Berry M.J. 1999. Regulation of human thioredoxin reductase expression and activity by 3′-untranslated region selenocysteine insertion sequence and mRNA instability elements. *J. Biol. Chem.* **274:** 25379–25385.

Gesteland R.F. and Atkins J.F. 1996. Recoding: Dynamic reprogramming of translation. *Annu. Rev. Biochem.* **65:** 741–768.

Gladyshev V.N. and Hatfield D.L. 1999. Selenocysteine-containing proteins in mammals. *J. Biomed. Sci.* **6:** 151–160.

Gladyshev V.N., Krause M., Xu X.M., Korotkov K.V., Kryukov G.V., Sun Q.A., Lee B.J., Wooton J.C., and Hatfield D.L. 1999. Selenocysteine-containing thioredoxin reductase in *C. elegans*. *Biochem. Biophys. Res. Commun.* **259:** 244–249.

Grundner-Culemann E., Martin G.W., Harney J.W., and Berry M.J. 1999. Two distinct SECIS structures capable of directing selenocysteine incorporation in eukaryotes. *RNA* **5:** 625–635.

Gu Q.P., Beilstein M.A., Barofsky E., Ream W., and Whanger P.D. 1999. Purification, characterization, and glutathione binding to selenoprotein W from monkey muscle. *Arch. Biochem. Biophys.* **361:** 25–33.

Hatfield D.L. and Portugal F.H. 1970. Seryl-tRNA in mammalian tissues: Chromatographic differences in brain and liver and a specific response to the codon, UGA. *Proc. Natl. Acad. Sci.* **67:** 1200–1206.

Heider J., Baron C., and Böck A. 1992. Coding from a distance: Dissection of the mRNA determinants required for the incorporation of selenocysteine into protein. *EMBO J.* **11:** 3759–3766.

Hentze M.W. and Kulozik A.E. 1999. A perfect message: RNA surveillance and nonsense-mediated decay. *Cell* **96:** 307–310.

Hill K.E., Lloyd R.S., and Burk R.F. 1993. Conserved nucleotide sequences in the open reading frame and 3′ untranslated region of selenoprotein P mRNA. *Proc. Natl. Acad. Sci.* **90:** 537–541.

Himeno S., Chittum H.S., and Burk R.F. 1996. Isoforms of selenoprotein P in rat plasma. *J. Biol. Chem.* **271:** 15769–15775.

Hubert N., Walczak R., Carbon P., and Krol A. 1996. A protein binds the selenocysteine insertion element in the 3′UTR of mammalian selenoprotein mRNAs. *Nucleic. Acids Res.* **24:** 464–469.

Huttenhofer A. and Böck A. 1998. RNA structures involved in selenoprotein synthesis. In *RNA structure and function* (ed. R. Simons and M. Grunberg-Manago), pp. 603–639. Cold Spring Harbor Laboratory Press, Cold Spring Harbor, New York.

Jung J.-E., Karoor V., Sandbaken M.G., Lee B.J., Ohama T., Gesteland R.F., Atkins J.F., Mullenbach G.T., Hill K.E., Wahba A.J., and Hatfield D.L. 1994. Utilization of selenocysteyl-tRNA$^{[Ser]Sec}$ and seryl-tRNA $^{[Ser]Sec}$ in protein synthesis. *J. Biol. Chem.* **269:** 29739–29745.

Koelman J.H., Peeters W.H.M., and Koudstaal-Hol C.H.M. 1973. Mercury-selenium correlations in marine mammals. *Nature* **245:** 385–386.

Kollmus H., Flohé L., and McCarthy J.E.G. 1996. Analysis of eukaryotic mRNA structures directing cotranslational incorporation of selenocysteine. *Nucleic Acids Res.* **24:** 1195–1201.

Kosta L., Byrne A.R., and Zelenko V. 1975. Correlation between selenium and mercury in man following exposure to inorganic mercury. *Nature* **254:** 238–239.

Kromayer M., Wilting R., Tormay P., and Böck A. 1996. Domain structure of the prokaryotic selenocysteine-specific elongation factor SELB. *J. Mol. Biol.* **262:** 413–420.

Kryukov G.V., Kryukov V.M., and Gladyshev V.N. 1999. New mammalian selenocysteine-containing proteins identified with an algorithm that searches for selenocysteine insertion sequence elements . *J. Biol. Chem.* **274:** 33888–33897.

Larsen P.R. and Berry M.J. 1995. Nutritional and hormonal regulation of thyroid hormone deiodinases. *Annu. Rev. Nutr.* **15:** 323–352.

Lee B.J., Worland P.J., Davis J.N., Stadtman T.C., and Hatfield D. 1989. Identification of a selenocysteyl-tRNA^{Ser} in mammalian cells that recognizes the nonsense codon, UGA. *J. Biol. Chem.* **264:** 9724–9727.

Lee B.J., Rajagopalan M., Kim Y.S., You K.H., Jacobson K.B., and Hatfield D. 1990. Selenocysteine tRNA$^{[Ser]Sec}$ gene is ubiquitous within the animal kingdom. *Mol. Cell Biol.* **10:** 1940–1949.

Leinfelder W., Forchhammer K., Zinoni F., Sawers G., Mandrand-Berthelot M.A., and Böck A. 1988. *Escherichia coli* genes whose products are involved in selenium metabolism. *J. Bacteriol.* **170:** 540–546.

Lesoon A., Mehta A., Singh R., Chisolm G., and Driscoll D.M. 1997. An RNA-binding protein recognizes a mammalian selenocysteine insertion sequence element required for cotranslational incorporation of selenocysteine. *Mol. Cell. Biol.* **17:** 1977–1985.

Low S.C., Harney J.W., and Berry M.J. 1995. Cloning and functional characterization of human selenophosphate synthetase, an essential component of selenoprotein synthesis. *J. Biol. Chem.* **270:** 21659–21664.

Maenpaa P.H. and Bernfield M.R. 1970. A specific hepatic transfer RNA for phosphoserine. *Proc. Natl. Acad. Sci.* **67:** 688–695.

Maquat L.E. 1995. When cells stop making sense: Effects of nonsense codons on RNA metabolism in vertebrate cells. *RNA* **1:** 453–465.

Maquat L.E., Kinniburgh A.J., Rachmilewitz E.A., and Ross J. 1981. Unstable β-globin mRNA in mRNA-deficient β-thalassemia. *Cell* **27:** 543–553.

Martin G.W., Harney J.W., and Berry M.J. 1996. Selenocysteine incorporation in eukaryotes: Insights into mechanism and efficiency from sequence, structure, and spacing proximity studies of the type 1 deiodinase SECIS element. *RNA* **2:** 171–182.

———. 1998. Functionality of mutations at conserved nucleotides in eukaryotic SECIS elements is determined by the identity of a single non-conserved nucleotide. *RNA* **4:** 65–73.

Moriarty P.M., Reddy C.C., and Maquat L.E. 1998. Selenium deficiency reduces the abundance of mRNA for Se-dependent glutathione peroxidase 1 by a UGA-dependent mechanism likely to be nonsense codon-mediated decay of cytoplasmic mRNA. *Mol. Cell. Biol.* **18:** 2932–2939.

Moustafa M.E., El-Saadani M.A., Kandeel K.M., Mansur D.B., Lee B.J., Hatfield D.L., and Diamond A.M. 1998. Overproduction of selenocysteine tRNA in Chinese hamster ovary cells following transfection of the mouse $tRNA^{[Ser]Sec}$ gene. *RNA* **4:** 1436–1443.

Nagy E. and Maquat L.E. 1998. A rule for termination-codon position within intron-containing genes: When nonsense affects RNA abundance. *Trends Biochem. Sci.* **23:** 198–199.

Negrutskii B.S., Shalak V.F., Kerjan P., El'skaya A.V., and Mirande M. 1999. Functional interaction of mammalian valyl-tRNA synthetase with elongation factor EF-1a in the complex with EF-1H. *J. Biol. Chem.* **274:** 4545–4550.

Rocher C., Lalanne J.L., and Chaudière J. 1992. Purification and properties of a recombinant sulfur analog of murine selenium-glutathione peroxidase. *Eur. J. Biochem.* **205:** 955–960.

Saedi M.S., Smith C.G., Frampton J., Chambers I., Harrison P.R., and Sunde R.A. 1988. Effects of selenium status on mRNA levels for glutathione peroxidase in rat liver. *Biochem. Biophys. Res. Comm.* **153:** 855–861.

Senkevich T.G., Bugert J.J., Shisler J.R., Koonin E.V., Darai G., and Moss B. 1996. Genome sequence of a tumorigenic poxvirus: Prediction of specific host response-evasion genes. *Science* **273:** 813–816.

Shen Q., Leonard J.L., and Newburger P.E. 1995. Structure and function of the selenium translation element in the 3′-untranslated region of human cellular glutathione peroxidase mRNA. *RNA* **1:** 519–525.

Shen Q., Wu R., Leonard J.L., and Newburger P.E. 1998. Identification and molecular cloning of a human selenocysteine insertion sequence-binding protein. A bifunctional role for DNA-binding protein B. *J. Biol. Chem.* **273:** 5443–5446.

Shisler J.L., Senkevich T.G., Berry M.J., and Moss B. 1998. UV-induced cell death blocked by a selenoprotein from a human dermatotropic poxvirus. *Science* **279:** 102–105.

Stadtman T.C. 1996. Selenocysteine. *Annu. Rev. Biochem.* **65:** 83–100.

Stapulionis R. and Deutscher M.P. 1995. A channeled tRNA cycle during mammalian protein synthesis. *Proc. Natl. Acad. Sci.* **92:** 7158–7161.

St. Germain D.L. and Galton V.A. 1997. The deiodinase family of selenoproteins. *Thyroid* **7:** 655–668.

Sunde R.A. 1994. Intracellular glutathione peroxidases — Structure, regulation, and function. In *Selenium in biology and human health* (ed. R.F. Burk), pp. 146–177. Springer–Verlag, New York.

Sunde R.A. and Evenson J.K. 1987. Serine incorporation into the selenocysteine moiety of glutathione peroxidase. *J. Biol. Chem.* **262:** 933–937.

Sunde R.A., Saedi M.S., Knight S.A.B., Smith C.G., and Evenson J.K. 1989. Regulation of expression of glutathione peroxidase by selenium. In *Selenium in biology and medicine* (ed. A. Wendel), vol. 4, pp. 8-13. Springer-Verlag, Heidelberg, Germany.

Sunde R.A., Thompson B.M., Palm M.D., Weiss S.L., Thompson K.M., and Evenson J.K. 1997. Selenium regulation of selenium-dependent glutathione peroxidases in animals and transfected CHO cells. *Biomed. Environ. Sci.* **10:** 346–355.

Suppmann S., Persson B.C., and Böck A. 1999. Dynamics and efficiency in vivo of UGA-directed selenocysteine insertion at the ribosome. *EMBO J.* **18:** 2284–2293.

Thermann R., Neu-Yilik G., Deters A., Frede U., Wehr K., Hagemeier C., Hentze M.W., and Kulozik A.E. 1998. Binary specification of nonsense codons by splicing and cytoplasmic translation. *EMBO J.* **17:** 3484–3494.

Tujebajeva R.M., Copeland P.R., Xu X.M., Carlson B.A., Harney J.W., Driscoll D.M., Hatfield D.L., and Berry M.J. 2000. Decoding apparatus for eukaryotic selenocysteine insertion. *EMBO J.* (in press).

Ursini F., Maiorino M., and Roveri A. 1997. Phospholipid hydroperoxide glutathione peroxidase (PHGPx): More than an antioxidant enzyme? *Biomed. Environ. Sci.*. **10:** 327–332.

Ursini F., Heim S., Kiess M., Maiorino M., Roveri A., Wissing J., and Flohe L. 1999. Dual function of the selenoprotein PHGPx during sperm maturation. *Science* **285:** 1393–1396.

Vendeland S.C., Beilstein M.A., Yeh J-Y., Ream W., and Whanger P.D. 1995. Rat skeletal muscle selenoprotein W: cDNA clone and mRNA modulation by dietary selenium. *Proc. Natl. Acad. Sci.* **92:** 8749–8753.

Walczak R., Westhof E., Carbon P., and Krol A. 1996. A novel RNA structural motif in the selenocysteine insertion element of eukaryotic selenoprotein mRNAs. *RNA* **2:** 367–379.

Warner G.S., Berry M.J., Moustafa M.E., Carlson B.A., Hatfield D.L., and Faust J.R. 2000. Inhibition of selenoprotein synthesis by selenocysteyl and tRNA$^{[Ser]Sec}$ lacking isopentenyl adenosine. *J. Biol. Chem.* (in press).

Watanabe K. and Osawa S. 1995. tRNA sequences and variations in the genetic code. In *tRNA: Structure, biosynthesis and function* (eds. D. Soll and U. RajBhandary), pp. 225–250. American Society for Microbiology, Washington, D.C.

Weiss S.L. and Sunde R.A. 1998. *cis*-acting elements are required for selenium regulation of glutathione peroxidase-1 mRNA levels. *RNA* **4:** 816–827.

Wilting R., Schorling S., Persson B.C., and Böck A. 1997. Selenoprotein synthesis in Archaea: Identification of an mRNA element of *Methanococcus jannaschii* probably directing selenocysteine insertion. *J. Mol. Biol.* **266:** 637–641.

Xu X.M., Zhou X., Carlson B.A., Kim L.K., Huh T.L., Lee B.J., and Hatfield D.L. 1999. The zebrafish genome contains two distinct selenocysteyl tRNA$^{[Ser]Sec}$ genes. *FEBS Lett.* **454:** 16–20.

Yan J. and Barrett J.N. 1998. Purification from bovine serum of a survival-promoting factor for cultured central neurons and its identification as selenoprotein-P. *J. Neurosci.* **18:** 8682–8691.

Yoneda S. and Suzuki K.T. 1997. Equimolar Hg-Se complex binds to selenoprotein P. *Biochem. Biophys. Res. Commun.* **231:** 7–11.

27

Influence of Polyadenylation-induced Translation on Metazoan Development and Neuronal Synaptic Function

Joel D. Richter
Department of Molecular Genetics and Microbiology
University of Massachusetts Medical School
Worcester, Massachusetts 01655

The oocytes of probably all metazoans contain a large store of dormant mRNAs that are destined for translation at subsequent stages of development. A substantial number of these molecules, generally referred to as masked or maternal mRNA, have relatively short poly(A) tails. When these tails lengthen in response to exogenous cues, translation ensues. Conversely, some translating mRNAs lose their poly(A) tails and, as a consequence, become translationally dormant. These dynamic changes in translation that are controlled by poly(A) addition and removal regulate early development. Although cytoplasmic polyadenylation has been known to be a hallmark of early animal development for some time, the extent to which it was used to modulate gene expression in late development or in adult tissues was unknown. It now appears that this regulatory process also takes place in the mammalian brain and may have important implications for synaptic function.

In this chapter, I focus on the molecular biology of cytoplasmic polyadenylation-induced translation. Studies that were published prior to 1996 will be discussed when necessary, but a general review of this topic prior to that date may be found in the first edition of this volume (Richter 1996; Wickens et al. 1996).

ESSENTIAL FEATURES OF A *XENOPUS* OOCYTE SIGNALING PATHWAY

Because much of the molecular biology of cytoplasmic polyadenylation is derived from work using *Xenopus* oocytes, it is important to outline

Translational Control of Gene Expression
© 2000 Cold Spring Harbor Laboratory Press 0-87969-568-4/00 $5 +. 00

some of the key features of the developmental process that makes these cells so useful for studying RNA processing and translational control. During the 3–6 months required for the oocytes to grow to their full size, they are arrested in prophase of meiosis I, which corresponds to G_2 of the cell cycle (Sagata 1997; Taieb et al. 1997). At this stage, mRNA is synthesized and stored in a translationally dormant form. Much of this masked mRNA is activated at several subsequent stages, most notably when the oocyte re-enters the meiotic divisions, or after fertilization (Davidson 1976). Re-entry into meiosis, also known as oocyte maturation, is stimulated by progesterone, which interacts with an unidentified cell-surface receptor to induce a rapid but transient decrease in cAMP. Eg2, a member of the Aurora/Ipl1p family of mitotic kinases (for review, see Bischoff and Plowman 1999) then becomes activated (Andresson and Ruderman 1998) and phosphorylates CPEB, a key RNA-binding protein that stimulates the cytoplasmic polyadenylation and translational activation of c-*mos* mRNA (Hake and Richter 1994; Mendez et al. 2000). Because oocytes contain no Mos protein, the only source of this critical factor is that which is translated de novo by polyadenylation (Sheets et al. 1995). Mos, a serine-threonine MAP kinase kinase kinase, activates the MAP kinase signaling pathway, which results in oocyte maturation. Whether MAP kinase indirectly activates the M-phase-promoting factor (also known as maturation promoting factor, or MPF), a heterodimer of cyclin B and the kinase cdc2, to effect maturation is unclear (Huang and Ferrell 1996; Fisher et al. 1999; Howard et al. 1999; for reviews, see Gebauer and Richter 1997; Sagata 1997). Irrespective of this, the phosphorylation of a number of substrates by active MAP kinase and/or cdc2 is responsible for many of the obvious manifestations of oocyte maturation, such as germinal vesicle breakdown (GVBD, i.e., nuclear envelope dissolution) and chromatin condensation (Fig. 1).

DEVELOPMENTAL BIOLOGY OF POLYADENYLATION-INDUCED TRANSLATION IN *XENOPUS*

The most distal portion of an mRNA 3′ UTR, usually about 20–30 nucleotides from the beginning of the poly(A) tail, contains the hexanucleotide AAUAAA, a sequence that is essential for cytoplasmic as well as nuclear polyadenylation. Because this element is present in almost all mRNAs, an additional mRNA-specific sequence is necessary for dictating which mRNAs undergo polyadenylation. This specificity sequence, the cytoplasmic polyadenylation element, or CPE, has the general structure UUUUUAU (Fox et al. 1989; McGrew et al. 1989). Although the

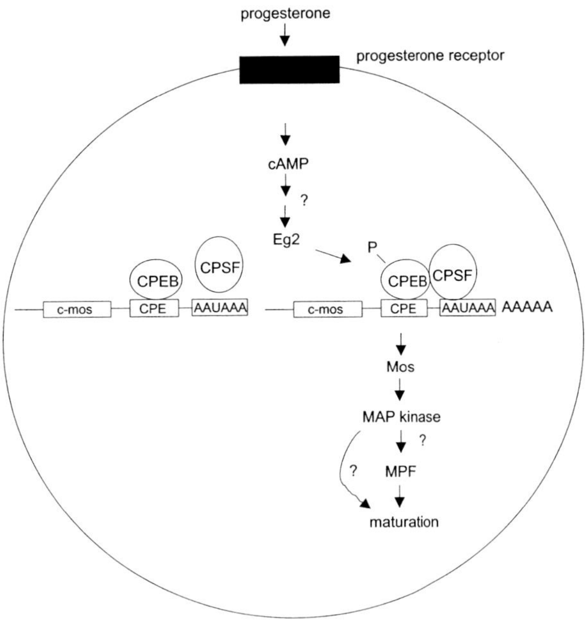

Figure 1 Key events during oocyte maturation in *Xenopus*. Progesterone interacts with a receptor on the surface of the oocyte membrane, which is followed by a decrease in cAMP and, through an unknown number of steps, the activation of the kinase Eg2. Eg2 phosphorylates CPEB, a CPE-specific binding protein, which is essential for the cytoplasmic polyadenylation and translational activation of *c-mos* mRNA. Mos activates the MAP kinase pathway, which culminates in oocyte maturation. Whether M-phase-promoting factor (MPF), a heterodimer of cyclin B and cdc2, is activated by the MAP kinase cascade, or another signaling pathway, is uncertain. CPSF (cleavage and polyadenylation specificity factor) interacts with the AAUAAA hexanucleotide (see also Fig. 3).

CPE usually resides about 15–50 nucleotides upstream of the hexanucleotide, these two sequences are sometimes separated by up to 100 nucleotides or even no nucleotides at all.

When dormant, mRNAs in the oocyte cytoplasm have poly(A) tails of usually fewer than 20–30 nucleotides, although they can be up to at least 90 nucleotides. Progesterone-induced maturation results in poly(A) tail growth to 100–150 nucleotides, but there is a heirarchy as to which mRNAs are first polyadenylated and translated (Sheets et al. 1994; Ballantyne et al. 1997; de Moor et al. 1997). For example, whereas c-*mos* mRNA is polyadenylated very early during maturation, cyclin A1 and B1 mRNAs are polyadenylated much later, at a time that is coincident with GVBD. Furthermore, the cyclin mRNAs have Mos response elements and require Mos activity to be polyadenylated (de Moor and Richter 1997). The polyadenylation of other mRNAs also exhibits a differential sensitivity to kinase activity; the mRNA encoding transcription factor D7 requires cdc2 kinase activity whereas the polyadenylation of another mRNA, H4, does not (Ballantyne et al. 1997). These observations demonstrate that the CPE and AAUAAA are not the only sequences involved in the regulation of polyadenylation.

Although Mos is clearly necessary for the induction of oocyte maturation in *Xenopus*, the extent to which this protein is sufficient for this process is in dispute. For example, Yew et al. (1992) injected Mos protein into oocytes incubated in the presence of cycloheximide and observed GVBD. Although this suggests that Mos is sufficient for maturation, other studies indicate that multiple factors may be involved in this process. Consider the experiments of Barkoff et al. (1998), who showed that the amputation of the 3′ end of c-*mos* mRNA (containing the CPE and AAUAAA) with an antisense oligonucleotide prevented progesterone-induced maturation. The annealing of a new "prosthetic" c-*mos* 3′UTR containing a poly(A) tail failed to induce Mos synthesis or maturation, but it did so when oocytes were incubated in progesterone. These experiments suggest that a factor(s), in addition to Mos, must be synthesized to promote maturation. Although the identity of the factor(s) is unknown, it could be p33(ringo) (Ferby et al. 1999), or speedy (Spy1) (Lenormand et al. 1999), both of which are maturation-inducing kinases that are synthesized de novo soon after oocytes are treated with progesterone.

A number of mRNAs undergo polyadenylation during oocyte maturation, but some do so only after fertilization. Many of these broadly fall into the Cl class of mRNAs that are expressed during the early cleavage stages (Paris et al. 1988). The polyadenylation of these mRNAs is regulated by the AAUAAA and an embryonic-type CPE, which is oligo $(U)_{12\text{-}27}$. (Simon et al. 1992, 1996; Simon and Richter 1994). Interestingly, interference with the cytoplasmic polyadenylation of one of these mRNAs, encoding the activin receptor, results in an embryo with apparent defects in body patterning (Simon et al. 1996).

The embryonic CPE is bound by ElrA (Wu et al. 1997), a member of the ELAV family of RNA-binding proteins (Good 1997). ElrA is present in oocytes and can bind the CPE even before maturation. However, precocious polyadenylation at this time is inhibited by other *cis* elements. The extent to which these repression elements occur in all mRNAs that are polyadenylated in embryos is unclear, but it should be noted that at least one additional injected mRNA is specifically polyadenylated in *Xenopus* embryos but not during oocyte maturation (Verrotti et al. 1996).

Recently, Xwnt-11 mRNA, which encodes a cell-signaling determinant, was found to undergo spatially restricted polyadenylation in *Xenopus* embryos (Schroeder et al. 1999). That is, while the mRNA is uniformly distributed along the dorsal–ventral axis, polyadenylation-induced translation occurs only in the dorsal region. This asymmetric expression of Xwnt-11, and possibly other mRNAs, could have important implications for establishment of the vertebrate body plan.

BIOCHEMISTRY OF CYTOPLASMIC POLYADENYLATION IN *XENOPUS*

There are now known to be six factors involved in cytoplasmic polyadenylation-induced translation in *Xenopus* oocytes, three of which we refer to as the "core factors." The first and most obvious of these factors is poly(A) polymerase (PAP). Alternative RNA processing is probably responsible for the multiple PAPs that have been cloned from *Xenopus* oocytes (Zhao and Manley 1996), but determining which one, or ones, catalyze cytoplasmic polyadenylation has not been clear-cut. For example, the PAP isoform identified by Ballantyne et al. (1995) is both nuclear and cytoplasmic in oocytes and closely resembles mammalian somatic PAP. Gebauer and Richter (1995), on the other hand, cloned a PAP isoform that is exclusively cytoplasmic. Are both forms involved in cytoplasmic polyadenylation? Unfortunately, no direct experiment has addressed this issue. However, an interesting corollary is that as somatic cells enter mitosis, PAP becomes multiply phosphorylated by cdc2 and is consequently inactivated (Colgan et al. 1996, 1998; Colgan and Manley 1997). PAP also becomes phosphorylated by cdc2 as oocytes mature, and presumably is inactivated at this time (Ballantyne et al. 1995; Colgan et al. 1996). Because mature oocytes and ovulated eggs are very proficient at cytoplasmic polyadenylation (Stebbins-Boaz and Richter 1994), the inactivation of PAP seems paradoxical. It is worth noting that the major cdc2 sites on PAP reside in the carboxy-terminal half of the protein, which is the portion that is missing in the isoform identified by Gebauer and Richter (1995). Thus, it is possible that this PAP isoform escapes cdc2-mediated inactivation because it lacks the sites of phosphorylation by this kinase.

The second factor involved in cytoplasmic polyadenylation is cleavage and polyadenylation specificity factor (CPSF), a four-subunit complex (sizes of ~160, 100, 70, and 30 kD) that has long been known to be essential for nuclear pre-mRNA cleavage and polyadenylation (Zhao et al. 1999). CPSF binds AAUAAA and probably facilitates PAP positioning on the end of the pre-mRNA. Bilger et al. (1994) demonstrated that purified mammalian CPSF could promote CPE-dependent polyadenylation in vitro, but it was the CPSF immunodepletion experiments by Dickson et al. (1999) that demonstrated a clear requirement for CPSF in cytoplasmic polyadenylation. Interestingly, because the 70-kD subunit has not been detected in the oocyte cytoplasm, Dickson et al. (1999) suggest that there may be a unique form of CPSF that promotes cytoplasmic polyadenylation.

The third core factor is CPEB, a CPE-specific binding protein that contains two RNA recognition motifs and a zinc finger, all of which contribute to RNA binding (Paris et al. 1991; Hake and Richter 1994; Hake et al. 1998). Egg extracts depleted of CPEB fail to polyadenylate RNA (Hake and Richter 1994), and the injection of CPEB antibody prevents cytoplasmic polyadenylation when oocytes are stimulated with progesterone. Because polyadenylation is necessary for Mos synthesis, such oocytes do not mature (Stebbins-Boaz et al. 1996). Although the precise function of CPEB is unclear, it may stabilize the binding of CPSF on the AAUAAA.

Although these three factors are central for cytoplasmic polyadenylation, they do not indicate how polyadenylation is initially activated. Because the known signaling events between the binding of progesterone to its presumed surface receptor and the polyadenylation of c-*mos* mRNA involve kinase activity, it seemed plausible that phosphorylation of one or more core factors was important for the activation of polyadenylation. As discussed previously, the PAP phosphorylation is involved in the inhibition, not the activation, of polyadenylation; it is unknown whether CPSF is phosphorylated during maturation. CPEB is phosphorylated at a time that is coincident with the polyadenylation of cyclin B1 and histone B4 mRNAs (Paris et al. 1991; Hake and Richter 1994), but because this is cdc-2-mediated, it cannot be responsible for the early polyadenylation of c-*mos* mRNA. What, then, triggers cytoplasmic polyadenylation?

Because cdc2-mediated phosphorylation induces a mobility shift in CPEB, SDS-PAGE has routinely been used to examine the timing of this posttranslational modification (see, e.g., de Moor and Richter 1997). However, when maturing oocytes were metabolically labeled with $^{32}P_i$, CPEB phosphorylation was clearly evident prior to cdc2 activation and prior to the mobility shift (Mendez et al. 2000). This suggested that CPEB undergoes two rounds of phosphorylation: an early round whose timing is consistent with an involvement in the activation of polyadenylation, and a late round that is mediated by cdc2. Mendez et al. (2000) subsequently showed that a single serine residue of CPEB, which resides in the motif LDS^{174}R, was phosphorylated prior to Mos synthesis and that this event was necessary for polyadenylation. Replacement of this serine with alanine prevented polyadenylation in mRNA injected oocytes, whereas replacement with an aspartate induced polyadenylation even in the absence of progesterone. Finally, these investigators demonstrated that the kinase Eg2 was responsible for the early phosphorylation of CPEB and suggested that this event was important for the recruitment of CPSF to the AAUAAA. Thus, Eg2 is the fourth factor directly involved in cytoplasmic polyadenylation.

TRANSLATIONAL ACTIVATION BY CYTOPLASMIC POLYADENYLATION

The two remaining factors are not involved in polyadenylation per se, but instead are translation factors that are probably regulated by cytoplasmic polyadenylation. Before describing these, however, it is necessary to recall that translational repression in *Xenopus* oocytes is mediated, at least in some cases, by 3′UTR elements. The translation of FGF receptor 1 mRNA, for example, is repressed by a 180-base sequence referred to as the translation inhibitory element (TIE). The TIE, whose activity is controlled by both Mos and cdc2 activity (Culp and Musci 1999), is bound by an unidentified 43-kD protein (Robbie et al. 1995; Culp and Musci 1998). In the case of cyclin B1 mRNA, it is, surprisingly, the CPE that mediates translational repression (de Moor and Richter 1999). Because efficient CPE-mediated repression required the presence of a 5′ cap on the mRNA, de Moor and Richter (1999) suggested that this structure, or cap-binding proteins, were the targets of the translational inhibitory machinery. The involvement of CPEB in the translation inhibition seemed likely since it was the only CPE-binding protein that could be detected. However, whether CPEB interacted directly with the cap or cap-binding proteins, or indirectly through other factors, was unknown.

Recent CPEB co-immunoprecipitation experiments have identified a new protein that is the most proximal factor involved in CPE-mediated translational repression. This protein, referred to as maskin, has a motif that is reminiscent of an eIF4E-binding domain. Indeed, an alignment of this maskin motif with the eIF4E-binding regions of other eIF4E interacting proteins (i.e., eIF4G and eIF4E-binding proteins, or BPs, Gingras et al. 1999) reveals a remarkable primary sequence conservation (Fig. 2). In addition to co-immunoprecipitation experiments, chromatography on histidine-tagged CPEB, ^{7m}GTP-Sepharose (i.e., cap column), and yeast two-hybrid analysis demonstrated that CPEB binds maskin, and maskin in turn binds eIF4E (Stebbins-Boaz et al. 1999). Two lines of evidence indicate that maskin inhibits translation: A peptide containing the maskin

```
Xenopus Maskin   EFKLATEADFLLAA
 Human 4E-BP 2   GTRIIYDRKFLLDR
 Human 4E-BP 3   GTRIIYDRKFLLEC
 Human 4E-BP 1   GTRIIYDRKFLMEC
 Human eIF4G I   EEKKRYDREFLLGF
     Fly eIF4G   SGKKQYDREQLLQL
   Yeast eIF4G   HVKYTYGPTFLLQF
Yeast Caf20p     --MIKYTIDELFQL
```

Figure 2 Alignment of the eIF4E-binding domain of maskin with the eIF4E-binding domains of eIF4G and other eIF4E-binding proteins. The position of the boxed Y (tyrosine) is conserved in all the denoted eIF4E-binding proteins except for maskin, which contains a T (threonine). The accession numbers for these proteins are: *Xenopus* maskin (AF200212), human eIF4E-BP2 (NM_004096), human eIF4E-BP3 (NM_003732), human eIF4E-BP1 (NM_004095), human eIF4GI (AF104913), *Drosophila* eIF4G (AF030155), *S. cerevisiae* eIF4G, also known as p150 Tif4631p (L16924), and *S. cerevisiae* Caf20p, which is probably equivalent to eIF4E-BP (X15731).

eIF4E-binding region competes with eIF4G for binding eIF4E, and the maskin–eIF4E complex is destroyed during oocyte maturation, a time when translation is activated. Thus, in the dormant state, maskin inhibits translation of CPE-containing mRNAs via an interaction with both CPEB and eIF4E. During oocyte maturation, cytoplasmic polyadenylation might lead to the dissolution of the maskin–eIF4E interaction, a necessary prerequisite for an eIF4E–eIF4G interaction and translation initiation. Consequently, maskin and eIF4E are two additional factors that are directly involved in polyadenylation-induced translation.

A covalent modification of the cap structure, which also appears to be regulated by cytoplasmic polyadenylation, enhances translational activity. The cap 0 structure (i.e., ^{7m}GpppG...) at the 5′ end of an mRNA is converted to cap I and cap II (i.e., methylation of the first and second ribose moieties that are immediately 3′ of the triphosphate bridge) when injected into oocytes that are stimulated to mature by progesterone. However, this cap ribose methylation occurs only if the injected mRNA undergo cytoplasmic polyadenylation (Kuge and Richter 1995). SIBA, an inhibitor of methyltransferase reactions, has no significant effect on cytoplasmic polyadenylation, but it prevents cap ribose methylation and translational activation (Kuge and Richter 1995). These results, plus the observation that an injected cap-I-containing mRNA has an elevated rate of translation compared to an identical mRNA containing cap 0 (Kuge et al. 1998), strongly indicate that cap-specific 2′-O-methylation potentiates translation in maturing oocytes. It should be borne in mind, however, that cap ribose methylation may involve only a subset of mRNAs that undergo cytoplasmic polyadenylation (Gillian-Daniel et al. 1998). Figure 3 depicts a model for the masking and unmasking of cyclin B1 mRNA during maturation.

MECHANISMS OF mRNA DEADENYLATION IN *XENOPUS* DEVELOPMENT

There are two phases of mRNA deadenylation in *Xenopus* development, the first of which occurs in maturing oocytes. At this stage, deadenylation is not directed by any particular *cis*-acting element, but instead occurs by default (Fox and Wickens 1990; Varnum and Wormington 1990). That is, if an mRNA does not contain a CPE to promote poly(A) addition, it is deadenylated. In the early embryo, however, specific *cis*-acting elements are required for mRNA-specific deadenylation, three of which have been described so far: a bipartite element in the 3′UTR of cdk2 mRNA (Stebbins-Boaz and Richter 1994), a 17-nucleotide EDEN (embryonic deadenylation) element in the 3′UTRs of Eg2 (the kinase that phosphorylates CPEB), Eg5 (a kinesin-like protein) and c-*mos* mRNAs (Legagneux et al. 1992, 1995; Bouvet et al. 1994; Audic et al. 1997), and the AU-rich element (AUUUA), which has been shown to function on a reporter RNA only (Voeltz and Steitz 1998).

Varnum et al. (1992) have demonstrated that factors from both the oocyte germinal vesicle and the cytoplasm must act in concert to effect default deadenylation after GVBD. Although the firm identity of both factors is not clear, one is obviously the important deadenylating enzyme (originally referred to as *dea*denylating *n*uclease, or DAN, but subsequently referred to as the *p*oly(*A*) specific *r*ibo*n*uclease, or PARN; Korner et al. 1998). This enzyme is also present in mammalian somatic cells (Korner and Wahle 1997) and can functionally substitute for the amphibian enzyme (Korner et al. 1998).

Recently, a factor involved in EDEN-directed deadenylation in embryos has been identified and cloned. This 50-kD protein, EDEN-BP, when immunodepleted from egg extracts, abolishes EDEN-dependent deadenylation but has no significant effect on default deadenylation (Paillard et al. 1998). Although the molecular function of EDEN-BP is unknown, one might reasonably suspect that it somehow recruits the PARN to the poly(A) tail.

Because oocytes contain a paucity of PAP, the long poly(A) tails of translating mRNAs in oocytes might be especially accessible to the PARN. On the basis of this reasoning, Wormington et al. (1996) sought to protect the poly(A) tail of ribosomal protein mRNA by overexpressing PAP

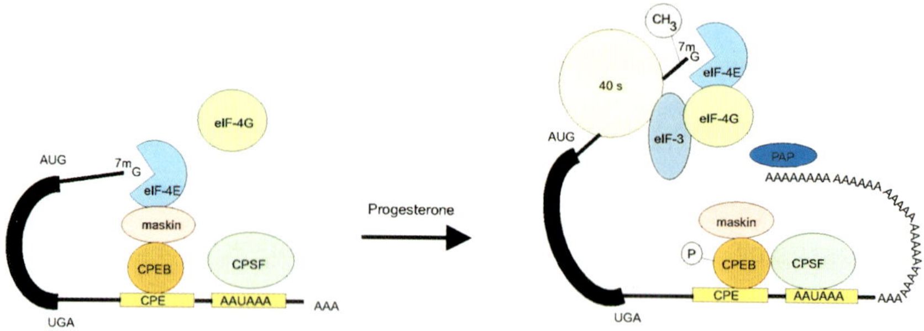

Figure 3 Model for masking and unmasking of mRNA. In oocytes prior to maturation, mRNAs such as the CPE-containing cyclin B1 are dormant. They are bound by CPEB, which in turn is bound by maskin, which in turn is bound by eIF4E. Because eIF4E is bound by maskin, it cannot interact with eIF4G at this time but probably still can bind the ^{7m}G cap. CPSF may or may not be bound to the hexanucleotide AAAAA at this time. AUG and UGA demarcate the coding region. Following progesterone stimulation, CPEB is phosphorylated by the kinase Eg2, which might stabilize the binding of CPSF on the mRNA. CPSF then recruits poly(A) polymerase to catalyze poly(A) elongation. Coincident with polyadenylation is the dissociation of maskin and eIF4E, which allows eIF4G to bind eIF4E. eIF4G and associated eIF-3 then help position the 40s ribosomal subunit on the mRNA. Polyadenylation-dependent, cap-specific 2′-O-methylation, which enhances translation, also occurs during maturation.

mRNA in oocytes. High levels of PAP not only prevented deadenylation of this mRNA, but translational inactivation as well. Thus, it seems very likely that translational silencing is the result of mRNA deadenylation.

POLYADENYLATION IN MOUSE OOCYTES

While investigating the expression of tissue-type plasminogen activator (tPA) mRNA in maturing mouse oocytes, Strickland et al. (1988) were the first to show that the 3′UTR was necessary for cytoplasmic polyadenylation. Subsequent experiments identified the *cis* elements necessary for this process: the U-rich adenylation control element (ACE), which resembles the CPE, and the hexanucleotide AAUAAA (Huarte et al. 1992). The sequences that drive polyadenylation in *Xenopus* and the mouse are fundamentally the same because *Xenopus* mRNAs are polyadenylated in injected mouse oocytes, and vice versa (Verrotti et al. 1996). In mouse oocytes, as in *Xenopus* oocytes, cytoplasmic polyadenylation of c-*mos* mRNA plays a critical role in meiotic progression (Gebauer et al. 1994).

In the mouse, the ACE/CPE appears to have three functions. The first, of course, is to drive cytoplasmic polyadenylation-induced translation (Vassalli et al. 1989). The second function is that of a deadenylation-promoting element in growing oocytes. That is, Huarte et al. (1992) posed the question of how tPA mRNA comes to be stored in the cytoplasm with a short poly(A) tail; unlike other pre-mRNAs, does it acquire a short poly(A) tail in the nucleus, or does it acquire the usual long poly(A) tail in the nucleus that subsequently undergoes shortening in the cytoplasm? Using PCR primers specific for intron and exon sequences to distinguish nuclear from cytoplasmic transcripts, plus the poly(A) tail, these investigators showed clearly that nuclear tPA pre-mRNA contains a normal long poly(A) tail in the nucleus that is subsequently shortened in the cytoplasm.

The third function of the ACE/CPE in mouse oocytes is that of a translational repression element. Stutz et al. (1997) injected a number of antisense oligodeoxynucleotides complementary to different portions of tPA mRNA and determined which would support RNase-H-directed strand scission. They reasoned that oligonucleotide annealing and RNA destruction would not occur if an RNA-bound protein occluded that particular region of the message. These investigators showed that oligonucleotides directed against the ACE/CPE and the AAUAAA did not lead to mRNA destruction in primary oocytes, but did lead to destruction in maturing oocytes. They suggested that proteins bound these regions to repress translation in oocytes, but that these proteins were removed from the mRNA during maturation to allow translation to proceed. The ACE/CPE-mediated translational regulation was subsequently shown to occur on an injected reporter mRNA, demonstrating that this element is sufficient to confer repression (Stutz et al. 1998). In many respects, then, these data are similar to those of de Moor and Richter (1999), who examined cyclin B1 mRNA in *Xenopus* oocytes. Whether CPEB is involved in tPA mRNA unmasking is unclear, but Stutz et al. (1998) do detect an ACE/CPE-binding protein of similar size (Gebauer and Richter 1996). The CPE represses translation of cyclin B1 mRNA in mouse oocytes, which clearly is bound by CPEB (Tay et al. 2000).

POLY(A) TAIL CHANGES IN INVERTEBRATES

Three invertebrate species have been used to examine translational control by polyadenylation: *Drosophila*, *C. elegans*, and *Spisula*, the surf clam. In flies, axis formation in the embryo is controlled by several mRNAs whose expression is regulated both spatially and temporally (Curtis et al. 1995; Wickens et al. 1996). The translation of three of these mRNAs, *bicoid*,

Toll, and *torso*, is controlled by cytoplasmic polyadenylation (Salles et al. 1994). Although there does not appear to be a small discrete *cis* element that directs cytoplasmic polyadenylation, as is the case in vertebrates, at least for *Toll* mRNA, an essential sequence is contained within a 192-nucleotide 3′UTR region (Schisa and Strickland 1998).

At least three factors are involved in cytoplasmic polyadenylation in *Drosophila*. Two of these, *grauzone* and *cortex*, were identified in genetic screens for female-sterile mutations that result in reduced Bicoid protein levels. Neither *bicoid* nor *Toll* mRNAs undergo cytoplasmic polyadenylation in these mutant embryos (Lieberfarb et al. 1996). Because the proteins encoded by *grauzone* or *cortex* are unknown, it is difficult to surmise whether they are involved in a developmental process upstream of polyadenylation or whether they encode factors that are an integral part of the polyadenylation apparatus. UV cross-linking has identified two proteins (101 and 89 kD) that interact with the *Toll* mRNA 192 polyadenylation element; however, the relationship, if any, of these with the *grauzone* or *cortex* genes is unknown (Schisa and Strickland 1998).

A protein that appears to be a direct component of the cytoplasmic polyadenylation machinery is Orb, the *Drosophila* homolog of CPEB. Orb not only binds *oskar* mRNA, which contains CPE-like sequences in its 3′UTR, but in flies that are mutant for *orb*, the poly(A) tail length of *oskar* mRNA is either reduced in weak alleles, or completely absent in strong alleles (Chang et al. 1999). These data are compelling, however, it should be noted that Lie and Macdonald (1999) have found that poly(A) tail-length changes do not accompany *oskar* mRNA translation.

Deadenylation, as well as polyadenylation, plays an important role in *Drosophila* embryogenesis. For example, one of the key factors that regulate axis formation is Nanos, whose mRNA resides predominantly in the posterior pole of the embryo. In this region, Nanos, acting in concert with the *pumilio* gene product, suppresses *hunchback* mRNA expression by deadenylation (Wreden et al. 1997), which is necessary for proper segmentation of the abdomen (Murata and Wharton 1995). The mechanism of deadenylation remains unclear, but it probably involves a ternary complex composed of Nanos, pumilio, and the nanos response element (NRE) that is present in the *hunchback* mRNA 3′UTR (Sonada and Wharton 1999).

As in the case of *Drosophila*, a CPEB homolog also regulates cytoplasmic polyadenylation in eggs of *Spisula*, the surf clam (Minshall et al. 1999; Walker et al. 1999). In this species, the CPEB-like protein, referred to as p82, not only promotes polyadenylation by interacting with a CPE, but also masks mRNA translation in a CPE-dependent manner. Whether this masking is direct, or indirect, as is the case with the CPEB–maskin–eIF4E complex in *Xenopus* oocytes, is unknown. However, the phosphorylation of p82, by either MAP kinase, cdc2, or both, leads to the destruction of the protein, which could be a necessary step for translational activation (Katsu et al. 1999; Minshall et al. 1999).

The signaling events that mediate sex determination in *C. elegans* include the translational control of a key mRNA, *tra-2*, whose repression is essential for XX hermaphrodites to become males (Goodwin and Evans 1997). The translational repression of this mRNA, which is correlated with a short poly(A) tail, is mediated by 3′UTR elements (Jan et al. 1997). These *cis* elements, referred to as TGEs, are the sites of regulation by the *laf-1* gene product, a negative regulator of *tra-2* mRNA expression. However, the laf-1 protein might act indirectly on TGEs because another protein, gld-1, directly binds this sequence (Jan et al. 1999).

CYTOPLASMIC POLYADENYLATION IN THE MAMMALIAN BRAIN: THE MOLECULAR BASIS OF SYNAPTIC PLASTICITY?

Axon–dendrite synaptic connections are the conduits for cell–cell information relay in the mammalian brain. Synapses are not static structures, but instead are modified in a use-dependent manner, which generally is referred to as synaptic plasticity. Such modifications can last from seconds to perhaps decades, and may constitute the molecular foundation of learning and memory (Wells et al. 2000). Thus, a typical neuron in the central nervous system is thought to use a molecular “tag” to “remember” which synapse is stimulated out of the many thousands of connections it may have. Accumulating evidence has indicated a link between long-term synaptic changes (e.g., the long-lasting phase of long-term potentiation, or L-LTP) and local synaptic protein synthesis (Huang 1999; Schuman 1999; Wells et al. 2000). Although this link was suggested by the observations that specific mRNAs (Crino and Eberwine 1996) and general translation factors (Tiedge and Brosius 1996) are present in dendrites and/or synapses, the more analytical studies of Kang and Schuman (1996) revealed a cause-and-effect relationship. These investigators found that the application of BDNF (brain-derived neurotrophic factor) to rat hippocampal slices induced activity in the synapto-dendritic compartment. However, this plasticity, though unaffected by the severing of the dendritic layer from the cell body by microdissection, was abrogated if the whole slice preparation, or just the dendritic layer, was treated with BDNF and cycloheximide. These results strongly imply that local protein synthesis, but not de novo transcription, is a key facilitator of synaptic plasticity.

A number of mechanisms might control translation at synapses, but the only one so far described is cytoplasmic polyadenylation. Wu et al. (1998) have shown that CPEB is present in the mammalian hippocampus, which is the part of the mammalian brain that is necessary for long-term memory storage, and the visual cortex, whose synapses are massively activated by light experience. Within the hippocampus, CPEB is present in the dendritic layer; in cultured hippocampal neurons, CPEB resides at synapses. Most importantly, CPEB co-fractionates with the postsynaptic density fraction, a proteinaceous matrix that immediately underlies the postsynaptic membrane. This region is likely to be the site of synaptic activity-dependent translational control (Wu et al. 1998).

Wu et al. suspected that one mRNA whose translation might be regulated by cytoplasmic polyadenylation encodes α-CaMKII (calmodulin-dependent protein kinase II). The α-CaMKII gene is essential for L-LTP and spatial learning (Silva et al. 1992a,b), the α-CaMKII mRNA is present in dendrites (Mayford et al. 1996), and α-CaMKII protein levels increase upon synaptic stimulation (Ouyang et al. 1999). Indeed, the α-CaMKII 3′ UTR contains CPE sequences, and the α-CaMKII mRNA undergoes polyadenylation and translational activation in the visual cortex following light-induced synaptic stimulation (Wu et al. 1998). Thus, cytoplasmic polyadenylation may play a critical role in synaptic plasticity and long-term memory storage. Figure 4 presents a working model for activity-dependent cytoplasmic polyadenylation at synapses. Prior to synaptic stimulation, CPEB, which is bound to translationally dormant CPE-containing mRNA, is anchored to the postsynaptic density. Following neurotransmitter release and signaling through the postsynaptic membrane, CPEB is phosphorylated, which by analogy with the *Xenopus* oocyte is essential for its activity. CPEB then induces mRNA-specific polyadenylation and translation, the result of which is a new protein that tags the stimulated synapse.

CONCLUDING REMARKS

Some of the most important advances in translational control by cytoplasmic polyadenylation have come in the last few years. Certainly, the identification of the CPE as a translational repression element in mice, clams, and frogs has given us new ways to think about how polyadenylation induces translation. The CPE-as-repressor observation led directly to the discovery of maskin, which bridges factors that bind the 5′ cap (eIF4E) and the 3′UTR (CPEB). The activity of maskin, a competitive inhibitor of eIF4G, tells us how translational dormancy is maintained and how polyadenylation may induce translation (by destroying the eIF4E-maskin interaction). Although the precise polyadenylation-inducing activity of CPEB remains elusive, the fact that a unique form of the AAUAAA-bind-

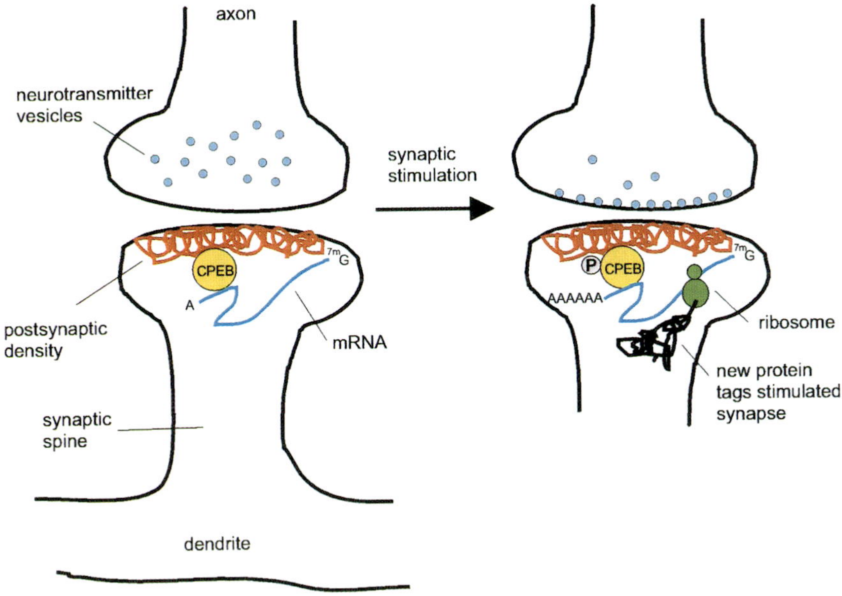

Figure 4 Model for local protein synthesis following synaptic stimulation. CPEB is associated with the postsynaptic density and is bound to quiescent, CPE-containing mRNA. Axonal activity results in neurotransmitter release and synaptic stimulation, which by analogy with the *Xenopus* oocyte, results in site-specific phosphorylation of CPEB. This modification is essential for polyadenylation-induced translation. The newly synthesized protein then tags the stimulated synapse, which distinguishes it from other nonstimulated synapses in the same neuron.

ing CPSF is involved certainly suggests an interaction between these two proteins. Another important advance is the observation that site-specific phosphorylation of CPEB by the kinase Eg2 activates polyadenylation. Thus, the upstream signaling pathway that culminates in polyadenylation may now be examined. Finally, the identification of CPEB in the brain and its possible role in learning and memory suggests whole new sets of experiments to perform. Perhaps the most essential of these experiments requires the generation of a mouse line with a hippocampal-specific knockout of the CPEB gene; it would be of great interest to determine whether such animals have defects in long-term memory storage.

ACKNOWLEDGMENTS

I thank the members of my laboratory, past and present, who have made such important contributions to some of the work reviewed here.

REFERENCES

Andresson T. and Ruderman J.V. 1998. The kinase Eg2 is a component of the *Xenopus* oocyte progesterone-activated signaling pathway. *EMBO J.* **17:** 5627–5637.

Audic Y., Omilli F., and Osborne H.B. 1997. Postfertilization deadenylation of mRNAs in *Xenopus laevis* embryos is sufficient to cause their degradation at the blastula stage. *Mol. Cell. Biol.* **17:** 209–218.

Ballantyne S., Daniel D.L., Jr., and Wickens M. 1997. A dependent pathway of cytoplasmic polyadenylation reactions linked to cell cycle control by c-*mos* and CDK1 activation. *Mol. Biol. Cell* **8:** 1633–1648.

Ballantyne S., Bilger A., Astrom J., Virtanen A., and Wickens M. 1995. Poly (A) polymerases in the nucleus and cytoplasm of frog oocytes: Dynamic changes during oocyte maturation and early development. *RNA* **1:** 64–78.

Barkoff A., Ballantyne S., and Wickens M. 1998. Meiotic maturation in *Xenopus* requires polyadenylation of multiple mRNAs. *EMBO J.* **17:** 3168–3175.

Bilger A., Fox C.A., Wahle E., and Wickens M. 1994. Nuclear polyadenylation factors recognize cytoplasmic polyadenylation elements. *Genes Dev.* **8:** 1106–1116.

Bischoff J.R. and Plowman G.D. 1999. The Aurora/Ipl1p kinase family: Regulators of chromosome segregation and cytokinesis. *Trends Cell Biol.* **9:** 454–459.

Bouvet P., Omilli F., Arlot-Bonnemains Y., Legagneux V., Roghi C., Bassez T., and Osborne H.B. 1994. The deadenylation conferred by the 3′ untranslated region of a developmentally controlled mRNA in *Xenopus* embryos is switched to polyadenylation by deletion of a short sequence element. *Mol. Cell. Biol.* **14:** 1893–1900.

Chang J.S., Tan L., and Schedl P. 1999. The *Drosophila* CPEB homolog, Orb, is required for Oskar protein expression in oocytes. *Dev. Biol.* **215:** 91–106.

Colgan D.F. and Manley J.L. 1997. Mechanism and regulation of mRNA polyadenylation. *Genes Dev.* **11:** 2755–2766.

Colgan D.F., Murthy K.G., Prives C., and Manley J.L. 1996. Cell-cycle related regulation of poly(A) polymerase by phosphorylation. *Nature* **384:** 282–285.

Colgan D.F., Murthy K.G., Zhao W., Prives C., and Manley J.L. 1998. Inhibition of poly(A) polymerase requires p34cdc2/cyclin B phosphorylation of multiple consensus and non-consensus sites. *EMBO J.* **17:** 1053–1062.

Crino P.B. and Eberwine J. 1996. Molecular characterization of the dendritic growth cone: Regulated mRNA transport and local protein synthesis. *Neuron* **17:** 1173–1187.

Culp P.A. and Musci T.J. 1998. Translational activation and cytoplasmic polyadenylation of FGF receptor-1 are independently regulated during *Xenopus* oocyte maturation. *Dev. Biol.* **193:** 63–76.

———. 1999. c-*mos* and cdc2 cooperate in the translational activation of fibroblast growth factor receptor-1 during *Xenopus* oocyte maturation. *Mol. Biol. Cell* **10:** 3567–3581.

Curtis D., Lehmann R., and Zamore P.D. 1995. Translational regulation in development. *Cell* **21:** 171–178.

Davidson E.H. 1976. *Gene activity in early development*. Academic Press, New York.

de Moor C.H. and Richter J.D. 1997. The Mos pathway regulates cytoplasmic polyadenylation in *Xenopus* oocytes. *Mol. Cell. Biol.* **17:** 6419–6426.

———. 1999. The CPE mediates masking and unmasking of cyclinB1 mRNA. *EMBO J.* **18:** 2294–2303.

Dickson K.S., Bilger A., Ballantyne S., and Wickens M.P. 1999. The cleavage and polyadenylation specificity factor in *Xenopus laevis* oocytes is a cytoplasmic factor involved in regulated polyadenylation. *Mol. Cell. Biol.* **19:** 5707–5717.

Ferby I., Blazquez M., Palmer A., Eritja R., and Nebreda A. 1999. A novel p34(cdc2)-binding and activating protein that is necessary and sufficient to trigger G(2)/M progression in *Xenopus* oocytes. *Genes Dev.* **13:** 2177–2189.

Fisher D.L., Brassac T., Galas S., and Doree M. 1999. Dissociation of MAP kinase activation and MPF activation in hormone-stimulated maturation of *Xenopus* oocytes. *Development* **126:** 4537–4546.

Fox C.A. and Wickens M. 1990. Poly(A) removal during oocyte maturation: A default reaction selectively prevented by specific sequences in the 3′ UTR of certain maternal mRNAs. *Genes Dev.* **4:** 2287–2298.

Fox C.A., Sheets M.D., and Wickens M.P. 1989. Poly(A) addition during maturation of frog oocytes: Distinct nuclear and cytoplasmic activities and regulation by the sequence UUUUUAU. *Genes Dev.* **3:** 2151–2162.

Gebauer F. and Richter J.D. 1995. Cloning and characterization of a *Xenopus* poly(A) polymerase. *Mol. Cell. Biol.* **15:** 1422–1430.

———. 1996. Mouse cytoplasmic polyadenylylation element binding protein: An evolutionarily conserved protein that interacts with the cytoplasmic polyadenylylation elements of c-*mos* mRNA. *Proc. Natl. Acad. Sci.* **93:** 14602–14607.

———. 1997. Synthesis and function of Mos: The control switch of vertebrate oocyte meiosis. *Bioessays* **19:** 23–28.

Gebauer F., Xu W., Cooper G.M., and Richter J.D. 1994. Translational control by cytoplasmic polyadenylation of c-*mos* mRNA is necessary for oocyte maturation in the mouse. *EMBO J.* **13:** 5712–5720.

Gillian-Daniel D., Gray N.K., Åström J., Barkoff A., and Wickens M. 1998. Modifications of the 5′ cap of mRNAs during *Xenopus* oocyte maturation: Independence from changes in poly(A) length and impact on translation. *Mol. Cell. Biol.* **18:** 6152–6163.

Gingras A.-C., Raught B., and Sonenberg N. 1999. eIF4E initiation factors: Effectors of mRNA recruitment to ribosomes and regulators of translation. *Annu. Rev. Biochem.* **68:** 913–963.

Good P.J. 1997. The role of elav-like genes, a conserved family encoding RNA-binding proteins, in growth and development. *Semin. Cell Dev. Biol.* **8:** 577–584.

Goodwin E.B. and Evans T.C. 1997. Translational control of development in *C. elegans*. *Semin. Cell Dev. Biol.* **8:** 551–559.

Hake L.E. and Richter J.D. 1994. CPEB is a specificity factor that mediates cytoplasmic polyadenylation during *Xenopus* oocyte maturation. *Cell* **79:** 617–627.

Hake L.E., Mendez R., and Richter J.D. 1998. Specificity of RNA binding by CPEB: Requirement for RNA recognition motifs and a novel zinc finger. *Mol. Cell. Biol.* **18:** 685–693.

Howard E.L., Charlesworth A., Welk J., and MacNicol A.M. 1999. The mitogen-activated protein kinase signaling pathway stimulates mos mRNA cytoplasmic polyadenylation during *Xenopus* oocyte maturation. *Mol. Cell. Biol.* **19:** 1990–1999.

Huang C.Y. and Ferrell J.E., Jr. 1996. Dependence of Mos-induced cdc2 activation on MAP kinase function in a cell-free system. *EMBO J.* **15:** 2169–2173.

Huang E.P. 1999. Synaptic plasticity: Regulated translation in dendrites. *Curr. Biol.* **9:** 168–170.

Huarte J., Stutz A., O′Connell M.L., Gubler P., Belin D., Darrow A.L., Strickland S., and Vassalli J.D. 1992. Transient translational silencing by reversible mRNA deadenylation. *Cell* **69:** 1021–1030.

Jan E., Motzny C.K., Graves L.E., and Goodwin E.B. 1999. The STAR protein, GLD-1, is a translational regulator of sexual identity in *Caenorhabditis elegans*. *EMBO J.* **18:** 258–269.

Jan E., Yoon J.W., Walterhouse D., Iannaccone P., and Goodwin E.B. 1997. Conservation of the *C. elegans tra-2* 3′UTR translational control. *EMBO J.* **16:** 6301–6313.

Kang H. and Schuman E.M. 1996. A requirement for local protein synthesis in neurotrophin-induced hippocampal synaptic plasticity. *Science* **273:** 1402–1406.

Katsu Y., Minshall N., Nagahama Y., and Standart N. 1999. Ca^{2+} is required for phosphorylation of clam p82/CPEB in vitro: Implications for dual and independent roles of MAP and cdc2 kinases. *Dev. Biol.* **209:** 186–199.

Korner C.G. and Wahle E. 1997. Poly(A) tail shortening by a mammalian poly(A)-specific 3′-exoribonuclease. *J. Biol. Chem.* **272:** 10448–10456.

Korner C.G., Wormington M., Muckenthaler M., Schneider S., Dehlin E., and Wahle E. 1998. The deadenylating nuclease (DAN) is involved in poly(A) tail removal during the meiotic maturation of *Xenopus* oocytes. *EMBO J.* **17:** 5427–5437.

Kuge H. and Richter J.D. 1995. Cytoplasmic 3′ poly(A) addition induces 5′ cap ribose methylation: Implications for translational control of maternal mRNA. *EMBO J.* **14:** 6301–6310.

Kuge H., Brownlee G.G., Gershon P.D., and Richter J.D. 1998. Cap ribose methylation of c-*mos* mRNA stimulates translation and oocyte maturation in *Xenopus laevis*. *Nucleic Acids Res.* **26:** 3208–3214.

Legagneux V., Omilli F., and Osborne H.B. 1995. Substrate-specific regulation of RNA deadenylation in *Xenopus* embryo and activated egg extracts. *RNA* **1:** 1001–1008.

Legagneux V., Bouvet P., Omilli F., Chevalier S., and Osborne H.B. 1992. Identification of RNA-binding proteins specific to *Xenopus* Eg maternal mRNAs: Association with the portion of Eg2 mRNA that promotes deadenylation in embryos. *Development* **116:** 1193–1202.

Lenormand J.L., Dellinger R.W., Knudsen K.E., Subramani S., and Donoghue D.J. 1999. Speedy: A novel cell cycle regulator of the G2/M transition. *EMBO J.* **18:** 1869–1877.

Lie Y.S. and Macdonald P.M. 1999. Translational regulation of *oskar* mRNA occurs independently of the cap and poly(A) tail in *Drosophila* ovarian extracts. *Dev. Biol.* **126:** 4989–4996.

Lieberfarb M.E., Chu T., Wreden C., Theurkauf W., Gergen J.P., and Strickland S. 1996. Mutations that perturb poly(A)-dependent maternal mRNA activation block the initiation of development. *Development* **122:** 579–588.

Mayford M., Baranes D., Podsypanina K., and Kandel E.R. 1996. The 3′-untranslated region of CaMKII alpha is a *cis*-acting signal for the localization and translation of mRNA in dendrites. *Proc. Natl. Acad. Sci.* **112:** 13250–13255.

McGrew L.L., Dworkin-Rastl E., Dworkin M.B., and Richter J.D. 1989. Poly(A) elongation during *Xenopus* oocyte maturation is required for translational recruitment and is mediated by a short sequence element. *Genes Dev.* **3:** 803–815.

Mendez R., Hake L.E., Andresson T., Littlepage L.E., Ruderman J.V., and Richter J.D. 2000. Phosphorylation of CPEB by Eg2 regulates c-ms mRNA translation. *Nature* **404:** 302–307.

Minshall N., Walker J., Dale M., and Standart N. 1999. Dual roles of p82, the clam CPEB homolog, in cytoplasmic polyadenylation and translational masking. *RNA* **5:** 27–38.

Murata Y. and Wharton R.P. 1995. Binding of pumilio to maternal *hunchback* mRNA is required for posterior patterning in *Drosophila* embryos. *Cell* **80:** 747–756.

Ouyang Y., Rosenstein A., Kreiman G., Schuman E.M., and Kennedy M.B. 1999. Tetanic stimulation leads to increased accumulation of Ca(2+)/calmodulin-dependent protein kinase II via dendritic protein synthesis in hippocampal neurons. *J. Neurosci.* **19:** 7823–7833.

Paillard L., Omilli F., Legagneux V., Bassez T., Maniey D., and Osborne H.B. 1998. EDEN and EDEN-BP, a *cis* element and an associated factor that mediate sequence-specific mRNA deadenylation in *Xenopus* embryos. *EMBO J.* **17:** 278–287.

Paris J., Swenson K., Piwnica-Worms H., and Richter J.D. 1991. Maturation-specific polyadenylation: *In vitro* activation by p34^{cdc2} and phosphorylation of a 58-kD CPE-binding protein. *Genes Dev.* **5:** 1697–1708.

Paris J., Osborne B., Couturier A., LeGuellec R., and Philippe M. 1988. Changes in the polyadenylation of specific stable RNAs during the early development of *Xenopus laevis. Gene* **72:** 169–176.

Richter J.D. 1996. Dynamics of poly(A) addition and removal during development. In *Translational control* (ed. J.W.B. Hershey et al.), pp. 481–503. Cold Spring Harbor Laboratory Press, Cold Spring Harbor, New York.

Robbie E.P., Peterson M., Amaya E., and Musci T.J. 1995. Temporal regulation of the *Xenopus* FGF receptor in development: A translational inhibitory element in the 3′ untranslated region. *Development* **121:** 1775–1785.

Sagata N. 1997. What does Mos do in oocytes and somatic cells? *Bioessays* **19:** 13–21.

Salles F.J., Lieberfarb M.E., Wreden C., Gergen J.P., and Strickland S. 1994. Coordinate initiation of *Drosophila* development by regulated polyadenylation of maternal messenger RNAs. *Science* **266:** 1996–1999.

Schisa J.A. and Strickland S. 1998. Cytoplasmic polyadenylation of *Toll* mRNA is required for dorsal-ventral patterning in *Drosophila* embryogenesis. *Development* **125:** 2995–3003.

Schroeder K.E., Condic M.L., Eisenberg L.M., and Yost H.J. 1999. Spatially regulated translation in embryos: Asymmetric expression of maternal wnt-11 along the dorsal-ventral axis in *Xenopus*. *Dev. Biol.* **214:** 288–297.

Schuman E.M. 1999. mRNA trafficking and local protein synthesis at the synapse. *Neuron* **23:** 645–648.

Sheets M.D., Wu M., and Wickens M. 1995. Polyadenylation of c-*mos* mRNA as a control point in *Xenopus* meiotic maturation. *Nature* **374:** 511–516.

Sheets M.D., Fox C.A., Hunt T., Vande Woude G., and Wickens M. 1994. The 3′-untranslated regions of c-*mos* and cyclin mRNAs stimulate translation by regulating cytoplasmic polyadenylation. *Genes Dev.* **8:** 926–938.

Silva A.J., Paylor R., Wehner J.M., and Tonegawa S. 1992a. Impaired spatial learning in alpha-calcium-calmodulin kinase II mutant mice. *Science* **257:** 206–211.

Silva A.J., Stevens C.F., Tonegawa S., and Wang Y. 19992b. Deficient hippocampal long-term potentiation in alpha-calcium-calmodulin kinase II mutant mice. *Science* **257:** 201–206.

Simon R. and Richter J.D. 1994. Further analysis of cytoplasmic polyadenylation in *Xenopus* embryos and identification of embryonic cytoplasmic polyadenylation element-binding proteins. *Mol. Cell. Biol.* **14:** 7867–7875.

Simon R., Tassan J.P., and Richter J.D. 1992. Translational control by poly(A) elongation during *Xenopus* development: Differential repression and enhancement by a novel cytoplasmic polyadenylation element. *Genes Dev.* **6:** 2580–2591.

Simon R., Wu L., and Richter J.D. 1996. Cytoplasmic polyadenylation of activin receptor mRNA and the control of pattern formation in *Xenopus* development. *Dev. Biol.* **179:** 239–250.

Sonoda J. and Wharton R.P. 1999. Recruitment of Nanos to *hunchback* mRNA by Pumilio. *Genes Dev.* **13:** 2704–2712.

Stebbins-Boaz B. and Richter J.D. 1994. Multiple sequence elements and a maternal mRNA product control cdk2 RNA polyadenylation and translation during early *Xenopus* development. *Mol. Cell. Biol.* **14:** 5870–5880.

Stebbins-Boaz B., Hake L.E., and Richter J.D. 1996. CPEB controls the cytoplasmic polyadenylation of cyclin, Cdk2 and c-*mos* mRNAs and is necessary for oocyte maturation in *Xenopus. EMBO J.* **15:** 2582–2592.

Stebbins-Boaz B., Cao Q., de Moor C.H., Mendez R., and Richter J.D. 1999. Maskin is a CPEB-associated factor that transiently interacts with eIF4E. *Mol. Cell* **4:** 1017–1027.

Strickland S., Huarte J., Belin D., Vassalli A., Rickles R.J., and Vassalli J.D. 1988. Antisense RNA directed against the 3′ noncoding region prevents dormant mRNA activation in mouse oocytes. *Science* **241:** 680–684.

Stutz A., Huarte J., Gubler P., Conne B., Belin D., and Vassalli J.D. 1997. *In vivo* antisense oligodeoxynucleotide mapping reveals masked regulatory elements in an mRNA dormant in mouse oocytes. *Mol. Cell. Biol.* **17:** 1759–1767.

Stutz A., Conne B., Huarte J., Gubler P., Volkel V., Flandin P., and Vassalli J.D. 1998. Masking, unmasking, and regulated polyadenylation cooperate in the translational control of a dormant mRNA in mouse oocytes. *Genes Dev.* **12:** 2535–2548.

Taieb F., Thibier C., and Jessus C. 1997. On cyclins, oocytes, and eggs. *Mol. Reprod. Dev.* **48**: 397-411.

Tay J., Hodgman R., and Richter J.D. 2000. The control of cyclin B1 mRNA translation during mouse oocyte maturation. *Dev. Biol.* **221:** 1–9.

Tiedge H. and Brosius J. 1996. Translational machinery in dendrites of hippocampal neurons in culture. *J. Neurosci.* **16:** 7171–7181.

Varnum S.M. and Wormington W.M. 1990. Deadenylation of maternal mRNAs during *Xenopus* oocyte maturation does not require specific *cis*-sequences: A default mechanism for translational control. *Genes Dev.* **4:** 2278–2286.

Varnum S.M., Hurney C.A., and Wormington W.M. 1992. Maturation-specific deadenylation in *Xenopus* oocytes requires nuclear and cytoplasmic factors. *Dev. Biol.* **153:** 283–290.

Vassalli J.D., Huarte J., Belin D., Gubler P., Vassalli A., O′Connell M.L., Parton L.A., R. Rickles J., and Strickland S. 1989. Regulated polyadenylation controls mRNA translation during meiotic maturation of mouse oocytes. *Genes Dev.* **3:** 2163–2171.

Verrotti A.C., Thompson S.R., Wreden C., Strickland S., and Wickens M. 1996. Evolutionary conservation of sequence elements controlling cytoplasmic polyadenylylation. *Proc. Natl. Acad. Sci.* **93:** 9027–9032.

Voeltz G.K. and Steitz J.A. 1998. AUUUA sequences direct mRNA deadenylation uncoupled from decay during *Xenopus* early development. *Mol. Cell. Biol.* **18:** 7537–7545.

Walker J., Carnall N., Hake L., Richter J., and Standart N. 1999. The clam 3′ UTR masking element-binding protein p82 is a member of the CPEB family. *RNA* **5:** 14–26.

Wells D.G., Richter J.D., and Fallon J.R. 2000. Molecular mechanisms for activity-regulated protein synthesis in the synapto-dendritic compartment. *Curr. Opin. Neurobiol.* **10:** 132–137.

Wickens M., Kimble J., and Strickland S. 1996. Translational control of developmental decisions. In *Translational control* (ed. J.W.B. Hershey et al.), pp. 441–450. Cold

Spring Harbor Laboratory Press, Cold Spring Harbor, New York.

Wormington W.M., Searfoss A.M., and Hurney C.A. 1996. Overexpression of poly(A) binding protein prevents maturation-specific deadenylation and translational inactivation in *Xenopus* oocytes. *EMBO J.* **15:** 900–909.

Wreden C., Verrotti A.C., Schisa J.A., Lieberfarb M.E., and Strickland S. 1997. Nanos and pumilio establish embryonic polarity in *Drosophila* by promoting posterior deadenylation of hunchback mRNA. *Development* **124:** 3015–3023.

Wu L., Good P.J., and Richter J.D. 1997. The 36-kilodalton embryonic-type cytoplasmic polyadenylation element-binding protein in *Xenopus laevis* is ElrA, a member of the ELAV family of RNA-binding proteins. *Mol. Cell. Biol.* **17:** 6402–6409.

Wu L., Wells D., Tay J., Mendis D., Abbot M.A., Barnitt A., Quinlan E., Heynen A., Fallon J.R., and Richter J.D. 1998. CPEB-mediated cytoplasmic polyadenylation and the regulation of experience-dependent translation of α-CaMKII mRNA at synapses. *Neuron* **21:** 1129–1139.

Yew N., Mellini M.L., and Vande Woude G.F. 1992. Meiotic initiation by the mos protein in *Xenopus. Nature* **355:** 649–652.

Zhao W. and Manley J.L. 1996. Complex alternative RNA processing generates an unexpected diversity of poly(A) polymerase isoforms. *Mol. Cell. Biol.* **16:** 2378–2386.

Zhao J., Hyman L., and Moore C. 1999. Formation of mRNA 3′ ends in eukaryotes: Mechanism, regulation, and interrelationships with other steps in mRNA synthesis. *Microbiol. Mol. Biol. Rev.* **63:** 405–445.

28

Interaction of mRNA Translation and mRNA Degradation in *Saccharomyces cerevisiae*

David C. Schwartz and Roy Parker[1]

Department of Molecular and Cellular Biology and

[1]Howard Hughes Medical Institute

University of Arizona

Tucson, Arizona 85721

BIOLOGICAL SIGNIFICANCE OF mRNA TURNOVER

It has become clear that mRNA turnover can play a major role in the control of gene expression. In eukaryotic cells, the decay rates of individual mRNAs vary by more than two orders of magnitude (Singer and Penman 1973; Spradling et al. 1975; Cabrera et al. 1984). There have also been a growing number of examples wherein the regulation of gene expression in response to an environmental cue is attained by altering the rates of degradation for specific mRNAs (for review, see Ross 1995). In addition, a specialized system of mRNA decay, referred to as mRNA surveillance, functions to ensure that mRNA biogenesis has been completed correctly by degrading aberrant transcripts (for reviews, see Chapters 29 and 30). Finally, specific mechanisms of mRNA degradation function in the gene-specific silencing induced by double-stranded RNA (for review, see Fire 1999). The numerous and diverse roles for mRNA degradation suggest that the modulation of mRNA stability is a critical step in the regulation of eukaryotic gene expression.

PATHWAYS OF mRNA DEGRADATION

Four distinct pathways of mRNA turnover have been identified in eukaryotic organisms (Fig. 1). In yeast, degradation begins by shortening of the 3′ poly(A) tail followed by removal of the 5′ cap structure (decapping), thus allowing 5′ to 3′ exonucleolytic digestion of the body of the tran-

Translational Control of Gene Expression
© 2000 Cold Spring Harbor Laboratory Press 0-87969-568-4/00 $5 + . 00

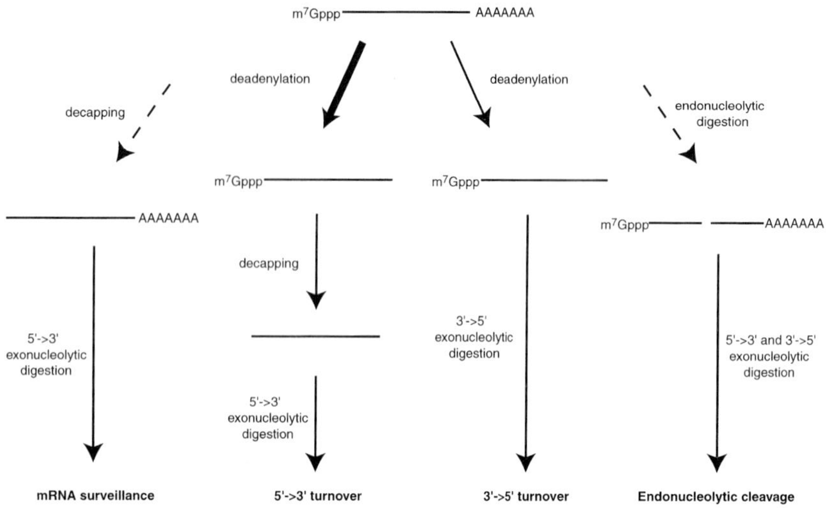

Figure 1 Four pathways of mRNA degradation. There are two pathways of mRNA degradation thought to occur on all mRNAs. The 5′→3′ pathway initiates with deadenylation, which leads to decapping followed by 5′→3′ exonucleolytic digestion. The 3′→5 pathway also initiates with deadenylation but is followed directly by 3′→5′ exonucleolytic digestion. Two alternative pathways of mRNA degradation appear to be mRNA-specific. The mRNA surveillance pathway degrades mRNAs the cell regards as aberrant. This pathway leads to rapid decapping followed by 5′→3′ exonucleolytic digestion. Finally, some mRNAs are known to contain *cis*-acting elements which allow recognition by specific endonucleases.

script (Muhlrad and Parker 1992; Decker and Parker 1993; Hsu and Stevens 1993; Muhlrad et al. 1994, 1995). Alternatively, mRNAs can be degraded 3′ to 5′ following deadenylation (Muhlrad et al. 1995; Anderson and Parker 1998). Both deadenylation leading to decapping or 3′ to 5′ degradation appear to be general pathways of turnover, since all mRNAs examined in yeast utilize one or both of these pathways. Two additional pathways of decay show more specificity in their mRNA substrates. For example, during mRNA surveillance, aberrant mRNAs are rapidly decapped and degraded 5′ to 3′ independently of deadenylation (Muhlrad and Parker 1994). Similarly, some vertebrate mRNAs contain specific recognition sites for endonucleases that initiate degradation of the mRNA (see, e.g., Bernstein et al. 1992; Nielsen and Christiansen 1992; Brown et al. 1993; Binder et al. 1994).

These mRNA degradation pathways have primarily been characterized in yeast, but there is evidence that these pathways are conserved in other organisms. In mammals, many mRNAs deadenylate as a first step in degradation (see, e.g., Wilson and Treisman 1988, Shyu et al. 1991), and molecules have been detected that could correspond to a deadenylated and decapped intermediate (Couttet et al. 1997). Moreover, several of the proteins involved in yeast mRNA decay, including the 5′ to 3′ ribonuclease, Xrn1p (Bashkirov et al. 1997), two proteins involved in decapping, Dcp1p and Dcp2p (Dunckley and Parker 1999; Tharun and Parker 1999), and the exosome complex, which is responsible for 3′ to 5′ degradation of the mRNA body (Anderson and Parker 1998; Allmang et al. 1999), are conserved in other eukaryotes.

TRANSLATION AND TURNOVER ARE INTERCONNECTED

It has been a long-standing observation that the processes of mRNA turnover and mRNA translation are interconnected. This is based on a variety of observations wherein changing the translation status of a mammalian or yeast mRNA by mutation in *cis*, or by the addition of translation inhibitors such as cycloheximide, alters the degradation rate of the mRNA (for review, see Jacobson and Peltz 1996). The identification of a discrete set of pathways for mRNA turnover has allowed insight into the relationship between translation and turnover. In this review chapter, we examine how translation interacts with the main pathway of deadenylation-dependent mRNA degradation, focusing primarily on work in yeast. An emerging theme with mechanistic implications is that the interplay between translation and turnover is achieved in part by the efficiency of translation initiation.

Interaction of Translation Initiation and Decapping

Several recent lines of experimental evidence lead to the conclusion that the decapping step of the mRNA turnover pathway is controlled by a competition between translation initiation factors and the decapping enzyme. This idea is based on the fact that the cap structure, in addition to being the site of decapping, also functions as a site of assembly for the cytoplasmic cap-binding complex (also termed eIF4F), which promotes translation initiation (for review, see Chapter 2). This dual nature of the cap structure has led to the hypothesis that decapping could be in competition with the cytoplasmic cap-binding complex for access to the cap (Decker and Parker 1994; Jacobson and Peltz 1996). In the simplest view, if the cap-binding complex binds to the cap structure, translation occurs; alternatively, if the decapping enzyme binds to the cap, the mRNA is degraded.

Mutations in Translation Initiation Factors Increase mRNA Decapping Rates

Evidence for a competition between translation initiation and decapping has come from analysis of mRNA turnover in yeast strains containing mutations in proteins in the translation initiation complex (Schwartz and Parker 1999). Mutations in components of the cap-binding complex, including eIF4E, eIF4G, and eIF4A, that strongly inhibit translation initiation also increase the rate of decapping of both the unstable MFA2 and stable PGK1 transcripts. Similarly, mutations in the Prt1 protein, a component of the eIF3 complex that interacts with the cap-binding complex to stabilize the 40S preinitiation complex on the mRNA (Chaudhuri et al. 1999), also increase the rate of decapping of both MFA2 and PGK1 transcripts (Schwartz and Parker 1999), and have been reported to accelerate the decay of the SSA1 and Ip mRNAs under specific conditions (Cereghino et al. 1995; Barnes 1998). Interestingly, partial loss-of-function mutations in the cap-binding protein, eIF4E, that lead to only a modest decrease in translation in vivo, have little effect on the degradation rate of the PGK1 mRNA in yeast (Linz et al. 1997). This observation suggests that in order to affect mRNA decay, the defect in the translation initiation complex assembly must be substantial. Together, these results indicate that decreasing translation initiation increases mRNA decapping rate, and support the idea that translation initiation and decapping are in competition for access to the cap.

Additional evidence for this model has come from the inhibition of translation of individual mRNAs. In one case, the insertion of a stem-loop into the 5′UTR of the stable PGK1 mRNA, which effectively blocks translation by preventing 40S scanning, leads to accelerated decapping (Muhlrad et al. 1995). Thus, in yeast, decreasing translation initiation either in *cis* or in *trans* can promote faster decapping. However, it is important to note that there are a number of examples from metazoan organisms where insertion of stem-loops into the 5′UTR of mRNAs acts to stabilize these transcripts (see, e.g., Aharon and Schneider 1993; Curatola et al. 1995).

In striking contrast to the stimulation of mRNA degradation under conditions of translation initiation inhibition, blocks to translation elongation stabilize mRNAs. For example, inhibiting translation elongation with cycloheximide greatly stabilizes several yeast mRNAs (Herrick et al. 1990; Beelman and Parker 1994). Similarly, mutations in factors required for the production of functional tRNAs lead to an apparent elongation block and also stabilize mRNAs (Peltz et al. 1992; Zuk et al. 1999). In the case of stabilization due to cycloheximide, the effect can be attributed to an inhibition of decapping (Beelman and Parker 1994), although the reason for this effect is currently unclear. One possible interpretation of these observations is that inhibiting translation elongation traps the translation initiation complex in a phase that is resistant to mRNA decapping.

The Cap-binding Protein Inhibits Dcp1p In Vitro

Further support of a competition between the cap-binding complex and the decapping enzyme for access to the cap has come from in vitro experiments with purified cap-binding protein (eIF4E) and decapping enzyme. The key observation is that purified eIF4E acts as an inhibitor of decapping by Dcp1p in vitro (D.C. Schwartz and R. Parker, in prep.). Two facts suggest that this inhibition occurs by binding of the eIF4E protein to the cap structure. First, the inhibition can be alleviated by addition of m^7GTP (D.C. Schwartz and R. Parker, in prep.), which has been shown to efficiently compete with capped mRNAs for eIF4E binding (Sonenberg et al. 1978). Second, a mutant eIF4E that fails to bind the cap structure in vitro also fails to block decapping by Dcp1p in vitro (D.C. Schwartz and R. Parker, in prep.). The inhibition of decapping by eIF4E protein strongly argues that dissociation of eIF4E from the cap structure is required before decapping can occur.

Poly(A)-binding Protein Inhibits Decapping and Promotes Translation Initiation

The status of the Pab1p association with the 3′ poly(A) tail also influences the rates of both translation initiation and mRNA turnover. The poly(A)-binding protein is known to bind the 3′ poly(A) tail and enhance translation initiation, at least in part, through interactions with the eIF4G and eIF4B components of the cap-binding complex (Tarun and Sachs 1995, 1996; Le et al. 1997). In addition, the poly(A)-binding protein is also required for the 3′ poly(A) tail to function as an inhibitor of decapping. mRNAs from strains containing a deletion of the PAB1 gene are decapped prior to deadenylation, showing that the Pab1p is necessary for linking of the two processes (Caponigro and Parker 1995; Coller et al. 1998). One simple model is that the interaction of Pab1p with translation initiation factors promotes translation and thereby inhibits decapping. However, this model is likely oversimplified since strains carrying mutations in translation initiation factors that lead to a severe defect in translation initiation still exhibit deadenylation-dependent mRNA decapping (Schwartz and Parker 1999). This strongly suggests that the ability of the poly(A) tail to inhibit decapping may include additional unidentified interactions.

Interaction between Translation Initiation and Deadenylation

Several lines of experimental evidence suggest that efficient translation initiation of yeast transcripts also slows the rate of deadenylation. For example, all mutations in translation initiation factors that lead to faster rates of decapping also lead to faster rates of deadenylation (Schwartz and Parker 1999). Similarly, inhibition of translation of the PGK1 mRNA with a secondary structure in its 5′UTR leads to 3- to 4-fold faster deadenylation rates (Muhlrad et al. 1995). Interestingly, the AU-rich instability elements found in many mammalian mRNAs are known to both promote deadenylation rates and inhibit translation (Kruys et al. 1987, 1989). In addition, translation inhibitors can effect the distribution of poly(A) tails lengths on the ferritin mRNA in mammalian cells (Muckenthaler et al. 1997). These observations suggest a significant relationship between the rates of translation and the rates of deadenylation.

There are two appealing models for how translation initiation rate could be coupled to deadenylation (Fig. 2). The first model is based on the observation that interaction of the poly(A)-binding protein with the cap-binding complex increases its affinity for poly(A) (Le et al. 1997). Thus, a stable assembly of initiation factors on the 5′ end of the transcript would stabilize Pab1p binding and thereby prevent deadenylation. In this hypothesis, the Pab1p is an inhibitor of deadenylation, which appears to be the case in vertebrate systems (Bernstein et al. 1989; Wormington et al. 1996; Dehlin et al. 2000) but may not be true in yeast (Sachs and Davis 1989; Caponigro and Parker 1995). Alternatively, in the second model, the cap structure of the mRNA, unoccupied by a translation initiation complex, would yield a higher intrinsic rate of deadenylation. This model is supported by the observation that, in vertebrate cells, the nuclease thought to be responsible for deadenylation is bound to, and is stimulated by a 5′ cap structure.

Model for the Interaction of Translation and Decapping

The above observations suggest a working hypothesis for the interaction between translation initiation and decapping (Fig. 2). In this view, a polyadenylated mRNA is protected from decapping because the assembly of Pab1p, along with the cap-binding complex, leads to a stable association of initiation components on the 5′ end of the mRNA and the cap structure is therefore not accessible to the decapping enzyme. Following the comple-

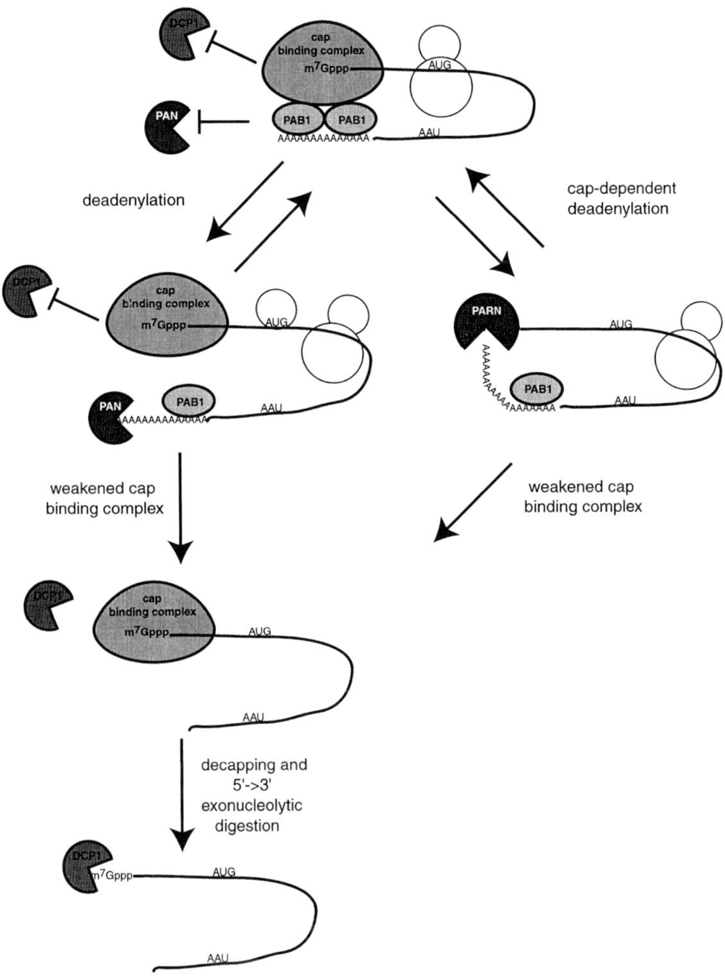

Figure 2 Translation regulates mRNA stability through the interaction of translation initiation factors with the 5′ and 3′ ends of the mRNA. mRNAs that are well translated are protected from degradation enzymes by the binding of translation initiation factors. Deadenylation may occur between rounds of translation initiation when there is a weakening of the 5′ and 3′ associated proteins. Loss of the poly(A) tail and the poly(A)-binding protein weaken the cap-binding complex, allowing decapping. An alternative mechanism of deadenylation is shown wherein deadenylation can occur via a cap-dependent poly(A) nuclease.

tion of deadenylation, the poly(A)-binding protein dissociates, leading to a weaker interaction between the cap-binding complex and the mRNA, and thereby promoting disassembly of this complex and allowing access by the decapping enzyme. In this hypothesis, the stabilizing effect of drugs or mutations that inhibit translation elongation can be interpreted as trapping the translation initiation complex in a state wherein the cap structure is protected from decapping. Alternatively, any block to translation initiation that prevents this assembly (whether through mutation or secondary structure) would promote disassembly of the translation initiation complex, leaving the cap structure readily accessible to the decapping enzyme.

SEQUENCE ELEMENTS THAT AFFECT mRNA DECAY CAN AFFECT TRANSLATION RATE

The model that mRNA decay is influenced by translation initiation predicts that *cis*-acting sequences which promote mRNA turnover may actually serve to repress translation initiation, thereby leading to increased rates of deadenylation and decapping. To date, many sequence elements have been identified in yeast that contribute to mRNA-specific rates of turnover. For instance, instability elements have been discovered in MFA2, MATα1, HIS3, STE3, SPO13, PPR1, SDH1, and HTB1 (Parker and Jacobson 1990; Xu et al. 1990; Heaton et al. 1992; Herrick and Jacobson 1992; Lombardo et al. 1992; Muhlrad and Parker 1992; Surosky and Esposito 1992; Pierrat et al. 1993). These elements can be found in the 5´UTR, the coding region, or the 3´UTR, and in all cases examined they accelerate the rates of both deadenylation and decapping. For example, mutation of the 3´UTR of the unstable MFA2 transcript leads to slower rates of deadenylation and decapping (Muhlrad and Parker 1992). Similarly, instability elements within the MATα1 coding region can also promote both deadenylation and decapping (Caponigro and Parker 1996). Finally, changes in the context of the translation start codon of the PGK1 coding region that increase the decay rate of the mRNA accelerate the rates of both decapping and deadenylation (LaGrandeur and Parker 1999). Importantly, the changes in the PGK1 start codon context that destabilize the transcript also reduce its translational efficiency. These observations suggest that instability elements may function by affecting translation initiation rate and thereby ultimately affect the rates of both deadenylation and decapping.

Another example of a *cis*-acting sequence affecting both translation initiation and decapping is found in the process of mRNA surveillance (see Chapter 29 Peltz and Jacobson). In this case, recognition of an early nonsense codon by the translation machinery leads to rapid decapping of the mRNA independent of prior deadenylation (Muhlrad and Parker 1994). Utilizing a reporter system where the levels of mRNA and translation efficiency could both be monitored, it was observed that a premature nonsense codon led to a decrease in the amount of protein produced per mRNA

(Muhlrad and Parker 1999). This occurred even when mRNA degradation was prevented by inactivation of the decapping enzyme. These observations argue that the recognition of a premature termination codon somehow leads to a decrease in translation initiation rates and suggest that decapping is a consequence of this change in translational status of the mRNA.

Several other instability elements have been proposed to affect translation initiation. For example, the decay rate of the Ip mRNA in yeast is regulated by carbon source in a manner requiring the 5′UTR (Cereghino et al. 1995). This 5′UTR has been proposed to promote decay by regulating the translation of the mRNA in different carbon sources (Scheffler et al. 1998), although the translation efficiency of the transcript has not yet been measured in each case. A related case occurs in the OLE1 mRNA, where the 5′UTR is required for the regulation of decay rate in response to specific fatty acids (Gonzalez and Martin 1996). A likely possibility is that the OLE1 5′UTR actually regulates translation rate under the different conditions. Some examples of this type of link between translation and decay exist in mammalian cells as well. For example, the AU-rich destabilizing elements found in many mammalian transcripts can in some cases promote degradation and also decrease mRNA translation (see, e.g., Kruys and Huez 1994).

A prediction made from the above data is that many *cis*-acting elements known to alter mRNA half-lives will also affect rates of translation. Sequences that stabilize mRNAs could do so by increasing the rates of translation; alternatively, sequences that destabilize mRNAs might be acting by repressing translation. The data suggest that regulation of translation initiation, and ultimately mRNA half-life, could be controlled at many points. Sequences located in the 3′UTR could both affect the ability of the Pab1p to bind the poly(A) tail and also influence the strength of the interactions between the Pab1p and the cap-binding complex. Second, sequences could alter the ability of the many RNA-binding proteins in the cap-binding complex, such as eIF4E, eIF4G (Pestova et al. 1996), and eIF4B, to assemble on the 5′ end of an mRNA. Finally, the sequences located in and around the AUG may be a last point at which translation initiation is regulated. The diverse ways in which translation initiation can be regulated would allow for multiple mechanisms to regulate mRNA decay rate.

TRANSLATION AND THE mRNA DECAY MACHINERY

As discussed above, the stability of the cap-binding complex appears to be a key modulator of mRNA deadenylation and decapping rates. At present, the interaction of the mRNA decay machinery and the translation machinery is not clear. However, several proteins are known to affect mRNA decay, including both degradative nucleases and proteins that promote mRNA decapping. In many cases, links from these proteins to the translation machinery are emerging. In the following sections we discuss the proteins involved in turnover and how they interact with translation.

Several nucleases involved in mRNA degradation have been identified. The enzyme responsible for both deadenylation-dependent and deadenylation-independent decapping is encoded by the DCP1 gene (Beelman et al. 1996; LaGrandeur and Parker 1998). In addition, Dcp1p-dependent decapping activity requires the Dcp2p, which is either required for production of active Dcp1p or could be a second subunit of the decapping enzyme required for function in vivo (Dunckley and Parker 1999). The 5′ to 3′ exoribonuclease that degrades mRNA following decapping is encoded by the XRN1 gene (Hsu and Stevens 1993; Muhlrad et al. 1994). Although not understood, efficient 5′ to 3′ mRNA digestion after decapping requires the eIF5A protein (Zuk and Jacobson 1998). Conversely, 3′ to 5′ degradation of the mRNA body appears to be performed by the exosome complex and the accessory proteins Ski2p, Ski3p, and Ski8p (Mitchell et al. 1997; Anderson and Parker 1998; Allmang et al. 1999). The identity of the nuclease(s) responsible for the cytoplasmic shortening of the poly(A) tail is not yet clear. A poly(A)-specific nuclease has been identified in yeast, encoded by the PAN2 gene, but this nuclease appears to function in the nuclear trimming of newly synthesized poly(A) tails (Brown and Sachs 1998). A promising candidate for the poly(A) nuclease in vertebrates has been identified by purification of a poly(A)-specific activity from HeLa cells, termed PARN or DAN (Körner and Wahle 1997). Antibody inhibition experiments suggest that this protein is responsible for cytoplasmic deadenylation in *Xenopus* oocytes (Körner et al. 1998). However, whether this is the major deadenylase in somatic cells remains to be determined. Moreover, examination of the yeast genome does not reveal any clear homologs of PARN/DAN, although there presumably is a functional homolog. For each of these degradative nucleases, their activity on individual mRNAs appears to be controlled by accessory or activator proteins.

Upf Proteins Activate mRNA Decapping in Response to Aberrant Translation Termination

One example of such activator proteins are the Upf proteins, which are required for the rapid deadenylation-independent decapping induced by aberrant translation termination (for discussion, see Chapters 29 and 30). In this case, genetic and biochemical evidence indicates that the Upf1, Upf2, and Upf3 proteins interact with the peptide release factors and affect the nature of the translation termination event (Cui et al. 1995; Lee and Culbertson 1995; Weng et al. 1996; Czaplinski et al. 1998). This suggests that differences in translation termination events can be communicated to the 5′ end to affect translation initiation and decapping. This is consistent with the observation that recognition of mRNAs as "nonsense-containing" leads to a down-regulation of translation efficiency (Fig. 3a) (Muhlrad and Parker 1999). Moreover, specific mutations in the translation initiation factor sui1/mof2 inhibit the process of mRNA surveillance (Cui et al. 1999). This suggests that Sui1p may be an important link between recognition of an aberrant mRNA at translation termination and signaling alterations in the translation initiation complex to allow decapping independent of deadenylation.

A Cytoplasmic Lsm Complex and Pat1/Mrt1p Activate Decapping on Normal mRNAs

Proteins involved in activating decapping on normal mRNAs include a complex of Sm-like proteins (referred to as Lsm proteins) and the associated Pat1/Mrt1 protein (Hatfield et al. 1996; Boeck et al. 1998; Tharun et al. 2000). Interestingly, mutations in these genes only partially inhibit decapping (Boeck et al. 1998; Tharun et al. 2000). This suggests that the Lsm proteins are not a required part of the decapping machinery but instead serve as activators of this process.

The Lsm proteins are members of a conserved family of proteins that contain a "Sm" motif and consist of nine proteins in yeast, referred to as Lsm1p to Lsm9p (Mayes et al. 1999; Salgado-Garrido et al. 1999). The Sm motif was first found in the canonical Sm proteins, which are known to be associated with snRNAs and to function in splicing (Cooper et al. 1995; Hermann et al. 1995; Seraphin 1995). Biochemical and structural analysis suggests that the Sm motif serves as a site of protein–protein interaction, which allows the Sm proteins to form a seven-member ring structure (Hermann et al. 1995; Kambach et al. 1999). Similar to the Sm proteins, the Lsm proteins are known to associate with each other and are a component of the U6 snRNP (Pannone et al. 1998; Mayes et al. 1999; Salgado-Garrido et al. 1999). Consistent with this function, inactivation of Lsm2 through Lsm8p leads to a defect in pre-mRNA splicing, presumably due to a destabilization of the U6 snRNA (Pannone et al. 1998; Mayes et al. 1999; Salgado-Garrido et al. 1999). Thus, mutations in several of the Lsm proteins affect both mRNA splicing and mRNA degradation.

Several observations suggest that the Lsm proteins are directly involved in decapping. First, two-hybrid analysis has identified several

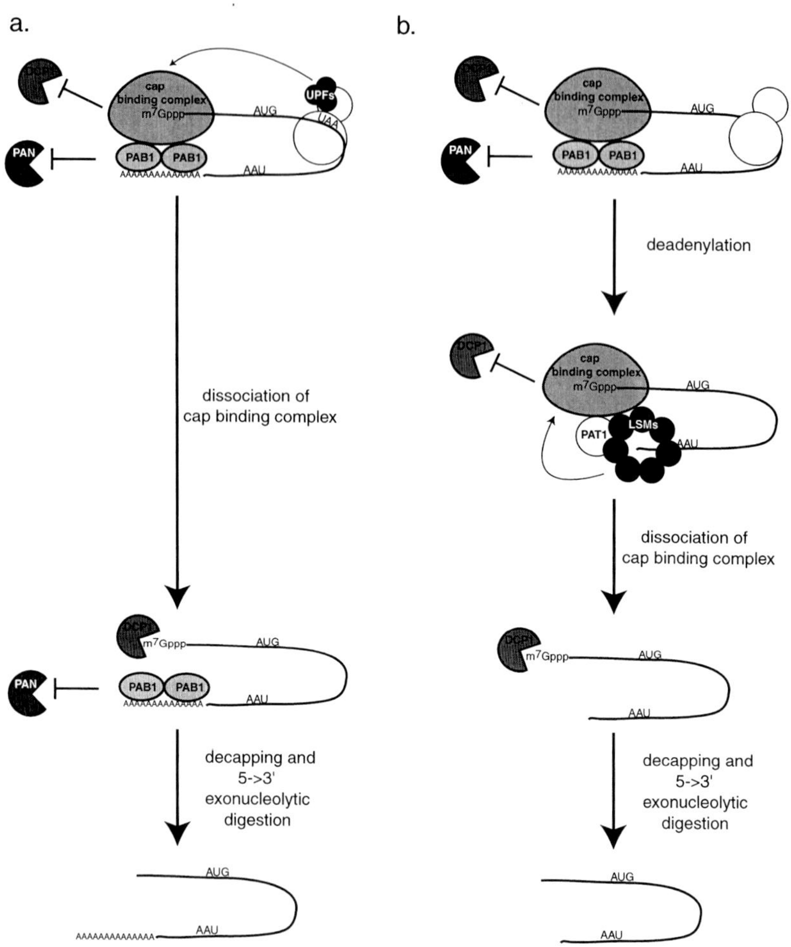

Figure 3 Regulation of decapping may occur by dissociation of the cap-binding complex. (*a*) Nonsense-mediated turnover of mRNAs is thought to occur via a signal sent from the UPF complex of proteins to the cap-binding complex. This signal causes dissociation of the cap-binding complex, allowing rapid decapping prior to deadenylation. (*b*) A similar mechanism might occur for all mRNAs. Once deadenylation has been completed, the LSM/PAT1 complex may bind to the 3′ end of an mRNA and signal dissociation of the cap-binding complex to allow decapping.

interactions between the Lsm proteins and Dcp1, Dcp2, Pat1/Mrt1, and Xrn1 (M. Fromont-Racine and J. Beggs, in prep.), all proteins required for mRNA degradation (Hsu and Stevens 1993; Muhlrad et al. 1994; Beelman et al. 1996; Hatfield et al. 1996; Dunckley and Parker 1999;

Tharun and Parker 1999). Second, the Lsm proteins co-immunoprecipitate with Dcp1p, Dcp2p, and Pat1/Mrt1p (Tharun et al. 2000). Third, immunolocalization of the Lsm1p indicates that this polypeptide is predominantly cytoplasmic. In contrast, the Lsm7p was found in both the nucleus and cytoplasm, consistent with its having a role in both mRNA decay and splicing (Tharun et al. 2000). Finally, Lsm proteins have been shown to directly interact with the U6 snRNA and with mRNA, suggesting that they may affect decapping by interacting directly with the substrate (Vidal et al. 1999; Tharun et al. 2000). These results argue that the Lsm proteins function in the cytoplasm to stimulate mRNA degradation.

An unresolved question is how the cytoplasmic Lsm complex affects mRNA degradation. One appealing model is that the Lsm complex, and the associated proteins, affect the dynamics of the translation initiation complex in some manner (Fig. 3b). This possibility is suggested by some of the phenotypes of lsm1Δ strains. Since Lsm1p does not function in splicing (Mayes et al. 1999), its phenotypes can be interpreted as reflecting the function of the cytoplasmic Lsm protein complex. Interestingly, lsm1Δ strains show a defect in mRNA degradation at all temperatures but are unable to grow at high temperatures (Mayes et al. 1999). This phenotype is similar to the phenotype of the pat1/mrt1Δ mutants (Hatfield et al. 1996; Wang et al. 1996) and suggests that these proteins affect some essential function, possibly translation. Consistent with that hypothesis, the Lsm proteins show two-hybrid interactions with three translation factors (M. Fromont-Racine and J. Beggs, in prep.). These include the γ subunit of eIF2, which delivers the initiator tRNA to the ribosome (Dorris et al. 1995); an eIF2B subunit (GCN3), which is involved in the GDP-GTP exchange of eIF2 (Bushman et al. 1993; see Chapter 5); and a ribosomal protein (RPS28), which is a component of the decoding site of the 40S subunit (Alksne and Warner 1993). It is striking that all three of these proteins are involved in the recognition of the AUG start codon in some manner, suggesting that the cytoplasmic Lsm protein complex may somehow function in the process of AUG recognition in a manner that affects mRNA decapping.

SUMMARY

It is clear that the process of translation initiation is central to determining the rate of mRNA decay in yeast. This hypothesis suggests that an understanding of decapping and its control will require determining the dynamics of the translation initiation complex and how that process interacts with the mRNA decay machinery. This view also suggests that insight will be gained from understanding how proteins affecting mRNA decay, such as the Lsm complex and the Pat1/Mrt1p, affect the different substeps of the translation process.

ACKNOWLEDGMENTS

The authors thank John Jacobs Anderson and Pat Hilleran for their helpful comments on the manuscript, Rhett Michelson for his critical evaluation of the figures, and to Anne Beudert for her help in the preparation of the manuscript. The authors' work has been supported by the Howard Hughes Medical Institute (R.P.) and from the National Institutes of Health (R.P. and D.C.S.).

REFERENCES

Aharon T. and Schneider R.J. 1993. Selective destabilization of short-lived mRNAs with the granulocyte-macrophage colony-stimulating factor AU-rich 3′ noncoding region is mediated by a cotranslational mechanism. *Mol. Cell. Biol.* **13:** 1971–1980.

Alksne L.E. and Warner J.R. 1993. A novel cloning strategy reveals the gene for the yeast homologue to *Escherichia coli* ribosomal protein S12. *J. Biol. Chem.* **268:** 10813–10819.

Allmang C., Petfalski E., Podtelejnikov A., Mann M., Tollervey D., and Mitchell P. 1999. The yeast exosome and human PM-Scl are related complexes of 3′→5′ exonucleases. *Genes Dev.* **13:** 2148–2158.

Anderson J.S.J. and Parker R. 1998. The 3′ to 5′ degradation of yeast mRNAs is a general mechanism for mRNA turnover that requires the SKI2 DEVH box protein and 3′ to 5′ exonucleases of the exosome complex. *EMBO J.* **17:** 1497–1506.

Barnes C.A. 1998. Upf1 and Upf2 proteins mediate normal yeast mRNA degradation when translation initiation is limited. *Nucleic Acids Res.* **26:** 2433–2441.

Bashkirov V.I., Scherthan H., Solinger J.A., Buerstedde J.M., and Heyer W.D. 1997. A mouse cytoplasmic exoribonuclease (mXRN1p) with preference for G4 tetraplex substrates. *J. Cell Biol.* **136:** 761–773.

Beelman C.A. and Parker R. 1994. Differential effects of translational inhibition in *cis* and in *trans* on the decay of the unstable yeast MFA2 mRNA. *J. Biol. Chem.* **269:** 9687–9692.

Beelman C.A., Stevens A., Caponigro G., LaGrandeur T.E., Hatfield L., Fortner D.M., and Parker R. 1996. An essential component of the decapping enzyme required for normal rates of mRNA turnover. *Nature* **382:** 642–646.

Bernstein P., Peltz S.W., and Ross J. 1989. The poly(A)-poly(A)-binding protein complex is a major determinant of mRNA stability in vitro. *Mol. Cell. Biol.* **9:** 659–670.

Bernstein P.L., Herrick D.J., Prokipcak R.D., and Ross J. 1992. Control of c-*myc* mRNA half-life in vitro by a protein capable of binding to a coding region stability determinant. *Genes Dev.* **6:** 642–654.

Binder R., Horowitz J.A., Basilion J.P., Koeller D.M., Klausner R.D., and Harford J.B. 1994. Evidence that the pathway of transferrin receptor mRNA degradation involves an endonucleolytic cleavage within the 3′UTR and does not involve poly(A) tail shortening. *EMBO J.* **13:** 1969–1980.

Boeck R., Lapeyre B., Brown C.E., and Sachs A.B. 1998. Capped mRNA degradation intermediates accumulate in the yeast spb8-2 mutant. *Mol. Cell. Biol.* **18:** 5062–5072.

Brown B.D., Zipkin I.D., and Harland R.M. 1993. Sequence-specific endonucleolytic cleavage and protection of mRNA in *Xenopus* and *Drosophila. Genes Dev.* **7:** 1620–1631.

Brown C.E. and Sachs A.B. 1998. Poly(A) tail length control in *Saccharomyces cerevisiae* occurs by message-specific deadenylation. *Mol. Cell. Biol.* **18:** 6548–6559.

Bushman J.L., Foiani M., Cigan A.M., Paddon C.J., and Hinnebusch A.G. 1993. Guanine nucleotide exchange factor for eukaryotic translation initiation factor 2 in *Saccharomyces cerevisiae:* Interactions between the essential subunits GCD2, GCD6, and GCD7 and the regulatory subunit GCN3. *Mol. Cell. Biol.* **13:** 4618–4631.

Cabrera C.V., Lee J.J., Ellison J.W., Britten R.J., and Davidson E.H. 1984. Regulation of cytoplasmic mRNA prevalence in sea urchin embryos. Rates of appearance and turnover for specific sequences. *J. Mol. Biol.* **174:** 85–111.

Caponigro G. and Parker R. 1995. Multiple functions for the poly(A)-binding protein in mRNA decapping and deadenylation in yeast. *Genes Dev.* **9:** 2421–2432.

———1996. mRNA turnover in yeast promoted by the MATα1 instability element. *Nucleic Acids Res.* **24:** 4304–4312.

Cereghino G.P., Atencio D.P., Saghbini M., Beiner J., and Scheffler I.E. 1995. Glucose-dependent turnover of the mRNAs encoding succinate dehydrogenase peptides in *Saccharomyces cerevisiae:* Sequence elements in the 5′ untranslated region of the Ip mRNA play a dominant role. *Mol. Biol. Cell.* **6:** 1125–1143.

Chaudhuri J., Chowdhury D., and Maitra U. 1999. Distinct functions of eukaryotic translation initiation factors eIF1A and eIF3 in the formation of the 40S ribosomal preinitiation complex. *J. Biol. Chem.* **274:** 17975–17980.

Coller J.M., Gray N.K., and Wickens M.P. 1998. mRNA stabilization by poly(A) binding protein is independent of poly(A) and requires translation. *Genes Dev.* **12:** 3226–3235.

Cooper M., Johnston L.H., and Beggs J.D. 1995. Identification and characterization of Uss1p (Sdb23p): A novel U6 snRNA-associated protein with significant similarity to core proteins of small nuclear ribonucleoproteins. *EMBO J.* **14:** 2066–2075.

Couttet P., Fromont-Racine M., Steel D., Pictet R., and Grange T. 1997. Messenger RNA deadenylation precedes decapping in mammalian cells. *Proc. Natl. Acad. Sci.* **94:** 5628–5633.

Cui Y., Hagan K.W., Zhang S., and Peltz S.W. 1995. Identification and characterization of genes that are required for the accelerated degradation of mRNAs containing a premature translational termination codon. *Genes Dev.* **9:** 423–436.

Cui Y., Gonzalez C.I., Kinzy T.G., Dinman J.D., and Peltz S.W. 1999. Mutations in the MOF2/SUI1 gene affect both translation and nonsense-mediated mRNA decay. *RNA* **5:** 794–804.

Curatola A.M., Nadal M.S., and Schneider R.J. 1995. Rapid degradation of AU-rich element (ARE) mRNAs is activated by ribosome transit and blocked by secondary structure at any position 5′ to the ARE. *Mol. Cell. Biol.* **15:** 6331–6340.

Czaplinski K., Ruiz-Echevarria M.J., Paushkin S.V., Han X., Weng Y., Perlick H.A., Dietz H.C., Ter-Avanesyan M.D., and Peltz S.W. 1998. The surveillance complex interacts with the translation release factors to enhance termination and degrade aberrant mRNAs. *Genes Dev.* **12:** 1665–1677.

Decker C.J. and Parker R. 1993. A turnover pathway for both stable and unstable mRNAs in yeast: Evidence for a requirement for deadenylation. *Genes Dev.* **7:** 1632–1643.

———. 1994. Mechanisms of mRNA degradation in eukaryotes. *Trends Biochem. Sci.* **19:** 336–340.

Dehlin E., Wormington M., Körner C. G., and Wahle E. 2000. Cap-dependent deadenylation of mRNA. *EMBO J.* **19:** 1079–1086.

Dorris D.R., Erickson F.L., and Hannig E.M. 1995. Mutations in GCD11, the structural gene for eIF-2γ in yeast, alter translational regulation of GCN4 and the selection of the start site for protein synthesis. *EMBO J.* **14:** 2239–2249.

Dunckley T. and Parker R. 1999. The DCP2 protein is required for mRNA decapping in *Saccharomyces cerevisiae* and contains a functional MutT motif. *EMBO J.* **18:** 5411–5422.

Fire A. 1999. RNA-triggered gene silencing. *Trends Genet.* **15:** 358–363.

Gonzalez C.I. and Martin C.E. 1996. Fatty acid-responsive control of mRNA stability. *J. Biol. Chem.* **271:** 25801–25809.

Hatfield L., Beelman C.A., Stevens A., and Parker R. 1996. Mutations in *trans*-acting factors affecting mRNA decapping in *Saccharomyces cerevisiae. Mol. Cell. Biol.* **16:** 5830–5838.

Heaton B., Decker C., Muhlrad D., Donahue J., Jacobson A., and Parker R. 1992. Analysis of chimeric mRNAs derived from the STE3 mRNA identified multiple regions within yeast mRNAs that modulate mRNA decay. *Nucleic Acids Res.* **20:** 5365–5373.

Hermann H., Fabrizio P., Raker V.A., Foulaki K., Hornig H., Brahms H., and Luhrmann R. 1995. snRNP Sm proteins share two evolutionarily conserved sequence motifs which are involved in Sm protein-protein interactions. *EMBO J.* **14:** 2076–2088.

Herrick D. and Jacobson A. 1992. A coding region segment is necessary, but not sufficient for rapid decay of the HIS3 mRNA in yeast. *Gene* **114:** 35–41.

Herrick D., Parker R., and Jacobson A. 1990. Identification and comparison of stable and unstable mRNAs in *Saccharomyces cerevisiae. Mol. Cell. Biol.* **10:** 2269–2284.

Hsu C.L. and Stevens A. 1993. Yeast cells lacking 5′→3′ exoribonuclease 1 contain mRNA species that are poly(A) deficient and partially lack the 5′ cap structure. *Mol. Cell. Biol.* **13:** 4826–4835.

Jacobson A. and Peltz S.W. 1996. Interrelationships of the pathways of mRNA decay and translation in eukaryotic cells. *Annu. Rev. Biochem.* **65:** 693–739.

Kambach C., Walke S., Young R., Avis J.M., de la Fortelle E., Raker V.A., Luhrmann R., Li J., and Nagai K. 1999. Crystal structures of two Sm protein complexes and their implications for the assembly of the spliceosomal snRNPs. *Cell* **96:** 375–387.

Körner C.G. and Wahle E. 1997. Poly(A) tail shortening by a mammalian poly(A)-specific 3′-exoribonuclease. *J. Biol. Chem.* **272:** 10448–10456.

Körner C.G., Wormington M., Muckenthaler M., Schneider S., Dehlin E., and Wahle E. 1998. The deadenylating nuclease (DAN) is involved in poly(A) tail removal during the meiotic maturation of *Xenopus* oocytes. *EMBO J.* **17:** 5427–5437.

Kruys V. and Huez G. 1994. Translational control of cytokine expression by 3′ UA-rich sequences. *Biochimie* **76:** 862–866.

Kruys V., Marinx O., Shaw G., Deschamps J., and Huez G. 1989. Translation blockade imposed by cytokine-derived UA-rich sequences. *Science* **245:** 852–855.

Kruys V., Wathelet M., Poupart P., Contreras R., Fiers W., Content J., and Huez G. 1987. The 3′ untranslated region of the human interferon-beta mRNA has an inhibitory effect on translation. *Proc. Natl. Acad. Sci.* **84:** 6030–6034.

LaGrandeur T.E. and Parker R. 1998. Isolation and characterization of Dcp1p, the yeast mRNA decapping enzyme. *EMBO J.* **17:** 1487–1496.

———. 1999. The *cis* acting sequences responsible for the differential decay of the unstable MFA2 and stable PGK1 transcripts in yeast include the context of the translational start codon. *RNA* **5:** 420–433.

Le H., Tanguay R.L., Balasta M.L., Wei C., Browning K.S., Metz A.M., Goss D.J., and Gallie D.R. 1997. Translation initiation factors eIF-iso4G and eIF-4B interact with the poly(A)-binding protein and increase its RNA binding activity. *J. Biol. Chem.* **272:** 16247–16255.

Lee B.S. and Culbertson M.R. 1995. Identification of an additional gene required for eukaryotic nonsense mRNA turnover. *Proc. Natl. Acad. Sci.* **92:** 10354–10358.

Linz B., Koloteva N., Vasilescu S., and McCarthy J.E. 1997. Disruption of ribosomal scanning on the 5′-untranslated region, and not restriction of translational initiation per se, modulates the stability of nonaberrant mRNAs in the yeast *Saccharomyces cerevisiae. J. Biol. Chem.* **272:** 9131–9140.

Lombardo A., Cereghino G.P., and Scheffler I.E. 1992. Control of mRNA turnover as a mechanism of glucose repression in *Saccharomyces cerevisiae. Mol. Cell. Biol.* **12:** 2941–2948.

Mayes A.E., Verdone L., Legrain P., and Beggs J.D. 1999. Characterization of Sm-like proteins in yeast and their association with U6 snRNA. *EMBO J.* **18:** 4321–4331.

Mitchell P., Petfalski E., Shevchenko A., Mann M., and Tollervey D. 1997. The exosome: A conserved eukaryotic RNA processing complex containing multiple 3′→5′ exoribonucleases. *Cell* **91:** 457–466.

Muckenthaler M., Gunkel N., Stripecke R., and Hentze M.W. 1997. Regulated poly(A) tail shortening in somatic cells mediated by cap-proximal translational repressor proteins and ribosome association. *RNA* **3:** 983–995.

Muhlrad D. and Parker R. 1992. Mutations affecting stability and deadenylation of the yeast MFA2 transcript. *Genes Dev.* **6:** 2100–2111.

———. 1994. Premature translational termination triggers mRNA decapping. *Nature* **370:** 578–581.

———. 1999. Recognition of yeast mRNAs as "nonsense containing" leads to both inhibition of mRNA translation and mRNA degradation: Implications for the control of mRNA decapping. *Mol. Biol. Cell.* **10:** 3971–3978.

Muhlrad D., Decker C.J., and Parker R. 1994. Deadenylation of the unstable mRNA encoded by the yeast MFA2 gene leads to decapping followed by 5′→3′ digestion of the transcript. *Genes Dev.* **8:** 855–866.

———. 1995. Turnover mechanisms of the stable yeast PGK1 mRNA. *Mol. Cell. Biol.* **15:** 2145–2156.

Nielsen F.C. and Christiansen J. 1992. Endonucleolysis in the turnover of insulin-like growth factor II mRNA. *J. Biol. Chem.* **267:** 19404–19411.

Pannone B.K., Xue D., and Wolin S.L. 1998. A role for the yeast La protein in U6 snRNP assembly: Evidence that the La protein is a molecular chaperone for RNA polymerase III transcripts. *EMBO J.* **17:** 7442–7453.

Parker R. and Jacobson A. 1990. Translation and a forty-two nucleotide segment within the coding region of the mRNA encoded by the MATα1 gene are involved in promoting mRNA decay in yeast. *Proc. Natl. Acad. Sci.* **87:** 2780–2784.

Peltz S.W., Donahue J.L., and Jacobson A. 1992. A mutation in the tRNA nucleotidyltransferase gene promotes stabilization of mRNAs in *Saccharomyces cerevisiae. Mol.*

Cell. Biol. **12:** 5778–5784.

Pestova T.V., Shatsky I.N., and Hellen C.U. 1996. Functional dissection of eukaryotic initiation factor 4F: The 4A subunit and the central domain of the 4G subunit are sufficient to mediate internal entry of 43S preinitiation complexes. *Mol. Cell. Biol.* **16:** 6870–6878.

Pierrat B., Lacroute F., and Losson R. 1993. The 5′ untranslated region of the PPR1 regulatory gene dictates rapid mRNA decay in yeast. *Gene* **131:** 43–51.

Ross J. 1995. mRNA stability in mammalian cells. *Microbiol. Rev.* **59:** 423–450.

Sachs A.B. and Davis R.W. 1989. The poly(A) binding protein is required for poly(A) shortening and 60S ribosomal subunit-dependent translation initiation. *Cell* **58:** 857–867.

Salgado-Garrido J., Bragado-Nilsson E., Kandels-Lewis S., and Seraphin B. 1999. Sm and Sm-like proteins assemble in two related complexes of deep evolutionary origin. *EMBO J.* **18:** 3451–3462.

Scheffler I.E., de la Cruz B.J., and Prieto S. 1998. Control of mRNA turnover as a mechanism of glucose repression in *Saccharomyces cerevisiae. Int. J. Biochem. Cell Biol.* **30:** 1175–1193.

Schwartz D.C. and Parker R. 1999. Mutations in translation initiation factors lead to increased rates of deadenylation and decapping of mRNAs in *Saccharomyces cerevisiae. Mol. Cell. Biol.* **19:** 5247–5256.

Seraphin B. 1995. Sm and Sm-like proteins belong to a large family: Identification of proteins of the U6 as well as the U1, U2, U4 and U5 snRNPs. *EMBO J.* **14:** 2089–2098.

Shyu A.B., Belasco J.G., and Greenberg M.E. 1991. Two distinct destabilizing elements in the c-*fos* message trigger deadenylation as a first step in rapid mRNA decay. *Genes Dev.* **5:** 221–231.

Singer R.H. and Penman S. 1973. Messenger RNA in HeLa cells: Kinetics of formation and decay. *J. Mol. Biol.* **78:** 321–334.

Sonenberg N., Morgan M.A., Merrick W.C., and Shatkin A.J. 1978. A polypeptide in eukaryotic initiation factors that crosslinks specifically to the 5′-terminal cap in mRNA. *Proc. Natl. Acad. Sci.* **75:** 4843–4847.

Spradling A., Hui H., and Penman S. 1975. Two very different components of messenger RNA in an insect cell line. *Cell* **4:** 131–137.

Surosky R.T. and Esposito R.E. 1992. Early meiotic transcripts are highly unstable in *Saccharomyces cerevisiae. Mol. Cell. Biol.* **12:** 3948–3958.

Tarun S.Z., Jr. and Sachs A.B. 1995. A common function for mRNA 5′ and 3′ ends in translation initiation in yeast. *Genes Dev.* **9:** 2997–3007.

———. 1996. Association of the yeast poly(A) tail binding protein with translation initiation factor eIF-4G. *EMBO J.* **15:** 7168–7177.

Tharun S. and Parker R. 1999. Analysis of mutations in the yeast mRNA decapping enzyme. *Genetics* **151:** 1273–1285.

Tharun S., He W., Mayes A.E., Lennertz P., Beggs J.D., and Parker R. 2000. Yeast Sm-like proteins function in mRNA decapping and decay. *Nature* **404:** 515–518.

Vidal V.P., Verdone L., Mayes A.E., and Beggs J.D. 1999. Characterization of U6 snRNA-protein interactions. *RNA* **5:** 1470–1481.

Wang X., Watt P.M., Louis E.J., Borts R.H., and Hickson I.D. 1996. Pat1: A topoisomerase II-associated protein required for faithful chromosome transmission in *Saccharomyces cerevisiae. Nucleic Acids Res.* **24:** 4791–4797.

Weng Y., Czaplinski K., and Peltz S.W. 1996. Identification and characterization of mutations in the UPF1 gene that affect nonsense suppression and the formation of the Upf protein complex but not mRNA turnover. *Mol. Cell. Biol.* **16:** 5491–5506.

Wilson T. and Treisman R. 1988. Removal of poly(A) and consequent degradation of c-*fos* mRNA facilitated by 3′ AU-rich sequences. *Nature* **336:** 396–399.

Wormington M., Searfoss A.M., and Hurney C.A. 1996. Overexpression of poly(A) binding protein prevents maturation-specific deadenylation and translational inactivation in *Xenopus* oocytes. *EMBO J.* **15:** 900–909.

Xu H., Johnson L., and Grunstein M. 1990. Coding and noncoding sequences at the 3′ end of yeast histone H2B mRNA confer cell cycle regulation. *Mol. Cell. Biol.* **10:** 2687–2694.

Zuk D. and Jacobson A. 1998. A single amino acid substitution in yeast eIF-5A results in mRNA stabilization. *EMBO J.* **17:** 2914–2925.

Zuk D., Belk J.P., and Jacobson A. 1999. Temperature-sensitive mutation in the *Saccharomyces cerevisiae* MRT4, GRC5, SLA2 and THS1 genes result in defects in mRNA turnover. *Genetics* **153:** 35–47.

29

Destabilization of Nonsense-containing Transcripts in *Saccharomyces cerevisiae*

Allan Jacobson
Department of Molecular Genetics and Microbiology
University of Massachusetts Medical School
Worcester, Massachusetts 01655

Stuart W. Peltz
Department of Molecular Genetics and Microbiology
Robert Wood Johnson Medical School
University of Medicine and Dentistry of New Jersey
Piscataway, New Jersey 08854

DESTABILIZATION OF ABERRANT mRNAS CONTRIBUTES TO THE ACCURACY OF GENE EXPRESSION

Gene expression is a highly accurate process in which the final products, i.e., proteins, contain no more than one incorrect amino acid per 10,000 inserted (Kurland 1992; Yarus 1992). This high level of accuracy underwrites nature's genetic systems and ensures that even very large polypeptides will be functional. Several cooperative and concurrently operating mechanisms are responsible for maintaining fidelity in the flow of genetic information, including those that monitor the integrity of RNA synthesis and processing, tRNA aminoacylation, codon:anticodon pairing, and peptide elongation (Yarus 1992; Chin and Pyle 1995; Freist et al. 1996; Jeon and Agarwal 1996; Ibba and Soll 1999; Yoshizawa et al. 1999). Because these mechanisms are incapable of rectifying errors attributable to mutations, other processes have evolved that rid the cell of improperly folded proteins and mRNAs that lack complete open reading frames (He et al. 1993; Cui et al. 1995; Wickner et al. 1999). The latter transcripts are typically those containing premature termination (or "nonsense") codons, and the process which ensures that these mRNAs do not accumulate as substrates for the translation apparatus has been dubbed nonsense-medi-

© 2000 Cold Spring Harbor Laboratory Press 0-87969-568-4/00 $5 +. 00

ated mRNA decay (NMD), or mRNA surveillance (Peltz et al. 1993a, 1994; Pulak and Anderson 1993).

Although shown to operate in a broad spectrum of eukaryotic organisms (for a discussion of this phenomenon in mammals, see Chapter 30), NMD has been most extensively studied in the yeast *Saccharomyces cerevisiae*, largely because of the facility of genetic analyses (Losson and Lacroute 1979; Leeds et al. 1991, 1992; Peltz et al. 1993a,b, 1994; Caponigro and Parker 1996; Jacobson and Peltz 1996; Ruiz-Echevarria et al. 1996; Weng et al. 1997; Culbertson 1999; Czaplinski et al. 1999). As discussed in this chapter, studies of yeast NMD have not only provided insight into cellular quality control mechanisms, but have also elucidated fundamental interrelationships between the pathways of mRNA translation and mRNA decay.

The phenomenon of nonsense-mediated mRNA decay is illustrated in Figure 1, in which decay rate measurements for two species of yeast *PGK1* mRNA are depicted. One of the *PGK1* transcripts is derived from the wild-type gene and the other from an allele harboring a nonsense mutation at nucleotide 361 (Peltz et al. 1993a). Half-lives for the two transcripts were determined by monitoring their relative abundance after the inhibition of transcription. It is evident from Figure 1 that the nonsense-containing mRNA has both a lower steady-state level (compare the relative mRNA abundance at t = 0) and a more rapid decay rate than its wild-type counterpart (wild-type mRNA, $t_{1/2}$ = 30–60 minutes; mutant mRNA, $t_{1/2}$ =3-5 minutes). Comparable destabilization and reduction in overall abundance have been observed in other nonsense-containing transcripts, including those expressed from the yeast *ADE2*, *ADE3*, *CAN1*, *CYC1*, *HIS3*, *HIS4*, *LEU2*, *MATα1*, *PGK1*, *TYR7*, *URA1*, and *URA3* genes (Losson and Lacroute 1979; Pelsy and Lacroute 1984; Peltz et al. 1993a; Hagan et al.

Figure 1 Destabilization of the *PGK1* mRNA by a premature nonsense codon. Decay of wild-type and nonsense-containing *PGK1* transcripts was monitored by Northern blotting, using RNA isolated at different times after the inhibition of transcription in a strain harboring a temperature-sensitive RNA polymerase II. (For details, see Peltz et al. 1993a).

1995; Yun and Sherman 1995; Zhang et al. 1995; Hennigan and Jacobson 1996, 1997, in prep.; Weng et al. 1996a,b; Maderazo et al. 2000; J. Belk and A. Jacobson, unpubl.). Although such transcripts are generally derived from genes in which a mutation has given rise to a premature nonsense codon, it is clear that the fate of the mRNAs would have been no different if their nonsense-containing status were attributable to errors in transcription, pre-mRNA splicing, or RNA editing, or caused by failure of the ribosome to maintain the normal reading frame.

MULTIPLE CLASSES OF TRANSCRIPTS ARE SUBSTRATES OF THE NONSENSE-MEDIATED mRNA DECAY PATHWAY

The existence of a pathway that promotes rapid decay of nonsense-containing mRNAs raised the question of whether the substrates of this pathway are restricted to aberrant mRNAs. It seemed unlikely that the normal function of the nonsense-mediated decay pathway was anticipatory; i.e., solely involved in the degradation of mRNAs derived from nonsense alleles, so additional substrates were sought. The standard approach was to identify mRNAs that were selectively stabilized in strains harboring mutations in one or more of the genes encoding factors essential for NMD. These studies have shown that, in addition to standard mRNAs with prematurely terminated open reading frames, the substrates of the NMD pathway include (see Fig. 2): (1) inefficiently spliced pre-mRNAs that enter the cytoplasm with their introns intact (He et al. 1993); (2) mRNAs in which the ribosome has bypassed the initiator AUG and commenced translation further downstream (Welch and Jacobson 1999); (3) some mRNAs containing upstream open reading frames (uORFs; Cui et al. 1995; Vilela et al. 1998, 1999); and (4) transcripts with extended 3′UTRs (Pulak and Anderson 1993; Muhlrad and Parker 1999; F. Sherman, pers. comm.).

Of these substrates, the standard nonsense-containing mRNAs, the intron-containing pre-mRNAs, and the mRNAs in which the ribosome has scanned downstream from the normal initiation codon can all be considered to be targets of a quality control system seeking to reduce the generation of potentially deleterious polypeptide fragments. However, the existence of the remaining classes of substrates (i.e., mRNAs with uORFs or extended 3′UTRs) indicates that the NMD pathway has additional regulatory capabilities. Studies of these substrates suggest that the NMD system may be a general regulator of translational termination efficiency (see below, Cross-talk between mRNA translation and turnover is mediated by

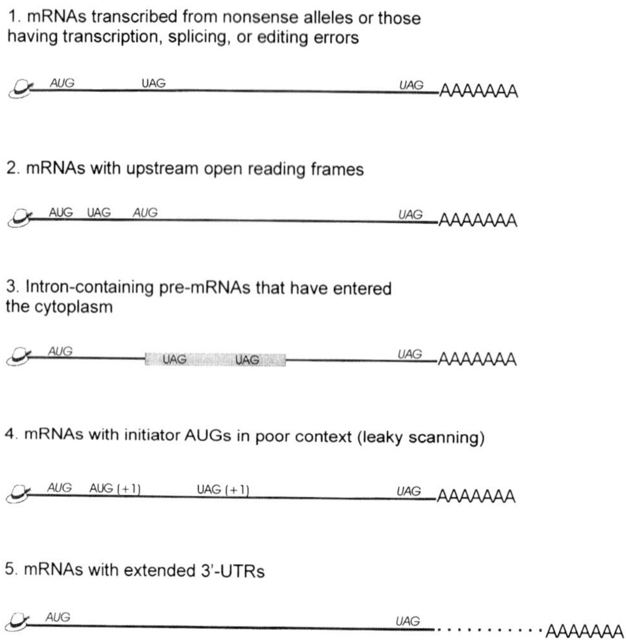

Figure 2 Substrates of the nonsense-mediated mRNA decay pathway. Initiation and termination codons that delineate the principal open reading frames in all classes of mRNA are indicated by italics. Termination codons likely to trigger nonsense-mediated mRNA decay are indicated (in a standard font) for substrate classes 1–4. For class 5, the dotted line indicates that the 3′UTR is considerably longer than its counterpart in the wild-type transcript.

the UPF/NMD factors; Weng et al., 1996a,b,1997; Ruiz-Echevarria et al. 1998b; Muhlrad and Parker 1999; Hilleren and Parker 1999; Maderazo et al. 2000) and that some mRNAs subject to NMD regulation encode factors that, in turn, regulate the abundance of other mRNAs (Leeds et al. 1991; Dahlseid et al. 1998; Lew et al. 1998; Lelivelt and Culbertson 1999; Welch and Jacobson 1999).

NMD REQUIRES TRANSLATION, DEADENYLATION-INDEPENDENT DECAPPING, AND SPECIFIC *TRANS*-ACTING FACTORS

Destabilization of Nonsense-containing mRNAs Requires Their Translation

The mere presence of a premature nonsense codon within an mRNA is not sufficient to promote its degradation by the NMD pathway. This conclusion follows from experiments showing that the normal rate of decay can be restored to a nonsense-containing mRNA if a nonsense-suppressing tRNA is coexpressed in the same cells as the nonsense-containing transcript (Losson and Lacroute 1979; Gozalbo and Hohmann 1990) or if the initiator AUG is deleted from the transcript (Ruiz-Echevarria et al. 1998a). It follows, therefore, that mRNA destabilization depends on recognition of the nonsense codon as such by the translational apparatus. This conclusion is consistent with several other observations, including those demonstrating (1) the association of nonsense-containing mRNAs with polysomes whose size reflects the relative position of the nonsense codon within the respective open reading frames (He et al. 1993); (2) the inhibition of NMD by drugs and mutations that block either translational initiation or elongation (Herrick et al. 1990; Peltz et al. 1992, 1997; Welch and Jacobson 1999; Zuk et al. 1999); (3) the destabilization of polysome-associated nonsense-containing mRNAs immediately after release from cycloheximide-promoted stabilization (Peltz et al. 1997); (4) the colocalization of factors required for NMD and polysomes (Peltz et al 1993b; Atkin et al. 1995,1997; Mangus and Jacobson 1999); and (5) the interaction of Upf1p, a factor essential for NMD (see below, Characterization of factors involved in controlling the NMD pathway; Leeds et al. 1991; Peltz et al. 1993a), with the polypeptide release factors eRF1 and eRF3 (Czaplinski et al. 1998) and the ribosome-associated factor Nmd3p (Belk et al. 1999). The importance of termination codon recognition to the process of NMD is further underscored by the observation that deletion of any of the genes encoding the three principal NMD regulatory factors (i.e., *UPF1*, *UPF2/NMD2*, or *UPF3*) promotes nonsense codon readthrough and phenotypic suppression (Leeds et al. 1992; Cui et al. 1995; Weng et al. 1996a,b; Maderazo et al. 2000; K. Czaplinski et al.; in prep.).

The concurrence of NMD and translation, as well as the association and involvement of principal NMD factors with the translation apparatus, implies that the degradation of yeast NMD substrates occurs in the cytoplasm. This conclusion is noteworthy because some experiments in mammalian cells indicate that certain nonsense-containing mRNAs are degraded while still in association with the nucleus (Maquat 1995; Jacobson and Peltz 1996; Chapter 30). This apparent conundrum can be resolved by postulating that degradation of nonsense-containing transcripts can be initiated as soon as they are accessible to the cytoplasmic translation apparatus and that the differences observed between mammalian cells and yeast simply reflect differences in the overall kinetics of mRNA transport, mRNP assembly, and translation initiation (Jacobson and Peltz 1996).

Nonsense-containing mRNAs Are Decapped without Prior Deadenylation

The first mechanistic insight into the yeast NMD pathway was the recognition that nonsense-containing mRNAs were shunted into a late step in the predominant pathway used for the degradation of most wild-type mRNAs (see Fig. 3). These mRNAs decay by a mechanism in which the initial nucleolytic event is the shortening of the poly(A) tail to an oligo(A) length of 10–15 nucleotides. After poly(A) shortening, transcripts are cleaved one or two nucleotides from their 5′ends by the *DCP1* protein (Stevens 1988; Beelman et al. 1996; LaGrandeur and Parker 1998), removing the 5′ cap structure. Decapped and deadenylated mRNAs are then digested exonucleolytically by the 5′ to 3′ exoribonuclease, Xrn1p (Decker and Parker 1993; Hsu and Stevens 1993; Muhlrad and Parker 1994). Degradation of nonsense-containing mRNAs, in contrast, is deadenylation-independent, entering the 5′→3′ pathway without prior poly(A) shortening. Transcripts recognized as nonsense-containing are decapped by Dcp1p while retaining full-length poly(A) tails and subsequently degraded 5′→3′ by the Xrn1p exonuclease (Fig. 3) (Muhlrad and Parker 1994; Hagan et al. 1995; Beelman et al. 1996; Hatfield et al. 1996).

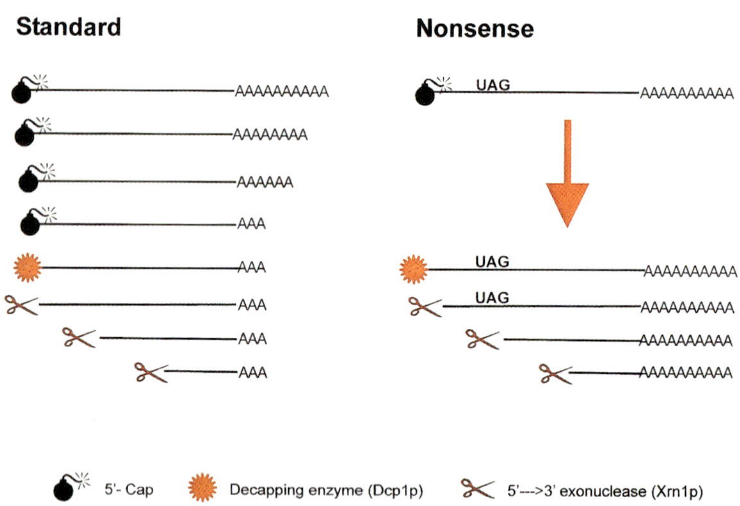

Figure 3 Decay pathways for wild-type and nonsense-containing mRNAs. (*Left*) Decay of wild-type ("standard") mRNAs by the 5′→3′ deadenylation dependent pathway. (*Right*) Decay of nonsense-containing ("nonsense") mRNAs by the 5′→3′ deadenylation-independent pathway. (For details, see Muhlrad and Parker 1994; Muhlrad et al. 1994, 1995.).

Interestingly, although nonsense-containing and wild-type mRNAs are both decapped by Dcp1p, there are differences in the modes by which the respective RNAs are treated by the decapping machinery. This is illustrated by two observations: (1) the identification of mutations that inhibit decapping of deadenylated wild-type mRNAs without affecting the decapping of nonsense-containing transcripts (Hatfield et al. 1996; Boeck et al. 1998; Tharun and Parker 1999) and (2) the ability of nonsense-containing *PGK1* mRNA to be decapped rapidly, whereas its deadenylated wild-type counterpart is decapped slowly (Muhlrad and Parker 1994; Muhlrad et al. 1995; Hilleren and Parker 1999). Since the susceptibility of deadenylated wild-type mRNAs to Dcp1p digestion is thought to arise in part from the elimination of poly(A) tail-promoted mRNA 5′/3′ interactions (Jacobson 1996; Jacobson and Peltz 1996; see Chapter 10 Sachs), these observations suggest that the recognition of premature nonsense codons can short-circuit these interactions (Hilleren and Parker 1999). This conclusion is reinforced by experiments showing that a premature nonsense codon will also destabilize an mRNA in which 5′/3′ interactions are established independent of a poly(A) tail (Coller et al. 1998).

Characterization of Factors Involved In Controlling the NMD Pathway

The powerful genetic system of *S. cerevisiae* has been particularly fruitful in the identification of genes whose products are involved in the decay of nonsense-containing mRNAs. Yeast mutants that affect this process were initially isolated in a screen for allosuppressors of the *his4-38* frameshift mutation (Culbertson et al. 1980). Subsequent screens that also identified components of this pathway included those seeking omnipotent suppressors, regulators of frameshifting or translation, suppressors of upstream initiation codons, or two-hybrid interactors with known factors (Hampsey et al. 1991; Pinto et al. 1992; Dinman and Wickner 1994; Cui et al. 1995, 1999a,b; He and Jacobson 1995; Lee et al. 1995; He et al. 1997; Welsh and Jacobson 1999). Analyses of the genes identified in all of these studies have demonstrated that mutations in the *UPF1*, *UPF2/NMD2*, *UPF3*, *PRT1*, *HRP1*, *MOF2*, *MOF5*, *MOF8*, and *DBP2* genes result in stabilization and increased accumulation of nonsense-containing mRNAs while having little or no effect on the abundance and stability of most wild-type transcripts (Leeds et al. 1991, 1992; Cui et al. 1995, 1999a,b; He and Jacobson 1995; Lee and Culbertson 1995; He et al. 1997; Gonzalez et al. 2000).

UPF1, *UPF2/NMD2*, and *UPF3* have been characterized extensively, in part for chronological reasons, but largely because these nonessential genes encode conserved central regulators of the NMD pathway. The *UPF1* gene encodes a protein that has a predicted mass of 109 kD, a cysteine- and histidine-rich region near its amino terminus that may comprise two Zn^{++}-fingers, and seven conserved motifs common to the members of helicase superfamily I (Altamura et al. 1992; Koonin 1992; Leeds et al. 1992). Upf1p has been purified from yeast cells and shown to possess RNA binding, as well as RNA-dependent ATPase and RNA helicase activities (Czaplinski et al. 1995; Weng et al. 1996a,b, 1998). *UPF2/NMD2* encodes an acidic protein with a predicted mass of 127 kD and no significant homologies with other polypeptides (Cui et al. 1995; He and Jacobson 1995). *UPF3* encodes a basic 45-kD protein harboring NLS- and NES-like segments, suggesting that it may shuttle between the nucleus and the cytoplasm (Lee and Culbertson 1995; Shirley et al. 1998).

Single or multiple mutations within *UPF1*, *UPF2/NMD2*, or *UPF3* yield similar mRNA decay phenotypes. For example, the respective individual, double, and triple disruptions all show comparable stabilization of the *CYH2* pre-mRNA, an endogenous nonsense-containing substrate (He et al. 1993, 1997; Cui et al. 1995). The similarity of the respective decay phenotypes indicates that all three gene products must be functionally related and act in a common pathway. Substantial support for this conclusion has been obtained from experiments on the biochemical and genetic interactions of these factors and their genes.

Upf1p, Upf2/Nmd2p, and Upf3p are all polysome-associated interacting proteins (Peltz et al. 1993b; Atkin et al. 1995, 1997; Weng et al. 1996a,b; He et al. 1997; Mangus and Jacobson 1999). Upf3p interacts with a central domain of Upf2/Nmd2p (He et al. 1997), and an interaction between a 157-amino acid carboxy-terminal segment of Upf2p/Nmd2p and the aminoterminus of Upf1p is required for mRNA decay (He et al. 1996). These results, and functional studies discussed below, indicate that these three proteins either interact sequentially, or function as a complex, to regulate translation termination as well as NMD.

CROSS-TALK BETWEEN mRNA TRANSLATION AND TURNOVER IS MEDIATED BY THE *UPF/NMD* FACTORS

Mutations in the *UPF/NMD* genes not only lead to the stabilization of nonsense-containing mRNAs, but they also result in a nonsense suppression phenotype (Leeds et al. 1992; Weng et al. 1996a,b). Nonsense suppression occurs when a near cognate tRNA (or a suppressor tRNA) effectively competes with the translation termination release factors for binding to the A-site of the ribosome while a nonsense codon is in residence at the decoding site. As a consequence, the nascent peptidyl-tRNA bond is not hydrolyzed and the polypeptide continues to be extended. Normally, a nonsense allele of an essential gene is the formal equivalent of a disruption; i.e., a null allele that leads to complete loss of gene function and inviability. Suppression of nonsense alleles occurs when the efficiency of termination codon readthrough is sufficient for restoration of normal growth.

Nonsense suppression in *upf/nmd* strains could simply be due to a combination of enhanced mRNA abundance and an inherent rate of nonsense codon readthrough that is sufficient to generate the minimal amount of protein required for function of the respective genes. Alternatively, the *UPF/NMD* proteins could be involved in modulating translation termination efficiencies directly. A role for these proteins in translation termination first became evident from the isolation of *upf1* alleles that separated the mRNA turnover and nonsense suppression phenotypes (Weng et al. 1997). One set of mutations inactivated mRNA decay but failed to allow suppression, whereas another set promoted mRNA decay but allowed suppression to occur (Weng et al. 1996a,b). Further evidence for independent suppression and mRNA decay activities of the *UPF/NMD* proteins was provided by experiments showing that (1) merely elevating the level of a nonsense-containing transcript was insufficient to promote suppression and (2) suppression of a *can1* nonsense allele in *upf2Δ/nmd2Δ* or *upf3Δ* strains could be reversed by overexpression of *UPF1* without concomitant changes in the level of the *can1* mRNA (Maderazo et al., 2000.). Because *upf* mutations do not affect polysome profiles (Leeds et al. 1991; He et al. 1993; Atkin et al. 1995, 1997), it is likely that the translational effects that promote suppression are not targeted to general initiation or elongation but, rather, to the premature termination event. This conclusion is supported by recent experiments demonstrating nonsense codon readthrough in the *can1-100* mRNA (Maderazo et al. 2000).

A principal role for Upf1p in the modulation of termination efficiency is also strongly indicated by experiments showing that yeast (and human) Upf1p interact with the translation termination factors eRF1 and eRF3 (Czaplinski et al. 1998). Interaction with both factors inhibits the ATPase activity of Upf1p, and interaction with eRF3 prevents formation of a Upf1p:RNA complex (Czaplinski et al. 1998). These observations, and the properties of the *upf1* alleles alluded to above, suggest that the ATPase/helicase and RNA-binding activities of Upf1p are required for function in mRNA decay but not in translation termination (Weng et al. 1996a,b, 1998). Moreover, since eRF3 inhibits Upf1p binding to RNA, eRF3 and RNA may compete for binding to Upf1p (Czaplinski et al. 1998). In turn, this suggests that factors such as ATP, which is capable of decreasing Upf1p affinity for RNA, might promote release factor binding (Weng et al. 1996a, 1998) or regulate Upf1p interaction with the release factors in the presence of competing RNAs (Czaplinski et al. 1998). The identification of a mutant form of Upf1p that can bind but not hydrolyze ATP, and which is active in translation termination but inactive in mRNA decay, suggests that ATP may be a cofactor that switches Upf1p between its translation termination and NMD activities (Weng et al. 1996a,b, 1998).

Strains harboring a *upf1Δ* mutation show higher levels of *can1* nonsense suppression than strains harboring either the *upf2Δ/nmd2Δ* or *upf3Δ* mutations, and the *upf1Δ* phenotype predominates in doubly mutant strains (Maderazo et al. 2000). These data, and the demonstration that overexpression of *UPF1* can compensate for *upf2Δ/nmd2Δ* and *upf3Δ* mutations but not vice versa, suggest that Upf2p/Nmd2p and Upf3p can regulate Upf1p activity, or that an imbalance in the concentration of Upf2p/Nmd2p and Upf3p can alter the efficiency of the translation termination factors (Maderazo et al. 2000; K. Czaplinski and S.W. Peltz, in prep.).

CIS-ACTING SEQUENCES REGULATE NMD

mRNA Destabilization Requires a Premature Nonsense Codon and a Downstream Element

All functional mRNAs contain a translation termination codon, yet they are not substrates for the nonsense-mediated mRNA decay pathway. What appears to distinguish a normal nonsense codon from one that promotes mRNA destabilization is its sequence context, or more precisely, the presence of specific sequences 3′ to the nonsense codon. A requirement for such downstream elements, or DSEs, was initially indicated by experiments demonstrating that (1) deletion of most of the *PGK1* protein coding region downstream from an early nonsense mutation reduced mRNA decay rates markedly and (2) reinsertion of a specific small segment of the *PGK1* coding region into the construct harboring the large deletion was sufficient to activate NMD (Peltz et al. 1993a). Characterization of a *PGK1* DSE, and sequences from other mRNAs with comparable activity, has established a weak sequence consensus (5′-YGCUGAUGYYYYY-3′) and shown that DSE-like sequences are present in at least one copy in the coding regions of most yeast mRNAs (Peltz et al. 1993a; Hagan et al. 1995; Yun and Sherman 1995; Zhang et al. 1997). Their presence within the coding regions of wild-type mRNAs of diverse decay rates suggests that DSEs are inactive unless preceded by an upstream termination codon

Further clarification of the role of the DSE has been obtained from studies of the yeast *GCN4* mRNA. The 5′UTR of this transcript contains four short uORFs whose cycles of translation initiation and termination regulate *GCN4* expression (see Chapter 5). The uORFs lack a 3′ DSE and the *GCN4* mRNA is thus relatively stable. However, insertion of a *PGK1* DSE 3′ to *GCN4* uORF1 is sufficient to trigger NMD (Ruiz-Echevarria and Peltz 1996). Additional experiments with the *GCN4* system have provided insight into DSE/nonsense codon interactions, demonstrating that (1) DSEs are only functional after at least one translation initiation/termination cycle has been completed, (2) although the initiation and termination phases of translation are required for NMD, the elongation phase is dispensable, (3) a DSE can activate the NMD pathway when it is located within ~150 nucleotides of the stop codon, and (4) DSEs are not functional if they are traversed by ribosomes (Ruiz-Echevarria and Peltz 1996; Ruiz-Echevarria et al.1998a).

The general occurrence of DSEs within mRNA coding regions is consistent with four of the known classes of NMD substrates: i.e., those with premature termination codons, uORFs, retained introns, and AUGs in poor context (see Fig. 2). However, in the case of mRNAs with extended 3′UTRs, the sequence downstream from the termination codon can be derived from intergenic sequences that are not normally transcribed (Zaret and Sherman 1982, 1984). This would suggest that DSEs are interspersed throughout the genome or that a DSE is simply a sequence 3′ to a termination codon that lacks some feature(s) of a normal 3′ UTR, and that it is the absence of that essential feature which triggers decay. In either case, control of the stability of an mRNA is likely to be the consequence of factors that interact with the DSE (see below, Roles of the factors and sequences: Surveillance complex or faux-UTR?).

STABILIZER ELEMENTS CAN INACTIVATE NONSENSE-MEDIATED mRNA DECAY

For many yeast mRNAs, the destabilizing effects of premature nonsense codons are position-dependent. Generally, nonsense codons located within the first two-thirds to three-quarters of the coding region accelerate an mRNA's decay rate up to 20-fold, whereas nonsense mutations located within the remaining portions of the coding region have little or no effect on mRNA decay (Losson and Lacroute 1979; Pelsy and Lacroute 1984; Peltz et al. 1993a; Hagan et al. 1995; Yun and Sherman 1995; Zhang et al. 1995; Hennigan and Jacobson 1996). One possible explanation for these position effects is that 3′-proximal nonsense mutations may lack functional DSEs. For at least the *PGK1* mRNA, this possibility was excluded by experiments showing that there was no effect on mRNA half-life when DSEs were inserted 3′ to nonsense mutations located within the final third of the coding region (Peltz et al. 1993a). Rather, these experiments suggested that prior to recognition of "late" *PGK1* nonsense codons the ribosome had traversed a sequence element that inactivated its ability to stimulate NMD. Such stabilizer elements (STEs) may thus be present in mRNAs as part of a mechanism that ensures efficient termination at the end of bona fide open reading frames. STE inactivation of NMD is apparently dominant to the effects of DSEs, since STEs preclude degradation of mRNAs with extended 3′UTRs (Peltz et al. 1993a). Unfortunately, comparisons of the STEs in several mRNAs (i.e., the regions separating destabilizing from non-destabilizing nonsense codons) have yet to reveal any significant conservation of sequence or secondary structure.

A different kind of STE has been identified in the 5′ leader region of the *GCN4* mRNA (Ruiz-Echevarria et al. 1998a). This element inactivates the NMD pathway when positioned downstream from a termination codon but still 5′ of a DSE. The *GCN4* STE is homologous to a comparably positioned sequence within the *YAP1* mRNA that also appears to have NMD-inactivating activity (M.J. Ruiz-Echevarria and S.W. Peltz, in prep.). The role of this class of STE has been clarified considerably by recent experiments showing that they bind the poly(U)-binding protein, Pub1p, and fail to function in Pub1p-deficient strains (M.J. Ruiz-Echevarria and S.W. Peltz, in prep.). These observations suggest that Pub1p bound 3′ to a premature termination codon can either antagonize activation of the decay apparatus or mimic the RNP context of a normal 3′UTR (see next section).

ROLES OF THE FACTORS AND SEQUENCES: SURVEILLANCE COMPLEX OR FAUX UTR?

As discussed in detail above, degradation of nonsense-containing mRNAs requires the activity of Upf1p, Upf2p/Nmd2p, and Upf3p. All three of these factors appear to regulate but not catalyze decapping of nonsense-containing transcripts (F. He and A. Jacobson, in prep.). How, then, do we reconcile decapping as well as translation termination functions for these factors with their known interactions and biochemical properties, and with the regulation of the decay pathway by DSEs and STEs? Do the translational functions of the factors dictate their mRNA decay functions, or vice versa, or are the decay and translation events distinct? Although there is, at present, no precise mechanistic understanding of the process,

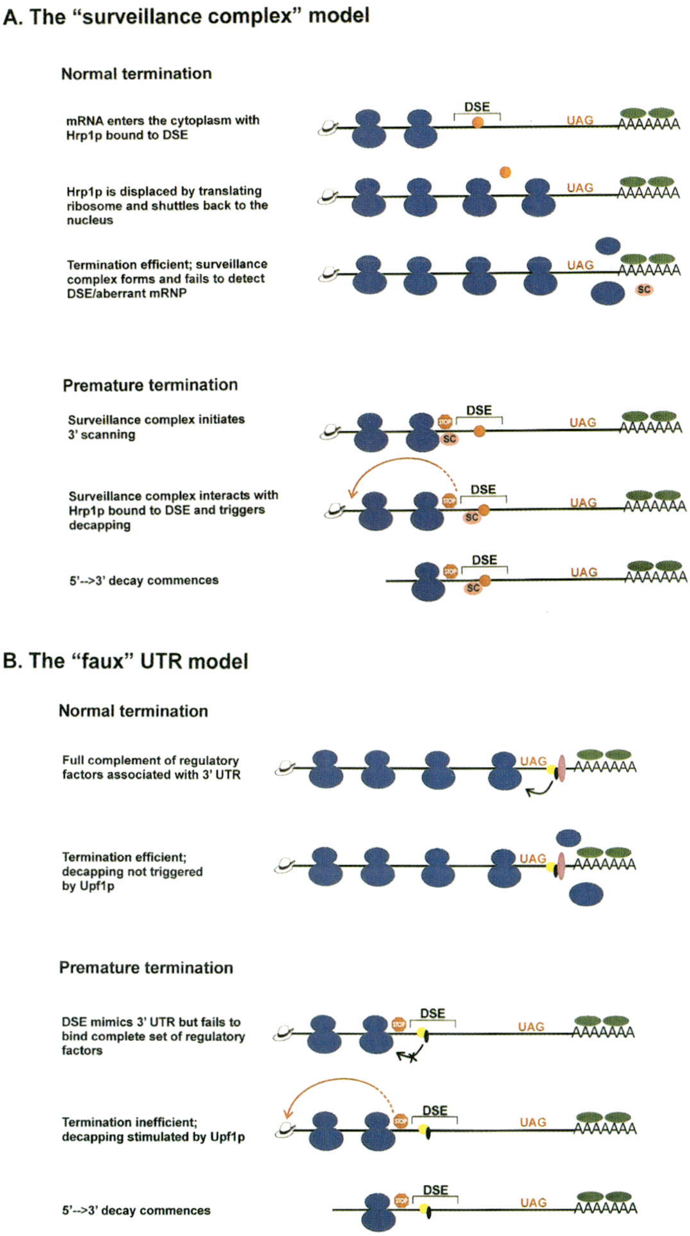

Figure 4 Two models for the mechanism of mRNA destabilization subsequent to the recognition of a premature nonsense codon. (*A*) Surveillance complex model. (*B*) Faux UTR model. (For details, see text; Czaplinski et al. 1999; D. Zuk and A. Jacobson, in prep.).

useful working models have been put forth. The two predominant (and related) models for the mechanism of NMD recognize that no direct link between Dcp1p and the *UPF/NMD* factors has been found and thus postulate that decay is a consequence of events occurring during or after translation termination. In one model, decay occurs in response to recognition of the DSE by a scanning complex of *UPF/NMD* factors (Czaplinski et al. 1998, 1999) and in the other, decay is triggered by the failure to terminate adjacent to a properly configured 3′UTR (Fig. 4) (Hilleren and Parker 1999; D. Zuk and A. Jacobson, in prep.).

The major premise of the first model is that an aberrant RNP formed as a consequence of the premature termination event is recognized by a surveillance complex, minimally comprising the *UPF/NMD* proteins (Czaplinski et al. 1998, 1999). In this model, assembly of the surveillance complex is triggered when the translation termination factors eRF1 and eRF3 bind the A-site of a ribosome paused at a termination codon, and interaction between the eRFs and the complex is thought to enhance termination factor activity (Czaplinski et al. 1998, 1999). Subsequent to hydrolysis of the peptidyl-tRNA bond, dissociation of the release factors is postulated to activate the RNA-binding and ATPase activities of Upf1p, allowing it to utilize these activities to monitor the RNP complex 3′ of the termination codon. A termination event is considered aberrant if a DSE-bound protein, 3′ of the termination codon, is recognized by the surveillance complex. The RNP that results from such recognition becomes a substrate for rapid decapping, and degradation of the body of the transcript ensues (Czaplinski et al. 1999; Gonzalez et al. 2000).

Two important issues that this model must consider are the nature of the RNA-binding proteins which interact with DSEs and the mechanism which prevents decay of most wild-type transcripts. Recent results suggest that the activities of Hrp1p, an RNA-binding protein which can interact with both Upf1p and the DSE, could address these issues (Gonzalez et al. 2000). Hrp1p shuttles between the nucleus and the cytoplasm (Kessler et al. 1997), and mutations in the RNA-binding domain of Hrp1p lead both to specific stabilization of nonsense-containing mRNAs and to loss of DSE and Upf1p interaction (Gonzalez et al. 2000). This suggests that Hrp1p may bind to newly synthesized transcripts but be displaced upon cytoplasmic remodeling of the mRNP that ensues when translation extends to the normal end of an open reading frame. Failure to translate an mRNA completely could thus lead to retention of bound Hrp1p, allowing an interaction between Hrp1p and the surveillance complex and the initiation of rapid mRNA degradation (Fig. 4) (Czaplinski et al. 1999; Gonzalez et al. 2000).

The second model considers the DSE to be a defective or "faux" 3′UTR and suggests that proper termination of translation only occurs in the context of interactions between a terminating ribosome and a specific RNP domain or set of factors localized 3′ to the stop codon (Bonetti et al. 1995; Hilleren and Parker 1999; D. Zuk et al., in prep.). Furthermore, it is suggested that the failure to terminate properly ultimately triggers decay, either because proper termination restructures the mRNP to a stable form or because an aberrant mode of ribosome release or recycling triggers decapping. In this general model, Upf1p is thought to play a direct role in termination, possibly using its ATPase and helicase activities to promote ribosome release or some form of conformational change among the components at the termination site (Hilleren and Parker 1999; D. Zuk and A. Jacobson, in prep.). If interactions with factors bound 3′ to the termination site dictate Upf1p's activity, then the efficiency of such release or conformational change may be drastically different with normal versus premature stop codons.

The second model raises the question of why the 3′UTR created by a premature termination codon should differ from that in a wild-type mRNA. Unlike model 1, which suggests that a nonsense-containing mRNA fails to evict a DSE-bound factor, model 2 suggests that the DSE is a "problem" because it lacks a factor (or factors) normally present on a legitimate 3′UTR. Two hypotheses have been put forth to explain the latter inadequacy: Either translation to the normal end of a coding region remodels an mRNP (Czaplinski et al. 1998, 1999; Hilleren and Parker 1999) or proximity to the poly(A) tail has a qualitative and/or quantitative influence on the nature of proteins bound to the UTR (D. Zuk and A. Jacobson, in prep.).

Either model also provides ample opportunities to explain the role of the STEs and other factors shown to play a role in NMD. Coding region STEs, like that of the *PGK1* mRNA (Peltz et al. 1993a), may alter the availability or activity of a ribosome-associated factor critical for efficient termination. Translocation through the STE could thus inactivate a component of the surveillance complex or lead to the ribosome association of a component that subsequently interacts with a factor normally bound to the 3′ UTR. Likewise, the STE characteristic of the *GCN4* leader may mimic a normal 3′UTR, leading to proper termination, or somehow inhibit 3′ scanning of the surveillance complex. Other *trans*-acting factors previously shown to be involved in NMD may provide a binding site on the ribosome for Upf1p (Nmd3p; Belk et al. 1999), participate in subunit dissociation at termination (Prt1p; Welch and Jacobson 1999), modulate translational fidelity (Mof2p; Cui et al. 1999a,b), or contribute an additional RNA helicase activity required for scanning of the surveillance complex, dissociation of the termination complex, or ribosome maturation (Dbp2p; A. Bond and A. Jacobson, in prep.). Finally, both models can be readily applied to all classes of substrates, except those with extended 3′UTRs (see Fig. 2). Application of model 1 to these transcripts requires that an extended 3′UTR harbor a DSE. For model 2 to accommodate these substrates, extension of the UTR must alter the cohort of factors that are proximal to the termination codon.

CONCLUDING REMARKS

The phenomenon of nonsense-mediated mRNA, originally recognized more than 20 years ago (Losson and Lacroute 1979), has only recently been studied intensively. As with so many biological phenomena, the details are just beginning to emerge, but those that have emerged were unanticipated. What was once thought to represent a trivial deprotection of mRNA by release of ribosomes has now come to illustrate the complex cross-talk between the pathways of mRNA decay and translation (Jacobson and Peltz 1996). Rather than being solely a mechanism for ridding the cell of aberrant transcripts, NMD is also now seen as a potential regulatory circuit for normal mRNAs (Dahlseid et al. 1998; Lew et al. 1998; Welch and Jacobson 1999). Factors once thought to be exclusively devoted to regulating mRNA decay have been shown to have key roles in regulating termination efficiency and translational fidelity (Leeds et al. 1992; Cui et al. 1996; Weng et al 1996a,b; Ruiz-Echevarria et al. 1998b; K. Czaplinski and S.W. Peltz; Maderazo et al. 2000). A related pathway in mammalian cells may provide the cell with broad flexibility in gene expression and may facilitate the progress of evolution (Hilleren and Parker 1999; see Chapter 30). Finally, and perhaps most importantly, a detailed understanding of the mechanisms of NMD may provide medicine with an unprecedented opportunity to treat hundreds of inherited diseases whose causative alleles are attributable to nonsense mutations (Howard et al. 1996; Bedwell et al. 1997; Barton-Davis et al. 1999).

ACKNOWLEDGMENTS

This work was supported by grants from the National Institutes of Health to A.J. (GM-27757) and S.W.P. (GM-48631) and by an American Heart Association Established Investigator Award to S.W.P. We are indebted to members of the Jacobson and Peltz labs, present and past, for the experiments and ideas that formed the underpinnings of much of this review. We also thank David Mangus and Dorit Zuk for helpful editorial suggestions.

REFERENCES

Altamura N., Groudinsky O., Dujardin G., and Slonimski P.P. 1992. *NAM7* nuclear gene encodes a novel member of a family of helicases with a Zn-ligand motif and is involved in mitochondrial functions in *Saccharomyces cerevisiae*. *J. Mol. Biol.* **224:** 575–587.

Atkin A.L., Altamura N., Leeds P., and Culbertson M.R. 1995. The majority of yeast *UPF1* co-localizes with polyribosomes in the cytoplasm. *Mol. Biol. Cell* **6:** 611–625.

Atkin A.L., Schenkman L.R., Eastham M., Dahlseid J.N., Lelivelt M.J., and Culbertson M.R. 1997. Relationship between yeast polyribosomes and the Upf proteins required for nonsense mRNA decay. *J. Biol. Chem.* **272:** 22163–22172.

Barton-Davis E.R., Cordier L., Shoturma D.I., Leland S.E., and Sweeney H.L. 1999. Amino-glycoside antibiotics restore dystrophin function to skeletal muscles of mdx mice. *J. Clin. Invest.* **104:** 375–381.

Bedwell D.M., Kaenjak A., Benos D.J., Bebok Z., Bubien J.K., Hong J., Tousson A., Clancy J.P., and Sorscher E.J. 1997. Suppression of a CFTR premature stop mutation in a bronchial epithelial cell line. *Nature Med.* **3:** 1280–1284.

Beelman C.A., Stevens A., Caponigro G., LaGrandeur T.E., Hatfield L., and Parker R. 1996. An essential component of the decapping enzyme required for normal rates of mRNA turnover. *Nature* **382:** 642–646.

Belk J.P., He F., and Jacobson A. 1999. Overexpression of truncated Nmd3p inhibits protein synthesis in yeast. *RNA* **5:** 1055–1070.

Boeck R., Lapeyre B., Brown C., and Sachs A.B. 1998. Capped mRNA degradation intermediates accumulate in the yeast *spb8-2* mutant. *Mol. Cell. Biol.* **18:** 5062–5072.

Bonetti B., Fu L., Moon J., and Bedwell D.M. 1995. The efficiency of translation termination is determined by a synergistic interplay between upstream and downstream sequences in *Saccharomyces cerevisiae*. *J. Mol. Biol.* **251:** 334–345.

Caponigro G. and Parker R. 1996. Mechanisms and control of mRNA turnover in *Saccharomyces cerevisiae*. *Microbiol. Rev.* **60:** 233–249.

Chin K. and Pyle A.M. 1995. Branch-point attack in group II introns is a highly reversible transesterification, providing a potential proofreading mechanism for 5′-splice site selection. *RNA* **1:** 391–406.

Coller J.M., Gray N.K., and Wickens M.P. 1998. mRNA stabilization by poly(A)-binding protein is independent of poly(A) and requires translation. *Genes Dev.* **12:** 3226–3235.

Cui Y., Dinman J.D., and Peltz S.W. 1996. *Mof4-1* is an allele of the *UPF1/IFS2* gene which affects both mRNA turnover and -1 ribosomal frameshifting efficiency. *EMBO J.* **15:** 5726–5736.

Cui Y., Dinman J.D., Kinzy T.G., and Peltz S.W. 1999a. The Mof2/Sui1 protein is a general monitor of translational accuracy. *Mol. Cell. Biol.* **18:** 1506–1516.

Cui Y., Hagan K.W., Zhang S., and Peltz S.W. 1995. Identification and characterization of genes that are required for the accelerated degradation of mRNAs containing a premature translational termination codon. *Genes Dev.* **9:** 423–436.

Cui Y., Gonzalez C.I., Kinzy T.G., Dinman J.D., and Peltz S.W. 1999b. Mutations in the *MOF2/SUI1* gene affect both translation and nonsense-mediated mRNA decay. *RNA* **5:** 794–804.

Culbertson M.R. 1999. RNA surveillance. Unforseen consequences for gene expression, inherited genetic disorders and cancer. *Trends Genet.* **15:** 74–80.

Culbertson M.R., Underbrink K.M., and Fink G.R. 1980. Frameshift suppression in *Saccharomyces cerevisiae*. II. Genetic properties of group II suppressors. *Genetics* **95:** 833–853.

Czaplinski K., Ruiz-Echevarria M.J., Paushkin S.V., Han X., Weng Y., Perlick H.A., Dietz H.C., Ter-Avanesyan M.D., and Peltz S.W. 1998. The surveillance complex interacts with the translation release factors to enhance termination and degrade aberrant mRNAs. *Genes Dev.* **12:** 1665–1677.

Czaplinski K., Weng Y., Hagan K.W., and Peltz S.W. 1995. Purification and characterization of the Upf1p: A factor involved in mRNA turnover. *RNA* **1:** 610–623.

Czaplinski K., Ruiz-Echevarria M.J., Gonzalez C.I., and Peltz S.W. 1999. Should we kill the messenger? The role of the surveillance complex in translation termination and mRNA turnover. *Bioessays* **21:** 685–696.

Dahlseid J.N., Puziss J., Shirley R.L., Atkin A.L., Hieter P., and Culbertson M. 1998. Accumulation of mRNA coding for the ctf13p kinetochore subunit of *Saccharomyces cerevisiae* depends on the same factors that promote rapid decay of nonsense mRNAs. *Genetics* **150:** 1019 1035.

Decker C.J. and Parker R. 1993. A turnover pathway for both stable and unstable mRNAs in yeast: Evidence for a requirement for deadenylation. *Genes Dev.* **7:** 1632–1643.

Dinman J.D. and Wickner R.B. 1994. Translational maintenance of frame: Mutants of *Saccharomyces cerevisiae* with altered -1 ribosomal frameshifting efficiencies. *Genetics* **136:** 75–86.

Freist W., Sternbach H., and Cramer F. 1996. Phenylalanyl-tRNA synthetase from yeast and its discrimination of 19 amino acids in aminoacylation of tRNA(Phe)-C-C-A and tRNA(Phe)-C-C-A(3′NH2). *Eur. J. Biochem.* **240:** 526–531.

Gonzalez C.I., Ruiz-Echevarria M.J., Vasudevan S., Henry M.F., and Peltz S.W. 2000. The yeast hnRNP-like protein Hrp1/Nab4 marks a transcript for nonsense-mediated mRNA decay. *Mol. Cell* **5:** 489–499.

Gozalbo D. and Hohmann S. 1990. Nonsense suppressors partially revert the decrease of the mRNA levels of a nonsense mutant allele in yeast. *Curr. Genet.* **17:** 77–79.

Hagan K.W., Ruiz-Echevarria M.J., Quan Y., and Peltz S.W. 1995. Characterization of *cis*-acting sequences and decay intermediates involved in nonsense-mediated mRNA turnover. *Mol. Cell. Biol.* **15:** 809–823.

Hampsey M.J., Na G., Pinto I., Ware D.E., and Berroteran R.W. 1991. Extragenic suppressors of a translation initiation defect in the *CYC1* gene of *Saccharomyces cerevisiae. Biochimie* **73:** 1445–1455.

Hatfield L., Beelman C.A., Stevens A., and Parker R. 1996. Mutations in *trans*-acting factors affecting mRNA decapping in *Saccharomyces cerevisiae. Mol. Cell. Biol.* **16:** 5830–5838.

He F. and Jacobson A. 1995. Identification of a novel component of the nonsense-mediated mRNA decay pathway using an interacting protein screen. *Genes Dev.* **9:** 437–454.

He F., Brown A.H., and Jacobson A. 1996. Interaction between Nmd2p and Upf1p is required for activity but not for dominant-negative inhibition of the nonsense-mediated mRNA decay pathway in yeast. *RNA* **2:** 153–170.

———. 1997. Upf1p, Nmd2p, and Upf3p are interacting components of the yeast nonsense-mediated mRNA decay pathway. *Mol. Cell. Biol.* **17:** 1580–1594.

He F., Peltz S.W., Donahue J.L., Rosbash M., and Jacobson A. 1993. Stabilization and ribosome association of unspliced pre-mRNAs in a yeast *upf1*- mutant *Proc. Natl. Acad. Sci.* **90:** 7034–7038.

Hennigan A.N. and Jacobson A. 1996. Functional mapping of the translation-dependent instability element of yeast *MATα1* mRNA. *Mol. Cell. Biol.* **16:** 3833–3843.

———. 1997. A genetic approach to mapping coding region determinants if mRNA instability in yeast. In *mRNA formation and function* (ed. J.D. Richter). pp. 149–161.

Academic Press, San Diego, California.

Herrick D., Parker R., and Jacobson A. 1990. Identification and comparison of stable and unstable mRNAs in the yeast *Saccharomyces cerevisiae. Mol. Cell. Biol.* **10:** 2269–2284.

Hilleren P. and Parker R. 1999. Mechanisms of mRNA surveillance in eukaryotes. *Annu. Rev. Genet.* **33:** 229–260.

Hinnebusch A.G. 1988. Mechanisms of gene regulation in the general control of amino acid biosynthesis in *Saccharomyces cerevisiae. Microbiol. Rev.* **52:** 248–273.

Howard M., Frizzell R.A., and Bedwell D.M. 1996. Aminoglycoside antibiotics restore CFTR function by overcoming premature stop mutations. *Nat. Med.* **2:** 467–469.

Hsu C.L. and Stevens A. 1993. Yeast cells lacking 5′→3′ exoribonuclease 1 contain mRNA species that are poly(A) deficient and partially lack the 5′ cap structure. *Mol. Cell. Biol.* **13:** 4826–4835.

Ibba M. and Soll D. 1999. Quality control mechanisms during translation. *Science* **286:** 1893–1897.

Jacobson A. 1996. Poly(A) metabolism and translation: The closed-loop model. In *Translational control* (ed. J.W.B. Hershey et al.), pp. 451–480. Cold Spring Harbor Laboratory Press, Cold Spring Harbor, New York.

Jacobson A. and Peltz S.W. 1996. Interrelationships of the pathways of mRNA decay and translation in eukaryotic cells. *Annu. Rev. Biochem.* **65:** 693–739.

Jeon C. and Agarwal K. 1996. Fidelity of RNA polymerase II transcription controlled by elongation factor TFIIS. *Proc. Natl. Acad. Sci.* **93:** 13677–13682.

Kessler M.M., Henry M.F., Shen E., Zhao J., Gross S., Silver P.A., and Moore C.L. 1997. Hrp1p, a sequence-specific RNA-binding protein that shuttles between the nucleus and the cytoplasm, is required for mRNA 3′-end formation in yeast. *Genes & Dev.* **11:** 2545–2556.

Koonin E.V. 1992. A new group of putative RNA helicases. *Trends Biochem. Sci.* **17:** 495–497.

Kurland C.G. 1992.Translational accuracy and the fitness of bacteria. *Ann. Rev. Genet.* **26:** 29–50.

Lee B.S. and Culbertson M.R. 1995. Identification of an additional gene required for eukaryotic nonsense mRNA turnover. *Proc. Natl. Acad. Sci.* **92:** 10354–10358.

Lee S.I., Umen J.G., and Varmus H.E. 1995. A genetic screen identifies cellular factors involved in retroviral -1 frameshifting. *Proc. Natl. Acad. Sci.* **92:** 6587–6591.

Leeds P., Peltz S.W., Jacobson A., and Culbertson M.R. 1991. The product of the yeast *UPF1* gene is required for rapid turnover of mRNAs containing a premature translational termination codon. *Genes Dev.* **5:** 2303–2314.

Leeds P., Wood J.M., Lee B.S., and Culbertson M.R. 1992. Gene products that promote mRNA turnover in *Saccharomyces cerevisiae. Mol. Cell. Biol.* **12:** 2165–2177.

LeGrandeur T.E. and Parker R. 1998. Isolation and characterization of Dcp1p, the yeast mRNA decapping enzyme. *EMBO J.* **17:** 1487–1496.

Lelivelt M.J. and Culberson M.R. 1999. Yeast Upf proteins required for RNA surveillance affect global expression of the yeast transcriptome. *Mol. Cell. Biol.* **10:** 6710–6719.

Lew J.E., Enomoto S., and Berman J. 1998. Telomere length regulation and telomeric chromatin require the nonsense-mediated mRNA decay pathway. *Mol. Cell. Biol.* **18:** 6121–6130.

Losson R. and Lacroute F. 1979. Interference of nonsense mutations with eukaryotic messenger RNA stability. *Proc. Natl. Acad. Sci.* **76:** 5134–5137.

Maderazo A.B., He F., Mangus D.A., and Jacobson A. 2000. Vpf1p control of nonsense mRNA translation is regulated by Nmd2p and Upf3p. *Mol. Cell. Biol.* **20:** 4591–4603.

Mangus D.A. and Jacobson A. 1999. Linking turnover and translation: Assessing the polyribosomal association of mRNA decay factors and degradative intermediates. *Methods* **17:** 28–37.

Maquat L.E. 1995. When cells stop making sense: Effects of nonsense codons on RNA metabolism in vertebrate cells. *RNA* **1:** 453–465.

Muhlrad D. and Parker R. 1994. Premature translational termination triggers mRNA decapping. *Nature* **340:** 578–581.

———. 1999. Aberrant mRNAs with extended 3′ UTRs are substrates for rapid degradation by mRNA surveillance. *RNA* **5:** 1299–1307.

Muhlrad D., Decker C.J., and Parker R. 1994. Deadenylation of the unstable mRNA encoded by the yeast *MFA2* gene leads to decapping followed by 5′→3′ digestion of the transcript. *Genes Dev.* **8:** 855–866.

———. 1995. Turnover mechanisms of the stable yeast *PGK1* mRNA. *Mol. Cell. Biol.* **15:** 2145–2156.

Pelsy F. and Lacroute F. 1984. Effect of ochre nonsense mutations on yeast URA1 mRNA stability. *Curr. Genet.* **8:** 277–282.

Peltz S.W., Brown A.H., and Jacobson A. 1993a. mRNA destabilization triggered by premature translational termination depends on at least three *cis*-acting sequence elements and one *trans*-acting factor. *Genes Dev.* **7:** 1737–1754.

Peltz S.W., Donahue J.L., and Jacobson A. 1992. A mutation in the tRNA nucleotidyltransferase gene promotes stabilization of mRNAs in *Saccharomyces cerevisiae. Mol. Cell. Biol.* **12:** 5778–5784.

Peltz S.W., He F., Welch E., and Jacobson A. 1994. Nonsense-mediated mRNA decay in yeast. *Prog. Nucleic Acid Res. Mol. Biol.* **47:** 271–298.

Peltz S.W., Welch E.M., Zhang S., Hagan K., Brown A.H., and Jacobson A. 1997. Polysome-associated mRNAs are substrates for the nonsense-mediated mRNA decay pathway in *Saccharomyces cerevisiae. RNA* **3:** 234–244.

Peltz S.W., Trotta C., He F., Brown A.H., Donahue J.L., Welch E., and Jacobson A. 1993b. Identification of the *cis*-acting sequences and *trans*-acting factors involved in nonsense-mediated mRNA decay. In *Protein synthesis and targeting in yeast* (ed. M. Tuite et al.), pp1–10.

Pinto I., Na J.G., Sherman F., and Hampsey M. 1992. *cis* and *trans*-acting suppressors of a translation initiation defect at the *cyc1* locus of *Saccharomyces cerevisiae. Genetics* **132:** 97–112.

Pulak R. and Anderson P. 1993. mRNA surveillance by the *Caenorhabditis elegans smg* genes. *Genes Dev.* **7:** 1885–1897.

Ruiz-Echevarria M.J. and Peltz S.W. 1996. Utilizing the *GCN4* leader region to investigate the role of the sequence determinants in nonsense-mediated mRNA decay. *EMBO J.* **15:** 2810–2819.

Ruiz-Echevarria M.J., Czaplinski K., and Peltz S.W. 1996. Making sense of nonsense in yeast. *Trends Biochem. Sci.* **21:** 433–438.

Ruiz-Echevarria M.J., Gonzalez C.I., and Peltz S.W. 1998a. Identifying the right stop: Determining how the surveillance complex recognizes and degrades an aberrant mRNA. *EMBO J.* **17:** 575–589.

Ruiz-Echevarria M.J., Yasenchak J.M., Han X., Dinman J.D., and Peltz S.W. 1998b. The Upf3 protein is a component of the surveillance complex that monitors both translation and mRNA turnover and affects viral propagation. *Proc. Natl. Acad. Sci.*. **95:** 8721–8726.

Shirley R.L., Lelivelt M.J., Schenkman L.R., Dahlseid J.N., and Culbertson M.R. 1998. A factor required for nonsense-mediated mRNA decay in yeast is exported from the nucleus to the cytoplasm by a nuclear export signal sequence. *J. Cell Sci.* **111:** 3129–3143.

Stevens A. 1988. mRNA-decapping enzyme from *Saccharomyces cerevisiae:* Purification and unique specificity for long RNA chains. *Mol. Cell. Biol.* **8:** 2005–2010.

Tharun S. and Parker R. 1999. Analysis of mutations in the yeast mRNA decapping enzyme. *Genetics* **151:** 1273–1285.

Vilela C., Linz B., Rodrigues-Pousada C., and McCarthy J.E. 1998. The yeast transcription factor genes *YAP1* and *YAP2* are subject to differential control at the levels of both translation and mRNA stability. *Nucleic Acids Res.* **26:** 1150–1159.

Vilela C., Ramirez C.V., Linz B., Rodrigues-Pousada C., and McCarthy J.E. 1999. Post-termination ribosome interactions with the 5′UTR modulate yeast mRNA stability. *EMBO J.* **18:** 3139–3152.

Welch E.M. and Jacobson A. 1999. An internal open reading frame triggers nonsense-mediated decay of the yeast *SPT10* mRNA. *EMBO J.* **18:** 6134–6145.

Weng Y., Czaplinski K., and Peltz S.W. 1996a. Identification and characterization of mutations in the *UPF1* gene that affect nonsense suppression and the formation of the Upf protein complex but not mRNA turnover. *Mol. Cell. Biol.* **16:** 5491–5506.

———. 1996b. Genetic and biochemical characterization of mutations in the ATPase and helicase regions of the Upfl protein. *Mol. Cell. Biol.* **16:** 5477–5590.

———. 1998. ATP is a cofactor of the Upf1p protein that modulates its translation termination and RNA binding activities. *RNA* **4:** 205–214.

Weng Y., Ruiz-Echevarria M.J., Zhang S., Cui Y., Czaplinski K., Dinman J.D., and Peltz S.W. 1997. Characterization of the nonsense-mediated mRNA decay pathway and its effect on modulating translation termination and programmed frameshifting. In *mRNA metabolism and post-transcriptional gene regulation* (eds. J.B. Harford and D.R. Morris), pp. 241–263. Wiley-Liss, New York.

Wickner S., Maurizi M.R., and Gottesman S. 1999. Posttranslational quality control: Folding, refolding, and degrading proteins. *Science* **286:** 1888–1893.

Yarus M. 1992. Proofreading, NTPases and translation: Successful increase in specificity. *Trends Biochem. Sci.* **17:** 171–174.

Yoshizawa S., Fourmy D., and Puglisi J.D. 1999. Recognition of the codon-anticodon helix by ribosomal RNA. *Science* **285:** 1722–1725.

Yun D.F. and Sherman F. 1995. Initiation of translation can occur only in a restricted region of the *CYC1* mRNA of *Saccharomyces cerevisiae. Mol. Cell. Biol.* **15:** 1021–1033.

Zaret K.S. and Sherman F. 1982. DNA sequence required for efficient transcription termination in yeast. *Cell* **28:** 563–573.

———. 1984. Mutationally altered 3′ ends of yeast *CYC1* mRNA affect transcript stability and translational efficiency. *J. Mol. Biol.* **177:** 107–135.

Zhang S., Ruiz-Echevarria M.J., Quan Y., and Peltz S.W. 1995. Identification and characterization of a sequence motif involved in nonsense mediated mRNA decay. *Mol. Cell. Biol.* **15:** 2231–2244.

Zhang S., Welch E.M., Hogan K., Brown A.H., Peltz S.W., and Jacobson A. 1997. Polysome-associated mRNAs are substrates for the nonsense-mediated mRNA decay pathway in *Saccharomyces cerevisiae. RNA* **3:** 234–244.

Zuk D., Belk J.P., and Jacobson A. 1999. Temperature-sensitive mutations in the yeast *MRT4, GRC5, SLA2,* and *THS1* genes result in defects in mRNA turnover. *Genetics* **153:** 35–47.

30

Nonsense-mediated RNA Decay in Mammalian Cells: A Splicing-dependent Means to Down-regulate the Levels of mRNAs That Prematurely Terminate Translation

Lynne E. Maquat
Department of Biochemistry and Biophysics
School of Medicine and Dentistry
University of Rochester
Rochester, New York 14642

Within all eukaryotic organisms that have been examined, including yeast, worms, flies, and mammals, there is a surveillance mechanism referred to as nonsense-mediated RNA decay (NMD). NMD provides a means to down-regulate aberrant gene expression by degrading RNAs harboring premature termination codons (PTCs). NMD also contributes, both directly and indirectly, to a proper balance of normal gene expression by degrading specific, naturally occurring transcripts. In this chapter, I describe the effects of PTCs on RNA metabolism in mammalian cells with the aim of presenting current mechanistic concepts and progress toward using PTC suppression as a treatment for human genetic diseases. Related studies of NMD in yeast are discussed in Chapter 29.

REASONS FOR NMD

PTCs (UGA, UAA, and UAG) can arise as a consequence of routine cellular processes that are occasionally programmed but are usually caused by errors. Errors permit molecular diversity, cellular adaptability, and improved organismal viability, but they do so at the cost of leading to nonproductive or deleterious gene expression most of the time. Processes by which the generation of PTCs is programmed include the posttran-

Translational Control of Gene Expression
© 2000 Cold Spring Harbor Laboratory Press 0-87969-568-4/00 $5 + .00

scriptional cytidine deamination of nuclear apolipoprotein (apo) B transcripts in enterocytes of the mammalian small intestine that results in conversion of codon 2153 within exon 26 from CAA to UAA (for review, see Chang et al. 1998). Processes by which PTCs arise as the consequence of errors include incomplete or inaccurate pre-mRNA splicing, which has the potential to generate multiple proteins from a single gene but more often generates either an intron-derived PTC or a frameshift that results in an exon-derived PTC (for review, see Maquat 1996). Other error-prone processes are the somatic rearrangements and hypermutations characterizing the immunoglobulin (Ig) and T-cell receptor (TCR) genes, which can generate a diverse set of receptors that recognize different proteins but more frequently result in gene inactivation by creating frameshift or nonsense mutations (for review, see Li and Wilkinson 1998). Additionally, PTCs can arise as the consequence of disease-causing translocations, deletions, insertions, or point mutations within germ-line or somatic DNA (for review, see Maquat 1996). Disease-causing mutations include TpG dinucleotides formed by deamination of methylated CpG dinucleotides (Hendrich et al. 1999 and references therein), 25% of which comprise a TGA sequence, as well as age-dependent deletions in sporadic Alzheimer's disease and other neurodegenerative pathologies (van Leeuwen et al. 1998). Notably, not all PTCs are recognized as nonsense, as exemplified by those UGA codons that encode the unusual amino acid selenocysteine within selenoprotein mRNAs such as glutathione peroxidase (GPx) 1 mRNA (for review, see Chapter 26). Nevertheless, the frequency with which the GPx1 UGA codon is routinely recognized as nonsense rather than encoding selenocysteine makes GPx1 mRNA a natural substrate for NMD (Moriarty et al. 1998).

A mechanism to degrade RNAs that harbor PTCs presumably exists as a means to preclude the synthesis of most, but not all (see below), encoded truncated proteins, which can be deleterious to cells by manifesting new or dominant-negative functions. For example, NMD reduces the steady-state level of PTC-containing mRNA and, as a consequence, the potentially deleterious proteins produced by the nonproductively rearranged Ig or TCR genes that characterize more than half of viable B or T lymphoblasts, respectively (for review, see Li and Wilkinson 1998). The benefits of NMD are illustrated by the disease-causing effects of PTCs within the final exon of the β-globin gene, which fail to elicit NMD and also happen to generate truncated proteins at near-normal levels. As opposed to PTCs within either of the other β-globin gene exons, which do elicit NMD, these PTCs result in ineffective erythropoiesis and a dominantly inherited form of thalassemia that is usually recessive (Kazazian et al. 1992; Hall and Thein 1994). NMD probably influences the metabolism of numerous natural substrates in mammalian cells in addition to GPx1 and other selenoprotein mRNAs. Screens of high-density oligonucleotide arrays have revealed that expression of ≥ 225 of the ~6000 genes of *Saccharomyces cerevisiae* is affected by inactivation of NMD (Lelivelt and Culbertson 1999). Furthermore, NMD has a demonstrated role in regulating telomeric length (Lew et al. 1998) and the levels of a kinetochore subunit (Dahlseid et al. 1998) in *S. cerevisiae* as well as certain alternatively spliced RNAs in *Caenorhabditis elegans* (Morrison et al. 1997). Although NMD is nonessential in lower eukaryotes, it is essential in mammals: Mouse embryos inactive in NMD are resorbed prior to 9.5 days postcoitus (S. Medghalchi and H. Dietz, pers. comm.), indicating that the failure to eliminate error-generated PTCs, and/or to properly regulate natural substrates of NMD, results in embryonic lethality.

A COMPARISON OF NUCLEUS-ASSOCIATED AND CYTOPLASMIC NMD

The most readily envisioned mechanism by which PTCs elicit NMD involves the degradation of polysome-associated mRNA during mRNA translation, as has been reported for *S. cerevisiae* (Zhang et al. 1997). Consistent with this view, results of nuclear run-on analyses indicate that mammalian NMD is posttranscriptional (for review, see Maquat 1995; Li and Wilkinson 1998; Hentze and Kulozik 1999). Whereas some mRNAs are subject to NMD in the cytoplasm, others are subject to NMD while nucleus-associated, i.e., before they have been completely transported from nuclei to the cytoplasm. Issues that remain to be resolved include what determines whether NMD is nucleus-associated or cytoplasmic and on which side of the nuclear envelope nucleus-associated NMD takes place. With regard to nucleus-associated NMD, it will be important to validate or disprove claims that recognition of particular PTCs takes place within particular pre-mRNAs so as to influence the efficiency and/or choice of splice-site utilization (for review, see Maquat 1995, 1996; Li and Wilkinson 1998; Frischmeyer and Dietz 1999; Hentze and Kulozik 1999).

mRNAs that are subject to cytoplasmic NMD, i.e., NMD after export to the cytoplasm has been completed, include (1) β^0-thalassemic β-globin mRNA in patient bone-marrow aspirates, the erythroid tissues of mice made transgenic for β^0-thalassemic alleles, or mouse erythroleukemia cells stably transfected with β^0-thalassemic alleles (Maquat et al. 1981; Lim et al. 1989, 1992; Lim and Maquat 1992; S. Sekularac and L.E. Maquat, unpubl.); (2) GPx1 mRNA regardless of the tissue source (Moriarty et al. 1998); and (3) mRNA for the α-subunit of β-hexosaminidase (HEXA) in lymphoblasts of patients with Tay-Sachs disease (K.S. Rajavel and E.F. Neufeld, pers. comm.). Consistent with the view that cytoplasmic NMD involves translation, NMD of HEXA mRNA has recently been shown to be inhibited by cycloheximide (K.S. Rajavel and E.F. Neufeld, pers. comm.). Although mRNAs are normally degraded by a pathway that requires deadenylation as an early step (Couttet et al. 1997; for review, see Chapter 28; Jacobson and Peltz 1996), NMD of β^0-thalassemic β-globin mRNAs in mouse erythroid cells generates classes of polyadenylated decay intermediates that are missing ~65–185 nucleotides from the mRNA 5′ end and, surprisingly, seem to have a cap-like structure in place of the missing nucleotides (Lim et al. 1989; Lim and Maquat 1992; S. Sekularac and L.E. Maquat, unpubl.). These data provide the sole evidence that NMD in mammals initiates at or near the mRNA 5′ end independently of appreciable deadenylation, as is the case for NMD in *S. cerevisiae* (Muhlrad and Parker 1994). The findings that full-length PTC-containing β-globin mRNA can be chased into the polyadenylated decay intermediates during a block in transcription with actinomycin D (Lim et al. 1992), and that a PTC reduces the cytoplasmic half-life of GPx1 mRNA produced from a regulatable promoter from 8–10 hours to 2.5 hours (Sun et al. 2000), indicate that cytoplasmic NMD is characteristic of steady-state mRNA.

Despite the readily envisioned mechanism of cytoplasmic NMD, most mammalian mRNAs examined to date are subject to NMD while nucleus-associated and, remarkably, become immune to NMD once released into the cytoplasm (for review, see Maquat 1995, 1996; Li and Wilkinson 1998; Hentze and Kulozik 1999). NMD intermediates have never been detected for any mRNA of this class despite attempts to slow the progression of ribonucleases by the insertion of one of a number of hairpin or poly(G) structures (L. Stephenson and L.E. Maquat, unpubl.), as has been done for *S. cerevisiae* (Muhlrad and Parker 1994). Nevertheless, there is no reason to believe that the nucleolytic activities responsible for nucleus-associated and cytoplasmic NMD differ. A PTC that elicits nucleus-associated NMD acts only in *cis* and, in the case of genes producing more than one transcript, affects only those transcripts harboring the PTC, making the possibility of feedback from the cytoplasm to the nucleus unlikely (for review, see Maquat 1995). PTC recognition during nucleus-associated NMD could be mediated by cytoplasmic ribosomes if the mRNA substrate were bound by ribosomes either during transit across the nuclear pore, as has been shown for the Balbiani ring granule of the insect *Chironomus tentans* (Mehlin et al. 1992), or after transit but prior to release into the cytoplasm (for review, see Maquat

1995, 1996; Hentze and Koluzik 1999). Alternatively, PTC recognition could be mediated within the nucleoplasm by a nuclear ribosome, ribosome-like complex, or novel molecule(s), in which case current perceptions of nuclear functions would have to be expanded (for review, see Li and Wilkinson 1998).

In support of a role for ribosomes, PTC recognition in nucleus-associated NMD is indistinguishable from cytoplasmic translation with respect to its sensitivity to translational effectors such as (1) a suppressor tRNA (Belgrader et al. 1993; Li et al. 1997); (2) ribosome-binding drugs, including anisomycin, cycloheximide, emetine, puromycin, and pactamycin (Qian et al. 1993; Menon and Neufeld 1994; Carter et al. 1995); (3) a secondary structure or iron-responsive element within the 5′-untranslated region of mRNAs that blocks translation initiation (Belgrader et al. 1993; Thermann et al. 1998); (4) poliovirus infection, which inactivates cap-dependent translation (Carter et al. 1995); and (5) translation reinitiation downstream of and in frame with a PTC (Zhang and Maquat 1997). By inference, in-depth studies of codon recognition in *Escherichia coli* also support a role for ribosomes in nucleus-associated NMD. For example, the high fidelity of translation in *E. coli* (about one error per 10^4 amino acids) despite the relatively low specificity of codon–anticodon interactions derives from interactions between the anticodon stem-loop region of tRNA and highly conserved regions of 16S rRNA (Yoshizawa et al. 1999 and references therein). Therefore, the 30S ribosomal subunit is critical for the molecular discrimination of correct versus incorrect codon–anticodon pairs and, by extrapolation, for the molecular definition of PTCs in *E. coli*. Moreover, correct codon-anticodon pairs trigger conformational changes that are transmitted to the 50S ribosomal subunit, where interactions between the aminoacylated acceptor stem of tRNA and 23S rRNA promote peptidyl transferase activity (Ban et al. 1999 and references therein). Considering the high degree to which ribosomes have been conserved throughout evolution, PTC recognition and translation termination in mammals probably also require both ribosomal subunits. Notably, translation termination and NMD appear to be inextricably linked (see below). Recent studies of *Xenopus laevis* oocytes indicate that arguments against the possibility of PTC recognition in the nucleoplasm cannot be based on the absence of nucleoplasmic tRNA aminoacylation. Although the bulk of tRNA aminoacylation takes place in the cytoplasm, all tRNAs can be aminoacylated in nuclei of *X. laevis* oocytes (Lund and Dahlberg 1998). Nuclear aminoacylation can be viewed as a mechanism for functional testing of newly made tRNAs, much as nucleus-associated NMD can be viewed as a mechanism for functional testing of newly made mRNAs. Arguments against the possibility of PTC recognition in the nucleoplasm, however, can be based on the absence of translationally active ribosomes. For example, detailed studies of *S. cerevisiae* indicate that nucleoplasmic ribosomes lack proteins characteristic of cytoplasmic ribosomes, at least one of which is required for subunit joining (for review, see Dick et al. 1997).

As would be predicted if translation were involved in nucleus-associated NMD, PTC recognition appears to take place after splicing, as evidenced by the ability of PTCs that span two exons (i.e., are interrupted by an intron) to elicit NMD as effectively as PTCs that reside entirely within a single exon (Carter et al. 1996; Zhang and Maquat 1996). Furthermore, the only established target for nucleus-associated NMD is fully spliced mRNA: Using the human c-*fos* promoter to elicit a burst and subsequent shutoff of triosephosphate isomerase (TPI) gene transcription, (1) PTCs did not affect the level of any intron within TPI pre-mRNA (Cheng and Maquat 1993); (2) the NMD of TPI mRNA was complete by 30 minutes after gene shutoff, a time at which the mRNA still copurified with nuclei (Belgrader et al. 1994); and (3) cytoplasmic TPI was immune to NMD despite being associated with polysomes (Cheng and Maquat 1993; Stephenson and Maquat 1996). These findings indicate that the target for nucleus-associated NMD is exclusively newly synthesized mRNA, and there must be a fundamental difference between the resulting cytoplasmic mRNA and cytoplasmic mRNA that is a substrate for NMD.

PTC-ASSOCIATED ALTERATIONS IN SPLICING

PTCs have also been associated with (1) skipping of a PTC-containing exon and, in some instances, one or more flanking exons; (2) retention of the intron that resides upstream or downstream from a PTC; or (3) use of one or more cryptic splice sites in the vicinity of a PTC (for review, see Maquat 1995; Valentine 1998; Hentze and Kulozik 1999). Most logically, PTC-associated alterations in splicing occur when a PTC changes the sequence of a *cis*-acting effector of splicing. In support of this view, (1) only particular PTCs within particular exons are associated with exon skipping, (2) missense mutations can also be associated with exon skipping, and (3) nonsense and missense mutations associated with exon skipping often reside in or around purine-rich or AC-rich sequences, either of which could potentially represent functional exon splicing enhancers (Valentine 1998; R.J. Kendzior and K.L. Beemon, pers. comm.). In fact, PTC-associated skipping of exon 5 of adenosine deaminase transcripts and exon 27 of dystrophin transcripts has been attributed to the disruption of an exon splice enhancer (Santisteban et al. 1995; Shiga et al. 1997). Additionally, a nonsense mutation as well as a missense mutation within exon 18 of the breast-cancer-associated BRCA1 gene have recently been shown to cause exon 18 skipping by disrupting the interaction of an exon splice enhancer with the serine-arginine-rich (SR) protein SF2/ASF, whereas a different nonsense mutation predicted to improve the exon splice enhancer–SF2/ASF interaction does not result in exon skipping (A. Krainer and H.-X. Liu, pers. comm.). In addition, skipping of exon 51 of fibrillin transcripts is induced not only by particular PTCs within the exon (Dietz et al. 1993; Dietz and Kendzior 1994), but also by a silent mutation within the exon (Liu et al. 1997). All of these findings suggest that exon skipping is attributable to disruption of an exon splice enhancer–splicing factor interaction rather than to disruption of the translational reading frame. This view is given additional support with data indicating that exon splice enhancers are surprisingly prevalent: 15–20% of random 18- or 20-mers promote splicing when substituted for an exon splice enhancer within different exon contexts (Liu et al. 1998; Schaal and Maniatis 1999).

Some investigators have proposed that PTCs can be effectors of either the efficiency or choice of splice-site usage by disrupting the translational reading frame, implying the existence of a nuclear mechanism for PTC recognition (Naeger et al. 1992; Dietz et al. 1993; Dietz and Kendzior 1994; Lozano et al. 1994; Gersappe and Pintel 1999; Gersappe et al. 1999). However, evidence that PTC recognition occurs by a mechanism distinct from conventional translation is provided by the findings that mutation of the translation initiation codon does not always affect PTC-associated intron retention in the case of minute virus of mouse RNA (Gersappe et al. 1999), and PTC-associated exon skipping is insensitive to anisomycin, puromycin, or cycloheximide in the case of fibrillin RNA (R.J. Kendzior and K.L. Beemon, pers. comm.).

DETERMINANTS OF THE CELLULAR SITE AND EXTENT OF NMD

The discovery that mRNAs are subject to NMD either while nucleus-associated or in the cytoplasm raises the intriguing issue of what determines the cellular site of NMD. Evidence indicates that determinants derive from conditions of expression as well as the particular mRNA. With regard to conditions of expression, although β-globin mRNA is subject to nucleus-associated NMD in every non-erythroid cell tested, it is subject to cytoplasmic NMD in erythroid cells (Baserga and Benz 1992;

Lim et al. 1992; Kugler et al. 1995; Thermann et al. 1998; Zhang et al. 1998a). With regard to mRNA-specific determinants, irrespective of the cell type tested, certain mRNAs (such as TPI mRNA) are subject to NMD while nucleus associated, whereas other mRNAs (such as GPx1 mRNA) are subject to NMD in the cytoplasm (Moriarty et al. 1998; Zhang et al. 1998a,b; J. Zhang and L.E. Maquat, unpubl.).

The finding that NMD reduces the abundance of some mRNAs <2-fold and other mRNAs ~100-fold raises the issue of what determines the extent of NMD. One determinant can be cell-type-specific, since the abundance of PTC-containing TCR-β mRNA is reduced 10- to 20-fold in T-lymphoma cells but an order of magnitude less in HeLa cells (Carter et al. 1996). Other determinants are undoubtedly gene-specific, since variations in the extent of NMD are evident for different mRNAs in the same cell type; in one remarkable case, the extent of NMD reflects the nature of the promoter driving RNA synthesis (Enssle et al. 1993; Kugler et al. 1995).

cis-ACTING SEQUENCES

The discovery of NMD raised the provocative issue of how PTCs that elicit NMD differ from normal termination codons, which generally do not elicit NMD. It is now clear that PTCs elicit NMD—either nucleus-associated or cytoplasmic—depending on their position relative to the 3′-most exon–exon junction within mRNA: only those termination codons located more than 50–55 nucleotides upstream of the 3′-most exon-exon junction elicit NMD (Cheng et al. 1994; Thermann et al. 1998; Zhang et al. 1998a,b; Sun et al. 2000). This rule for termination codon position is consistent with the observation that most normal termination codons reside within the final exon and, therefore, do not reside upstream of an exon-exon junction (Hawkins 1988). Remarkably, a survey of genes having one or two 3′-untranslated exons, i.e., normal termination codons followed by one or two exon–exon junctions, revealed that the normal termination codon in 99 out of 101 of these genes resides less than 50 nucleotides upstream of the 3′-most exon–exon junction (Nagy and Maquat 1998). This finding corroborates the rule, since normal termination codons are not expected to elicit NMD. Transcripts from the human HLA 6.09 gene and chicken Gnot gene, the two genes with a normal termination codon that does not follow the rule, either break the rule or are natural substrates for NMD. TCR-β transcripts may also be an exception to the rule, since PTCs as few as 8–10 nucleotides upstream of the 3′-most exon–exon junction elicit NMD (Carter et al. 1996). Even so, the junction must be generated by a spliceable intron in order to function in NMD, since deletion of all introns downstream from a PTC essentially eliminates NMD, and insertion of one or more introns at least 50 nucleotides downstream from a normal termination elicits NMD (Carter et al. 1996; Thermann et al. 1998).

Conceivably, the process of splicing alters mRNP structure so as to deposit one or more proteins required for NMD. This implies that nuclear splicing and cytoplasmic translation are mechanistically linked in the case of cytoplasmic NMD and, possibly, nucleus-associated NMD, depending on the cellular site of PTC recognition (Fig. 1). Notably, an impact of nuclear pre-mRNA metabolism on cytoplasmic mRNA translation has also been reported for maternal mRNA of *X. laevis* ooctyes (Matsumoto et al. 1998 and references therein). Consistent with the idea that splicing is important to NMD, transcripts that derive from intron-less counterparts of either the GPx1 or TPI gene, which normally contain one or more introns, are no longer susceptible to NMD (Moriarty et al. 1997; Zhang et al. 1998b; see below). It will be interesting to determine whether transcripts from naturally intron-less genes, such as the *hsp70* gene, are subject to NMD and, if they are, to assess the nature of the destabilizing element.

A PTC within exon 1 of the TPI gene fails to elicit NMD when all introns but the 3′-most intron (intron 6) are deleted so that only the 3′-most exon–exon junction is generated by splicing (Zhang et al. 1998b; Sun and Maquat 2000). This finding suggests that exon–exon junctions other than the 3′-most normally function in NMD elicited by PTCs located near the 5′ end of TPI mRNA. By implication, splicing-dependent alterations in mRNP are not restricted to the 3′-most junction. Whereas this finding also suggests that NMD requires a minimal distance between a PTC and a downstream junction, studies of other transcripts indicate that distances as large as several kilobases support NMD (X. Li and L.E. Maquat, unpubl.; Y. Ishigaki and L.E. Maquat, unpubl.). In contrast, evidence exists that NMD in *S. cerevisiae*, an organism characterized by mostly intron-less genes, does require a minimal distance between a PTC and a so-called downstream destabilizing element (Ruiz-Echevarria et al. 1998; for review, see Chapter 29). The yeast downstream element can be thought of as the functional equivalent of a mammalian exon–exon junction, even though it is not generated by splicing. Also analogous to the yeast downstream destabilizing element are ill-defined sequences within certain mammalian mRNAs that can function in NMD on behalf of an exon–exon junction as long as another splicing-dependent junction is present (Zhang et al. 1998a,b).

Considering the role of exon–exon junctions in NMD, introns could be either positive effectors of gene expression that preclude NMD when located upstream of, or less than 50–55 nucleotides downstream from, a termination codon, or negative effectors of gene expression that elicit

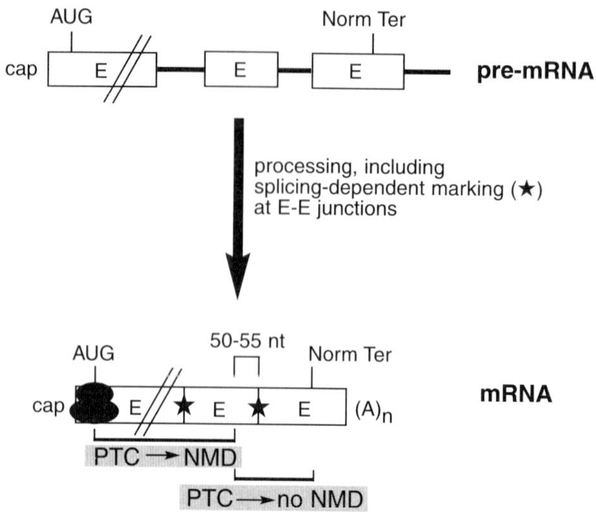

Figure 1 Model for NMD: How a PTC decreases the half-life of nucleus-associated or cytoplasmic mRNA. Pre-mRNA splicing in the nucleus leaves marks (★) at the exon–exon junctions of product mRNA. If a PTC resides more than 50–55 nucleotides upstream of the 3′-most junction, then the mRNA will be subject to NMD (PTC→NMD). However, if a PTC resides less than 50–55 nucleotides upstream or downstream of the 3′-most junction, then the mRNA will not be subject to NMD (PTC→no NMD). Depending on undefined mRNA features or conditions of production, NMD takes place in association with nuclei or in the cytoplasm. Since evidence indicates that PTC recognition involves translation by ribosomes ( ) regardless of the cellular site of NMD, nucleus-associated NMD is most readily envisioned taking place during mRNA export to the cytoplasm, at a point where mRNA copurifies with nuclei but has access to translationally active cytoplasmic ribosomes. Alternatively, nucleus-associated NMD could take place within the nucleoplasm, in which case PTC recognition could be mediated by a nuclear ribosome or ribosome-like molecule. For those mRNAs subject to nucleus-associated NMD, a fraction escapes NMD and is exported to the cytoplasm where it is immune to NMD, possibly because of a change in mRNP structure during export. In contrast, those mRNAs subject to cytoplasmic NMD are degraded after export to the cytoplasm. Currently, it is unclear what differentiates mRNAs subject to nucleus-associated NMD from mRNAs subject to cytoplasmic NMD. AUG, E, and Norm Ter specify, respectively, the initiation codon, an exon, and the normal termination codon.

NMD when located more than 50-55 nucleotides downstream from a termination codon (Hilleren and Parker 1999). The first possibility predicts that intron-less counterparts of normally intron-containing genes would produce transcripts that are subject to NMD, and the second possibility predicts that such transcripts would be immune to NMD. In favor of the second possibility, expression of an intron-less version of the GPx1 gene under conditions of selenium deficiency or when the TGA Sec codon is converted to a TAA nonsense codon fails to increase the efficiency of NMD despite increasing the efficiency with which the Sec codon is recognized as nonsense (Moriarty et al. 1997). In contrast, expression of the natural intron-containing GPx1 gene under identical circumstances does increase the efficiency of NMD (Moriarty et al. 1998). Also in favor of the second possibility, expression of the intron-less GPx1 gene is not affected when NMD is inhibited by overexpression of a dominant-negative version of human Upf1p (Sun and Maquat 2000), a protein required for NMD (see below).

To date, there is no evidence for the existence of *cis*-acting sequences in mammalian cells that preclude NMD, possibly analogous to the stablizing elements of the GCN4 and YAP1 transcripts of *S. cerevisiae* that preclude NMD when situated between a termination codon and a downstream destabilizing element (Ruiz-Echevarria et al. 1998; for review, see Chapter 29). However, indications that such an element could exist derive from those transcripts that are predicted to be subject to NMD but, in fact, are not. For example, mice engineered to harbor a PTC at codon 2153 of exon 26 within both apo B alleles, so as to produce PTC-containing apo B transcripts identical to the products of posttranscriptional editing, do not generate an abnormally low level of apo B mRNA (Farese et al. 1992). In contrast, mice engineered to harbor a PTC located within an exon downstream from exon 26 do produce PTC-containing mRNA at an abnormally low level (Kim et al. 1998). In theory, the inability of the PTC at codon 2153 to elicit NMD could be due to either the existence of some type of stabilizing element between the PTC and the downstream exon–exon junction or the large (>7000 nucleotides) distance between the PTC and the downstream exon–exon.

TRANS-ACTING FACTORS

NMD requires at least three *cis*-acting sequences: an initiation codon, a PTC, and an appropriately positioned downstream destabilizing element, which is usually an exon–exon junction generated by the process of pre-mRNA splicing. It follows that *trans*-acting factors required for NMD should minimally consist of translational factors, splicing or splicing-related factors that mark exon–exon junctions, and mRNA decay factors. Notably, a PTC positioned immediately downstream from the initiation codon can elicit NMD, indicating that peptide-bond formation and, therefore, the associated factors are not required for NMD (Zhang and Maquat 1997).

On the basis of sequence similarities to the so-called Upf proteins of *S. cerevisiae* and SMG proteins of *C. elegans* that function in NMD, at least three mammalian factors likely to function in NMD have been identified. The human (h) factors have been called hUpf1p, hUpf2p, and hUpf3p after their *S. cerevisiae* counterparts, each of which plays a different role in aspects of translation (for review, see Chapter 29). Notably, hUpf1p (also called RENT1 by Perlick et al. [1996] or HUPF1 by Applequist et al. [1997]) is related to *C. elegans* SMG-2 (Page et al. 1999), hUpf2p (G. Serin and L.E. Maquat, unpubl.) is related to *C. elegans* SMG-3 (S. Kuchma and P. Anderson, pers. comm.), and hUpf3p (G. Serin and L.E. Maquat, unpubl.) is related to *C. elegans* SMG-4 (R. Aronoff and J. Hodgkin, pers. comm.). hUpf1p is a group 1 RNA helicase, and the only one of the three hUpf factors demonstrated so far to function in NMD (Sun et al. 1998). It is a phosphoprotein, as is SMG-2 (Page et al. 1999), that is equally distributed among polysomal, sub-polysomal, and polysome-free fractions of exponentially growing HeLa cells, capable of hydrolyzing ATP and binding single-stranded RNA, and present at ~3×10^6 copies per cell (M. Pal and L.E. Maquat, unpubl.). hUpf1p interacts with purified release factors RF1 and RF3 when synthesized in vitro (Czaplinski et al. 1998) and modulates translation termination when produced as a hybrid human/yeast protein in yeast (Czaplinski et al. 1998). These and other data indicate that hUpf1p, like its yeast counterpart, increases the efficiency of translation termination in addition to being essential for NMD. Considering that SMG-3 and SMG-4 are required for SMG-2 phosphorylation in *C. elegans* (Page et al. 1999), it is likely that hUpf2p and hUpf3p are required for hUpf1p phosphorylation. Human orthologs to SMG-1, a phosphatidylinositol 3-kinase-related kinase, and the several SMG factors known to be required for SMG-2 dephosphorylation (Page et al. 1999) have yet to be discovered.

With regard to splicing or splicing-related factors that could function in NMD, novel cross-linking and RNase protection strategies have been developed that identify proteins that bind only to mRNAs generated by splicing in HeLa-cell nuclear extract (Le Hir et al. 2000 and unpubl.). Results indicate that a stable ~335-kD complex of proteins forms 20–24 nucleotides upstream of mRNA exon–exon junctions as a direct consequence of splicing. Glycerol gradient fractionation demonstrated that the proteins remain associated with mRNA after its release from the spliceosome, indicating that they are bona fide constituents of mRNP, at least in vitro. To date, immunoprecipitation experiments have identified five proteins: SRm160, a nuclear-matrix-associated splicing coactivator; DEK, another splicing coactivator; RNPS1, a general splicing activator; and Y14 and REF, both of which associate with the mRNA export factor TAP. Future studies of these and other proteins that comprise the complex will assay for a functional role in NMD.

Factors required for the mRNA decay aspect of NMD in *S. cerevisiae* include Dcp1p, a decapping enzyme (LaGrandeur and Parker 1998) and Xrn1p, a cytoplasmic 5′ to 3′ exonuclease (for review, see Jacobson and Peltz 1996), both of which are also involved in the decay of nonsense-free mRNAs (for review, see Chapters 28 and 29). Despite the existence of mammalian orthologs to Xrn1p (Bashkirov et al. 1997) and of mammalian activities comparable to Dcp1p (see, e.g., Coutts and Brawerman 1993), their likely role in NMD has yet to be investigated.

CLINICAL APPROACHES TO ABROGATING NMD AND PTC-ASSOCIATED DISEASE PHENOTYPES

Approximately one-third of inherited genetic disorders and many cancers are thought to be attributable to frameshift or nonsense mutations that generate PTCs (Frischmeyer and Dietz 1999). As remarkable examples, PTCs are generated by at least 90% of mutations associated with Duchenne muscular dystrophy, familial adenomatous polyposis, hereditary desmoid disease, ataxia telangiectasia, hereditary breast and ovarian cancer, and polycystic kidney disease, and at least 75% of mutations associated with Emery-Dreifuss muscular dystrophy, Fanconi anemia, and hereditary non-polyposis colorectal cancer (for review, see Hogervorst 1997). The prevalence of PTCs in these and other diseases has prompted the development of a quick and easy protein truncation test that is applicable to any gene from which cDNA can be synthesized (Hogervorst 1997). In this test, cDNA to the transcript region under analysis is synthesized using a specific downstream (antisense) DNA primer, and the cDNA is subsequently amplified using an upstream (sense) DNA primer that has a bacteriophage RNA polymerase promoter sequence and translation initiation codon and the downstream primer. Protein synthesized in vitro using the resulting double-stranded DNA and a coupled transcription–translation reaction will be smaller than normal if the region under analysis harbors a PTC.

In theory, diseases caused by a nonsense mutation could be treated by suppressing PTC recognition so that functional full-length protein is made at a level sufficient to abrogate the disease phenotype. For this to happen, the efficiency and type of amino acid incorporation at the PTC would have to allow for an adequate level of protein function. To date, PTC suppression has been achieved in cultured cells and mice that model a human disease by using aminoglycoside antibiotics, suppressor tRNAs, or antisense oligoribonucleotides. For example, a nonsense mutation in the cystic fibrosis transmembrane conductance regulator gene (CFTR) was suppressed by growing a defective bronchial epithelial cell line in the presence of the aminoglycoside gentamicin: Gentamicin, which binds to a specific site in ribosomal RNA and allows for translation through a termination codon, restored the level of CFTR expression to 10–20% of normal and, in so doing, restored cAMP-activated chloride transport (Howard et al. 1996; Bedwell et al. 1997). Gentamicin has also been used to suppress a nonsense mutation within exon 23 of the dystrophin gene of *mdx* mice, which exhibit hallmarks of human Duchenne muscular dystrophy, so as to elicit the production of dystrophin in the cell membrane of all striated muscles examined and protect against muscle injury (Barton-Davis et al. 1999). In other studies, the direct injection of *mdx* mice with plasmid DNA encoding a serine tRNA suppressor resulted in dystrophin-positive fibers (Li et al. 1997). Studies performed in parallel involving the co-injection of the suppressor tRNA gene and a PTC-containing chloramphenicol acetyl transferase gene into isolated rat heart revealed a suppression efficiency of 5.5% (Li et al. 1997). In experiments that employed primary cultures of *mdx* myoblasts, delivery of a 2′-O-methyl oligoribonucleotide complementary to the 3′ splice site upstream of the PTC-containing dystrophin exon resulted in skipping of the PTC-containing exon and six downstream exons so that an in-frame dystrophin mRNA was generated (Dunckley et al. 1998). Since patients with comparable in-frame deletions show relatively mild myopathic symptoms, antisense therapy may be useful in abrogating the deleterious effect of a number of PTC-containing dystrophin transcripts that derive from either nonsense or frameshift mutations.

CONCLUSIONS

Studies of NMD in mammalian cells have uncovered remarkable links between pre-mRNA splicing and mRNA translation and have identified translation termination as a determinant of mRNA stability. Future studies should elucidate how splicing-dependent mRNA–protein interactions and translational factors collaborate to elicit NMD. It is hoped that a better understanding of NMD together with more comprehensive analyses of drug-mediated PTC suppression will lead to the development of therapeutic protocols for a number of PTC-generated genetic disorders.

ACKNOWLEDGMENTS

I thank many colleagues for communicating results prior to publication and apologize for often referencing reviews rather than primary reports due to space constraints. Thanks also to Anand Gersappe for help preparing the manuscript, Ben Blencowe and David Bedwell for helpful conversations, and members of the Maquat lab, especially Yasuhito Ishigaki, for comments on the manuscript. Work cited from this laboratory is currently supported by National Institutes of Health grants DK-33938 and GM-59614 to L.E.M.

REFERENCES

Applequist S.E., Selg M., Raman C., and Jäck H.M. 1997. Cloning and characterization of HUPF1, a human homolog of the *Saccharomyces cerevisiae* nonsense mRNA-reducing UPF1 protein. *Nucleic Acids Res.* **25:** 814–821.

Ban N., Nissen P., Hansen J., Capel M., Moore P.B., and Steitz T.A. 1999. Placement of protein and RNA structures into a 5Å-resolution map of the 50S ribosomal subunit. *Nature* **400:** 841–847.

Barton-Davis E.R., Cordier L., Shoturma D.I., Leland S.E., and Sweeney H.L. 1999. Aminoglycoside antibiotics restore dystrophin function to skeletal muscles of mdx mice. *J. Clin. Invest.* **104:** 375–381.

Baserga S.J. andBenz E.J., Jr. 1992. β-Globin nonsense mutation: Deficient accumulation of mRNA occurs despite normal cytoplasmic stability. *Proc. Natl. Acad. Sci.* **89:** 2935–2939.

Bashkirov V.I., Scherthan H., Solinger J.A., Buerstedde J.-M., and Heyer W.-D. 1997. A mouse cytoplasmic exoribonuclease (mXRN1p) with preference for G4 tetraplex substrates. *J. Cell Biol.* **136:** 761–773.

Bedwell D.M., Kaenjak A., Benos D.J., Bebok Z., Bubien J.K., Hong J., Tousson A., Clancy J.P., and Sorscher E.J. 1997. Suppression of a CFTR premature stop mutation in a bronchial epithelial cell line. *Nat. Med.* **3:** 1280–1284.

Belgrader P., Cheng J., and Maquat L.E. 1993. Evidence to implicate translation by ribosomes in the mechanism by which nonsense codons reduce the nuclear level of human triosephosphate isomerase mRNA. *Proc. Natl. Acad. Sci.* **90:** 482–486.

Belgrader P., Cheng J., Zhou X., Stephenson L.S., and Maquat L.E. 1994. Mammalian nonsense codons can be *cis*-effectors of nuclear mRNA half-life. *Mol. Cell. Biol.* **14:** 8219–8228.

Carter M.S., Li S., and Wilkinson M.F. 1996. A splicing-dependent regulatory mechanism that detects translation signals. *EMBO J.* **15:** 5965–5975.

Carter M.S., Doskow J., Morris P., Li S., Nhim R.P., Sandstedt S., and Wilkinson M.F. 1995. A regulatory mechanism that detects premature nonsense codons in T-cell receptor transcripts in vivo is reversed by protein synthesis inhibitors in vitro. *J. Biol. Chem.* **270:** 28995–29003.

Chang B.H.-J., Lau P.P., and Chan L. 1998. Apolipoprotein B mRNA editing. In *Modification and editing of RNA.* (ed. H. Grosjean and R. Benne) pp. 325–342. ASM Press, Washington, DC.

Cheng J. and Maquat L.E. 1993. Nonsense codons can reduce the abundance of nuclear mRNA without affecting the abundance of pre-mRNA or the half-life of cytoplasmic mRNA. *Mol. Cell. Biol.* **13:** 1892–1902.

Cheng J., Belgrader P., Zhou X., and Maquat L.E. 1994. Introns are *cis* effectors of the nonsense-codon-mediated reduction in nuclear mRNA abundance. *Mol. Cell. Biol.* **14:** 6317–6325.

Couttet P., Fromont-Racine M., Steel D., Pictet R., and Grange T. 1997. Messenger RNA deadenylation precedes decapping in mammalian cells. *Proc. Natl. Acad. Sci.* **94:** 5628–5633.

Coutts M. and Brawerman G. 1993. A 5′ exoribonuclease from cytoplasmic extracts of mouse sarcoma 180 ascites cells. *Biochim. Biophys. Acta* **1173:** 57–62.

Czaplinski K., Ruiz-Echevarría M.J., Paushkin S.V., Han X., Weng Y., Perlick H.A., Dietz H.C., Ter Avanesyan M.D., and Peltz S.W. 1998. The surveillance complex interacts with the translation release factors to enhance termination and degrade aberrant mRNAs. *Genes Dev.* **12:** 1665–1677.

Dahlseid J.N., Puziss J., Shirley R.L., Atkin A.L., Hieter P., and Culbertson M.R. 1998. Accumulation of mRNA coding for the Ctf13p kinetochore subunit of *Saccharomyces cerevisiae* depends on the same factors that promote rapid decay of nonsense mRNAs. *Genetics* **150:** 1019–1035.

Dick F.A., Eisinger D.P., and Trumpower B.L. 1997. Exchangeability of Qsr1p, a large ribosomal subunit protein required for subunit joining, suggests a novel translational regulatory mechanism. *FEBS Lett.* **419:** 1–3.

Dietz H.C. and Kendzior R.J., Jr. 1994. Maintenance of an open reading frame as an additional level of scrutiny during splice site selection. Nat. Genet. **8:** 183–188.

Dietz H.C., Valle D., Francomano C.A., Kendzior R.J., Jr., Pyeritz R.E., and Cutting G.R. 1993. The skipping of constitutive exons in vivo induced by nonsense mutations. *Science* **259:** 680–683.

Dunckley M.G., Manoharan M., Villiet P., Eperon I.C., and Dickson G. 1998. Modification of splicing in the dystrophin gene in cultured *Mdx* muscle cells by antisense oligoribonucleotides. *Hum. Mol. Genet.* **7:** 1083–1090.

Enssle J., Kugler W., Hentze M.W., and Kulozik A.E. 1993. Determination of mRNA fate by different RNA polymerase II promoters. *Proc. Natl. Acad. Sci.* **90:** 10091–10095.

Farese R.V., Linton M.F., and Young S.G. 1992. Apolipoprotein B gene mutations affecting cholesterol levels. *J. Intern. Med.* **231:** 643–652.

Frischmeyer P.A. and Dietz H.C. 1999. Nonsense-mediated mRNA decay in health and disease. *Hum. Mol. Genet.* **8:** 1893–1900.

Gersappe A. and Pintel D.J. 1999. A premature termination codon interferes with the nuclear function of an exon splicing enhancer in an open reading frame-dependent manner. *Mol. Cell. Biol.* **19:** 1640–1650.

Gersappe A., Burger L., and Pintel D.J. 1999. A premature termination codon in either exon of minute virus of mice P4 promoter-generated pre-mRNA can inhibit nuclear splicing of the intervening intron in an open reading frame-dependent manner. *J. Biol. Chem.* **274:** 22452–22458.

Hall G.W. and Thein S. 1994. Nonsense codon mutations in the terminal exon of the β-globin gene are not associated with a reduction in β-globin mRNA accumulation: A mechanism for the phenotype of dominant β-thalassemia. *Blood* **83:** 2031–2037.

Hawkins J.D. 1988. A survey on intron and exon lengths. *Nucleic Acids Res.* **16:** 9893–9908.

Hendrich B., Hardeland U., Ng H.H., Jiricny J., and Bird A. 1999. The thymine glycosylase MBD4 can bind to the product of deamination at methylated CpG sites. *Nature* **401:** 301–304.

Hentze M.W. and Kulozik A.E. 1999. A perfect message: RNA surveillance and nonsense-mediated decay. *Cell* **96:** 307–310.

Hilleren P. and Parker R. 1999. mRNA surveillance in eukaryotes: Kinetic proofreading of proper translation termination as assessed by mRNP domain organization? *RNA* **5:** 711–719.

Hogervorst F.B.L. 1997. The protein truncation test. *Promega Notes Magazine* **62:** 7–10.

Howard M., Frizell R.A., and Bedwell D.M. 1996. Aminoglycoside antibiotics restore CFTR function by overcoming premature stop mutations. *Nat. Med.* **2:** 467–469.

Jacobson A. and Peltz S.W. 1996. Interrelationships of the pathways of mRNA decay and translation in eukaryotic cells. *Annu. Rev. Biochem.* **65:** 693–739.

Kazazian H.H., Jr., Dowling C.E., Hurwitz R.L., Coleman M., Stopeck A., and Adams J.G., III. 1992. Dominant thalassemia-like phenotypes associated with mutations in exon 3 of the β-globin gene. *Blood* **79:** 3014–3018.

Kim E., Cham C.M., Vénaint M.M., Ambroziak P., and Young S.G. 1998. Dual mechanisms for the low plasma levels of truncated apolipoprotein B proteins in familial hypobetalipoproteinemia. *J. Clin. Invest.* **101:** 1468–1477.

Kugler W., Enssle J., Hentze M.W., and Kulozik A.E. 1995. Nuclear degradation of nonsense mutated β-globin mRNA: A post-transcriptional mechanism to protect heterozygotes from severe clinical manifestations of β-thalassemia? *Nucleic. Acids Res.* **23:** 413–418.

LaGrandeur T.E. and Parker R. 1998. Isolation and characterization of Dcp1p, the yeast mRNA decapping enzyme. *EMBO J.* **17:** 1487–1496.

Le Hir H., Moore M.J., and Maquat L.E. 2000. Pre-mRNA splicing alters mRNP composition: Evidence for stable association of proteins at exon-exon junctions. *Genes Dev.* **14:** 1098–1108.

Lelivelt M.J. and Culbertson M.R. 1999. Yeast Upf proteins required for RNA surveillance affect global expression of the yeast transcriptome. *Mol. Cell. Biol.* **19:** 6710–6719.

Lew J.E., Enomoto S., and Berman J. 1998. Telomere length regulation and telomeric chromatin require the nonsense–mediated mRNA decay pathway. *Mol. Cell. Biol.* **18:** 6121–6130.

Li S. and Wilkinson M.F. 1998. Nonsense surveillance in lymphocytes? *Immunity* **8:** 135–141.

Li S., Leonard D., and Wilkinson M.F. 1997. T cell receptor (TCR) mini-gene mRNA expression regulated by nonsense codons: A nuclear-associated translation-like mechanism. *J. Exp. Med.* **185:** 985–992.

Lim S.-K. and Maquat L.E. 1992. Human β-globin mRNAs that harbor a nonsense codon are degraded in murine erythroid tissues to intermediates lacking regions of exon I or exons I and II that have a cap-like structure at the 5′ termini. *EMBO J.* **11:** 3271–3278.

Lim S.K., Sigmund C.D., Gross K.W., and Maquat L.E. 1992. Nonsense codons in human β-globin mRNA result in the production of mRNA degradation products. *Mol. Cell.*

Biol. **12:** 1149–1161.

Lim S., Mullins J.J., Chen C.M., Gross K.W., and Maquat L.E. 1989. Novel metabolism of several β^0-thalassemic β-globin mRNAs in the erythroid tissues of transgenic mice. *EMBO J.* **8:** 2613–2619.

Liu H.X., Zhang M., and Krainer A.R. 1998. Identification of functional exonic splicing enhancer motifs recognized by individual SR proteins. *Genes Dev.* **12:** 1998–2012.

Liu W., Qian C., and Francke U. 1997. Silent mutation induces exon skipping in fibrillin-1 gene in Marfan syndrome. *Nat. Genet.* **16:** 328–329.

Lozano F., Maertzdorf B., Pannell R., and Milstein C. 1994. Low cytoplasmic mRNA levels of immunoglobulin kappa light chain genes containing nonsense codons correlate with inefficient splicing. *EMBO J.* **13**: 4617–4622.

Lund E. and Dahlberg J.E. 1998. Proofreading and aminoacylation of tRNAs before export from the nucleus. *Science* **282:** 2082–2085.

Maquat L.E. 1995. When cells stop making sense: Effects of nonsense codons on RNA metabolism in vertebrate cells. *RNA* **1:** 453–465.

———. 1996. Defects in RNA splicing and the consequence of shortened translational reading frames. *Am. J. Hum. Genet.* **59:** 279–286.

Maquat L.E., Kinniburgh A.J., Rachmilewitz E.A., and Ross J. 1981. Unstable β-globin mRNA in mRNA-deficient β^0- thalassemia. *Cell* **27:** 543–553.

Matsumoto K., Wassarman K.M., and Wolffe A.P. 1998. Nuclear history of a pre-mRNA determines the translational activity of cytoplasmic mRNA. *EMBO J.* **17:** 2107–2121.

Mehlin H., Daneholt B., and Skoglund U. 1992. Translocation of a specific premessenger ribonucleoprotein particle through the nuclear pore studied with electron microscope tomography. *Cell* **69:** 605–613.

Menon K.P. and Neufeld E.F. 1994. Evidence for degradation of mRNA encoding alpha-L-iduronidase in Hurler fibroblasts with premature termination alleles. *Cell. Mol. Biol.* **40:** 999–1005.

Moriarty P.M., Reddy C.C., and Maquat L.E. 1997. The presence of an intron within the rat gene for selenium-dependent glutathione peroxidase 1 is not required to protect nuclear RNA from UGA-mediated decay. *RNA* **3:** 1369–1373.

———. 1998. Selenium deficiency reduces the abundance of mRNA for Se-dependent glutathione peroxidase 1 by a UGA-dependent mechanism likely to be nonsense codon-mediated decay of cytoplasmic mRNA. *Mol. Cell. Biol.* **18:** 2932–2939.

Morrison M., Harris K.S., and Roth M.B. 1997. *smg* mutants affect the expression of alternatively spliced SR protein mRNAs in *Caenorhabditis elegans*. *Proc. Natl. Acad. Sci.* **94:** 9782–9785.

Muhlrad D. and Parker R. 1994. Premature translational termination triggers mRNA decapping. *Nature* **370:** 578–581.

Naeger L.K., Schoborg R.V., Zhao Q., Tullis G.E., and Pintel D.J. 1992. Nonsense mutations inhibit splicing of MVM RNA in *cis* when they interrupt the reading frame of either exon of the final spliced product. *Genes Dev.* **6:** 1107–1119.

Nagy E. and Maquat L.E. 1998. A rule for termination-codon position within intron-containing genes: When nonsense affects RNA abundance. *Trends Biochem. Sci.* **23:** 198–199.

Page M.F., Carr B., Anders K.R., Grimson A., and Anderson P. 1999. SMG-2 is a phosphorylated protein required for mRNA surveillance in *Caenorhabditis elegans* and related to Upf1p in yeast. *Mol. Cell. Biol.* **19:** 5943–5951.

Perlick H.A., Medghalchi S.M., Spencer F.A., Kendzior R.J., Jr., and Dietz H.C. 1996.

Mammalian orthologues of a yeast regulator of nonsense transcript stability. *Proc. Natl. Acad. Sci.* **93:** 10928–10932.

Qian L., Theodor L., Carter M., Vu M.N., Sasaki A.W., and Wilkinson M.F. 1993. T cell receptor-β mRNA splicing: Regulation of unusual splicing intermediates. *Mol. Cell. Biol.* **13:** 1686–1696.

Ruiz-Echevarría M.J., Gonzalez C.I., and Peltz S.W. 1998. Identifying the right stop: Determining how the surveillance complex recognizes and degrades an aberrant mRNA. *EMBO J.* **17:** 575–589.

Santisteban I., Arredondo-Vega F.X., Kelly S., Loubser M., Meydan N., Roifman C., Howell P.L., Bowen T., Weinberg K.I., and Schroeder M.L., et al. 1995. Three new adenosine deaminase mutations that define a splicing enhancer and cause severe and partial phenotypes: Implications for evolution of a CpG hotspot and expression of a transduced ADA cDNA. *Hum. Mol. Genet.* **4:** 2081–2087.

Schaal T.D. and Maniatis T. 1999. Selection and characterization of pre-mRNA splicing enhancers: Identification of novel SR protein-specific enhancer sequences. *Mol. Cell. Biol.* **19:** 1705–1719.

Shiga N., Takeshima Y., Sakamoto H., Inoue K., Yokota Y., Yokoyama M., and Matsuo M. 1997. Disruption of the splicing enhancer sequence within exon 27 of the dystrophin gene by a nonsense mutation induces partial skipping of the exon and is responsible for Becker muscular dystrophy. *J. Clin. Invest.* **100:** 2204–2210.

Stephenson L.S. and Maquat L.E. 1996. Cytoplasmic mRNA for human triosephosphate isomerase is immune to nonsense-mediated decay despite forming polysomes. *Biochimie* **78:** 1043–1048.

Sun X. and Maquat L.E. 2000. mRNA surveillance in mammalian cells: The relationship between introns and translation termination. *RNA* **6:** 1–8.

Sun X., Moriarty P.M., and Maquat L.E. 2000. Nonsense-mediated decay of glutathione peroxidase 1 mRNA in the cytoplasm depends on the intron position. *EMBO J.* (in press).

Sun X., Perlick H.A., Dietz H.C., and Maquat L.E. 1998. A mutated human homologue to yeast Upf1 protein has a dominant-negative effect on the decay of nonsense-containing mRNAs in mammalian cells. *Proc. Natl. Acad. Sci.* **95:** 10009–10014.

Thermann R., Neu-Yilik G., Deters A., Frede U., Wehr K., Hagemeier C., Hentze M.W., and Kulozik A.E. 1998. Binary specification of nonsense codons by splicing and cytoplasmic translation. *EMBO J.* **17:** 3484–3494.

Valentine C.R. 1998. The association of nonsense codons with exon skipping. *Mutat. Res.* **411:** 87–117.

van Leeuwen F.W., de Kleijn D.P.V., van den Hurk H.H., Neubauer A., Sonnemans M.A.F., Sluijs J.A., Koycu S., Ramdjielal R.D.J., Salehi A., Martens G.J.M., Grosveld F.G., Burbach J.P.H., and Hol E.M. 1998. Frameshift mutants of β amyloid precursor protein and ubiquitin-B in Alzheimer's and Down patients. *Science* **279:** 242–247.

Yoshizawa S., Fourmy D., and Puglisi J.D. 1999. Recognition of the codon-anticodon helix by ribosomal RNA. *Science* **285:** 1722–1725.

Zhang J. and Maquat L.E. 1996. Evidence that the decay of nucleus-associated nonsense mRNA for human triosephosphate isomerase involves nonsense codon recognition after splicing. *RNA* **2:** 235–243.

———. 1997. Evidence that translation reinitiation abrogates nonsense-mediated mRNA decay in mammalian cells. *EMBO J.* **16:** 826–833.

Zhang J., Sun X., Qian Y., and Maquat L.E. 1998a. Intron function in the nonsense-mediated decay of β-globin mRNA: Indications that pre-mRNA splicing in the nucleus can influence mRNA translation in the cytoplasm. *RNA* **4:** 801–815.

Zhang J., Sun X., Qian Y., LaDuca J.P., and Maquat L.E. 1998b. At least one intron is required for the nonsense-mediated decay of triosephosphate isomerase mRNA: A possible link between nuclear splicing and cytoplasmic translation. *Mol. Cell. Biol.* **18**: 5272–5283.

Zhang S., Welch E.M., Hogan K., Brown A.H., Peltz S.W., and Jacobson A. 1997. Polysome-associated mRNAs are substrates for the nonsense-mediated mRNA decay pathway in *Saccharomyces cerevisiae*. *RNA* **3:** 234–244.

31
Translation Initiation on Picornavirus RNA

Graham J. Belsham
BBSRC Institute for Animal Health
Pirbright, Woking
Surrey, United Kingdom

Richard J. Jackson
Department of Biochemistry
University of Cambridge
Cambridge, United Kingdom

The animal picornaviruses are a large family of viruses which is currently subdivided into six genera (Table 1). They are non-enveloped viruses with a positive-strand RNA genome some 8000 nucleotides in length. Replication occurs entirely within the cytoplasm, as witnessed by the fact that they can replicate efficiently in enucleated cells (Follett et al. 1975). There is a 3′ poly(A) tail that is encoded rather than being added in a template-independent process by a poly(A) polymerase. At the 5′ end there is a covalently linked small virus-encoded protein, VPg (also known as 3B). It is believed that soon after entry into the infected cell, this protein is cleaved off by cellular activities, so that the RNA is effectively translated as an uncapped mRNA. Certainly, VPg is not essential for infectivity, since full length in vitro synthesized RNA (without VPg) is infectious. There is a single open reading frame coding for a polyprotein (Fig. 1), which is rapidly processed to give the individual capsid proteins and viral nonstructural proteins.

The common features of picornavirus 5′-untranslated regions (UTRs) are that they are very long (610 to >1200 nucleotides depending on the species); they have numerous AUG triplets, most of which are not conserved even between different isolates of the same virus (Pöyry et al. 1992); and they are predicted to fold into complex secondary structures. These characteristics, together with the fact that the RNA is not capped, imply that the RNAs cannot possibly be translated efficiently by the con-

Translational Control of Gene Expression
© 2000 Cold Spring Harbor Laboratory Press 0-87969-568-4/00 $5 + . 00

Table 1 Picornavirus classification

Picornaviridae genus	Representative species[a]	eIF4G cleavage
Enterovirus	PV, CV-A, CV-B	yes (2A-protease)
Rhinovirus	HRV	yes (2A-protease)
Aphthovirus	FMDV, ERAV	yes (L-protease)
Cardiovirus	EMCV, TMEV, MV	no
Hepatovirus	HAV	no
Parechovirus	EV-22, EV-23	no

[a]Abbreviations: (PV) Poliovirus; (CV) Coxsackie virus; (HRV) human rhinovirus (>100 serotypes); (FMDV) foot-and-mouth disease virus; (ERAV) equine rhinitis A virus; (EMCV) encephalomyocarditis virus; (TMEV) Theilers murine encephalomyelitis virus; (MV) mengovirus; (HAV) hepatitis A virus; (EV) echovirus.

ventional scanning mechanism. Initiation by direct internal ribosome entry was proven by showing that the insertion of a picornavirus 5′UTR between the two cistrons of a laboratory-constructed dicistronic mRNA leads to dramatic enhancement of expression of the downstream cistron (Jang et al. 1988; Pelletier and Sonenberg 1988). Thus, the concept of the "internal ribosome entry segment/site" (IRES), *cis*-acting RNA elements that direct internal ribosome entry, was born.

Picornaviruses are usually cytopathic. Most of them actively shut off host-cell mRNA translation, some of them very rapidly. Invariably this shutoff is caused by modification of the host-cell translation machinery (Fig. 1) and not through destruction of host-cell mRNA. However, there are some differences between different virus species as to the mechanism of this shutoff. Despite these differences, however, the fact that they do cause host-cell shutoff means that the IRES-dependent mechanism needs to facilitate efficient translation of the viral RNA both in the early stages of the infection and after the shut-off.

Many picornaviruses exhibit considerable tissue tropism. In part this is due to the tissue distribution of the relevant receptor, but this may not be the total explanation. For example, poliovirus neurovirulent strains, Sabin vaccine strains, and a recently constructed chimeric virus in which part of the poliovirus IRES was replaced by the rhinovirus equivalent (Gromeier et al. 1996, 1999) all have the same capsid proteins (and thus all use the same receptor), and they all replicate efficiently in HeLa cells, yet only the first of them will propagate in neuronal cells. Since these viruses differ only in their IRESs, this suggests that not only is viral RNA translation dependent on cellular proteins that differ between cell types, but that even IRESs which are quite similar to each other have different requirements for these host factors.

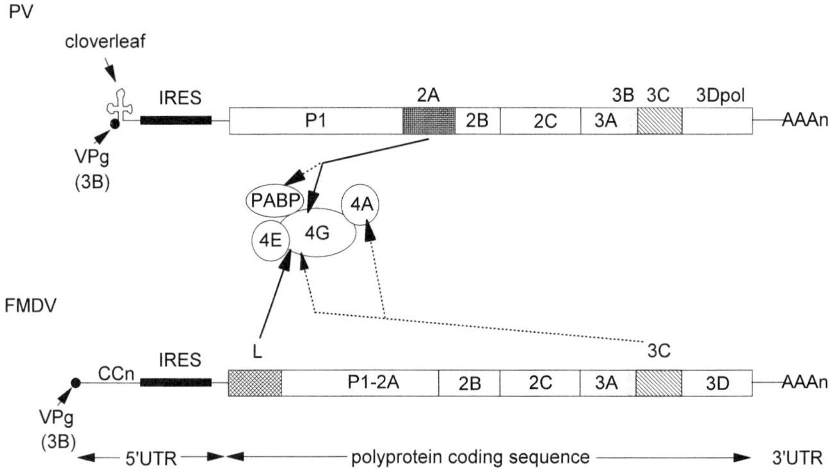

Figure 1 Schematic representation of picornavirus genome structure. The genomes of poliovirus (PV) and foot-and-mouth disease virus (FMDV) are illustrated. Virus-encoded proteins mentioned in the text are indicated above the genomes. Rapid cleavage of eIF4G is induced by PV 2A and by FMDV L-proteases (*solid lines*). The slower cleavage of eIF4A and eIF4G by FMDV 3C and of PABP by PV 2A are indicated by dashed lines. The RNA structures termed the cloverleaf and the IRES are also labeled (see text for details). The presence of a poly(C) tract in the FMDV genome and poly(A) tracts at the 3′ termini of both RNAs is indicated. Abbreviations: (4E), eIF4E; (4A), eIF4A; (4G), eIF4G; (3Dpol), the viral RNA polymerase.

The discussion in this chapter focuses on the mechanism by which the viral RNAs are translated, the mechanism of shutoff of host-cell mRNA translation, and the possible mechanism of down-regulation of viral RNA translation necessary to allow negative-strand RNA synthesis.

TRANSLATION OF PICORNAVIRUS RNAs BY INTERNAL RIBOSOME ENTRY

Characteristics of Picornavirus IRESs

Assay of deletion mutants in the standard dicistronic mRNA assay shows that picornavirus IRESs are typically about 450 nucleotides long. The extreme 5′-proximal part of the viral genome does not generally score as part of the IRES, although mutations in this 5′-proximal region of poliovirus RNA can inhibit IRES activity by up to fivefold (Simoes and Sarnow 1991). It has often been thought that the extreme 5′-proximal sequences are exclusively RNA replication signals and that the downstream IRES is exclusively concerned with translation. However, more recent evidence suggests some overlap or crosstalk: Not only can the interaction of proteins with the 5′-proximal "RNA replication signal" influence translation efficiency (Simoes and Sarnow 1991; Gamarnik and Andino 1998), but some RNA replication signals seem to reside within the IRES (Borman et al. 1994).

By the criteria of sequence conservation, the picornavirus IRESs can be divided into one minor and two major groups (for review, see Jackson and Kaminski 1995): (1) hepatitis A virus (the minor group); (2) cardio- and aphthoviruses, to which should be added the more recently defined parechoviruses (Table 1); and (3) entero- and rhinoviruses. The only feature all three types of IRES share in common is that at the 3′ end they have an ~25-nucleotide segment that starts with a tract of some 10 pyrimidines, followed by a G-poor sequence and ending with an AUG. Elsewhere the three types of IRES differ in sequence and predicted structure, but the two main groups share two small motifs in common: a GNRA tetraloop (stem-loop A), and a C-rich bulge (loop B), each located in similar sequence and structure background in the two types of IRES (see below).

A model has been proposed (Fig. 2A) that attempts to provide a unified mechanism applicable, with small variations in detail, to all picornavirus IRESs belonging to the two major groups (Jackson and Kaminski 1995). In essence this model proposes (1) that most of the ~450 nucleotides of the IRES is required by virtue of its secondary (and presumably tertiary) structure, and the critical primary sequence elements are relatively short and few in number; (2) that the actual ribosome entry site is invariably at an AUG triplet located at the 3′ end of the IRES, some 20–25 nucleotides downstream from the start of the pyrimidine-rich tract; but (3) that the events which follow ribosome entry at this site differ between the different viruses. In this and the following section, the evidence in support of these propositions is reviewed, after which the roles of canonical translation initiation factors and other cellular RNA-binding proteins are discussed.

Within each group, the conservation of predicted secondary structure is considerably greater than conservation of primary sequence. The high degree of genetic drift in these RNA viruses generates a large number of variant sequences, but these show patterns of covariance that result in conservation of secondary structure. Since the primary sequence motifs that are critical for IRES activity are likely to be highly conserved between closely related species, the phylogenetic comparisons suggest that these motifs are quite short, are mainly located in unpaired regions of the IRES, are highly dispersed within the IRES, and occur at higher den-

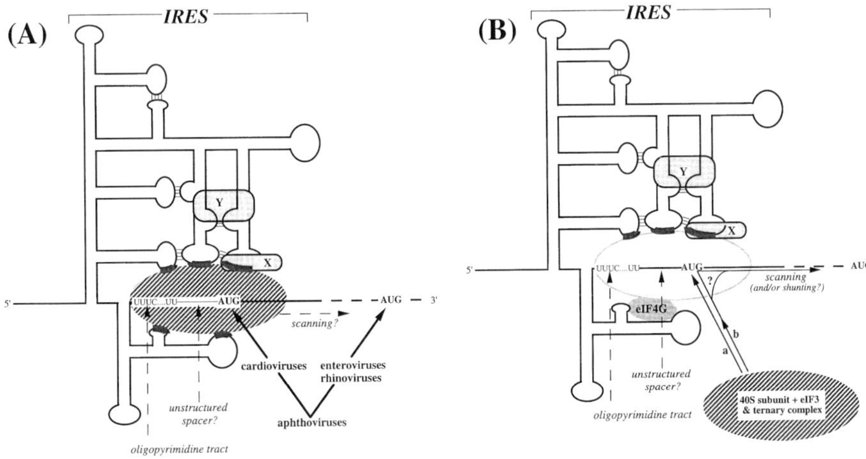

Figure 2 General model for internal initiation of translation on picornavirus IRESs: (*A*) as proposed previously (Jackson 1996), and (*B*) revised as discussed in the text. (*A*) The IRES consists of a number of base-paired stem-loops, with possible tertiary structure interactions between the loops and bulges. This structure presents several quite short unpaired primary sequence motifs, denoted by thickened lines, in the appropriate three-dimensional spatial organization for internal ribosome entry. These motifs may be binding sites for specific RNA-binding proteins, which are then recognized by the initiating ribosome (*shaded oval*), or they may be recognized directly by the ribosome and associated initiation factors. Two possible roles for cellular *trans*-acting factors are shown. Protein X binds at a specific site and directs internal initiation via specific interactions with the initiating 40S subunit. Protein Y binds to specific sequences and thereby promotes or stabilizes the appropriate three-dimensional structure of the IRES. (*B*) The revised model takes into account that eIF4G binds to a specific site in the IRES, and, together with the associated eIF4A (not shown), serves to recruit the 40S subunit primed for initiation by binding eIF3 and the eIF2/Met-$tRNA_i$/GTP ternary complex. The model posits that direct interactions between the 40S subunit and the IRES also contribute to ribosome recruitment. Two routes of internal initiation are suggested. In *a*, representative of the EMCV IRES, the 40S subunit enters and initiates translation at the AUG downstream from the oligopyrimidine tract. In *b*, representative of the entero-/rhinovirus IRESs, 40S subunit entry may be at the equivalent AUG or possibly at sites just downstream of it, but all the ribosomes are then transferred to the next AUG downstream, most probably mainly by linear scanning, although some ribosome shunting cannot be ruled out (see text). In the case of the FMDV IRES, up to one-third of the ribosomes follow route *a*, and the others route *b*.

sity toward the 3′ end of the segment (Figs. 3 and 4) (Jackson and Kaminski 1995). Validating the presumption that these conserved motifs are important will require extensive mutagenesis, which has so far been done only on a limited scale.

The results of mutation of the GNRA tetraloop mentioned above suggest that although some highly conserved residues are indeed essential, others may not be. Extensive mutagenesis of this loop (stem-loop A) in the FMDV IRES (Fig. 3) showed that only sequences conforming to GNRA were permissive to high IRES activity (Lopez de Quinto and Martinez-Salas 1997). In an alternative approach, which promises to be a powerful general method of determining which residues are essential for IRES activity, Robertson et al. (1999) randomized the tetraloop in the EMCV IRES to generate a pool of 256 sequences, made a preliminary selection of the winners in a transfection assay, then cloned these out and assayed them individually either in vitro or in transfected cells. The selected sequences conformed to a pattern of RNRA, and reasonably high activity was found even with a pyrimidine in the 5′ position (Robertson et al. 1999). Thus in both types of IRES, the 3′ A residue of the GNRA motif is absolutely essential, but the 5′ G does not seem to be very critical for the EMCV IRES, despite the fact that it is conserved in all known cardiovirus IRESs.

Defective IRES elements from PV, FMDV, and EMCV, containing point mutations or deletions, have been complemented *in trans*, by coexpression in the transfected cells of their respective wild-type IRES elements (Stone et al. 1993; Drew and Belsham 1994; van der Velden et al. 1995; Roberts and Belsham 1997; Tang et al. 1999). The complementation is quite efficient, requires the production of positive sense RNA, and does not involve recombination at either the RNA or DNA level. It has only been observed within cells expressing high levels of RNA transcripts derived from transfected plasmid DNAs via infection with recombinant vaccinia virus vTF7-3 expressing T7 RNA polymerase, and appears to be a special feature of picornavirus IRESs, since no complementation has been observed between mutant and wild-type hepatitis C virus IRESs in the same system (Tang et al. 1999). It is postulated that complementation may involve RNA–RNA interactions between different domains of the IRES, and such interactions between FMDV IRES domains have recently been demonstrated using gel shift assays (Ramos and Martinez-Salas 1999).

Where Is the Ribosome Entry Site and What Determines the Events That Immediately Follow Ribosome Entry?

These questions are most easily addressed in the case of the cardiovirus IRESs. In the commonly studied EMCV R strain internal initiation occurs mostly at the 11th AUG (Fig. 3), with some initiation also at AUG-12

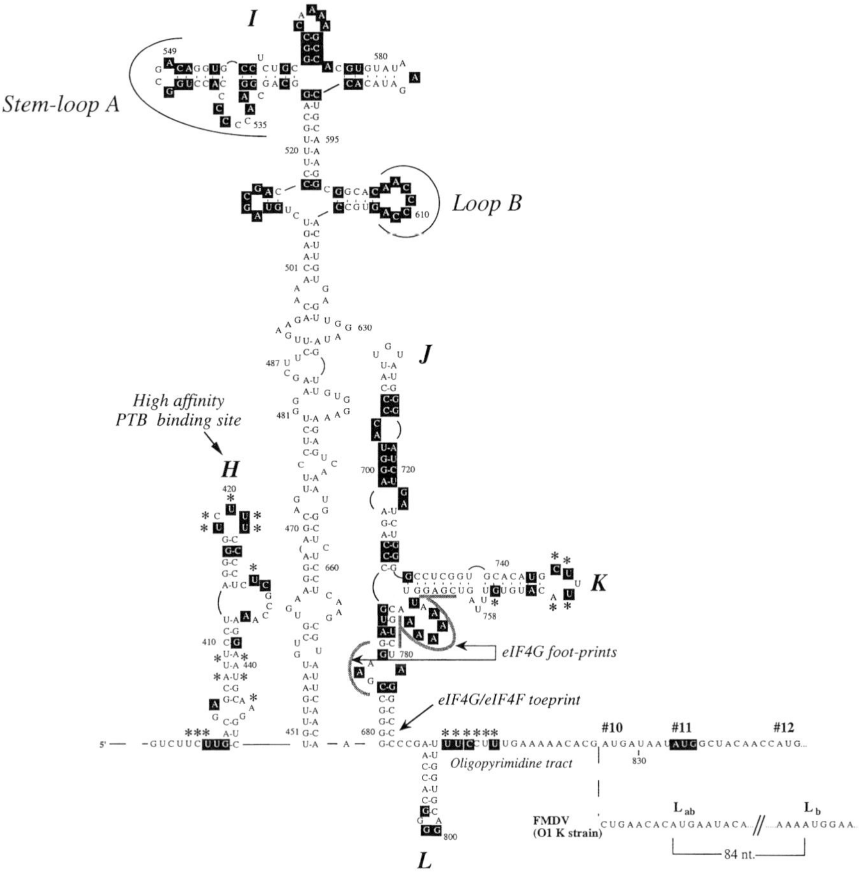

Figure 3 Conservation of sequence and structure among the cardio-/aphthovirus IRESs. The sequence shown is that of EMCV strain R (with the sequence around the initiation sites of FMDV strain O1K shown for comparison), and the structure is based on that of Pilipenko et al. (1989a). Sequences conserved in all cardiovirus, FMDV and echovirus 22 (a parechovirus) IRESs are blacked out, but as neither equine rhinitis virus IRESs nor any IRES sequence published since 1996 has been included in the analysis, the pattern of conservation should be regarded as illustrative rather than definitive. The various subdomains mentioned in the text are designated H–L (subdomains A–G are not part of the IRES). Stem-loop A and loop B are two structural features shared with entero-/rhinovirus IRESs (see Fig. 4). The residues protected when PTB binds to the IRES are denoted by asterisks (Kolupaeva et al. 1996). The eIFG/eIF4F toeprint position is shown, and the residues protected when eIF4G binds to the IRES are indicated by thickened shaded curves (Pestova et al. 1996b; Kolupaeva et al. 1998).

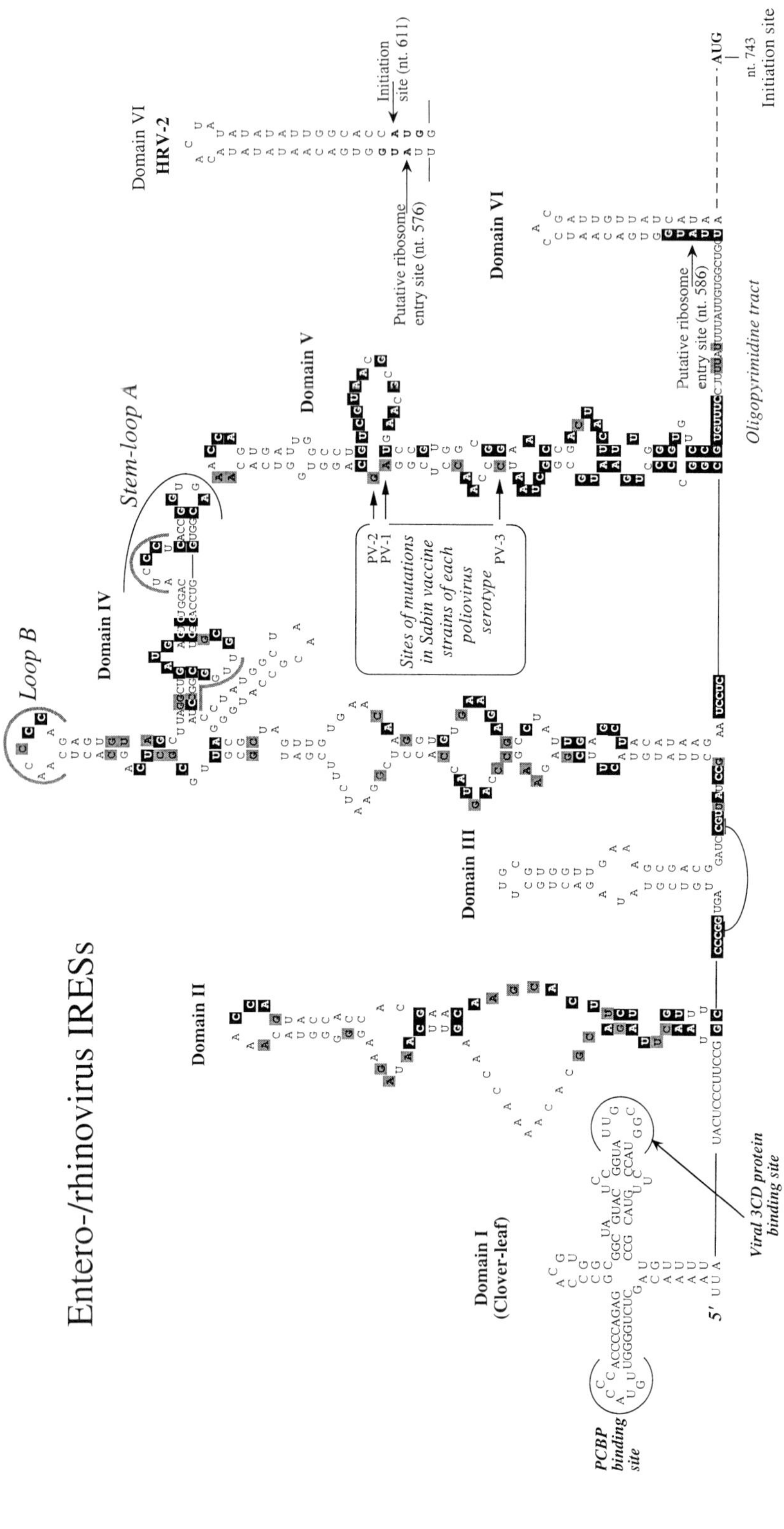

Figure 4 (*See facing page for legend.*).

located 12 nucleotides farther downstream, but almost none at AUG-10 just 8 nucleotides upstream of AUG-11 (Kaminski et al. 1990, 1994). Since AUG-10 is used quite efficiently if the IRES is amputated to generate an mRNA translated by the scanning mechanism, these results show that the IRES must direct the ribosome to land directly at or very close to AUG-11, rather than landing at an upstream site and scanning to this AUG (Kaminski et al. 1990, 1994). It is not known whether the ribosomes that initiate at AUG-12 reached this site by landing first at AUG-11 and then scanning on, or whether they accessed AUG-12 without ever "seeing" AUG-11.

The FMDV IRES is clearly related to the cardiovirus IRESs, but there are two functional initiation sites in all FMDV strains (Fig. 3): one (the L_{ab} site) analogous to EMCV AUG-11 at the 3′ end of the IRES, some 22 nucleotides downstream from the start of the oligopyrimidine tract, and the other (L_b site) in the same reading frame but 84 nucleotides downstream (Sangar et al. 1987; Belsham 1992). Looking for close parallels with the EMCV IRES, it was suggested that all ribosomes first enter at the upstream L_{ab} site, but that only a small proportion (20–33%, depending on the conditions) actually initiate there, whereas the rest scan to the L_b site. The entero-/rhinovirus IRESs represent a more extreme version of FMDV. There is very little (Ohlmann and Jackson 1999), or perhaps even no, initiation (Pestova et al. 1994) at the AUG (nucleotide 586 in poliovirus type 1) at the 3′ end of the IRES (Fig. 4), and all initiation occurs at the next AUG, ~40 nucleotides downstream in rhinoviruses and ~160 nucleotides in enteroviruses (AUG_{743} in poliovirus type 1).

Figure 4 Conservation of sequence and structure among the entero-/rhinovirus IRESs. The sequence shown is that of poliovirus type 1 (with the sequence of Domain VI of human rhinovirus 2 as an inset), and the structure is based on an amalgamation of Pilipenko et al. (1989b), Andino et al. (1990), and Pöyry et al. (1992). Blacked-out residues are absolutely conserved, whereas shaded residues are conserved in more than 80% of the sequences analyzed, but as the analysis does not include any IRES sequence published after 1995, the pattern of conservation should be regarded as illustrative rather than definitive. In addition, this analysis was been confined to the IRES itself, and thus did not include Domain I and sequences downstream from the putative ribosome entry site AUG. Stem-loop A and loop B are two structural features shared with cardio-/aphthovirus IRESs (see Fig. 3). The binding sites of PCBP and viral 3CD to the cloverleaf (Domain I) are shown (Gamarnik and Andino 1997), and the residues in the top of Domain IV which are protected when PCBP-2 binds to the IRES are indicated by thickened shaded curves (Gamarnik and Andino 2000).

The first question raised by these findings is, Why is the AUG at the 3′ end of the IRES utilized inefficiently (FMDV L_{ab}) or not at all (PV AUG_{586})? Context is often invoked as the explanation, and it is true that upgrading the context of this AUG can dramatically increase the frequency of initiation at this site (Pestova et al. 1994; Lopez de Quinto and Martinez-Salas 1999; Ohlmann and Jackson 1999), although, curiously, this increase is not accompanied by a quantitatively equivalent decrease in initiation at the downstream AUG (AUG_{743} in PV-1 or the L_b AUG in FMDV). Moreover, with an EMCV–FMDV chimeric IRES there was much more initiation at the L_{ab} site and less at the L_b site as compared with a wild-type FMDV IRES, although the context of the L_{ab} site was identical in both IRESs (Ohlmann and Jackson 1999). Thus, although context may have some influence, it cannot be the whole explanation.

The next question is whether efficient initiation at the FMDV L_b site or the authentic entero-/rhinovirus initiation sites is strictly dependent on the AUG at the 3′ end of the IRES, which would imply that this AUG might be an obligatory ribosome entry site. For FMDV the answer appears to be negative, because although mutation of the L_b initiation codon is lethal, mutation of the upstream L_{ab} AUG, the putative obligatory entry site, has only a marginal influence on infectivity (Cao et al. 1995). Thus, ribosomes must be able to access the L_b site without having entered at an upstream AUG. It is not known whether this means that they can access the L_b site directly, or whether they can enter at any codon (not necessarily an AUG) in the approximate position of the L_{ab} initiation site.

For the entero-/rhinovirus IRESs the evidence is more controversial as to whether the (almost) silent AUG at the 3′ end of the IRES is essential. In one study, an AUG located close to nucleotide 586 of PV-1 was essential for infectivity; mutants that lacked an AUG in this position generated pseudo-revertants which acquired one, either by point mutation or by deletions that brought a previously more distant AUG to the appropriate location (Pilipenko et al. 1992). On the other hand, mutation of the equivalent AUG in poliovirus type 2 resulted in just a small plaque phenotype (rather than quasi-lethality) and reduced IRES-dependent translation of linked reporter cistrons by no more than 60–70% (Pelletier et al. 1988; Meerovitch et al. 1991; Nicholson et al. 1991).

Assuming ribosome entry is at some upstream site, although not necessarily at an AUG, the final question is whether the transfer to the downstream AUG, the FMDV L_b site or the unique authentic initiation site in entero-/rhinoviruses, is by strictly linear scanning. If this were the case, insertion of AUGs or hairpin structures between AUG_{586} and AUG_{743} in PV-1 RNA, or between the FMDV L_{ab} and L_b sites, would be expected to reduce initiation at AUG_{743} and the L_b site, respectively. This is indeed what is observed, but the magnitude of the reduction seems surprisingly small in some experiments (Pelletier and Sonenberg 1988; Belsham 1992; Hellen et al. 1994; Lopez de Quinto and Martinez-Salas 1999). Similarly, antisense oligonucleotides targeted to the region downstream of the FMDV L_{ab} site had a lesser effect on initiation at the L_b site than would be expected from a strictly linear scanning model (Lopez de Quinto and Martinez-Salas 1999). It is true that no AUG triplet has ever been found in the ~160-nucleotide segment between the putative entry site AUG and the authentic initiation codon in any natural poliovirus isolate (Pöyry et al. 1992), but insertion of an AUG into this region actually resulted in a small plaque phenotype rather than lethality (Kuge et al. 1989a,b).

To summarize, the enthusiasm to produce a unified model that can embrace all the major types of picornavirus IRES may have resulted in a less critical evaluation of the evidence than should have been the case. There is little doubt that binding at an upstream site and transfer by linear scanning can explain how the *majority* of ribosomes access the L_b site of FMDV and the authentic initiation codon of enteroviruses, but there remain doubts as to whether some ribosomes take a different route, and whether entry necessarily has to be at an AUG triplet.

A further problem, peculiar to the entero-/rhinovirus IRESs, is that the putative ribosome entry site AUG is at the base of a stem-loop (Fig. 4). This is conserved in all viruses of this group except bovine enterovirus, yet deletions that destroy it by removing the 3′ side have little influence on infectivity (Kuge and Nomoto 1987), implying that it is neither an impediment nor a positive determinant for ribosome entry. It is intuitive that this stem-loop must be opened, if not to allow silent ribosome entry at PV-1 nucleotide 586, then at least for initiation at the authentic start site in rhinovirus RNAs (Fig. 4) and for the initiation at PV-1 AUG_{586}, which can be induced by upgrading its context (Pestova et al. 1994; Ohlmann and Jackson 1999). It also seems intuitive that the melting of this stem must be due to a focused unwinding localized to this region, rather than the result of a general unwinding of the 5′UTR.

Role of Canonical Initiation Factors in Initiation on Picornavirus IRESs

One weakness of the model depicted in Figure 2A is that it says nothing about the role of canonical initiation factors in the mechanism of internal initiation. In fact, internal initiation on the two main classes of picornavirus IRES is believed to require all the canonical initiation factors needed for the scanning mechanism, except that eIF4E seems entirely redundant and the complete eIF4G polypeptide is not required (the hepatitis A virus IRES appears to be an exception to both of these rules.) Here again, it is the EMCV IRES that has provided the most information, mainly through investigation of initiation complex formation with 40S subunits and highly purified or recombinant initiation factors, assayed by sucrose gradients or toeprinting (Pestova et al. 1996a,b). Initiation complex formation absolutely required eIF2, 3, 4A, and either the complete eIF4F complex, or recombinant fragments of eIF4G that included the central domain, which has the eIF3 interaction site and one of the eIF4A-binding sites (Lamphear et al. 1995; Imataka and Sonenberg 1997). There was also a partial requirement for eIF4B, which increased the yield of initiation complexes by about twofold.

Toeprinting assays revealed that the central domain of eIF4G binds to the J-K domain of the EMCV IRES fairly close to the correct initiation site (Fig. 3). This toeprint was identical to that obtained with the eIF4F holoenzyme complex (Pestova et al. 1996a,b; Kolupaeva et al. 1998) and was uninfluenced by the presence or absence of eIF4A, eIF4B, and ATP. In UV-crosslinking assays with the eIF4F holoenzyme complex and the whole EMCV IRES, all three eIF4F polypeptides, even eIF4E, were crosslinked to the IRES (Pestova et al. 1996b). With recombinant subdomains of eIF4G and the same probe, there seemed to be cooperativity of binding (or of cross-linking) of eIF4A, eIF4B, and the central domain of eIF4G, but not the extreme carboxy-terminal domain (Pestova et al. 1996b). Although this assay was done with the entire EMCV IRES, the presumption is that the binding of all three polypeptides was to the J-K domain, an assumption which is supported by the fact that eIF4B in crude reticulocyte lysates can be crosslinked by UV irradiation to the J domain of the FMDV IRES in the presence of ATP (Meyer et al. 1995).

It is not known which part of eIF4G actually contacts the J-K domain. It was proposed that both isoforms of yeast eIF4G have an RNA-binding motif of the RNP-1/RNP-2 family (Goyer et al. 1993), and a similar motif is found in the central domain of mammalian eIF4GI and eIF4GII (Gradi et al. 1998b; Gingras et al. 1999). However, the putative RNP-1 and RNP-2 motifs are somewhat noncanonical, and the spacing between them is greater than is typical. There is no direct evidence that this is an active RNA-binding domain, or that it is responsible for the binding of eIF4G to the EMCV IRES.

It seems likely that the binding of the central domain of eIF4G (together with eIF4A and 4B) to the J-K domain of the EMCV IRES serves to deliver the 40S subunit to the nearby initiation site via the eIF4G/eIF3/40S interaction relay. However, this cannot be the whole secret of internal initiation, otherwise the isolated J-K domain of the IRES would be sufficient to direct internal initiation, which is certainly not the case. The upstream H and I domains (Fig. 3) must play an essential role, which is unlikely to be related to eIF4G binding, since fairly drastic mutations in stem-loop I have no effect on the eIF4G/J-K domain interactions. It is not known whether domains H and I bind other initiation factors, or whether they are the site of ribosome–IRES interactions. However, if the latter is the case, the interactions must be rather weak, since binding of the 40S subunits to the IRES in the absence of initiation factors cannot be detected (Pestova et al. 1996a).

As discussed in the preceding section, internal initiation on the EMCV IRES involves very little, if any, ribosome scanning. Nevertheless, singular eIF4A is required even if initiation complex formation is driven by the eIF4F complex complete with its bound eIF4A subunit (Pestova et al. 1996a). It also requires ATP hydrolysis, albeit lower ATP concentrations than are needed for scanning-dependent initiation (Jackson 1991). The ATP requirement can be substituted by dATP with reasonable efficiency (Pestova et al. 1996b), which correlates with the fact that in RNA unwinding assays dATP can support the helicase activity of eIF4A (Rozen et al. 1990). The requirement for eIF4A, probably as part of eIF4F, is further confirmed by the fact that translation dependent on the EMCV IRES is as sensitive to inhibition by dominant negative eIF4A mutants as is initiation dependent on the conventional scanning mechanism (Pause et al. 1994a). In fact, some long-standing data suggest that translation of EMCV RNA may need *more* eIF4A than does conventional scanning-dependent translation initiation. Partial fractionation of a cell-free system from ascites cells decreased its capacity to translate EMCV RNA to a greater extent than globin mRNA. A factor that selectively restored EMCV RNA translation was purified; originally named IF_{EMC}, it subsequently turned out to be eIF4A (Wigle and Smith 1973).

This requirement for eIF4A and ATP hydrolysis despite the fact that there is probably no scanning often causes some surprise, since the function of eIF4A has often been considered to be related to scanning. However, as discussed elsewhere in this volume (see Chapter 4), results with dominant negative eIF4A mutants have questioned whether singular eIF4A plays a significant role in translation initiation. It is suggested, instead, that the main influence of eIF4A is exerted as a subunit of eIF4F holoenzyme complex, i.e. in association with eIF4G (Pause et al. 1994a). Thus it seems likely that the binding of eIF4G (and associated eIF4A) to the J-K domain of the EMCV IRES not only delivers the 40S subunit to the nearby initiation site via the eIF4G/eIF3/40S subunit interaction relay, but also focuses the helicase activity of the associated eIF4A to unwind the region around the initiation site.

It is not known whether the central domain of eIF4G also binds to entero-/rhinovirus IRESs, and if so, to what site. A feature of the eIF4G-binding site in the EMCV IRES is an A-rich bulge (Fig. 3); binding of eIF4G strongly protects these A residues against chemical reagents (Kolupaeva et al. 1998). The entero-/rhinovirus IRESs have no obvious equivalent to this A-rich bulge. Nevertheless, the expectation is that eIF4G will be found to bind at a specific site in the entero-/rhinovirus IRESs. Certainly, the poliovirus and EMCV IRESs resemble each other with respect to their sensitivity to inhibition by dominant negative eIF4A mutants (Pause et al. 1994a). Consequently, the model depicted in Figure 2A needs modifying, or clarifying (see Fig. 2B), to convey that an important aspect of internal initiation, although not the complete explanation, is the site-specific binding of eIF4G to the IRES that, via the eIF4G/eIF3/40S subunit relay, helps to deliver the ribosomal subunit to the internal entry site and directs the action of the associated eIF4A helicase to this region.

Identification of Cellular RNA-binding Proteins Required for Picornavirus IRES Function

It has been known for over 20 years that poliovirus RNA translation in rabbit reticulocyte lysates is inefficient and inaccurate (especially at high RNA concentration), but can be rescued by addition of HeLa cell fractions (Brown and Ehrenfeld 1979; Dorner et al. 1984). Similar results were reported subsequently for other enteroviruses, and especially for rhinovirus RNAs (Borman et al. 1993, 1995). In addition, supplementation with liver extracts, but not HeLa cell cytoplasm, stimulated the activity of the HAV IRES in reticulocyte lysate systems (Glass and Summers 1993). Thus, it is clear that the activity of at least some IRESs requires other cellular proteins in addition to canonical initiation factors, which has aroused considerable interest as a possible explanation for the tissue tropism of the viruses.

In the early 1990s, UV-crosslinking reactions were widely used in attempts to identify which proteins bind specifically to the IRESs, in the hope that these would turn out to be required for the activity of the IRES. It is now recognized that there is no substitute for a functional assay; there are several examples of proteins that bind to an IRES apparently at a specific site, yet the binding has no effect on IRES activity (Kaminski and

Jackson 1998; Walter et al. 1999). The autoantigen La was the first cellular factor to be identified by a functional assay, although initial identification was by virtue of the fact that it binds to the region of the poliovirus IRES representing the putative ribosome entry site (AUG_{586} in PV-1, Fig. 4). Addition of La to rabbit reticulocyte lysates was shown to give a fair stimulation of poliovirus RNA translation, and to improve the accuracy of selection of the correct initiation site (Meerovitch et al. 1993; Svitkin et al. 1994). However, the concentrations of La required for this effect were significantly higher than would be present when poliovirus RNA translation in reticulocyte lysates is rescued by addition of HeLa cell extracts. It is still not clear why such high concentrations are needed. Thus, it can be argued that La is not a physiologically relevant stimulator of enterovirus IRES activity, but at high concentrations it may mimic the physiological factors. It is noteworthy, however, that poliovirus infection results in the redistribution of La from the nucleus to the cytoplasm (Meerovitch et al. 1993) following poliovirus 3C protease-mediated removal of the extreme carboxyl terminus of La (Shiroki et al. 1999).

A strictly functional in vitro translation assay was used to purify those factors which are necessary for rhinovirus IRES activity and which are present in reticulocyte lysates at relatively low abundance relative to HeLa cell extracts (Hunt and Jackson 1999; Hunt et al. 1999). Two HeLa cell factors were isolated: When tested individually, each stimulated rhinovirus IRES activity in the reticulocyte lysate, and together their effects were at least additive and more often synergistic (Hunt and Jackson 1999; Hunt et al. 1999). One of them proved to be polypyrimidine tract-binding protein (PTB) (Hunt and Jackson 1999), a protein whose normal role is believed to be regulation of alternative splicing (Valcarcel and Gebauer 1997). It is located predominantly in the nucleus, although it does have a significant cytoplasmic presence (Ghetti et al. 1992). The other factor was *unr*, associated with a novel GH-WD repeat protein (Hunt et al. 1999). *unr*, which gets its name from the fact it is encoded by a gene located just *u*pstream of *N-r*as, is a single-stranded nucleic-acid-binding protein, that is fairly ubiquitously expressed in mammalian tissues and is largely cytoplasmic (Boussadia et al. 1993; Jacquemin-Sablon et al. 1994).

The rhinovirus, rather than poliovirus, IRES was used for these purification assays because it gives a lower background of IRES activity in the unsupplemented reticulocyte lysate, yet the close structural similarity of the IRESs gave reason to suppose that the same factors might be required for both IRESs. This turned out not to be true. Although addition of PTB to reticulocyte lysates stimulates the poliovirus IRES, *unr* has little effect, and although relatively low concentrations of PTB and *unr* stimulate HRV IRES activity as effectively as HeLa cell extract, the same is not true for the poliovirus IRES (Hunt et al. 1999). Thus, it seems that there must be at least one other, as yet unidentified, protein needed for maximum poliovirus IRES activity. It will be difficult to identify this using the reticulocyte lysate system as a functional assay, since, unlike the HRV IRES, the poliovirus IRES works moderately efficiently in reticulocyte lysates at the low RNA concentrations favorable for monitoring purification of *trans*-acting factors. A more promising approach may be that developed by Gamarnik and Andino (1996), who found that the poliovirus IRES functions very inefficiently in Stage VI *Xenopus* oocytes, unless HeLa cell cytoplasm was coinjected or HeLa cell mRNA had been previously injected. The HeLa cell activity behaved as a single entity of around 300 kD, which does not correspond to any of the proteins discussed above.

Another protein found to be necessary for the activity of the entero- and rhinovirus IRESs is poly(rC) binding protein-2 (PCBP-2), a cytoplasmic RNA-binding protein with three KH domains that is relatively abundant not only in HeLa cell extracts but also in reticulocyte lysates. This was originally identified as a protein that binds to stem-loop IV (Fig. 4) of the poliovirus IRES (Blyn et al. 1996). When HeLa cell extracts were depleted of PCBPs by an affinity column procedure, they lost the capacity to support translation dependent on the IRESs of poliovirus, coxsackie B virus, and rhinovirus, but EMCV or FMDV IRES activity was unaffected. Recombinant PCBP-2, but not the closely related PCBP-1 (>80% amino acid identity), restored the activity of the entero-/rhinovirus IRESs (Blyn et al. 1997; Walter et al. 1999).

Cardiovirus IRESs function very efficiently not only in rabbit reticulocyte lysates but also in *Xenopus* oocytes (Laskey et al. 1972; Gamarnik and Andino 1996). Nevertheless, the fact that they have a much higher affinity for PTB than does the rhinovirus IRES, coupled with the knowledge that reticulocyte lysates do contain some PTB (Borman et al. 1993), albeit at lower concentrations than HeLa cells, raised the speculation that PTB might actually play a role in the function of the EMCV IRES. This idea seemed to gain some credibility when it was shown that depleting reticulocyte lysates of PTB by an affinity column approach had little effect on scanning-dependent translation but resulted in loss of IRES-dependent translation, which could be restored by addition of low concentrations of recombinant PTB (Kaminski et al. 1995). However, when the IRES of another cardiovirus, TMEV, was tested in the same system, no evidence of PTB dependence was seen (Kaminski et al. 1995), which seemed counterintuitive given the close relationship between EMCV and

TMEV. Closer study revealed that this difference was due to the fact that whereas the TMEV construct had a truly wild-type IRES and a reporter that started with viral coding sequences, the EMCV IRES used for this work had acquired an additional A-residue in the A-rich bulge in the J-K domain (Fig. 3) and was linked to a heterologous reporter. Thus, the TMEV IRES could be rendered highly dependent on PTB if all viral coding sequences were eliminated and the A-rich bulge was enlarged by one residue, and the EMCV IRES became PTB-independent when a wild-type IRES was linked to viral polyprotein coding sequences as reporter (Kaminski and Jackson 1998).

What Is the Role of Cellular RNA-binding Proteins in Internal Initiation?

It is particularly striking that these manipulations to the EMCV and TMEV IRESs make absolutely no difference to the binding of PTB, at least to the high-affinity site that is stem-loop H (Fig. 3). Thus, two types of RNA transcript can be obtained, a wild-type IRES linked to viral coding sequences, or an IRES with an enlarged A-bulge linked to a heterologous reporter. Both RNAs bind PTB with comparable affinity, yet in one case this binding has no influence on IRES activity and in the other it is almost absolutely essential for such activity (Kaminski and Jackson 1998). This seems incompatible with the idea that PTB binding directly promotes 40S subunit entry, and is much more consistent with the notion that the binding of PTB may, in special circumstances, help to stabilize the appropriate three-dimensional structure of the IRES. It is therefore pertinent that PTB has four RRMs in the monomer and probably exists in solution as a dimer (Pérez et al. 1997), which suggests that a PTB dimer could interact with the IRES at several different points, and thereby stabilize the higher-order structure. Footprinting has indeed shown that in addition to the high affinity site (stem-loop H), PTB interacts with several other sites (Kolupaeva et al. 1996), that appear to be quite widely dispersed in the IRES (Fig. 3). Moreover, in circumstances in which cardiovirus IRES activity is dependent on PTB, all four RRMs seem to be essential. The carboxy-terminal half of PTB, with two RRMs, binds equally tightly to the high-affinity site on the IRES yet cannot support internal initiation and actually inhibits the action of full-length PTB (Kaminski et al. 1995), which suggests that it is the multipoint contact with the IRES that is critical for stimulation by full-length PTB.

In a similar vein, it is worth noting that PCBP-2 has three copies of the KH domain that is thought to be the RNA-binding surface, and again exists as a dimer (Gamarnik and Andino 1997). As for *unr*, it is not known whether this exists as a monomer or dimer. It belongs to the cold-shock-domain family of single-stranded nucleic-acid-binding proteins (Graumann and Marahiel 1998), but it is the only member of this family with as many as five cold-shock domains, all of which share a unique amino acid sequence signature (Hunt et al. 1999). Thus, both *unr* and PCBP-2 proteins also have the potential for multipoint interaction with the IRES that could help in the establishment and maintenance of the appropriate higher-order structure. Although La is not generally considered to have multiple RNA-binding domains, it does dimerize, and there is evidence that dimerization is necessary for stimulation of poliovirus IRES activity (Craig et al. 1997).

These considerations suggest that it is much more likely that the role of these cellular RNA-binding proteins is to help to maintain or attain the appropriate three-dimensional structure of the IRES, rather than acting directly to recruit the 40S subunit in much the same way as the central domain of eIF4G bound to the J-K domain of the EMCV IRES serves as a ribosome recruitment factor. In short, their role is much more like that of protein Y in the model shown in Figure 2, than of protein X. Regrettably, there is no indication that the tissue tropism of the viruses can be explained by the tissue distribution of the protein factors that have been identified so far as necessary for IRES activity. In particular, it is still not known why the single site mutation that is the critical difference between the IRESs of poliovirus Sabin vaccine strains and their wild-type neurovirulent parents (Fig. 4) restricts IRES activity in neuronal cells but not in HeLa cells.

MECHANISMS AND CONSEQUENCES OF VIRUS-INDUCED PERTURBATION OF THE TRANSLATION MACHINERY

eIF4G Cleavage in Host-cell Shutoff

Infection with entero-/rhinoviruses or FMDV results in a fairly rapid inhibition of host-cell protein synthesis that is accompanied by cleavage of eIF4G into an amino-terminal one-third fragment, which has the eIF4E-binding site, and a carboxy-terminal two-thirds fragment that includes the eIF3 and both eIF4A interaction sites (Lamphear et al. 1995; Imataka and Sonenberg 1997). The resulting separation of the cap-binding function from the helicase and eIF3 interaction domains results in an inhibition of cap-dependent host-cell mRNA translation, but does not block viral IRES-dependent translation (discussed in more detail below). However, in the presence of inhibitors of viral RNA replication, the correlation between eIF4G cleavage and inhibition of host-cell protein synthesis was seen to break down, since complete eIF4G cleavage was observed and yet host-cell mRNA translation was partially maintained (Bonneau and Sonenberg 1987; Perez and Carrasco 1992). These experiments were done at a time when it was believed that there is a single species of eIF4G. The subsequent discovery of a functional homolog (eIF4GII) of eIF4G (now designated as eIF4GI) with only 46% identity to eIF4GI appears to provide an explanation for the discrepancy (Gradi et al. 1998b). The cleavage of eIF4GI occurs early after infection by entero-/rhinoviruses and is still observed when viral RNA replication is blocked, whereas cleavage of eIF4GII occurs more slowly, especially in the presence of replication inhibitors. Thus, the complete shutoff of cellular protein synthesis correlates with cleavage of both eIF4GI and eIF4GII, and loss of all intact eIF4G (Gradi et al. 1998a).

Mutation of poliovirus 2A (Bernstein et al. 1985) results in a greatly delayed and incomplete shutoff of host-cell mRNA translation, and the same is true if FMDV L-protease is deleted (Piccone et al. 1995), although in this case the delay is less acute since FMDV 3C protease also cleaves eIF4G (Belsham et al. 2000). However, although these results indicate that eIF4G cleavage is certainly triggered by the 2A or L-proteases, there is still controversy over whether the cleavage is direct or indirect. There is no doubt that the viral proteases *can* cleave eIF4G directly (Lamphear et al. 1993; Liebig et al. 1993; Kirchweger et al. 1994; Bovee et al. 1998a). What is in dispute, however, is whether direct cleavage is what happens in the infected cell. One reason for questioning whether cleavage is direct is the fact that when a purification of the eIF4G cleavage activity from poliovirus-infected cells was attempted, the cleavage activity did not copurify with the 2A protein (Bovee et al. 1998b). Another reason is that direct cleavage of eIF4G in vitro using recombinant 2A proteases expressed in bacteria requires high levels of protease (Bovee et al. 1998a). In contrast, cleavage in the infected cell probably requires comparatively little protease, given that cleavage can occur within 2–3 hours postinfection even if RNA replication has been blocked by guanidine, and thus the only source of 2A is translation of the input RNA (Gradi et al. 1998a).

It is likely that a similar discrepancy exists with respect to FMDV L-protease. When L-protease is expressed by in vitro translation in rabbit reticulocyte lysate, an extremely potent cleaving activity is generated (Ohlmann et al. 1995), such that if it is diluted 100-fold into fresh lysate, all the eIF4GI in the fresh lysate is cleaved within 10 minutes. On the basis of the claims as to how much protein is synthesized in commercially available reticulocyte lysates, it is likely that complete cleavage of eIF4G occurs with as little as 10–20 ng/ml L-protease. In contrast, concentrations in the range of 1–10 μg/ml of bacterially expressed recombinant L-protease generally have to be used to achieve a similarly rapid cleavage (Ohlmann et al. 1996; Borman et al. 1997b). Cleavage of eIF4G in vivo is also likely to require only very low protease concentrations, since it occurs even when cells are electroporated with a replication-incompetent derivative of a FMDV replicon (Belsham et al. 2000).

Thus, despite the fact that recombinant viral proteases can cleave eIF4G directly, there is reason to question whether this is what happens in the infected cell. Furthermore, in view of the observation that the eIF4G cleaving activity from poliovirus-infected cells does not copurify with 2A (Bovee et al. 1998b), an alternative explanation is that the viral proteases activate latent cellular proteases, and it is mostly these that cleave eIF4G.

Effect of Picornavirus Proteases on Viral IRES Activity

Since many picornaviruses induce cleavage of eIF4G early in infection, it is not surprising that the activity of their IRESs is not inhibited by this cleavage (Borman et al. 1995, 1997a,b; Roberts et al. 1998), even in the case of EMCV, which does not induce eIF4G cleavage. The one notable exception to this generalization is the hepatitis A virus IRES, which appears to require intact eIF4G (Borman and Kean 1997), and perhaps also eIF4E. Apart from this exception, there can be little doubt that the eIF4G requirement for picornavirus IRES activity can be fully supported by just the carboxy-terminal two-thirds fragment, and at least in some cases by just the central one-third domain (Ohlmann et al. 1996; Pestova et al. 1996b).

More intriguing and more controversial are reports that picornavirus IRES activity can actually be stimulated by viral proteases in certain situations. When stimulation is observed, the hierarchy is generally HRV•PV>>EMCV>FMDV, but the absolute magnitude varies according to the system. In HeLa cell transfections, or in rabbit reticulocyte lysates, only the entero-/rhinovirus IRESs are stimulated significantly (Hambidge and Sarnow 1992; Borman et al. 1995, 1997a). At the other extreme, in Neuro 2A or especially NRK cells, coexpression of the proteases stimulated all the IRESs tested, even the FMDV IRES (Borman et al. 1997a; Roberts et al. 1998).

There are two schools of thought about this stimulation by viral proteases. One is that it is due to cleavage of eIF4G, implying that the carboxy-terminal two-thirds cleavage product must somehow be more efficient than the uncleaved factor in promoting IRES-dependent translation, for unknown reasons. In support of this argument, addition of the carboxy-terminal cleavage product stimulated HRV IRES activity in vitro (Borman et al. 1997b), although the stimulation (less than twofold) was much less than the sevenfold achieved by adding viral protease. In transfection assays, coexpression of the carboxy-terminal two-thirds fragment of eIF4G stimulated EMCV IRES activity fourfold in COS cells (Yamanaka et al. 1997) and twofold in HeLa cells (Imataka and Sonenberg 1997). However, in other transfection assays, no stimulation of IRES activity was seen when the carboxy-terminal two-thirds fragment or the central domain of eIF4G was coexpressed in BHK cells (Roberts et al. 1998). Moreover, coexpression of an eIF4E-binding protein (4E-BP2; Pause et al. 1994b) did not stimulate the IRESs, although it did inhibit cap-dependent translation as expected, which shows that stimulation of the IRESs by coexpression of viral proteases cannot be just a secondary consequence of inhibition of competing translation of capped mRNAs. Thus, in view of the failure of coexpression of the carboxy-terminal two-thirds fragment of eIF4G to mimic the effect of the viral proteases, it was suggested that the stimulation of IRES activity is due to the cleavage of some other, as yet unidentified, cellular protein (Roberts et al. 1998).

Mechanism of Host-cell Shutoff in Cardiovirus Infection

Infection by cardioviruses does not result in cleavage of eIF4G, yet these viruses do shutoff host-cell protein synthesis, albeit in a more gradual and progressive manner than is seen with entero-/rhinoviruses or FMDV. The degree of shutoff is profoundly influenced by the tonicity of the medium (Alonso and Carrasco 1982). Host-cell mRNA translation is rescued in hypotonic conditions and strongly suppressed in hypertonic media. The main explanation for the shutoff during EMCV infection is the dephosphorylation of 4E-BPs by some unknown mechanism. Dephosphorylated 4E-BPs bind eIF4E, preventing the entry of eIF4E into the eIF4F complex, and hence the dephosphorylation results in inhibition of translation of capped cellular mRNAs but not viral RNA (Pause et al. 1994b). The evidence for this hypothesis is that the kinetics of dephosphorylation of 4E-BP1 parallel the inhibition of host-cell mRNA translation in EMCV-infected cells (Gingras et al. 1996), and rapamycin, which inhibits 4E-BP1 phosphorylation and therefore induces dephosphorylation, accelerated and augmented the inhibition of cellular protein synthesis (Beretta et al. 1996). Dephosphorylation of 4E-BP1 was also seen in poliovirus-infected cells but is unlikely to make a significant contribution to host-cell shutoff, since it occurred too late in the infectious cycle (Gingras et al. 1996).

Other Virus-induced Modifications to the Translation Machinery

In addition to eIF4G, some other components of the translation machinery become cleaved during picornavirus infection, although relatively late in the infectious cycle. One example is the cleavage of eIF4A by the FMDV 3C protease, which occurs too late to have a significant impact on host-cell shutoff, but could have a negative effect on viral protein synthesis and thereby serve to direct the newly synthesized plus-strand viral RNA toward encapsidation (Belsham et al. 2000).

Another protein cleaved during infection by poliovirus or coxsackie B viruses is poly(A) binding protein (PABP). Both the 2A and 3C proteases have been implicated in this cleavage (Joachims et al. 1999; Kerekette et al. 1999). Whereas cleavage of eIF4GI occurs rather suddenly during the progression of an infection, PABP cleavage is more progressive and gradual. Some 50% of PABP is still intact at the time when host-cell shutoff is complete. Nevertheless, the authors claim that other correlates between PABP cleavage and shutoff indicate that this event could be at least a contributor to host-cell shutoff, even if not the complete explanation. PABP binds to the 3′ poly(A) tails of cellular mRNAs but also interacts with the extreme amino terminus of eIF4G (Imataka et al. 1998), so that when eIF4G, as part of the eIF4F holoenzyme complex, is bound to the 5′ cap, the mRNA is effectively circularized. This is believed to enhance the efficiency of initiation, or reinitiation, particularly under conditions of strong competition between mRNAs (Preiss and Hentze 1998; Preiss et al. 1998). Thus, disruption of these interactions could, in principle, augment shutoff of host-cell mRNA translation. However, the enterovirus proteases cleave PABP near its caboxyl terminus (Joachims et al. 1999; Kerekette et al. 1999), and cleavage at this site would not be expected to prevent the binding of PABP to poly(A), nor its interaction with eIF4G (Imataka et al. 1998), and thus would not disrupt circularization of host-cell mRNAs.

Whether PABP cleavage has any influence on viral RNA translation is not known, for the simple reason that there are no data on whether the encoded poly(A) tail of picornaviruses has a significant influence on IRES-dependent translation. In principle, PABP/poly(A)-dependent circularization of viral RNA could be maintained throughout the whole infectious cycle in the case of cardioviruses. With entero-/rhinoviruses and FMDV, it could only occur in the early stages; the cleavage of eIF4G would disrupt any such circularization, and the carboxy-terminal two-thirds fragment of eIF4G would no longer be tethered to the mRNA via interaction with PABP bound to the poly(A) tail.

HOST-CELL mRNAs RESISTANT TO TRANSLATIONAL SHUTOFF

The translation of a few host-cell mRNAs persists for some time following the general shutoff of host-cell mRNA translation by poliovirus infection. The first of these to be identified was immunoglobulin heavy-chain-binding protein (BiP) mRNA, also known as glucose-regulated protein 78 (Sarnow 1989). Synthesis of BiP actually increased for a short period around the time of global shutoff, but later it too was shutoff at a time when viral protein synthesis was still in full flood or even increasing. These observations led to the demonstration that BiP mRNA could be translated by internal ribosome entry (Macejak and Sarnow 1991). More recently, cDNA microarray technology has been used to show that at least 200 different species of cellular mRNA (out of 7000 screened) remain polysome-associated just after shutoff of global host-cell protein synthesis (Johannes et al.1999; Carter et al., Chapter 19). It would be interesting to know whether this cohort includes mRNAs coding for antiviral response proteins, which might have specifically evolved to be resistant to the virus-induced shutoff.

Although it is likely that this group of cellular mRNAs includes many that are translated by an IRES-dependent mechanism, this is not necessarily true for all of them. This experiment is really testing whether translation of a given mRNA can occur under conditions of no, or very low, eIF4F holoenzyme activity. Some mRNAs, such as late adenovirus mRNAs (Dolph et al. 1988), vaccinia virus mRNAs (Mulder et al. 1998), or heat-shock protein mRNAs (Castrillo and Carrasco 1987), would score positive in this assay, yet they are all almost certainly translated by a cap-dependent mechanism that requires eIF4F, even if only at very low concentration, as discussed elsewhere in this volume (see Chapters 4 and 17). On the other hand, hepatitis A virus RNA would score negative in this test, even though it is almost unquestionably translated by an IRES-dependent mechanism, albeit one which requires intact eIF4F holoenzyme complex (Borman et al. 1995; Borman and Kean 1997).

SWITCHING BETWEEN TRANSLATION AND MINUS-STRAND RNA SYNTHESIS

With all positive-strand RNA viruses there is the problem that the input RNA has to serve both as a template for translation by ribosomes in a 5′–3′ direction, and also as a template for minus-strand RNA synthesis by viral RNA polymerase moving in the 3′–5′ direction. It is intuitive that these two processes are mutually incompatible, and, indeed, in an in vitro system where both events could potentially occur in competition, translation won outright, and there was no RNA synthesis whatsoever (Gamarnik and Andino 1998). Thus, translation initiation must be temporarily halted in order to allow negative-strand synthesis, and some insights into how this might happen in the poliovirus system are starting to emerge.

The 5′-proximal sequences, which exist as a cloverleaf structure (Fig. 4), bind the viral 3CD protein, an uncleaved precursor of 3C protease and 3D RNA polymerase (Andino et al. 1990, 1993). The actual RNA-binding entity is the 3C moiety, since mutations in the cloverleaf can be compensated by mutations in 3C (Andino et al. 1990). There is evidence that this binding of 3CD to the cloverleaf (Fig. 4) inhibits translation dependent on the poliovirus IRES (Gamarnik and Andino 1998) by an indirect mechanism involving changes in the status of PCBP interaction with the 5′UTR (Gamarnik and Andino 2000).

A complex of 3CD and PCBP forms at the cloverleaf (Andino et al. 1993; Gamarnik and Andino 1997; Parsley et al. 1997), and PCBP also binds to an internal site within the IRES (Fig. 4), stem-loop IV (Blyn et al. 1996, 1997; Gamarnik and Andino 1997; Walter et al. 1999). It is the amino-terminal of the three KH domains of PCBP that is involved in the binding to both sites (Silvera et al. 1999), and as PCBPs dimerize, it is possible, but not proven, that a PCBP dimer binds to both sites simultaneously, in effect bridging between these two parts of the 5′UTR. The binding of PCBP to the 5′UTR is necessary for IRES activity (Blyn et al. 1997; Gamarnik and Andino 1997; Walter et al. 1999), and by inference it is the binding to stem-loop IV that is critical for this (Gamarnik and Andino 2000).

The binding of 3CD to the cloverleaf results in a sixfold increase in the affinity of PCBP for this structure, coupled with a loss of detectable contacts between PCBP and stem-loop IV (Gamarnik and Andino 2000). Hence, the IRES is inactivated, permitting negative-strand RNA synthesis (Gamarnik and Andino 1998). One possible explanation for how increased affinity of PCBP binding at one site results in loss of contacts at the other is simple competition between the two sites for PCBP binding (Gamarnik and Andino 2000). Alternatively, if a PCBP dimer bridges between the cloverleaf and stem-loop IV, it is possible that changes in the status of interaction at one of the sites could disrupt binding to the other. The inhibition of translation would be readily reversible by dissociation of the 3CD, either as such, or after the protease activity of 3C has cleaved the 3CD, to generate separate 3C and 3D, neither of which binds strongly to the cloverleaf (Andino et al. 1993). This mechanism is likely to apply to other enteroviruses and rhinoviruses, which all have a 5′ cloverleaf and seem to require PCBP for IRES activity (Walter et al. 1999). In contrast, it could not apply exactly in this form to cardioviruses and FMDV, since these IRESs do not require PCBP (Walter et al. 1999).

CONCLUDING REMARKS

Quite apart from their intrinsic importance and interest, picornaviruses have contributed enormously to our understanding of eukaryotic translation mechanisms in general. If it were not for picornaviruses, the discovery of internal initiation of translation would have been greatly delayed. Another important contribution is the insight into the domain structure of eIF4G that has been gained from studying the cleavage of eIF4G by viral proteases. It is quite remarkable what complex and intricate mechanisms are embodied in just 8000 nucleotides of RNA with only ~7000 nucleotides of coding sequence. This should be a sobering thought for those who appear to believe that when the complete 3 billion base pair sequence of the human genome is known, all of human biology will become immediately obvious.

ACKNOWLEDGMENTS

We gratefully acknowledge the contributions of our colleagues and financial support from the BBSRC (G.J.B.) and the Wellcome Trust (R.J.J.).

REFERENCES

Alonso M.A. and Carrasco L. 1982. Reversion by hypotonic medium of the shut off of protein synthesis induced by encephalomyocarditis virus. *J. Virol.* **37:** 535–540.

Andino R., Rieckhof G.E., and Baltimore D. 1990. A functional ribonucleoprotein complex forms around the 5′ end of poliovirus RNA. *Cell* **63:** 369–380.

Andino R., Rieckhof G.E., Achacoso P.L., and Baltimore D. 1993. Poliovirus RNA synthesis utilizes an RNP complex formed around the 5′-end of viral RNA. *EMBO J.* **12:** 3587–3598.

Belsham G.J. 1992. Dual initiation sites of protein synthesis on foot-and-mouth disease virus RNA are selected following internal entry and scanning of ribosomes in vivo. *EMBO J.* **11:** 1105–1110.

Belsham G.J., McInerney G.M., and Ross-Smith N. 2000. Foot-and-mouth disease virus 3C protease induces the cleavage of translation initiation factors eIF4A and eIF4G within infected cells. *J. Virol.***74:** 272–280.

Beretta L., Svitkin Y.V., and Sonenberg N. 1996. Rapamycin stimulates viral protein synthesis and augments the shutoff of host cell protein synthesis upon picornavirus infection. *J. Virol.* **70:** 8993–8996.

Bernstein H.D., Sonenberg N., and Baltimore D. 1985. Poliovirus mutant that does not selectively inhibit host-cell protein synthesis. *Mol. Cell. Biol.* **5:** 2913–2923.

Blyn L.B., Towner J.S., Semler B.L., and Ehrenfeld E. 1997. Requirement of poly(rC) binding protein 2 for translation of poliovirus RNA. *J. Virol.* **71:** 6243–6246.

Blyn L.B., Swiderek K.M., Richards O., Stahl D.C. , Semler B.L., and Ehrenfeld E. 1996. Poly(rC) binding protein 2 binds to stem-loop IV of the poliovirus RNA 5′ noncoding region: Identification by automated liquid chromatography-tandem mass spectrometry. *Proc. Natl. Acad. Sci.* **93:** 11115–11120.

Bonneau A.-M. and Sonenberg N. 1987. Proteolysis of the p220 component of the cap-binding protein complex is not sufficient for complete inhibition of host cell protein synthesis after poliovirus infection. *J. Virol.* **61:** 986–991.

Borman A.M. and Kean K.M. 1997. Intact eukaryotic initiation factor 4G is required for hepatitis A virus internal initiation of translation. *Virology* **237:** 129–136.

Borman A.M., Deliat F.G., and Kean K.M. 1994. Sequences within the poliovirus internal ribosome entry segment control viral RNA replication. *EMBO J.* **13:** 3149–3157.

Borman A.M., Bailly J.-L., Girard M., and Kean K.M. 1995. Picornavirus internal ribosome entry segments-comparison of translation efficiency and the requirements for optimal internal initiation of translation in vitro. *Nucleic Acids Res.* **23:** 3656–3663.

Borman A.M., Howell M.T., Patton J.G., and Jackson R.J. 1993. The involvement of a spliceosome component in internal initiation of human rhinovirus RNA translation. *J. Gen. Virol.* **74:** 1775–1788.

Borman A.M., Le Mercier P., Girard M., and Kean K.M. 1997a. Comparison of picornaviral IRES-driven internal initiation in cultured cells of different origins. *Nucleic Acids Res.* **25:** 925–932.

Borman A.M., Kirchweger R., Ziegler E., Rhoads R.E., Skern T., and Kean K.M. 1997b. eIF4G and its proteolytic cleavage products: Effect on initiation of protein synthesis from capped, uncapped and IRES-containing mRNAs. *RNA* **3:** 186–196.

Boussadia O., Jacquemin-Sablon H., and Dautry F. 1993. Exon skipping in the expression of the gene immediately upstream of N-*ras* (*unr*/*NRU*). *Biochim. Biophys. Acta* **1172:** 64–72.

Bovee M.L., Lamphear B.J., Rhoads R.E., and Lloyd R.E. 1998a. Direct cleavage of eIF4G by poliovirus 2A protease is inefficient in vitro. *Virology* **245:** 241–249.

Bovee M.L., Marissen W.E., Zamora M., and Lloyd R.E. 1998b. The predominant eIF4G-specific cleavage activity in poliovirus HeLa cells is distinct from 2A protease. *Virology* **245:** 229–240.

Brown B. and Ehrenfeld E. 1979. Translation of poliovirus RNA in vitro: Changes in cleavage pattern and initiation sites by ribosomal salt wash. *Virology* **97:** 396–405.

Cao X.M., Bergmann I.E., Fullkrug R., and Beck E. 1995. Functional analysis of the two alternative translation initiation sites of foot-and-mouth-disease virus. *J. Virol.* **69:**

560–563.

Castrillo J.L. and Carrasco L. 1987. Adenovirus late protein synthesis is resistant to the inhibition of translation induced by poliovirus. *J. Biol. Chem.* **262:** 7328–7334.

Craig A.W.B., Svitkin Y.V., Lee H.S., Belsham G.J., and Sonenberg N. 1997. The La autoantigen contains a dimerization domain that is essential for enhancing translation. *Mol. Cell. Biol.* **17:** 163–169.

Dolph P.J., Racaniello V., Villamarin A., Palladion F., and Schneider R.J. 1988. The adenovirus tripartite leader eliminates the requirement for cap binding protein during translation initiation. *J. Virol.* **62:** 2059–2066.

Dorner A.J., Semler B.L., Jackson R.J., Hanecak R., Duprey E., and Wimmer E. 1984. In vitro translation of poliovirus RNA: Utilization of internal initiation sites in reticulocyte lysate. *J. Virol.* **50:** 507–514.

Drew J. and Belsham G.J. 1994. Trans complementation of defective foot-and-mouth disease virus internal ribosome entry site elements. *J. Virol.* **68:** 697–703.

Follett E.A.C., Pringle C.R., and Pennington T.H. 1975 Virus development in enucleate cells: Echovirus, poliovirus, pseudorabies virus, reovirus, respiratory syncytial virus and Semliki forest virus. *J. Gen. Virol.* **26:** 183–196.

Gamarnik A.V. and Andino R. 1996. Replication of poliovirus in *Xenopus* oocytes requires two human factors. *EMBO J.* **15:** 5988–5998.

———. 1997. Two functional complexes formed by KH domain containing proteins with the 5′ noncoding region of poliovirus RNA. *RNA* **3:** 882–892.

———. 1998. Switch from translation to RNA replication in a positive stranded RNA virus. *Genes Dev.* **12:** 2293–2304.

———. 2000. Interactions of viral protein 3CD and poly(rC) binding protein with the 5′ untranslated region of the poliovirus genome. *J. Virol.* **74:** 2219–2226.

Ghetti A., Piñol-Roma S., Michael W.M., Morandi C., and Dreyfuss G. 1992. hnRNP I, the polypyrimidine tract binding protein: Distinct nuclear localization and association with hnRNPs. *Nucleic Acids Res.* **14:** 3671–3678.

Gingras A.-C., Raught B., and Sonenberg N. 1999. eIF4 initiation factors: Effectors of mRNA recruitment to ribosomes and regulators of translation. *Annu. Rev. Biochem.* **68:** 913–963.

Gingras A.-C., Svitkin Y., Belsham G.J., Pause A., and Sonenberg N. 1996. Activation of the translational suppressor 4E-BP1 following infection with encephalomyocarditis virus and poliovirus. *Proc. Natl. Acad. Sci.* **93:** 5578–5583.

Glass M.J. and Summers D.F. 1993. Identification of a *trans*-acting activity from liver that stimulates hepatitis A virus translation in vitro. *Virology* **193:**1047–1050.

Goyer C., Altmann M., Lee H.S., Blanc A., Deshmukh M., Woolford J.L., Trachsel H., and Sonenberg N. 1993. *TIF4631* and *TIF4632:* Two yeast genes encoding the high molecular weight subunits of cap-binding protein complex (eukaryotic initiation factor 4F) contain an RNA recognition motif-like sequence and carry out an essential function. *Mol. Cell. Biol.* **13:** 4860–4874.

Gradi A., Svitkin Y.V., Imataka H., and Sonenberg N. 1998a. Proteolysis of human eukaryotic translation initiation factor eIFGII, but not eIF4GI, coincides with the shutoff of host protein synthesis after poliovirus infection. *Proc. Natl. Acad. Sci.* **95:** 11089–11094.

Gradi A., Imataka H., Svitkin Y.V., Rom E., Raught B., Morino S., and Sonenberg N. 1998b. A novel functional human eukaryotic translation initiation factor 4G. *Mol. Cell. Biol.* **18:** 334–342.

Graumann P.L. and Marahiel M.A. 1998. A superfamily of proteins that contain the cold-shock domain. *Trends Biochem. Sci.* **23:** 286–290.

Gromeier M., Alexander L., and Wimmer E. 1996. Internal ribosome entry site substitution eliminates neurovirulence in intergenic poliovirus recombinants. *Proc. Natl. Acad. Sci.* **93:** 2370–2375.

Gromeier M., Bossert B., Arita M., Nomoto A., and Wimmer E. 1999. Dual stem loops within the poliovirus internal ribosome entry site control neurovirulence. *J. Virol.* **73:** 958–964.

Hambidge S.J. and Sarnow P. 1992. Translational enhancement of the poliovirus 5′ noncoding region mediated by virus-encoded polypeptide 2A. *Proc. Natl. Acad. Sci.* **89:** 10272–10276.

Hellen C.U.T., Pestova T.V., and Wimmer E. 1994. Effect of mutations downstream of the internal ribosome entry site on initiation of poliovirus protein synthesis. *J. Virol.* **68:** 6312–6322.

Hunt S.L. and Jackson R.J. 1999. Polypyrimidine tract binding protein (PTB) is necessary, but not sufficient, for efficient internal initiation of translation of human rhinovirus-2 RNA. *RNA* **5:** 344–359.

Hunt S.L., Hsuan J.J., Totty N., and Jackson R.J. 1999. unr, a cellular cytoplasmic RNA-binding protein with five cold-shock domains, is required for internal initiation of translation of human rhinovirus RNA. *Genes. Dev.* **13:** 437–448.

Imataka H. and Sonenberg N. 1997. Human eukaryotic initiation factor 4G (eIF4G) possesses two separate and independent binding sites for eIF4A. *Mol. Cell. Biol.* **17:** 6940–6947.

Imataka H., Gradi A., and Sonenberg N. 1998. A newly identified N-terminal amino acid sequence of human eIF4G binds poly(A)-binding protein and functions in poly(A)-dependent translation. *EMBO J.* **17:** 7480–7489.

Jackson R.J. 1991. The ATP requirement for initiation of eukaryotic translation varies according to the mRNA species. *Eur. J. Biochem.* **200:** 285–294.

———. 1996. A comparative view of initiation site selection mechanisms. In *Translational control* (ed. J.W.B. Hershey et al.), pp. 71–112. Cold Spring Harbor Laboratory Press, Cold Spring Harbor, New York.

Jackson R.J. and Kaminski A. 1995. Internal initiation of translation in eukaryotes: The picornavirus paradigm and beyond. *RNA* **1:** 985–1000.

Jacquemin-Sablon H., Triqueneaux G., Deschamps S., le Maire M., Doniger J., and Dautry F. 1994. Nucleic acid binding and intracellular localization of unr, a protein with five cold shock domains. *Nucleic Acids Res.* **22:** 2643–2650.

Jang S.K., Krausslich H.-G., Nicklin M.J.H., Duke G.M., Palmenberg A.C., and Wimmer E. 1988. A segment of the 5′ nontranslated region of encephalomyocarditis virus RNA directs internal entry of ribosomes during in vitro translation. *J. Virol.* **62:** 2636–2643.

Joachims M., van Breugel P.C., and Lloyd R.E. 1999. Cleavage of poly(A)-binding protein by enterovirus proteases concurrent with inhibition of translation in vitro. *J. Virol.* **73:** 718–727.

Johannes G., Carter M.S., Eisen M.B., Brown P.O., and Sarnow P. 1999. Identification of eukaryotic mRNAs that are translated at reduced cap binding complex eIF4F concentrations using a cDNA microarray. *Proc. Natl. Acad. Sci.* **96:** 13118–13123.

Kaminski A. and Jackson R.J. 1998. The polypyrimidine tract binding protein (PTB) requirement for internal initiation of translation of cardiovirus RNAs is conditional rather than absolute. *RNA* **4:** 626–638.

Kaminski A., Belsham G.J., and Jackson R.J. 1994. Translation of encephalomyocarditis virus RNA; parameters influencing the selection of the internal initiation site. *EMBO J.* **13:** 1673–1681.

Kaminski A., Howell M.T., and Jackson R.J. 1990. Initiation of encephalomyocarditis virus RNA translation: The authentic initiation site is not selected by a scanning mechanism. *EMBO J.* **9:** 3753–3759.

Kaminski A., Hunt S.L., Patton J.G., and Jackson R.J. 1995. Direct evidence that polypyrimidine tract binding protein (PTB) is essential for internal initiation of encephalomyocarditis virus RNA translation. *RNA* **1:** 924–938.

Kerekette V., Keiper B.D., Badorff C., Cai A., Knowlton K.U., and Rhoads R.E. 1999. Cleavage of poly(A) binding protein by coxsackie 2A protease in vitro and in vivo: Another mechanism for host protein synthesis shutoff? *J. Virol.* **73:** 709–717.

Kirchweger R., Ziegler E., Lamphear B.J., Waters D., Liebig H.D., Sommergruber W., Sobrino F., Hohenadl C., Blaas D., Rhoads R.E., and Skern T. 1994. Foot-and-mouth disease virus leader proteinase: Purification of the Lb form and determination of its cleavage site on eIF-4 gamma. *J. Virol.* **61:** 2711–2718.

Kolupaeva V.G., Hellen C.U.T., and Shatsky I.N. 1996. Structural analysis of the interaction of the pyrimidine tract binding protein with the internal ribosome entry segment of encephalomyocarditis virus and foot-and-mouth disease virus. *RNA* **2:** 1199–1212.

Kolupaeva V.G., Pestova T.V., Hellen C.U.T., and Shatsky I.N. 1998. Translation eukaryotic initiation factor 4G recognizes a specific structural element within the internal ribosome entry site of encephalomyocarditis virus RNA. *J. Biol. Chem.* **273:** 18599–18604.

Kuge S. and Nomoto A. 1987. Construction of viable deletion and insertion mutants of the Sabin strain of type 1 poliovirus: Function of the 5′ noncoding sequence in viral replication. *J. Virol.* **61:** 1478–1487.

Kuge S., Kawamura N., and Nomoto A. 1989a. Genetic variation occurring on the genome of an in vitro insertion mutant of poliovirus type 1. *J. Virol.* **63:** 1069–1075.

———. 1989b. Strong inclination toward transition mutations in nucleotide substitutions by poliovirus replicase. *J. Mol. Biol.* **207:** 175–182.

Lamphear B.J., Kirchweger R., Skern T., and Rhoads R.E. 1995. Mapping of functional domains in eukaryotic protein synthesis initiation factor 4G (eIF4G) with picornaviral proteases. *J. Biol. Chem.* **270:** 21975–21983.

Lamphear B.J., Yan R.Q., Yang F., Waters D., Liebig H.D., Klump H., Kuechler E., Skern T., and Rhoads R.E. 1993. Mapping the cleavage site in protein synthesis initiation factor eIF-4γ of the 2A proteases from human coxsackievirus and rhinovirus. *J. Biol. Chem.* **268:** 19200–19203.

Laskey R.A., Gurdon J.B., and Crawford L.V. 1972. Translation of encephalomyocarditis viral RNA in oocytes of *Xenopus laevis*. *Proc. Natl. Acad. Sci* **69:** 3665–3669.

Liebig H.D., Ziegler E., Yan R., Hartmuth K., Klump H., Kowalski H., Blaas D., Sommergruber W., Frasel L., Lamphear B., Rhoads R., Kuechler E., and Skern T. 1993. Purification of 2 picornaviral 2A proteinases: Interaction with eIF4γ and influence on *in vitro* translation. *Biochemistry* **32:** 7581–7588.

Lopez de Quinto S. and Martinez-Salas E. 1997. Conserved structural motifs located in distal loops of aphthovirus internal ribosome entry site domain 3 are required for internal initiation of translation. *J. Virol.* **71:** 4171–4175.

———. 1999. Involvement of the aphthovirus RNA region located between the two functional AUGs in start codon selection. *Virology* **255:** 324–336.

Macejak D.G. and Sarnow P. 1991. Internal initiation of translation mediated by the 5′ leader of a cellular mRNA. *Nature* **353:** 90–94.

Meerovitch K., Nicholson R., and Sonenberg N. 1991. In vitro mutational analysis of *cis*-acting RNA translational elements within the poliovirus type 2 5′ untranslated region. *J. Virol.* **65:** 5895–5901.

Meerovitch K., Svitkin Y.V., Lee H.S., Lejbkowicz F., Kenan D.J., Chan E.K.L., Agol V.I., Keene J.D., and Sonenberg N. 1993. La autoantigen enhances and corrects aberrant translation of poliovirus RNA in reticulocyte lysate. *J. Virol.* **67:** 3798–3807.

Meyer K., Petersen A., Niepmann M., and Beck E. 1995. Interaction of eukaryotic initiation factor eIF4B with a picornavirus internal translation site. *J. Virol.* **69:** 2819–2824.

Mulder J., Robertson M.E.M., Seamons R.A., and Belsham G.J. 1998. Vaccinia virus protein synthesis has a low requirement for the intact translation initiation factor eIF4F, the cap-binding complex, within infected cells. *J. Virol.* **72:** 8813–8819.

Nicholson R., Pelletier J., Le S.-Y., and Sonenberg N. 1991. Structural and functional analysis of the ribosome landing pad of poliovirus type 2: In vivo translation studies. *J. Virol.* **65:** 5886–5894.

Ohlmann T. and Jackson R.J. 1999. The properties of chimaeric picornavirus IRESes show that discrimination between internal initiation sites is influenced by the identity of the IRES and not just the context of the AUG codon. *RNA* **5:** 764–778.

Ohlmann T., Rau M., Morley S.J., and Pain V.M. 1995. Proteolytic cleavage of initiation factor eIF-4γ in the reticulocyte lysate inhibits translation of capped mRNAs but enhances that of uncapped mRNAs. *Nucleic Acids Res.* **23:** 334–340.

Ohlmann T., Rau M., Pain V.M., and Morley S.J. 1996. The C-terminal domain of eukaryotic protein synthesis initiation factor (eIF) 4G is sufficient to support cap-independent translation in the absence of eIF4E. *EMBO J.* **15:** 1371–1382.

Parsley T.B., Towner J.S., Blyn L.B., Ehrenfeld E., and Semler B.L. 1997. Poly (rC) binding protein 2 forms a ternary complex with the 5′-terminal sequences of poliovirus RNA and the viral 3CD proteinase. *RNA* **3:** 1124–1134.

Pause A., Methot N., Svitkin Y.V., Merrick W.C., and Sonenberg N. 1994a. Dominant negative mutants of mammalian translation initiation factor eIF-4A define a critical role for eIF-4F in cap-dependent and cap-independent initiation of translation. *EMBO J.* **13:** 1205–1215.

Pause A., Belsham G.J., Gingras A.-C., Donze O., Lin T.-A., Lawrence J.C., Jr., and Sonenberg N. 1994b. Insulin dependent stimulation of protein synthesis by phosphorylation of a novel regulator of 5′-cap function. *Nature* **371:** 762–767.

Pelletier J. and Sonenberg N. 1988. Internal initiation of translation of eukaryotic mRNA directed by a sequence derived from poliovirus RNA. *Nature* **334:** 320–325.

Pelletier J., Flynn M.E., Kaplan G., Racaniello V., and Sonenberg N. 1988. Mutational analysis of upstream AUG codons of poliovirus RNA. *J. Virol.* **62:** 4486–4492.

Pérez L. and Carrasco L. 1992. Lack of direct correlation between p220 cleavage and shut-off of host translation after poliovirus infection. *Virology* **189:** 178–186.

Pérez I., McAfee J.G., and Patton J.G. 1997. Multiple RRMs contribute to RNA binding specificity and affinity for polypyrimidine tract binding protein. *Biochemistry* **36:** 11881–11890.

Pestova T.V., Hellen C.U.T., and Shatsky I.N. 1996a. Canonical eukaryotic initiation-factors determine initiation of translation by internal ribosomal entry. *Mol. Cell. Biol.* **16:** 6859–6869.

Pestova T.V., Shatsky I.N., and Hellen C.U.T. 1996b. Functional dissection of eukaryotic initiation-factor 4F: The 4A subunit and the central domain of the 4G subunit are sufficient to mediate internal entry of 43S preinitiation complexes. *Mol. Cell. Biol.* **16:** 6870–6878.

Pestova T.V., Hellen C.U.T., and Wimmer E. 1994. A conserved AUG triplet in the 5′ non-translated region of poliovirus can function as an initiation codon in vitro and in vivo. *Virology* **204:** 729–737.

Piccone M.E., Rieder E., Mason P.W., and Grubman M.J. 1995. The foot-and-mouth disease leader proteinase gene is not required for viral replication. *J. Virol.* **69:** 5376–5382.

Pilipenko E.V., Blinov V.M., Chernov B.K., Dmitrieva T.M., and Agol V.I. 1989a. Conservation of the secondary structure elements of the 5′-untranslated region of cardio- and aphthoviruses. *Nucleic Acids Res.* **17:** 5701–5711.

Pilipenko E.V., Blinov V.M., Romanova L.I., Sinyakov A.N., Maslova S.V., and Agol V.I. 1989b. Conserved structural domains in the 5′-untranslated region of picornaviral genomes. *Virology* **168:** 201–209.

Pilipenko E.V., Gmyl A.P., Maslova S.V., Svitkin Y.V., Sinyakov A.N., and Agol V.I. 1992. Prokaryotic-like *cis* elements in the cap-independent internal initiation of translation on picornavirus RNA. *Cell* **68:** 119–131.

Pöyry T., Kinnunen L., and Hovi T. 1992. Genetic variation in vivo and proposed functional domains of the 5′ noncoding region of poliovirus RNA. *J. Virol.* **66:** 5313–5319.

Preiss T. and Hentze M.W. 1998. Dual function of the messenger RNA cap structure in poly(A)-tail-promoted translation in yeast. *Nature* **392:** 516–520.

Preiss T., Muckenthaler M., and Hentze M.W. 1998. Poly(A) tail promoted translation in yeast: Implications for translational control. *RNA* **4:** 1321–1331.

Ramos R. and Martinez-Salas E. 1999. Long-range RNA interactions between structural domains of the aphthovirus internal ribosome entry site (IRES). *RNA* **5:** 1374–1383.

Roberts L.O. and Belsham G.J. 1997. Complementation of defective picornavirus internal ribosome entry site (IRES) elements by the co-expression of fragments of the IRES. *Virology* **227:** 53–62.

Roberts L.O., Seamons R.A., and Belsham G.J. 1998. Recognition of picornavirus internal ribosome entry sites within cells; influence of cellular and viral proteins. *RNA* **4:** 520–529.

Robertson M.E.M., Seamons R.A., and Belsham G.J. 1999. A selection system for functional internal ribosome entry site (IRES) elements: Analysis of the requirement for a conserved GNRA tetraloop in the encephalomyocarditis virus IRES. *RNA* **5:** 1167–1179.

Rozen F., Edery I., Meerovitch K., Dever T.E., Merrick W.C., and Sonenberg N. 1990. Bidirectional RNA helicase activity of eucaryotic initiation factors 4A and 4F. *Mol. Cell. Biol.* **10:** 1134–1144.

Sangar D.V., Newton S.E., Rowlands D.J., and Clarke B.E. 1987. All foot and mouth disease serotypes initiate protein synthesis at two separate AUGs. *Nucleic Acids Res.* **15:** 3305–3315.

Sarnow P. 1989. Translation of glucose regulated protein 78/immunoglobulin heavy chain binding protein mRNA is increased in poliovirus-infected cells at a time when cap-dependent translation of cellular mRNAs is inhibited. *Proc. Natl. Acad. Sci.* **86:** 5795–5799.

Shiroki K., Isoyama T., Kuge S., Ishii T., Ohmi S., Hata S., Suzuki K., Takasaki Y., and Nomoto A. 1999. Intracellular redistribution of truncated La protein produced by poliovirus 3Cpro-mediated cleavage. *J. Virol.* **73:** 2193–2200.

Silvera D., Gamarnik A.V., and Andino R. 1999. The N-terminal K-homology domain of the poly(rC) binding protein is a major determinant for binding to the poliovirus 5′-untranslated region and acts as an inhibitor of viral translation. *J. Biol. Chem.* **274:** 38163–38170.

Simoes E.A.F. and Sarnow P. 1991. An RNA hairpin at the extreme 5′ end of the poliovirus RNA genome modulates viral translation in human cells. *J. Virol.* **65:** 913–921.

Stone D., Almond J.W., Brangwyn J.K., and Belsham G.J. 1993. Trans-complementation of cap-independent translation directed by poliovirus 5′ non-coding region deletion mutants: Evidence for RNA/RNA interactions. *J. Virol.* **67:** 6215–6223.

Svitkin Y.V., Meerovitch K., Lee H.S., Dholakia J.N., Kenan D.J., Agol V.I., and Sonenberg N. 1994. Internal translation initiation on poliovirus RNA: Further characterization of La function in poliovirus translation in vitro. *J. Virol.* **68:** 1544–1550.

Tang S.X., Collier A.J., and Elliott R.M. 1999. Alterations to both the primary and predicted secondary structure of stem-loop IIIc of the hepatitis C virus 5′ untranslated region (5′UTR) lead to mutants severely defective in translation which cannot be complemented in trans by the wild-type 5′UTR sequence. *J. Virol.* **73:** 2359–2364.

Valcarcel J. and Gebauer F. 1997. Post-transcriptional regulation: The dawn of PTB. *Curr. Biol.* **7:** R705–R708.

van der Velden A., Kaminski A., Jackson R.J., and Belsham G.J. 1995. Defective point mutants of encephalomyocarditis virus internal ribosome entry site can be complemented *in trans*. *Virology* **214:** 82–90.

Walter B.L., Nguyen J.H.C., Ehrenfeld E., and Semler B.L. 1999. Differential utilization of poly (rC) binding protein 2 in translation directed by picornavirus IRES elements. *RNA* **5:** 1570–1585.

Wigle D.T. and Smith A.E. 1973. Specificity in initiation in a fractionated mammalian cell-free system. *Nat. New Biol.* **242:** 136–140.

Yamanaka S., Poksay K.S., Arnold K.S., and Innerarity T.L. 1997. A novel translational repressor mRNA is edited extensively in livers containing tumors caused by the transgene expression of the apoB mRNA editing enzyme. *Genes Dev.* **11:** 321–333.

32

Adenovirus Inhibition of Cellular Protein Synthesis and Preferential Translation of Viral mRNAs

Robert J. Schneider
Department of Microbiology
New York University School of Medicine
New York, New York 10016

Human adenovirus dominates host cell protein synthesis, implementing sophisticated approaches that assure the exclusive translation of viral mRNAs while simultaneously suppressing those of the cell. Adenovirus also blocks the host cell antiviral response that would normally inhibit viral translation. Consequently, adenovirus–host cell interactions at the level of translation are complex. The current understanding of translational control in adenovirus-infected cells is discussed here, with a particular emphasis on establishing likely mechanisms for translational dominance of the host cell. A more detailed discussion of specific features of infection by adenovirus, and of host cell–virus interactions, can be found elsewhere (Schneider and Shenk 1987; Mathews and Shenk 1991; Schneider 1996; Chapter 8).

OVERVIEW OF ADENOVIRUS INFECTION AND GENE EXPRESSION

Adenoviruses infect a variety of tissues in humans and animals. The virus contains a double-stranded linear DNA genome, typically 36 kb in size. Adenovirus gene expression is organized into early and late phases, corresponding to expression of genes prior to, or subsequent to, viral DNA replication, respectively. Gene products synthesized during the early phase of infection are associated with functions typically required for viral DNA replication, suppression of host immune recognition, and activation of late viral genes. Late gene products encode structural and nonstructural polypeptides that are required in large amounts for assembly of viral capsids. A detailed understanding of adenovirus biology (and the

© 2000 Cold Spring Harbor Laboratory Press 0-87969-568-4/00 $5 +. 00

discussion herein) is restricted largely to human viruses, particularly serotypes 2, 5, 7, 9, and 12.

Early adenovirus gene expression takes place in the nucleus using cellular RNA polymerases, and can be detected within an hour of infection. Early gene transcription initiates from six different transcription units distributed throughout the genome, located on both DNA strands. Viral early gene expression gives rise to a low abundance of viral mRNAs and proteins in the cell and slightly stimulates cellular protein synthesis during this phase (Feigenblum and Schneider 1996; Gingras and Sonenberg 1997). Late adenovirus gene expression begins anywhere from 10 to 18 hours after infection of HeLa cells, although this can vary widely in different cell types. Most late adenovirus transcription is initiated from the major late promoter (MLP), which synthesizes a single long precursor RNA corresponding to 80% of the adenovirus genome. Five families of adenovirus late mRNAs (L1–L5) are derived from the MLP primary late transcript by differential splicing and polyadenylation (Fig. 1A). Abundant amounts of mRNAs that encode pIX and pIVa2 proteins are also transcribed during the late phase, but from promoters independent of the MLP. Because many of the early transcription units are in opposition to the late transcription unit, double-stranded RNA (dsRNA) is apparently formed after adenovirus enters late phase, generated by the pairing of early and late mRNAs (Maran and Mathews 1988).

All mRNAs derived from the MLP possess a common 5′ noncoding region 212 nucleotides in length called the tripartite leader, produced by the splicing of three small exons (Berget et al.1977; Moore et al. 1987). The tripartite leader has two known functions: (1) It is essential for the exclusive translation of late adenovirus mRNAs during late phase and (2) it promotes the export of mRNAs to the cytoplasm. Adenovirus also encodes two small, RNA-polymerase-III-transcribed products known as virion-associated or VA RNAs I and II (for review, see Mathews and Shenk 1991). Exclusive translation of late viral mRNAs and shutoff of cell protein synthesis occurs several hours after adenovirus enters the late phase of infection. The infectious cycle typically lasts about 30 hours, culminating in dramatic cytopathic effects (CPE) (see, e.g., Zhang and Schneider 1994).

ADENOVIRUS BLOCKS CELLULAR AND INTERFERON-INDUCED ANTIVIRAL RESPONSES AT THE TRANSLATIONAL LEVEL

Cellular and Interferon α/β Antiviral Translational Responses

Infection of animal cells with many different viruses induces intracellular antiviral responses and extracellular interferon-mediated antiviral activi-

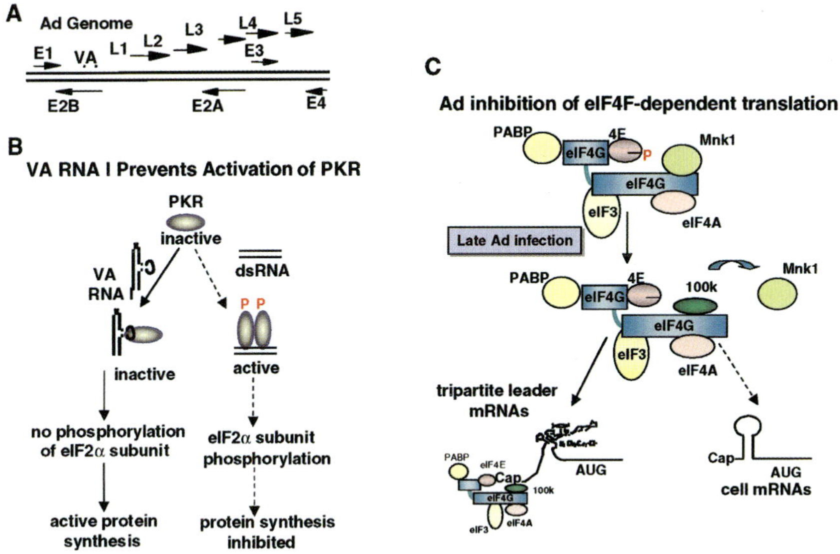

Figure 1 Model for translational control during the late phase of adenovirus infection. (*A*) Adenovirus dsDNA genome showing early transcription units E1–E4, and the family of late transcripts L1–L5. (*B*) Inhibition of PKR activation by VA RNA I. Activation of PKR by dsRNA occurs by autophosphorylation, probably by bridging a preformed PKR dimer, or by bringing two PKR proteins together. Activated PKR in turn phosphorylates the α subunit of eIF2, phosphorylated eIF2α tightly binds eIF2B, sequestering it in an inactive complex, blocking initiation. VA RNA I competitively binds PKR, preventing activation by dsRNA. (*C*) Inhibition of cell protein synthesis and preferential translation of late adenovirus mRNAs. Shown is the eIF4F cap-binding complex with associated eIF3, poly(A)-binding protein (PABP), and the eIF4E kinase, Mnk1. The adenovirus late L4-100K protein blocks phosphorylation of eIF4E, promoting dissociation of the eIF4E kinase, Mnk1. Cellular mRNA translation is therefore suppressed. The tripartite leader 5′ noncoding region on late adenovirus mRNAs permits translation when levels of phosphorylated eIF4E are low, either by efficiently recruiting the small amounts of active (eIF4E phosphorylated) eIF4F, or by utilizing eIF4E/eIF4F containing the viral 100K protein.

ties at the level of protein synthesis (see Chapter 8). One response comprises the 2′–5′-oligoadenylate synthetase(OAS)/nuclease system. The 2′–5′ OAS enzymes are activated by dsRNA, synthesize oligonucleotides with the structure ppp(A2′p5′)nA, which in turn activate endoribonuclease (RNase) L, which then degrades both cellular and viral RNAs. A second response is activation of the protein kinase PKR by dsRNA, which in turn phosphorylates the α subunit of translation factor eIF2 and inhibits initiation (see Chapters 2, 5, and 13).

Adenovirus Prevents Cellular and Interferon Antiviral Translation Responses

Adenovirus prevents activation of cellular antiviral activities mediated by PKR. Adenoviruses express one or two VA RNA genes (depending on the serotype) (Ma and Mathews 1993). The VA RNAs are small (about 160 nucleotides) and highly structured with no coding capacity (Mathews and Shenk 1991). VA RNAs I and II of human group C adenoviruses 2 and 5 are the best studied. VA RNA I, the major species, accumulates rapidly during late infection to $\sim 10^8$ copies per cell, whereas VA RNA II levels are 10- to 40-fold lower (Soderlund et al. 1976). VA RNAs are mostly cytoplasmic, with some bound to polyribosomes (Schneider et al. 1984; Katze et al. 1987; Kostura and Mathews 1989). The primary function of VA RNA I is to block activation of PKR during virus infection (Fig. 1B). An adenovirus mutant deleted of the VA RNA I gene (Ad5*dl*331), but not VA RNA II, is significantly impaired in virus growth (Thimmappaya et al. 1982). The defect results from global inhibition of viral and cellular protein synthesis (Thimmappaya et al. 1982), at an early step in initiation (Schneider et al. 1984; Reichel et al. 1985), corresponding to elevated phosphorylation of the α subunit of eIF2 by PKR, possibly activated by viral dsRNA (Maran and Mathews 1988). Inactivation of the initiation factor eIF2 occurs through sequestration of eIF2B (Schneider et al. 1984; Reichel et al. 1985; O'Malley et al. 1986a,b), the GTP to GDP exchange factor for eIF2 (see Chapters 2 and 5). The central domain region of VA RNA I directly binds PKR, through its dsRNA-binding motif (dsRBM), blocking interaction with authentic dsRNA (Fig. 1B) (Mellits and Mathews 1988; Kostura and Mathews 1989; Furtado et al. 1989; Ghadge et al. 1991; Clarke et al. 1994).

VA RNA I may possess other activities in addition to inhibition of PKR. It is reported to stabilize ribosome-associated reporter mRNAs in transiently transfected cells (Bhat et al. 1989; Strijker et al. 1989), to enhance the cytoplasmic abundance of mRNAs (Svensson and Akusjarvi 1985), and to bind the RNA-specific adenosine-deaminase (ADAR) enzyme, blocking its ability to catalyze C-6 deamination of adenosine to inosine in dsRNA (Lei et al. 1998), a process known as RNA editing. Clearly, VA RNA I possesses important activities apart from the ability to block PKR activation. In fact, VA RNA I was found to enhance the levels of transfected reporter mRNA and protein expression in cells lacking

PKR (Terenzi et al. 1999), although the mechanism is obscure. The potential roles of these responses in late adenovirus translational control have not been investigated.

An adenovirus mutant lacking both VA RNAs I and II is more sensitive to activation of PKR during infection and to inhibition by interferon-α (Bhat and Thimmappaya 1984, 1985; Bhat et al. 1985), although VA RNA II is a poor inhibitor of PKR activation (Kitajewski et al. 1986; Ma and Mathews 1993). Thus, VA RNA II likely has a different primary function from VA RNA I. RNA helicase A, which is involved in transcriptional activation (Nakajima et al. 1997), and possibly in RNA export from the nucleus (Tang et al. 1997), binds strongly and specifically to VA RNA II (Liao et al. 1998). VA RNA II is also associated with the p90 subunit of the transcription factor NF-AT (nuclear factor of activated T cells) (Liao et al. 1998), about which little is known. It is not yet known whether VA RNA II influences the activity of these factors in vivo, although it does impair RNA helicase activity in vitro (Liao et al. 1998).

SHUTOFF OF CELL PROTEIN SYNTHESIS DURING LATE ADENOVIRUS INFECTION

Evidence for Translational Control in Late Adenovirus-infected Cells

As adenovirus enters the late phase of infection, the cytoplasmic accumulation of cellular rRNAs and mRNAs is significantly inhibited (Beltz and Flint 1979). Two early viral proteins, identified by their respective transcription units as the region E1B-55k and E4-34k proteins, form a complex that overrides the block in RNA transport, and selectively exports adenovirus mRNAs (Babiss and Ginsberg 1984; Halbert et al. 1984). The adenovirus tripartite leader 5′ noncoding region, which is found on all late viral mRNAs transcribed from the MLP, further enhances nuclear export of newly transcribed mRNAs (Huang and Flint 1998). Cell mRNAs remain stable, capped and polyadenylated, and can be efficiently translated in vitro despite the inhibition of their translation in vivo (Thimmappaya et al. 1982). Certain cellular mRNAs, such as β-tubulin, escape the viral block to cellular mRNA export but are still excluded from translation during late infection (Moore et al. 1987). Consequently, >95% of translating mRNAs are late adenovirus mRNAs, although they represent less than 20% of the cytoplasmic pool (Schneider 1996). There is, therefore, selective translation inhibition of cellular mRNAs and exclusive translation of adenovirus late mRNAs in infected cells.

Mechanism for Inhibition of Cellular mRNA Translation during Adenovirus Late Infection

Adenovirus Inhibition of Cell Protein Synthesis Is Associated with Dephosphorylation of eIF4E

The tripartite leader permits mRNA translation in cells that are superinfected by poliovirus (Castrillo and Carrasco 1987; Dolph et al. 1988), which blocks eIF4F-dependent mRNA translation through proteolytic degradation of initiation factor eIF4G (see Chapters 8, 19, and 31). Thus, the tripartite leader reduces the dependence of adenovirus late mRNAs on eIF4F activity. In late adenovirus-infected cells, eIF4F is not inactivated by proteolysis of eIF4G (Dolph et al. 1988), but by dephosphorylation of >95% of eIF4E (Huang and Schneider 1991), which is coincident with inhibition of cell mRNA translation (Zhang et al. 1994). The importance of eIF4E phosphorylation in translation is reviewed in Chapter 6. Dephosphorylation of eIF4E likely impairs most cap-dependent protein synthesis by reducing the functional abundance of eIF4F complexes that promote 40S ribosome–mRNA interaction. Studies suggested that adenovirus enters the late phase of infection and synthesizes one or more late gene products to specifically block eIF4E phosphorylation and host cell protein synthesis (Zhang et al. 1994). Moreover, several drugs that inhibit adenovirus shutoff of cell protein synthesis also block dephosphorylation of eIF4E (Huang and Schneider 1990; Feigenblum et al. 1998). The dephosphorylation of eIF4E in adenovirus-infected cells is therefore tightly coupled to inhibition of cell mRNA translation.

Adenovirus L4-100k Protein Is Central to Late Viral Translational Control

The 100K protein is a late adenovirus polypeptide encoded by the L4 transcription unit in very large amounts and is involved in morphogenesis of the viral particle (Oosterom-Dragon and Ginsberg 1981). The 100K protein can also be photo-UV crosslinked to both cellular and viral mRNAs (Adam and Dreyfuss 1987), implying a possible role in translational control. Genetic evidence has linked the 100K protein to a role in promoting late adenovirus mRNA translation (Hayes et al. 1990; Riley and Flint 1993). An adenovirus containing a specific temperature sensitive (ts) mutation in 100K protein (Ad2*ts*1) is reduced in its ability to carry out late viral protein synthesis at restrictive temperature. It was originally reported that Ad2*ts*1 still prevents cellular protein synthesis at the restrictive temperature during late infection (Hayes et al. 1990), although cell mRNA translation was not strongly impaired by wild-type virus in this study. More recent analysis (described below) confirmed that Ad2*ts*1 at restrictive temperature does not efficiently translate late adenovirus mRNAs, and in addition, does not strongly block cell protein synthesis.

In vivo phosphorylation of eIF4E is carried out by the protein kinase Mnk1 (Fukunaga and Hunter 1997; Waskiewicz et al. 1997), which only efficiently phosphorylates eIF4E in vivo when both are bound to eIF4G, eIF4E at the amino terminus and Mnk1 at the carboxyl terminus (Pyronnet et al. 1999; Waskiewicz et al. 1999; see Chapters 2 and 6). The block in phosphorylation of eIF4E brought about by adenovirus does not involve sequestration or removal of eIF4E from eIF4F by the eIF4E-binding proteins, 4E-BP1 or BP2. In fact, the adenovirus early E1A gene product induces phosphorylation of 4E-BP proteins during the early phase of infection, which blocks their binding to eIF4E (Feigenblum and Schneider 1996; Gingras and Sonenberg 1997). Instead, adenovirus specifically displaces Mnk1 from eIF4F complexes during late infection, concomitant with inhibition of eIF4E phosphorylation and host cell mRNA translation (Fig. 1C) (Cuesta et al. 2000). eIF4E, eIF4A, eIF4G, and PABP all remain associated in a cap-binding complex that contains the adenovirus L4-100K protein. The 100K protein binds the carboxy-terminal fragment of eIF4G, apparently at or near the site normally occupied by Mnk1, implicating 100K protein in direct competitive displacement of the kinase. Thus, 100K protein is involved in specifically blocking the association of Mnk1 with eIF4G, leading to inhibition of cell protein synthesis. 100K protein may also be involved in enhanced expression of adenovirus late mRNAs by ribosome shunting during late infection (Q. Xi and R.J. Schneider, unpubl.).

SELECTIVE TRANSLATION OF ADENOVIRUS LATE mRNAs

The tripartite leader is essential for selective translation of mRNAs following inhibition of cell protein synthesis (Logan and Shenk 1984). Structural and mutational studies of the tripartite leader showed that it contains an extensive unstructured 5′ end, followed by several moderately stable hairpins possessing large single-stranded loops (Zhang et al. 1989; Dolph et al. 1990). Both the unstructured 5′ end and the hairpin structures are important for tripartite leader translation during late adenovirus infection or during heat shock of cells, when eIF4E phosphorylation is blocked, as well as in poliovirus-infected cells, when eIF4G is cleaved (Dolph et al. 1988, 1990; Yueh and Schneider 1996). The tripartite leader requires both a cap for translation and eIF4E (Dolph et al. 1990; Thomas et al. 1992). The tripartite leader does not function as an internal ribosome entry site (IRES) (Dolph et al. 1990; Jang et al. 1990).

Tripartite Leader Promotes Selective Translation by Ribosome Shunting

The tripartite leader directs a novel form of translation initiation known as ribosome shunting. Ribosome shunting has been described for the cauliflower mosaic virus 35S mRNA (Futterer et al. 1993), adenovirus late mRNAs (Yueh and Schneider 1996, 2000), the Sendai virus Y mRNA (Curran and Kolakofsky 1988; Latorre et al. 1998), and papillomavirus E1 mRNA (Remm et al. 1999). The general mechanism for ribosome shunting is very poorly understood but involves loading of 40S ribosome subunits to the 5′ end of the mRNA, possibly limited scanning, followed by direct translocation of 40S subunits to the downstream initiation codon, directed by the "shunting elements". 40S ribosome shunting is predicted to decrease the dependence of mRNAs for eIF4F during initiation by reducing the need for mRNA unwinding activity. The tripartite leader is blocked in its ability to direct translation at the downstream AUG codon only if strong secondary structure, or an AUG codon, is inserted within the first 80 nucleotides of the 5′ noncoding region (Yueh and Schneider 1996). If inserted thereafter, translation is directed by the tripartite leader at normal levels. Shunting occurs to a limited distance downstream from the tripartite leader, as the AUG cannot be located farther than ~160 nucleotides from the 3′ end of the leader. The limitation in translation of 40S ribosome subunits is consistent with the location of adenovirus late AUG codons, which are within 35 nucleotides of the 3′ end of the tripartite leader. The RNA "shunting elements" include hairpin structures downstream from nucleotide 80. In uninfected cells, the tripartite leader directs translation by conventional 5′ scanning of 40S ribosome subunits and by 40S ribosome shunting at roughly equal levels. However, ribosome initiation by both scanning and shunting are incompatible on the same tripartite leader mRNA, probably because scanning ribosomes disrupt the conformation of the shunting elements, thereby abolishing their ability to direct ribosomes to shunt (Yueh and Schneider 1996). Consequently, in late adenovirus-infected cells, the tripartite leader directs translation solely by ribosome shunting (Yueh and Schneider 1996, 2000), suggesting that dephosphorylation of eIF4E and/or inactivation of eIF4F is a switch, converting the tripartite leader to initiate translation exclusively by shunting.

Within each of the three tripartite leader exons is a conserved complementarity to the 3′ end of 18S ribosomal RNA (rRNA), a component of the 40S ribosome. Mutagenesis showed that the unstructured 5′ end of the tripartite leader is essential for translation during late adenovirus infection, when the levels of phosphorylated eIF4E and functional eIF4F are very low (Dolph et al. 1990; Yueh and Schneider 1996). Later analysis also identified the complementarities to the 3′ hairpin of 18S rRNA as vital for shunting as well (Yueh and Schneider 2000). Individual mutation of each rRNA complementarity had no impact on either the shunting or scanning mode of translation initiation by the tripartite leader (Yueh and Schneider 2000). Paired deletion of rRNA complementarities showed that exons 2 and 3 are particularly critical, in that their combined deletion reduces shunting translation by ~20-fold without significantly decreasing scanning initiation. Thus, the 18S rRNA complementarities are involved in an unknown manner in strongly promoting ribosome shunting in the tripartite leader, and there is a redundancy and a hierarchy in their function.

The molecular mechanism by which 40S ribosome shunting is directed by the tripartite leader remains to be elucidated. However, two molecular models can be envisioned based on the available data (Fig. 2). The "dissociating model" for ribosome shunting proposes that 40S ribosome

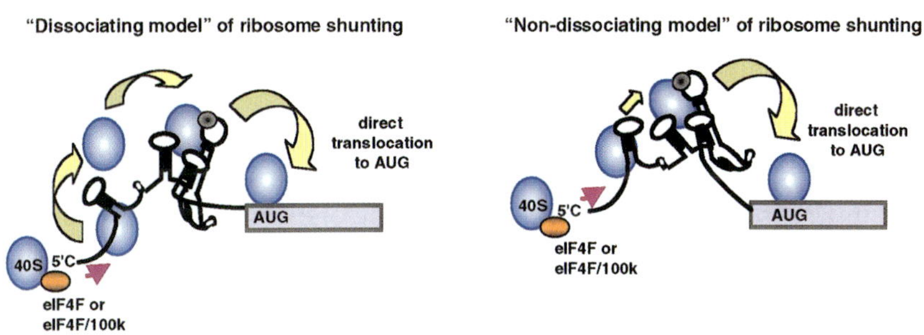

Figure 2 Model for ribosome shunting directed by the tripartite leader during late adenovirus infection. In the "dissociating model" of ribosome shunting, 40S ribosome subunits are thought to be displaced from the tripartite leader due to the inability to unwind the RNA in the absence of normal eIF4F activity. In the "non-dissociating model," 40S ribosome subunits are thought to stall on the mRNA but do not dissociate. The shunting elements would then recover dissociated ribosome subunits, or directly interact with tethered 40S ribosomes, promoting direct translocation to the downstream AUG. 40S ribosomes might interact directly through 18S rRNA:tripartite leader RNA interaction, or via initiation factors recruited to the shunting elements.

subunits are loaded onto the 5′ unstructured end of the tripartite leader, either by 5′ scanning or by direct placement from the cap. However, due to an intrinsically high off-rate resulting from the stability of the internal hairpin structures and the loss of eIF4F/RNA unwinding activity, a significant number of 40S subunits dissociate from the tripartite leader. The shunting element might then recover these ribosomal subunits by direct binding to the RNA shunting elements through a tripartite leader RNA:18S rRNA hairpin interaction. Alternatively, the shunting element might bind the very limited amount of native eIF4F or eIF4F/100K complex, or specific translation factors such as eIF4G or eIF3, which in turn would recruit 40S subunits to the shunting elements. The translocation step might involve direct placement of 40S subunits to the AUG, or upstream of the AUG, followed by limited scanning. The "non-dissociating model" of ribosome shunting proposes that 40S ribosome subunits are loaded onto the 5′ end of the tripartite leader as above, but rather than dissociate, they are blocked from proceeding into the body of the tripartite leader due to dephosphorylation of eIF4E and loss of eIF4F mRNA-unwinding activity. Tethered 40S subunits would then interact with the shunting elements, possibly by RNA:RNA interactions, or by initiation factors recruited to the shunting elements.

CONCLUDING REMARKS

The mechanisms by which adenovirus inhibits cellular protein synthesis and selectively translates its family of late mRNAs are beginning to emerge. Future research must now describe, in molecular detail, the processes by which translation of cellular mRNAs is inhibited and that of viral mRNAs is promoted. The identity of one viral gene product is known, the L4-100K protein, which is involved in the preferential translation of adenovirus late mRNAs. Additional work may disclose the involvement of other viral proteins, and possibly the VA RNAs as well. The precise function of the 100K protein in the dissociation of Mnk1 from eIF4F complexes needs to be clarified, and whether this inhibits the function of the complex or alters it so that it is specific for adenovirus late mRNA translation by ribosome shunting. Finally, the understanding of ribosome shunting in general, and as it occurs on late adenovirus mRNAs in particular, is presently only rudimentary. The molecular mechanism of ribosome shunting needs to be more fully characterized, including the roles of canonical cellular translation factors and possibly noncanonical factors.

ACKNOWLEDGMENTS

I thank R. Cuesta for thoughtful discussions. The author's work described in this review was supported by a grant from the National Institutes of Health (CA-42357).

REFERENCES

Adam S.A. and Dreyfuss G. 1987. Adenovirus proteins associated with mRNA and hnRNA in infected Hela cells. *J. Virol.* **61:** 3276–3283.

Babiss L.E. and Ginsberg H.S. 1984. Adenovirus type 5 early region 1b gene product is required for efficient shutoff of host protein synthesis. *J. Virol.* **50:** 202–212.

Beltz G.A. and Flint S.J. 1979. Inhibition of HeLa cell protein synthesis during adenovirus infection. *J. Mol. Biol.* **131:** 353–373.

Berget S.M., Moore C., and Sharp P. 1977. Spliced segments at the 5′ terminus of Ad2 late mRNA. *Proc. Natl. Acad. Sci.* **74:** 3171–3175.

Bhat R.A. and Thimmappaya B. 1984. Adenovirus mutants with DNA sequence perturbations in the intragenic promoter of VAI RNA gene allow the enhanced transcription of VAII RNA gene in HeLa cells. *Nucleic Acids Res.* **12:** 7377–7388.

———. 1985. Construction and analysis of additional adenovirus substitution mutants confirm the complementation of VAI RNA function by two small RNAs encoded by Epstein-Barr virus. *J. Virol.* **56:** 750–756.

Bhat R.A., Domer P.H., and Thimmappaya B. 1985. Structural requirements of adenovirus VAI RNA for its translation enhancement function. *Mol. Cell. Biol.* **5:** 187–196.

Bhat R.A., Furtado M.R., and Thimmappaya B. 1989. Efficient expression of small polymerase III genes from a novel simian virus 40 vector and their effect on viral gene expression. *Nucleic Acids Res.* **17:** 1159–1176.

Castrillo J.L. and Carrasco L. 1987. Adenovirus late protein synthesis is resistant to the inhibition of translation induced by poliovirus. *J. Biol. Chem.* **262:** 7328–7334.

Clarke P.A., Pe'ery T., Ma Y., and Mathews M.B. 1994. Structural features of adenovirus 2 virus-associated RNA required for binding to the protein kinase DAI. *Nucleic Acids Res.* **22:** 4364–4374.

Cuesta R., Xi Q., and Schneider R.J. 2000. Adenovirus-specific translation by displacement of kinase Mnk1 from cap-initiation complex eIF4F. *EMBO J.* **19:** 3465–3474.

Curran J. and Kolakofsky D. 1988. Scanning independent ribosomal initiation of the Sendai virus X protein. *EMBO J.* **7:** 2869–2874.

Dolph P.J., Huang J., and Schneider R.J. 1990. Translation by the adenovirus tripartite leader: Elements which determine independence from cap-binding protein complex. *J. Virol.* **64:** 2669–2677.

Dolph P.J., Racaniello V., Villamarin A., Palladino F., and Schneider R.J. 1988. The adenovirus tripartite leader eliminates the requirement for cap binding protein during translation initiation. *J. Virol.* **62:** 2059–2066.

Feigenblum D. and Schneider R.J. 1996. Cap-binding protein (eukaryotic initiation factor 4E) and 4E-inactivating protein BP-1 independently regulate cap-dependent translation. *Mol. Cell. Biol.* **16:** 5450–5457.

Feigenblum D., Walker R., and Schneider R.J. 1998. Adenovirus induction of an interferon-regulatory factor during entry into the late phase of infection. *J. Virol.* **72:** 9257–9266.

Fukunaga R. and Hunter T. 1997. MNK1, a new MAP kinase-activated protein kinase, isolated by a novel expression screening method for identifying protein kinase substrates. *EMBO J.* **16:** 1921–1933.

Furtado M.R., Subramanian S., Bhat R.A., Wowlkes D.M., Safer B., and Thimmappaya B. 1989. Functional dissection of adenovirus VA1 RNA. *J. Virol.* **63:** 3423–3434.

Futterer J., Kiss-Laszlo Z., and Hohn T. 1993. Nonlinear ribosome migration on cauliflower mosaic virus 35S RNA. *Cell* **73:** 789–802.

Ghadge G.D., Swaminathan S., Katze M.G., and Thimmappaya B. 1991. Binding of the adenovirus VAI RNA to the interferon induced 68 kDa protein kinase correlates with function. *Proc. Natl. Acad. Sci.* **88:** 7140–7144.

Gingras A.C. and Sonenberg N. 1997. Adenovirus infection inactivates the translational inhibitors 4E-BP1 and 4E-BP2. *Virology* **237:** 182–186.

Halbert D.N., Cutt J.R., and Shenk T. 1984. Adenovirus early region 4 encodes functions required for efficient DNA replication, late gene expression and host cell shutoff. *J. Virol.* **56:** 250–257.

Hayes B.W., Telling G.C., Myat M.M., Williams J.F., and Flint S.J. 1990. The adenovirus L4 100 kilodalton protein is necessary for efficient translation of viral late mRNA species. *J. Virol.* **64:** 2732–2742.

Huang J. and Schneider R.J. 1990. Adenovirus inhibition of cellular protein synthesis is prevented by the drug 2-aminopurine. *Proc. Natl. Acad. Sci.* **87:** 7115–7119.

———. 1991. Adenovirus inhibition of cellular protein synthesis involves inactivation of cap binding protein. *Cell* **65:** 271–280.

Huang W. and Flint S.J. 1998. The tripartite leader sequence of subgroup C adenovirus major late mRNAs can increase the efficiency of mRNA transport. *J. Virol.* **72:** 225–235.

Jang S.K., Pestova T.V., Hellen C.U.T., Witherall G.W., and Wimmer E. 1990. Cap-independent translation of picornaviral RNAs: Structure and function of internal ribosome entry site. *Enzyme* **44:** 292–309.

Katze M.G., DeCorato D., Safer B., Galabru J., and Hovanessian A.G. 1987. Adenovirus VA1 RNA complexes with the 68,000 M_r protein kinase to regulate its autophosphorylation and activity. *EMBO J.* **6:** 689–697.

Kitajewski J., Schneider R.J., Safer B., and Shenk T. 1986. An adenovirus mutant unable to express VA1 RNA displays different growth responses and sensitivity to interferon in various host cell lines. *Mol. Cell. Biol.* **6:** 4493–4498.

Kostura M. and Mathews M.B. 1989. Purification and activation of the double-stranded RNA dependent eIF-2 kinase DAI. *Mol. Cell. Biol.* **9:** 1576–1586.

Latorre P., Kolakofsky D., and Curran J. 1998. Sendai virus Y proteins are initiated by a ribosomal shunt. *Mol. Cell. Biol.* **18:** 5021–5031.

Lei M., Liu Y., and Samuel C.E. 1998. Adenovirus VAI RNA antagonizes the RNA-editing activity of the ADAR adenosine deaminase. *Virology* **245:** 188–196.

Liao H.J., Kobayashi R., and Mathews M.B. 1998. Activities of adenovirus virus-associated RNAs: Purification and characterization of RNA binding proteins. *Proc. Natl. Acad. Sci.* **95:** 8514–8519.

Logan J. and Shenk T. 1984. Adenovirus tripartite leader sequence enhances translation of mRNAs late after infection. *Proc. Natl. Acad. Sci.* **81:** 3655–3659.

Ma Y. and Mathews M.B. 1993. Comparative analysis of the structure and function of adenovirus virus-associated RNAs. *J. Virol.* **67:** 6605–6617.

Maran A. and Mathews M.B. 1988. Characterization of the double-stranded RNA implicated in inhibition of protein synthesis in cells infected with a mutant adenovirus defective for VA RNA I. *Virology* **164:** 106–113.

Mathews M.B. and Shenk T. 1991. Adenovirus virus-associated RNA and translational control. *J. Virol.* **65:** 5657–5662.

Mellits K.H. and Mathews M.B. 1988. Effects of mutations in stem and loop regions on the structure and function of adenovirus VA RNA I. *EMBO J.* **7:** 2849–2859.

Moore M., Schaack J., Baim S.B., Morimoto R.I., and Shenk T. 1987. Induced heat shock mRNAs escape the nucleocytoplasmic transport block in adenovirus-infected HeLa cells. *Mol. Cell. Biol.* **7:** 4505–4512.

Nakajima T., Uchida C., Anderson S.F., Lee C.G., Hurwitz J., Parvin J.D., and Montminy M. 1997. RNA helicase A mediates association of CBP with RNA polymerase II. *Cell* **90:** 1107–1112.

O'Malley R.P., Mariano T.M., Siekierka J., and Mathews M.B. 1986a. A mechanism for the control of protein synthesis by adenovirus VA RNA I. *Cell* **44:** 391–400.

O'Malley R.P., Mariano T.M., Siekierka J., Merrick W.C., Reichel P.A., and Mathews M.B. 1986b. The control of protein synthesis by adenovirus VA RNA. *Cancer Cells* **4:** 291–301.

Oosterom-Dragon E.A. and Ginsberg H.S. 1981. Characterization of two temperature sensitive mutants of type 5 adenovirus with mutations in the 100,000k dalton protein gene. *J. Virol.* **40:** 491–500.

Pyronnet S., Imataka H., Gingras A.C., Fukunaga R., Hunter T., and Sonenberg N. 1999. Human eukaryotic translation initiation factor 4G (eIF4G) recruits Mnk1 to phosphorylate eIF4E. *EMBO J.* **18:** 270–279.

Reichel P.A., Merrick W.C., Siekierka J., and Mathews M.B. 1985. Regulation of a protein initiation factor by adenovirus associated RNA I. *Nature* **313:** 196–200.

Remm M., Remm A., and Ustav M. 1999. Human papillomavirus type 18 E1 protein is translated from polycistronic mRNA by a discontinuous scanning mechanism. *J. Virol.* **73:** 3062–3070.

Riley D. and Flint S.J. 1993. RNA-binding properties of a translational activator, the adenovirus L4 100-kilodalton protein. *J. Virol.* **67:** 3586–3595.

Schneider R.J. 1996. Adenovirus and vaccinia virus translational control. In *Translational control* (ed. J.W.B. Hershey et al.), pp. 575–605. Cold Spring Harbor Laboratory Press, Cold Spring Harbor, New York.

Schneider R.J. and Shenk T. 1987. Impact of virus infection on host cell protein synthesis. *Annu. Rev. Biochem.* **56:** 317–332.

Schneider R.J., Weinberger C., and Shenk T. 1984. Adenovirus VA1 RNA facilitates the initiation of translation in virus infected cells. *Cell* **37:** 291–298.

Soderlund H., Petterson U., Vennstrom B., Philipson L., and Mathews M.B. 1976. A new species of virus coded low molecular weight RNA from cells infected with adenovirus type 2. *Cell* **7:** 585–593.

Strijker R., Fritz D.T., and Levinson A.D. 1989. Adenovirus VAI-RNA regulates gene expression by controlling stability of ribosome bound RNAs. *EMBO J.* **8:** 2669–2675.

Svensson C. and Akusjarvi G. 1985. Adenovirus VAI RNA mediates a translational stimulation which is not restricted to the viral mRNAs. *EMBO J.* **4:** 957–964.

Tang H., Gaietta G.M., Fischer W.H., Ellisman M.H., and Wong-Staal F. 1997. A cellular cofactor for the constitutive transport element of type D retrovirus. *Science* **276:** 1412–1415.

Terenzi F., deVeer M.J., Ying H., Restifo N.P., Williams B.R., and Silverman R.H. 1999.

The antiviral enzymes PKR and RNase L suppress gene expression from viral and non-viral based vectors. *Nucleic Acids Res.* **27:** 4369–4375.

Thimmappaya B., Weinberger C., Schneider R.J., and Shenk T. 1982. Adenovirus VA1 RNA is required for efficient translation of viral mRNA at late times after infection. *Cell* **31:** 543–551.

Thomas A.M., Scheper G.C., Kleijn M., DeBoer M., and Voorma H.O. 1992. Dependence of the adenovirus tripartite leader on the p220 subunit of eukaryotic initiation factor 4F during in vitro translation. *Eur. J. Biochem.* **207:** 471–477.

Waskiewicz A.J., Flynn A., Proud C.G., and Cooper J.A. 1997. Mitogen-activated protein kinases activate the serine/threonine kinases Mnk1 and Mnk2. *EMBO J.* **16:** 1909–1920.

Waskiewicz A.J., Johnson J.C., Penn B., Mahalingham M., Kimball S.R., and Cooper J.A. 1999. Phosphorylation of the cap-binding protein eukaryotic translation initiation factor 4E by protein kinase Mnk1 in vivo. *Mol. Cell. Biol.* **19:** 1871–1880.

Yueh A. and Schneider R.J. 1996. Selective translation by ribosome jumping in adenovirus infected and heat shocked cells. *Genes Dev.* **10:** 1557–1567.

———. 2000. Translation by ribosome shunting on adenovirus and Hsp70 mRNAs facilitated by complementarity to 18S rRNA. *Genes Dev.* **14:** 414–421.

Zhang Y. and Schneider R.J. 1994. Adenovirus inhibition of cell translation facilitates release of virus particles and enhances degradation of the cytokeratin network. *J. Virol.* **68:** 2544–2555.

Zhang Y., Dolph P.J., and Schneider R.J. 1989. Secondary structure analysis of adenovirus tripartite leader. *J. Biol. Chem.* **264:** 10679–10684.

Zhang Y., Feigenblum D., and Schneider R.J. 1994. A late adenovirus factor induces eIF-4E dephosphorylation and inhibition of cell protein synthesis. *J. Virol.* **68:** 7040–7050.

33

Reovirus Translational Control

Aaron J. Shatkin
Center for Advanced Biotechnology and Medicine
Piscataway, New Jersey 08854

THE REOVIRIDAE FAMILY

Reoviridae are cytolytic parasites that infect a broad spectrum of insect, plant, and vertebrate species, including humans (Nibert et al. 1996). They can readily initiate replication because they contain, as virally encoded components, all the enzymes necessary to produce functional viral mRNAs. However, like other viruses, *Reoviridae* require the host translational machinery to complete the replicative cycle, and control mechanisms operate in infected cells to facilitate a shift from cellular to viral protein synthesis. Several regulatory events have been suggested to explain how this is accomplished, but even the intensively studied continued translation of reoviral mRNAs in the face of initiation shutoff by the interferon-induced, double-stranded (ds) RNA-activated protein kinase (PKR; see Chapters 8 and 13) remains incompletely understood and controversial.

The prototype dsRNA-containing reoviruses were isolated from mammals including humans and characterized in the 1950s and 1960s (Fields 1996). Later discoveries of other dsRNA viruses from mammals, fish, birds, insects, plants, and diverse organisms including fungi and bacteria demonstrated that viruses with segmented dsRNA genomes are widely distributed and successfully established in nature. The *Reoviridae* family of metazoan dsRNA viruses has been divided into nine genera based on host range and a genome segment number between 10 and 12. They include the Orthoreoviruses (10), Orbiviruses (10), Rotaviruses (11), and Coltiviruses (12) which all replicate in humans; the Phytoviruses (12), Oryzaviruses (10), and Fijiviruses (10) of plants; Aquareoviruses (11) of fish and shellfish; and Cypoviruses (10) that grow in insects.

© 2000 Cold Spring Harbor Laboratory Press 0-87969-568-4/00 $5 +. 00

In addition to the common features of a dsRNA genome consisting of 10–12 unique segments enclosed within a capsid shell that is usually multilayered, *Reoviridae* are all nonenveloped, icosahedral, about 70–85-nm particles that replicate in the cytoplasm of infected cells. Several of the virion-associated structural proteins also function as enzymes (Yue and Shatkin 1998), initiating replication by transcribing each of the duplex segments end to end to produce mRNAs that also serve as templates for genome formation and are m7G-capped at the 5′ end (Furuichi et al. 1975) but not 3′-polyadenylated (Stoltzfus et al. 1973). Mammalian reovirus (Orthoreovirus) mRNAs synthesized by viral cores in vitro (Shatkin and Sipe 1968) proved invaluable not only for analyzing viral-specific characteristics such as protein coding assignments, but also for discovering and deciphering general aspects of eukaryotic gene expression, e.g., the structure, synthesis, and translational functions of mRNA caps (Furuichi and Shatkin 2000); the activities of cap-binding proteins in protein synthesis (Sonenberg et al. 1978; Gingras et al. 1999), and the scanning mechanism of translation initiation (Kozak and Shatkin 1978; Kozak 1999).

In this chapter, I focus on the mammalian reoviruses, their inhibitory effects on cell metabolism, and the interplay between host defenses and viral countermeasures that in infected cells results in preferential synthesis of viral proteins. Included among these translation-related phenomena are the shutoff of host protein synthesis, selective translation of viral mRNAs, PKR activation and its inhibition by the viral dsRNA-binding protein σ3, other viral protein interactions that modulate initiation, Ras signaling and PKR down-regulation, and the control of bicistronic mRNA expression during elongation by ribosome pausing.

STRUCTURE AND FUNCTION OF MAMMALIAN REOVIRUSES

The Ten Genomic dsRNAs and Cognate mRNAs

The reovirus genus includes three immunologically distinct serotypes. Virions of each type contain ten essential dsRNAs that can be separated by polyacrylamide gel electrophoresis into three large (L, ~3.9 kbp), three medium (M, ~2.3 kbp), and four small (S,~1.2–1.4 kbp) species (Shatkin et al. 1968). The gel migration patterns are similar but serotype-distinct and thus have provided an important basis for gene mapping by analysis of virus reassortants derived by mixed infections with different serotypes and containing a reassorted combination of the ten essential dsRNAs (Ramig and Fields 1983). All the reovirus dsRNAs have been cloned, sequenced, and characterized, e.g., with respect to evolutionary relatedness and coding assignments. With the exception of the bicistronic S1 RNA, reovirus mRNAs each contain a single functional open reading frame (ORF). In all but the weakly translated m1 mRNA which starts at the second AUG (Roner et al. 1993), initiation is from the AUG proximal to the conserved 5′-terminal capped sequence, m^7GpppG^mCUA (Shatkin and Kozak 1983). The 5′-untranslated regions (UTRs) of reovirus mRNAs are relatively short (in serotype 3, 12–32 nucleotides), consistent with direct and efficient translation. All the 3′-UTRs (35-184 nucleotides) terminate in the sequence ...UCAUC-OH (Darzynkiewicz and Shatkin 1980; Antczak et al. 1982), demonstrating that polyadenylation is not essential for eukaryotic mRNA translation.

The Eight Virion-associated and Three Nonstructural Proteins

The coding assignments of the reovirus type 3 genome segments and some properties of the corresponding proteins are summarized in Table 1. The three largest (λ) polypeptides interact in cores to produce the capped transcripts. The RNA-dependent RNA polymerase catalytic site probably resides in λ3, which is located at each of the 12 icosahedral five-fold axes or vertices (Starnes and Joklik 1993; Nibert et al. 1996). Polypeptide λ2 forms pentameric spikes surrounding the vertices and extending from the core to the virus surface. These spikes may be channels for extrusion of nascent transcripts (Gillies et al. 1971; Dryden et al. 1993), consistent

Table 1 Reovirus type 3 proteins

Gene	Protein	Amino acids	Location	Copies	Functions
L1	λ3	1267	core	12	RNA polymerase
L2	λ2	1290	outer shell	60	guanylyltransferase, methyltransferase
L3	λ1	1233	core	120	ATPase/helicase, RNA 5′-triphosphatase
M1	μ2	736	core	12 (?)	dsRNA synthesis
M2	μ1	708	outer shell	600	assembly/penetration
M3	μNS	719	nonstructural		binds ssRNA
S1	σ1	455	outer shell	36	cell attachment, DNA inhibitor serotype determinant
	σ1s	120	nonstructural		cytopathogenesis(?)
S2	σ2	418	core	120–180	binds dsRNA
S3	σNS	366	nonstructural		binds ssRNA
S4	σ3	365	outer shell	600	binds dsRNA translation control

with the enzymatic roles of λ2 in transcript 5′-end guanylylation and methylation (Seliger et al. 1987; Fausnaugh and Shatkin 1990; Luongo et al. 1998) to form the cap structure, m^7GpppG^m (Furuichi et al. 1976). Polypeptide λ1 binds dsRNA, contains NTPase activity, and likely corresponds to the energy-dependent helicase required to unwind the duplex template RNAs during reovirus transcription (Bisaillon et al. 1997; Nibert 1998). It probably also provides the RNA 5′-triphosphatase that converts pppG to ppG ends, an essential step in mRNA capping.

Two of the three middle-sized polypeptides are structural (Table 1). Although the exact function(s) of μ2 remains undefined, mutant analyses demonstrated that it influences reovirus pathogenesis and is required for dsRNA synthesis (Brown 1998). Major polypeptide μ1 is present mostly as proteolytic cleavage products μ1C and the amino-terminal fragment μ1N (Nibert et al. 1996). In association with another major outer shell polypeptide (σ3), μ1/μ1C promotes virion stability. The amino-terminal glycine of μ1 is uniquely N-myristoylated (Nibert et al. 1991), suggesting a role in virus penetration into cells, virion maturation, or possibly other membrane-related events (Tosteson et al. 1993). The third M class polypeptide, μNS, is nonstructural, forms oligomeric complexes, and binds single-stranded (ss) RNA (Antczak and Joklik 1992; Nibert et al. 1996). It may participate in the assorting of template plus strands during assembly of viral nucleoprotein complexes that synthesize minus strands and are destined to become infectious virions. Interaction of μNS with viral mRNA may also stimulate translation, like rotavirus nonstructural protein NSP3 (Piron et al. 1998), which also forms oligomers (Mattion et al. 1992; Piron et al. 1999).

The four S genome segments code for two nonstructural proteins and three virion components. The largest, σ1, forms a trimeric "lollypop" structure at each of the virion vertices and initiates infection by attaching to cell-surface receptors (Lee and Gilmore 1998). In addition to cell tropism, it specifies virus serotype, consistent with the S1 genome sequences being the least conserved among all the segments of the three different serotypes (Sharpe and Fields 1982a; Cashdollar et al. 1985). In contrast, the overall organization of S1 is highly conserved, assuring the production of a second reading frame protein, σ1s, in all three serotypes. Core polypeptide σ2 binds weakly to dsRNA and has been implicated in RNA synthesis (Nibert et al. 1996), whereas the nonstructural polypeptide σNS, like μNS, binds ssRNA strongly and may also be involved in assorting viral transcripts into morphogenetic complexes (Gillian and Nibert 1998). The S4 segment-encoded polypeptide σ3 is a major stabilizing component of the outer shell, and its assembly onto subviral particles that are actively synthesizing mRNA blocks further transcription. This property is being exploited in a particle "recoating genetics" approach to defining determinants involved in reovirus entry into cells (Jane-Valbuena et al. 1999) and may help elucidate other σ3 properties including dsRNA binding to block PKR and facilitate viral protein synthesis.

Inhibition of Host Macromolecular Synthesis in Reovirus-infected Cells

Structural studies (Nibert 1998) have helped define the cellular effects of the proteins in intact reovirus and in infectious subviral particles (missing σ3) and noninfectious cores (missing μ1, σ1, σ3) prepared by proteolytic treatment of purified virus (but analogous to particles formed during virus entry and replication). Cell attachment protein σ1 and dsRNA-binding protein σ3 are key outer shell proteins that affect cellular synthetic processes. An early consequence of reovirus type 3 (or type 2 but not 1) attachment to mouse L cells is the inhibition of DNA synthesis. It does not require virus replication and can occur even in the absence of viral entry (Shaw and Cox 1973). Replicons active at the time of virus attachment, and during presumptive second messenger signaling from the cell surface to the nucleus, apparently are completed, but new initiation sites are not activated (Hand and Tamm 1974). As a consequence, infected cells accumulate in the G_1 phase of the cell cycle, apparently due to inactivation of p21*ras* and a decrease in Raf phosphorylation and mitogen-activated protein kinase activity (Saragovi et al. 1999).

Cellular RNA and protein synthesis can also decrease in response to reovirus infection (Kudo and Graham 1965; Zweerink and Joklik 1970; Sharpe and Fields 1982b). In mouse L cells the magnitude of inhibition varies with virus serotype and strain (type 1 being the least inhibitory, but noninhibitory strains have also been described [see Schmechel et al. 1997]), multiplicity of infection (greater inhibition at higher inputs that increase the rate of multiplication), and time after infection (increased inhibition at late times, consistent with increasing effects of accumulating viral products). An early study showed that the inhibitory effects on RNA and protein formation both segregate with the S4 genome segment (Sharpe and Fields 1982b), implicating the major outer shell protein σ3. Despite the co-segregation of S4 and the translational shutoff phenotype, σ3 can reverse the translational inhibition of infection by binding dsRNA activators of PKR (Jacobs and Langland 1998). This indirect sparing action of σ3 on host-cell translational capacity may be an essential countermeasure to the antiviral defense elicited in the host by activation of PKR.

HOST RESPONSES TO REOVIRUS INFECTION

Activation of Interferon-induced PKR and Initiation Shutoff

Infection by reoviruses is among the many external stimuli of cells that can induce the synthesis of interferon (IFN), leading to the antiviral state. Induction of IFN and its inhibitory effects on reovirus replication are dependent on both the cell origin and virus serotype (e.g., in mouse L cells type 3 is a more effective inducer than type 1 and is more sensitive to IFN; Samuel 1998). PKR plays a critical role in this host defense process. Its synthesis is induced by enhanced transcription at an interferon-responsive promoter (Kuhen and Samuel 1997), and although no "free" dsRNA has been detected in reovirus-infected cells, stem-loop duplex regions in viral mRNA, notably S1 transcripts (Bischoff and Samuel 1989), can activate PKR. The autophosphorylated and activated PKR then phosphorylates the α subunit of the heterotrimeric initiation factor eIF2. Consequently, the guanine nucleotide exchange factor eIF2B, which is present in limiting quantities, binds tightly to the phosphorylated eIF2/GDP derived from 40S ribosome-bound eIF2/Met-$tRNA_i$/GTP ternary complexes when 60S subunits join, a process energized by GTP hydrolysis. Recycling of eIF2 catalyzed by eIF2B is thus prevented, and shutoff of translation initiation ensues (see Chapters 2, 5, and 13).

Apoptosis Induction by PKR

Another defensive measure with the potential to restrict virus spread in the infected animal is apoptosis. However, cell culture and mouse CNS studies suggest that apoptosis is ineffective in limiting reovirus replication and cytopathogenesis (Oberhaus et al. 1998). Strikingly, PKR is also a key player in virus-induced apoptosis. Exactly how the increased expression of PKR leads to programmed death is not clear, but activation of caspase 3 and events upstream of the inhibitory action of Bcl-2 are apparently involved (Lee et al. 1997). Phosphorylation of eIF2 (Srivastava et al. 1998) and activation of NF-κB (Gil et al. 1999) and other transcription factors (Cuddihy et al. 1999) also induce apoptosis in virus-infected cells by various pathways, as reviewed recently (Kaufman 1999a,b).

Reovirus-induced apoptosis occurs early in infection and is linked to, and possibly coordinated with, inhibition of host DNA synthesis; i.e., both properties segregate with the S1 genome segment and are similarly altered in infections with different virus serotypes (Oberhaus et al. 1998). Calpain, a Ca-dependent cysteine protease, recently was implicated in reovirus-induced apoptosis (Debiasi et al. 1999). PKR and/or other eIF2 kinases

(Brostrom and Brostrom 1998; Kaufman 1999a) and, possibly as a consequence, NF-κB (Kaufman 1999a,b) can also be activated by agents that mobilize intracellular Ca, suggesting a central role of this ion in multiple pathways leading to programmed cell death in reovirus-infected cells.

VIRAL COUNTERMEASURES

The suicidal responses of cells to infection may have evolved to prevent virus spread in the organism by limiting pathogenic effects to initially infected cells. However, viruses clearly are highly successful parasites and have developed various countermeasures to block PKR and overcome the antiviral state (see Chapter 8 Pe´ery and Mathews). These include eIF2α decoy (Beattie et al. 1991) and dsRNA-binding (Yue and Shatkin 1998; Hatada et al. 1999) proteins as well as dsRNA mimics that compete for common binding sites on PKR (Mathews and Shenk 1991; Gunnery et al. 1992). In addition, PKR autophosphorylation is an intermolecular process that can be prevented by heterodimerization with virus-induced proteins (Cosentino et al. 1995; Benkirane et al. 1997; Gale et al. 1999; Melville et al. 1999; Taylor et al. 1999). Other blocking mechanisms include PKR degradation (Black et al. 1993) and dephosphorylation of eIF2 (and/or PKR) (He et al. 1998). A significant sequence homology between the dsRNA-binding region of reovirus σ3 and the regulatory A subunit of protein phosphatase (PP2A) has been noted (Miller and Samuel 1992), and the partially purified phosphatase has been shown to dephosphorylate the α subunit of eIF2 (Chen et al. 1989). This raises the intriguing possibility that σ3 binds and stimulates the PP2A catalytic subunit with reversal of the PKR effects on translation by viral (σ3-containing) polysomes (Miller and Samuel 1992). It would also be of interest to know whether PP2A can act on other transiently phosphorylated translational regulatory elements such as 4E-BP1.

RNA Binding by Reovirus σ3 and Rotavirus NSP3 Proteins

Outer capsid protein σ3 was initially implicated in reovirus translational control by the report that host protein synthesis is selectively shut off in mouse L cells infected with type 2 and, to a lesser extent, type 3 virus (but not type 1) and that the inhibitory phenotype segregates with the S4 genomic segment (Sharpe and Fields 1982b). Many subsequent studies are consistent with the idea that σ3 binds sequence-independently and sequesters dsRNA activators of PKR in infected cells. The eIF2 activity and translation initiation capacity spared as a consequence could then be effectively competed for by viral mRNAs.

Recent reviews have detailed the supporting evidence (much of it indirect) for this hypothesis, starting with the early demonstration that the PKR inhibitory activity in extracts of reovirus-infected cells copurified with σ3 (Imani and Jacobs 1988) and that reporter CAT expression increased in cells cotransfected with S4 cDNA (Giantini and Shatkin 1989). In addition, σ3 effectively replaced PKR inhibitors such as VA RNA (see Chapter 17) in both adenovirus infections and plasmid transfections (Lloyd and Shatkin 1992). Although the amino acid sequences of σ3 are more than 90% conserved in the three reovirus serotypes, type 1 S4 was the least stimulatory in CAT reporter-cotransfected cells, reminiscent of the serotype-dependent effects on host translation in infected L cells (Seliger et al. 1992). Replacement of *E3L* by reovirus S4 (both encode a viral dsRNA-binding protein, see Chapter 35) in recombinant vaccinia virus resulted in a wild-type phenotype by several tests including interferon resistance, failure to induce apoptosis, and prevention of PKR activation (Jacobs and Langland 1998). Mutant analyses indicated further that basic residues in two motifs in the carboxy-terminal domain of σ3 are important for dsRNA binding, and this requirement correlated with *trans*-complementation. Similar rescue of *E3L*$^-$ virus was observed with porcine rotavirus NSP3, consistent with dsRNA-binding activity and PKR inhibition (Langland et al. 1994). An alternative explanation for the selective translation of viral mRNAs in reovirus-infected cells was based on reports that late viral transcripts are uncapped and utilize a cap-independent initiation mechanism (Skup and Millward 1980; Skup et al. 1981), but this was not generally confirmed (Detjen et al. 1982; Lemieux et al. 1984; Munoz et al. 1985) and remains an intriguing hypothesis.

A novel mechanism by which a *Reoviridae* protein, rotavirus nonstructural protein NSP3, may promote preferential translation of rotavirus mRNAs in infected cells has recently been reported. NSP3 binds sequence-specifically to rotavirus mRNAs at the conserved 3′-terminal end (Poncet et al. 1994). It also interacts with initiation factor eIF4G, leading to disruption of complexes that contain poly(A)-binding protein (Piron et al. 1998) and apparently can circularize, and stimulate the translation of, capped and polyadenylated cellular mRNAs (Tarun and Sachs 1996; Le et al. 1997; Imataka et al. 1998). As a consequence of the presence of NSP3, translation of cellular mRNAs would be diminished in infected cells and utilization of capped but nonpolyadenylated viral mRNAs increased by interactions with new complexes containing NSP3 and cap-binding initiation factors (Piron et al. 1998, 1999).

Modulation of σ3 by μ1 Protein

The outer shell of reovirus consists mainly of an equivalent number of copies (600) of σ3 and μ1, present mostly in the μ1C cleaved form. Binding to μ1 depends on the presence of the zinc finger in the amino-terminal domain of σ3 (Shepard et al. 1996), and processing of μ1 to μ1C requires myristoylation as well as σ3 interaction (Tillotson and Shatkin 1992). Although the μ1-cleaving protease has not been identified, the conserved protease-like sequence in σ3 is not required (Mabrouk and Lemay 1994), and it seems more likely that an altered conformation of μ1 resulting from interaction with σ3 facilitates autoproteolysis. In this regard, it is interesting that a conserved GDSG motif, containing the catalytic triad serine and required for the autocatalytic cleavage of Sindbis core protein and several other serine proteinases (Choi et al. 1991), is also present in μ1.

Stable tetracycline *trans*-activator-controlled expression of type 3 σ3 in HeLa cells resulted in no obvious phenotype (Yue and Shatkin 1996), but in retrospect, it would have been useful to compare S4 genes derived from several different virus strains. The viral protein was distributed in both nuclear and cytoplasmic compartments. Coexpression of μ1/μ1C, which was cytoplasmic, resulted in the colocalization of σ3 in the cytoplasm, consistent with stable complex formation. Although different domains in σ3 bind dsRNA and μ1/μ1C, interaction with μ1/μ1C abrogated dsRNA binding and increased the protease sensitivity of σ3 (Shepard et al. 1995). In addition, coexpression of μ1/μ1C prevented the stimulatory effects of σ3 on reporter CAT mRNA translation (Tillotson and Shatkin 1992). These results demonstrate that viral structural proteins can regulate protein synthesis by interactions that alter conformation and modulate binding to PKR activators. However, the mechanism of preferential translation of viral mRNAs in infected cells remains enigmatic, as does the selective stimulation of plasmid-encoded reporter CAT and dihydrofolate reductase synthesis by σ3 and VA RNA in transfected cells. One obvious explanation may be concentration of PKR inhibitors at highly specific sites where viral or plasmid-encoded mRNAs are being translated.

Intracellular Localization of σ3 and PKR Resistance of Viral mRNAs

Schmechel et al. (1997) have reported a correlation between the distribution of σ3 in the cytoplasm of L cells infected with different reovirus strains and the capacity to shut off host protein synthesis. At a high multiplicity of infection with type 3 Dearing strain, σ3 was present uniformly throughout the cytoplasm, and cellular mRNA translation continued almost unabated. However, in cells infected with type 3 Abney strain or type 2 Jones strain, which both dramatically down-regulate host protein synthesis, σ3 was concentrated in cytoplasmic inclusions that probably correspond to sites of virus replication. Host mRNA translation was decreased strongly (84%) or weakly (25%) at 20 hours after infection with Jones (inhibitory) and Dearing (noninhibitory) strains, respectively, but the cells in each case contained similar amounts of total S4 mRNA and σ3 protein. Thus, viral mRNA abundance alone cannot account for the strain-specific, selective production of viral proteins in infected cells. In addition, mixed infection with noninhibitory (Dearing) and inhibitory (Jones) strains indicated that the translation sparing phenotype is largely dominant, consistent with localization of σ3 to putative sites of viral mRNA translation and presumably in a form available to inhibit PKR (i.e., not in complexes with μ1/μ1C). This model helps explain selective translation, but studies are needed to assess the PKR phosphorylation status and activity of viral polysomes in the cytoplasmic inclusion sites. Immunofluorescence and immunoprecipitation studies of cells infected with inhibitory and noninhibitory strains indicated that μ1/μ1C is also distributed in cytoplasmic clusters in both cases but with a greater proportion of "free" σ3 available for dsRNA binding in noninhibitory strains, consistent with the σ3 localization model of preferential mRNA translation.

RAS SIGNALING AND PKR INHIBITION

Although reoviruses are found in nearly all mammals, many mammalian cells in culture are relatively resistant to reovirus and become susceptible after transformation (Hashiro et al. 1977; Duncan et al. 1978). These findings followed earlier observations that mouse ascites tumor cells contain in the cytoplasm an oncolytic 70-nm virus (Bennette 1960; Nelson and Tarnowski 1960). Recent studies have implicated the Ras signaling pathway in transformed cell susceptibility to reovirus (Strong et al. 1998). In several parental NIH-3T3 cell lines, viral mRNAs were synthesized but only weakly translated due to PKR activation, and virus yields were correspondingly low. In transformed mouse cell lines containing activated *ras*, PKR was apparently not phosphorylated (or was dephosphorylated [P.W.K. Lee, pers. comm.]; perhaps by σ3-activated PP2A?) and both viral mRNA translation and replication were enhanced. Precisely how Ras signaling down-regulates PKR remains to be defined, but a correlation has been observed between ERK activation and susceptibility to reovirus infection (P.W.K. Lee, pers. comm.).

BICISTRONIC S1 mRNA: INITIATION, ELONGATION, AND RIBOSOME PAUSING

All three reovirus serotypes contain in the S1 RNA two different, overlapping, and functional ORFs (Fajardo and Shatkin 1990a). Rotavirus segment 11 also apparently codes for two proteins from different ORFs (Mattion et al. 1991). In type 3 reovirus s1 mRNA, synthesis of the minor but important 49-kD cell attachment protein (viral hemagglutinin, σ1) begins 13 nucleotides from the 5′ cap and terminates 36 residues from the 3′-terminal C. The second ORF starts at the second AUG (nucleotide 71) and generates a 14-kD nonstructural protein, σ1s. This arrangement to conserve a +1 ORF within a larger ORF clearly requires strong selection to retain sense codons throughout both frames.

Although σ1s is expressed in infected cells (and apparently at levels similar to σ1; Fajardo and Shatkin 1990b), it is not required for virus replication or apoptosis induction (Rodgers et al. 1998), and its function, perhaps as a modulator of reovirus cytopathogenesis (Fajardo and Shatkin 1990c), remains to be discovered. Studies of the relative levels of σ1s and σ1 made in S1 cDNA-transfected COS cells have provided insights into a novel regulation by inter-ORF interference during elongation (Fajardo and Shatkin 1990a). In these studies, initiation at the first AUG influenced translational starts at the second as predicted by the modified scanning model (Kozak 1999). For example, conversion of the σ1 AUG to UUG or introduction of a downstream terminator nearby both increased σ1s expression severalfold. On the other hand, although σ1s was decreased when sequences surrounding the first AUG were adjusted to conform more closely to the consensus for initiation (GCC GCC ACC AUGG), σ1 levels remained unchanged, suggesting that its synthesis was limited by elongation rather than initiation.

Codon usage analysis predicted abundant expression of σ1s, whereas the presence of many rarely used codons in the σ1 ORF, including the overlapping region, indicated slow progression for σ1-synthesizing ribosomes. Thus, ribosomes slowed or paused in the larger ORF would interfere with passage of normally rapidly translating ribosomes in the σ1s ORF. This concept was tested directly with s1 and other viral mRNAs using a primer extension assay. The results demonstrated nonuniform distribution of translating ribosomes, including stacking at initiation and termination sites as well as pausing or arrested elongation (Doohan and Samuel 1992).

CONCLUDING REMARKS

Reovirus has been crystallized recently (S.C. Harrison, pers. comm.), and X-ray studies promise exciting new insights into how the viral structural proteins accomplish a myriad of functions. They include not only attachment to cells, replication of viral RNAs and proteins, and their assembly into infectious progeny, but also shutoff of host gene expression and preferential translation of viral mRNAs. Susceptibility to reovirus infection is linked in some way to the Ras signaling pathway, and virus attachment at the cell surface is sufficient to elicit a block in DNA replication and cell division. Despite the importance of these downstream events, the reovirus receptor still remains to be elucidated.

A key host defense measure in reovirus (and other) infections is activation of PKR. This protein kinase is likely a common element in several cytopathic effects besides the shutoff of host protein synthesis. Polypeptide σ3 has been shown to bind dsRNA, to counteract PKR, and to promote the selective translation of viral mRNAs, but more direct evidence is needed for (or against) this as a mechanism of control in infected cells. In addition to PKR and eIF2, it may be of interest to determine the phosphorylation levels of other initiation factors and translational regulatory proteins such as 4E-BP1 in reovirus infections. The possible modulating effects of σ3 and other viral proteins on protein phosphatases also need to be explored.

Finally, reovirus mRNAs presumably are removed from the translation process when they serve as templates for the synthesis of genomic dsRNAs. Packaging of the dsRNA segments during virus maturation is an accurate and highly ordered but perplexing process (Joklik 1998). Development of a reverse genetics system (Roner 1999) that makes possible the routine engineering of specific mutations into the reovirus genome would facilitate a better understanding of many replication events including virus assembly. Bacterial dsRNA virus ϕ6, which contains three polycistronic genome segments, has been assembled in vitro (Mindich 1999) and may provide a model for reoviruses.

REFERENCES

Antczak J.B. and Joklik W.K. 1992. Reovirus genome segment assortment into progeny genomes studied by the use of monoclonal antibodies directed against reovirus proteins. *Virology* **187:** 760–776.

Antczak J.B., Chmelo R., Pickup D.J., and Joklik W.K. 1982. Sequences at both termini of the ten genes of reovirus serotype 3 (strain Dearing). *Virology* **121:** 307–319.

Beattie E., Tattaglia J., and Paoletti E. 1991. Vaccinia virus-encoded eIF-2α homologue abrogates the antiviral effect of interferon. *Virology* **183:** 419–422.

Benkirane M., Neuveut C., Chun R.F., Smith S.M., Samuel C.E., Gatignol A., and Jeang K.T. 1997. Oncogenic potential of TAR RNA binding protein TRBP and its regulatory interaction with RNA-dependent protein kinase PKR. *EMBO J.* **16:** 611–624.

Bennette J.G. 1960. Isolation of a non-pathogenic tumour-destroying virus from mouse ascites. *Nature* **187:** 72–73.

Bisaillon M., Bergeron J., and Lemay G. 1997. Characterization of the nucleoside triphosphate phosphohydrolase and helicase activities of the reovirus λ1 protein. *J. Biol. Chem.* **272:** 18298–18303.

Bischoff J.R. and Samuel C.E. 1989. Mechanism of interferon action. Activation of the P1/eIF-2 alpha protein kinase by individual reovirus s-class mRNAs: s1 mRNA is a potent activator relative to s4 mRNA. *Virology* **172:** 106–115.

Black T.L., Barber G.N., and Katze M.G. 1993. Degradation of the interferon-induced 68,000-M(r) protein kinase by poliovirus requires RNA. *J. Virol.* **67:** 791–800.

Brostrom C.O. and Brostrom M.A. 1998. Regulation of translational initiation during cellular response to stress. *Prog. Nucleic Acid Res. Mol. Biol.* **58:** 79–125.

Brown E.B. 1998. Reovirus M1 gene expression. *Curr. Top. Microbiol. Immunol.* **233/I:** 197–213.

Cashdollar L.W., Chmelo R.A., Wiener J.R., and Joklik W.K. 1985. Sequences of the S1 genes of the three serotypes of reovirus. *Proc. Natl. Acad. Sci.* **82:** 24–28.

Chen S.-C., Kramer G., and Hardesty B. 1989. Isolation and partial characterization of an Mr 60,000 subunit of a type 2A phosphatase from rabbit reticulocytes. *J. Biol. Chem.* **264:** 7267–7275.

Choi H.K., Tong L., Minor W., Dumas P., Boege U., Rossmann M.G., and Wengler G. 1991. Structure of Sindbis virus core protein reveals a chymotrypsin-like serine proteinase and the organization of the virion. *Nature* **354:** 37–43.

Cosentino G.P., Venkatesan S., Serluca F.C., Green S.R., Mathews M.B., and Sonenberg N. 1995. Double-stranded-RNA-dependent protein kinase and TAR RNA-binding protein form homo- and heterodimers in vivo. *Proc. Natl. Acad. Sci.* **92:** 9445–9449.

Cuddihy A.R., Li S., Tam N.W., Wong A.H., Taya Y., Abraham N., Bell J.C., and Koromilas A.E. 1999. Double-stranded-RNA-activated protein kinase PKR enhances transcriptional activation by tumor suppressor p53. *Mol. Cell. Biol.* **19:** 2475–2484.

Darzynkiewicz E. and Shatkin A.J. 1980. Assignment of reovirus mRNA ribosome binding sites to virion genome segments by nucleotide squence analyses. *Nucleic Acids Res.* **8:** 337–350.

Debiasi R.L., Squier M.K.T., Pike B., Wynes M., Dermody T.S., Cohen J.J., and Tyler K.L. 1999. Reovirus-induced apoptosis is preceded by increased cellular calpain activity and is blocked by calpain inhibitors. *J. Virol.* **73:** 695–701.

Detjen B.M., Walden W.E., and Thach R.E. 1982. Translational specificity in reovirus-infected mouse fibroblasts. *J. Biol. Chem.* **257:** 9855–9860.

Doohan J.P. and Samuel C.E. 1992. Biosynthesis of reovirus-specified polypeptides: Ribosome pausing during the translation of reovirus S1 mRNA. *Virology* **182:** 409–425.

Dryden K.A., Wang G., Yeager M., Nibert M.L., Coombs K.M., Furlong D.B., Fields B.N., and Baker T.S. 1993. Early steps in reovirus infection are associated with dramatic changes in supramolecular structure and protein conformation: Analysis of virions and subviral particles by cryoelectron microscopy and image reconstruction. *J. Cell Biol.* **122:** 1023–1041.

Duncan M.R., Stanish S.M., and Cox D.C. 1978. Differential sensitivity of normal and transformed human cells to reovirus infection. *J. Virol.* **28:** 444–449.

Fajardo J.E. and Shatkin A.J. 1990a. Translation of bicistronic viral mRNA in transfected cells: Regulation at the level of elongation. *Proc. Natl. Acad. Sci.* **87:** 328–332.

———. 1990b. Effects of elongation on the translation of a reovirus bicistronic mRNA. *Enzyme* **44:** 235–243.

———. 1990c. Expression of the two reovirus S1 gene products in transfected mammalian cells. *Virology* **178:** 223–231.

Fausnaugh J. and Shatkin A.J. 1990. Active site localization in a viral mRNA capping enzyme. *J. Biol. Chem.* **265:** 7669–7672.

Fields B.N. 1996. Reoviridae. In *Fields virology* (ed. B.N. Fields et al.), pp. 1553–1555. Lippincott-Raven, Philadelphia, Pennsylvania.

Furuichi Y. and Shatkin A.J. 2000. Viral and cellular mRNA capping: Past and prospects. *Adv. Virus Res.* **55:** 135–184.

Furuichi Y., Morgan M., Muthukrishnan S., and Shatkin A.J. 1975. Reovirus messenger RNA contains a methylated, blocked 5′-terminal structure: $m^7G(5')ppp(5')G^mpCp$-. *Proc. Natl. Acad. Sci.* **72:** 362–366.

Furuichi Y., Muthukrishnan S., Tomasz J., and Shatkin A.J. 1976. Mechanism of formation of reovirus mRNA 5′-terminal blocked and methylated sequence, $m^7GpppG^{mp}C$. *J. Biol. Chem.* **251:** 5043–5053.

Gale M., Kwieciszewski B., Dossett M., Nakao H., and Katze M.G. 1999. Antiapoptotic and oncogenic potentials of hepatitis C virus are linked to interferon resistance by viral repression of the PKR protein kinase. *J. Virol.* **73:** 6506–6516.

Giantini M. and Shatkin A.J. 1989. Stimulation of chloramphenicol acetyltransferase mRNA translation by reovirus capsid polypeptide σ3 in cotransfected COS cells. *J. Virol.* **63:** 2415–2421.

Gil J., Alcami J., and Esteban M. 1999. Induction of apoptosis by double-stranded-RNA-dependent protein kinase (PKR) involves the alpha subunit of eukaryotic translation initiation factor 2 and NF-κB. *Mol. Cell. Biol.* **19:** 4653–4663.

Gillian A.L. and Nibert M.L. 1998. Amino terminus of reovirus nonstructural protein sigma NS is important for ssRNA binding and nucleoprotein complex formation. *Virology* **240:** 1–11.

Gillies S., Bullivant S., and Bellamy A.R. 1971. Viral RNA polymerases: Electron microscopy of reovirus reaction cores. *Science* **174:** 694–696.

Gingras A.-C., Raught B., and Sonenberg N. 1999. EIF4 initiation factors: Effectors of mRNA recruitment to ribosomes and regulators of translation. *Annu. Rev. Biochem.* **68:** 913–963.

Gunnery S., Green S.R., and Mathews M.B. 1992. Tat-responsive region RNA of human immunodeficiency virus type 1 stimulates protein synthesis in vivo and in vitro: Relationship between structure and function. *Proc. Natl. Acad. Sci.* **89:** 11557–11561.

Hand R. and Tamm I. 1974. Initiation of DNA replication in mammalian cells and its inhibition by reovirus infection. *J. Mol. Biol.* **82:** 175–183.

Hashiro G., Loh P.C., and Yau J.T. 1977. The preferential cytotoxicity of reovirus for certain transformed cell lines. *Arch. Virol.* **54:** 307–315.

Hatada E., Saito S., and Fukuda R. 1999. Mutant influenza viruses with a defective NS1 protein cannot block the activation of PKR in infected cells. *J. Virol.* **73:** 2425–2433.

He B., Gross M., and Roizman B. 1998. The gamma134.5 protein of herpes simplex virus 1 has the structural and functional attributes of a protein phosphatase 1 regulatory subunit and is present in a high molecular weight complex with the enzyme in infected cells. *J. Biol. Chem.* **273:** 20737–20743.

Imani F. and Jacobs B.L. 1988. Inhibitory activity for the interferon-induced protein kinase is associated with the reovirus serotype 1 sigma 3 protein. *Proc. Natl. Acad. Sci.* **85:** 7887–7891.

Imataka H., Gradi A., and Sonenberg N. 1998. A newly identified N-terminal amino acid sequence of human eIF4G binds poly(A)-binding protein and functions in poly(A)-dependent translation. *EMBO J.* **17:** 7480–7489.

Jacobs B.L. and Langland J.O. 1998. Reovirus σ3 protein: dsRNA binding and inhibition of RNA-activated protein kinase. *Curr. Top. Microbiol. Immunol.* **233/I:** 185–196.

Jane-Valbuena J., Nibert M.L., Spencer S.M., Walker S.B., Baker T.S., Chen Y., Centonze V.E., and Schiff L.A. 1999. Reovirus virion-like particles obtained by recoating infectious subvirion particles with baculovirus-expressed σ3 protein: An approach for analyzing σ3 functions during virus entry. *J. Virol.* **73:** 2963–2973.

Joklik W.K. 1998. Assembly of the reovirus genome. *Curr. Top. Microbiol. Immunol.* **233/I:** 57–68.

Kaufman R.J. 1999a. Stress signaling from the lumen of the endoplasmic reticulum: Coordination of gene transcriptional and translational controls. *Genes Dev.* **13:** 1211–1233.

———. 1999b. Double-stranded RNA-activated protein kinase mediates virus-induced apoptosis: A new role for an old actor. *Proc. Natl. Acad. Sci.* **96:** 11693–11695.

Kozak M. 1999. Initiation of translation in prokaryotes and eukaryotes. *Gene* **234:** 187–208.

Kozak M. and Shatkin A.J. 1978. Migration of 40S ribosomal subunits on messenger RNA in the presence of edeine. *J. Biol. Chem.* **253:** 6568–6577.

Kudo H. and Graham A.F. 1965. Synthesis of reovirus ribonucleic acid in L cells. *J. Bacteriol.* **90:** 936–945.

Kuhen K.L. and Samuel C.E. 1997. Isolation of the interferon-inducible RNA-dependent protein kinase Pkr promoter and identification of a novel DNA element within the 5′-flanking region of human and mouse Pkr genes. *Virology* **227:** 119–130.

Langland J.O., Pettiford S., Jiang B., and Jacobs B.L. 1994. Products of the porcine group C rotavirus NSP3 gene bind specifically to double-stranded RNA and inhibit activation of the interferon-induced protein kinase PKR. *J. Virol.* **68:** 3821–3829.

Le H., Tanguay R.L., Balasta M.L., Wei C.-C., Browning K.S., Metz A.M., Goss D.J., and Gallie D.R. 1997. Translation initiation factors eIF-iso4G and eIF-4B interact with the poly(A)-binding protein and increase its RNA binding activity. *J. Biol. Chem.* **272:** 16247–16255.

Lee P.W.K. and Gilmore R. 1998. Reovirus cell attachment protein σ1: Structure-function relationships and biogenesis. *Curr. Top. Microbiol. Immunol.* **233/I:** 137–153.

Lee S.B., Rodriguez D., Rodriguez J.R., and Esteban M. 1997. The apoptosis pathway triggered by the interferon-induced protein kinase PKR requires the third basic domain, initiates upstream of Bcl-2, and involves ICE-like proteases. *Virology* **231:** 81–88.

Lemieux R., Zarbl H., and Millward S. 1984. mRNA discrimination in extracts from uninfected and reovirus-infected L-cells. *J. Virol.* **51:** 215–222.

Lloyd R.M. and Shatkin A.J. 1992. Translational stimulation by reovirus polypeptide σ3: Substitution for VAI RNA and inhibition of phosphorylation of the α subunit of eukaryotic initiation factor 2. *J. Virol.* **66:** 6878–6884.

Luongo C.L., Contreras C.M., Farsetta D.L., and Nibert M.L. 1998. Binding site for S-adenosyl-L-methionine in a central region of mammalian reovirus lambda2 protein. Evidence for activities in mRNA cap methylation. *J. Biol. Chem.* **273:** 23773–23780.

Mabrouk T. and Lemay G. 1994. The sequence similarity of reovirus σ3 protein to picornaviral proteases is unrelated to its role in μ1 viral protein cleavage. *Virology* **202:** 615–620.

Mathews M.B. and Shenk T. 1991. Adenovirus virus-associated RNA and translational control. *J. Virol.* **65:** 5657–5662.

Mattion N.M., Cohen J., Aponte C., and Estes M.K. 1992. Characterization of an oligomerization domain and RNA-binding properties on rotavirus nonstructural protein NS34. *Virology* **190:** 68–83.

Mattion N.M., Mitchell D.B., Both G.W., and Estes M.K. 1991. Expression of rotavirus proteins encoded by alternative open reading frames of genome segment 11. *Virology* **181:** 295–304.

Melville M.W., Tan S.L., Wambach M., Song J., Morimoto R.I., and Katze M.G. 1999. The cellular inhibitor of the PKR protein kinase, P58(IPK), is an influenza virus-activated co-chaperone that modulates heat shock protein 70 activity. *J. Biol. Chem.* **274:** 3797–3803.

Miller J.E. and Samuel C.E. 1992. Proteolytic cleavage of the reovirus sigma 3 protein results in enhanced double-stranded RNA-binding activity: Identification of a repeated basic amino acid motif within the C-terminal binding region. *J. Virol.* **66:** 5347–5356.

Mindich L. 1999. Reverse genetics of dsRNA bacteriophage ϕ6. *Adv. Virus Res.* **53:** 341–353.

Munoz A., Alonso M.A., and Carrasco L. 1985. The regulation of translation in reovirus-infected cells. *J. Gen. Virol.* **66:** 2161–2170.

Nelson J.B. and Tarnowski G.S. 1960. An oncolytic virus recovered from Swiss mice during passage of an ascites tumour. *Nature* **188:** 866–867.

Nibert M.L. 1998. Structure of mammalian orthoreovirus particles. *Curr. Top. Microbiol. Immunol.* **233/I:** 1–30.

Nibert M.L., Schiff L.A., and Fields B.N. 1991. Mammalian reoviruses contain a myristoylated structural protein. *J. Virol.* **88:** 1960–1967.

———. 1996. Reoviruses and their replication. In *Fields virology* (ed. B.N. Fields et al.), pp. 1557–1596. Lippincott-Raven, Philadelphia, Pennsylvania.

Oberhaus S.M., Dermody T.S., and Tyler K.L. 1998. Apoptosis and the cytopathic effects of reovirus. *Curr. Top. Microbiol. Immunol.* **233/II:** 23–49.

Piron M., Delaunay T., Grosclaude J., and Poncet D. 1999. Identification of the RNA-binding, dimerization, and eIF4GI-binding domains of rotavirus nonstructural protein NSP3. *J. Virol.* **73:** 5411–5421.

Piron M., Vende P., Cohen J., and Poncet D. 1998. Rotavirus RNA-binding protein NSP3 interacts with eIF4GI and evicts the poly(A) binding protein from eIF4F. *EMBO J.* **17:** 5811–5821.

Poncet D., Laurent S., and Cohen J. 1994. Four nucleotides are the minimal requirement for RNA recognition by rotavirus non-structural protein NSP3. *EMBO J.* **13:** 4165–4173.

Ramig R.F. and Fields B.N. 1983. Genetics of reoviruses. In *The reoviridae* (ed. W.K. Joklik), pp.197–228. Plenum Press, New York.

Rodgers S.E., Connolly J.L., Chappell J.D., and Dermody T.S. 1998. Reovirus growth in cell culture does not require the full complement of viral proteins: Identification of a σ1s-null mutant. *J. Virol.* **72:** 8597–8604.

Roner M.R. 1999. Rescue systems for dsRNA viruses of higher organisms. *Adv. Virus Res.* **53:** 355–367.

Roner M.R., Roner L.A., and Joklik W.K. 1993. Translation of reovirus RNA species m1 can initiate at either of the first two in-frame initiation codons. *Proc. Natl. Acad. Sci.* **90:** 8947–8951.

Samuel C.E. 1998. Reoviruses and the interferon system. *Curr. Top. Microbiol. Immunol.* **233/II:** 125–145.

Saragovi H.U., Rebai N., Di Guglielmo G.M., Macleod R., Sheng J., Rubin D.H., and Greene M.I. 1999. A G1 cell cycle arrest induced by ligands of the reovirus type 3 receptor is secondary to inactivation of p21ras and mitogen-activated protein kinase. *DNA Cell Biol.* **18:** 763–770.

Schmechel S., Chute M., Skinner P., Anderson R., and Schiff L. 1997. Preferential translation of reovirus mRNA by a σ3-dependent mechanism. *Virology* **232:** 62–73.

Seliger L.S., Zheng K., and Shatkin A.J. 1987. Complete nucleotide sequence of reovirus L2 gene and deduced amino acid sequence of viral mRNA guanylyltransferase. *J. Biol. Chem.* **262:** 16289–16293.

Seliger L.S., Giantini M., and Shatkin A.J. 1992. Translational effects and sequence comparisons of the three serotypes of the reovirus S4 gene. *Virology* **187:** 202–210.

Sharpe A.H. and Fields B.N. 1982a. Reovirus inhibition of cellular DNA synthesis: Role of the S1 gene. *J. Virol.* **38:** 380–392.

———. 1982b. Reovirus inhibition of cellular RNA and protein synthesis: Role of the S4 gene. *Virology* **122:** 381–391.

Shatkin A.J. and Kozak M. 1983. Biochemical aspects of reovirus transcription and translation. In *The reoviridae* (ed. W.K. Joklik), pp.79–106. Plenum Press, New York.

Shatkin A.J. and Sipe J.D. 1968. RNA polymerase activity in purified reoviruses. *Proc. Natl. Acad. Sci.* **61:** 1462–1469.

Shatkin A.J., Sipe J.D., and Loh P.C. 1968. Separation of ten reovirus genome fragments by polyacrylamide gel electrophoresis. *J. Virol.* **2:** 986–991.

Shaw J.E. and Cox D.C. 1973. Early inhibition of cellular DNA synthesis by high multiplicities of infectious and UV-irradiated reovirus. *J. Virol.* **12:** 704–710.

Shepard D.A., Ehnstrom J.G., and Schiff L.A. 1995. Association of reovirus outer capsid protein σ3 and μ1 causes a conformational change that renders σ3 protease sensitive. *J. Virol.* **69:** 8180–8184.

Shepard D.A., Ehnstrom J.G., Skinner P.J., and Schiff L.A. 1996. Mutations in the zinc-binding motif of the reovirus capsid protein σ3 eliminate its ability to associate with capsid protein μ1. *J. Virol.* **70:** 2065–2068.

Skup D. and Millward S. 1980. Reovirus-induced modification of cap dependent translation in infected L cells. *Proc. Natl. Acad. Sci.* **77:** 152–156.

Skup D., Zarbl H., and Millward S. 1981. Regulation of translation in L-cells infected with reovirus. *J. Mol. Biol.* **151:** 35–55.

Sonenberg N., Morgan M.A., Merrick W.C., and Shatkin A.J. 1978. A polypeptide in eukaryotic initiation factors that crosslinks specifically to the 5′-terminal cap in mRNA. *Proc. Natl. Acad. Sci.* **75:** 4843–4847.

Srivastava S.P., Kumar K.U., and Kaufman R.J. 1998. Phosphorylation of eukaryotic translation initiation factor 2 mediates apoptosis in response to activation of the double-stranded RNA-dependent protein kinase. *J. Biol. Chem.* **273:** 2416–2423.

Starnes M.C. and Joklik W.K. 1993. Reovirus protein λ3 is a poly(C)-dependent poly(G) polymerase. *Virology* **193:** 356–366.

Stoltzfus C.M., Shatkin A.J., and Banerjee A.K. 1973. Absence of polyadenylic acid from reovirus messenger RNA. *J. Biol. Chem.* **248:** 7993–7998.

Strong J.E., Coffey M.C., Tang D., Sabinin P., and Lee P.W.K. 1998. The molecular basis of viral oncolysis: Usurpation of the Ras signaling pathway by reovirus. *EMBO J.* **17:** 3351–3362.

Tarun S.Z. and Sachs A.B. 1996. Association of the yeast poly(A) tail binding protein with translation factor eIF4G. *EMBO J.* **15:** 7168–7177.

Taylor D.R., Shi S.T., Romano P.R., Barber G.N., and Lai M.M. 1999. Inhibition of the interferon-inducible protein kinase PKR by HCV E2 protein. *Science* **285:** 107–110.

Tillotson L. and Shatkin A.J. 1992. Reovirus polypeptide σ3 and N-terminal myristoylation of polypeptide μ1 are required for site-specific cleavage to μ1C in transfected cells. *J. Virol.* **66:** 2180–2186.

Tosteson M.T., Nibert M.L., and Fields B.N. 1993. Ion channels induced in lipid bilayers by subvirion particles of the nonenveloped mammalian reoviruses. *Proc. Natl. Acad. Sci.* **90:** 10549–10552.

Yue Z. and Shatkin A.J. 1996. Regulated, stable expression and nuclear presence of reovirus double-stranded RNA-binding protein σ3 in HeLa cells. *J. Virol.* **70:** 3497–3501.

———. 1998. Enzymatic and control functions of reovirus structural proteins. *Curr. Top. Microbiol. Immunol.* **233/I:** 31–56.

Zweerink H.J. and Joklik W.K. 1970. Studies on the intracellular synthesis of reovirus-specified proteins. *Virology* **41:** 501–518.

34

Translational Reprogramming during Influenza Virus Infection

Seng-Lai Tan and Michael G. Katze
Department of Microbiology
University of Washington School of Medicine
Seattle, Washington 98195

Michael J. Gale, Jr.
Department of Microbiology
University of Texas Southwestern Medical Center
Dallas, Texas 75235

Influenza virus infection is one of the most common human infectious diseases and a worldwide health problem. It is also one of the most dreadful threats for a recurring pandemic, responsible for more than 20 million deaths worldwide in 1918–1919 (the "Spanish" flu). Indeed, the recent outbreaks of H5 influenza viruses in Hong Kong in 1997–1998 gave the world a frightening feeling of déjà vu. Apart from being a medically important viral pathogen, influenza virus serves as an attractive system for studying translational control because it severely impairs the ability of the host to initiate protein synthesis on cellular mRNAs, a phenomenon known as host shutoff. Yet this highly cytopathic virus is able to maintain selective and efficient translation of its own mRNAs in a cap-dependent manner. This requires the infected cell to be translationally proficient so that the processes involved in the protein synthetic pathway remain intact. Furthermore, like many animal viruses, influenza virus has to overcome the host innate antiviral response, of which one important mediator, namely the PKR protein kinase (see Chapters 8 and 13), exerts its effect by arresting global protein synthesis at the translation initiation step. To accomplish these goals, influenza virus has evolved diverse and intricate strategies, ranging from the inhibition of cellular mRNA translation to the preferential translation of viral mRNAs to the inactivation of the PKR protein kinase (Table 1).

Translational Control of Gene Expression
© 2000 Cold Spring Harbor Laboratory Press 0-87969-568-4/00 $5 +. 00

Table 1 Translational regulatory events of influenza virus-infected cells

Strategy	Reference
Host shutoff of cellular protein synthesis	Lazarowitz et al. (1971); Skehel (1972); Katze and Krug 1984
Inhibition of cellular mRNA transport and degradation of mRNAs in the nucleus	Katze and Krug (1984)
Inhibition of cellular mRNA translation at initiation and elongation stages	Katze et al. (1986a); Garfinkel and Katze (1992)
Dephosphorylation of eukaryotic initiation factor 4E	Feigenblum and Schneider (1993)
Cap-dependent, selective translation of influenza virus mRNAs	Katze and Krug (1984); Katze et al. (1986a); Alonso-Caplen et al. (1988); Garfinkel and Katze (1992, 1993)
Innate ability of influenza virus mRNAs for preferential translation	Alonso-Caplen et al. (1988); Garfinkel and Katze (1992, 1993)
Viral 5′UTR-mediated preferential translation of influenza virus mRNAs	Garfinkel and Katze (1993); Park and Katze (1995); Park et al. (1999)
Temporal regulation of influenza virus protein synthesis	Yamanaka et al. (1988, 1991); Enami et al. (1994)
Inhibition of PKR	Katze et al. (1984, 1986b, 1988); Lee et al. (1990, 1992)
PKR Inhibition by P58^{IPK}	Barber et al. (1994); Lee et al. (1994)
PKR Inhibition by NS1	Lu et al. (1995); Tan and Katze (1998); Hatada et al. (1999)

STRATEGIES OF HOST SHUTOFF BY INFLUENZA VIRUS

Like other cytopathic viruses, influenza virus dramatically perturbs the normal synthesis of host macromolecules. However, influenza virus is different from most non-oncogenic RNA viruses in that its replicative cycle includes both nuclear and cytoplasmic phases. The evolution of the virus's translational strategies is likely to be dictated by a different set of selective pressures (Krug et al. 1989). Moreover, the genetic coding capacity of influenza virus is limited by its small negative-segmented RNA genome size, which encodes no more than ten proteins: three polymerase proteins, PB1, PB2, and PA; two glycoproteins, hemagglutinin (HA) and neuraminidase (NA); two matrix proteins, M1 and M2; two nonstructural proteins, NS1 and NS2; and the nucleocapsid protein, NP. Thus, it is almost intuitive that such a successful virus must have evolved a plethora of ingenious schemes to ensure selective and efficient translation of viral mRNAs.

Earlier studies of influenza translational control have focused on the mechanism underlying the drastic host shutoff because of its potential implication in the virus pathogenic killing of infected cells. These studies revealed that transcription of cellular mRNA is decreased in infected cells, although at a modest rate (Katze and Krug 1984, 1990). Furthermore, degradation of cellular mRNAs is only evident late after infection (Inglis 1982; Beloso et al. 1992). In addition, newly synthesized cellular mRNAs containing short (~12 nucleotide) poly(A) tails are retained in the nucleus (Katze et al. 1984), a process that may be mediated by the influenza virus nonstructural protein NS1. Recent studies demonstrated that NS1 can interact with two cellular 3′ processing proteins, the cellular 30-kD subunit of the cleavage and polyadenylation specificity factor, CPSF (Nemeroff et al. 1998), and the poly(A)-binding protein II (PABII) of the cellular 3′-end processing machinery (Chen et al. 1999). Apart from directly associating with each other, the 30-kD CPSF and PABII proteins also bind to nonoverlapping regions of the NS1 protein in vitro. Thus, it is possible that NS1 binding to CPSF and PABII may disrupt 3′-end formation of cellular pre-mRNAs, resulting in their nuclear retention.

Cellular mRNAs are also degraded in the nucleus during influenza virus infection, which is thought to be induced by the cleavage of the 5′ ends of cellular RNA polymerase II transcripts by the viral cap-dependent endonuclease (Katze and Krug 1984). Influenza-induced aberrant host pre-mRNA splicing results in unspliced cellular transcripts that may be rapidly degraded (Lu et al. 1994). Thus, the possible contribution of this anomalous splicing to host shutoff cannot be excluded. Indeed, such a mechanism has been proposed for the shutoff of cellular protein synthesis by herpes simplex virus (HSV) (Hardy and Sandri-Goldin 1994; Sandri-Goldin et al. 1995). In this regard, yet another role has been proposed for the viral NS1 protein, which has been demonstrated to inhibit cellular pre-mRNA splicing (Lu et al. 1994). Furthermore, NS1 interacts with a novel cellular protein, termed NS1-BP, that is capable of promoting pre-mRNA splicing (Wolff et al. 1998). This interaction may account for the splicing-inhibitory effect of NS1, which in turn could trigger host shutoff by influenza. Given the multiple functions assigned to NS1 in mediating host shutoff, it is remarkable that a recombinant influenza virus that lacks the *NS1* gene is still able to replicate to high titer within cells (Garcia-Sastre et al. 1998).

Indeed, the above events cannot completely account for the dramatic cessation of cellular protein synthesis because preexisting cytoplasmic cellular mRNAs are stable and functional when tested in cell-free translation systems in vitro (Garfinkel and Katze 1992). Moreover, both cellular and viral mRNAs remain associated with polysomes that are cytoskeleton-bound (Katze et al. 1989), suggesting that other mechanisms may be at work. Finally, there is evidence indicating that cellular mRNA translation is inhibited at both the initiation and elongation stages in the infected cell (Katze et al. 1986a; Garfinkel and Katze 1992). The inhibition of cellular mRNA translational initiation may be a consequence of influenza-induced limitation of functional eIF4E translation factor (see discussion below). It has also been shown that influenza virus impairs ribosome transit on a number of cellular mRNAs, slowing the runoff of ribosomes and thereby leading to the accumulation of polysomes. Although little progress has been made since these observations to determine the exact molecular mechanisms for these events, it is clear that influenza virus deploys multiple schemes to induce host shutoff (Table 1).

SELECTIVE TRANSLATION OF VIRAL mRNAS BY INFLUENZA VIRUS

The mechanism by which influenza virus selectively translates its mRNAs is beginning to be understood. While poliovirus and adenovirus mRNAs contain the internal ribosome entry site (IRES) element of poliovirus and the tripartite leader of adenovirus, respectively (see Chapters 31 and 32), influenza virus mRNAs do not appear to possess any extensive secondary structures. Instead, all influenza virus mRNAs contain a short, conserved 5′-untranslated region (UTR) with no upstream AUGs (Katze and Krug 1990). Yet, this short linear 5′UTR is capable of directing viral mRNA translation with a remarkably high efficiency (Garfinkel and Katze 1992, 1993). Importantly, a cellular mRNA artificially created to carry the 5′UTR of the influenza viral nucleoprotein (NP) gene is also translated efficiently and selectively, demonstrating that the viral 5′UTR alone is sufficient to recruit the cellular translational machinery. That the viral 5′UTR contains an innate ability to promote preferential translation of viral mRNAs is further supported by the observation that the process, although cap dependent, appears to be independent of the availabilty of functional eIF4E translation factor (Feigenblum and Schneider 1993). Present in limiting amounts within the cell, eIF4E is an important translation initiation factor that binds both cellular and viral mRNAs and is required for assembly of the 5′ cap-binding complex, eIF4F (see Chapter 6). During influenza virus infection, the phosphorylation levels of eIF4E are diminished, which can lead to a significant decrease in the cap-binding activity (Feigenblum and Schneider 1993). This dephosphorylation of eIF4E corresponds to a decrease in the level of host mRNA translation with little or no effect on translation of viral mRNAs, suggesting that it may contribute to the inhibition of cellular mRNA translation in influenza virus-infected cells. Moreover, influenza virus mRNAs are efficiently translated in adenovirus-infected cells, where eIF4E is severely dephosphorylated. Although these observations have not been followed up, they suggest that the 5′UTR of influenza viral mRNAs possess features that render them less susceptible to limitations in the availability of eIF4F and associated helicase activity imposed by eIF4E dephosphorylation. Indeed, recent studies demonstrate that the 5′UTR region functions to recruit both cellular and influenza viral proteins to aid in selective translation of viral mRNAs.

Viral 5′UTR Recruits Cellular Protein grsf-1 for Selective Translation

The involvement of *trans*-acting factors in the translational control of influenza virus mRNAs was suggested by the finding that several cellular proteins interact with specific regions of the 5′UTRs of viral but not cellular mRNAs, as revealed by gel mobility shift and UV cross-linking analysis (Park and Katze 1995). One of these 5′UTR-interacting factors was identified as the cellular protein GRSF-1 in a yeast three-hybrid screen (Park et al. 1999). Originally cloned by using a Northwestern strategy and a labeled G-rich RNA element as probe, the GRSF-1 protein is predominantly localized in the cytoplasm and contains three RNA-recognition motifs (RRMs), categorizing it as a member of the RRM-containing protein superfamily (Qian and Wilusz 1994). The specificity of GRSF-1 was demonstrated by its binding to the 5′ UTR of the viral NP gene, but not to NP 5′UTR mutants or cellular mRNA 5′UTRs (Park et al. 1999). Recombinant GRSF-1 specifically stimulates translation of a NP 5′ UTR-driven template in cell-free translation systems. Furthermore, the translational efficiency of NP 5′UTR-driven templates is markedly reduced in GRSF-1-depleted HeLa cell extracts but is restored by the addition of recombinant GRSF-1 in a concentration-dependent fashion. Finally, competition experiments using NP 5′UTR sequences demonstrated a requirement for GRSF-1-binding in the translation of viral, but not cellular, mRNA. Taken together, these results strongly suggest a specific interaction between GRSF-1 and the influenza virus 5′UTR, and that the recruitment of GRSF-1 to the 5′UTR facilitates the selective translation of viral mRNAs in infected cells. The precise mechanism by which GRSF-1 affects the overall translation efficiency of influenza virus mRNAs in the infected cell is not known. However, it should be mentioned that since the action of GRSF-1 was tested on viral 5′UTR-driven chimera templates in a cell-free translation system, it is essential to perform parallel experiments using bona fide viral mRNAs. Ultimately, to prove its biological function it will be necessary to examine the selective translatability of viral mRNAs in cells that lack functional endogenous GRSF-1.

Role of NS1 Protein in Preferential Translation of Select Viral mRNA

NS1, the only nonstructural protein of influenza virus, is an RNA-binding protein that inhibits both the nuclear export of poly(A)-containing mRNA and the splicing of pre-mRNA (Fortes et al. 1994; Lu et al. 1994; Qiu and Krug 1994). In addition to binding to poly(A) mRNA and a stem-bulge region in U6 small nuclear RNA, the NS1 protein also binds to double-stranded (ds) RNA (Hatada and Fukuda 1992). There is evidence that NS1 is also involved in the selective translation of viral mRNAs. Enami and colleagues (Enami et al. 1994) showed in an in vivo transfection assay that the expression of the matrix (M) 5′UTR- or NP 5′UTR-driven reporter gene is enhanced in the presence of NS1, whereas that of the NS1 5′UTR-driven gene is not. Ortin and colleagues confirmed the involvement of NS1 and the viral 5′UTR in the translational enhancement of M and NP mRNAs in a cotransfection analysis (de la Luna et al. 1995). Polysome analysis of the reporter mRNAs expressed in the presence or absence of NS1 indicates that the stimulation occurs at the level of translation initiation. Consistent with its role in the translational enhancement of viral mRNAs, NS1 is present in the polysomes during virus infection (Lazarowitz et al. 1971) and can bind to the 5′UTRs of M and NP mRNAs (Park and Katze 1995). However, it is essential to validate the role of NS1 in translational regulation by using a more direct approach, such as an in vitro system that can discriminate between cellular and viral mRNA translation.

Considering these properties, it is possible that NS1 may simply recruit specific translation initiation factors to viral mRNAs for selective translation. NS1 can form a complex with eIF4G in extracts from influenza virus-infected cells as shown by coimmunoprecipitation and affinity chromatography studies (Aragon et al. 2000). In vitro binding assay and deletion analysis demonstrated that the interaction is likely to be mediated by RNA-independent protein–protein interaction. NS1 may recruit eIF4G to the 5′UTR of viral mRNAs for translation initiation, which may in turn also contribute the host shutoff due to competition with cellular mRNAs for eIF4G. Equally possible is that NS1 binding competes with eIF4G interaction with either eIF4E and/or poly(A)-binding proteins, thereby disrupting the normal translation of cellular mRNAs. However, the exact mechanism by which NS1 mediates viral mRNA translation remains to be determined. It might be interesting to examine whether NS1 interacts with GRSF-1 to synergistically promote translation of viral mRNAs. In this regard, it should be mentioned that the two proteins bind to distinct regions with viral 5′UTR (Park and Katze 1995; Park et al. 1999).

The NS1 protein, despite its small size, has a number of cellular protein partners apart from eIF4G. These include a human homolog of the porcine 17-β-estradiol dehydrogenase (Wolff et al. 1996), the 30-kD subunit of the CPSF factor (Nemeroff et al. 1998), the poly(A)-binding protein II (Chen et al. 1999), a novel human 70-kD protein (NS1-BP) (Wolff et al. 1998), and the human homolog of Staufen protein (hStaufen) (Falcon et al. 1999). Whether these proteins play any role in selective translation of viral mRNAs remains to be determined. It is interesting to note that hStaufen is an RNA-binding protein that is associated with polysomes. Furthermore, the *Drosophilia melanogaster* Staufen protein has been shown to colocalize specifically with certain mRNAs during early development of the fly (St Johnston et al. 1991). It is speculated that the interaction of NS1 with hStaufen may localize viral mRNAs to precise sites in the cytoplasm where NS1 exerts its regulatory function, although this has not been shown experimentally.

MODULATION OF A TRANSLATIONAL INITIATION CHECKPOINT DURING INFLUENZA VIRUS INFECTION

To induce a dramatic shutoff of host protein synthesis, but yet maintain selective and efficient cap-dependent translation of its own mRNAs, influenza virus needs to ensure that the cell remains translationally competent during infection. Furthermore, like many animal viruses, influenza virus has to overcome the host innate antiviral defense mediated by the interferon (IFN) response system. Clearly, given its limited genetic coding capacity, influenza virus must capitalize on existing cellular pathways to accomplish these goals. In the following sections, we discuss how influenza virus has adeptly recruited the cellular stress response pathway to modulate a key mediator of the IFN response, namely the PKR protein kinase.

Recruitment of a Novel Molecular Co-chaperone Pathway

PKR (*p*rotein *k*inase, *R*NA-activated) mediates the cellular IFN-induced antiviral response, at least in part, through its ability to repress protein synthesis (see Chapters 5, 8, and 13). PKR is activated upon binding dsRNA or highly structured RNA molecules, a process that is associated with kinase dimerization and autophosphorylation. Viral RNAs are potent activators of PKR, as they are produced at high levels during an acute infection and frequently contain double-stranded structural elements. In addition, many viruses synthesize dsRNA as part of their replicative process. Once activated, PKR catalyzes the phosphorylation of the α subunit of eukaryotic initiation factor 2 (eIF2α), on serine 51 (S51), leading to an inhibition of translation initiation. Thus, PKR shuts down protein synthesis in a virally infected cell to prevent or limit the spread of virus infection.

The increasing number of different viruses that use distinct mechanisms to down-regulate PKR attests to the importance of the protein kinase as part of the antiviral response (see Chapters 8 and 13). Unlike the strategies used by other viruses, influenza virus utilizes a cellular protein to block PKR activity. This protein, termed $P58^{IPK}$ (*i*nhibitor of *p*rotein *k*inase), contains nine tetratricopeptide repeat (TPR) motifs, arranged in tandem, which comprise over 60% of the $P58^{IPK}$ sequence (Lee et al. 1994). In addition, the carboxyl terminus of $P58^{IPK}$ contains a region homologous to the J-domain of the DnaJ family of proteins. In uninfected cells, $P58^{IPK}$ appears to form an inactive complex with its own inhibitor(s). In response to activating stimuli, such as viral infection or other cellular stresses, $P58^{IPK}$ is released from its inhibitor. As a result, $P58^{IPK}$ is now free to repress PKR activity, apparently by blocking both the autophosphorylation of PKR and the subsequent phosphorylation of eIF2α by preexisting activated PKR molecules (Lee et al. 1990, 1992). Consistent with its requirement for inhibition of kinase activity in vivo (Tang et al. 1996), the sixth TPR motif (TPR6) of $P58^{IPK}$ acts as a docking site for PKR interaction (Gale et al. 1996; Polyak et al. 1996).

The eukaryotic DnaJ homolog, Hsp40, is a negative regulator of $P58^{IPK}$ (Melville et al. 1997). Influenza virus probably activates $P58^{IPK}$ by promoting the dissociation of Hsp40 from $P58^{IPK}$ during infection (Melville et al. 1999). $P58^{IPK}$ and Hsp40 interact with each other in a direct manner, as determined with the use of purified proteins. $P58^{IPK}$ also forms a complex with the molecular chaperone Hsp70, requiring both Hsp40 and ATP. The dependence on Hsp40 was confirmed in vivo by means of a yeast two-hybrid approach that measured $P58^{IPK}$ and Hsp70 interaction in the presence or absence of Hsp40. Furthermore, the binding is localized to the ATPase domain of Hsp70. It is not clear, however, whether Hsp40 promotes complex formation through its interaction with $P58^{IPK}$ (via protein conformational change) or acts as a molecular bridge between $P58^{IPK}$ and Hsp70. Another inhibitor of $P58^{IPK}$, the cellular protein $P52^{rIPK}$ (Gale et al. 1998), has recently been identified. Although we do not know which region of $P58^{IPK}$ mediates Hsp40 binding, the seventh TPR motif (TPR7) is required for the interaction with $P52^{rIPK}$ (Gale et al. 1998). In support of the role of $P52^{rIPK}$, functional analysis in yeast showed that a TPR7 deletion mutant ($P58^{IPK}\Delta$TPR7) is a more effective inhibitor of PKR than wild-type $P58^{IPK}$. However, the stimulus that triggers the release of $P52^{rIPK}$ from $P58^{IPK}$ has yet to be identified.

How Does $P58^{IPK}$ Regulate the PKR Pathway?

Initially, it was thought that $P58^{IPK}$ might function as a pseudosubstrate inhibitor of PKR because the central region of $P58^{IPK}$ has limited homology with eIF2α, including the conserved S51 that is phosphorylated by PKR (Tang et al. 1996). However, mutation of this conserved serine residue in $P58^{IPK}$ does not abrogate its ability to inhibit PKR. In fact, $P58^{IPK}$ likely operates as a cochaperone capable of stimulating Hsp70 via its J-domain to modulate the protein conformation of PKR (Fig. 1). As discussed in previous sections, $P58^{IPK}$ is likely held in an inactive complex with Hsp40, possibly along with $P52^{rIPK}$, Hsp70, and PKR, as well as other unidentified proteins. In addition to stabilizing $P58^{IPK}$ interaction with Hsp70, Hsp40 may also function as an inhibitor by blocking or competing with $P58^{IPK}$ for stimulating Hsp70. In response to a stress event, such as heat shock or influenza virus infection, Hsp40 dissociates from $P58^{IPK}$, although the mechanism of the release is unclear. It is possible that Hsp40 is recruited to protein-folding pathways during heat shock (e.g., for refolding denatured proteins) or virus infection (e.g., for viral replication or packaging). Alternatively, Hsp40 may be subjected to some posttranslational modification during a stress event such that it can no longer bind $P58^{IPK}$. At any rate, Hsp40-free $P58^{IPK}$ is presumably now "activated" and thus is able to stimulate Hsp70 to inactivate PKR, possibly by inducing a protein conformational change. Perhaps the liberation of Hsp40 alleviates competition between the Hsp40 and $P58^{IPK}$ J-domains for Hsp70 binding, allowing the J-domain of $P58^{IPK}$ to stimulate Hsp70. Indeed, $P58^{IPK}$ is capable of stimulating the ATPase activity of Hsp70, which promotes the refolding activity of Hsp70 (Melville et al. 1999).

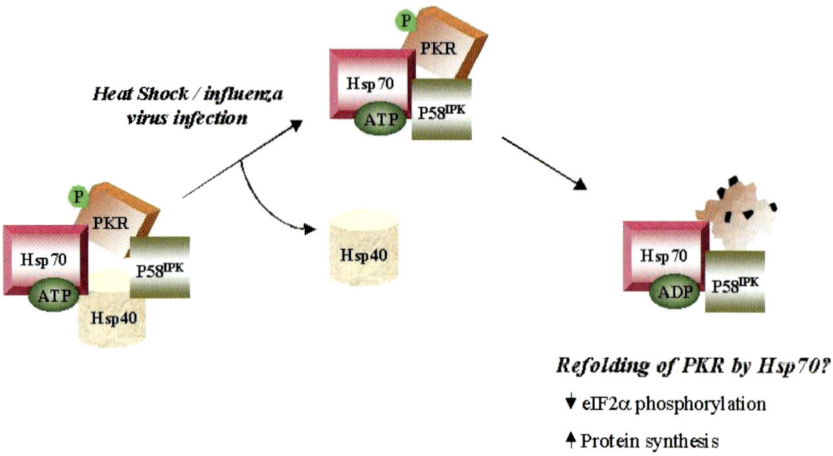

Figure 1 Model for P58^{IPK} regulation of PKR. P58^{IPK} functions as a cochaperone that stimulates Hsp70 via its J-domain to modulate the protein conformation of PKR. In the absence of stress signals, P58^{IPK} is held in an inactive complex with Hsp40, Hsp70, and PKR. In response to a stress event, such as heat shock or influenza virus infection, Hsp40 dissociates from P58^{IPK}. Hsp40-free P58^{IPK} stimulates the ATPase activity of Hsp70, which promotes the refolding activity of Hsp70. PKR may be refolded and therefore inactivated by Hsp70. See text for further details.

Does the NS1 Protein Also Function as an Inhibitor of PKR?

Given their capability of binding to dsRNA, it was hypothesized that NS1 and PKR could compete in influenza virus-infected cells for dsRNA binding. NS1 can indeed inhibit the activation of PKR and, as a result, the phosphorylation of eIF2α in vitro (Lu et al. 1995). Furthermore, the NS1 protein also blocks the inhibition of translation caused by dsRNA-mediated activation of PKR in reticulocyte lysate extracts. An inactive mutant of NS1, which lacks a functional RNA-binding domain, is unable to inhibit PKR. Although NS1 and PKR can form a specific complex in vitro, presumably through an RNA-mediated bridging mechanism (Tan and Katze 1998; Falcon et al. 1999), evidence for an in vivo association has not been demonstrated. Nevertheless, evidence supporting the role of NS1 in modulating PKR function in vivo is provided by studies of mutant influenza viruses with a defective NS1 protein (Hatada et al. 1999). These variant viruses do not block the activation of PKR in infected cells, leading to enhanced phosphorylation of eIF2α and suppression of translation. Furthermore, the level of phosphorylation of PKR and eIF2α correlates with the defect in virus protein synthesis. Thus, NS1 may confer translational competence to influenza virus by removing the translational blockade imposed by PKR.

However, a recent study of a recombinant influenza virus that lacks the *NS1* gene, termed delNS1, suggests that NS1 may not play a direct role in translational control and viral replication: rather, it may function to block the antiviral effects of the host IFN system (Garcia-Sastre et al. 1998). Although delNS1 replicates to high titer within cells deficient in IFN-signaling pathways, its replication is severely limited in cells in which IFN-signaling remains intact. Furthermore, delNS1 replicates to lethal titers in mice with a target deletion in the *STAT1* gene, which renders cells unable to respond to IFN, whereas IFN-competent control mice effectively suppress delNS1 replication.

PKR AS A LINK BETWEEN TRANSLATION CONTROL AND APOPTOSIS: HOW DOES INFLUENZA VIRUS CAPITALIZE?

Recently, it has become increasingly clear that PKR is capable of mediating apoptosis (for review, see Tan and Katze 1999). PKR-dependent apoptosis might be mediated, at least in part, through phosphorylation of eIF2α, suggesting a cross-talk between translation initiation and apoptosis. Influenza virus infection is known to induce cellular apoptosis (Takizawa et al. 1993; Fesq et al. 1994). Furthermore, influenza-induced apoptosis can be overcome by inhibition of endogenous PKR function (Takizawa et al. 1996), thus suggesting a relationship between PKR inhibition and apoptosis during influenza virus infection. As discussed earlier, influenza virus infection activates a novel host molecular chaperone protein, $P58^{IPK}$, to inhibit PKR. In addition, the virus may also utilize NS1 to sequester dsRNA activators from activating PKR. NIH-3T3 cell lines overexpressing $P58^{IPK}$ are resistant to dsRNA- and TNFα-induced apoptosis, highlighting $P58^{IPK}$ as a possible novel apoptotic antagonist; namely, one that acts through the PKR pathway (Tang et al. 1999). However, the role of NS1 as a PKR inhibitor is less straightforward, because expression of NS1 was found to be necessary and sufficient to induce apoptosis in Madin-Darby canine kidney cells (Schultz-Cherry et al. 1998). Furthermore, both the apoptotis-inducing and PKR inhibitory effects of NS1 require the RNA-binding domain of NS1 to be intact (Schultz-Cherry et al. 1998; Tan and Katze 1998). Because previous studies of the PKR inhibitory effect of NS1 were carried out largely by in vitro assays, it is possible that the NS1 protein may mediate apoptosis by activating PKR under physiological conditions. In this regard, it should be mentioned that a newly discovered cellular dsRNA-binding protein, PACT, whose interaction with PKR is mediated by the dsRNA-binding motifs of both proteins, is capable of activating kinase activity and promoting apoptosis (Patel and Sen 1998; G. Sen, pers. comm.).

CONCLUDING REMARKS AND FUTURE PERSPECTIVES

As discussed above, a number of mechanisms have been proposed for host shutoff and selective viral mRNA translation by influenza, but several important questions remain unanswered. For example, we do not yet know which viral protein(s) is required for mediating host shutoff, a question that can be addressed by analysis of mammalian cell lines stably expressing individual viral or cellular genes (Fodor et al. 1999), or by infection with gene knock-out recombinant influenza viruses (Neumann et al. 1999). The role of NS1 in both host shutoff and selective viral mRNA translation and the significance of its interaction with the various cellular proteins will need to be further explored in the context of virus infection. Since the *NS1* gene of influenza A virus is now amenable to genetic manipulation, it should be possible to test the various roles assigned to NS1 by mutating the binding sites for individual NS1-binding partners within the NS1 protein and examining the consequences for their proposed functions. Finally, the mechanism by which influenza virus induces eIF4E dephosphorylation during shutoff or inhibits cellular mRNA translation at the elongation step should be investigated further.

How does GRSF-1 mediate selective translation of influenza virus mRNAs? GRSF-1 may chaperone viral mRNAs to a translationally competent compartment, such as the cytoskeleton (Katze et al. 1989), where components of the translation apparatus are abundant or remain intact during infection. Alternatively, GRSF-1 may function by recruiting translation initiation factors that are involved in ribosome loading. The identification of such interacting factors could help decipher the mechanism of GRSF-1-mediated selective translation. A potential candidate for a GRSF-1-interacting protein is the influenza virus NS1 protein. NS1 is known to stimulate the translation of certain viral mRNAs and has the ability to bind to both viral 5′UTRs and eIF4G. Thus, it will be interesting to test whether GRSF-1 interacts with NS1 and/or eIF4G, and whether this interaction synergistically enhances translation of viral mRNAs.

Influenza virus appear to use two strategies, a cellular factor ($P58^{IPK}$) and a viral protein (NS1), to evade the PKR-mediated translational block. However, several questions remain to be addressed in future studies: (1) How does influenza virus activate the $P58^{IPK}$/PKR regulatory pathway? (2) Would cells devoid of $P58^{IPK}$ be more susceptible to influenza virus infection and replication? (3) Do NS1 and $P58^{IPK}$ work in a synergistic manner to inhibit PKR function and enhance translation of viral mRNAs and/or inhibit apoptosis? Although recombinant influenza viruses lacking the *NS1* gene (Egorov et al. 1998; Garcia-Sastre et al. 1998) or containing a defective *NS1* gene (Hatada et al. 1999) are now available, definitive proof of the physiological role of $P58^{IPK}$ will have to be obtained from the generation and analysis of mice or cell lines devoid of the $P58^{IPK}$ gene.

In addition, the mechanism of the release of $P58^{IPK}$ by Hsp40 in influenza virus-infected cells needs to be addressed. It may be that Hsp40 is recruited to another site in the cell or is modified in some way. Curiously, this dissociation is also observed in cells recovering from heat shock (Melville et al. 1999). It is tempting to speculate that $P58^{IPK}$ activation may be part of a typical stress response, and influenza virus infection may simply trigger this response. Influenza virus infection of lung tissue mimics the cellular stress induced by oxidative stress (Choi et al. 1996). Similarly, the conditions under which $P52^{rIPK}$ and $P58^{IPK}$ interact within the cell, as well as the consequences of this interaction in vivo, will need to be identified. Also, how does Hsp70 recognize PKR as a substrate in the context of bound $P58^{IPK}$? The answer may be provided by the recent observation that $P58^{IPK}$ blocks dimerization of PKR (Tan et al. 1998). Monomerization of PKR by $P58^{IPK}$ may expose hydrophobic domains on PKR to Hsp70, which binds and subsequently refolds the kinase. In this regard, monomerization of the *E. coli* P1 replication initiator RepA requires a functional DnaK–DnaJ system (Wickner et al. 1991, 1992), thus implicating involvement of a J-domain member in dissociation of a protein complex. Refolding by Hsp70 may induce a defective conformation of PKR for binding activator dsRNA, kinase activity, substrate recognition, or dimerization.

Finally, it is expected that a few remaining cellular mRNAs, whose protein products play an important role in the replication of the virus in the host, would be continuously translated during compromised cellular protein synthesis. It will be important to identify such mRNAs and determine how they evade influenza virus-induced host shutoff, which may provide further insights into cellular mRNA translation. DNA microarray technology has expanded our views on global gene expression at the transcriptional level, comparing the profile between diseased (including viral infection) versus non-diseased states (Lockhart et al. 1996; Schena et al. 1998; Zhu et al. 1998; Brown and Botstein 1999; Geiss et al. 2000). Coupled with polysome distribution analysis, DNA microarrays have now also been applied to the identification of translationally regulated mRNAs in virus-infected cells (Johannes et al. 1999; Zong et al. 1999; see Chapter 19). Furthermore, with the rapid development of sophisticated protein complex analytical techniques, such as tandem mass spectrometry (Blyn et al. 1996; Gygi et al. 1999a,b; Link et al. 1999), we should be able to identify mRNAs and their encoded gene products that are preferentially translated during influenza virus infection. These emerging technologies could add a new dimension to our exploration of the complexity in translational regulation in virus-infected cells.

Note Added in Proof

The involvement of NS1 in PKR inhibition (Bergmann et al. 2000) and translational stimulation of the M1 protein (Enami and Enami 2000) is further supported by recent studies using NS1-defective influenza virus mutants.

ACKNOWLEDGMENTS

We thank members of the Katze laboratory, past and present, for their contributions to our work. We are also grateful to those individuals who have collaborated with us over the years. We thank G. Sen, J. Ortin, and A. Nieto for sharing unpublished results, and M. Korth and S. Tareen for editorial assistance. Work in the Katze laboratory is supported by National Institutes of Health grants AI-22646, RR-00166, and AI-41629; Ribogene, Inc.; and a grant from Gustavus & Louise Pfeiffer Research Foundation.

REFERENCES

Alonso-Caplen F.V., Katze M.G., and Krug R.M. 1988. Efficient transcription, not translation, is dependent on adenovirus tripartite leader sequences at late times of infection. *J. Virol.* **62:** 1606–1616.

Aragon T., de la Luna S., Novoa I., Carracso L., Ortin J., and Nieto A. 2000. Translation factor eIF4GI is a cellular target for NS1 protein, a translational activator of influenza virus. *Mol. Cell. Biol.* (in press).

Barber G.N., Thompson S., Lee T.G., Strom T., Jagus R., Darveau A., and Katze M.G. 1994. The 58-kilodalton inhibitor of the interferon-induced double-stranded RNA-activated protein kinase is a tetratricopeptide repeat protein with oncogenic properties. *Proc. Natl. Acad. Sci.* **91:** 4278–4282.

Beloso A., Martinez C., Valcarcel J., Santaren J.F., and Ortin J. 1992. Degradation of cellular mRNA during influenza virus infection: Its possible role in protein synthesis shutoff. *J. Gen. Virol.* **73:** 575–581.

Bergmann M., Garcia-Sastre A., Carnero E., Pehamberger H., Wolff K., Palese P., and Muster T. 2000. Influenza virus NSI protein counteracts PKR-mediated inhibition of replication. *J. Virol.* **74:** 6203–6206.

Blyn L.B., Swiderek K.M., Richards O., Stahl D.C., Semler B.L., and Ehrenfeld E. 1996. Poly(rC) binding protein 2 binds to stem-loop IV of the poliovirus RNA 5′ noncoding region: Identification by automated liquid chromatography-tandem mass spectrometry. *Proc. Natl. Acad. Sci.* **93:** 11115–11120.

Brown P.O. and Botstein D. 1999. Exploring the new world of the genome with DNA microarrays. *Nat. Genet.* **21:** 33–37.

Chen Z., Li Y., and Krug R.M. 1999. Influenza A virus NS1 protein targets poly(A)-binding protein II of the cellular 3′-end processing machinery. *EMBO J.* **18:** 2273–2283.

Choi A.M., Knobil K., Otterbein S.L., Eastman D.A., and Jacoby D.B. 1996. Oxidant stress responses in influenza virus pneumonia: Gene expression and transcription factor activation. *Am. J. Physiol.* **271:** 383–391.

de la Luna S., Fortes P., Beloso A., and Ortin J. 1995. Influenza virus NS1 protein enhances the rate of translation initiation of viral mRNAs. *J. Virol.* **68:** 2427–2433.

Egorov A., Brandt S., Sereinig S., Romanova J., Ferko B., Katinger D., Grassauer A., Alexandrova G., Katinger H., and Muster T. 1998. Transfectant influenza A viruses with long deletions in the NS1 protein grow efficiently in Vero cells. *J. Virol.* **72:** 6437–6441.

Enami M. and Enami K. 2000. Characterization of influenza virus NS1 protein by using a novel helper-virus-free reverse genetic system. *J. Virol.* **74:** 5556–5561.

Enami K., Sato T.A., Nakada S., and Enami M. 1994. Influenza virus NS1 protein stimulates translation of the M1 protein. *J. Virol.* **68:** 1432–1437.

Falcon A.M, P. Fortes P., R.M. Marion R.M., Beloso A., and Ortin J. 1999. Interaction of influenza virus NS1 protein and the human homologue of Staufen in vivo and in vitro. *Nucleic Acids Res.* **27:** 2241–2247.

Feigenblum D. and Schneider R.J. 1993. Modification of eukaryotic initiation factor 4F during infection by influenza virus. *J. Virol.* **67:** 3027–3035.

Fesq H., Bacher M., Nain M., and Gemsa D. 1994. Programmed cell death (apoptosis) in human monocytes infected by influenza virus. *Immunobiology* **190:** 175–182.

Fodor E., Devenish L., Engelhardt O.G., Palese P., Brownlee G.G., and Garcia-Sastre A. 1999. Rescue of influenza A virus from recombinant DNA. *J. Virol.* **73:** 9679–9682.

Fortes P., Beloso A., and Ortin J. 1994. Influenza virus NS1 protein inhibits pre-mRNA splicing and blocks mRNA nucleocytoplasmic transport. *EMBO J.* **13:** 704–712.

Gale M., Jr., Tan S.-L., Wambach M., and Katze M.G. 1996. Interaction of the interferon-induced PKR protein kinase with inhibitory proteins $P58^{IPK}$ and vaccinia virus K3L is mediated by unique domains: Implications for kinase regulation. *Mol. Cell. Biol.* **16:** 4172–4181.

Gale M., Jr., Blakely C.M., Hopkins D.A., Melville M.W., Wambach M., Romano P.R., and Katze M.G. 1998. Regulation of interferon-induced protein kinase PKR: Modulation of $P58^{IPK}$ inhibitory function by a novel protein, $P52^{rIPK}$. *Mol. Cell. Biol.* **18:** 859–871.

Garcia-Sastre A., Egorov A., Matassov D., Brandt S., Levy D.E., Durbin J.E., Palese P., and Muster T. 1998. Influenza A virus lacking the NS1 gene replicates in interferon-deficient systems. *Virology* **252:** 324–330.

Garfinkel M.S. and Katze M.G. 1992. Translational control by influenza virus: Selective and cap-dependent translation of viral mRNAs in infected cells. *J. Biol. Chem.* **267:** 9383–9390.

———. 1993. Translational control by influenza virus: Selective translation is mediated by sequences within the viral mRNA 5′-untranslated region. *J. Biol. Chem.* **268:** 22223–22226.

Geiss G.K., Bumgarner R.E., An M., Agy M.B., van′t Wout A., Hammersmark E., Carter V., Upchurch D., Mullins J.I., and Katze M.G. 2000. Large-scale monitoring of host cell gene expression during HIV-1 infection using cDNA microarrays. *Virology* **266:** 8–16.

Gygi S.P., Rochon Y., Franza B.R., and Aebersold R. 1999a. Correlation between protein and mRNA abundance in yeast. *Mol. Cell. Biol.* **19:** 1720–1730.

Gygi S.P., Han D.K.M., Gingras A.-C., Sonenberg N., and Aebersold R. 1999b. Protein analysis by mass spectrometry and sequence database searching: Tools for cancer research in the post-genomic era. *Electrophoresis* **20:** 310–319.

Hardy W.R. and Sandri-Goldin R.M. 1994. Herpes simplex virus inhibits host cell splicing, and regulatory protein ICP27 is required for this effect. *J. Virol.* **68:** 7790–7799.

Hatada E. and Fukuda R. 1992. Binding of influenza A virus NS1 protein to dsRNA in vitro. *J. Gen. Virol.* **73:** 3325–3329.

Hatada E., Saito S., and Fukuda R. 1999. Mutant influenza viruses with a defective NS1 protein cannot block the activation of PKR in infected cells. *J. Virol.* **73:** 2425–2433.

Inglis S.C. 1982. Inhibition of host protein synthesis and degradation of cellular mRNAs during infection by influenza and herpes simplex virus. *Mol. Cell. Biol.* **2:** 1644–1648.

Johannes G., Carter M.S., Eisen M.B., Brown P.O., and Sarnow P. 1999. Identification of eukaryotic mRNAs that are translated at reduced cap binding complex eIF4F concentrations using a cDNA microarray. *Proc. Natl. Acad. Sci.* **96:** 13118–13123.

Katze M.G. and Krug R.M. 1984. Metabolism and expression of RNA polymerase II transcripts in influenza virus-infected cells. *Mol. Cell. Biol.* **4:** 2198–2206.

———. 1990. Translational control in influenza virus-infected cells. *Enzyme* **44:** 265–277.

Katze M.G., Chen Y.T., and Krug R.M. 1984. Nuclear-cytoplasmic transport and VAI RNA-independent translation of influenza viral messenger RNAs in late adenovirus-infected cells. *Cell* **37:** 483–490.

Katze M.G., DeCorato D., and Krug R.M. 1986a. Cellular mRNA translation is blocked at both initiation and elongation after infection by influenza virus or adenovirus. *J. Virol.* **60:** 1027–1039.

Katze M.G., Lara J., and Wambach M. 1989. Nontranslated cellular mRNAs are associated with cytoskeletal framework in influenza virus or adenovirus infected cells. *Virology* **169:** 312–322.

Katze M.G., Detjen B.M., Safer B., and Krug R.M. 1986b. Translational control by influenza virus: Suppression of the kinase that phosphorylates the alpha subunit of initiation factor eIF-2 and selective translation of influenza viral mRNAs. *Mol. Cell. Biol.* **6:** 1741–1750.

Katze M.G., Tomita J., Black T., Krug R.M., Safer B., and Hovanessian A. 1988. Influenza virus regulates protein synthesis during infection by repressing autophosphorylation and activity of the cellular 68,000-M_r protein kinase. *J. Virol.* **62:** 3710–3717.

Krug R.M., Alonso-Caplen F., Julkunen I., and Katze M.G. 1989. Expression and replication of the influenza virus genome. In Influenza viruses (ed. R.M. Krug), pp. 89–152. Press, New York.

Lazarowitz S.G., Compans R.W., and Choppin P.W. 1971. Influenza virus structural and nonstructural proteins in infected cells and their plasma membrane. *Virology* **46:** 830–843.

Lee T.G., Tomita J., Hovanessian A.G., and Katze M.G. 1990. Purification and partial characterization of a cellular inhibitor of the interferon-induced protein kinase of M_r 68,000 from influenza virus-infected cells. *Proc. Natl. Acad. Sci.* **87:** 6208–6212.

———. 1992. Characterization and regulation of the 58,000-dalton cellular inhibitor of the interferon-induced, dsRNA-activated protein kinase. *J. Biol. Chem.* **267:** 14238–14243.

Lee T.G., Tang N., Thompson S., Miller J., and Katze M.G. 1994. The 58,000-dalton cellular inhibitor of the interferon-induced double-stranded RNA-activated protein kinase (PKR) is a member of the tetratricopeptide repeat family of proteins. *Mol. Cell. Biol.* **14:** 2331–2342.

Link A.J., Eng J., Schieltz D.M., Carmack E., Mize G.J., Morris D.R., Garvik B.M., and Yates J.R. 1999. Direct analysis of protein complexes using mass spectrometry. *Nat. Biotechnol.* **17:** 676–682.

Lockhart D.J., Dong H., Byrne M.C., Follettie M.T., Gallo M.V., Chee M.S., Mittmann M.,

Wang C., Kobayashi M., Horton H., and Brown E.L. 1996. Expression monitoring by hybridization to high-density oligonucleotide arrays. *Nat. Biotechnol.* **14:** 1675–1680.

Lu Y., Qian X.-Y., and Krug R.M. 1994. The influenza virus NS1 protein: A novel inhibitor of pre-mRNA splicing. *Genes Dev.* **8:** 1817–1828.

Lu Y., Wambach M., Katze M.G., and Krug R.M. 1995. Binding of the influenza virus NS1 protein to double-stranded RNA inhibits the activation of the protein kinase that phosphorylates the eIF-2 translation initiation factor. *Virology* **214:** 222–228.

Melville M.W., Hansen W.J., Freeman B.C., Welch W.J., and Katze M.G. 1997. The molecular chaperone hsp40 regulates the activity of $P58^{IPK}$, the cellular inhibitor of PKR. *Proc. Natl. Acad. Sci.* **94:** 97–102.

Melville M.W., Tan S.-L., Wambach M., Song J., Morimoto R.I., and Katze M.G. 1999. The cellular inhibitor of the PKR protein kinase, $P58^{IPK}$, is an influenza virus-activated co-chaperone that modulates heat shock protein 70 activity. *J. Biol. Chem.* **274:** 3797–3803.

Nemeroff M.E., Barabino S.M., Li Y., Keller W., and Krug R.M. 1998. Influenza virus NS1 protein interacts with the cellular 30-kDa subunit of CPSF and inhibits 3′ end formation of cellular pre-mRNAs. *Mol. Cell* **1:** 991–1000.

Neumann G., Watanabe T., Ito H., Watanabe S., Goto H., Gao P., Hughes M., Perez D.R., Donis R., Hoffmann E., Hobom G., and Kawaoka Y. 1999. Generation of influenza A viruses entirely from cloned cDNAs. *Proc. Natl. Acad. Sci.* **96:** 9345–9350.

Park Y.-W. and Katze M.G. 1995. Translational control by influenza virus: Identification of *cis*-acting sequences and *trans*-acting factors which may regulate selective viral mRNA translation. *J. Biol. Chem.* **270:** 28433–28439.

Park Y.-W., Wilusz J., and Katze M.G. 1999. Regulation of eukaryotic protein synthesis: Selective influenza viral mRNA translation is mediated by the cellular RNA-binding protein GRSF-1. *Proc. Natl. Acad. Sci.* **96:** 6694–6699.

Patel R.K. and Sen G.C. 1998. PACT, a protein activator of the interferon-induced protein kinase, PKR. *EMBO J.* **17:** 4379–4390.

Polyak S.J., Tang N.M., Wambach M., Barber G.N., and Katze M.G. 1996. The p58 cellular inhibitor complexes with the interferon-induced, double-stranded RNA-dependent protein kinase, PKR, to regulate its autophosphorylation and activity. *J. Biol. Chem.* **271:** 1702–1707.

Qian Z. and Wilusz J. 1994. GRSF-1: A $poly(A)^+$ mRNA binding protein which interacts with a conserved G-rich element. *Nucleic Acids Res.* **22:** 2334–2343.

Qiu Y. and Krug R.M. 1994. The influenza virus NS1 protein is a poly(A)-binding protein that inhibits nuclear export of mRNAs containing poly(A). *J. Virol.* **68:** 2425–2432.

Sandri-Goldin R.M., Hibbard M.K., and Hardwicke M.A. 1995. The C-terminal repressor region of herpes simplex virus type 1 ICP27 is required for the redistribution of small nuclear ribonucleoprotein particles and splicing factor SC35; however, these alterations are not sufficient to inhibit host cell splicing. *J. Virol.* **69:** 6063–6076.

Schena M., Heller R.A., Theriault T.P., Konrad K., Lachenmeier E., and Davis R.W. 1998. Microarrays: Biotechnology's discovery platform for functional genomics. *Trends Biotechnol.* **16:** 301–306.

Schultz-Cherry S., Krug R.M., and Hinshaw V.S. 1998. Induction of apoptosis by influenza virus. *Semin. Virol.* **8:** 491-495.

Skehel J.J. 1972. Polypeptide synthesis in influenza virus-infected cells. *Virology* **49:** 23–36.

St Johnston D., Beuchle D., and Nüsslein-Volhard C. 1991. Staufen, a gene required to localize maternal RNAs in the Drosophila egg. *Cell* **66:** 51–63.

Takizawa T., Ohashi K., and Nakanishi Y. 1996. Possible involvement of double-stranded

RNA-activated protein kinase in cell death by influenza virus infection. *J. Virol.* **70:** 8128–8132.

Takizawa T., Matsukawa S., Higuchi Y., Nakamura S., Nakanishi Y., and Fukuda R. 1993. Induction of programmed cell death (apoptosis) by influenza virus infection in tissue culture cells. *J. Gen. Virol.* **74:** 2347–2355.

Tan S.-L. and Katze M.G. 1998. Biochemical and genetic evidence for complex formation between the influenza A virus NS1 protein and the interferon-induced PKR protein kinase. *J. Interferon Cytokine Res.* **18:** 757–766.

———. 1999. The emerging role of the interferon-induced PKR protein kinase as an apoptotic effector: A new face of death? *J. Interferon Cytokine Res.* **19:** 543–554.

Tan S.-L., Gale M.J., Jr., and Katze M.G. 1998. Double-stranded RNA-independent dimerization of interferon-induced protein kinase PKR and inhibition of dimerization by the cellular P58^{IPK} inhibitor. *Mol. Cell. Biol.* **18:** 2431–2443.

Tang N.M., Ho C.Y., and Katze M.G. 1996. The 58-kDa cellular inhibitor of the double stranded RNA-dependent protein kinase requires the tetratricopeptide repeat 6 and DnaJ motifs to stimulate protein synthesis in vivo. *J. Biol. Chem.* **271:** 28660–28866.

Tang N.M., Korth M.J., Gale M., Jr., Wambach M., Der S.D., Bandyopadhyay S.K., Williams B.R., and Katze M.G. 1999. Inhibition of double-stranded RNA- and tumor necrosis factor alpha-mediated apoptosis by tetratricopeptide repeat protein and cochaperone P58^{IPK}. *Mol. Cell. Biol.* **19:** 4757–4765.

Wickner S., Hoskins J., and McKenney K. 1991. Monomerization of RepA dimers by heat shock proteins activates binding to DNA replication origin. *Proc. Natl. Acad. Sci.* **88:** 7903–7907.

Wickner S., Skowyra D., Hoskins J., and McKenney K. 1992. DnaJ, DnaK, and GrpE heat shock proteins are required in oriP1 DNA replication solely at the RepA monomerization step. *Proc. Natl. Acad. Sci.* **89:** 10345–10249.

Wolff T., O'Neill R.E., and Palese P. 1996. Interaction cloning of NS1-I, a human protein that binds to the nonstructural NS1 proteins of influenza A and B viruses. *J. Virol.* **70:** 5363–5372.

———. 1998. NS1-binding protein (NS1-BP): A novel human protein that interacts with influenza A virus nonstructural NS1 protein is relocalized in the nuclei of infected cells. *J. Virol.* **72:** 7170–7180.

Yamanaka K., Nagata K., and Ishihama A. 1988. Translational regulation of influenza virus mRNAs. *Virus Genes* **2:** 19–30.

———. 1991. Temporal control for translation of influenza virus mRNAs. *Arch. Virol.* **120:** 33–42.

Zhu H., Cong J.-P., Mamtora G., Gingeras T., and Shenk T. 1998. Cellular gene expression altered by human cytomegalovirus: Global monitoring with oligonucleotide arrays. *Proc. Natl. Acad. Sci.* **95:** 14470–14475.

Zong Q., Schummer M., Hood L., and Morris D.R. 1999. Messenger RNA translation state: The second dimension of high-throughput expression screening. *Proc. Natl. Acad. Sci.* **96:** 10632–10636.

35

Translational Control in Poxvirus-infected Cells

Bertram L. Jacobs
Department of Microbiology
Program in Molecular and Cellular Biology
Arizona State University
Tempe, Arizona 85287-2701

As might be expected, regulation of gene expression occurs at many levels in poxvirus-infected cells. Regulation of the translational machinery primarily involves control of the double-stranded (ds) RNA-activated regulators of translation, PKR and the 2′,5′ oligoadenylate system; the selective translation of viral mRNA in some cells; and the selective inhibition of host protein synthesis. These aspects of translational control of gene expression in poxvirus-infected cells are covered in this chapter.

THE POXVIRIDAE: A BRIEF OVERVIEW

Classification and replication of poxviruses has been reviewed recently (Fields et al. 1996). The poxviridae family is made up of large viruses with a (ds)DNA genome 130–300 kbp in length. Poxviral genomes code for more than 100 polypeptides, with the central portion coding for proteins required for virus replication, and the termini coding for proteins involved in virus/host interactions. The family is subdivided into the chordopoxviridae and entomopoxviridae, which infect chordates and arthropods, respectively. The chordopoxviridae are separated into eight recognized genera, based on morphology, antigenic cross-reactivity, and host range. The largest genus is the orthopoxviruses, which contains closely related viruses infecting buffalo, camel, monkey, mouse (ectromelia), rabbit, raccoon, vole, cattle, and humans (smallpox). Vaccinia virus, the type virus of the orthopoxviruses, is the virus that was used to immunize against smallpox in the 20th century. Whereas the original virus used for smallpox "vaccination" by Jenner was derived from a poxvirus that infect-

Translational Control of Gene Expression
© 2000 Cold Spring Harbor Laboratory Press 0-87969-568-4/00 $5 + .00

ed cattle (the terms vaccinia and vaccination come from the Latin, *vacca*, for cow), the virus we now call vaccinia virus is clearly distinct from either cowpox or smallpox viruses, and the derivation of vaccinia virus is at present unclear. Vaccinia virus has no known natural host. More distantly related are the parapoxviruses (e.g., ORF virus), which infect sheep and occasionally humans; the avipoxviruses, which are arthropod-borne viruses of birds; the capripoxviruses, which are arthropod-borne viruses of ungulates; the suipoxviruses, the leporipoxviruses (myxoma and rabbit fibroma viruses), which cause immunosuppressive diseases in rabbits; and the mollusci- and yatapoxviruses, which cause benign tumors in humans and primates, respectively.

Poxviruses replicate entirely in the cytoplasm of infected cells, and thus code for much of the machinery required for viral transcription and DNA replication. For vaccinia virus, early RNAs are synthesized by a virion DNA-dependent RNA polymerase. Early mRNAs are capped and 3′-polyadenylated by viral enzymes and have precise 5′ and 3′ ends, due to precise initiation and termination of transcription. There is a profound inhibition of host protein synthesis during the early phase of virus replication. Early gene expression leads to the transcription of intermediate genes, whose products, along with DNA replication, are required for late gene expression. Late transcripts are also capped and 3′-polyadenylated, but contain 5′ poly(A) leaders from 5–30 nucleotides long just upstream from the initiator AUG codon, as well. Late mRNAs are heterogeneous at their 3′ ends due to imprecise termination of transcription. These long late run-on transcripts can be complementary to transcripts read from the opposite strand of the viral genome and are thought to be the source of dsRNA in infected cells.

Synthesis of dsRNA in vaccinia-virus-infected cells appears to be under genetic control. Conditional lethal mutations in the A18R gene lead to synthesis of increased levels of dsRNA during infection under restrictive conditions (Bayliss and Condit 1993). Under these conditions the viral DNA-dependent RNA polymerase has increased processivity compared to wild-type polymerase (Xiang et al. 1998). Mutations in A18R are suppressed by mutations in the G2R gene (Condit et al. 1996) that decrease accumulation of dsRNA (Black and Condit 1996) by decreasing processivity of the polymerase complex (Black and Condit 1996). Mutations in G2R are themselves suppressed by the anti-poxvirus drug, isatin-beta-thiosemicarbazone (IBT), which increases processivity of the viral polymerase (Hassett and Condit 1994). Thus, the vaccinia virus system is unique in that accumulation of viral dsRNA in an infected cell can be modified depending on the genetic background of the virus used.

POXVIRUSES AND THE INTERFERON SYSTEM: SKIF AND E3L

Vaccinia virus was one of the earliest viruses to be characterized as resistant, in at least some cell lines in culture, to the antiviral effects of interferon (IFN) (Youngner et al. 1972). In addition, coinfection of several interferon-sensitive viruses, such as vesicular stomatitis virus (VSV, a rhabdovirus) and encephalomyocarditis virus (EMCV, a picornavirus), with vaccinia virus rescues the interferon-sensitive viruses from the antiviral effects of IFN (Thacore and Younger 1973; Whitaker-Dowling and Younger 1986). For VSV, rescue occurs with a restoration of the translation of viral mRNA into protein (Whitaker-Dowling and Younger 1983). These results suggest that vaccinia virus makes a *trans*-acting factor or factors that block the antiviral effects of IFN.

Vaccinia virus was also one of the earliest viruses to be shown to induce the synthesis of an inhibitor of the IFN-inducible eIF2α kinase, PKR (Paez and Esteban 1984; Rice and Kerr 1984; Whitaker-Dowling and Younger 1984). This inhibitory activity was originally called SKIF, for *s*pecific *k*inase *i*nhibitory *f*actor. Addition of an optimal concentration of dsRNA fails to activate PKR in extracts of vaccinia-virus-infected cells. This inhibitory activity is overcome by adding 10- to 100-fold excess dsRNA, suggesting that the inhibitor interacts with the dsRNA activator. The interaction appears to be noncatalytic, in that it does not require preincubation of extracts from infected cells with dsRNA or PKR in order to inhibit activation (Whitaker-Dowling and Younger 1984). These results suggest that SKIF may be a dsRNA-binding protein that sequesters the activator dsRNA away from PKR. Such a vaccinia-virus-specific dsRNA-binding protein was indeed found, and shown to copurify with SKIF activity (Watson et al. 1991). Partial amino acid sequence analysis of this dsRNA-binding protein demonstrates that it is encoded by the E3L gene. Recombinant E3L protein has SKIF activity, confirming that E3L encodes a potent inhibitor of PKR (Chang et al. 1992).

To determine the function of E3L in the context of a virus infection, E3L was deleted from vaccinia virus (Denzler and Jacobs 1994). Replication of VVΔE3L is sensitive to pretreatment of rabbit kidney RK-13 cells with IFNβ. VVΔE3L also has a severe host-range phenotype in that it does not replicate in human HeLa, and monkey kidney COS, CV-1, or BSC-40 cells, even in the absence of IFN treatment (Beattie et al. 1996). VVΔE3L-infection also induces apoptosis in HeLa cells in an IFN-independent manner (Lee and Esteban 1994; Kibler et al. 1997) and in IFN-treated RK-13 cells (Kibler et al. 1997). Infection with VVΔE3L leads to activation of PKR and increased phosphorylation of eIF2α (Beattie et al. 1995a,b) in HeLa cells irrespective of IFN-treatment, and in IFN-treated RK-13 cells. E3L can also inhibit PKR activation and eIF2α phosphorylation mediated by mammalian PKR in yeast cells, and it can rescue yeast cells from the growth inhibitory activity of PKR (Romano et al. 1998).

The E3L gene encodes two related proteins, p20 and p25 (Chang et al. 1992; Yuwen et al. 1993). The entire E3L open reading frame is 190 codons long. p25 is the product of the entire E3L open reading frame, whereas p20 is the product of initiation of translation at the 2nd in-frame AUG codon (the 38th codon of the open reading frame), likely due to leaky scanning (Yuwen et al. 1993). p25 is usually made in at least 5-fold excess over p20 (Chang and Jacobs 1993; Yuwen et al. 1993). Nonetheless, the replication in cells in culture of virus expressing only p20 is indistinguishable from that of wild-type virus (Shors et al. 1997).

As originally predicted for SKIF, the products of the cloned E3L gene bind specifically to dsRNA (Chang et al. 1992). The E3L gene products consist of amino-terminal and carboxy-terminal domains, separated by a trypsin-sensitive spacer region (Ho and Shuman 1996b). Both p20 and p25 bind tightly (K_D of 7–9 nM) and specifically to dsRNA, but not ssRNA, ssDNA, B-form dsDNA, or RNA:DNA hybrids (Chang et al. 1992; Chang and Jacobs 1993; Ho and Shuman 1996a,b). The carboxy-terminal domain, which has similarity to several other dsRNA-binding proteins (Fierro-Monti and Mathews 2000), is necessary for binding to dsRNA (Chang and Jacobs 1993; Ho and Shuman 1996a). This domain is necessary and sufficient for IFN resistance, a normal host range, and inhibition of apoptosis in virus-infected cells in culture (Kibler et al. 1997; Shors et al. 1997). The carboxy-terminal domain is necessary but not sufficient for inhibition of mammalian PKR expressed in yeast (Romano et al. 1998). Yeast expressing PKR exhibit a slow growth phenotype. This phenotype can be rescued by expression of E3L, and rescue requires a functional dsRNA-binding domain (Romano et al. 1998). E3L function in virus can be complemented by other dsRNA-binding proteins, such as the unrelated reovirus σ3 protein (Beattie et al. 1995b), mammalian TRBP (Park et al. 1994), the 69-amino-acid group C rotavirus p8 protein (Langland et al. 1994), and the *Escherichia coli* dsRNA-specific ribonuclease, RNase III (Shors and Jacobs 1997). These results further suggest that the major function of E3L-encoded proteins in a vaccinia-virus-infected cell in culture is to bind dsRNA.

The E3L gene products are early proteins, synthesized as early as 0.5 hour postinfection, with synthesis decreasing by 4–6 hours postinfection (Watson et al. 1991). However, the E3L gene products function at late times postinfection, with both viral and host protein synthesis ceasing in restrictive or IFN-treated cells by 2–3 hours postinfection (Chang et al. 1995). The shutoff of host and viral protein synthesis in cells infected with VVΔE3L can be completely prevented by treatment with cytosine arabinoside or hydroxyurea (H.-W. Chang and B.L. Jacobs, unpubl.), both of which block entry into the late phase of virus replication. Thus, it is likely that E3L proteins are binding to dsRNA that accumulates from symmetrical viral transcripts at late, but not early, times postinfection.

In addition to acting as an inhibitor of PKR, the E3L gene products act as inhibitors of the 2′,5′ oligoadenylate (2′,5′A) synthetase pathway. The 2′,5′A synthetase family of enzymes is activated by binding dsRNA. Active 2′,5′A synthetase catalyzes polymerization of ATP in a novel 2′,5′ linkage. Oligomers of 2′5′A activate a latent ribonuclease, RNase L, which subsequently cleaves both viral and cellular single-stranded (ss)RNA, including rRNA. Infection of several cells with virus deleted of E3L leads to rRNA degradation typical of activation of the 2′,5′A synthetase/RNase L pathway (Beattie et al. 1995b). Furthermore, the E3L gene is required to prevent activation of the 2′,5′A synthetase and RNase L in cells transiently expressing these proteins from plasmids (Rivas et al. 1998).

The E3L gene products exist as oligomers, both in extracts from infected cells (Watson et al. 1991) and in extracts of *E. coli* engineered to produce E3L gene products (Ho and Shuman 1996b). The ability to oligomerize maps to the carboxy-terminal 90 amino acids of E3L (Ho and Shuman 1996b), coincident with the dsRNA-binding domain of E3L-encoded proteins. Small deletions at the carboxyl terminus of E3L (Δ7C) prevent dimerization (H.-W. Chang and B.L. Jacobs, unpubl.), and decrease the affinity for dsRNA by 1000-fold (Chang and Jacobs 1993). VVE3LΔ7C has an intermediate phenotype. It is IFN-resistant in RK-13 cells, but has a narrow host range (Shors et al. 1997).

Even though vaccinia virus replicates in the cytoplasm of infected cells, an intact nucleus is required for replication. The E3L gene products are the only vaccinia virus gene products known to localize to both the nucleus and cytoplasm of infected cells (Yuwen et al. 1993; Chang et al. 1995). Sequences at the amino terminus of E3L, between residues 38 and 54, are necessary for accumulation of E3L products in the nucleus, and fusion of E3L residues 1–54 with blue fluorescent protein (BFP) drives BFP accumulation in the nucleus (K. Perkins and B.L. Jacobs, unpubl.). Virus containing deletion mutants of E3L that accumulate solely in the cytoplasm have a wild-type phenotype in cells in culture (Chang et al. 1995). On the contrary, fusion of an SV-40 T-antigen nuclear localization signal onto E3LΔ83N (E3LΔ83N/NLS), which drives accumulation solely in the nucleus, yields a virus with a restricted host range that is sensitive to the antiviral effects of IFN (K. Perkins and B.L. Jacobs, unpubl.). These results suggest that cytoplasmic but not nuclear accumulation of the E3L gene products is required for efficient replication in cells in culture.

The amino-terminal 70 amino acids of E3L share significant sequence homology with the eukaryotic RNA-editing enzyme ADAR1, which catalyzes the deamination of adenosine residues that are present in dsRNA, or in secondary structures of predominantly ssRNA (Patterson et al. 1995). ADAR1 has also been isolated in a screen for Z-DNA binding proteins (Herbert et al. 1997). The region of homology between E3L and ADAR1 is sufficient for Z-DNA binding of ADAR1 (Liu et al. 1998), and p25 has been shown to bind Z-DNA (A. Herbert, pers. comm.). Viruses containing mutations of E3L that destroy Z-DNA binding have a broad host range and are IFN-resistant in cells in culture (K. Perkins and B.L. Jacobs, unpubl.), suggesting that Z-DNA binding is dispensable for replication in cells in culture.

The function of the amino terminus of E3L has proved a conundrum. On the one hand, the amino terminus is as well conserved as the carboxyl terminus between vaccinia virus E3L and the distantly related ORF virus E3L. On the other hand, the amino terminus is deleted in the more closely related myxoma virus E3L. The amino-terminal 45% of E3L, upstream of the dsRNA-binding domain, is not essential for replication of vaccinia virus in several different cell lines in culture (Kibler et al. 1997; Shors et al. 1997). This domain is, however, essential for rescue of yeast from the growth inhibitory effects of mammalian PKR (Romano et al. 1998). In fact, E3L mutated at W66, in the amino-terminal domain, fails to rescue yeast cells from the effects of PKR (Romano et al. 1998), but virus containing a mutation at this site is fully wild type in cells in culture (K. Perkins and B.L. Jacobs, unpubl.). The amino terminus of E3L proteins has been reported to directly interact with the catalytic domain of PKR, suggesting that this interaction may be required for function of E3L proteins in yeast (Romano et al. 1998).

To address this conundrum, we have begun to ask whether the amino terminus of E3L plays a role in infection in an animal model. The WR strain of vaccinia virus is highly pathogenic in inbred mice, causing lethal encephalitis at 4–8 days post-intranasal inoculation. Deletion of E3L yields a virus that is at least 10,000-fold less pathogenic than wild-type WR-VV (T. Brandt and B.L. Jacobs, unpubl.). The entire E3L gene appears to be required for pathogenesis, since pathogenesis is drastically reduced by mutations in either the carboxy-terminal dsRNA-binding domain or the amino-terminal Z-DNA-binding/nuclear localization/PKR-binding domain (T. Brandt and B.J. Jacobs, unpubl.). We have not yet determined whether PKR-binding, Z-DNA-binding, or accumulation in the nucleus is necessary for pathogenesis in this animal model.

As noted above, vaccinia virus infection rescues both VSV and encephalomyocarditis virus (EMCV) from the antiviral effects of IFN. The E3L gene is both necessary and sufficient for rescue of VSV (Shors et al. 1998). Virus deleted of E3L fails to rescue VSV from the antiviral effects of IFN treatment, and cloned E3L can rescue VSV during transient transfection experiments. Rescue appears to be dependent on binding to dsRNA, since rescue with mutants of E3L correlates with their ability to bind dsRNA. Surprisingly, cloned E3L fails to rescue EMCV from the effects of IFN, and virus deleted of E3L can still rescue EMCV, suggesting that EMCV rescue depends on a second vaccinia virus IFN-resistance gene (Shors et al. 1998), the K3L gene. The domain structure that has been described in this section for E3L-encoded proteins is illustrated in Figure 1.

POXVIRUSES AND THE INTERFERON SYSTEM: K3L

Vaccinia virus encodes a second PKR inhibitor, the K3L gene product (Davies et al. 1992). K3L was first identified during the project to sequence the entire vaccinia virus genome (Beattie et al. 1991). The K3L gene encodes an 88-amino-acid protein that shares sequence homology with the amino-terminal 30% of eIF2α (~30% sequence identity and 50% sequence similarity), one of the key substrates for PKR. This region of

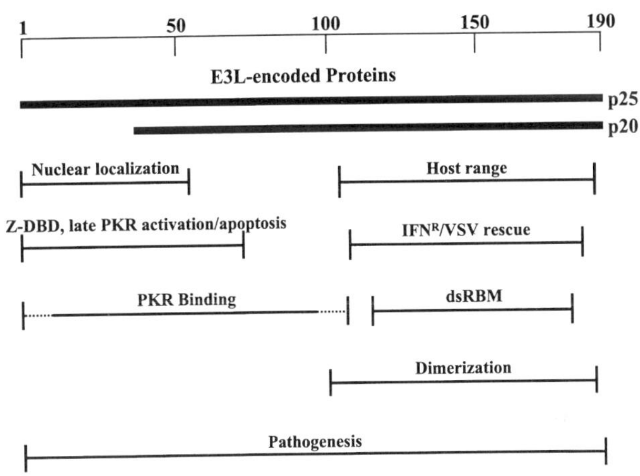

Figure 1 Domain structure for the vaccinia virus E3L-encoded proteins, p25 and p20. (IFN^R) Interferon resistance; (Z-DBD) Z-DNA-binding domain; (dsRBM) dsRNA-binding motif.

eIF2α contains the site of phosphorylation by PKR, although the site of phosphorylation is not well conserved in vaccinia virus K3L.

Recombinant K3L inhibits eIF2α phosphorylation in infected (Beattie et al. 1995a) and transfected cells (Davies et al. 1992) and decreases autophosphorylation of PKR (Davies et al. 1992; Carroll et al. 1993). K3L proteins are not themselves phosphorylated by PKR in these reactions (Davies et al. 1992). K3L-encoded proteins coprecipitate with human PKR, using anti-PKR monoclonal antibodies (Carroll et al. 1993). Both K3L and eIF2α bind to a similar or overlapping region in the catalytic domain of PKR (residues 367–415) (Gale et al. 1997; Sharp et al. 1997). These results suggest that K3L proteins act as pseudosubstrates, binding to PKR in competition with authentic eIF2, without themselves being phosphorylated. K3L protein also inhibits the three other known eIF2α kinases, the heme-regulated kinase (Carroll et al. 1993), the yeast GCN2 kinase (Qian et al. 1996), and the endoplasmic-reticulum-resident kinase, PEK (Sood et al. 2000). In yeast, K3L decreases eIF2α phosphorylation mediated by active GCN2 and prevents induction of the general amino acid control pathway by amino acid starvation (Qian et al. 1996). These results suggest that K3L proteins are mimicking a conserved interaction between eIF2 and eIF2α kinases.

The carboxy-terminal 10 amino acids of K3L have the greatest homology with eIF2α. Deletion and point mutation analysis suggests that this region is necessary for binding to and inhibition of PKR (Kawagishi-Kobayashi et al. 1997; Sharp et al. 1997). The highly conserved sequence KGYID appears to be critical for interaction with PKR. Mutations that increase the ability of K3L to inhibit PKR map to the central part of K3L and increase the similarity between K3L and eIF2α near the phosphorylation site on eIF2α (Kawagishi-Kobayashi et al. 1997).

Although vaccinia virus deleted for E3L has a strong host range and IFN-sensitive phenotype in cells in culture, virus deleted for K3L has at best a subtle phenotype. VVΔK3L replicates efficiently in most cell lines tested and is partially restricted only in BHK cells (J.O. Langland and B.L. Jacobs, unpubl.). Furthermore, replication of VVΔK3L is IFN-resistant in several cell lines (J.O. Langland and B.L. Jacobs, unpubl.) and is IFN-sensitive only in mouse L929 cells (Beattie et al. 1995a). It is possible that K3L functions to inhibit PKR in those cells in which function of E3L is overcome with saturating amounts of dsRNA. In vitro, E3L proteins appear to be more efficient inhibitors of PKR than K3L proteins (Davies et al. 1993). VVΔK3L does induce activation of PKR and phosphorylation of eIF2α in the semi-restrictive BHK cell line (J.O. Langland and B.L. Jacobs, unpubl.), and eIF2α phosphorylation in IFN-treated mouse L929 cells (Beattie et al. 1995a). In addition, replication of VVΔK3L is hypersensitive to treatment of RK-13 cells with IBT (J.O. Langland and B.L. Jacobs, unpubl.). IBT is an anti-poxvirus drug that leads to accumulation of excess dsRNA in treated cells (Bayliss and Condit 1993) and thus may be able to saturate E3L proteins. Interestingly, infection of IFN-treated mouse L929 cells with VVΔK3L leads to inhibition of both viral and cellular protein synthesis by 0.5 hour postinfection (Beattie et al. 1995a), a time well before viral dsRNA is thought to accumulate from late symmetrical transcription. Thus, the nature of the activator, and perhaps even of the kinase, in L929 cells that leads to phosphorylation of eIF2α during infection with VVΔK3L is unclear.

The K3L gene appears to be involved in rescue of EMCV but not VSV from the antiviral effects of IFN (Shors et al. 1998). K3L expressed from a plasmid at least partially rescues EMCV from the effects of IFN but has no effect on replication of VSV in IFN-treated cells. VVΔK3L does partially rescue EMCV from the effects of IFN, suggesting that the K3L gene is necessary for full rescue of EMCV but that some other gene product may be involved as well. The other gene product is likely not the product of the E3L gene, since cloned E3L does not increase rescue of EMCV by cloned K3L.

POXVIRUSES AND THE INTERFERON SYSTEM: INTERFERON SENSITIVITY OF POXVIRUSES

Despite having two IFN-resistance genes, vaccinia virus has instances in which replication is sensitive to IFN or to overexpression of PKR or the 2′,5′A synthetase/RNase L pathway. Vaccinia virus replication in CEF cells is sensitive to treatment with chicken IFNβ (Bodo et al. 1972; Rosel and Jungwirth 1983). Virus replication is blocked at an early stage. Early mRNAs are synthesized at normal, or in fact, increased rates, but are rapidly degraded in IFN-treated CEF cells (Grun et al. 1987). However, no degradation of rRNA, characteristic of activation of the 2′,5′ A synthetase/RNase L pathway, was detected (Grun et al. 1987). Some, but not all, heterologous cellular RNAs expressed from the viral thymidine kinase locus are also turned over rapidly in IFN treated CEF cells, suggesting that flanking viral sequences impart IFN-sensitivity to normally IFN-resistant cellular mRNAs (Grun et al. 1991; Degen et al. 1992).

Replication of vaccinia virus is also sensitive to treatment of BHK cells with recombinant human IFNαA/D (J.O. Langland and B.L. Jacobs, unpubl.). IFN treatment leads to an inhibition of both viral and host protein synthesis. Infection of IFN-treated BHK cells leads to increased eIF2α phosphorylation, but no detectable rRNA degradation indicative of activation of the 2′,5′A synthetase /RNase L system (J.O. Langland and B.L. Jacobs, unpubl.).

Perhaps most interestingly, replication of vaccinia virus is sensitive to IFN treatment of suspension but not monolayer cultures of L929 cells (Esteban et al. 1986). Infection of IFN-treated suspension cultures of L929 cells leads to an early inhibition of both viral and host protein synthesis. Early viral RNAs are synthesized at increased rates but are rapidly degraded, similar to infection of IFN-treated CEF cells. rRNA was also degraded at early times postinfection, suggestive of activation of the 2′,5′A synthetase/RNase L pathway. Finally, overexpression of either PKR, or 2′,5′A synthetase and RNase L from vaccinia virus recombinants, leads to inhibition of virus replication (Lee et al. 1994; Diaz-Guerra et al. 1997). Thus, although E3L and K3L are powerful inhibitors of IFN action, there are clear instances where their effects can be overcome. However, in none of the systems yet identified is a mechanistic explanation available.

LARGE DNA VIRUSES AND THE INTERFERON SYSTEM: EVOLUTIONARY CONSIDERATIONS

The E3L gene is highly conserved throughout the poxvirus family. Among the orthopoxviruses, vaccinia virus (Goebel et al. 1990), variola virus (Shchelkunov et al. 1993), cowpox, rabbit pox, and ectromelia viruses all contain highly conserved E3L genes (>90% identity among vaccinia, variola, and ectromelia viruses) or make antigenically cross-reactive proteins of the same relative molecular weights (p20 and p25) (B.L. Jacobs, unpubl.). An E3L homolog is also present in the parapoxvirus, ORF (McInnes et al. 1998), and in myxoma and rabbit fibroma viruses (accession # AAF14917 and AAF17911, respectively). The ORF virus E3L homolog is only distantly related to the orthopoxvirus E3Ls (~30% sequence identity) but shows clear similarity in the dsRNA-binding and the amino-terminal ADAR1 homology domains, with little similarity in between. Interestingly, ORF E3L likely encodes only one protein, a p25 equivalent, since the first AUG is in an excellent context for initiation of translation, and since there is an isoleucine codon in the position where the downstream initiation of translation occurs in the orthopoxvirus E3Ls. The myxoma virus E3L gene is truncated at its 5′-end compared to the orthopoxvirus E3Ls, and is predicted to encode a 115-amino-acid protein of M_r=12,600. A dsRNA-binding protein of this size has been detected in extracts from myxoma-virus-infected cells (J.O.

Langland and B.L. Jacobs, unpubl.). Homology between myxoma virus E3L and vaccinia virus E3L is reasonably high (45% identity) but is limited to the dsRNA-binding domain. Spacing between E3L and the next upstream gene (E2L) in myxoma virus suggests that the Z-DNA-binding motif has either been deleted from myxoma virus, or inserted into ORF and the orthopoxvirus genomes. In comparing the dsRNA-binding domains, myxoma virus E3L and orthopoxvirus E3L are more similar to each other than they are to ORF virus E3L. This suggests that a deletion event has occurred in the amino terminus of the myxoma virus E3L after divergence of the myxoma/orthopoxvirus lineage from the ORF virus lineage. The dsRNA-binding domain of the E3L-like genes is most closely related to the cellular enzyme ADAR1. Since the E3Ls share a Z-DNA-binding motif with ADAR1 and have a closely related dsRNA-binding motif, it is likely that the poxvirus E3L gene arose by capture of a cellular ADAR-like gene. Of the poxviruses whose genomes have been sequenced in their entirety, only canarypox (J. Tartaglia, pers. comm.) and molluscum contagiosum (Senkevich et al. 1996) viruses do not contain an E3L-like gene. Molluscum contagiosum can be successfully treated with IFN (Nelson et al. 1995). Schematic structures of the various E3L-related proteins are shown in Figure 2.

Conservation of K3L is less widespread than for E3L. Whereas a K3L-like open reading frame is present in vaccinia (Goebel et al. 1990), cowpox (accession # CAA58597) and variola (Shchelkunov et al. 1993) viruses, the K3L-like genes in ectromelia (M. Buller, pers. comm.) and monkeypox viruses (S. Shchelkunov, pers. comm.) have in-frame stop codons that would almost certainly produce an inactive protein. Myxoma and rabbit fibroma viruses contain K3L-like genes (accession # AAF15043.1 and # AAF17892.1, respectively) but the highly conserved sequence, KGYID, which is necessary for K3L function in yeast

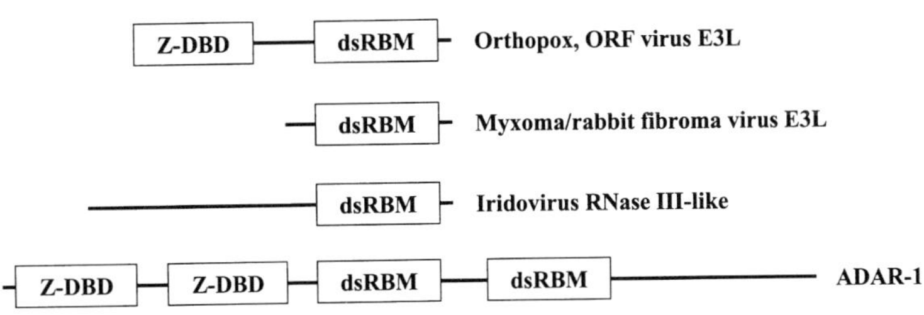

Figure 2 Schematic structure of E3L-related proteins. (Z-DBD) Z-DNA-binding domain; (dsRBM) dsRNA-binding motif; (ADAR-1) adenosine deaminase that acts on RNA-1.

(Kawagishi-Kobayashi et al. 1997; Sharp et al. 1997), is not conserved in the myxoma and rabbit fibroma genes. A K3L-like gene that contains the highly conserved KGYID sequence is, however, present in the distantly related suipox (Massung et al. 1993) and yatapoxviruses (accession # AAD46178). Again, as for E3L, neither canarypox nor molluscum contagiosum viruses contain a K3L-like gene. Schematic structures of the various K3L-related proteins are shown in Figure 3.

The iridoviruses are large DNA viruses that share many features of replication with the poxviruses, including cytoplasmic transcription and DNA synthesis. Although the iridoviruses do not contain identifiable E3L-like genes, they do contain genes with similarity to the dsRNA-specific ribonuclease, RNase III (see Fig. 2) (Kutish et al. 1996). Since *Escherichia coli* RNase III can at least partially complement deletion of E3L in vaccinia virus (Shors and Jacobs 1997), it is possible that the iridovirus RNase III-like gene can act as an inhibitor of dsRNA-mediated events in iridovirus-infected cells. Iridoviruses that infect fish (lymphocystis disease virus; Tidona and Darai 1997), insects (chilo iridescent virus; Bahr et al. 1997), and paramecium (paramecium bursaria chlorella virus 1; Kutish et al. 1996) have RNase III-like genes. In no case has the iridovirus RNase III-like gene been shown to encode a functional enzyme.

Iridoviruses also contain an eIF2α homolog (Yu et al. 1999). At 259 amino acids, the iridovirus homolog is much larger than the poxvirus K3L-encoded proteins (see Fig. 3). The iridovirus eIF2α homolog shows the greatest similarity in its amino terminus with eukaryotic eIF2α pro-

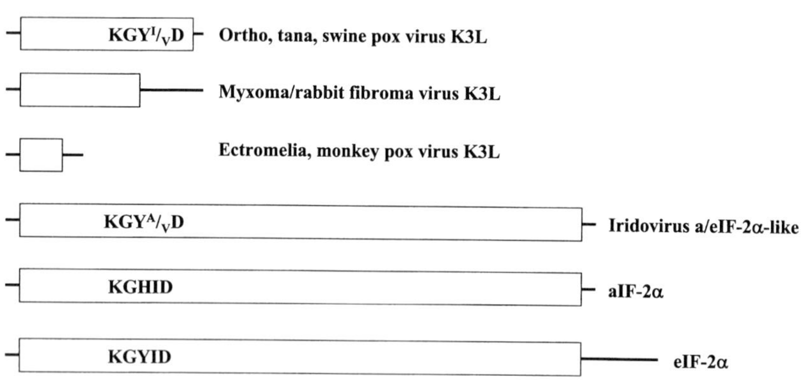

Figure 3 Schematic structure of K3L-related proteins. Boxed sequences show conservation among the eIF2α proteins. KGYID (or related variants) is the highly conserved sequence that is required for K3L function. (aIF-2α) Archaebacterial eIF2α-like protein.

teins, but shows greater similarity through its entire length with archaebacterial eIF2-like translation initiation factors. The eIF2α kinase interaction sequences, identified for poxvirus K3L-like proteins, are well conserved in the iridovirus a/eIF2α-homologs. Thus, it is unclear whether the iridovirus homolog is acting as a eIF2α kinase inhibitor, or given its large size, as an alternative eIF2α-like translation initiation factor.

HOST RANGE

The two vaccinia virus IFN-resistance genes, E3L and K3L, also function as host-range genes: E3L allows replication in human HeLa cells (Beattie et al. 1996), and K3L allows replication in BHK cells (J.O. Langland and B.L. Jacobs, unpubl.). Interestingly, several other poxvirus host-range genes function by preserving protein synthesis in certain infected cells. The vaccinia virus K1L gene is necessary for replication in RK-13 cells (Perkus et al. 1990), whereas C7L is necessary for replication in hamster Dede cells (Perkus et al. 1990; Oguiura et al. 1993) and the cowpox CP77 gene (C9L) is necessary for replication in Chinese hamster ovary (CHO) cells (Spehner et al. 1988). Either an intact K1L, C7L, or CP77 gene is required for vaccinia virus replication in cells of human origin (Perkus et al. 1990).

RK-13 cells infected with VVΔK1L synthesize early mRNAs, but the synthesis of both host proteins and early viral proteins is inhibited (Ramsey-Ewing and Moss 1996). The CP77 gene from cowpox can complement deletion of K1L (Perkus et al. 1990), restoring translation of early mRNA in RK-13 cells (Ramsey-Ewing and Moss 1996). CP77 seems to work at a later stage in the viral life cycle, since infection of RK-13 cells with VVΔK1Lcp77+ leads to a rapid inhibition of both host and early viral protein synthesis, followed by a resurrection of early viral protein synthesis at 6 hours postinfection (Ramsey-Ewing and Moss 1996). Correspondingly, infection with virus lacking a CP77 gene leads to early viral protein synthesis, intermediate viral mRNA synthesis, but no intermediate protein synthesis in CHO cells (Ramsey-Ewing and Moss 1995). The molecular mechanism by which these host-range genes support translation is unclear.

INHIBITION OF HOST PROTEIN SYNTHESIS

Poxvirus infection leads to a profound inhibition of the synthesis of host proteins at early times postinfection. Infection at high multiplicity of infection (moi) leads to an inhibition of host cell protein synthesis within minutes (Ben-Hamida and Beaud 1978), whereas infection at an moi of

1–10 usually leads to an inhibition of protein synthesis by 1–2 hours postinfection. At least some of the inhibition of synthesis of host proteins is due to the rapid degradation of cellular mRNA in a vaccinia-virus-infected cell (Rice and Roberts 1983). However, some inhibition of host translation appears to occur prior to degradation of host mRNA. Treatment of cells with low concentrations of actinomycin D that inhibit transcription of viral mRNA do not prevent shutoff of protein synthesis (Person and Beaud 1980). UV-irradiation experiments suggest that 50–100-nucleotide long transcripts, accumulation of which is not inhibited by low concentrations of actinomycin D, could be involved in shutoff of protein synthesis in infected cells (Bablanian et al. 1981). The effects of chain-terminating inhibitors of transcription on inhibition of translation are controversial. Cordecypin (3′-deoxyadenosine) prevented the shutoff of protein synthesis in HeLa cells (Bablanian et al. 1978), but did not affect shutoff of protein synthesis in Ehrlich ascites tumor (EAT) cells (Person and Beaud 1980). The contradictory results with cordecypin have led one group of researchers to suggest that a virion protein is likely involved in shutoff of protein synthesis in EAT cells, and another group has suggested that short polyadenylated RNAs are likely responsible for shutoff of protein synthesis in HeLa cells.

Shutoff of protein synthesis in EAT cells occurs primarily at the step of initiation of translation (Person and Beaud 1978), concomitant with an inhibition of 40S-methionyl-tRNA complex formation in extracts from virus-infected cells (Person et al. 1980). Again, in EAT cells, inhibition of protein synthesis is resistant to treatment of cells with actinomycin-D or cordecypin, and to UV-irradiation of virus (Person and Beaud 1980). These results suggest that neither RNA nor viral protein synthesis is needed for inhibition of protein synthesis in EAT cells, suggesting that a pre-existing virion component is likely responsible for inhibition. Incubation of protein-synthesizing systems with virion cores in vitro leads to a rapid inhibition of protein synthesis and a rapid breakdown of polysomes, consistent with an inhibition of initiation of translation (Ben-Hamida and Beaud 1978). Inhibition in this system occurs at a virion core to ribosome ratio of 1:100 (Ben-Hamida and Beaud 1978). An 11-kD virion phosphoprotein that is released from nontranscribing cores copurified with the translation inhibitory activity. Interestingly, transcribing cores appear not to release this translation inhibitory activity (Lobo et al. 1986). The 11-kD protein inhibited protein synthesis and 40S-methionyl-tRNA complex formation at a 1:1 protein:ribosome ratio (Person-Fernandez and Beaud 1986). This protein appears to inhibit translation in vitro of viral mRNA as well as cellular mRNA. A newly synthesized viral protein has been hypothesized to be necessary to rescue viral protein synthesis in infected cells from the effects of the inhibitor. A phosphorylated form of the 11-kD virion protein is also found associated with ribosomes in infected cells (Sagot and Beaud 1979).

In HeLa cells, where cordecypin treatment has been reported to prevent inhibition of host protein synthesis, small virus-induced polyadenylated RNAs (100–500 nucleotides long), named POLADS, have been suggested as the inhibitor of host protein synthesis (Lu and Bablanian 1996). These RNAs are synthesized by transcribing viral cores in vitro, in cells infected with vaccinia virus (Su and Bablanian 1990), and to a greater extent in cells infected with UV-treated vaccinia virus (Bablanian and Banerjee 1981), probably by the rather promiscuous viral poly(A) polymerase. POLADS have been reported to preferentially inhibit translation in vitro of cellular mRNA compared to viral mRNA (Bablanian and Banerjee 1986). Inhibition of translation can be overcome by adding oligo(dT) (Bablanian Banerjee 1986) or purified poly(A)-binding protein (PABp) to reactions (Su and Bablanian 1990), suggesting that POLADS compete for a limiting amount of PABp. These results also suggest that translation of viral mRNA is less sensitive to limitations in PABp than is translation of cellular mRNA. Both viral and nuclear and cytoplasmic cellular RNAs are found in the pool of small poly(A) RNAs at early and late times postinfection (Lu and Bablanian 1996). Many RNAs that do not normally contain poly(A), such as U2 RNA and tRNA, are found in the pool of POLADS after infection. Thus, the unique viral or cellular sequences at the 5′end of the POLADs seem irrelevant to inhibition of translation in vitro (Lu and Bablanian 1996).

The precise role of either of these pathways in inhibition of host protein synthesis in infected cells is at present unclear. The only genetic information implicating either of these pathways in events in infected cells is the observation that a temperature-sensitive mutant of VV that does not shut off host protein synthesis at restrictive temperature also produces decreased amounts of POLADS at restrictive temperature (Cacoullos and Bablanian 1993). Further genetic analysis of inhibition of host protein synthesis would likely be illuminating.

CONCLUDING REMARKS

Perhaps no viruses demonstrate the intimate interactions between virus and host better than the poxviruses. In this chapter, I have highlighted the role that regulation of protein synthesis plays in resistance to the antiviral effects of IFN, and in allowing poxviruses to infect a wide range of cells.

In addition to the significant resources that poxviruses have invested to maintain and control protein synthesis in infected cells, poxviruses produce numerous cytokine inhibitors, complement inhibitors, and apoptosis inhibitors that allow this family of viruses to parry many of the host's attempts to protect itself from these viruses. How these varied viral products interact in the infected cell and in an infected organism to regulate the host environment remains to be explained. Future research is required to define the precise molecular mechanisms involved in the control and maintenance of protein synthesis in the infected cell. Likely the unraveling of these events will tell us as much about how cells regulate protein synthesis as it will about how these viruses regulate protein synthesis.

ACKNOWLEDGMENTS

The author thanks Jim Tartaglia, Mark Buller, Alan Herbert, and Sergei Shchelkunov for sharing information prior to publication; Jeff Langland and Chandra Mitnick for help with figures; and Karen Kibler for a critical review of the manuscript. The work described from the author's lab was supported by National Institutes of Health grant CA48654.

REFERENCES

Bablanian R. and Banerjee A.K. 1986. Poly(riboadenylic acid) preferentially inhibits in vitro translation of cellular mRNAs compared with vaccinia virus mRNAs: Possible role in vaccinia virus cytopathology. *Proc. Natl. Acad. Sci.* **83:** 1290–1294.

Bablanian R., Coppola G., Scribani S., and Esteban M. 1981. Inhibition of protein synthesis by vaccinia virus. IV. The role of low-molecular-weight viral RNA in the inhibition of protein synthesis. *Virology* **112:** 13–24.

Bablanian R., Esteban M., Baxt B., and Sonnabend J.A. 1978. Studies on the mechanisms of vaccina virus cytopathic effects. I. Inhibition of protein synthesis in infected cells is associated with virus-induced RNA synthesis. *J. Gen. Virol.* **39:** 391–402.

Bahr U., Tidona C.A., and Darai G. 1997. The DNA sequence of Chilo iridescent virus between the genome coordinates 0.101 and 0.391: Similarities in coding strategy between insect and vertebrate iridoviruses. *Virus Genes* **15:** 235–245.

Bayliss C.D. and Condit R.C. 1993. Temperature-sensitive mutants in the vaccinia virus A18R gene increase double-stranded RNA synthesis as a result of aberrant viral transcription. *Virology* **194:** 254–262.

Beattie E., Paoletti E., and Tartaglia J. 1995a. Distinct patterns of IFN sensitivity observed in cells infected with vaccinia K3L- and E3L-mutant viruses. *Virology* **210:** 254–263.

Beattie E., Tartaglia J., and Paoletti E. 1991. Vaccinia virus-encoded eIF2α homologue abrogates the antiviral effect of interferon. *Virology* **183:** 419–422.

Beattie E., Denzler K.L., Tartaglia J., Perkus M.E., Paoletti E., and Jacobs B.L. 1995b. Reversal of the interferon-sensitive phenotype of a vaccinia virus lacking E3L by expression of the reovirus S4 gene. *J. Virol.* **69:** 499–505.

Beattie E., Kaufman E., Martinez H., Perkus M., Jacobs B.L., Paoletti E., and Tartaglia J. 1996. Host range restriction of vaccinia virus E3L-specific deletion mutants. *Virus Genes* **12:** 89–94.

Ben-Hamida F. and Beaud G. 1978. In vitro inhibition of protein synthesis by purified cores from vaccinia virus. *Proc. Natl. Acad. Sci.* **75:** 175–179.

Black E.P. and Condit R.C. 1996. Phenotypic characterization of mutants in vaccinia virus gene G2R, a putative transcription elongation factor. *J. Virol.* **70:** 47–54.

Bodo G., Scheirer W., Suh M., Schultze B., Horak I., and Jungwirth C. 1972. Protein synthesis in pox-infected cells treated with interferon. *Virology* **50:** 140–147.

Cacoullos N. and Bablanian R. 1993. Role of polyadenylated RNA sequences (POLADS) in vaccinia virus infection: Correlation between accumulation of POLADS and extent of shut-off in infected cells. *Cell. Mol. Biol. Res.* **39:** 657–664.

Carroll K., Elroy-Stein O., Moss B., and Jagus R. 1993. Recombinant vaccinia virus K3L gene product prevents activation of double-stranded RNA-dependent, initiation factor 2 alpha-specific protein kinase. *J. Biol. Chem.* **268:** 12837–12842.

Chang H.-W. and Jacobs B.L. 1993. Identification of a conserved motif that is necessary for binding of the vaccinia virus E3L gene products to double-stranded RNA. *Virology* **194:** 537–547.

Chang H.-W., Uribe L.H., and Jacobs B.L. 1995. Rescue of vaccinia virus lacking the E3L gene by mutants of E3L. *J. Virol.* **69:** 6605–6608.

Chang H.-W., Watson J., and Jacobs B.L. 1992. The vaccinia virus E3L gene encodes a double-stranded RNA-binding protein with inhibitory activity for the interferon-induced protein kinase. *Proc. Natl. Acad. Sci.* **89:** 4825–4829.

Condit R.C., Xiang Y., and Lewis J.I. 1996. Mutation of vaccinia virus gene G2R causes suppression of gene A18R ts mutants: Implications for control of transcription. *Virology* **220:** 10–19.

Davies M.V., Chang H.W., Jacobs B.L., and Kaufman R.J. 1993. The E3L and K3L vaccinia virus gene products stimulate translation through inhibition of the double-stranded RNA-dependent protein kinase by different mechanisms. *J. Virol.* **67:** 1688–1692.

Davies M.V., Elroy-Stein O., Jagus R., Moss B., and Kaufman R.J. 1992. The vaccinia virus K3L gene product potentiates translation by inhibiting double-stranded-RNA-activated protein kinase and phosphorylation of the alpha subunit of eukaryotic initiation factor 2. *J. Virol.* **66:** 1943–1950.

Degen H.J., Blum D., Grun J., and Jungwirth C. 1992. Expression of authentic vaccinia virus-specific and inserted viral and cellular genes under control of an early vaccinia virus promoter is regulated post-transcriptionally in interferon-treated chick embryo fibroblasts. *Virology* **188:** 114–121.

Denzler K.L. and Jacobs B.L. 1994. Site-directed mutagenic analysis of reovirus sigma 3 protein binding to dsRNA. *Virology* **204:** 190–199.

Diaz-Guerra M., Rivas C., and Esteban M. 1997. Inducible expression of the 2-5A synthetase/RNase L system results in inhibition of vaccinia virus replication. *Virology* **227:** 220–228.

Esteban M., Benavente J., and Paez E. 1986. Mode of sensitivity and resistance of vaccinia virus replication to interferon. *J. Gen. Virol.* **67:** 801–808.

Fields B.N., Knipe D.M., and Howley P.M., Eds. 1996. *Virology*. Lippincott-Raven, Philadelphia, Pennsylvania.

Fierro-Monti I. and Mathews M.B. 2000. Proteins binding to duplexed RNA: One motif, multiple functions. *Trends Biol. Sci.* **25:** 241–246.

Gale M.J., Jr., Korth M.J., Tang N.M., Tan S.L., Hopkins D.A., Dever T.E., Polyak S.J., Gretch D.R., and Katze M.G. 1997. Evidence that hepatitis C virus resistance to interferon is mediated through repression of the PKR protein kinase by the nonstructural 5A protein. *Virology* **230:** 217–227.

Goebel S.J., Johnson G.P., Perkus M.E., Davis S.W., Winslow J.P., and Paoletti E. 1990. The complete DNA sequence of vaccinia virus. *Virology* **179:** 247–266.

Grun J., Kroon E., Zoller B., Krempien U., and Jungwirth C. 1987. Reduced steady-state levels of vaccinia virus-specific early mRNAs in interferon-treated chick embryo fibroblasts. *Virology* **158:** 28–33.

Grun J., Redmann-Muller I., Blum D., Degen H.J., Doenecke D., Zentgraf H.W., and Jungwirth C. 1991. Regulation of histone H5 and H1 zero gene expression under the control of vaccinia virus-specific sequences in interferon-treated chick embryo fibroblasts. *Virology* **180:** 535–542.

Hassett D.E. and Condit R.C. 1994. Targeted construction of temperature-sensitive mutations in vaccinia virus by replacing clustered charged residues with alanine. *Proc. Natl. Acad. Sci.* **91:** 4554–4558.

Herbert A., Alfken J., Kim Y.G., Mian I.S., Nishikura K., and Rich A. 1997. A Z-DNA binding domain present in the human editing enzyme, double-stranded RNA adenosine deaminase. *Proc. Natl. Acad. Sci.* **94:** 8421–8426.

Ho C.K. and Shuman S. 1996a. Mutational analysis of the vaccinia virus E3 protein defines amino acid residues involved in E3 binding to double-stranded RNA. *J. Virol.* **70:** 2611–2614.

———. 1996b. Physical and functional characterization of the double-stranded RNA binding protein encoded by the vaccinia virus E3 gene. *Virology* **217:** 272–284.

Kawagishi-Kobayashi M., Silverman J.B., Ung T.L., and Dever T.E. 1997. Regulation of the protein kinase PKR by the vaccinia virus pseudosubstrate inhibitor K3L is dependent on residues conserved between the K3L protein and the PKR substrate eIF2α. *Mol. Cell. Biol.* **17:** 4146–4158.

Kibler K.V., Shors T., Perkins K.B., Zeman C.C., Banaszak M.P., Biesterfeldt J., Langland J.O., and Jacobs B.L. 1997. Double-stranded RNA is a trigger for apoptosis in vaccinia virus-infected cells. *J. Virol.* **71:** 1992–2003.

Kutish G.F., Li Y., Lu Z., Furuta M., Rock D.L., and Van Etten J.L. 1996. Analysis of 76 kb of the chlorella virus PBCV-1 330-kb genome: Map positions 182 to 258. *Virology* **223:** 303–317.

Langland J.O., Pettiford S.M., Jiang B., and Jacobs B.L. 1994. Products of the porcine NSP3 gene bind specifically to double-stranded RNA and inhibit activation of the interferon-induced protein kinase, PKR. *J. Virol.* **68:** 3821–3829.

Lee S.B. and Esteban M. 1994. The interferon-induced double-stranded RNA-activated protein kinase induces apoptosis. *Virology* **199:** 491–496.

Lee S.B., Green S.R., Mathews M.B., and Esteban M. 1994. Activation of the double-stranded RNA (dsRNA)-activated human protein kinase in vivo in the absence of its dsRNA binding domain. *Proc. Natl. Acad. Sci.* **91:** 10551–10555.

Liu Y., Herbert A., Rich A., and Samuel C.E. 1998. Double-stranded RNA-specific adenosine deaminase: Nucleic acid binding properties. *Methods* **15:** 199–205.

Lobo D.S., Rebello M.A., and Moussatche N. 1986. Core transcription restores in vitro inhibition of protein synthesis induced by vaccinia virus. *Biochim. Biophys. Acta* **868:** 183–189.

Lu C. and Bablanian R. 1996. Characterization of small nontranslated polyadenylylated

RNAs in vaccinia virus-infected cells. *Proc. Natl. Acad. Sci.* **93:** 2037–2042.

Massung R.F., Jayarama V., and Moyer R.W. 1993. DNA sequence analysis of conserved and unique regions of swinepox virus: Identification of genetic elements supporting phenotypic observations including a novel G protein-coupled receptor homologue. *Virology* **197:** 511–528.

McInnes C.J., Wood A.R., and Mercer A.A. 1998. Orf virus encodes a homolog of the vaccinia virus interferon-resistance gene E3L. *Virus Genes* **17:** 107–115.

Nelson M.R., Chard S., and Barton S.E. 1995. Intralesional interferon for the treatment of recalcitrant molluscum contagiosum in HIV antibody positive individuals — A preliminary report. *Int. J. STD AIDS* **6:** 351–352.

Oguiura N., Spehner D., and Drillien R. 1993. Detection of a protein encoded by the vaccinia virus C7L open reading frame and study of its effect on virus multiplication in different cell lines. *J. Gen. Virol.* **74:**1409–1413.

Paez E. and Esteban M. 1984. Resistance of vaccinia virus to interferon is related to an interference phenomenon between the virus and the interferon system. *Virology* **134:** 12–28.

Park H., Davies M.V., Langland J.O., Chang H.-W., Nam Y.S., Tartaglia J., Paoletti E., Jacobs B.L., Kaufman R.J., and Venkatesan S. 1994. A cellular protein that binds several structured viral RNAs is an inhibitor of the interferon induced PKR protein kinase in vitro and in vivo. *Proc. Natl. Acad. Sci.* **91:** 4713–4717.

Patterson J.B., Samuel C.E., Patterson J.B., Thomis D.C., Hans S.L., and Samuel C.E. 1995. Expression and regulation by interferon of a double-stranded-RNA-specific adenosine deaminase from human. *Mol. Cell. Biol.* **15:** 5376–5388.

Perkus M.E., Goebel S.J., Davis S.W., Johnson G.P., Limbach K., Norton E.K., and Paoletti E. 1990. Vaccinia virus host range genes. *Virology* **179:** 276–286.

Person A. and Beaud G. 1978. Inhibition of host protein synthesis in vaccinia virus-infected cells in the presence of cordycepin (3´0-deoxyadenosine). *J. Virol.* **25:** 11–18.

———. 1980. Shut-off of host protein synthesis in vaccinia-virus-infected cells exposed to cordycepin. A study in vitro. *Eur. J. Biochem.* **103:** 85–93.

Person A., Ben-Hamida F., and Beaud G. 1980. Inhibition of 40S–Met–tRNAfMet ribosomal initiation complex formation by vaccinia virus. *Nature* **287:** 355–357.

Person-Fernandez A. and Beaud G. 1986. Purification and characterization of a protein synthesis inhibitor associated with vaccinia virus. *J. Biol. Chem.* **261:** 8283–8289.

Qian W., Zhu S., Sobolev A.Y., and Wek R.C. 1996. Expression of vaccinia virus K3L protein in yeast inhibits eukaryotic initiation factor-2 kinase GCN2 and the general amino acid control pathway. *J. Biol. Chem.* **271:** 13202–13207.

Ramsey-Ewing A.L. and Moss B. 1995. Restriction of vaccinia virus replication in CHO cells occurs at the stage of viral intermediate protein synthesis. *Virology* **206:** 984–993.

———. 1996. Complementation of a vaccinia virus host-range K1L gene deletion by the nonhomologous CP77 gene. *Virology* **222:** 75–86.

Rice A.P. and Kerr I.M. 1984. Interferon-mediated, double-stranded RNA-dependent protein kinase is inhibited in extracts from vaccinia virus-infected cells. *J. Virol.* **50:** 229–236.

Rice A.P. and Roberts B.E. 1983. Vaccinia virus induces cellular mRNA degradation. *J. Virol.* **47:** 529–539.

Rivas C., Gil J., Melkova Z., Esteban M., and Diaz-Guerra M. 1998. Vaccinia virus E3L protein is an inhibitor of the interferon (i.f.n.)- induced 2-5A synthetase enzyme. *Virology* **243:** 406–414.

Romano P.R., Zhang F., Tan S.L., Garcia-Barrio M.T., Katze M.G., Dever T.E., and Hinnebusch A.G. 1998. Inhibition of double-stranded RNA-dependent protein kinase PKR by vaccinia virus E3: Role of complex formation and the E3 N-terminal domain. *Mol. Cell. Biol.* **18:** 7304–7316.

Rosel J. and Jungwirth C. 1983. Isolation of early viral proteins from poxvirus-infected chick embryo fibroblasts by DNA-cellulose chromatography and inhibition of their synthesis by chicken interferon. *Eur. J. Biochem.* **132:** 361–367.

Sagot J. and Beaud G. 1979. Phosphorylation in vivo of a vaccinia-virus structural protein found associated with the ribosomes from infected cells. *Eur. J. Biochem.* **98:** 131–140.

Senkevich T.G., Bugert J.J., Sisler J.R., Koonin E.V., Darai G., and Moss B. 1996. Genome sequence of a human tumorigenic poxvirus: Prediction of specific host response-evasion genes. *Science* **273:** 813–816.

Sharp T.V., Witzel J.E., and Jagus R. 1997. Homologous regions of the alpha subunit of eukaryotic translational initiation factor 2 (eIF2α) and the vaccinia virus K3L gene product interact with the same domain within the dsRNA-activated protein kinase (PKR). *Eur. J. Biochem.* **250:** 85–91.

Shchelkunov S.N., Blinov V.M., and Sandakhchiev L.S. 1993. Genes of variola and vaccinia viruses necessary to overcome the host protective mechanisms. *FEBS Lett.* **319:** 80–83.

Shors S.T., Beattie E., Paoletti E., Tartaglia J., and Jacobs B.L. 1998. Role of the vaccinia virus E3L and K3L gene products in rescue of VSV and EMCV from the effects of IFN-α. *J. Interferon Cytokine Res.* **18:** 721–729.

Shors T. and Jacobs B.L. 1997. Complementation of deletion of the vaccinia virus E3L gene by the *Escherichia coli* RNase III gene. *Virology* **227:** 77–87.

Shors T., Kibler K.V., Perkins K.B., Seidler-Wulff R., Banaszak M.P., and Jacobs B.L. 1997. Complementation of vaccinia virus deleted of the E3L gene by mutants of E3L. *Virology* **239:** 269–276.

Simpson D.A. and Condit R.C. 1995. Vaccinia virus gene A18R encodes an essential DNA helicase. *J. Virol.* **69:** 6131–6139.

Sood R., Porter A.C., Ma K., Quilliam L.A., and Wek R.C. 2000. Pancreatic eukaryotic initiation factor-2α kinase (PEK) homologues in humans, *Drosophila melanogaster* and *Caenorhabditis elegans* that mediate translational control in response to endoplasmic reticulum stress. *Biochem. J.* **346:** 281–293.

Spehner D., Gillard S., Drillien R., and Kirn A. 1988. A cowpox virus gene required for multiplication in Chinese hamster ovary cells. *J. Virol.* **62:** 1297–1304.

Su M.J. and Bablanian R. 1990. Polyadenylated RNA sequences from vaccinia virus-infected cells selectively inhibit translation in a cell-free system: Structural properties and mechanism of inhibition. *Virology* **179:** 679–693.

Thacore H.R. and Youngner J.S. 1973. Rescue of vesicular stomatitis virus from interferon-induced resistance by superinfection with vaccinia virus. II. Effect of UV-inactivated vaccinia and metabolic inhibitors. *Virology* **56:** 512–522.

Tidona C.A. and Darai G. 1997. The complete DNA sequence of lymphocystis disease virus. *Virology* **230:** 207–216.

Watson J., Chang H.-W., and Jacobs B.L. 1991. Characterization of a vaccinia virus-induced dsRNA-binding protein that may be the inhibitor of the dsRNA-dependent protein kinase. *Virology* **185:** 206–216.

Whitaker-Dowling P.A. and Youngner J.S. 1983. Vaccinia rescue of VSV from interferon-induced resistance: Reversal of translation block and inhibition of protein kinase activity. *Virology* **131:** 128–136.

———. 1984. Characterization of a specific kinase inhibitory factor produced by vaccinia virus which inhibits the interferon-induced protein kinase activity. *Virology* **137:** 171–181.

———. 1986. Vaccinia-mediated rescue of encephalomyocarditis virus from the inhibitory effects of interferon. *Virology* **152:** 50–57.

Xiang Y., Simpson D.A., Spiegel J., Zhou A., Silverman R.H., and Condit R.C. 1998. The vaccinia virus A18R DNA helicase is a postreplicative negative transcription elongation factor. *J. Virol.* **72:** 7012–7023.

Youngner J.S., Thacore H.R., and Kelley M.E. 1972. Sensitivity of ribonucleic acid and deoxyribonucleic acid viruses to different species of interferon in cell cultures. *J. Virol.* **10:** 171–178.

Yu Y.X., Bearzotti M., Vende P., Ahne W., and Bremont M. 1999. Partial mapping and sequencing of a fish iridovirus genome reveals genes homologous to the frog virus 3 p31, p40 and human eIF2α. *Virus Res.* **63:** 53–63.

Yuwen H., Cox J.H., Yewdell J.W., Bennink J.R., and Moss B. 1993. Nuclear localization of a double-stranded RNA-binding protein encoded by the vaccinia virus E3L gene. *Virology* **195:** 732–744.

36

Nontranslational Functions of Components of the Translational Apparatus

Terri Goss Kinzy

Robert Wood Johnson Medical School
Department of Molecular Genetics & Microbiology
University of Medicine & Dentistry of New Jersey
Piscataway, New Jersey 08854

Emanuel Goldman

Department of Microbiology & Molecular Genetics
New Jersey Medical School
University of Medicine & Dentistry of New Jersey
Newark, New Jersey 07103

A growing number of studies have identified new and often nontranslational functions for many components of the translational apparatus, including tRNAs, aminoacyl-tRNA synthetases, ribosomal proteins, and initiation and elongation factors (Fig. 1). Although initially the sheer abundance of some of these components made their identification in novel functions suspect, increasingly detailed biochemical and genetic studies have established the multifunctional nature of these molecules. The studies include classic and elegant examples of viral systems that recruit host factors for their replication and maintenance, as well as cellular processes that adapt these abundant cellular components to new and perhaps related functions. Furthermore, the increasing recognition that some translational components reside in previously unexpected locations may serve to link the regulation of gene expression and the quality of protein synthesis with new functions of the translational apparatus itself (Fig. 1).

Although beyond the scope of this chapter, it is a fascinating evolutionary question as to how these components came to perform other functions in addition to their translational roles. It seems likely that at least some viruses evolved to take advantage of these preexisting abundant components (see Chapter 8), but in cases where cellular functions are

Translational Control of Gene Expression
© 2000 Cold Spring Harbor Laboratory Press 0-87969-568-4/00 $5 + . 00

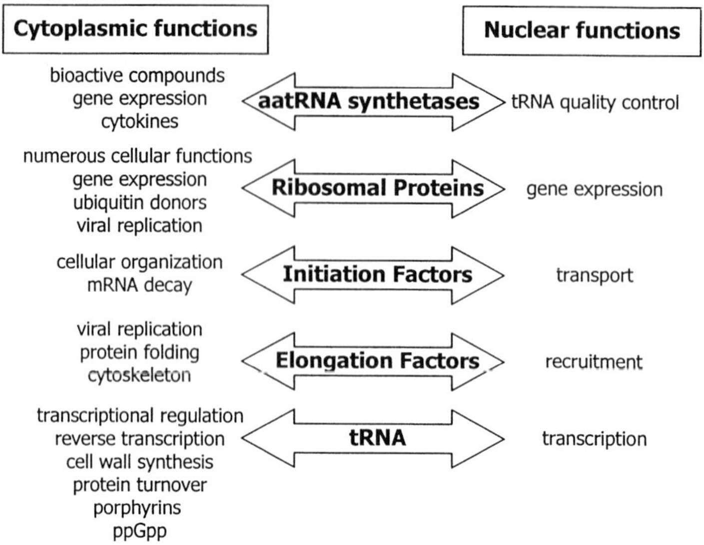

Figure 1 Summary of nontranslational functions of components of the translational apparatus. In addition to roles in protein synthesis, translational components perform both well characterized and novel but less well understood roles in many viral and cellular processes. These processes occur in prokaryotes, and in both the cytoplasm and nucleus of eukaryotic cells, as indicated.

involved, it is not so obvious whether the translational or the other function came first. In this chapter, we briefly review the classic examples of nontranslational functions of components of the protein synthetic machinery while focusing on the newly emerging themes of nontranslational functions, and potential explanations of the varied and important roles these factors play in addition to their critical roles in protein synthesis.

tRNA

The primary function of this ubiquitous class of molecule is to serve the crucial decoding role as the linker between the information encoded in nucleic acid (according to the rules of the genetic code) and the amino acids specified in protein. Perhaps because of their ubiquity, there is also a variety of nontranslational functions in which tRNAs have been implicated, including roles in replication, mutagenesis, transcription, regulation, cell wall synthesis, protein turnover, and even heme biosynthesis. These are briefly summarized below.

Primers for Reverse Transcriptase

The discovery that specific tRNAs were packaged in retrovirus particles and commonly used as primers for DNA synthesis by retroviral reverse transcriptases (for review, see Marquet et al. 1995; Arts and Le Grice 1998) generated an enormous amount of interest because of the worldwide HIV epidemic. In the case of HIV-1, $tRNA_3^{Lys}$ is specifically used (for review, see Litvak et al. 1994). This involves interactions of both the template with the 3′ end of the tRNA, and the enzyme with the 5′ end of the tRNA (Zakharova et al. 1995). In contrast, some yeast retrotransposons utilize other parts of the tRNA molecule to interact with template, such as the anticodon stem-loop in the case of Ty5 (Ke et al. 1999).

Bacterial Gene Regulation

tRNA has been shown to have fundamental roles in several bacterial regulatory systems. In addition to direct effects on transcription in *Bacillus* species, there are also indirect effects on *Escherichia coli* transcription mediated by uncharged tRNA entering the A-site of the ribosome, in competition with EF1A (previously EF-Tu)-bound charged tRNA, under conditions when the uncharged tRNA is in large excess.

Bacterial Stringent Control (ppGpp Synthesis)

Guanosine tetraphosphate (ppGpp), a pleiotropic regulator in *E. coli* and other bacteria, is synthesized in large amounts in response to amino acid or energy deprivation (for review, see Gallant 1979; Lamond and Travers 1985; Cashel et al. 1996). This is called the "stringent response"; mutants that fail to synthesize ppGpp are called "relaxed." Basal levels of ppGpp may also be involved in growth rate regulation, but this role remains controversial. During a stringent response, ppGpp inhibits transcription of genes encoding stable RNAs (rRNA and tRNA) and mRNAs for ribosomal proteins, while activating transcription of many amino acid biosynthetic operons; it also increases the degradation of proteins and the accuracy of protein synthesis, among other activities. During amino acid starvation, synthesis of ppGpp is mediated by the enzyme stringent factor (the product of the *relA* gene) on ribosomes (discussed below) and requires a codon–anticodon interaction with cognate uncharged tRNA in the A-site of the ribosome (Haseltine and Block 1973; Rojiani et al. 1989; for review, see Goldman and Jakubowski 1990). During energy starvation, ppGpp is synthesized by another enzyme, the product of the *spoT* gene; it does not appear that components of the translational apparatus are involved in this (Gentry and Cashel 1995); however, the *spoT* enzyme also has a ppGpp hydrolase activity that has been proposed to be inhibited directly by uncharged tRNA (Murray and Bremer 1996).

Attenuation of Transcription of Bacterial Amino Acid Biosynthetic Operons

In *E. coli* and some other bacteria, transcription of most amino acid biosynthetic operons is controlled by an "attenuation" mechanism, in which the mRNA contains a leader region that can assume alternate conformations, one of which provides a signal for intrinsic (rho-independent) transcription termination. Within the leader region lies a short peptide-encoding sequence that contains multiple tandem codons specifying the same amino acid which the enzymes encoded by that operon would produce (for review, see Yanofsky 1981). When the amino acid in question is present in excess, the leader peptide is translated to completion, and the leader RNA assumes the transcription–termination conformation. When the amino acid in question is limiting, however, uncharged tRNA is believed to enter the ribosome when it is trying to translate the multiple tandem codons, stalling the ribosome, thus allowing the leader RNA to assume an alternate (antitermination) conformation so that transcription of the operon proceeds. This process can be described as "translational control of transcription termination." Mutations in enzymes that modify certain tRNAs in their anticodon loops (e.g., the *trpX* mutation for isopentenyl and the *hisT* mutation for pseudouracil), as well as mutations in tRNA itself or in a cognate aminoacyl-tRNA synthetase (Yanofsky and Soll 1977), have been shown to affect the attenuation mechanism, supporting the model of participation of uncharged tRNA. Thus, as in stringent control, the role of tRNA in this process apparently involves uncharged tRNA competing for the ribosomal A-site. This mechanism also represents a variation on the theme of autoregulation, since excess amino acid, via its cognate tRNA, inhibits synthesis of mRNA for enzymes that would produce more of the same amino acid.

Transcription of Bacillus *Genes Encoding aa-tRNA Synthetases*

In *Bacillus subtilis* and some other gram-positive bacteria, alternate leader RNA structures also govern termination or antitermination of transcription for most aminoacyl-tRNA synthetase genes (for review, see Henkin 1996). However, in this mechanism, there are no leader peptide-encoding regions; instead, the leader RNAs themselves carry an appropriately positioned codon-like triplet, which pairs with the anticodon of uncharged cognate tRNA to trigger an antitermination structure in the leader RNA. When the tRNA is charged, this interaction results instead in a transcription termination structure (for review, see Henkin 1994). This is also a variation on the theme of autoregulation, because when charged tRNA is present in excess, transcription of mRNA for the very enzyme that would charge more of that tRNA is inhibited.

Eukaryotic Transcription

A $tRNA^{Ile}$ has been reported to be involved in transcription by silkworm RNA polymerase III (Dunstan et al. 1994a). Originally called TFIIIR (although no longer considered a transcription factor per se), this highly specific tRNA isoacceptor appears to act by blocking transcriptional inhibition that would otherwise be caused by low-frequency cleavage of the DNA template for class III genes (Dunstan et al. 1994b).

Biosynthesis of Porphyrins

In plants and bacteria, Glu-$tRNA^{Glu}$ is a precursor in the biosynthetic pathway of porphyrins, which in turn are precursors of chlorophyll and heme (for review, see Jahn et al. 1992). In the first step of the pathway, Glu-tRNA reacts with a unique glutamyl-tRNA reductase, converting glutamic acid to glutamate semialdehyde, releasing uncharged tRNA. The semialdehyde is then converted to 5-aminolevulinic acid, eight molecules of which are used to make the porphyrin ring.

Bacterial Cell Wall Synthesis

The involvement of tRNAs in bacterial cell wall, and possibly cell membrane, synthesis has been a neglected area of study for many years now, but elegant early work demonstrated that several different tRNAs (in different species) transferred their attached amino acids to interpeptide bridges in peptidoglycan (for review, see Strominger 1969; Soffer 1974). Curiously, at least some of these tRNAs, for example, $tRNA^{Gly}$ in *Staphylococcus aureus*, appeared to be nonfunctional in protein synthesis in vitro, and this is one area that might be of interest for study with currently available tools. Of less certain significance are observations that in some bacteria, lysine and/or alanine can be transferred by tRNA to phosphatidyl glycerol (for review, see Soffer 1974), a major constituent of membranes. The function of these amino acid transfers remains unclear.

Aminoacyl-tRNA Protein Transferases

These enzymes catalyze transfer of amino acids from specific aa-tRNAs directly to amino-terminal residues of proteins and peptides in all species examined (for review, see Deutch et al. 1978). The function of these reactions remained mysterious for many years until it was recognized that they were an important component of protein degradation pathways (Gonda et al. 1989). In eukaryotes, arginyl-tRNA protein transferases deposit arginine at the amino terminus of proteins, a primary destabilizing signal for the "N-end rule" in ubiquitin-mediated proteolysis. In bacteria, the Leu-Phe-tRNA protein transferase is required for protein turnover according to the bacterial version of the N-end rule of proteolysis (Tobias et al. 1991).

Bacterial Mutator Gene

In *E. coli*, an anticodon mutation that makes $tRNA^{Gly}$ a missense suppressor of aspartate codons also functions as a mutator gene (Slupska et al. 1996; Murphy and Humayun 1997). Strains containing this mutant $tRNA^{Gly}$, which mistranslate aspartic acid codons as glycine at a low frequency, show dramatically increased frequencies of mutation compared to wild type. Although the mechanism by which this tRNA mutation leads to elevated mutagenesis remains to be elucidated, it has been proposed that "translational stress" is responsible (Murphy and Humayun 1997; Ren et al. 1999).

AMINOACYL-tRNA SYNTHETASES

A growing number of functions for aminoacyl-tRNA synthetases (aaRSs), aside from the activation of amino acids for protein synthesis, have been the subject of many recent studies. The heterogeneity in the aaRS sequences and structures, and the presence of subdomains not clearly ascribed to functions in aminoacylation, have helped fuel studies in this area.

Bacterial Gene Regulation

Whereas the expression of many aminoacyl-tRNA synthetases is regulated by the level of uncharged tRNA by a transcriptional antiterminator mechanism (see above; for review, see Henkin 1994; Putzer et al. 1995; Condon et al. 1996), other modes of regulation dependent on the synthetase itself have emerged. Synthetases, like ribosomal proteins (discussed below), have evolved in at least some cases to help autoregulate their own synthesis. The known examples are best described in prokaryotic systems. This includes the repression of transcription by direct binding of the *E. coli* AlaRS to a promoter element (i.e., an "operator") of the gene (Putney and Schimmel 1981). This may not be an isolated case, as subsequent studies have shown that PheRS from *Thermus thermophilus* has a helix-loop-helix motif and is capable of binding DNA (Lechler and Kruetzer 1998). Although the biological relevance remains to be determined, the possibility of other functions for PheRS has been previously raised based on its altered expression patterns in human cell lines (Sen et al. 1997). Synthetases are also capable of binding RNAs besides tRNA, as shown by the translational repression activity of the *E. coli* ThrRS binding to its own mRNA (Springer et al. 1985). The recent structural analysis of the ThrRS-tRNA^{Thr} cocrystal predicts that the two stem-loop structures of the *thrS* gene transcript recognize the anticodon-binding domains of each monomer of ThrRS (Sankaranarayanan et al. 1999).

Eukaryotic Gene Regulation

Some group I introns, while self splicing in vitro, are aided in vivo by aaRSs (for review, see Dujardin and Herbert 1997). Genetic studies have demonstrated that the *cyt18* gene of *Neurospora crassa* encodes the mitochondrial TyrRS and affects the splicing of several mitochondrial group I introns. The *Saccharomyces cerevisiae NAM2* gene encoding mitochondrial LeuRS is involved in excision of the aI4 and bI4 introns. A yeast genetic screen also identified a temperature-sensitive mutation in the *THS1* gene encoding ThrRS that affects mRNA decay, although the mechanism of this effect and the specificity to this aaRS will require further characterization (Zuk et al. 1999).

Cytokine-like Domains and Association with Growth Factors

The unique amino- and carboxy-terminal extensions of metazoan aaRSs present yet another potential domain for additional functions. The human TyrRS carboxy-terminal domain, although not essential for aminoacylation, has 51% sequence identity with human endothelial monocyte-activating peptide II (EMAP II; Wakasugi and Schimmel 1999). The carboxy-terminal domain of TyrRS showed properties similar to human EMAP II. More surprisingly, the amino-terminal fragment showed activities of a leukocyte chemoattractant and bound the IL-8 receptor type A while still functioning as an aaRS. The TyrRS was also secreted, unlike other control aaRSs tested, and cleavage to produce the growth factors was linked to events in apoptosis. This link may have other important implications since the Pro-EMAP II molecule has been shown to associate with the multienzyme aaRS complex (Quevillon et al. 1997) and to stimulate the activity of the ArgRS (Park et al. 1999). Thus, a new and exciting area of research into the multifunctional nature of tRNA-synthetases has been established.

Generation of Bioactive Molecules

Synthetases can also contribute to the generation of novel molecules in the cell. An example which has been studied in depth is that of Met-tRNA synthetases, from all species, which edit homocysteine (Hcy) to homocysteine-thiolactone (Jakubowski 1990, 1991; Jakubowski and Goldman 1992, 1993). MetRSs activate Hcy to Hcy-AMP, but before transfer to $tRNA^{Met}$, convert the activated Hcy to the cyclic Hcy-thiolactone (releasing AMP). This highly reactive molecule diffuses from cells and can homocysteinylate lysine residues in proteins (Jakubowski 1999), which has been proposed as a mechanism for the involvement of homocysteine as an independent high-risk factor in the etiology of atherosclerosis (Jakubowski 1997).

aa-RS Aminoacylation of tRNA-like Molecules

In several plant viral RNAs, tRNA-like structures are found at the 3´ ends of the genomes, and specific amino acids can be attached to them by cognate aa-tRNA syntetases (Haenni et al. 1982; Florentz and Giegé 1995). There have also been isolated reports of this phenomenon in some animal viruses (Salomon and Littauer 1974; Lindley and Stebbing 1977). In at least some cases, the tRNA mimicry extends to interaction of the viral RNAs with other components including elongation factors (for review, see Giegé et al. 1998). These tRNA-like structures appear to be involved as origins in viral RNA replication, and in some but not all instances, aminoacylation is required (see Giegé 1996). Other functions for aminoacylation of viral RNA 3´ ends have also been proposed.

aa-RS-like Molecules

Although it is outside the scope of this review, it is clear that a growing number of aaRS-like molecules are important in regulation of gene expression and amino acid biosynthesis. Examples such as the GCN2 kinase (Hinnebusch 1997 and Chapter 5) and the HisRS-like protein involved in histidine biosynthesis (Sissler et al. 1999) attest to the ability of the cell to recruit aaRS motifs to perform multiple roles.

RIBOSOMAL PROTEINS

The bifunctional nature of certain ribosomal proteins is perhaps the most advanced example of the recruitment of components of the translational apparatus to other functions within the cell. Some of these functions have been found for other components, such as the aminoacyl tRNA-synthetases described above. Many examples, both well developed and still speculative, have been discussed recently in several excellent reviews (Wool 1996a,b; Wool et al. 1996; Nomura et al. 1984). Below are some representative examples of the many types of functions attributed to ribosomal proteins. Some ribosomal proteins show motifs that give clues to potential secondary functions, such as zinc finger, leucine zipper, and helix-turn-helix motifs and ubiquitin domains. In particular, similar functions have been found for prokaryotic and eukaryotic ribosomal proteins, indicating trends conserved throughout evolution.

Prokaryotic

S1 of E. coli *Is a Component of RNA Phage RNA-dependent RNA Replicases*

The RNA phage of *E. coli* (MS2 group and Qβ) encode an RNA-dependent RNA replicase protein, but this is only one subunit in the mature enzyme, which contains three additional host subunits, all derived from the translational apparatus: protein S1, from the 30S ribosomal subunit, and elongation factors EF1A and EF1B (also called EF-Tu and EF-Ts). Curiously, the roles of these translational components in the RNA replicase have some similarity to their roles in translation: S1 is only needed for initiation of RNA synthesis (Schuppli et al. 1998); EF-Tu and EF-Ts remain in the elongating replicase complex (Blumenthal et al. 1976), apparently needed to stabilize the complex (Blumenthal 1980). S1 has had a checkered history as a ribosomal protein. At ~60 kD, it is more than twice as large as all the other bacterial ribosomal proteins and was originally incorrectly thought to be dispensable for protein synthesis. It is now regarded as a kind of quasi-initiation factor (Moll et al. 1998), apparently leaving the ribosome after initiation (Laughrea and Moore 1977; for review, see Subramanian 1983).

S10 of E. coli *Is a Component of Transcription Antitermination Complexes*

Bacteriophage λ and host *E. coli*, both use transcription antitermination as regulatory mechanisms. Unique λ-specified proteins N and Q are transcription antiterminators in λ morphogenesis; in the case of N, signals in the transcript ("nut" sites) direct several host proteins ("nus" proteins) to join in an antitermination complex (for review, see Roberts 1993), including ribosomal protein S10 of the 30S subunit. This was first recognized when a mutant in a host factor required for N function turned out to have an altered S10 protein (Friedman et al. 1981). *E. coli* uses essentially the same transcript signals and set of proteins, including S10 but obviously without N, for antitermination of transcription of rRNA and some other genes (for review, see Condon et al. 1995). It is not known whether the S10 used in antitermination complexes is derived from degraded 30S subunits or from tightly regulated synthesis in excess of the required amounts for balanced assembly of new 30S particles. The *E. coli* system, lacking N protein, is somewhat less efficient in antitermination than the λ system, suggesting that λ evolved to take advantage of and improve upon the preexisting *E. coli* components.

L11 of E. coli *Is the relC Product, Required for Synthesis of ppGpp*

As discussed in the section "Bacterial stringent control" (above), guanosine tetraphosphate (ppGpp) is synthesized on ribosomes in large amounts during amino acid limitation. In addition to the enzyme responsible for this synthesis (stringent factor, the product of the *relA* gene), a ribosomal protein in the 50S subunit, L11, is required (Parker et al. 1976). Mutants of L11 are unable to make ppGpp during amino acid limitation; because they cannot mount a stringent response, these mutants are "relaxed" and the locus is also known as *relC*. The activity of L11 in this process apparently occurs while it is a part of the 50S subunit rather than a free molecule.

Many E. coli *Ribosomal Proteins Are Involved in Autoregulation, as Translational Repressors*

Messenger RNAs for many ribosomal proteins contain structures that mimic the structures to which the ribosomal proteins bind during ribosome biogenesis. When excess ribosomal proteins (for example, L10) bind to these structures in mRNA, they act as translational repressors, thereby preventing further production of ribosomal proteins (for review, see Nomura et al. 1984). Not all ribosomal proteins act as translational repressors, but many of the genes for ribosomal proteins are clustered in multigene operons, such that the reading frames for the respective ribosomal proteins are translationally coupled. Thus, translational repression of the first gene of the operon is sufficient to shut down translation of the entire operon.

E. coli Ribosomal Protein L4 Is Involved in Transcriptional Attenuation of the S10 Operon

In a similar theme to translational autoregulation, excess ribosomal protein L4 interacts with the transcription complex transcribing the leader RNA of the S10 operon (which includes the gene for L4). This facilitates an intrinsic (rho-independent) termination of transcription at the end of the leader RNA, whereas in the absence of L4 interaction, transcription of the operon continues (Zengel and Lindahl 1996; for review, see Zengel and Lindahl 1994).

Eukaryotic Ribosomal Proteins

Ribosomal Proteins and Viral Functions

Consistent with both prokaryotic ribosomes and many translational components, eukaryotic ribosomal proteins also interact with viral targets. The Epstein-Barr virus short noncoding EBER-1 RNA is associated with rpL22 (Toczyski et al. 1994). Unlike the previously described cases, however, it may be that this is a cellular response to prevent the EBER-1 RNA from inactivating PKR. The link between rpL22 function and PKR inhibition, as well as the oncogenic nature of this virus, dominant negative PKR mutants, and the observation that the rpL22 gene is a translocation site in some leukemia patients is an intriguing but ongoing question (Mathews 1996; Chapters 8 and 13).

Regulation of Gene Expression by Ribosomal Proteins

The classic example of ribosomal proteins regulating their own gene expression, originally shown in bacteria, has proven to be conserved in eukaryotes as well. Yeast has provided an excellent genetic system to both identify these functions and to test the mechanisms of action. At the transcriptional level, the *RPS20* (previously *URP2*) gene product shows a genetic interaction with RNA polymerase III as a high-copy suppressor (Denmat et al. 1994). Although the biochemical mechanism remains unknown, the resemblance to the *E. coli* S10 protein in sequence and function lends credence to the conservation of this regulatory function. Posttranscriptional regulation in yeast has taken advantage of the RNA-binding properties of ribosomal proteins to regulate mRNA splicing of the genes encoding rpL30 (formerly rpL32; Li et al. 1996), and rpS14 (Fewell and Woolford 1999), translation of the rpL30 mRNA (Li et al. 1996), and stability of the rpL4 (formerly L20) mRNA (Presutti et al. 1991). These regulatory mechanisms are further conserved in metazoans for rpL4 of *Xenopus* (Bozzoni et al. 1984) and rpS14 of humans (Tasheva and Roufa 1995).

Eukaryotic Ribosomal Protein Functions in Divergent Cellular Processes

Eukaryotic ribosomal proteins also appear to extend their roles to even more divergent functions. A large number of preliminary observations, including association with proteins of known function and genetic studies, indicates that our understanding of the breadth of these processes is just beginning. Ribosomal proteins have potential roles in processes such as DNA repair (mammalian S3 and *Drosophila* S3 and P0), transformation (p53 binding by rpL5-5S rRNA), development (such as in minute mutations of *Drosophila*), nuclear envelope disassembly (rpS8), bone induction (rpL32), and iron binding (P2) (for review, see Wool 1996a,b; Wool et al. 1996). Studies of the effects of alterations in rpS3a expression on transformation and apoptosis (Naora et al. 1998) and changes in ribosomal protein expression associated with some cancers (Hershey, Chapter 20) will continue to raise the question as to what effects are a direct consequence of new and nontranslational functions of the ribosomal proteins. Yeast genetic screens are further extending the processes that may be linked to specific ribosomal proteins. For example, rpL30 (formerly rpL32) was identified in a screen for factors that disrupt telomeric silencing when overexpressed (Singer et al. 1998), and a mutation in rpL9 affects mRNA turnover (Zuk et al. 1999). Although this provides new leads to potential roles for translation apparatus components in these pathways, more work is required to determine the mechanism by which these phenotypes occur and to rule out the possible indirect mechanisms that may result from altered gene expression when ribosomal subunits are limiting.

Ubiquitin Fusions of Ribosomal Proteins

Some eukaryotic ribosomal proteins donate a specific functional domain to the cell. The rpS31, rpL40A, and rpL40B are ubiquitin fusions in yeast (Finley et al. 1989), as is rpS27a in humans (Redman and Rechsteiner 1989). This adaptation benefits the individual ribosomal protein in its translational function, as the ubiquitin domain is required for efficient incorporation of the ribosomal protein tail into the ribosome and may aid in folding. The cell also receives a large pool of ubiquitin for conjugation to other proteins, since ribosomal proteins are highly expressed.

TRANSLATION FACTORS

A number of soluble protein factors function in the initiation and elongation of protein synthesis in prokaryotes and eukaryotes, (for review, see

Chapters 2, 3, 5, 6, and 10). Some of these factors, however, have roles which are either a link to other processes or perhaps novel and separate functions. Many observations were initially met with skepticism, especially for the highly abundant factors such as eEF1A, but continued work has demonstrated many of the interactions to be specific. Work is now addressing the functional implications of whether these interactions affect protein synthesis or perhaps define functions completely unrelated to the roles of these proteins in the translational process.

Initiation Factors

Roles of Initiation Factors in Postinitiation Events in Gene Expression

A classic example is the identification of mutations in several translation initiation factors that alter mRNA decay. Examples such as mutations in the Sui1p subunit of eIF3 (Cui et al. 1999) and eIF5A (Zuk and Jacobson 1998), which are discussed along with others by Jacobson and Peltz (Chapter 29), link processes expected to show some cross-communication. The *mof2-1* allele of *SUI1*, however, also affects programmed ribosomal frameshifting, and may link not only mRNA decay and the maintenance of the translational reading frame, but also perhaps both the initiation and elongation steps of protein synthesis (Cui et al. 1998).

Multiple Functions Assigned to eIF3

Initiation factor eIF3, the largest of the mammalian initiation factors, has multiple subunits contributing to alternative functions including yeast p16 (*SUI1/MOF2*) described above. The mammalian eIF3 complex consists of at least 11 subunits (see Chapter 2 Hershey and Merrick), many of which contain sequence motifs conserved in other proteins, particularly in the multisubunit COP9 signalosome complex and 19S regulatory particle (RP) of the proteasome. The mammalian p36 subunit is a WD40 repeat protein identical to TRIP-1, a substrate for phosphorylation by the TGF-β type II receptor (Asano et al. 1997a). The p40 and p47 subunits share the MPN motif found in the 2 Mov-34 family member subunits of the RP and 2 COP9 subunits (Asano et al. 1997b; Aravind and Ponting 1998; Glickman et al. 1998; Hofmann and Bucher 1998; Wei et al. 1998). The p48, p110, and p170 subunits show PCI motifs (*p*roteasome-*C*OP9-*I*nitiation motifs), also found in 6 COP9 subunits and 5 RP subunits

(Aravind and Ponting 1998; Glickman et al. 1998, Hofmann and Bucher 1998; Wei et al. 1998). The finding that three multiprotein complexes share significant sequence conservation may reflect a shared evolutionary ancestry or motifs common to large multiprotein complexes. Further work is required to determine whether other hypotheses that these complexes may interact with similar substrates or that eIF3 and COP9 may also interact with the 20S core proteasome complex are correct and may define a broader biological role for these complexes.

Multiple Functions Assigned to eIF5A

The eIF5A protein, in addition to its effects on mRNA decay noted above, has been implicated in a wide range of functions which may, however, share a link to nuclear transport events. eIF5A is one of several proteins proposed to function as a cellular cofactor for HIV-1 Rev (Bevec et al. 1996). This may be related to other studies that have linked eIF5A function to nuclear transport. *TIF51A* gene was identified as a high-copy suppressor of a TATA-binding protein mutation in a screen that also yielded Kap114, an importin/karyopherin protein (Morehouse et al. 1999). Localization of eIF5A at nuclear-preassociated intranuclear filaments in mammalian cells and the interaction with the CRM1 nuclear export receptor also fit with these findings (Elfgang et al. 1999; Rosorius et al. 1999). Furthermore, following retinoic acid stimulation, tissue transglutaminase II is activated to bind membranes and increase lipid turnover as well as bind eIF5A (Singh et al. 1998). It remains to be determined how these functions are linked to each other and to proposed functions of eIF5A in protein synthesis.

Metazoan-specific Functions for eIF6

eIF6 is another factor potentially involved in multiple functions. Human eIF6 was simultaneously identified as a factor that affects 40S and 60S subunit association (Si et al. 1997) and as a β4 integrin-binding protein (Biffo et al. 1997). Characterization of the yeast homolog demonstrated that loss of eIF6/p27BBP function affects 60S ribosomal subunit biogenesis, whereas the human protein shows nucleolar localization and association with the basal membrane (Sanvito et al. 1999; Si and Maitra 1999). The lack of β4 integrin in yeast has led to the hypothesis that metazoans may have adapted eIF6 to a unique function.

Elongation Factors

Elongation Factor Function in Phage and Viral Replication

As described above in "S1 of *E. coli* is a component of RNA phage RNA-dependent RNA replicases," EF1A (formerly EF-Tu) and EF1B (formerly EF-Ts) are also components of RNA phage RNA-dependent RNA replicases. It is perhaps a conserved association, since it has now been demonstrated that eukaryotic eEF1A (formerly EF-1α) binds the vesicular stomatitis virus RNA polymerase, an association that is enhanced by the eEF1Bα (formerly EF-1β) and eEF1Bγ (formerly EF-1γ) subunits (Das et al. 1998). The major head protein of bacteriophage T4 binds EF1A, which is proposed to play a dual role in head assembly and triggering inactivation of EF1A (Bingham et al. 2000). eEF1A is also associated with the HIV type I gag polyprotein, an interaction that may assist in the release of viral RNA from polyribosomes for packaging (Cimarelli and Luban 1999) and the 3′stem-loops of West Nile virus and turnip yellow mosaic virus RNAs (Joshi et al. 1986; Blackwell and Brinton 1997). It should be noted that eEF1A has also been found in complexes with several cellular nucleic acids as in mRNP complexes (Greenberg and Slobin 1987), in association with the eEF1Bαβγ subunits in a tRNA–vigilin complex (Kruse et al. 1998), and in recognition of chromium and transplatin-damaged DNA (Wang et al. 1997). The eEF1Bβ subunit (formerly EF-1δ) has also been reported to associate with viral factors such as the herpes simplex virus 1 α regulatory protein ICP0 (Kaxaguchi et al. 1997) and the second coding intron of HIV-1 Tat (Xiao et al. 1998). Since the function of eEF1Bβ itself remains less well understood and appears metazoan-specific, the consequences of these associations may reflect either novel or translation-related functions.

eEF1A Interactions with Cytoskeletal Components

The initial observation that *Dictyostelium* eEF1A can bind and bundle actin is a classic example of a unique secondary function of the translational apparatus (Yang et al. 1990; for review, see Condeelis 1995). Since that time, this interaction has been carefully dissected to determine the effect of eEF1A on actin polymerization (Murray et al. 1996) and the effect of the association with actin to inhibit both aa-tRNA binding and nucleotide hydrolysis (Liu et al. 1996; Edmonds et al. 1998). Although the full implications of this interaction are still not known, several observations support its physiological relevance. First, the translational apparatus is clearly subject to a higher-order structure, and actin plays a role in maintaining efficient protein synthesis (Stapulionis et al. 1997). Second, yeast eEF1A also interacts with Bni1p, a downstream target of Rho1p that can regulate reorganization of the actin cytoskeleton (Umikawa et al. 1998). Third, in *S. cerevisiae*, overexpression of eEF1A (Carr-Schmid et al. 1999), like overexpression of actin and known actin-binding proteins such as Fimbrin (Sac6p) and ABP1 (Liu et al. 1992; Sandrock et al. 1999), severely inhibits cell growth. In *Schizosaccharomyces pombe*, higher-level overexpression of eEF1A is lethal and affects cell morphology; however, growth is restored by co-overexpression of actin (Suda et al. 1999). Further work is required to understand the functional implications of these observations. Reports of eEF1A both severing (Shiina et al. 1994) and stabilizing (Moore et al. 1998) microtubules offer additional possible links to the cytoskeleton, although the apparent contradictory nature of these findings will require more analysis. Additionally, the reported association of eEF1A with the mitotic apparatus may prove another structural or function role (Ohta et al. 1990). Since the eEF1Bγ subunit is also a tubulin-binding protein (Janssen and Moller 1988), multiple subunits of eEF1 may have a cytoskeletal connection.

Elongation Factor Roles in Protein Folding and Degradation

Prokaryotic EF1A (EF-Tu) has been shown to have potential chaperone-like activity on several substrates (Kudlicki et al. 1997; Caldas et al. 1998). At the other extreme, eukaryotic eEF1A was identified as a factor involved in ubiquitin-dependent degradation of some proteins, a function that could also be performed by EF1A (Gonen et al. 1994). It is interesting to speculate how these functions could occur cotranslationally or sense cellular stress. Several other protein–protein interactions involving eEF1A include calmodulin binding and activation of several protein kinases and may be another mechanism to sense or respond to cellular conditions (Yang et al. 1993; Durso and Cyr 1994; Kaur and Ruben 1994; Yang and Boss 1994; Wang and Poovaiah 1999).

NUCLEAR LOCALIZATION OF TRANSLATIONAL COMPONENTS

In the course of studies of the translational apparatus, reports have surfaced of nuclear localization of translation initiation (Lejbkowicz et al. 1992; Lang et al. 1994; Lobo et al. 1997) and elongation (Minella et al. 1996; Sanders et al. 1996) factors and aminoacyl-tRNA synthetases (Kisselev and Wolfson 1994; Popenko et al. 1994). Whereas ribosomal proteins and tRNAs spend a portion of their lives becoming activated and assembled in the nucleus, initiation and elongation factors as well as aminoacyl-tRNA synthetases were considered mislocalized. However, several recent studies have led to a new consideration of the implications of nuclear localization of these proteins. eEF1A has been shown to bind the Zpr1p protein, a zinc finger protein translocated from the cytoplasm to the nucleus upon treatment of cells with mitogens (Gangwani et al. 1998). This provides a new framework in which to consider the possibility that translational components are recruited to the nucleus under specific conditions to perform as yet unknown roles in the cell. It has recently been demonstrated that tRNAs are aminoacylated in *Xenopus* oocyte nuclei, a process that may serve as a "quality control" point to proofread tRNA processing (Arts et al. 1998; Lund and Dahlberg 1998). Furthermore, many *S. cerevisiae* aaRSs have classic nuclear-localization signals (Schimmel and Wang 1999). The aa-tRNA-binding protein eEF1A also appears to affect tRNA export. Overexpression of eEF1A in yeast suppresses the lethality of a synthetic lethal strain lacking the *LOS1* gene product (the yeast Xpo-t homolog), and mutations in eEF1A can also affect tRNA export (Grosshans et al. 2000). Although models put forth by the authors do not make this an obligatory step, it stimulates an attractive extension of the channeling of the protein synthesis machinery (Stapulionis and Deutscher 1995) and will continue to extend the growing lists of nontranslational functions of the protein synthetic apparatus.

ACKNOWLEDGMENTS

We thank Anne Carr-Schmid, Raj Munshi, and Louis Valente for comments on the manuscript and John L. Woolford Jr. and William C. Merrick for helpful discussions. T.G.K. is supported by a grant from the National Institutes of Health (RO1 GM-57483). E.G. was supported by grants from the National Science Foundation (MCB-9513127 and -9981879).

REFERENCES

Aravind L. and Ponting C.P. 1998. Homologues of 26S proteasome subunits are regulators of transcription and translation. *Protein Sci.* **7:** 1250–1254.

Arts E.J. and Le Grice S.F. 1998. Interaction of retroviral reverse transcriptase with template-primer duplexes during replication. *Prog. Nucleic Acid Res. Mol. Biol.* **58:** 339–393.

Arts G.-J., Kuersten S., Romby P., Ehresmann B., and Mattaj I.W. 1998. The role of exportin-t in selective nuclear export of mature tRNAs. *EMBO J.* **17:** 7430–7441.

Asano K., Kinzy T.G., Merrick W.C., and Hershey J.W.B. 1997a. Conservation and diversity of eukaryotic translation initiation factor eIF3. *J. Biol. Chem.* **272:** 1101–1109.

Asano K., Vornlocher H.P., Richter-Cook N.J., Merrick W.C., Hinnebusch A.G., and Hershey J.W.B. 1997b. Structure of cDNAs encoding human eukaryotic initiation factor 3 subunits. Possible roles in RNA binding and macromolecular assembly. *J. Biol. Chem.* **272:** 27042–27052.

Bevec D., Jaksche H., Oft M., Wohl T., Himmelspach M., Pacher A., Schebesta M., Koettnitz K., Dobrovnik M., Csonga R., Lottspeich F., and Hauber J. 1996. Inhibition of HIV-1 replication in lymphocytes by mutants of the rev cofactor eIF-5A. *Science* **271:** 1858–1860.

Biffo S., Sanvito F., Costa S., Preve L., Pignatelli R., Spindardi L., and Marchisio P.C. 1997. Isolation of a novel beta4 integrin-binding protein (p27(BBP)) highly expressed in epithelial cells. *J. Biol. Chem.* **272:** 30314–30321.

Bingham R., Ekunwe S.I., Falk S., Snyder L., and Kleanthous C. 2000. The major head protein of bacteriophage T4 binds specifically to elongation factor Tu. *J. Biol. Chem.* **275:** (in press.)

Blackwell J.L. and Brinton M.A. 1997. Translation elongation factor-1 alpha interacts with the 3′ stem-loop region of West Nile virus genomic RNA. *J. Virol.* **71:** 6433–6444.

Blumenthal T. 1980. Interaction of host-coded and virus-coded polypeptides in RNA phage replication. *Proc. R. Soc. Lond. B Biol. Sci.* **210:** 321–335.

Blumenthal T., Young R.A., and Brown S. 1976. Function and structure in phage Qβ RNA replicase. Association of EF-Tu-Ts with the other enzyme subunits. *J. Biol. Chem.* **251:** 2740–2743.

Bozzoni I., Fragapane P., Annesi F., Pierandrei-Amaldi P., Amaldi F., and Beccari E. 1984. Expression of two *Xenopus laevis* ribosomal protein genes in injected frog oocytes: A specific splicing block interferes with the L1 RNA maturation. *J. Mol. Biol.* **180:** 987–1005.

Caldas T.D., El Yaagoubi A., and Richarme G. 1998. Chaperone properties of bacterial elongation factor EF-Tu. *J. Biol. Chem.* **273:** 11478–11482.

Carr-Schmid A., Valente L., Loik V.I., Williams T., Starita L.M., and Kinzy T.G. 1999. Mutations in elongation factor 1β, a guanine nucleotide exchange factor, enhance translational fidelity. *Mol. Cell. Biol.* **19:** 5257–5266.

Cashel M., Gentry D.R., Hernandez V.J., and Vinella D. 1996. The stringent response. In Escherichia coli *and* Salmonella: *Cellular and molecular biology*, 2nd edition (ed. F.C. Neidhardt et al.), pp. 1458–1496. American Society for Microbiology, Washington, D.C.

Cimarelli A. and Luban J. 1999. Translation elongation factor 1-alpha interacts specifically with the human immunodeficiency virus type 1 gag polyprotein. *J. Virol.* **73:** 5388–5401.

Condeelis J. 1995. Elongation factor 1α, translation and the cytoskeleton. *Trends Biochem. Sci.* **20:** 169–170.

Condon C., Grunberg-Manago M., and Putzer H. 1996. Aminoacyl-tRNA synthetase gene regulation in *Bacillus subtilis*. *Biochimie* **78:** 381–389.

Condon C., Squires C., and Squires C.L. 1995. Control of rRNA transcription in *Escherichia coli*. *Microbiol. Rev.* **59:** 623–645.

Cui Y., Dinman J.D., Kinzy T.G., and Peltz S.W. 1998. The Mof2/Sui1 protein is general monitor of translational accuracy. *Mol. Cell. Biol.* **18:** 1506–1516.

Cui Y., Gonzales C.I., Kinzy T.G., Dinman J.D., and Peltz S.W. 1999. Mutations in the *MOF2/SUI1* gene affect both translation and nonsense-mediated mRNA decay. *RNA* **5:** 794–804.

Das T., Mathur M., Gupta A.K., Janssen G.M., and Banerjee A.K. 1998. RNA polymerase of vesicular stomatitis virus specifically associates with translation elongation factor-1 αβγ for its activity. *Proc. Natl. Acad. Sci.* **95:** 1449–1454.

Denmat S.H.-L., Sipiczki M., and Thuriaux P. 1994. Suppression of yeast RNA polymerase II mutations by the *URP2* gene encoding a protein homologous to the mammalian ribosomal protein S20. *J. Mol. Biol.* **240:** 1–7.

Deutch C.E., Scarpulla R.C., and Soffer R.L. 1978. Posttranslational NH2-terminal aminoacylation. *Curr. Top. Cell Regul.* **13:** 1–28.

Dujardin G. and Herbert C.J. 1997. Aminoacyl tRNA synthetases involved in group I intron splicing. In *Ribosomal RNA and group I introns* (ed. R. Green and R. Schroeder), pp. 179–198. R.G. Landes, Austin, Texas.

Dunstan H.M., Young L.S., and Sprague K.U. 1994a. TFIIIR is an isoleucine tRNA. *Mol. Cell. Biol.* **14:** 3588–3595.

———. 1994b. tRNA(IleIAU) (TFIIIR) plays an indirect role in silkworm class III transcription in vitro and inhibits low-frequency DNA cleavage (erratum in *Mol. Cell. Biol.* [1994] **14:** 5617). *Mol. Cell. Biol.* **14:** 3596–3603.

Durso N.A. and Cyr R.J. 1994. A calmodulin-sensitive interaction between microtubules and a higher plant homolog of elongation factor-1 alpha. *Plant Cell* **6:** 893–905.

Edmonds B.T., Bell A., Wyckoff J., Condeelis J., and Leyh T.S. 1998. The effect of F-actin on the binding and hydrolysis of guanine nucleotide by *Dictyostelium* elongation factor 1A. *J. Biol. Chem.* **273:** 10288–10295.

Elfgang C., Rosorius O., Hofer L., Jaksche H., Hauber J., and Bevec D. 1999. Evidence for specific nucleocytoplasmic transport pathways used by leucine-rich nuclear export signals. *Proc. Natl. Acad. Sci.* **96:** 6229–6234.

Fewell S.W. and Woolford J.L. 1999. Ribosomal protein S14 of *Saccharomyces cerevisiae* regulates its expression by binding to *RPS14B* pre-mRNA and to 18S rRNA. *Mol. Cell. Biol.* **19:** 826–834.

Finley D., Bartel B., and Varshavsky A. 1989. The tails of ubiquitin precursors are ribosomal proteins whose fusion to ubiquitin facilitates ribosome biogenesis. *Nature* **338:** 394–401.

Florentz C. and Giegé R. 1995. tRNA-like structures in plant viral RNAs. In *tRNA: Structure, biosynthesis, and function* (ed. D. Söll and U.L. RajBhandary), pp. 141–163. ASM Press, Washington, D.C.

Friedman D.I., Schauer A.T., Baumann M.R., Baron L.S., and Adhya S.L. 1981. Evidence that ribosomal protein S10 participates in control of transcription termination. *Proc. Natl. Acad. Sci.* **78:** 1115–1118.

Gallant J.A. 1979. Stringent control in *E. coli*. *Annu. Rev. Genet.* **13:** 393–415.

Gangwani L., Mikrut M., Galcheva-Gargova Z., and Davis R.J. 1998. Interaction of ZPR1 with translation elongation factor-1α in proliferating cells. *J. Cell Biol.* **143:** 1471–1484.

Gentry D.R. and Cashel M. 1995. Cellular localization of the *Escherichia coli SpoT* protein. *J. Bacteriol.* **177:** 3890–3893.

Giegé R. 1996. Commentary: Interplay of tRNA-like structures from plant viral RNAs with partners of the translation and replication machineries. *Proc. Natl. Acad. Sci.* **93:** 12078–12081.

Giegé R., Frugier M., and Rudinger J. 1998. tRNA mimics (erratum in *Curr. Opin. Struct. Biol.* [1998] **8:** 812). *Curr. Opin. Struct. Biol.* **8:** 286–293.

Glickman M.H., Rusin D.M., Coux O., Wefes I., Pfeifer G., Cjeka Z., Baumeister W., Fried V.A., and Finley D. 1998. A subcomplex of the proteasome regulatory particle required for ubiquitin-conjugate degradation and related to the COP9-signalosome and eIF3. *Cell* **94:** 615–623.

Goldman E. and Jakubowski H. 1990. Uncharged tRNA, protein synthesis, and the bacterial stringent response. *Mol. Microbiol.* **4 :** 2035–2040.

Gonda D.K., Bachmair A., Wunning I., Tobias J.W., Lane W.S., and Varshavsky A. 1989. Universality and structure of the N-end rule. *J. Biol. Chem.* **264:** 16700–16712.

Gonen H., Smith C.E., Siegel N.R., Kahana C., Merrick W.C., Chakraburtty K., Schwartz A.L., and Ciechanover A. 1994. Protein synthesis elongation factor EF-1α is essential for ubiquitin-dependent degradation of certain N^{α}-acetylated proteins and may be substituted for by the bacterial elongation factor EF-Tu. *Proc. Natl. Acad. Sci.* **91:** 7649–7652.

Greenberg J.R. and Slobin L.I. 1987. Eukaryotic elongation factor Tu is present in mRNA-protein complexes. *FEBS Lett.* **224:** 54–58.

Grosshans H., Hurt E., and Simos G. 2000. An aminoacylation-dependent nuclear tRNA export pathway in yeast. *Genes Dev.* **14:** 830–840.

Haenni A.L., Joshi S., and Chapeville F. 1982. tRNA-like structures in the genomes of RNA viruses. *Prog. Nucleic Acid Res. Mol. Biol.* **27:** 85–104.

Haseltine W.A. and Block R. 1973. Synthesis of guanosine tetra and pentaphosphate requires the presence of a codon-specific, uncharged transfer ribonucleic acid in the acceptor site of ribosomes. *Proc. Natl. Acad. Sci.* **70:** 1564–1568.

Henkin T.M. 1994. tRNA-directed transcription antitermination. *Mol. Microbiol.* **13:** 381–387.

———. 1996. Control of transcription termination in prokaryotes. *Annu. Rev. Genet.* **30:** 35–57.

Hinnebusch A.G. 1997. Translational regulation of yeast *GCN4*. *J. Biol. Chem.* **272:** 21661–21664.

Hofmann K. and Bucher P. 1998. The PCI domain: A common theme in three multiprotein complexes. *Trends Biochem. Sci.* **23:** 204–205.

Jahn D., Verkamp E., and Söll D. 1992. Glutamyl-transfer RNA: A precursor of heme and chlorophyll biosynthesis. *Trends Biochem. Sci.* **17:** 215–218.

Jakubowski H. 1990. Proofreading in vivo: Editing of homocysteine by methionyl-tRNA synthetase in *Escherichia coli*. *Proc. Natl. Acad. Sci.* **87:** 4504–4508.

———. 1991. Proofreading in vivo: Editing of homocysteine by methionyl-tRNA synthetase in the yeast *Saccharomyces cerevisiae*. *EMBO J.* **10:** 593–598.

———. 1997. Metabolism of homocysteine thiolactone in human cell cultures. Possible mechanism for pathological consequences of elevated homocysteine levels. *J. Biol. Chem.* **272:** 1935–1942.

———. 1999. Protein homocysteinylation: Possible mechanism underlying pathological consequences of elevated homocysteine levels. *FASEB. J.* **13:** 2277–2283.

Jakubowski H. and Goldman E. 1992. Editing of errors in selection of amino acids for protein synthesis. *Microbiol. Rev.* **56:** 412–429.

———. 1993. Synthesis of homocysteine thiolactone by methionyl-tRNA synthetase in cultured mammalian cells. *FEBS Lett.* **317:** 237–240.

Janssen G.M. and Moller W. 1988. Elongation factor 1$\beta\gamma$ from *Artemia*: Purification and properties of its subunits. *Eur. J. Biochem.* **171:** 119–129.

Joshi R.L., Ravel J.M., and Haenni A. 1986. Interaction of turnip yellow mosaic virus Val-tRNA with eukaryotic elongation factor EF-1α. Search for a function. *EMBO J.* **5:** 1143–1148.

Kaur K.J. and Ruben L. 1994. Protein translation elongation factor-1 alpha from

Trypanosoma brucei binds calmodulin. *J. Biol. Chem.* **269:** 23045–23050.

Kaxaguchi Y., Bruni R., and Roizman B. 1997. Interaction of herpes simplex virus 1 α regulatory protein ICP0 with elongation factor 1δ: ICP0 affects translational machinery. *J. Virol.* **71:** 1019–1024.

Ke N., Gao X., Keeney J.B., Boeke J.D., and Voytas D.F. 1999. The yeast retrotransposon Ty5 uses the anticodon stem-loop of the initiator methionine tRNA as a primer for reverse transcription. *RNA* **5:** 929–938.

Kisselev L.L. and Wolfson A.D. 1994. Aminoacyl-tRNA synthetases from higher eukaryotes. *Prog. Nucleic Acid Res. Mol. Biol.* **48:** 83–142.

Kruse C., Grunweller A., Willkomm D.K., Pfeiffer T., Hartmann R.K. and Muller P.K. 1998. tRNA is entrapped in similar, but distinct, nuclear and cytoplasmic ribonucleoprotein complexes, both of which contain vigilin and elongation factor 1 alpha. *Biochem. J.* **329:** 615–621.

Kudlicki W., Coffman A., Kramer G., and Hardesty B. 1997. Renaturation of rhodanese by translation elongation factor (EF) Tu. Protein refolding by EF-Tu flexing. *J. Biol. Chem.* **272:** 32206–32210.

Lamond A.I. and Travers A.A. 1985. Stringent control of bacterial transcription. *Cell* **41:** 6–8.

Lang V., Zanchin N.I., Lunsdorf H., Tuite M., and McCarthy J.E. 1994. Initiation factor eIF-4E of *Saccharomyces cerevisiae*. Distribution within the cell, binding to mRNA, and consequences of its overproduction. *J. Biol. Chem.* **269:** 6117–6123.

Laughrea M. and Moore P.B. 1977. Physical properties of ribosomal protein S1 and its interaction with the 30 S ribosomal subunit of *Escherichia coli*. *J. Mol. Biol.* **112:** 399–421.

Lechler A. and Kreutzer R. 1998. The phenylalanyl-tRNA synthetase specifically binds DNA. *J. Mol. Biol.* **278:** 897–901.

Lejbkowicz F., Goyer C., Darveau A., Neron S., Lemieux R., and Sonenberg N. 1992. A fraction of the mRNA 5′ cap-binding protein, eukaryotic initiation factor 4E, localizes to the nucleus. *Proc. Natl. Acad. Sci.* **89:** 9612–9616.

Li B., Vilardell J., and Warner J.R. 1996. An RNA structure involved in feedback regulation of splicing and of translation is critical for biological fitness. *Proc. Natl. Acad. Sci.* **93:** 1596–1600.

Lindley I.G. and Stebbing N. 1977. Aminoacylation of encephalomyocarditis virus RNA. *J. Gen. Virol.* **34:** 177–182.

Litvak S., Sarih-Cottin L., Fournier M., Andreola M., and Tarrago-Litvak L. 1994. Priming of HIV replication by tRNA(Lys3): Role of reverse transcriptase. *Trends Biochem. Sci.* **19:** 114–118.

Liu G., Tang J., Edmonds B.T., Murray J., Levin S., and Condeelis J. 1996. F-actin sequesters elongation factor 1α from interaction with aminoacyl-tRNA in a pH-dependent reaction. *J. Cell Biol.* **135:** 953–963.

Liu H., Krizek J., and Bretscher A. 1992. Construction of a *GAL1*-regulated yeast cDNA expression library and its application to the identification of genes whose overexpression causes lethality in yeast. *Genetics* **132:** 665–673.

Lobo M.V., Alonso F.J., Rodriguez S., Alcazar A., Martin E., Munoz F., Santander R.G., Salinas M., and Fando J.L. 1997. Localization of eukaryotic initiation factor 2 in neuron primary cultures and established cell lines. *Histochem. J.* **29:** 453–468

Lund E. and Dahlberg J.E. 1998. Proofreading and aminoacylation of tRNAs before export from the nucleus. *Science* **282:** 2082–2085.

Marquet R., Isel C., Ehresmann C., and Ehresmann B. 1995. tRNAs as primer of reverse transcriptases. *Biochimie* **77:** 113–124.

Mathews M.B. 1996. Interactions between viruses and the cellular machinery for protein synthesis. In *Translational control* (ed. J.W.B. Hershey et al.), pp. 505–548. Cold Spring Harbor Laboratory Press, Cold Spring Harbor, New York.

Minella O., Mulner-Lorillon O., De Smedt V., Hourdez S., Cormier P., and Belle R. 1996. Major intracellular localization of elongation factor-1. *Cell. Mol. Biol.* **42:** 805–810.

Moll I., Resch A., and Blasi U. 1998. Discrimination of 5′-terminal start codons by translation initiation factor 3 is mediated by ribosomal protein S1. *FEBS Lett.* **436:** 213–217.

Moore R.C., Durso N.A., and Cyr R.J. 1998. Elongation factor-1alpha stabilizes microtubules in a calcium/calmodulin-dependent manner. *Cell Motil. Cytoskelet.* **41:** 168–180.

Morehouse H., Buratowski R.M., Silver P.A., and Buratowski S. 1999. The importin/karyopherin Kap114 mediates the nuclear import of TATA-binding protein. *Proc. Natl. Acad. Sci.* **96:** 12542–12547.

Murphy H.S. and Humayun M.Z. 1997. *Escherichia coli* cells expressing a mutant glyV (glycine tRNA) gene have a UVM-constitutive phenotype: Implications for mechanisms underlying the mutA or mutC mutator effect (erratum in *J. Bacteriol.* [1998] **180:** 2271). *J. Bacteriol.* **179:** 7507–7514.

Murray K.D. and Bremer H. 1996. Control of *spoT*-dependent ppGpp synthesis and degradation in *Escherichia coli*. *J. Mol. Biol.* **259:** 41–57.

Murray J.W., Edmonds B.T., Liu G., and Condeelis J. 1996. Bundling of actin filaments by elongation factor 1α inhibits polymerization at filament ends. *J. Cell Biol.* **135:** 1309–1321.

Naora H., Takai I., Adachi M., and Naora H. 1998. Altered cellular responses by varying expression of a ribosomal protein gene: Sequential coordination of enhancement and suppression of ribosomal protein S3a gene expression induces apoptosis. *J. Cell Biol.* **141:** 741–753.

Nomura M., Gourse R., and Baughman G. 1984. Regulation of the synthesis of ribosomes and ribosomal components. *Annu. Rev. Biochem.* **53:** 75–117.

Ohta K., Toriyama M., Miyazaki M., Murofushi H., Hosoda S., Endo S., and Sakai H. 1990. The mitotic apparatus-associated 51-kDa protein from sea urchin eggs is a GTP-binding protein and is immunologically related to yeast polypeptide elongation factor 1 alpha. *J. Biol. Chem.* **265:** 3240–3247.

Park S.G., Jung K.H., Lee J.S., Jo Y.J., Motegi H., Kim S., and Shiba K. 1999. Precursor of pro-apoptotic cytokine modulates aminoacylation activity of tRNA synthetase. *J. Biol. Chem.* **274:** 16673–16676.

Parker J., Watson R.J., and Friesen J.D. 1976. A relaxed mutant with an altered ribosomal protein L11. *Mol. Gen. Genet.* **144:** 111–114.

Popenko V.I., Ivanova J.L., Cherny N.E., Filonenko V.V., Beresten S.F., Wolfson A.D., and Kisselev L.L. 1994. Compartmentalization of certain components of the protein synthesis apparatus in mammalian cells. *Eur. J. Cell Biol.* **65:** 60–69.

Presutti C., Caifre S.-A., and Bozzoni I. 1991. The ribosomal protein L2 in *S. cerevisiae* controls the level of accumulation of its own mRNA. *EMBO J.* **10:** 2215–2221.

Putney S.D. and Schimmel P. 1981. An aminoacyl-tRNA synthetase binds to a specific DNA sequence and regulates its gene expression. *Nature* **306:** 441–447.

Putzer H., Grunberg-Manago M., and Springer M. 1995. Bacterial aminoacyl-tRNA synthetases: Genes and regulation of expression. In *tRNA: Structure, biosynthesis and function* (ed. D. Söll and U. RajBhandary), pp. 293–333. ASM Press, Washington, D.C.

Quevillon S., Agou F., Robinson J.C., and Mirande M. 1997. The p43 component of the mammalian multi-synthetase complex is likely to be the precursor of the endothelial monocyte-activating polypeptide II cytokine. *J. Biol. Chem.* **272:** 32573–32579.

Redman K.L. and Rechsteiner M. 1989. Identification of the long ubiquitin extension as ribosomal protein S27a. *Nature* **338:** 438–440.

Ren L., Al Mamun A.A., and Humayun M.Z. 1999. The mutA mistranslator tRNA-induced mutator phenotype requires *recA* and *recB* genes, but not the derepression of *lexA*-regulated functions. *Mol. Microbiol.* **32:** 607–615.

Roberts J.W. 1993. RNA and protein elements of *E. coli* and lambda transcription antitermination complexes. *Cell* **72:** 653–655.

Rojiani M.V., Jakubowski H., and Goldman E. 1989. Effect of variation of charged and uncharged tRNA(Trp) levels on ppGpp synthesis in *E. coli*. *J. Bacteriol.* **171:** 6493–6502.

Rosorius O., Reichart B., Kratzer F., Heger P., Dabauvalle M.C., and Hauber J. 1999. Nuclear pore localization and nucleocytoplasmic transport of eIF-5A; evidence for direct interaction with the export receptor CRM1. *J. Cell Sci.* **112:** 2369–2380.

Salomon R. and Littauer U.Z. 1974. Enzymatic acylation of histidine to mengovirus RNA. *Nature* **249:** 32–34.

Sanders J., Brandsma M., Janssen G.M., Dijk J., and Moller W. 1996. Immunofluorescence studies of human fibroblasts demonstrate the presence of the complex of elongation factor-1 beta gamma delta in the endoplasmic reticulum. *J. Cell Sci.* **109:** 1113–1117.

Sandrock T.M., Brower S.M., Toenjes K.A., and Adams A.E.M. 1999. Suppressor analysis of fimbrin (Sac6p) overexpression in yeast. *Genetics* **151:** 1287–1297.

Sankaranarayanan R., Dock-Bregeon A.C., Romby P., Caillet J., Springer M., Rees B., Ehresmann C., Ehresmann B., and Moras D. 1999. The structure of threonyl-tRNA synthetase-tRNA(Thr) complex enlightens its repressor activity and reveals an essential zinc ion in the active site. *Cell* **97:** 371–381.

Sanvito F., Piatti S., Villa A., Bossi M., Lucchini G., Marchisio P.C., and Biffo S. 1999. The β4 integrin interactor $p27^{BBP/eIF6}$ is an essential nuclear matrix protein involved in 60S ribosomal subunit assembly. *J. Cell Biol.* **144:** 823–837.

Schimmel P. and Wang C.-C. 1999. Getting tRNA synthetases into the nucleus. *Trends Biochem. Sci.* **24:** 127–128.

Schuppli D., Miranda G., Qiu S., and Weber H. 1998. A branched stem-loop structure in the M-site of bacteriophage Qβ RNA is important for template recognition by Qβ replicase holoenzyme. *J. Mol. Biol.* **283:** 585–593.

Sen S., Zhou H., Ripmaster T., Hittelman W.N., Schimmel P., and White R.A. 1997. Expression of a gene encoding a tRNA synthetase-like protein is enhanced in tumorigenic human myeloid leukemia cells and is cell cycle stage- and differentiation-dependent. *Proc. Natl. Acad. Sci.* **94:** 6164–6169.

Shiina N., Gotoh Y., Kubomura N., Iwamatsu A., and Nishida E. 1994. Microtubule severing by elongation factor 1α. *Science* **266:** 282–285.

Si K. and Maitra U. 1999. The *Saccharomyces cerevisiae* homologue of mammalian translation initiation factor 6 does not function as a translation initiation factor. *Mol. Cell. Biol.* **19:** 1416–1426.

Si K., Chaudhuri J., Chevesich J., and Maitra U. 1997. Molecular cloning and functional expression of a human cDNA encoding translation initiation factor 6. *Proc. Natl. Acad. Sci.* **94:** 14285–14290.

Singer M.S., Kahana A., Wolf A.J., Meisinger L.L., Peterson S.E., Goggin C., Mahowald M., and Gottschling D.E. 1998. Identification of high-copy disruptors of telomeric silencing in *Saccharomyces cerevisiae*. *Genetics* **150:** 613–632.

Singh U.S., Li Q., and Cerione R. 1998. Identification of the eukaryotic initiation factor 5A as a retinoic acid-stimulated cellular binding partner for tissue transglutaminase II. *J. Biol. Chem.* **273:** 1946–1950.

Sissler M., Delorme C., Bond J., Ehrlich S.D., Renault P., and Francklyn C. 1999. An aminoacyl-tRNA synthetase paralog with a catalytic role on histidine biosynthesis. *Proc. Natl. Acad. Sci.* **96:** 8985–8990.

Slupska M.M., Baikalov C., Lloyd R., and Miller J.H. 1996. Mutator tRNAs are encoded by the *Escherichia coli* mutator genes *mutA* and *mutC:* A novel pathway for mutagenesis. *Proc. Natl. Acad. Sci.* **93:** 4380–4385.

Soffer R.L. 1974. Aminoacyl-tRNA transferases. *Adv. Enzymol.* **40:** 91–139.

Springer M., Plumbridge J.A., Butler J.S., Graffe M., Dondon J., Mayaux J.F., Fayat G., Lestienne P., Blanquet S., and Grunberg-Manago M. 1985. Autogenous control of *Escherichia coli* threonyl-tRNA synthetase expression *in vivo*. *J. Mol. Biol.* **185:** 93–104.

Stapulionis R. and Deutscher M.P. 1995. A channeled tRNA cycle during mammalian protein synthesis. *Proc. Natl. Acad. Sci.* **92:** 7158–7161.

Stapulionis R., Kolli S., and Deutscher M.P. 1997. Efficient mammalian protein synthesis requires an intact F-actin system. *J. Biol. Chem.* **272:** 24980–24986.

Strominger J.L. 1969. Penicillin-sensitive enzymatic reactions in bacterial cell wall synthesis. *Harvey Lect.* **64:** 179–213.

Subramanian A.R. 1983. Structure and functions of ribosomal protein S1. *Prog. Nucleic Acid Res. Mol. Biol.* **28:** 101–142.

Suda M., Fukui M., Sogabe Y., Sate K., Morimatsu A., Arai R., Motegi F., Miyakawa T., Mabuchi I., and Hirata D. 1999. Overproduction of elongation factor 1α, an essential translational component, causes aberrant cell morphology by affecting the control of growth polarity in fission yeast. *Genes Cells* **4:** 517–527.

Tasheva E.S. and Roufa D.J. 1995. Regulation of human RPS14 transcription by intronic antisense and ribosomal protein S14. *Genes Dev.* **9:** 304–316.

Tobias J.W., Shrader T.E., Rocap G., and Varshavsky A. 1991. The N-end rule in bacteria. *Science* **254:** 1374–1377.

Toczyski D.P., Matera A.G., Ward D.C., and Steitz J.A. 1994. The Epstein-Barr virus (EBV) small RNA EBER1 binds and relocalizes ribosomal protein L22 in EBV-infected human B lympocytes. *Proc. Natl. Acad. Sci.* **91:** 3463–3467.

Umikawa M., Tanaka K., Kamei T., Shimizu K., Imamura H., Sasaki T., and Takai Y. 1998. Interaction of Rho1p target Bni1p with F-actin-binding elongation factor 1α: Implications in Rho1p-regulated reorganization of the actin cytoskeleton in *Saccharomyces cerevisiae*. *Oncogene* **16:** 2011–2016.

Wakasugi K. and Schimmel P. 1999. Two distinct cytokines released from a human aminoacyl-tRNA synthetase. *Science* **284:** 147–150.

Wang J.F., Engelsberg B.N., Johnson S.W., Witmer C., Merrick W.C., Rozmiarek H., and Billings P.C. 1997. DNA binding activity of the mammalian translation elongation complex: Recognition of chromium- and transplatin-damaged DNA. *Arch. Toxicol.* **71:** 450–454.

Wang W. and Poovaiah B.W. 1999. Interaction of plant chimeric calcium/calmodulin-dependent protein kinase with a homolog of eukaryotic elongation factor-1alpha. *J. Biol. Chem.* **274:** 12001–12008.

Wei N., Tsuge T., Serino G., Dohmae N., Takio K., Matsui M., and Deng X.W. 1998. The COP9 complex is conserved between plants and mammals and is related to the 26S proteasome regulatory complex. *Curr. Biol.* **8:** 919–922.

Wool I.G. 1996a. Extraribosomal functions of ribosomal proteins. *Trends Biochem. Sci.* **21:** 164–165.

———. 1996b. Extraribosomal functions of ribosomal proteins. In *Ribosomal RNA and group I introns* (ed. R. Green and R. Schroeder), pp. 153–178. R.G. Landes, Austin, Texas.

Wool I.G., Chan Y.-L., and Gluck A. 1996. Mammalian ribosomes: The structure and the evolution of the proteins. In *Translational control* (ed. J.W.B. Hershey et al.), pp. 685–732. Cold Spring Harbor Laboratory Press, Cold Spring Harbor, NewYork.

Xiao H., Neuveut C., Benkirane M., and Jeang K.T. 1998. Interaction of the second coding intron of Tat with human EF-1 delta delineates a mechanism for HIN-1-mediated shut-off of host mRNA translation. *Biochem. Biophys. Res. Commun.* **244:** 384–389.

Yang F., Demma F., Warren V., Dharmawardhane S., and Condeelis J. 1990. Identification of an actin-binding protein from *Dictyostelium* as elongation factor 1α. *Nature* **347:** 494–496.

Yang W. and Boss W.F. 1994. Regulation of phosphatidylinositol 4-kinase by the protein activator PIK-A49. Activation requires phosphorylation of PIK-A49. *J. Biol. Chem.* **269:** 3852–3857.

Yang W., Burkhart W., Cavallius J., Merrick W.C., and Boss W.F. 1993. Purification and characterization of a phosphatidylinositol 4-kinase activator in carrot cells. *J. Biol. Chem.* **268**: 392–398.

Yanofsky C. 1981. Attenuation in the control of expression of bacterial operons. *Nature* **289:** 751–758.

Yanofsky C. and Soll L. 1977. Mutations affecting $tRNA^{Trp}$ and its charging and their effect on regulation of transcription termination at the attenuator of the tryptophan operon. *J. Mol. Biol.* **113:** 663–677.

Zakharova O.D., Tarrago-Litvak L., Maksakova G., Andreola M.L., Dufour E., Litvak S., and Nevinsky G.A. 1995. High-affinity interaction of human immunodeficiency virus type-1 reverse transcriptase with partially complementary primers. *Eur. J. Biochem.* **233:** 856–863.

Zengel J.M. and Lindahl L. 1994. Diverse mechanisms for regulating ribosomal protein synthesis in *Escherichia coli*. *Prog. Nucleic Acid Res. Mol. Biol.* **47:** 331–370.

———. 1996. A hairpin structure upstream of the terminator hairpin required for ribosomal protein L4-mediated attenuation control of the S10 operon of *Escherichia coli*. *J. Bacteriol.* **178:** 2383–2387.

Zuk D. and Jacobson A. 1998. A single amino acid substitution in yeast eIF-5A results in mRNA stabilization. *EMBO J.* **17:** 2914–2925.

Zuk D., Belk J.P., and Jacobson A. 1999. Temperature sensitive mutations in the *Saccharomyces cerevisiae MRT4, GRC5, SLA2* and *THS1* genes result in defects in mRNA turnover. *Genetics* **153:** 35–47.

Index